Da ns
and Networking

McGraw-Hill Forouzan Networking Series

Titles by Behrouz A. Forouzan:

Data Communications and Networking
TCP/IP Protocol Suite
Computer Networks: A Top-Down Approach
Cryptography and Network Security

GLOBAL EDITION

Data Communications and Networking

FIFTH EDITION

Behrouz A. Forouzan

GLOBAL EDITION: DATA COMMUNICATIONS AND NETWORKING, FIFTH EDITION

Published by McGraw-Hill, a business unit of The McGraw-Hill Companies, Inc., 1221 Avenue of the Americas, New York, NY 10020.

Some ancillaries, including electronic and print components, may not be available to customers outside the United States.

This book is printed on acid-free paper.

1 2 3 4 5 6 7 8 9 0 DOC/DOC 1 0 9 8 7 6 5 4 3 2

ISBN 978-0-07-131586-9
MHID 0-07-131586-1

Cover Image: © *Science Photo Library RF/Getty Images*

www.mhhe.com

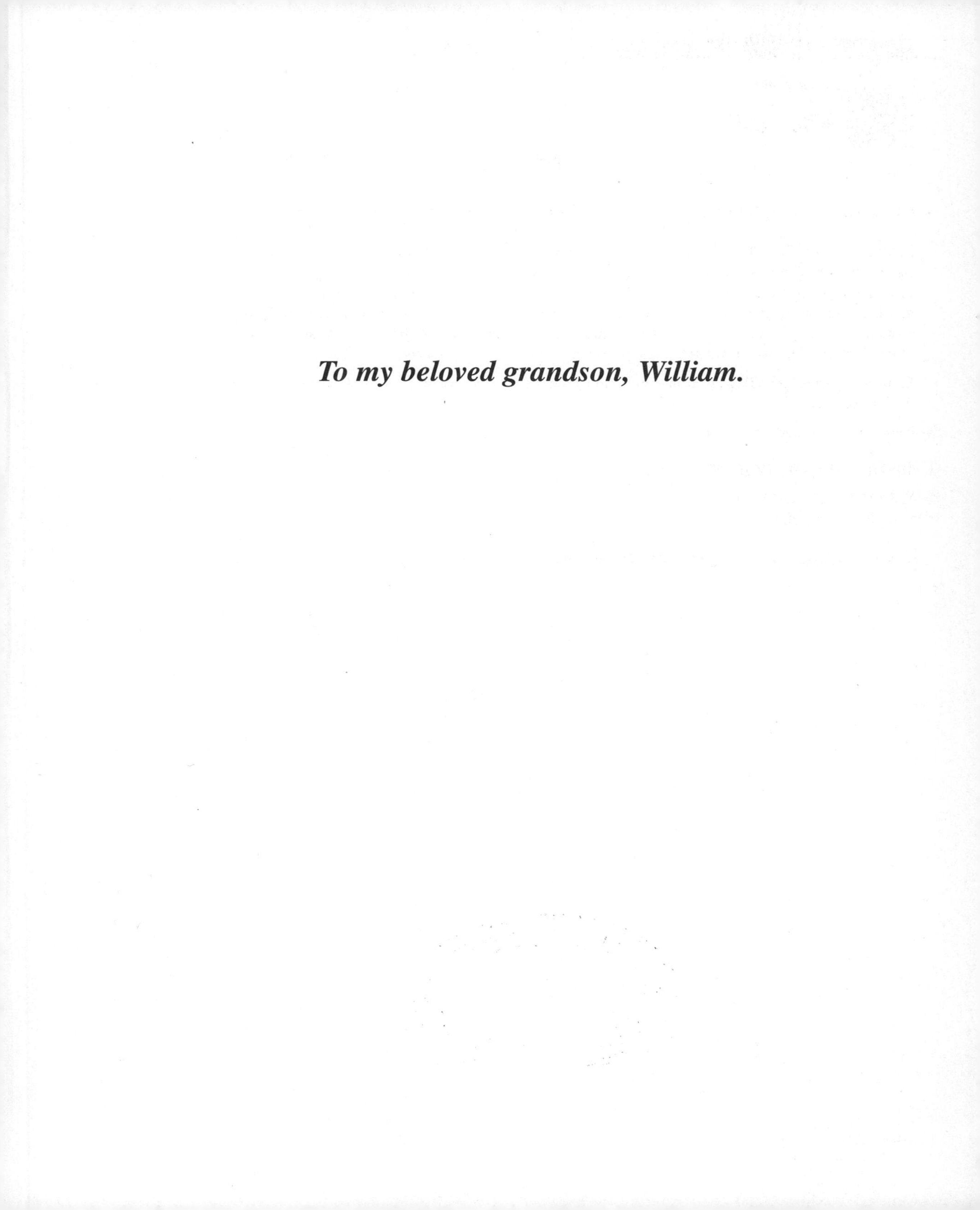

To my beloved grandson, William.

BRIEF CONTENTS

CONTENTS

PART IV: Network Layer 509

PREFACE

Technologies related to data communication and networking may be the fastest growing in our culture today. The appearance of some new social networking applications every year is a testimony to this claim. People use the Internet more and more every day. They use the Internet for research, shopping, airline reservations, checking the latest news and weather, and so on.

In this Internet-oriented society, specialists need be trained to run and manage the Internet, part of the Internet, or an organization's network that is connected to the Internet. This book is designed to help students understand the basics of data communications and networking in general and the protocols used in the Internet in particular.

Features

Although the main goal of the book is to teach the principles of networking, it is designed to teach these principles using the following goals:

Protocol Layering

The book is designed to teach the principles of networking by using the protocol layering of the Internet and the TCP/IP protocol suite. Some of the networking principles may have been duplicated in some of these layers, but with their own special details. Teaching these principles using protocol layering is beneficial because these principles are repeated and better understood in relation to each layer. For example, although *addressing* is an issue that is applied to four layers of the TCP/IP suite, each layer uses a different addressing format for different purposes. In addition, addressing has a different domain in each layer. Another example is *framing and packetizing,* which is repeated in several layers, but each layer treats the principle differently.

Bottom-Up Approach

This book uses a bottom-up approach. Each layer in the TCP/IP protocol suite is built on the services provided by the layer below. We learn how bits are moving at the physical layer before learning how some programs exchange messages at the application layer.

Changes in the Fifth Edition

I have made several categories of changes in this edition.

Changes in the Organization

Although the book is still made of seven parts, the contents and order of chapters have been changed. Some chapters have been combined, some have been moved, some are new. Sometimes part of a chapter is eliminated because the topic is deprecated. The following shows the relationship between chapters in the fourth and fifth editions.

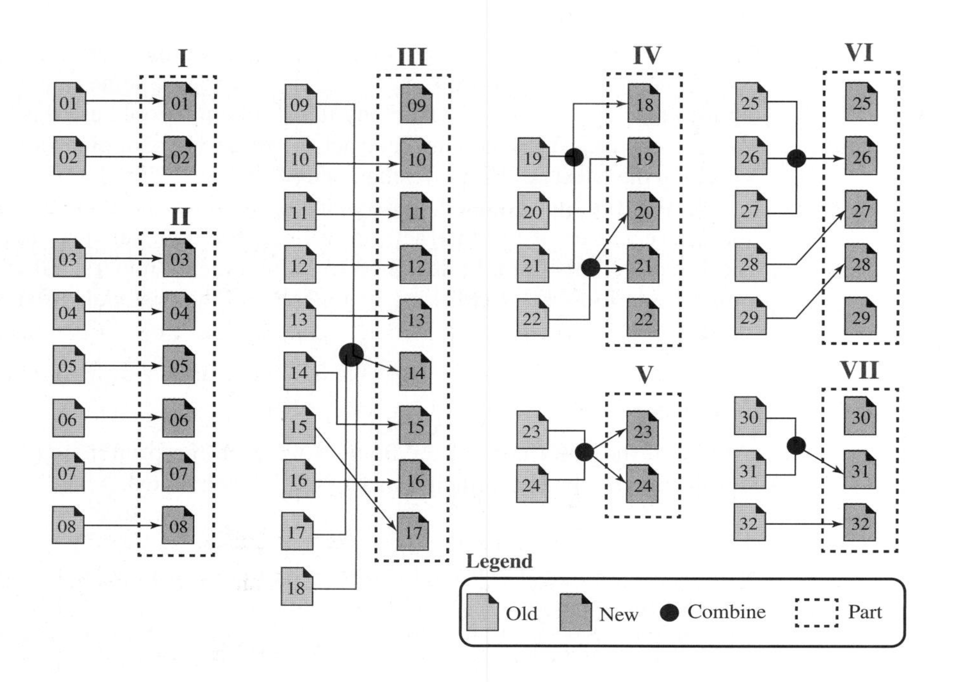

- Some chapters have been combined into one chapter. Chapters 9, 17, and 18 are combined into one chapter because some topics in each chapter have been deprecated. Chapters 19 and 21 are combined into Chapter 18. Chapters 25, 26, and 27 are also combined into one chapter because the topics are related to each other. Chapters 30 and 31 are also combined because they cover the same issue.
- Some chapters have been split into two chapters because of content augmentation. For example, Chapter 22 is split into Chapters 20 and 21.
- Some chapters have been first combined, but then split for better organization. For example, Chapters 23 and 24 are first combined and then split into two chapters again.
- Some chapters have been moved to better fit in the organization of the book. Chapter 15 now becomes Chapter 17. Chapters 28 and 29 now become Chapters 27 and 28.
- Some chapters have been moved to fit better in the sequence. For example, Chapter 15 has become Chapter 17 to cover more topics.
- Some chapters are new. Chapter 9 is an introduction to the data-link layer. Chapter 25 is an introduction to the application layer and includes socket-interface programming in C and Java. Chapter 30 is almost new. It covers QoS, which was part of other chapters in the previous edition.

New and Augmented Materials

Although the contents of each chapter have been updated, some new materials have also been added to this edition:

- *Peer-to-Peer paradigm* has been added as a new chapter (Chapter 29).
- *Quality of service* (QoS) has been augmented and added as a new chapter (Chapter 30).
- Chapter 10 is augmented to include the *forward error correction.*
- WiMAX, as the wireless access network, has been added to Chapter 16.
- The coverage of the transport-layer protocol has been augmented (Chapter 23).
- Socket-interface programming in Java has been added to Chapter 25.
- Chapter 28, on multimedia, has been totally revised and augmented.
- Contents of unicast and multicast routing (Chapters 20 and 21) have gone through a major change and have been augmented.
- The next generation IP is augmented and now belongs to Chapter 22.

Changes in the End-Chapter Materials

The end-chapter materials have gone through a major change:

- The practice set is augmented; it has many new problems in some appropriate chapters.
- Lab assignments have been added to some chapters to allow students to see some data in motion.
- Some applets have been posted on the book website to allow students to see some problems and protocols in action.
- Some programming assignments allow the students to write some programs to solve problems.

Extra Materials

Some extra materials, which could not be fit in the contents and volume of the book, have been posted on the book website for further study.

New Organization

This edition is divided into seven parts, which reflects the structure of the Internet model.

Part One: Overview

The first part gives a general overview of data communications and networking. Chapter 1 covers introductory concepts needed for the rest of the book. Chapter 2 introduces the Internet model.

Part Two: Physical Layer

The second part is a discussion of the physical layer of the Internet model. It is made of six chapters. Chapters 3 to 6 discuss telecommunication aspects of the physical layer.

Chapter 7 introduces the transmission media, which, although not part of the physical layer, is controlled by it. Chapter 8 is devoted to switching, which can be used in several layers.

Part Three: Data-Link Layer

The third part is devoted to the discussion of the data-link layer of the Internet model. It is made of nine chapters. Chapter 9 introduces the data-link layer. Chapter 10 covers error detection and correction, which can also be used in some other layers. Chapters 11 and 12 discuss issues related to two sublayers in the data-link layer. Chapters 13 and 14 discuss wired networks. Chapters 15 and 16 discuss wireless networks. Chapter 17 shows how networks can be combined to create larger or virtual networks.

Part Four: Network Layer

The fourth part is devoted to the discussion of the network layer of the Internet model. Chapter 18 introduces this layer and discusses the network-layer addressing. Chapter 19 discusses the protocols in the current version. Chapters 20 and 21 are devoted to routing (unicast and multicast). Chapter 22 introduces the next generation protocol.

Part Five: Transport Layer

The fifth part is devoted to the discussion of the transport layer of the Internet model. Chapter 23 gives an overview of the transport layer and discusses the services and duties of this layer. Chapter 24 discusses the transport-layer protocols in the Internet: UDP, TCP, and SCTP.

Part Six: Application Layer

Chapter 25 introduces the application layer and discusses some network programming in both C and Java. Chapter 26 discusses most of the standard client-server programming in the Internet. Chapter 27 discusses network management. Chapter 28 is devoted to the multimedia, an issue which is very hot today. Finally, Chapter 29 is an introduction to the peer-to-peer paradigm, a trend which is on the rise in the today's Internet.

Part Seven: Topics Related to All Layers

The last part of the book discusses the issues that belong to some or all layers. Chapter 30 discusses the quality of service. Chapters 31 and 32 discuss security.

Appendices

The appendices (available online at http://www.mhhe.com/forouzan) are intended to provide a quick reference or review of materials needed to understand the concepts discussed in the book. There are eight appendices that can be used by the students for reference and study:

- Appendix A: Unicode
- Appendix B: Positional Numbering System
- Appendix C: HTML,CSS, XML, and XSL
- Appendix D: A Touch of Probability
- Appendix E: Mathematical Review

- ❑ Appendix F: 8B/6T Code
- ❑ Appendix G: Miscellaneous Information
- ❑ Appendix H: Telephone History

References

The book contains a list of references for further reading.

Glossary and Acronyms

The book contains an extensive glossary and a list of acronyms for finding the corresponding term quickly.

Pedagogy

Several pedagogical features of this text are designed to make it particularly easy for students to understand data communication and networking.

Visual Approach

The book presents highly technical subject matter without complex formulas by using a balance of text and figures. More than 830 figures accompanying the text provide a visual and intuitive opportunity for understanding the material. Figures are particularly important in explaining networking concepts. For many students, these concepts are more easily grasped visually than verbally.

Highlighted Points

I have repeated important concepts in boxes for quick reference and immediate attention.

Examples and Applications

Whenever appropriate, I have included examples that illustrate the concepts introduced in the text. Also, I have added some real-life applications throughout each chapter to motivate students.

End-of-Chapter Materials

Each chapter ends with a set of materials that includes the following:

Key Terms

The new terms used in each chapter are listed at the end of the chapter and their definitions are included in the glossary.

Recommended Reading

This section gives a brief list of references relative to the chapter. The references can be used to quickly find the corresponding literature in the reference section at the end of the book.

Summary

Each chapter ends with a summary of the material covered by that chapter. The summary glues the important materials together to be seen in one shot.

Practice Set

Each chapter includes a practice set designed to reinforce salient concepts and encourage students to apply them. It consists of three parts: quizzes, questions, and problems.

Quizzes

Quizzes, which are posted on the book website, provide quick concept checking. Students can take these quizzes to check their understanding of the materials. The feedback to the students' responses is given immediately.

Questions

This section contains simple questions about the concepts discussed in the book. Answers to the odd-numbered questions are posted on the book website to be checked by the student. There are more than 630 questions at the ends of chapters.

Problems

This section contains more difficult problems that need a deeper understanding of the materials discussed in the chapter. I strongly recommend that the student try to solve all of these problems. Answers to the odd-numbered problems are also posted on the book website to be checked by the student. There are more than 600 problems at the ends of chapters.

Simulation Experiments

Network concepts and the flow and contents of the packets can be better understood if they can be analyzed in action. Some chapters include a section to help students experiment with these. This section is divided into two parts: applets and lab assignments.

Applets

Java applets are interactive experiments that are created by the authors and posted on the website. Some of these applets are used to better understand the solutions to some problems; others are used to better understand the network concepts in action.

Lab Assignments

Some chapters include lab assignments that use Wireshark simulation software. The instructions for downloading and using Wireshark are given in Chapter 1. In some other chapters, there are a few lab assignments that can be used to practice sending and receiving packets and analyzing their contents.

Programming Assignments

Some chapters also include programming assignments. Writing a program about a process or procedure clarifies many subtleties and helps the student better understand the concept behind the process. Although the student can write and test programs in any computer language she or he is comfortable with, the solutions are given in Java language at the book website for the use of professors.

Audience

This book is written for both academic and professional audiences. The book can be used as a self-study guide for interested professionals. As a textbook, it can be used for a one-semester or one-quarter course. It is designed for the last year of undergraduate study or the first year of graduate study. Although some problems at the end of the chapters require some knowledge of probability, the study of the text needs only general mathematical knowledge taught in the first year of college.

Instruction Resources

The book contains complete instruction resources that can be downloaded from the book site **http://www.mhhe.com/forouzan**. They include:

Presentations

The site includes a set of colorful and animated PowerPoint presentations for teaching the course.

Solutions to Practice Set

Solutions to all questions and problems are provided on the book website for the use of professors who teach the course.

Solution to Programming Assignments

Solutions to programming assignments are also provided on the book website. The programs are mostly in Java language.

Student Resources

The book contains complete student resources that can be downloaded from the book website **http://www.mhhe.com/forouzan**. They include:

Quizzes

There are quizzes at the end of each chapter that can be taken by the students. Students are encouraged to take these quizzes to test their general understanding of the materials presented in the corresponding chapter.

Solution to Odd-Numbered Practice Set

Solutions to all odd-numbered questions and problems are provided on the book website for the use of students.

Lab Assignments

The descriptions of lab assignments are also included in the student resources.

Applets

There are some applets for each chapter. Applets can either show the solution to some examples and problems or show some protocols in action. It is strongly recommended that students activate these applets.

Extra Materials

Students can also access the extra materials at the book website for further study.

How to Use the Book

The chapters in the book are organized to provide a great deal of flexibility. I suggest the following:

- Materials provided in Part I are essential for understanding the rest of the book.
- Part II (physical layer) is essential to understand the rest of the book, but the professor can skip this part if the students already have the background in engineering and the physical layer.
- Parts III to VI are based on the Internet model. They are required for understanding the use of the networking principle in the Internet.
- Part VII (QoS and Security) is related to all layers of the Internet mode. It can be partially or totally skipped if the students will be taking a course that covers these materials.

Website

The McGraw-Hill website contains much additional material, available at **www.mhhe.com/forouzan.** As students read through *Data Communications and Networking,* they can go online to take self-grading quizzes. They can also access lecture materials such as PowerPoint slides, and get additional review from animated figures from the book. Selected solutions are also available over the Web. The solutions to odd-numbered problems are provided to students, and instructors can use a password to access the complete set of solutions.

McGraw-Hill Create™

Craft your teaching resources to match the way you teach! With McGraw-Hill Create, www.mcgrawhillcreate.com, you can easily rearrange chapters, combine material from other content sources, and quickly upload content you have written like your course syllabus or teaching notes. Find the content you need in Create by searching through thousands of leading McGraw-Hill textbooks. Arrange your book to fit your teaching style. Create even allows you to personalize your book's appearance by selecting the cover and adding your name, school, and course information. Order a Create book and you'll receive a complimentary print review copy in 3–5 business days or a complimentary electronic review copy (eComp) via email in minutes. Go to www.mcgrawhillcreate.com today and register to experience how McGraw-Hill Create empowers you to teach *your* students *your* way.

Electronic Textbook Option

This text is offered through CourseSmart for both instructors and students. CourseSmart is an online resource where students can purchase the complete text online at almost half the cost of a traditional text. Purchasing the eTextbook allows students to take advantage of CourseSmart's web tools for learning, which include full text search, notes and highlighting, and email tools for sharing notes between classmates. To learn more about CourseSmart options, contact your sales representative or visit www.CourseSmart.com.

Acknowledgments

It is obvious that the development of a book of this scope needs the support of many people. I would like to acknowledge the contributions from peer reviewers to the development of the book. These reviewers are:

Tricha Anjali, Illinois Institute of Technology
Yoris A. Au, University of Texas at San Antonio
Randy J. Fortier, University of Windsor
Tirthankar Ghosh, Saint Cloud State University
Lawrence Hill, Rochester Institute of Technology
Ezzat Kirmani, Saint Cloud State University
Robert Koeneke, University of Central Florida
Mike O'Dell, University of Texas at Arlington

Special thanks go to the staff of McGraw-Hill. Raghu Srinivasan, the publisher, proved how a proficient publisher can make the impossible, possible. Melinda Bilecki, the developmental editor, gave help whenever I needed it. Jane Mohr, the project manager, guided us through the production process with enormous enthusiasm. I also thank Dheeraj Chahal, full-service project manager, Brenda A. Rolwes, the cover designer, and Kathryn DiBernardo, the copy editor.

Behrouz A. Forouzan
Los Angeles, California, United States of America
January 2012

TRADE MARK

Throughout the text we have used several trademarks. Rather than insert a trademark symbol with each mention of the trademark name, we acknowledge the trademarks here and state that they are used with no intention of infringing upon them. Other product names, trademarks, and registered trademarks are the property of their respective owners.

PART I

Overview

In the first part of the book, we discuss some general ideas related to both data communications and networking. This part lays the plan for the rest of the book. The part is made of two chapters that prepare the reader for the long journey ahead.

CHAPTER 1

Introduction

Data communications and networking have changed the way we do business and the way we live. Business decisions have to be made ever more quickly, and the decision makers require immediate access to accurate information. Why wait a week for that report from Europe to arrive by mail when it could appear almost instantaneously through computer networks? Businesses today rely on computer networks and internetworks.

Data communication and networking have found their way not only through business and personal communication, they have found many applications in political and social issues. People have found how to communicate with other people in the world to express their social and political opinions and problems. Communities in the world are not isolated anymore.

But before we ask how quickly we can get hooked up, we need to know how networks operate, what types of technologies are available, and which design best fills which set of needs.

This chapter paves the way for the rest of the book. It is divided into five sections.

- The first section introduces data communications and defines their components and the types of data exchanged. It also shows how different types of data are represented and how data is flowed through the network.
- The second section introduces networks and defines their criteria and structures. It introduces four different network topologies that are encountered throughout the book.
- The third section discusses different types of networks: LANs, WANs, and internetworks (internets). It also introduces the Internet, the largest internet in the world. The concept of switching is also introduced in this section to show how small networks can be combined to create larger ones.
- The fourth section covers a brief history of the Internet. The section is divided into three eras: early history, the birth of the Internet, and the issues related to the Internet today. This section can be skipped if the reader is familiar with this history.
- The fifth section covers standards and standards organizations. The section covers Internet standards and Internet administration. We refer to these standards and organizations throughout the book.

1.1 DATA COMMUNICATIONS

When we communicate, we are sharing information. This sharing can be local or remote. Between individuals, local communication usually occurs face to face, while remote communication takes place over distance. The term ***telecommunication,*** which includes telephony, telegraphy, and television, means communication at a distance (*tele* is Greek for "far"). The word ***data*** refers to information presented in whatever form is agreed upon by the parties creating and using the data.

Data communications are the exchange of data between two devices via some form of transmission medium such as a wire cable. For data communications to occur, the communicating devices must be part of a communication system made up of a combination of hardware (physical equipment) and software (programs). The effectiveness of a data communications system depends on four fundamental characteristics: delivery, accuracy, timeliness, and jitter.

1. **Delivery.** The system must deliver data to the correct destination. Data must be received by the intended device or user and only by that device or user.
2. **Accuracy.** The system must deliver the data accurately. Data that have been altered in transmission and left uncorrected are unusable.
3. **Timeliness.** The system must deliver data in a timely manner. Data delivered late are useless. In the case of video and audio, timely delivery means delivering data as they are produced, in the same order that they are produced, and without significant delay. This kind of delivery is called *real-time* transmission.
4. **Jitter.** Jitter refers to the variation in the packet arrival time. It is the uneven delay in the delivery of audio or video packets. For example, let us assume that video packets are sent every 30 ms. If some of the packets arrive with 30-ms delay and others with 40-ms delay, an uneven quality in the video is the result.

1.1.1 Components

A data communications system has five components (see Figure 1.1).

Figure 1.1 *Five components of data communication*

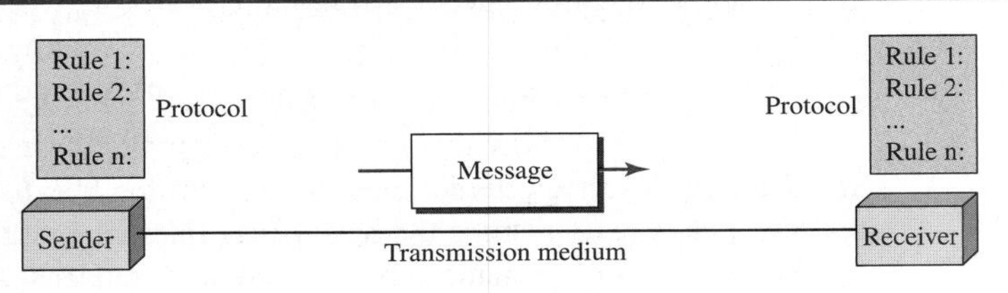

1. **Message.** The **message** is the information (data) to be communicated. Popular forms of information include text, numbers, pictures, audio, and video.
2. **Sender.** The **sender** is the device that sends the data message. It can be a computer, workstation, telephone handset, video camera, and so on.

3. **Receiver.** The **receiver** is the device that receives the message. It can be a computer, workstation, telephone handset, television, and so on.
4. **Transmission medium.** The **transmission medium** is the physical path by which a message travels from sender to receiver. Some examples of transmission media include twisted-pair wire, coaxial cable, fiber-optic cable, and radio waves.
5. **Protocol.** A protocol is a set of rules that govern data communications. It represents an agreement between the communicating devices. Without a protocol, two devices may be connected but not communicating, just as a person speaking French cannot be understood by a person who speaks only Japanese.

1.1.2 Data Representation

Information today comes in different forms such as text, numbers, images, audio, and video.

Text

In data communications, text is represented as a bit pattern, a sequence of bits (0s or 1s). Different sets of bit patterns have been designed to represent text symbols. Each set is called a **code,** and the process of representing symbols is called coding. Today, the prevalent coding system is called **Unicode,** which uses 32 bits to represent a symbol or character used in any language in the world. The **American Standard Code for Information Interchange (ASCII),** developed some decades ago in the United States, now constitutes the first 127 characters in Unicode and is also referred to as **Basic Latin.** Appendix A includes part of the Unicode.

Numbers

Numbers are also represented by bit patterns. However, a code such as ASCII is not used to represent numbers; the number is directly converted to a binary number to simplify mathematical operations. Appendix B discusses several different numbering systems.

Images

Images are also represented by bit patterns. In its simplest form, an image is composed of a matrix of pixels (picture elements), where each pixel is a small dot. The size of the pixel depends on the *resolution.* For example, an image can be divided into 1000 pixels or 10,000 pixels. In the second case, there is a better representation of the image (better resolution), but more memory is needed to store the image.

After an image is divided into pixels, each pixel is assigned a bit pattern. The size and the value of the pattern depend on the image. For an image made of only black-and-white dots (e.g., a chessboard), a 1-bit pattern is enough to represent a pixel.

If an image is not made of pure white and pure black pixels, we can increase the size of the bit pattern to include gray scale. For example, to show four levels of gray scale, we can use 2-bit patterns. A black pixel can be represented by 00, a dark gray pixel by 01, a light gray pixel by 10, and a white pixel by 11.

There are several methods to represent color images. One method is called **RGB,** so called because each color is made of a combination of three primary colors: *r*ed, *g*reen, and *b*lue. The intensity of each color is measured, and a bit pattern is assigned to it. Another method is called **YCM,** in which a color is made of a combination of three other primary colors: *y*ellow, *c*yan, and *m*agenta.

Audio

Audio refers to the recording or broadcasting of sound or music. Audio is by nature different from text, numbers, or images. It is continuous, not discrete. Even when we use a microphone to change voice or music to an electric signal, we create a continuous signal. We will learn more about audio in Chapter 26.

Video

Video refers to the recording or broadcasting of a picture or movie. Video can either be produced as a continuous entity (e.g., by a TV camera), or it can be a combination of images, each a discrete entity, arranged to convey the idea of motion. We will learn more about video in Chapter 26.

1.1.3 Data Flow

Communication between two devices can be simplex, half-duplex, or full-duplex as shown in Figure 1.2.

Figure 1.2 *Data flow (simplex, half-duplex, and full-duplex)*

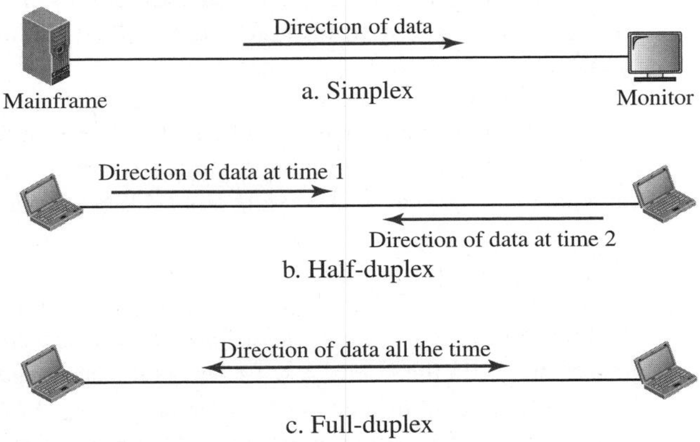

Direction of data

Mainframe

a. Simplex

Monitor

Direction of data at time 1

Direction of data at time 2

b. Half-duplex

Direction of data all the time

c. Full-duplex

Simplex

In **simplex mode,** the communication is unidirectional, as on a one-way street. Only one of the two devices on a link can transmit; the other can only receive (see Figure 1.2a).

Keyboards and traditional monitors are examples of simplex devices. The keyboard can only introduce input; the monitor can only accept output. The simplex mode can use the entire capacity of the channel to send data in one direction.

Half-Duplex

In **half-duplex mode,** each station can both transmit and receive, but not at the same time. When one device is sending, the other can only receive, and vice versa (see Figure 1.2b).

The half-duplex mode is like a one-lane road with traffic allowed in both directions. When cars are traveling in one direction, cars going the other way must wait. In a half-duplex transmission, the entire capacity of a channel is taken over by whichever of the two devices is transmitting at the time. Walkie-talkies and CB (citizens band) radios are both half-duplex systems.

The half-duplex mode is used in cases where there is no need for communication in both directions at the same time; the entire capacity of the channel can be utilized for each direction.

Full-Duplex

In **full-duplex mode** (also called *duplex*), both stations can transmit and receive simultaneously (see Figure 1.2c).

The full-duplex mode is like a two-way street with traffic flowing in both directions at the same time. In full-duplex mode, signals going in one direction share the capacity of the link with signals going in the other direction. This sharing can occur in two ways: Either the link must contain two physically separate transmission paths, one for sending and the other for receiving; or the capacity of the channel is divided between signals traveling in both directions.

One common example of full-duplex communication is the telephone network. When two people are communicating by a telephone line, both can talk and listen at the same time.

The full-duplex mode is used when communication in both directions is required all the time. The capacity of the channel, however, must be divided between the two directions.

1.2 NETWORKS

A **network** is the interconnection of a set of devices capable of communication. In this definition, a device can be a **host** (or an *end system* as it is sometimes called) such as a large computer, desktop, laptop, workstation, cellular phone, or security system. A device in this definition can also be a **connecting device** such as a router, which connects the network to other networks, a switch, which connects devices together, a modem (modulator-demodulator), which changes the form of data, and so on. These devices in a network are connected using wired or wireless transmission media such as cable or air. When we connect two computers at home using a plug-and-play router, we have created a network, although very small.

1.2.1 Network Criteria

A network must be able to meet a certain number of criteria. The most important of these are performance, reliability, and security.

Performance

Performance can be measured in many ways, including transit time and response time. Transit time is the amount of time required for a message to travel from one device to another. Response time is the elapsed time between an inquiry and a response. The performance of a network depends on a number of factors, including the number of users, the type of transmission medium, the capabilities of the connected hardware, and the efficiency of the software.

Performance is often evaluated by two networking metrics: **throughput** and **delay.** We often need more throughput and less delay. However, these two criteria are often contradictory. If we try to send more data to the network, we may increase throughput but we increase the delay because of traffic congestion in the network.

Reliability

In addition to accuracy of delivery, network **reliability** is measured by the frequency of failure, the time it takes a link to recover from a failure, and the network's robustness in a catastrophe.

Security

Network **security** issues include protecting data from unauthorized access, protecting data from damage and development, and implementing policies and procedures for recovery from breaches and data losses.

1.2.2 Physical Structures

Before discussing networks, we need to define some network attributes.

Type of Connection

A network is two or more devices connected through links. A link is a communications pathway that transfers data from one device to another. For visualization purposes, it is simplest to imagine any link as a line drawn between two points. For communication to occur, two devices must be connected in some way to the same link at the same time. There are two possible types of connections: point-to-point and multipoint.

Point-to-Point

A **point-to-point connection** provides a dedicated link between two devices. The entire capacity of the link is reserved for transmission between those two devices. Most point-to-point connections use an actual length of wire or cable to connect the two ends, but other options, such as microwave or satellite links, are also possible (see Figure 1.3a). When we change television channels by infrared remote control, we are establishing a point-to-point connection between the remote control and the television's control system.

Multipoint

A **multipoint** (also called **multidrop**) **connection** is one in which more than two specific devices share a single link (see Figure 1.3b).

Figure 1.3 *Types of connections: point-to-point and multipoint*

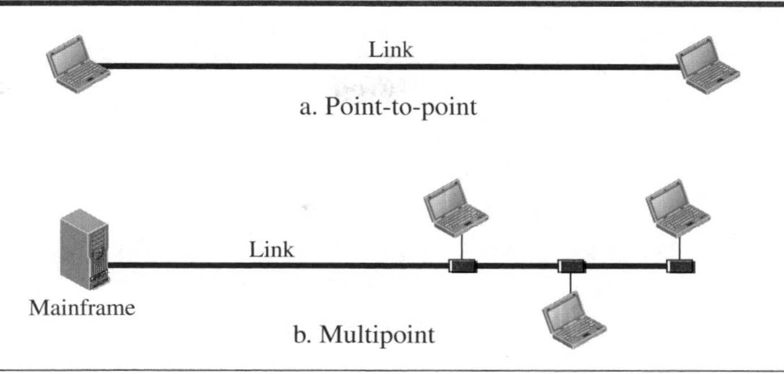

In a multipoint environment, the capacity of the channel is shared, either spatially or temporally. If several devices can use the link simultaneously, it is a *spatially shared* connection. If users must take turns, it is a *timeshared* connection.

Physical Topology

The term ***physical topology*** refers to the way in which a network is laid out physically. Two or more devices connect to a link; two or more links form a topology. The topology of a network is the geometric representation of the relationship of all the links and linking devices (usually called ***nodes***) to one another. There are four basic topologies possible: mesh, star, bus, and ring.

Mesh Topology

In a **mesh topology,** every device has a dedicated point-to-point link to every other device. The term *dedicated* means that the link carries traffic only between the two devices it connects. To find the number of physical links in a fully connected mesh network with n nodes, we first consider that each node must be connected to every other node. Node 1 must be connected to $n - 1$ nodes, node 2 must be connected to $n - 1$ nodes, and finally node n must be connected to $n - 1$ nodes. We need $n (n - 1)$ physical links. However, if each physical link allows communication in both directions (duplex mode), we can divide the number of links by 2. In other words, we can say that in a mesh topology, we need $\boldsymbol{n (n - 1) / 2}$ duplex-mode links. To accommodate that many links, every device on the network must have $n - 1$ input/output (I/O) ports (see Figure 1.4) to be connected to the other $n - 1$ stations.

A mesh offers several advantages over other network topologies. First, the use of dedicated links guarantees that each connection can carry its own data load, thus eliminating the traffic problems that can occur when links must be shared by multiple devices. Second, a mesh topology is robust. If one link becomes unusable, it does not incapacitate the entire system. Third, there is the advantage of privacy or security. When every message travels along a dedicated line, only the intended recipient sees it. Physical boundaries prevent other users from gaining access to messages. Finally, point-to-point links make fault identification and fault isolation easy. Traffic can be routed to avoid links with suspected problems. This facility enables the network manager to discover the precise location of the fault and aids in finding its cause and solution.

Figure 1.4 *A fully connected mesh topology (five devices)*

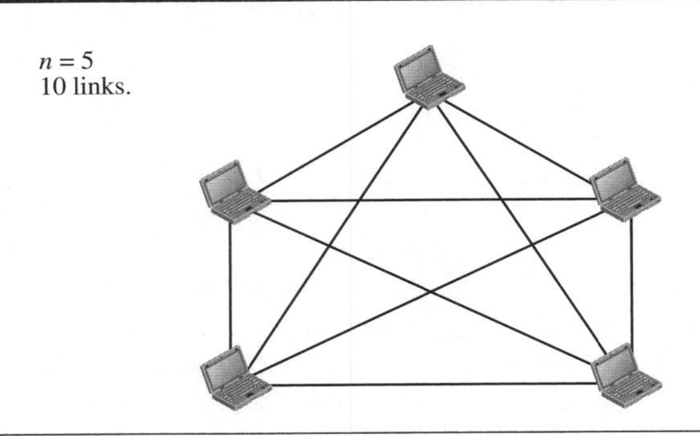

The main disadvantages of a mesh are related to the amount of cabling and the number of I/O ports required. First, because every device must be connected to every other device, installation and reconnection are difficult. Second, the sheer bulk of the wiring can be greater than the available space (in walls, ceilings, or floors) can accommodate. Finally, the hardware required to connect each link (I/O ports and cable) can be prohibitively expensive. For these reasons a mesh topology is usually implemented in a limited fashion, for example, as a backbone connecting the main computers of a hybrid network that can include several other topologies.

One practical example of a mesh topology is the connection of telephone regional offices in which each regional office needs to be connected to every other regional office.

Star Topology

In a **star topology,** each device has a dedicated point-to-point link only to a central controller, usually called a ***hub.*** The devices are not directly linked to one another. Unlike a mesh topology, a star topology does not allow direct traffic between devices. The controller acts as an exchange: If one device wants to send data to another, it sends the data to the controller, which then relays the data to the other connected device (see Figure 1.5) .

Figure 1.5 *A star topology connecting four stations*

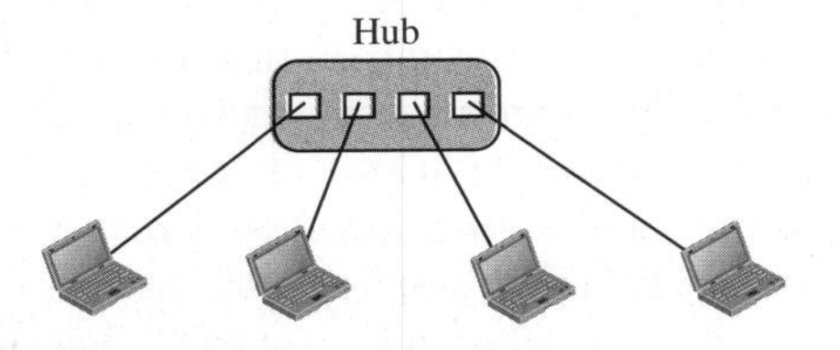

A star topology is less expensive than a mesh topology. In a star, each device needs only one link and one I/O port to connect it to any number of others. This factor also makes it easy to install and reconfigure. Far less cabling needs to be housed, and additions, moves, and deletions involve only one connection: between that device and the hub.

Other advantages include robustness. If one link fails, only that link is affected. All other links remain active. This factor also lends itself to easy fault identification and fault isolation. As long as the hub is working, it can be used to monitor link problems and bypass defective links.

One big disadvantage of a star topology is the dependency of the whole topology on one single point, the hub. If the hub goes down, the whole system is dead.

Although a star requires far less cable than a mesh, each node must be linked to a central hub. For this reason, often more cabling is required in a star than in some other topologies (such as ring or bus).

The star topology is used in local-area networks (LANs), as we will see in Chapter 13. High-speed LANs often use a star topology with a central hub.

Bus Topology

The preceding examples all describe point-to-point connections. A **bus topology,** on the other hand, is multipoint. One long cable acts as a **backbone** to link all the devices in a network (see Figure 1.6).

Figure 1.6 *A bus topology connecting three stations*

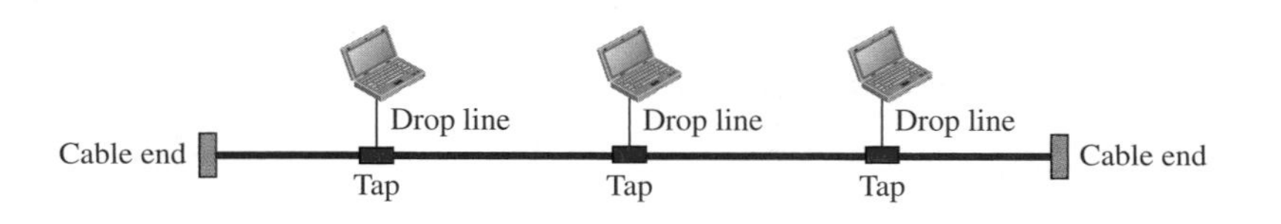

Nodes are connected to the bus cable by drop lines and taps. A drop line is a connection running between the device and the main cable. A tap is a connector that either splices into the main cable or punctures the sheathing of a cable to create a contact with the metallic core. As a signal travels along the backbone, some of its energy is transformed into heat. Therefore, it becomes weaker and weaker as it travels farther and farther. For this reason there is a limit on the number of taps a bus can support and on the distance between those taps.

Advantages of a bus topology include ease of installation. Backbone cable can be laid along the most efficient path, then connected to the nodes by drop lines of various lengths. In this way, a bus uses less cabling than mesh or star topologies. In a star, for example, four network devices in the same room require four lengths of cable reaching all the way to the hub. In a bus, this redundancy is eliminated. Only the backbone cable stretches through the entire facility. Each drop line has to reach only as far as the nearest point on the backbone.

Disadvantages include difficult reconnection and fault isolation. A bus is usually designed to be optimally efficient at installation. It can therefore be difficult to add new devices. Signal reflection at the taps can cause degradation in quality. This degradation can be controlled by limiting the number and spacing of devices connected to a given length of cable. Adding new devices may therefore require modification or replacement of the backbone.

In addition, a fault or break in the bus cable stops all transmission, even between devices on the same side of the problem. The damaged area reflects signals back in the direction of origin, creating noise in both directions.

Bus topology was the one of the first topologies used in the design of early local-area networks. Traditional Ethernet LANs can use a bus topology, but they are less popular now for reasons we will discuss in Chapter 13.

Ring Topology

In a **ring topology,** each device has a dedicated point-to-point connection with only the two devices on either side of it. A signal is passed along the ring in one direction, from device to device, until it reaches its destination. Each device in the ring incorporates a repeater. When a device receives a signal intended for another device, its repeater regenerates the bits and passes them along (see Figure 1.7).

Figure 1.7 *A ring topology connecting six stations*

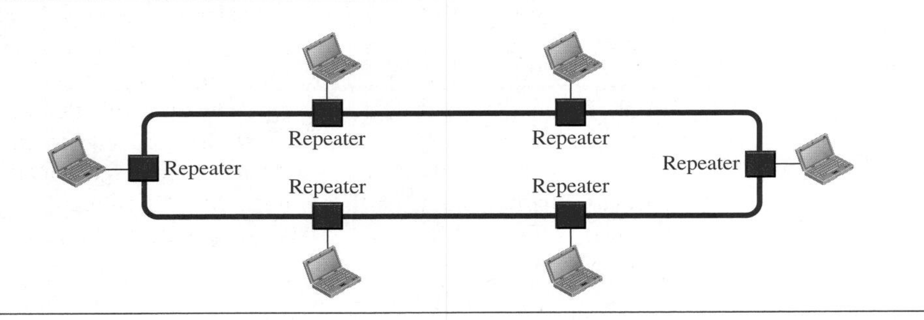

A ring is relatively easy to install and reconfigure. Each device is linked to only its immediate neighbors (either physically or logically). To add or delete a device requires changing only two connections. The only constraints are media and traffic considerations (maximum ring length and number of devices). In addition, fault isolation is simplified. Generally, in a ring a signal is circulating at all times. If one device does not receive a signal within a specified period, it can issue an alarm. The alarm alerts the network operator to the problem and its location.

However, unidirectional traffic can be a disadvantage. In a simple ring, a break in the ring (such as a disabled station) can disable the entire network. This weakness can be solved by using a dual ring or a switch capable of closing off the break.

Ring topology was prevalent when IBM introduced its local-area network, Token Ring. Today, the need for higher-speed LANs has made this topology less popular.

1.3 NETWORK TYPES

After defining networks in the previous section and discussing their physical structures, we need to discuss different types of networks we encounter in the world today. The criteria of distinguishing one type of network from another is difficult and sometimes confusing. We use a few criteria such as size, geographical coverage, and ownership to make this distinction. After discussing two types of networks, LANs and WANs, we define switching, which is used to connect networks to form an internetwork (a network of networks).

1.3.1 Local Area Network

A **local area network** (**LAN**) is usually privately owned and connects some hosts in a single office, building, or campus. Depending on the needs of an organization, a LAN can be as simple as two PCs and a printer in someone's home office, or it can extend throughout a company and include audio and video devices. Each host in a LAN has an identifier, an address, that uniquely defines the host in the LAN. A packet sent by a host to another host carries both the source host's and the destination host's addresses.

In the past, all hosts in a network were connected through a common cable, which meant that a packet sent from one host to another was received by all hosts. The intended recipient kept the packet; the others dropped the packet. Today, most LANs use a smart connecting switch, which is able to recognize the destination address of the packet and guide the packet to its destination without sending it to all other hosts. The switch alleviates the traffic in the LAN and allows more than one pair to communicate with each other at the same time if there is no common source and destination among them. Note that the above definition of a LAN does not define the minimum or maximum number of hosts in a LAN. Figure 1.8 shows a LAN using either a common cable or a switch.

Figure 1.8 *An isolated LAN in the past and today*

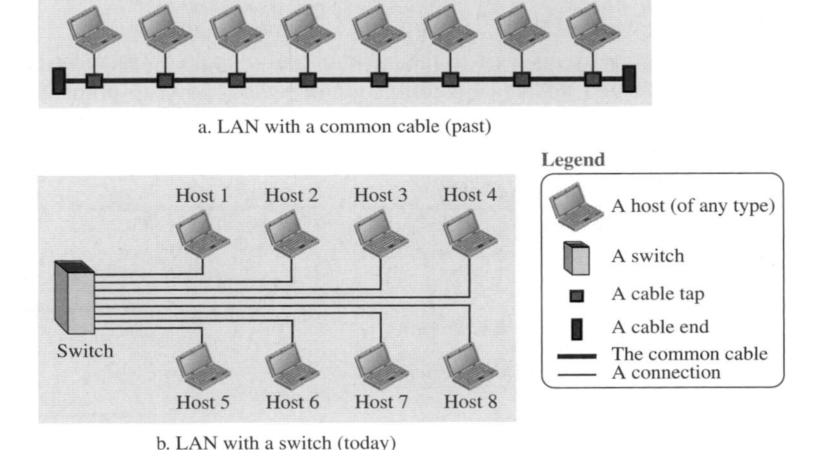

LANs are discussed in more detail in Part III of the book.

When LANs were used in isolation (which is rare today), they were designed to allow resources to be shared between the hosts. As we will see shortly, LANs today are connected to each other and to WANs (discussed next) to create communication at a wider level.

1.3.2 Wide Area Network

A **wide area network (WAN)** is also an interconnection of devices capable of communication. However, there are some differences between a LAN and a WAN. A LAN is normally limited in size, spanning an office, a building, or a campus; a WAN has a wider geographical span, spanning a town, a state, a country, or even the world. A LAN interconnects hosts; a WAN interconnects connecting devices such as switches, routers, or modems. A LAN is normally privately owned by the organization that uses it; a WAN is normally created and run by communication companies and leased by an organization that uses it. We see two distinct examples of WANs today: point-to-point WANs and switched WANs.

Point-to-Point WAN

A point-to-point WAN is a network that connects two communicating devices through a transmission media (cable or air). We will see examples of these WANs when we discuss how to connect the networks to one another. Figure 1.9 shows an example of a point-to-point WAN.

Figure 1.9 *A point-to-point WAN*

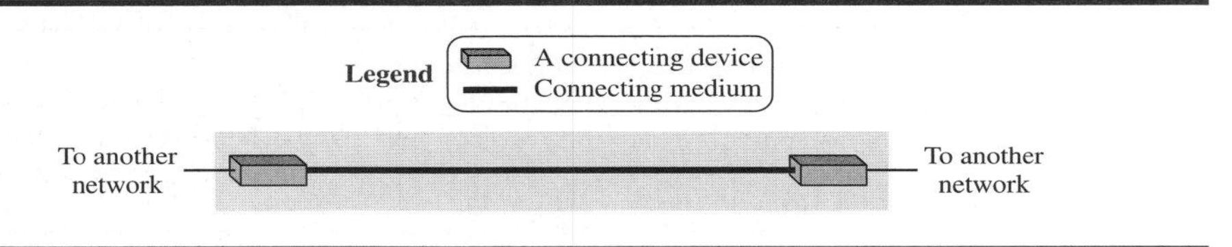

Switched WAN

A switched WAN is a network with more than two ends. A switched WAN, as we will see shortly, is used in the backbone of global communication today. We can say that a switched WAN is a combination of several point-to-point WANs that are connected by switches. Figure 1.10 shows an example of a switched WAN.

Figure 1.10 *A switched WAN*

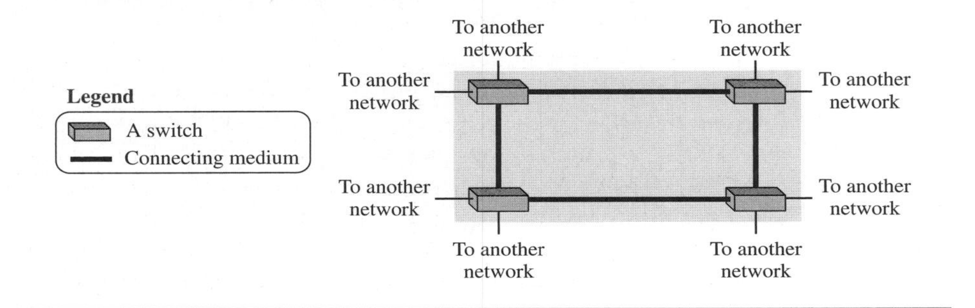

WANs are discussed in more detail in Part II of the book.

Internetwork

Today, it is very rare to see a LAN or a WAN in isolation; they are connected to one another. When two or more networks are connected, they make an **internetwork,** or **internet.** As an example, assume that an organization has two offices, one on the east coast and the other on the west coast. Each office has a LAN that allows all employees in the office to communicate with each other. To make the communication between employees at different offices possible, the management leases a point-to-point dedicated WAN from a service provider, such as a telephone company, and connects the two LANs. Now the company has an internetwork, or a private internet (with lowercase *i*). Communication between offices is now possible. Figure 1.11 shows this internet.

Figure 1.11 *An internetwork made of two LANs and one point-to-point WAN*

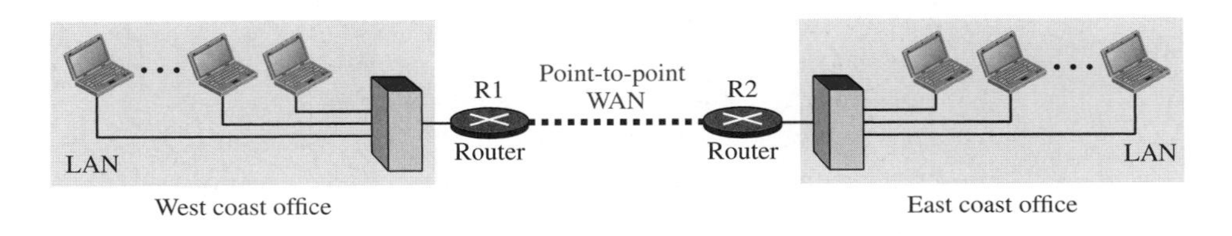

When a host in the west coast office sends a message to another host in the same office, the router blocks the message, but the switch directs the message to the destination. On the other hand, when a host on the west coast sends a message to a host on the east coast, router R1 routes the packet to router R2, and the packet reaches the destination.

Figure 1.12 (see next page) shows another internet with several LANs and WANs connected. One of the WANs is a switched WAN with four switches.

1.3.3 Switching

An internet is a **switched network** in which a switch connects at least two links together. A switch needs to forward data from a network to another network when required. The two most common types of switched networks are circuit-switched and packet-switched networks. We discuss both next.

Circuit-Switched Network

In a **circuit-switched network,** a dedicated connection, called a circuit, is always available between the two end systems; the switch can only make it active or inactive. Figure 1.13 shows a very simple switched network that connects four telephones to each end. We have used telephone sets instead of computers as an end system because circuit switching was very common in telephone networks in the past, although part of the telephone network today is a packet-switched network.

In Figure 1.13, the four telephones at each side are connected to a switch. The switch connects a telephone set at one side to a telephone set at the other side. The thick

Figure 1.12 *A heterogeneous network made of four WANs and three LANs*

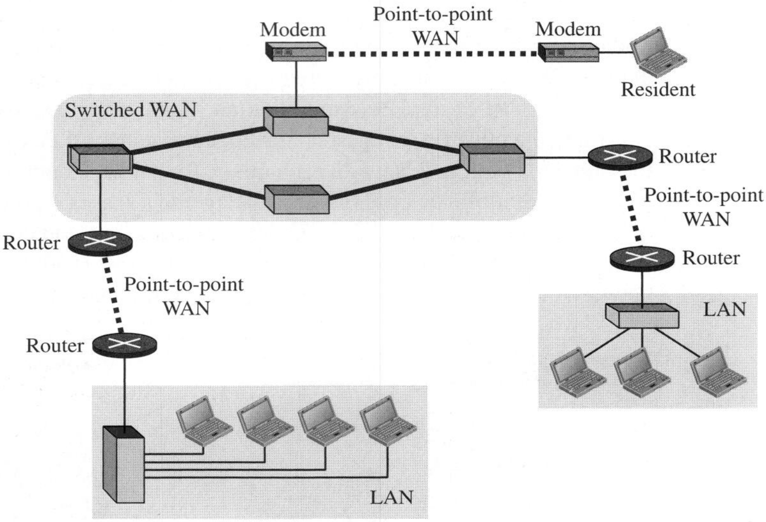

Figure 1.13 *A circuit-switched network*

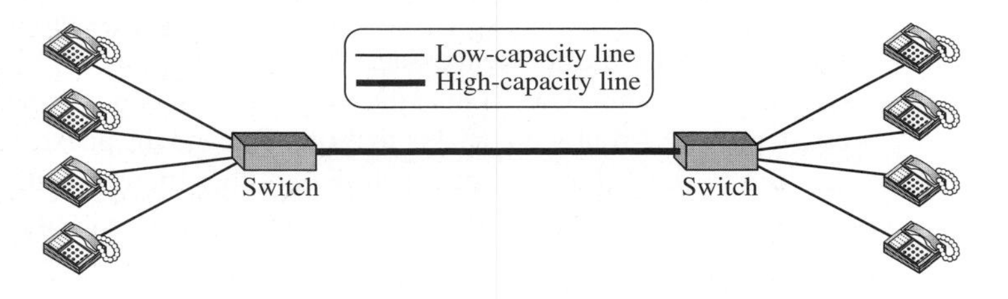

line connecting two switches is a high-capacity communication line that can handle four voice communications at the same time; the capacity can be shared between all pairs of telephone sets. The switches used in this example have forwarding tasks but no storing capability.

Let us look at two cases. In the first case, all telephone sets are busy; four people at one site are talking with four people at the other site; the capacity of the thick line is fully used. In the second case, only one telephone set at one side is connected to a telephone set at the other side; only one-fourth of the capacity of the thick line is used. This means that a circuit-switched network is efficient only when it is working at its full capacity; most of the time, it is inefficient because it is working at partial capacity. The reason that we need to make the capacity of the thick line four times the capacity of each voice line is that we do not want communication to fail when all telephone sets at one side want to be connected with all telephone sets at the other side.

Packet-Switched Network

In a computer network, the communication between the two ends is done in blocks of data called **packets.** In other words, instead of the continuous communication we see between two telephone sets when they are being used, we see the exchange of individual data packets between the two computers. This allows us to make the switches function for both storing and forwarding because a packet is an independent entity that can be stored and sent later. Figure 1.14 shows a small packet-switched network that connects four computers at one site to four computers at the other site.

Figure 1.14 *A packet-switched network*

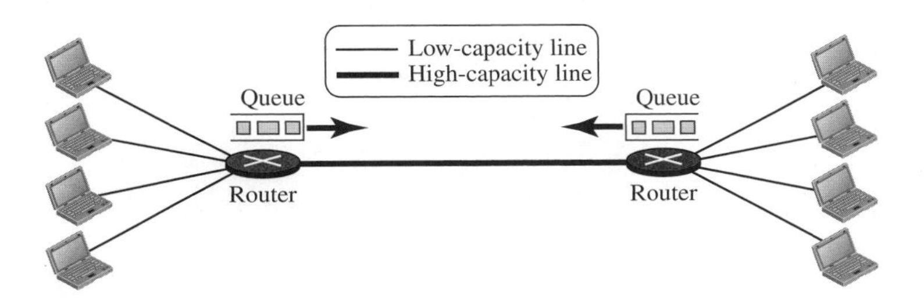

A router in a packet-switched network has a queue that can store and forward the packet. Now assume that the capacity of the thick line is only twice the capacity of the data line connecting the computers to the routers. If only two computers (one at each site) need to communicate with each other, there is no waiting for the packets. However, if packets arrive at one router when the thick line is already working at its full capacity, the packets should be stored and forwarded in the order they arrived. The two simple examples show that a packet-switched network is more efficient than a circuit-switched network, but the packets may encounter some delays.

In this book, we mostly discuss packet-switched networks. In Chapter 18, we discuss packet-switched networks in more detail and discuss the performance of these networks.

1.3.4 The Internet

As we discussed before, an internet (note the lowercase *i*) is two or more networks that can communicate with each other. The most notable internet is called the **Internet** (uppercase *I*), and is composed of thousands of interconnected networks. Figure 1.15 shows a conceptual (not geographical) view of the Internet.

The figure shows the Internet as several backbones, provider networks, and customer networks. At the top level, the *backbones* are large networks owned by some communication companies such as Sprint, Verizon (MCI), AT&T, and NTT. The backbone networks are connected through some complex switching systems, called *peering points*. At the second level, there are smaller networks, called *provider networks*, that use the services of the backbones for a fee. The provider networks are connected to backbones and sometimes to other provider networks. The *customer networks* are

Figure 1.15 *The Internet today*

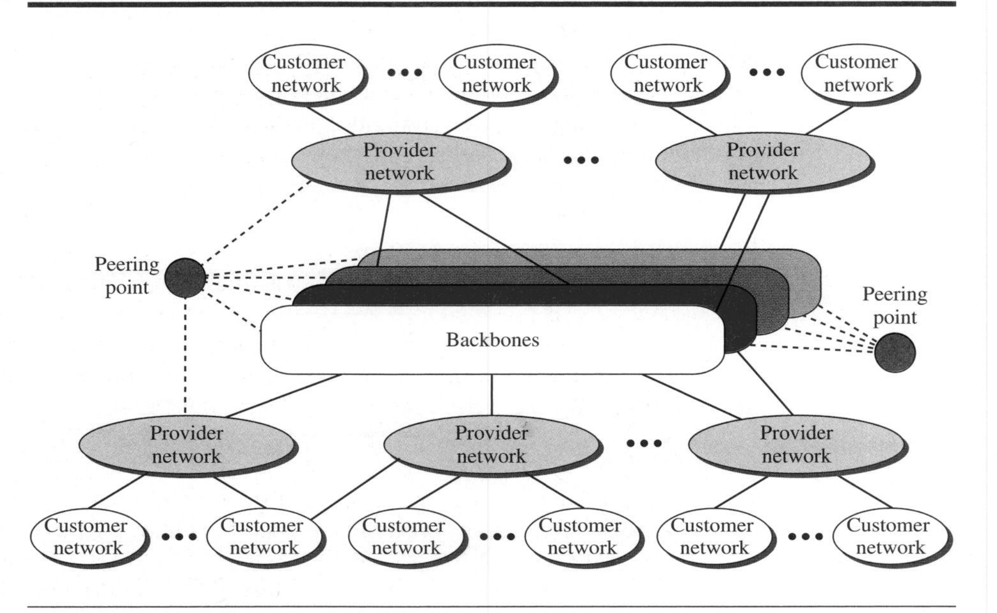

networks at the edge of the Internet that actually use the services provided by the Internet. They pay fees to provider networks for receiving services.

Backbones and provider networks are also called **Internet Service Providers (ISPs).** The backbones are often referred to as *international ISPs;* the provider networks are often referred to as *national* or *regional ISPs*.

1.3.5 Accessing the Internet

The Internet today is an internetwork that allows any user to become part of it. The user, however, needs to be physically connected to an ISP. The physical connection is normally done through a point-to-point WAN. In this section, we briefly describe how this can happen, but we postpone the technical details of the connection until Chapters 14 and 16.

Using Telephone Networks

Today most residences and small businesses have telephone service, which means they are connected to a telephone network. Since most telephone networks have already connected themselves to the Internet, one option for residences and small businesses to connect to the Internet is to change the voice line between the residence or business and the telephone center to a point-to-point WAN. This can be done in two ways.

- ❑ ***Dial-up service.*** The first solution is to add to the telephone line a modem that converts data to voice. The software installed on the computer dials the ISP and imitates making a telephone connection. Unfortunately, the dial-up service is very slow, and when the line is used for Internet connection, it cannot be used for telephone (voice) connection. It is only useful for small residences. We discuss dial-up service in Chapter 14.

- ***DSL Service.*** Since the advent of the Internet, some telephone companies have upgraded their telephone lines to provide higher speed Internet services to residences or small businesses. The DSL service also allows the line to be used simultaneously for voice and data communication. We discuss DSL in Chapter 14.

Using Cable Networks

More and more residents over the last two decades have begun using cable TV services instead of antennas to receive TV broadcasting. The cable companies have been upgrading their cable networks and connecting to the Internet. A residence or a small business can be connected to the Internet by using this service. It provides a higher speed connection, but the speed varies depending on the number of neighbors that use the same cable. We discuss the cable networks in Chapter 14.

Using Wireless Networks

Wireless connectivity has recently become increasingly popular. A household or a small business can use a combination of wireless and wired connections to access the Internet. With the growing wireless WAN access, a household or a small business can be connected to the Internet through a wireless WAN. We discuss wireless access in Chapter 16.

Direct Connection to the Internet

A large organization or a large corporation can itself become a local ISP and be connected to the Internet. This can be done if the organization or the corporation leases a high-speed WAN from a carrier provider and connects itself to a regional ISP. For example, a large university with several campuses can create an internetwork and then connect the internetwork to the Internet.

1.4 INTERNET HISTORY

Now that we have given an overview of the Internet, let us give a brief history of the Internet. This brief history makes it clear how the Internet has evolved from a private network to a global one in less than 40 years.

1.4.1 Early History

There were some communication networks, such as telegraph and telephone networks, before 1960. These networks were suitable for constant-rate communication at that time, which means that after a connection was made between two users, the encoded message (telegraphy) or voice (telephony) could be exchanged. A computer network, on the other hand, should be able to handle *bursty* data, which means data received at variable rates at different times. The world needed to wait for the packet-switched network to be invented.

Birth of Packet-Switched Networks

The theory of packet switching for bursty traffic was first presented by Leonard Kleinrock in 1961 at MIT. At the same time, two other researchers, Paul Baran at Rand Institute and Donald Davies at National Physical Laboratory in England, published some papers about packet-switched networks.

ARPANET

In the mid-1960s, mainframe computers in research organizations were stand-alone devices. Computers from different manufacturers were unable to communicate with one another. The **Advanced Research Projects Agency (ARPA)** in the Department of Defense (DOD) was interested in finding a way to connect computers so that the researchers they funded could share their findings, thereby reducing costs and eliminating duplication of effort.

In 1967, at an Association for Computing Machinery (ACM) meeting, ARPA presented its ideas for the **Advanced Research Projects Agency Network (ARPANET),** a small network of connected computers. The idea was that each host computer (not necessarily from the same manufacturer) would be attached to a specialized computer, called an *interface message processor* (IMP). The IMPs, in turn, would be connected to each other. Each IMP had to be able to communicate with other IMPs as well as with its own attached host.

By 1969, ARPANET was a reality. Four nodes, at the University of California at Los Angeles (UCLA), the University of California at Santa Barbara (UCSB), Stanford Research Institute (SRI), and the University of Utah, were connected via the IMPs to form a network. Software called the *Network Control Protocol* (NCP) provided communication between the hosts.

1.4.2 Birth of the Internet

In 1972, Vint Cerf and Bob Kahn, both of whom were part of the core ARPANET group, collaborated on what they called the *Internetting Project*. They wanted to link dissimilar networks so that a host on one network could communicate with a host on another. There were many problems to overcome: diverse packet sizes, diverse interfaces, and diverse transmission rates, as well as differing reliability requirements. Cerf and Kahn devised the idea of a device called a *gateway* to serve as the intermediary hardware to transfer data from one network to another.

TCP/IP

Cerf and Kahn's landmark 1973 paper outlined the protocols to achieve end-to-end delivery of data. This was a new version of NCP. This paper on transmission control protocol (TCP) included concepts such as encapsulation, the datagram, and the functions of a gateway. A radical idea was the transfer of responsibility for error correction from the IMP to the host machine. This ARPA Internet now became the focus of the communication effort. Around this time, responsibility for the ARPANET was handed over to the Defense Communication Agency (DCA).

In October 1977, an internet consisting of three different networks (ARPANET, packet radio, and packet satellite) was successfully demonstrated. Communication between networks was now possible.

Shortly thereafter, authorities made a decision to split TCP into two protocols: **Transmission Control Protocol (TCP)** and **Internet Protocol (IP).** IP would handle datagram routing while TCP would be responsible for higher level functions such as segmentation, reassembly, and error detection. The new combination became known as TCP/IP.

In 1981, under a Defence Department contract, UC Berkeley modified the UNIX operating system to include TCP/IP. This inclusion of network software along with a popular operating system did much for the popularity of internetworking. The open (non-manufacturer-specific) implementation of the Berkeley UNIX gave every manufacturer a working code base on which they could build their products.

In 1983, authorities abolished the original ARPANET protocols, and TCP/IP became the official protocol for the ARPANET. Those who wanted to use the Internet to access a computer on a different network had to be running TCP/IP.

MILNET

In 1983, ARPANET split into two networks: **Military Network (MILNET)** for military users and ARPANET for nonmilitary users.

CSNET

Another milestone in Internet history was the creation of CSNET in 1981. **Computer Science Network (CSNET)** was a network sponsored by the National Science Foundation (NSF). The network was conceived by universities that were ineligible to join ARPANET due to an absence of ties to the Department of Defense. CSNET was a less expensive network; there were no redundant links and the transmission rate was slower.

By the mid-1980s, most U.S. universities with computer science departments were part of CSNET. Other institutions and companies were also forming their own networks and using TCP/IP to interconnect. The term *Internet,* originally associated with government-funded connected networks, now referred to the connected networks using TCP/IP protocols.

NSFNET

With the success of CSNET, the NSF in 1986 sponsored the **National Science Foundation Network (NSFNET),** a backbone that connected five supercomputer centers located throughout the United States. Community networks were allowed access to this backbone, a T-1 line (see Chapter 6) with a 1.544-Mbps data rate, thus providing connectivity throughout the United States. In 1990, ARPANET was officially retired and replaced by NSFNET. In 1995, NSFNET reverted back to its original concept of a research network.

ANSNET

In 1991, the U.S. government decided that NSFNET was not capable of supporting the rapidly increasing Internet traffic. Three companies, IBM, Merit, and Verizon, filled the void by forming a nonprofit organization called Advanced Network & Services (ANS) to build a new, high-speed Internet backbone called **Advanced Network Services Network (ANSNET).**

1.4.3 Internet Today

Today, we witness a rapid growth both in the infrastructure and new applications. The Internet today is a set of pier networks that provide services to the whole world. What has made the Internet so popular is the invention of new applications.

World Wide Web

The 1990s saw the explosion of Internet applications due to the emergence of the World Wide Web (WWW). The Web was invented at CERN by Tim Berners-Lee. This invention has added the commercial applications to the Internet.

Multimedia

Recent developments in the multimedia applications such as voice over IP (telephony), video over IP (Skype), view sharing (YouTube), and television over IP (PPLive) has increased the number of users and the amount of time each user spends on the network. We discuss multimedia in Chapter 28.

Peer-to-Peer Applications

Peer-to-peer networking is also a new area of communication with a lot of potential. We introduce some peer-to-peer applications in Chapter 29.

1.5 STANDARDS AND ADMINISTRATION

In the discussion of the Internet and its protocol, we often see a reference to a standard or an administration entity. In this section, we introduce these standards and administration entities for those readers that are not familiar with them; the section can be skipped if the reader is familiar with them.

1.5.1 Internet Standards

An **Internet standard** is a thoroughly tested specification that is useful to and adhered to by those who work with the Internet. It is a formalized regulation that must be followed. There is a strict procedure by which a specification attains Internet standard status. A specification begins as an Internet draft. An **Internet draft** is a working document (a work in progress) with no official status and a six-month lifetime. Upon recommendation from the Internet authorities, a draft may be published as a **Request for Comment (RFC).** Each RFC is edited, assigned a number, and made available to all interested parties. RFCs go through maturity levels and are categorized according to their requirement level.

Maturity Levels

An RFC, during its lifetime, falls into one of six *maturity levels:* proposed standard, draft standard, Internet standard, historic, experimental, and informational (see Figure 1.16).

- ❑ ***Proposed Standard.*** A proposed standard is a specification that is stable, well understood, and of sufficient interest to the Internet community. At this level, the specification is usually tested and implemented by several different groups.

Figure 1.16 *Maturity levels of an RFC*

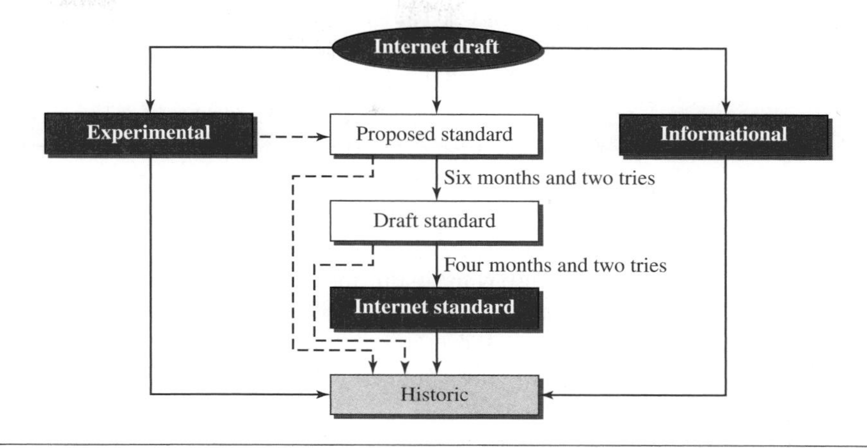

- ***Draft Standard.*** A proposed standard is elevated to draft standard status after at least two successful independent and interoperable implementations. Barring difficulties, a draft standard, with modifications if specific problems are encountered, normally becomes an Internet standard.
- ***Internet Standard.*** A draft standard reaches Internet standard status after demonstrations of successful implementation.
- ***Historic.*** The historic RFCs are significant from a historical perspective. They either have been superseded by later specifications or have never passed the necessary maturity levels to become an Internet standard.
- ***Experimental.*** An RFC classified as experimental describes work related to an experimental situation that does not affect the operation of the Internet. Such an RFC should not be implemented in any functional Internet service.
- ***Informational.*** An RFC classified as informational contains general, historical, or tutorial information related to the Internet. It is usually written by someone in a non-Internet organization, such as a vendor.

Requirement Levels

RFCs are classified into five *requirement levels:* required, recommended, elective, limited use, and not recommended.

- ***Required.*** An RFC is labeled *required* if it must be implemented by all Internet systems to achieve minimum conformance. For example, IP and ICMP (Chapter 19) are required protocols.
- ***Recommended.*** An RFC labeled recommended is not required for minimum conformance; it is recommended because of its usefulness. For example, FTP (Chapter 26) and TELNET (Chapter 26) are recommended protocols.
- ***Elective.*** An RFC labeled elective is not required and not recommended. However, a system can use it for its own benefit.

- ❑ ***Limited Use.*** An RFC labeled limited use should be used only in limited situations. Most of the experimental RFCs fall under this category.
- ❑ ***Not Recommended.*** An RFC labeled not recommended is inappropriate for general use. Normally a historic (deprecated) RFC may fall under this category.

RFCs can be found at http://www.rfc-editor.org.

1.5.2 Internet Administration

The Internet, with its roots primarily in the research domain, has evolved and gained a broader user base with significant commercial activity. Various groups that coordinate Internet issues have guided this growth and development. Appendix G gives the addresses, e-mail addresses, and telephone numbers for some of these groups. Figure 1.17 shows the general organization of Internet administration.

Figure 1.17 *Internet administration*

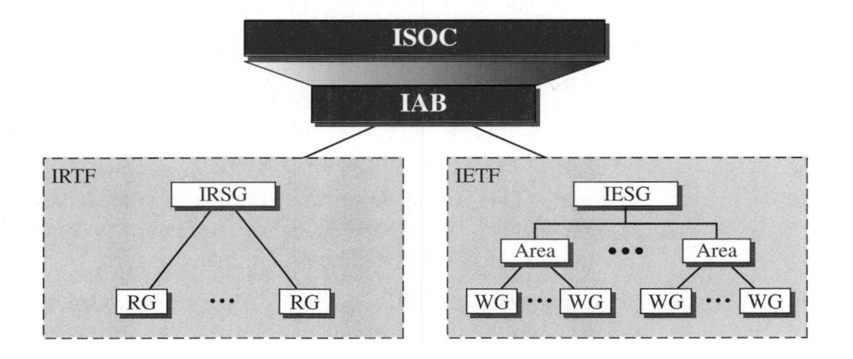

ISOC

The **Internet Society (ISOC)** is an international, nonprofit organization formed in 1992 to provide support for the Internet standards process. ISOC accomplishes this through maintaining and supporting other Internet administrative bodies such as IAB, IETF, IRTF, and IANA (see the following sections). ISOC also promotes research and other scholarly activities relating to the Internet.

IAB

The **Internet Architecture Board (IAB)** is the technical advisor to the ISOC. The main purposes of the IAB are to oversee the continuing development of the TCP/IP Protocol Suite and to serve in a technical advisory capacity to research members of the Internet community. IAB accomplishes this through its two primary components, the Internet Engineering Task Force (IETF) and the Internet Research Task Force (IRTF). Another responsibility of the IAB is the editorial management of the RFCs, described earlier. IAB is also the external liaison between the Internet and other standards organizations and forums.

IETF

The **Internet Engineering Task Force (IETF)** is a forum of working groups managed by the Internet Engineering Steering Group (IESG). IETF is responsible for identifying operational problems and proposing solutions to these problems. IETF also develops and reviews specifications intended as Internet standards. The working groups are collected into areas, and each area concentrates on a specific topic. Currently nine areas have been defined. The areas include applications, protocols, routing, network management next generation (IPng), and security.

IRTF

The **Internet Research Task Force (IRTF)** is a forum of working groups managed by the Internet Research Steering Group (IRSG). IRTF focuses on long-term research topics related to Internet protocols, applications, architecture, and technology.

1.6 END-CHAPTER MATERIALS

1.6.1 Recommended Reading

For more details about subjects discussed in this chapter, we recommend the following books. The items enclosed in brackets [. . .] refer to the reference list at the end of the book.

Books

The introductory materials covered in this chapter can be found in [Sta04] and [PD03]. [Tan03] also discusses standardization.

1.6.2 Key Terms

Advanced Network Services Network (ANSNET)
Advanced Research Projects Agency (ARPA)
Advanced Research Projects Agency Network (ARPANET)
American Standard Code for Information Interchange (ASCII)
audio
backbone
Basic Latin
bus topology
circuit-switched network
code
Computer Science Network (CSNET)
data
data communications
delay
full-duplex mode
half-duplex mode
hub
image
internet
Internet
Internet Architecture Board (IAB)
Internet draft
Internet Engineering Task Force (IETF)
Internet Research Task Force (IRTF)
Internet Service Provider (ISP)
Internet Society (ISOC)
Internet standard
internetwork
local area network (LAN)
mesh topology
message

- Military Network (MILNET)
- multipoint or multidrop connection
- National Science Foundation Network (NSFNET)
- network
- node
- packet
- packet-switched network
- performance
- physical topology
- point-to-point connection
- protocol
- Request for Comment (RFC)
- RGB
- ring topology
- simplex mode
- star topology
- switched network
- TCP/IP protocol suite
- telecommunication
- throughput
- Transmission Control Protocol/ Internet Protocol (TCP/IP)
- transmission medium
- Unicode
- video
- wide area network (WAN)
- YCM

1.6.3 Summary

Data communications are the transfer of data from one device to another via some form of transmission medium. A data communications system must transmit data to the correct destination in an accurate and timely manner. The five components that make up a data communications system are the message, sender, receiver, medium, and protocol. Text, numbers, images, audio, and video are different forms of information. Data flow between two devices can occur in one of three ways: simplex, half-duplex, or full-duplex.

A network is a set of communication devices connected by media links. In a point-to-point connection, two and only two devices are connected by a dedicated link. In a multipoint connection, three or more devices share a link. Topology refers to the physical or logical arrangement of a network. Devices may be arranged in a mesh, star, bus, or ring topology.

A network can be categorized as a local area network or a wide area network. A LAN is a data communication system within a building, plant, or campus, or between nearby buildings. A WAN is a data communication system spanning states, countries, or the whole world. An internet is a network of networks. The Internet is a collection of many separate networks.

The Internet history started with the theory of packet switching for bursty traffic. The history continued when The ARPA was interested in finding a way to connect computers so that the researchers they funded could share their findings, resulting in the creation of ARPANET. The Internet was born when Cerf and Kahn devised the idea of a device called a *gateway* to serve as the intermediary hardware to transfer data from one network to another. The TCP/IP protocol suite paved the way for creation of today's Internet. The invention of WWW, the use of multimedia, and peer-to-peer communication helps the growth of the Internet.

An Internet standard is a thoroughly tested specification. An Internet draft is a working document with no official status and a six-month lifetime. A draft may be published as a Request for Comment (RFC). RFCs go through maturity levels and are categorized according to their requirement level. The Internet administration has evolved with the Internet. ISOC promotes research and activities. IAB is the technical advisor to the ISOC. IETF is a forum of working groups responsible for operational problems. IRTF is a forum of working groups focusing on long-term research topics.

1.7 PRACTICE SET

1.7.1 Quizzes

A set of interactive quizzes for this chapter can be found on the book website. It is strongly recommended that the student take the quizzes to check his/her understanding of the materials before continuing with the practice set.

1.7.2 Questions

Q1-1. Why are protocols needed?

Q1-2. What are the two types of line configuration?

Q1-3. Name the four basic network topologies, and cite an advantage of each type.

Q1-4. Explain the difference between a required RFC and a recommended RFC.

Q1-5. When a resident uses a dial-up or DLS service to connect to the Internet, what is the role of the telephone company?

Q1-6. What is an internet? What is the Internet?

Q1-7. What are the advantages of a multipoint connection over a point-to-point one?

Q1-8. What is the first principle we discussed in this chapter for protocol layering that needs to be followed to make the communication bidirectional?

Q1-9. Identify the five components of a data communications system.

Q1-10. What is the difference between half-duplex and full-duplex transmission modes?

Q1-11. Categorize the four basic topologies in terms of line configuration.

Q1-12. What are the three criteria necessary for an effective and efficient network?

Q1-13. How many point-to-point WANs are needed to connect n LANs if each LAN should be able to directly communicate with any other LAN?

Q1-14. For n devices in a network, what is the number of cable links required for a mesh, ring, bus, and star topology?

Q1-15. Explain the difference between an Internet draft and a proposed standard.

Q1-16. When we use local telephones to talk to a friend, are we using a circuit-switched network or a packet-switched network?

Q1-17. What are some of the factors that determine whether a communication system is a LAN or WAN?

Q1-18. In a LAN with a link-layer switch (Figure 1.8b), Host 1 wants to send a message to Host 3. Since communication is through the link-layer switch, does the switch need to have an address? Explain.

Q1-19. Explain the difference between the duties of the IETF and IRTF.

1.7.3 Problems

P1-1. In the bus topology in Figure 1.6, what happens if one of the stations is unplugged?

P1-2. Compare the telephone network and the Internet. What are the similarities? What are the differences?

P1-3. When a party makes a local telephone call to another party, is this a point-to-point or multipoint connection? Explain the answer.

P1-4. A color image uses 24 bits to represent a pixel. What is the maximum number of different colors that can be represented?

P1-5. What is the maximum number of characters or symbols that can be represented by Unicode?

P1-6. For each of the following four networks, discuss the consequences if a connection fails.

a. Seven devices arranged in a mesh topology
b. Seven devices arranged in a star topology (not counting the hub)
c. Seven devices arranged in a bus topology
d. Seven devices arranged in a ring topology

P1-7. Assume eight devices are arranged in a mesh topology. How many cables are needed? How many ports are needed for each device?

P1-8. Performance is inversely related to delay. When we use the Internet, which of the following applications are more sensitive to delay?

a. Sending an e-mail
b. Copying a file
c. Surfing the Internet

P1-9. We have two computers connected by an Ethernet hub at home. Is this a LAN or a WAN? Explain the reason.

P1-10. In the ring topology in Figure 1.7, what happens if one of the stations is unplugged?

1.8 SIMULATION EXPERIMENTS

1.8.4 Applets

One of the ways to show the network protocols in action or visually see the solution to some examples is through the use of interactive animation. We have created some Java applets to show some of the main concepts discussed in this chapter. It is strongly recommended that the students activate these applets on the book website and carefully examine the protocols in action. However, note that applets have been created only for some chapters, not all (see the book website).

1.8.5 Lab Assignments

Experiments with networks and network equipment can be done using at least two methods. In the first method, we can create an isolated networking laboratory and use networking hardware and software to simulate the topics discussed in each chapter. We can create an internet and send and receive packets from any host to another. The flow of packets can be observed and the performance can be measured. Although the first method is more effective and more instructional, it is expensive to implement and not all institutions are ready to invest in such an exclusive laboratory.

In the second method, we can use the Internet, the largest network in the world, as our virtual laboratory. We can send and receive packets using the Internet. The existence of some free-downloadable software allows us to capture and examine the packets exchanged. We can analyze the packets to see how theoretical aspects of networking are put into action. Although the second method may not be as effective as the first method, in that we cannot control and change the packet routes to see how the Internet behaves, the method is much cheaper to implement. It does not need a physical networking lab; it can be implemented using our desktop or laptop. The required software is also free to download.

There are many programs and utilities available for Windows and UNIX operating systems that allow us to sniff, capture, trace, and analyze packets that are exchanged between our computer and the Internet. Some of these, such as *Wireshark* and *Ping-Plotter,* have graphical user interface (GUI); others, such as *traceroute, nslookup*, *dig*, *ipconfig,* and *ifconfig,* are network administration command-line utilities. Any of these programs and utilities can be a valuable debugging tool for network administrators and educational tool for computer network students.

In this book, we mostly use Wireshark for lab assignments, although we occasionally use other tools. It captures live packet data from a network interface and displays them with detailed protocol information. Wireshark, however, is a passive analyzer. It only "measures" things from the network without manipulating them; it doesn't send packets on the network or perform other active operations. Wireshark is not an intrusion detection tool either. It does not give warning about any network intrusion. It, nevertheless, can help network administrators or network security engineers to figure out what is going on inside a network and to troubleshoot network problems. In addition to being an indispensable tool for network administrators and security engineers, Wireshark is a valuable tool for protocol developers, who may use it to debug protocol implementations, and a great educational tool for computer networking students who can use it to see details of protocol operations in real time. However, note that we can use lab assignments only with a few chapters.

Lab1-1. In this lab assignment we learn how to download and install Wireshark. The instructions for downloading and installing the software are posted on the book website in the lab section for Chapter 1. In this document, we also discuss the general idea behind the software, the format of its window, and how to use it. The full study of this lab prepares the student to use Wireshark in the lab assignments for other chapters.

CHAPTER 2

Network Models

The second chapter is a preparation for the rest of the book. The next five parts of the book is devoted to one of the layers in the TCP/IP protocol suite. In this chapter, we first discuss the idea of network models in general and the TCP/IP protocol suite in particular.

Two models have been devised to define computer network operations: the TCP/IP protocol suite and the OSI model. In this chapter, we first discuss a general subject, protocol layering, which is used in both models. We then concentrate on the TCP/IP protocol suite, on which the book is based. The OSI model is briefly discuss for comparison with the TCP/IP protocol suite.

- The first section introduces the concept of protocol layering using two scenarios. The section also discusses the two principles upon which the protocol layering is based. The first principle dictates that each layer needs to have two opposite tasks. The second principle dictates that the corresponding layers should be identical. The section ends with a brief discussion of logical connection between two identical layers in protocol layering. Throughout the book, we need to distinguish between logical and physical connections.
- The second section discusses the five layers of the TCP/IP protocol suite. We show how packets in each of the five layers (physical, data-link, network, transport, and application) are named. We also mention the addressing mechanism used in each layer. Each layer of the TCP/IP protocol suite is a subject of a part of the book. In other words, each layer is discussed in several chapters; this section is just an introduction and preparation.
- The third section gives a brief discussion of the OSI model. This model was never implemented in practice, but a brief discussion of the model and its comparison with the TCP/IP protocol suite may be useful to better understand the TCP/IP protocol suite. In this section we also give a brief reason for the OSI model's lack of success.

2.1 PROTOCOL LAYERING

We defined the term *protocol* in Chapter 1. In data communication and networking, a protocol defines the rules that both the sender and receiver and all intermediate devices need to follow to be able to communicate effectively. When communication is simple, we may need only one simple protocol; when the communication is complex, we may need to divide the task between different layers, in which case we need a protocol at each layer, or **protocol layering.**

2.1.1 Scenarios

Let us develop two simple scenarios to better understand the need for protocol layering.

First Scenario

In the first scenario, communication is so simple that it can occur in only one layer. Assume Maria and Ann are neighbors with a lot of common ideas. Communication between Maria and Ann takes place in one layer, face to face, in the same language, as shown in Figure 2.1.

Figure 2.1 *A single-layer protocol*

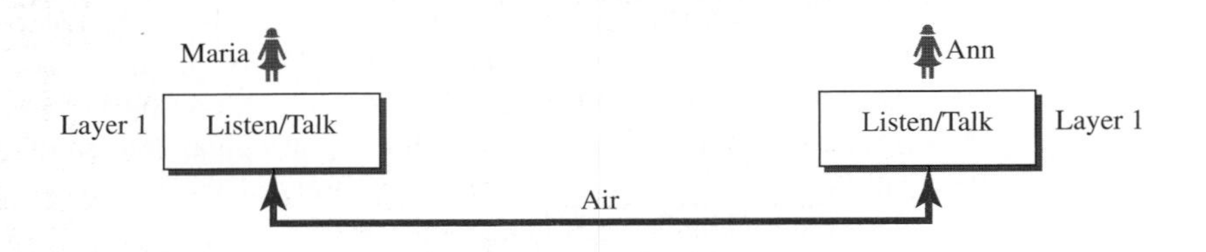

Even in this simple scenario, we can see that a set of rules needs to be followed. First, Maria and Ann know that they should greet each other when they meet. Second, they know that they should confine their vocabulary to the level of their friendship. Third, each party knows that she should refrain from speaking when the other party is speaking. Fourth, each party knows that the conversation should be a dialog, not a monolog: both should have the opportunity to talk about the issue. Fifth, they should exchange some nice words when they leave.

We can see that the protocol used by Maria and Ann is different from the communication between a professor and the students in a lecture hall. The communication in the second case is mostly monolog; the professor talks most of the time unless a student has a question, a situation in which the protocol dictates that she should raise her hand and wait for permission to speak. In this case, the communication is normally very formal and limited to the subject being taught.

Second Scenario

In the second scenario, we assume that Ann is offered a higher-level position in her company, but needs to move to another branch located in a city very far from Maria. The two friends still want to continue their communication and exchange ideas because they have come up with an innovative project to start a new business when they both retire. They decide to continue their conversation using regular mail through the post office. However, they do not want their ideas to be revealed by other people if the letters are intercepted. They agree on an encryption/decryption technique. The sender of the letter encrypts it to make it unreadable by an intruder; the receiver of the letter decrypts it to get the original letter. We discuss the encryption/decryption methods in Chapter 31, but for the moment we assume that Maria and Ann use one technique that makes it hard to decrypt the letter if one does not have the key for doing so. Now we can say that the communication between Maria and Ann takes place in three layers, as shown in Figure 2.2. We assume that Ann and Maria each have three machines (or robots) that can perform the task at each layer.

Figure 2.2 *A three-layer protocol*

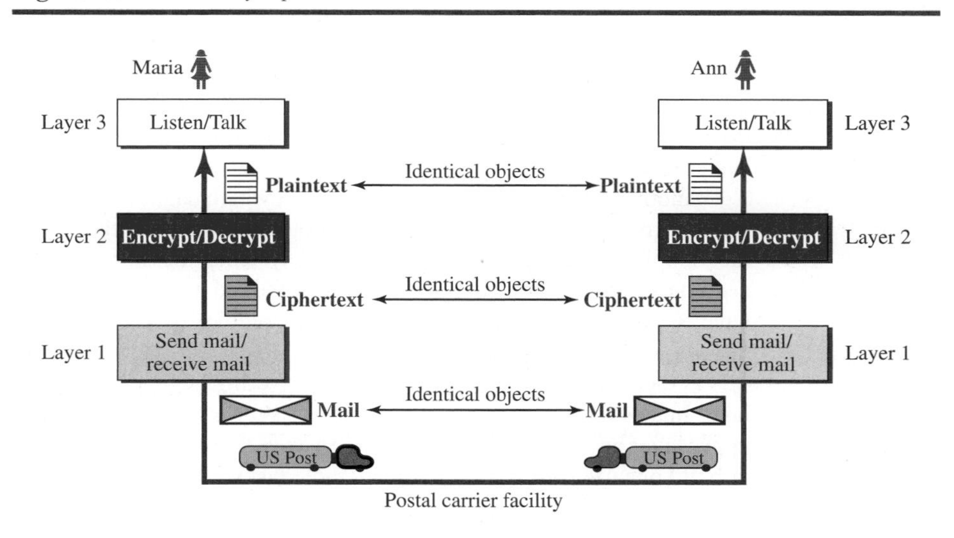

Let us assume that Maria sends the first letter to Ann. Maria talks to the machine at the third layer as though the machine is Ann and is listening to her. The third layer machine listens to what Maria says and creates the plaintext (a letter in English), which is passed to the second layer machine. The second layer machine takes the plaintext, encrypts it, and creates the ciphertext, which is passed to the first layer machine. The first layer machine, presumably a robot, takes the ciphertext, puts it in an envelope, adds the sender and receiver addresses, and mails it.

At Ann's side, the first layer machine picks up the letter from Ann's mail box, recognizing the letter from Maria by the sender address. The machine takes out the ciphertext from the envelope and delivers it to the second layer machine. The second layer machine decrypts the message, creates the plaintext, and passes the plaintext to the third-layer machine. The third layer machine takes the plaintext and reads it as though Maria is speaking.

Protocol layering enables us to divide a complex task into several smaller and simpler tasks. For example, in Figure 2.2, we could have used only one machine to do the job of all three machines. However, if Maria and Ann decide that the encryption/decryption done by the machine is not enough to protect their secrecy, they would have to change the whole machine. In the present situation, they need to change only the second layer machine; the other two can remain the same. This is referred to as *modularity.* Modularity in this case means independent layers. A layer (module) can be defined as a black box with inputs and outputs, without concern about how inputs are changed to outputs. If two machines provide the same outputs when given the same inputs, they can replace each other. For example, Ann and Maria can buy the second layer machine from two different manufacturers. As long as the two machines create the same ciphertext from the same plaintext and vice versa, they do the job.

One of the advantages of protocol layering is that it allows us to separate the services from the implementation. A layer needs to be able to receive a set of services from the lower layer and to give the services to the upper layer; we don't care about how the layer is implemented. For example, Maria may decide not to buy the machine (robot) for the first layer; she can do the job herself. As long as Maria can do the tasks provided by the first layer, in both directions, the communication system works.

Another advantage of protocol layering, which cannot be seen in our simple examples but reveals itself when we discuss protocol layering in the Internet, is that communication does not always use only two end systems; there are intermediate systems that need only some layers, but not all layers. If we did not use protocol layering, we would have to make each intermediate system as complex as the end systems, which makes the whole system more expensive.

Is there any disadvantage to protocol layering? One can argue that having a single layer makes the job easier. There is no need for each layer to provide a service to the upper layer and give service to the lower layer. For example, Ann and Maria could find or build one machine that could do all three tasks. However, as mentioned above, if one day they found that their code was broken, each would have to replace the whole machine with a new one instead of just changing the machine in the second layer.

2.1.2 Principles of Protocol Layering

Let us discuss two principles of protocol layering.

First Principle

The first principle dictates that if we want bidirectional communication, we need to make each layer so that it is able to perform two opposite tasks, one in each direction. For example, the third layer task is to listen (in one direction) and *talk* (in the other direction). The second layer needs to be able to encrypt and decrypt. The first layer needs to send and receive mail.

Second Principle

The second principle that we need to follow in protocol layering is that the two objects under each layer at both sites should be identical. For example, the object under layer 3 at both sites should be a plaintext letter. The object under layer 2 at both sites should be a ciphertext letter. The object under layer 1 at both sites should be a piece of mail.

2.1.3 Logical Connections

After following the above two principles, we can think about logical connection between each layer as shown in Figure 2.3. This means that we have layer-to-layer communication. Maria and Ann can think that there is a logical (imaginary) connection at each layer through which they can send the object created from that layer. We will see that the concept of logical connection will help us better understand the task of layering we encounter in data communication and networking.

Figure 2.3 *Logical connection between peer layers*

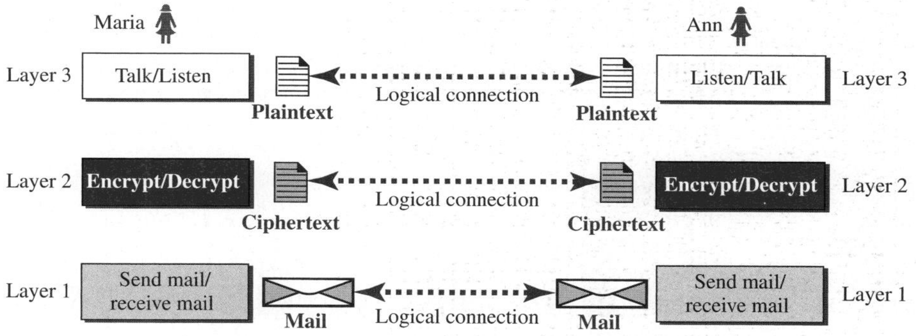

2.2 TCP/IP PROTOCOL SUITE

Now that we know about the concept of protocol layering and the logical communication between layers in our second scenario, we can introduce the TCP/IP (Transmission Control Protocol/Internet Protocol). TCP/IP is a protocol suite (a set of protocols organized in different layers) used in the Internet today. It is a hierarchical protocol made up of interactive modules, each of which provides a specific functionality. The term *hierarchical* means that each upper level protocol is supported by the services provided by one or more lower level protocols. The original TCP/IP protocol suite was defined as four software layers built upon the hardware. Today, however, TCP/IP is thought of as a five-layer model. Figure 2.4 shows both configurations.

2.2.1 Layered Architecture

To show how the layers in the TCP/IP protocol suite are involved in communication between two hosts, we assume that we want to use the suite in a small internet made up of three LANs (links), each with a link-layer switch. We also assume that the links are connected by one router, as shown in Figure 2.5.

Figure 2.4 *Layers in the TCP/IP protocol suite*

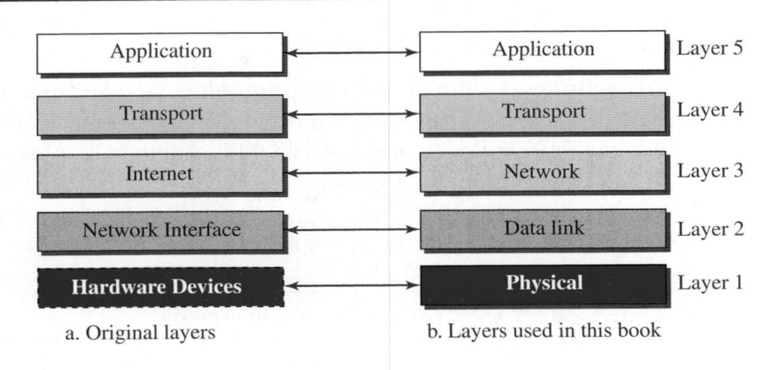

Figure 2.5 *Communication through an internet*

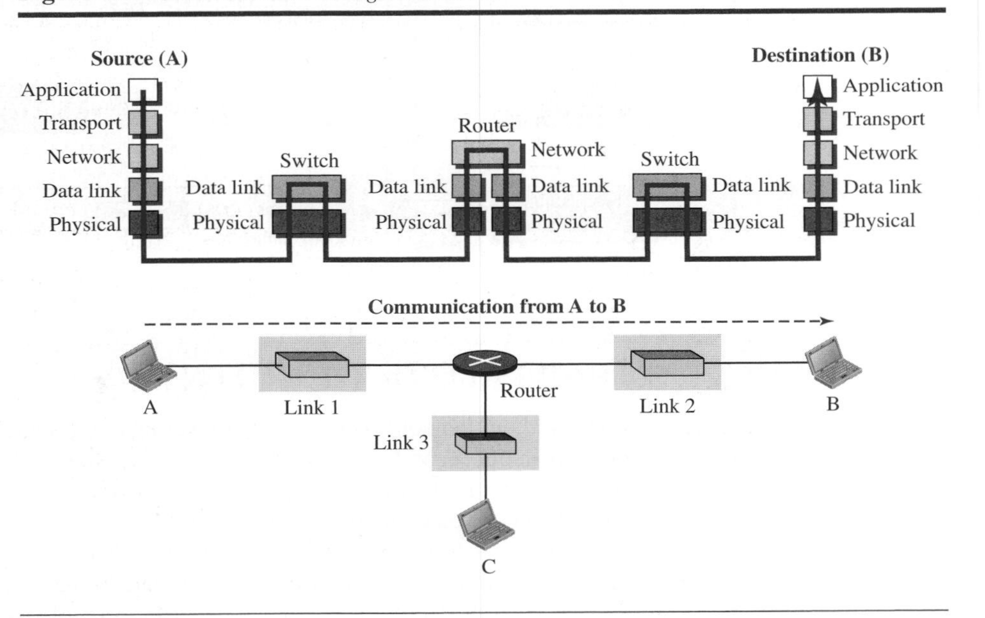

Let us assume that computer A communicates with computer B. As the figure shows, we have five communicating devices in this communication: source host (computer A), the link-layer switch in link 1, the router, the link-layer switch in link 2, and the destination host (computer B). Each device is involved with a set of layers depending on the role of the device in the internet. The two hosts are involved in all five layers; the source host needs to create a message in the application layer and send it down the layers so that it is physically sent to the destination host. The destination host needs to receive the communication at the physical layer and then deliver it through the other layers to the application layer.

The router is involved in only three layers; there is no transport or application layer in a router as long as the router is used only for routing. Although a router is always involved in one network layer, it is involved in n combinations of link and physical layers in which n is the number of links the router is connected to. The reason is that each link may use its own data-link or physical protocol. For example, in the above figure, the router is involved in three links, but the message sent from source A to destination B is involved in two links. Each link may be using different link-layer and physical-layer protocols; the router needs to receive a packet from link 1 based on one pair of protocols and deliver it to link 2 based on another pair of protocols.

A link-layer switch in a link, however, is involved only in two layers, data-link and physical. Although each switch in the above figure has two different connections, the connections are in the same link, which uses only one set of protocols. This means that, unlike a router, a link-layer switch is involved only in one data-link and one physical layer.

2.2.2 Layers in the TCP/IP Protocol Suite

After the above introduction, we briefly discuss the functions and duties of layers in the TCP/IP protocol suite. Each layer is discussed in detail in the next five parts of the book. To better understand the duties of each layer, we need to think about the logical connections between layers. Figure 2.6 shows logical connections in our simple internet.

Figure 2.6 *Logical connections between layers of the TCP/IP protocol suite*

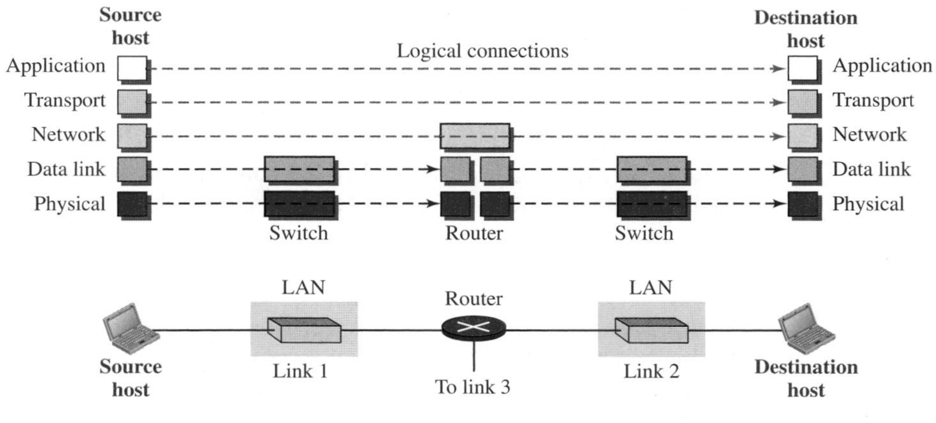

Using logical connections makes it easier for us to think about the duty of each layer. As the figure shows, the duty of the application, transport, and network layers is end-to-end. However, the duty of the data-link and physical layers is hop-to-hop, in which a hop is a host or router. In other words, the domain of duty of the top three layers is the internet, and the domain of duty of the two lower layers is the link.

Another way of thinking of the logical connections is to think about the data unit created from each layer. In the top three layers, the data unit (packets) should not be changed by any router or link-layer switch. In the bottom two layers, the packet created by the host is changed only by the routers, not by the link-layer switches.

Figure 2.7 shows the second principle discussed previously for protocol layering. We show the identical objects below each layer related to each device.

Figure 2.7 *Identical objects in the TCP/IP protocol suite*

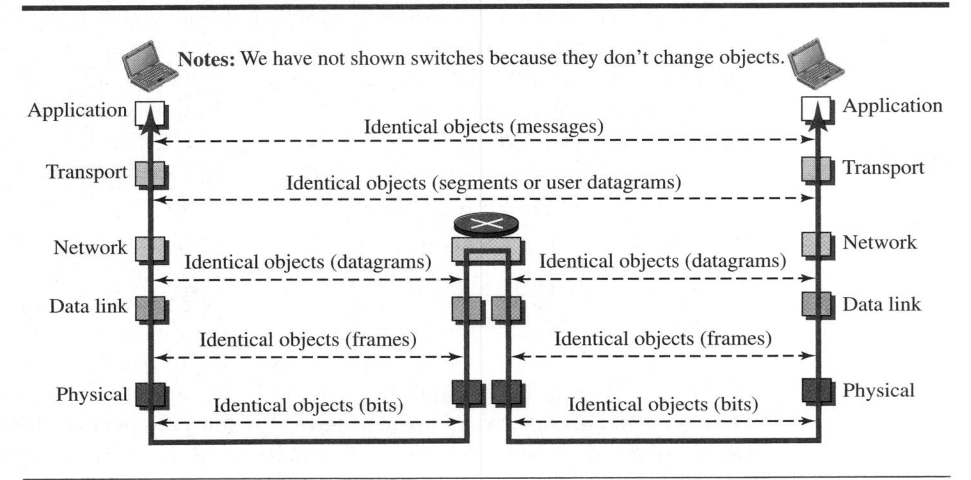

Note that, although the logical connection at the network layer is between the two hosts, we can only say that identical objects exist between two hops in this case because a router may fragment the packet at the network layer and send more packets than received (see fragmentation in Chapter 19). Note that the link between two hops does not change the object.

2.2.3 Description of Each Layer

After understanding the concept of logical communication, we are ready to briefly discuss the duty of each layer. Our discussion in this chapter will be very brief, but we come back to the duty of each layer in next five parts of the book.

Physical Layer

We can say that the physical layer is responsible for carrying individual bits in a frame across the link. Although the physical layer is the lowest level in the TCP/IP protocol suite, the communication between two devices at the physical layer is still a logical communication because there is another, hidden layer, the transmission media, under the physical layer. Two devices are connected by a transmission medium (cable or air). We need to know that the transmission medium does not carry bits; it carries electrical or optical signals. So the bits received in a frame from the data-link layer are transformed and sent through the transmission media, but we can think that the logical unit between two physical layers in two devices is a *bit*. There are several protocols that transform a bit to a signal. We discuss them in Part II when we discuss the physical layer and the transmission media.

Data-link Layer

We have seen that an internet is made up of several links (LANs and WANs) connected by routers. There may be several overlapping sets of links that a datagram can travel from the host to the destination. The routers are responsible for choosing the *best* links. However, when the next link to travel is determined by the router, the data-link layer is responsible for taking the datagram and moving it across the link. The link can be a wired LAN with a link-layer switch, a wireless LAN, a wired WAN, or a wireless WAN. We can also have different protocols used with any link type. In each case, the data-link layer is responsible for moving the packet through the link.

TCP/IP does not define any specific protocol for the data-link layer. It supports all the standard and proprietary protocols. Any protocol that can take the datagram and carry it through the link suffices for the network layer. The data-link layer takes a datagram and encapsulates it in a packet called a *frame*.

Each link-layer protocol may provide a different service. Some link-layer protocols provide complete error detection and correction, some provide only error correction. We discuss wired links in Chapters 13 and 14 and wireless links in Chapters 15 and 16.

Network Layer

The network layer is responsible for creating a connection between the source computer and the destination computer. The communication at the network layer is host-to-host. However, since there can be several routers from the source to the destination, the routers in the path are responsible for choosing the best route for each packet. We can say that the network layer is responsible for host-to-host communication and routing the packet through possible routes. Again, we may ask ourselves why we need the network layer. We could have added the routing duty to the transport layer and dropped this layer. One reason, as we said before, is the separation of different tasks between different layers. The second reason is that the routers do not need the application and transport layers. Separating the tasks allows us to use fewer protocols on the routers.

The network layer in the Internet includes the main protocol, Internet Protocol (IP), that defines the format of the packet, called a datagram at the network layer. IP also defines the format and the structure of addresses used in this layer. IP is also responsible for routing a packet from its source to its destination, which is achieved by each router forwarding the datagram to the next router in its path.

IP is a connectionless protocol that provides no flow control, no error control, and no congestion control services. This means that if any of theses services is required for an application, the application should rely only on the transport-layer protocol. The network layer also includes unicast (one-to-one) and multicast (one-to-many) routing protocols. A routing protocol does not take part in routing (it is the responsibility of IP), but it creates forwarding tables for routers to help them in the routing process.

The network layer also has some auxiliary protocols that help IP in its delivery and routing tasks. The Internet Control Message Protocol (ICMP) helps IP to report some problems when routing a packet. The Internet Group Management Protocol (IGMP) is another protocol that helps IP in multitasking. The Dynamic Host Configuration Protocol (DHCP) helps IP to get the network-layer address for a host. The Address Resolution Protocol (ARP) is a protocol that helps IP to find the link-layer address of a host or a router when its network-layer address is given. ARP is discussed in Chapter 9, ICMP in Chapter 19, and IGMP in Chapter 21.

Transport Layer

The logical connection at the transport layer is also end-to-end. The transport layer at the source host gets the message from the application layer, encapsulates it in a transport-layer packet (called a *segment* or a *user datagram* in different protocols) and sends it, through the logical (imaginary) connection, to the transport layer at the destination host. In other words, the transport layer is responsible for giving services to the application layer: to get a message from an application program running on the source host and deliver it to the corresponding application program on the destination host. We may ask why we need an end-to-end transport layer when we already have an end-to-end application layer. The reason is the separation of tasks and duties, which we discussed earlier. The transport layer should be independent of the application layer. In addition, we will see that we have more than one protocol in the transport layer, which means that each application program can use the protocol that best matches its requirement.

As we said, there are a few transport-layer protocols in the Internet, each designed for some specific task. The main protocol, Transmission Control Protocol (TCP), is a connection-oriented protocol that first establishes a logical connection between transport layers at two hosts before transferring data. It creates a logical pipe between two TCPs for transferring a stream of bytes. TCP provides flow control (matching the sending data rate of the source host with the receiving data rate of the destination host to prevent overwhelming the destination), error control (to guarantee that the segments arrive at the destination without error and resending the corrupted ones), and congestion control to reduce the loss of segments due to congestion in the network. The other common protocol, User Datagram Protocol (UDP), is a connectionless protocol that transmits user datagrams without first creating a logical connection. In UDP, each user datagram is an independent entity without being related to the previous or the next one (the meaning of the term *connectionless*). UDP is a simple protocol that does not provide flow, error, or congestion control. Its simplicity, which means small overhead, is attractive to an application program that needs to send short messages and cannot afford the retransmission of the packets involved in TCP, when a packet is corrupted or lost. A new protocol, Stream Control Transmission Protocol (SCTP) is designed to respond to new applications that are emerging in the multimedia. We will discuss UDP, TCP, and SCTP in Chapter 24.

Application Layer

As Figure 2.6 shows, the logical connection between the two application layers is end-to-end. The two application layers exchange *messages* between each other as though there were a bridge between the two layers. However, we should know that the communication is done through all the layers.

Communication at the application layer is between two *processes* (two programs running at this layer). To communicate, a process sends a request to the other process and receives a response. Process-to-process communication is the duty of the application layer. The application layer in the Internet includes many predefined protocols, but a user can also create a pair of processes to be run at the two hosts. In Chapter 25, we explore this situation.

The Hypertext Transfer Protocol (HTTP) is a vehicle for accessing the World Wide Web (WWW). The Simple Mail Transfer Protocol (SMTP) is the main protocol used in electronic mail (e-mail) service. The File Transfer Protocol (FTP) is used for transferring files from one host to another. The Terminal Network (TELNET) and Secure Shell (SSH) are used for accessing a site remotely. The Simple Network Management Protocol (SNMP) is used by an administrator to manage the Internet at global and local levels. The Domain Name System (DNS) is used by other protocols to find the network-layer address of a computer. The Internet Group Management Protocol (IGMP) is used to collect membership in a group. We discuss most of these protocols in Chapter 26 and some in other chapters.

2.2.4 Encapsulation and Decapsulation

One of the important concepts in protocol layering in the Internet is encapsulation/decapsulation. Figure 2.8 shows this concept for the small internet in Figure 2.5.

Figure 2.8 *Encapsulation/Decapsulation*

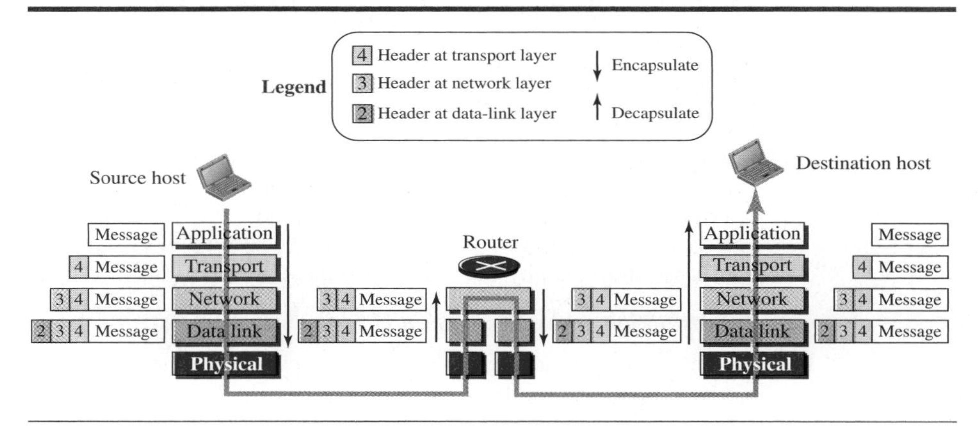

We have not shown the layers for the link-layer switches because no encapsulation/decapsulation occurs in this device. In Figure 2.8, we show the encapsulation in the source host, decapsulation in the destination host, and encapsulation and decapsulation in the router.

Encapsulation at the Source Host

At the source, we have only encapsulation.

1. At the application layer, the data to be exchanged is referred to as a *message*. A message normally does not contain any header or trailer, but if it does, we refer to the whole as the message. The message is passed to the transport layer.
2. The transport layer takes the message as the payload, the load that the transport layer should take care of. It adds the transport layer header to the payload, which contains the identifiers of the source and destination application programs that

want to communicate plus some more information that is needed for the end-to-end delivery of the message, such as information needed for flow, error control, or congestion control. The result is the transport-layer packet, which is called the *segment* (in TCP) and the *user datagram* (in UDP). The transport layer then passes the packet to the network layer.

3. The network layer takes the transport-layer packet as data or payload and adds its own header to the payload. The header contains the addresses of the source and destination hosts and some more information used for error checking of the header, fragmentation information, and so on. The result is the network-layer packet, called a *datagram*. The network layer then passes the packet to the data-link layer.
4. The data-link layer takes the network-layer packet as data or payload and adds its own header, which contains the link-layer addresses of the host or the next hop (the router). The result is the link-layer packet, which is called a *frame*. The frame is passed to the physical layer for transmission.

Decapsulation and Encapsulation at the Router

At the router, we have both decapsulation and encapsulation because the router is connected to two or more links.

1. After the set of bits are delivered to the data-link layer, this layer decapsulates the datagram from the frame and passes it to the network layer.
2. The network layer only inspects the source and destination addresses in the datagram header and consults its forwarding table to find the next hop to which the datagram is to be delivered. The contents of the datagram should not be changed by the network layer in the router unless there is a need to fragment the datagram if it is too big to be passed through the next link. The datagram is then passed to the data-link layer of the next link.
3. The data-link layer of the next link encapsulates the datagram in a frame and passes it to the physical layer for transmission.

Decapsulation at the Destination Host

At the destination host, each layer only decapsulates the packet received, removes the payload, and delivers the payload to the next-higher layer protocol until the message reaches the application layer. It is necessary to say that decapsulation in the host involves error checking.

2.2.5 Addressing

It is worth mentioning another concept related to protocol layering in the Internet, *addressing*. As we discussed before, we have logical communication between pairs of layers in this model. Any communication that involves two parties needs two addresses: source address and destination address. Although it looks as if we need five pairs of addresses, one pair per layer, we normally have only four because the physical layer does not need addresses; the unit of data exchange at the physical layer is a bit, which definitely cannot have an address. Figure 2.9 shows the addressing at each layer.

As the figure shows, there is a relationship between the layer, the address used in that layer, and the packet name at that layer. At the application layer, we normally use names to define the site that provides services, such as *someorg.com*, or the e-mail

Figure 2.9 *Addressing in the TCP/IP protocol suite*

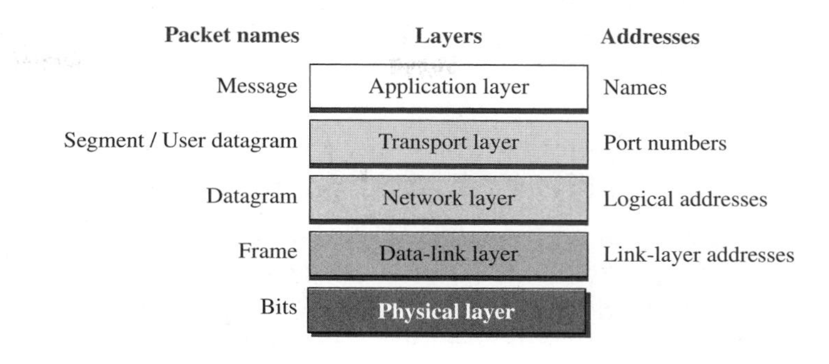

address, such as *somebody@coldmail.com.* At the transport layer, addresses are called port numbers, and these define the application-layer programs at the source and destination. Port numbers are local addresses that distinguish between several programs running at the same time. At the network-layer, the addresses are global, with the whole Internet as the scope. A network-layer address uniquely defines the connection of a device to the Internet. The link-layer addresses, sometimes called MAC addresses, are locally defined addresses, each of which defines a specific host or router in a network (LAN or WAN). We will come back to these addresses in future chapters.

2.2.6 Multiplexing and Demultiplexing

Since the TCP/IP protocol suite uses several protocols at some layers, we can say that we have multiplexing at the source and demultiplexing at the destination. Multiplexing in this case means that a protocol at a layer can encapsulate a packet from several next-higher layer protocols (one at a time); demultiplexing means that a protocol can decapsulate and deliver a packet to several next-higher layer protocols (one at a time). Figure 2.10 shows the concept of multiplexing and demultiplexing at the three upper layers.

Figure 2.10 *Multiplexing and demultiplexing*

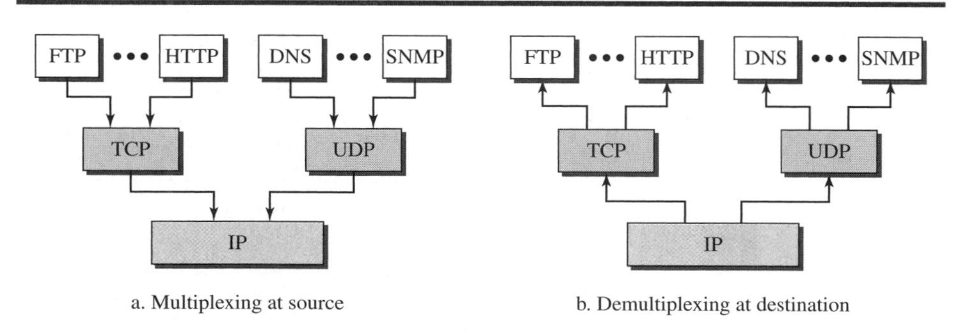

To be able to multiplex and demultiplex, a protocol needs to have a field in its header to identify to which protocol the encapsulated packets belong. At the transport layer, either UDP or TCP can accept a message from several application-layer protocols. At the network layer, IP can accept a segment from TCP or a user datagram from UDP. IP can also accept a packet from other protocols such as ICMP, IGMP, and so on. At the data-link layer, a frame may carry the payload coming from IP or other protocols such as ARP (see Chapter 9).

2.3 THE OSI MODEL

Although, when speaking of the Internet, everyone talks about the TCP/IP protocol suite, this suite is not the only suite of protocols defined. Established in 1947, the **International Organization for Standardization (ISO)** is a multinational body dedicated to worldwide agreement on international standards. Almost three-fourths of the countries in the world are represented in the ISO. An ISO standard that covers all aspects of network communications is the **Open Systems Interconnection (OSI) model.** It was first introduced in the late 1970s.

ISO is the organization; OSI is the model.

An *open system* is a set of protocols that allows any two different systems to communicate regardless of their underlying architecture. The purpose of the OSI model is to show how to facilitate communication between different systems without requiring changes to the logic of the underlying hardware and software. The OSI model is not a protocol; it is a model for understanding and designing a network architecture that is flexible, robust, and interoperable. The OSI model was intended to be the basis for the creation of the protocols in the OSI stack.

The OSI model is a layered framework for the design of network systems that allows communication between all types of computer systems. It consists of seven separate but related layers, each of which defines a part of the process of moving information across a network (see Figure 2.11).

Figure 2.11 *The OSI model*

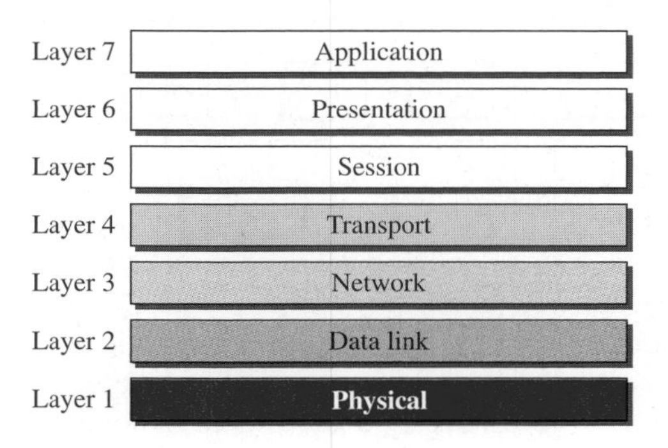

2.3.1 OSI versus TCP/IP

When we compare the two models, we find that two layers, session and presentation, are missing from the TCP/IP protocol suite. These two layers were not added to the TCP/IP protocol suite after the publication of the OSI model. The application layer in the suite is usually considered to be the combination of three layers in the OSI model, as shown in Figure 2.12.

Figure 2.12 *TCP/IP and OSI model*

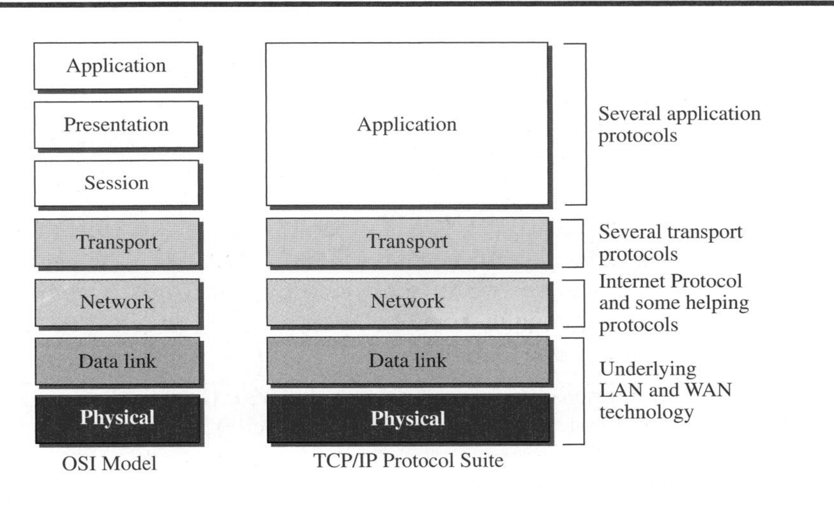

Two reasons were mentioned for this decision. First, TCP/IP has more than one transport-layer protocol. Some of the functionalities of the session layer are available in some of the transport-layer protocols. Second, the application layer is not only one piece of software. Many applications can be developed at this layer. If some of the functionalities mentioned in the session and presentation layers are needed for a particular application, they can be included in the development of that piece of software.

2.3.2 Lack of OSI Model's Success

The OSI model appeared after the TCP/IP protocol suite. Most experts were at first excited and thought that the TCP/IP protocol would be fully replaced by the OSI model. This did not happen for several reasons, but we describe only three, which are agreed upon by all experts in the field. First, OSI was completed when TCP/IP was fully in place and a lot of time and money had been spent on the suite; changing it would cost a lot. Second, some layers in the OSI model were never fully defined. For example, although the services provided by the presentation and the session layers were listed in the document, actual protocols for these two layers were not fully defined, nor were they fully described, and the corresponding software was not fully developed. Third, when OSI was implemented by an organization in a different application, it did not show a high enough level of performance to entice the Internet authority to switch from the TCP/IP protocol suite to the OSI model.

2.4 END-CHAPTER MATERIALS

2.4.1 Recommended Reading

For more details about subjects discussed in this chapter, we recommend the following books, and RFCs. The items enclosed in brackets refer to the reference list at the end of the book.

Books and Papers

Several books and papers give a thorough coverage about the materials discussed in this chapter: [Seg 98], [Lei et al. 98], [Kle 04], [Cer 89], and [Jen et al. 86].

RFCs

Two RFCs in particular discuss the TCP/IP suite: RFC 791 (IP) and RFC 817 (TCP). In future chapters we list different RFCs related to each protocol in each layer.

2.4.2 Key Terms

International Organization for Standardization (ISO)
Open Systems Interconnection (OSI) model
protocol layering

2.4.3 Summary

A protocol is a set of rules that governs communication. In protocol layering, we need to follow two principles to provide bidirectional communication. First, each layer needs to perform two opposite tasks. Second, two objects under each layer at both sides should be identical. In a protocol layering, we need to distinguish between a logical connection and a physical connection. Two protocols at the same layer can have a logical connection; a physical connection is only possible through the physical layers.

TCP/IP is a hierarchical protocol suite made of five layers: physical, data link, network, transport, and application. The physical layer coordinates the functions required to transmit a bit stream over a physical medium. The data-link layer is responsible for delivering data units from one station to the next without errors. The network layer is responsible for the source-to-destination delivery of a packet across multiple network links. The transport layer is responsible for the process-to-process delivery of the entire message. The application layer enables the users to access the network.

Four levels of addresses are used in an internet following the TCP/IP protocols: physical (link) addresses, logical (IP) addresses, port addresses, and specific addresses. The physical address, also known as the link address, is the address of a node as defined by its LAN or WAN. The IP address uniquely defines a host on the Internet. The port address identifies a process on a host. A specific address is a user-friendly address.

Another model that defines protocol layering is the Open Systems Interconnection (OSI) model. Two layers in the OSI model, session and presentation, are missing from the TCP/IP protocol suite. These two layers were not added to the TCP/IP protocol suite after the publication of the OSI model. The application layer in the suite is usually considered to be the combination of three layers in the OSI model. The OSI model did not replace the TCP/IP protocol suite because it was completed when TCP/IP was fully in place and because some layers in the OSI model were never fully defined.

2.5 PRACTICE SET

2.5.1 Quizzes

A set of interactive quizzes for this chapter can be found on the book website. It is strongly recommended that the student take the quizzes to check his/her understanding of the materials before continuing with the practice set.

2.5.2 Questions

Q2-1. What are the types of addresses (identifiers) used in each of the following layers?

a. application layer **b.** network layer **c.** data-link layer

Q2-2. Assume we want to connect two isolated hosts together to let each host communicate with the other. Do we need a link-layer switch between the two? Explain.

Q2-3. Can you explain why we did not mention multiplexing/demultiplexing services for the application layer?

Q2-4. Which of the following data units has an application-layer message plus the header from layer 4?

a. a frame **b.** a user datagram **c.** a bit

Q2-5. What is the first principle we discussed in this chapter for protocol layering that needs to be followed to make the communication bidirectional?

Q2-6. If a port number is 16 bits (2 bytes), what is the minimum header size at the transport layer of the TCP/IP protocol suite?

Q2-7. Which of the following data units is decapsulated from a user datagram?

a. a datagram **b.** a segment **c.** a message

Q2-8. In the TCP/IP protocol suite, what are the identical objects at the sender and the receiver sites when we think about the logical connection at the application layer?

Q2-9. A host communicates with another host using the TCP/IP protocol suite. What is the unit of data sent or received at each of the following layers?

a. application layer **b.** network layer **c.** data-link layer

Q2-10. When we say that the transport layer multiplexes and demultiplexes application-layer messages, do we mean that a transport-layer protocol can combine several messages from the application layer in one packet? Explain.

Q2-11. List some application-layer protocols mentioned in this chapter.

Q2-12. Which of the following data units is encapsulated in a frame?

a. a user datagram **b.** a datagram **c.** a segment

Q2-13. A router connects three links (networks). How many of each of the following layers can the router be involved with?

d. physical layer **e.** data-link layer **f.** network layer

Q2-14. Which layers of the TCP/IP protocol suite are involved in a link-layer switch?

Q2-15. If there is a single path between the source host and the destination host, do we need a router between the two hosts?

2.5.3 Problems

P2-1. Protocol layering can be found in many aspects of our lives such as air travelling. Imagine you make a round-trip to spend some time on vacation at a resort. You need to go through some processes at your city airport before flying. You also need to go through some processes when you arrive at the resort airport. Show the protocol layering for the round trip using some layers such as baggage checking/claiming, boarding/unboarding, takeoff/landing.

P2-2. The presentation of data is becoming more and more important in today's Internet. Some people argue that the TCP/IP protocol suite needs to add a new layer to take care of the presentation of data. If this new layer is added in the future, where should its position be in the suite? Redraw Figure 2.4 to include this layer.

P2-3. In an internet, we change the LAN technology to a new one. Which layers in the TCP/IP protocol suite need to be changed?

P2-4. Assume that an application-layer protocol is written to use the services of UDP. Can the application-layer protocol use the services of TCP without change?

P2-5. Answer the following questions about Figure 2.2 when the communication is from Maria to Ann:

a. What is the service provided by layer 1 to layer 2 at Maria's site?

b. What is the service provided by layer 1 to layer 2 at Ann's site?

P2-6. Assume a system uses five protocol layers. If the application program creates a message of 150 bytes and each layer (including the fifth and the first) adds a header of 20 bytes to the data unit, what is the efficiency (the ratio of application-layer bytes to the number of bytes transmitted) of the system?

P2-7. Assume a private internet uses three different protocols at the data-link layer (L1, L2, and L3). Redraw Figure 2.10 with this assumption. Can we say that, in the data-link layer, we have demultiplexing at the source node and multiplexing at the destination node?

P2-8. Match the following to one or more layers of the TCP/IP protocol suite:

a. route determination

b. connection to transmission media

c. providing services for the end user

P2-9. Assume we have created a packet-switched internet. Using the TCP/IP protocol suite, we need to transfer a huge file. What are the advantage and disadvantage of sending large packets?

P2-10. In Figure 2.10, when the IP protocol decapsulates the transport-layer packet, how does it know to which upper-layer protocol (UDP or TCP) the packet should be delivered?

P2-11. Assume that the number of hosts connected to the Internet at year 2010 is five hundred million. If the number of hosts increases only 20 percent per year, what is the number of hosts in year 2020?

P2-12. Answer the following questions about Figure 2.2 when the communication is from Maria to Ann:

a. What is the service provided by layer 2 to layer 3 at Maria's site?

b. What is the service provided by layer 2 to layer 3 at Ann's site?

P2-13. Match the following to one or more layers of the TCP/IP protocol suite:

a. creating user datagrams

b. responsibility for handling frames between adjacent nodes

c. transforming bits to electromagnetic signals

P2-14. Assume that a private internet requires that the messages at the application layer be encrypted and decrypted for security purposes. If we need to add some information about the encryption/decryption process (such as the algorithms used in the process), does it mean that we are adding one layer to the TCP/IP protocol suite? Redraw the TCP/IP layers (Figure 2.4 part b) if you think so.

P2-15. Using the internet in Figure 1.11 (Chapter 1) in the text, show the layers of the TCP/IP protocol suite and the flow of data when two hosts, one on the west coast and the other on the east coast, exchange messages.

PART 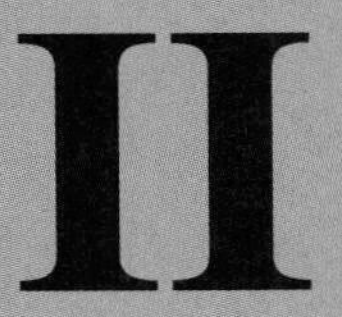

Physical Layer

In the second part of the book, we discuss the physical layer, including the transmission media that is connected to the physical layer. The part is made of six chapters. The first introduces the entities involved in the physical layer. The next two chapters cover transmission. The following chapter discusses how to use the available bandwidth. The transmission media alone occupy all of the next chapter. Finally, the last chapter discusses switching, which can occur in any layer, but we introduce the topic in this part of the book.

CHAPTER 3

Introduction to Physical Layer

One of the major functions of the physical layer is to move data in the form of electromagnetic signals across a transmission medium. Whether you are collecting numerical statistics from another computer, sending animated pictures from a design workstation, or causing a bell to ring at a distant control center, you are working with the transmission of **data** across network connections.

Generally, the data usable to a person or application are not in a form that can be transmitted over a network. For example, a photograph must first be changed to a form that transmission media can accept. Transmission media work by conducting energy along a physical path. For transmission, data needs to be changed to **signals.**

This chapter is divided into six sections:

- The first section shows how data and signals can be either analog or digital. Analog refers to an entity that is continuous; digital refers to an entity that is discrete.
- The second section shows that only periodic analog signals can be used in data communication. The section discusses simple and composite signals. The attributes of analog signals such as period, frequency, and phase are also explained.
- The third section shows that only nonperiodic digital signals can be used in data communication. The attributes of a digital signal such as bit rate and bit length are discussed. We also show how digital data can be sent using analog signals. Baseband and broadband transmission are also discussed in this section.
- The fourth section is devoted to transmission impairment. The section shows how attenuation, distortion, and noise can impair a signal.
- The fifth section discusses the data rate limit: how many bits per second we can send with the available channel. The data rates of noiseless and noisy channels are examined and compared.
- The sixth section discusses the performance of data transmission. Several channel measurements are examined including bandwidth, throughput, latency, and jitter. Performance is an issue that is revisited in several future chapters.

3.1 DATA AND SIGNALS

Figure 3.1 shows a scenario in which a scientist working in a research company, Sky Research, needs to order a book related to her research from an online bookseller, Scientific Books.

Figure 3.1 *Communication at the physical layer*

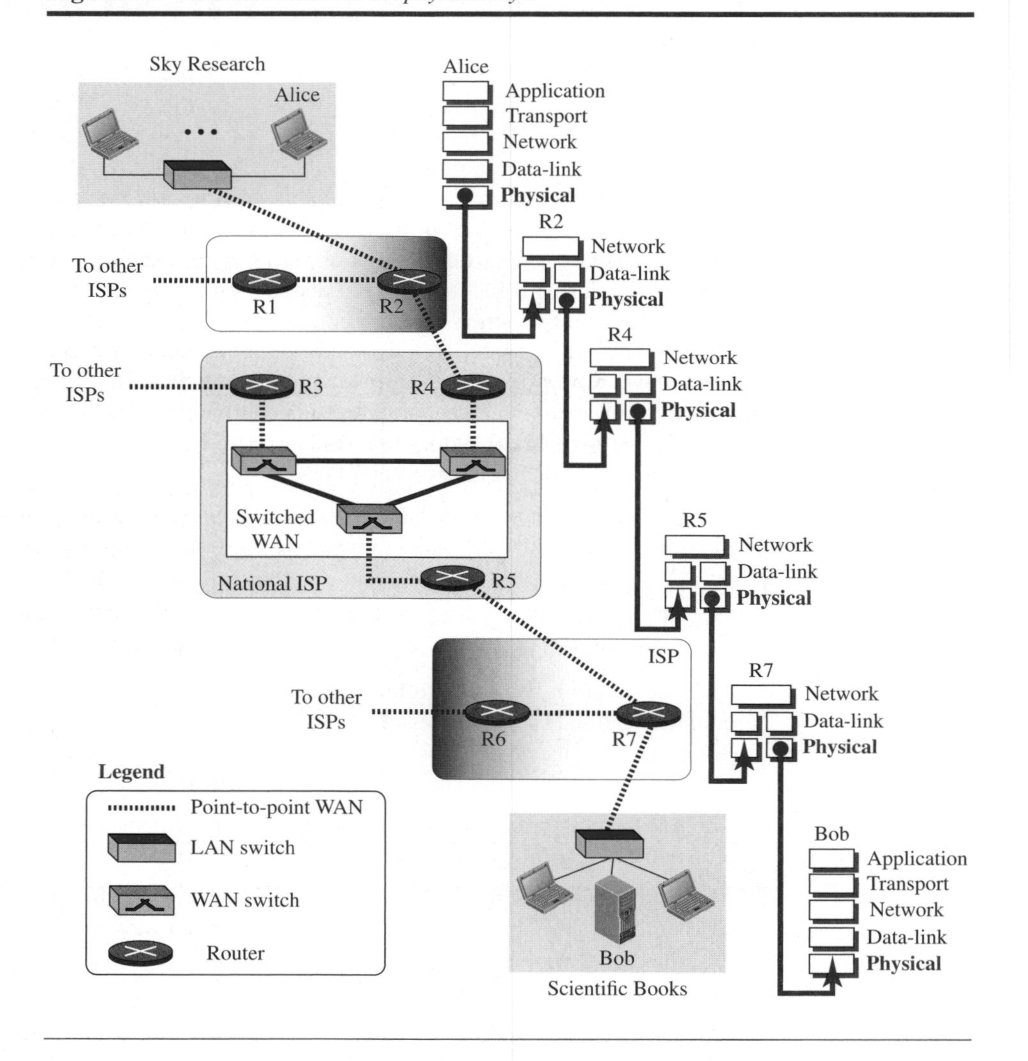

We can think of five different levels of communication between Alice, the computer on which our scientist is working, and Bob, the computer that provides online service. Communication at application, transport, network, or data-link is *logical*; communication at the physical layer is *physical*. For simplicity, we have shown only host-to-router, router-to-router, and router-to-host, but the switches are also involved in the physical communication.

Although Alice and Bob need to exchange *data*, communication at the physical layer means exchanging *signals*. Data need to be transmitted and received, but the media have to change data to signals. Both data and the signals that represent them can be either **analog** or **digital** in form.

3.1.1 Analog and Digital Data

Data can be analog or digital. The term **analog data** refers to information that is continuous; **digital data** refers to information that has discrete states. For example, an analog clock that has hour, minute, and second hands gives information in a continuous form; the movements of the hands are continuous. On the other hand, a digital clock that reports the hours and the minutes will change suddenly from 8:05 to 8:06.

Analog data, such as the sounds made by a human voice, take on continuous values. When someone speaks, an analog wave is created in the air. This can be captured by a microphone and converted to an analog signal or sampled and converted to a digital signal.

Digital data take on discrete values. For example, data are stored in computer memory in the form of 0s and 1s. They can be converted to a digital signal or modulated into an analog signal for transmission across a medium.

3.1.2 Analog and Digital Signals

Like the data they represent, **signals** can be either analog or digital. An **analog signal** has infinitely many levels of intensity over a period of time. As the wave moves from value *A* to value *B*, it passes through and includes an infinite number of values along its path. A **digital signal,** on the other hand, can have only a limited number of defined values. Although each value can be any number, it is often as simple as 1 and 0.

The simplest way to show signals is by plotting them on a pair of perpendicular axes. The vertical axis represents the value or strength of a signal. The horizontal axis represents time. Figure 3.2 illustrates an analog signal and a digital signal. The curve representing the analog signal passes through an infinite number of points. The vertical lines of the digital signal, however, demonstrate the sudden jump that the signal makes from value to value.

Figure 3.2 *Comparison of analog and digital signals*

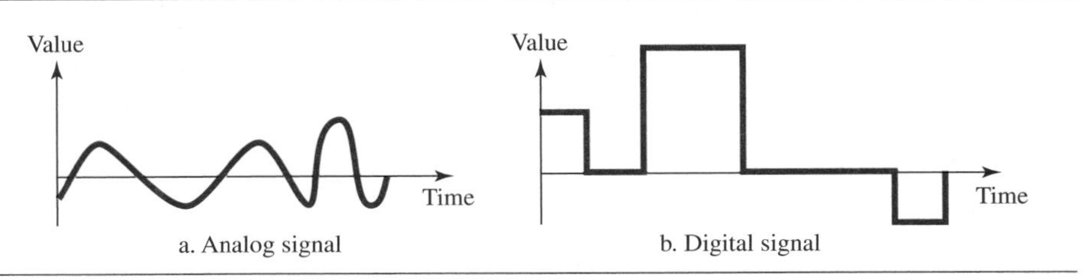

3.1.3 Periodic and Nonperiodic

Both analog and digital signals can take one of two forms: *periodic* or *nonperiodic* (sometimes referred to as *aperiodic;* the prefix *a* in Greek means "non").

A **periodic signal** completes a pattern within a measurable time frame, called a **period,** and repeats that pattern over subsequent identical periods. The completion of one full pattern is called a **cycle.** A **nonperiodic signal** changes without exhibiting a pattern or cycle that repeats over time.

Both analog and digital signals can be periodic or nonperiodic. In data communications, we commonly use periodic analog signals and nonperiodic digital signals, as we will see in future chapters.

In data communications, we commonly use periodic analog signals and nonperiodic digital signals.

3.2 PERIODIC ANALOG SIGNALS

Periodic analog signals can be classified as simple or composite. A simple periodic analog signal, a **sine wave,** cannot be decomposed into simpler signals. A composite periodic analog signal is composed of multiple sine waves.

3.2.1 Sine Wave

The sine wave is the most fundamental form of a periodic analog signal. When we visualize it as a simple oscillating curve, its change over the course of a cycle is smooth and consistent, a continuous, rolling flow. Figure 3.3 shows a sine wave. Each cycle consists of a single arc above the time axis followed by a single arc below it.

Figure 3.3 *A sine wave*

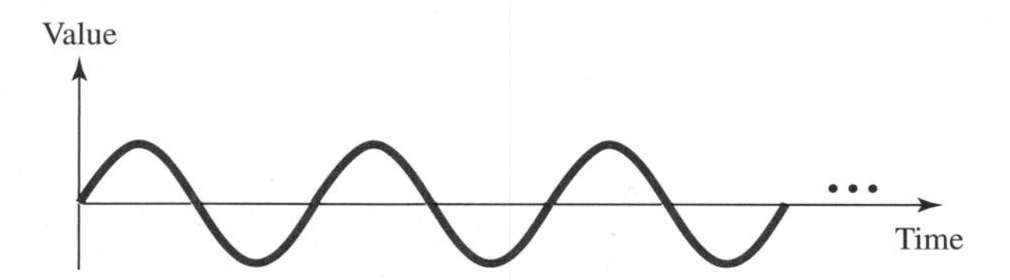

We discuss a mathematical approach to sine waves in Appendix E.

A sine wave can be represented by three parameters: the *peak amplitude,* the *frequency,* and the *phase*. These three parameters fully describe a sine wave.

Peak Amplitude

The **peak amplitude** of a signal is the absolute value of its highest intensity, proportional to the energy it carries. For electric signals, peak amplitude is normally measured in *volts*. Figure 3.4 shows two signals and their peak amplitudes.

Figure 3.4 *Two signals with the same phase and frequency, but different amplitudes*

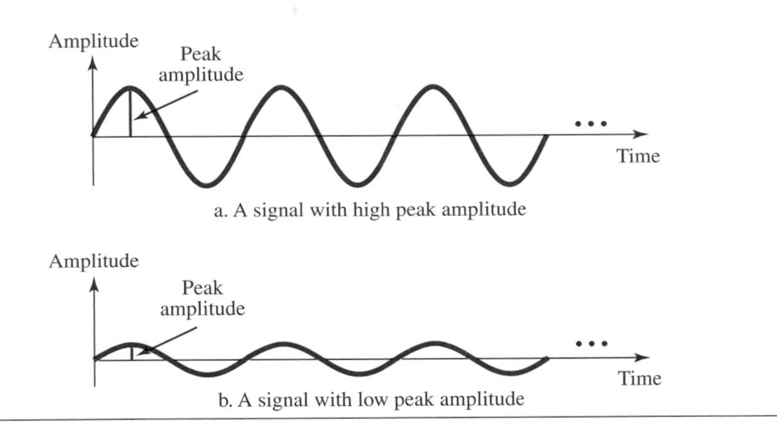

Example 3.1

The power in your house can be represented by a sine wave with a peak amplitude of 155 to 170 V. However, it is common knowledge that the voltage of the power in U.S. homes is 110 to 120 V. This discrepancy is due to the fact that these are root mean square (rms) values. The signal is squared and then the average amplitude is calculated. The peak value is equal to $2^{1/2} \times$ rms value.

Example 3.2

The voltage of a battery is a constant; this constant value can be considered a sine wave, as we will see later. For example, the peak value of an AA battery is normally 1.5 V.

Period and Frequency

Period refers to the amount of time, in seconds, a signal needs to complete 1 cycle. **Frequency** refers to the number of periods in 1 s. Note that period and frequency are just one characteristic defined in two ways. Period is the inverse of frequency, and frequency is the inverse of period, as the following formulas show.

$$f = \frac{1}{T} \quad \textbf{and} \quad T = \frac{1}{f}$$

Frequency and period are the inverse of each other.

Figure 3.5 shows two signals and their frequencies. Period is formally expressed in seconds. Frequency is formally expressed in **Hertz (Hz),** which is cycle per second. Units of period and frequency are shown in Table 3.1.

Figure 3.5 *Two signals with the same amplitude and phase, but different frequencies*

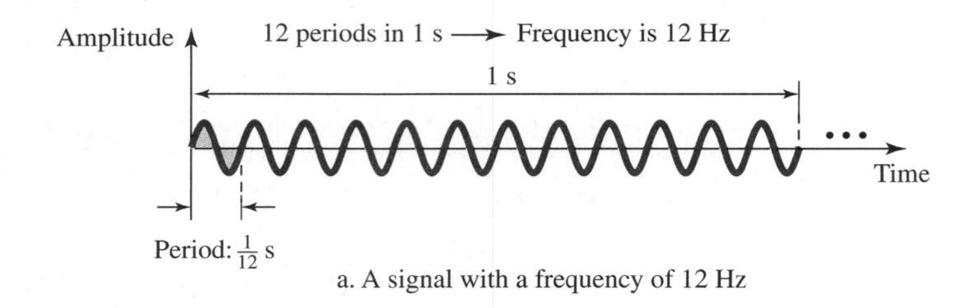

a. A signal with a frequency of 12 Hz

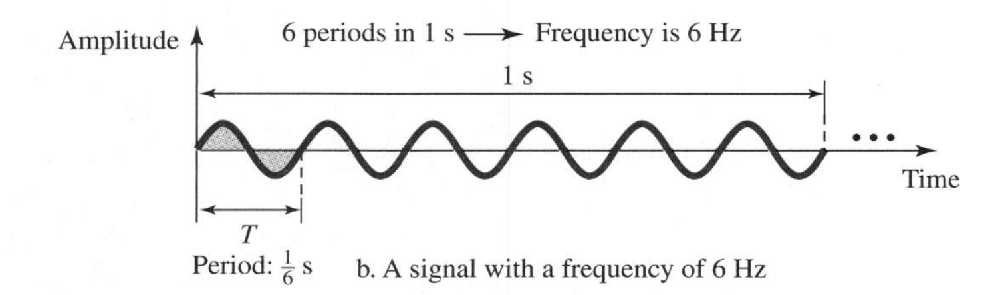

b. A signal with a frequency of 6 Hz

Table 3.1 *Units of period and frequency*

Period		*Frequency*	
Unit	*Equivalent*	*Unit*	*Equivalent*
Seconds (s)	1 s	Hertz (Hz)	1 Hz
Milliseconds (ms)	10^{-3} s	Kilohertz (kHz)	10^3 Hz
Microseconds (μs)	10^{-6} s	Megahertz (MHz)	10^6 Hz
Nanoseconds (ns)	10^{-9} s	Gigahertz (GHz)	10^9 Hz
Picoseconds (ps)	10^{-12} s	Terahertz (THz)	10^{12} Hz

Example 3.3

The power we use at home has a frequency of 60 Hz (50 Hz in Europe). The period of this sine wave can be determined as follows:

$$T = \frac{1}{f} = \frac{1}{60} = 0.0166 \text{ s} = 0.0166 \times 10^3 \text{ ms} = 16.6 \text{ ms}$$

This means that the period of the power for our lights at home is 0.0116 s, or 16.6 ms. Our eyes are not sensitive enough to distinguish these rapid changes in amplitude.

Example 3.4

Express a period of 100 ms in microseconds.

Solution

From Table 3.1 we find the equivalents of 1 ms (1 ms is 10^{-3} s) and 1 s (1 s is 10^6 μs). We make the following substitutions:

$$100 \text{ ms} = 100 \times 10^{-3} \text{ s} = 100 \times 10^{-3} \times 10^6 \ \mu\text{s} = 10^2 \times 10^{-3} \times 10^6 \ \mu\text{s} = 10^5 \ \mu\text{s}$$

Example 3.5

The period of a signal is 100 ms. What is its frequency in kilohertz?

Solution

First we change 100 ms to seconds, and then we calculate the frequency from the period (1 Hz = 10^{-3} kHz).

$$100 \text{ ms} = 100 \times 10^{-3} \text{ s} = 10^{-1} \text{ s}$$

$$f = \frac{1}{T} = \frac{1}{10^{-1}} \text{Hz} = 10 \text{ Hz} = 10 \times 10^{-3} \text{ kHz} = 10^{-2} \text{ kHz}$$

More About Frequency

We already know that frequency is the relationship of a signal to time and that the frequency of a wave is the number of cycles it completes in 1 s. But another way to look at frequency is as a measurement of the rate of change. Electromagnetic signals are oscillating waveforms; that is, they fluctuate continuously and predictably above and below a mean energy level. A 40-Hz signal has one-half the frequency of an 80-Hz signal; it completes 1 cycle in twice the time of the 80-Hz signal, so each cycle also takes twice as long to change from its lowest to its highest voltage levels. Frequency, therefore, though described in cycles per second (hertz), is a general measurement of the rate of change of a signal with respect to time.

Frequency is the rate of change with respect to time. Change in a short span of time means high frequency. Change over a long span of time means low frequency.

If the value of a signal changes over a very short span of time, its frequency is high. If it changes over a long span of time, its frequency is low.

Two Extremes

What if a signal does not change at all? What if it maintains a constant voltage level for the entire time it is active? In such a case, its frequency is zero. Conceptually, this idea is a simple one. If a signal does not change at all, it never completes a cycle, so its frequency is 0 Hz.

But what if a signal changes instantaneously? What if it jumps from one level to another in no time? Then its frequency is infinite. In other words, when a signal changes instantaneously, its period is zero; since frequency is the inverse of period, in this case, the frequency is 1/0, or infinite (unbounded).

If a signal does not change at all, its frequency is zero.
If a signal changes instantaneously, its frequency is infinite.

3.2.2 Phase

The term **phase,** or phase shift, describes the position of the waveform relative to time 0. If we think of the wave as something that can be shifted backward or forward along the time axis, phase describes the amount of that shift. It indicates the status of the first cycle.

Phase describes the position of the waveform relative to time 0.

Phase is measured in degrees or radians [360° is 2π rad; 1° is 2π/360 rad, and 1 rad is 360/(2π)]. A phase shift of 360° corresponds to a shift of a complete period; a phase shift of 180° corresponds to a shift of one-half of a period; and a phase shift of 90° corresponds to a shift of one-quarter of a period (see Figure 3.6).

Figure 3.6 *Three sine waves with the same amplitude and frequency, but different phases*

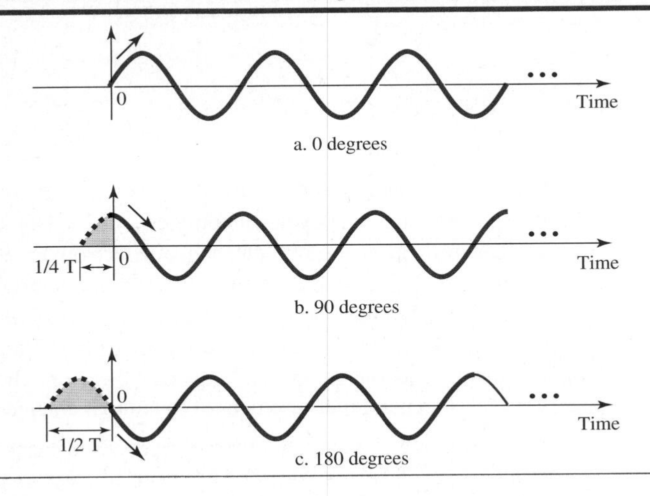

Looking at Figure 3.6, we can say that

a. A sine wave with a phase of 0° starts at time 0 with a zero amplitude. The amplitude is increasing.

b. A sine wave with a phase of 90° starts at time 0 with a peak amplitude. The amplitude is decreasing.

c. A sine wave with a phase of 180° starts at time 0 with a zero amplitude. The amplitude is decreasing.

Another way to look at the phase is in terms of shift or offset. We can say that

a. A sine wave with a phase of 0° is not shifted.

b. A sine wave with a phase of 90° is shifted to the left by $\frac{1}{4}$ cycle. However, note that the signal does not really exist before time 0.

c. A sine wave with a phase of 180° is shifted to the left by $\frac{1}{2}$ cycle. However, note that the signal does not really exist before time 0.

Example 3.6

A sine wave is offset $\frac{1}{6}$ cycle with respect to time 0. What is its phase in degrees and radians?

Solution

We know that 1 complete cycle is 360°. Therefore, $\frac{1}{6}$ cycle is

$$\frac{1}{6} \times 360 = 60° = 60 \times \frac{2\pi}{360} \text{ rad} = \frac{\pi}{3} \text{ rad} = 1.046 \text{ rad}$$

3.2.3 Wavelength

Wavelength is another characteristic of a signal traveling through a transmission medium. Wavelength binds the period or the frequency of a simple sine wave to the **propagation speed** of the medium (see Figure 3.7).

Figure 3.7 *Wavelength and period*

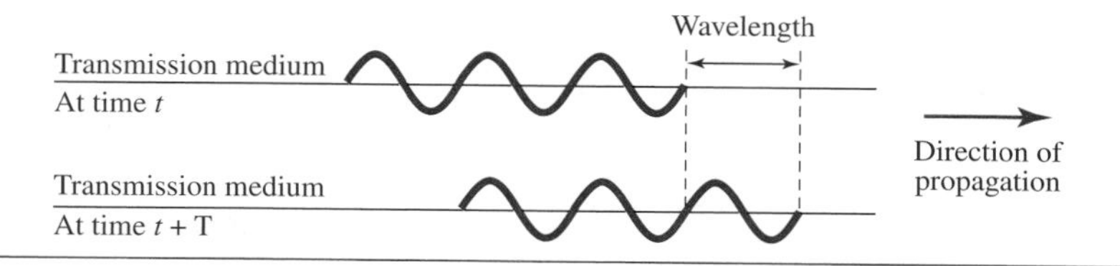

While the frequency of a signal is independent of the medium, the wavelength depends on both the frequency and the medium. Wavelength is a property of any type of signal. In data communications, we often use wavelength to describe the transmission of light in an optical fiber. The wavelength is the distance a simple signal can travel in one period.

Wavelength can be calculated if one is given the propagation speed (the speed of light) and the period of the signal. However, since period and frequency are related to each other, if we represent wavelength by λ, propagation speed by c (speed of light), and frequency by f, we get

$$\textbf{Wavelength} = \textbf{(propagation speed)} \times \textbf{period} = \frac{\textbf{propagation speed}}{\textbf{frequency}}$$

$$\lambda = \frac{c}{f}$$

The propagation speed of electromagnetic signals depends on the medium and on the frequency of the signal. For example, in a vacuum, light is propagated with a speed of 3×10^8 m/s. That speed is lower in air and even lower in cable.

The wavelength is normally measured in micrometers (microns) instead of meters. For example, the wavelength of red light (frequency $= 4 \times 10^{14}$) in air is

$$\lambda = \frac{c}{f} = \frac{3 \times 10^8}{4 \times 10^{14}} = 0.75 \times 10^{-6} \text{ m} = 0.75 \ \mu\text{m}$$

In a coaxial or fiber-optic cable, however, the wavelength is shorter (0.5 μm) because the propagation speed in the cable is decreased.

3.2.4 Time and Frequency Domains

A sine wave is comprehensively defined by its amplitude, frequency, and phase. We have been showing a sine wave by using what is called a **time-domain** plot. The time-domain plot shows changes in signal amplitude with respect to time (it is an amplitude-versus-time plot). Phase is not explicitly shown on a time-domain plot.

To show the relationship between amplitude and frequency, we can use what is called a **frequency-domain** plot. A frequency-domain plot is concerned with only the peak value and the frequency. Changes of amplitude during one period are not shown. Figure 3.8 shows a signal in both the time and frequency domains.

Figure 3.8 *The time-domain and frequency-domain plots of a sine wave*

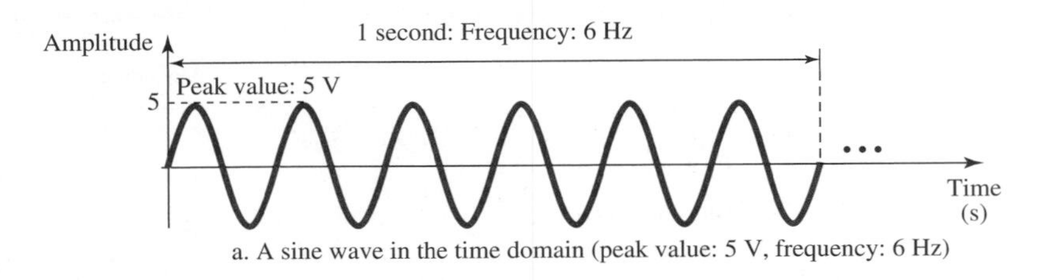

a. A sine wave in the time domain (peak value: 5 V, frequency: 6 Hz)

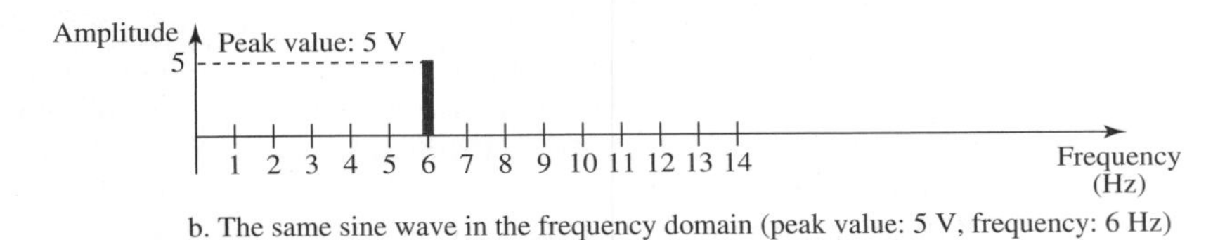

b. The same sine wave in the frequency domain (peak value: 5 V, frequency: 6 Hz)

It is obvious that the frequency domain is easy to plot and conveys the information that one can find in a time domain plot. The advantage of the frequency domain is that we can immediately see the values of the frequency and peak amplitude. A complete sine wave is represented by one spike. The position of the spike shows the frequency; its height shows the peak amplitude.

A complete sine wave in the time domain can be represented by one single spike in the frequency domain.

Example 3.7

The frequency domain is more compact and useful when we are dealing with more than one sine wave. For example, Figure 3.9 shows three sine waves, each with different amplitude and frequency. All can be represented by three spikes in the frequency domain.

Figure 3.9 *The time domain and frequency domain of three sine waves*

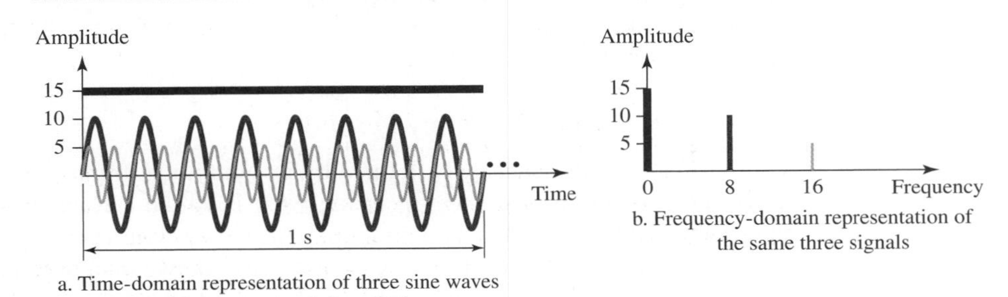

a. Time-domain representation of three sine waves with frequencies 0, 8, and 16

b. Frequency-domain representation of the same three signals

3.2.5 Composite Signals

So far, we have focused on simple sine waves. Simple sine waves have many applications in daily life. We can send a single sine wave to carry electric energy from one place to another. For example, the power company sends a single sine wave with a frequency of 60 Hz to distribute electric energy to houses and businesses. As another example, we can use a single sine wave to send an alarm to a security center when a burglar opens a door or window in the house. In the first case, the sine wave is carrying energy; in the second, the sine wave is a signal of danger.

If we had only one single sine wave to convey a conversation over the phone, it would make no sense and carry no information. We would just hear a buzz. As we will see in Chapters 4 and 5, we need to send a composite signal to communicate data. A **composite signal** is made of many simple sine waves.

A single-frequency sine wave is not useful in data communications; we need to send a composite signal, a signal made of many simple sine waves.

In the early 1900s, the French mathematician Jean-Baptiste Fourier showed that any composite signal is actually a combination of simple sine waves with different frequencies, amplitudes, and phases. **Fourier analysis** is discussed in Appendix E; for our purposes, we just present the concept.

According to Fourier analysis, any composite signal is a combination of simple sine waves with different frequencies, amplitudes, and phases. Fourier analysis is discussed in Appendix E.

A composite signal can be periodic or nonperiodic. A periodic composite signal can be decomposed into a series of simple sine waves with discrete frequencies—frequencies that have integer values (1, 2, 3, and so on). A nonperiodic composite signal can be decomposed into a combination of an infinite number of simple sine waves with continuous frequencies, frequencies that have real values.

If the composite signal is periodic, the decomposition gives a series of signals with discrete frequencies; if the composite signal is nonperiodic, the decomposition gives a combination of sine waves with continuous frequencies.

Example 3.8

Figure 3.10 shows a periodic composite signal with frequency *f*. This type of signal is not typical of those found in data communications.We can consider it to be three alarm systems, each with a different frequency. The analysis of this signal can give us a good understanding of how to decompose signals.

It is very difficult to manually decompose this signal into a series of simple sine waves. However, there are tools, both hardware and software, that can help us do the job. We are not concerned about how it is done; we are only interested in the result. Figure 3.11 shows the result of decomposing the above signal in both the time and frequency domains.

The amplitude of the sine wave with frequency *f* is almost the same as the peak amplitude of the composite signal. The amplitude of the sine wave with frequency 3*f* is one-third of that of

Figure 3.10 *A composite periodic signal*

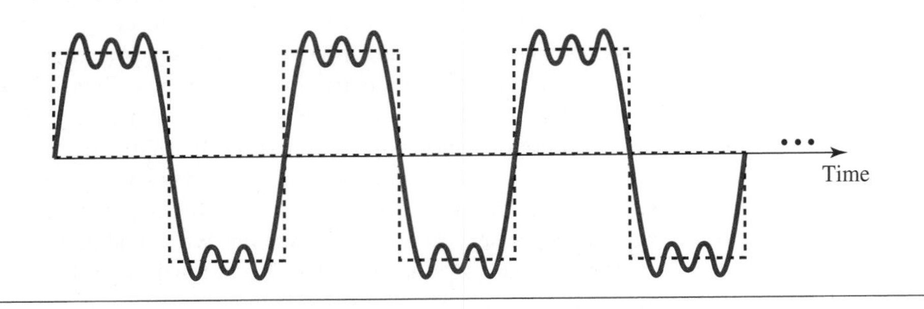

Figure 3.11 *Decomposition of a composite periodic signal in the time and frequency domains*

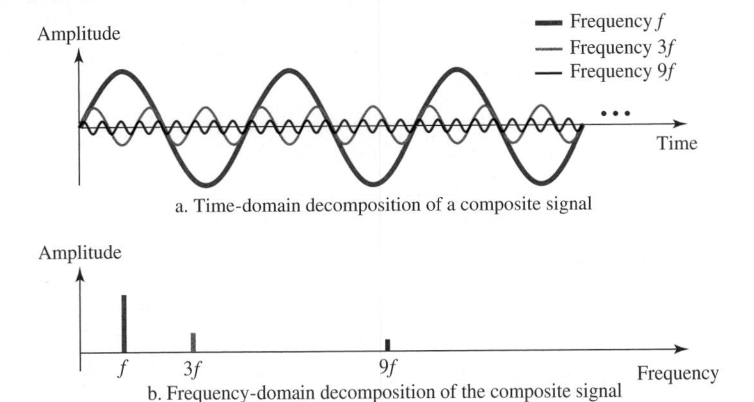

the first, and the amplitude of the sine wave with frequency $9f$ is one-ninth of the first. The frequency of the sine wave with frequency f is the same as the frequency of the composite signal; it is called the **fundamental frequency,** or first **harmonic.** The sine wave with frequency $3f$ has a frequency of 3 times the fundamental frequency; it is called the third harmonic. The third sine wave with frequency $9f$ has a frequency of 9 times the fundamental frequency; it is called the ninth harmonic.

Note that the frequency decomposition of the signal is discrete; it has frequencies f, $3f$, and $9f$. Because f is an integral number, $3f$ and $9f$ are also integral numbers. There are no frequencies such as $1.2f$ or $2.6f$. The frequency domain of a periodic composite signal is always made of discrete spikes.

Example 3.9

Figure 3.12 shows a nonperiodic composite signal. It can be the signal created by a microphone or a telephone set when a word or two is pronounced. In this case, the composite signal cannot be periodic, because that implies that we are repeating the same word or words with exactly the same tone.

Figure 3.12 *The time and frequency domains of a nonperiodic signal*

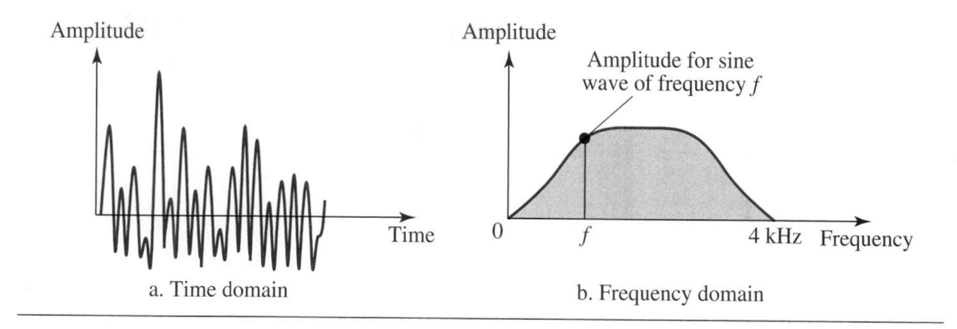

In a time-domain representation of this composite signal, there are an infinite number of simple sine frequencies. Although the number of frequencies in a human voice is infinite, the range is limited. A normal human being can create a continuous range of frequencies between 0 and 4 kHz.

Note that the frequency decomposition of the signal yields a continuous curve. There are an infinite number of frequencies between 0.0 and 4000.0 (real values). To find the amplitude related to frequency f, we draw a vertical line at f to intersect the envelope curve. The height of the vertical line is the amplitude of the corresponding frequency.

3.2.6 Bandwidth

The range of frequencies contained in a composite signal is its **bandwidth.** The bandwidth is normally a difference between two numbers. For example, if a composite signal contains frequencies between 1000 and 5000, its bandwidth is 5000 – 1000, or 4000.

The bandwidth of a composite signal is the difference between the highest and the lowest frequencies contained in that signal.

Figure 3.13 shows the concept of bandwidth. The figure depicts two composite signals, one periodic and the other nonperiodic. The bandwidth of the periodic signal contains all integer frequencies between 1000 and 5000 (1000, 1001, 1002, . . .). The bandwidth of the nonperiodic signals has the same range, but the frequencies are continuous.

Example 3.10

If a periodic signal is decomposed into five sine waves with frequencies of 100, 300, 500, 700, and 900 Hz, what is its bandwidth? Draw the spectrum, assuming all components have a maximum amplitude of 10 V.

Solution

Let f_h be the highest frequency, f_l the lowest frequency, and B the bandwidth. Then

$$B = f_h - f_l = 900 - 100 = 800 \text{ Hz}$$

The spectrum has only five spikes, at 100, 300, 500, 700, and 900 Hz (see Figure 3.14).

Figure 3.13 *The bandwidth of periodic and nonperiodic composite signals*

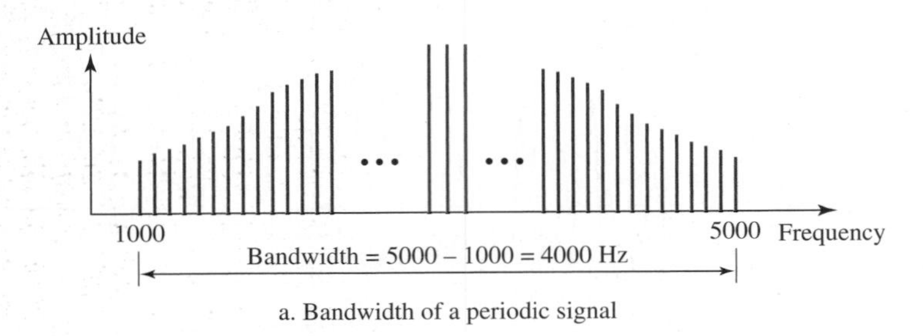

a. Bandwidth of a periodic signal

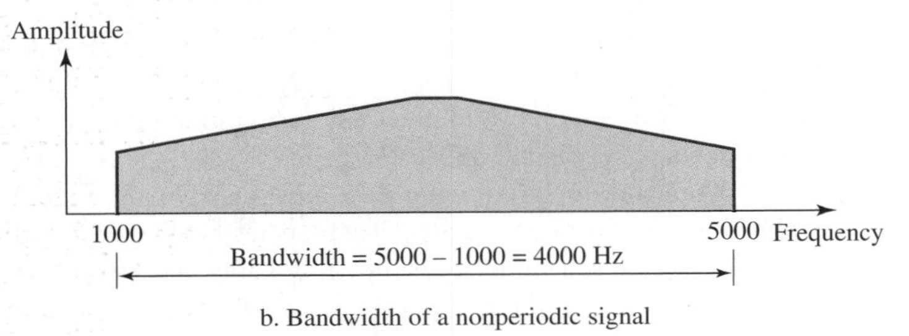

b. Bandwidth of a nonperiodic signal

Figure 3.14 *The bandwidth for Example 3.10*

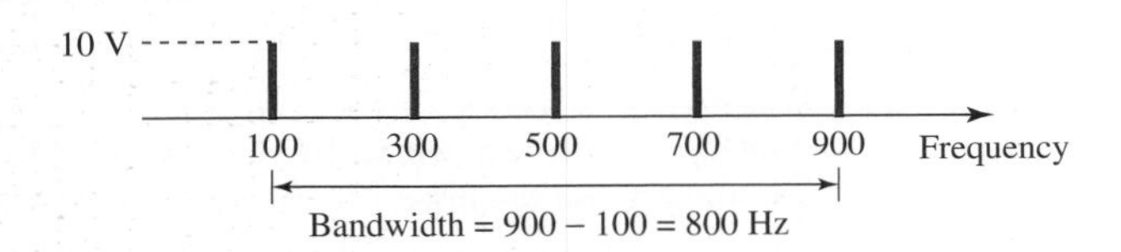

Example 3.11

A periodic signal has a bandwidth of 20 Hz. The highest frequency is 60 Hz. What is the lowest frequency? Draw the spectrum if the signal contains all frequencies of the same amplitude.

Solution

Let f_h be the highest frequency, f_l the lowest frequency, and B the bandwidth. Then

$$B = f_h - f_l \longrightarrow 20 = 60 - f_l \longrightarrow f_l = 60 - 20 = 40 \text{ Hz}$$

The spectrum contains all integer frequencies. We show this by a series of spikes (see Figure 3.15).

Example 3.12

A nonperiodic composite signal has a bandwidth of 200 kHz, with a middle frequency of 140 kHz and peak amplitude of 20 V. The two extreme frequencies have an amplitude of 0. Draw the frequency domain of the signal.

Figure 3.15 *The bandwidth for Example 3.11*

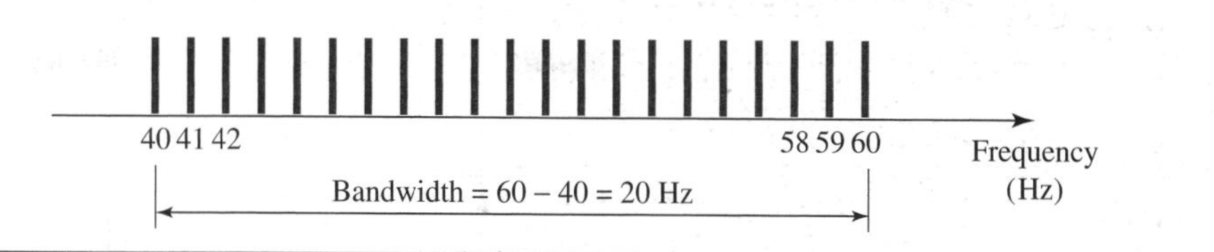

Solution

The lowest frequency must be at 40 kHz and the highest at 240 kHz. Figure 3.16 shows the frequency domain and the bandwidth.

Figure 3.16 *The bandwidth for Example 3.12*

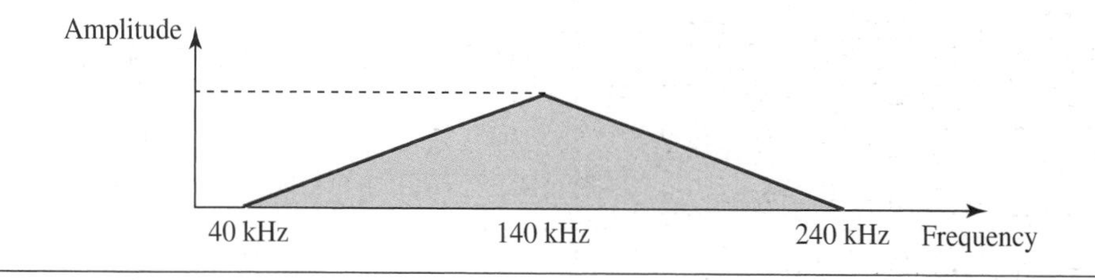

Example 3.13

An example of a nonperiodic composite signal is the signal propagated by an AM radio station. In the United States, each AM radio station is assigned a 10-kHz bandwidth. The total bandwidth dedicated to AM radio ranges from 530 to 1700 kHz. We will show the rationale behind this 10-kHz bandwidth in Chapter 5.

Example 3.14

Another example of a nonperiodic composite signal is the signal propagated by an FM radio station. In the United States, each FM radio station is assigned a 200-kHz bandwidth. The total bandwidth dedicated to FM radio ranges from 88 to 108 MHz. We will show the rationale behind this 200-kHz bandwidth in Chapter 5.

Example 3.15

Another example of a nonperiodic composite signal is the signal received by an old-fashioned analog black-and-white TV. A TV screen is made up of pixels (picture elements) with each pixel being either white or black. The screen is scanned 30 times per second. (Scanning is actually 60 times per second, but odd lines are scanned in one round and even lines in the next and then interleaved.) If we assume a resolution of 525 × 700 (525 vertical lines and 700 horizontal lines), which is a ratio of 3:4, we have 367,500 pixels per screen. If we scan the screen 30 times per second, this is 367,500 × 30 = 11,025,000 pixels per second. The worst-case scenario is alternating black and white pixels. In this case, we need to represent one color by the minimum amplitude and the other color by the maximum amplitude. We can send 2 pixels per cycle. Therefore, we need 11,025,000 / 2 = 5,512,500 cycles per second, or Hz. The bandwidth needed is 5.5124 MHz.

This worst-case scenario has such a low probability of occurrence that the assumption is that we need only 70 percent of this bandwidth, which is 3.85 MHz. Since audio and synchronization signals are also needed, a 4-MHz bandwidth has been set aside for each black and white TV channel. An analog color TV channel has a 6-MHz bandwidth.

3.3 DIGITAL SIGNALS

In addition to being represented by an analog signal, information can also be represented by a digital signal. For example, a 1 can be encoded as a positive voltage and a 0 as zero voltage. A digital signal can have more than two levels. In this case, we can send more than 1 bit for each level. Figure 3.17 shows two signals, one with two levels and the other with four. We send 1 bit per level in part a of the figure and 2 bits per level in part b of the figure. In general, if a signal has L levels, each level needs $\log_2 L$ bits. For this reason, we can send $\log_2 4 = 2$ bits in part b.

Figure 3.17 *Two digital signals: one with two signal levels and the other with four signal levels*

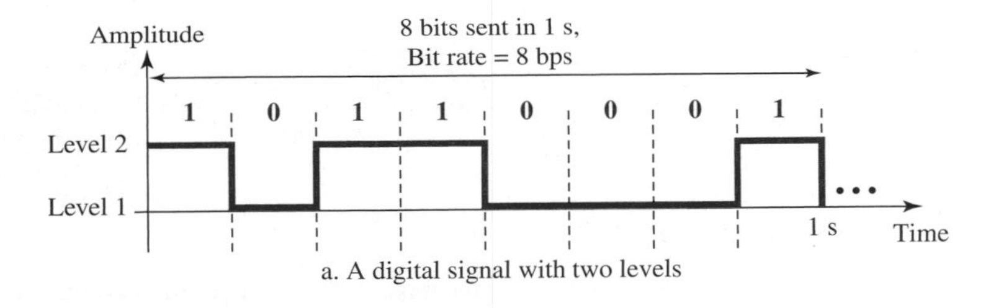

a. A digital signal with two levels

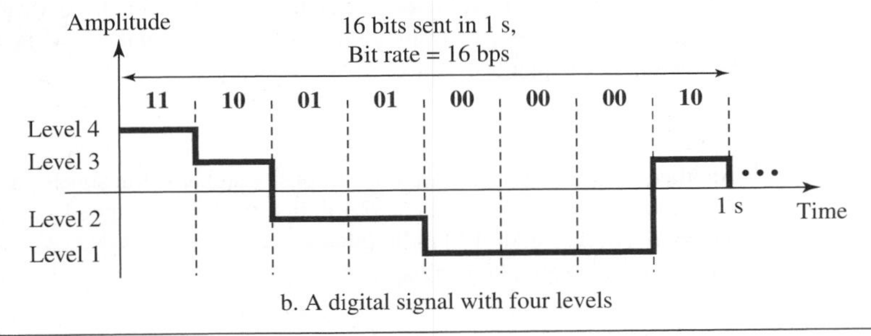

b. A digital signal with four levels

Example 3.16

A digital signal has eight levels. How many bits are needed per level? We calculate the number of bits from the following formula. Each signal level is represented by 3 bits.

$$\text{Number of bits per level} = \log_2 8 = 3$$

Example 3.17

A digital signal has nine levels. How many bits are needed per level? We calculate the number of bits by using the formula. Each signal level is represented by 3.17 bits. However, this answer is not realistic. The number of bits sent per level needs to be an integer as well as a power of 2. For this example, 4 bits can represent one level.

3.3.1 Bit Rate

Most digital signals are nonperiodic, and thus period and frequency are not appropriate characteristics. Another term—*bit rate* (instead of *frequency*)—is used to describe digital signals. The **bit rate** is the number of bits sent in 1s, expressed in **bits per second (bps).** Figure 3.17 shows the bit rate for two signals.

Example 3.18

Assume we need to download text documents at the rate of 100 pages per second. What is the required bit rate of the channel?

Solution

A page is an average of 24 lines with 80 characters in each line. If we assume that one character requires 8 bits, the bit rate is

$$100 \times 24 \times 80 \times 8 = 1{,}536{,}000 \text{ bps} = 1.536 \text{ Mbps}$$

Example 3.19

A digitized voice channel, as we will see in Chapter 4, is made by digitizing a 4-kHz bandwidth analog voice signal. We need to sample the signal at twice the highest frequency (two samples per hertz). We assume that each sample requires 8 bits. What is the required bit rate?

Solution

The bit rate can be calculated as

$$2 \times 4000 \times 8 = 64{,}000 \text{ bps} = 64 \text{ kbps}$$

Example 3.20

What is the bit rate for high-definition TV (HDTV)?

Solution

HDTV uses digital signals to broadcast high quality video signals. The HDTV screen is normally a ratio of 16 : 9 (in contrast to 4 : 3 for regular TV), which means the screen is wider. There are 1920 by 1080 pixels per screen, and the screen is renewed 30 times per second. Twenty-four bits represents one color pixel. We can calculate the bit rate as

$$1920 \times 1080 \times 30 \times 24 = 1{,}492{,}992{,}000 \approx 1.5 \text{ Gbps}$$

The TV stations reduce this rate to 20 to 40 Mbps through compression.

3.3.2 Bit Length

We discussed the concept of the wavelength for an analog signal: the distance one cycle occupies on the transmission medium. We can define something similar for a digital signal: the bit length. The **bit length** is the distance one bit occupies on the transmission medium.

$$\text{Bit length} = \text{propagation speed} \times \text{bit duration}$$

3.3.3 Digital Signal as a Composite Analog Signal

Based on Fourier analysis (See Appendix E), a digital signal is a composite analog signal. The bandwidth is infinite, as you may have guessed. We can intuitively come up with this concept when we consider a digital signal. A digital signal, in the time domain, comprises connected vertical and horizontal line segments. A vertical line in the time domain means a frequency of infinity (sudden change in time); a horizontal line in the time domain means a frequency of zero (no change in time). Going from a frequency of zero to a frequency of infinity (and vice versa) implies all frequencies in between are part of the domain.

Fourier analysis can be used to decompose a digital signal. If the digital signal is periodic, which is rare in data communications, the decomposed signal has a frequency-domain representation with an infinite bandwidth and discrete frequencies. If the digital signal is nonperiodic, the decomposed signal still has an infinite bandwidth, but the frequencies are continuous. Figure 3.18 shows a periodic and a nonperiodic digital signal and their bandwidths.

Figure 3.18 *The time and frequency domains of periodic and nonperiodic digital signals*

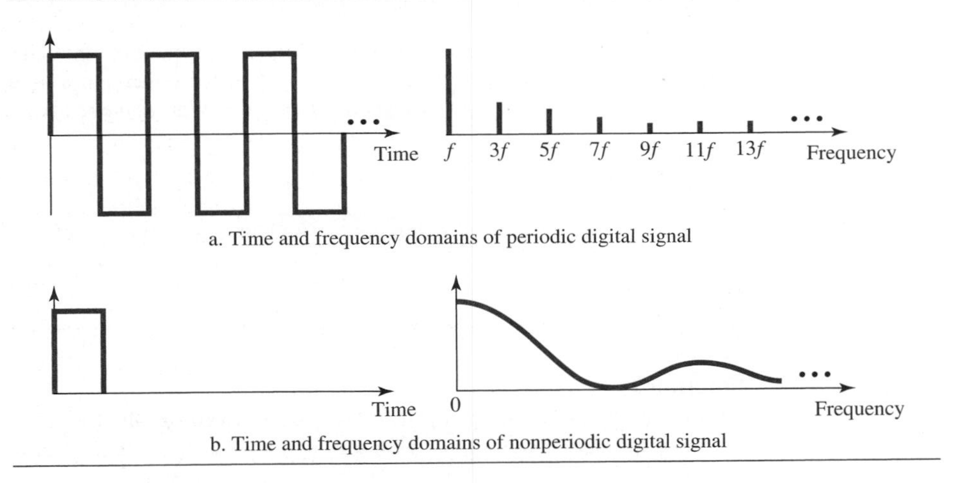

Note that both bandwidths are infinite, but the periodic signal has discrete frequencies while the nonperiodic signal has continuous frequencies.

3.3.4 Transmission of Digital Signals

The previous discussion asserts that a digital signal, periodic or nonperiodic, is a composite analog signal with frequencies between zero and infinity. For the remainder of the discussion, let us consider the case of a nonperiodic digital signal, similar to the ones we encounter in data communications. The fundamental question is, How can we send a digital signal from point *A* to point *B*? We can transmit a digital signal by using one of two different approaches: baseband transmission or broadband transmission (using modulation).

A digital signal is a composite analog signal with an infinite bandwidth.

Baseband Transmission

Baseband transmission means sending a digital signal over a channel without changing the digital signal to an analog signal. Figure 3.19 shows **baseband** transmission.

Figure 3.19 *Baseband transmission*

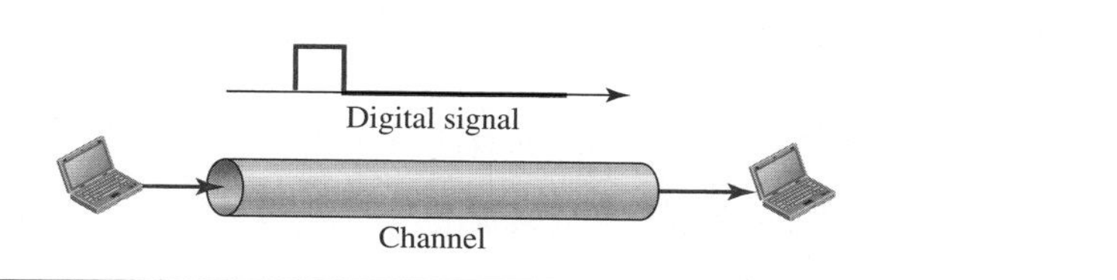

Baseband transmission requires that we have a **low-pass channel,** a channel with a bandwidth that starts from zero. This is the case if we have a dedicated medium with a bandwidth constituting only one channel. For example, the entire bandwidth of a cable connecting two computers is one single channel. As another example, we may connect several computers to a bus, but not allow more than two stations to communicate at a time. Again we have a low-pass channel, and we can use it for baseband communication. Figure 3.20 shows two low-pass channels: one with a narrow bandwidth and the other with a wide bandwidth. We need to remember that a low-pass channel with infinite bandwidth is ideal, but we cannot have such a channel in real life. However, we can get close.

Figure 3.20 *Bandwidths of two low-pass channels*

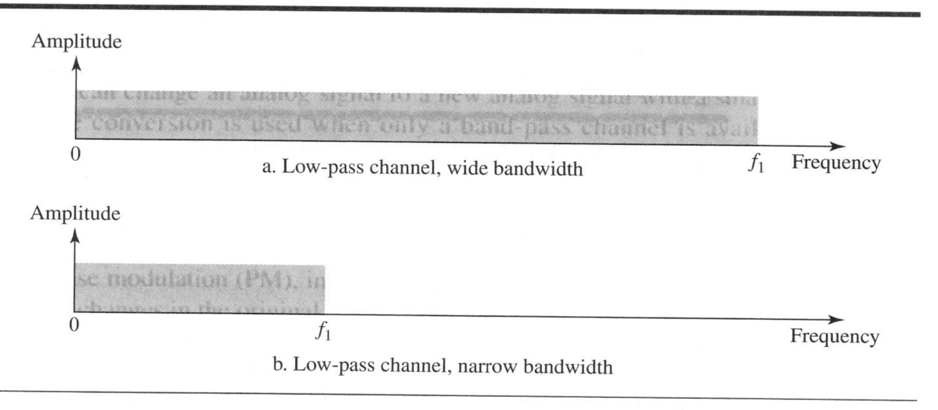

Let us study two cases of a baseband communication: a low-pass channel with a wide bandwidth and one with a limited bandwidth.

Case 1: Low-Pass Channel with Wide Bandwidth

If we want to preserve the exact form of a nonperiodic digital signal with vertical segments vertical and horizontal segments horizontal, we need to send the entire spectrum, the continuous range of frequencies between zero and infinity. This is possible if we have a dedicated medium with an infinite bandwidth between the sender and receiver that preserves the exact amplitude of each component of the composite signal. Although this may be possible inside a computer (e.g., between CPU and memory), it is not possible between two devices. Fortunately, the amplitudes of the frequencies at the border of the bandwidth are so small that they can be ignored. This means that if we have a medium, such as a coaxial or fiber optic cable, with a very wide bandwidth, two stations can communicate by using digital signals with very good accuracy, as shown in Figure 3.21. Note that f_1 is close to zero, and f_2 is very high.

Figure 3.21 *Baseband transmission using a dedicated medium*

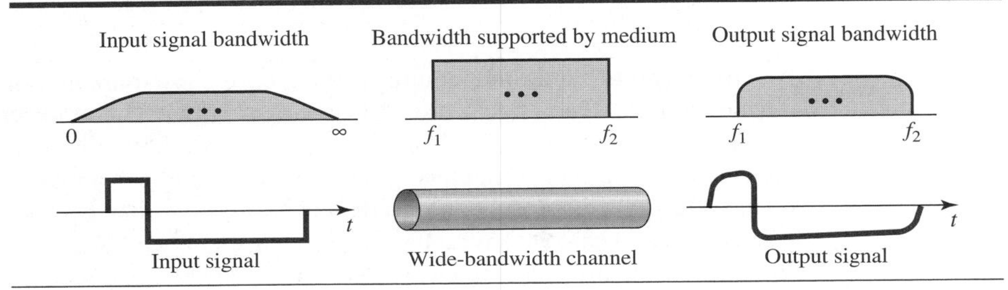

Although the output signal is not an exact replica of the original signal, the data can still be deduced from the received signal. Note that although some of the frequencies are blocked by the medium, they are not critical.

Baseband transmission of a digital signal that preserves the shape of the digital signal is possible only if we have a low-pass channel with an infinite or very wide bandwidth.

Example 3.21

An example of a dedicated channel where the entire bandwidth of the medium is used as one single channel is a LAN. Almost every wired LAN today uses a dedicated channel for two stations communicating with each other. In a bus topology LAN with multipoint connections, only two stations can communicate with each other at each moment in time (timesharing); the other stations need to refrain from sending data. In a star topology LAN, the entire channel between each station and the hub is used for communication between these two entities. We study LANs in Chapter 13.

Case 2: Low-Pass Channel with Limited Bandwidth

In a low-pass channel with limited bandwidth, we approximate the digital signal with an analog signal. The level of approximation depends on the bandwidth available.

Rough Approximation

Let us assume that we have a digital signal of bit rate N. If we want to send analog signals to roughly simulate this signal, we need to consider the worst case, a maximum number of changes in the digital signal. This happens when the signal carries the sequence 01010101 . . . or the sequence 10101010. . . . To simulate these two cases, we need an analog signal of frequency $f = N/2$. Let 1 be the positive peak value and 0 be the negative peak value. We send 2 bits in each cycle; the frequency of the analog signal is one-half of the bit rate, or $N/2$. However, just this one frequency cannot make all patterns; we need more components. The maximum frequency is $N/2$. As an example of this concept, let us see how a digital signal with a 3-bit pattern can be simulated by using analog signals. Figure 3.22 shows the idea. The two similar cases (000 and 111) are simulated with a signal with frequency $f = 0$ and a phase of 180° for 000 and a phase of 0° for 111. The two worst cases (010 and 101) are simulated with an analog signal with frequency $f = N/2$ and phases of 180° and 0°. The other four cases can only be simulated with an analog signal with $f = N/4$ and phases of 180°, 270°, 90°, and 0°. In other words, we need a channel that can handle frequencies 0, $N/4$, and $N/2$. This rough approximation is referred to as using the first harmonic ($N/2$) frequency. The required bandwidth is

$$\textbf{Bandwidth} = \frac{N}{2} - 0 = \frac{N}{2}$$

Figure 3.22 *Rough approximation of a digital signal using the first harmonic for worst case*

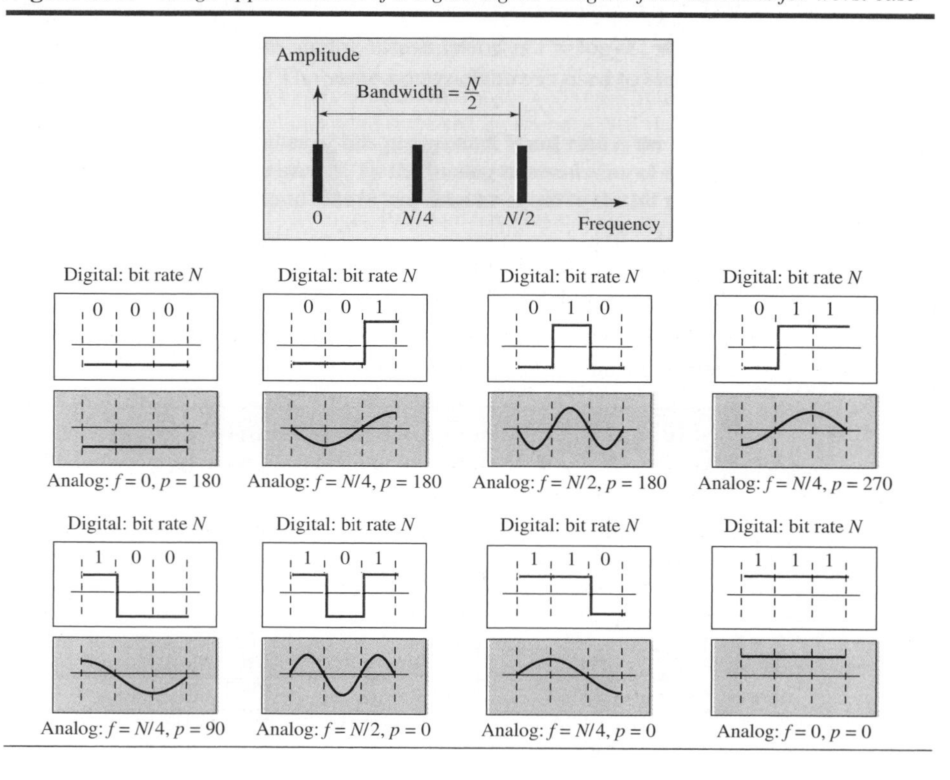

Better Approximation

To make the shape of the analog signal look more like that of a digital signal, we need to add more harmonics of the frequencies. We need to increase the bandwidth. We can increase the bandwidth to $3N/2$, $5N/2$, $7N/2$, and so on. Figure 3.23 shows the effect of

Figure 3.23 *Simulating a digital signal with first three harmonics*

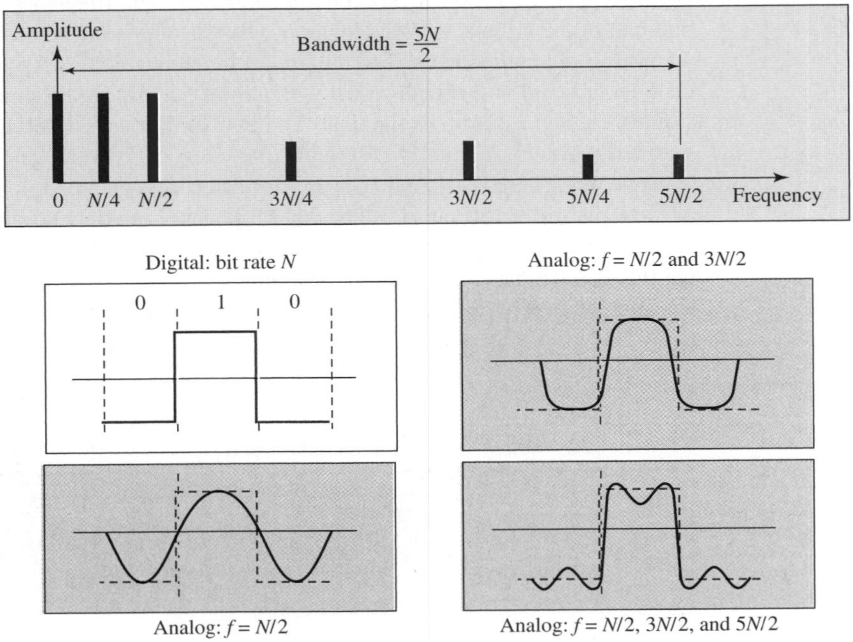

this increase for one of the worst cases, the pattern 010. Note that we have shown only the highest frequency for each harmonic. We use the first, third, and fifth harmonics. The required bandwidth is now $5N/2$, the difference between the lowest frequency 0 and the highest frequency $5N/2$. As we emphasized before, we need to remember that the required bandwidth is proportional to the bit rate.

In baseband transmission, the required bandwidth is proportional to the bit rate; if we need to send bits faster, we need more bandwidth.

By using this method, Table 3.2 shows how much bandwidth we need to send data at different rates.

Table 3.2 *Bandwidth requirements*

Bit Rate	*Harmonic 1*	*Harmonics 1, 3*	*Harmonics 1, 3, 5*
n = 1 kbps	B = 500 Hz	B = 1.5 kHz	B = 2.5 kHz
n = 10 kbps	B = 5 kHz	B = 15 kHz	B = 25 kHz
n = 100 kbps	B = 50 kHz	B = 150 kHz	B = 250 kHz

Example 3.22

What is the required bandwidth of a low-pass channel if we need to send 1 Mbps by using baseband transmission?

Solution

The answer depends on the accuracy desired.

a. The minimum bandwidth, a rough approximation, is B = bit rate /2, or 500 kHz. We need a low-pass channel with frequencies between 0 and 500 kHz.

b. A better result can be achieved by using the first and the third harmonics with the required bandwidth $B = 3 \times 500$ kHz = 1.5 MHz.

c. A still better result can be achieved by using the first, third, and fifth harmonics with $B = 5 \times 500$ kHz = 2.5 MHz.

Example 3.23

We have a low-pass channel with bandwidth 100 kHz. What is the maximum bit rate of this channel?

Solution

The maximum bit rate can be achieved if we use the first harmonic. The bit rate is 2 times the available bandwidth, or 200 kbps.

Broadband Transmission (Using Modulation)

Broadband transmission or modulation means changing the digital signal to an analog signal for transmission. Modulation allows us to use a **bandpass channel**—a channel with a bandwidth that does not start from zero. This type of channel is more available than a low-pass channel. Figure 3.24 shows a bandpass channel.

Figure 3.24 *Bandwidth of a bandpass channel*

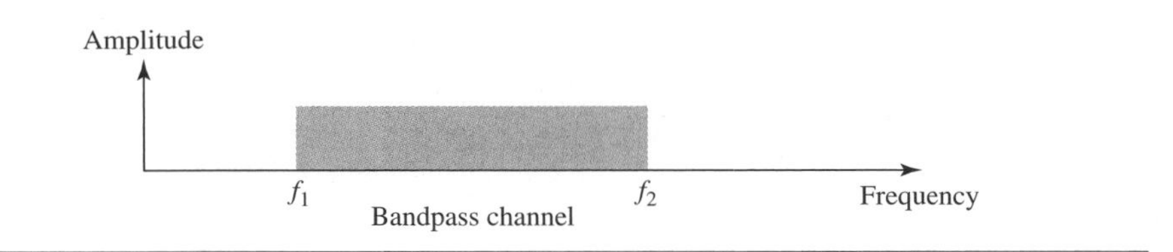

Note that a low-pass channel can be considered a bandpass channel with the lower frequency starting at zero.

Figure 3.25 shows the modulation of a digital signal. In the figure, a digital signal is converted to a composite analog signal. We have used a single-frequency analog signal (called a carrier); the amplitude of the carrier has been changed to look like the digital signal. The result, however, is not a single-frequency signal; it is a composite signal, as we will see in Chapter 5. At the receiver, the received analog signal is converted to digital, and the result is a replica of what has been sent.

If the available channel is a bandpass channel, we cannot send the digital signal directly to the channel; we need to convert the digital signal to an analog signal before transmission.

Figure 3.25 *Modulation of a digital signal for transmission on a bandpass channel*

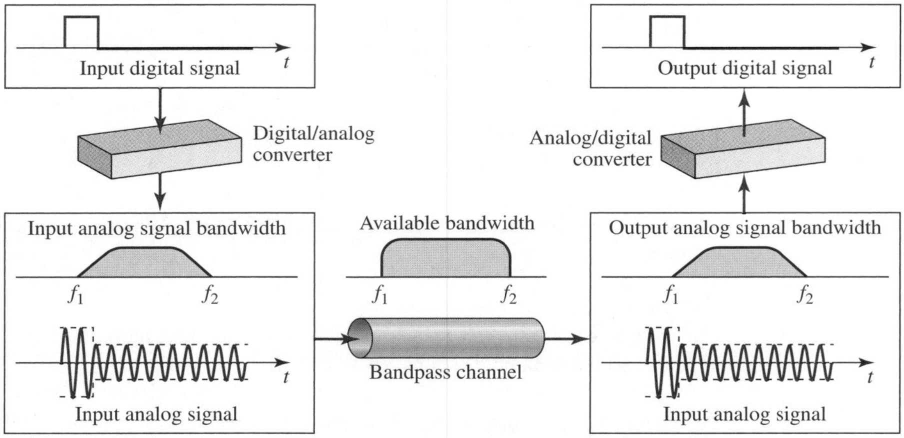

Example 3.24

An example of broadband transmission using modulation is the sending of computer data through a telephone subscriber line, the line connecting a resident to the central telephone office. These lines, installed many years ago, are designed to carry voice (analog signal) with a limited bandwidth (frequencies between 0 and 4 kHz). Although this channel can be used as a low-pass channel, it is normally considered a bandpass channel. One reason is that the bandwidth is so narrow (4 kHz) that if we treat the channel as low-pass and use it for baseband transmission, the maximum bit rate can be only 8 kbps. The solution is to consider the channel a bandpass channel, convert the digital signal from the computer to an analog signal, and send the analog signal. We can install two converters to change the digital signal to analog and vice versa at the receiving end. The converter, in this case, is called a *modem* (*mo*dulator/*dem*odulator), which we discuss in detail in Chapter 5.

Example 3.25

A second example is the digital cellular telephone. For better reception, digital cellular phones convert the analog voice signal to a digital signal (see Chapter 16). Although the bandwidth allocated to a company providing digital cellular phone service is very wide, we still cannot send the digital signal without conversion. The reason is that we only have a bandpass channel available between caller and callee. For example, if the available bandwidth is W and we allow 1000 couples to talk simultaneously, this means the available channel is W/1000, just part of the entire bandwidth. We need to convert the digitized voice to a composite analog signal before sending. The digital cellular phones convert the analog audio signal to digital and then convert it again to analog for transmission over a bandpass channel.

3.4 TRANSMISSION IMPAIRMENT

Signals travel through transmission media, which are not perfect. The imperfection causes signal impairment. This means that the signal at the beginning of the medium is not the same as the signal at the end of the medium. What is sent is not what is received. Three causes of impairment are attenuation, distortion, and noise (see Figure 3.26).

Figure 3.26 *Causes of impairment*

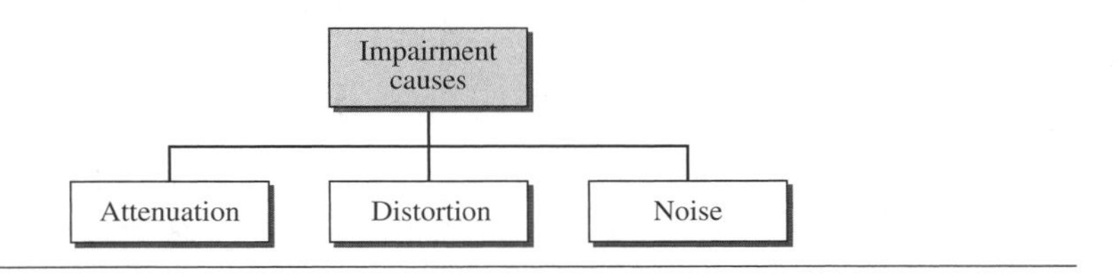

3.4.1 Attenuation

Attenuation means a loss of energy. When a signal, simple or composite, travels through a medium, it loses some of its energy in overcoming the resistance of the medium. That is why a wire carrying electric signals gets warm, if not hot, after a while. Some of the electrical energy in the signal is converted to heat. To compensate for this loss, amplifiers are used to amplify the signal. Figure 3.27 shows the effect of attenuation and amplification.

Figure 3.27 *Attenuation*

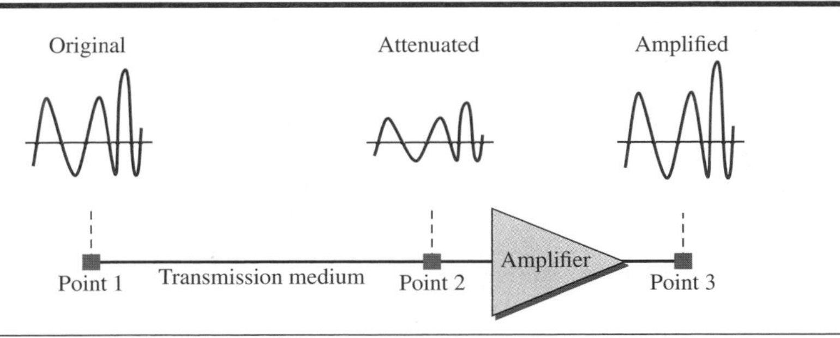

Decibel

To show that a signal has lost or gained strength, engineers use the unit of the decibel. The **decibel (dB)** measures the relative strengths of two signals or one signal at two different points. Note that the decibel is negative if a signal is attenuated and positive if a signal is amplified.

$$\mathbf{dB = 10 \log_{10} \frac{P_2}{P_1}}$$

Variables P_1 and P_2 are the powers of a signal at points 1 and 2, respectively. Note that some engineering books define the decibel in terms of voltage instead of power. In this case, because power is proportional to the square of the voltage, the formula is $\text{dB} = 20 \log_{10} (V_2/V_1)$. In this text, we express dB in terms of power.

Example 3.26

Suppose a signal travels through a transmission medium and its power is reduced to one-half. This means that $P_2 = \frac{1}{2} P_1$. In this case, the attenuation (loss of power) can be calculated as

$$10 \log_{10} \frac{P_2}{P_1} = 10 \log_{10} \frac{0.5P_1}{P_1} = 10 \log_{10} 0.5 = 10(-0.3) = -3 \text{ dB}$$

A loss of 3 dB (–3 dB) is equivalent to losing one-half the power.

Example 3.27

A signal travels through an amplifier, and its power is increased 10 times. This means that $P_2 = 10P_1$. In this case, the amplification (gain of power) can be calculated as

$$10 \log_{10} \frac{P_2}{P_1} = 10 \log_{10} \frac{10P_1}{P_1} = 10 \log_{10} 10 = 10(1) = 10 \text{ dB}$$

Example 3.28

One reason that engineers use the decibel to measure the changes in the strength of a signal is that decibel numbers can be added (or subtracted) when we are measuring several points (cascading) instead of just two. In Figure 3.28 a signal travels from point 1 to point 4. The signal is attenuated by the time it reaches point 2. Between points 2 and 3, the signal is amplified. Again, between points 3 and 4, the signal is attenuated. We can find the resultant decibel value for the signal just by adding the decibel measurements between each set of points.

Figure 3.28 *Decibels for Example 3.28*

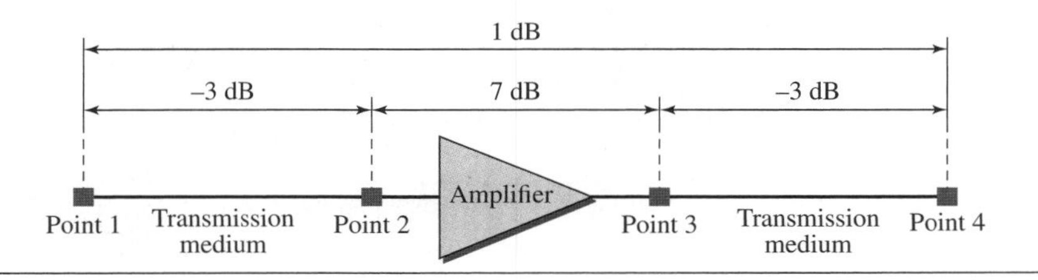

In this case, the decibel value can be calculated as

$$\text{dB} = -3 + 7 - 3 = +1$$

The signal has gained in power.

Example 3.29

Sometimes the decibel is used to measure signal power in milliwatts. In this case, it is referred to as dB_m and is calculated as $\text{dB}_\text{m} = 10 \log_{10} P_m$, where P_m is the power in milliwatts. Calculate the power of a signal if its $\text{dB}_\text{m} = -30$.

Solution

We can calculate the power in the signal as

$$\text{dB}_\text{m} = 10 \log_{10} \longrightarrow dB_m = -30 \longrightarrow \log_{10} P_m = -3 \longrightarrow P_m = 10^{-3} \text{ mW}$$

Example 3.30

The loss in a cable is usually defined in decibels per kilometer (dB/km). If the signal at the beginning of a cable with −0.3 dB/km has a power of 2 mW, what is the power of the signal at 5 km?

Solution

The loss in the cable in decibels is $5 \times (-0.3) = -1.5$ dB. We can calculate the power as

$$\text{dB} = 10 \log_{10} (P_2 / P_1) = -1.5 \quad \longrightarrow \quad (P_2 / P_1) = 10^{-0.15} = 0.71$$

$$P_2 = 0.71 P_1 = 0.7 \times 2 \text{ mW} = 1.4 \text{ mW}$$

3.4.2 Distortion

Distortion means that the signal changes its form or shape. Distortion can occur in a composite signal made of different frequencies. Each signal component has its own propagation speed (see the next section) through a medium and, therefore, its own delay in arriving at the final destination. Differences in delay may create a difference in phase if the delay is not exactly the same as the period duration. In other words, signal components at the receiver have phases different from what they had at the sender. The shape of the composite signal is therefore not the same. Figure 3.29 shows the effect of distortion on a composite signal.

Figure 3.29 *Distortion*

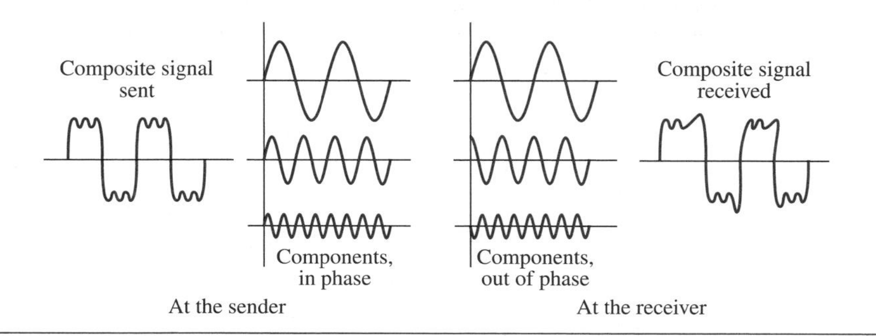

3.4.3 Noise

Noise is another cause of impairment. Several types of noise, such as thermal noise, induced noise, crosstalk, and impulse noise, may corrupt the signal. Thermal noise is the random motion of electrons in a wire, which creates an extra signal not originally sent by the transmitter. Induced noise comes from sources such as motors and appliances. These devices act as a sending antenna, and the transmission medium acts as the receiving antenna. Crosstalk is the effect of one wire on the other. One wire acts as a sending antenna and the other as the receiving antenna. Impulse noise is a spike (a signal with high energy in a very short time) that comes from power lines, lightning, and so on. Figure 3.30 shows the effect of noise on a signal. We discuss error in Chapter 10.

Figure 3.30 *Noise*

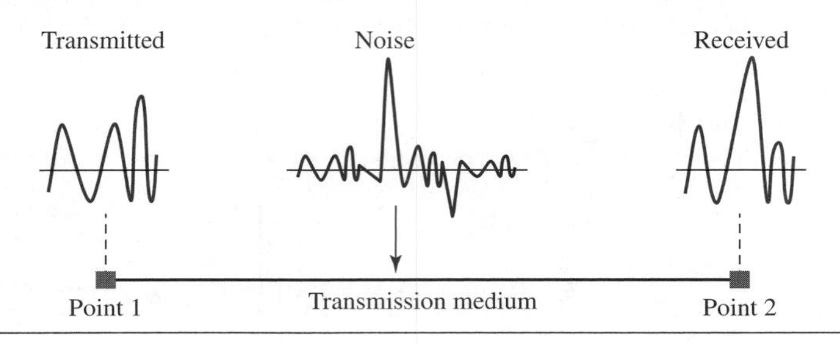

Signal-to-Noise Ratio (SNR)

As we will see later, to find the theoretical bit rate limit, we need to know the ratio of the signal power to the noise power. The **signal-to-noise ratio** is defined as

$$\mathbf{SNR} = \frac{\textbf{average signal power}}{\textbf{average noise power}}$$

We need to consider the average signal power and the average noise power because these may change with time. Figure 3.31 shows the idea of SNR.

Figure 3.31 *Two cases of SNR: a high SNR and a low SNR*

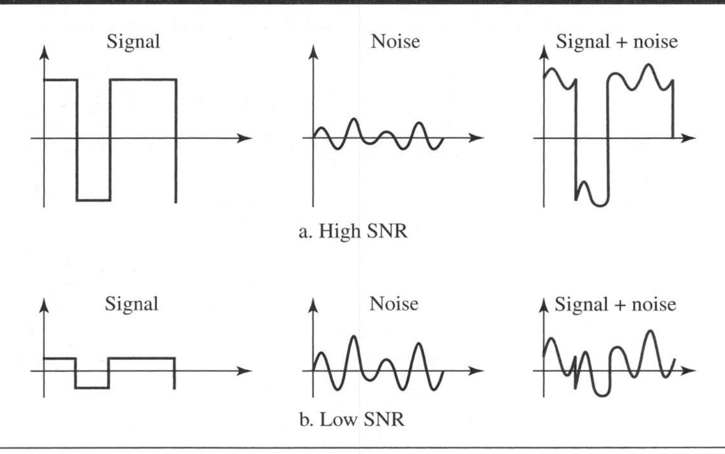

SNR is actually the ratio of what is wanted (signal) to what is not wanted (noise). A high SNR means the signal is less corrupted by noise; a low SNR means the signal is more corrupted by noise.

Because SNR is the ratio of two powers, it is often described in decibel units, SNR_{dB}, defined as

$$\mathbf{SNR_{dB} = 10 \log_{10} SNR}$$

Example 3.31

The power of a signal is 10 mW and the power of the noise is 1 μW; what are the values of SNR and SNR_{dB}?

Solution

The values of SNR and SNR_{dB} can be calculated as follows:

$$SNR = (10{,}000\ \mu w) / (1\ \mu w) = 10{,}000 \qquad SNR_{dB} = 10 \log_{10} 10{,}000 = 10 \log_{10} 10^4 = 40$$

Example 3.32

The values of SNR and SNR_{dB} for a noiseless channel are

$$SNR = (\text{signal power}) / 0 = \infty \longrightarrow SNR_{dB} = 10 \log_{10} \infty = \infty$$

We can never achieve this ratio in real life; it is an ideal.

3.5 DATA RATE LIMITS

A very important consideration in data communications is how fast we can send data, in bits per second, over a channel. Data rate depends on three factors:

1. The bandwidth available
2. The level of the signals we use
3. The quality of the channel (the level of noise)

Two theoretical formulas were developed to calculate the data rate: one by Nyquist for a noiseless channel, another by Shannon for a noisy channel.

3.5.1 Noiseless Channel: Nyquist Bit Rate

For a noiseless channel, the **Nyquist bit rate** formula defines the theoretical maximum bit rate

$$\text{BitRate} = 2 \times \text{bandwidth} \times \log_2 L$$

In this formula, bandwidth is the bandwidth of the channel, L is the number of signal levels used to represent data, and BitRate is the bit rate in bits per second.

According to the formula, we might think that, given a specific bandwidth, we can have any bit rate we want by increasing the number of signal levels. Although the idea is theoretically correct, practically there is a limit. When we increase the number of signal levels, we impose a burden on the receiver. If the number of levels in a signal is just 2, the receiver can easily distinguish between a 0 and a 1. If the level of a signal is 64, the receiver must be very sophisticated to distinguish between 64 different levels. In other words, increasing the levels of a signal reduces the reliability of the system.

Increasing the levels of a signal may reduce the reliability of the system.

Example 3.33

Does the Nyquist theorem bit rate agree with the intuitive bit rate described in baseband transmission?

Solution

They match when we have only two levels. We said, in baseband transmission, the bit rate is 2 times the bandwidth if we use only the first harmonic in the worst case. However, the Nyquist formula is more general than what we derived intuitively; it can be applied to baseband transmission and modulation. Also, it can be applied when we have two or more levels of signals.

Example 3.34

Consider a noiseless channel with a bandwidth of 3000 Hz transmitting a signal with two signal levels. The maximum bit rate can be calculated as

$$\text{BitRate} = 2 \times 3000 \times \log_2 2 = 6000 \text{ bps}$$

Example 3.35

Consider the same noiseless channel transmitting a signal with four signal levels (for each level, we send 2 bits). The maximum bit rate can be calculated as

$$\text{BitRate} = 2 \times 3000 \times \log_2 4 = 12{,}000 \text{ bps}$$

Example 3.36

We need to send 265 kbps over a noiseless channel with a bandwidth of 20 kHz. How many signal levels do we need?

Solution

We can use the Nyquist formula as shown:

$$265{,}000 = 2 \times 20{,}000 \times \log_2 L \longrightarrow \log_2 L = 6.625 \longrightarrow L = 2^{6.625} = 98.7 \text{ levels}$$

Since this result is not a power of 2, we need to either increase the number of levels or reduce the bit rate. If we have 128 levels, the bit rate is 280 kbps. If we have 64 levels, the bit rate is 240 kbps.

3.5.2 Noisy Channel: Shannon Capacity

In reality, we cannot have a noiseless channel; the channel is always noisy. In 1944, Claude Shannon introduced a formula, called the **Shannon capacity,** to determine the theoretical highest data rate for a noisy channel:

$$\text{Capacity} = \text{bandwidth} \times \log_2(1 + \text{SNR})$$

In this formula, bandwidth is the bandwidth of the channel, SNR is the signal-to-noise ratio, and capacity is the capacity of the channel in bits per second. Note that in the Shannon formula there is no indication of the signal level, which means that no matter how many levels we have, we cannot achieve a data rate higher than the capacity of the channel. In other words, the formula defines a characteristic of the channel, not the method of transmission.

Example 3.37

Consider an extremely noisy channel in which the value of the signal-to-noise ratio is almost zero. In other words, the noise is so strong that the signal is faint. For this channel the capacity C is calculated as

$$C = B \log_2 (1 + \text{SNR}) = B \log_2(1 + 0) = B \log_2 1 = B \times 0 = 0$$

This means that the capacity of this channel is zero regardless of the bandwidth. In other words, we cannot receive any data through this channel.

Example 3.38

We can calculate the theoretical highest bit rate of a regular telephone line. A telephone line normally has a bandwidth of 3000 Hz (300 to 3300 Hz) assigned for data communications. The signal-to-noise ratio is usually 3162. For this channel the capacity is calculated as

$$C = B \log_2 (1 + \text{SNR}) = 3000 \log_2(1 + 3162) = 3000 \times 11.62 = 34{,}860 \text{ bps}$$

This means that the highest bit rate for a telephone line is 34.860 kbps. If we want to send data faster than this, we can either increase the bandwidth of the line or improve the signal-to-noise ratio.

Example 3.39

The signal-to-noise ratio is often given in decibels. Assume that $\text{SNR}_{\text{dB}} = 36$ and the channel bandwidth is 2 MHz. The theoretical channel capacity can be calculated as

$$\text{SNR}_{\text{dB}} = 10 \log_{10}\text{SNR} \longrightarrow \text{SNR} = 10^{\text{SNR}_{\text{dB}}/10} \longrightarrow \text{SNR} = 10^{3.6} = 3981$$

$$C = B \log_2(1 + \text{SNR}) = 2 \times 10^6 \times \log_2 3982 = 24 \text{ Mbps}$$

Example 3.40

When the SNR is very high, we can assume that SNR + 1 is almost the same as SNR. In these cases, the theoretical channel capacity can be simplified to $C = B \times \text{SNR}_{\text{dB}}$. For example, we can calculate the theoretical capacity of the previous example as

$$C = 2 \text{ MHz} \times (36 / 3) = 24 \text{ Mbps}$$

3.5.3 Using Both Limits

In practice, we need to use both methods to find the limits and signal levels. Let us show this with an example.

Example 3.41

We have a channel with a 1-MHz bandwidth. The SNR for this channel is 63. What are the appropriate bit rate and signal level?

Solution

First, we use the Shannon formula to find the upper limit.

$$C = B \log_2(1 + \text{SNR}) = 10^6 \log_2(1 + 63) = 10^6 \log_2 64 = 6 \text{ Mbps}$$

The Shannon formula gives us 6 Mbps, the upper limit. For better performance we choose something lower, 4 Mbps, for example. Then we use the Nyquist formula to find the number of signal levels.

$$4 \text{ Mbps} = 2 \times 1 \text{ MHz} \times \log_2 L \longrightarrow L = 4$$

**The Shannon capacity gives us the upper limit;
the Nyquist formula tells us how many signal levels we need.**

3.6 PERFORMANCE

Up to now, we have discussed the tools of transmitting data (signals) over a network and how the data behave. One important issue in networking is the performance of the network—how good is it? We discuss quality of service, an overall measurement of network performance, in greater detail in Chapter 30. In this section, we introduce terms that we need for future chapters.

3.6.1 Bandwidth

One characteristic that measures network performance is bandwidth. However, the term can be used in two different contexts with two different measuring values: bandwidth in hertz and bandwidth in bits per second.

Bandwidth in Hertz

We have discussed this concept. Bandwidth in hertz is the range of frequencies contained in a composite signal or the range of frequencies a channel can pass. For example, we can say the bandwidth of a subscriber telephone line is 4 kHz.

Bandwidth in Bits per Seconds

The term *bandwidth* can also refer to the number of bits per second that a channel, a link, or even a network can transmit. For example, one can say the bandwidth of a Fast Ethernet network (or the links in this network) is a maximum of 100 Mbps. This means that this network can send 100 Mbps.

Relationship

There is an explicit relationship between the bandwidth in hertz and bandwidth in bits per second. Basically, an increase in bandwidth in hertz means an increase in bandwidth in bits per second. The relationship depends on whether we have baseband transmission or transmission with modulation. We discuss this relationship in Chapters 4 and 5.

In networking, we use the term *bandwidth* in two contexts.

- **The first, *bandwidth in hertz,* refers to the range of frequencies in a composite signal or the range of frequencies that a channel can pass.**
- **The second, *bandwidth in bits per second,* refers to the speed of bit transmission in a channel or link.**

Example 3.42

The bandwidth of a subscriber line is 4 kHz for voice or data. The bandwidth of this line for data transmission can be up to 56,000 bps using a sophisticated modem to change the digital signal to analog.

Example 3.43

If the telephone company improves the quality of the line and increases the bandwidth to 8 kHz, we can send 112,000 bps by using the same technology as mentioned in Example 3.42.

3.6.2 Throughput

The **throughput** is a measure of how fast we can actually send data through a network. Although, at first glance, bandwidth in bits per second and throughput seem the same, they are different. A link may have a bandwidth of B bps, but we can only send T bps through this link with T always less than B. In other words, the bandwidth is a potential measurement of a link; the throughput is an actual measurement of how fast we can send data. For example, we may have a link with a bandwidth of 1 Mbps, but the devices connected to the end of the link may handle only 200 kbps. This means that we cannot send more than 200 kbps through this link.

Imagine a highway designed to transmit 1000 cars per minute from one point to another. However, if there is congestion on the road, this figure may be reduced to 100 cars per minute. The bandwidth is 1000 cars per minute; the throughput is 100 cars per minute.

Example 3.44

A network with bandwidth of 10 Mbps can pass only an average of 12,000 frames per minute with each frame carrying an average of 10,000 bits. What is the throughput of this network?

Solution

We can calculate the throughput as

$$\textbf{Throughput} = (12{,}000 \times 10{,}000) / 60 = 2 \text{ Mbps}$$

The throughput is almost one-fifth of the bandwidth in this case.

3.6.3 Latency (Delay)

The **latency** or delay defines how long it takes for an entire message to completely arrive at the destination from the time the first bit is sent out from the source. We can say that latency is made of four components: propagation time, transmission time, queuing time and processing delay.

Latency = propagation time + transmission time + queuing time + processing delay

Propagation Time

Propagation time measures the time required for a bit to travel from the source to the destination. The propagation time is calculated by dividing the distance by the propagation speed.

Propagation time = Distance / (Propagation Speed)

The propagation speed of electromagnetic signals depends on the medium and on the frequency of the signal. For example, in a vacuum, light is propagated with a speed of 3×10^8 m/s. It is lower in air; it is much lower in cable.

Example 3.45

What is the propagation time if the distance between the two points is 12,000 km? Assume the propagation speed to be 2.4×10^8 m/s in cable.

Solution

We can calculate the propagation time as

$$\textbf{Propagation time} = (12{,}000 \times 10{,}000) / (2.4 \times 2^8) = 50 \text{ ms}$$

The example shows that a bit can go over the Atlantic Ocean in only 50 ms if there is a direct cable between the source and the destination.

Transmission Time

In data communications we don't send just 1 bit, we send a message. The first bit may take a time equal to the propagation time to reach its destination; the last bit also may take the same amount of time. However, there is a time between the first bit leaving the sender and the last bit arriving at the receiver. The first bit leaves earlier and arrives earlier; the last bit leaves later and arrives later. The **transmission time** of a message depends on the size of the message and the bandwidth of the channel.

$$\textbf{Transmission time} = (\text{Message size}) / \text{Bandwidth}$$

Example 3.46

What are the propagation time and the transmission time for a 2.5-KB (kilobyte) message (an e-mail) if the bandwidth of the network is 1 Gbps? Assume that the distance between the sender and the receiver is 12,000 km and that light travels at 2.4×10^8 m/s.

Solution

We can calculate the propagation and transmission time as

$$\textbf{Propagation time} = (12{,}000 \times 1000) / (2.4 \times 10^8) = 50 \text{ ms}$$

$$\textbf{Transmission time} = (2500 \times 8) / 10^9 = 0.020 \text{ ms}$$

Note that in this case, because the message is short and the bandwidth is high, the dominant factor is the propagation time, not the transmission time. The transmission time can be ignored.

Example 3.47

What are the propagation time and the transmission time for a 5-MB (megabyte) message (an image) if the bandwidth of the network is 1 Mbps? Assume that the distance between the sender and the receiver is 12,000 km and that light travels at 2.4×10^8 m/s.

Solution

We can calculate the propagation and transmission times as

$$\textbf{Propagation time} = (12{,}000 \times 1000) / (2.4 \times 10^8) = 50 \text{ ms}$$

$$\textbf{Transmission time} = (5{,}000{,}000 \times 8) / 10^6 = 40 \text{ s}$$

Note that in this case, because the message is very long and the bandwidth is not very high, the dominant factor is the transmission time, not the propagation time. The propagation time can be ignored.

Queuing Time

The third component in latency is the **queuing time**, the time needed for each intermediate or end device to hold the message before it can be processed. The queuing time is not a fixed factor; it changes with the load imposed on the network. When there is heavy traffic on the network, the queuing time increases. An intermediate device, such as a router, queues the arrived messages and processes them one by one. If there are many messages, each message will have to wait.

3.6.4 Bandwidth-Delay Product

Bandwidth and delay are two performance metrics of a link. However, as we will see in this chapter and future chapters, what is very important in data communications is the product of the two, the bandwidth-delay product. Let us elaborate on this issue, using two hypothetical cases as examples.

- **Case 1.** Figure 3.32 shows case 1.

Figure 3.32 *Filling the link with bits for case 1*

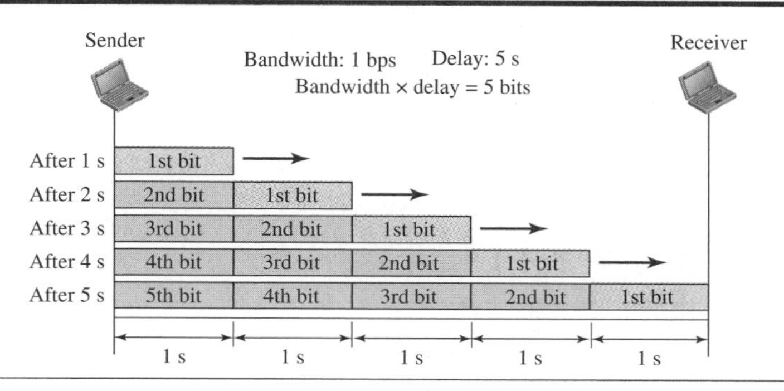

Let us assume that we have a link with a bandwidth of 1 bps (unrealistic, but good for demonstration purposes). We also assume that the delay of the link is 5 s (also unrealistic). We want to see what the bandwidth-delay product means in this case. Looking at the figure, we can say that this product 1×5 is the maximum number of bits that can fill the link. There can be no more than 5 bits at any time on the link.

- **Case 2.** Now assume we have a bandwidth of 5 bps. Figure 3.33 shows that there can be maximum $5 \times 5 = 25$ bits on the line. The reason is that, at each second, there are 5 bits on the line; the duration of each bit is 0.20 s.

The above two cases show that the product of bandwidth and delay is the number of bits that can fill the link. This measurement is important if we need to send data in bursts and wait for the acknowledgment of each burst before sending the next one. To use the maximum capability of the link, we need to make the size of our burst 2 times the product

Figure 3.33 *Filling the link with bits in case 2*

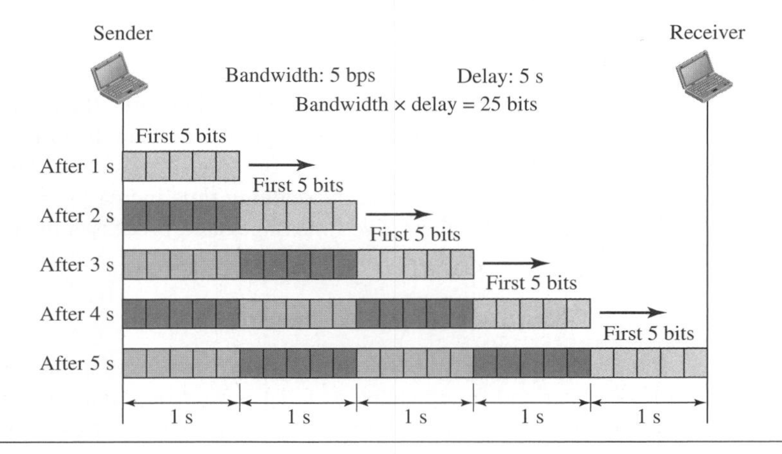

of bandwidth and delay; we need to fill up the full-duplex channel (two directions). The sender should send a burst of data of (2 × bandwidth × delay) bits. The sender then waits for receiver acknowledgment for part of the burst before sending another burst. The amount 2 × bandwidth × delay is the number of bits that can be in transition at any time.

The bandwidth-delay product defines the number of bits that can fill the link.

Example 3.48

We can think about the link between two points as a pipe. The cross section of the pipe represents the bandwidth, and the length of the pipe represents the delay. We can say the volume of the pipe defines the bandwidth-delay product, as shown in Figure 3.34.

Figure 3.34 *Concept of bandwidth-delay product*

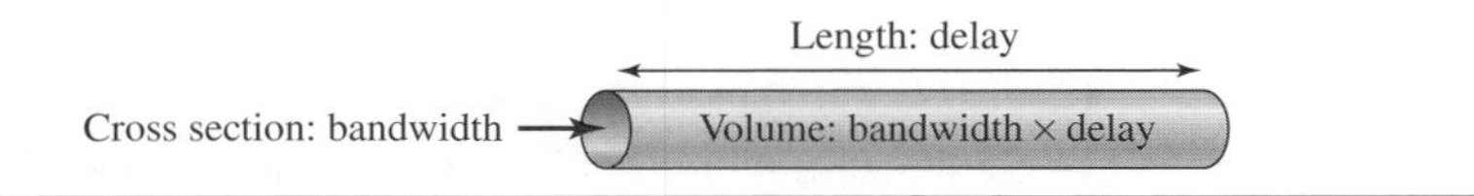

3.6.5 Jitter

Another performance issue that is related to delay is **jitter.** We can roughly say that jitter is a problem if different packets of data encounter different delays and the application using the data at the receiver site is time-sensitive (audio and video data, for example). If the delay for the first packet is 20 ms, for the second is 45 ms, and for the third is 40 ms, then the real-time application that uses the packets endures jitter. We discuss jitter in greater detail in Chapter 28.

3.7 END-CHAPTER MATERIALS

3.7.1 Recommended Reading

For more details about subjects discussed in this chapter, we recommend the following books. The items in brackets [. . .] refer to the reference list at the end of the text.

Books

Data and signals are discussed in [Pea92]. [Cou01] gives excellent coverage of signals. More advanced materials can be found in [Ber96]. [Hsu03] gives a good mathematical approach to signaling. Complete coverage of Fourier Analysis can be found in [Spi74]. Data and signals are discussed in [Sta04] and [Tan03].

3.7.2 Key Terms

analog
analog data
analog signal
attenuation
bandpass channel
bandwidth
baseband transmission
bit length
bit rate
bits per second (bps)
broadband transmission
composite signal
cycle
data
decibel (dB)
digital
digital data
digital signal
distortion
Fourier analysis
frequency
frequency-domain
fundamental frequency
harmonic
Hertz (Hz)
jitter
latency
low-pass channel
noise
nonperiodic signal
Nyquist bit rate
peak amplitude
period
periodic signal
phase
processing delay
propagation speed
propagation time
queuing time
Shannon capacity
signal
signal-to-noise ratio (SNR)
sine wave
throughput
time-domain
transmission time
wavelength

3.7.3 Summary

Data must be transformed to electromagnetic signals to be transmitted. Data can be analog or digital. Analog data are continuous and take continuous values. Digital data have discrete states and take discrete values. Signals can be analog or digital. Analog signals can have an infinite number of values in a range; digital signals can have only a limited number of values.

In data communications, we commonly use periodic analog signals and nonperiodic digital signals. Frequency and period are the inverse of each other. Frequency is the rate of change with respect to time. Phase describes the position of the waveform relative to time 0. A complete sine wave in the time domain can be represented by one single spike in the frequency domain. A single-frequency sine wave is not useful in data communications; we need to send a composite signal, a signal made of many simple sine waves. According to Fourier analysis, any composite signal is a combination of simple sine waves with different frequencies, amplitudes, and phases. The bandwidth of a composite signal is the difference between the highest and the lowest frequencies contained in that signal.

A digital signal is a composite analog signal with an infinite bandwidth. Baseband transmission of a digital signal that preserves the shape of the digital signal is possible only if we have a low-pass channel with an infinite or very wide bandwidth. If the available channel is a bandpass channel, we cannot send a digital signal directly to the channel; we need to convert the digital signal to an analog signal before transmission.

For a noiseless channel, the Nyquist bit rate formula defines the theoretical maximum bit rate. For a noisy channel, we need to use the Shannon capacity to find the maximum bit rate. Attenuation, distortion, and noise can impair a signal. Attenuation is the loss of a signal's energy due to the resistance of the medium. Distortion is the alteration of a signal due to the differing propagation speeds of each of the frequencies that make up a signal. Noise is the external energy that corrupts a signal. The bandwidth-delay product defines the number of bits that can fill the link.

3.8 PRACTICE SET

3.8.1 Quizzes

A set of interactive quizzes for this chapter can be found on the book website. It is strongly recommended that the student take the quizzes to check his/her understanding of the materials before continuing with the practice set.

3.8.2 Questions

Q3-1. How can a composite signal be decomposed into its individual frequencies?

Q3-2. Distinguish between a low-pass channel and a band-pass channel.

Q3-3. What is the relationship between period and frequency?

Q3-4. Is the frequency domain plot of an alarm system discrete or continuous?

Q3-5. Why do optical signals used in fiber optic cables have a very short wave length?

Q3-6. We send a digital signal from one station on a LAN to another station. Is this baseband or broadband transmission?

Q3-7. We send a voice signal from a microphone to a recorder. Is this baseband or broadband transmission?

Q3-8. What does the Shannon capacity have to do with communications?

Q3-9. What does the Nyquist theorem have to do with communications?

Q3-10. Name three types of transmission impairment.

Q3-11. Is the frequency domain plot of a voice signal discrete or continuous?

Q3-12. What does the amplitude of a signal measure? What does the frequency of a signal measure? What does the phase of a signal measure?

Q3-13. Distinguish between baseband transmission and broadband transmission.

Q3-14. Can we say whether a signal is periodic or nonperiodic by just looking at its frequency domain plot? How?

Q3-15. We modulate several voice signals and send them through the air. Is this baseband or broadband transmission?

3.8.3 Problems

P3-1. What is the length of a bit in a channel with a propagation speed of 2×10^8 m/s if the channel bandwidth is

a. 10 Mbps? **b.** 100 Mbps? **c.** 1 Gbps?

P3-2. What is the frequency of the signal in Figure 3.35?

Figure 3.35 *Problem P3-2*

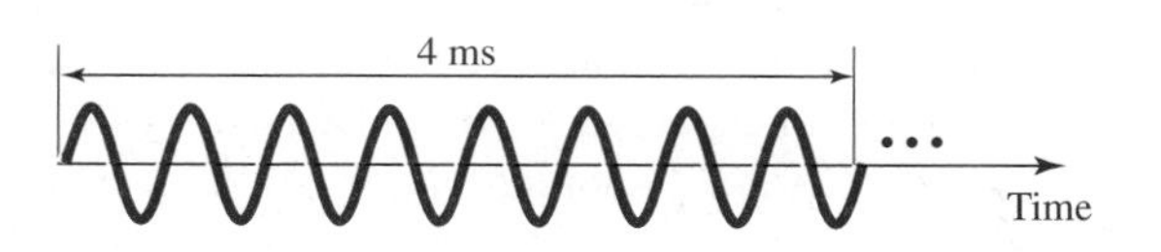

P3-3. The light of the sun takes approximately eight minutes to reach the earth. What is the distance between the sun and the earth?

P3-4. A computer monitor has a resolution of 1600 by 800 pixels. If each pixel uses 1024 colors, how many bits are needed to send the complete contents of a screen?

P3-5. A line has a signal-to-noise ratio of 2000 and a bandwidth of 5000 KHz. What is the maximum data rate supported by this line?

P3-6. A signal has a wavelength of 1 μm in air. How far can the front of the wave travel during 500 periods?

P3-7. What is the bit rate for each of the following signals?

a. A signal in which 1 bit lasts 0.001 s

b. A signal in which 1 bit lasts 2 ms

c. A signal in which 10 bits last 20 μs

P3-8. If the bandwidth of the channel is 5 Kbps, how long does it take to send a frame of 1,000,000 bits out of this device?

P3-9. A signal travels from point A to point B. At point A, the signal power is 100 W. At point B, the power is 80 W. What is the attenuation in decibels?

P3-10. The attenuation of a signal is –10 dB. What is the final signal power if it was originally10 W?

P3-11. What is the bit rate for the signal in Figure 3.36?

Figure 3.36 *Problem P3-11*

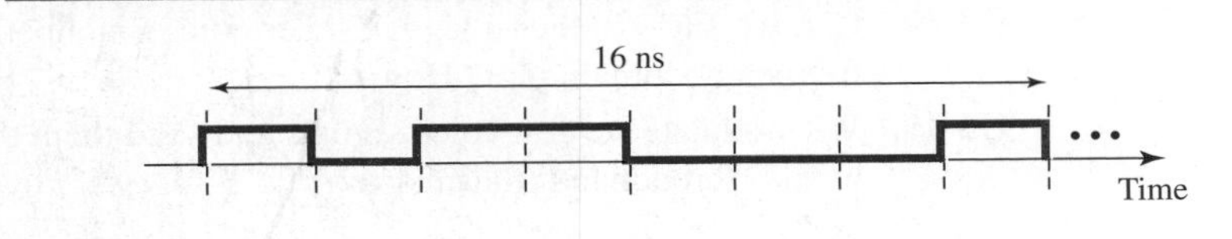

P3-12. A device is sending out data at the rate of 1000 bps.

a. How long does it take to send out 10 bits?

b. How long does it take to send out a single character (8 bits)?

c. How long does it take to send a file of 100,000 characters?

P3-13. What is the phase shift for the following?

a. A sine wave with the maximum amplitude at time zero

b. A sine wave with maximum amplitude after 1/4 cycle

c. A sine wave with zero amplitude after 3/4 cycle and increasing

P3-14. Which signal has a wider bandwidth, a sine wave with a frequency of 100 Hz or a sine wave with a frequency of 200 Hz?

P3-15. A file contains 3 million bytes. How long does it take to download this file using a 56-Kbps channel? 1-Mbps channel?

P3-16. How many bits can fit on a link with a 2 ms delay if the bandwidth of the link is

a. 1 Mbps? **b.** 10 Mbps? **c.** 100 Mbps?

P3-17. What is the theoretical capacity of a channel in each of the following cases?

a. Bandwidth: 30 KHz SNR_{dB} ∇ 40

b. Bandwidth: 100 KHz SNR_{dB} ∇ 4

c. Bandwidth: 1 MHz SNR_{dB} ∇ 20

P3-18. A TV channel has a bandwidth of 6 MHz. If we send a digital signal using one channel, what are the data rates if we use one harmonic, three harmonics, and five harmonics?

P3-19. A signal with 200 milliwatts power passes through 20 devices, each with an average noise of 2 microwatts. What is the SNR? What is the SNRdB?

P3-20. What is the bandwidth of a signal that can be decomposed into five sine waves with frequencies at 0, 20, 50, 100, and 200 Hz? All peak amplitudes are the same. Draw the bandwidth.

P3-21. What is the bandwidth of the composite signal shown in Figure 3.37?

Figure 3.37 *Problem P3-21*

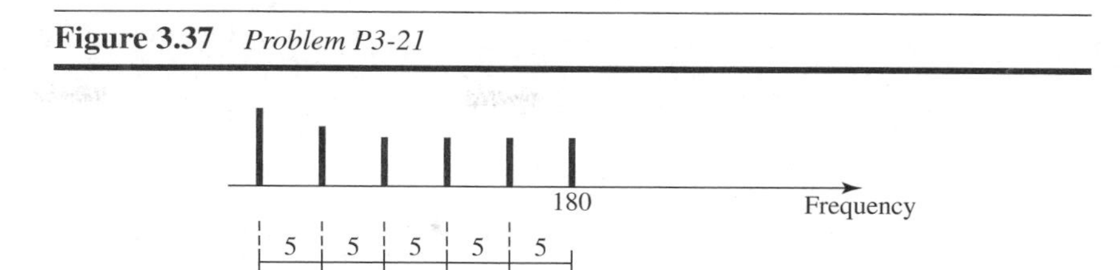

P3-22. Given the following periods, calculate the corresponding frequencies.

a. 5 s **b.** 12 μs **c.** 220 ns

P3-23. A signal has passed through three cascaded amplifiers, each with a 4 dB gain. What is the total gain? How much is the signal amplified?

P3-24. If the peak voltage value of a signal is 20 times the peak voltage value of the noise, what is the SNR? What is the SNR_{dB}?

P3-25. A periodic composite signal with a bandwidth of 2000 Hz is composed of two sine waves. The first one has a frequency of 100 Hz with a maximum amplitude of 20 V; the second one has a maximum amplitude of 5 V. Draw the bandwidth.

P3-26. We measure the performance of a telephone line (4 KHz of bandwidth). When the signal is 10 V, the noise is 5 mV. What is the maximum data rate supported by this telephone line?

P3-27. A nonperiodic composite signal contains frequencies from 10 to 30 KHz. The peak amplitude is 10 V for the lowest and the highest signals and is 30 V for the 20-KHz signal. Assuming that the amplitudes change gradually from the minimum to the maximum, draw the frequency spectrum.

P3-28. We need to upgrade a channel to a higher bandwidth. Answer the following questions:

a. How is the rate improved if we double the bandwidth?

b. How is the rate improved if we double the SNR?

P3-29. We have a channel with 4 KHz bandwidth. If we want to send data at 100 Kbps, what is the minimum SNR_{dB}? What is the SNR?

P3-30. A periodic composite signal contains frequencies from 10 to 30 KHz, each with an amplitude of 10 V. Draw the frequency spectrum.

P3-31. Given the frequencies listed below, calculate the corresponding periods.

a. 24 Hz **b.** 8 MHz **c.** 140 KHz

P3-32. What is the transmission time of a packet sent by a station if the length of the packet is 1 million bytes and the bandwidth of the channel is 200 Kbps?

P3-33. What is the total delay (latency) for a frame of size 5 million bits that is being sent on a link with 10 routers each having a queuing time of 2 μs and a processing time of 1 μs. The length of the link is 2000 Km. The speed of light inside the link is 2×10^8 m/s. The link has a bandwidth of 5 Mbps. Which component of the total delay is dominant? Which one is negligible?

3.9 SIMULATION EXPERIMENTS

3.9.1 Applets

We have created some Java applets to show some of the main concepts discussed in this chapter. It is strongly recommended that the students activate these applets on the book website and carefully examine the protocols in action.

CHAPTER 4

Digital Transmission

A computer network is designed to send information from one point to another. This information needs to be converted to either a digital signal or an analog signal for transmission. In this chapter, we discuss the first choice, conversion to digital signals; in Chapter 5, we discuss the second choice, conversion to analog signals.

We discussed the advantages and disadvantages of digital transmission over analog transmission in Chapter 3. In this chapter, we show the schemes and techniques that we use to transmit data digitally. First, we discuss **digital-to-digital conversion** techniques, methods which convert digital data to digital signals. Second, we discuss **analog-to-digital conversion** techniques, methods which change an analog signal to a digital signal. Finally, we discuss **transmission modes.** We have divided this chapter into three sections:

- The first section discusses digital-to-digital conversion. Line coding is used to convert digital data to a digital signal. Several common schemes are discussed. The section also describes block coding, which is used to create redundancy in the digital data before they are encoded as a digital signal. Redundancy is used as an inherent error detecting tool. The last topic in this section discusses scrambling, a technique used for digital-to-digital conversion in long-distance transmission.
- The second section discusses analog-to-digital conversion. Pulse code modulation is described as the main method used to sample an analog signal. Delta modulation is used to improve the efficiency of the pulse code modulation.
- The third section discusses transmission modes. When we want to transmit data digitally, we need to think about parallel or serial transmission. In parallel transmission, we send multiple bits at a time; in serial transmission, we send one bit at a time.

4.1 DIGITAL-TO-DIGITAL CONVERSION

In Chapter 3, we discussed data and signals. We said that data can be either digital or analog. We also said that signals that represent data can also be digital or analog. In this section, we see how we can represent digital data by using digital signals. The conversion involves three techniques: line coding, block coding, and scrambling. Line coding is always needed; block coding and scrambling may or may not be needed.

4.1.1 Line Coding

Line coding is the process of converting digital data to digital signals. We assume that data, in the form of text, numbers, graphical images, audio, or video, are stored in computer memory as sequences of bits (see Chapter 1). Line coding converts a sequence of bits to a digital signal. At the sender, digital data are encoded into a digital signal; at the receiver, the digital data are recreated by decoding the digital signal. Figure 4.1 shows the process.

Figure 4.1 *Line coding and decoding*

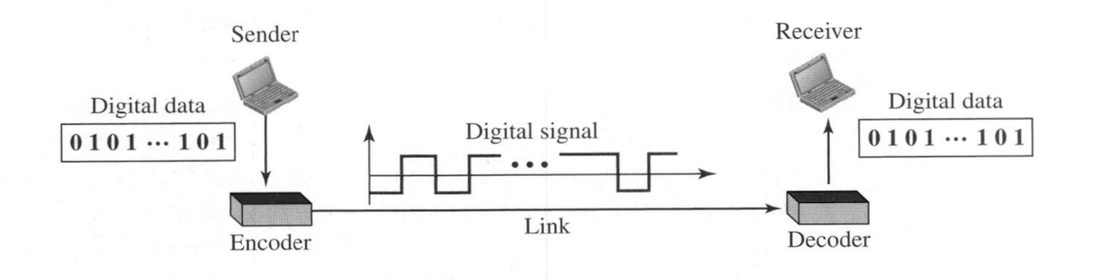

Characteristics

Before discussing different line coding schemes, we address their common characteristics.

Signal Element Versus Data Element

Let us distinguish between a **data element** and a **signal element.** In data communications, our goal is to send data elements. A data element is the smallest entity that can represent a piece of information: this is the bit. In digital data communications, a signal element carries data elements. A signal element is the shortest unit (timewise) of a digital signal. In other words, data elements are what we need to send; signal elements are what we can send. Data elements are being carried; signal elements are the carriers.

We define a ratio r which is the number of data elements carried by each signal element. Figure 4.2 shows several situations with different values of r.

In part a of the figure, one data element is carried by one signal element ($r = 1$). In part b of the figure, we need two signal elements (two transitions) to carry each data element ($r = \frac{1}{2}$). We will see later that the extra signal element is needed to guarantee synchronization. In part c of the figure, a signal element carries two data elements ($r = 2$).

Figure 4.2 *Signal element versus data element*

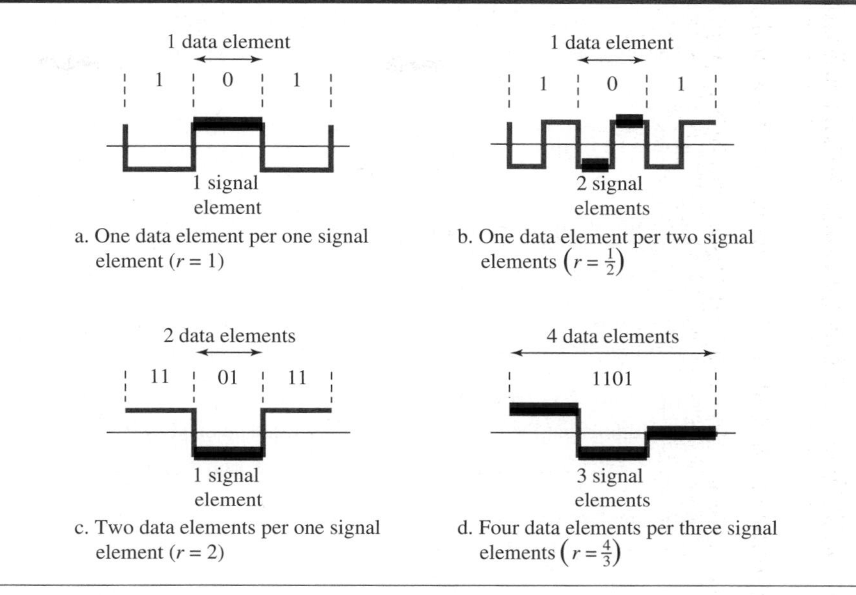

Finally, in part d, a group of 4 bits is being carried by a group of three signal elements ($r = 4/3$). For every line coding scheme we discuss, we will give the value of r.

An analogy may help here. Suppose each data element is a person who needs to be carried from one place to another. We can think of a signal element as a vehicle that can carry people. When $r = 1$, it means each person is driving a vehicle. When $r > 1$, it means more than one person is travelling in a vehicle (a carpool, for example). We can also have the case where one person is driving a car and a trailer ($r = 1/2$).

Data Rate Versus Signal Rate

The **data rate** defines the number of data elements (bits) sent in 1s. The unit is bits per second (bps). The **signal rate** is the number of signal elements sent in 1s. The unit is the baud. There are several common terminologies used in the literature. The data rate is sometimes called the **bit rate;** the signal rate is sometimes called the **pulse rate,** the **modulation rate,** or the **baud rate.**

One goal in data communications is to increase the data rate while decreasing the signal rate. Increasing the data rate increases the speed of transmission; decreasing the signal rate decreases the bandwidth requirement. In our vehicle-people analogy, we need to carry more people in fewer vehicles to prevent traffic jams. We have a limited *bandwidth* in our transportation system.

We now need to consider the relationship between data rate (N) and signal rate (S)

$$S = N/r$$

in which r has been previously defined. This relationship, of course, depends on the value of r. It also depends on the data pattern. If we have a data pattern of all 1s or all 0s, the signal rate may be different from a data pattern of alternating 0s and 1s. To derive a formula for the relationship, we need to define three cases: the worst, best, and average. The worst case is when we need the maximum signal rate; the best case is when we need the minimum. In data communications, we are usually interested in the average case. We can formulate the relationship between data rate and signal rate as

$$S_{ave} = c \times N \times (1/r) \quad \textbf{baud}$$

where N is the data rate (bps); c is the case factor, which varies for each case; S is the number of signal elements per second; and r is the previously defined factor.

Example 4.1

A signal is carrying data in which one data element is encoded as one signal element ($r = 1$). If the bit rate is 100 kbps, what is the average value of the baud rate if c is between 0 and 1?

Solution

We assume that the average value of c is 1/2. The baud rate is then

$$S = c \times N \times (1 / r) = 1/2 \times 100{,}000 \times (1/1) = 50{,}000 = 50 \text{ kbaud}$$

Bandwidth

We discussed in Chapter 3 that a digital signal that carries information is nonperiodic. We also showed that the bandwidth of a nonperiodic signal is continuous with an infinite range. However, most digital signals we encounter in real life have a bandwidth with finite values. In other words, the bandwidth is theoretically infinite, but many of the components have such a small amplitude that they can be ignored. The effective bandwidth is finite. From now on, when we talk about the bandwidth of a digital signal, we need to remember that we are talking about this effective bandwidth.

Although the actual bandwidth of a digital signal is infinite, the effective bandwidth is finite.

We can say that the baud rate, not the bit rate, determines the required bandwidth for a digital signal. If we use the transportation analogy, the number of vehicles, not the number of people being carried, affects the traffic. More changes in the signal mean injecting more frequencies into the signal. (Recall that frequency means change and change means frequency.) The bandwidth reflects the range of frequencies we need. There is a relationship between the baud rate (signal rate) and the bandwidth. Bandwidth is a complex idea. When we talk about the bandwidth, we normally define a range of frequencies. We need to know where this range is located as well as the values of the lowest and the highest frequencies. In addition, the amplitude (if not the phase) of each component is an important issue. In other words, we need more information about the bandwidth than just its value; we need a diagram of the bandwidth. We will show the bandwidth for most schemes we discuss in the chapter. For the moment, we can say that the bandwidth (range of frequencies) is proportional to the signal rate (baud rate). The minimum bandwidth can be given as

$$B_{min} = c \times N \times (1 / r)$$

We can solve for the maximum data rate if the bandwidth of the channel is given.

$$N_{max} = (1 / c) \times B \times r$$

Example 4.2

The maximum data rate of a channel (see Chapter 3) is $N_{max} = 2 \times B \times \log_2 L$ (defined by the Nyquist formula). Does this agree with the previous formula for N_{max}?

Solution

A signal with L levels actually can carry $\log_2 L$ bits per level. If each level corresponds to one signal element and we assume the average case ($c = 1/2$), then we have

$$N_{max} = (1/c) \times B \times r = 2 \times B \times \log_2 L$$

Baseline Wandering

In decoding a digital signal, the receiver calculates a running average of the received signal power. This average is called the ***baseline.*** The incoming signal power is evaluated against this baseline to determine the value of the data element. A long string of 0s or 1s can cause a drift in the baseline (**baseline wandering**) and make it difficult for the receiver to decode correctly. A good line coding scheme needs to prevent baseline wandering.

DC Components

When the voltage level in a digital signal is constant for a while, the spectrum creates very low frequencies (results of Fourier analysis). These frequencies around zero, called DC (direct-current) *components,* present problems for a system that cannot pass low frequencies or a system that uses electrical coupling (via a transformer). We can say that DC component means 0/1 parity that can cause base-line wondering. For example, a telephone line cannot pass frequencies below 200 Hz. Also a long-distance link may use one or more transformers to isolate different parts of the line electrically. For these systems, we need a scheme with no **DC component.**

Self-synchronization

To correctly interpret the signals received from the sender, the receiver's bit intervals must correspond exactly to the sender's bit intervals. If the receiver clock is faster or slower, the bit intervals are not matched and the receiver might misinterpret the signals. Figure 4.3 shows a situation in which the receiver has a shorter bit duration. The sender sends 10110001, while the receiver receives 110111000011.

A **self-synchronizing** digital signal includes timing information in the data being transmitted. This can be achieved if there are transitions in the signal that alert the receiver to the beginning, middle, or end of the pulse. If the receiver's clock is out of synchronization, these points can reset the clock.

Example 4.3

In a digital transmission, the receiver clock is 0.1 percent faster than the sender clock. How many extra bits per second does the receiver receive if the data rate is 1 kbps? How many if the data rate is 1 Mbps?

Solution

At 1 kbps, the receiver receives 1001 bps instead of 1000 bps.

Figure 4.3 *Effect of lack of synchronization*

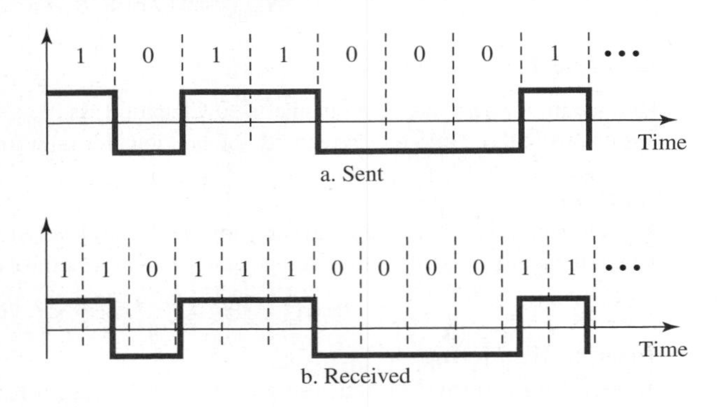

1000 bits sent	→	1001 bits received	→	1 extra bps

At 1 Mbps, the receiver receives 1,001,000 bps instead of 1,000,000 bps.

1,000,000 bits sent	→	1,001,000 bits received	→	1000 extra bps

Built-in Error Detection

It is desirable to have a built-in error-detecting capability in the generated code to detect some or all of the errors that occurred during transmission. Some encoding schemes that we will discuss have this capability to some extent.

Immunity to Noise and Interference

Another desirable code characteristic is a code that is immune to noise and other interferences. Some encoding schemes that we will discuss have this capability.

Complexity

A complex scheme is more costly to implement than a simple one. For example, a scheme that uses four signal levels is more difficult to interpret than one that uses only two levels.

4.1.2 Line Coding Schemes

We can roughly divide line coding schemes into five broad categories, as shown in Figure 4.4.

There are several schemes in each category. We need to be familiar with all schemes discussed in this section to understand the rest of the book. This section can be used as a reference for schemes encountered later.

Unipolar Scheme

In a **unipolar** scheme, all the signal levels are on one side of the time axis, either above or below.

Figure 4.4 *Line coding schemes*

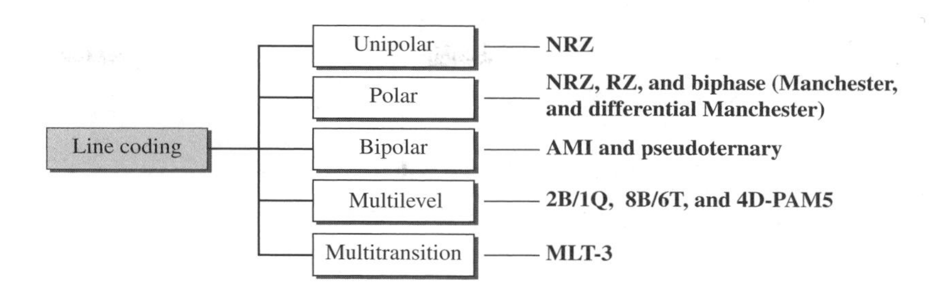

NRZ (Non-Return-to-Zero)

Traditionally, a unipolar scheme was designed as a **non-return-to-zero (NRZ)** scheme in which the positive voltage defines bit 1 and the zero voltage defines bit 0. It is called NRZ because the signal does not return to zero at the middle of the bit. Figure 4.5 shows a unipolar NRZ scheme.

Figure 4.5 *Unipolar NRZ scheme*

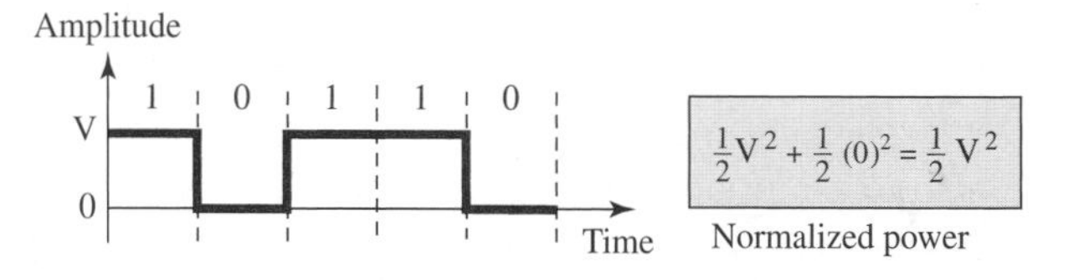

Compared with its polar counterpart (see the next section), this scheme is very costly. As we will see shortly, the normalized power (the power needed to send 1 bit per unit line resistance) is double that for polar NRZ. For this reason, this scheme is normally not used in data communications today.

Polar Schemes

In **polar** schemes, the voltages are on both sides of the time axis. For example, the voltage level for 0 can be positive and the voltage level for 1 can be negative.

Non-Return-to-Zero (NRZ)

In **polar NRZ** encoding, we use two levels of voltage amplitude. We can have two versions of polar NRZ: NRZ-L and NRZ-I, as shown in Figure 4.6. The figure also shows the value of *r*, the average baud rate, and the bandwidth. In the first variation, NRZ-L (**NRZ-Level**), the level of the voltage determines the value of the bit. In the second variation, NRZ-I (**NRZ-Invert**), the change or lack of change in the level of the voltage determines the value of the bit. If there is no change, the bit is 0; if there is a change, the bit is 1.

Figure 4.6 *Polar NRZ-L and NRZ-I schemes*

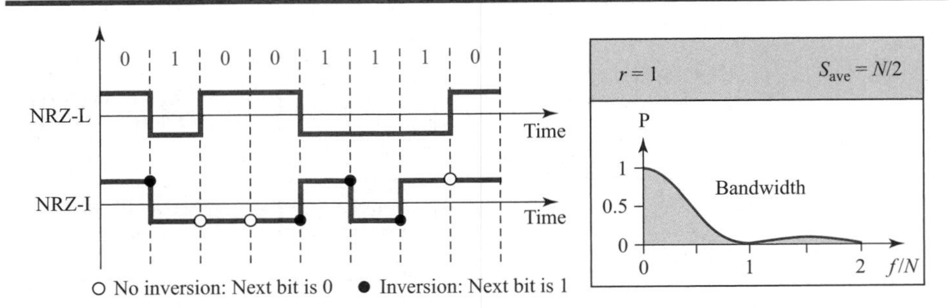

In NRZ-L the level of the voltage determines the value of the bit. In NRZ-I the inversion or the lack of inversion determines the value of the bit.

Let us compare these two schemes based on the criteria we previously defined. Although baseline wandering is a problem for both variations, it is twice as severe in NRZ-L. If there is a long sequence of 0s or 1s in NRZ-L, the average signal power becomes skewed. The receiver might have difficulty discerning the bit value. In NRZ-I this problem occurs only for a long sequence of 0s. If somehow we can eliminate the long sequence of 0s, we can avoid baseline wandering. We will see shortly how this can be done.

The synchronization problem (sender and receiver clocks are not synchronized) also exists in both schemes. Again, this problem is more serious in NRZ-L than in NRZ-I. While a long sequence of 0s can cause a problem in both schemes, a long sequence of 1s affects only NRZ-L.

Another problem with NRZ-L occurs when there is a sudden change of polarity in the system. For example, if twisted-pair cable is the medium, a change in the polarity of the wire results in all 0s interpreted as 1s and all 1s interpreted as 0s. NRZ-I does not have this problem. Both schemes have an average signal rate of $N/2$ Bd.

NRZ-L and NRZ-I both have an average signal rate of $N/2$ Bd.

Let us discuss the bandwidth. Figure 4.6 also shows the normalized bandwidth for both variations. The vertical axis shows the power density (the power for each 1 Hz of bandwidth); the horizontal axis shows the frequency. The bandwidth reveals a very serious problem for this type of encoding. The value of the power density is very high around frequencies close to zero. This means that there are DC components that carry a high level of energy. As a matter of fact, most of the energy is concentrated in frequencies between 0 and $N/2$. This means that although the average of the signal rate is $N/2$, the energy is not distributed evenly between the two halves.

NRZ-L and NRZ-I both have a DC component problem.

Example 4.4

A system is using NRZ-I to transfer 10-Mbps data. What are the average signal rate and minimum bandwidth?

Solution

The average signal rate is $S = N/2 = 500$ kbaud. The minimum bandwidth for this average baud rate is $B_{min} = S = 500$ kHz.

Return-to-Zero (RZ)

The main problem with NRZ encoding occurs when the sender and receiver clocks are not synchronized. The receiver does not know when one bit has ended and the next bit is starting. One solution is the **return-to-zero (RZ)** scheme, which uses three values: positive, negative, and zero. In RZ, the signal changes not between bits but during the bit. In Figure 4.7 we see that the signal goes to 0 in the middle of each bit. It remains there until the beginning of the next bit. The main disadvantage of RZ encoding is that it requires two signal changes to encode a bit and therefore occupies greater bandwidth. The same problem we mentioned, a sudden change of polarity resulting in all 0s interpreted as 1s and all 1s interpreted as 0s, still exists here, but there is no DC component problem. Another problem is the complexity: RZ uses three levels of voltage, which is more complex to create and discern. As a result of all these deficiencies, the scheme is not used today. Instead, it has been replaced by the better-performing Manchester and differential Manchester schemes (discussed next).

Figure 4.7 *Polar RZ scheme*

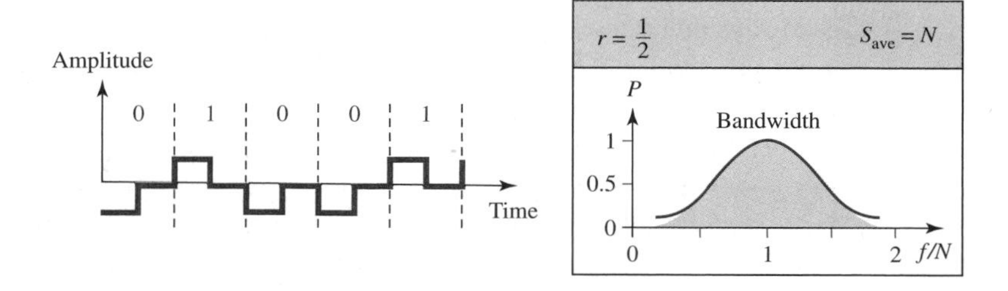

Biphase: Manchester and Differential Manchester

The idea of RZ (transition at the middle of the bit) and the idea of NRZ-L are combined into the **Manchester** scheme. In Manchester encoding, the duration of the bit is divided into two halves. The voltage remains at one level during the first half and moves to the other level in the second half. The transition at the middle of the bit provides synchronization. **Differential Manchester,** on the other hand, combines the ideas of RZ and NRZ-I. There is always a transition at the middle of the bit, but the bit values are determined at the beginning of the bit. If the next bit is 0, there is a transition; if the next bit is 1, there is none. Figure 4.8 shows both Manchester and differential Manchester encoding.

The Manchester scheme overcomes several problems associated with NRZ-L, and differential Manchester overcomes several problems associated with NRZ-I. First, there

Figure 4.8 *Polar biphase: Manchester and differential Manchester schemes*

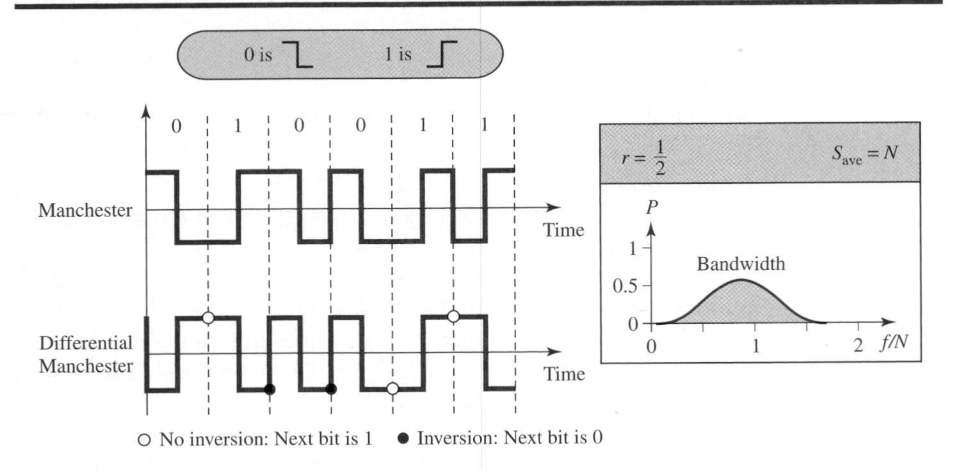

In Manchester and differential Manchester encoding, the transition at the middle of the bit is used for synchronization.

is no baseline wandering. There is no DC component because each bit has a positive and negative voltage contribution. The only drawback is the signal rate. The signal rate for Manchester and differential Manchester is double that for NRZ. The reason is that there is always one transition at the middle of the bit and maybe one transition at the end of each bit. Figure 4.8 shows both Manchester and differential Manchester encoding schemes. Note that Manchester and differential Manchester schemes are also called **biphase** schemes.

The minimum bandwidth of Manchester and differential Manchester is 2 times that of NRZ.

Bipolar Schemes

In **bipolar** encoding (sometimes called ***multilevel binary***), there are three voltage levels: positive, negative, and zero. The voltage level for one data element is at zero, while the voltage level for the other element alternates between positive and negative.

In bipolar encoding, we use three levels: positive, zero, and negative.

AMI and Pseudoternary

Figure 4.9 shows two variations of bipolar encoding: AMI and pseudoternary. A common bipolar encoding scheme is called bipolar **alternate mark inversion (AMI).** In the term *alternate mark inversion,* the word *mark* comes from telegraphy and means 1. So AMI means alternate 1 inversion. A neutral zero voltage represents binary 0. Binary

1s are represented by alternating positive and negative voltages. A variation of AMI encoding is called **pseudoternary** in which the 1 bit is encoded as a zero voltage and the 0 bit is encoded as alternating positive and negative voltages.

Figure 4.9 *Bipolar schemes: AMI and pseudoternary*

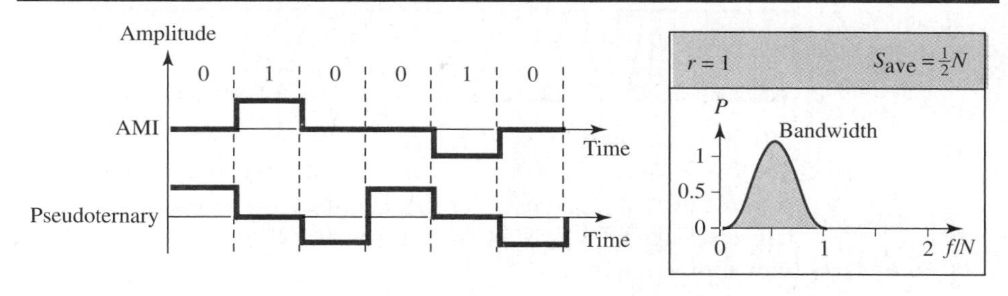

The bipolar scheme was developed as an alternative to NRZ. The bipolar scheme has the same signal rate as NRZ, but there is no DC component. The NRZ scheme has most of its energy concentrated near zero frequency, which makes it unsuitable for transmission over channels with poor performance around this frequency. The concentration of the energy in bipolar encoding is around frequency $N/2$. Figure 4.9 shows the typical energy concentration for a bipolar scheme.

One may ask why we do not have a DC component in bipolar encoding. We can answer this question by using the Fourier transform, but we can also think about it intuitively. If we have a long sequence of 1s, the voltage level alternates between positive and negative; it is not constant. Therefore, there is no DC component. For a long sequence of 0s, the voltage remains constant, but its amplitude is zero, which is the same as having no DC component. In other words, a sequence that creates a constant zero voltage does not have a DC component.

AMI is commonly used for long-distance communication, but it has a synchronization problem when a long sequence of 0s is present in the data. Later in the chapter, we will see how a scrambling technique can solve this problem.

Multilevel Schemes

The desire to increase the data rate or decrease the required bandwidth has resulted in the creation of many schemes. The goal is to increase the number of bits per baud by encoding a pattern of m data elements into a pattern of n signal elements. We only have two types of data elements (0s and 1s), which means that a group of m data elements can produce a combination of 2^m data patterns. We can have different types of signal elements by allowing different signal levels. If we have L different levels, then we can produce L^n combinations of signal patterns. If $2^m = L^n$, then each data pattern is encoded into one signal pattern. If $2^m < L^n$, data patterns occupy only a subset of signal patterns. The subset can be carefully designed to prevent baseline wandering, to provide synchronization, and to detect errors that occurred during data transmission. Data encoding is not possible if $2^m > L^n$ because some of the data patterns cannot be encoded.

The code designers have classified these types of coding as *mBnL*, where *m* is the length of the binary pattern, *B* means binary data, *n* is the length of the signal pattern, and *L* is the number of levels in the signaling. A letter is often used in place of *L*: *B* (binary) for $L = 2$, *T* (ternary) for $L = 3$, and *Q* (quaternary) for $L = 4$. Note that the first two letters define the data pattern, and the second two define the signal pattern.

In *mBnL* schemes, a pattern of *m* data elements is encoded as a pattern of *n* signal elements in which $2^m \leq L^n$.

2B1Q

The first *mBnL* scheme we discuss, **two binary, one quaternary (2B1Q),** uses data patterns of size 2 and encodes the 2-bit patterns as one signal element belonging to a four-level signal. In this type of encoding $m = 2$, $n = 1$, and $L = 4$ (quaternary). Figure 4.10 shows an example of a 2B1Q signal.

Figure 4.10 *Multilevel: 2B1Q scheme*

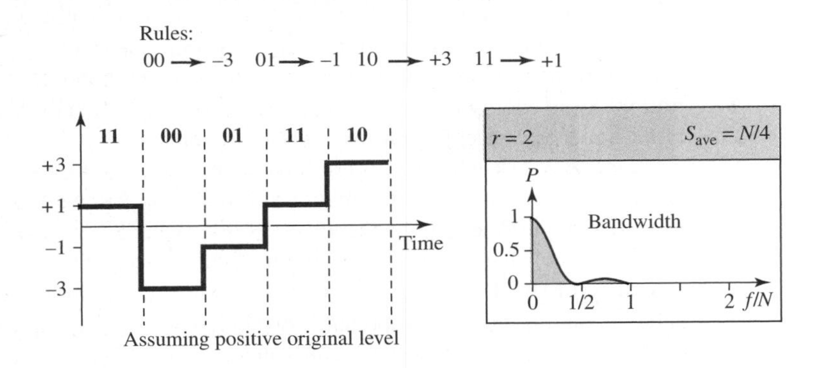

The average signal rate of 2B1Q is $S = N/4$. This means that using 2B1Q, we can send data 2 times faster than by using NRZ-L. However, 2B1Q uses four different signal levels, which means the receiver has to discern four different thresholds. The reduced bandwidth comes with a price. There are no redundant signal patterns in this scheme because $2^2 = 4^1$.

The 2B1Q scheme is used in DSL (Digital Subscriber Line) technology to provide a high-speed connection to the Internet by using subscriber telephone lines (see Chapter 14).

8B6T

A very interesting scheme is **eight binary, six ternary (8B6T).** This code is used with 100BASE-4T cable, as we will see in Chapter 13. The idea is to encode a pattern of 8 bits as a pattern of six signal elements, where the signal has three levels (ternary). In this type of scheme, we can have $2^8 = 256$ different data patterns and $3^6 = 729$ different signal patterns. The mapping table is shown in Appendix F. There are $729 - 256 = 473$ redundant signal elements that provide synchronization and error detection. Part of the redundancy is also used to provide DC balance. Each signal pattern has a weight of 0 or +1 DC values. This means that there is no pattern with the weight −1. To make the whole stream DC-balanced, the sender keeps track of the weight. If two groups of weight 1 are encountered one after another, the first one is sent as is, while the next one is totally inverted to give a weight of −1.

Figure 4.11 shows an example of three data patterns encoded as three signal patterns. The three possible signal levels are represented as −, 0, and +. The first 8-bit pattern 00010001 is encoded as the signal pattern − 0 − 0 + + with weight 0; the second 8-bit pattern 01010011 is encoded as − + − + + 0 with weight +1. The third 8-bit pattern 01010000 should be encoded as + − − + 0 + with weight +1. To create DC balance, the sender inverts the actual signal. The receiver can easily recognize that this is an inverted pattern because the weight is −1. The pattern is inverted before decoding.

Figure 4.11 *Multilevel: 8B6T scheme*

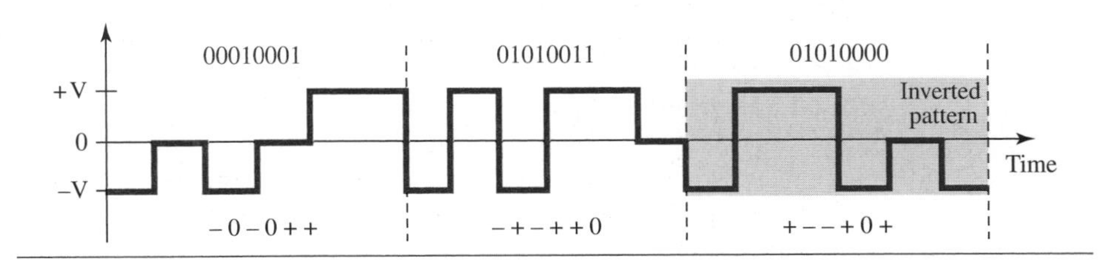

The average signal rate of the scheme is theoretically $S_{ave} = \frac{1}{2} \times N \times \frac{6}{8}$; in practice the minimum bandwidth is very close to $6N/8$.

4D-PAM5

The last signaling scheme we discuss in this category is called **four-dimensional five-level pulse amplitude modulation (4D-PAM5).** The 4D means that data is sent over four wires at the same time. It uses five voltage levels, such as −2, −1, 0, 1, and 2. However, one level, level 0, is used only for forward error detection (discussed in Chapter 10). If we assume that the code is just one-dimensional, the four levels create something similar to 8B4Q. In other words, an 8-bit word is translated to a signal element of four different levels. The worst signal rate for this imaginary one-dimensional version is $N \times 4/8$, or $N/2$.

The technique is designed to send data over four channels (four wires). This means the signal rate can be reduced to $N/8$, a significant achievement. All 8 bits can be fed into a wire simultaneously and sent by using one signal element. The point here is that the four signal elements comprising one signal group are sent simultaneously in a four-dimensional setting. Figure 4.12 shows the imaginary one-dimensional and the actual four-dimensional implementation. Gigabit LANs (see Chapter 13) use this technique to send 1-Gbps data over four copper cables that can handle 125 Mbaud. This scheme has a lot of redundancy in the signal pattern because 2^8 data patterns are matched to 4^4 = 256 signal patterns. The extra signal patterns can be used for other purposes such as error detection.

Figure 4.12 *Multilevel: 4D-PAM5 scheme*

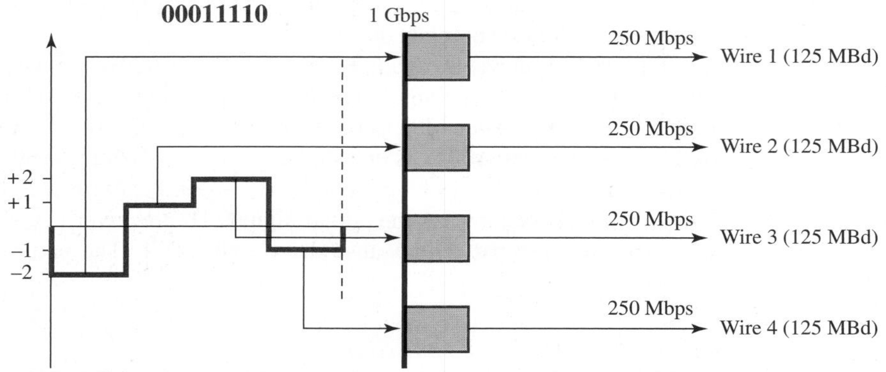

Multitransition: MLT-3

NRZ-I and differential Manchester are classified as differential encoding but use two transition rules to encode binary data (no inversion, inversion). If we have a signal with more than two levels, we can design a differential encoding scheme with more than two transition rules. MLT-3 is one of them. The **multiline transmission, three-level (MLT-3) scheme uses** three levels (+*V*, 0, and −*V*) and three transition rules to move between the levels.

1. If the next bit is 0, there is no transition.
2. If the next bit is 1 and the current level is not 0, the next level is 0.
3. If the next bit is 1 and the current level is 0, the next level is the opposite of the last nonzero level.

The behavior of MLT-3 can best be described by the state diagram shown in Figure 4.13. The three voltage levels (−*V*, 0, and +*V*) are shown by three states (ovals). The transition from one state (level) to another is shown by the connecting lines. Figure 4.13 also shows two examples of an MLT-3 signal.

One might wonder why we need to use MLT-3, a scheme that maps one bit to one signal element. The signal rate is the same as that for NRZ-I, but with greater complexity (three levels and complex transition rules). It turns out that the shape of the signal in this scheme helps to reduce the required bandwidth. Let us look at the worst-case scenario, a sequence of 1s. In this case, the signal element pattern +*V* 0 −*V* 0 is repeated every 4 bits. A nonperiodic signal has changed to a periodic signal with the period equal to 4 times the bit duration. This worst-case situation can be simulated as an analog signal with a frequency one-fourth of the bit rate. In other words, the signal rate for MLT-3 is one-fourth the bit rate. This makes MLT-3 a suitable choice when we need to send 100 Mbps on a copper wire that cannot support more than 32 MHz (frequencies above this level create electromagnetic emissions). MLT-3 and LANs are discussed in Chapter 13.

Figure 4.13 *Multitransition: MLT-3 scheme*

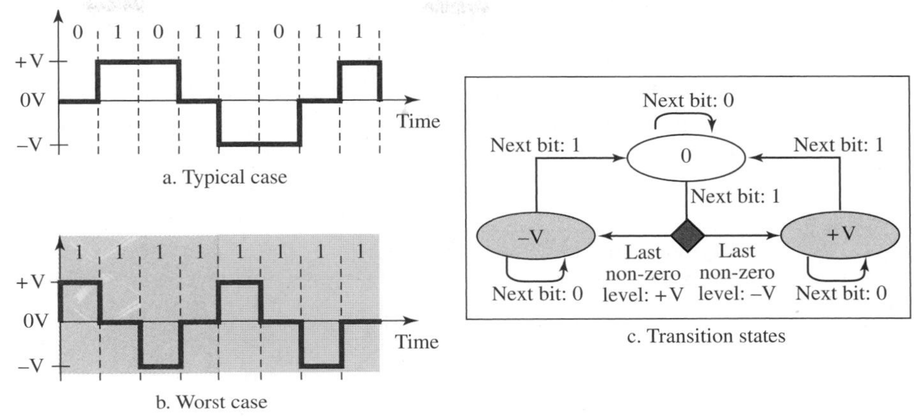

a. Typical case

b. Worst case

c. Transition states

Summary of Line Coding Schemes

We summarize in Table 4.1 the characteristics of the different schemes discussed.

Table 4.1 *Summary of line coding schemes*

Category	*Scheme*	*Bandwidth (average)*	*Characteristics*
Unipolar	NRZ	$B = N/2$	Costly, no self-synchronization if long 0s or 1s, DC
Polar	NRZ-L	$B = N/2$	No self-synchronization if long 0s or 1s, DC
	NRZ-I	$B = N/2$	No self-synchronization for long 0s, DC
	Biphase	$B = N$	Self-synchronization, no DC, high bandwidth
Bipolar	AMI	$B = N/2$	No self-synchronization for long 0s, DC
Multilevel	2B1Q	$B = N/4$	No self-synchronization for long same double bits
	8B6T	$B = 3N/4$	Self-synchronization, no DC
	4D-PAM5	$B = N/8$	Self-synchronization, no DC
Multitransition	MLT-3	$B = N/3$	No self-synchronization for long 0s

4.1.3 Block Coding

We need redundancy to ensure synchronization and to provide some kind of inherent error detecting. Block coding can give us this redundancy and improve the performance of line coding. In general, **block coding** changes a block of *m* bits into a block of *n* bits, where *n* is larger than *m*. Block coding is referred to as an *mB/nB* encoding technique.

Block coding is normally referred to as *mB/nB* coding; it replaces each *m*-bit group with an *n*-bit group.

The slash in block encoding (for example, 4B/5B) distinguishes block encoding from multilevel encoding (for example, 8B6T), which is written without a slash. Block coding normally involves three steps: division, substitution, and combination. In the division step, a sequence of bits is divided into groups of m bits. For example, in 4B/5B encoding, the original bit sequence is divided into 4-bit groups. The heart of block coding is the substitution step. In this step, we substitute an m-bit group with an n-bit group. For example, in 4B/5B encoding we substitute a 4-bit group with a 5-bit group. Finally, the n-bit groups are combined to form a stream. The new stream has more bits than the original bits. Figure 4.14 shows the procedure.

Figure 4.14 *Block coding concept*

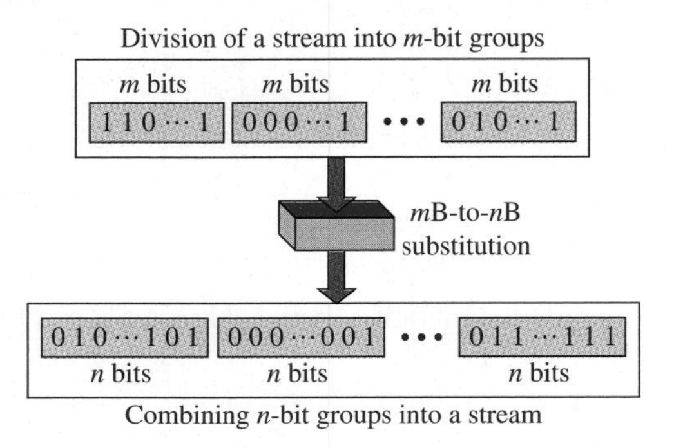

4B/5B

The **four binary/five binary (4B/5B)** coding scheme was designed to be used in combination with NRZ-I. Recall that NRZ-I has a good signal rate, one-half that of the biphase, but it has a synchronization problem. A long sequence of 0s can make the receiver clock lose synchronization. One solution is to change the bit stream, prior to encoding with NRZ-I, so that it does not have a long stream of 0s. The 4B/5B scheme achieves this goal. The block-coded stream does not have more that three consecutive 0s, as we will see later. At the receiver, the NRZ-I encoded digital signal is first decoded into a stream of bits and then decoded to remove the redundancy. Figure 4.15 shows the idea.

Figure 4.15 *Using block coding 4B/5B with NRZ-I line coding scheme*

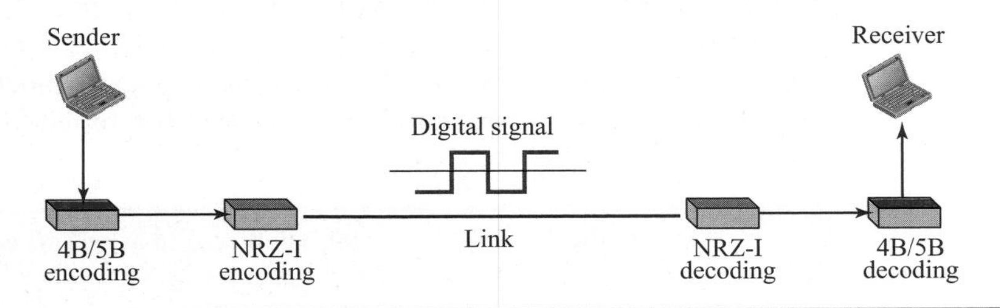

In 4B/5B, the 5-bit output that replaces the 4-bit input has no more than one leading zero (left bit) and no more than two trailing zeros (right bits). So when different groups are combined to make a new sequence, there are never more than three consecutive 0s. (Note that NRZ-I has no problem with sequences of 1s.) Table 4.2 shows the corresponding pairs used in 4B/5B encoding. Note that the first two columns pair a 4-bit group with a 5-bit group. A group of 4 bits can have only 16 different combinations while a group of 5 bits can have 32 different combinations. This means that there are 16 groups that are not used for 4B/5B encoding. Some of these unused groups are used for control purposes; the others are not used at all. The latter provide a kind of error detection. If a 5-bit group arrives that belongs to the unused portion of the table, the receiver knows that there is an error in the transmission.

Table 4.2 *4B/5B mapping codes*

Data Sequence	*Encoded Sequence*	*Control Sequence*	*Encoded Sequence*
0000	11110	Q (Quiet)	00000
0001	01001	I (Idle)	11111
0010	10100	H (Halt)	00100
0011	10101	J (Start delimiter)	11000
0100	01010	K (Start delimiter)	10001
0101	01011	T (End delimiter)	01101
0110	01110	S (Set)	11001
0111	01111	R (Reset)	00111
1000	10010		
1001	10011		
1010	10110		
1011	10111		
1100	11010		
1101	11011		
1110	11100		
1111	11101		

Figure 4.16 shows an example of substitution in 4B/5B coding. 4B/5B encoding solves the problem of synchronization and overcomes one of the deficiencies of NRZ-I. However, we need to remember that it increases the signal rate of NRZ-I. The redundant bits add 20 percent more baud. Still, the result is less than the biphase scheme which has a signal rate of 2 times that of NRZ-I. However, 4B/5B block encoding does not solve the DC component problem of NRZ-I. If a DC component is unacceptable, we need to use biphase or bipolar encoding.

Example 4.5

We need to send data at a 1-Mbps rate. What is the minimum required bandwidth, using a combination of 4B/5B and NRZ-I or Manchester coding?

Solution

First 4B/5B block coding increases the bit rate to 1.25 Mbps. The minimum bandwidth using NRZ-I is $N/2$ or 625 kHz. The Manchester scheme needs a minimum bandwidth of 1 MHz. The

Figure 4.16 *Substitution in 4B/5B block coding*

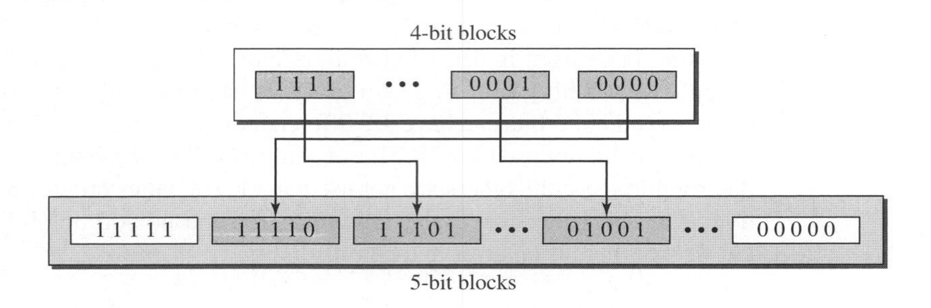

first choice needs a lower bandwidth, but has a DC component problem; the second choice needs a higher bandwidth, but does not have a DC component problem.

8B/10B

The **eight binary/ten binary (8B/10B)** encoding is similar to 4B/5B encoding except that a group of 8 bits of data is now substituted by a 10-bit code. It provides greater error detection capability than 4B/5B. The 8B/10B block coding is actually a combination of 5B/6B and 3B/4B encoding, as shown in Figure 4.17.

Figure 4.17 *8B/10B block encoding*

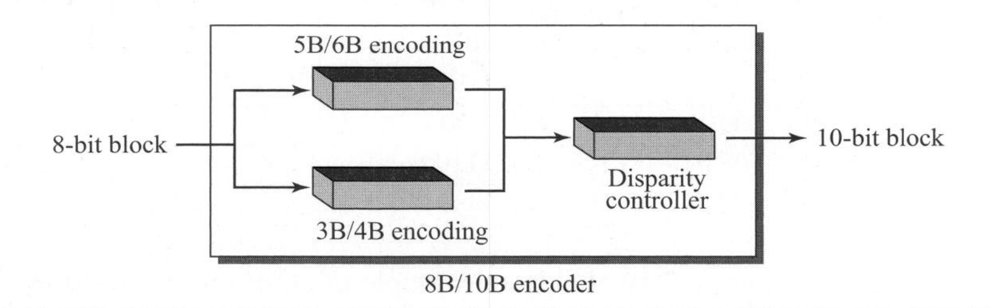

The five most significant bits of a 10-bit block are fed into the 5B/6B encoder; the three least significant bits are fed into a 3B/4B encoder. The split is done to simplify the mapping table. To prevent a long run of consecutive 0s or 1s, the code uses a disparity controller which keeps track of excess 0s over 1s (or 1s over 0s). If the bits in the current block create a disparity that contributes to the previous disparity (either direction), then each bit in the code is complemented (a 0 is changed to a 1 and a 1 is changed to a 0). The coding has $2^{10} - 2^8 = 768$ redundant groups that can be used for disparity checking and error detection. In general, the technique is superior to 4B/5B because of better built-in error-checking capability and better synchronization.

4.1.4 Scrambling

Biphase schemes that are suitable for dedicated links between stations in a LAN are not suitable for long-distance communication because of their wide bandwidth requirement. The combination of block coding and NRZ line coding is not suitable for long-distance encoding either, because of the DC component. Bipolar AMI encoding, on the other hand, has a narrow bandwidth and does not create a DC component. However, a long sequence of 0s upsets the synchronization. If we can find a way to avoid a long sequence of 0s in the original stream, we can use bipolar AMI for long distances. We are looking for a technique that does not increase the number of bits and does provide synchronization. We are looking for a solution that substitutes long zero-level pulses with a combination of other levels to provide synchronization. One solution is called **scrambling.** We modify part of the AMI rule to include scrambling, as shown in Figure 4.18. Note that scrambling, as opposed to block coding, is done at the same time as encoding. The system needs to insert the required pulses based on the defined scrambling rules. Two common scrambling techniques are B8ZS and HDB3.

Figure 4.18 *AMI used with scrambling*

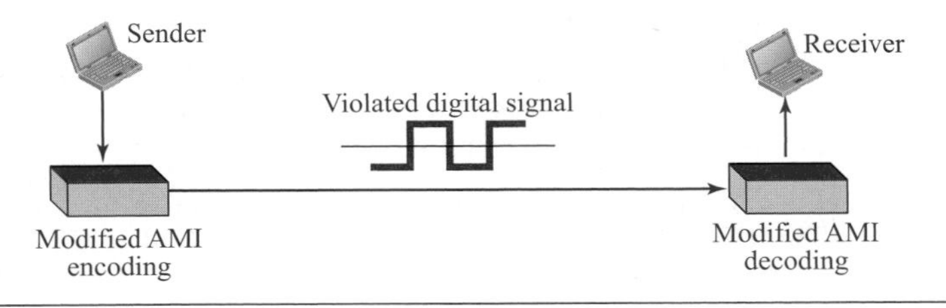

B8ZS

Bipolar with 8-zero substitution (B8ZS) is commonly used in North America. In this technique, eight consecutive zero-level voltages are replaced by the sequence **000VB0VB.** The V in the sequence denotes *violation;* this is a nonzero voltage that breaks an AMI rule of encoding (opposite polarity from the previous). The B in the sequence denotes *bipolar,* which means a nonzero level voltage in accordance with the AMI rule. There are two cases, as shown in Figure 4.19.

Figure 4.19 *Two cases of B8ZS scrambling technique*

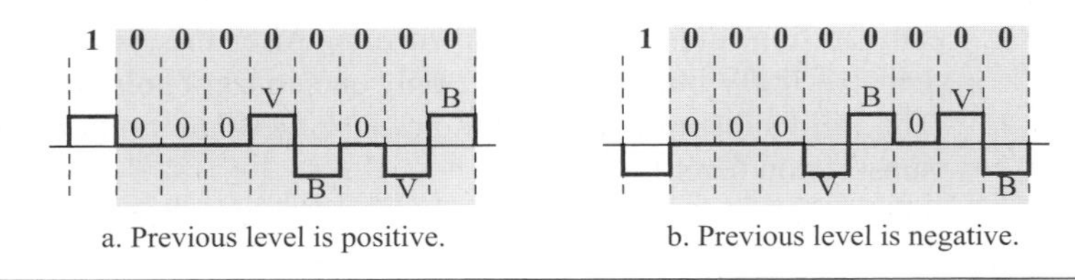

Note that the scrambling in this case does not change the bit rate. Also, the technique balances the positive and negative voltage levels (two positives and two negatives), which means that the DC balance is maintained. Note that the substitution may change the polarity of a 1 because, after the substitution, AMI needs to follow its rules.

B8ZS substitutes eight consecutive zeros with 000VB0VB.

One more point is worth mentioning. The letter V (violation) or B (bipolar) here is relative. The V means the same polarity as the polarity of the previous nonzero pulse; B means the polarity opposite to the polarity of the previous nonzero pulse.

HDB3

High-density bipolar 3-zero (HDB3) is commonly used outside of North America. In this technique, which is more conservative than B8ZS, four consecutive zero-level voltages are replaced with a sequence of **000V** or **B00V.** The reason for two different substitutions is to maintain the even number of nonzero pulses after each substitution. The two rules can be stated as follows:

1. If the number of nonzero pulses after the last substitution is odd, the substitution pattern will be **000V**, which makes the total number of nonzero pulses even.
2. If the number of nonzero pulses after the last substitution is even, the substitution pattern will be **B00V**, which makes the total number of nonzero pulses even.

Figure 4.20 shows an example.

Figure 4.20 *Different situations in HDB3 scrambling technique*

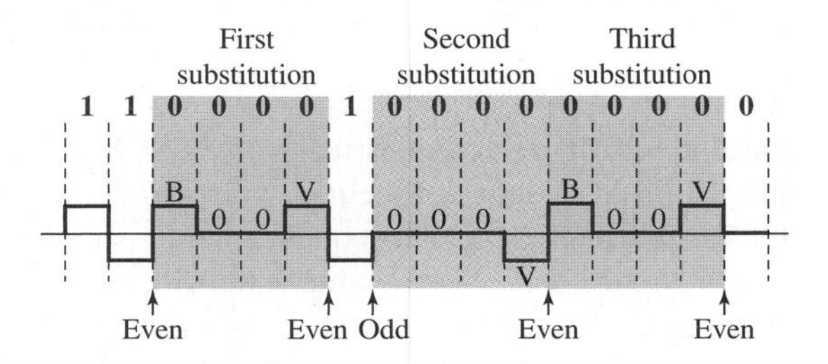

There are several points we need to mention here. First, before the first substitution, the number of nonzero pulses is even, so the first substitution is B00V. After this substitution, the polarity of the 1 bit is changed because the AMI scheme, after each substitution, must follow its own rule. After this bit, we need another substitution, which is 000V because we have only one nonzero pulse (odd) after the last substitution. The third substitution is B00V because there are no nonzero pulses after the second substitution (even).

HDB3 substitutes four consecutive zeros with 000V or B00V depending on the number of nonzero pulses after the last substitution.

4.2 ANALOG-TO-DIGITAL CONVERSION

The techniques described in Section 4.1 convert digital data to digital signals. Sometimes, however, we have an analog signal such as one created by a microphone or camera. We have seen in Chapter 3 that a digital signal is superior to an analog signal. The tendency today is to change an analog signal to digital data. In this section we describe two techniques, pulse code modulation and delta modulation. After the digital data are created (digitization), we can use one of the techniques described in Section 4.1 to convert the digital data to a digital signal.

4.2.1 Pulse Code Modulation (PCM)

The most common technique to change an analog signal to digital data (**digitization**) is called **pulse code modulation (PCM).** A PCM encoder has three processes, as shown in Figure 4.21.

Figure 4.21 *Components of PCM encoder*

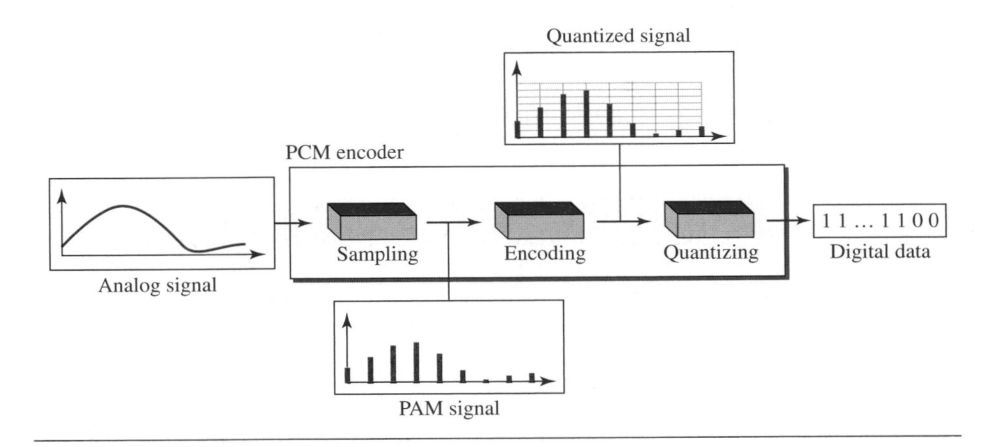

1. The analog signal is sampled.
2. The sampled signal is quantized.
3. The quantized values are encoded as streams of bits.

Sampling

The first step in PCM is **sampling.** The analog signal is sampled every T_s s, where T_s is the sample interval or period. The inverse of the sampling interval is called the ***sampling rate*** or ***sampling frequency*** and denoted by f_s, where $f_s = 1/T_s$. There are three sampling methods—ideal, natural, and flat-top—as shown in Figure 4.22.

In ideal sampling, pulses from the analog signal are sampled. This is an ideal sampling method and cannot be easily implemented. In natural sampling, a high-speed switch is turned on for only the small period of time when the sampling occurs. The result is a sequence of samples that retains the shape of the analog signal. The most

Figure 4.22 *Three different sampling methods for PCM*

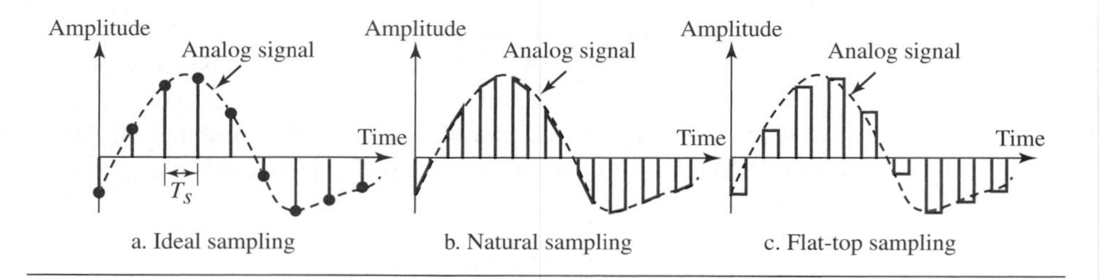

common sampling method, called ***sample and hold,*** however, creates flat-top samples by using a circuit.

The sampling process is sometimes referred to as **pulse amplitude modulation (PAM).** We need to remember, however, that the result is still an analog signal with nonintegral values.

Sampling Rate

One important consideration is the sampling rate or frequency. What are the restrictions on T_s? This question was elegantly answered by Nyquist. According to the **Nyquist theorem,** to reproduce the original analog signal, one necessary condition is that the *sampling rate* be at least twice the highest frequency in the original signal.

> **According to the Nyquist theorem, the sampling rate must be at least 2 times the highest frequency contained in the signal.**

We need to elaborate on the theorem at this point. First, we can sample a signal only if the signal is band-limited. In other words, a signal with an infinite bandwidth cannot be sampled. Second, the sampling rate must be at least 2 times the highest frequency, not the bandwidth. If the analog signal is low-pass, the bandwidth and the highest frequency are the same value. If the analog signal is bandpass, the bandwidth value is lower than the value of the maximum frequency. Figure 4.23 shows the value of the sampling rate for two types of signals.

Example 4.6

For an intuitive example of the Nyquist theorem, let us sample a simple sine wave at three sampling rates: $f_s = 4f$ (2 times the Nyquist rate), $f_s = 2f$ (Nyquist rate), and $f_s = f$ (one-half the Nyquist rate). Figure 4.24 shows the sampling and the subsequent recovery of the signal.

It can be seen that sampling at the Nyquist rate can create a good approximation of the original sine wave (part a). Oversampling in part b can also create the same approximation, but it is redundant and unnecessary. Sampling below the Nyquist rate (part c) does not produce a signal that looks like the original sine wave.

Example 4.7

As an interesting example, let us see what happens if we sample a periodic event such as the revolution of a hand of a clock. The second hand of a clock has a period of 60 s. According to the

Figure 4.23 *Nyquist sampling rate for low-pass and bandpass signals*

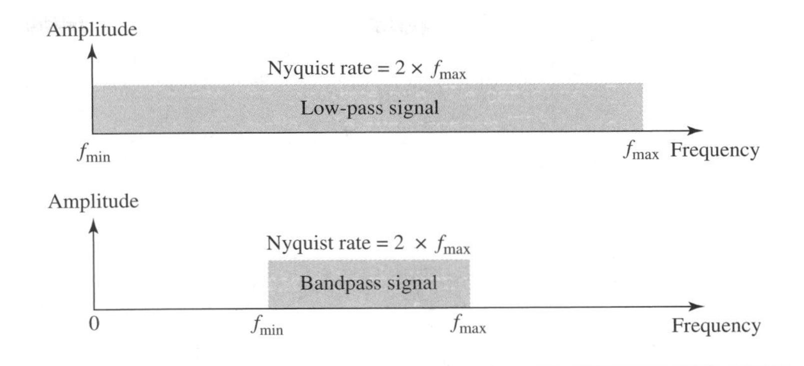

Figure 4.24 *Recovery of a sampled sine wave for different sampling rates*

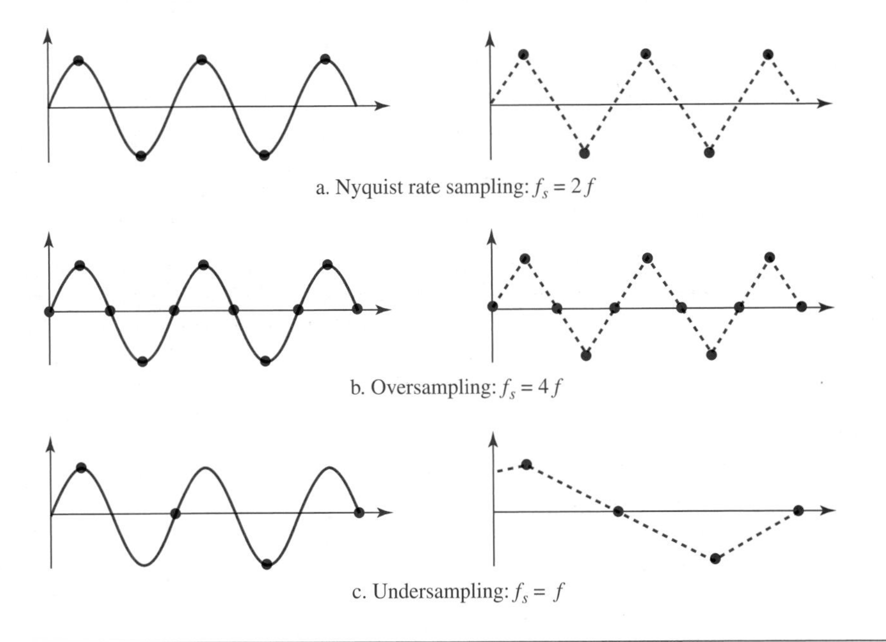

Nyquist theorem, we need to sample the hand (take and send a picture) every 30 s ($T_s = \frac{1}{2}T$ or $f_s = 2f$). In Figure 4.25a, the sample points, in order, are 12, 6, 12, 6, 12, and 6. The receiver of the samples cannot tell if the clock is moving forward or backward. In part b, we sample at double the Nyquist rate (every 15 s). The sample points, in order, are 12, 3, 6, 9, and 12. The clock is moving forward. In part c, we sample below the Nyquist rate ($T_s = \frac{3}{4}T$ or $f_s = \frac{4}{3}f$). The sample

Figure 4.25 *Sampling of a clock with only one hand*

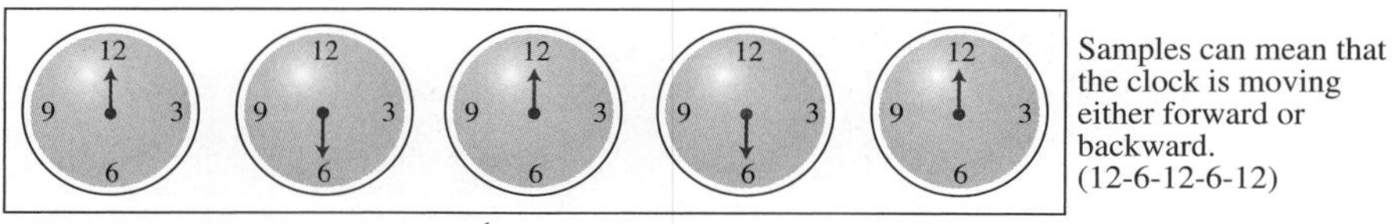

a. Sampling at Nyquist rate: $T_s = T\frac{1}{2}$

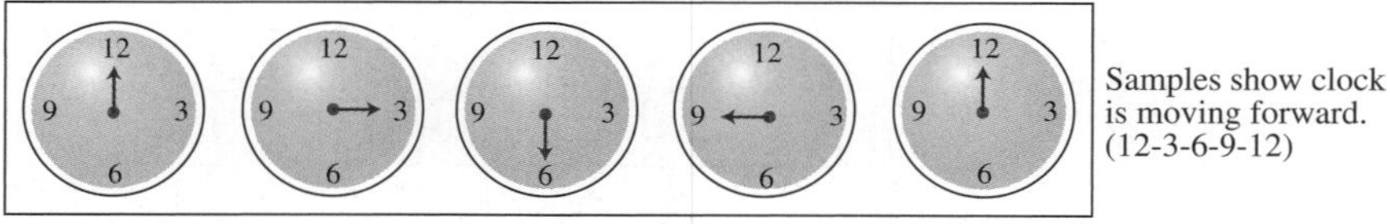

b. Oversampling (above Nyquist rate): $T_s = T\frac{1}{4}$

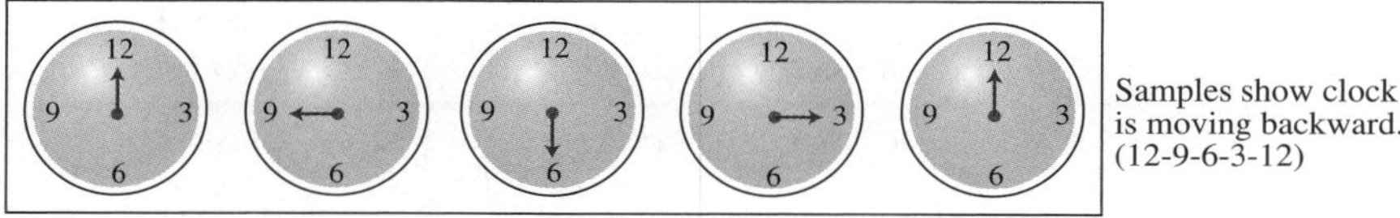

c. Undersampling (below Nyquist rate): $T_s = T\frac{3}{4}$

points, in order, are 12, 9, 6, 3, and 12. Although the clock is moving forward, the receiver thinks that the clock is moving backward.

Example 4.8

An example related to Example 4.7 is the seemingly backward rotation of the wheels of a forward-moving car in a movie. This can be explained by undersampling. A movie is filmed at 24 frames per second. If a wheel is rotating more than 12 times per second, the undersampling creates the impression of a backward rotation.

Example 4.9

Telephone companies digitize voice by assuming a maximum frequency of 4000 Hz. The sampling rate therefore is 8000 samples per second.

Example 4.10

A complex low-pass signal has a bandwidth of 200 kHz. What is the minimum sampling rate for this signal?

Solution

The bandwidth of a low-pass signal is between 0 and *f*, where *f* is the maximum frequency in the signal. Therefore, we can sample this signal at 2 times the highest frequency (200 kHz). The sampling rate is therefore 400,000 samples per second.

Example 4.11

A complex bandpass signal has a bandwidth of 200 kHz. What is the minimum sampling rate for this signal?

Solution

We cannot find the minimum sampling rate in this case because we do not know where the bandwidth starts or ends. We do not know the maximum frequency in the signal.

Quantization

The result of sampling is a series of pulses with amplitude values between the maximum and minimum amplitudes of the signal. The set of amplitudes can be infinite with nonintegral values between the two limits. These values cannot be used in the encoding process. The following are the steps in quantization:

1. We assume that the original analog signal has instantaneous amplitudes between V_{min} and V_{max}.
2. We divide the range into L zones, each of height Δ (delta).

$$\Delta = \frac{V_{max} - V_{min}}{L}$$

3. We assign quantized values of 0 to $L - 1$ to the midpoint of each zone.
4. We approximate the value of the sample amplitude to the quantized values.

As a simple example, assume that we have a sampled signal and the sample amplitudes are between −20 and +20 V. We decide to have eight levels ($L = 8$). This means that $\Delta = 5$ V. Figure 4.26 shows this example.

We have shown only nine samples using ideal sampling (for simplicity). The value at the top of each sample in the graph shows the actual amplitude. In the chart, the first row is the normalized value for each sample (actual amplitude/Δ). The quantization process selects the quantization value from the middle of each zone. This means that the normalized quantized values (second row) are different from the normalized amplitudes. The difference is called the *normalized error* (third row). The fourth row is the quantization code for each sample based on the quantization levels at the left of the graph. The encoded words (fifth row) are the final products of the conversion.

Quantization Levels

In the previous example, we showed eight quantization levels. The choice of L, the number of levels, depends on the range of the amplitudes of the analog signal and how accurately we need to recover the signal. If the amplitude of a signal fluctuates between two values only, we need only two levels; if the signal, like voice, has many amplitude values, we need more quantization levels. In audio digitizing, L is normally chosen to be 256; in video it is normally thousands. Choosing lower values of L increases the quantization error if there is a lot of fluctuation in the signal.

Quantization Error

One important issue is the error created in the quantization process. (Later, we will see how this affects high-speed modems.) Quantization is an approximation process. The

Figure 4.26 *Quantization and encoding of a sampled signal*

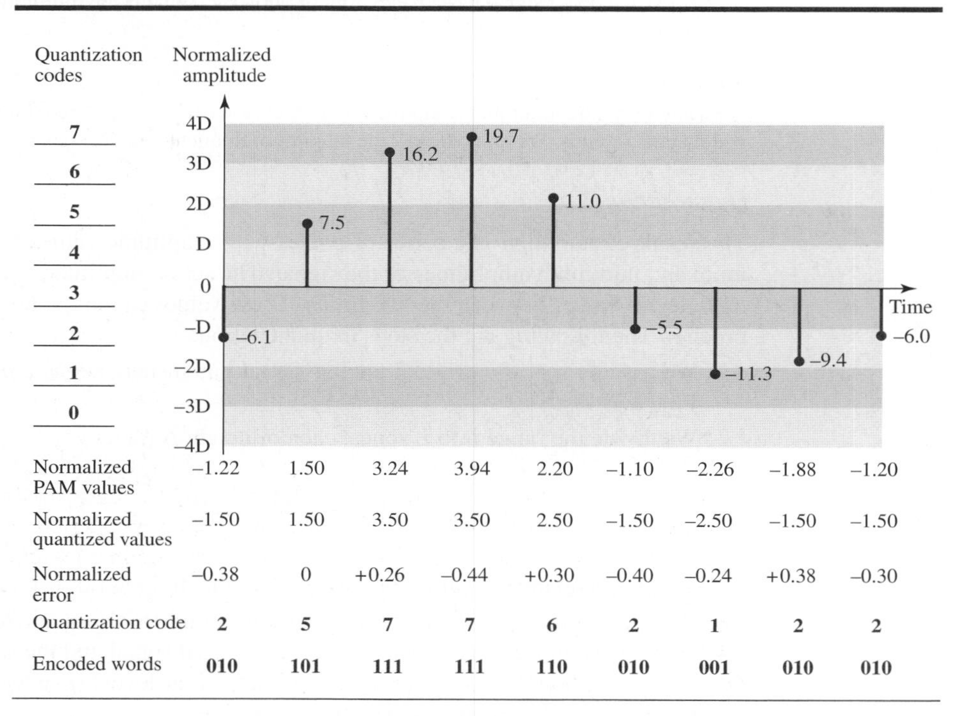

Normalized PAM values	–1.22	1.50	3.24	3.94	2.20	–1.10	–2.26	–1.88	–1.20
Normalized quantized values	–1.50	1.50	3.50	3.50	2.50	–1.50	–2.50	–1.50	–1.50
Normalized error	–0.38	0	+0.26	–0.44	+0.30	–0.40	–0.24	+0.38	–0.30
Quantization code	**2**	**5**	**7**	**7**	**6**	**2**	**1**	**2**	**2**
Encoded words	**010**	**101**	**111**	**111**	**110**	**010**	**001**	**010**	**010**

input values to the quantizer are the real values; the output values are the approximated values. The output values are chosen to be the middle value in the zone. If the input value is also at the middle of the zone, there is no quantization error; otherwise, there is an error. In the previous example, the normalized amplitude of the third sample is 3.24, but the normalized quantized value is 3.50. This means that there is an error of +0.26. The value of the error for any sample is less than $\Delta/2$. In other words, we have $-\Delta/2 \le \text{error} \le \Delta/2$.

The quantization error changes the signal-to-noise ratio of the signal, which in turn reduces the upper limit capacity according to Shannon.

It can be proven that the contribution of the **quantization error** to the SNR_{dB} of the signal depends on the number of quantization levels L, or the bits per sample n_b, as shown in the following formula:

$$\mathbf{SNR_{dB} = 6.02}\boldsymbol{n_b}\mathbf{ + 1.76 \quad dB}$$

Example 4.12

What is the SNR_{dB} in the example of Figure 4.26?

Solution

We can use the formula to find the quantization. We have eight levels and 3 bits per sample, so $SNR_{dB} = 6.02(3) + 1.76 = 19.82$ dB. Increasing the number of levels increases the SNR.

Example 4.13

A telephone subscriber line must have an SNR_{dB} above 40. What is the minimum number of bits per sample?

Solution

We can calculate the number of bits as

$$\mathbf{SNR_{dB} = 6.02n_b + 1.76 = 40 \quad \rightarrow \quad n = 6.35}$$

Telephone companies usually assign 7 or 8 bits per sample.

Uniform Versus Nonuniform Quantization

For many applications, the distribution of the instantaneous amplitudes in the analog signal is not uniform. Changes in amplitude often occur more frequently in the lower amplitudes than in the higher ones. For these types of applications it is better to use nonuniform zones. In other words, the height of Δ is not fixed; it is greater near the lower amplitudes and less near the higher amplitudes. Nonuniform quantization can also be achieved by using a process called **companding and expanding.** The signal is companded at the sender before conversion; it is expanded at the receiver after conversion. *Companding* means reducing the instantaneous voltage amplitude for large values; expanding is the opposite process. Companding gives greater weight to strong signals and less weight to weak ones. It has been proved that nonuniform quantization effectively reduces the SNR_{dB} of quantization.

Encoding

The last step in PCM is encoding. After each sample is quantized and the number of bits per sample is decided, each sample can be changed to an n_b-bit code word. In Figure 4.26 the encoded words are shown in the last row. A quantization code of 2 is encoded as 010; 5 is encoded as 101; and so on. Note that the number of bits for each sample is determined from the number of quantization levels. If the number of quantization levels is L, the number of bits is $n_b = \log_2 L$. In our example L is 8 and n_b is therefore 3. The bit rate can be found from the formula

$$\textbf{Bit rate = sampling rate} \times \textbf{number of bits per sample} = f_s \times n_b$$

Example 4.14

We want to digitize the human voice. What is the bit rate, assuming 8 bits per sample?

Solution

The human voice normally contains frequencies from 0 to 4000 Hz. So the sampling rate and bit rate are calculated as follows:

$$\textbf{Sampling rate} = 4000 \times 2 = 8000 \textbf{ samples/s}$$

$$\textbf{Bit rate} = 8000 \times 8 = 64{,}000 \textbf{ bps} = 64 \textbf{ kbps}$$

Original Signal Recovery

The recovery of the original signal requires the PCM decoder. The decoder first uses circuitry to convert the code words into a pulse that holds the amplitude until the next pulse. After the staircase signal is completed, it is passed through a low-pass filter to smooth the staircase signal into an analog signal. The filter has the same cutoff frequency as the original signal at the sender. If the signal has been sampled at (or greater than) the Nyquist sampling rate and if there are enough quantization levels, the original signal will be recreated. Note that the maximum and minimum values of the original signal can be achieved by using amplification. Figure 4.27 shows the simplified process.

Figure 4.27 *Components of a PCM decoder*

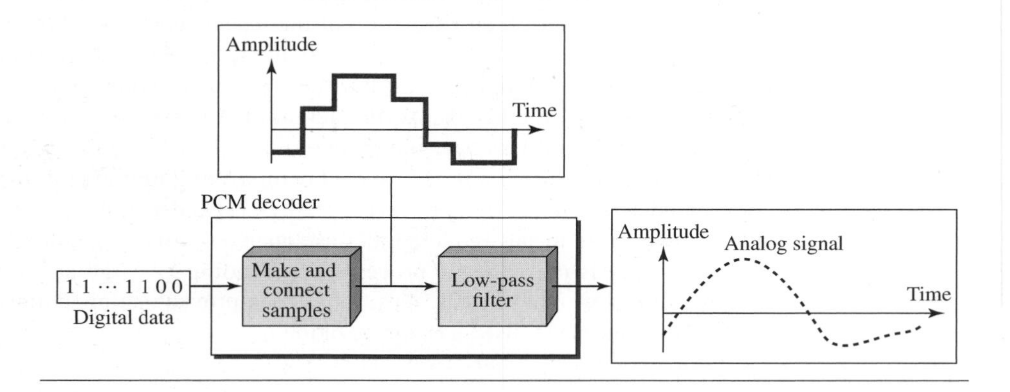

PCM Bandwidth

Suppose we are given the bandwidth of a low-pass analog signal. If we then digitize the signal, what is the new minimum bandwidth of the channel that can pass this digitized signal? We have said that the minimum bandwidth of a line-encoded signal is $B_{\text{min}} = c \times N \times (1/r)$. We substitute the value of N in this formula:

$$B_{\text{min}} = c \times N \times \frac{1}{r} = c \times n_b \times f_s \times \frac{1}{r} = c \times n_b \times 2 \times B_{\text{analog}} \times \frac{1}{r}$$

When $1/r = 1$ (for a NRZ or bipolar signal) and $c = (1/2)$ (the average situation), the minimum bandwidth is

$$B_{\text{min}} = n_b \times B_{\text{analog}}$$

This means the minimum bandwidth of the digital signal is n_b times greater than the bandwidth of the analog signal. This is the price we pay for digitization.

Example 4.15

We have a low-pass analog signal of 4 kHz. If we send the analog signal, we need a channel with a minimum bandwidth of 4 kHz. If we digitize the signal and send 8 bits per sample, we need a channel with a minimum bandwidth of 8×4 kHz = 32 kHz.

Maximum Data Rate of a Channel

In Chapter 3, we discussed the Nyquist theorem, which gives the data rate of a channel as $N_{max} = 2 \times B \times \log_2 L$. We can deduce this rate from the Nyquist sampling theorem by using the following arguments.

1. We assume that the available channel is low-pass with bandwidth B.
2. We assume that the digital signal we want to send has L levels, where each level is a signal element. This means $r = 1/\log_2 L$.
3. We first pass the digital signal through a low-pass filter to cut off the frequencies above B Hz.
4. We treat the resulting signal as an analog signal and sample it at $2 \times B$ samples per second and quantize it using L levels. Additional quantization levels are useless because the signal originally had L levels.
5. The resulting bit rate is $N = f_s \times n_b = 2 \times B \times \log_2 L$. This is the maximum bandwidth.

$$N_{max} = 2 \times B \times \log_2 L \quad \text{bps}$$

Minimum Required Bandwidth

The previous argument can give us the minimum bandwidth if the data rate and the number of signal levels are fixed. We can say

$$B_{min} = \frac{N}{(2 \times \log_2) L} \quad \text{Hz}$$

4.2.2 Delta Modulation (DM)

PCM is a very complex technique. Other techniques have been developed to reduce the complexity of PCM. The simplest is **delta modulation.** PCM finds the value of the signal amplitude for each sample; DM finds the change from the previous sample. Figure 4.28 shows the process. Note that there are no code words here; bits are sent one after another.

Figure 4.28 *The process of delta modulation*

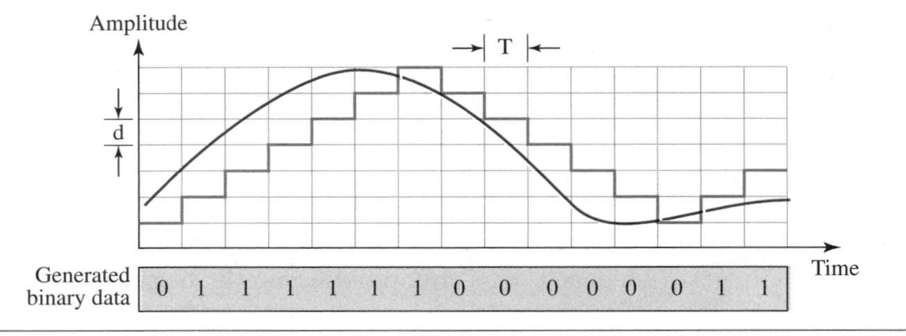

Modulator

The modulator is used at the sender site to create a stream of bits from an analog signal. The process records the small positive or negative changes, called delta δ. If the delta is positive, the process records a 1; if it is negative, the process records a 0. However, the process needs a base against which the analog signal is compared. The modulator builds a second signal that resembles a staircase. Finding the change is then reduced to comparing the input signal with the gradually made staircase signal. Figure 4.29 shows a diagram of the process.

Figure 4.29 *Delta modulation components*

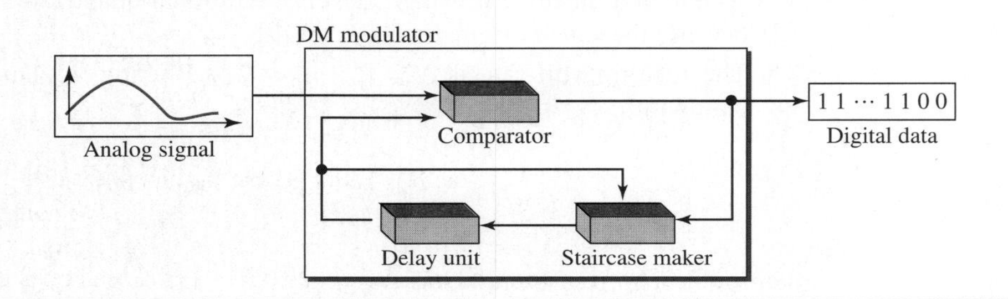

The modulator, at each sampling interval, compares the value of the analog signal with the last value of the staircase signal. If the amplitude of the analog signal is larger, the next bit in the digital data is 1; otherwise, it is 0. The output of the comparator, however, also makes the staircase itself. If the next bit is 1, the staircase maker moves the last point of the staircase signal δ up; if the next bit is 0, it moves it δ down. Note that we need a delay unit to hold the staircase function for a period between two comparisons.

Demodulator

The demodulator takes the digital data and, using the staircase maker and the delay unit, creates the analog signal. The created analog signal, however, needs to pass through a low-pass filter for smoothing. Figure 4.30 shows the schematic diagram.

Adaptive DM

A better performance can be achieved if the value of δ is not fixed. In **adaptive delta modulation,** the value of δ changes according to the amplitude of the analog signal.

Quantization Error

It is obvious that DM is not perfect. Quantization error is always introduced in the process. The quantization error of DM, however, is much less than that for PCM.

Figure 4.30 *Delta demodulation components*

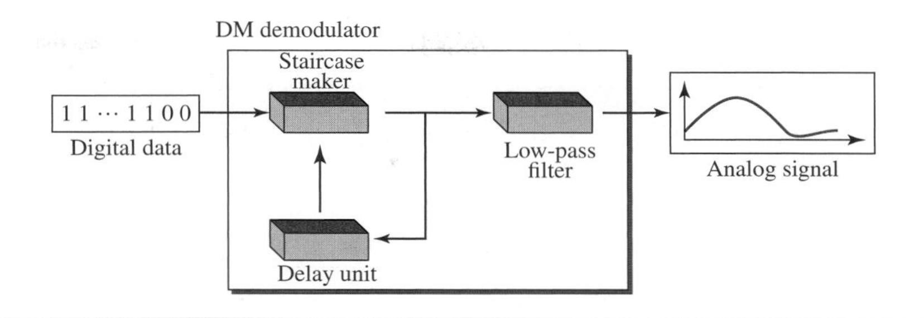

4.3 TRANSMISSION MODES

Of primary concern when we are considering the transmission of data from one device to another is the wiring, and of primary concern when we are considering the wiring is the data stream. Do we send 1 bit at a time; or do we group bits into larger groups and, if so, how? The transmission of binary data across a link can be accomplished in either parallel or serial mode. In parallel mode, multiple bits are sent with each clock tick. In serial mode, 1 bit is sent with each clock tick. While there is only one way to send parallel data, there are three subclasses of serial transmission: asynchronous, synchronous, and isochronous (see Figure 4.31).

Figure 4.31 *Data transmission and modes*

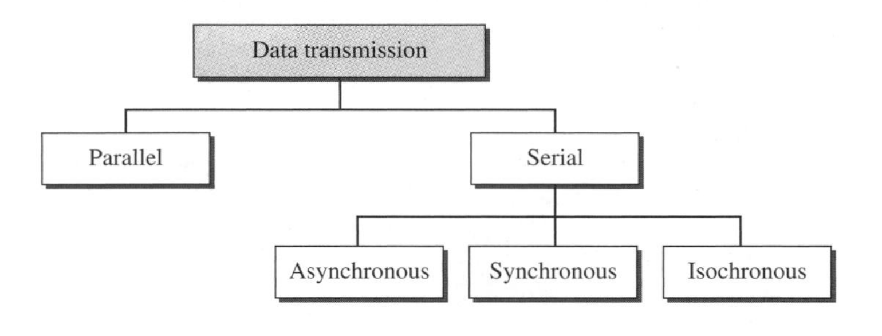

4.3.1 Parallel Transmission

Binary data, consisting of 1s and 0s, may be organized into groups of n bits each. Computers produce and consume data in groups of bits much as we conceive of and use spoken language in the form of words rather than letters. By grouping, we can send data n bits at a time instead of 1. This is called ***parallel transmission.***

The mechanism for parallel transmission is a conceptually simple one: Use n wires to send n bits at one time. That way each bit has its own wire, and all n bits of one group can be transmitted with each clock tick from one device to another. Figure 4.32 shows how parallel transmission works for $n = 8$. Typically, the eight wires are bundled in a cable with a connector at each end.

Figure 4.32 *Parallel transmission*

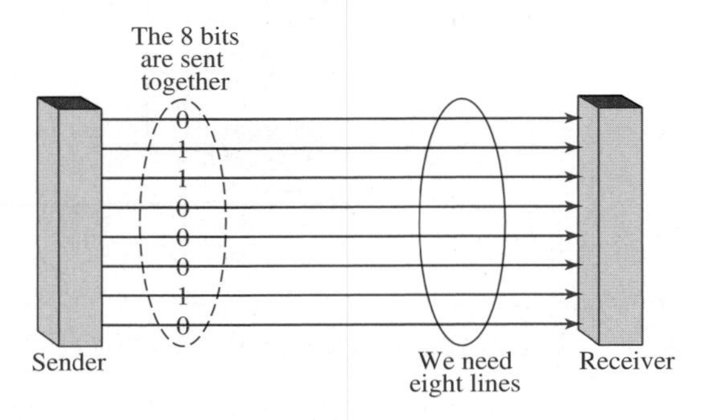

The advantage of parallel transmission is speed. All else being equal, parallel transmission can increase the transfer speed by a factor of n over serial transmission. But there is a significant disadvantage: cost. Parallel transmission requires n communication lines (wires in the example) just to transmit the data stream. Because this is expensive, parallel transmission is usually limited to short distances.

4.3.2 Serial Transmission

In **serial transmission** one bit follows another, so we need only one communication channel rather than n to transmit data between two communicating devices (see Figure 4.33).

Figure 4.33 *Serial transmission*

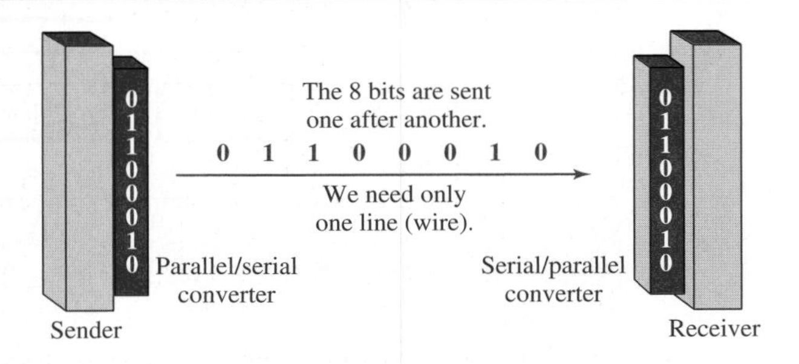

The advantage of serial over parallel transmission is that with only one communication channel, serial transmission reduces the cost of transmission over parallel by roughly a factor of n.

Since communication within devices is parallel, conversion devices are required at the interface between the sender and the line (parallel-to-serial) and between the line and the receiver (serial-to-parallel).

Serial transmission occurs in one of three ways: asynchronous, synchronous, and isochronous.

Asynchronous Transmission

Asynchronous transmission is so named because the timing of a signal is unimportant. Instead, information is received and translated by agreed upon patterns. As long as those patterns are followed, the receiving device can retrieve the information without regard to the rhythm in which it is sent. Patterns are based on grouping the bit stream into bytes. Each group, usually 8 bits, is sent along the link as a unit. The sending system handles each group independently, relaying it to the link whenever ready, without regard to a timer.

Without synchronization, the receiver cannot use timing to predict when the next group will arrive. To alert the receiver to the arrival of a new group, therefore, an extra bit is added to the beginning of each byte. This bit, usually a 0, is called the **start bit.** To let the receiver know that the byte is finished, 1 or more additional bits are appended to the end of the byte. These bits, usually 1s, are called **stop bits.** By this method, each byte is increased in size to at least 10 bits, of which 8 bits is information and 2 bits or more are signals to the receiver. In addition, the transmission of each byte may then be followed by a gap of varying duration. This gap can be represented either by an idle channel or by a stream of additional stop bits.

In asynchronous transmission, we send 1 start bit (0) at the beginning and 1 or more stop bits (1s) at the end of each byte. There may be a gap between bytes.

The start and stop bits and the gap alert the receiver to the beginning and end of each byte and allow it to synchronize with the data stream. This mechanism is called *asynchronous* because, at the byte level, the sender and receiver do not have to be synchronized. But within each byte, the receiver must still be synchronized with the incoming bit stream. That is, some synchronization is required, but only for the duration of a single byte. The receiving device resynchronizes at the onset of each new byte. When the receiver detects a start bit, it sets a timer and begins counting bits as they come in. After n bits, the receiver looks for a stop bit. As soon as it detects the stop bit, it waits until it detects the next start bit.

Asynchronous here means "asynchronous at the byte level," but the bits are still synchronized; their durations are the same.

Figure 4.34 is a schematic illustration of asynchronous transmission. In this example, the start bits are 0s, the stop bits are 1s, and the gap is represented by an idle line rather than by additional stop bits.

The addition of stop and start bits and the insertion of gaps into the bit stream make asynchronous transmission slower than forms of transmission that can operate

Figure 4.34 *Asynchronous transmission*

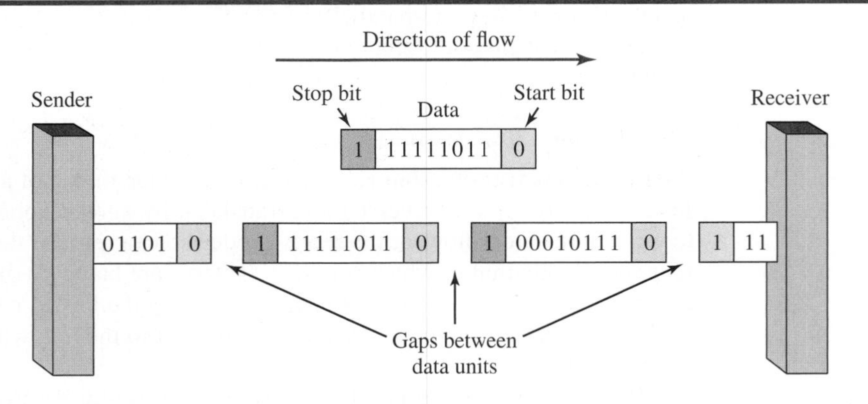

without the addition of control information. But it is cheap and effective, two advantages that make it an attractive choice for situations such as low-speed communication. For example, the connection of a keyboard to a computer is a natural application for asynchronous transmission. A user types only one character at a time, types extremely slowly in data processing terms, and leaves unpredictable gaps of time between characters.

Synchronous Transmission

In **synchronous transmission,** the bit stream is combined into longer "frames," which may contain multiple bytes. Each byte, however, is introduced onto the transmission link without a gap between it and the next one. It is left to the receiver to separate the bit stream into bytes for decoding purposes. In other words, data are transmitted as an unbroken string of 1s and 0s, and the receiver separates that string into the bytes, or characters, it needs to reconstruct the information.

In synchronous transmission, we send bits one after another without start or stop bits or gaps. It is the responsibility of the receiver to group the bits.

Figure 4.35 gives a schematic illustration of synchronous transmission. We have drawn in the divisions between bytes. In reality, those divisions do not exist; the sender puts its data onto the line as one long string. If the sender wishes to send data in separate bursts, the gaps between bursts must be filled with a special sequence of 0s and 1s that means *idle*. The receiver counts the bits as they arrive and groups them in 8-bit units.

Without gaps and start and stop bits, there is no built-in mechanism to help the receiving device adjust its bit synchronization midstream. Timing becomes very important, therefore, because the accuracy of the received information is completely dependent on the ability of the receiving device to keep an accurate count of the bits as they come in.

Figure 4.35 *Synchronous transmission*

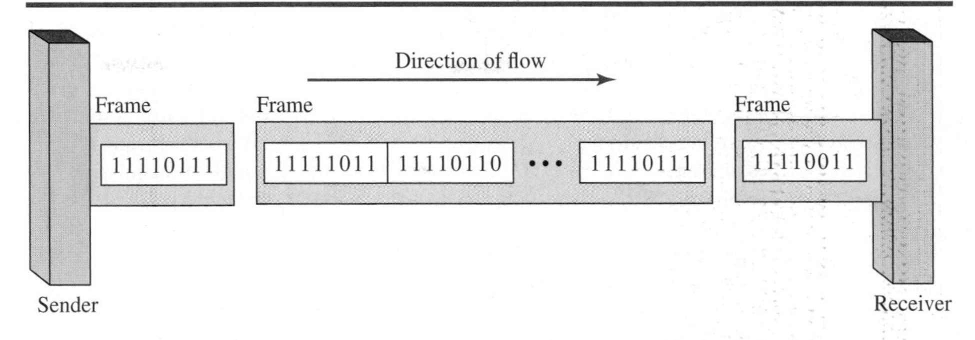

The advantage of synchronous transmission is speed. With no extra bits or gaps to introduce at the sending end and remove at the receiving end, and, by extension, with fewer bits to move across the link, synchronous transmission is faster than asynchronous transmission. For this reason, it is more useful for high-speed applications such as the transmission of data from one computer to another. Byte synchronization is accomplished in the data-link layer.

We need to emphasize one point here. Although there is no gap between characters in synchronous serial transmission, there may be uneven gaps between frames.

Isochronous

In real-time audio and video, in which uneven delays between frames are not acceptable, synchronous transmission fails. For example, TV images are broadcast at the rate of 30 images per second; they must be viewed at the same rate. If each image is sent by using one or more frames, there should be no delays between frames. For this type of application, synchronization between characters is not enough; the entire stream of bits must be synchronized. The **isochronous transmission** guarantees that the data arrive at a fixed rate.

4.4 END-CHAPTER MATERIALS

4.4.1 Recommended Reading

For more details about subjects discussed in this chapter, we recommend the following books. The items in brackets [...] refer to the reference list at the end of the text.

Books

Digital to digital conversion is discussed in [Pea92], [Cou01], and [Sta04]. Sampling is discussed in [Pea92], [Cou01], and [Sta04]. [Hsu03] gives a good mathematical approach to modulation and sampling. More advanced materials can be found in [Ber96].

4.4.2 Key Terms

adaptive delta modulation
alternate mark inversion (AMI)
analog-to-digital conversion
asynchronous transmission
baseline
baseline wandering
baud rate
biphase
bipolar
bipolar with 8-zero substitution (B8ZS)
bit rate
block coding
companding and expanding
data element
data rate
DC component
delta modulation (DM)
differential Manchester
digital-to-digital conversion
digitization
eight binary/ten binary (8B/10B)
eight-binary, six-ternary (8B6T)
four binary/five binary (4B/5B)
four dimensional, five-level pulse amplitude modulation (4D-PAM5)
high-density bipolar 3-zero (HDB3)
isochronous transmission
line coding
Manchester
modulation rate
multilevel binary
multiline transmission, three-level (MLT-3)
non-return-to-zero (NRZ)
non-return-to-zero, invert (NRZ-I)
non-return-to-zero, level (NRZ-L)
Nyquist theorem
parallel transmission
polar
pseudoternary
pulse amplitude modulation (PAM)
pulse code modulation (PCM)
pulse rate
quantization
quantization error
return-to-zero (RZ)
sample and hold
sampling
sampling rate
scrambling
self-synchronizing
serial transmission
signal element
signal rate
start bit
stop bit
synchronous transmission
transmission mode
two-binary, one quaternary (2B1Q)
unipolar

4.4.3 Summary

Digital-to-digital conversion involves three techniques: line coding, block coding, and scrambling. Line coding is the process of converting digital data to a digital signal. We can roughly divide line coding schemes into five broad categories: unipolar, polar, bipolar, multilevel, and multitransition. Block coding provides redundancy to ensure synchronization and inherent error detection. Block coding is normally referred to as mB/nB coding; it replaces each m-bit group with an n-bit group. Scrambling provides synchronization without increasing the number of bits. Two common scrambling techniques are B8ZS and HDB3.

The most common technique to change an analog signal to digital data (digitization) is called pulse code modulation (PCM). The first step in PCM is sampling. The analog signal is sampled every T_s second, where T_s is the sample interval or period. The inverse of the sampling interval is called the *sampling rate* or *sampling frequency* and denoted by f_s, where $f_s = 1/T_s$. There are three sampling methods—ideal, natural, and flat-top. According to the *Nyquist theorem,* to reproduce the original analog signal, one necessary condition is that the *sampling rate* be at least twice the highest frequency in the original signal. Other sampling techniques have been developed to reduce the complexity of PCM. The simplest is delta modulation. PCM finds the value of the signal amplitude for each sample; DM finds the change from the previous sample.

While there is only one way to send parallel data, there are three subclasses of serial transmission: asynchronous, synchronous, and isochronous. In asynchronous transmission, we send 1 start bit (0) at the beginning and 1 or more stop bits (1s) at the end of each byte. In synchronous transmission, we send bits one after another without start or stop bits or gaps. It is the responsibility of the receiver to group the bits. The isochronous mode provides synchronization for the entire stream of bits. In other words, it guarantees that the data arrive at a fixed rate.

4.5 PRACTICE SET

4.5.1 Quizzes

A set of interactive quizzes for this chapter can be found on the book website. It is strongly recommended that the student take the quizzes to check his/her understanding of the materials before continuing with the practice set.

4.5.2 Questions

Q4-1. List five line coding schemes discussed in this book.

Q4-2. Define the characteristics of a self-synchronizing signal.

Q4-3. Define scrambling and give its purpose.

Q4-4. Compare and contrast PCM and DM.

Q4-5. Distinguish between data rate and signal rate.

Q4-6. Distinguish between a signal element and a data element.

Q4-7. What are the differences between parallel and serial transmission?

Q4-8. Define block coding and give its purpose.

Q4-9. List three techniques of digital-to-digital conversion.

Q4-10. List three different techniques in serial transmission and explain the differences.

Q4-11. Define a DC component and its effect on digital transmission.

Q4-12. Define baseline wandering and its effect on digital transmission.

4.5.3 Problems

P4-1. How many invalid (unused) code sequences can we have in 5B/6B encoding? How many in 3B/4B encoding?

P4-2. In a digital transmission, the sender clock is 0.3 percent faster than the receiver clock. How many extra bits per second does the sender send if the data rate is 1 Mbps?

P4-3. What is the maximum data rate of a channel with a bandwidth of 200 KHz if we use four levels of digital signaling.

P4-4. The input stream to a 4B/5B block encoder is

0100 0000 0000 0000 0000 0001

Answer the following questions:

a. What is the output stream?

b. What is the length of the longest consecutive sequence of 0s in the input?

c. What is the length of the longest consecutive sequence of 0s in the output?

P4-5. What is the Nyquist sampling rate for each of the following signals?

a. A low-pass signal with bandwidth of 300 KHz?

b. A band-pass signal with bandwidth of 300 KHz if the lowest frequency is 100 KHz?

P4-6. We want to transmit 1200 characters with each character encoded as 8 bits.

a. Find the number of transmitted bits for synchronous transmission.

b. Find the number of transmitted bits for asynchronous transmission.

c. Find the redundancy percent in each case.

P4-7. We have a baseband channel with a 2-MHz bandwidth. What is the data rate for this channel if we use each of the following line coding schemes?

a. NRZ-L **b.** Manchester **c.** MLT-3 **d.** 2B1Q

P4-8. We have sampled a low-pass signal with a bandwidth of 300 KHz using 1024 levels of quantization.

a. Calculate the bit rate of the digitized signal.

b. Calculate the SNR_{dB} for this signal.

c. Calculate the PCM bandwidth of this signal.

P4-9. Find the 8-bit data stream for each case depicted in Figure 4.36.

Figure 4.36 *Problem P4-9*

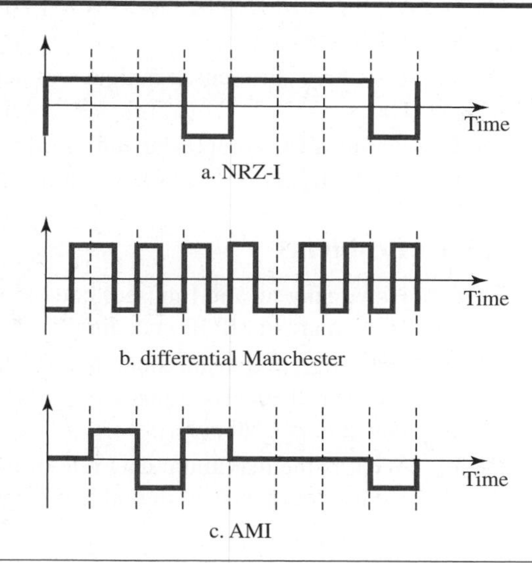

P4-10. Draw the graph of the NRZ-I scheme using each of the following data streams, assuming that the last signal level has been positive. From the graphs, guess the bandwidth for this scheme using the average number of changes in the signal level. Compare your guess with the corresponding entry in Table 4.1.

a. 00000000 **b.** 11111111 **c.** 01010101 **d.** 00110011

P4-11. Calculate the value of the signal rate for each case in Figure 4.2 if the data rate is 1 Mbps and c = 1/2.

P4-12. An analog signal has a bandwidth of 20 KHz. If we sample this signal and send it through a 30 Kbps channel, what is the SNR_{dB}?

P4-13. A Manchester signal has a data rate of 100 Kbps. Using Figure 4.8, calculate the value of the normalized energy (P) for frequencies at 0 Hz, 50 KHz, 100 KHz.

P4-14. What is the result of scrambling the sequence 11100000000000 using each of the following scrambling techniques? Assume that the last non-zero signal level has been positive.

a. B8ZS

b. HDB3 (The number of nonzero pulses is odd after the last substitution.)

P4-15. Draw the graph of the NRZ-L scheme using each of the following data streams, assuming that the last signal level has been positive. From the graphs, guess the bandwidth for this scheme using the average number of changes in the signal level. Compare your guess with the corresponding entry in Table 4.1.

a. 00000000 **b.** 11111111 **c.** 01010101 **d.** 00110011

P4-16. Repeat Problem P4-15 for the MLT-3 scheme, but use the following data streams.

a. 00000000 **b.** 11111111 **c.** 01010101 **d.** 00011000

P4-17. Repeat Problem P4-15 for the 2B1Q scheme, but use the following data streams.

a. 0000000000000000

b. 1111111111111111

c. 0101010101010101

d. 0011001100110011

P4-18. Repeat Problem P4-15 for the differential Manchester scheme.

P4-19. Repeat Problem P4-15 for the Manchester scheme.

P4-20. An NRZ-I signal has a data rate of 120 Kbps. Using Figure 4.6, calculate the value of the normalized energy (P) for frequencies at 0 Hz, 50 KHz, and 100 KHz.

4.6 SIMULATION EXPERIMENTS

4.6.1 Applets

We have created some Java applets to show some of the main concepts discussed in this chapter. It is strongly recommended that the students activate these applets on the book website and carefully examine the protocols in action.

CHAPTER 5

Analog Transmission

In Chapter 3, we discussed the advantages and disadvantages of digital and analog transmission. We saw that while digital transmission is very desirable, a low-pass channel is needed. We also saw that analog transmission is the only choice if we have a bandpass channel. Digital transmission was discussed in Chapter 4; we discuss analog transmission in this chapter.

Converting digital data to a bandpass analog signal is traditionally called digital-to-analog conversion. Converting a low-pass analog signal to a bandpass analog signal is traditionally called analog-to-analog conversion. In this chapter, we discuss these two types of conversions in two sections:

- The first section discusses digital-to-analog conversion. The section shows how we can change digital data to an analog signal when a band-pass channel is available. The first method described is called amplitude shift keying (ASK), in which the amplitude of a carrier is changed using the digital data. The second method described is called frequency shift keying (FSK), in which the frequency of a carrier is changed using the digital data. The third method described is called phase shift keying (PSK), in which the phase of a carrier signal is changed to represent digital data. The fourth method described is called quadrature amplitude modulation (QAM), in which both amplitude and phase of a carrier signal are changed to represent digital data.
- The second section discusses analog-to-analog conversion. The section shows how we can change an analog signal to a new analog signal with a smaller bandwidth. The conversion is used when only a band-pass channel is available. The first method is called amplitude modulation (AM), in which the amplitude of a carrier is changed based on the changes in the original analog signal. The second method is called frequency modulation (FM), in which the phase of a carrier is changed based on the changes in the original analog signal. The third method is called phase modulation (PM), in which the phase of a carrier signal is changed to show the changes in the original signal.

5.1 DIGITAL-TO-ANALOG CONVERSION

Digital-to-analog conversion is the process of changing one of the characteristics of an analog signal based on the information in digital data. Figure 5.1 shows the relationship between the digital information, the digital-to-analog modulating process, and the resultant analog signal.

Figure 5.1 *Digital-to-analog conversion*

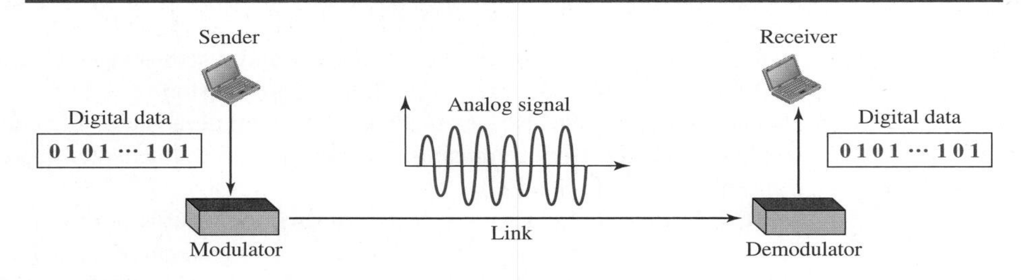

As discussed in Chapter 3, a sine wave is defined by three characteristics: amplitude, frequency, and phase. When we vary any one of these characteristics, we create a different version of that wave. So, by changing one characteristic of a simple electric signal, we can use it to represent digital data. Any of the three characteristics can be altered in this way, giving us at least three mechanisms for modulating digital data into an analog signal: **amplitude shift keying (ASK), frequency shift keying (FSK),** and **phase shift keying (PSK).** In addition, there is a fourth (and better) mechanism that combines changing both the amplitude and phase, called **quadrature amplitude modulation (QAM).** QAM is the most efficient of these options and is the mechanism commonly used today (see Figure 5.2).

Figure 5.2 *Types of digital-to-analog conversion*

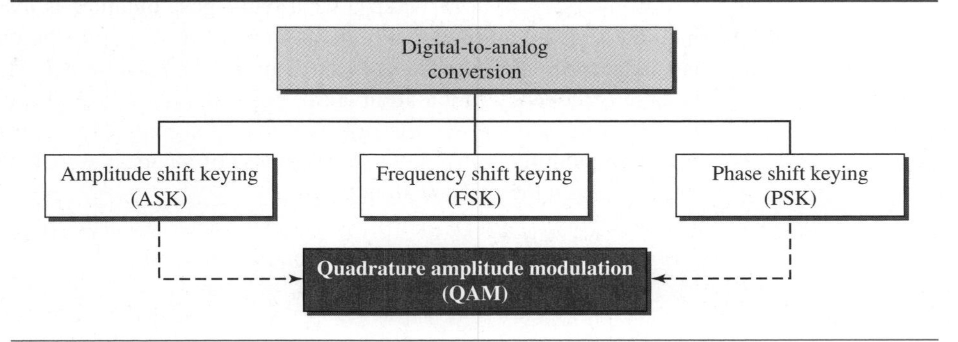

5.1.1 Aspects of Digital-to-Analog Conversion

Before we discuss specific methods of digital-to-analog modulation, two basic issues must be reviewed: bit and baud rates and the carrier signal.

Data Element Versus Signal Element

In Chapter 4, we discussed the concept of the data element versus the signal element. We defined a data element as the smallest piece of information to be exchanged, the bit. We also defined a signal element as the smallest unit of a signal that is constant. Although we continue to use the same terms in this chapter, we will see that the nature of the signal element is a little bit different in analog transmission.

Data Rate Versus Signal Rate

We can define the data rate (bit rate) and the signal rate (baud rate) as we did for digital transmission. The relationship between them is

$$S = N \times \frac{1}{r} \quad \text{baud}$$

where N is the data rate (bps) and r is the number of data elements carried in one signal element. The value of r in analog transmission is $r = \log_2 L$, where L is the number of different signal elements. The same nomenclature is used to simplify the comparisons.

Bit rate is the number of bits per second. Baud rate is the number of signal elements per second. In the analog transmission of digital data, the baud rate is less than or equal to the bit rate.

The same analogy we used in Chapter 4 for bit rate and baud rate applies here. In transportation, a baud is analogous to a vehicle, and a bit is analogous to a passenger. We need to maximize the number of people per car to reduce the traffic.

Example 5.1

An analog signal carries 4 bits per signal element. If 1000 signal elements are sent per second, find the bit rate.

Solution

In this case, $r = 4$, $S = 1000$, and N is unknown. We can find the value of N from

$$S = N \times (1/r) \quad \text{or} \quad N = S \times r = 1000 \times 4 = 4000 \text{ bps}$$

Example 5.2

An analog signal has a bit rate of 8000 bps and a baud rate of 1000 baud. How many data elements are carried by each signal element? How many signal elements do we need?

Solution

In this example, $S = 1000$, $N = 8000$, and r and L are unknown. We first find the value of r and then the value of L.

$$S = N \times 1/r \longrightarrow r = N / S = 8000 / 10{,}000 = 8 \text{ bits/baud}$$

$$r = \log_2 L \longrightarrow L = 2^r = 2^8 = 256$$

Bandwidth

The required bandwidth for analog transmission of digital data is proportional to the signal rate except for FSK, in which the difference between the carrier signals needs to be added. We discuss the bandwidth for each technique.

Carrier Signal

In analog transmission, the sending device produces a high-frequency signal that acts as a base for the information signal. This base signal is called the **carrier signal** or *carrier frequency.* The receiving device is tuned to the frequency of the carrier signal that it expects from the sender. Digital information then changes the carrier signal by modifying one or more of its characteristics (amplitude, frequency, or phase). This kind of modification is called modulation (shift keying).

5.1.2 Amplitude Shift Keying

In amplitude shift keying, the amplitude of the carrier signal is varied to create signal elements. Both frequency and phase remain constant while the amplitude changes.

Binary ASK (BASK)

Although we can have several levels (kinds) of signal elements, each with a different amplitude, ASK is normally implemented using only two levels. This is referred to as *binary amplitude shift keying* or *on-off keying* (OOK). The peak amplitude of one signal level is 0; the other is the same as the amplitude of the carrier frequency. Figure 5.3 gives a conceptual view of binary ASK.

Figure 5.3 *Binary amplitude shift keying*

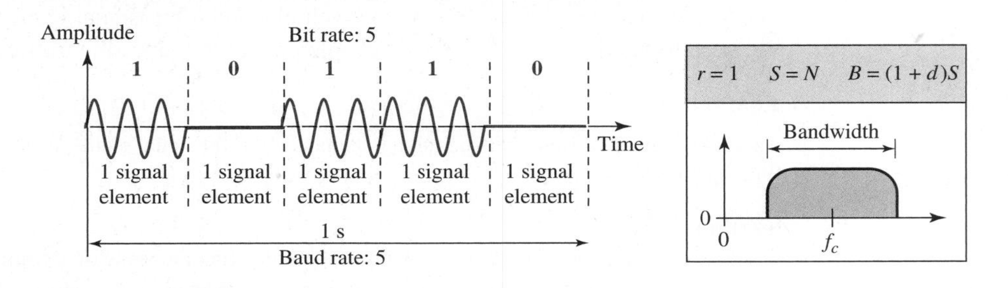

Bandwidth for ASK

Figure 5.3 also shows the bandwidth for ASK. Although the carrier signal is only one simple sine wave, the process of modulation produces a nonperiodic composite signal. This signal, as was discussed in Chapter 3, has a continuous set of frequencies. As we expect, the bandwidth is proportional to the signal rate (baud rate). However, there is normally another factor involved, called *d*, which depends on the modulation and filtering process. The value of *d* is between 0 and 1. This means that the bandwidth can be expressed as shown, where *S* is the signal rate and the *B* is the bandwidth.

$$B = (1 + d) \times S$$

The formula shows that the required bandwidth has a minimum value of S and a maximum value of $2S$. The most important point here is the location of the bandwidth. The middle of the bandwidth is where f_c, the carrier frequency, is located. This means if we have a bandpass channel available, we can choose our f_c so that the modulated signal occupies that bandwidth. This is in fact the most important advantage of digital-to-analog conversion. We can shift the resulting bandwidth to match what is available.

Implementation

The complete discussion of ASK implementation is beyond the scope of this book. However, the simple ideas behind the implementation may help us to better understand the concept itself. Figure 5.4 shows how we can simply implement binary ASK.

Figure 5.4 *Implementation of binary ASK*

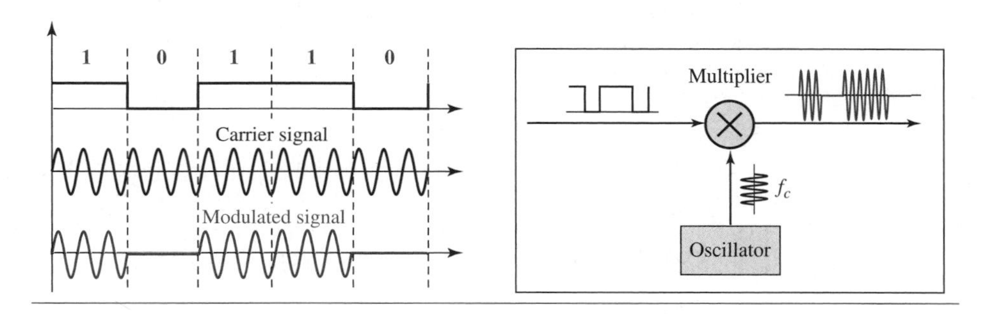

If digital data are presented as a unipolar NRZ (see Chapter 4) digital signal with a high voltage of 1 V and a low voltage of 0 V, the implementation can achieved by multiplying the NRZ digital signal by the carrier signal coming from an oscillator. When the amplitude of the NRZ signal is 1, the amplitude of the carrier frequency is held; when the amplitude of the NRZ signal is 0, the amplitude of the carrier frequency is zero.

Example 5.3

We have an available bandwidth of 100 kHz which spans from 200 to 300 kHz. What are the carrier frequency and the bit rate if we modulated our data by using ASK with $d = 1$?

Solution

The middle of the bandwidth is located at 250 kHz. This means that our carrier frequency can be at $f_c = 250$ kHz. We can use the formula for bandwidth to find the bit rate (with $d = 1$ and $r = 1$).

$$B = (1 + d) \times S = 2 \times N \times (1/r) = 2 \times N = 100 \text{ kHz} \longrightarrow N = 50 \text{ kbps}$$

Example 5.4

In data communications, we normally use full-duplex links with communication in both directions. We need to divide the bandwidth into two with two carrier frequencies, as shown in Figure 5.5. The figure shows the positions of two carrier frequencies and the bandwidths. The available bandwidth for each direction is now 50 kHz, which leaves us with a data rate of 25 kbps in each direction.

Figure 5.5 *Bandwidth of full-duplex ASK used in Example 5.4*

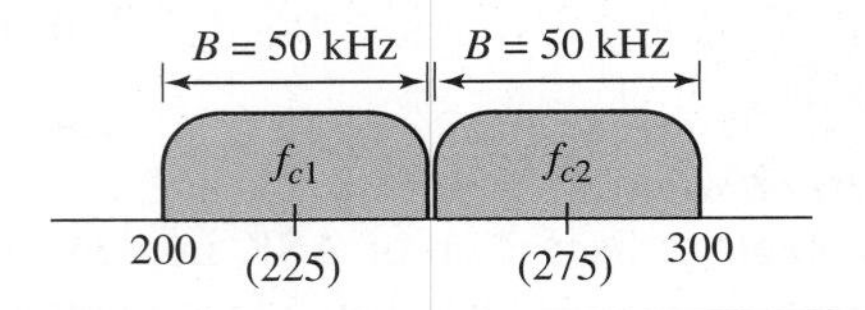

Multilevel ASK

The above discussion uses only two amplitude levels. We can have multilevel ASK in which there are more than two levels. We can use 4, 8, 16, or more different amplitudes for the signal and modulate the data using 2, 3, 4, or more bits at a time. In these cases, $r = 2$, $r = 3$, $r = 4$, and so on. Although this is not implemented with pure ASK, it is implemented with QAM (as we will see later).

5.1.3 Frequency Shift Keying

In frequency shift keying, the frequency of the carrier signal is varied to represent data. The frequency of the modulated signal is constant for the duration of one signal element, but changes for the next signal element if the data element changes. Both peak amplitude and phase remain constant for all signal elements.

Binary FSK (BFSK)

One way to think about binary FSK (or BFSK) is to consider two carrier frequencies. In Figure 5.6, we have selected two carrier frequencies, f_1 and f_2. We use the first carrier if the data element is 0; we use the second if the data element is 1. However, note that this is an unrealistic example used only for demonstration purposes. Normally the carrier frequencies are very high, and the difference between them is very small.

Figure 5.6 *Binary frequency shift keying*

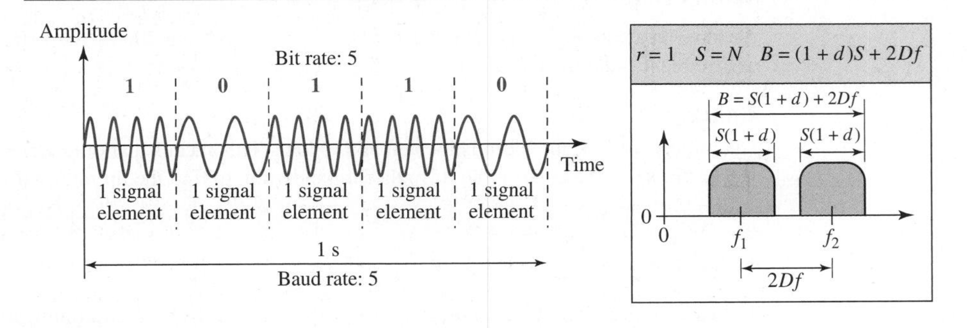

As Figure 5.6 shows, the middle of one bandwidth is f_1 and the middle of the other is f_2. Both f_1 and f_2 are Δ_f apart from the midpoint between the two bands. The difference between the two frequencies is $2\Delta_f$.

Bandwidth for BFSK

Figure 5.6 also shows the bandwidth of FSK. Again the carrier signals are only simple sine waves, but the modulation creates a nonperiodic composite signal with continuous frequencies. We can think of FSK as two ASK signals, each with its own carrier frequency (f_1 or f_2). If the difference between the two frequencies is $2\Delta_f$, then the required bandwidth is

$$B = (1 + d) \times S + 2\Delta\phi$$

What should be the minimum value of $2\Delta_f$? In Figure 5.6, we have chosen a value greater than $(1 + d)S$. It can be shown that the minimum value should be at least S for the proper operation of modulation and demodulation.

Example 5.5

We have an available bandwidth of 100 kHz which spans from 200 to 300 kHz. What should be the carrier frequency and the bit rate if we modulated our data by using FSK with $d = 1$?

Solution

This problem is similar to Example 5.3, but we are modulating by using FSK. The midpoint of the band is at 250 kHz. We choose $2\Delta_f$ to be 50 kHz; this means

$$B = (1 + d) \times S + 2\Delta_f = 100 \longrightarrow 2S = 50 \text{ kHz} \longrightarrow S = 25 \text{ kbaud} \longrightarrow N = 25 \text{ kbps}$$

Compared to Example 5.3, we can see the bit rate for ASK is 50 kbps while the bit rate for FSK is 25 kbps.

Implementation

There are two implementations of BFSK: noncoherent and coherent. In noncoherent BFSK, there may be discontinuity in the phase when one signal element ends and the next begins. In coherent BFSK, the phase continues through the boundary of two signal elements. Noncoherent BFSK can be implemented by treating BFSK as two ASK modulations and using two carrier frequencies. Coherent BFSK can be implemented by using one *voltage-controlled oscillator* (VCO) that changes its frequency according to the input voltage. Figure 5.7 shows the simplified idea behind the second implementation. The input to the oscillator is the unipolar NRZ signal. When the amplitude of NRZ is zero, the oscillator keeps its regular frequency; when the amplitude is positive, the frequency is increased.

Multilevel FSK

Multilevel modulation (MFSK) is not uncommon with the FSK method. We can use more than two frequencies. For example, we can use four different frequencies f_1, f_2, f_3, and f_4 to send 2 bits at a time. To send 3 bits at a time, we can use eight frequencies. And so on. However, we need to remember that the frequencies need to be $2\Delta_f$ apart. For the proper operation of the modulator and demodulator, it can be shown that the minimum value of $2\Delta_f$ needs to be S. We can show that the bandwidth is

$$B = (1 + d) \times S + (L - 1)2\Delta_f \longrightarrow B = L \times S$$

Figure 5.7 *Implementation of BFSK*

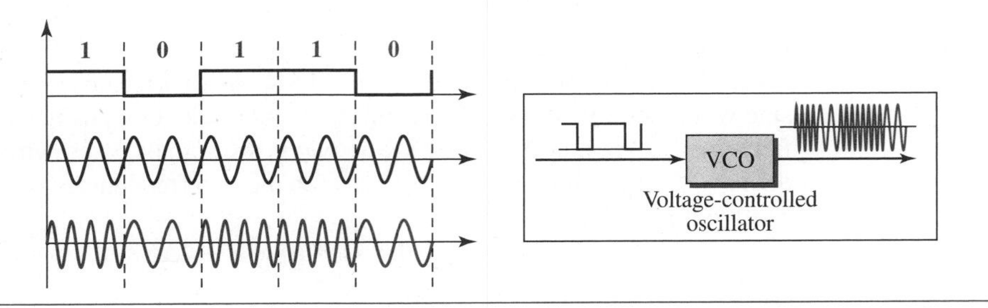

Note that MFSK uses more bandwidth than the other techniques; it should be used when noise is a serious issue.

Example 5.6

We need to send data 3 bits at a time at a bit rate of 3 Mbps. The carrier frequency is 10 MHz. Calculate the number of levels (different frequencies), the baud rate, and the bandwidth.

Solution

We can have $L = 2^3 = 8$. The baud rate is $S = 3$ MHz/3 = 1 Mbaud. This means that the carrier frequencies must be 1 MHz apart ($2\Delta_f$ = 1 MHz). The bandwidth is $B = 8 \times 1 = 8$ MHz. Figure 5.8 shows the allocation of frequencies and bandwidth.

Figure 5.8 *Bandwidth of MFSK used in Example 5.6*

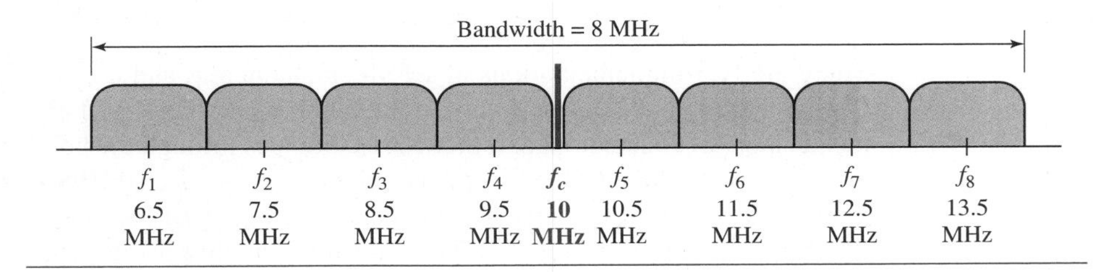

5.1.4 Phase Shift Keying

In phase shift keying, the phase of the carrier is varied to represent two or more different signal elements. Both peak amplitude and frequency remain constant as the phase changes. Today, PSK is more common than ASK or FSK. However, we will see shortly that QAM, which combines ASK and PSK, is the dominant method of digital-to-analog modulation.

Binary PSK (BPSK)

The simplest PSK is binary PSK, in which we have only two signal elements, one with a phase of 0°, and the other with a phase of 180°. Figure 5.9 gives a conceptual view of PSK. Binary PSK is as simple as binary ASK with one big advantage—it is less susceptible to noise. In ASK, the criterion for bit detection is the amplitude of the

Figure 5.9 *Binary phase shift keying*

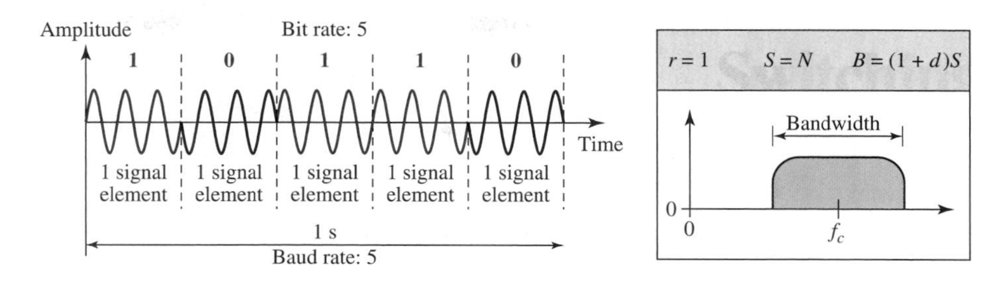

signal; in PSK, it is the phase. Noise can change the amplitude easier than it can change the phase. In other words, PSK is less susceptible to noise than ASK. PSK is superior to FSK because we do not need two carrier signals. However, PSK needs more sophisticated hardware to be able to distinguish between phases.

Bandwidth

Figure 5.9 also shows the bandwidth for BPSK. The bandwidth is the same as that for binary ASK, but less than that for BFSK. No bandwidth is wasted for separating two carrier signals.

Implementation

The implementation of BPSK is as simple as that for ASK. The reason is that the signal element with phase 180° can be seen as the complement of the signal element with phase 0°. This gives us a clue on how to implement BPSK. We use the same idea we used for ASK but with a polar NRZ signal instead of a unipolar NRZ signal, as shown in Figure 5.10. The polar NRZ signal is multiplied by the carrier frequency; the 1 bit (positive voltage) is represented by a phase starting at 0°; the 0 bit (negative voltage) is represented by a phase starting at 180°.

Figure 5.10 *Implementation of BASK*

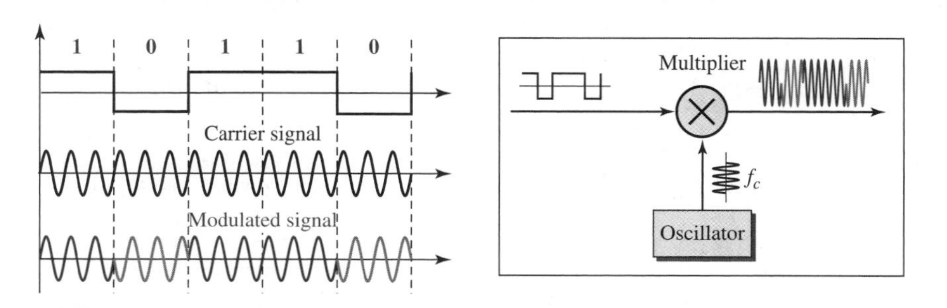

Quadrature PSK (QPSK)

The simplicity of BPSK enticed designers to use 2 bits at a time in each signal element, thereby decreasing the baud rate and eventually the required bandwidth. The scheme is called *quadrature PSK* or *QPSK* because it uses two separate BPSK modulations; one is in-phase, the other quadrature (out-of-phase). The incoming bits are first passed through a serial-to-parallel conversion that sends one bit to one modulator and the next bit to the other modulator. If the duration of each bit in the incoming signal is T, the duration of each bit sent to the corresponding BPSK signal is $2T$. This means that the bit to each BPSK signal has one-half the frequency of the original signal. Figure 5.11 shows the idea.

Figure 5.11 *QPSK and its implementation*

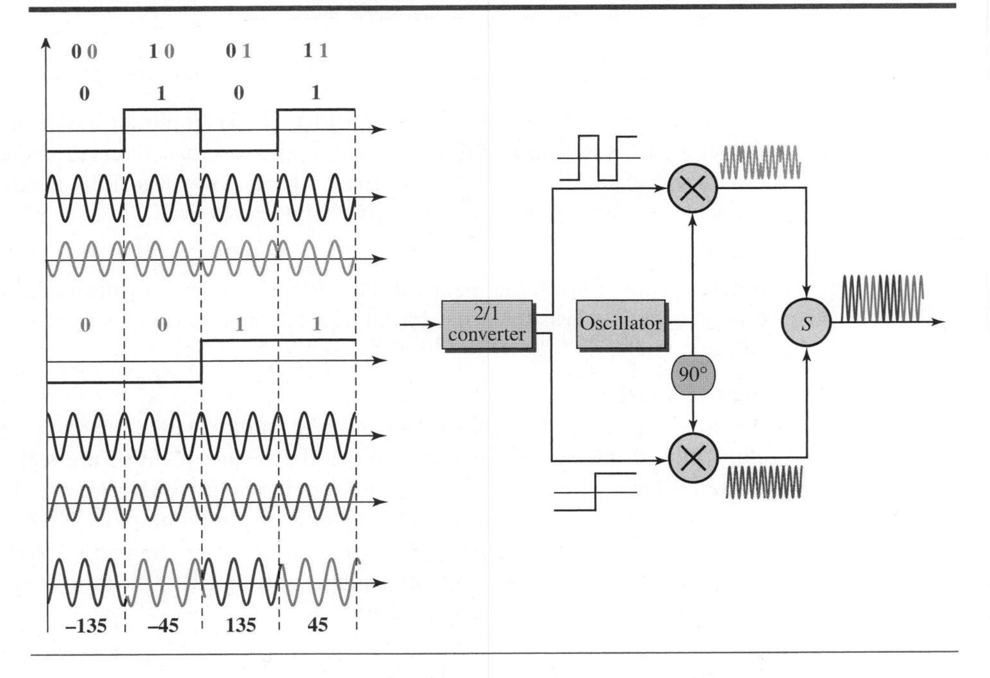

The two composite signals created by each multiplier are sine waves with the same frequency, but different phases. When they are added, the result is another sine wave, with one of four possible phases: 45°, −45°, 135°, and −135°. There are four kinds of signal elements in the output signal ($L = 4$), so we can send 2 bits per signal element ($r = 2$).

Example 5.7

Find the bandwidth for a signal transmitting at 12 Mbps for QPSK. The value of $d = 0$.

Solution

For QPSK, 2 bits are carried by one signal element. This means that $r = 2$. So the signal rate (baud rate) is $S = N \times (1/r) = 6$ Mbaud. With a value of $d = 0$, we have $B = S = 6$ MHz.

Constellation Diagram

A **constellation diagram** can help us define the amplitude and phase of a signal element, particularly when we are using two carriers (one in-phase and one quadrature). The diagram is useful when we are dealing with multilevel ASK, PSK, or QAM (see next section). In a constellation diagram, a signal element type is represented as a dot. The bit or combination of bits it can carry is often written next to it.

The diagram has two axes. The horizontal *X* axis is related to the in-phase carrier; the vertical *Y* axis is related to the quadrature carrier. For each point on the diagram, four pieces of information can be deduced. The projection of the point on the *X* axis defines the peak amplitude of the in-phase component; the projection of the point on the *Y* axis defines the peak amplitude of the quadrature component. The length of the line (vector) that connects the point to the origin is the peak amplitude of the signal element (combination of the *X* and *Y* components); the angle the line makes with the *X* axis is the phase of the signal element. All the information we need can easily be found on a constellation diagram. Figure 5.12 shows a constellation diagram.

Figure 5.12 *Concept of a constellation diagram*

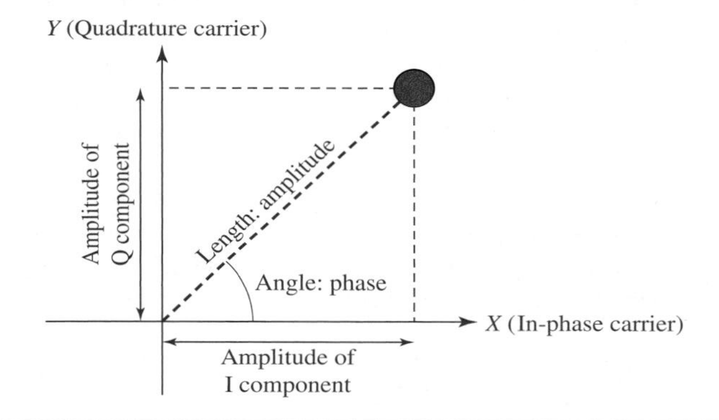

Example 5.8

Show the constellation diagrams for ASK (OOK), BPSK, and QPSK signals.

Solution

Figure 5.13 shows the three constellation diagrams. Let us analyze each case separately:

Figure 5.13 *Three constellation diagrams*

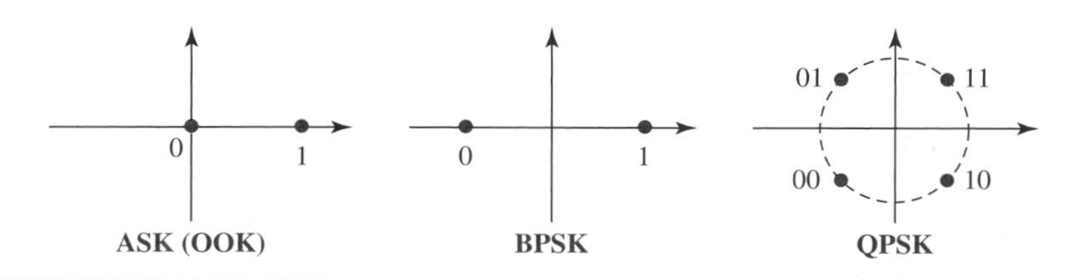

- For ASK, we are using only an in-phase carrier. Therefore, the two points should be on the *X* axis. Binary 0 has an amplitude of 0 V; binary 1 has an amplitude of 1 V (for example). The points are located at the origin and at 1 unit.

- BPSK also uses only an in-phase carrier. However, we use a polar NRZ signal for modulation. It creates two types of signal elements, one with amplitude 1 and the other with amplitude −1. This can be stated in other words: BPSK creates two different signal elements, one with amplitude 1 V and in phase and the other with amplitude 1 V and 180° out of phase.
- QPSK uses two carriers, one in-phase and the other quadrature. The point representing 11 is made of two combined signal elements, both with an amplitude of 1 V. One element is represented by an in-phase carrier, the other element by a quadrature carrier. The amplitude of the final signal element sent for this 2-bit data element is $2^{1/2}$, and the phase is 45°. The argument is similar for the other three points. All signal elements have an amplitude of $2^{1/2}$, but their phases are different (45°, 135°, −135°, and −45°). Of course, we could have chosen the amplitude of the carrier to be $1/(2^{1/2})$ to make the final amplitudes 1 V.

5.1.5 Quadrature Amplitude Modulation

PSK is limited by the ability of the equipment to distinguish small differences in phase. This factor limits its potential bit rate. So far, we have been altering only one of the three characteristics of a sine wave at a time; but what if we alter two? Why not combine ASK and PSK? The idea of using two carriers, one in-phase and the other quadrature, with different amplitude levels for each carrier is the concept behind **quadrature amplitude modulation (QAM).**

Quadrature amplitude modulation is a combination of ASK and PSK.

The possible variations of QAM are numerous. Figure 5.14 shows some of these schemes. Figure 5.14a shows the simplest 4-QAM scheme (four different signal element types) using a unipolar NRZ signal to modulate each carrier. This is the same mechanism we used for ASK (OOK). Part b shows another 4-QAM using polar NRZ, but this is exactly the same as QPSK. Part c shows another QAM-4 in which we used a signal with two positive levels to modulate each of the two carriers. Finally, Figure 5.14d shows a 16-QAM constellation of a signal with eight levels, four positive and four negative.

Figure 5.14 *Constellation diagrams for some QAMs*

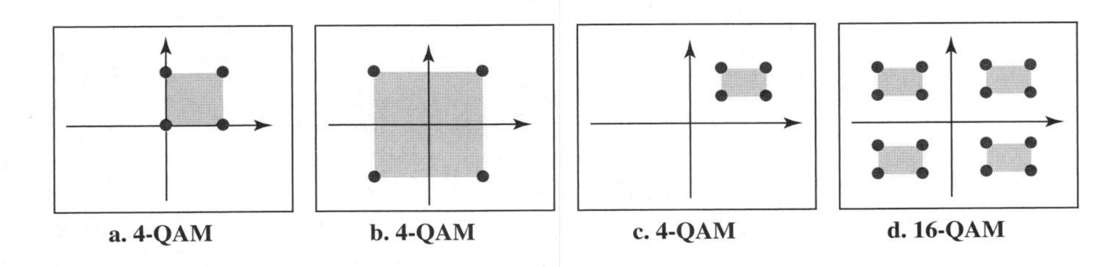

Bandwidth for QAM

The minimum bandwidth required for QAM transmission is the same as that required for ASK and PSK transmission. QAM has the same advantages as PSK over ASK.

5.2 ANALOG-TO-ANALOG CONVERSION

Analog-to-analog conversion, or analog modulation, is the representation of analog information by an analog signal. One may ask why we need to modulate an analog signal; it is already analog. Modulation is needed if the medium is bandpass in nature or if only a bandpass channel is available to us. An example is radio. The government assigns a narrow bandwidth to each radio station. The analog signal produced by each station is a low-pass signal, all in the same range. To be able to listen to different stations, the low-pass signals need to be shifted, each to a different range.

Analog-to-analog conversion can be accomplished in three ways: **amplitude modulation (AM), frequency modulation (FM),** and **phase modulation (PM).** FM and PM are usually categorized together. See Figure 5.15.

Figure 5.15 *Types of analog-to-analog modulation*

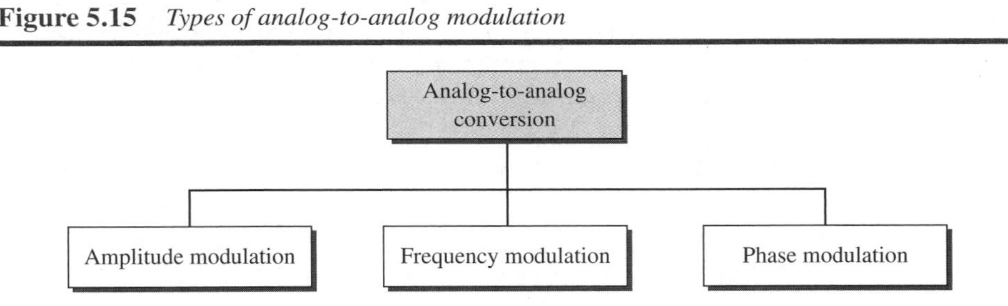

5.2.1 Amplitude Modulation (AM)

In AM transmission, the carrier signal is modulated so that its amplitude varies with the changing amplitudes of the modulating signal. The frequency and phase of the carrier remain the same; only the amplitude changes to follow variations in the information. Figure 5.16 shows how this concept works. The modulating signal is the envelope of the carrier. As Figure 5.16 shows, AM is normally implemented by using a simple multiplier because the amplitude of the carrier signal needs to be changed according to the amplitude of the modulating signal.

AM Bandwidth

Figure 5.16 also shows the bandwidth of an AM signal. The modulation creates a bandwidth that is twice the bandwidth of the modulating signal and covers a range centered on the carrier frequency. However, the signal components above and below the carrier frequency carry exactly the same information. For this reason, some implementations discard one-half of the signals and cut the bandwidth in half.

Figure 5.16 *Amplitude modulation*

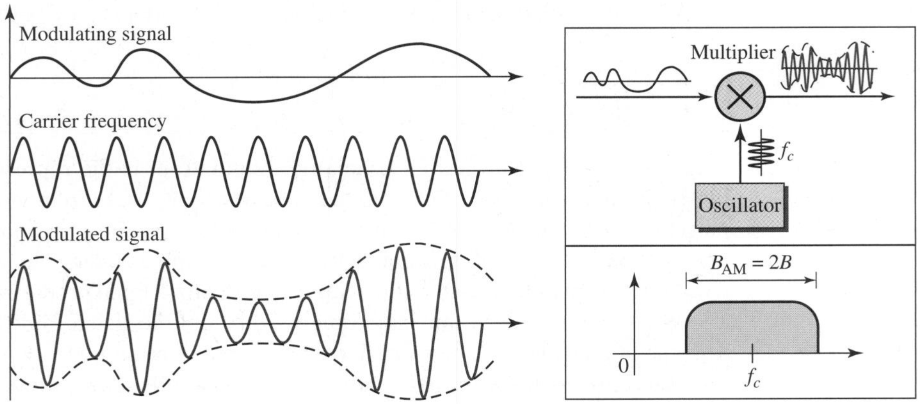

The total bandwidth required for AM can be determined from the bandwidth of the audio signal: $B_{AM} = 2B$.

Standard Bandwidth Allocation for AM Radio

The bandwidth of an audio signal (speech and music) is usually 5 kHz. Therefore, an AM radio station needs a bandwidth of 10 kHz. In fact, the Federal Communications Commission (FCC) allows 10 kHz for each AM station.

AM stations are allowed carrier frequencies anywhere between 530 and 1700 kHz (1.7 MHz). However, each station's carrier frequency must be separated from those on either side of it by at least 10 kHz (one AM bandwidth) to avoid interference. If one station uses a carrier frequency of 1100 kHz, the next station's carrier frequency cannot be lower than 1110 kHz (see Figure 5.17).

Figure 5.17 *AM band allocation*

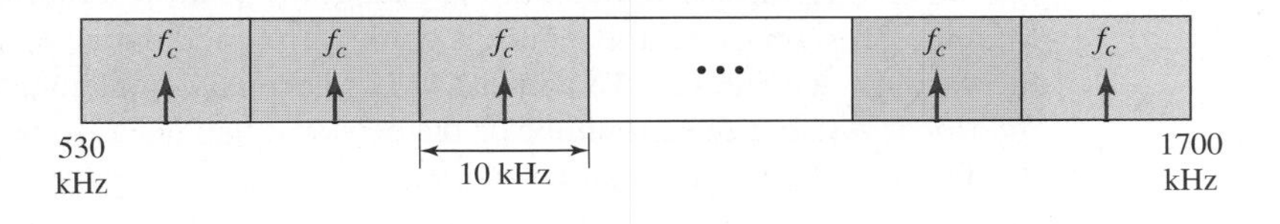

5.2.2 Frequency Modulation (FM)

In FM transmission, the frequency of the carrier signal is modulated to follow the changing voltage level (amplitude) of the modulating signal. The peak amplitude and phase of the carrier signal remain constant, but as the amplitude of the information signal changes, the frequency of the carrier changes correspondingly. Figure 5.18 shows the relationships of the modulating signal, the carrier signal, and the resultant FM signal.

As Figure 5.18 shows, FM is normally implemented by using a voltage-controlled oscillator as with FSK. The frequency of the oscillator changes according to the input voltage which is the amplitude of the modulating signal.

FM Bandwidth

Figure 5.18 also shows the bandwidth of an FM signal. The actual bandwidth is difficult to determine exactly, but it can be shown empirically that it is several times that of the analog signal or $2(1 + \beta)B$ where β is a factor that depends on modulation technique with a common value of 4.

The total bandwidth required for FM can be determined from the bandwidth of the audio signal: $B_{FM} = 2(1 \times \beta)B$.

Figure 5.18 *Frequency modulation*

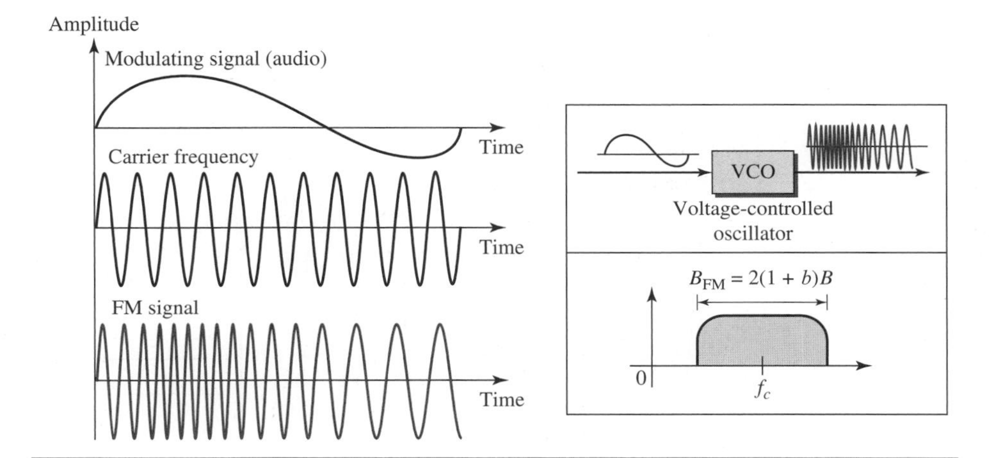

Standard Bandwidth Allocation for FM Radio

The bandwidth of an audio signal (speech and music) broadcast in stereo is almost 15 kHz. The FCC allows 200 kHz (0.2 MHz) for each station. This mean $\beta = 4$ with some extra guard band. FM stations are allowed carrier frequencies anywhere between 88 and 108 MHz. Stations must be separated by at least 200 kHz to keep their bandwidths from overlapping. To create even more privacy, the FCC requires that in a given area, only alternate bandwidth allocations may be used. The others remain unused to prevent any possibility of two stations interfering with each other. Given 88 to 108 MHz as a range, there are 100 potential FM bandwidths in an area, of which 50 can operate at any one time. Figure 5.19 illustrates this concept.

5.2.3 Phase Modulation (PM)

In PM transmission, the phase of the carrier signal is modulated to follow the changing voltage level (amplitude) of the modulating signal. The peak amplitude and frequency

Figure 5.19 *FM band allocation*

of the carrier signal remain constant, but as the amplitude of the information signal changes, the phase of the carrier changes correspondingly. It can be proved mathematically (see Appendix E) that PM is the same as FM with one difference. In FM, the instantaneous change in the carrier frequency is proportional to the amplitude of the modulating signal; in PM the instantaneous change in the carrier frequency is proportional to the derivative of the amplitude of the modulating signal. Figure 5.20 shows the relationships of the modulating signal, the carrier signal, and the resultant PM signal.

Figure 5.20 *Phase modulation*

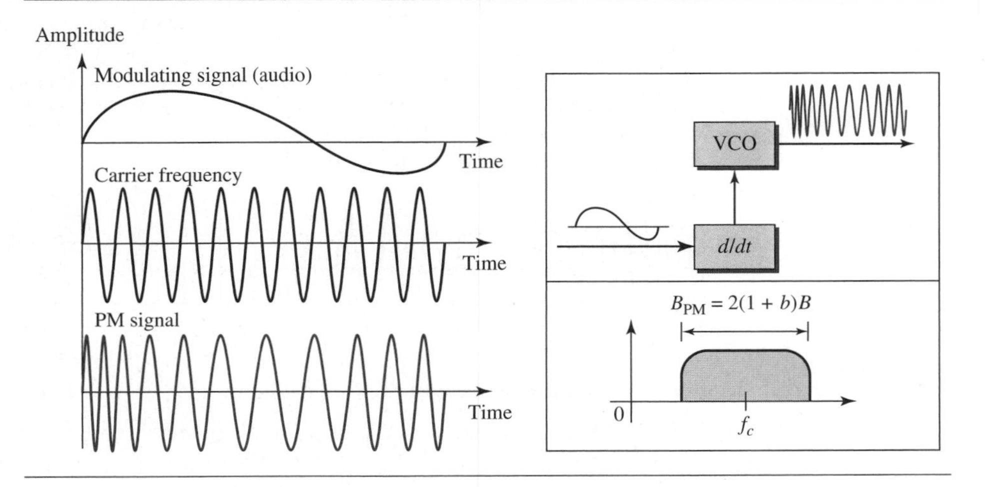

As Figure 5.20 shows, PM is normally implemented by using a voltage-controlled oscillator along with a derivative. The frequency of the oscillator changes according to the derivative of the input voltage, which is the amplitude of the modulating signal.

PM Bandwidth

Figure 5.20 also shows the bandwidth of a PM signal. The actual bandwidth is difficult to determine exactly, but it can be shown empirically that it is several times that of the analog signal. Although the formula shows the same bandwidth for FM and PM, the value of β is lower in the case of PM (around 1 for narrowband and 3 for wideband).

The total bandwidth required for PM can be determined from the bandwidth and maximum amplitude of the modulating signal: $B_{PM} = 2(1 + \beta)B$.

5.3 END-CHAPTER MATERIALS

5.3.1 Recommended Reading

For more details about subjects discussed in this chapter, we recommend the following books. The items in brackets [. . .] refer to the reference list at the end of the text.

Books

Digital-to-analog conversion is discussed in [Pea92], [Cou01], and [Sta04]. Analog-to-analog conversion is discussed in [Pea92], Chapter 5 of [Cou01], [Sta04]. [Hsu03] gives a good mathematical approach to all materials discussed in this chapter. More advanced materials can be found in [Ber96].

5.3.2 Key Terms

amplitude modulation (AM)
amplitude shift keying (ASK)
analog-to-analog conversion
carrier signal
constellation diagram
digital-to-analog conversion
frequency modulation (FM)
frequency shift keying (FSK)
phase modulation (PM)
phase shift keying (PSK)
quadrature amplitude modulation (QAM)

5.3.3 Summary

Digital-to-analog conversion is the process of changing one of the characteristics of an analog signal based on the information in the digital data. Digital-to-analog conversion can be accomplished in several ways: amplitude shift keying (ASK), frequency shift keying (FSK), and phase shift keying (PSK). Quadrature amplitude modulation (QAM) combines ASK and PSK. In amplitude shift keying, the amplitude of the carrier signal is varied to create signal elements. Both frequency and phase remain constant while the amplitude changes. In frequency shift keying, the frequency of the carrier signal is varied to represent data. The frequency of the modulated signal is constant for the duration of one signal element, but changes for the next signal element if the data element changes. Both peak amplitude and phase remain constant for all signal elements. In phase shift keying, the phase of the carrier is varied to represent two or more different signal elements. Both peak amplitude and frequency remain constant as the phase changes. A constellation diagram shows us the amplitude and phase of a signal element, particularly when we are using two carriers (one in-phase and one quadrature). Quadrature amplitude modulation (QAM) is a combination of ASK and PSK. QAM uses two carriers, one in-phase and the other quadrature, with different amplitude levels for each carrier. Analog-to-analog conversion is the representation of analog information by an analog signal. Conversion is needed if the medium is bandpass in nature or if only a bandpass bandwidth is available to us.

Analog-to-analog conversion can be accomplished in three ways: amplitude modulation (AM), frequency modulation (FM), and phase modulation (PM). In AM transmission, the carrier signal is modulated so that its amplitude varies with the changing amplitudes of the modulating signal. The frequency and phase of the carrier remain the same; only the amplitude changes to follow variations in the information. In FM transmission, the frequency of the carrier signal is modulated to follow the changing voltage level (amplitude) of the modulating signal. The peak amplitude and phase of the carrier signal remain constant, but as the amplitude of the information signal changes, the frequency of the carrier changes correspondingly. In PM transmission, the phase of the carrier signal is modulated to follow the changing voltage level (amplitude) of the modulating signal. The peak amplitude and frequency of the carrier signal remain constant, but as the amplitude of the information signal changes, the phase of the carrier changes correspondingly.

5.4 PRACTICE SET

5.4.1 Quizzes

A set of interactive quizzes for this chapter can be found on the book website. It is strongly recommended that the student take the quizzes to check his/her understanding of the materials before continuing with the practice set.

5.4.2 Questions

Q5-1. Define *digital-to-analog conversion.*

Q5-2. Define *carrier signal* and explain its role in analog transmission.

Q5-3. Which characteristics of an analog signal are changed to represent the lowpass analog signal in each of the following analog-to-analog conversions?

a. AM **b.** FM **c.** PM

Q5-4. Which of the three analog-to-analog conversion techniques (AM, FM, or PM) is the most susceptible to noise? Defend your answer.

Q5-5. Define *analog transmission.*

Q5-6. Which characteristics of an analog signal are changed to represent the digital signal in each of the following digital-to-analog conversions?

d. ASK **e.** FSK **f.** PSK **g.** QAM

Q5-7. What are the two components of a signal when the signal is represented on a constellation diagram? Which component is shown on the horizontal axis? Which is shown on the vertical axis?

Q5-8. Define *analog-to-analog conversion.*

Q5-9. Which of the four digital-to-analog conversion techniques (ASK, FSK, PSK or QAM) is the most susceptible to noise? Defend your answer.

Q5-10. Define *constellation diagram* and explain its role in analog transmission.

5.4.3 Problems

P5-1. What is the required bandwidth for the following cases if we need to send 6000 bps? Let d = 1.

a. ASK

b. FSK with $2\Delta f$ = 4 KHz

c. QPSK

d. 16-QAM

P5-2. A cable company uses one of the cable TV channels (with a bandwidth of 6 MHz) to provide digital communication for each resident. What is the available data rate for each resident if the company uses a 64-QAM technique?

P5-3. Find the bandwidth for the following situations if we need to modulate a 5-KHz voice.

a. AM **b.** FM ($\beta = 5$) **c.** PM ($\beta = 1$)

P5-4. Find the total number of channels in the corresponding band allocated by FCC.

a. AM **b.** FM

P5-5. What is the number of bits per baud for the following techniques?

a. ASK with four different amplitudes

b. FSK with eight different frequencies

c. PSK with four different phases

d. QAM with a constellation of 128 points

P5-6. Draw the constellation diagram for the following:

a. ASK, with peak amplitude values of 1 and 3

b. BPSK, with a peak amplitude value of 2

c. QPSK, with a peak amplitude value of 3

d. 8-QAM with two different peak amplitude values, 1 and 3, and four different phases

P5-7. Calculate the baud rate for the given bit rate and type of modulation.

a. 3000 bps, FSK

b. 2000 bps, ASK

c. 4000 bps, QPSK

d. 36,000 bps, 64-QAM

P5-8. Calculate the bit rate for the given baud rate and type of modulation.

a. 1000 baud, FSK

b. 1000 baud, ASK

c. 1000 baud, BPSK

d. 1000 baud, 16-QAM

P5-9. A corporation has a medium with a 1-MHz bandwidth (lowpass). The corporation needs to create 10 separate independent channels each capable of sending at least 10 Mbps. The company has decided to use QAM technology. What is the minimum number of bits per baud for each channel? What is the number of points in the constellation diagram for each channel? Let d = 0.

P5-10. How many bits per baud can we send in each of the following cases if the signal constellation has one of the following number of points?

a. 4 **b.** 8 **c.** 2 **d.** 1024

P5-11. Draw the constellation diagram for the following cases. Find the peak amplitude value for each case and define the type of modulation (ASK, FSK, PSK, or QAM). The numbers in parentheses define the values of I and Q respectively.

a. Two points at (2, 0) and (3, 0)

b. Two points at (3, 0) and (−3, 0)

c. Four points at (2, 2), (−2, 2), (−2, −2), and (2, −2)

d. Two points at (0, 2) and (0, −2)

P5-12. The telephone line has 4 KHz bandwidth. What is the maximum number of bits we can send using each of the following techniques? Let d = 0.

a. ASK **b.** QPSK **c.** 16-QAM **d.** 64-QAM

5.5 SIMULATION EXPERIMENTS

5.5.1 Applets

We have created some Java applets to show some of the main concepts discussed in this chapter. It is strongly recommended that the students activate these applets on the book website and carefully examine the protocols in action.

CHAPTER 6

Bandwidth Utilization: Multiplexing and Spectrum Spreading

In real life, we have links with limited bandwidths. The wise use of these bandwidths has been, and will be, one of the main challenges of electronic communications. However, the meaning of *wise* may depend on the application. Sometimes we need to combine several low-bandwidth channels to make use of one channel with a larger bandwidth. Sometimes we need to expand the bandwidth of a channel to achieve goals such as privacy and antijamming. In this chapter, we explore these two broad categories of bandwidth utilization: multiplexing and spectrum spreading. In multiplexing, our goal is efficiency; we combine several channels into one. In spectrum spreading, our goals are privacy and antijamming; we expand the bandwidth of a channel to insert redundancy, which is necessary to achieve these goals.

This chapter is divided into two sections:

- The first section discusses multiplexing. The first method described in this section is called *frequency-division multiplexing* (FDM), which means to combine several analog signals into a single analog signal. The second method is called *wavelength-division multiplexing* (WDM), which means to combine several optical signals into one optical signal.The third method is called *time-division multiplexing* (TDM), which allows several digital signals to share a channel in time.
- The second section discusses spectrum spreading, in which we first spread the bandwidth of a signal to add redundancy for the purpose of more secure transmission before combining different channels. The first method described in this section is called *frequency hopping spread spectrum* (FHSS), in which different modulation frequencies are used in different periods of time. The second method is called *direct sequence spread spectrum* (DSSS), in which a single bit in the original signal is changed to a sequence before transmission.

6.1 MULTIPLEXING

Whenever the bandwidth of a medium linking two devices is greater than the bandwidth needs of the devices, the link can be shared. **Multiplexing** is the set of techniques that allow the simultaneous transmission of multiple signals across a single data link. As data and telecommunications use increases, so does traffic. We can accommodate this increase by continuing to add individual links each time a new channel is needed; or we can install higher-bandwidth links and use each to carry multiple signals. As described in Chapter 7, today's technology includes high-bandwidth media such as optical fiber and terrestrial and satellite microwaves. Each has a bandwidth far in excess of that needed for the average transmission signal. If the bandwidth of a link is greater than the bandwidth needs of the devices connected to it, the bandwidth is wasted. An efficient system maximizes the utilization of all resources; bandwidth is one of the most precious resources we have in data communications.

In a multiplexed system, n lines share the bandwidth of one link. Figure 6.1 shows the basic format of a multiplexed system. The lines on the left direct their transmission streams to a **multiplexer (MUX),** which combines them into a single stream (many-to-one). At the receiving end, that stream is fed into a **demultiplexer (DEMUX),** which separates the stream back into its component transmissions (one-to-many) and directs them to their corresponding lines. In the figure, the word **link** refers to the physical path. The word **channel** refers to the portion of a link that carries a transmission between a given pair of lines. One link can have many (n) channels.

Figure 6.1 *Dividing a link into channels*

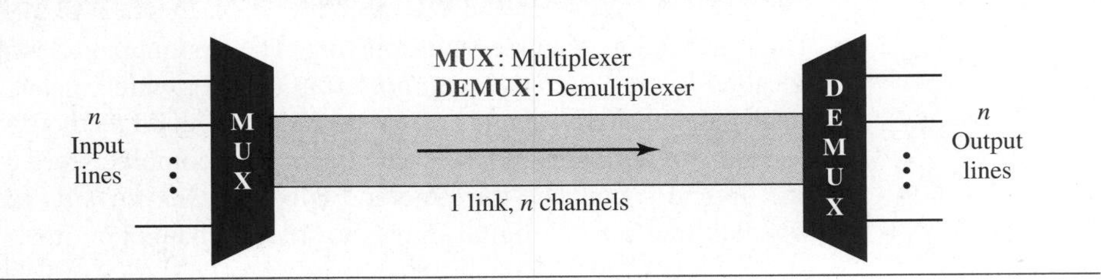

There are three basic multiplexing techniques: frequency-division multiplexing, wavelength-division multiplexing, and time-division multiplexing. The first two are techniques designed for analog signals, the third, for digital signals (see Figure 6.2).

Figure 6.2 *Categories of multiplexing*

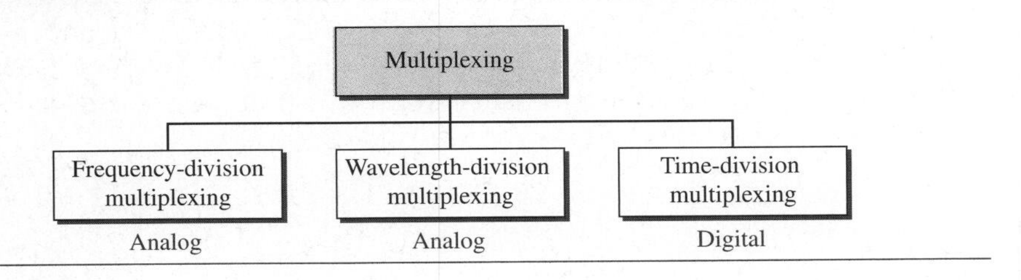

Although some textbooks consider carrier division multiple access (CDMA) as a fourth multiplexing category, we discuss CDMA as an access method (see Chapter 12).

6.1.1 Frequency-Division Multiplexing

Frequency-division multiplexing (FDM) is an analog technique that can be applied when the bandwidth of a link (in hertz) is greater than the combined bandwidths of the signals to be transmitted. In FDM, signals generated by each sending device modulate different carrier frequencies. These modulated signals are then combined into a single composite signal that can be transported by the link. Carrier frequencies are separated by sufficient bandwidth to accommodate the modulated signal. These bandwidth ranges are the channels through which the various signals travel. Channels can be separated by strips of unused bandwidth—**guard bands**—to prevent signals from overlapping. In addition, carrier frequencies must not interfere with the original data frequencies.

Figure 6.3 gives a conceptual view of FDM. In this illustration, the transmission path is divided into three parts, each representing a channel that carries one transmission.

Figure 6.3 *Frequency-division multiplexing*

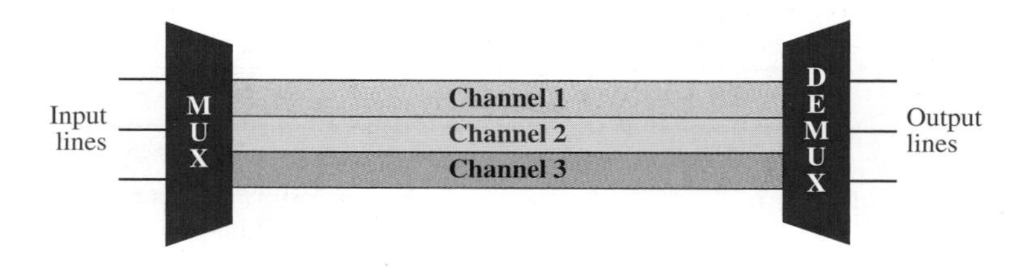

We consider FDM to be an analog multiplexing technique; however, this does not mean that FDM cannot be used to combine sources sending digital signals. A digital signal can be converted to an analog signal (with the techniques discussed in Chapter 5) before FDM is used to multiplex them.

FDM is an analog multiplexing technique that combines analog signals.

Multiplexing Process

Figure 6.4 is a conceptual illustration of the multiplexing process. Each source generates a signal of a similar frequency range. Inside the multiplexer, these similar signals modulate different carrier frequencies (f_1, f_2, and f_3). The resulting modulated signals are then combined into a single composite signal that is sent out over a media link that has enough bandwidth to accommodate it.

Figure 6.4 *FDM process*

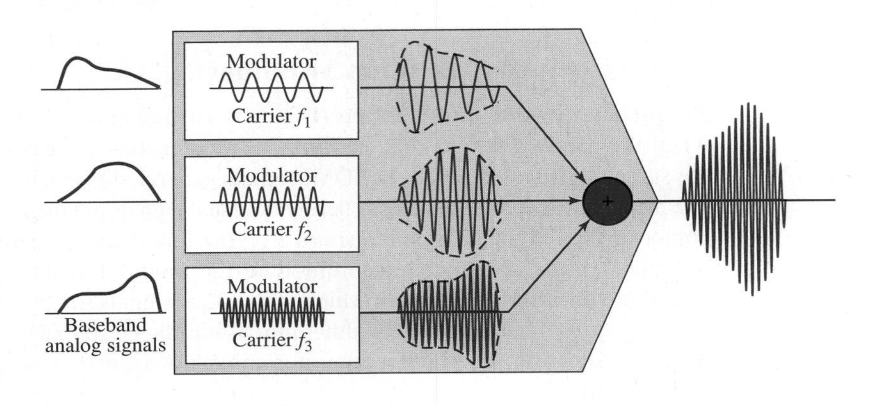

Demultiplexing Process

The demultiplexer uses a series of filters to decompose the multiplexed signal into its constituent component signals. The individual signals are then passed to a demodulator that separates them from their carriers and passes them to the output lines. Figure 6.5 is a conceptual illustration of demultiplexing process.

Figure 6.5 *FDM demultiplexing example*

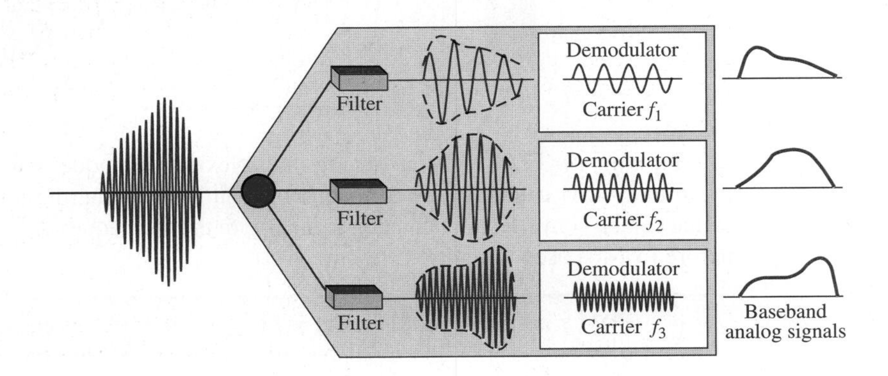

Example 6.1

Assume that a voice channel occupies a bandwidth of 4 kHz. We need to combine three voice channels into a link with a bandwidth of 12 kHz, from 20 to 32 kHz. Show the configuration, using the frequency domain. Assume there are no guard bands.

Solution

We shift (modulate) each of the three voice channels to a different bandwidth, as shown in Figure 6.6. We use the 20- to 24-kHz bandwidth for the first channel, the 24- to 28-kHz bandwidth

Figure 6.6 *Example 6.1*

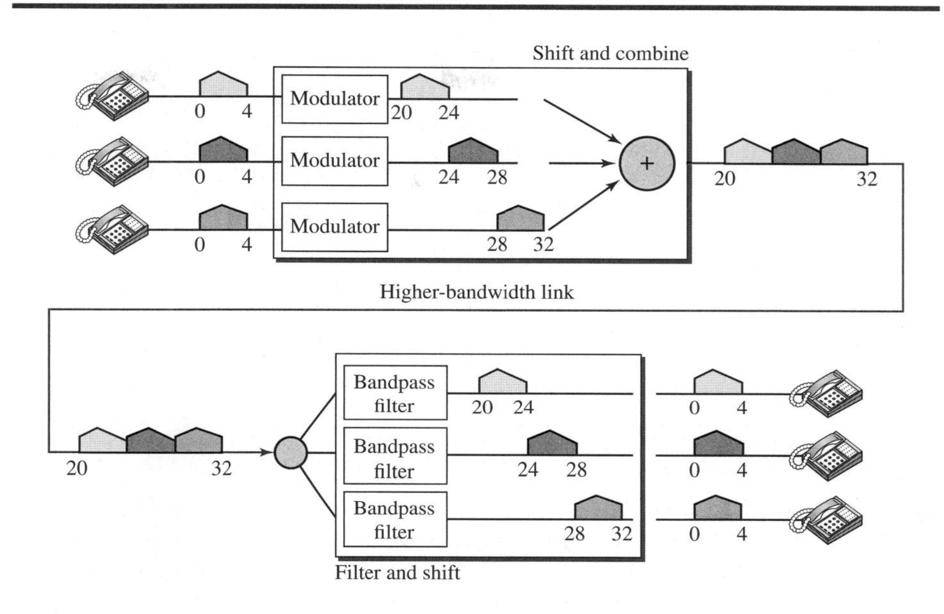

for the second channel, and the 28- to 32-kHz bandwidth for the third one. Then we combine them as shown in Figure 6.6. At the receiver, each channel receives the entire signal, using a filter to separate out its own signal. The first channel uses a filter that passes frequencies between 20 and 24 kHz and filters out (discards) any other frequencies. The second channel uses a filter that passes frequencies between 24 and 28 kHz, and the third channel uses a filter that passes frequencies between 28 and 32 kHz. Each channel then shifts the frequency to start from zero.

Example 6.2

Five channels, each with a 100-kHz bandwidth, are to be multiplexed together. What is the minimum bandwidth of the link if there is a need for a guard band of 10 kHz between the channels to prevent interference?

Solution

For five channels, we need at least four guard bands. This means that the required bandwidth is at least $5 \times 100 + 4 \times 10 = 540$ kHz, as shown in Figure 6.7.

Example 6.3

Four data channels (digital), each transmitting at 1 Mbps, use a satellite channel of 1 MHz. Design an appropriate configuration, using FDM.

Solution

The satellite channel is analog. We divide it into four channels, each channel having a 250-kHz bandwidth. Each digital channel of 1 Mbps is modulated so that each 4 bits is modulated to 1 Hz. One solution is 16-QAM modulation. Figure 6.8 shows one possible configuration.

Figure 6.7 *Example 6.2*

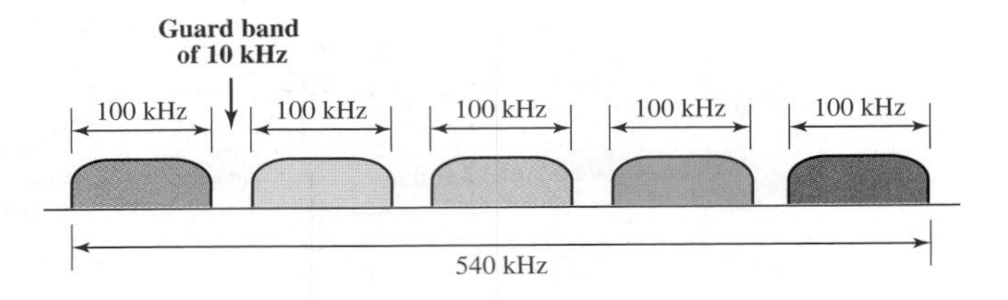

Figure 6.8 *Example 6.3*

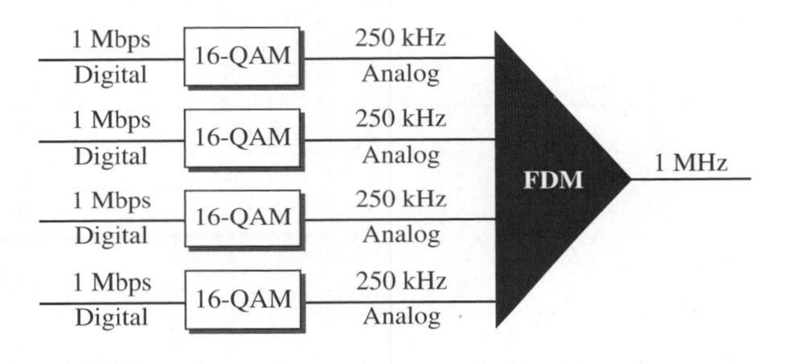

The Analog Carrier System

To maximize the efficiency of their infrastructure, telephone companies have traditionally multiplexed signals from lower-bandwidth lines onto higher-bandwidth lines. In this way, many switched or leased lines can be combined into fewer but bigger channels. For analog lines, FDM is used.

One of these hierarchical systems used by telephone companies is made up of groups, supergroups, master groups, and jumbo groups (see Figure 6.9).

In this **analog hierarchy,** 12 voice channels are multiplexed onto a higher-bandwidth line to create a **group.** A group has 48 kHz of bandwidth and supports 12 voice channels.

At the next level, up to five groups can be multiplexed to create a composite signal called a **supergroup.** A supergroup has a bandwidth of 240 kHz and supports up to 60 voice channels. Supergroups can be made up of either five groups or 60 independent voice channels.

At the next level, 10 supergroups are multiplexed to create a **master group.** A master group must have 2.40 MHz of bandwidth, but the need for guard bands between the supergroups increases the necessary bandwidth to 2.52 MHz. Master groups support up to 600 voice channels.

Finally, six master groups can be combined into a **jumbo group.** A jumbo group must have 15.12 MHz (6×2.52 MHz) but is augmented to 16.984 MHz to allow for guard bands between the master groups.

Figure 6.9 *Analog hierarchy*

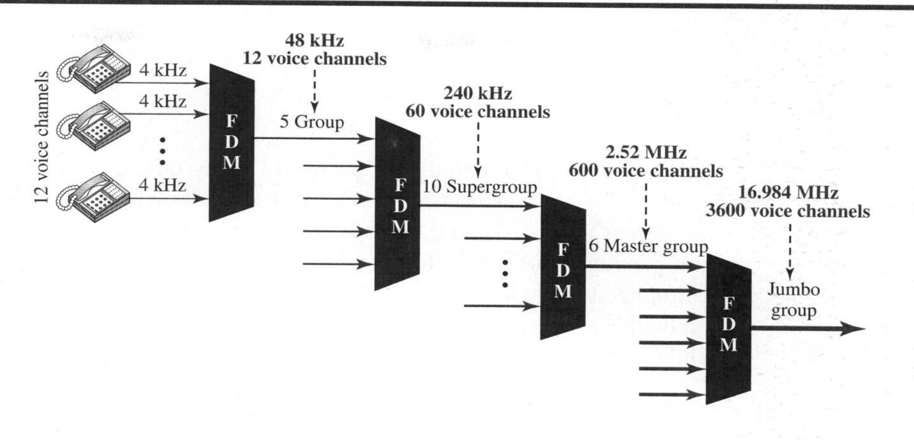

Other Applications of FDM

A very common application of FDM is AM and FM radio broadcasting. Radio uses the air as the transmission medium. A special band from 530 to 1700 kHz is assigned to AM radio. All radio stations need to share this band. As discussed in Chapter 5, each AM station needs 10 kHz of bandwidth. Each station uses a different carrier frequency, which means it is shifting its signal and multiplexing. The signal that goes to the air is a combination of signals. A receiver receives all these signals, but filters (by tuning) only the one which is desired. Without multiplexing, only one AM station could broadcast to the common link, the air. However, we need to know that there is no physical multiplexer or demultiplexer here. As we will see in Chapter 12, multiplexing is done at the data-link layer.

The situation is similar in FM broadcasting. However, FM has a wider band of 88 to 108 MHz because each station needs a bandwidth of 200 kHz.

Another common use of FDM is in television broadcasting. Each TV channel has its own bandwidth of 6 MHz.

The first generation of cellular telephones (See Chapter 16) also uses FDM. Each user is assigned two 30-kHz channels, one for sending voice and the other for receiving. The voice signal, which has a bandwidth of 3 kHz (from 300 to 3300 Hz), is modulated by using FM. Remember that an FM signal has a bandwidth 10 times that of the modulating signal, which means each channel has 30 kHz (10×3) of bandwidth. Therefore, each user is given, by the base station, a 60-kHz bandwidth in a range available at the time of the call.

Example 6.4

The Advanced Mobile Phone System (AMPS) uses two bands. The first band of 824 to 849 MHz is used for sending, and 869 to 894 MHz is used for receiving. Each user has a bandwidth of 30 kHz in each direction. The 3-kHz voice is modulated using FM, creating 30 kHz of modulated signal. How many people can use their cellular phones simultaneously?

Solution

Each band is 25 MHz. If we divide 25 MHz by 30 kHz, we get 833.33. In reality, the band is divided into 832 channels. Of these, 42 channels are used for control, which means only 790 channels are available for cellular phone users. We discuss AMPS in greater detail in Chapter 16.

Implementation

FDM can be implemented very easily. In many cases, such as radio and television broadcasting, there is no need for a physical multiplexer or demultiplexer. As long as the stations agree to send their broadcasts to the air using different carrier frequencies, multiplexing is achieved. In other cases, such as the cellular telephone system, a base station needs to assign a carrier frequency to the telephone user. There is not enough bandwidth in a cell to permanently assign a bandwidth range to every telephone user. When a user hangs up, her or his bandwidth is assigned to another caller.

6.1.2 Wavelength-Division Multiplexing

Wavelength-division multiplexing (WDM) is designed to use the high-data-rate capability of fiber-optic cable. The optical fiber data rate is higher than the data rate of metallic transmission cable, but using a fiber-optic cable for a single line wastes the available bandwidth. Multiplexing allows us to combine several lines into one.

WDM is conceptually the same as FDM, except that the multiplexing and demultiplexing involve optical signals transmitted through fiber-optic channels. The idea is the same: We are combining different signals of different frequencies. The difference is that the frequencies are very high.

Figure 6.10 gives a conceptual view of a WDM multiplexer and demultiplexer. Very narrow bands of light from different sources are combined to make a wider band of light. At the receiver, the signals are separated by the demultiplexer.

Figure 6.10 *Wavelength-division multiplexing*

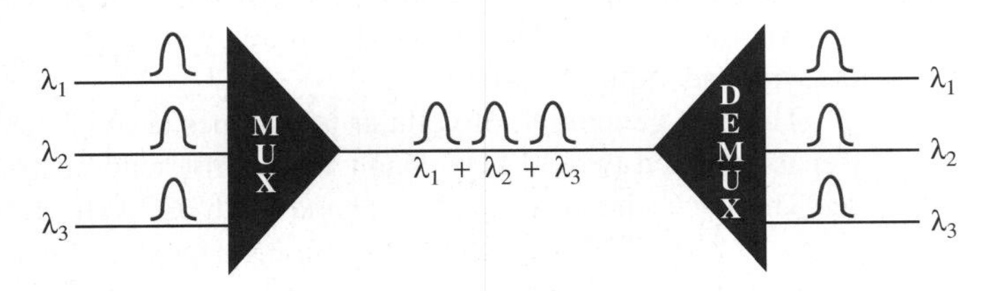

WDM is an analog multiplexing technique to combine optical signals.

Although WDM technology is very complex, the basic idea is very simple. We want to combine multiple light sources into one single light at the multiplexer and do the reverse at the demultiplexer. The combining and splitting of light sources are easily handled by a prism. Recall from basic physics that a prism bends a beam of light based on the angle of incidence and the frequency. Using this technique, a multiplexer can be made to combine several input beams of light, each containing a narrow band of frequencies, into one output beam of a wider band of frequencies. A demultiplexer can also be made to reverse the process. Figure 6.11 shows the concept.

Figure 6.11 *Prisms in wavelength-division multiplexing and demultiplexing*

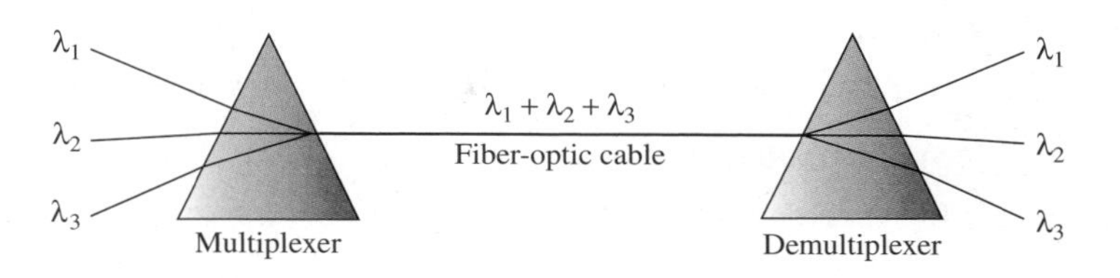

One application of WDM is the SONET network, in which multiple optical fiber lines are multiplexed and demultiplexed. We discuss SONET in Chapter 14.

A new method, called **dense WDM (DWDM),** can multiplex a very large number of channels by spacing channels very close to one another. It achieves even greater efficiency.

6.1.3 Time-Division Multiplexing

Time-division multiplexing (TDM) is a digital process that allows several connections to share the high bandwidth of a link. Instead of sharing a portion of the bandwidth as in FDM, time is shared. Each connection occupies a portion of time in the link. Figure 6.12 gives a conceptual view of TDM. Note that the same link is used as in FDM; here, however, the link is shown sectioned by time rather than by frequency. In the figure, portions of signals 1, 2, 3, and 4 occupy the link sequentially.

Figure 6.12 *TDM*

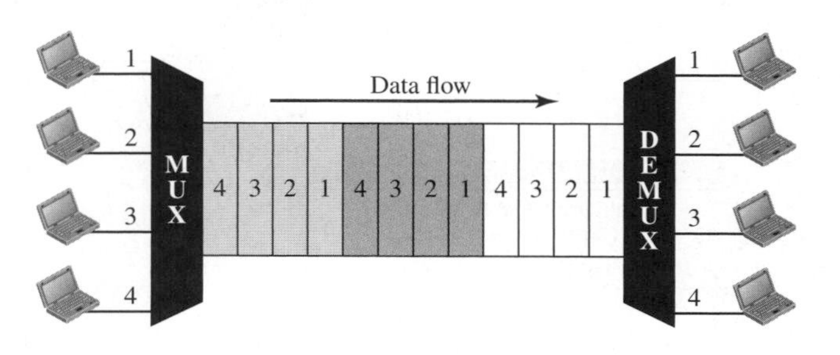

Note that in Figure 6.12 we are concerned with only multiplexing, not switching. This means that all the data in a message from source 1 always go to one specific destination, be it 1, 2, 3, or 4. The delivery is fixed and unvarying, unlike switching.

We also need to remember that TDM is, in principle, a digital multiplexing technique. Digital data from different sources are combined into one timeshared link. However, this does not mean that the sources cannot produce analog data; analog data can be sampled, changed to digital data, and then multiplexed by using TDM.

TDM is a digital multiplexing technique for combining several low-rate channels into one high-rate one.

We can divide TDM into two different schemes: synchronous and statistical. We first discuss **synchronous TDM** and then show how **statistical TDM** differs.

Synchronous TDM

In synchronous TDM, each input connection has an allotment in the output even if it is not sending data.

Time Slots and Frames

In synchronous TDM, the data flow of each input connection is divided into units, where each input occupies one input time slot. A unit can be 1 bit, one character, or one block of data. Each input unit becomes one output unit and occupies one output time slot. However, the duration of an output time slot is n times shorter than the duration of an input time slot. If an input time slot is T s, the output time slot is T/n s, where n is the number of connections. In other words, a unit in the output connection has a shorter duration; it travels faster. Figure 6.13 shows an example of synchronous TDM where n is 3.

Figure 6.13 *Synchronous time-division multiplexing*

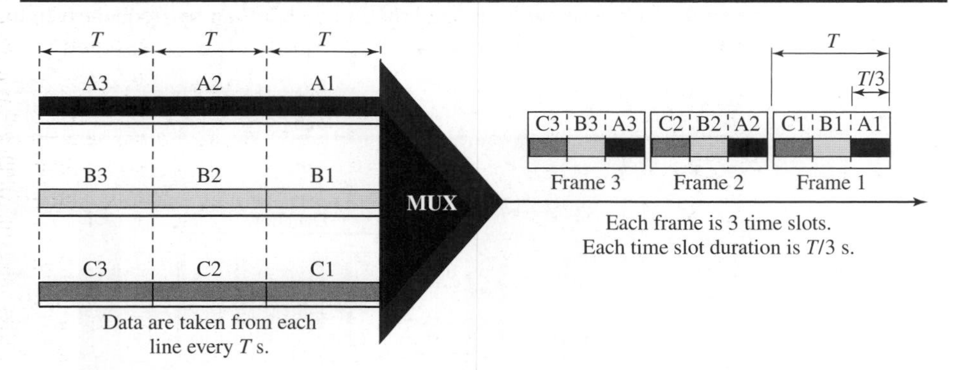

In synchronous TDM, a round of data units from each input connection is collected into a frame (we will see the reason for this shortly). If we have n connections, a frame is divided into n time slots and one slot is allocated for each unit, one for each input line. If the duration of the input unit is T, the duration of each slot is T/n and the duration of each frame is T (unless a frame carries some other information, as we will see shortly).

The data rate of the output link must be n times the data rate of a connection to guarantee the flow of data. In Figure 6.13, the data rate of the link is 3 times the data rate of a connection; likewise, the duration of a unit on a connection is 3 times that of the time slot (duration of a unit on the link). In the figure we represent the data prior to multiplexing as 3 times the size of the data after multiplexing. This is just to convey the idea that each unit is 3 times longer in duration before multiplexing than after.

In synchronous TDM, the data rate of the link is n times faster, and the unit duration is n times shorter.

Time slots are grouped into frames. A frame consists of one complete cycle of time slots, with one slot dedicated to each sending device. In a system with n input lines, each frame has n slots, with each slot allocated to carrying data from a specific input line.

Example 6.5

In Figure 6.13, the data rate for each input connection is 1 kbps. If 1 bit at a time is multiplexed (a unit is 1 bit), what is the duration of

1. each input slot,
2. each output slot, and
3. each frame?

Solution

We can answer the questions as follows:

1. The data rate of each input connection is 1 kbps. This means that the bit duration is 1/1000 s or 1 ms. The duration of the input time slot is 1 ms (same as bit duration).
2. The duration of each output time slot is one-third of the input time slot. This means that the duration of the output time slot is 1/3 ms.
3. Each frame carries three output time slots. So the duration of a frame is $3 \times 1/3$ ms, or 1 ms. The duration of a frame is the same as the duration of an input unit.

Example 6.6

Figure 6.14 shows synchronous TDM with a data stream for each input and one data stream for the output. The unit of data is 1 bit. Find (1) the input bit duration, (2) the output bit duration, (3) the output bit rate, and (4) the output frame rate.

Figure 6.14 *Example 6.6*

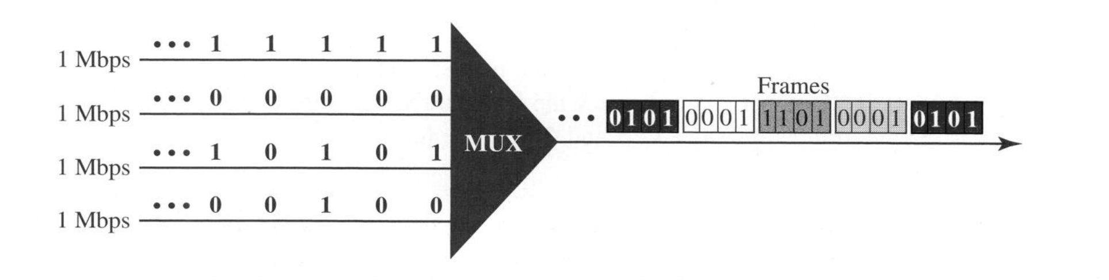

Solution

We can answer the questions as follows:

1. The input bit duration is the inverse of the bit rate: 1/1 Mbps = 1 μs.
2. The output bit duration is one-fourth of the input bit duration, or 1/4 μs.

3. The output bit rate is the inverse of the output bit duration, or 1/4 μs, or 4 Mbps. This can also be deduced from the fact that the output rate is 4 times as fast as any input rate; so the output rate = 4 × 1 Mbps = 4 Mbps.
4. The frame rate is always the same as any input rate. So the frame rate is 1,000,000 frames per second. Because we are sending 4 bits in each frame, we can verify the result of the previous question by multiplying the frame rate by the number of bits per frame.

Example 6.7

Four 1-kbps connections are multiplexed together. A unit is 1 bit. Find (1) the duration of 1 bit before multiplexing, (2) the transmission rate of the link, (3) the duration of a time slot, and (4) the duration of a frame.

Solution

We can answer the questions as follows:

1. The duration of 1 bit before multiplexing is 1/1 kbps, or 0.001 s (1 ms).
2. The rate of the link is 4 times the rate of a connection, or 4 kbps.
3. The duration of each time slot is one-fourth of the duration of each bit before multiplexing, or 1/4 ms or 250 μs. Note that we can also calculate this from the data rate of the link, 4 kbps. The bit duration is the inverse of the data rate, or 1/4 kbps or 250 μs.
4. The duration of a frame is always the same as the duration of a unit before multiplexing, or 1 ms. We can also calculate this in another way. Each frame in this case has four time slots. So the duration of a frame is 4 times 250 μs, or 1 ms.

Interleaving

TDM can be visualized as two fast-rotating switches, one on the multiplexing side and the other on the demultiplexing side. The switches are synchronized and rotate at the same speed, but in opposite directions. On the multiplexing side, as the switch opens in front of a connection, that connection has the opportunity to send a unit onto the path. This process is called **interleaving.** On the demultiplexing side, as the switch opens in front of a connection, that connection has the opportunity to receive a unit from the path.

Figure 6.15 shows the interleaving process for the connection shown in Figure 6.13. In this figure, we assume that no switching is involved and that the data from the first connection at the multiplexer site go to the first connection at the demultiplexer. We discuss switching in Chapter 8.

Example 6.8

Four channels are multiplexed using TDM. If each channel sends 100 bytes/s and we multiplex 1 byte per channel, show the frame traveling on the link, the size of the frame, the duration of a frame, the frame rate, and the bit rate for the link.

Solution

The multiplexer is shown in Figure 6.16. Each frame carries 1 byte from each channel; the size of each frame, therefore, is 4 bytes, or 32 bits. Because each channel is sending 100 bytes/s and a frame carries 1 byte from each channel, the frame rate must be 100 frames per second. The

Figure 6.15 *Interleaving*

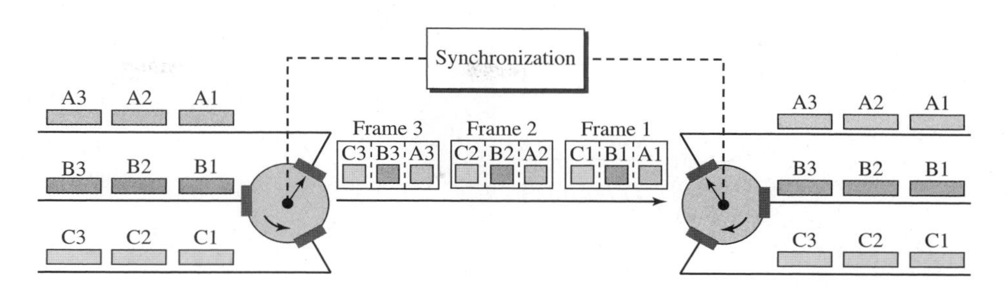

duration of a frame is therefore 1/100 s. The link is carrying 100 frames per second, and since each frame contains 32 bits, the bit rate is 100×32, or 3200 bps. This is actually 4 times the bit rate of each channel, which is $100 \times 8 = 800$ bps.

Figure 6.16 *Example 6.8*

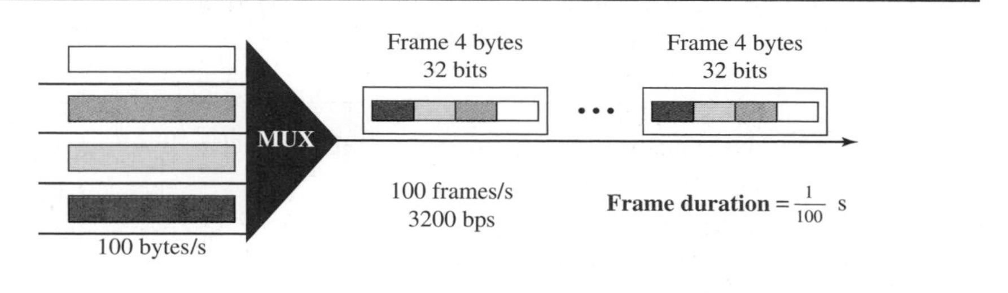

Example 6.9

A multiplexer combines four 100-kbps channels using a time slot of 2 bits. Show the output with four arbitrary inputs. What is the frame rate? What is the frame duration? What is the bit rate? What is the bit duration?

Solution

Figure 6.17 shows the output for four arbitrary inputs. The link carries 50,000 frames per second since each frame contains 2 bits per channel. The frame duration is therefore 1/50,000 s or 20 μs. The frame rate is 50,000 frames per second, and each frame carries 8 bits; the bit rate is $50{,}000 \times 8 = 400{,}000$ bits or 400 kbps. The bit duration is 1/400,000 s, or 2.5 μs. Note that the frame duration is 8 times the bit duration because each frame is carrying 8 bits.

Empty Slots

Synchronous TDM is not as efficient as it could be. If a source does not have data to send, the corresponding slot in the output frame is empty. Figure 6.18 shows a case in which one of the input lines has no data to send and one slot in another input line has discontinuous data.

The first output frame has three slots filled, the second frame has two slots filled, and the third frame has three slots filled. No frame is full. We learn in the next section

Figure 6.17 *Example 6.9*

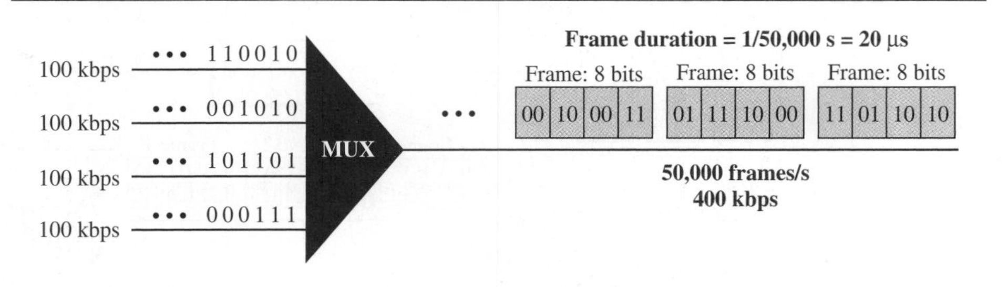

Figure 6.18 *Empty slots*

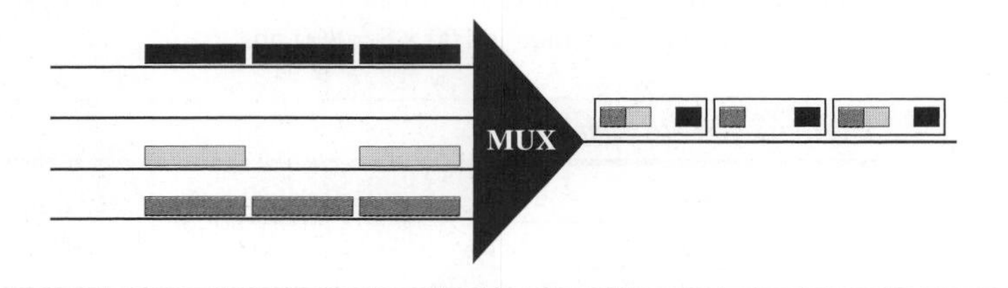

that statistical TDM can improve the efficiency by removing the empty slots from the frame.

Data Rate Management

One problem with TDM is how to handle a disparity in the input data rates. In all our discussion so far, we assumed that the data rates of all input lines were the same. However, if data rates are not the same, three strategies, or a combination of them, can be used. We call these three strategies **multilevel multiplexing, multiple-slot allocation,** and **pulse stuffing.**

Multilevel Multiplexing Multilevel multiplexing is a technique used when the data rate of an input line is a multiple of others. For example, in Figure 6.19, we have two inputs of 20 kbps and three inputs of 40 kbps. The first two input lines can be multiplexed together to provide a data rate equal to the last three. A second level of multiplexing can create an output of 160 kbps.

Figure 6.19 *Multilevel multiplexing*

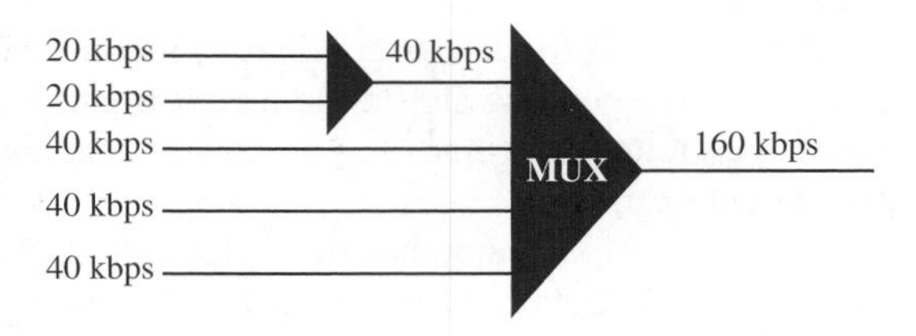

Multiple-Slot Allocation Sometimes it is more efficient to allot more than one slot in a frame to a single input line. For example, we might have an input line that has a data rate that is a multiple of another input. In Figure 6.20, the input line with a 50-kbps data rate can be given two slots in the output. We insert a demultiplexer in the line to make two inputs out of one.

Figure 6.20 *Multiple-slot multiplexing*

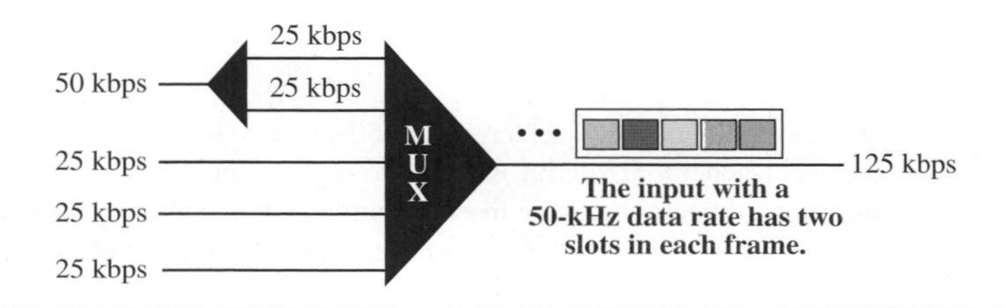

Pulse Stuffing Sometimes the bit rates of sources are not multiple integers of each other. Therefore, neither of the above two techniques can be applied. One solution is to make the highest input data rate the dominant data rate and then add dummy bits to the input lines with lower rates. This will increase their rates. This technique is called ***pulse stuffing,*** *bit padding,* or *bit stuffing.* The idea is shown in Figure 6.21. The input with a data rate of 46 is pulse-stuffed to increase the rate to 50 kbps. Now multiplexing can take place.

Figure 6.21 *Pulse stuffing*

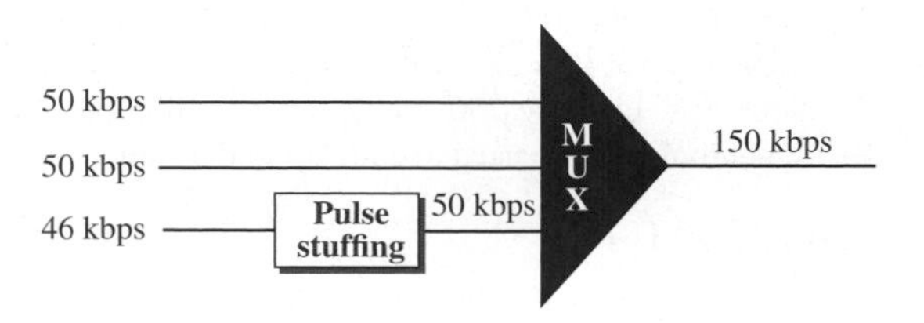

Frame Synchronizing

The implementation of TDM is not as simple as that of FDM. Synchronization between the multiplexer and demultiplexer is a major issue. If the multiplexer and the demultiplexer are not synchronized, a bit belonging to one channel may be received by the wrong channel. For this reason, one or more synchronization bits are usually added to the beginning of each frame. These bits, called **framing bits,** follow a pattern, frame to frame, that allows the demultiplexer to synchronize with the incoming stream so that it can separate the time slots accurately. In most cases, this synchronization information consists of 1 bit per frame, alternating between 0 and 1, as shown in Figure 6.22.

Figure 6.22 *Framing bits*

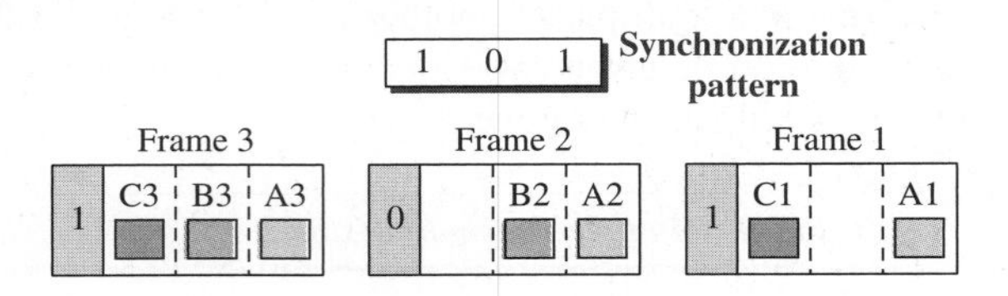

Example 6.10

We have four sources, each creating 250 characters per second. If the interleaved unit is a character and 1 synchronizing bit is added to each frame, find (1) the data rate of each source, (2) the duration of each character in each source, (3) the frame rate, (4) the duration of each frame, (5) the number of bits in each frame, and (6) the data rate of the link.

Solution

We can answer the questions as follows:

1. The data rate of each source is $250 \times 8 = 2000$ bps = 2 kbps.
2. Each source sends 250 characters per second; therefore, the duration of a character is 1/250 s, or 4 ms.
3. Each frame has one character from each source, which means the link needs to send 250 frames per second to keep the transmission rate of each source.
4. The duration of each frame is 1/250 s, or 4 ms. Note that the duration of each frame is the same as the duration of each character coming from each source.
5. Each frame carries 4 characters and 1 extra synchronizing bit. This means that each frame is $4 \times 8 + 1 = 33$ bits.
6. The link sends 250 frames per second, and each frame contains 33 bits. This means that the data rate of the link is 250×33, or 8250 bps. Note that the bit rate of the link is greater than the combined bit rates of the four channels. If we add the bit rates of four channels, we get 8000 bps. Because 250 frames are traveling per second and each contains 1 extra bit for synchronizing, we need to add 250 to the sum to get 8250 bps.

Example 6.11

Two channels, one with a bit rate of 100 kbps and another with a bit rate of 200 kbps, are to be multiplexed. How this can be achieved? What is the frame rate? What is the frame duration? What is the bit rate of the link?

Solution

We can allocate one slot to the first channel and two slots to the second channel. Each frame carries 3 bits. The frame rate is 100,000 frames per second because it carries 1 bit from the first channel. The frame duration is 1/100,000 s, or 10 ms. The bit rate is 100,000 frames/s $\times$ 3 bits per frame, or 300 kbps. Note that because each frame carries 1 bit from the first channel, the bit rate for the first channel is preserved. The bit rate for the second channel is also preserved because each frame carries 2 bits from the second channel.

Digital Signal Service

Telephone companies implement TDM through a hierarchy of digital signals, called ***digital signal (DS) service*** or *digital hierarchy.* Figure 6.23 shows the data rates supported by each level.

Figure 6.23 *Digital hierarchy*

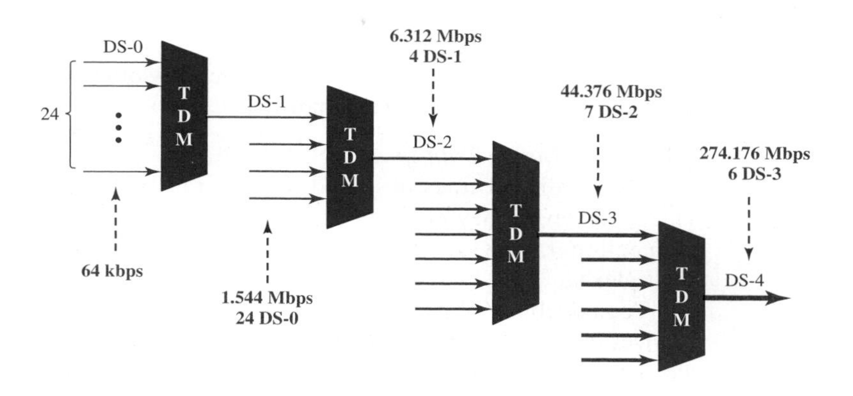

- **DS-0** is a single digital channel of 64 kbps.
- **DS-1** is a 1.544-Mbps service; 1.544 Mbps is 24 times 64 kbps plus 8 kbps of overhead. It can be used as a single service for 1.544-Mbps transmissions, or it can be used to multiplex 24 DS-0 channels or to carry any other combination desired by the user that can fit within its 1.544-Mbps capacity.
- **DS-2** is a 6.312-Mbps service; 6.312 Mbps is 96 times 64 kbps plus 168 kbps of overhead. It can be used as a single service for 6.312-Mbps transmissions; or it can be used to multiplex 4 DS-1 channels, 96 DS-0 channels, or a combination of these service types.
- **DS-3** is a 44.376-Mbps service; 44.376 Mbps is 672 times 64 kbps plus 1.368 Mbps of overhead. It can be used as a single service for 44.376-Mbps transmissions; or it can be used to multiplex 7 DS-2 channels, 28 DS-1 channels, 672 DS-0 channels, or a combination of these service types.
- **DS-4** is a 274.176-Mbps service; 274.176 is 4032 times 64 kbps plus 16.128 Mbps of overhead. It can be used to multiplex 6 DS-3 channels, 42 DS-2 channels, 168 DS-1 channels, 4032 DS-0 channels, or a combination of these service types.

T Lines

DS-0, DS-1, and so on are the names of services. To implement those services, the telephone companies use **T lines** (T-1 to T-4). These are lines with capacities precisely matched to the data rates of the DS-1 to DS-4 services (see Table 6.1). So far only T-1 and T-3 lines are commercially available.

Table 6.1 *DS and T line rates*

Service	*Line*	*Rate (Mbps)*	*Voice Channels*
DS-1	T-1	1.544	24
DS-2	T-2	6.312	96
DS-3	T-3	44.736	672
DS-4	T-4	274.176	4032

The T-1 line is used to implement DS-1; T-2 is used to implement DS-2; and so on. As you can see from Table 6.1, DS-0 is not actually offered as a service, but it has been defined as a basis for reference purposes.

T Lines for Analog Transmission
T lines are digital lines designed for the transmission of digital data, audio, or video. However, they also can be used for analog transmission (regular telephone connections), provided the analog signals are first sampled, then time-division multiplexed.

The possibility of using T lines as analog carriers opened up a new generation of services for the telephone companies. Earlier, when an organization wanted 24 separate telephone lines, it needed to run 24 twisted-pair cables from the company to the central exchange. (Remember those old movies showing a busy executive with 10 telephones lined up on his desk? Or the old office telephones with a big fat cable running from them? Those cables contained a bundle of separate lines.) Today, that same organization can combine the 24 lines into one T-1 line and run only the T-1 line to the exchange. Figure 6.24 shows how 24 voice channels can be multiplexed onto one T-1 line. (Refer to Chapter 4 for PCM encoding.)

Figure 6.24 *T-1 line for multiplexing telephone lines*

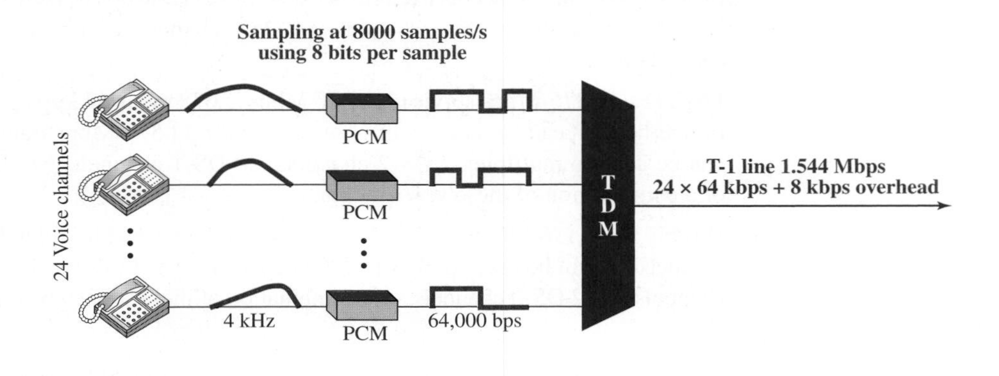

The T-1 Frame As noted above, DS-1 requires 8 kbps of overhead. To understand how this overhead is calculated, we must examine the format of a 24-voice-channel frame.

The frame used on a T-1 line is usually 193 bits divided into 24 slots of 8 bits each plus 1 extra bit for synchronization ($24 \times 8 + 1 = 193$); see Figure 6.25. In other words,

Figure 6.25 *T-1 frame structure*

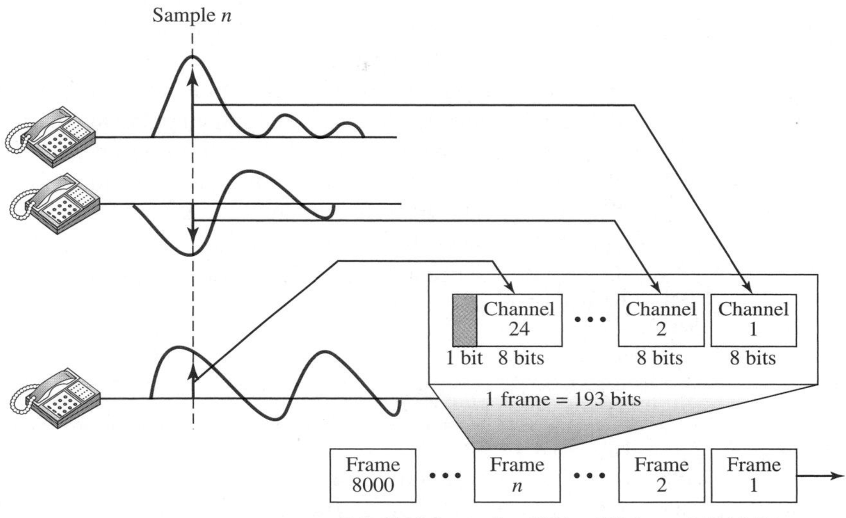

each slot contains one signal segment from each channel; 24 segments are interleaved in one frame. If a T-1 line carries 8000 frames, the data rate is 1.544 Mbps ($193 \times 8000 =$ 1.544 Mbps)—the capacity of the line.

E Lines

Europeans use a version of T lines called **E lines.** The two systems are conceptually identical, but their capacities differ. Table 6.2 shows the E lines and their capacities.

Table 6.2 *E line rates*

Line	*Rate (Mbps)*	*Voice Channels*
E-1	2.048	30
E-2	8.448	120
E-3	34.368	480
E-4	139.264	1920

More Synchronous TDM Applications

Some second-generation cellular telephone companies use synchronous TDM. For example, the digital version of cellular telephony divides the available bandwidth into 30-kHz bands. For each band, TDM is applied so that six users can share the band. This means that each 30-kHz band is now made of six time slots, and the digitized voice signals of the users are inserted in the slots. Using TDM, the number of telephone users in each area is now 6 times greater. We discuss second-generation cellular telephony in Chapter 16.

Statistical Time-Division Multiplexing

As we saw in the previous section, in synchronous TDM, each input has a reserved slot in the output frame. This can be inefficient if some input lines have no data to send. In statistical time-division multiplexing, slots are dynamically allocated to improve bandwidth efficiency. Only when an input line has a slot's worth of data to send is it given a slot in the output frame. In statistical multiplexing, the number of slots in each frame is less than the number of input lines. The multiplexer checks each input line in round-robin fashion; it allocates a slot for an input line if the line has data to send; otherwise, it skips the line and checks the next line.

Figure 6.26 shows a synchronous and a statistical TDM example. In the former, some slots are empty because the corresponding line does not have data to send. In the latter, however, no slot is left empty as long as there are data to be sent by any input line.

Figure 6.26 *TDM slot comparison*

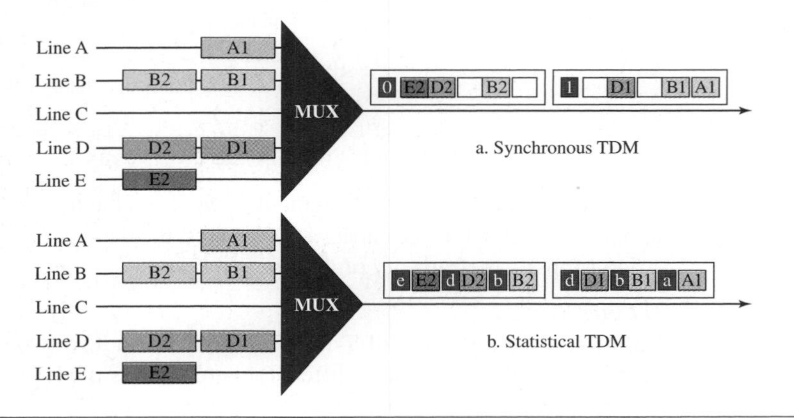

Addressing

Figure 6.26 also shows a major difference between slots in synchronous TDM and statistical TDM. An output slot in synchronous TDM is totally occupied by data; in statistical TDM, a slot needs to carry data as well as the address of the destination. In synchronous TDM, there is no need for addressing; synchronization and preassigned relationships between the inputs and outputs serve as an address. We know, for example, that input 1 always goes to input 2. If the multiplexer and the demultiplexer are synchronized, this is guaranteed. In statistical multiplexing, there is no fixed relationship between the inputs and outputs because there are no preassigned or reserved slots. We need to include the address of the receiver inside each slot to show where it is to be delivered. The addressing in its simplest form can be *n* bits to define *N* different output lines with $n = \log_2 N$. For example, for eight different output lines, we need a 3-bit address.

Slot Size

Since a slot carries both data and an address in statistical TDM, the ratio of the data size to address size must be reasonable to make transmission efficient. For example, it would be inefficient to send 1 bit per slot as data when the address is 3 bits. This would mean an overhead of 300 percent. In statistical TDM, a block of data is usually many bytes while the address is just a few bytes.

No Synchronization Bit

There is another difference between synchronous and statistical TDM, but this time it is at the frame level. The frames in statistical TDM need not be synchronized, so we do not need synchronization bits.

Bandwidth

In statistical TDM, the capacity of the link is normally less than the sum of the capacities of each channel. The designers of statistical TDM define the capacity of the link based on the statistics of the load for each channel. If on average only x percent of the input slots are filled, the capacity of the link reflects this. Of course, during peak times, some slots need to wait.

6.2 SPREAD SPECTRUM

Multiplexing combines signals from several sources to achieve bandwidth efficiency; the available bandwidth of a link is divided between the sources. In **spread spectrum (SS),** we also combine signals from different sources to fit into a larger bandwidth, but our goals are somewhat different. Spread spectrum is designed to be used in wireless applications (LANs and WANs). In these types of applications, we have some concerns that outweigh bandwidth efficiency. In wireless applications, all stations use air (or a vacuum) as the medium for communication. Stations must be able to share this medium without interception by an eavesdropper and without being subject to jamming from a malicious intruder (in military operations, for example).

To achieve these goals, spread spectrum techniques add redundancy; they spread the original spectrum needed for each station. If the required bandwidth for each station is B, spread spectrum expands it to B_{SS}, such that $B_{SS} >> B$. The expanded bandwidth allows the source to wrap its message in a protective envelope for a more secure transmission. An analogy is the sending of a delicate, expensive gift. We can insert the gift in a special box to prevent it from being damaged during transportation, and we can use a superior delivery service to guarantee the safety of the package.

Figure 6.27 shows the idea of spread spectrum. Spread spectrum achieves its goals through two principles:

1. The bandwidth allocated to each station needs to be, by far, larger than what is needed. This allows redundancy.

2. The expanding of the original bandwidth B to the bandwidth B_{ss} must be done by a process that is independent of the original signal. In other words, the spreading process occurs after the signal is created by the source.

Figure 6.27 *Spread spectrum*

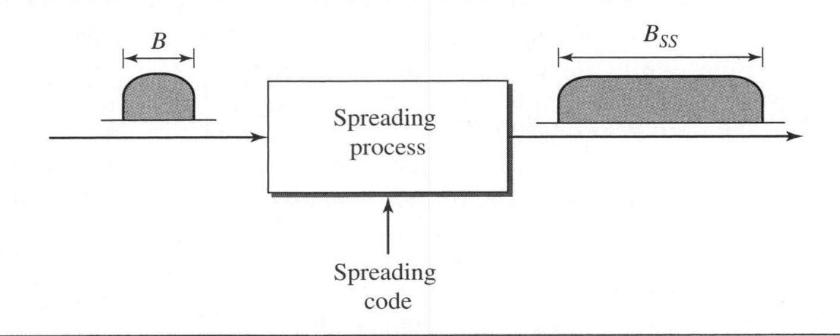

After the signal is created by the source, the spreading process uses a spreading code and spreads the bandwidth. The figure shows the original bandwidth B and the spread bandwidth B_{SS}. The spreading code is a series of numbers that look random, but are actually a pattern.

There are two techniques to spread the bandwidth: frequency hopping spread spectrum (FHSS) and direct sequence spread spectrum (DSSS).

6.2.1 Frequency Hopping Spread Spectrum

The **frequency hopping spread spectrum (FHSS)** technique uses M different carrier frequencies that are modulated by the source signal. At one moment, the signal modulates one carrier frequency; at the next moment, the signal modulates another carrier frequency. Although the modulation is done using one carrier frequency at a time, M frequencies are used in the long run. The bandwidth occupied by a source after spreading is $B_{FHSS} >> B$.

Figure 6.28 shows the general layout for FHSS. A **pseudorandom code generator,** called ***pseudorandom noise*** **(PN),** creates a k-bit pattern for every **hopping period** T_h. The frequency table uses the pattern to find the frequency to be used for this hopping period and passes it to the frequency synthesizer. The frequency synthesizer creates a carrier signal of that frequency, and the source signal modulates the carrier signal.

Suppose we have decided to have eight hopping frequencies. This is extremely low for real applications and is just for illustration. In this case, M is 8 and k is 3. The pseudorandom code generator will create eight different 3-bit patterns. These are mapped to eight different frequencies in the frequency table (see Figure 6.29).

The pattern for this station is 101, 111, 001, 000, 010, 011, 100. Note that the pattern is pseudorandom; it is repeated after eight hoppings. This means that at hopping period 1, the pattern is 101. The frequency selected is 700 kHz; the source signal modulates this carrier frequency. The second k-bit pattern selected is 111, which selects the 900-kHz carrier; the eighth pattern is 100, and the frequency is 600 kHz. After eight hoppings, the pattern repeats, starting from 101 again. Figure 6.30 shows how the signal

Figure 6.28 *Frequency hopping spread spectrum (FHSS)*

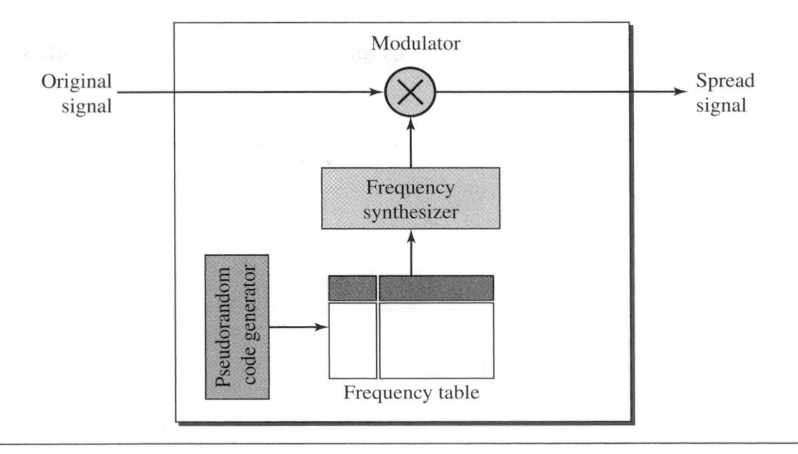

Figure 6.29 *Frequency selection in FHSS*

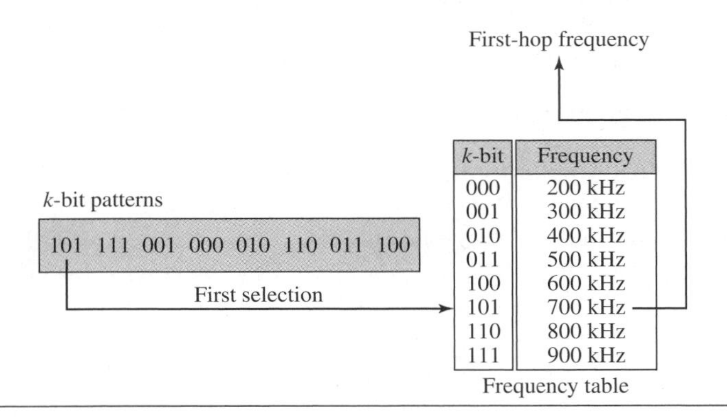

hops around from carrier to carrier. We assume the required bandwidth of the original signal is 100 kHz.

It can be shown that this scheme can accomplish the previously mentioned goals. If there are many k-bit patterns and the hopping period is short, a sender and receiver can have privacy. If an intruder tries to intercept the transmitted signal, she can only access a small piece of data because she does not know the spreading sequence to quickly adapt herself to the next hop. The scheme also has an antijamming effect. A malicious sender may be able to send noise to jam the signal for one hopping period (randomly), but not for the whole period.

Bandwidth Sharing

If the number of hopping frequencies is M, we can multiplex M channels into one by using the same B_{ss} bandwidth. This is possible because a station uses just one frequency in each hopping period; $M - 1$ other frequencies can be used by $M - 1$ other stations. In

Figure 6.30 *FHSS cycles*

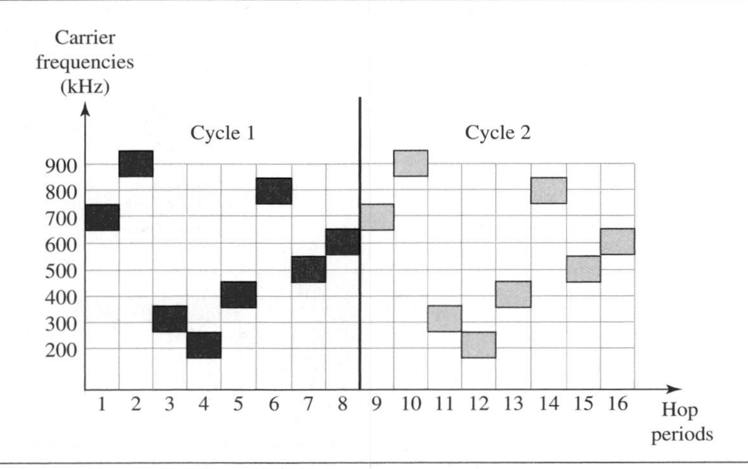

other words, M different stations can use the same B_{SS} if an appropriate modulation technique such as multiple FSK (MFSK) is used. FHSS is similar to FDM, as shown in Figure 6.31.

Figure 6.31 *Bandwidth sharing*

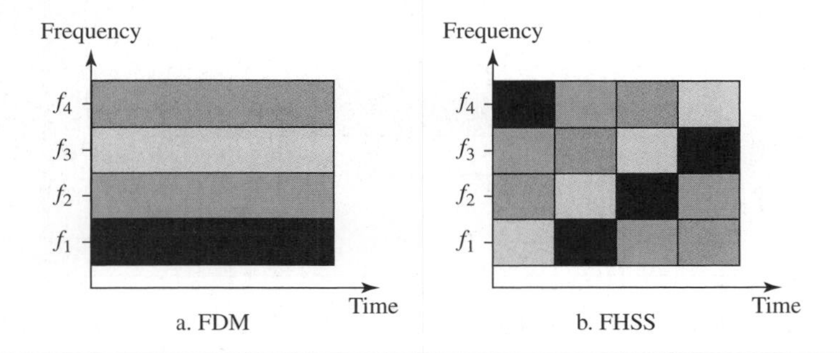

Figure 6.31 shows an example of four channels using FDM and four channels using FHSS. In FDM, each station uses $1/M$ of the bandwidth, but the allocation is fixed; in FHSS, each station uses $1/M$ of the bandwidth, but the allocation changes hop to hop.

6.2.2 Direct Sequence Spread Spectrum

The **direct sequence spread spectrum (DSSS)** technique also expands the bandwidth of the original signal, but the process is different. In DSSS, we replace each data bit with n bits using a spreading code. In other words, each bit is assigned a code of n bits, called ***chips,*** where the chip rate is n times that of the data bit. Figure 6.32 shows the concept of DSSS.

Figure 6.32 *DSSS*

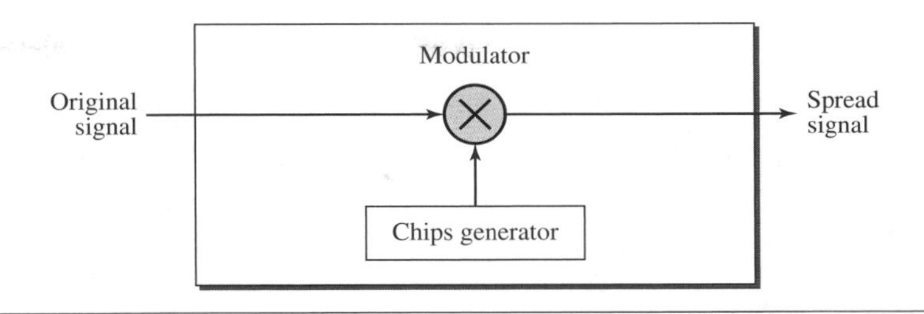

As an example, let us consider the sequence used in a wireless LAN, the famous **Barker sequence,** where n is 11. We assume that the original signal and the chips in the chip generator use polar NRZ encoding. Figure 6.33 shows the chips and the result of multiplying the original data by the chips to get the spread signal.

Figure 6.33 *DSSS example*

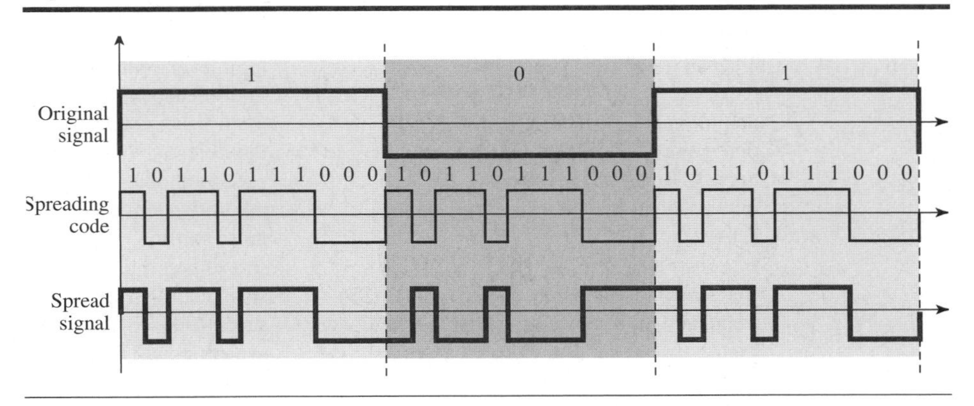

In Figure 6.33, the spreading code is 11 chips having the pattern 10110111000 (in this case). If the original signal rate is N, the rate of the spread signal is $11N$. This means that the required bandwidth for the spread signal is 11 times larger than the bandwidth of the original signal. The spread signal can provide privacy if the intruder does not know the code. It can also provide immunity against interference if each station uses a different code.

Bandwidth Sharing

Can we share a bandwidth in DSSS as we did in FHSS? The answer is no and yes. If we use a spreading code that spreads signals (from different stations) that cannot be combined and separated, we cannot share a bandwidth. For example, as we will see in Chapter 15, some wireless LANs use DSSS and the spread bandwidth cannot be shared. However, if we use a special type of sequence code that allows the combining and separating of spread signals, we can share the bandwidth. As we will see in

Chapter 16, a special spreading code allows us to use DSSS in cellular telephony and share a bandwidth among several users.

6.3 END-CHAPTER MATERIALS

6.3.1 Recommended Reading

For more details about subjects discussed in this chapter, we recommend the following books. The items in brackets [...] refer to the reference list at the end of the text.

Books

Multiplexing is discussed in [Pea92]. [Cou01] gives excellent coverage of TDM and FDM. More advanced materials can be found in [Ber96]. Multiplexing is discussed in [Sta04]. A good coverage of spread spectrum can be found in [Cou01] and [Sta04].

6.3.2 Key Terms

analog hierarchy
Barker sequence
channel
chip
demultiplexer (DEMUX)
dense WDM (DWDM)
digital signal (DS) service
direct sequence spread spectrum (DSSS)
E line
framing bit
frequency hopping spread spectrum (FHSS)
frequency-division multiplexing (FDM)
group
guard band
hopping period
interleaving
jumbo group
link
master group
multilevel multiplexing
multiple-slot allocation
multiplexer (MUX)
multiplexing
pseudorandom code generator
pseudorandom noise (PN)
pulse stuffing
spread spectrum (SS)
statistical TDM
supergroup
synchronous TDM
T line
time-division multiplexing (TDM)
wavelength-division multiplexing (WDM)

6.3.3 Summary

Bandwidth utilization is the use of available bandwidth to achieve specific goals. Efficiency can be achieved by using multiplexing; privacy and antijamming can be achieved by using spreading.

Multiplexing is the set of techniques that allow the simultaneous transmission of multiple signals across a single data link. In a multiplexed system, n lines share the bandwidth of one link. The word *link* refers to the physical path. The word *channel* refers to the portion of a link that carries a transmission. There are three basic multiplexing techniques: frequency-division multiplexing, wavelength-division multiplexing, and time-division multiplexing. The first two are techniques designed for analog signals, the third, for digital signals. Frequency-division multiplexing (FDM) is an analog technique that can be applied when the bandwidth of a link (in hertz) is greater than the combined bandwidths of the signals to be transmitted. Wavelength-division multiplexing (WDM) is designed to use the high bandwidth capability of fiber-optic cable. WDM is an analog multiplexing technique to combine optical signals. Time-division multiplexing (TDM) is a digital process that allows several connections to share the high bandwidth of a link. TDM is a digital multiplexing technique for combining several low-rate channels into one high-rate one. We can divide TDM into two different schemes: synchronous or statistical. In synchronous TDM, each input connection has an allotment in the output even if it is not sending data. In statistical TDM, slots are dynamically allocated to improve bandwidth efficiency.

In spread spectrum (SS), we combine signals from different sources to fit into a larger bandwidth. Spread spectrum is designed to be used in wireless applications in which stations must be able to share the medium without interception by an eavesdropper and without being subject to jamming from a malicious intruder. The frequency hopping spread spectrum (FHSS) technique uses M different carrier frequencies that are modulated by the source signal. At one moment, the signal modulates one carrier frequency; at the next moment, the signal modulates another carrier frequency. The direct sequence spread spectrum (DSSS) technique expands the bandwidth of a signal by replacing each data bit with n bits using a spreading code. In other words, each bit is assigned a code of n bits, called chips.

6.4 PRACTICE SET

6.4.1 Quizzes

A set of interactive quizzes for this chapter can be found on the book website. It is strongly recommended that the student take the quizzes to check his/her understanding of the materials before continuing with the practice set.

6.4.2 Questions

Q6-1. Distinguish between synchronous and statistical TDM.

Q6-2. Define spread spectrum and its goal. List the two spread spectrum techniques discussed in this chapter.

Q6-3. Define FHSS and explain how it achieves bandwidth spreading.

Q6-4. Define the digital hierarchy used by telephone companies and list different levels of the hierarchy.

Q6-5. Describe the goals of multiplexing.

Q6-6. Which of the three multiplexing techniques is (are) used to combine analog signals? Which of the three multiplexing techniques is (are) used to combine digital signals?

Q6-7. Which of the three multiplexing techniques is common for fiber-optic links? Explain the reason.

Q6-8. Distinguish between multilevel TDM, multiple-slot TDM, and pulse-stuffed TDM.

Q6-9. Distinguish between a link and a channel in multiplexing.

Q6-10. List three main multiplexing techniques mentioned in this chapter.

Q6-11. Define the analog hierarchy used by telephone companies and list different levels of the hierarchy.

Q6-12. Define DSSS and explain how it achieves bandwidth spreading.

6.4.3 Problems

P6-1. We need to use synchronous TDM and combine 20 digital sources, each of 200 Kbps. Each output slot carries 2 bit from each digital source, but one extra bit is added to each frame for synchronization. Answer the following questions:

a. What is the size of an output frame in bits?

b. What is the output frame rate?

c. What is the duration of an output frame?

d. What is the output data rate?

e. What is the efficiency of the system (ratio of useful bits to the total bits)?

P6-2. Four channels, two with a bit rate of 300 Kbps and two with a bit rate of 150 kbps, are to be multiplexed using multiple-slot TDM with no synchronization bits. Answer the following questions:

a. What is the size of a frame in bits?

b. What is the frame rate?

c. What is the duration of a frame?

d. What is the data rate?

P6-3. In the analog hierarchy of Figure 6.9, find the overhead (extra bandwidth for guard band or control) in each hierarchy level (group, supergroup, master group, and jumbo group).

P6-4. We have 12 sources, each creating 500 8-bit characters per second. Since only some of these sources are active at any moment, we use statistical TDM to combine these sources using character interleaving. Each frame carries 6 slots at a time, but we need to add 4-bit addresses to each slot. Answer the following questions:

a. What is the size of an output frame in bits?

b. What is the output frame rate?

c. What is the duration of an output frame?

d. What is the output data rate?

P6-5. Figure 6.34 shows a demultiplexer in a synchronous TDM. If the input slot is 16 bits long (no framing bits), what is the bit stream in each output? The bits arrive at the demultiplexer as shown by the arrows.

P6-6. An FHSS system uses a 4-bit PN sequence. If the bit rate of the PN is 64 bits per second, answer the following questions:

a. What is the total number of possible channels?

b. What is the time needed to finish a complete cycle of PN?

Figure 6.34 *Problem P6-5*

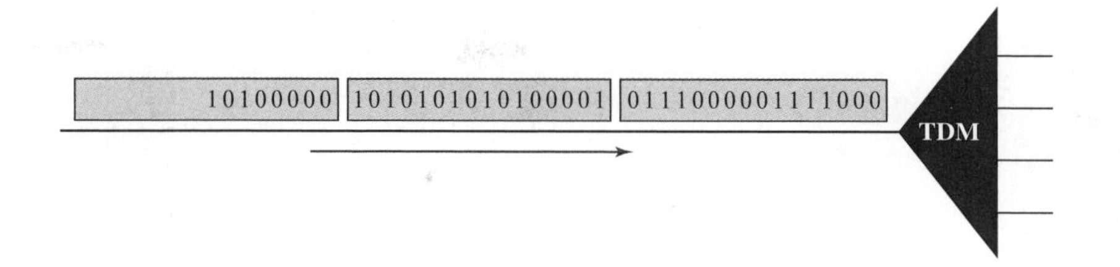

P6-7. A pseudorandom number generator uses the following formula to create a random series:

$$N_{i+1} = (5 + 7N_i) \bmod 17 - 1$$

In which N_i defines the current random number and N_{i+1} defines the next random number. The term *mod* means the value of the remainder when dividing $(5 + 7N_i)$ by 17. Show the sequence created by this generator to be used for spread spectrum.

P6-8. Answer the following questions about the digital hierarchy in Figure 6.23:

a. What is the overhead (number of extra bits) in the DS-1 service?

b. What is the overhead (number of extra bits) in the DS-2 service?

c. What is the overhead (number of extra bits) in the DS-3 service?

d. What is the overhead (number of extra bits) in the DS-4 service?

P6-9. What is the minimum number of bits in a PN sequence if we use FHSS with a channel bandwidth of $B = 8$ KHz and $B_{SS} = 100$ KHz?

P6-10. We need to use synchronous TDM and combine 40 digital sources, each of 100 Kbps. Each output slot carries 1 bit from each digital source, but one extra bit is added to each frame for synchronization. Answer the following questions:

a. What is the size of an output frame in bits?

b. What is the output frame rate?

c. What is the duration of an output frame?

d. What is the output data rate?

e. What is the efficiency of the system (ratio of useful bits to the total bits)?

P6-11. Assume that a voice channel occupies a bandwidth of 4 KHz. We need to multiplex 12 voice channels with guard bands of 500 Hz using FDM. Calculate the required bandwidth.

P6-12. Figure 6.35 shows a multiplexer in a synchronous TDM system. Each output slot is only 10 bits long (3 bits taken from each input plus 1 framing bit). What is the output stream? The bits arrive at the multiplexer as shown by the arrows.

P6-13. Ten sources, six with a bit rate of 200 kbps and four with a bit rate of 400 kbps, are to be combined using multilevel TDM with no synchronizing bits. Answer the following questions about the final stage of the multiplexing:

a. What is the size of a frame in bits?

b. What is the frame rate?

Figure 6.35 *Problem P6-12*

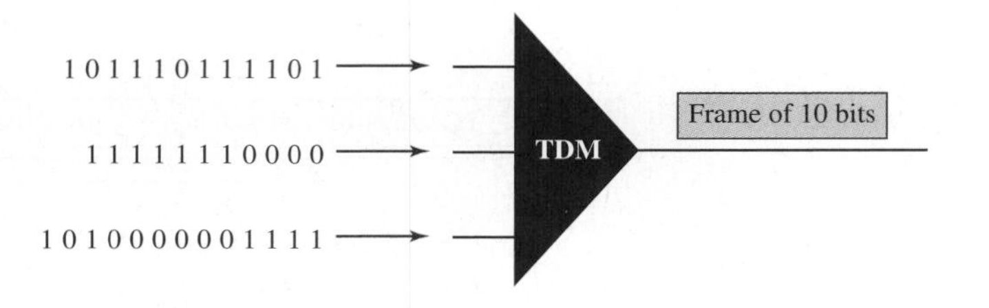

c. What is the duration of a frame?

d. What is the data rate?

P6-14. Answer the following questions about a T-1 line:

a. What is the duration of a frame?

b. What is the overhead (number of extra bits per second)?

P6-15. Show the contents of the five output frames for a synchronous TDM multiplexer that combines four sources sending the following characters. Note that the characters are sent in the same order that they are typed. The third source is silent.

a. Source 1 message: HELLO

b. Source 2 message: HI

c. Source 3 message:

d. Source 4 message: BYE

P6-16. We need to transmit 100 digitized voice channels using a passband channel of 30 KHz. What should be the ratio of bits/Hz if we use no guard band?

P6-17. Two channels, one with a bit rate of 190 kbps and another with a bit rate of 180 kbps, are to be multiplexed using pulse-stuffing TDM with no synchronization bits. Answer the following questions:

a. What is the size of a frame in bits?

b. What is the frame rate?

c. What is the duration of a frame?

d. What is the data rate?

P6-18. We have a digital medium with a data rate of 20 Mbps. How many 64-kbps voice channels can be carried by this medium if we use DSSS with the Barker sequence?

6.5 SIMULATION EXPERIMENTS

6.5.1 Applets

We have created some Java applets to show some of the main concepts discussed in this chapter. It is strongly recommended that the students activate these applets on the book website and carefully examine the protocols in action.

CHAPTER 7

Transmission Media

We discussed many issues related to the physical layer in Chapters 3 through 6. In this chapter, we discuss transmission media. We definitely need transmission media to conduct signals from the source to the destination. However, the media can be wired or wireless.

This chapter is divided into three sections:

- The first section introduces the transmission media and defines its position in the Internet model. It shows that we can classify transmission media into two broad categories: guided and unguided media.
- The second section discusses guided media. The first part describes twisted-pair cables and their characteristics and applications. The second part describes coaxial cables and their characteristics and applications. Finally, the third part describes fiber-optic cables and their characteristics and applications.
- The third section discusses unguided media. The first part describes radio waves and their characteristics and applications. The second part describes microwaves and their characteristics and applications. Finally, the third part describes infrared waves and their characteristics and applications.

7.1 INTRODUCTION

Transmission media are actually located below the physical layer and are directly controlled by the physical layer. We could say that transmission media belong to layer zero. Figure 7.1 shows the position of transmission media in relation to the physical layer.

Figure 7.1 *Transmission medium and physical layer*

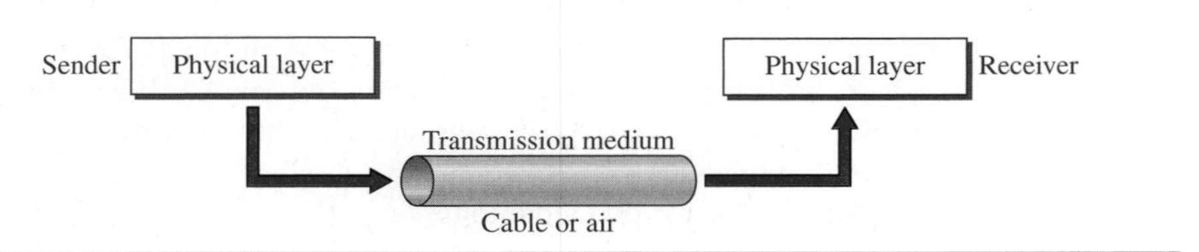

A **transmission medium** can be broadly defined as anything that can carry information from a source to a destination. For example, the transmission medium for two people having a dinner conversation is the air. The air can also be used to convey the message in a smoke signal or semaphore. For a written message, the transmission medium might be a mail carrier, a truck, or an airplane.

In data communications the definition of the information and the transmission medium is more specific. The transmission medium is usually free space, metallic cable, or fiber-optic cable. The information is usually a signal that is the result of a conversion of data from another form.

The use of long-distance communication using electric signals started with the invention of the telegraph by Morse in the 19th century. Communication by telegraph was slow and dependent on a metallic medium.

Extending the range of the human voice became possible when the telephone was invented in 1869. Telephone communication at that time also needed a metallic medium to carry the electric signals that were the result of a conversion from the human voice. The communication was, however, unreliable due to the poor quality of the wires. The lines were often noisy and the technology was unsophisticated.

Wireless communication started in 1895 when Hertz was able to send high-frequency signals. Later, Marconi devised a method to send telegraph-type messages over the Atlantic Ocean.

We have come a long way. Better metallic media have been invented (twisted-pair and coaxial cables, for example). The use of optical fibers has increased the data rate incredibly. Free space (air, vacuum, and water) is used more efficiently, in part due to the technologies (such as modulation and multiplexing) discussed in the previous chapters.

As discussed in Chapter 3, computers and other telecommunication devices use signals to represent data. These signals are transmitted from one device to another in the form of electromagnetic energy, which is propagated through transmission media.

Electromagnetic energy, a combination of electric and magnetic fields vibrating in relation to each other, includes power, radio waves, infrared light, visible light, ultraviolet light, and X, gamma, and cosmic rays. Each of these constitutes a portion of the **electromagnetic spectrum.** Not all portions of the spectrum are currently usable for telecommunications, however. The media to harness those that are usable are also limited to a few types.

In telecommunications, transmission media can be divided into two broad categories: guided and unguided. Guided media include twisted-pair cable, coaxial cable, and fiber-optic cable. Unguided medium is free space. Figure 7.2 shows this taxonomy.

Figure 7.2 *Classes of transmission media*

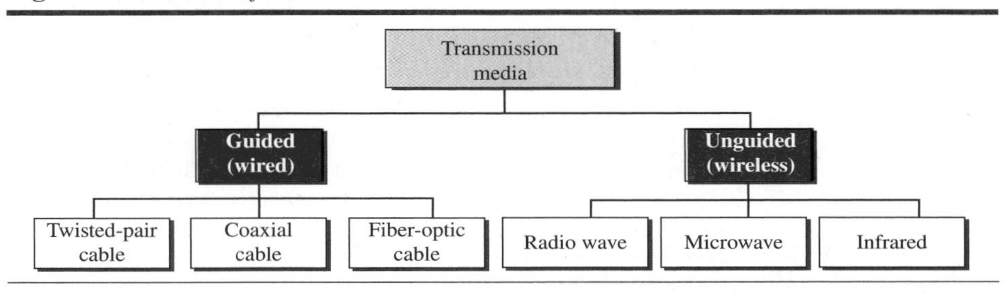

7.2 GUIDED MEDIA

Guided media, which are those that provide a conduit from one device to another, include **twisted-pair cable, coaxial cable,** and **fiber-optic cable.** A signal traveling along any of these media is directed and contained by the physical limits of the medium. Twisted-pair and coaxial cable use metallic (copper) conductors that accept and transport signals in the form of electric current. **Optical fiber** is a cable that accepts and transports signals in the form of light.

7.2.1 Twisted-Pair Cable

A twisted pair consists of two conductors (normally copper), each with its own plastic insulation, twisted together, as shown in Figure 7.3.

Figure 7.3 *Twisted-pair cable*

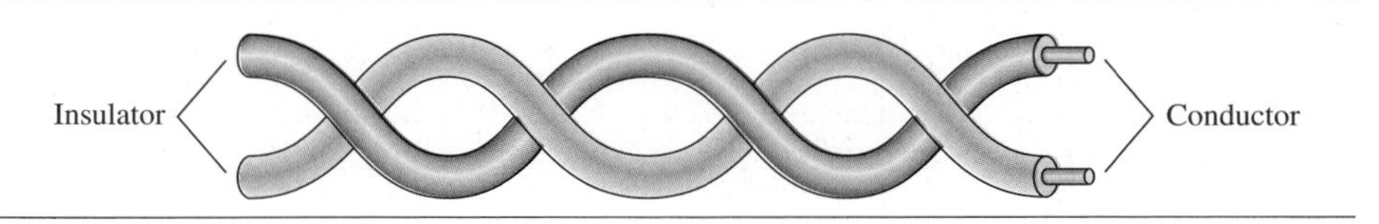

One of the wires is used to carry signals to the receiver, and the other is used only as a ground reference. The receiver uses the difference between the two.

In addition to the signal sent by the sender on one of the wires, interference (noise) and crosstalk may affect both wires and create unwanted signals.

If the two wires are parallel, the effect of these unwanted signals is not the same in both wires because they are at different locations relative to the noise or crosstalk sources (e.g., one is closer and the other is farther). This results in a difference at the receiver.

By twisting the pairs, a balance is maintained. For example, suppose in one twist, one wire is closer to the noise source and the other is farther; in the next twist, the reverse is true. Twisting makes it probable that both wires are equally affected by external influences (noise or crosstalk). This means that the receiver, which calculates the difference between the two, receives no unwanted signals. The unwanted signals are mostly canceled out. From the above discussion, it is clear that the number of twists per unit of length (e.g., inch) has some effect on the quality of the cable.

Unshielded Versus Shielded Twisted-Pair Cable

The most common twisted-pair cable used in communications is referred to as ***unshielded twisted-pair*** **(UTP).** IBM has also produced a version of twisted-pair cable for its use, called ***shielded twisted-pair*** **(STP).** STP cable has a metal foil or braided-mesh covering that encases each pair of insulated conductors. Although metal casing improves the quality of cable by preventing the penetration of noise or crosstalk, it is bulkier and more expensive. Figure 7.4 shows the difference between UTP and STP. Our discussion focuses primarily on UTP because STP is seldom used outside of IBM.

Figure 7.4 *UTP and STP cables*

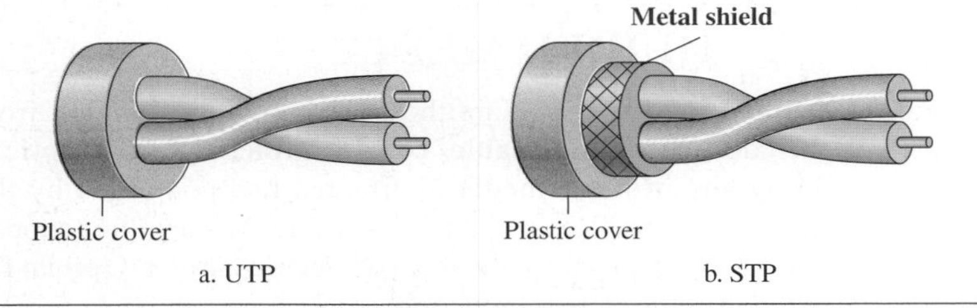

Categories

The Electronic Industries Association (EIA) has developed standards to classify unshielded twisted-pair cable into seven categories. Categories are determined by cable quality, with 1 as the lowest and 7 as the highest. Each EIA category is suitable for specific uses. Table 7.1 shows these categories.

Table 7.1 *Categories of unshielded twisted-pair cables*

Category	*Specification*	*Data Rate (Mbps)*	*Use*
1	Unshielded twisted-pair used in telephone	< 0.1	Telephone
2	Unshielded twisted-pair originally used in T lines	2	T-1 lines
3	Improved CAT 2 used in LANs	10	LANs
4	Improved CAT 3 used in Token Ring networks	20	LANs
5	Cable wire is normally 24 AWG with a jacket and outside sheath	100	LANs

Table 7.1 *Categories of unshielded twisted-pair cables (continued)*

Category	*Specification*	*Data Rate (Mbps)*	*Use*
5E	An extension to category 5 that includes extra features to minimize the crosstalk and electromagnetic interference	125	LANs
6	A new category with matched components coming from the same manufacturer. The cable must be tested at a 200-Mbps data rate.	200	LANs
7	Sometimes called *SSTP (shielded screen twisted-pair)*. Each pair is individually wrapped in a helical metallic foil followed by a metallic foil shield in addition to the outside sheath. The shield decreases the effect of crosstalk and increases the data rate.	600	LANs

Connectors

The most common UTP connector is **RJ45** (RJ stands for registered jack), as shown in Figure 7.5. The RJ45 is a keyed connector, meaning the connector can be inserted in only one way.

Figure 7.5 *UTP connector*

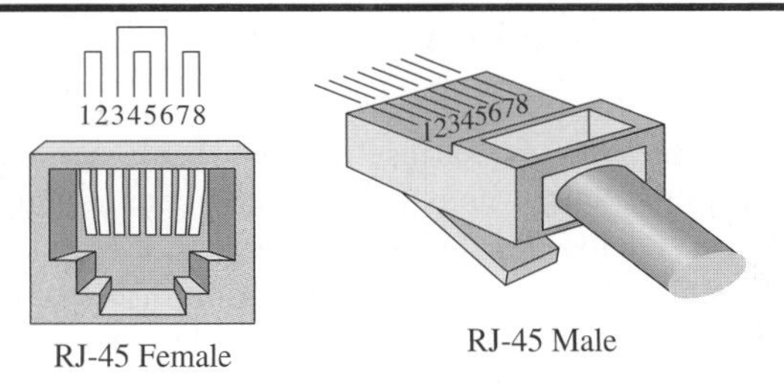

Performance

One way to measure the performance of twisted-pair cable is to compare attenuation versus frequency and distance. A twisted-pair cable can pass a wide range of frequencies. However, Figure 7.6 shows that with increasing frequency, the attenuation, measured in decibels per kilometer (dB/km), sharply increases with frequencies above 100 kHz. Note that ***gauge*** is a measure of the thickness of the wire.

Applications

Twisted-pair cables are used in telephone lines to provide voice and data channels. The local loop—the line that connects subscribers to the central telephone office—commonly consists of unshielded twisted-pair cables. We discuss telephone networks in Chapter 14.

The DSL lines that are used by the telephone companies to provide high-data-rate connections also use the high-bandwidth capability of unshielded twisted-pair cables. We discuss DSL technology in Chapter 14.

Figure 7.6 *UTP performance*

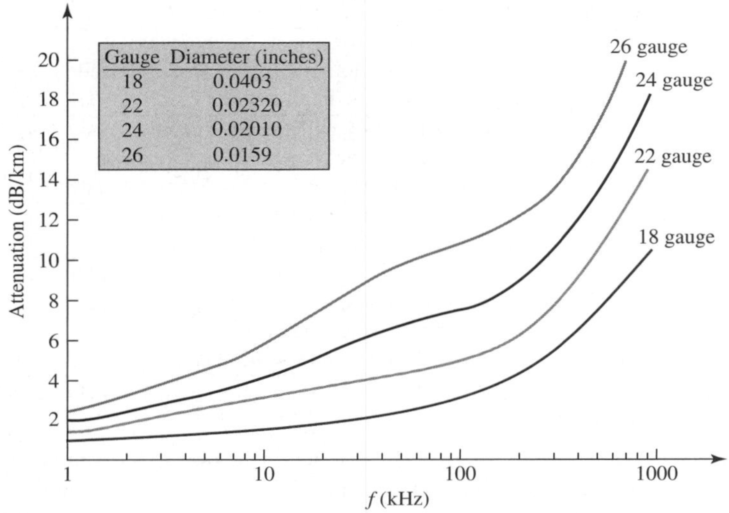

Gauge	Diameter (inches)
18	0.0403
22	0.02320
24	0.02010
26	0.0159

Local-area networks, such as 10Base-T and 100Base-T, also use twisted-pair cables. We discuss these networks in Chapter 13.

7.2.2 Coaxial Cable

Coaxial cable (or *coax*) carries signals of higher frequency ranges than those in twisted-pair cable, in part because the two media are constructed quite differently. Instead of having two wires, coax has a central core conductor of solid or stranded wire (usually copper) enclosed in an insulating sheath, which is, in turn, encased in an outer conductor of metal foil, braid, or a combination of the two. The outer metallic wrapping serves both as a shield against noise and as the second conductor, which completes the circuit. This outer conductor is also enclosed in an insulating sheath, and the whole cable is protected by a plastic cover (see Figure 7.7).

Figure 7.7 *Coaxial cable*

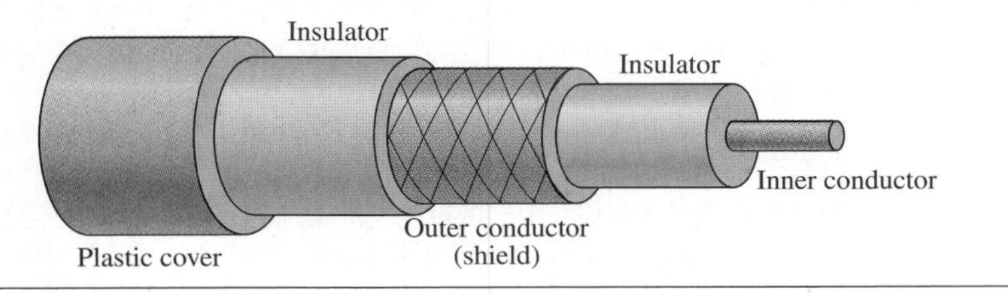

Coaxial Cable Standards

Coaxial cables are categorized by their **Radio Government (RG)** ratings. Each RG number denotes a unique set of physical specifications, including the wire gauge of the inner conductor, the thickness and type of the inner insulator, the construction of the shield, and the size and type of the outer casing. Each cable defined by an RG rating is adapted for a specialized function, as shown in Table 7.2.

Table 7.2 *Categories of coaxial cables*

Category	*Impedance*	*Use*
RG-59	75 Ω	Cable TV
RG-58	50 Ω	Thin Ethernet
RG-11	50 Ω	Thick Ethernet

Coaxial Cable Connectors

To connect coaxial cable to devices, we need coaxial connectors. The most common type of connector used today is the **Bayonet Neill-Concelman (BNC)** connector. Figure 7.8 shows three popular types of these connectors: the BNC connector, the BNC T connector, and the BNC terminator.

Figure 7.8 *BNC connectors*

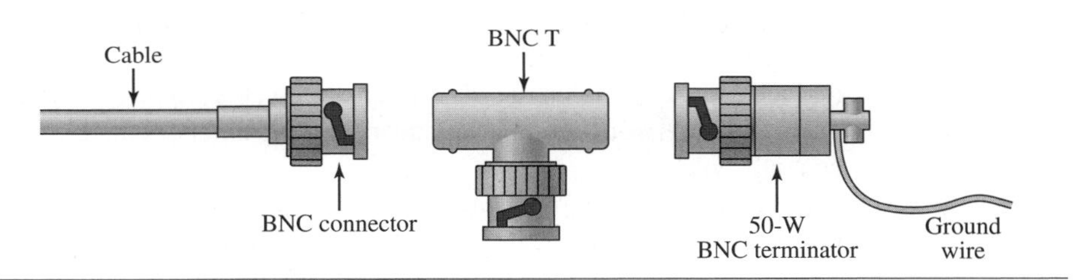

The BNC connector is used to connect the end of the cable to a device, such as a TV set. The BNC T connector is used in Ethernet networks (see Chapter 13) to branch out to a connection to a computer or other device. The BNC terminator is used at the end of the cable to prevent the reflection of the signal.

Performance

As we did with twisted-pair cable, we can measure the performance of a coaxial cable. We notice in Figure 7.9 that the attenuation is much higher in coaxial cable than in twisted-pair cable. In other words, although coaxial cable has a much higher bandwidth, the signal weakens rapidly and requires the frequent use of repeaters.

Applications

Coaxial cable was widely used in analog telephone networks where a single coaxial network could carry 10,000 voice signals. Later it was used in digital telephone networks where a single coaxial cable could carry digital data up to 600 Mbps. However, coaxial cable in telephone networks has largely been replaced today with fiber-optic cable.

Figure 7.9 *Coaxial cable performance*

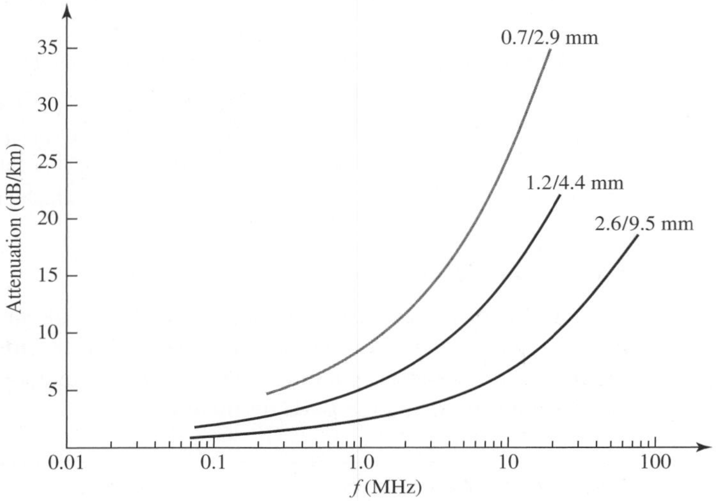

Cable TV networks (see Chapter 14) also use coaxial cables. In the traditional cable TV network, the entire network used coaxial cable. Later, however, cable TV providers replaced most of the media with fiber-optic cable; hybrid networks use coaxial cable only at the network boundaries, near the consumer premises. Cable TV uses RG-59 coaxial cable.

Another common application of coaxial cable is in traditional Ethernet LANs (see Chapter 13). Because of its high bandwidth, and consequently high data rate, coaxial cable was chosen for digital transmission in early Ethernet LANs. The 10Base-2, or Thin Ethernet, uses RG-58 coaxial cable with BNC connectors to transmit data at 10 Mbps with a range of 185 m. The 10Base5, or Thick Ethernet, uses RG-11 (thick coaxial cable) to transmit 10 Mbps with a range of 5000 m. Thick Ethernet has specialized connectors.

7.2.3 Fiber-Optic Cable

A fiber-optic cable is made of glass or plastic and transmits signals in the form of light. To understand optical fiber, we first need to explore several aspects of the nature of light.

Light travels in a straight line as long as it is moving through a single uniform substance. If a ray of light traveling through one substance suddenly enters another substance (of a different density), the ray changes direction. Figure 7.10 shows how a ray of light changes direction when going from a more dense to a less dense substance.

As the figure shows, if the **angle of incidence** *I* (the angle the ray makes with the line perpendicular to the interface between the two substances) is less than the **critical angle,** the ray **refracts** and moves closer to the surface. If the angle of incidence is equal to the critical angle, the light bends along the interface. If the angle is greater than the critical angle, the ray **reflects** (makes a turn) and travels again in the denser

Figure 7.10 *Bending of light ray*

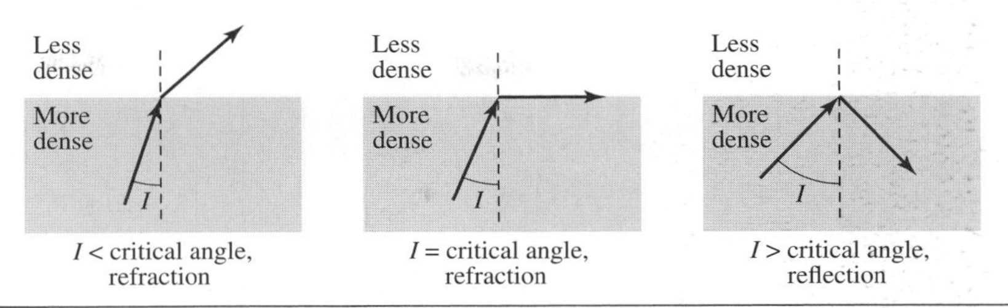

substance. Note that the critical angle is a property of the substance, and its value differs from one substance to another.

Optical fibers use reflection to guide light through a channel. A glass or plastic **core** is surrounded by a **cladding** of less dense glass or plastic. The difference in density of the two materials must be such that a beam of light moving through the core is reflected off the cladding instead of being refracted into it. See Figure 7.11.

Figure 7.11 *Optical fiber*

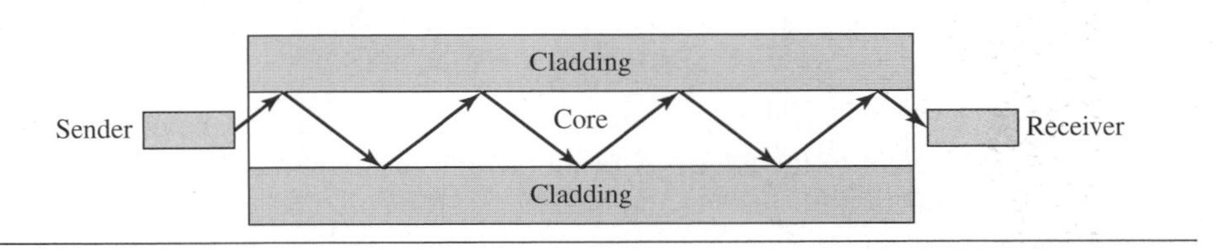

Propagation Modes

Current technology supports two modes (multimode and single mode) for propagating light along optical channels, each requiring fiber with different physical characteristics. Multimode can be implemented in two forms: step-index or graded-index (see Figure 7.12).

Figure 7.12 *Propagation modes*

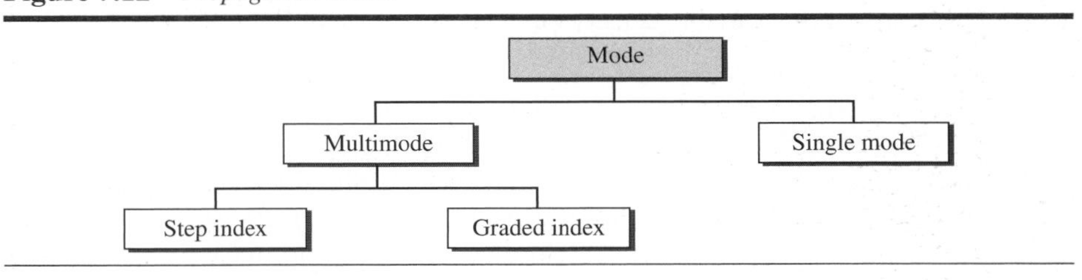

Multimode

Multimode is so named because multiple beams from a light source move through the core in different paths. How these beams move within the cable depends on the structure of the core, as shown in Figure 7.13.

Figure 7.13 *Modes*

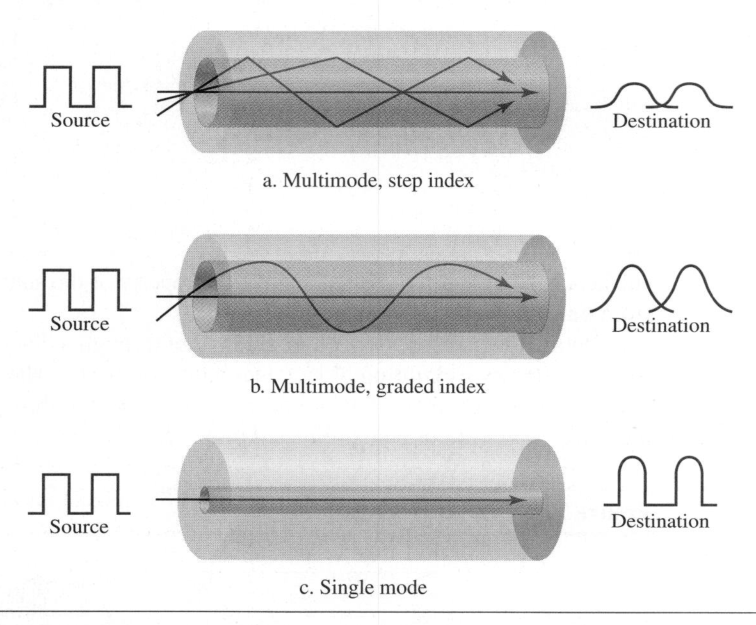

In **multimode step-index fiber,** the density of the core remains constant from the center to the edges. A beam of light moves through this constant density in a straight line until it reaches the interface of the core and the cladding. At the interface, there is an abrupt change due to a lower density; this alters the angle of the beam's motion. The term *step-index* refers to the suddenness of this change, which contributes to the distortion of the signal as it passes through the fiber.

A second type of fiber, called **multimode graded-index fiber,** decreases this distortion of the signal through the cable. The word *index* here refers to the index of refraction. As we saw above, the index of refraction is related to density. A graded-index fiber, therefore, is one with varying densities. Density is highest at the center of the core and decreases gradually to its lowest at the edge. Figure 7.13 shows the impact of this variable density on the propagation of light beams.

Single-Mode

Single-mode uses step-index fiber and a highly focused source of light that limits beams to a small range of angles, all close to the horizontal. The **single-mode fiber** itself is manufactured with a much smaller diameter than that of multimode fiber, and with substantially lower density (index of refraction). The decrease in density results in a critical angle that is close enough to 90° to make the propagation of beams almost horizontal. In this case, propagation of different beams is almost identical, and delays are negligible. All the beams arrive at the destination "together" and can be recombined with little distortion to the signal (see Figure 7.13).

Fiber Sizes

Optical fibers are defined by the ratio of the diameter of their core to the diameter of their cladding, both expressed in micrometers. The common sizes are shown in Table 7.3. Note that the last size listed is for single-mode only.

Table 7.3 *Fiber types*

Type	*Core (μm)*	*Cladding (μm)*	*Mode*
50/125	50.0	125	Multimode, graded index
62.5/125	62.5	125	Multimode, graded index
100/125	100.0	125	Multimode, graded index
7/125	7.0	125	Single mode

Cable Composition

Figure 7.14 shows the composition of a typical fiber-optic cable. The outer jacket is

Figure 7.14 *Fiber construction*

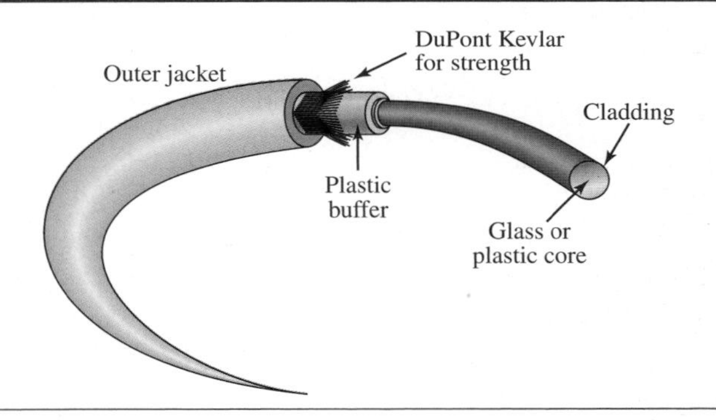

made of either PVC or Teflon. Inside the jacket are Kevlar strands to strengthen the cable. Kevlar is a strong material used in the fabrication of bulletproof vests. Below the Kevlar is another plastic coating to cushion the fiber. The fiber is at the center of the cable, and it consists of cladding and core.

Fiber-Optic Cable Connectors

There are three types of connectors for fiber-optic cables, as shown in Figure 7.15. The **subscriber channel (SC) connector** is used for cable TV. It uses a push/pull locking system. The **straight-tip (ST) connector** is used for connecting cable to networking devices. It uses a bayonet locking system and is more reliable than SC. **MT-RJ** is a connector that is the same size as RJ45.

Performance

The plot of attenuation versus wavelength in Figure 7.16 shows a very interesting phenomenon in fiber-optic cable. Attenuation is flatter than in the case of twisted-pair cable and coaxial cable. The performance is such that we need fewer (actually one-tenth as many) repeaters when we use fiber-optic cable.

Figure 7.15 *Fiber-optic cable connectors*

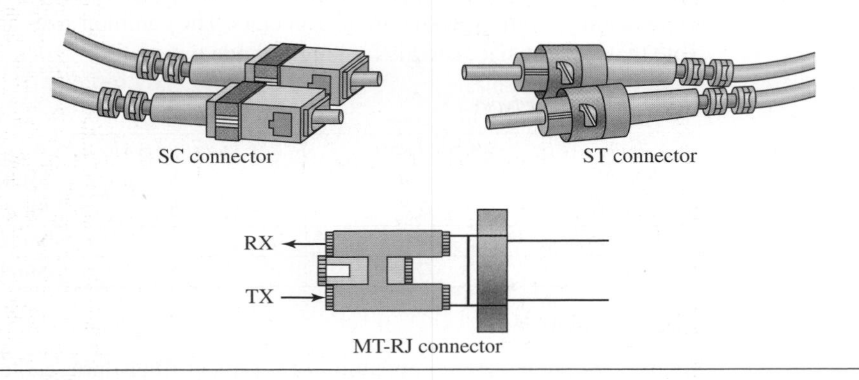

Figure 7.16 *Optical fiber performance*

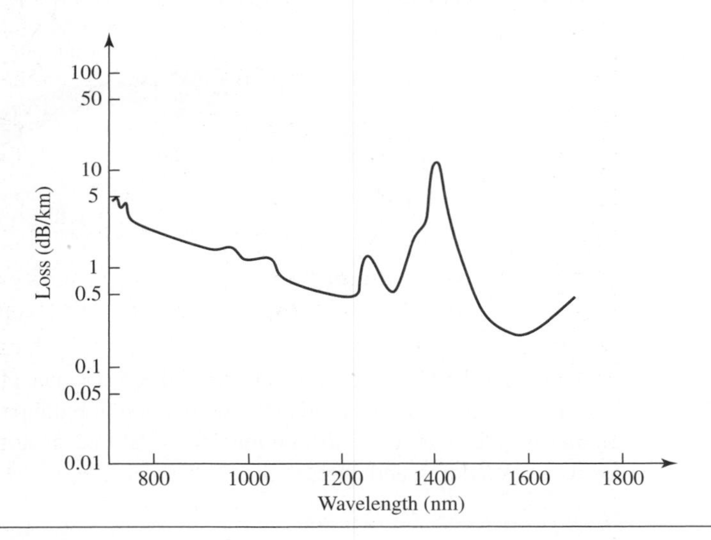

Applications

Fiber-optic cable is often found in backbone networks because its wide bandwidth is cost-effective. Today, with wavelength-division multiplexing (WDM), we can transfer data at a rate of 1600 Gbps. The SONET network that we discuss in Chapter 14 provides such a backbone.

Some cable TV companies use a combination of optical fiber and coaxial cable, thus creating a hybrid network. Optical fiber provides the backbone structure while coaxial cable provides the connection to the user premises. This is a cost-effective configuration since the narrow bandwidth requirement at the user end does not justify the use of optical fiber.

Local-area networks such as 100Base-FX network (Fast Ethernet) and 1000Base-X also use fiber-optic cable.

Advantages and Disadvantages of Optical Fiber

Advantages

Fiber-optic cable has several advantages over metallic cable (twisted-pair or coaxial).

- ❑ **Higher bandwidth.** Fiber-optic cable can support dramatically higher bandwidths (and hence data rates) than either twisted-pair or coaxial cable. Currently, data rates and bandwidth utilization over fiber-optic cable are limited not by the medium but by the signal generation and reception technology available.
- ❑ **Less signal attenuation.** Fiber-optic transmission distance is significantly greater than that of other guided media. A signal can run for 50 km without requiring regeneration. We need repeaters every 5 km for coaxial or twisted-pair cable.
- ❑ **Immunity to electromagnetic interference.** Electromagnetic noise cannot affect fiber-optic cables.
- ❑ **Resistance to corrosive materials.** Glass is more resistant to corrosive materials than copper.
- ❑ **Light weight.** Fiber-optic cables are much lighter than copper cables.
- ❑ **Greater immunity to tapping.** Fiber-optic cables are more immune to tapping than copper cables. Copper cables create antenna effects that can easily be tapped.

Disadvantages

There are some disadvantages in the use of optical fiber.

- ❑ **Installation and maintenance.** Fiber-optic cable is a relatively new technology. Its installation and maintenance require expertise that is not yet available everywhere.
- ❑ **Unidirectional light propagation.** Propagation of light is unidirectional. If we need bidirectional communication, two fibers are needed.
- ❑ **Cost.** The cable and the interfaces are relatively more expensive than those of other guided media. If the demand for bandwidth is not high, often the use of optical fiber cannot be justified.

7.3 UNGUIDED MEDIA: WIRELESS

Unguided medium transport electromagnetic waves without using a physical conductor. This type of communication is often referred to as ***wireless communication.*** Signals are normally broadcast through free space and thus are available to anyone who has a device capable of receiving them.

Figure 7.17 shows the part of the electromagnetic spectrum, ranging from 3 kHz to 900 THz, used for wireless communication.

Unguided signals can travel from the source to the destination in several ways: ground propagation, sky propagation, and line-of-sight propagation, as shown in Figure 7.18.

Figure 7.17 *Electromagnetic spectrum for wireless communication*

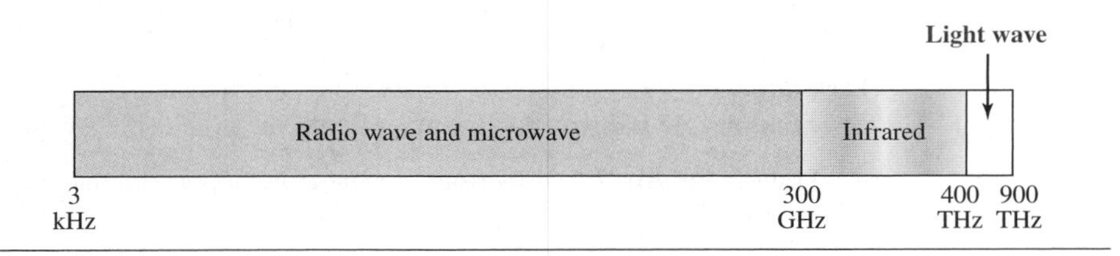

Figure 7.18 *Propagation methods*

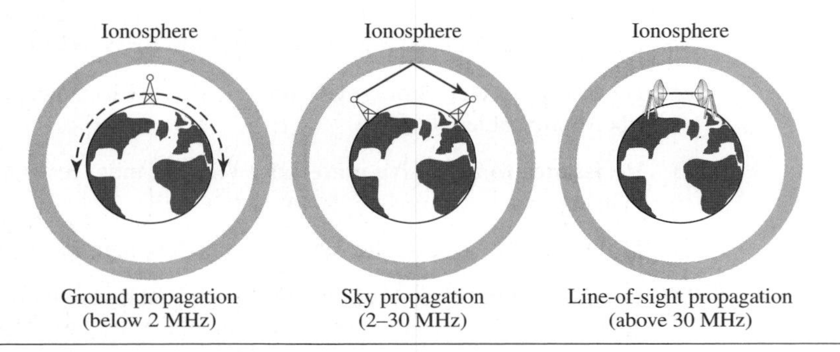

In **ground propagation,** radio waves travel through the lowest portion of the atmosphere, hugging the earth. These low-frequency signals emanate in all directions from the transmitting antenna and follow the curvature of the planet. Distance depends on the amount of power in the signal: The greater the power, the greater the distance. In **sky propagation,** higher-frequency radio waves radiate upward into the ionosphere (the layer of atmosphere where particles exist as ions) where they are reflected back to earth. This type of transmission allows for greater distances with lower output power. In **line-of-sight propagation,** very high-frequency signals are transmitted in straight lines directly from antenna to antenna. Antennas must be directional, facing each other, and either tall enough or close enough together not to be affected by the curvature of the earth. Line-of-sight propagation is tricky because radio transmissions cannot be completely focused.

The section of the electromagnetic spectrum defined as radio waves and microwaves is divided into eight ranges, called *bands,* each regulated by government authorities. These bands are rated from *very low frequency* (VLF) to *extremely high frequency* (EHF). Table 7.4 lists these bands, their ranges, propagation methods, and some applications.

Table 7.4 *Bands*

Band	*Range*	*Propagation*	*Application*
very low frequency (VLF)	3–30 kHz	Ground	Long-range radio navigation
low frequency (LF)	30–300 kHz	Ground	Radio beacons and navigational locators

Table 7.4 *Bands (continued)*

Band	*Range*	*Propagation*	*Application*
middle frequency (MF)	300 kHz–3 MHz	Sky	AM radio
high frequency (HF)	3–30 MHz	Sky	Citizens band (CB), ship/aircraft
very high frequency (VHF)	30–300 MHz	Sky and line-of-sight	VHF TV, FM radio
ultrahigh frequency (UHF)	300 MHz–3 GHz	Line-of-sight	UHF TV, cellular phones, paging, satellite
superhigh frequency (SF)	3–30 GHz	Line-of-sight	Satellite
extremely high frequency (EHF)	30–300 GHz	Line-of-sight	Radar, satellite

We can divide wireless transmission into three broad groups: radio waves, microwaves, and infrared waves.

7.3.1 Radio Waves

Although there is no clear-cut demarcation between radio waves and microwaves, electromagnetic waves ranging in frequencies between 3 kHz and 1 GHz are normally called **radio waves;** waves ranging in frequencies between 1 and 300 GHz are called **microwaves.** However, the behavior of the waves, rather than the frequencies, is a better criterion for classification.

Radio waves, for the most part, are omnidirectional. When an antenna transmits radio waves, they are propagated in all directions. This means that the sending and receiving antennas do not have to be aligned. A sending antenna sends waves that can be received by any receiving antenna. The omnidirectional property has a disadvantage, too. The radio waves transmitted by one antenna are susceptible to interference by another antenna that may send signals using the same frequency or band.

Radio waves, particularly those waves that propagate in the sky mode, can travel long distances. This makes radio waves a good candidate for long-distance broadcasting such as AM radio.

Radio waves, particularly those of low and medium frequencies, can penetrate walls. This characteristic can be both an advantage and a disadvantage. It is an advantage because, for example, an AM radio can receive signals inside a building. It is a disadvantage because we cannot isolate a communication to just inside or outside a building. The radio wave band is relatively narrow, just under 1 GHz, compared to the microwave band. When this band is divided into subbands, the subbands are also narrow, leading to a low data rate for digital communications.

Almost the entire band is regulated by authorities (e.g., the FCC in the United States). Using any part of the band requires permission from the authorities.

Omnidirectional Antenna

Radio waves use **omnidirectional antennas** that send out signals in all directions. Based on the wavelength, strength, and the purpose of transmission, we can have several types of antennas. Figure 7.19 shows an omnidirectional antenna.

Figure 7.19 *Omnidirectional antenna*

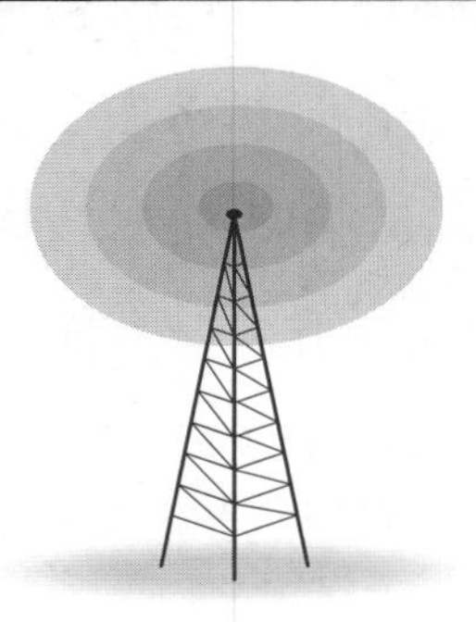

Applications

The omnidirectional characteristics of radio waves make them useful for multicasting, in which there is one sender but many receivers. AM and FM radio, television, maritime radio, cordless phones, and paging are examples of multicasting.

Radio waves are used for multicast communications, such as radio and television, and paging systems.

7.3.2 Microwaves

Electromagnetic waves having frequencies between 1 and 300 GHz are called microwaves. Microwaves are unidirectional. When an antenna transmits microwaves, they can be narrowly focused. This means that the sending and receiving antennas need to be aligned. The unidirectional property has an obvious advantage. A pair of antennas can be aligned without interfering with another pair of aligned antennas. The following describes some characteristics of microwave propagation:

- Microwave propagation is line-of-sight. Since the towers with the mounted antennas need to be in direct sight of each other, towers that are far apart need to be very tall. The curvature of the earth as well as other blocking obstacles do not allow two short towers to communicate by using microwaves. Repeaters are often needed for long-distance communication.
- Very high-frequency microwaves cannot penetrate walls. This characteristic can be a disadvantage if receivers are inside buildings.
- The microwave band is relatively wide, almost 299 GHz. Therefore wider subbands can be assigned, and a high data rate is possible.
- Use of certain portions of the band requires permission from authorities.

Unidirectional Antenna

Microwaves need **unidirectional antennas** that send out signals in one direction. Two types of antennas are used for microwave communications: the parabolic dish and the horn (see Figure 7.20).

Figure 7.20 *Unidirectional antennas*

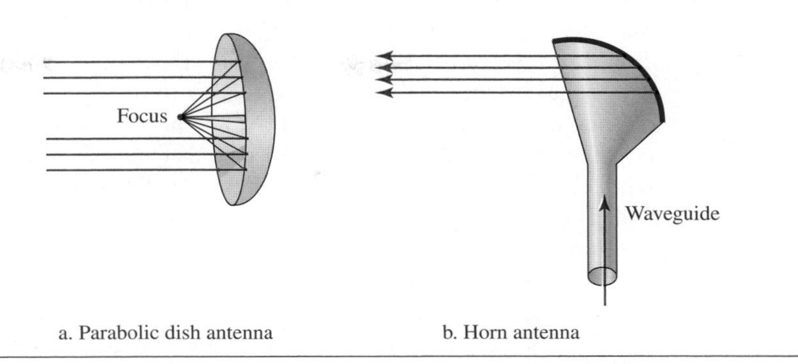

a. Parabolic dish antenna   b. Horn antenna

A **parabolic dish antenna** is based on the geometry of a parabola: Every line parallel to the line of symmetry (line of sight) reflects off the curve at angles such that all the lines intersect in a common point called the focus. The parabolic dish works as a funnel, catching a wide range of waves and directing them to a common point. In this way, more of the signal is recovered than would be possible with a single-point receiver.

Outgoing transmissions are broadcast through a horn aimed at the dish. The microwaves hit the dish and are deflected outward in a reversal of the receipt path.

A **horn antenna** looks like a gigantic scoop. Outgoing transmissions are broadcast up a stem (resembling a handle) and deflected outward in a series of narrow parallel beams by the curved head. Received transmissions are collected by the scooped shape of the horn, in a manner similar to the parabolic dish, and are deflected down into the stem.

Applications

Microwaves, due to their unidirectional properties, are very useful when unicast (one-to-one) communication is needed between the sender and the receiver. They are used in cellular phones (Chapter 16), satellite networks (Chapter 16), and wireless LANs (Chapter 15).

Microwaves are used for unicast communication such as cellular telephones, satellite networks, and wireless LANs.

7.3.3 Infrared

Infrared waves, with frequencies from 300 GHz to 400 THz (wavelengths from 1 mm to 770 nm), can be used for short-range communication. Infrared waves, having high frequencies, cannot penetrate walls. This advantageous characteristic prevents interference between one system and another; a short-range communication system in one room cannot be affected by another system in the next room. When we use our infrared remote control, we do not interfere with the use of the remote by our neighbors. However, this same characteristic makes infrared signals useless for long-range communication. In addition, we cannot use infrared waves outside a building because the sun's rays contain infrared waves that can interfere with the communication.

Applications

The infrared band, almost 400 THz, has an excellent potential for data transmission. Such a wide bandwidth can be used to transmit digital data with a very high data rate. The *Infrared Data Association* (IrDA), an association for sponsoring the use of infrared waves, has established standards for using these signals for communication between devices such as keyboards, mice, PCs, and printers. For example, some manufacturers provide a special port called the **IrDA port** that allows a wireless keyboard to commnicate with a PC. The standard originally defined a data rate of 75 kbps for a distance up to 8 m. The recent standard defines a data rate of 4 Mbps.

Infrared signals defined by IrDA transmit through line of sight; the IrDA port on the keyboard needs to point to the PC for transmission to occur.

Infrared signals can be used for short-range communication in a closed area using line-of-sight propagation

7.4 END-CHAPTER MATERIALS

7.4.1 Recommended Reading

For more details about subjects discussed in this chapter, we recommend the following books. The items in brackets [. . .] refer to the reference list at the end of the text.

Books

Transmission media is discussed in [GW04], [Sta04], and [Tan03]. [SSS05] gives full coverage of transmission media.

7.4.2 Key Terms

angle of incidence
Bayonet Neill-Concelman (BNC)
cladding
coaxial cable
core
critical angle
electromagnetic spectrum
fiber-optic cable
gauge
ground propagation
guided media
horn antenna
infrared wave
IrDA port
line-of-sight propagation
microwave
MT-RJ
multimode graded-index fiber
multimode step-index fiber
omnidirectional antenna
optical fiber
parabolic dish antenna
Radio Government (RG) rating
radio wave
reflection
refraction
RJ45
shielded twisted-pair (STP)
single-mode fiber
sky propagation
straight-tip (ST) connector
subscriber channel (SC) connector
transmission medium
twisted-pair cable
unguided medium
unidirectional antenna
unshielded twisted-pair (UTP)
wireless communication

7.4.3 Summary

Transmission media are actually located below the physical layer and are directly controlled by the physical layer. We could say that transmission media belong to layer zero.

A guided medium provides a physical conduit from one device to another. Twisted-pair cable consists of two insulated copper wires twisted together. Twisted-pair cable is used for voice and data communications. Coaxial cable consists of a central conductor and a shield. Coaxial cable is used in cable TV networks and traditional Ethernet LANs. Fiber-optic cables are composed of a glass or plastic inner core surrounded by cladding, all encased in an outside jacket. Fiber-optic transmission is becoming increasingly popular due to its noise resistance, low attenuation, and high-bandwidth capabilities. Fiber-optic cable is used in backbone networks, cable TV networks, and Fast Ethernet networks.

Unguided media (free space) transport electromagnetic waves without the use of a physical conductor. Wireless data are transmitted through ground propagation, sky propagation, and line-of-sight propagation.Wireless waves can be classified as radio waves, microwaves, or infrared waves. Radio waves are omnidirectional; microwaves are unidirectional. Microwaves are used for cellular phone, satellite, and wireless LAN communications. Infrared waves are used for short-range communications such as those between a PC and a peripheral device. They can also be used for indoor LANs.

7.5 PRACTICE SET

7.5.1 Quizzes

A set of interactive quizzes for this chapter can be found on the book website. It is strongly recommended that the student take the quizzes to check his/her understanding of the materials before continuing with the practice set.

7.5.2 Questions

Q7-1. What is the function of the twisting in twisted-pair cable?

Q7-2. Name the two major categories of transmission media.

Q7-3. What is the purpose of cladding in an optical fiber?

Q7-4. What is the difference between omnidirectional waves and unidirectional waves?

Q7-5. How does sky propagation differ from line-of-sight propagation?

Q7-6. Name the advantages of optical fiber over twisted-pair and coaxial cable.

Q7-7. How do guided media differ from unguided media?

Q7-8. What is refraction? What is reflection?

Q7-9. What is the position of the transmission media in the OSI or the Internet model?

Q7-10. What are the three major classes of guided media?

7.5.3 Problems

P7-1. Using Figure 7.6, tabulate the attenuation (in dB) of an 18-gauge UTP for the indicated frequencies and distances.

Table 7.5 *Attenuation for 18-gauge UTP*

Distance	*dB at 1 KHz*	*dB at 10 KHz*	*dB at 100 KHz*
1 Km			
10 Km			
15 Km			
20 Km			

P7-2. Use the results of Problem P7-1 to infer that the bandwidth of a UTP cable decreases with an increase in distance.

P7-3. Calculate the bandwidth of the light for the following wavelength ranges (assume a propagation speed of 2×10^8 m):

a. 1000 to 1200 nm **b.** 1000 to 1400 nm

P7-4. Using Figure 7.9, tabulate the attenuation (in dB) of a 2.6/9.5 mm coaxial cable for the indicated frequencies and distances.

Table 7.6 *Attenuation for 2.6/9.5 mm coaxial cable*

Distance	*dB at 1 KHz*	*dB at 10 KHz*	*dB at 100 KHz*
1 Km			
10 Km			
15 Km			
20 Km			

P7-5. Use the results of Problem P7-4 to infer that the bandwidth of a coaxial cable decreases with the increase in distance.

P7-6. A light signal is travelling through a fiber. What is the delay in the signal if the length of the fiber-optic cable is 5 m, 500 m, and 1 Km (assume a propagation speed of 2×10^8 m)?

P7-7. Using Figure 7.16, tabulate the attenuation (in dB) of an optical fiber for the indicated wavelength and distances.

Table 7.7 *Attenuation for optical fiber*

Distance	*dB at 800 nm*	*dB at 1000 nm*	*dB at 1200 nm*
1 Km			
10 Km			
15 Km			
20 Km			

P7-8. If the power at the beginning of a 1 Km 2.6/9.5 mm coaxial cable is 300 mw, what is the power at the end for frequencies 1 KHz, 10 KHz, and 100 KHz? Use the results of Problem P7-4.

P7-9. If the power at the beginning of a 1 Km 18-gauge UTP is 300 mw, what is the power at the end for frequencies 1 KHz, 10 KHz, and 100 KHz? Use the results of Problem P7-1.

P7-10. The horizontal axes in Figures 7.6 and 7.9 represent frequencies. The horizontal axis in Figure 7.16 represents wavelength. Can you explain the reason? If the propagation speed in an optical fiber is 2×10^8 m, can you change the units in the horizontal axis to frequency? Should the vertical-axis units be changed too? Should the curve be changed too?

P7-11. A beam of light moves from one medium to another medium with less density. The critical angle is 60°. Do we have refraction or reflection for each of the following incident angles? Show the bending of the light ray in each case.

a. 40° **b.** 60° **c.** 80°

CHAPTER 8

Switching

Switching is a topic that can be discussed at several layers. We have switching at the physical layer, at the data-link layer, at the network layer, and even logically at the application layer (message switching). We have decided to discuss the general idea behind switching in this chapter, the last chapter related to the physical layer. We particularly discuss circuit-switching, which occurs at the physical layer. We introduce the idea of packet-switching, which occurs at the data-link and network layers, but we postpone the details of these topics until the appropriate chapters. Finally, we talk about the physical structures of the switches and routers.

This chapter is divided into four sections:

- The first section introduces switching. It mentions three methods of switching: circuit switching, packet switching, and message switching. The section then defines the switching methods that can occur in some layers of the Internet model.
- The second section discusses circuit-switched networks. It first defines three phases in these types of networks. It then describes the efficiency of these networks. The section also discusses the delay in circuit-switched networks.
- The third section briefly discusses packet-switched networks. It first describes datagram networks, listing their characteristics and advantages. The section then describes virtual circuit networks, explaining their features and operations. We will discuss packet-switched networks in more detail in Chapter 18.
- The last section discusses the structure of a switch. It first describes the structure of a circuit switch. It then explains the structure of a packet switch.

8.1 INTRODUCTION

A network is a set of connected devices. Whenever we have multiple devices, we have the problem of how to connect them to make one-to-one communication possible. One solution is to make a point-to-point connection between each pair of devices (a mesh topology) or between a central device and every other device (a star topology). These methods, however, are impractical and wasteful when applied to very large networks. The number and length of the links require too much infrastructure to be cost-efficient, and the majority of those links would be idle most of the time. Other topologies employing multipoint connections, such as a bus, are ruled out because the distances between devices and the total number of devices increase beyond the capacities of the media and equipment.

A better solution is **switching.** A switched network consists of a series of interlinked nodes, called ***switches.*** Switches are devices capable of creating temporary connections between two or more devices linked to the switch. In a switched network, some of these nodes are connected to the end systems (computers or telephones, for example). Others are used only for routing. Figure 8.1 shows a switched network.

Figure 8.1 *Switched network*

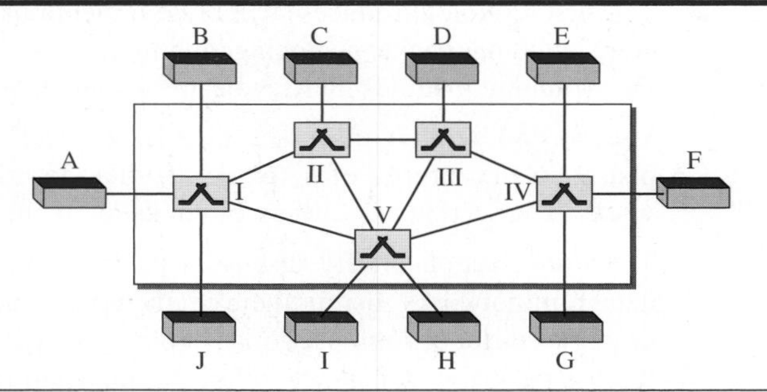

The **end systems** (communicating devices) are labeled A, B, C, D, and so on, and the switches are labeled I, II, III, IV, and V. Each switch is connected to multiple links.

8.1.1 Three Methods of Switching

Traditionally, three methods of switching have been discussed: **circuit switching**, **packet switching**, and **message switching**. The first two are commonly used today. The third has been phased out in general communications but still has networking applications. Packet switching can further be divided into two subcategories—virtual-circuit approach and datagram approach—as shown in Figure 8.2. In this chapter, we discuss only circuit switching and packet switching; message switching is more conceptual than practical.

Figure 8.2 *Taxonomy of switched networks*

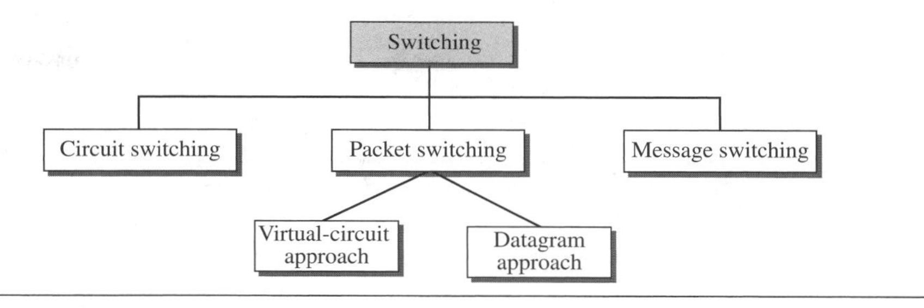

8.1.2 Switching and TCP/IP Layers

Switching can happen at several layers of the TCP/IP protocol suite.

Switching at Physical Layer

At the physical layer, we can have only circuit switching. There are no packets exchanged at the physical layer. The switches at the physical layer allow signals to travel in one path or another.

Switching at Data-Link Layer

At the data-link layer, we can have packet switching. However, the term *packet* in this case means *frames* or *cells*. Packet switching at the data-link layer is normally done using a virtual-circuit approach.

Switching at Network Layer

At the network layer, we can have packet switching. In this case, either a virtual-circuit approach or a datagram approach can be used. Currently the Internet uses a datagram approach, as we see in Chapter 18, but the tendency is to move to a virtual-circuit approach.

Switching at Application Layer

At the application layer, we can have only message switching. The communication at the application layer occurs by exchanging messages. Conceptually, we can say that communication using e-mail is a kind of message-switched communication, but we do not see any network that actually can be called a message-switched network.

8.2 CIRCUIT-SWITCHED NETWORKS

A **circuit-switched network** consists of a set of switches connected by physical links. A connection between two stations is a dedicated path made of one or more links. However, each connection uses only one dedicated channel on each link. Each link is normally divided into n channels by using FDM or TDM, as discussed in Chapter 6.

A circuit-switched network is made of a set of switches connected by physical links, in which each link is divided into *n* channels.

Figure 8.3 shows a trivial circuit-switched network with four switches and four links. Each link is divided into *n* (*n* is 3 in the figure) channels by using FDM or TDM.

Figure 8.3 *A trivial circuit-switched network*

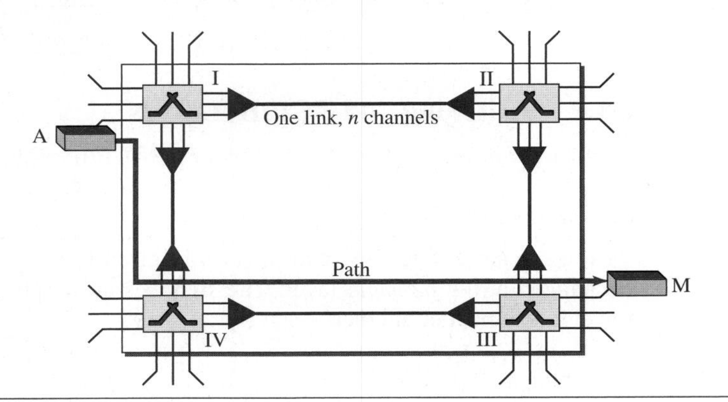

We have explicitly shown the multiplexing symbols to emphasize the division of the link into channels even though multiplexing can be implicitly included in the switch fabric.

The end systems, such as computers or telephones, are directly connected to a switch. We have shown only two end systems for simplicity. When end system A needs to communicate with end system M, system A needs to request a connection to M that must be accepted by all switches as well as by M itself. This is called the **setup phase;** a circuit (channel) is reserved on each link, and the combination of circuits or channels defines the dedicated path. After the dedicated path made of connected circuits (channels) is established, the **data-transfer phase** can take place. After all data have been transferred, the circuits are torn down.

We need to emphasize several points here:

- Circuit switching takes place at the physical layer.
- Before starting communication, the stations must make a reservation for the resources to be used during the communication. These resources, such as channels (bandwidth in FDM and time slots in TDM), switch buffers, switch processing time, and switch input/output ports, must remain dedicated during the entire duration of data transfer until the **teardown phase.**
- Data transferred between the two stations are not packetized (physical layer transfer of the signal). The data are a continuous flow sent by the source station and received by the destination station, although there may be periods of silence.

- There is no addressing involved during data transfer. The switches route the data based on their occupied band (FDM) or time slot (TDM). Of course, there is end-to-end addressing used during the setup phase, as we will see shortly.

In circuit switching, the resources need to be reserved during the setup phase; the resources remain dedicated for the entire duration of data transfer until the teardown phase.

Example 8.1

As a trivial example, let us use a circuit-switched network to connect eight telephones in a small area. Communication is through 4-kHz voice channels. We assume that each link uses FDM to connect a maximum of two voice channels. The bandwidth of each link is then 8 kHz. Figure 8.4 shows the situation. Telephone 1 is connected to telephone 7; 2 to 5; 3 to 8; and 4 to 6. Of course the situation may change when new connections are made. The switch controls the connections.

Figure 8.4 *Circuit-switched network used in Example 8.1*

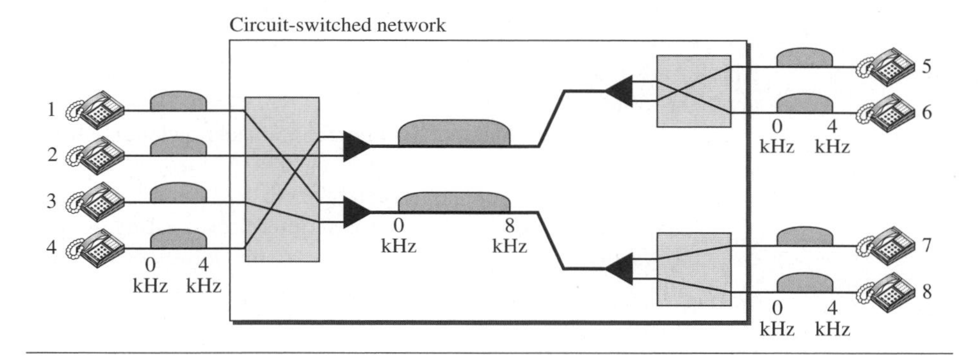

Example 8.2

As another example, consider a circuit-switched network that connects computers in two remote offices of a private company. The offices are connected using a T-1 line leased from a communication service provider. There are two 4×8 (4 inputs and 8 outputs) switches in this network. For each switch, four output ports are folded into the input ports to allow communication between computers in the same office. Four other output ports allow communication between the two offices. Figure 8.5 shows the situation.

8.2.1 Three Phases

The actual communication in a circuit-switched network requires three phases: connection setup, data transfer, and connection teardown.

Setup Phase

Before the two parties (or multiple parties in a conference call) can communicate, a dedicated circuit (combination of channels in links) needs to be established. The end systems are normally connected through dedicated lines to the switches, so connection setup

Figure 8.5 *Circuit-switched network used in Example 8.2*

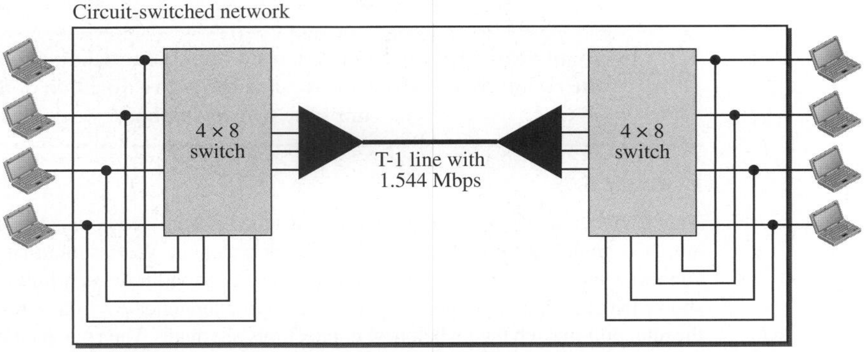

means creating dedicated channels between the switches. For example, in Figure 8.3, when system A needs to connect to system M, it sends a setup request that includes the address of system M, to switch I. Switch I finds a channel between itself and switch IV that can be dedicated for this purpose. Switch I then sends the request to switch IV, which finds a dedicated channel between itself and switch III. Switch III informs system M of system A's intention at this time.

In the next step to making a connection, an acknowledgment from system M needs to be sent in the opposite direction to system A. Only after system A receives this acknowledgment is the connection established.

Note that end-to-end addressing is required for creating a connection between the two end systems. These can be, for example, the addresses of the computers assigned by the administrator in a TDM network, or telephone numbers in an FDM network.

Data-Transfer Phase

After the establishment of the dedicated circuit (channels), the two parties can transfer data.

Teardown Phase

When one of the parties needs to disconnect, a signal is sent to each switch to release the resources.

8.2.2 Efficiency

It can be argued that circuit-switched networks are not as efficient as the other two types of networks because resources are allocated during the entire duration of the connection. These resources are unavailable to other connections. In a telephone network, people normally terminate the communication when they have finished their conversation. However, in computer networks, a computer can be connected to another computer even if there is no activity for a long time. In this case, allowing resources to be dedicated means that other connections are deprived.

8.2.3 Delay

Although a circuit-switched network normally has low efficiency, the delay in this type of network is minimal. During data transfer the data are not delayed at each switch; the resources are allocated for the duration of the connection. Figure 8.6 shows the idea of delay in a circuit-switched network when only two switches are involved.

Figure 8.6 *Delay in a circuit-switched network*

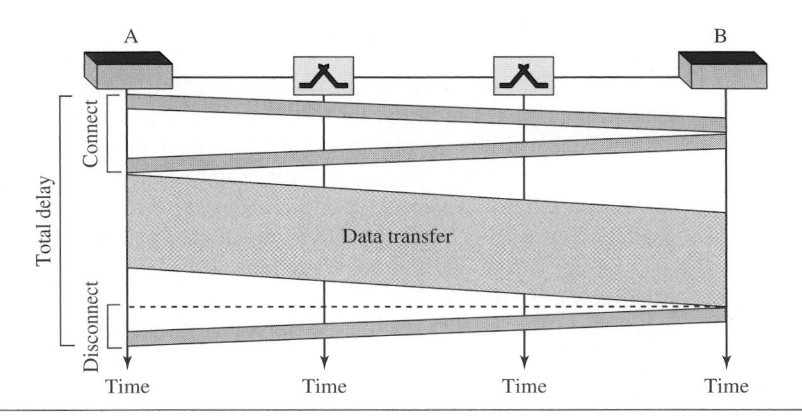

As Figure 8.6 shows, there is no waiting time at each switch. The total delay is due to the time needed to create the connection, transfer data, and disconnect the circuit. The delay caused by the setup is the sum of four parts: the propagation time of the source computer request (slope of the first gray box), the request signal transfer time (height of the first gray box), the propagation time of the acknowledgment from the destination computer (slope of the second gray box), and the signal transfer time of the acknowledgment (height of the second gray box). The delay due to data transfer is the sum of two parts: the propagation time (slope of the colored box) and data transfer time (height of the colored box), which can be very long. The third box shows the time needed to tear down the circuit. We have shown the case in which the receiver requests disconnection, which creates the maximum delay.

8.3 PACKET SWITCHING

In data communications, we need to send messages from one end system to another. If the message is going to pass through a **packet-switched network**, it needs to be divided into packets of fixed or variable size. The size of the packet is determined by the network and the governing protocol.

In packet switching, there is no resource allocation for a packet. This means that there is no reserved bandwidth on the links, and there is no scheduled processing time for each packet. Resources are allocated on demand. The allocation is done on a first-come, first-served basis. When a switch receives a packet, no matter what the source or destination is, the packet must wait if there are other packets being processed. As with other systems in our daily life, this lack of reservation may create delay. For example, if we do not have a reservation at a restaurant, we might have to wait.

> **In a packet-switched network, there is no resource reservation; resources are allocated on demand.**

We can have two types of packet-switched networks: datagram networks and virtual-circuit networks.

8.3.1 Datagram Networks

In a **datagram network**, each packet is treated independently of all others. Even if a packet is part of a multipacket transmission, the network treats it as though it existed alone. Packets in this approach are referred to as ***datagrams***.

Datagram switching is normally done at the network layer. We briefly discuss datagram networks here as a comparison with circuit-switched and virtual-circuit-switched networks. In Chapter 18 of this text, we go into greater detail.

Figure 8.7 shows how the datagram approach is used to deliver four packets from station A to station X. The switches in a datagram network are traditionally referred to as routers. That is why we use a different symbol for the switches in the figure.

Figure 8.7 *A datagram network with four switches (routers)*

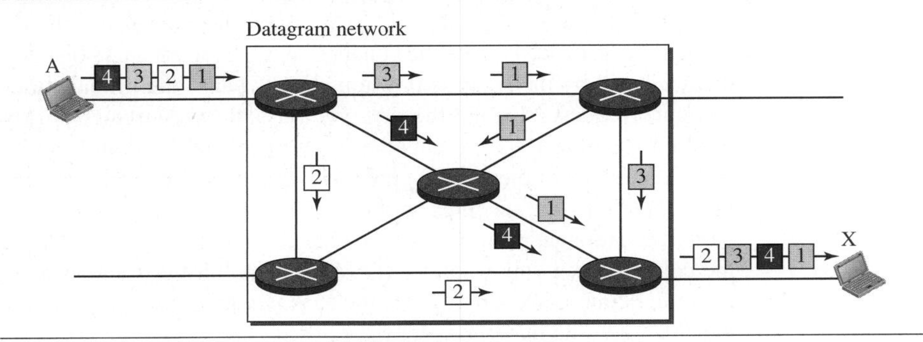

In this example, all four packets (or datagrams) belong to the same message, but may travel different paths to reach their destination. This is so because the links may be involved in carrying packets from other sources and do not have the necessary bandwidth available to carry all the packets from A to X. This approach can cause the datagrams of a transmission to arrive at their destination out of order with different delays between the packets. Packets may also be lost or dropped because of a lack of resources. In most protocols, it is the responsibility of an upper-layer protocol to reorder the datagrams or ask for lost datagrams before passing them on to the application.

The datagram networks are sometimes referred to as *connectionless networks*. The term *connectionless* here means that the switch (packet switch) does not keep information about the connection state. There are no setup or teardown phases. Each packet is treated the same by a switch regardless of its source or destination.

Routing Table

If there are no setup or teardown phases, how are the packets routed to their destinations in a datagram network? In this type of network, each switch (or packet switch) has a routing table which is based on the destination address. The routing tables are dynamic and are updated periodically. The destination addresses and the corresponding forwarding output ports are recorded in the tables. This is different from the table of a circuit-switched network (discussed later) in which each entry is created when the setup phase is completed and deleted when the teardown phase is over. Figure 8.8 shows the routing table for a switch.

Figure 8.8 *Routing table in a datagram network*

Destination address	Output port
1232	1
4150	2
⋮	⋮
9130	3

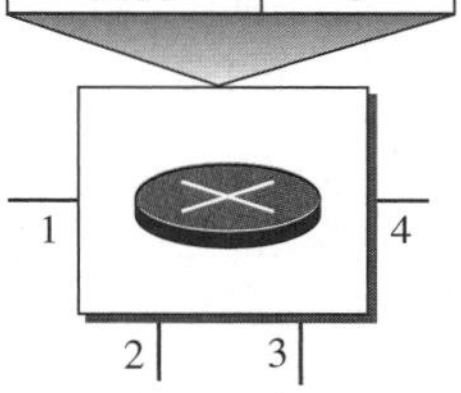

A switch in a datagram network uses a routing table that is based on the destination address.

Destination Address

Every packet in a datagram network carries a header that contains, among other information, the destination address of the packet. When the switch receives the packet, this destination address is examined; the routing table is consulted to find the corresponding port through which the packet should be forwarded. This address, unlike the address in a virtual-circuit network, remains the same during the entire journey of the packet.

The destination address in the header of a packet in a datagram network remains the same during the entire journey of the packet.

Efficiency

The efficiency of a datagram network is better than that of a circuit-switched network; resources are allocated only when there are packets to be transferred. If a source sends a packet and there is a delay of a few minutes before another packet can be sent, the resources can be reallocated during these minutes for other packets from other sources.

Delay

There may be greater delay in a datagram network than in a virtual-circuit network. Although there are no setup and teardown phases, each packet may experience a wait at a switch before it is forwarded. In addition, since not all packets in a message necessarily travel through the same switches, the delay is not uniform for the packets of a message. Figure 8.9 gives an example of delay in a datagram network for one packet.

Figure 8.9 *Delay in a datagram network*

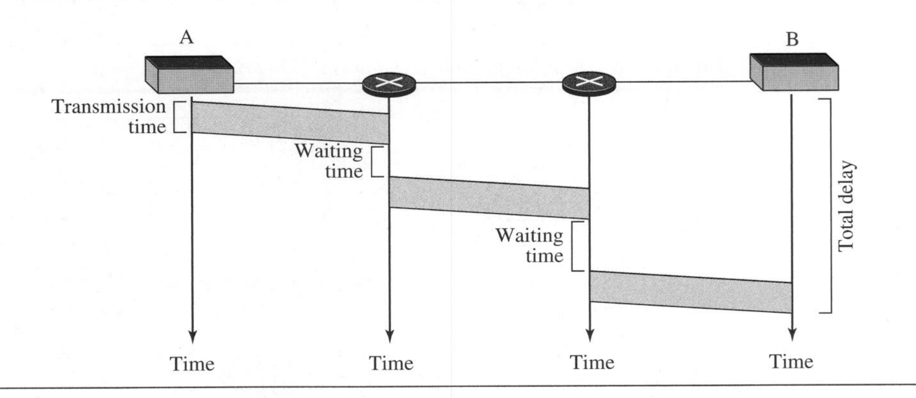

The packet travels through two switches. There are three transmission times ($3T$), three propagation delays (slopes 3τ of the lines), and two waiting times ($w_1 + w_2$). We ignore the processing time in each switch. The total delay is

$$\textbf{Total delay} = 3T + 3\tau + w_1 + w_2$$

8.3.2 Virtual-Circuit Networks

A **virtual-circuit network** is a cross between a circuit-switched network and a datagram network. It has some characteristics of both.

1. As in a circuit-switched network, there are setup and teardown phases in addition to the data transfer phase.
2. Resources can be allocated during the setup phase, as in a circuit-switched network, or on demand, as in a datagram network.
3. As in a datagram network, data are packetized and each packet carries an address in the header. However, the address in the header has local jurisdiction (it defines what the next switch should be and the channel on which the packet is being carried), not end-to-end jurisdiction. The reader may ask how the intermediate switches know where to send the packet if there is no final destination address carried by a packet. The answer will be clear when we discuss virtual-circuit identifiers in the next section.
4. As in a circuit-switched network, all packets follow the same path established during the connection.

5. A virtual-circuit network is normally implemented in the data-link layer, while a circuit-switched network is implemented in the physical layer and a datagram network in the network layer. But this may change in the future.

Figure 8.10 is an example of a virtual-circuit network. The network has switches that allow traffic from sources to destinations. A source or destination can be a computer, packet switch, bridge, or any other device that connects other networks.

Figure 8.10 *Virtual-circuit network*

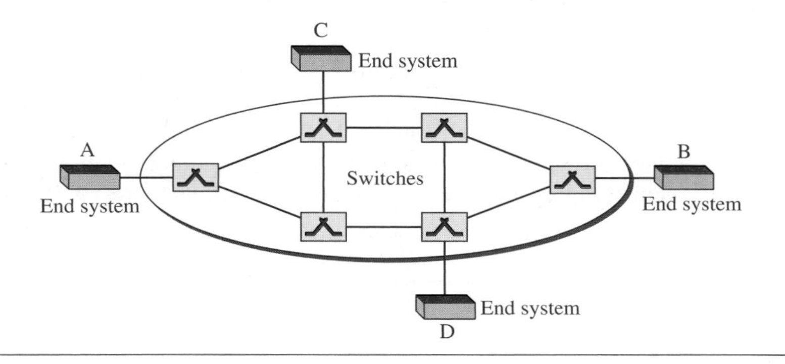

Addressing

In a virtual-circuit network, two types of addressing are involved: global and local (virtual-circuit identifier).

Global Addressing

A source or a destination needs to have a global address—an address that can be unique in the scope of the network or internationally if the network is part of an international network. However, we will see that a global address in virtual-circuit networks is used only to create a virtual-circuit identifier, as discussed next.

Virtual-Circuit Identifier

The identifier that is actually used for data transfer is called the ***virtual-circuit identifier* (VCI)** or the ***label.*** A VCI, unlike a global address, is a small number that has only switch scope; it is used by a frame between two switches. When a frame arrives at a switch, it has a VCI; when it leaves, it has a different VCI. Figure 8.11 shows how the VCI in a data frame changes from one switch to another. Note that a VCI does not need to be a large number since each switch can use its own unique set of VCIs.

Figure 8.11 *Virtual-circuit identifier*

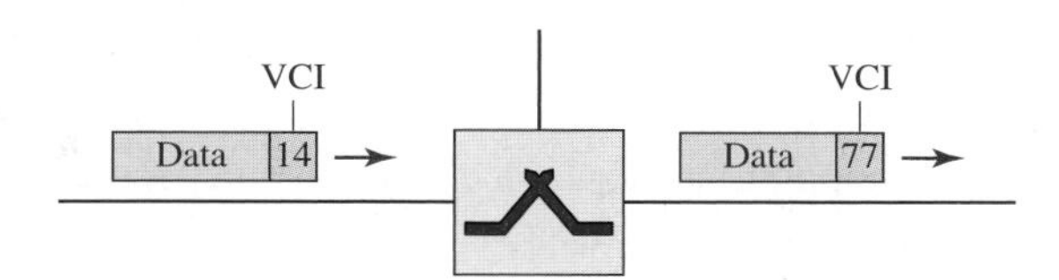

Three Phases

As in a circuit-switched network, a source and destination need to go through three phases in a virtual-circuit network: setup, data transfer, and teardown. In the setup phase, the source and destination use their global addresses to help switches make table entries for the connection. In the teardown phase, the source and destination inform the switches to delete the corresponding entry. Data transfer occurs between these two phases. We first discuss the data-transfer phase, which is more straightforward; we then talk about the setup and teardown phases.

Data-Transfer Phase

To transfer a frame from a source to its destination, all switches need to have a table entry for this virtual circuit. The table, in its simplest form, has four columns. This means that the switch holds four pieces of information for each virtual circuit that is already set up. We show later how the switches make their table entries, but for the moment we assume that each switch has a table with entries for all active virtual circuits. Figure 8.12 shows such a switch and its corresponding table.

Figure 8.12 *Switch and tables in a virtual-circuit network*

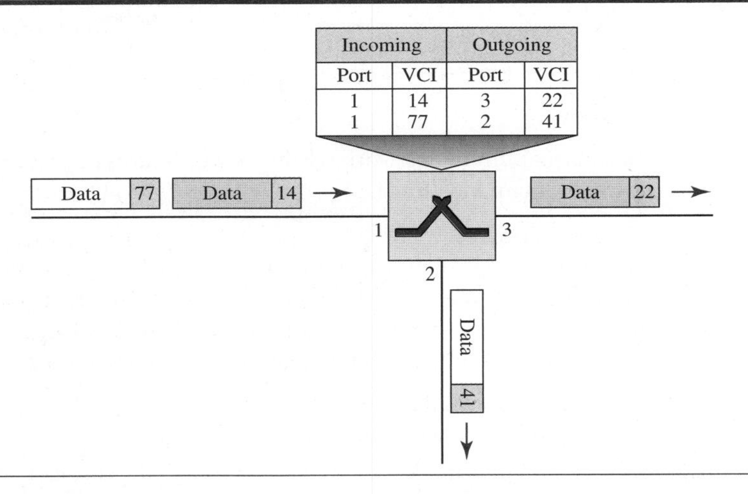

Figure 8.12 shows a frame arriving at port 1 with a VCI of 14. When the frame arrives, the switch looks in its table to find port 1 and a VCI of 14. When it is found, the switch knows to change the VCI to 22 and send out the frame from port 3.

Figure 8.13 shows how a frame from source A reaches destination B and how its VCI changes during the trip. Each switch changes the VCI and routes the frame.

The data-transfer phase is active until the source sends all its frames to the destination. The procedure at the switch is the same for each frame of a message. The process creates a virtual circuit, not a real circuit, between the source and destination.

Setup Phase

In the setup phase, a switch creates an entry for a virtual circuit. For example, suppose source A needs to create a virtual circuit to B. Two steps are required: the setup request and the acknowledgment.

Figure 8.13 *Source-to-destination data transfer in a virtual-circuit network*

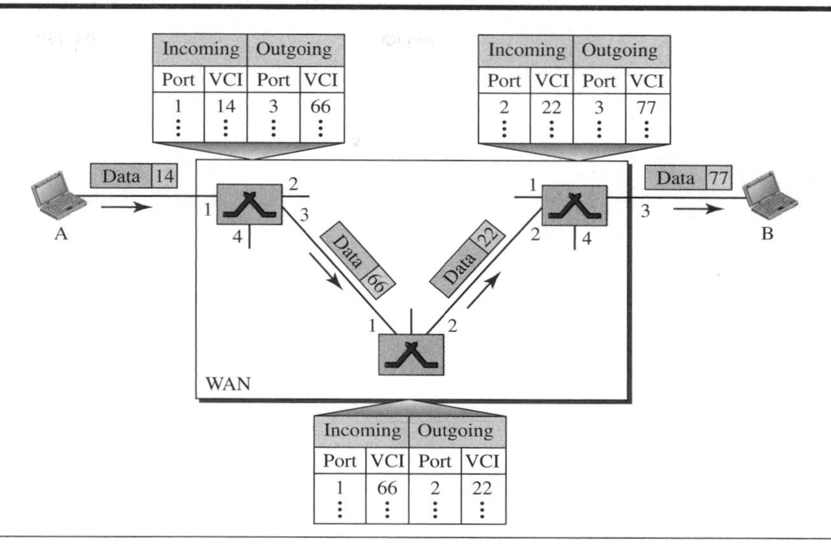

Setup Request

A setup request frame is sent from the source to the destination. Figure 8.14 shows the process.

Figure 8.14 *Setup request in a virtual-circuit network*

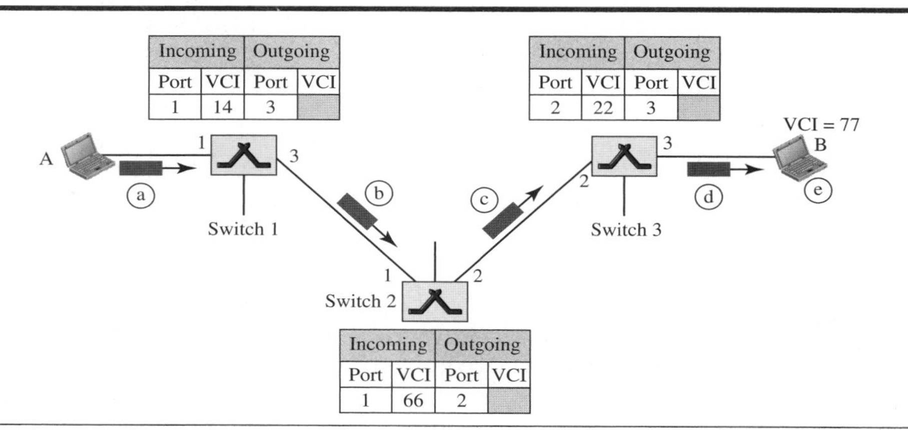

a. Source A sends a setup frame to switch 1.

b. Switch 1 receives the setup request frame. It knows that a frame going from A to B goes out through port 3. How the switch has obtained this information is a point covered in future chapters. The switch, in the setup phase, acts as a packet switch; it has a routing table which is different from the switching table. For the moment, assume that it knows the output port. The switch creates an entry in its table for this virtual circuit, but it is only able to fill three of the four columns. The switch assigns the incoming port (1) and chooses an available incoming VCI (14) and the outgoing port (3). It does not yet know the outgoing VCI, which will be found during the acknowledgment step. The switch then forwards the frame through port 3 to switch 2.

c. Switch 2 receives the setup request frame. The same events happen here as at switch 1; three columns of the table are completed: in this case, incoming port (1), incoming VCI (66), and outgoing port (2).

d. Switch 3 receives the setup request frame. Again, three columns are completed: incoming port (2), incoming VCI (22), and outgoing port (3).

e. Destination B receives the setup frame, and if it is ready to receive frames from A, it assigns a VCI to the incoming frames that come from A, in this case 77. This VCI lets the destination know that the frames come from A, and not other sources.

Acknowledgment

A special frame, called the *acknowledgment frame,* completes the entries in the switching tables. Figure 8.15 shows the process.

Figure 8.15 *Setup acknowledgment in a virtual-circuit network*

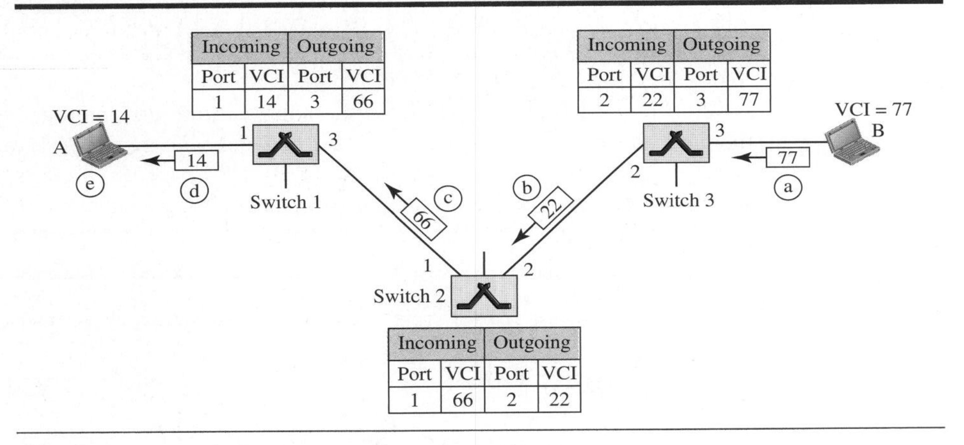

a. The destination sends an acknowledgment to switch 3. The acknowledgment carries the global source and destination addresses so the switch knows which entry in the table is to be completed. The frame also carries VCI 77, chosen by the destination as the incoming VCI for frames from A. Switch 3 uses this VCI to complete the outgoing VCI column for this entry. Note that 77 is the incoming VCI for destination B, but the outgoing VCI for switch 3.

b. Switch 3 sends an acknowledgment to switch 2 that contains its incoming VCI in the table, chosen in the previous step. Switch 2 uses this as the outgoing VCI in the table.

c. Switch 2 sends an acknowledgment to switch 1 that contains its incoming VCI in the table, chosen in the previous step. Switch 1 uses this as the outgoing VCI in the table.

d. Finally switch 1 sends an acknowledgment to source A that contains its incoming VCI in the table, chosen in the previous step.

e. The source uses this as the outgoing VCI for the data frames to be sent to destination B.

Teardown Phase

In this phase, source A, after sending all frames to B, sends a special frame called a *teardown request*. Destination B responds with a teardown confirmation frame. All switches delete the corresponding entry from their tables.

Efficiency

As we said before, resource reservation in a virtual-circuit network can be made during the setup or can be on demand during the data-transfer phase. In the first case, the delay for each packet is the same; in the second case, each packet may encounter different delays. There is one big advantage in a virtual-circuit network even if resource allocation is on demand. The source can check the availability of the resources, without actually reserving it. Consider a family that wants to dine at a restaurant. Although the restaurant may not accept reservations (allocation of the tables is on demand), the family can call and find out the waiting time. This can save the family time and effort.

In virtual-circuit switching, all packets belonging to the same source and destination travel the same path, but the packets may arrive at the destination with different delays if resource allocation is on demand.

Delay in Virtual-Circuit Networks

In a virtual-circuit network, there is a one-time delay for setup and a one-time delay for teardown. If resources are allocated during the setup phase, there is no wait time for individual packets. Figure 8.16 shows the delay for a packet traveling through two switches in a virtual-circuit network.

Figure 8.16 *Delay in a virtual-circuit network*

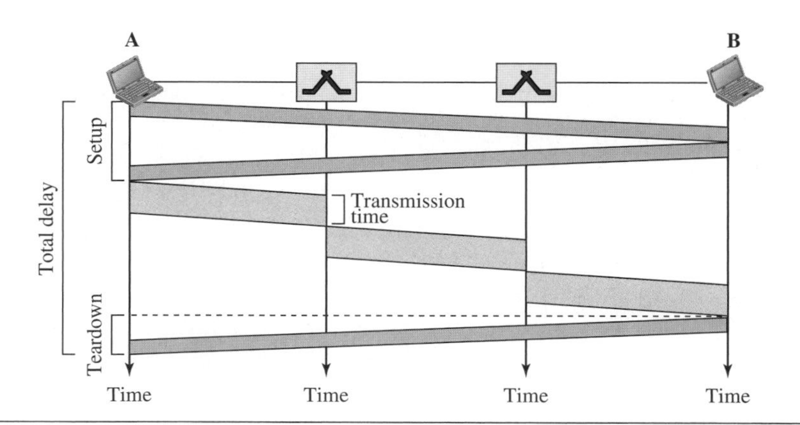

The packet is traveling through two switches (routers). There are three transmission times ($3T$), three propagation times (3τ), data transfer depicted by the sloping lines, a setup delay (which includes transmission and propagation in two directions), and a teardown delay (which includes transmission and propagation in one direction). We ignore the processing time in each switch. The total delay time is

Total delay + $3T$ + 3τ + setup delay + teardown delay

Circuit-Switched Technology in WANs

As we will see in Chapter 14, virtual-circuit networks are used in switched WANs such as ATM networks. The data-link layer of these technologies is well suited to the virtual-circuit technology.

Switching at the data-link layer in a switched WAN is normally implemented by using virtual-circuit techniques.

8.4 STRUCTURE OF A SWITCH

We use switches in circuit-switched and packet-switched networks. In this section, we discuss the structures of the switches used in each type of network.

8.4.1 Structure of Circuit Switches

Circuit switching today can use either of two technologies: the space-division switch or the time-division switch.

Space-Division Switch

In **space-division switching,** the paths in the circuit are separated from one another spatially. This technology was originally designed for use in analog networks but is used currently in both analog and digital networks. It has evolved through a long history of many designs.

Crossbar Switch

A **crossbar switch** connects n inputs to m outputs in a grid, using electronic micro-switches (transistors) at each **crosspoint** (see Figure 8.17). The major limitation of this design is the number of crosspoints required. To connect n inputs to m outputs using a

Figure 8.17 *Crossbar switch with three inputs and four outputs*

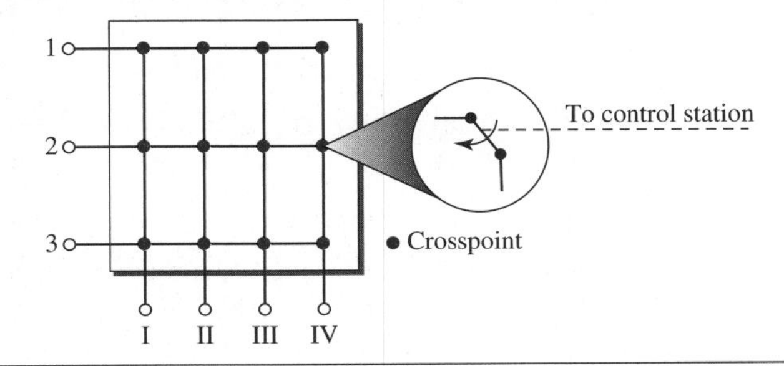

crossbar switch requires $n \times m$ crosspoints. For example, to connect 1000 inputs to 1000 outputs requires a switch with 1,000,000 crosspoints. A crossbar switch [?] with this number of crosspoints is impractical. Such a switch is also inefficient because statistics show that, in practice, fewer than 25 percent of the crosspoints are in use at any given time. The rest are idle.

Multistage Switch

The solution to the limitations of the crossbar switch is the **multistage switch,** which combines crossbar switches in several (normally three) stages, as shown in Figure 8.18. In a single crossbar switch, only one row or column (one path) is active for any connection. So we need $N \times N$ crosspoints. If we can allow multiple paths inside the switch, we can decrease the number of crosspoints. Each crosspoint in the middle stage can be accessed by multiple crosspoints in the first or third stage.

Figure 8.18 *Multistage switch*

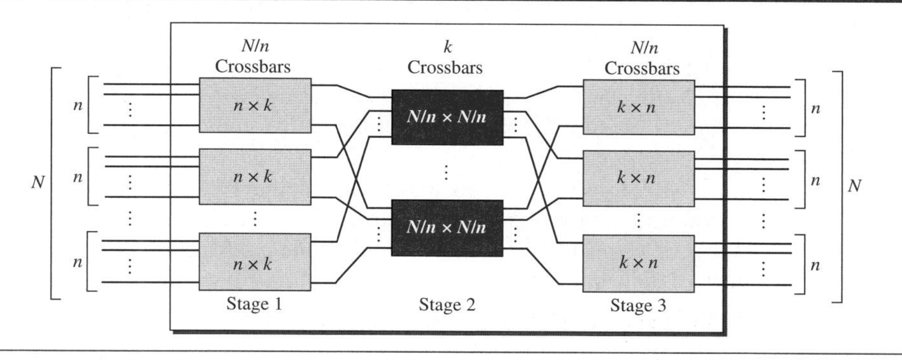

To design a three-stage switch, we follow these steps:

1. We divide the N input lines into groups, each of n lines. For each group, we use one crossbar of size $n \times k$, where k is the number of crossbars in the middle stage. In other words, the first stage has N/n crossbars of $n \times k$ crosspoints.
2. We use k crossbars, each of size $(N/n) \times (N/n)$ in the middle stage.
3. We use N/n crossbars, each of size $k \times n$ at the third stage.

We can calculate the total number of crosspoints as follows:

$$\frac{N}{n}(n \times k) + k\left(\frac{N}{n} \times \frac{N}{n}\right) + \frac{N}{n}(k \times n) = 2kN + k\left(\frac{N}{n}\right)^2$$

In a three-stage switch, the total number of crosspoints is

$$2kN + k\left(\frac{N}{n}\right)^2$$

which is much smaller than the number of crosspoints in a single-stage switch (N^2).

Example 8.3

Design a three-stage, 200×200 switch ($N = 200$) with $k = 4$ and $n = 20$.

Solution

In the first stage we have N/n or 10 crossbars, each of size 20×4. In the second stage, we have 4 crossbars, each of size 10×10. In the third stage, we have 10 crossbars, each of size 4×20. The total number of crosspoints is $2kN + k(N/n)^2$, or 2000 crosspoints. This is 5 percent of the number of crosspoints in a single-stage switch ($200 \times 200 = 40,000$).

The multistage switch in Example 8.3 has one drawback—**blocking** during periods of heavy traffic. The whole idea of multistage switching is to share the crosspoints in the middle-stage crossbars. Sharing can cause a lack of availability if the resources are limited and all users want a connection at the same time. *Blocking* refers to times when one input cannot be connected to an output because there is no path available between them—all the possible intermediate switches are occupied.

In a single-stage switch, blocking does not occur because every combination of input and output has its own crosspoint; there is always a path. (Cases in which two inputs are trying to contact the same output do not count. That path is not blocked; the output is merely busy.) In the multistage switch described in Example 8.3, however, only four of the first 20 inputs can use the switch at a time, only four of the second 20 inputs can use the switch at a time, and so on. The small number of crossbars at the middle stage creates blocking.

In large systems, such as those having 10,000 inputs and outputs, the number of stages can be increased to cut down on the number of crosspoints required. As the number of stages increases, however, possible blocking increases as well. Many people have experienced blocking on public telephone systems in the wake of a natural disaster when the calls being made to check on or reassure relatives far outnumber the regular load of the system.

Clos investigated the condition of nonblocking in multistage switches and came up with the following formula. In a nonblocking switch, the number of middle-stage switches must be at least $2n - 1$. In other words, we need to have $k \geq 2n - 1$.

Note that the number of crosspoints is still smaller than that in a single-stage switch. Now we need to minimize the number of crosspoints with a fixed N by using the Clos criteria. We can take the derivative of the equation with respect to n (the only variable) and find the value of n that makes the result zero. This n must be equal to or greater than $(N/2)^{1/2}$. In this case, the total number of crosspoints is greater than or equal to $4N [(2N)^{1/2} - 1]$. In other words, the minimum number of crosspoints according to the Clos criteria is proportional to $N^{3/2}$.

According to Clos criterion: $n = (N/2)^{1/2}$ *and* $k \geq 2n - 1$
Total number of crosspoints $\geq 4N [(2N)^{1/2} - 1]$

Example 8.4

Redesign the previous three-stage, 200×200 switch, using the Clos criteria with a minimum number of crosspoints.

Solution

We let $n = (200/2)^{1/2}$, or $n = 10$. We calculate $k = 2n - 1 = 19$. In the first stage, we have 200/10, or 20, crossbars, each with 10×19 crosspoints. In the second stage, we have 19 crossbars, each with 10 × 10 crosspoints. In the third stage, we have 20 crossbars each with 19 × 10 crosspoints. The total number of crosspoints is 20(10 × 19) + 19(10 × 10) + 20(19 × 10) = 9500. If we use a single-stage switch, we need 200 × 200 = 40,000 crosspoints. The number of crosspoints in this three-stage switch is 24 percent that of a single-stage switch. More points are needed than in Example 8.3 (5 percent). The extra crosspoints are needed to prevent blocking.

A multistage switch that uses the Clos criteria and a minimum number of crosspoints still requires a huge number of crosspoints. For example, to have a 100,000 input/output switch, we need something close to 200 million crosspoints (instead of 10 billion). This means that if a telephone company needs to provide a switch to connect 100,000 telephones in a city, it needs 200 million crosspoints. The number can be reduced if we accept blocking. Today, telephone companies use time-division switching or a combination of space- and time-division switches, as we will see shortly.

Time-Division Switch

Time-division switching uses time-division multiplexing (TDM) inside a switch. The most popular technology is called the **time-slot interchange (TSI).**

Time-Slot Interchange

Figure 8.19 shows a system connecting four input lines to four output lines. Imagine that each input line wants to send data to an output line according to the following pattern: (1 → 3), (2 → 4), (3 → 1), and (4 → 2), in which the arrow means "to."

Figure 8.19 *Time-slot interchange*

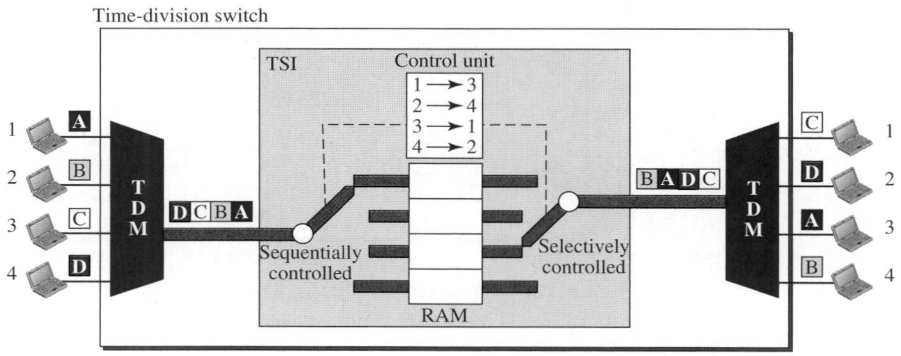

The figure combines a TDM multiplexer, a TDM demultiplexer, and a TSI consisting of random access memory (RAM) with several memory locations. The size of each location is the same as the size of a single time slot. The number of locations is the same as the number of inputs (in most cases, the numbers of inputs and outputs are equal). The RAM fills up with incoming data from time slots in the order received. Slots are then sent out in an order based on the decisions of a control unit.

Time- and Space-Division Switch Combinations

When we compare space-division and time-division switching, some interesting facts emerge. The advantage of space-division switching is that it is instantaneous. Its disadvantage is the number of crosspoints required to make space-division switching acceptable in terms of blocking.

The advantage of time-division switching is that it needs no crosspoints. Its disadvantage, in the case of TSI, is that processing each connection creates delays. Each time slot must be stored by the RAM, then retrieved and passed on.

In a third option, we combine space-division and time-division technologies to take advantage of the best of both. Combining the two results in switches that are optimized both physically (the number of crosspoints) and temporally (the amount of delay). Multistage switches of this sort can be designed as **time-space-time (TST) switches.**

Figure 8.20 shows a simple TST switch that consists of two time stages and one space stage and has 12 inputs and 12 outputs. Instead of one time-division switch, it divides the inputs into three groups (of four inputs each) and directs them to three time-slot interchanges. The result is that the average delay is one-third of what would result from using one time-slot interchange to handle all 12 inputs.

Figure 8.20 *Time-space-time switch*

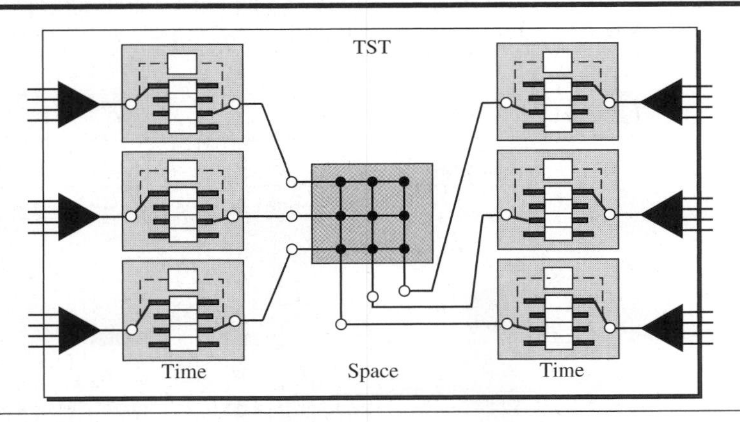

The last stage is a mirror image of the first stage. The middle stage is a space-division switch (crossbar) that connects the TSI groups to allow connectivity between all possible input and output pairs (e.g., to connect input 3 of the first group to output 7 of the second group).

8.4.2 Structure of Packet Switches

A switch used in a packet-switched network has a different structure from a switch used in a circuit-switched network.We can say that a packet switch has four components: **input ports, output ports,** the **routing processor,** and the **switching fabric,** as shown in Figure 8.21.

Figure 8.21 *Packet switch components*

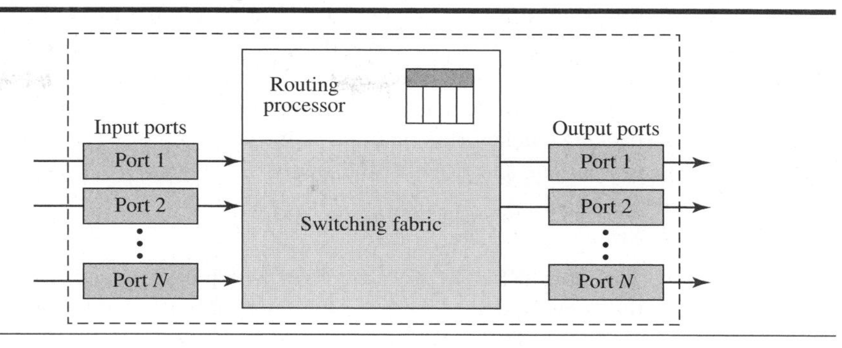

Input Ports

An input port performs the physical and data-link functions of the packet switch. The bits are constructed from the received signal. The packet is decapsulated from the frame. Errors are detected and corrected. The packet is now ready to be routed by the network layer. In addition to a physical-layer processor and a data-link processor, the input port has buffers (queues) to hold the packet before it is directed to the switching fabric. Figure 8.22 shows a schematic diagram of an input port.

Figure 8.22 *Input port*

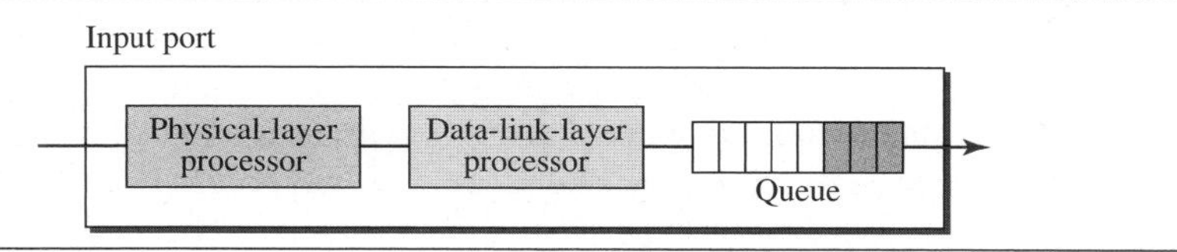

Output Port

The output port performs the same functions as the input port, but in the reverse order. First the outgoing packets are queued, then the packet is encapsulated in a frame, and finally the physical-layer functions are applied to the frame to create the signal to be sent on the line. Figure 8.23 shows a schematic diagram of an output port.

Figure 8.23 *Output port*

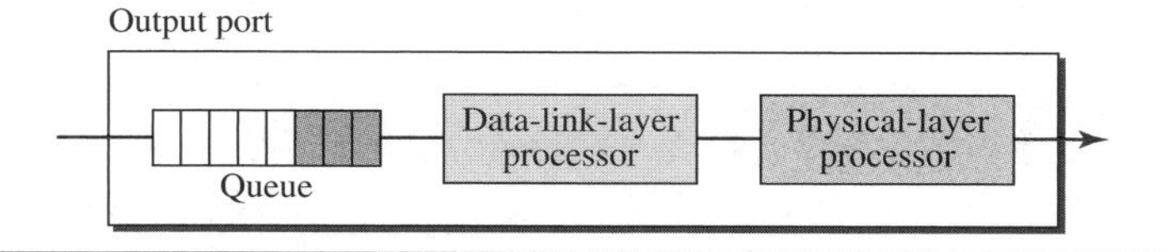

Routing Processor

The routing processor performs the functions of the network layer. The destination address is used to find the address of the next hop and, at the same time, the output port number from which the packet is sent out. This activity is sometimes referred to as **table lookup** because the routing processor searches the routing table. In the newer packet switches, this function of the routing processor is being moved to the input ports to facilitate and expedite the process.

Switching Fabrics

The most difficult task in a packet switch is to move the packet from the input queue to the output queue. The speed with which this is done affects the size of the input/output queue and the overall delay in packet delivery. In the past, when a packet switch was actually a dedicated computer, the memory of the computer or a bus was used as the switching fabric. The input port stored the packet in memory; the output port retrieved the packet from memory. Today, packet switches are specialized mechanisms that use a variety of switching fabrics. We briefly discuss some of these fabrics here.

Crossbar Switch

The simplest type of switching fabric is the crossbar switch, discussed in the previous section.

Banyan Switch

A more realistic approach than the crossbar switch is the **banyan switch** (named after the banyan tree). A banyan switch is a multistage switch with microswitches at each stage that route the packets based on the output port represented as a binary string. For n inputs and n outputs, we have $\log_2 n$ stages with $n/2$ microswitches at each stage. The first stage routes the packet based on the high-order bit of the binary string. The second stage routes the packet based on the second high-order bit, and so on. Figure 8.24 shows a banyan switch with eight inputs and eight outputs. The number of stages is $\log_2(8) = 3$.

Figure 8.24 *A banyan switch*

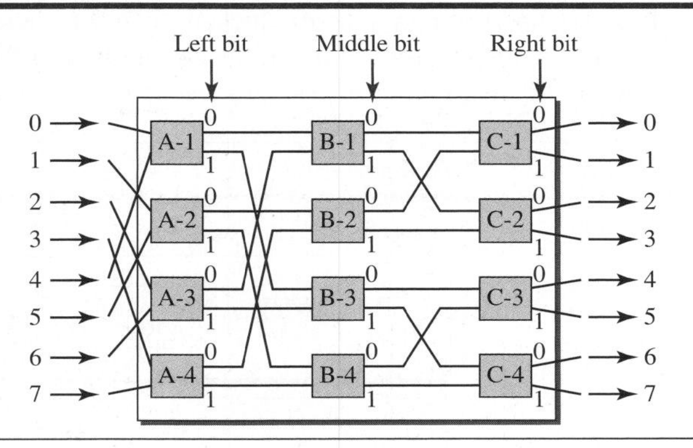

Figure 8.25 shows the operation. In part a, a packet has arrived at input port 1 and must go to output port 6 (110 in binary). The first microswitch (A-2) routes the packet

Figure 8.25 *Examples of routing in a banyan switch*

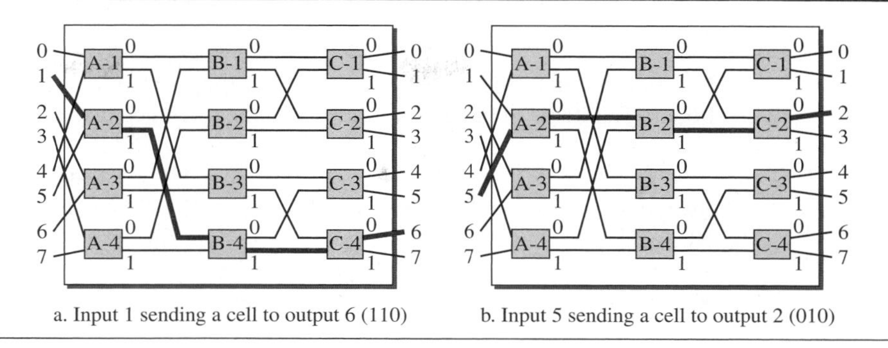

based on the first bit (1), the second microswitch (B-4) routes the packet based on the second bit (1), and the third microswitch (C-4) routes the packet based on the third bit (0). In part b, a packet has arrived at input port 5 and must go to output port 2 (010 in binary). The first microswitch (A-2) routes the packet based on the first bit (0), the second microswitch (B-2) routes the packet based on the second bit (1), and the third microswitch (C-2) routes the packet based on the third bit (0).

Batcher-Banyan Switch The problem with the banyan switch is the possibility of internal collision even when two packets are not heading for the same output port. We can solve this problem by sorting the arriving packets based on their destination port.

K. E. Batcher designed a switch that comes before the banyan switch and sorts the incoming packets according to their final destinations. The combination is called the **Batcher-banyan switch.** The sorting switch uses hardware merging techniques, but we do not discuss the details here. Normally, another hardware module called a **trap** is added between the Batcher switch and the banyan switch (see Figure 8.26) The trap module prevents duplicate packets (the packets with the same output destination) from passing to the banyan switch simultaneously. Only one packet for each destination is allowed at each tick; if there is more than one, they wait for the next tick.

Figure 8.26 *Batcher-banyan switch*

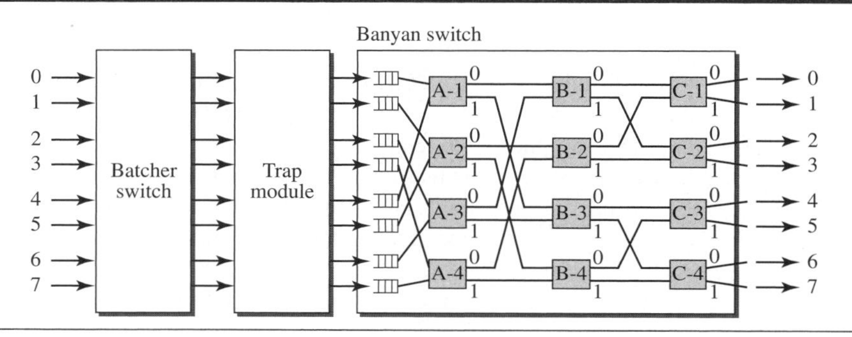

8.5 END-CHAPTER MATERIALS

8.5.1 Recommended Reading

For more details about subjects discussed in this chapter, we recommend the following books. The items in brackets [. . .] refer to the reference list at the end of the text.

Books

Switching is discussed in [Sta04] and [GW04]. Circuit-switching is fully discussed in [BEL01].

8.5.2 Key terms

banyan switch
Batcher-banyan switch
blocking
circuit switching
circuit-switched network
crossbar switch
crosspoint
data-transfer phase
datagram
datagram network
end system
input port
message switching
multistage switch
output port
packet switching
packet-switched network
routing processor
setup phase
space-division switching
switch
switching
switching fabric
table lookup
teardown phase
time-division switching
time-slot interchange (TSI)
time-space-time (TST) switch
trap
virtual-circuit identifier (VCI)
virtual-circuit network

8.5.3 Summary

A switched network consists of a series of interlinked nodes, called *switches*. Traditionally, three methods of switching have been important: circuit switching, packet switching, and message switching.

We can divide today's networks into three broad categories: circuit-switched networks, packet-switched networks, and message-switched networks. Packet-switched networks can also be divided into two subcategories: virtual-circuit networks and datagram networks. A circuit-switched network is made of a set of switches connected by physical links, in which each link is divided into n channels. Circuit switching takes place at the physical layer. In circuit switching, the resources need to be reserved during the setup phase; the resources remain dedicated for the entire duration of the data-transfer phase until the teardown phase.

In packet switching, there is no resource allocation for a packet. This means that there is no reserved bandwidth on the links, and there is no scheduled processing time for each packet. Resources are allocated on demand. In a datagram network, each packet is treated independently of all others. Packets in this approach are referred to as datagrams. There are no setup or teardown phases. A virtual-circuit network is a cross between a circuit-switched network and a datagram network. It has some characteristics of both. Circuit switching uses either of two technologies: the space-division switch or the time-division switch. A switch in a packet-switched network has a different structure from a switch used in a circuit-switched network. We can say that a packet switch has four types of components: input ports, output ports, a routing processor, and switching fabric.

8.6 PRACTICE SET

8.6.1 Quizzes

A set of interactive quizzes for this chapter can be found on the book website. It is strongly recommended that the student take the quizzes to check his/her understanding of the materials before continuing with the practice set.

8.6.2 Questions

Q8-1. Compare and contrast the two major categories of circuit switches.

Q8-2. What is TSI and what is its role in time-division switching?

Q8-3. Describe the need for switching and define a switch.

Q8-4. Compare and contrast a circuit-switched network and a packet-switched network.

Q8-5. What is the role of the address field in a packet traveling through a datagram network?

Q8-6. List four major components of a packet switch and their functions.

Q8-7. What are the two approaches to packet switching?

Q8-8. What is the role of the address field in a packet traveling through a virtual-circuit network?

Q8-9. Compare space-division and time-division switches.

Q8-10. List the three traditional switching methods. Which are the most common today?

8.6.3 Problems

P8-1. Transmission of information in any network involves end-to-end addressing and sometimes local addressing (such as VCI). Table 8.1 shows the types of networks and the addressing mechanism used in each of them.

Table 8.1 *P8-1*

Network	*Setup*	*Data Transfer*	*Teardown*
Circuit-switched	End-to-end		End-to-end
Datagram		End-to-end	
Virtual-circuit	End-to-end	Local	End-to-end

Answer the following questions:

a. Why does a circuit-switched network need end-to-end addressing during the setup and teardown phases? Why are no addresses needed during the data transfer phase for this type of network?

b. Why does a datagram network need only end-to-end addressing during the data transfer phase, but no addressing during the setup and teardown phases?

c. Why does a virtual-circuit network need addresses during all three phases?

P8-2. We need a three-stage time-space-time switch with $N = 100$. We use 10 TSIs at the first and third stages and 4 crossbars at the middle stage.

a. Draw the configuration diagram.

b. Calculate the total number of crosspoints.

c. Calculate the total number of memory locations we need for the TSIs.

P8-3. Answer the following questions:

a. Can a routing table in a datagram network have two entries with the same destination address? Explain.

b. Can a switching table in a virtual-circuit network have two entries with the same input port number? With the same output port number? With the same incoming VCIs? With the same outgoing VCIs? With the same incoming values (port, VCI)? With the same outgoing values (port, VCI)?

P8-4. Figure 8.26 shows a switch in a virtual-circuit network.

Figure 8.26 *Problem P8-4*

Incoming		Outgoing	
Port	VCI	Port	VCI
1	14	3	22
2	71	4	41
2	92	1	45
3	58	2	43
3	78	2	70
4	56	3	11

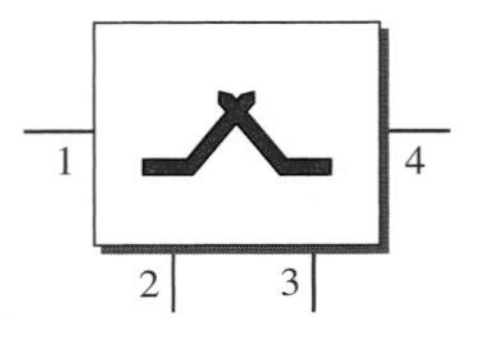

Find the output port and the output VCI for packets with the following input port and input VCI addresses:

a. Packet 1: 3, 78

b. Packet 2: 2, 92

c. Packet 3: 4, 56

d. Packet 4: 2, 71

P8-5. Figure 8.27 shows a switch (router) in a datagram network.

Figure 8.27 *Problem P8-5*

Destination address	Output port
1233	3
1456	2
3255	1
4470	4
7176	2
8766	3
9144	2

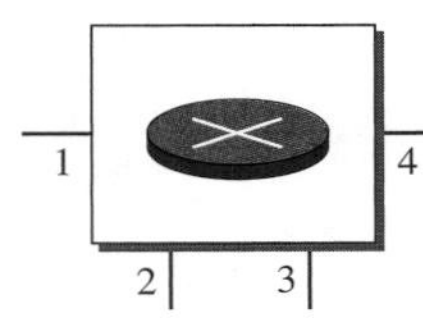

Find the output port for packets with the following destination addresses:

e. Packet 1: 7176
f. Packet 2: 1233
g. Packet 3: 8766
h. Packet 4: 9144

P8-6. The minimum number of columns in a datagram network is two; the minimum number of columns in a virtual-circuit network is four. Can you explain the reason? Is the difference related to the type of addresses carried in the packets of each network?

P8-7. We need to have a space-division switch with 1000 inputs and outputs. What is the total number of crosspoints in each of the following cases?

a. Using a single crossbar.

b. Using a multi-stage switch based on the Clos criteria.

P8-8. We need a three-stage space-division switch with $N = 100$. We use 10 crossbars at the first and third stages and 4 crossbars at the middle stage.

a. Draw the configuration diagram.

b. Calculate the total number of crosspoints.

c. Find the possible number of simultaneous connections.

d. Find the possible number of simultaneous connections if we use a single crossbar (100×100).

e. Find the blocking factor, the ratio of the number of connections in part c and in part d.

P8-9. A path in a digital circuit-switched network has a data rate of 1 Mbps. The exchange of 1000 bits is required for the setup and teardown phases. The distance between two parties is 5000 km. Answer the following questions if the propagation speed is 2×10^8 m:

a. What is the total delay if 1000 bits of data are exchanged during the data-transfer phase?

b. What is the total delay if 100,000 bits of data are exchanged during the data-transfer phase?

c. What is the total delay if 1,000,000 bits of data are exchanged during the data-transfer phase?

d. Find the delay per 1000 bits of data for each of the above cases and compare them. What can you infer?

P8-10. It is obvious that a router or a switch needs to search to find information in the corresponding table. The searching in a routing table for a datagram network is based on the destination address; the searching in a switching table in a virtual-circuit network is based on the combination of incoming port and incoming VCI. Explain the reason and define how these tables must be ordered (sorted) based on these values.

P8-11. An entry in the switching table of a virtual-circuit network is normally created during the setup phase and deleted during the teardown phase. In other words, the entries in this type of network reflect the current connections, the activity in the network. In contrast, the entries in a routing table of a datagram network do not depend on the current connections; they show the configuration of the network and how any packet should be routed to a final destination. The entries may remain the same even if there is no activity in the network. The routing tables, however, are updated if there are changes in the network. Can you explain the reason for these two different characteristics? Can we say that a virtual-circuit is a *connection-oriented* network and a datagram network is a *connectionless* network because of the above characteristics?

P8-12. We mentioned that two types of networks, datagram and virtual-circuit, need a routing or switching table to find the output port from which the information belonging to a destination should be sent out, but a circuit-switched network has no need for such a table. Give the reason for this difference.

P8-13. Consider an $n \times k$ crossbar switch with n inputs and k outputs.

a. Can we say that the switch acts as a multiplexer if $n > k$?

b. Can we say that the switch acts as a demultiplexer if $n < k$?

P8-14. Five equal-size datagrams belonging to the same message leave for the destination one after another. However, they travel through different paths as shown in Table 8.2.

Table 8.2 *P8-14*

Datagram	*Path Length*	*Visited Switches*
1	3200 km	1, 3, 5
2	11,700 km	1, 2, 5
3	12,200 km	1, 2, 3, 5
4	10,200 km	1, 4, 5
5	10,700 km	1, 4, 3, 5

We assume that the delay for each switch (including waiting and processing) is 3, 10, 20, 7, and 20 ms respectively. Assuming that the propagation speed is 2×10^8 m, find the order the datagrams arrive at the destination and the delay for each. Ignore any other delays in transmission.

P8-15. Repeat Problem 8-2 if we use 6 crossbars at the middle stage.

P8-16. Redesign the configuration of Problem 8-2 using the Clos criteria.

8.7 SIMULATION EXPERIMENTS

8.7.1 Applets

We have created some Java applets to show some of the main concepts discussed in this chapter. It is strongly recommended that the students activate these applets on the book website and carefully examine the protocols in action.

PART 

Data-Link Layer

In the third part of the book, we discuss the data-link layer in nine chapters. General topics are covered in Chapters 9 to 12. Wired networks are covered in Chapters 13 and 14. Wireless networks are covered in Chapters 15 and 16. Finally, we show how to connect LANs in Chapter 17.

CHAPTER 9

Introduction to Data-Link Layer

The TCP/IP protocol suite does not define any protocol in the data-link layer or physical layer. These two layers are territories of networks that when connected make up the Internet. These networks, wired or wireless, provide services to the upper three layers of the TCP/IP suite. This may give us a clue that there are several standard protocols in the market today. For this reason, we discuss the data-link layer in several chapters. This chapter is an introduction that gives the general idea and common issues in the data-link layer that relate to all networks.

- The first section introduces the data-link layer. It starts with defining the concept of links and nodes. The section then lists and briefly describes the services provided by the data-link layer. It next defines two categories of links: point-to-point and broadcast links. The section finally defines two sublayers at the data-link layer that will be elaborated on in the next few chapters.
- The second section discusses link-layer addressing. It first explains the rationale behind the existence of an addressing mechanism at the data-link layer. It then describes three types of link-layer addresses to be found in some link-layer protocols. The section discusses the Address Resolution Protocol (ARP), which maps the addresses at the network layer to addresses at the data-link layer. This protocol helps a packet at the network layer find the link-layer address of the next node for delivery of the frame that encapsulates the packet. To show how the network layer helps us to find the data-link-layer addresses, a long example is included in this section that shows what happens at each node when a packet is travelling through the Internet.

9.1 INTRODUCTION

The Internet is a combination of networks glued together by connecting devices (routers or switches). If a packet is to travel from a host to another host, it needs to pass through these networks. Figure 9.1 shows the same scenario we discussed in Chapter 3, but we are now interested in communication at the data-link layer. Communication at the data-link layer is made up of five separate logical connections between the data-link layers in the path.

Figure 9.1 *Communication at the data-link layer*

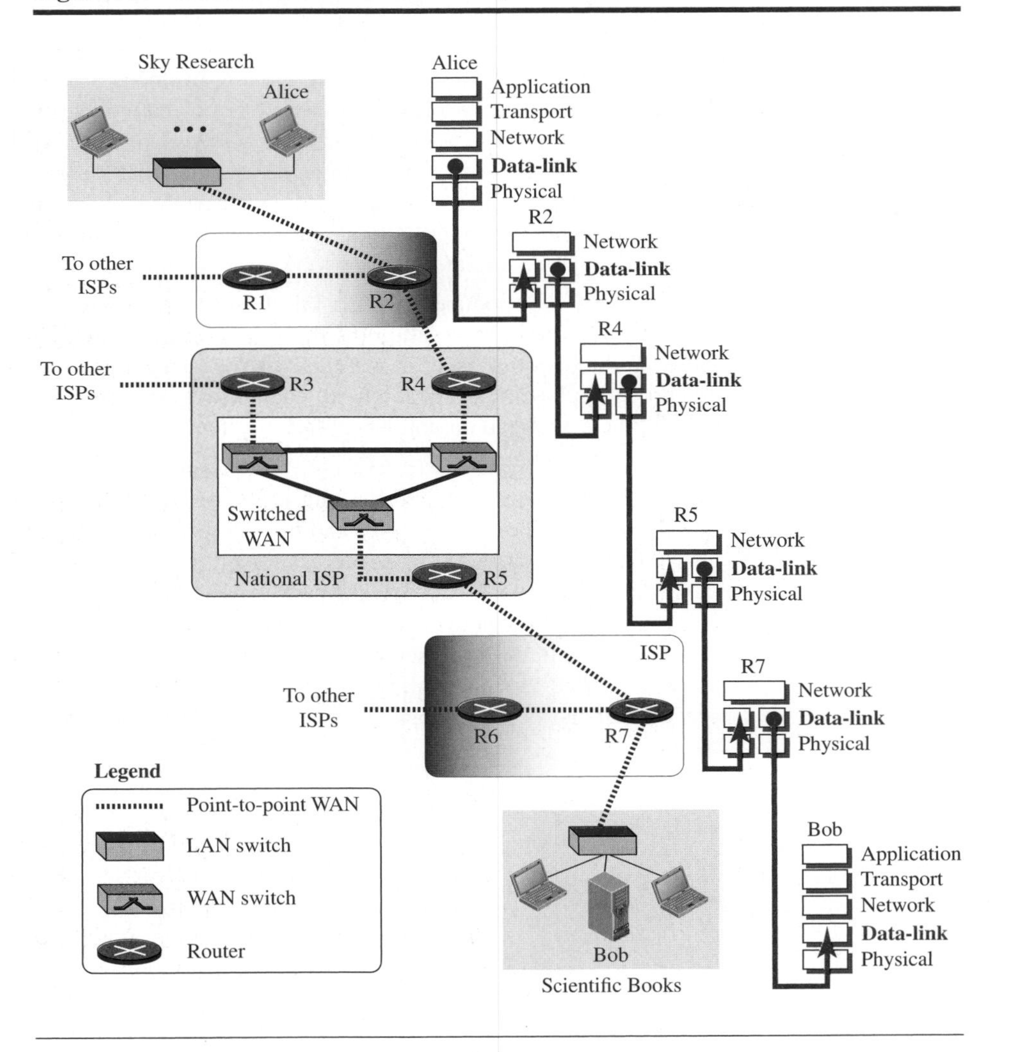

The data-link layer at Alice's computer communicates with the data-link layer at router R2. The data-link layer at router R2 communicates with the data-link layer at router R4, and so on. Finally, the data-link layer at router R7 communicates with the data-link layer at Bob's computer. Only one data-link layer is involved at the source or the destination, but two data-link layers are involved at each router. The reason is that Alice's and Bob's computers are each connected to a single network, but each router takes input from one network and sends output to another network. Note that although switches are also involved in the data-link-layer communication, for simplicity we have not shown them in the figure.

9.1.1 Nodes and Links

Communication at the data-link layer is node-to-node. A data unit from one point in the Internet needs to pass through many networks (LANs and WANs) to reach another point. Theses LANs and WANs are connected by routers. It is customary to refer to the two end hosts and the routers as ***nodes*** and the networks in between as ***links.*** Figure 9.2 is a simple representation of links and nodes when the path of the data unit is only six nodes.

Figure 9.2 *Nodes and Links*

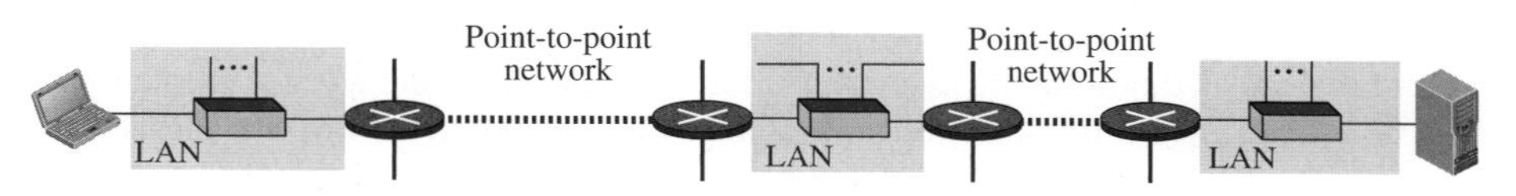

a. A small part of the Internet

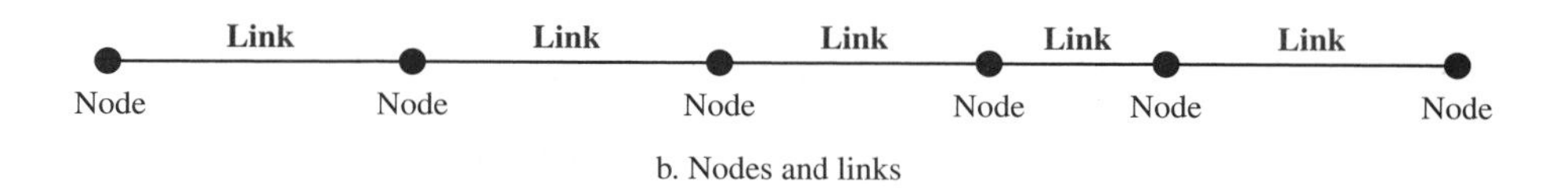

b. Nodes and links

The first node is the source host; the last node is the destination host. The other four nodes are four routers. The first, the third, and the fifth links represent the three LANs; the second and the fourth links represent the two WANs.

9.1.2 Services

The data-link layer is located between the physical and the network layers. The data-link layer provides services to the network layer; it receives services from the physical layer. Let us discuss services provided by the data-link layer.

The duty scope of the data-link layer is node-to-node. When a packet is travelling in the Internet, the data-link layer of a node (host or router) is responsible for delivering a datagram to the next node in the path. For this purpose, the data-link layer of the sending node needs to encapsulate the datagram received from the network in a frame, and the data-link layer of the receiving node needs to decapsulate the datagram from the frame. In other words, the data-link layer of the source host needs only to encapsulate, the data-link layer of the destination host needs to decapsulate, but each intermediate node needs to both encapsulate and decapsulate. One may ask why we need encapsulation and decapsulation at each intermediate node. The reason is that each link may be using a different protocol with a different frame format. Even if one link and the next are using the same protocol, encapsulation and decapsulation are needed because the link-layer addresses are normally different. An analogy may help in this case. Assume a person needs to travel from her home to her friend's home in another city. The traveller can use three transportation tools. She can take a taxi to go to the train station in her own city, then travel on the train from her own city to the city where her friend lives, and finally reach her friend's home using another taxi. Here we have a source node, a destination node, and two intermediate nodes. The traveller needs to get into the taxi at the source node, get out of the taxi and get into the train at the first intermediate node (train station in the city where she lives), get out of the train and get into another taxi at the second intermediate node (train station in the city where her friend lives), and finally get out of the taxi when she arrives at her destination. A kind of encapsulation occurs at the source node, encapsulation and decapsulation occur at the intermediate nodes, and decapsulation occurs at the destination node. Our traveller is the same, but she uses three transporting tools to reach the destination.

Figure 9.3 shows the encapsulation and decapsulation at the data-link layer. For simplicity, we have assumed that we have only one router between the source and destination. The datagram received by the data-link layer of the source host is encapsulated in a frame. The frame is logically transported from the source host to the router. The frame is decapsulated at the data-link layer of the router and encapsulated at another frame. The new frame is logically transported from the router to the destination host. Note that, although we have shown only two data-link layers at the router, the router actually has three data-link layers because it is connected to three physical links.

Figure 9.3 *A communication with only three nodes*

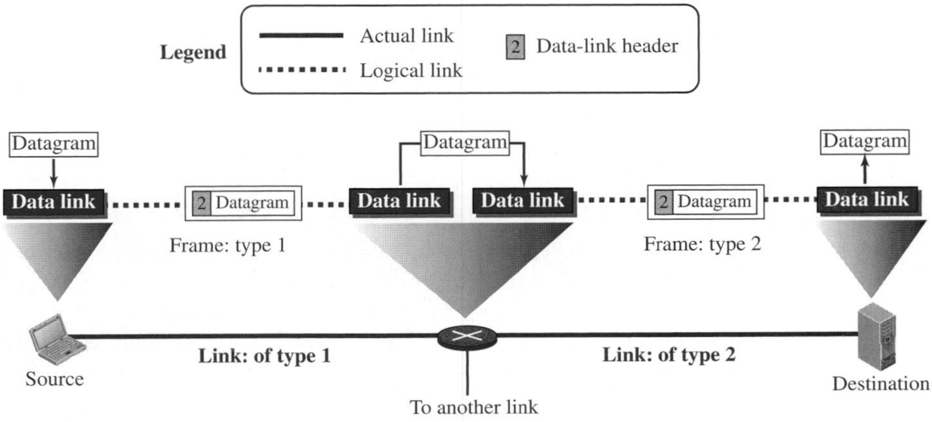

With the contents of the above figure in mind, we can list the services provided by a data-link layer as shown below.

Framing

Definitely, the first service provided by the data-link layer is **framing**. The data-link layer at each node needs to encapsulate the datagram (packet received from the network layer) in a **frame** before sending it to the next node. The node also needs to decapsulate the datagram from the frame received on the logical channel. Although we have shown only a header for a frame, we will see in future chapters that a frame may have both a header and a trailer. Different data-link layers have different formats for framing.

A packet at the data-link layer is normally called a *frame*.

Flow Control

Whenever we have a producer and a consumer, we need to think about flow control. If the producer produces items that cannot be consumed, accumulation of items occurs. The sending data-link layer at the end of a link is a producer of frames; the receiving data-link layer at the other end of a link is a consumer. If the rate of produced frames is higher than the rate of consumed frames, frames at the receiving end need to be buffered while waiting to be consumed (processed). Definitely, we cannot have an unlimited buffer size at the receiving side. We have two choices. The first choice is to let the receiving data-link layer drop the frames if its buffer is full. The second choice is to let the receiving data-link layer send a feedback to the sending data-link layer to ask it to stop or slow down. Different data-link-layer protocols use different strategies for flow control. Since flow control also occurs at the transport layer, with a higher degree of importance, we discuss this issue in Chapter 23 when we talk about the transport layer.

Error Control

At the sending node, a frame in a data-link layer needs to be changed to bits, transformed to electromagnetic signals, and transmitted through the transmission media. At the receiving node, electromagnetic signals are received, transformed to bits, and put together to create a frame. Since electromagnetic signals are susceptible to error, a frame is susceptible to error. The error needs first to be detected. After detection, it needs to be either corrected at the receiver node or discarded and retransmitted by the sending node. Since error detection and correction is an issue in every layer (node-to-node or host-to-host), we have dedicated all of Chapter 10 to this issue.

Congestion Control

Although a link may be congested with frames, which may result in frame loss, most data-link-layer protocols do not directly use a congestion control to alleviate congestion, although some wide-area networks do. In general, congestion control is considered an issue in the network layer or the transport layer because of its end-to-end nature. We will discuss congestion control in the network layer and the transport layer in later chapters.

9.1.3 Two Categories of Links

Although two nodes are physically connected by a transmission medium such as cable or air, we need to remember that the data-link layer controls how the medium is used. We can have a data-link layer that uses the whole capacity of the medium; we can also have a data-link layer that uses only part of the capacity of the link. In other words, we can have a *point-to-point link* or a *broadcast link*. In a point-to-point link, the link is dedicated to the two devices; in a broadcast link, the link is shared between several pairs of devices. For example, when two friends use the traditional home phones to chat, they are using a point-to-point link; when the same two friends use their cellular phones, they are using a broadcast link (the air is shared among many cell phone users).

9.1.4 Two Sublayers

To better understand the functionality of and the services provided by the link layer, we can divide the data-link layer into two sublayers: **data link control (DLC)** and **media access control (MAC).** This is not unusual because, as we will see in later chapters, LAN protocols actually use the same strategy. The data link control sublayer deals with all issues common to both point-to-point and broadcast links; the media access control sublayer deals only with issues specific to broadcast links. In other words, we separate these two types of links at the data-link layer, as shown in Figure 9.4.

Figure 9.4 *Dividing the data-link layer into two sublayers*

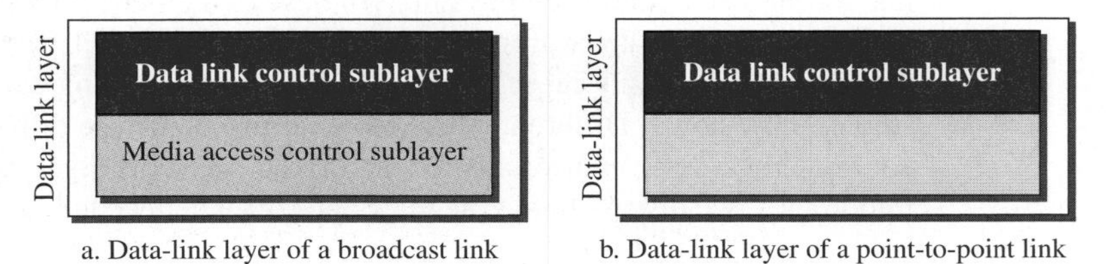

a. Data-link layer of a broadcast link   b. Data-link layer of a point-to-point link

We discuss the DLC and MAC sublayers later, each in a separate chapter. In addition, we discuss the issue of error detection and correction, a duty of the data-link and other layers, also in a separate chapter.

9.2 LINK-LAYER ADDRESSING

The next issue we need to discuss about the data-link layer is the link-layer addresses. In Chapter 18, we will discuss IP addresses as the identifiers at the network layer that define the exact points in the Internet where the source and destination hosts are connected. However, in a connectionless internetwork such as the Internet we cannot make a datagram reach its destination using only IP addresses. The reason is that each datagram in the Internet, from the same source host to the same destination host, may take a different path. The source and destination IP addresses define the two ends but cannot define which links the datagram should pass through.

We need to remember that the IP addresses in a datagram should not be changed. If the destination IP address in a datagram changes, the packet never reaches its destination; if the source IP address in a datagram changes, the destination host or a router can never communicate with the source if a response needs to be sent back or an error needs to be reported back to the source (see ICMP in Chapter 19).

The above discussion shows that we need another addressing mechanism in a connectionless internetwork: the link-layer addresses of the two nodes. A *link-layer address* is sometimes called a *link address*, sometimes a *physical address*, and sometimes a *MAC address*. We use these terms interchangeably in this book.

Since a link is controlled at the data-link layer, the addresses need to belong to the data-link layer. When a datagram passes from the network layer to the data-link layer, the datagram will be encapsulated in a frame and two data-link addresses are added to the frame header. These two addresses are changed every time the frame moves from one link to another. Figure 9.5 demonstrates the concept in a small internet.

Figure 9.5 *IP addresses and link-layer addresses in a small internet*

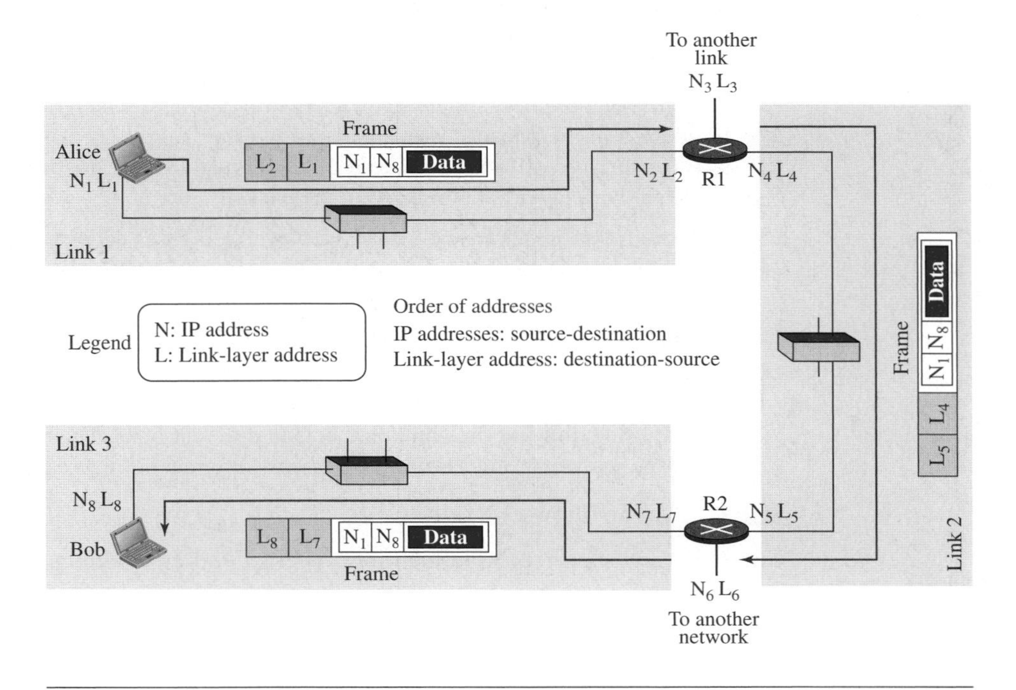

In the internet in Figure 9.5, we have three links and two routers. We also have shown only two hosts: Alice (source) and Bob (destination). For each host, we have shown two addresses, the IP addresses (N) and the link-layer addresses (L). Note that a router has as many pairs of addresses as the number of links the router is connected to. We have shown three frames, one in each link. Each frame carries the same datagram with the same source and destination addresses (**N1** and **N8**), but the link-layer addresses of the frame change from link to link. In link 1, the link-layer addresses are L_1 and L_2. In link 2, they are L_4 and L_5. In link 3, they are L_7 and L_8. Note that the IP addresses and the link-layer addresses are not in the same order. For IP addresses, the source address comes before the destination address; for link-layer addresses, the destination address comes before the source. The datagrams and frames are designed in this way, and we follow the design. We may raise several questions:

- If the IP address of a router does not appear in any datagram sent from a source to a destination, why do we need to assign IP addresses to routers? The answer is that in some protocols a router may act as a sender or receiver of a datagram. For example, in routing protocols we will discuss in Chapters 20 and 21, a router is a sender or a receiver of a message. The communications in these protocols are between routers.
- Why do we need more than one IP address in a router, one for each interface? The answer is that an interface is a connection of a router to a link. We will see that an IP address defines a point in the Internet at which a device is connected. A router with *n* interfaces is connected to the Internet at *n* points. This is the situation of a house at the corner of a street with two gates; each gate has the address related to the corresponding street.
- How are the source and destination IP addresses in a packet determined? The answer is that the host should know its own IP address, which becomes the source IP address in the packet. As we will discuss in Chapter 26, the application layer uses the services of DNS to find the destination address of the packet and passes it to the network layer to be inserted in the packet.
- How are the source and destination link-layer addresses determined for each link? Again, each hop (router or host) should know its own link-layer address, as we discuss later in the chapter. The destination link-layer address is determined by using the Address Resolution Protocol, which we discuss shortly.
- What is the size of link-layer addresses? The answer is that it depends on the protocol used by the link. Although we have only one IP protocol for the whole Internet, we may be using different data-link protocols in different links. This means that we can define the size of the address when we discuss different link-layer protocols.

9.2.1 Three Types of addresses

Some link-layer protocols define three types of addresses: unicast, multicast, and broadcast.

Unicast Address

Each host or each interface of a router is assigned a unicast address. Unicasting means one-to-one communication. A frame with a unicast address destination is destined only for one entity in the link.

Example 9.1

As we will see in Chapter 13, the unicast link-layer addresses in the most common LAN, Ethernet, are 48 bits (six bytes) that are presented as 12 hexadecimal digits separated by colons; for example, the following is a link-layer address of a computer.

A3:34:45:11:92:F1

Multicast Address

Some link-layer protocols define multicast addresses. Multicasting means one-to-many communication. However, the jurisdiction is local (inside the link).

Example 9.2

As we will see in Chapter 13, the multicast link-layer addresses in the most common LAN, Ethernet, are 48 bits (six bytes) that are presented as 12 hexadecimal digits separated by colons. The second digit, however, needs to be an even number in hexadecimal. The following shows a multicast address:

A2:34:45:11:92:F1

Broadcast Address

Some link-layer protocols define a broadcast address. Broadcasting means one-to-all communication. A frame with a destination broadcast address is sent to all entities in the link.

Example 9.3

As we will see in Chapter 13, the broadcast link-layer addresses in the most common LAN, Ethernet, are 48 bits, all 1s, that are presented as 12 hexadecimal digits separated by colons. The following shows a broadcast address:

FF:FF:FF:FF:FF:FF

9.2.2 Address Resolution Protocol (ARP)

Anytime a node has an IP datagram to send to another node in a link, it has the IP address of the receiving node. The source host knows the IP address of the default router. Each router except the last one in the path gets the IP address of the next router by using its forwarding table. The last router knows the IP address of the destination host. However, the IP address of the next node is not helpful in moving a frame through a link; we need the link-layer address of the next node. This is the time when the **Address Resolution Protocol (ARP)** becomes helpful. The ARP protocol is one of the auxiliary protocols defined in the network layer, as shown in Figure 9.6. It belongs to the network layer, but we discuss it in this chapter because it maps an IP address to a logical-link address. ARP accepts an IP address from the IP protocol, maps the address to the corresponding link-layer address, and passes it to the data-link layer.

Figure 9.6 *Position of ARP in TCP/IP protocol suite*

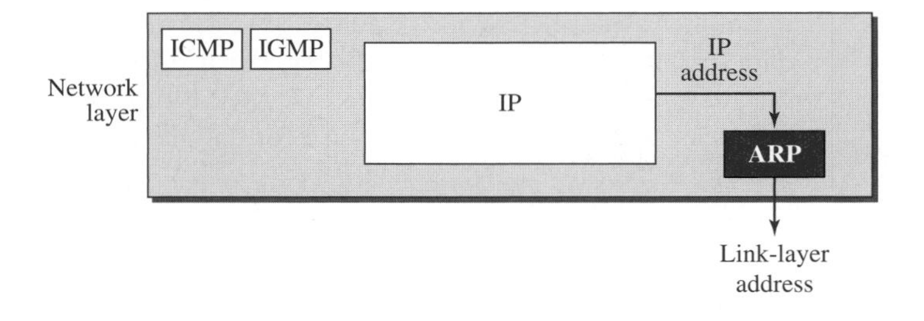

Anytime a host or a router needs to find the link-layer address of another host or router in its network, it sends an ARP request packet. The packet includes the link-layer and IP addresses of the sender and the IP address of the receiver. Because the sender does not know the link-layer address of the receiver, the query is broadcast over the link using the link-layer broadcast address, which we discuss for each protocol later (see Figure 9.7).

Figure 9.7 *ARP operation*

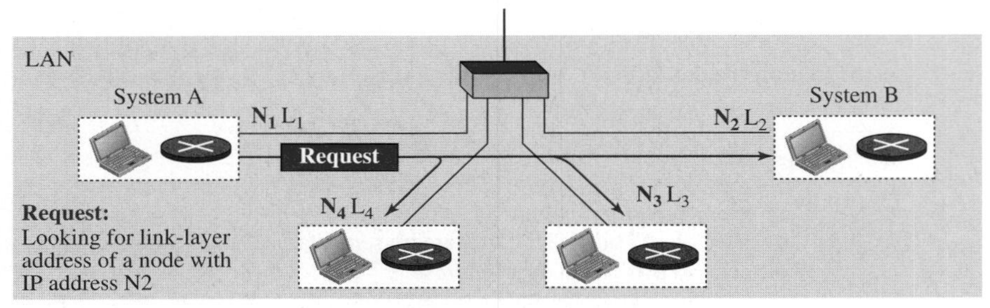

a. ARP request is broadcast

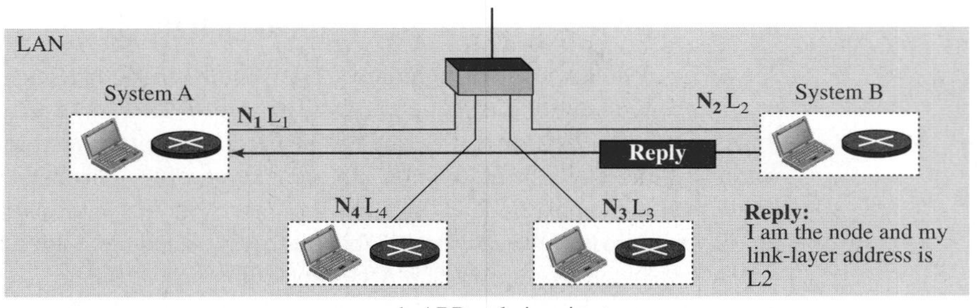

b. ARP reply is unicast

Every host or router on the network receives and processes the ARP request packet, but only the intended recipient recognizes its IP address and sends back an ARP response packet. The response packet contains the recipient's IP and link-layer addresses. The packet is unicast directly to the node that sent the request packet.

In Figure 9.7a, the system on the left (A) has a packet that needs to be delivered to another system (B) with IP address **N2**. System A needs to pass the packet to its data-link layer for the actual delivery, but it does not know the physical address of the recipient. It uses the services of ARP by asking the ARP protocol to send a broadcast ARP request packet to ask for the physical address of a system with an IP address of **N2**.

This packet is received by every system on the physical network, but only system B will answer it, as shown in Figure 9.7b. System B sends an ARP reply packet that includes its physical address. Now system A can send all the packets it has for this destination using the physical address it received.

Caching

A question that is often asked is this: If system A can broadcast a frame to find the link-layer address of system B, why can't system A send the datagram for system B using a broadcast frame? In other words, instead of sending one broadcast frame (ARP request), one unicast frame (ARP response), and another unicast frame (for sending the datagram), system A can encapsulate the datagram and send it to the network. System B receives it and keep it; other systems discard it.

To answer the question, we need to think about the efficiency. It is probable that system A has more than one datagram to send to system B in a short period of time. For example, if system B is supposed to receive a long e-mail or a long file, the data do not fit in one datagram.

Let us assume that there are 20 systems connected to the network (link): system A, system B, and 18 other systems. We also assume that system A has 10 datagrams to send to system B in one second.

a. Without using ARP, system A needs to send 10 broadcast frames. Each of the 18 other systems need to receive the frames, decapsulate the frames, remove the datagram and pass it to their network-layer to find out the datagrams do not belong to them.This means processing and discarding 180 broadcast frames.

b. Using ARP, system A needs to send only one broadcast frame. Each of the 18 other systems need to receive the frames, decapsulate the frames, remove the ARP message and pass the message to their ARP protocol to find that the frame must be discarded. This means processing and discarding only 18 (instead of 180) broadcast frames. After system B responds with its own data-link address, system A can store the link-layer address in its cache memory. The rest of the nine frames are only unicast. Since processing broadcast frames is expensive (time consuming), the first method is preferable.

Packet Format

Figure 9.8 shows the format of an ARP packet. The names of the fields are self-explanatory. The *hardware type* field defines the type of the link-layer protocol; Ethernet is given the type 1. The *protocol type* field defines the network-layer protocol: IPv4 protocol is $(0800)_{16}$. The source hardware and source protocol addresses are variable-length fields defining the link-layer and network-layer addresses of the sender. The destination hardware address and destination protocol address fields define the receiver link-layer and network-layer addresses. An ARP packet is encapsulated directly into a data-link frame. The frame needs to have a field to show that the payload belongs to the ARP and not to the network-layer datagram.

Example 9.4

A host with IP address **N1** and MAC address **L1** has a packet to send to another host with IP address **N2** and physical address **L2** (which is unknown to the first host). The two hosts are on the same network. Figure 9.9 shows the ARP request and response messages.

Figure 9.8 *ARP packet*

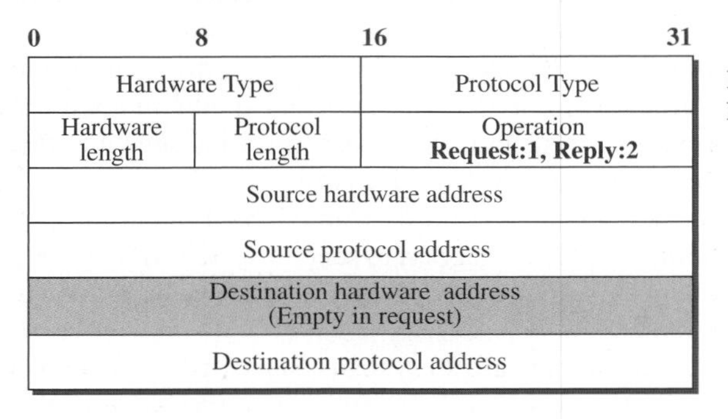

Figure 9.9 *Example 9.4*

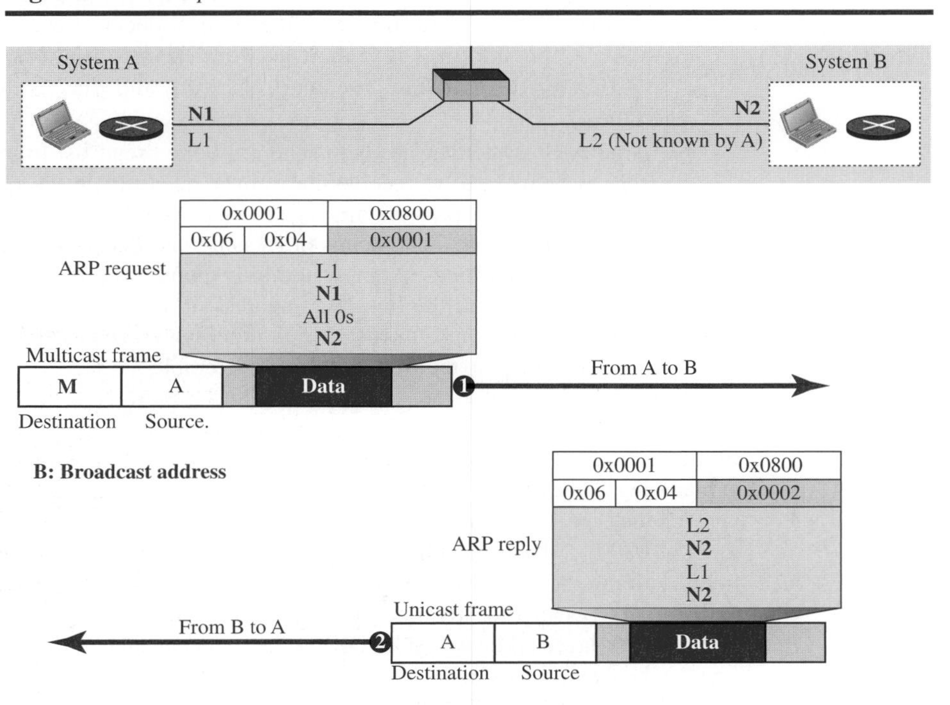

9.2.3 An Example of Communication

To show how communication is done at the data-link layer and how link-layer addresses are found, let us go through a simple example. Assume Alice needs to send a datagram to Bob, who is three nodes away in the Internet. How Alice finds the network-layer address of Bob is what we discover in Chapter 26 when we discuss DNS. For the moment, assume that Alice knows the network-layer (IP) address of Bob. In other words, Alice's host is given the data to be sent, the IP address of Bob, and the

IP address of Alice's host (each host needs to know its IP address). Figure 9.10 shows the part of the internet for our example.

Figure 9.10 *The internet for our example*

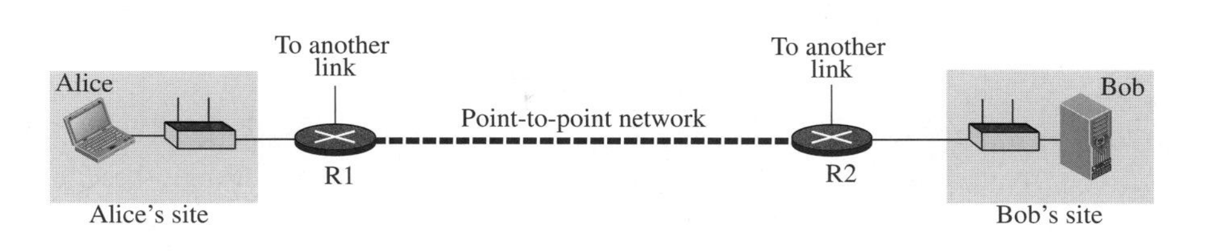

Activities at Alice's Site

We will use symbolic addresses to make the figures more readable. Figure 9.11 shows what happens at Alice's site.

Figure 9.11 *Flow of packets at Alice's computer*

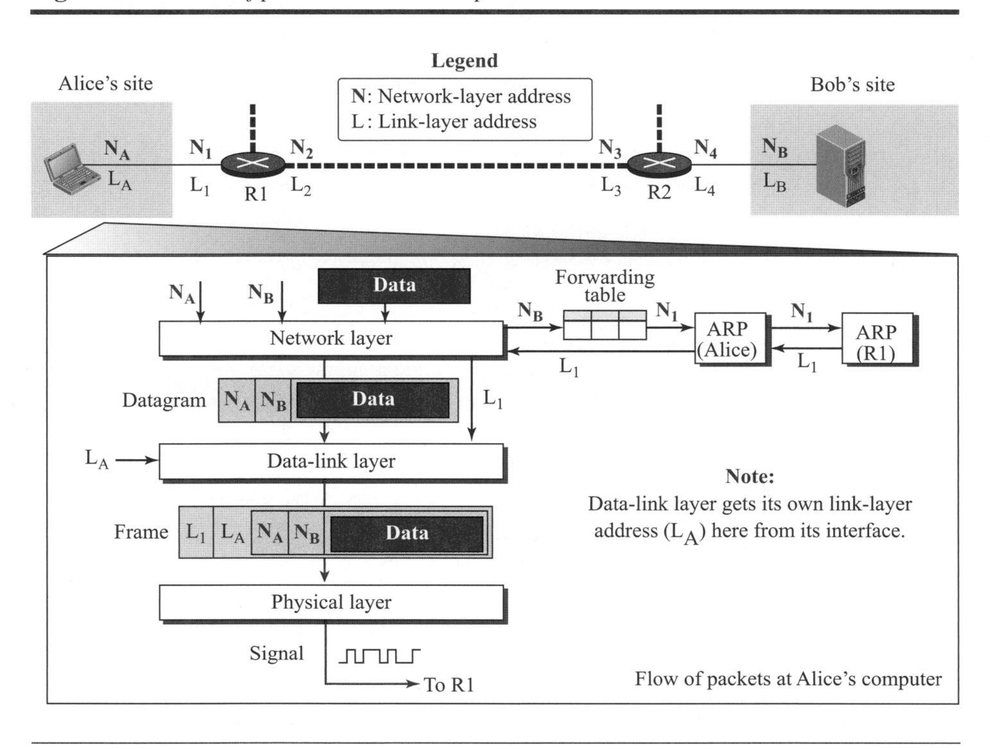

The network layer knows it's given $\mathbf{N_A}$, $\mathbf{N_B}$, and the packet, but it needs to find the link-layer address of the next node. The network layer consults its routing table and tries to find which router is next (the default router in this case) for the destination $\mathbf{N_B}$. As we will discuss in Chapter 18, the routing table gives $\mathbf{N_1}$, but the network layer needs to find the link-layer address of router R1. It uses its ARP to find the link-layer address $\mathbf{L_1}$. The network layer can now pass the datagram with the link-layer address to the data-link layer.

The data-link layer knows its own link-layer address, $\mathbf{L_A}$. It creates the frame and passes it to the physical layer, where the address is converted to signals and sent through the media.

Activities at Router R1

Now let us see what happens at Router R1. Router R1, as we know, has only three lower layers. The packet received needs to go up through these three layers and come down. Figure 9.12 shows the activities.

Figure 9.12 *Flow of activities at router R1*

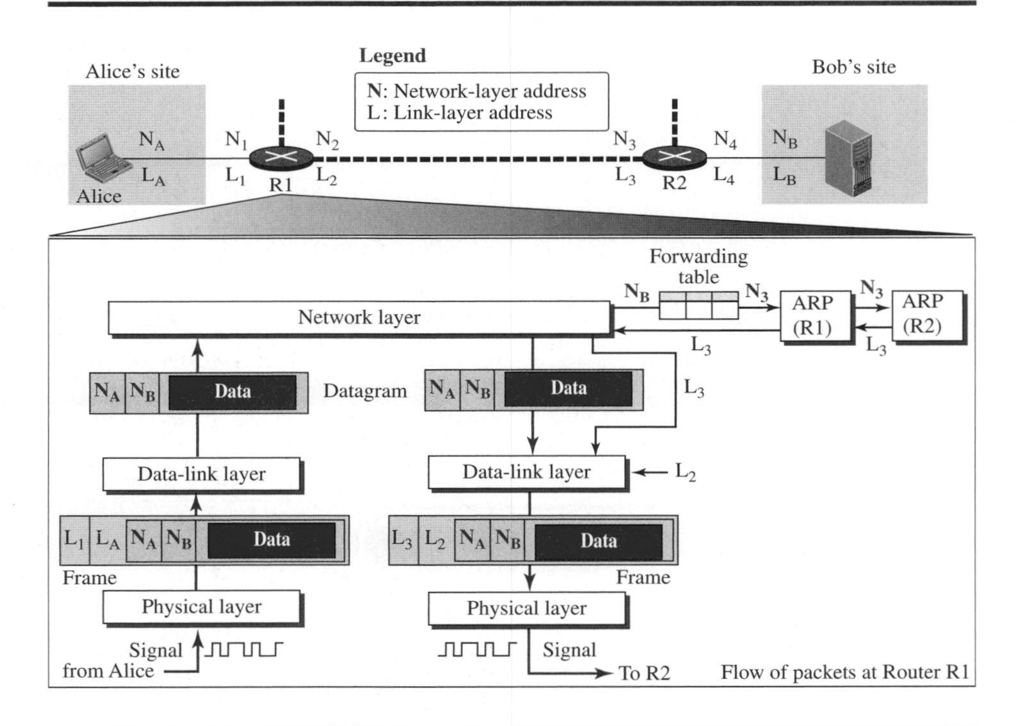

At arrival, the physical layer of the left link creates the frame and passes it to the data-link layer. The data-link layer decapsulates the datagram and passes it to the network layer. The network layer examines the network-layer address of the datagram and finds that the datagram needs to be delivered to the device with IP address $\mathbf{N_B}$. The network layer consults its routing table to find out which is the next node (router) in the path to $\mathbf{N_B}$. The forwarding table returns $\mathbf{N_3}$. The IP address of router R2 is in the same link with R1. The network layer now uses the ARP to find the link-layer address of this router, which comes up as $\mathbf{L_3}$. The network layer passes the datagram and $\mathbf{L_3}$ to the data-link layer belonging to the link at the right side. The link layer encapsulates the datagram, adds **L3** and **L2** (its own link-layer address), and passes the frame to the physical layer. The physical layer encodes the bits to signals and sends them through the medium to R2.

Activities at Router R2

Activities at router R2 are almost the same as in R1, as shown in Figure 9.13.

Figure 9.13 *Activities at router R2.*

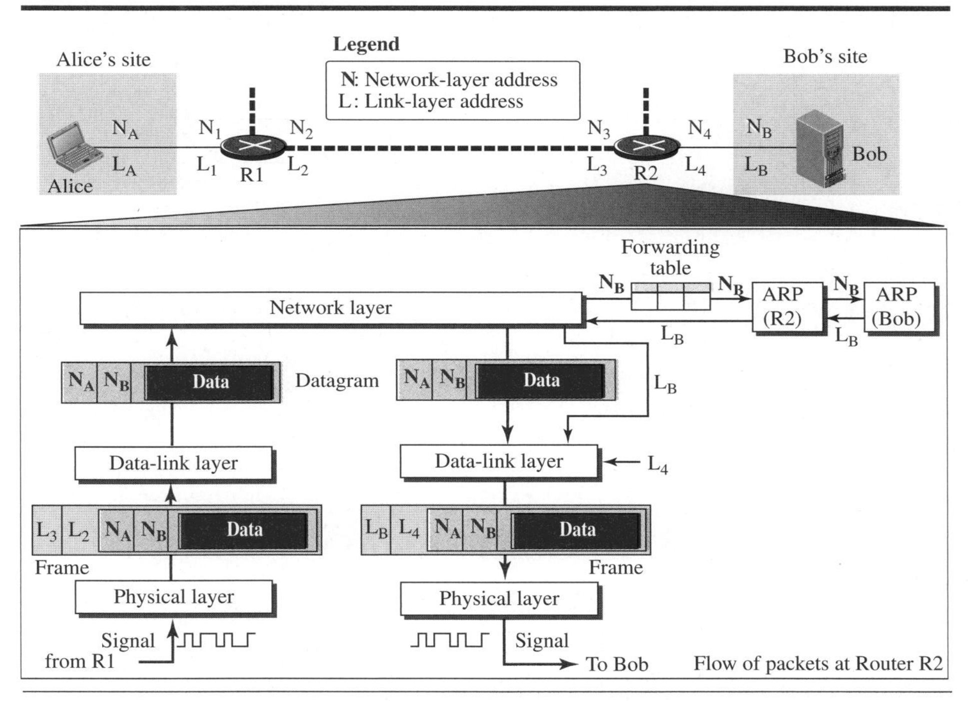

Activities at Bob's Site

Now let us see what happens at Bob's site. Figure 9.14 shows how the signals at Bob's site are changed to a message. At Bob's site there are no more addresses or mapping needed. The signal received from the link is changed to a frame. The frame is passed to the data-link layer, which decapsulates the datagram and passes it to the network layer. The network layer decapsulates the message and passes it to the transport layer.

Changes in Addresses

This example shows that the source and destination network-layer addresses, NA and NB, have not been changed during the whole journey. However, all four network-layer addresses of routers R1 and R2 (N1, N2, N3, and N4) are needed to transfer a datagram from Alice's computer to Bob's computer.

Figure 9.14 *Activities at Bob's site*

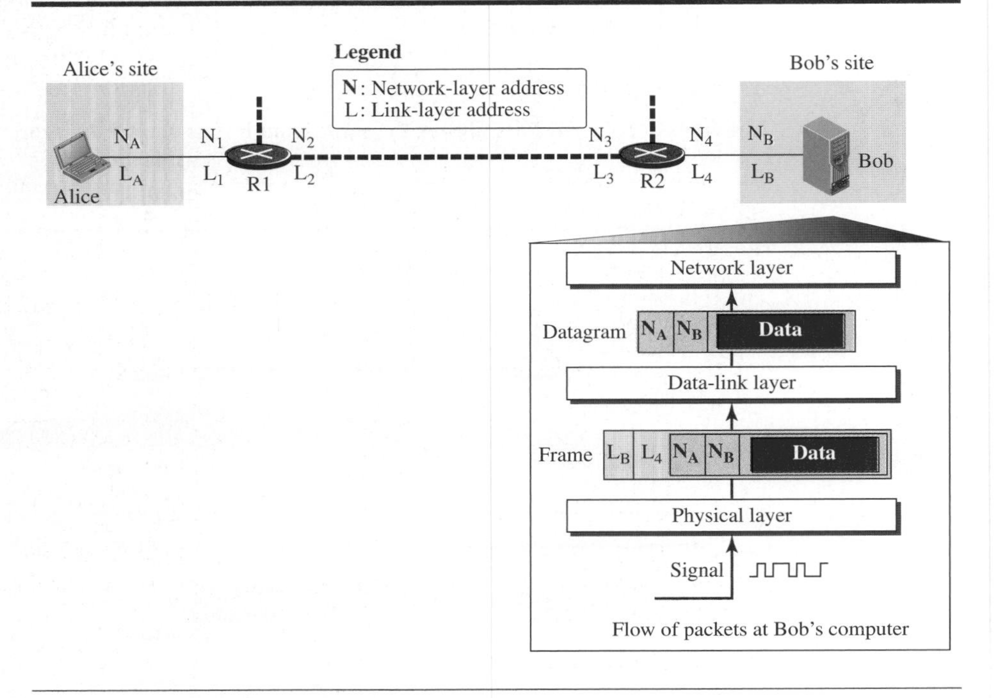

9.3 END-CHAPTER MATERIALS

9.3.1 Recommended Reading

For more details about subjects discussed in this chapter, we recommend the following books. The items in brackets [...] refer to the reference list at the end of the text.

Books

Several books discuss link-layer issues. Among them we recommend [Ham 80], [Zar 02], [Ror 96], [Tan 03], [GW 04], [For 03], [KMK 04], [Sta 04], [Kes 02], [PD 03], [Kei 02], [Spu 00], [KCK 98], [Sau 98], [Izz 00], [Per 00], and [WV 00].

9.3.2 Key Terms

Address Resolution Protocol (ARP)
data link control (DLC)
frame
framing
links
media access control (MAC)
nodes

9.3.3 Summary

The Internet is made of many hosts, networks, and connecting devices such as routers. The hosts and connecting devices are referred to as *nodes;* the networks are referred to as *links*. A path in the Internet from a source host to a destination host is a set of nodes and links through which a packet should travel.

The data-link layer is responsible for the creation and delivery of a frame to another node, along the link. It is responsible for packetizing (framing), flow control, error control, and congestion control along the link. Two data-link layers at the two ends of a link coordinate to deliver a frame from one node to the next.

As with any delivery between a source and destination in which there are many paths, we need two types of addressing. The end-to-end addressing defines the source and destination; the link-layer addressing defines the addresses of the nodes that the packet should pass through. To avoid including the link-layer addresses of all of these nodes in the frame, the Address Resolution Protocol (ARP) was devised to map an IP address to its corresponding link-layer address. When a packet is at one node ready to be sent to the next, the forwarding table finds the IP address of the next node and ARP find its link-layer address.

9.4 PRACTICE SET

9.4.1 Quizzes

A set of interactive quizzes for this chapter can be found on the book website. It is strongly recommended that the student take the quizzes to check his/her understanding of the materials before continuing with the practice set.

9.4.2 Questions

Q9-1. Why is it better not to change an end-to-end address from the source to the destination?

Q9-2. Assume we have an isolated link (not connected to any other link) such as a private network in a company. Do we still need addresses in both the network layer and the data-link layer? Explain.

Q9-3. Can two hosts in two different networks have the same link-layer address? Explain.

Q9-4. Distinguish between a point-to-point link and a broadcast link.

Q9-5. Why does a host or a router need to run the ARP program all of the time in the background?

Q9-6. When we talk about the broadcast address in a link, do we mean sending a message to all hosts and routers in the link or to all hosts and routers in the Internet? In other words, does a broadcast address have a local jurisdiction or a universal jurisdiction? Explain.

Q9-7. Distinguish between communication at the network layer and communication at the data-link layer.

Q9-8. In Figure 9.9, why is the destination hardware address of the frame from A to B a broadcast address?

Q9-9. What is the size of an ARP packet when the protocol is IPv4 and the hardware is Ethernet?

Q9-10. Is the size of the ARP packet fixed? Explain.

Q9-11. In Figure 9.9, how does system A know what the link-layer address of system B is when it receives the ARP reply?

Q9-12. How many IP addresses and how many link-layer addresses should a router have when it is connected to five links?

Q9-13. In Figure 9.9, why is the destination hardware address all 0s in the ARP request message?

Q9-14. Why does a router normally have more than one interface?

9.4.3 Problems

P9-1. Assume the network in Figure 9.7 does not support broadcasting. What do you suggest for sending the ARP request in this network?

P9-2. Assume we have an internet (a private small internet) in which all hosts are connected in a mesh topology. Do we need both network and data-link layers in this internet?

P9-3. Assume we have an internet (a private small internet) in which all hosts are connected in a mesh topology. Do we need routers in this internet? Explain.

P9-4. In Figure 9.7, do you think that system A should first check its cache for mapping from N2 to L2 before even broadcasting the ARP request?

P9-5. In Figure 9.7, assume system B is not running the ARP program. What would happen?

P9-6. In Figures 9.11 to 9.13, both the forwarding table and ARP are doing a kind of mapping. Show the difference between them by listing the input and output of mapping for a forwarding table and ARP.

P9-7. In Figure 9.5, assume Link 2 is broken. How can Alice communicate with Bob?

P9-8. Assume Alice is travelling from 2020 Main Street in Los Angeles to 1432 American Boulevard in Chicago. If she is travelling by air from Los Angeles Airport to Chicago Airport,

a. find the end-to-end addresses in this scenario.

b. find the link-layer addresses in this scenario.

P9-9. In the previous problem, assume Alice cannot find a direct flight from the Los Angeles to the Chicago. If she needs to change flights in Denver,

a. find the end-to-end addresses in this scenario.

b. find the link-layer addresses in this scenario.

P9-10. When we send a letter using the services provided by the post office, do we use an end-to-end address? Does the post office necessarily use an end-to-end address to deliver the mail? Explain.

P9-11. Is the current Internet using circuit-switching or packet-switching at the data-link layer? Explain.

P9-12. In Figure 9.5, show the process of frame change in routers R1 and R2.

P9-13. Explain why we do not need the router in Figure 9.15.

Figure 9.15 *Problem 9-13*

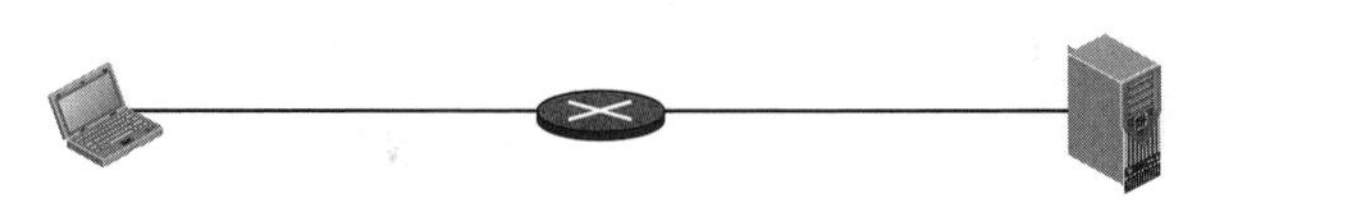

P9-14. Explain why we may need a router in Figure 9.16.

Figure 9.16 *Problem 9-14*

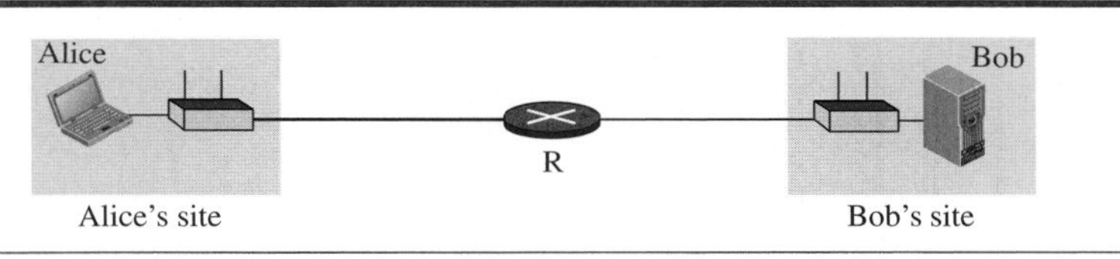

P9-15. Figure 9.7 shows a system as either a host or a router. What would be the actual entity (host or router) of system A and B in each of the following cases:

a. If the link is the first one in the path?

b. If the link is the middle one in the path?

c. If the link is the last one in the path?

d. If there is only one link in the path (local communication)?

CHAPTER 10

Error Detection and Correction

Networks must be able to transfer data from one device to another with acceptable accuracy. For most applications, a system must guarantee that the data received are identical to the data transmitted. Any time data are transmitted from one node to the next, they can become corrupted in passage. Many factors can alter one or more bits of a message. Some applications require a mechanism for detecting and correcting **errors.**

Some applications can tolerate a small level of error. For example, random errors in audio or video transmissions may be tolerable, but when we transfer text, we expect a very high level of accuracy.

At the data-link layer, if a frame is corrupted between the two nodes, it needs to be corrected before it continues its journey to other nodes. However, most link-layer protocols simply discard the frame and let the upper-layer protocols handle the retransmission of the frame. Some multimedia applications, however, try to correct the corrupted frame.

This chapter is divided into five sections.

- The first section introduces types of errors, the concept of redundancy, and distinguishes between error detection and correction.
- The second section discusses block coding. It shows how error can be detected using block coding and also introduces the concept of Hamming distance.
- The third section discusses cyclic codes. It discusses a subset of cyclic code, CRC, that is very common in the data-link layer. The section shows how CRC can be easily implemented in hardware and represented by polynomials.
- The fourth section discusses checksums. It shows how a checksum is calculated for a set of data words. It also gives some other approaches to traditional checksum.
- The fifth section discusses forward error correction. It shows how Hamming distance can also be used for this purpose. The section also describes cheaper methods to achieve the same goal, such as XORing of packets, interleaving chunks, or compounding high and low resolutions packets.

10.1 INTRODUCTION

Let us first discuss some issues related, directly or indirectly, to error detection and correction.

10.1.1 Types of Errors

Whenever bits flow from one point to another, they are subject to unpredictable changes because of **interference.** This interference can change the shape of the signal. The term ***single-bit error*** means that only 1 bit of a given data unit (such as a byte, character, or packet) is changed from 1 to 0 or from 0 to 1. The term ***burst error*** means that 2 or more bits in the data unit have changed from 1 to 0 or from 0 to 1. Figure 10.1 shows the effect of a single-bit and a burst error on a data unit.

Figure 10.1 *Single-bit and burst error*

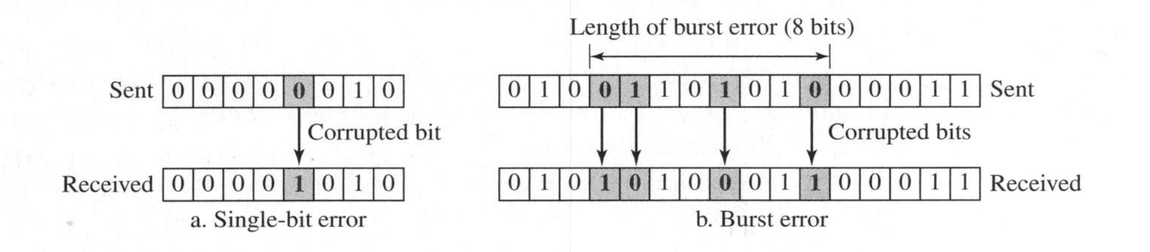

A burst error is more likely to occur than a single-bit error because the duration of the noise signal is normally longer than the duration of 1 bit, which means that when noise affects data, it affects a set of bits. The number of bits affected depends on the data rate and duration of noise. For example, if we are sending data at 1 kbps, a noise of 1/100 second can affect 10 bits; if we are sending data at 1 Mbps, the same noise can affect 10,000 bits.

10.1.2 Redundancy

The central concept in detecting or correcting errors is **redundancy**. To be able to detect or correct errors, we need to send some extra bits with our data. These redundant bits are added by the sender and removed by the receiver. Their presence allows the receiver to detect or correct corrupted bits.

10.1.3 Detection versus Correction

The correction of errors is more difficult than the detection. In **error detection**, we are only looking to see if any error has occurred. The answer is a simple yes or no. We are not even interested in the number of corrupted bits. A single-bit error is the same for us as a burst error. In **error correction**, we need to know the exact number of bits that are corrupted and, more importantly, their location in the message. The number of errors and the size of the message are important factors. If we need to correct a single error in an 8-bit data unit, we need to consider eight possible error locations; if we need to correct two errors in a data unit of the same size, we need to consider 28 (permutation of 8 by 2) possibilities. You can imagine the receiver's difficulty in finding 10 errors in a data unit of 1000 bits.

10.1.4 Coding

Redundancy is achieved through various coding schemes. The sender adds redundant bits through a process that creates a relationship between the redundant bits and the actual data bits. The receiver checks the relationships between the two sets of bits to detect errors. The ratio of redundant bits to data bits and the robustness of the process are important factors in any coding scheme.

We can divide coding schemes into two broad categories: **block coding** and **convolution coding**. In this book, we concentrate on block coding; convolution coding is more complex and beyond the scope of this book.

10.2 BLOCK CODING

In block coding, we divide our message into blocks, each of k bits, called ***datawords***. We add r redundant bits to each block to make the length $n = k + r$. The resulting n-bit blocks are called ***codewords***. How the extra r bits are chosen or calculated is something we will discuss later. For the moment, it is important to know that we have a set of datawords, each of size k, and a set of codewords, each of size of n. With k bits, we can create a combination of 2^k datawords; with n bits, we can create a combination of 2^n codewords. Since $n > k$, the number of possible codewords is larger than the number of possible datawords. The block coding process is one-to-one; the same dataword is always encoded as the same codeword. This means that we have $2^n - 2^k$ codewords that are not used. We call these codewords invalid or illegal. The trick in error detection is the existence of these invalid codes, as we discuss next. If the receiver receives an invalid codeword, this indicates that the data was corrupted during transmission.

10.2.1 Error Detection

How can errors be detected by using block coding? If the following two conditions are met, the receiver can detect a change in the original codeword.

1. The receiver has (or can find) a list of valid codewords.
2. The original codeword has changed to an invalid one.

Figure 10.2 shows the role of block coding in error detection. The sender creates codewords out of datawords by using a generator that applies the rules and procedures of encoding (discussed later). Each codeword sent to the receiver may change during transmission. If the received codeword is the same as one of the valid codewords, the word is accepted; the corresponding dataword is extracted for use. If the received codeword is not valid, it is discarded. However, if the codeword is corrupted during transmission but the received word still matches a valid codeword, the error remains undetected.

Figure 10.2 *Process of error detection in block coding*

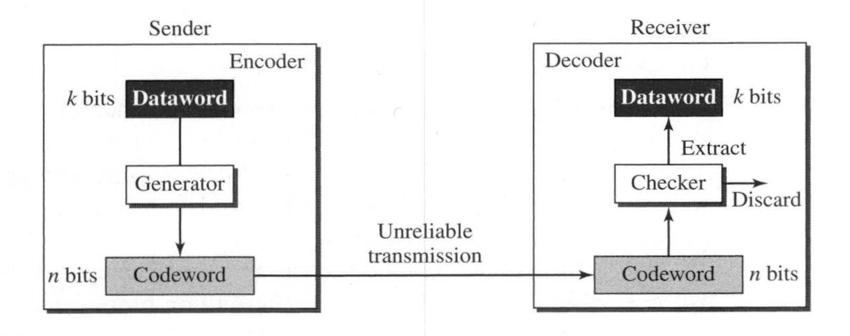

Example 10.1

Let us assume that $k = 2$ and $n = 3$. Table 10.1 shows the list of datawords and codewords. Later, we will see how to derive a codeword from a dataword.

Table 10.1 *A code for error detection in Example 10.1*

Dataword	*Codeword*	*Dataword*	*Codeword*
00	000	10	101
01	011	11	110

Assume the sender encodes the dataword 01 as 011 and sends it to the receiver. Consider the following cases:

1. The receiver receives 011. It is a valid codeword. The receiver extracts the dataword 01 from it.
2. The codeword is corrupted during transmission, and 111 is received (the leftmost bit is corrupted). This is not a valid codeword and is discarded.
3. The codeword is corrupted during transmission, and 000 is received (the right two bits are corrupted). This is a valid codeword. The receiver incorrectly extracts the dataword 00. Two corrupted bits have made the error undetectable.

An error-detecting code can detect only the types of errors for which it is designed; other types of errors may remain undetected.

Hamming Distance

One of the central concepts in coding for error control is the idea of the Hamming distance. The **Hamming distance** between two words (of the same size) is the number of differences between the corresponding bits. We show the Hamming distance between two words x and y as $d(x, y)$. We may wonder why Hamming distance is important for error detection. The reason is that the Hamming distance between the received codeword and the sent codeword is the number of bits that are corrupted during transmission. For example, if the codeword 00000 is sent and 01101 is received, 3 bits are in error and the Hamming distance between the two is $d(00000, 01101) = 3$. In other words, if the Hamming distance between the sent and the received codeword is not zero, the codeword has been corrupted during transmission.

The Hamming distance can easily be found if we apply the XOR operation (⊕) on the two words and count the number of 1s in the result. Note that the Hamming distance is a value greater than or equal to zero.

The Hamming distance between two words is the number of differences between corresponding bits.

Example 10.2

Let us find the Hamming distance between two pairs of words.

1. The Hamming distance $d(000, 011)$ is 2 because $(000 \oplus 011)$ is 011 (two 1s).
2. The Hamming distance $d(10101, 11110)$ is 3 because $(10101 \oplus 11110)$ is 01011 (three 1s).

Minimum Hamming Distance for Error Detection

In a set of codewords, the **minimum Hamming distance** is the smallest Hamming distance between all possible pairs of codewords. Now let us find the minimum Hamming distance in a code if we want to be able to detect up to s errors. If s errors occur during transmission, the Hamming distance between the sent codeword and received codeword is s. If our system is to detect up to s errors, the minimum distance between the valid codes must be $(s + 1)$, so that the received codeword does not match a valid codeword. In other words, if the minimum distance between all valid codewords is $(s + 1)$, the received codeword cannot be erroneously mistaken for another codeword. The error will be detected. We need to clarify a point here: Although a code with $d_{min} = s + 1$ may be able to detect more than s errors in some special cases, only s or fewer errors are guaranteed to be detected.

To guarantee the detection of up to s errors in all cases, the minimum Hamming distance in a block code must be $d_{min} = s + 1$.

We can look at this criteria geometrically. Let us assume that the sent codeword x is at the center of a circle with radius s. All received codewords that are created by 0 to s errors are points inside the circle or on the perimeter of the circle. All other valid codewords must be outside the circle, as shown in Figure 10.3. This means that d_{min} must be an integer greater than s or $d_{min} = s + 1$.

Example 10.3

The minimum Hamming distance for our first code scheme (Table 10.1) is 2. This code guarantees detection of only a single error. For example, if the third codeword (101) is sent and one error occurs, the received codeword does not match any valid codeword. If two errors occur, however, the received codeword may match a valid codeword and the errors are not detected.

Example 10.4

A code scheme has a Hamming distance $d_{min} = 4$. This code guarantees the detection of up to three errors ($d = s + 1$ or $s = 3$).

Figure 10.3 *Geometric concept explaining* d_{min} *in error detection*

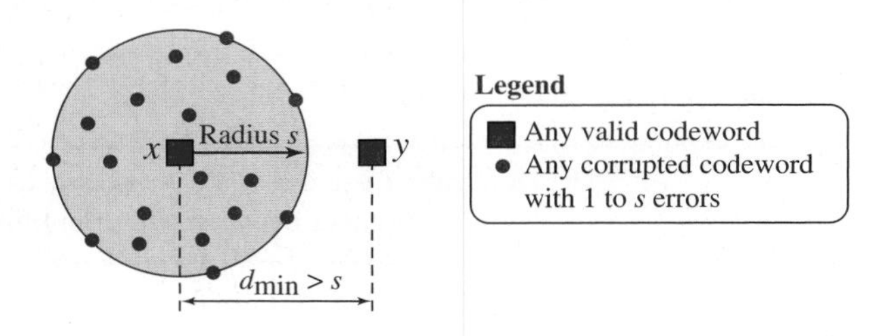

Linear Block Codes

Almost all block codes used today belong to a subset of block codes called ***linear block codes***. The use of nonlinear block codes for error detection and correction is not as widespread because their structure makes theoretical analysis and implementation difficult. We therefore concentrate on linear block codes. The formal definition of linear block codes requires the knowledge of abstract algebra (particularly Galois fields), which is beyond the scope of this book. We therefore give an informal definition. For our purposes, a linear block code is a code in which the exclusive OR (addition modulo-2) of two valid codewords creates another valid codeword.

Example 10.5

The code in Table 10.1 is a linear block code because the result of XORing any codeword with any other codeword is a valid codeword. For example, the XORing of the second and third codewords creates the fourth one.

Minimum Distance for Linear Block Codes

It is simple to find the minimum Hamming distance for a linear block code. The minimum Hamming distance is the number of 1s in the nonzero valid codeword with the smallest number of 1s.

Example 10.6

In our first code (Table 10.1), the numbers of 1s in the nonzero codewords are 2, 2, and 2. So the minimum Hamming distance is $d_{min} = 2$.

Parity-Check Code

Perhaps the most familiar error-detecting code is the **parity-check code.** This code is a linear block code. In this code, a k-bit dataword is changed to an n-bit codeword where $n = k + 1$. The extra bit, called the *parity bit,* is selected to make the total number of 1s in the codeword even. Although some implementations specify an odd number of 1s, we discuss the even case. The minimum Hamming distance for this category is $d_{min} = 2$, which means that the code is a single-bit error-detecting code. Our first code (Table 10.1) is a parity-check code ($k = 2$ and $n = 3$). The code in Table 10.2 is also a parity-check code with $k = 4$ and $n = 5$.

Table 10.2 *Simple parity-check code C(5, 4)*

Dataword	*Codeword*	*Dataword*	*Codeword*
0000	**00000**	1000	**10001**
0001	**00011**	1001	**10010**
0010	**00101**	1010	**10100**
0011	**00110**	1011	**10111**
0100	**01001**	1100	**11000**
0101	**01010**	1101	**11011**
0110	**01100**	1110	**11101**
0111	**01111**	1111	**11110**

Figure 10.4 shows a possible structure of an encoder (at the sender) and a decoder (at the receiver).

Figure 10.4 *Encoder and decoder for simple parity-check code*

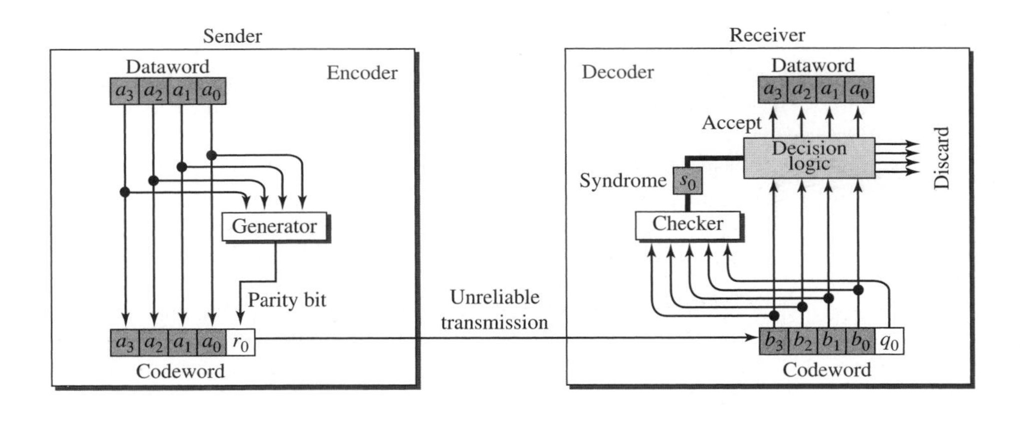

The calculation is done in **modular arithmetic** (see Appendix E). The encoder uses a generator that takes a copy of a 4-bit dataword (a_0, a_1, a_2, and a_3) and generates a parity bit r_0. The dataword bits and the parity bit create the 5-bit codeword. The parity bit that is added makes the number of 1s in the codeword even. This is normally done by adding the 4 bits of the dataword (modulo-2); the result is the parity bit. In other words,

$$r_0 = a_3 + a_2 + a_1 + a_0 \quad \textbf{(modulo-2)}$$

If the number of 1s is even, the result is 0; if the number of 1s is odd, the result is 1. In both cases, the total number of 1s in the codeword is even.

The sender sends the codeword, which may be corrupted during transmission. The receiver receives a 5-bit word. The checker at the receiver does the same thing as the generator in the sender with one exception: The addition is done over all 5 bits. The result, which is called the ***syndrome,*** is just 1 bit. The syndrome is 0 when the number of 1s in the received codeword is even; otherwise, it is 1.

$$s_0 = b_3 + b_2 + b_1 + b_0 + q_0 \qquad \textbf{(modulo-2)}$$

The syndrome is passed to the decision logic analyzer. If the syndrome is 0, there is no detectable error in the received codeword; the data portion of the received codeword is accepted as the dataword; if the syndrome is 1, the data portion of the received codeword is discarded. The dataword is not created.

Example 10.7

Let us look at some transmission scenarios. Assume the sender sends the dataword 1011. The codeword created from this dataword is 10111, which is sent to the receiver. We examine five cases:

1. No error occurs; the received codeword is 10111. The syndrome is 0. The dataword 1011 is created.
2. One single-bit error changes a_1. The received codeword is 10**0**11. The syndrome is 1. No dataword is created.
3. One single-bit error changes r_0. The received codeword is 1011**0**. The syndrome is 1. No dataword is created. Note that although none of the dataword bits are corrupted, no dataword is created because the code is not sophisticated enough to show the position of the corrupted bit.
4. An error changes r_0 and a second error changes a_3. The received codeword is **0**011**0**. The syndrome is 0. The dataword 0011 is created at the receiver. Note that here the dataword is wrongly created due to the syndrome value. The simple parity-check decoder cannot detect an even number of errors. The errors cancel each other out and give the syndrome a value of 0.
5. Three bits—a_3, a_2, and a_1—are changed by errors. The received codeword is **010**11. The syndrome is 1. The dataword is not created. This shows that the simple parity check, guaranteed to detect one single error, can also find any odd number of errors.

A parity-check code can detect an odd number of errors.

10.3 CYCLIC CODES

Cyclic codes are special linear block codes with one extra property. In a **cyclic code,** if a codeword is cyclically shifted (rotated), the result is another codeword. For example, if 1011000 is a codeword and we cyclically left-shift, then 0110001 is also a codeword. In this case, if we call the bits in the first word a_0 to a_6, and the bits in the second word b_0 to b_6, we can shift the bits by using the following:

$$b_1 = a_0 \quad b_2 = a_1 \quad b_3 = a_2 \quad b_4 = a_3 \quad b_5 = a_4 \quad b_6 = a_5 \quad b_0 = a_6$$

In the rightmost equation, the last bit of the first word is wrapped around and becomes the first bit of the second word.

10.3.1 Cyclic Redundancy Check

We can create cyclic codes to correct errors. However, the theoretical background required is beyond the scope of this book. In this section, we simply discuss a subset of cyclic codes called the **cyclic redundancy check (CRC)**, which is used in networks such as LANs and WANs.

Table 10.3 shows an example of a CRC code. We can see both the linear and cyclic properties of this code.

Table 10.3 *A CRC code with C(7, 4)*

Dataword	*Codeword*	*Dataword*	*Codeword*
0000	0000000	1000	1000101
0001	0001011	1001	1001110
0010	0010110	1010	1010011
0011	0011101	1011	1011000
0100	0100111	1100	1100010
0101	0101100	1101	1101001
0110	0110001	1110	1110100
0111	0111010	1111	1111111

Figure 10.5 shows one possible design for the encoder and decoder.

Figure 10.5 *CRC encoder and decoder*

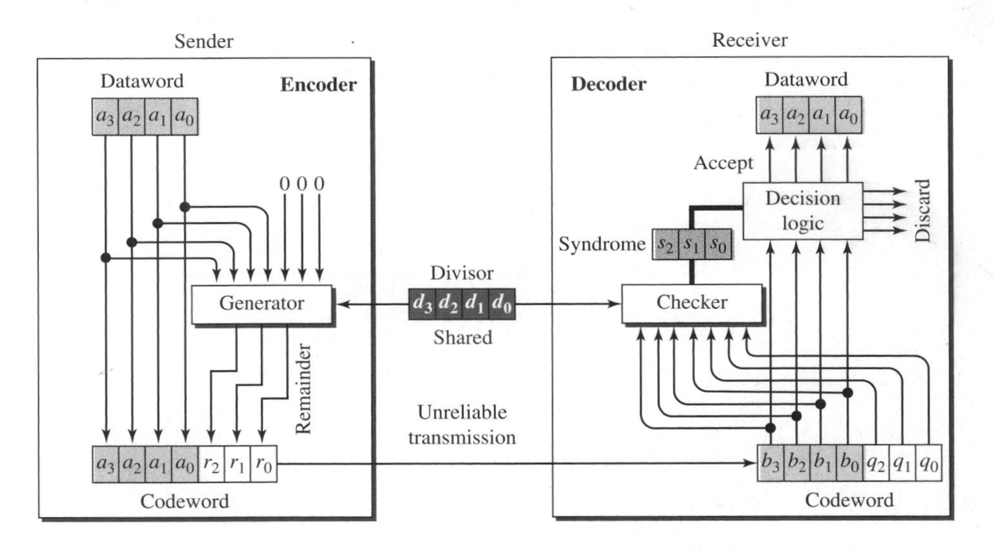

In the encoder, the dataword has k bits (4 here); the codeword has n bits (7 here). The size of the dataword is augmented by adding $n - k$ (3 here) 0s to the right-hand side of the word. The n-bit result is fed into the generator. The generator uses a divisor of size $n - k + 1$ (4 here), predefined and agreed upon. The generator divides the augmented dataword by the divisor (modulo-2 division). The quotient of the division is discarded; the remainder ($r_2r_1r_0$) is appended to the dataword to create the codeword.

The decoder receives the codeword (possibly corrupted in transition). A copy of all n bits is fed to the checker, which is a replica of the generator. The remainder produced by the checker is a syndrome of $n - k$ (3 here) bits, which is fed to the decision logic analyzer. The analyzer has a simple function. If the syndrome bits are all 0s, the 4 leftmost bits of the codeword are accepted as the dataword (interpreted as no error); otherwise, the 4 bits are discarded (error).

Encoder

Let us take a closer look at the encoder. The encoder takes a dataword and augments it with $n - k$ number of 0s. It then divides the augmented dataword by the divisor, as shown in Figure 10.6.

Figure 10.6 *Division in CRC encoder*

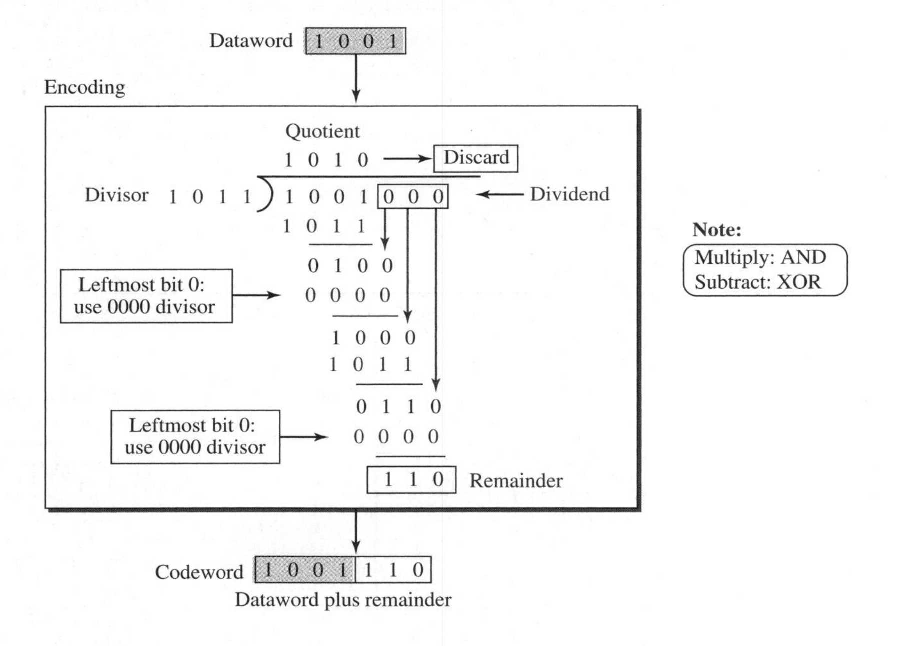

The process of modulo-2 binary division is the same as the familiar division process we use for decimal numbers. However, addition and subtraction in this case are the same; we use the XOR operation to do both.

As in decimal division, the process is done step by step. In each step, a copy of the divisor is XORed with the 4 bits of the dividend. The result of the XOR operation (remainder) is 3 bits (in this case), which is used for the next step after 1 extra bit is pulled down to make it 4 bits long. There is one important point we need to remember in this type of division. If the leftmost bit of the dividend (or the part used in each step) is 0, the step cannot use the regular divisor; we need to use an all-0s divisor.

When there are no bits left to pull down, we have a result. The 3-bit remainder forms the **check bits** (r_2, r_1, and r_0). They are appended to the dataword to create the codeword.

Decoder

The codeword can change during transmission. The decoder does the same division process as the encoder. The remainder of the division is the syndrome. If the syndrome is all 0s, there is no error with a high probability; the dataword is separated from the received codeword and accepted. Otherwise, everything is discarded. Figure 10.7 shows two cases: The left-hand figure shows the value of the syndrome when no error has occurred; the syndrome is 000. The right-hand part of the figure shows the case in which there is a single error. The syndrome is not all 0s (it is 011).

Figure 10.7 *Division in the CRC decoder for two cases*

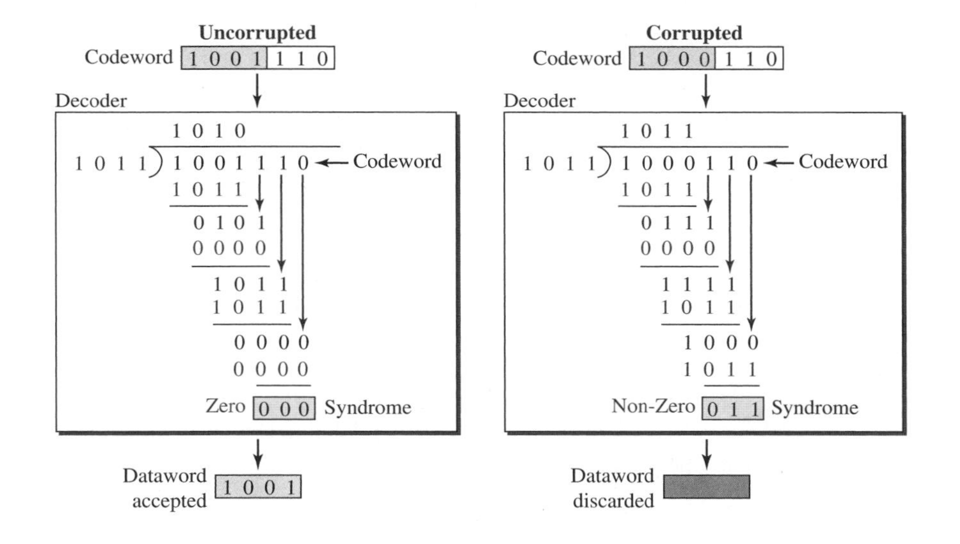

Divisor

We may be wondering how the divisor 1011 is chosen. This depends on the expectation we have from the code. We will show some standard divisors later in the chapter (Table 10.4) after we discuss polynomials.

10.3.2 Polynomials

A better way to understand cyclic codes and how they can be analyzed is to represent them as polynomials. Again, this section is optional.

A pattern of 0s and 1s can be represented as a **polynomial** with coefficients of 0 and 1. The power of each term shows the position of the bit; the coefficient shows the value of the bit. Figure 10.8 shows a binary pattern and its polynomial representation. In Figure 10.8a we show how to translate a binary pattern into a polynomial; in Figure 10.8b we show how the polynomial can be shortened by removing all terms with zero coefficients and replacing x^1 by x and x^0 by 1.

Figure 10.8 shows one immediate benefit; a 7-bit pattern can be replaced by three terms. The benefit is even more conspicuous when we have a polynomial such as

Figure 10.8 *A polynomial to represent a binary word*

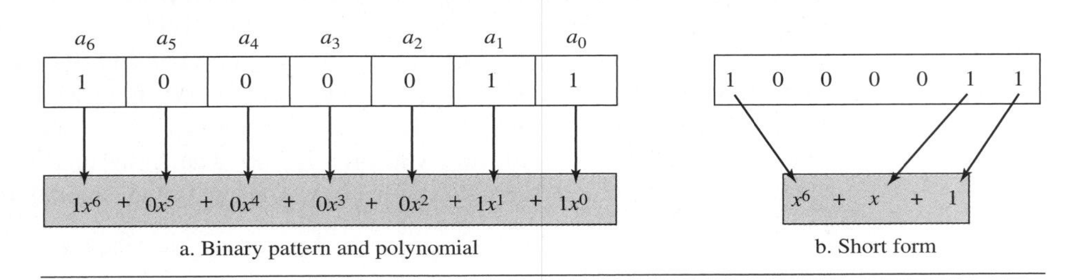

$x^{23} + x^3 + 1$. Here the bit pattern is 24 bits in length (three 1s and twenty-one 0s) while the polynomial is just three terms.

Degree of a Polynomial

The degree of a polynomial is the highest power in the polynomial. For example, the degree of the polynomial $x^6 + x + 1$ is 6. Note that the degree of a polynomial is 1 less than the number of bits in the pattern. The bit pattern in this case has 7 bits.

Adding and Subtracting Polynomials

Adding and subtracting polynomials in mathematics are done by adding or subtracting the coefficients of terms with the same power. In our case, the coefficients are only 0 and 1, and adding is in modulo-2. This has two consequences. First, addition and subtraction are the same. Second, adding or subtracting is done by combining terms and deleting pairs of identical terms. For example, adding $x^5 + x^4 + x^2$ and $x^6 + x^4 + x^2$ gives just $x^6 + x^5$. The terms x^4 and x^2 are deleted. However, note that if we add, for example, three polynomials and we get x^2 three times, we delete a pair of them and keep the third.

Multiplying or Dividing Terms

In this arithmetic, multiplying a term by another term is very simple; we just add the powers. For example, $x^3 \times x^4$ is x^7. For dividing, we just subtract the power of the second term from the power of the first. For example, x^5/x^2 is x^3.

Multiplying Two Polynomials

Multiplying a polynomial by another is done term by term. Each term of the first polynomial must be multiplied by all terms of the second. The result, of course, is then simplified, and pairs of equal terms are deleted. The following is an example:

$$
\begin{aligned}
(x^5 + x^3 + x^2 + x)(x^2 + x + 1) &= x^7 + x^6 + x^5 + x^5 + x^4 + x^3 + x^4 + x^3 + x^2 + x^3 + x^2 + x \\
&= x^7 + x^6 + x^3 + x
\end{aligned}
$$

Dividing One Polynomial by Another

Division of polynomials is conceptually the same as the binary division we discussed for an encoder. We divide the first term of the dividend by the first term of the divisor to get the first term of the quotient. We multiply the term in the quotient by the divisor and subtract the result from the dividend. We repeat the process until the dividend degree is less than the divisor degree. We will show an example of division later in this chapter.

Shifting

A binary pattern is often shifted a number of bits to the right or left. Shifting to the left means adding extra 0s as rightmost bits; shifting to the right means deleting some rightmost bits. Shifting to the left is accomplished by multiplying each term of the polynomial by x^m, where m is the number of shifted bits; shifting to the right is accomplished by dividing each term of the polynomial by x^m. The following shows shifting to the left and to the right. Note that we do not have negative powers in the polynomial representation.

Shifting left 3 bits: 10011 becomes 10011000 $\quad x^4 + x + 1$ becomes $x^7 + x^4 + x^3$
Shifting right 3 bits: 10011 becomes 10 $\quad x^4 + x + 1$ becomes x

When we augmented the dataword in the encoder of Figure 10.6, we actually shifted the bits to the left. Also note that when we concatenate two bit patterns, we shift the first polynomial to the left and then add the second polynomial.

10.3.3 Cyclic Code Encoder Using Polynomials

Now that we have discussed operations on polynomials, we show the creation of a codeword from a dataword. Figure 10.9 is the polynomial version of Figure 10.6. We can see that the process is shorter. The dataword 1001 is represented as $x^3 + 1$. The divisor 1011 is represented as $x^3 + x + 1$. To find the augmented dataword, we have left-shifted the dataword 3 bits (multiplying by x^3). The result is $x^6 + x^3$. Division is straightforward. We divide the first term of the dividend, x^6, by the first term of the divisor, x^3. The first term of the quotient is then x^6/x^3, or x^3. Then we multiply x^3 by the divisor and subtract (according to our previous definition of subtraction) the result from the dividend. The result is x^4, with a degree greater than the divisor's degree; we continue to divide until the degree of the remainder is less than the degree of the divisor.

Figure 10.9 *CRC division using polynomials*

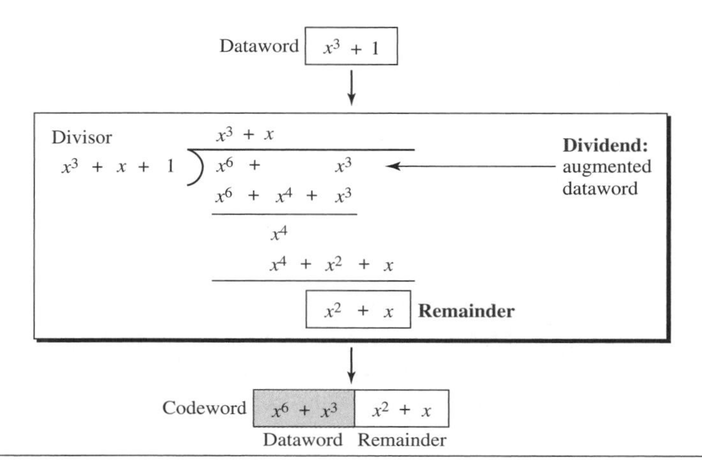

It can be seen that the polynomial representation can easily simplify the operation of division in this case, because the two steps involving all-0s divisors are not needed here. (Of course, one could argue that the all-0s divisor step can also be eliminated in binary division.) In a polynomial representation, the divisor is normally referred to as the ***generator polynomial*** $t(x)$.

The divisor in a cyclic code is normally called the *generator polynomial* or simply the *generator.*

10.3.4 Cyclic Code Analysis

We can analyze a cyclic code to find its capabilities by using polynomials. We define the following, where $f(x)$ is a polynomial with binary coefficients.

Dataword: $d(x)$ **Codeword:** $c(x)$ **Generator:** $g(x)$ **Syndrome:** $s(x)$ **Error:** $e(x)$

If $s(x)$ is not zero, then one or more bits is corrupted. However, if $s(x)$ is zero, either no bit is corrupted or the decoder failed to detect any errors. (Note that ¦ means divide).

In a cyclic code,

1. If $s(x)$ ¦ 0, one or more bits is corrupted.
2. If $s(x) = 0$, either
   a. No bit is corrupted, or
   b. Some bits are corrupted, but the decoder failed to detect them.

In our analysis we want to find the criteria that must be imposed on the generator, $g(x)$ to detect the type of error we especially want to be detected. Let us first find the relationship among the sent codeword, error, received codeword, and the generator. We can say

$$\textbf{Received codeword} = c(x) + e(x)$$

In other words, the received codeword is the sum of the sent codeword and the error. The receiver divides the received codeword by $g(x)$ to get the syndrome. We can write this as

$$\frac{\textbf{Received codeword}}{g(x)} = \frac{c(x)}{g(x)} + \frac{e(x)}{g(x)}$$

The first term at the right-hand side of the equality has a remainder of zero (according to the definition of codeword). So the syndrome is actually the remainder of the second term on the right-hand side. If this term does not have a remainder (syndrome = 0), either $e(x)$ is 0 or $e(x)$ is divisible by $g(x)$. We do not have to worry about the first case (there is no error); the second case is very important. Those errors that are divisible by $g(x)$ are not caught.

In a cyclic code, those $e(x)$ errors that are divisible by $g(x)$ are not caught.

Let us show some specific errors and see how they can be caught by a well-designed $g(x)$.

Single-Bit Error

What should the structure of $g(x)$ be to guarantee the detection of a single-bit error? A single-bit error is $e(x) = x^i$, where i is the position of the bit. If a single-bit error is caught, then x^i is not divisible by $g(x)$. (Note that when we say *not divisible,* we mean that there is a remainder.) If $g(x)$ has at least two terms (which is normally the case) and the coefficient of x^0 is not zero (the rightmost bit is 1), then $e(x)$ cannot be divided by $g(x)$.

If the generator has more than one term and the coefficient of x^0 is 1, all single-bit errors can be caught.

Example 10.8

Which of the following $g(x)$ values guarantees that a single-bit error is caught? For each case, what is the error that cannot be caught?

a. $x + 1$

b. x^3

c. 1

Solution

a. No x^i can be divisible by $x + 1$. In other words, $x^i/(x + 1)$ always has a remainder. So the syndrome is nonzero. Any single-bit error can be caught.

b. If i is equal to or greater than 3, x^i is divisible by $g(x)$. The remainder of x^i/x^3 is zero, and the receiver is fooled into believing that there is no error, although there might be one. Note that in this case, the corrupted bit must be in position 4 or above. All single-bit errors in positions 1 to 3 are caught.

c. All values of i make x^i divisible by $g(x)$. No single-bit error can be caught. In addition, this $g(x)$ is useless because it means the codeword is just the dataword augmented with $n - k$ zeros.

Two Isolated Single-Bit Errors

Now imagine there are two single-bit isolated errors. Under what conditions can this type of error be caught? We can show this type of error as $e(x) = x^j + x^i$. The values of i and j define the positions of the errors, and the difference $j - i$ defines the distance between the two errors, as shown in Figure 10.10.

Figure 10.10 *Representation of two isolated single-bit errors using polynomials*

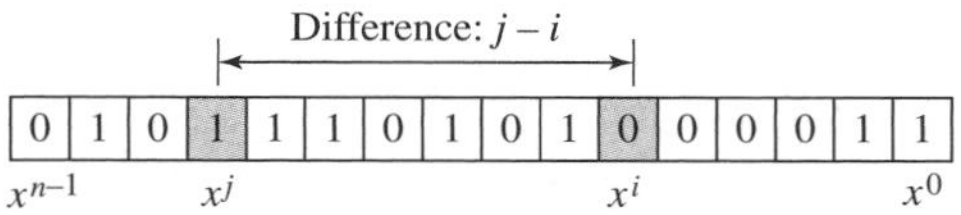

Difference: $j - i$

0	1	0	1	1	1	0	1	0	1	0	0	0	0	1	1
x^{n-1}			x^j							x^i					x^0

We can write $e(x) = x^i(x^{j-i} + 1)$. If $g(x)$ has more than one term and one term is x^0, it cannot divide x^i, as we saw in the previous section. So if $g(x)$ is to divide $e(x)$, it must divide $x^{j-i} + 1$. In other words, $g(x)$ must not divide $x^t + 1$, where t is between 0 and $n - 1$. However, $t = 0$ is meaningless and $t = 1$ is needed, as we will see later. This means t should be between 2 and $n - 1$.

If a generator cannot divide $x^t + 1$ (t between 0 and $n - 1$), then all isolated double errors can be detected.

Example 10.9

Find the status of the following generators related to two isolated, single-bit errors.

a. $x + 1$

b. $x^4 + 1$

c. $x^7 + x^6 + 1$

d. $x^{15} + x^{14} + 1$

Solution

a. This is a very poor choice for a generator. Any two errors next to each other cannot be detected.

b. This generator cannot detect two errors that are four positions apart. The two errors can be anywhere, but if their distance is 4, they remain undetected.

c. This is a good choice for this purpose.

d. This polynomial cannot divide any error of type $x^t + 1$ if t is less than 32,768. This means that a codeword with two isolated errors that are next to each other or up to 32,768 bits apart can be detected by this generator.

Odd Numbers of Errors

A generator with a factor of $x + 1$ can catch all odd numbers of errors. This means that we need to make $x + 1$ a factor of any generator. Note that we are not saying that the generator itself should be $x + 1$; we are saying that it should have a factor of $x + 1$. If it is only $x + 1$, it cannot catch the two adjacent isolated errors (see the previous section). For example, $x^4 + x^2 + x + 1$ can catch all odd-numbered errors since it can be written as a product of the two polynomials $x + 1$ and $x^3 + x^2 + 1$.

A generator that contains a factor of $x + 1$ can detect all odd-numbered errors.

Burst Errors

Now let us extend our analysis to the burst error, which is the most important of all. A burst error is of the form $e(x) = (x^j + \cdots + x^i)$. Note the difference between a burst error and two isolated single-bit errors. The first can have two terms or more; the second can only have two terms. We can factor out x^i and write the error as $x^i(x^{j-i} + \cdots + 1)$. If our generator can detect a single error (minimum condition for a generator), then it cannot divide x^i. What we should worry about are those generators that divide $x^{j-i} + \cdots + 1$. In other words, the remainder of $(x^{j-i} + \cdots + 1)/(x^r + \cdots + 1)$ must not be zero. Note that the denominator is the generator polynomial. We can have three cases:

1. If $j - i < r$, the remainder can never be zero. We can write $j - i = L - 1$, where L is the length of the error. So $L - 1 < r$ or $L < r + 1$ or L ð r. This means all burst errors with length smaller than or equal to the number of check bits r will be detected.
2. In some rare cases, if $j - i = r$, or $L = r + 1$, the syndrome is 0 and the error is undetected. It can be proved that in these cases, the probability of undetected burst error of length $r + 1$ is $(1/2)^{r-1}$. For example, if our generator is $x^{14} + x^3 + 1$, in which $r = 14$, a burst error of length $L = 15$ can slip by undetected with the probability of $(1/2)^{14-1}$ or almost 1 in 10,000.
3. In some rare cases, if $j - i > r$, or $L > r + 1$, the syndrome is 0 and the error is undetected. It can be proved that in these cases, the probability of undetected burst error of length greater than $r + 1$ is $(1/2)^r$. For example, if our generator is $x^{14} + x^3 + 1$, in which $r = 14$, a burst error of length greater than 15 can slip by undetected with the probability of $(1/2)^{14}$ or almost 1 in 16,000 cases.

- ❑ **All burst errors with $L \le r$ will be detected.**
- ❑ **All burst errors with $L = r + 1$ will be detected with probability $1 - (1/2)^{r-1}$.**
- ❑ **All burst errors with $L > r + 1$ will be detected with probability $1 - (1/2)^r$.**

Example 10.10

Find the suitability of the following generators in relation to burst errors of different lengths.

a. $x^6 + 1$
b. $x^{18} + x^7 + x + 1$
c. $x^{32} + x^{23} + x^7 + 1$

Solution

a. This generator can detect all burst errors with a length less than or equal to 6 bits; 3 out of 100 burst errors with length 7 will slip by; 16 out of 1000 burst errors of length 8 or more will slip by.

b. This generator can detect all burst errors with a length less than or equal to 18 bits; 8 out of 1 million burst errors with length 19 will slip by; 4 out of 1 million burst errors of length 20 or more will slip by.

c. This generator can detect all burst errors with a length less than or equal to 32 bits; 5 out of 10 billion burst errors with length 33 will slip by; 3 out of 10 billion burst errors of length 34 or more will slip by.

Summary

We can summarize the criteria for a good polynomial generator:

A good polynomial generator needs to have the following characteristics:

1. **It should have at least two terms.**
2. **The coefficient of the term x^0 should be 1.**
3. **It should not divide $x^t + 1$, for t between 2 and $n - 1$.**
4. **It should have the factor $x + 1$.**

Standard Polynomials

Some standard polynomials used by popular protocols for CRC generation are shown in Table 10.4 along with the corresponding bit pattern.

Table 10.4 *Standard polynomials*

Name	*Polynomial*	*Used in*
CRC-8	$x^8 + x^2 + x + 1$ 100000111	ATM header
CRC-10	$x^{10} + x^9 + x^5 + x^4 + x^2 + 1$ 11000110101	ATM AAL
CRC-16	$x^{16} + x^{12} + x^5 + 1$ 10001000000100001	HDLC
CRC-32	$x^{32} + x^{26} + x^{23} + x^{22} + x^{16} + x^{12} + x^{11} + x^{10} + x^8 + x^7 + x^5 + x^4 + x^2 + x + 1$ 100000100110000010001110110110111	LANs

10.3.5 Advantages of Cyclic Codes

We have seen that cyclic codes have a very good performance in detecting single-bit errors, double errors, an odd number of errors, and burst errors. They can easily be implemented in hardware and software. They are especially fast when implemented in hardware. This has made cyclic codes a good candidate for many networks.

10.3.6 Other Cyclic Codes

The cyclic codes we have discussed in this section are very simple. The check bits and syndromes can be calculated by simple algebra. There are, however, more powerful polynomials that are based on abstract algebra involving Galois fields. These are beyond the scope of this book. One of the most interesting of these codes is the **Reed-Solomon code** used today for both detection and correction.

10.3.7 Hardware Implementation

One of the advantages of a cyclic code is that the encoder and decoder can easily and cheaply be implemented in hardware by using a handful of electronic devices. Also, a hardware implementation increases the rate of check bit and syndrome bit calculation. In this section, we try to show, step by step, the process. The section, however, is optional and does not affect the understanding of the rest of the chapter.

Divisor

Let us first consider the divisor. We need to note the following points:

1. The divisor is repeatedly XORed with part of the dividend.
2. The divisor has $n - k + 1$ bits which either are predefined or are all 0s. In other words, the bits do not change from one dataword to another. In our previous example, the divisor bits were either 1011 or 0000. The choice was based on the leftmost bit of the part of the augmented data bits that are active in the XOR operation.

3. A close look shows that only $n - k$ bits of the divisor are needed in the XOR operation. The leftmost bit is not needed because the result of the operation is always 0, no matter what the value of this bit. The reason is that the inputs to this XOR operation are either both 0s or both 1s. In our previous example, only 3 bits, not 4, are actually used in the XOR operation.

Using these points, we can make a fixed (hardwired) divisor that can be used for a cyclic code if we know the divisor pattern. Figure 10.11 shows such a design for our previous example. We have also shown the XOR devices used for the operation.

Figure 10.11 *Hardwired design of the divisor in CRC*

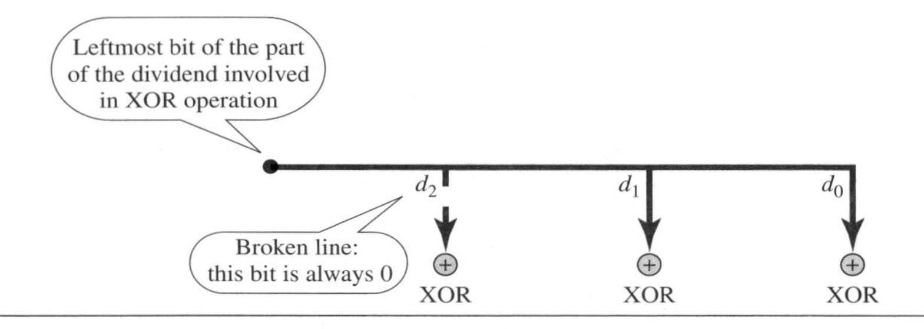

Note that if the leftmost bit of the part of the dividend to be used in this step is 1, the divisor bits ($d_2d_1d_0$) are 011; if the leftmost bit is 0, the divisor bits are 000. The design provides the right choice based on the leftmost bit.

Augmented Dataword

In our paper-and-pencil division process in Figure 10.6, we show the augmented dataword as fixed in position with the divisor bits shifting to the right, 1 bit in each step. The divisor bits are aligned with the appropriate part of the augmented dataword. Now that our divisor is fixed, we need instead to shift the bits of the augmented dataword to the left (opposite direction) to align the divisor bits with the appropriate part. There is no need to store the augmented dataword bits.

Remainder

In our previous example, the remainder is 3 bits ($n - k$ bits in general) in length. We can use three **registers** (single-bit storage devices) to hold these bits. To find the final remainder of the division, we need to modify our division process. The following is the step-by-step process that can be used to simulate the division process in hardware (or even in software).

1. We assume that the remainder is originally all 0s (000 in our example).
2. At each time click (arrival of 1 bit from an augmented dataword), we repeat the following two actions:
   a. We use the leftmost bit to make a decision about the divisor (011 or 000).
   b. The other 2 bits of the remainder and the next bit from the augmented dataword (total of 3 bits) are XORed with the 3-bit divisor to create the next remainder.

Figure 10.12 shows this simulator, but note that this is not the final design; there will be more improvements.

Figure 10.12 *Simulation of division in CRC encoder*

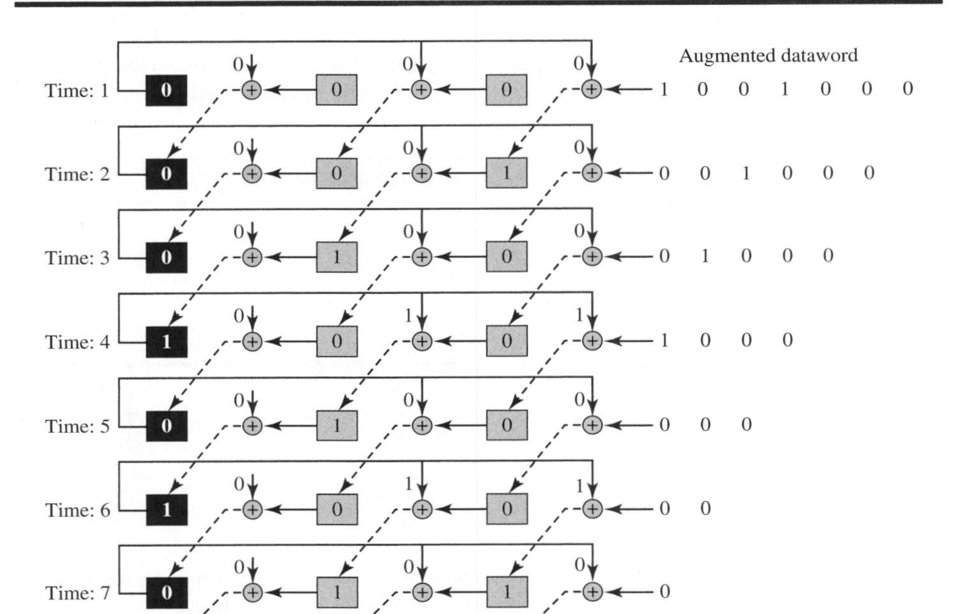

At each clock tick, shown as different times, one of the bits from the augmented dataword is used in the XOR process. If we look carefully at the design, we have seven steps here, while in the paper-and-pencil method we had only four steps. The first three steps have been added here to make each step equal and to make the design for each step the same. Steps 1, 2, and 3 push the first 3 bits to the remainder registers; steps 4, 5, 6, and 7 match the paper-and-pencil design. Note that the values in the remainder register in steps 4 to 7 exactly match the values in the paper-and-pencil design. The final remainder is also the same.

The above design is for demonstration purposes only. It needs simplification to be practical. First, we do not need to keep the intermediate values of the remainder bits; we need only the final bits. We therefore need only 3 registers instead of 24. After the XOR operations, we do not need the bit values of the previous remainder. Also, we do not need 21 XOR devices; two are enough because the output of an XOR operation in which one of the bits is 0 is simply the value of the other bit. This other bit can be used as the output. With these two modifications, the design becomes tremendously simpler and less expensive, as shown in Figure 10.13.

Figure 10.13 *The CRC encoder design using shift registers*

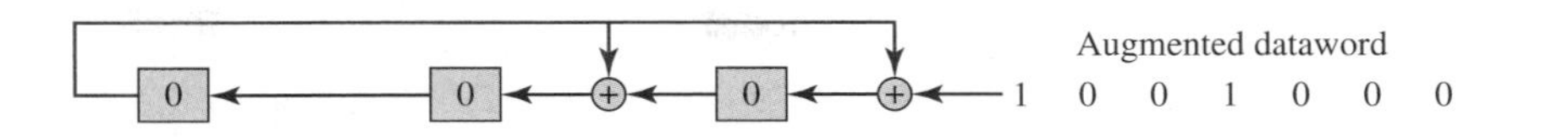

We need, however, to make the registers shift registers. A 1-bit shift register holds a bit for a duration of one clock time. At a time click, the shift register accepts the bit at its input port, stores the new bit, and displays it on the output port. The content and the output remain the same until the next input arrives. When we connect several 1-bit shift registers together, it looks as if the contents of the register are shifting.

General Design

A general design for the encoder and decoder is shown in Figure 10.14.

Figure 10.14 *General design of encoder and decoder of a CRC code*

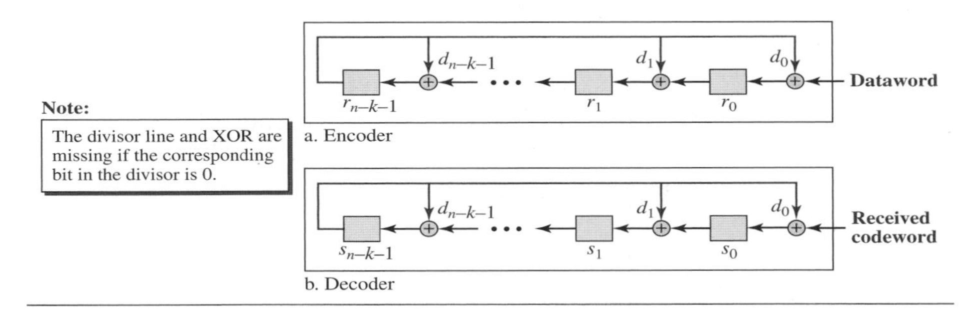

Note that we have $n - k$ 1-bit shift registers in both the encoder and decoder. We have up to $n - k$ XOR devices, but the divisors normally have several 0s in their pattern, which reduces the number of devices. Also note that, instead of augmented datawords, we show the dataword itself as the input because after the bits in the dataword are all fed into the encoder, the extra bits, which all are 0s, do not have any effect on the rightmost XOR. Of course, the process needs to be continued for another $n - k$ steps before the check bits are ready. This fact is one of the criticisms of this design. Better schemes have been designed to eliminate this waiting time (the check bits are ready after k steps), but we leave this as a research topic for the reader. In the decoder, however, the entire codeword must be fed to the decoder before the syndrome is ready.

10.4 CHECKSUM

Checksum is an error-detecting technique that can be applied to a message of any length. In the Internet, the checksum technique is mostly used at the network and transport layer rather than the data-link layer. However, to make our discussion of error-detecting techniques complete, we discuss the checksum in this chapter.

At the source, the message is first divided into m-bit units. The generator then creates an extra m-bit unit called the ***checksum,*** which is sent with the message. At the destination, the checker creates a new checksum from the combination of the message and sent checksum. If the new checksum is all 0s, the message is accepted; otherwise, the message is discarded (Figure 10.15). Note that in the real implementation, the checksum unit is not necessarily added at the end of the message; it can be inserted in the middle of the message.

Figure 10.15 *Checksum*

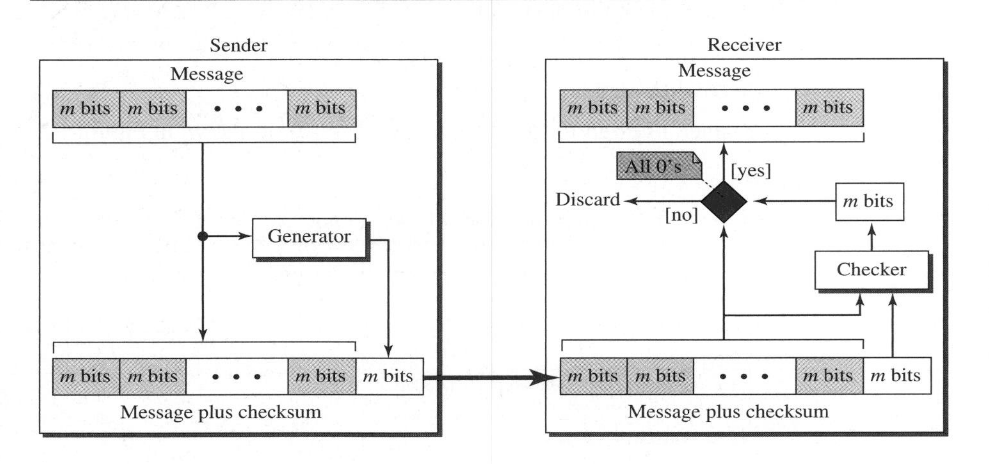

10.4.1 Concept

The idea of the traditional checksum is simple. We show this using a simple example.

Example 10.11

Suppose the message is a list of five 4-bit numbers that we want to send to a destination. In addition to sending these numbers, we send the sum of the numbers. For example, if the set of numbers is (7, 11, 12, 0, 6), we send (7, 11, 12, 0, 6, **36**), where 36 is the sum of the original numbers. The receiver adds the five numbers and compares the result with the sum. If the two are the same, the receiver assumes no error, accepts the five numbers, and discards the sum. Otherwise, there is an error somewhere and the message is not accepted.

One's Complement Addition

The previous example has one major drawback. Each number can be written as a 4-bit word (each is less than 15) except for the sum. One solution is to use **one's complement** arithmetic. In this arithmetic, we can represent unsigned numbers between 0 and $2^m - 1$ using only m bits. If the number has more than m bits, the extra leftmost bits need to be added to the m rightmost bits (wrapping).

Example 10.12

In the previous example, the decimal number 36 in binary is $(100100)_2$. To change it to a 4-bit number we add the extra leftmost bit to the right four bits as shown below.

$$\mathbf{(10)_2 + (0100)_2 = (0110)_2 \rightarrow (6)_{10}}$$

Instead of sending 36 as the sum, we can send 6 as the sum (7, 11, 12, 0, 6, **6**). The receiver can add the first five numbers in one's complement arithmetic. If the result is 6, the numbers are accepted; otherwise, they are rejected.

Checksum

We can make the job of the receiver easier if we send the complement of the sum, the checksum. In one's complement arithmetic, the complement of a number is found by completing all bits (changing all 1s to 0s and all 0s to 1s). This is the same as subtracting the number from $2^m - 1$. In one's complement arithmetic, we have two 0s: one positive and one negative, which are complements of each other. The positive zero has all m bits set to 0; the negative zero has all bits set to 1 (it is $2^m - 1$). If we add a number with its complement, we get a negative zero (a number with all bits set to 1). When the receiver adds all five numbers (including the checksum), it gets a negative zero. The receiver can complement the result again to get a positive zero.

Example 10.13

Let us use the idea of the checksum in Example 10.12. The sender adds all five numbers in one's complement to get the sum = 6. The sender then complements the result to get the checksum = 9, which is 15 − 6. Note that $6 = (0110)_2$ and $9 = (1001)_2$; they are complements of each other. The sender sends the five data numbers and the checksum (7, 11, 12, 0, 6, **9**). If there is no corruption in transmission, the receiver receives (7, 11, 12, 0, 6, **9**) and adds them in one's complement to get 15. The sender complements 15 to get 0. This shows that data have not been corrupted. Figure 10.16 shows the process.

Figure 10.16 *Example 10.13*

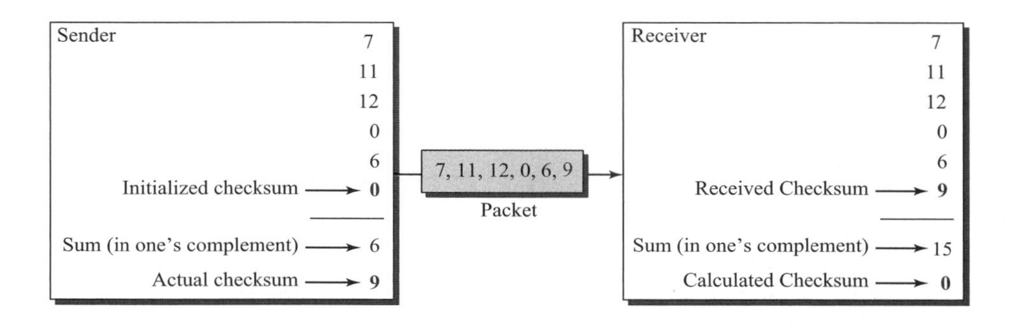

Internet Checksum

Traditionally, the Internet has used a 16-bit checksum. The sender and the receiver follow the steps depicted in Table 10.5. The sender or the receiver uses five steps.

Table 10.5 *Procedure to calculate the traditional checksum*

Sender	*Receiver*
1. The message is divided into 16-bit words.	1. The message and the checksum are received.
2. The value of the checksum word is initially set to zero.	2. The message is divided into 16-bit words.
3. All words including the checksum are added using one's complement addition.	3. All words are added using one's complement addition.
4. The sum is complemented and becomes the checksum.	4. The sum is complemented and becomes the new checksum.
5. The checksum is sent with the data.	5. If the value of the checksum is 0, the message is accepted; otherwise, it is rejected.

Algorithm

We can use the flow diagram of Figure 10.17 to show the algorithm for calculation of the checksum. A program in any language can easily be written based on the algorithm. Note that the first loop just calculates the sum of the data units in two's complement; the second loop wraps the extra bits created from the two's complement calculation to simulate the calculations in one's complement. This is needed because almost all computers today do calculation in two's complement.

Figure 10.17 *Algorithm to calculate a traditional checksum*

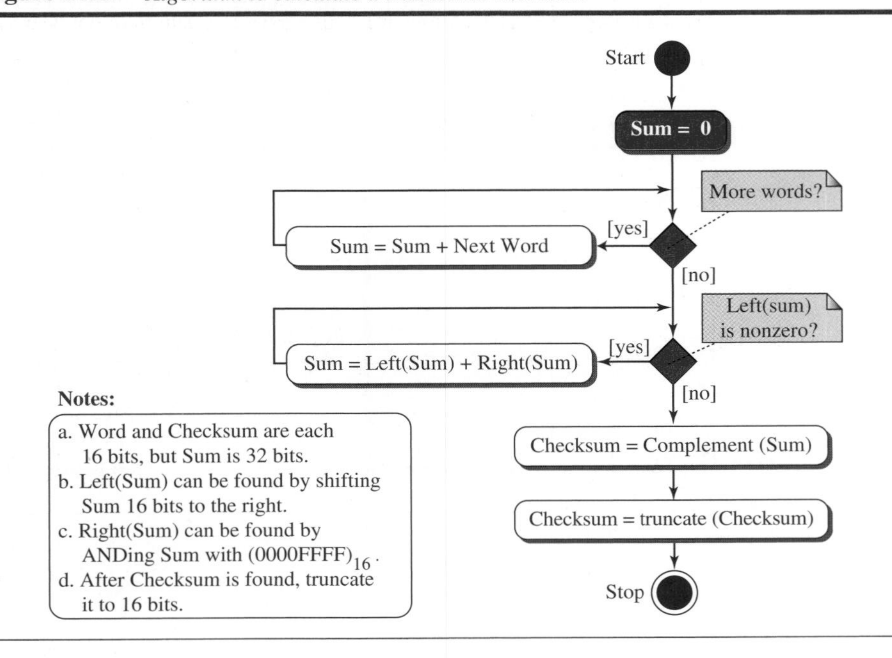

Performance

The traditional checksum uses a small number of bits (16) to detect errors in a message of any size (sometimes thousands of bits). However, it is not as strong as the CRC in error-checking capability. For example, if the value of one word is incremented and the value of another word is decremented by the same amount, the two errors cannot be detected because the sum and checksum remain the same. Also, if the values of several words are incremented but the sum and the checksum do not change, the errors are not detected. Fletcher and Adler have proposed some weighted checksums that eliminate the first problem. However, the tendency in the Internet, particularly in designing new protocols, is to replace the checksum with a CRC.

10.4.2 Other Approaches to the Checksum

As mentioned before, there is one major problem with the traditional checksum calculation. If two 16-bit items are transposed in transmission, the checksum cannot catch this error. The reason is that the traditional checksum is not weighted: it treats each data item equally. In other words, the order of data items is immaterial to the calculation. Several approaches have been used to prevent this problem. We mention two of them here: Fletcher and Adler.

Fletcher Checksum

The Fletcher checksum was devised to weight each data item according to its position. Fletcher has proposed two algorithms: 8-bit and 16-bit. The first, 8-bit Fletcher, calculates on 8-bit data items and creates a 16-bit checksum. The second, 16-bit Fletcher, calculates on 16-bit data items and creates a 32-bit checksum.

The 8-bit Fletcher is calculated over data octets (bytes) and creates a 16-bit checksum. The calculation is done modulo 256 (2^8), which means the intermediate results are divided by 256 and the remainder is kept. The algorithm uses two accumulators, L and R. The first simply adds data items together; the second adds a weight to the calculation. There are many variations of the 8-bit Fletcher algorithm; we show a simple one in Figure 10.18.

The 16-bit Fletcher checksum is similar to the 8-bit Fletcher checksum, but it is calculated over 16-bit data items and creates a 32-bit checksum. The calculation is done modulo 65,536.

Adler Checksum

The Adler checksum is a 32-bit checksum. Figure 10.19 shows a simple algorithm in flowchart form. It is similar to the 16-bit Fletcher with three differences. First, calculation is done on single bytes instead of 2 bytes at a time. Second, the modulus is a prime number (65,521) instead of 65,536. Third, L is initialized to 1 instead of 0. It has been proved that a prime modulo has a better detecting capability in some combinations of data.

Figure 10.18 *Algorithm to calculate an 8-bit Fletcher checksum*

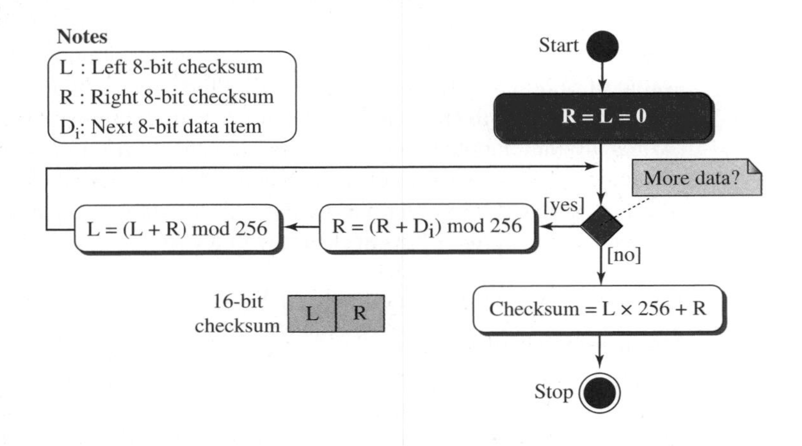

Figure 10.19 *Algorithm to calculate an Adler checksum*

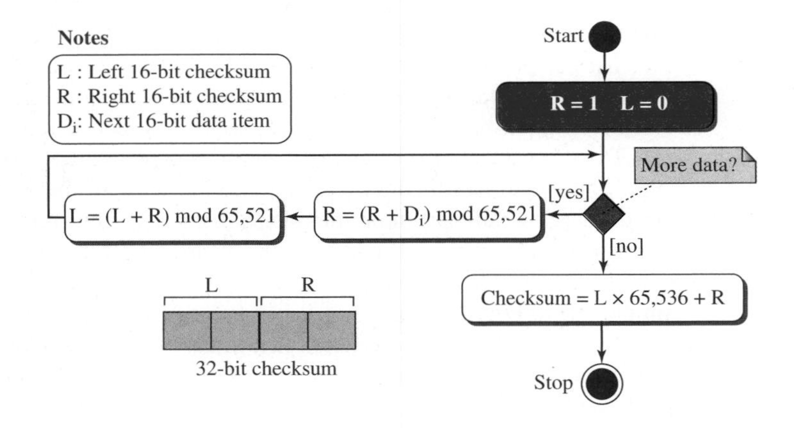

To see the behavior of the different checksum algorithms, check some of the applets for this chapter at the book website.

10.5 FORWARD ERROR CORRECTION

We discussed error detection and retransmission in the previous sections. However, retransmission of corrupted and lost packets is not useful for real-time multimedia transmission because it creates an unacceptable delay in reproducing: we need to wait until the lost or corrupted packet is resent. We need to correct the error or reproduce the packet immediately. Several schemes have been designed and used in this case that are collectively referred to as **forward error correction** (**FEC**) techniques. We briefly discuss some of the common techniques here.

10.5.1 Using Hamming Distance

We earlier discussed the Hamming distance for error detection. We said that to detect s errors, the minimum Hamming distance should be $d_{min} = s + 1$. For error detection, we definitely need more distance. It can be shown that to detect t errors, we need to have $d_{min} = 2t + 1$. In other words, if we want to correct 10 bits in a packet, we need to make the minimum hamming distance 21 bits, which means a lot of redundant bits need to be sent with the data. To give an example, consider the famous BCH code. In this code, if data is 99 bits, we need to send 255 bits (extra 156 bits) to correct just 23 possible bit errors. Most of the time we cannot afford such a redundancy. We give some examples of how to calculate the required bits in the practice set. Figure 10.20 shows the geometrical representation of this concept.

Figure 10.20 *Hamming distance for error correction*

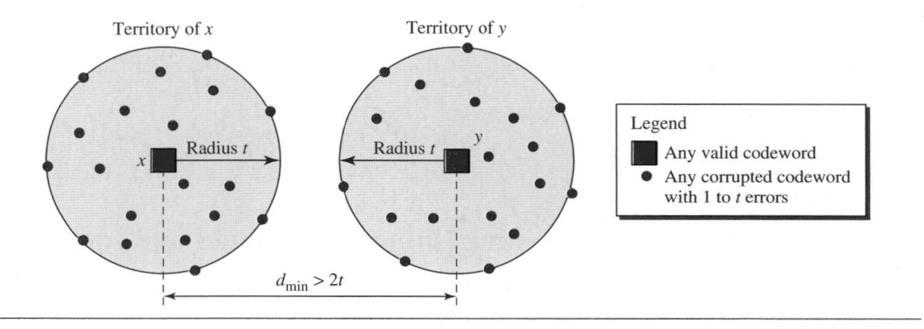

10.5.2 Using XOR

Another recommendation is to use the property of the exclusive OR operation as shown below.

$$\mathbf{R} = \mathbf{P}_1 \oplus \mathbf{P}_2 \oplus \ldots \oplus \mathbf{P}_i \oplus \ldots \oplus \mathbf{P}_N \quad \rightarrow \quad \mathbf{P}_i = \mathbf{P}_1 \oplus \mathbf{P}_2 \oplus \ldots \oplus \mathbf{R} \oplus \ldots \oplus \mathbf{P}_N$$

In other words, if we apply the exclusive OR operation on N data items (P_1 to P_N), we can recreate any of the data items by exclusive-ORing all of the items, replacing the one to be created by the result of the previous operation (R). This means that we can divide a packet into N chunks, create the exclusive OR of all the chunks and send $N + 1$ chunks. If any chunk is lost or corrupted, it can be created at the receiver site. Now the question is what should the value of N be. If $N = 4$, it means that we need to send 25 percent extra data and be able to correct the data if only one out of four chunks is lost.

10.5.3 Chunk Interleaving

Another way to achieve FEC in multimedia is to allow some small chunks to be missing at the receiver. We cannot afford to let all the chunks belonging to the same packet be missing; however, we can afford to let one chunk be missing in each packet. Figure 10.21 shows that we can divide each packet into 5 chunks (normally the number is much larger). We can then create data chunk by chunk (horizontally), but combine the chunks into packets vertically. In this case, each packet sent carries a chunk from several original packets. If the packet is lost, we miss only one chunk in each packet, which is normally acceptable in multimedia communication.

Figure 10.21 *Interleaving*

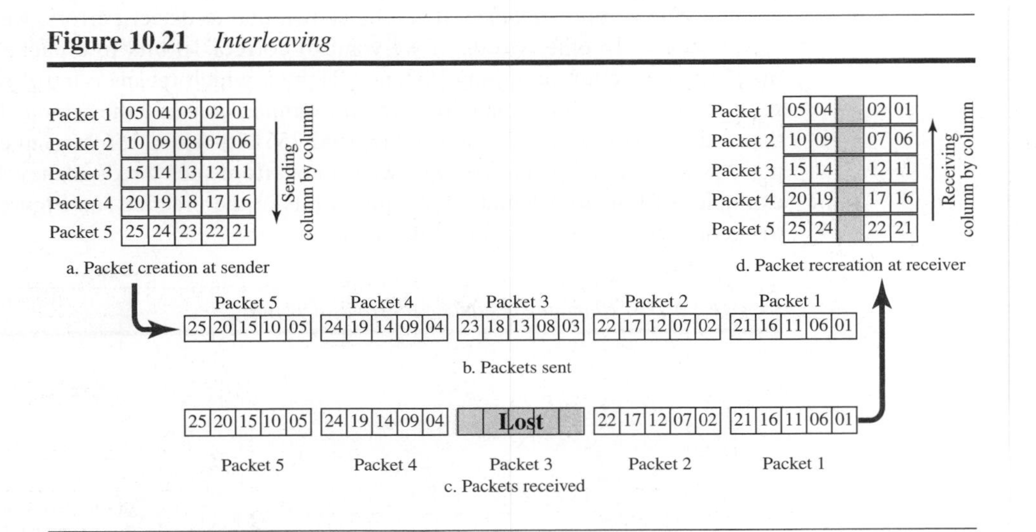

10.5.4 Combining Hamming Distance and Interleaving

Hamming distance and interleaving can be combined. We can first create n-bit packets that can correct t-bit errors. Then we interleave m rows and send the bits column by column. In this way, we can automatically correct burst errors up to $m \times t$-bit errors.

10.5.5 Compounding High- and Low-Resolution Packets

Still another solution is to create a duplicate of each packet with a low-resolution redundancy and combine the redundant version with the next packet. For example, we can create four low-resolution packets out of five high-resolution packets and send them as shown in Figure 10.22. If a packet is lost, we can use the low-resolution version from the next packet. Note that the low-resolution section in the first packet is empty. In this method, if the last packet is lost, it cannot be recovered, but we use the low-resolution version of a packet if the lost packet is not the last one. The audio and video reproduction does not have the same quality, but the lack of quality is not recognized most of the time.

Figure 10.22 *Compounding high- and low-resolution packets*

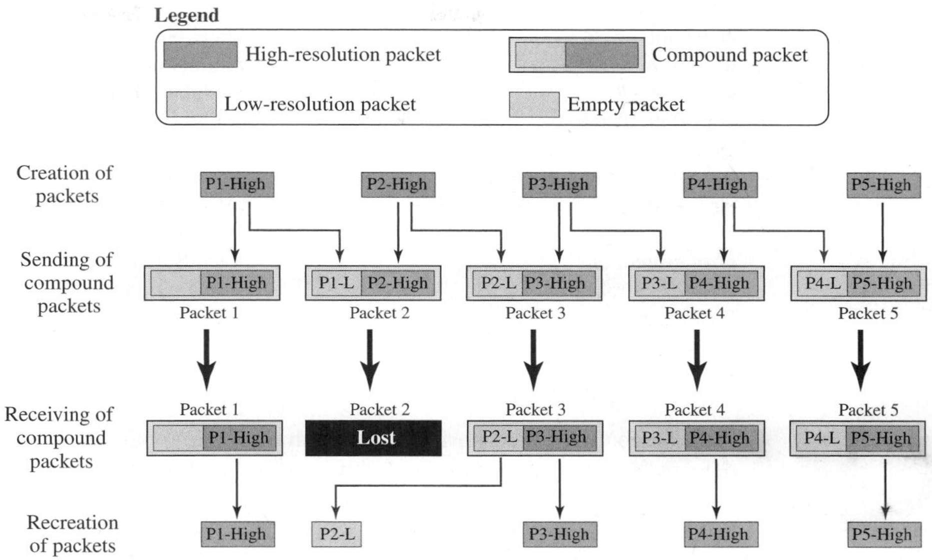

10.6 END-CHAPTER MATERIALS

10.6.1 Recommended Reading

For more details about subjects discussed in this chapter, we recommend the following books and RFCs. The items in brackets [...] refer to the reference list at the end of the text.

Books

Several excellent books discuss link-layer issues. Among them we recommend [Ham 80], [Zar 02], [Ror 96], [Tan 03], [GW 04], [For 03], [KMK 04], [Sta 04], [Kes 02], [PD 03], [Kei 02], [Spu 00], [KCK 98], [Sau 98], [Izz 00], [Per 00], and [WV 00].

RFCs

A discussion of the use of the checksum in the Internet can be found in RFC 1141.

10.6.2 Key Terms

block coding
burst error
check bit
checksum
codeword
convolution coding
cyclic code
cyclic redundancy check (CRC)
dataword
error
error correction
error detection
forward error correction (FEC)
generator polynomial
Hamming distance
interference
linear block code
minimum Hamming distance
modular arithmetic
one's complement
parity-check code
polynomial
redundancy
register
Reed-Solomon code
single-bit error
syndrome

10.6.3 Summary

Data can be corrupted during transmission. Some applications require that errors be detected and corrected. In a single-bit error, only one bit in the data unit has changed. A burst error means that two or more bits in the data unit have changed. To detect or correct errors, we need to send extra (redundant) bits with data. There are two main methods of error correction: forward error correction and correction by retransmission.

We can divide coding schemes into two broad categories: block coding and convolution coding. In coding, we need to use modulo-2 arithmetic. Operations in this arithmetic are very simple; addition and subtraction give the same results. We use the XOR (exclusive OR) operation for both addition and subtraction. In block coding, we divide our message into blocks, each of k bits, called *datawords*. We add r redundant bits to each block to make the length $n = k + r$. The resulting n-bit blocks are called *codewords*.

In block coding, errors be detected by using the following two conditions:

a. The receiver has (or can find) a list of valid codewords.

b. The original codeword has changed to an invalid one.

The Hamming distance between two words is the number of differences between corresponding bits. The minimum Hamming distance is the smallest Hamming distance between all possible pairs in a set of words. To guarantee the detection of up to s errors in all cases, the minimum Hamming distance in a block code must be $d_{min} = s + 1$. To guarantee correction of up to t errors in all cases, the minimum Hamming distance in a block code must be $d_{min} = 2t + 1$.

In a linear block code, the exclusive OR (XOR) of any two valid codewords creates another valid codeword.

A simple parity-check code is a single-bit error-detecting code in which $n = k + 1$ with $d_{min} = 2$. A simple parity-check code can detect an odd number of errors.

All Hamming codes discussed in this book have $d_{min} = 3$. The relationship between m and n in these codes is $n = 2m - 1$.

Cyclic codes are special linear block codes with one extra property. In a cyclic code, if a codeword is cyclically shifted (rotated), the result is another codeword. A category of cyclic codes called the cyclic redundancy check (CRC) is used in networks such as LANs and WANs.

A pattern of 0s and 1s can be represented as a polynomial with coefficients of 0 and 1. Traditionally, the Internet has been using a 16-bit checksum, which uses one's complement arithmetic. In this arithmetic, we can represent unsigned numbers between 0 and $2^n - 1$ using only n bits.

10.7 PRACTICE SET

10.7.1 Quizzes

A set of interactive quizzes for this chapter can be found on the book website. It is strongly recommended that the student take the quizzes to check his/her understanding of the materials before continuing with the practice set.

10.7.2 Questions

Q10-1. In a block code, a dataword is 20 bits and the corresponding codeword is 25 bits. What are the values of k, r, and n according to the definitions in the text? How many redundant bits are added to each dataword?

Q10-2. In CRC, if the dataword is 5 bits and the codeword is 8 bits, how many 0s need to be added to the dataword to make the dividend? What is the size of the remainder? What is the size of the divisor?

Q10-3. Can the value of a traditional checksum be all 0s (in binary)? Defend your answer.

Q10-4. If we want to be able to detect two-bit errors, what should be the minimum Hamming distance?

Q10-5. A category of error detecting (and correcting) code, called the ***Hamming code,*** is a code in which $d_{min} = 3$. This code can detect up to two errors (or correct one single error). In this code, the values of n, k, and r are related as: $n = 2^r - 1$ and $k = n - r$. Find the number of bits in the dataword and the codewords if r is 3.

Q10-6. In a codeword, we add two redundant bits to each 8-bit data word. Find the number of

a. valid codewords. **b.** invalid codewords.

Q10-7. In CRC, which of the following generators (divisors) guarantees the detection of a single bit error?

a. 101 **b.** 100 **c.** 1

Q10-8. What is the definition of a linear block code?

Q10-9. How does a single-bit error differ from a burst error?

Q10-10. Show how the Fletcher algorithm (Figure 10.18) attaches weights to the data items when calculating the checksum.

Q10-11. In CRC, we have chosen the generator 1100101. What is the probability of detecting a burst error of length

a. 5? **b.** 7? **c.** 10?

Q10-12. In CRC, which of the following generators (divisors) guarantees the detection of an odd number of errors?

a. 10111 **b.** 101101 **c.** 111

Q10-13. What is the minimum Hamming distance?

Q10-14. Assume we are sending data items of 16-bit length. If two data items are swapped during transmission, can the traditional checksum detect this error? Explain.

Q10-15. Show how the Adler algorithm (Figure 10.19) attaches weights to the data items when calculating the checksum.

10.7.3 Problems

P10-1. Using the code in Table 10.2, what is the dataword if each of the following codewords is received?

a. 01011 **b.** 11111 **c.** 00000 **d.** 11011

P10-2. Assume we want to send a dataword of two bits using FEC based on the Hamming distance. Show how the following list of datawords/codewords can automatically correct up to a one-bit error in transmission.

00 → 00000 01→ 01011 10 → 10101 11 → 11110

P10-3. We can create a general formula for correcting any number of errors (m) in a codeword of size (n). Develop such a formula. Use the combination of n objects taking x objects at a time.

P10-4. Answer the following questions:

a. What is the polynomial representation of 101110?

b. What is the result of shifting 101110 three bits to the left?

c. Repeat part b using polynomials.

d. What is the result of shifting 101110 four bits to the right?

e. Repeat part d using polynomials.

P10-5. Traditional checksum calculation needs to be done in one's complement arithmetic. Computers and calculators today are designed to do calculations in two's complement arithmetic. One way to calculate the traditional checksum is to add the numbers in two's complement arithmetic, find the quotient and remainder of dividing the result by 2^{16}, and add the quotient and the remainder to get the sum in one's complement. The checksum can be found by subtracting the sum from $2^{16} - 1$. Use the above method to find the checksum of the following four numbers: 43,689, 64,463, 45,112, and 59,683.

P10-6. Given the dataword 101001111 and the divisor 10111, show the generation of the CRC codeword at the sender site (using binary division).

P10-7. Manually simulate the Fletcher algorithm (Figure 10.18) to calculate the checksum of the following bytes: $(2B)_{16}$, $(3F)_{16}$, $(6A)_{16}$, and $(AF)_{16}$. Also show that the result is a weighted checksum.

P10-8. In Table 10.1, the sender sends dataword 10. A 3-bit burst error corrupts the codeword. Can the receiver detect the error? Defend your answer.

P10-9. A simple parity-check bit, which is normally added at the end of the word (changing a 7-bit ASCII character to a byte), cannot detect even numbers of errors. For example, two, four, six, or eight errors cannot be detected in this way. A better solution is to organize the characters in a table and create row and column parities. The bit in the row parity is sent with the byte, the column parity is sent as an extra byte (Figure 10.23).

Figure 10.23 *P10-9*

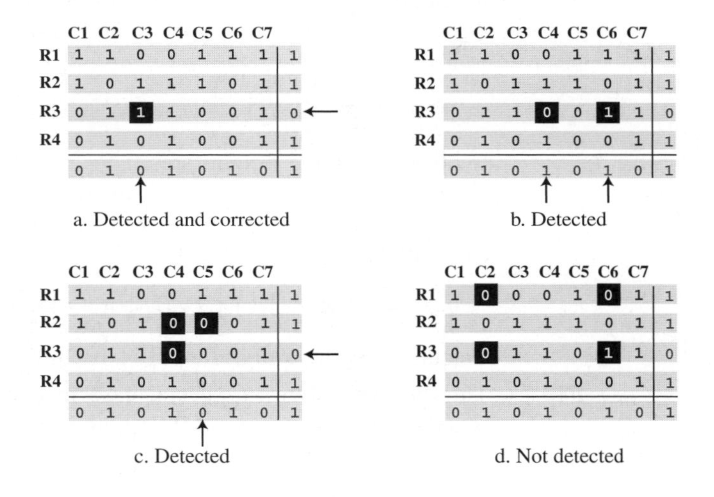

Show how the following errors can be detected:

a. An error at (R3, C3).

b. Two errors at (R3, C4) and (R3, C6).

c. Three errors at (R2, C4), (R2, C5), and (R3, C4).

d. Four errors at (R1, C2), (R1, C6), (R3, C2), and (R3, C6).

P10-10. Although it can be formally proved that the code in Table 10.3 is both linear and cyclic, use only two tests to partially prove the fact:

a. Test the cyclic property on codeword 0101100.

b. Test the linear property on codewords 0010110 and 1111111.

P10-11. Assume we need to create codewords that can automatically correct a one-bit error. What should the number of redundant bits (r) be, given the number of bits in the dataword (k)? Remember that the codeword needs to be $n = k + r$ bits, called $C(n, k)$. After finding the relationship, find the number of bits in r if k is 1, 2, 5, 50, or 1000.

P10-12. Referring to the CRC-8 polynomial in Table 10.7, answer the following questions:

a. Does it detect a single error? Defend your answer.

b. Does it detect a burst error of size 6? Defend your answer.

c. What is the probability of detecting a burst error of size 9?

d. What is the probability of detecting a burst error of size 15?

P10-13. One of the examples of a weighted checksum is the ISBN-10 code we see printed on the back cover of some books. In ISBN-10, there are 9 decimal digits that define the country, the publisher, and the book. The tenth (rightmost) digit is a checksum digit. The code, $D_1D_2D_3D_4D_5D_6D_7D_8D_9C$, satisfies the following.

$$[(10 \times D_1) + (9 \times D_2) + (8 \times D_3) + \ldots + (2 \times D_9) + (1 \times C)] \bmod 11 = 0$$

In other words, the weights are 10, 9, . . .,1. If the calculated value for C is 10, one uses the letter X instead. By replacing each weight w with its complement in modulo 11 arithmetic $(11 - w)$, it can be shown that the check digit can be calculated as shown below.

$$C = [(1 \times D_1) + (2 \times D_2) + (3 \times D_3) + \ldots + (9 \times D_9)] \bmod 11$$

Calculate the check digit for ISBN-10: **0-07-296775-C**.

P10-14. This problem shows a special case in checksum handling. A sender has two data items to send: $(4567)_{16}$ and $(BA98)_{16}$. What is the value of the checksum?

P10-15. What is the Hamming distance for each of the following codewords?

a. d (10000, 00000)

b. d (10101, 10000)

c. d (00000, 11111)

d. d (00000, 00000)

P10-16. Prove that the code represented by the following codewords is not linear. You need to find only one case that violates the linearity.

{(00000), (01011), (10111), (11111)}

P10-17. In the interleaving approach to FEC, assume each packet contains 10 samples from a sampled piece of music. Instead of loading the first packet with the first 10 samples, the second packet with the second 10 samples, and so on, the sender loads the first packet with the odd-numbered samples of the first 20 samples, the second packet with the even-numbered samples of the first 20 samples, and so on. The receiver reorders the samples and plays them. Now assume that the third packet is lost in transmission. What will be missed at the receiver site?

P10-18. Manually simulate the Adler algorithm (Figure 10.19) to calculate the checksum of the following words: $(FBFF)_{16}$ and $(EFAA)_{16}$. Also show that the result is a weighted checksum.

P10-19. Referring to the CRC-32 polynomial in Table 10.4, answer the following questions:

a. Does it detect a single error? Defend your answer.

b. Does it detect a burst error of size 16? Defend your answer.

c. What is the probability of detecting a burst error of size 33?

d. What is the probability of detecting a burst error of size 55?

P10-20. An ISBN-13 code, a new version of ISBN-10, is another example of a weighted checksum with 13 digits, in which there are 12 decimal digits defining the book and the last digit is the checksum digit. The code, $D_1D_2D_3D_4D_5D_6D_7D_8D_9D_{10}D_{11}D_{12}C$, satisfies the following.

$$[(1 \times D_1) + (3 \times D_2) + (1 \times D_3) + \ldots + (3 \times D_{12}) + (1 \times C)] \bmod 10 = 0$$

In other words, the weights are 1 and 3 alternately. Using the above description, calculate the check digit for ISBN-13: **978-0-07-296775-C**.

P10-21. Apply the following operations on the corresponding polynomials:

a. $(x^3 + x^2 + x + 1) + (x^4 + x^2 + x + 1)$

b. $(x^3 + x^2 + x + 1) - (x^4 + x^2 + x + 1)$

c. $(x^3 + x^2) \times (x^4 + x^2 + x + 1)$

d. $(x^3 + x^2 + x + 1) / (x^2 + 1)$

P10-22. Assume a packet is made only of four 16-bit words $(A7A2)_{16}$, $(CABF)_{16}$, $(903A)_{16}$, and $(A123)_{16}$. Manually simulate the algorithm in Figure 10.17 to find the checksum.

P10-23. Which of the following CRC generators guarantee the detection of a single bit error?

a. $x^3 + x + 1$ **b.** $x^4 + x^2$ **c.** 1 **d.** $x^2 + 1$

P10-24. Assume that the probability that a bit in a data unit is corrupted during transmission is p. Find the probability that x number of bits are corrupted in an n-bit data unit for each of the following cases.

a. $n = 8, x = 1, p = 0.2$

b. $n = 16, x = 3, p = 0.3$

c. $n = 32, x = 10, p = 0.4$

P10-25. Exclusive-OR (XOR) is one of the most used operations in the calculation of codewords. Apply the exclusive-OR operation on the following pairs of patterns. Interpret the results.

a. (10001) ⊕ (10001) **b.** (11100) ⊕ (00000) **c.** (10011) ⊕ (11111)

P10-26. In problem P10-11, we tried to find the number of bits to be added to a dataword to correct a single-bit error. If we need to correct more than one bit, the number of redundant bits increases. What should the number of redundant bits (r) be to automatically correct one or two bits (not necessarily contiguous) in a dataword of size k? After finding the relationship, find the number of bits in r if k is 1, 2, 5, 50, or 1000.

P10-27. What is the maximum effect of a 2-ms burst of noise on data transmitted at the following rates?

a. 1500 bps **b.** 12 kbps **c.** 100 kbps **d.** 100 Mbps

P10-28. In Figure 10.22, assume we have 100 packets. We have created two sets of packets with high and low resolutions. Each high-resolution packet carries on average 700 bits. Each low-resolution packet carries on average 400 bits. How many extra bits are we sending in this scheme for the sake of FEC? What is the percentage of overhead?

P10-29. Referring to the CRC-8 in Table 5.4, answer the following questions:

a. Does it detect a single error? Defend your answer.

b. Does it detect a burst error of size 6? Defend your answer.

c. What is the probability of detecting a burst error of size 9?

d. What is the probability of detecting a burst error of size 15?

P10-30. Assuming even parity, find the parity bit for each of the following data units.

a. 1001011 **b.** 0001100 **c.** 1000000 **d.** 1110111

10.8 SIMULATION EXPERIMENTS

10.8.1 Applets

We have created some Java applets to show some of the main concepts discussed in this chapter. It is strongly recommended that the students activate these applets on the book website and carefully examine the protocols in action.

10.9 PROGRAMMING ASSIGNMENTS

For each of the following assignments, write a program in the programming language you are familiar with.

Prg10-1. A program to simulate the calculation of CRC.

Prg10-2. A program to simulate the calculation of traditional checksum.

Prg10-3. A program to simulate the calculation of Fletcher checksum.

Prg10-4. A program to simulate the calculation of Adler checksum.

CHAPTER 11

Data Link Control (DLC)

As we discussed in Chapter 9, the data-link layer is divided into two sublayers. In this chapter, we discuss the upper sublayer of the data-link layer (DLC). The lower sublayer, multiple access control (MAC) will be discussed in Chapter 12. We have already discussed error detection and correction, an issue that is encountered in several layers, in Chapter 10.

This chapter is divided into four sections.

- The first section discusses the general services provided by the DLC sublayer. It first describes framing and two types of frames used in this sublayer. The section then discusses flow and error control. Finally, the section explains that a DLC protocol can be either connectionless or connection-oriented.
- The second section discusses some simple and common data-link protocols that are implemented at the DLC sublayer. The section first describes the Simple Protocol. It then explains the Stop-and-Wait Protocol.
- The third section introduces HDLC, a protocol that is the basis of all common data-link protocols in use today such as PPP and Ethernet. The section first talks about configurations and transfer modes. It then describes framing and three different frame formats used in this protocol.
- The fourth section discusses PPP, a very common protocol for point-to-point access. It first introduces the services provided by the protocol. The section also describes the format of the frame in this protocol. It then describes the transition mode in the protocol using an FSM. The section finally explains multiplexing in PPP.

11.1 DLC SERVICES

The **data link control (DLC)** deals with procedures for communication between two adjacent nodes—node-to-node communication—no matter whether the link is dedicated or broadcast. Data link control functions include *framing* and *flow and error control*. In this section, we first discuss framing, or how to organize the bits that are carried by the physical layer. We then discuss flow and error control.

11.1.1 Framing

Data transmission in the physical layer means moving bits in the form of a signal from the source to the destination. The physical layer provides bit synchronization to ensure that the sender and receiver use the same bit durations and timing. We discussed the physical layer in Part II of the book.

The data-link layer, on the other hand, needs to pack bits into frames, so that each frame is distinguishable from another. Our postal system practices a type of framing. The simple act of inserting a letter into an envelope separates one piece of information from another; the envelope serves as the delimiter. In addition, each envelope defines the sender and receiver addresses, which is necessary since the postal system is a many-to-many carrier facility.

Framing in the data-link layer separates a message from one source to a destination by adding a sender address and a destination address. The destination address defines where the packet is to go; the sender address helps the recipient acknowledge the receipt.

Although the whole message could be packed in one frame, that is not normally done. One reason is that a frame can be very large, making flow and error control very inefficient. When a message is carried in one very large frame, even a single-bit error would require the retransmission of the whole frame. When a message is divided into smaller frames, a single-bit error affects only that small frame.

Frame Size

Frames can be of fixed or variable size. In *fixed-size framing,* there is no need for defining the boundaries of the frames; the size itself can be used as a delimiter. An example of this type of framing is the ATM WAN, which uses frames of fixed size called *cells*. We discuss ATM in Chapter 14.

Our main discussion in this chapter concerns *variable-size framing,* prevalent in local-area networks. In variable-size framing, we need a way to define the end of one frame and the beginning of the next. Historically, two approaches were used for this purpose: a character-oriented approach and a bit-oriented approach.

Character-Oriented Framing

In *character-oriented (or byte-oriented) framing,* data to be carried are 8-bit characters from a coding system such as ASCII (see Appendix A). The header, which normally carries the source and destination addresses and other control information, and the trailer, which carries error detection redundant bits, are also multiples of 8 bits. To separate one frame from the next, an 8-bit (1-byte) **flag** is added at the beginning and the end of a frame. The flag, composed of protocol-dependent special characters, signals the start or end of a frame. Figure 11.1 shows the format of a frame in a character-oriented protocol.

Figure 11.1 *A frame in a character-oriented protocol*

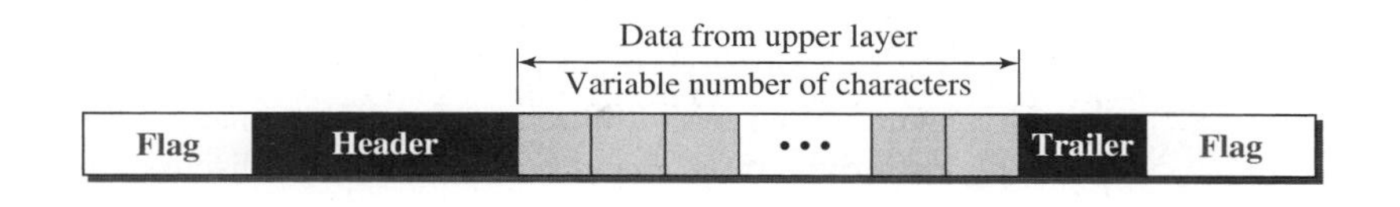

Character-oriented framing was popular when only text was exchanged by the data-link layers. The flag could be selected to be any character not used for text communication. Now, however, we send other types of information such as graphs, audio, and video; any character used for the flag could also be part of the information. If this happens, the receiver, when it encounters this pattern in the middle of the data, thinks it has reached the end of the frame. To fix this problem, a byte-stuffing strategy was added to character-oriented framing. In **byte stuffing** (or character stuffing), a special byte is added to the data section of the frame when there is a character with the same pattern as the flag. The data section is stuffed with an extra byte. This byte is usually called the *escape character (ESC)* and has a predefined bit pattern. Whenever the receiver encounters the ESC character, it removes it from the data section and treats the next character as data, not as a delimiting flag. Figure 11.2 shows the situation.

Figure 11.2 *Byte stuffing and unstuffing*

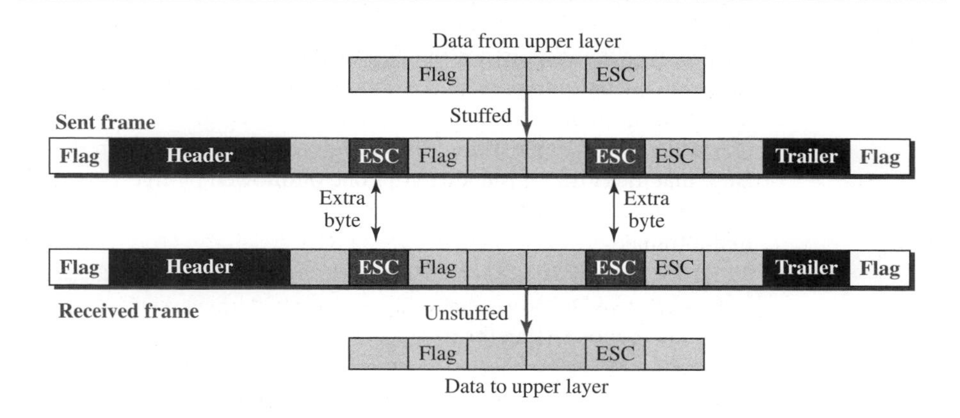

Byte stuffing is the process of adding one extra byte whenever there is a flag or escape character in the text.

Byte stuffing by the escape character allows the presence of the flag in the data section of the frame, but it creates another problem. What happens if the text contains one or more escape characters followed by a byte with the same pattern as the flag? The receiver removes the escape character, but keeps the next byte, which is incorrectly interpreted as the end of the frame. To solve this problem, the escape characters that are part of the text must also be marked by another escape character. In other words, if the escape character is part of the text, an extra one is added to show that the second one is part of the text.

Character-oriented protocols present another problem in data communications. The universal coding systems in use today, such as Unicode, have 16-bit and 32-bit characters that conflict with 8-bit characters. We can say that, in general, the tendency is moving toward the bit-oriented protocols that we discuss next.

Bit-Oriented Framing

In *bit-oriented framing,* the data section of a frame is a sequence of bits to be interpreted by the upper layer as text, graphic, audio, video, and so on. However, in addition to headers (and possible trailers), we still need a delimiter to separate one frame from the other. Most protocols use a special 8-bit pattern flag, 01111110, as the delimiter to define the beginning and the end of the frame, as shown in Figure 11.3.

Figure 11.3 *A frame in a bit-oriented protocol*

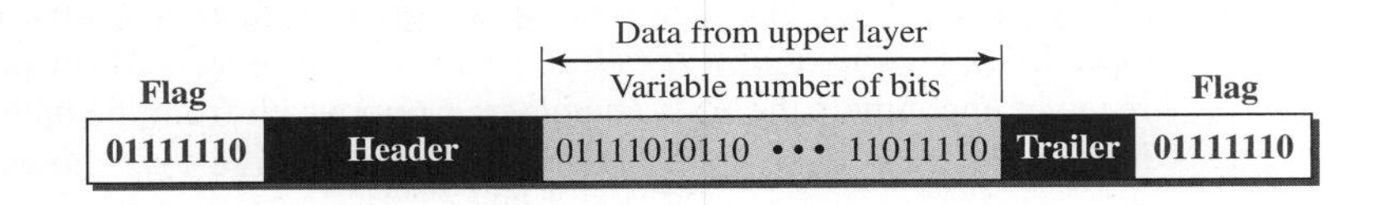

This flag can create the same type of problem we saw in the character-oriented protocols. That is, if the flag pattern appears in the data, we need to somehow inform the receiver that this is not the end of the frame. We do this by stuffing 1 single bit (instead of 1 byte) to prevent the pattern from looking like a flag. The strategy is called **bit stuffing.** In bit stuffing, if a 0 and five consecutive 1 bits are encountered, an extra 0 is added. This extra stuffed bit is eventually removed from the data by the receiver. Note that the extra bit is added after one 0 followed by five 1s regardless of the value of the next bit. This guarantees that the flag field sequence does not inadvertently appear in the frame.

Bit stuffing is the process of adding one extra 0 whenever five consecutive 1s follow a 0 in the data, so that the receiver does not mistake the pattern 0111110 for a flag.

Figure 11.4 shows bit stuffing at the sender and bit removal at the receiver. Note that even if we have a 0 after five 1s, we still stuff a 0. The 0 will be removed by the receiver.

This means that if the flaglike pattern 01111110 appears in the data, it will change to 011111010 (stuffed) and is not mistaken for a flag by the receiver. The real flag 01111110 is not stuffed by the sender and is recognized by the receiver.

Figure 11.4 *Bit stuffing and unstuffing*

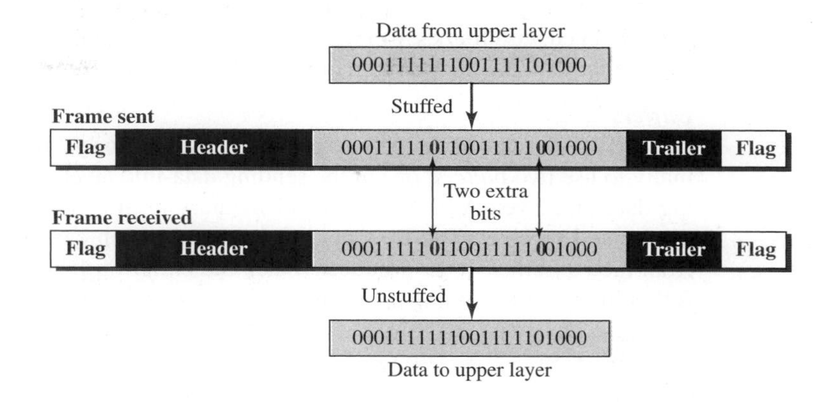

11.1.2 Flow and Error Control

We briefly defined flow and error control in Chapter 9; we elaborate on these two issues here. One of the responsibilities of the data-link control sublayer is flow and error control at the data-link layer.

Flow Control

Whenever an entity produces items and another entity consumes them, there should be a balance between production and consumption rates. If the items are produced faster than they can be consumed, the consumer can be overwhelmed and may need to discard some items. If the items are produced more slowly than they can be consumed, the consumer must wait, and the system becomes less efficient. Flow control is related to the first issue. We need to prevent losing the data items at the consumer site.

In communication at the data-link layer, we are dealing with four entities: network and data-link layers at the sending node and network and data-link layers at the receiving node. Although we can have a complex relationship with more than one producer and consumer (as we will see in Chapter 23), we ignore the relationships between networks and data-link layers and concentrate on the relationship between two data-link layers, as shown in Figure 11.5.

Figure 11.5 *Flow control at the data-link layer*

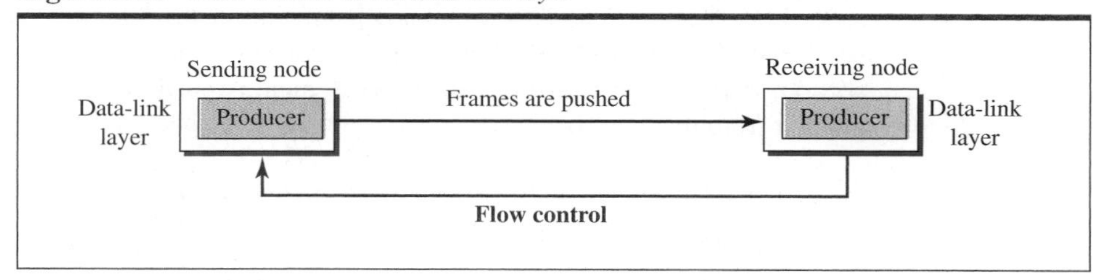

The figure shows that the data-link layer at the sending node tries to push frames toward the data-link layer at the receiving node. If the receiving node cannot process and deliver the packet to its network at the same rate that the frames arrive, it becomes overwhelmed with frames. Flow control in this case can be feedback from the receiving node to the sending node to stop or slow down pushing frames.

Buffers

Although flow control can be implemented in several ways, one of the solutions is normally to use two *buffers*; one at the sending data-link layer and the other at the receiving data-link layer. A buffer is a set of memory locations that can hold packets at the sender and receiver. The flow control communication can occur by sending signals from the consumer to the producer. When the buffer of the receiving data-link layer is full, it informs the sending data-link layer to stop pushing frames.

Example 11.1

The above discussion requires that the consumers communicate with the producers on two occasions: when the buffer is full and when there are vacancies. If the two parties use a buffer with only one slot, the communication can be easier. Assume that each data-link layer uses one single memory slot to hold a frame. When this single slot in the receiving data-link layer is empty, it sends a note to the network layer to send the next frame.

Error Control

Since the underlying technology at the physical layer is not fully reliable, we need to implement error control at the data-link layer to prevent the receiving node from delivering corrupted packets to its network layer. Error control at the data-link layer is normally very simple and implemented using one of the following two methods. In both methods, a CRC is added to the frame header by the sender and checked by the receiver.

- In the first method, if the frame is corrupted, it is silently discarded; if it is not corrupted, the packet is delivered to the network layer. This method is used mostly in wired LANs such as Ethernet.
- In the second method, if the frame is corrupted, it is silently discarded; if it is not corrupted, an acknowledgment is sent (for the purpose of both flow and error control) to the sender.

Combination of Flow and Error Control

Flow and error control can be combined. In a simple situation, the acknowledgment that is sent for flow control can also be used for error control to tell the sender the packet has arrived uncorrupted. The lack of acknowledgment means that there is a problem in the sent frame. We show this situation when we discuss some simple protocols in the next section. A frame that carries an acknowledgment is normally called an *ACK* to distinguish it from the data frame.

11.1.3 Connectionless and Connection-Oriented

A DLC protocol can be either connectionless or connection-oriented. We will discuss this issue very briefly here, but we return to this topic in the network and transport layer.

Connectionless Protocol

In a connectionless protocol, frames are sent from one node to the next without any relationship between the frames; each frame is independent. Note that the term *connectionless* here does not mean that there is no physical connection (transmission medium) between the nodes; it means that there is no *connection* between frames. The frames are not numbered and there is no sense of ordering. Most of the data-link protocols for LANs are connectionless protocols.

Connection-Oriented Protocol

In a connection-oriented protocol, a logical connection should first be established between the two nodes (setup phase). After all frames that are somehow related to each other are transmitted (transfer phase), the logical connection is terminated (teardown phase). In this type of communication, the frames are numbered and sent in order. If they are not received in order, the receiver needs to wait until all frames belonging to the same set are received and then deliver them in order to the network layer. Connection-oriented protocols are rare in wired LANs, but we can see them in some point-to-point protocols, some wireless LANs, and some WANs.

11.2 DATA-LINK LAYER PROTOCOLS

Traditionally four protocols have been defined for the data-link layer to deal with flow and error control: Simple, Stop-and-Wait, Go-Back-N, and Selective-Repeat. Although the first two protocols still are used at the data-link layer, the last two have disappeared. We therefore briefly discuss the first two protocols in this chapter, in which we need to understand some wired and wireless LANs. We postpone the discussion of all four, in full detail, to Chapter 23, where we discuss the transport layer.

The behavior of a data-link-layer protocol can be better shown as a **finite state machine (FSM).** An FSM is thought of as a machine with a finite number of states. The machine is always in one of the states until an *event* occurs. Each event is associated with two reactions: defining the list (possibly empty) of actions to be performed and determining the next state (which can be the same as the current state). One of the states must be defined as the initial state, the state in which the machine starts when it turns on. In Figure 11.6, we show an example of a machine using FSM. We have used rounded-corner rectangles to show states, colored text to show events, and regular black text to show actions. A horizontal line is used to separate the event from the actions, although later we replace the horizontal line with a slash. The arrow shows the movement to the next state.

The figure shows a machine with three states. There are only three possible events and three possible actions. The machine starts in state I. If event 1 occurs, the machine performs actions 1 and 2 and moves to state II. When the machine is in state II, two events may occur. If event 1 occurs, the machine performs action 3 and remains in the same state, state II. If event 3 occurs, the machine performs no action, but move to state I.

Figure 11.6 *Connectionless and connection-oriented service represented as FSMs*

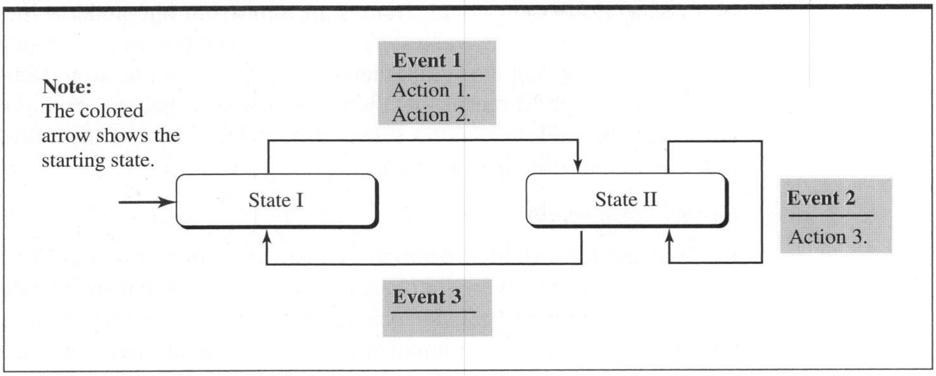

Note:
The colored arrow shows the starting state.

Event 1
Action 1.
Action 2.

State I

State II

Event 2
Action 3.

Event 3

11.2.1 Simple Protocol

Our first protocol is a **simple protocol** with neither flow nor error control. We assume that the receiver can immediately handle any frame it receives. In other words, the receiver can never be overwhelmed with incoming frames. Figure 11.7 shows the layout for this protocol.

Figure 11.7 *Simple protocol*

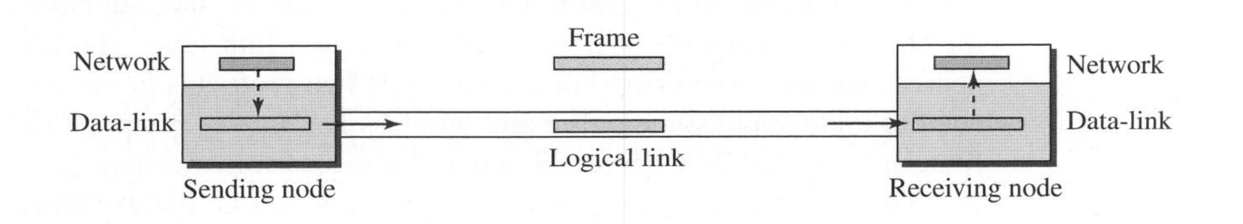

The data-link layer at the sender gets a packet from its network layer, makes a frame out of it, and sends the frame. The data-link layer at the receiver receives a frame from the link, extracts the packet from the frame, and delivers the packet to its network layer. The data-link layers of the sender and receiver provide transmission services for their network layers.

FSMs

The sender site should not send a frame until its network layer has a message to send. The receiver site cannot deliver a message to its network layer until a frame arrives. We can show these requirements using two FSMs. Each FSM has only one state, the *ready state*. The sending machine remains in the ready state until a request comes from the process in the network layer. When this event occurs, the sending machine encapsulates the message in a frame and sends it to the receiving machine. The receiving machine remains in the ready state until a frame arrives from the sending machine. When this event occurs, the receiving machine decapsulates the message out of the frame and delivers it to the process at the network layer. Figure 11.8 shows the FSMs for the simple protocol. We'll see more in Chapter 23, which uses this protocol.

Figure 11.8 *FSMs for the simple protocol*

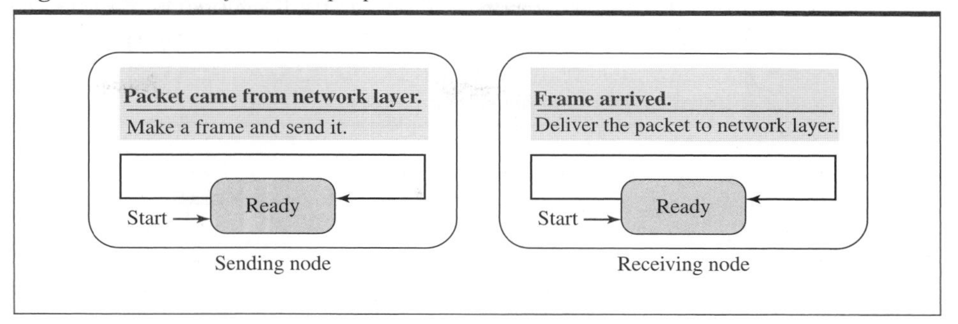

Example 11.2

Figure 11.9 shows an example of communication using this protocol. It is very simple. The sender sends frames one after another without even thinking about the receiver.

Figure 11.9 *Flow diagram for Example 11.2*

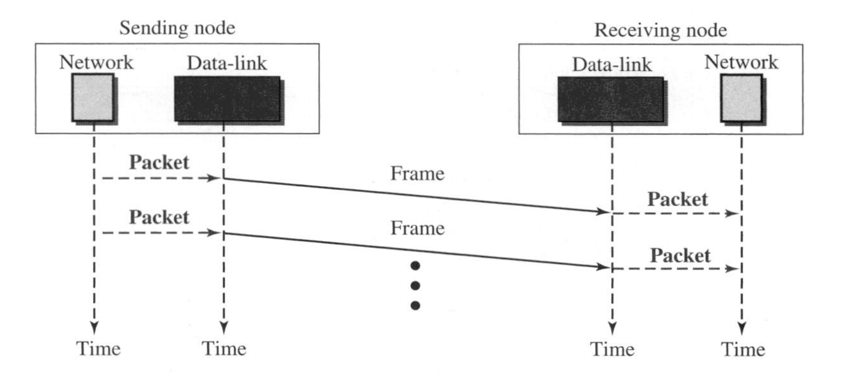

11.2.2 Stop-and-Wait Protocol

Our second protocol is called the **Stop-and-Wait protocol,** which uses both flow and error control. We show a primitive version of this protocol here, but we discuss the more sophisticated version in Chapter 23 when we have learned about sliding windows. In this protocol, the sender sends one frame at a time and waits for an acknowledgment before sending the next one. To detect corrupted frames, we need to add a CRC (see Chapter 10) to each data frame. When a frame arrives at the receiver site, it is checked. If its CRC is incorrect, the frame is corrupted and silently discarded. The silence of the receiver is a signal for the sender that a frame was either corrupted or lost. Every time the sender sends a frame, it starts a timer. If an acknowledgment arrives before the timer expires, the timer is stopped and the sender sends the next frame (if it has one to send). If the timer expires, the sender resends the previous frame, assuming that the frame was either lost or corrupted. This means that the sender needs to keep a copy of the frame until its acknowledgment arrives. When the corresponding acknowledgment arrives, the sender discards the copy and sends the next frame if it is ready. Figure 11.10 shows the outline for the Stop-and-Wait protocol. Note that only one frame and one acknowledgment can be in the channels at any time.

Figure 11.10 *Stop-and-Wait protocol*

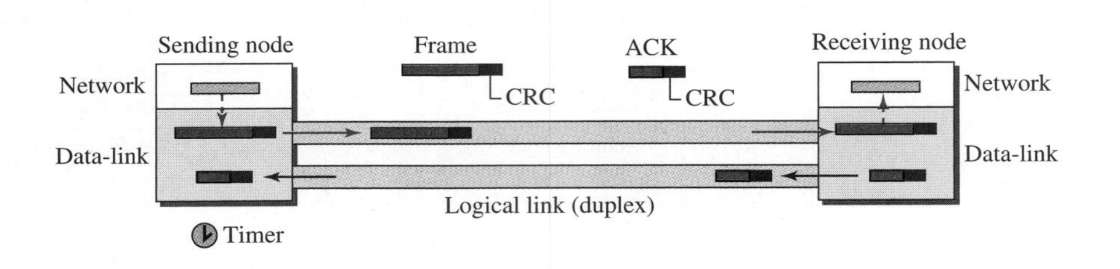

FSMs

Figure 11.11 shows the FSMs for our primitive Stop-and-Wait protocol.

Figure 11.11 *FSM for the Stop-and-Wait protocol*

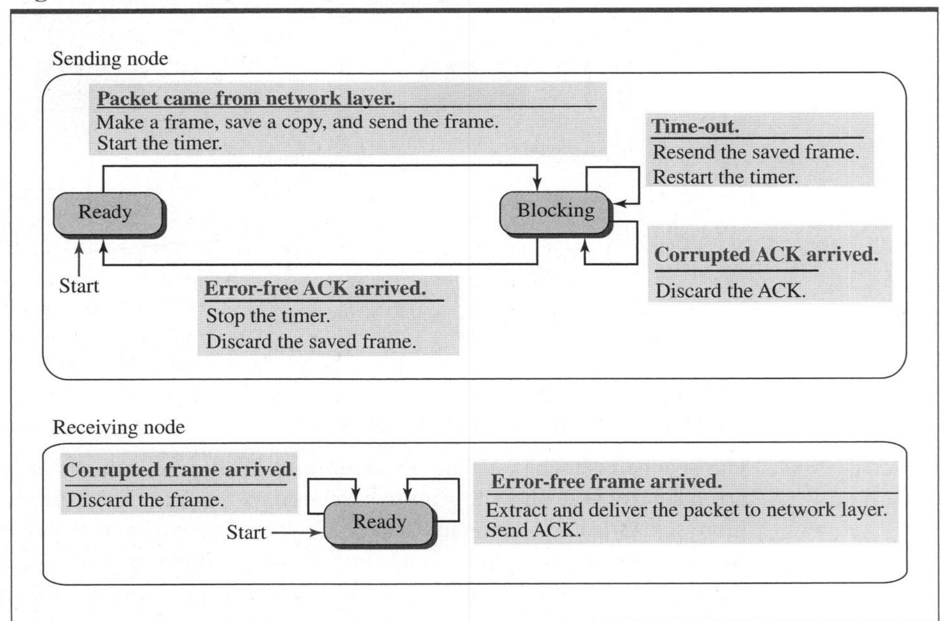

We describe the sender and receiver states below.

Sender States

The sender is initially in the ready state, but it can move between the ready and blocking state.

- ❑ ***Ready State.*** When the sender is in this state, it is only waiting for a packet from the network layer. If a packet comes from the network layer, the sender creates a frame, saves a copy of the frame, starts the only timer and sends the frame. The sender then moves to the blocking state.
- ❑ ***Blocking State.*** When the sender is in this state, three events can occur:
  a. If a time-out occurs, the sender resends the saved copy of the frame and restarts the timer.
  b. If a corrupted ACK arrives, it is discarded.
  c. If an error-free ACK arrives, the sender stops the timer and discards the saved copy of the frame. It then moves to the ready state.

Receiver

The receiver is always in the *ready* state. Two events may occur:

a. If an error-free frame arrives, the message in the frame is delivered to the network layer and an ACK is sent.

b. If a corrupted frame arrives, the frame is discarded.

Example 11.3

Figure 11.12 shows an example. The first frame is sent and acknowledged. The second frame is sent, but lost. After time-out, it is resent. The third frame is sent and acknowledged, but the acknowledgment is lost. The frame is resent. However, there is a problem with this scheme. The network layer at the receiver site receives two copies of the third packet, which is not right. In the next section, we will see how we can correct this problem using sequence numbers and acknowledgment numbers.

Sequence and Acknowledgment Numbers

We saw a problem in Example 11.3 that needs to be addressed and corrected. Duplicate packets, as much as corrupted packets, need to be avoided. As an example, assume we are ordering some item online. If each packet defines the specification of an item to be ordered, duplicate packets mean ordering an item more than once. To correct the problem in Example 11.3, we need to add **sequence numbers** to the data frames and **acknowledgment numbers** to the ACK frames. However, numbering in this case is very simple. Sequence numbers are 0, 1, 0, 1, 0, 1, . . . ; the acknowledgment numbers can also be 1, 0, 1, 0, 1, 0, ... In other words, the sequence numbers start with 0, the acknowledgment numbers start with 1. An acknowledgment number always defines the sequence number of the next frame to receive.

Example 11.4

Figure 11.13 shows how adding sequence numbers and acknowledgment numbers can prevent duplicates. The first frame is sent and acknowledged. The second frame is sent, but lost. After time-out, it is resent. The third frame is sent and acknowledged, but the acknowledgment is lost. The frame is resent.

FSMs with Sequence and Acknowledgment Numbers

We can change the FSM in Figure 11.11 to include the sequence and acknowledgment numbers, but we leave this as a problem at the end of the chapter.

Figure 11.12 *Flow diagram for Example 11.3*

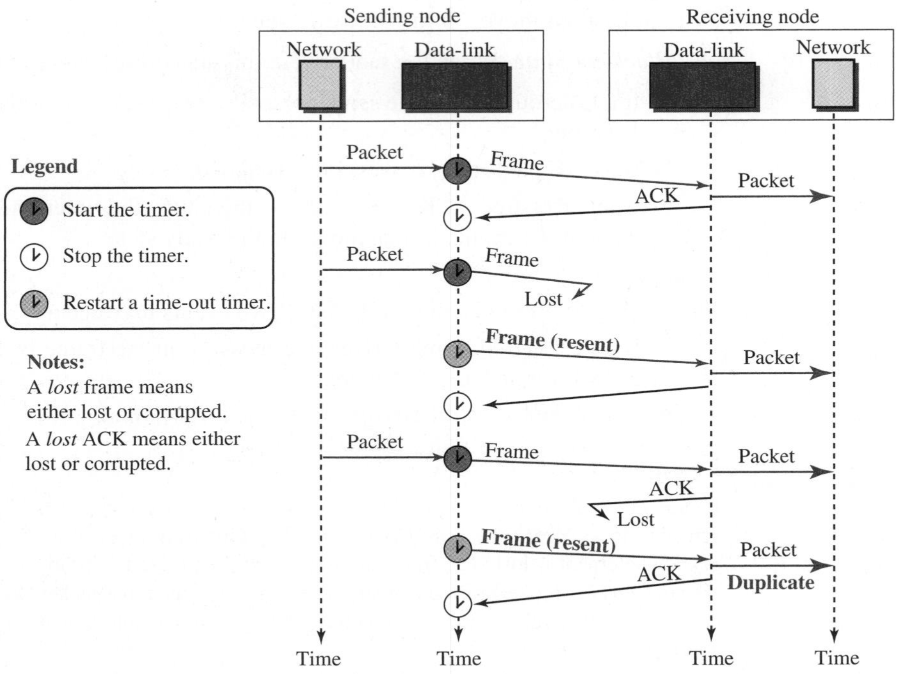

11.2.3 Piggybacking

The two protocols we discussed in this section are designed for unidirectional communication, in which data is flowing only in one direction although the acknowledgment may travel in the other direction. Protocols have been designed in the past to allow data to flow in both directions. However, to make the communication more efficient, the data in one direction is piggybacked with the acknowledgment in the other direction. In other words, when node A is sending data to node B, Node A also acknowledges the data received from node B. Because piggybacking makes communication at the data-link layer more complicated, it is not a common practice. We discuss two-way communication and piggybacking in more detail in Chapter 23.

11.3 HDLC

High-level Data Link Control (HDLC) is a bit-oriented protocol for communication over point-to-point and multipoint links. It implements the Stop-and-Wait protocol we discussed earlier. Although this protocol is more a theoretical issue than practical, most of the concept defined in this protocol is the basis for other practical protocols such as PPP, which we discuss next, or the Ethernet protocol, which we discuss in wired LANs (Chapter 13), or in wireless LANs (Chapter 15).

Figure 11.13 *Flow diagram for Example 11.4*

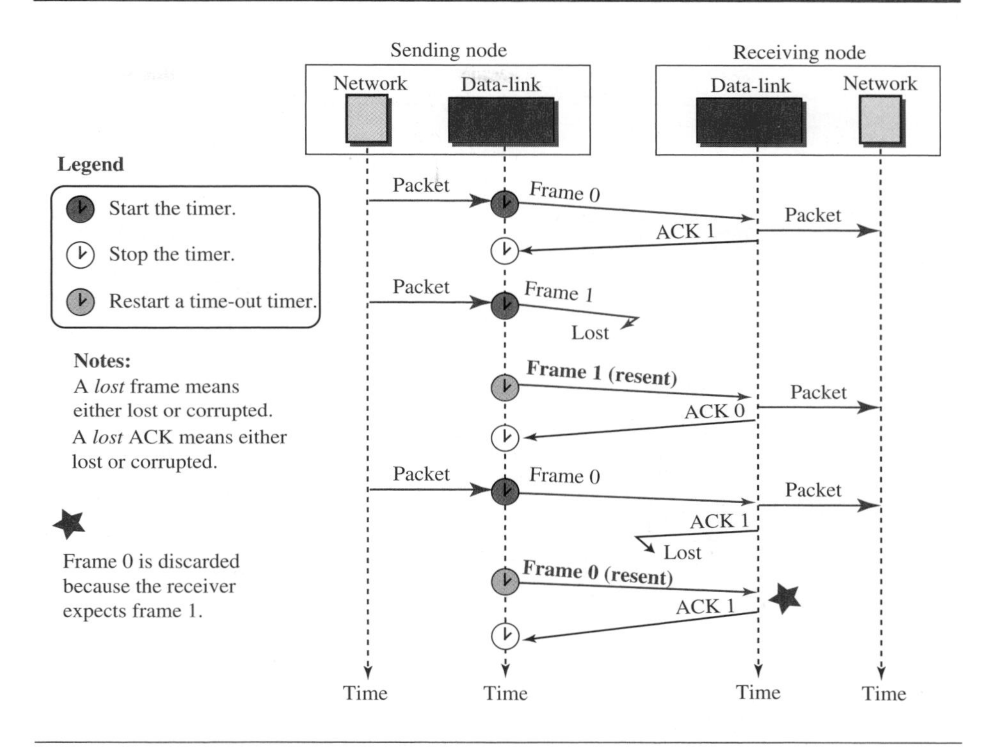

11.3.1 Configurations and Transfer Modes

HDLC provides two common transfer modes that can be used in different configurations: *normal response mode (NRM)* and *asynchronous balanced mode (ABM)*. In *normal response mode (NRM)*, the station configuration is unbalanced. We have one primary station and multiple secondary stations. A *primary station* can send commands; a *secondary station* can only respond. The NRM is used for both point-to-point and multipoint links, as shown in Figure 11.14.

In ABM, the configuration is balanced. The link is point-to-point, and each station can function as a primary and a secondary (acting as peers), as shown in Figure 11.15. This is the common mode today.

11.3.2 Framing

To provide the flexibility necessary to support all the options possible in the modes and configurations just described, HDLC defines three types of frames: *information frames (I-frames)*, *supervisory frames (S-frames)*, and *unnumbered frames (U-frames)*. Each type of frame serves as an envelope for the transmission of a different type of message. I-frames are used to data-link user data and control information relating to user data (piggybacking). S-frames are used only to transport control information. U-frames are reserved for system management. Information carried by U-frames is intended for managing the

Figure 11.14 *Normal response mode*

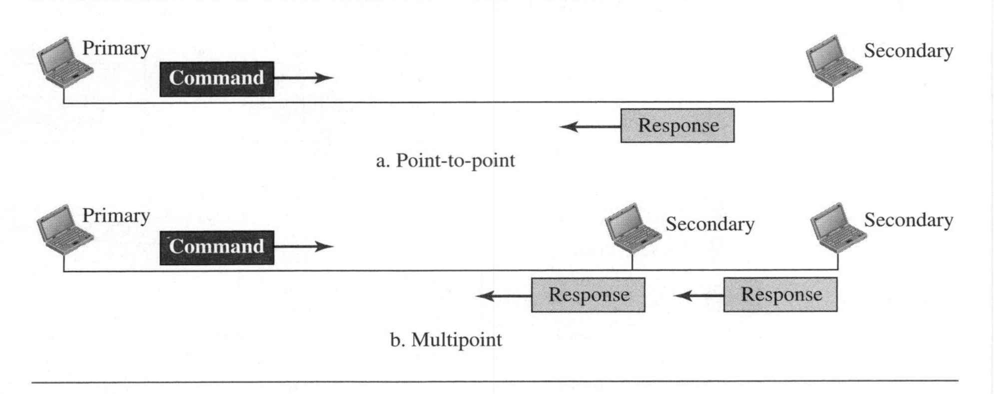

Figure 11.15 *Asynchronous balanced mode*

link itself. Each frame in HDLC may contain up to six fields, as shown in Figure 11.16: a beginning flag field, an address field, a control field, an information field, a frame check sequence (FCS) field, and an ending flag field. In multiple-frame transmissions, the ending flag of one frame can serve as the beginning flag of the next frame.

Figure 11.16 *HDLC frames*

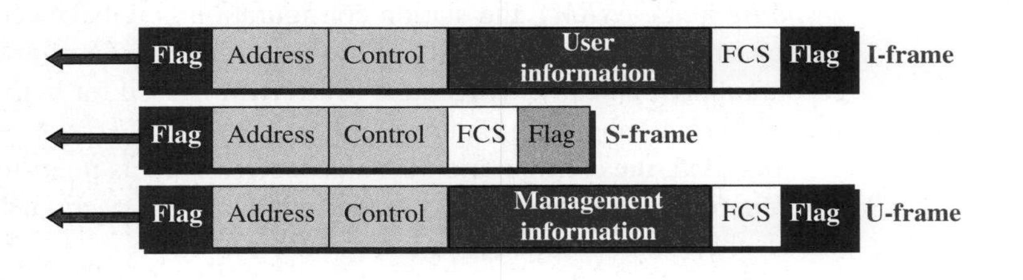

Let us now discuss the fields and their use in different frame types.

- ***Flag field.*** This field contains synchronization pattern 01111110, which identifies both the beginning and the end of a frame.
- ***Address field.*** This field contains the address of the secondary station. If a primary station created the frame, it contains a *to* address. If a secondary station creates the frame, it contains a *from* address. The address field can be one byte or several bytes long, depending on the needs of the network.

- ***Control field.*** The control field is one or two bytes used for flow and error control. The interpretation of bits are discussed later.
- ***Information field.*** The information field contains the user's data from the network layer or management information. Its length can vary from one network to another.
- ***FCS field.*** The frame check sequence (FCS) is the HDLC error detection field. It can contain either a 2- or 4-byte CRC.

The control field determines the type of frame and defines its functionality. So let us discuss the format of this field in detail. The format is specific for the type of frame, as shown in Figure 11.17.

Figure 11.17 *Control field format for the different frame types*

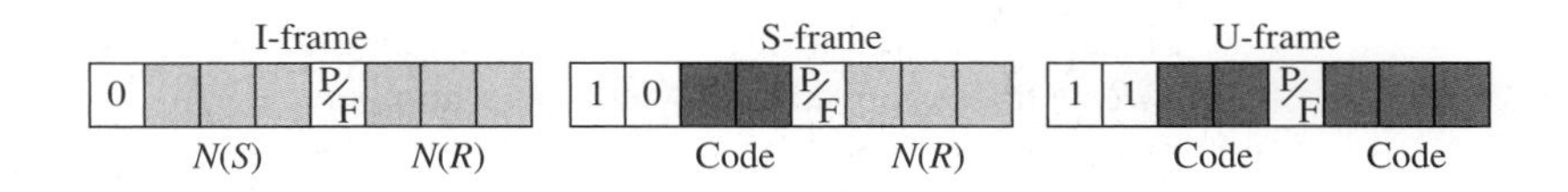

Control Field for I-Frames

I-frames are designed to carry user data from the network layer. In addition, they can include flow- and error-control information (piggybacking). The subfields in the control field are used to define these functions. The first bit defines the type. If the first bit of the control field is 0, this means the frame is an I-frame. The next 3 bits, called *N*(*S*), define the sequence number of the frame. Note that with 3 bits, we can define a sequence number between 0 and 7. The last 3 bits, called *N*(*R*), correspond to the acknowledgment number when piggybacking is used. The single bit between *N*(*S*) and *N*(*R*) is called the P/F bit. The P/F field is a single bit with a dual purpose. It has meaning only when it is set (bit = 1) and can mean poll or final. It means *poll* when the frame is sent by a primary station to a secondary (when the address field contains the address of the receiver). It means *final* when the frame is sent by a secondary to a primary (when the address field contains the address of the sender).

Control Field for S-Frames

Supervisory frames are used for flow and error control whenever piggybacking is either impossible or inappropriate. S-frames do not have information fields. If the first 2 bits of the control field are 10, this means the frame is an S-frame. The last 3 bits, called *N*(*R*), correspond to the acknowledgment number (ACK) or negative acknowledgment number (NAK), depending on the type of S-frame. The 2 bits called *code* are used to define the type of S-frame itself. With 2 bits, we can have four types of S-frames, as described below:

- ***Receive ready (RR).*** If the value of the code subfield is 00, it is an RR S-frame. This kind of frame acknowledges the receipt of a safe and sound frame or group of frames. In this case, the value of the *N*(*R*) field defines the acknowledgment number.

- **Receive not ready (RNR).** If the value of the code subfield is 10, it is an RNR S-frame. This kind of frame is an RR frame with additional functions. It acknowledges the receipt of a frame or group of frames, and it announces that the receiver is busy and cannot receive more frames. It acts as a kind of congestion-control mechanism by asking the sender to slow down. The value of *N*(*R*) is the acknowledgment number.
- **Reject (REJ).** If the value of the code subfield is 01, it is an REJ S-frame. This is a NAK frame, but not like the one used for Selective Repeat ARQ. It is a NAK that can be used in Go-Back-*N* ARQ to improve the efficiency of the process by informing the sender, before the sender timer expires, that the last frame is lost or damaged. The value of *N*(*R*) is the negative acknowledgment number.
- **Selective reject (SREJ).** If the value of the code subfield is 11, it is an SREJ S-frame. This is a NAK frame used in Selective Repeat ARQ. Note that the HDLC Protocol uses the term *selective reject* instead of *selective repeat*. The value of *N*(*R*) is the negative acknowledgment number.

Control Field for U-Frames

Unnumbered frames are used to exchange session management and control information between connected devices. Unlike S-frames, U-frames contain an information field, but one used for system management information, not user data. As with S-frames, however, much of the information carried by U-frames is contained in codes included in the control field. U-frame codes are divided into two sections: a 2-bit prefix before the P/F bit and a 3-bit suffix after the P/F bit. Together, these two segments (5 bits) can be used to create up to 32 different types of U-frames.

Control Field for U-Frames

Unnumbered frames are used to exchange session management and control information between connected devices. Unlike S-frames, U-frames contain an information field, but one used for system management information, not user data. As with S-frames, however, much of the information carried by U-frames is contained in codes included in the control field. U-frame codes are divided into two sections: a 2-bit prefix before the P/F bit and a 3-bit suffix after the P/F bit. Together, these two segments (5 bits) can be used to create up to 32 different types of U-frames.

Example 11.5

Figure 11.18 shows how U-frames can be used for connection establishment and connection release. Node A asks for a connection with a set asynchronous balanced mode (SABM) frame; node B gives a positive response with an unnumbered acknowledgment (UA) frame. After these two exchanges, data can be transferred between the two nodes (not shown in the figure). After data transfer, node A sends a DISC (disconnect) frame to release the connection; it is confirmed by node B responding with a UA (unnumbered acknowledgment).

Example 11.6

Figure 11.19 shows two exchanges using piggybacking. The first is the case where no error has occurred; the second is the case where an error has occurred and some frames are discarded.

Figure 11.18 *Example of connection and disconnection*

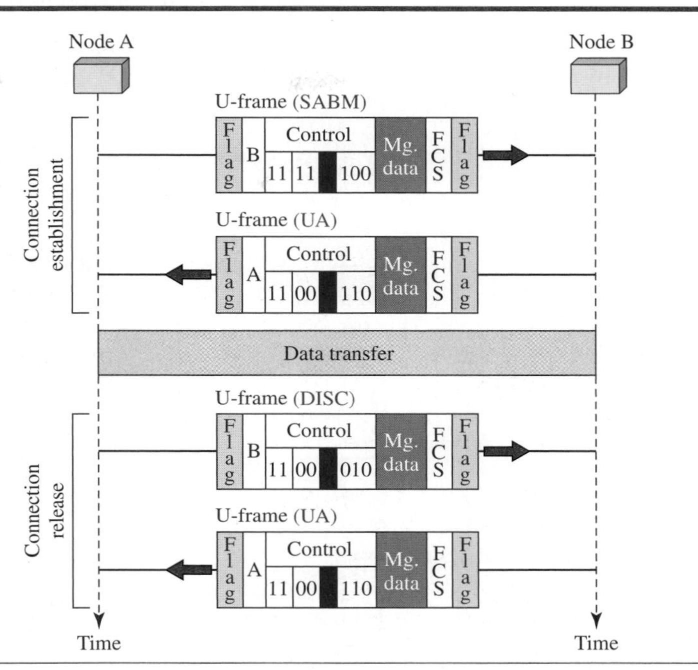

11.4 POINT-TO-POINT PROTOCOL (PPP)

One of the most common protocols for point-to-point access is the **Point-to-Point Protocol (PPP).** Today, millions of Internet users who need to connect their home computers to the server of an Internet service provider use PPP. The majority of these users have a traditional modem; they are connected to the Internet through a telephone line, which provides the services of the physical layer. But to control and manage the transfer of data, there is a need for a point-to-point protocol at the data-link layer. PPP is by far the most common.

11.4.1 Services

The designers of PPP have included several services to make it suitable for a point-to-point protocol, but have ignored some traditional services to make it simple.

Services Provided by PPP

PPP defines the format of the frame to be exchanged between devices. It also defines how two devices can negotiate the establishment of the link and the exchange of data. PPP is designed to accept payloads from several network layers (not only IP). Authentication is also provided in the protocol, but it is optional. The new version of PPP, called *Multilink PPP,* provides connections over multiple links. One interesting feature of PPP is that it provides network address configuration. This is particularly useful when a home user needs a temporary network address to connect to the Internet.

Figure 11.19 *Example of piggybacking with and without error*

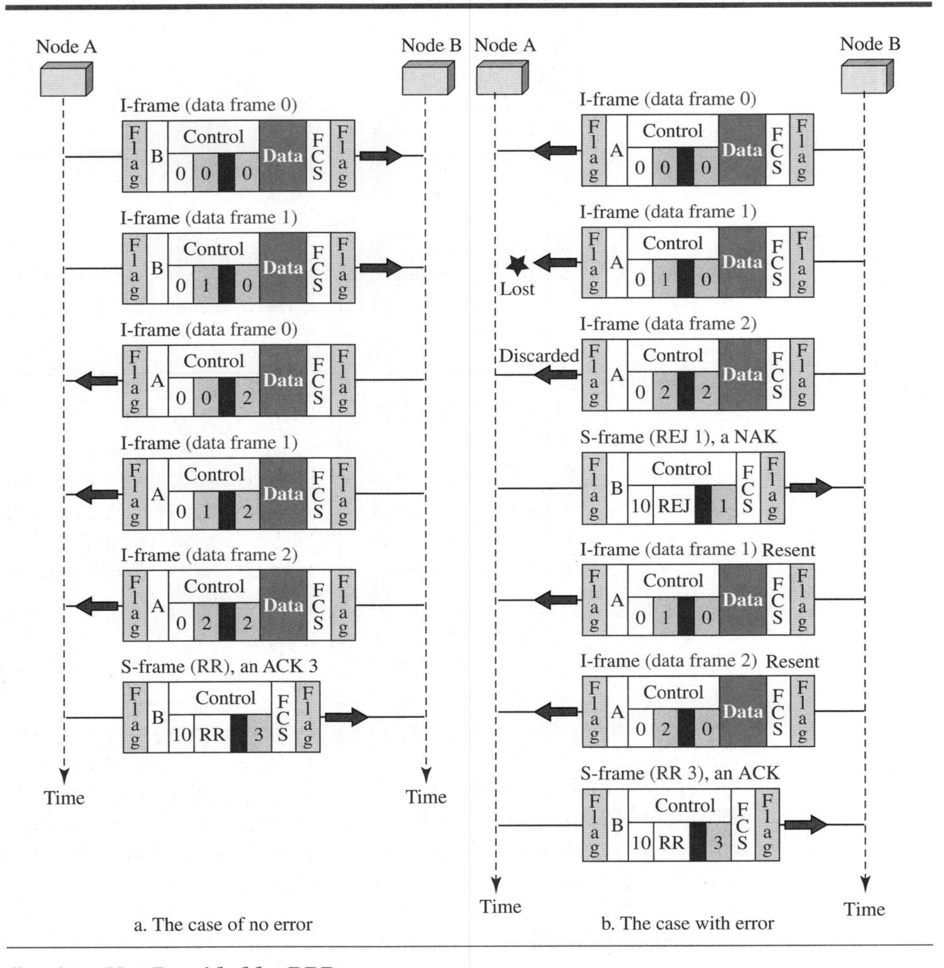

Services Not Provided by PPP

PPP does not provide flow control. A sender can send several frames one after another with no concern about overwhelming the receiver. PPP has a very simple mechanism for error control. A CRC field is used to detect errors. If the frame is corrupted, it is silently discarded; the upper-layer protocol needs to take care of the problem. Lack of error control and sequence numbering may cause a packet to be received out of order. PPP does not provide a sophisticated addressing mechanism to handle frames in a multipoint configuration.

11.4.2 Framing

PPP uses a character-oriented (or byte-oriented) frame. Figure 11.20 shows the format of a PPP frame. The description of each field follows:

- ***Flag.*** A PPP frame starts and ends with a 1-byte flag with the bit pattern 01111110.

Figure 11.20 *PPP frame format*

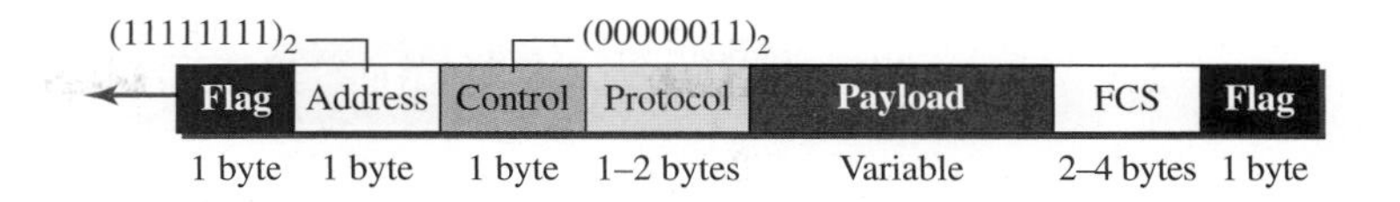

- ***Address.*** The address field in this protocol is a constant value and set to 11111111 (broadcast address).
- ***Control.*** This field is set to the constant value 00000011 (imitating unnumbered frames in HDLC). As we will discuss later, PPP does not provide any flow control. Error control is also limited to error detection.
- ***Protocol.*** The protocol field defines what is being carried in the data field: either user data or other information. This field is by default 2 bytes long, but the two parties can agree to use only 1 byte.
- ***Payload field.*** This field carries either the user data or other information that we will discuss shortly. The data field is a sequence of bytes with the default of a maximum of 1500 bytes; but this can be changed during negotiation. The data field is byte-stuffed if the flag byte pattern appears in this field. Because there is no field defining the size of the data field, padding is needed if the size is less than the maximum default value or the maximum negotiated value.
- ***FCS.*** The frame check sequence (FCS) is simply a 2-byte or 4-byte standard CRC.

Byte Stuffing

Since PPP is a byte-oriented protocol, the flag in PPP is a byte that needs to be escaped whenever it appears in the data section of the frame. The escape byte is 01111101, which means that every time the flaglike pattern appears in the data, this extra byte is stuffed to tell the receiver that the next byte is not a flag. Obviously, the escape byte itself should be stuffed with another escape byte.

11.4.3 Transition Phases

A PPP connection goes through phases which can be shown in a *transition phase* diagram (see Figure 11.21). The transition diagram, which is an FSM, starts with the *dead* state. In this state, there is no active carrier (at the physical layer) and the line is quiet. When one of the two nodes starts the communication, the connection goes into the *establish* state. In this state, options are negotiated between the two parties. If the two parties agree that they need authentication (for example, if they do not know each other), then the system needs to do authentication (an extra step); otherwise, the parties can simply start communication. The link-control protocol packets, discussed shortly, are used for this purpose. Several packets may be exchanged here. Data transfer takes place in the *open* state. When a connection reaches this state, the exchange of data packets can be started. The connection remains in this state until one of the endpoints wants to terminate the connection. In this case, the system goes to the *terminate* state. The system remains in this state until the carrier (physical-layer signal) is dropped, which moves the system to the *dead* state again.

Figure 11.21 *Transition phases*

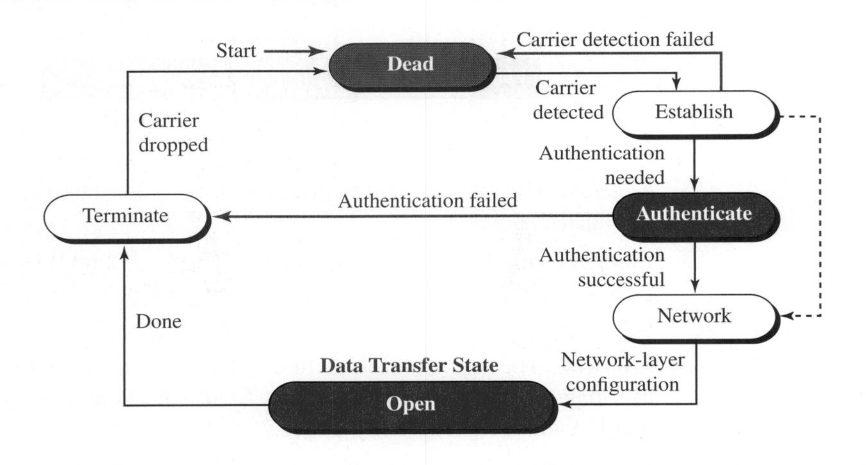

11.4.4 Multiplexing

Although PPP is a link-layer protocol, it uses another set of protocols to establish the link, authenticate the parties involved, and carry the network-layer data. Three sets of protocols are defined to make PPP powerful: the Link Control Protocol (LCP), two Authentication Protocols (APs), and several Network Control Protocols (NCPs). At any moment, a PPP packet can carry data from one of these protocols in its data field, as shown in Figure 11.22. Note that there are one LCP, two APs, and several NCPs. Data may also come from several different network layers.

Link Control Protocol

The **Link Control Protocol (LCP)** is responsible for establishing, maintaining, configuring, and terminating links. It also provides negotiation mechanisms to set options between the two endpoints. Both endpoints of the link must reach an agreement about the options before the link can be established. See Figure 11.21.

Figure 11.22 *Multiplexing in PPP*

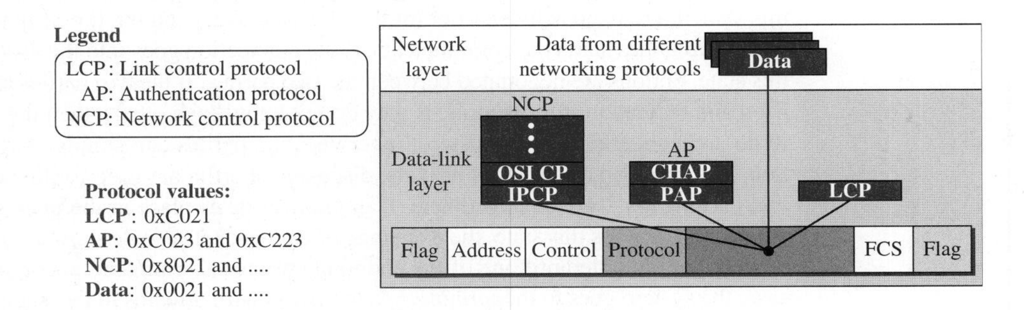

All LCP packets are carried in the payload field of the PPP frame with the protocol field set to C021 in hexadecimal (see Figure 11.23).

Figure 11.23 *LCP packet encapsulated in a frame*

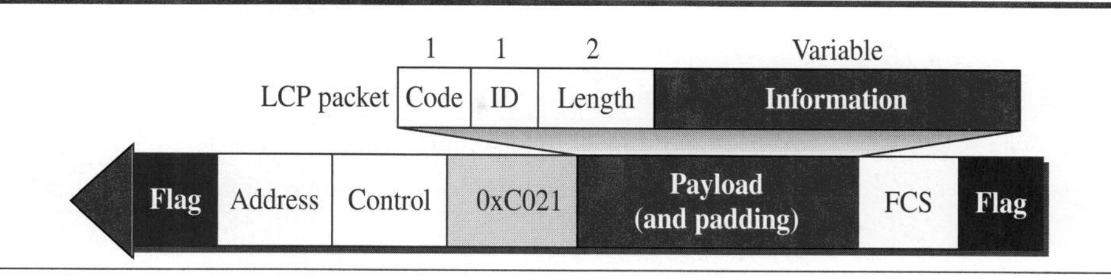

The code field defines the type of LCP packet. There are 11 types of packets, as shown in Table 11.1.

Table 11.1 *LCP packets*

Code	Packet Type	Description
0x01	Configure-request	Contains the list of proposed options and their values
0x02	Configure-ack	Accepts all options proposed
0x03	Configure-nak	Announces that some options are not acceptable
0x04	Configure-reject	Announces that some options are not recognized
0x05	Terminate-request	Request to shut down the line
0x06	Terminate-ack	Accept the shutdown request
0x07	Code-reject	Announces an unknown code
0x08	Protocol-reject	Announces an unknown protocol
0x09	Echo-request	A type of hello message to check if the other end is alive
0x0A	Echo-reply	The response to the echo-request message
0x0B	Discard-request	A request to discard the packet

There are three categories of packets. The first category, comprising the first four packet types, is used for link configuration during the establish phase. The second category, comprising packet types 5 and 6, is used for link termination during the termination phase. The last five packets are used for link monitoring and debugging.

The ID field holds a value that matches a request with a reply. One endpoint inserts a value in this field, which will be copied into the reply packet. The length field defines the length of the entire LCP packet. The information field contains information, such as options, needed for some LCP packets.

There are many options that can be negotiated between the two endpoints. Options are inserted in the information field of the configuration packets. In this case, the information field is divided into three fields: option type, option length, and option data. We list some of the most common options in Table 11.2.

Table 11.2 *Common options*

Option	*Default*
Maximum receive unit (payload field size)	1500
Authentication protocol	None
Protocol field compression	Off
Address and control field compression	Off

Authentication Protocols

Authentication plays a very important role in PPP because PPP is designed for use over dial-up links where verification of user identity is necessary. *Authentication* means validating the identity of a user who needs to access a set of resources. PPP has created two protocols for authentication: Password Authentication Protocol and Challenge Handshake Authentication Protocol. Note that these protocols are used during the authentication phase.

PAP

The **Password Authentication Protocol (PAP)** is a simple authentication procedure with a two-step process:

a. The user who wants to access a system sends an authentication identification (usually the user name) and a password.

b. The system checks the validity of the identification and password and either accepts or denies connection.

Figure 11.24 shows the three types of packets used by PAP and how they are actually exchanged. When a PPP frame is carrying any PAP packets, the value of the protocol field is 0xC023. The three PAP packets are authenticate-request, authenticate-ack, and authenticate-nak. The first packet is used by the user to send the user name and password. The second is used by the system to allow access. The third is used by the system to deny access.

CHAP

The **Challenge Handshake Authentication Protocol (CHAP)** is a three-way handshaking authentication protocol that provides greater security than PAP. In this method, the password is kept secret; it is never sent online.

a. The system sends the user a challenge packet containing a challenge value, usually a few bytes.

b. The user applies a predefined function that takes the challenge value and the user's own password and creates a result. The user sends the result in the response packet to the system.

c. The system does the same. It applies the same function to the password of the user (known to the system) and the challenge value to create a result. If the

Figure 11.24 *PAP packets encapsulated in a PPP frame*

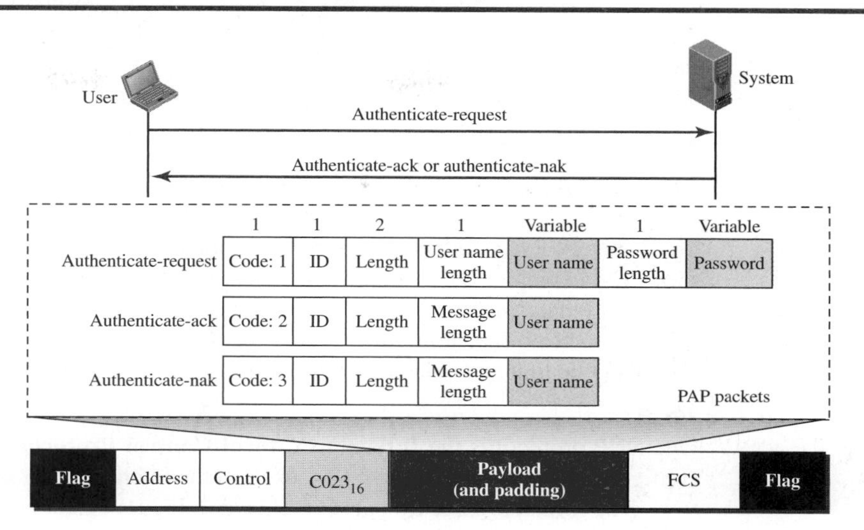

result created is the same as the result sent in the response packet, access is granted; otherwise, it is denied. CHAP is more secure than PAP, especially if the system continuously changes the challenge value. Even if the intruder learns the challenge value and the result, the password is still secret. Figure 11.25 shows the packets and how they are used.

Figure 11.25 *CHAP packets encapsulated in a PPP frame*

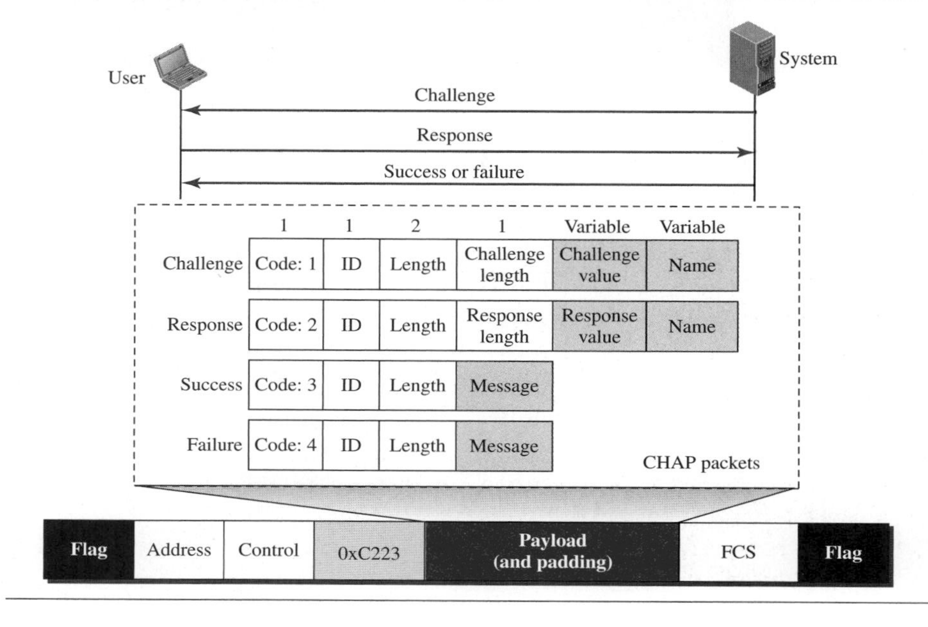

CHAP packets are encapsulated in the PPP frame with the protocol value C223 in hexadecimal. There are four CHAP packets: challenge, response, success, and failure. The first packet is used by the system to send the challenge value. The second is used by the user to return the result of the calculation. The third is used by the system to allow access to the system. The fourth is used by the system to deny access to the system.

Network Control Protocols

PPP is a multiple-network-layer protocol. It can carry a network-layer data packet from protocols defined by the Internet, OSI, Xerox, DECnet, AppleTalk, Novel, and so on. To do this, PPP has defined a specific Network Control Protocol for each network protocol. For example, IPCP (Internet Protocol Control Protocol) configures the link for carrying IP data packets. Xerox CP does the same for the Xerox protocol data packets, and so on. Note that none of the NCP packets carry network-layer data; they just configure the link at the network layer for the incoming data.

IPCP

One NCP protocol is the **Internet Protocol Control Protocol (IPCP).** This protocol configures the link used to carry IP packets in the Internet. IPCP is especially of interest to us. The format of an IPCP packet is shown in Figure 11.26. Note that the value of the protocol field in hexadecimal is 8021.

Figure 11.26 *IPCP packet encapsulated in PPP frame*

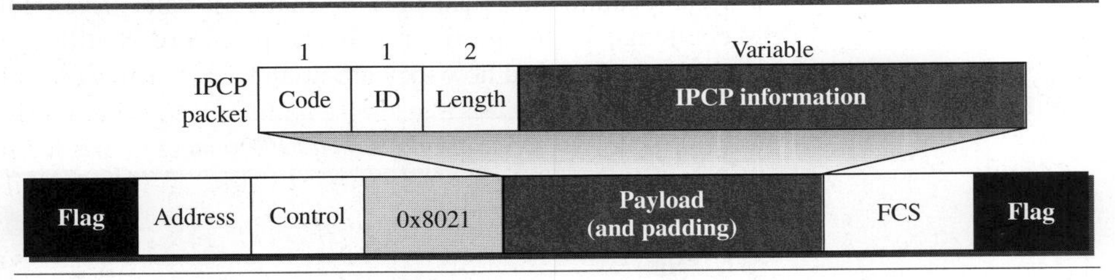

IPCP defines seven packets, distinguished by their code values, as shown in Table 11.3.

Table 11.3 *Code value for IPCP packets*

Code	*IPCP Packet*
0x01	Configure-request
0x02	Configure-ack
0x03	Configure-nak
0x04	Configure-reject
0x05	Terminate-request
0x06	Terminate-ack
0x07	Code-reject

Other Protocols

There are other NCP protocols for other network-layer protocols. The OSI Network Layer Control Protocol has a protocol field value of 8023; the Xerox NS IDP Control Protocol has a protocol field value of 8025; and so on.

Data from the Network Layer

After the network-layer configuration is completed by one of the NCP protocols, the users can exchange data packets from the network layer. Here again, there are different protocol fields for different network layers. For example, if PPP is carrying data from the IP network layer, the field value is 0021 (note that the three rightmost digits are the same as for IPCP). If PPP is carrying data from the OSI network layer, the value of the protocol field is 0023, and so on. Figure 11.27 shows the frame for IP.

Figure 11.27 *IP datagram encapsulated in a PPP frame*

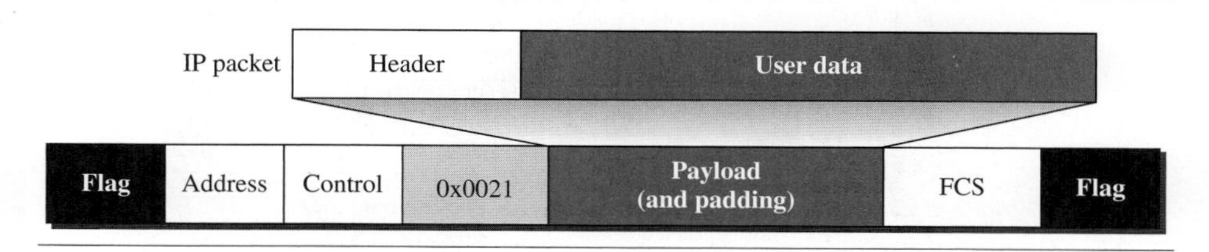

Multilink PPP

PPP was originally designed for a single-channel point-to-point physical link. The availability of multiple channels in a single point-to-point link motivated the development of Multilink PPP. In this case, a logical PPP frame is divided into several actual PPP frames. A segment of the logical frame is carried in the payload of an actual PPP frame, as shown in Figure 11.28. To show that the actual PPP frame is carrying a fragment of a logical PPP frame, the protocol field is set to $(003d)_{16}$. This new development adds complexity. For example, a sequence number needs to be added to the actual PPP frame to show a fragment's position in the logical frame.

Figure 11.28 *Multilink PPP*

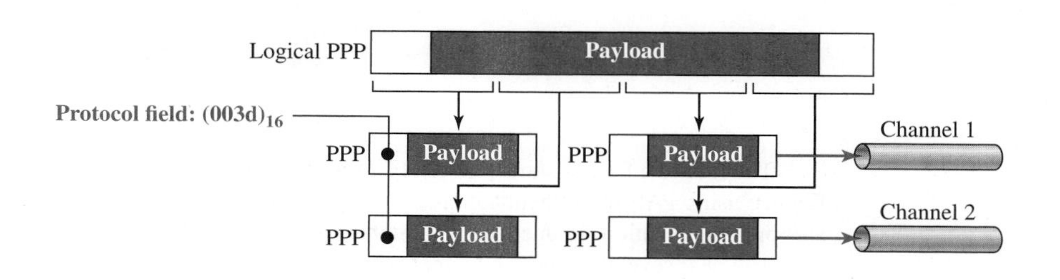

Example 11.7

Let us go through the phases followed by a network layer packet as it is transmitted through a PPP connection. Figure 11.29 shows the steps. For simplicity, we assume unidirectional movement of data from the user site to the system site (such as sending an e-mail through an ISP).

Figure 11.29 *An example*

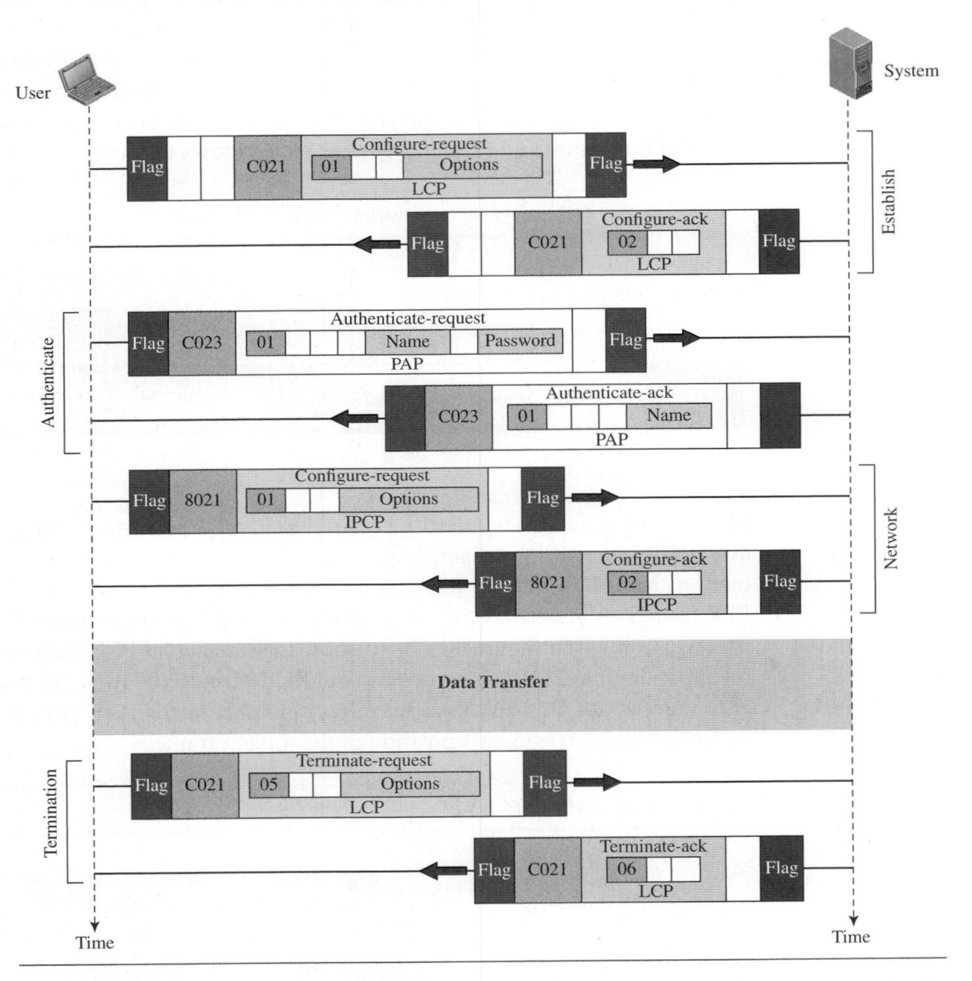

The first two frames show link establishment. We have chosen two options (not shown in the figure): using PAP for authentication and suppressing the address control fields. Frames 3 and 4 are for authentication. Frames 5 and 6 establish the network layer connection using IPCP.

The next several frames show that some IP packets are encapsulated in the PPP frame. The system (receiver) may have been running several network layer protocols, but it knows that the incoming data must be delivered to the IP protocol because the NCP protocol used before the data transfer was IPCP.

After data transfer, the user then terminates the data-link connection, which is acknowledged by the system. Of course the user or the system could have chosen to terminate the network-layer IPCP and keep the data-link layer running if it wanted to run another NCP protocol.

11.5 END-CHAPTER MATERIALS

11.5.1 Recommended Reading

For more details about subjects discussed in this chapter, we recommend the following books. The items in brackets [...] refer to the reference list at the end of the text.

Books

Several books discuss link-layer issues. Among them we recommend [Ham 80], [Zar 02], [Ror 96], [Tan 03], [GW 04], [For 03], [KMK 04], [Sta 04], [Kes 02], [PD 03], [Kei 02], [Spu 00], [KCK 98], [Sau 98], [Izz 00], [Per 00], and [WV 00].

11.5.2 Key Terms

acknowledgment number
bit stuffing
byte stuffing
Challenge Handshake Authentication Protocol (CHAP)
data link control (DLC)
finite state machine (FSM)
flag
High-level Data Link Control (HDLC)
Internet Protocol Control Protocol (IPCP)
Link Control Protocol (LCP)
Password Authentication Protocol (PAP)
piggybacking
Point-to-Point Protocol (PPP)
sequence number
Simple Protocol
Stop-and-Wait Protocol

11.5.3 Summary

Data link control deals with the design and procedures for communication between two adjacent nodes: node-to-node communication. Framing in the data-link layer separates one packet from another. In fixed-size framing, there is no need for defining the boundaries of frames; in variable-size framing, we need a delimiter (flag) to define the boundary of two frames. Variable-size framing uses two categories of protocols: byte-oriented (or character-oriented) and bit-oriented. In a byte-oriented protocol, the data section of a frame is a sequence of bytes; in a bit-oriented protocol, the data section of a frame is a sequence of bits. In byte-oriented protocols, we use byte stuffing; in bit-oriented protocols, we use bit stuffing.

Another duty of DLC is flow and error control. At the data-link layer, flow control means creating a balance between the frames sent by a node and the frames that can be handled by the next node. Error control at the data-link layer is normally implemented very simply. Corrupted frames are silently discarded; uncorrupted frames are accepted with or without sending acknowledgments to the sender.

A DLC protocol can be either connectionless or connection-oriented. In a connectionless protocol, frames are sent from one node to the next without any relationship between the frames; each frame is independent. In a connection-oriented protocol, a logical connection should first be established between the two nodes before sending the data frames. After all related frames are transmitted, the logical connection is terminated.

Data-link protocols have been designed to handle communication between two nodes. We discussed two protocols in this chapter. In the Simple Protocol, there is no flow and error control. In the Stop-and-Wait Protocol, there are both flow and error controls, but communication is a frame at a time.

High-level Data Link Control (HDLC) is a bit-oriented protocol for communication over point-to-point and multipoint links. It implements the Stop-and-Wait protocol. It is the basis of many protocols in practice today. HDLC defines three types of frames: information frames, supervisory frames, and unnumbered frames. The informational frames are used to carry data frames. Supervisory frames are used only to transport control information for flow and error control. Unnumbered frames are reserved for system management and provide connection-oriented service.

One of the most common protocols for point-to-point access is the Point-to-Point Protocol (PPP). PPP uses only one type of frame, but allows multiplexing of different payloads to achieve a kind of connection-oriented service authentication. Encapsulating different packets in a frame allows PPP to move to different states to provide necessary services.

11.6 PRACTICE SET

11.6.1 Quizzes

A set of interactive quizzes for this chapter can be found on the book website. It is strongly recommended that the student take the quizzes to check his/her understanding of the materials before continuing with the practice set.

11.6.2 Questions

Q11-1. Define *framing* and give the reason it is needed.

Q11-2. Define *piggybacking* and its benefit.

Q11-3. In PPP, we normally talk about *user* and *system* instead of *sending and receiving nodes;* explain the reason.

Q11-4. Explain why flags are needed when we use variable-size frames.

Q11-5. In Figure 11.9, we show the packet path as a horizontal line, but the frame path as a diagonal line. Can you explain the reason?

Q11-6. Compare Figure 11.6 and Figure 11.21. If both are FSMs, why are there no event/action pairs in the second?

Q11-7. In Figure 11.11, do the ready and blocking states use the same timer? Explain.

Q11-8. In Figure 11.12, explain why we need a timer at the sending site, but none at the receiving site.

Q11-9. In a bit-oriented protocol, should we first unstuff the extra bits and then remove the flags or reverse the process?

Q11-10. In a byte-oriented protocol, should we first unstuff the extra bytes and then remove the flags or reverse the process?

Q11-11. Compare and contrast byte-stuffing and bit-stuffing.

Q11-12. In Figure 11.20, explain why we need only one address field. Explain why the address is set to the predefined value of $(11111111)_2$.

Q11-13. In Example 11.4 (Figure 11.13) how many frames are in transit at the same time?

Q11-14. Compare and contrast byte-oriented and bit-oriented protocols.

Q11-15. Compare the flag byte and the escape byte in PPP. Are they the same? Explain.

Q11-16. In the traditional Ethernet protocol (Chapter 13), the frames are sent with the CRC. If the frame is corrupted, the receiving node just discards it. Is this an example of a Simple Protocol or the Stop-and-Wait Protocol? Explain.

Q11-17. In the Stop-and-Wait Protocol, assume that the sender has only one slot in which to keep the frame to send or the copy of the sent frame. What happens if the network layer delivers a packet to the data-link layer at this moment?

Q11-18. Compare and contrast flow control and error control.

Q11-19. In Figure 11.16, which frame type can be used for acknowledgment?

Q11-20. Compare and contrast HDLC with PPP.

Q11-21. Assume a new character-oriented protocol is using the 16-bit Unicode as the character set. What should the size of the flag be in this protocol?

Q11-22. Explain why there is no need for CRC in the Simple Protocol.

Q11-23. Does the duplex communication in Figure 11.10 necessarily mean we need two separate media between the two nodes? Explain.

Q11-24. In Example 11.3 (Figure 11.12) how many frames are in transit at the same time?

11.6.3 Problems

P11-1. Assume PPP is in the established phase; show payload encapsulated in the frame.

P11-2. Using the following specifications, draw a finite state machine with three states (I, II, and III), six events, and four actions:

a. If the machine is in state I, two events can occur. If event 1 occurs, the machine moves to state III. If event 3 occurs, the machine performs actions 2 and 4 and moves to state II.

b. If the machine is in state II, two events can occur. If event 4 occurs, the machine remains in state II. If event 6 occurs, the machine performs actions 1 and 2 and moves to state III.

c. If the machine is in state III, three events can occur. If event 2 occurs, the machine remains in state III. If event 6 occurs, the machine performs actions 2, 3, 4, and 5 moves to state I. If event 4 occurs, the machine performs actions 1 and 2 and moves to state I.

P11-3. Redraw Figure 11.12 using the following scenario:

a. The first frame is sent and acknowledged.

b. The second frame is sent and acknowledged, but the acknowledgment is lost.

c. The second frame is resent, but it is timed-out.

d. The second frame is resent and acknowledged.

P11-4. In Example 11.4 (Figure 11.13), assume the round trip time for a frame is 50 milliseconds. Explain what will happen if we set the time-out in each of the following cases.

a. 45 milliseconds **b.** 55 milliseconds **c.** 50 milliseconds

P11-5. Byte-stuff the following frame payload in which E is the escape byte, F is the flag byte, and D is a data byte other than an escape or a flag character.

D	E	D	D	E	D	D	E	F	D	F	D

P11-6. In Figure 11.11, show what happens in each of the following cases:

a. The receiver is in the ready state and a packet comes from the network layer.

b. The receiver is in the ready state and a corrupted frame arrives.

c. The receiver is in the ready state and an acknowledgment arrives.

P11-7. In Figure 11.11, show what happens in each of the following cases:

a. The sender is at the ready state and an error-free ACK arrives.

b. The sender is at the blocking state and a time-out occurs.

c. The sender is at the ready state and a time-out occurs.

P11-8. Redraw Figure 11.21 with the system not using authentication.

P11-9. Using the following specifications, draw a finite state machine with three states (I, II, and III), five events, and six actions:

a. If the machine is in state I, two events can occur. If event 1 occurs, the machine moves to state II. If event 2 occurs, the machine performs actions 1 and 2 and moves to state III.

b. If the machine is in state II, two events can occur. If event 3 occurs, the machine remains in state II. If event 4 occurs, the machine moves to state III.

c. If the machine is in state III, three events can occur. If event 2 occurs, the machine remains in state III. If event 3 occurs, the machine performs actions 1, 2, 4, and 5 moves to state II. If event 5 occurs, the machine performs actions 1, 2, and 6 and moves to state I.

P11-10. Redraw Figure 11.2 using the following scenario:

a. Frame 0 is sent, but lost.

b. Frame 0 is resent and acknowledged.

c. Frame 1 is sent and acknowledged, but the acknowledgment is lost.

d. Frame 1 is resent and acknowledged.

P11-11. Bit-stuff the following frame payload:

00011111000001111101000111111011110000111

P11-12. Assume the only computer in the residence uses PPP to communicate with the ISP. If the user sends 10 network-layer packets to ISP, how many frames are exchanged in each of the following cases:

a. Using no authentication?

b. Using PAP for authentication?

c. Using CHAP for authentication?

P11-13. Assume PPP is in the authentication phase, show payload exchanged between the nodes if PPP is using

a. PAP

b. CHAP

P11-14. Unstuff the following frame payload:

```
000111111000001011101110100111011111001101111
```

P11-15. Redraw Figure 11.11 using a variable to hold the one-bit sequence number and a variable to hold the one-bit acknowledgment number.

P11-16. Unstuff the following frame payload in which E is the escape byte, F is the flag byte, and D is a data byte other than an escape or a flag character.

P11-17. Assume we change the Stop-and-Wait Protocol to include a NAK (negative feedback), which is used only when a corrupted frame arrives and is discarded. Redraw Figure 11.9 to show this change.

P11-18. Redraw Figure 11.10 using piggybacking.

11.7 SIMULATION EXPERIMENTS

11.7.1 Applets

We have created some Java applets to show some of the main concepts discussed in this chapter. It is strongly recommended that the students activate these applets on the book website and carefully examine the protocols in action.

11.8 PROGRAMMING ASSIGNMENTS

For each of the following assignments, write a program in the programming language you are familiar with.

Prg11-1. Write and test a program that simulates the byte stuffing and byte unstuffing as shown in Figure 11.2.

Prg11-2. Write and test a program that simulates the bit stuffing and bit unstuffing as shown in Figure 11.4.

CHAPTER 12

Media Access Control (MAC)

When nodes or stations are connected and use a common link, called a *multipoint* or *broadcast link,* we need a multiple-access protocol to coordinate access to the link. The problem of controlling the access to the medium is similar to the rules of speaking in an assembly. The procedures guarantee that the right to speak is upheld and ensure that two people do not speak at the same time, do not interrupt each other, do not monopolize the discussion, and so on. Many protocols have been devised to handle access to a shared link. All of these protocols belong to a sublayer in the data-link layer called *media access control (MAC).* We categorize them into three groups, as shown in Figure 12.1.

Figure 12.1 *Taxonomy of multiple-access protocols*

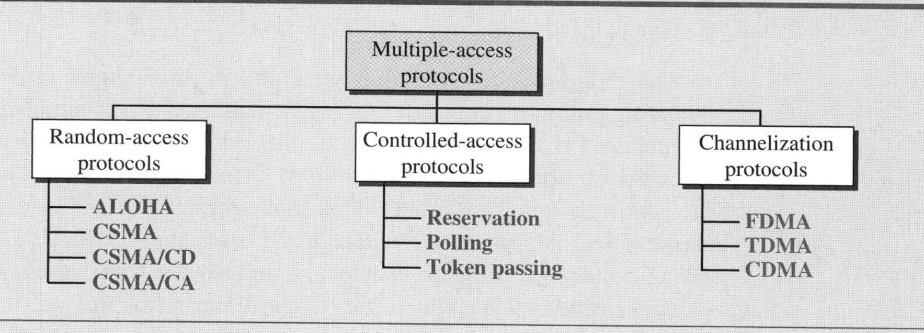

This chapter is divided into three sections:

- The first section discusses random-access protocols. Four protocols, ALOHA, CSMA, CSMA/CD, and CSMA/CA, are described in this section. These protocols are mostly used in LANs and WANs, which we discuss in future chapters.
- The second section discusses controlled-access protocols. Three protocols, reservation, polling, and token-passing, are described in this section. Some of these protocols are used in LANs, but others have some historical value.
- The third section discusses channelization protocols. Three protocols, FDMA, TDMA, and CDMA are described in this section. These protocols are used in cellular telephony, which we discuss in Chapter 16.

12.1 RANDOM ACCESS

In **random-access** or **contention** methods, no station is superior to another station and none is assigned control over another. At each instance, a station that has data to send uses a procedure defined by the protocol to make a decision on whether or not to send. This decision depends on the state of the medium (idle or busy). In other words, each station can transmit when it desires on the condition that it follows the predefined procedure, including testing the state of the medium.

Two features give this method its name. First, there is no scheduled time for a station to transmit. Transmission is random among the stations. That is why these methods are called *random access*. Second, no rules specify which station should send next. Stations compete with one another to access the medium. That is why these methods are also called *contention* methods.

In a random-access method, each station has the right to the medium without being controlled by any other station. However, if more than one station tries to send, there is an access conflict—***collision***—and the frames will be either destroyed or modified. To avoid access conflict or to resolve it when it happens, each station follows a procedure that answers the following questions:

- When can the station access the medium?
- What can the station do if the medium is busy?
- How can the station determine the success or failure of the transmission?
- What can the station do if there is an access conflict?

The random-access methods we study in this chapter have evolved from a very interesting protocol known as *ALOHA*, which used a very simple procedure called ***multiple access (MA).*** The method was improved with the addition of a procedure that forces the station to sense the medium before transmitting. This was called *carrier sense multiple access* (*CSMA*). This method later evolved into two parallel methods: *carrier sense multiple access with collision detection (CSMA/CD),* which tells the station what to do when a collision is detected, and *carrier sense multiple access with collision avoidance (CSMA/CA),* which tries to avoid the collision.

12.1.1 ALOHA

ALOHA, the earliest random access method, was developed at the University of Hawaii in early 1970. It was designed for a radio (wireless) LAN, but it can be used on any shared medium.

It is obvious that there are potential collisions in this arrangement. The medium is shared between the stations. When a station sends data, another station may attempt to do so at the same time. The data from the two stations collide and become garbled.

Pure ALOHA

The original ALOHA protocol is called ***pure ALOHA.*** This is a simple but elegant protocol. The idea is that each station sends a frame whenever it has a frame to send (multiple access). However, since there is only one channel to share, there is the possibility of collision between frames from different stations. Figure 12.2 shows an example of frame collisions in pure ALOHA.

Figure 12.2 *Frames in a pure ALOHA network*

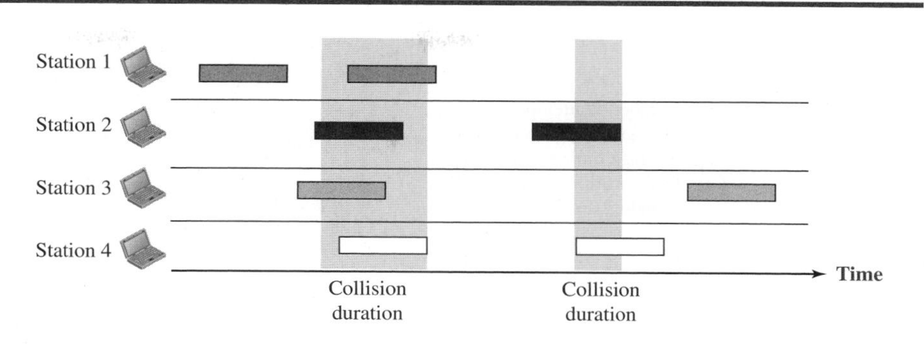

There are four stations (unrealistic assumption) that contend with one another for access to the shared channel. The figure shows that each station sends two frames; there are a total of eight frames on the shared medium. Some of these frames collide because multiple frames are in contention for the shared channel. Figure 12.2 shows that only two frames survive: one frame from station 1 and one frame from station 3. We need to mention that even if one bit of a frame coexists on the channel with one bit from another frame, there is a collision and both will be destroyed. It is obvious that we need to resend the frames that have been destroyed during transmission.

The pure ALOHA protocol relies on acknowledgments from the receiver. When a station sends a frame, it expects the receiver to send an acknowledgment. If the acknowledgment does not arrive after a time-out period, the station assumes that the frame (or the acknowledgment) has been destroyed and resends the frame.

A collision involves two or more stations. If all these stations try to resend their frames after the time-out, the frames will collide again. Pure ALOHA dictates that when the time-out period passes, each station waits a random amount of time before resending its frame. The randomness will help avoid more collisions. We call this time the *backoff time* T_B.

Pure ALOHA has a second method to prevent congesting the channel with retransmitted frames. After a maximum number of retransmission attempts K_{max}, a station must give up and try later. Figure 12.3 shows the procedure for pure ALOHA based on the above strategy.

The time-out period is equal to the maximum possible round-trip propagation delay, which is twice the amount of time required to send a frame between the two most widely separated stations ($2 \times T_p$). The backoff time T_B is a random value that normally depends on K (the number of attempted unsuccessful transmissions). The formula for T_B depends on the implementation. One common formula is the ***binary exponential backoff.*** In this method, for each retransmission, a multiplier $R = 0$ to $2^K - 1$ is randomly chosen and multiplied by T_p (maximum propagation time) or T_{fr} (the average time required to send out a frame) to find T_B. Note that in this procedure, the range of the random numbers increases after each collision. The value of K_{max} is usually chosen as 15.

Figure 12.3 *Procedure for pure ALOHA protocol*

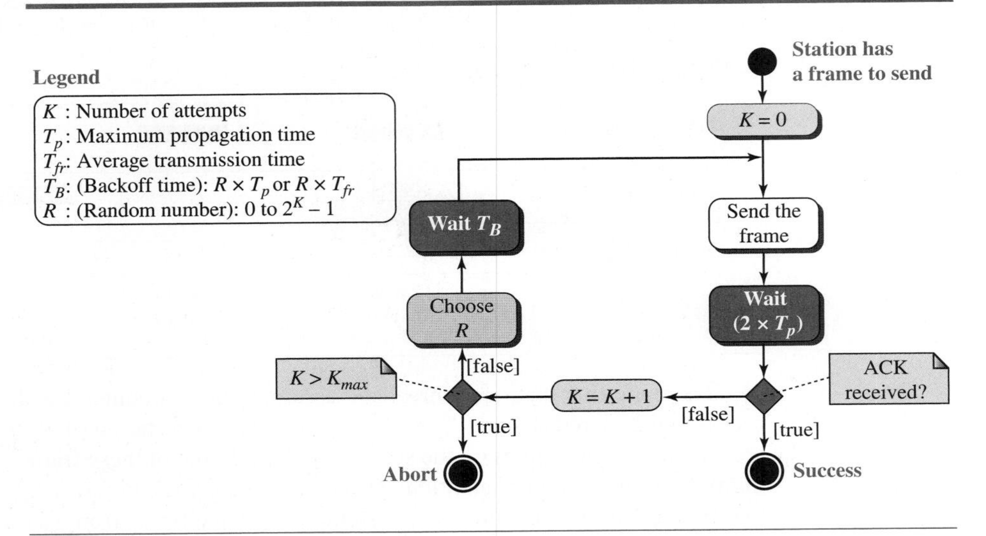

Example 12.1

The stations on a wireless ALOHA network are a maximum of 600 km apart. If we assume that signals propagate at 3×10^8 m/s, we find $T_p = (600 \times 10^3) / (3 \times 10^8) = 2$ ms. For $K = 2$, the range of R is $\{0, 1, 2, 3\}$. This means that T_B can be 0, 2, 4, or 6 ms, based on the outcome of the random variable R.

Vulnerable time

Let us find the ***vulnerable time,*** the length of time in which there is a possibility of collision. We assume that the stations send fixed-length frames with each frame taking T_{fr} seconds to send. Figure 12.4 shows the vulnerable time for station B.

Figure 12.4 *Vulnerable time for pure ALOHA protocol*

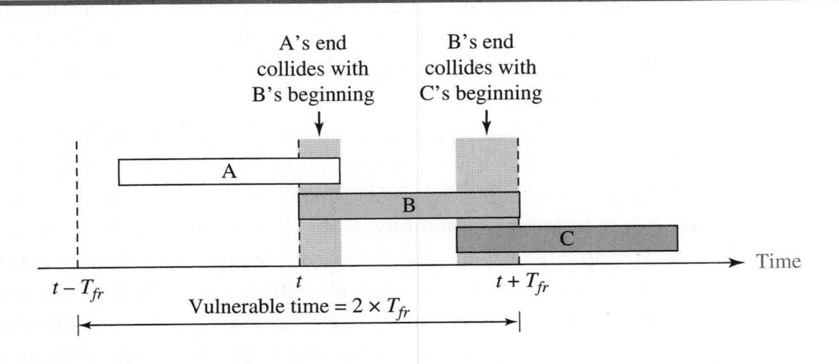

Station B starts to send a frame at time t. Now imagine station A has started to send its frame after $t - T_{fr}$. This leads to a collision between the frames from station B and station A. On the other hand, suppose that station C starts to send a frame before time $t + T_{fr}$. Here, there is also a collision between frames from station B and station C.

Looking at Figure 12.4, we see that the vulnerable time during which a collision may occur in pure ALOHA is 2 times the frame transmission time.

$$\textbf{Pure ALOHA vulnerable time} = 2 \times T_{fr}$$

Example 12.2

A pure ALOHA network transmits 200-bit frames on a shared channel of 200 kbps. What is the requirement to make this frame collision-free?

Solution

Average frame transmission time T_{fr} is 200 bits/200 kbps or 1 ms. The vulnerable time is 2×1 ms = 2 ms. This means no station should send later than 1 ms before this station starts transmission and no station should start sending during the period (1 ms) that this station is sending.

Throughput

Let us call G the average number of frames generated by the system during one frame transmission time. Then it can be proven that the average number of successfully transmitted frames for pure ALOHA is $S = G \times e^{-2G}$. The maximum throughput S_{max} is 0.184, for $G = 1/2$. (We can find it by setting the derivative of S with respect to G to 0; see Exercises.) In other words, if one-half a frame is generated during one frame transmission time (one frame during two frame transmission times), then 18.4 percent of these frames reach their destination successfully. We expect $G = 1/2$ to produce the maximum throughput because the vulnerable time is 2 times the frame transmission time. Therefore, if a station generates only one frame in this vulnerable time (and no other stations generate a frame during this time), the frame will reach its destination successfully.

The throughput for pure ALOHA is $S = G \times e^{-2G}$.
The maximum throughput $S_{max} = 1/(2e) = 0.184$ when $G = (1/2)$.

Example 12.3

A pure ALOHA network transmits 200-bit frames on a shared channel of 200 kbps. What is the throughput if the system (all stations together) produces

a. 1000 frames per second?

b. 500 frames per second?

c. 250 frames per second?

Solution

The frame transmission time is 200/200 kbps or 1 ms.

a. If the system creates 1000 frames per second, or 1 frame per millisecond, then $G = 1$. In this case $S = G \times e^{-2G} = 0.135$ (13.5 percent). This means that the throughput is $1000 \times 0.135 = 135$ frames. Only 135 frames out of 1000 will probably survive.

b. If the system creates 500 frames per second, or 1/2 frames per millisecond, then $G = 1/2$. In this case $S = G \times e^{-2G} = 0.184$ (18.4 percent). This means that the throughput is $500 \times 0.184 = 92$ and that only 92 frames out of 500 will probably survive. Note that this is the *maximum* throughput case, percentagewise.

c. If the system creates 250 frames per second, or 1/4 frames per millisecond, then $G = 1/4$. In this case $S = G \times e^{-2G} = 0.152$ (15.2 percent). This means that the throughput is $250 \times 0.152 = 38$. Only 38 frames out of 250 will probably survive.

Slotted ALOHA

Pure ALOHA has a vulnerable time of $2 \times T_{fr}$. This is so because there is no rule that defines when the station can send. A station may send soon after another station has started or just before another station has finished. Slotted ALOHA was invented to improve the efficiency of pure ALOHA.

In **slotted ALOHA** we divide the time into slots of T_{fr} seconds and force the station to send only at the beginning of the time slot. Figure 12.5 shows an example of frame collisions in slotted ALOHA.

Figure 12.5 *Frames in a slotted ALOHA network*

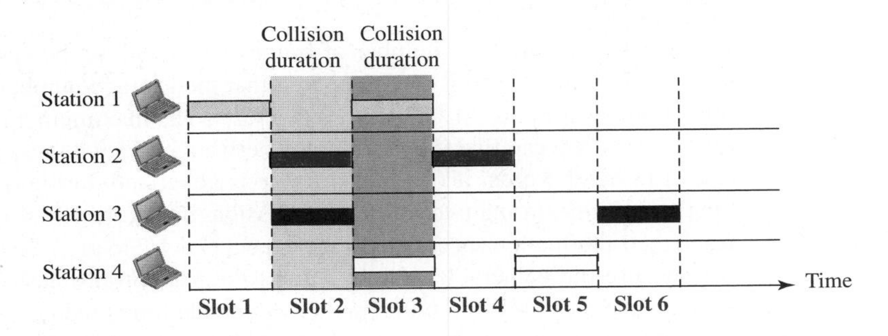

Because a station is allowed to send only at the beginning of the synchronized time slot, if a station misses this moment, it must wait until the beginning of the next time slot. This means that the station which started at the beginning of this slot has already finished sending its frame. Of course, there is still the possibility of collision if two stations try to send at the beginning of the same time slot. However, the vulnerable time is now reduced to one-half, equal to T_{fr}. Figure 12.6 shows the situation.

Figure 12.6 *Vulnerable time for slotted ALOHA protocol*

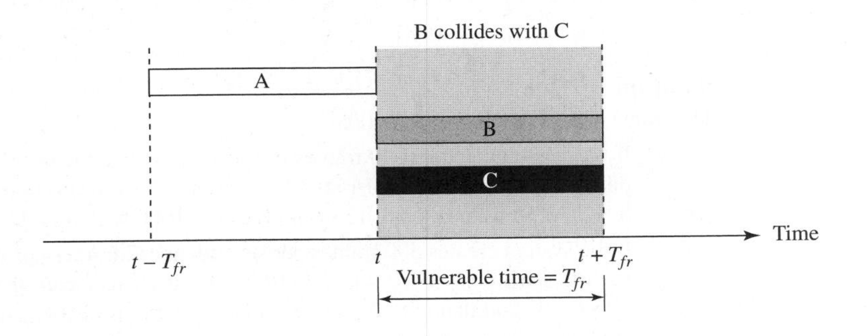

$$\textbf{Slotted ALOHA vulnerable time} = T_{fr}$$

Throughput

It can be proven that the average number of successful transmissions for slotted ALOHA is $S = G \times e^{-G}$. The maximum throughput S_{max} is 0.368, when $G = 1$. In other words, if one frame is generated during one frame transmission time, then 36.8 percent of these frames reach their destination successfully. We expect $G = 1$ to produce maximum throughput because the vulnerable time is equal to the frame transmission time. Therefore, if a station generates only one frame in this vulnerable time (and no other station generates a frame during this time), the frame will reach its destination successfully.

The throughput for slotted ALOHA is $S = G \times e^{-G}$.
The maximum throughput $S_{max} = 0.368$ when $G = 1$.

Example 12.4

A slotted ALOHA network transmits 200-bit frames using a shared channel with a 200-kbps bandwidth. Find the throughput if the system (all stations together) produces

a. 1000 frames per second.
b. 500 frames per second.
c. 250 frames per second.

Solution

This situation is similar to the previous exercise except that the network is using slotted ALOHA instead of pure ALOHA. The frame transmission time is 200/200 kbps or 1 ms.

a. In this case G is 1. So $S = G \times e^{-G} = 0.368$ (36.8 percent). This means that the throughput is $1000 \times 0.0368 = 368$ frames. Only 368 out of 1000 frames will probably survive. Note that this is the maximum throughput case, percentagewise.

b. Here G is 1/2. In this case $S = G \times e^{-G} = 0.303$ (30.3 percent). This means that the throughput is $500 \times 0.0303 = 151$. Only 151 frames out of 500 will probably survive.

c. Now G is 1/4. In this case $S = G \times e^{-G} = 0.195$ (19.5 percent). This means that the throughput is $250 \times 0.195 = 49$. Only 49 frames out of 250 will probably survive.

12.1.2 CSMA

To minimize the chance of collision and, therefore, increase the performance, the CSMA method was developed. The chance of collision can be reduced if a station senses the medium before trying to use it. **Carrier sense multiple access (CSMA)** requires that each station first listen to the medium (or check the state of the medium) before sending. In other words, CSMA is based on the principle "sense before transmit" or "listen before talk."

CSMA can reduce the possibility of collision, but it cannot eliminate it. The reason for this is shown in Figure 12.7, a space and time model of a CSMA network. Stations are connected to a shared channel (usually a dedicated medium).

The possibility of collision still exists because of propagation delay; when a station sends a frame, it still takes time (although very short) for the first bit to reach every station and for every station to sense it. In other words, a station may sense the medium and find it idle, only because the first bit sent by another station has not yet been received.

Figure 12.7 *Space/time model of a collision in CSMA*

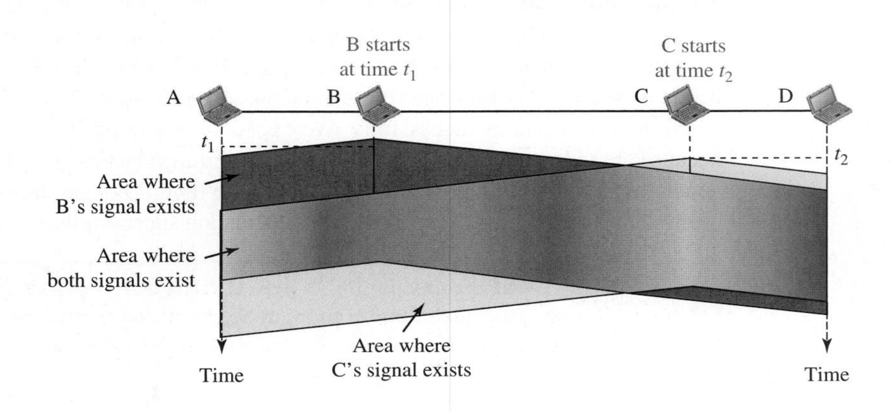

At time t_1, station B senses the medium and finds it idle, so it sends a frame. At time t_2 ($t_2 > t_1$), station C senses the medium and finds it idle because, at this time, the first bits from station B have not reached station C. Station C also sends a frame. The two signals collide and both frames are destroyed.

Vulnerable Time

The vulnerable time for CSMA is the ***propagation time*** T_p. This is the time needed for a signal to propagate from one end of the medium to the other. When a station sends a frame and any other station tries to send a frame during this time, a collision will result. But if the first bit of the frame reaches the end of the medium, every station will already have heard the bit and will refrain from sending. Figure 12.8 shows the worst case. The leftmost station, A, sends a frame at time t_1, which reaches the rightmost station, D, at time $t_1 + T_p$. The gray area shows the vulnerable area in time and space.

Figure 12.8 *Vulnerable time in CSMA*

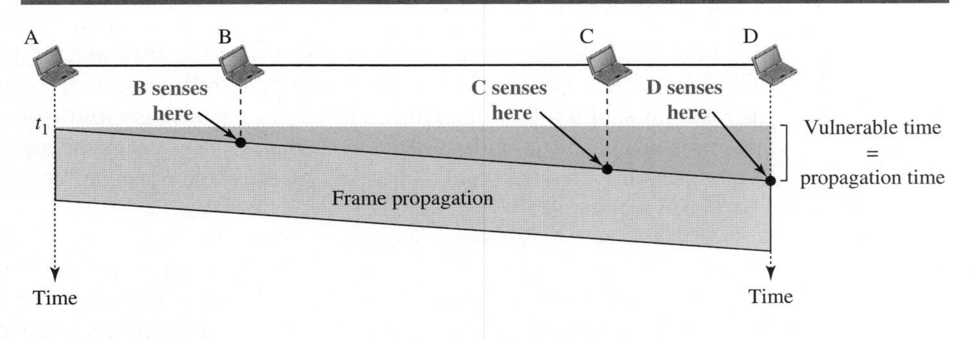

Persistence Methods

What should a station do if the channel is busy? What should a station do if the channel is idle? Three methods have been devised to answer these questions: the **1-persistent method,** the **nonpersistent method,** and the ***p*-persistent method.** Figure 12.9 shows the behavior of three persistence methods when a station finds a channel busy.

Figure 12.9 *Behavior of three persistence methods*

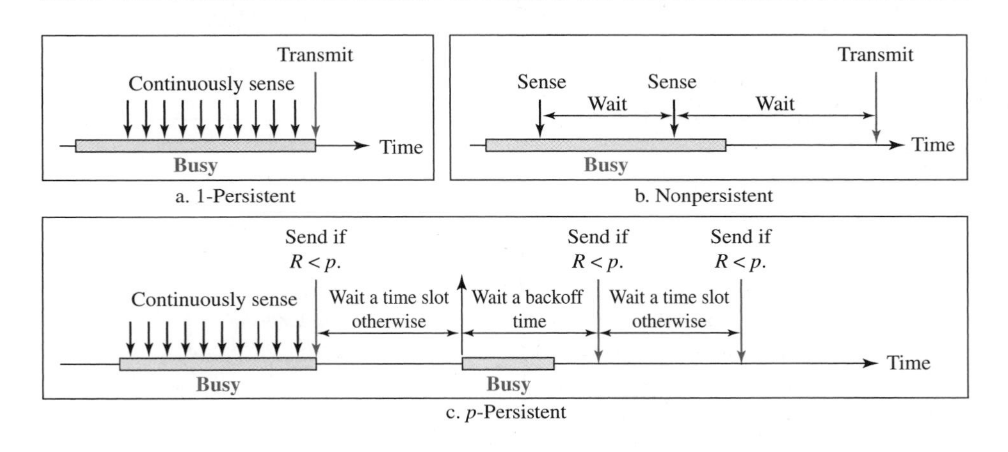

Figure 12.10 shows the flow diagrams for these methods.

1-Persistent

The *1-persistent method* is simple and straightforward. In this method, after the station finds the line idle, it sends its frame immediately (with probability 1). This method has the highest chance of collision because two or more stations may find the line idle and send their frames immediately. We will see later that Ethernet uses this method.

Nonpersistent

In the *nonpersistent method,* a station that has a frame to send senses the line. If the line is idle, it sends immediately. If the line is not idle, it waits a random amount of time and then senses the line again. The nonpersistent approach reduces the chance of collision because it is unlikely that two or more stations will wait the same amount of time and retry to send simultaneously. However, this method reduces the efficiency of the network because the medium remains idle when there may be stations with frames to send.

p-*Persistent*

The p-*persistent method* is used if the channel has time slots with a slot duration equal to or greater than the maximum propagation time. The p-persistent approach combines the advantages of the other two strategies. It reduces the chance of collision and improves efficiency. In this method, after the station finds the line idle it follows these steps:

1. With probability p, the station sends its frame.

Figure 12.10 *Flow diagram for three persistence methods*

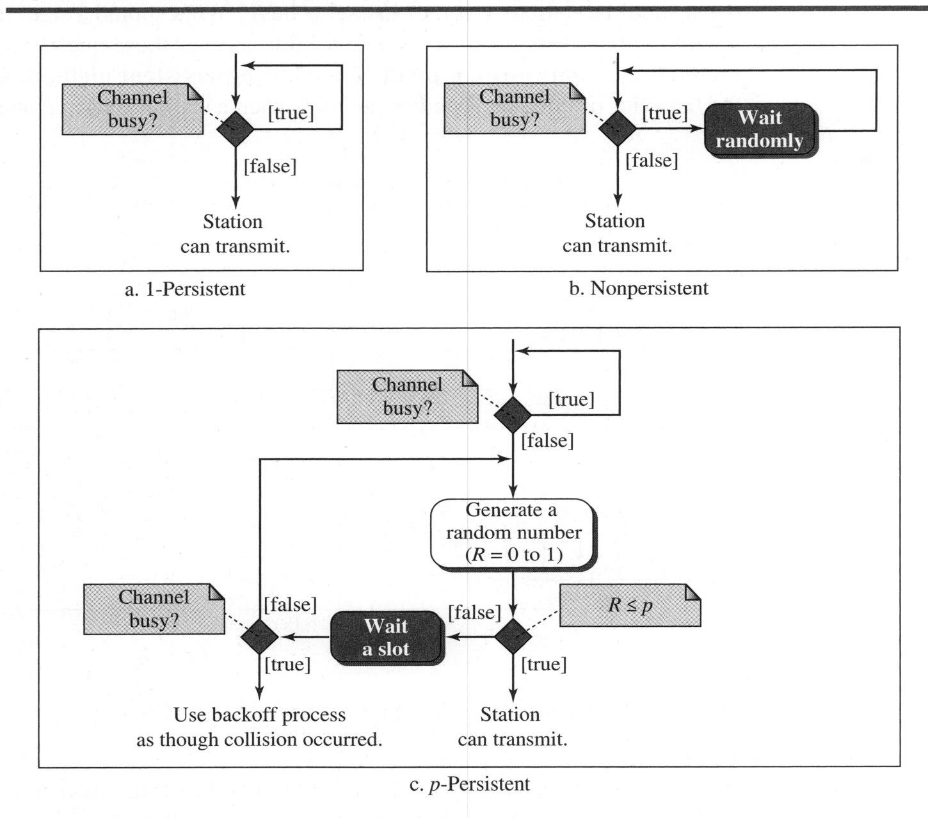

2. With probability $q = 1 - p$, the station waits for the beginning of the next time slot and checks the line again.
   a. If the line is idle, it goes to step 1.
   b. If the line is busy, it acts as though a collision has occurred and uses the back-off procedure.

12.1.3 CSMA/CD

The CSMA method does not specify the procedure following a collision. **Carrier sense multiple access with collision detection (CSMA/CD)** augments the algorithm to handle the collision.

In this method, a station monitors the medium after it sends a frame to see if the transmission was successful. If so, the station is finished. If, however, there is a collision, the frame is sent again.

To better understand CSMA/CD, let us look at the first bits transmitted by the two stations involved in the collision. Although each station continues to send bits in the frame until it detects the collision, we show what happens as the first bits collide. In Figure 12.11, stations A and C are involved in the collision.

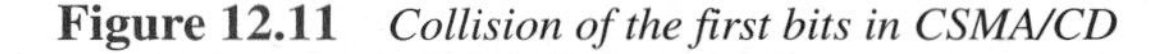

Figure 12.11 *Collision of the first bits in CSMA/CD*

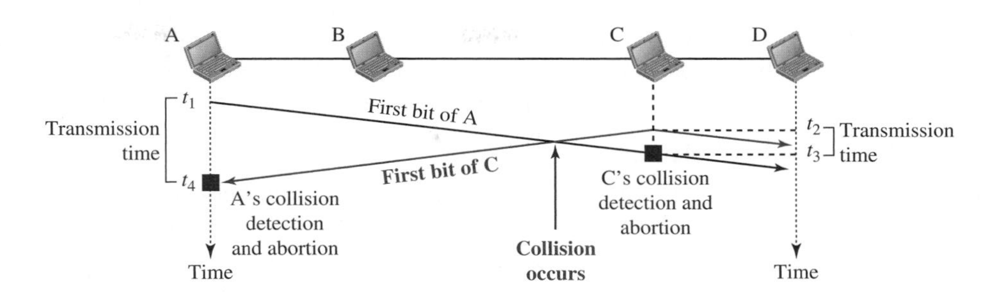

At time t_1, station A has executed its persistence procedure and starts sending the bits of its frame. At time t_2, station C has not yet sensed the first bit sent by A. Station C executes its persistence procedure and starts sending the bits in its frame, which propagate both to the left and to the right. The collision occurs sometime after time t_2. Station C detects a collision at time t_3 when it receives the first bit of A's frame. Station C immediately (or after a short time, but we assume immediately) aborts transmission. Station A detects collision at time t_4 when it receives the first bit of C's frame; it also immediately aborts transmission. Looking at the figure, we see that A transmits for the duration $t_4 - t_1$; C transmits for the duration $t_3 - t_2$.

Now that we know the time durations for the two transmissions, we can show a more complete graph in Figure 12.12.

Figure 12.12 *Collision and abortion in CSMA/CD*

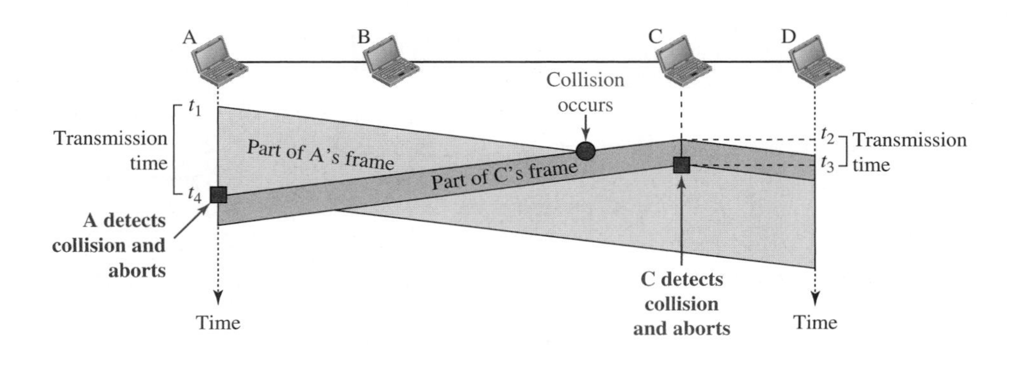

Minimum Frame Size

For CSMA/CD to work, we need a restriction on the frame size. Before sending the last bit of the frame, the sending station must detect a collision, if any, and abort the transmission. This is so because the station, once the entire frame is sent, does not keep a copy of the frame and does not monitor the line for collision detection. Therefore, the frame transmission time T_{fr} must be at least two times the maximum propagation time T_p. To understand the reason, let us think about the worst-case scenario. If the two stations involved in a collision are the maximum distance apart, the signal from the first takes time T_p to reach the second, and the effect of the collision takes another time T_P to reach the first. So the requirement is that the first station must still be transmitting after $2T_p$.

Example 12.5

A network using CSMA/CD has a bandwidth of 10 Mbps. If the maximum propagation time (including the delays in the devices and ignoring the time needed to send a jamming signal, as we see later) is 25.6 μs, what is the minimum size of the frame?

Solution

The minimum frame transmission time is $T_{fr} = 2 \times T_p = 51.2$ μs. This means, in the worst case, a station needs to transmit for a period of 51.2 μs to detect the collision. The minimum size of the frame is 10 Mbps × 51.2 μs = 512 bits or 64 bytes. This is actually the minimum size of the frame for Standard Ethernet, as we will see later in the chapter.

Procedure

Now let us look at the flow diagram for CSMA/CD in Figure 12.13. It is similar to the one for the ALOHA protocol, but there are differences.

Figure 12.13 *Flow diagram for the CSMA/CD*

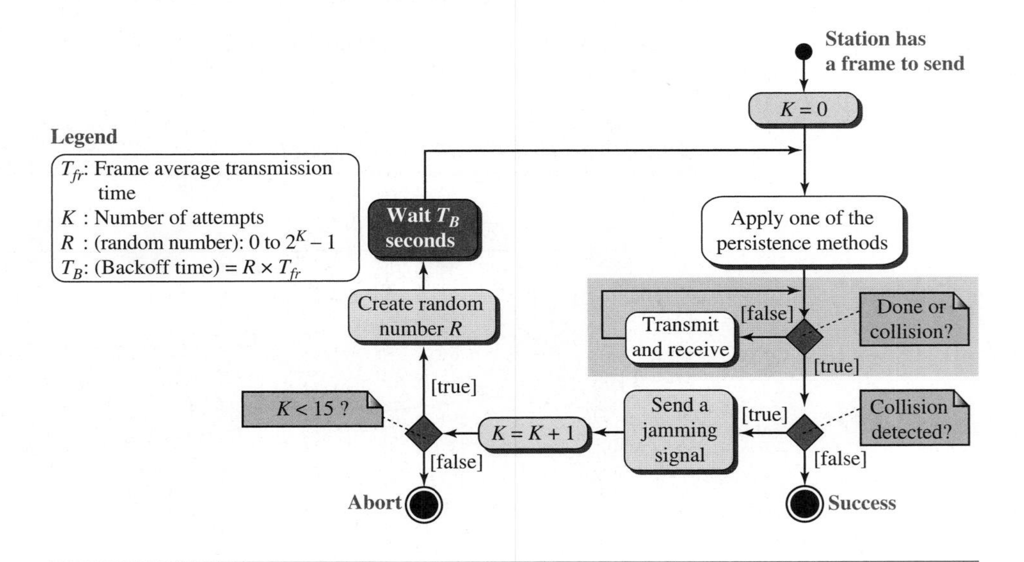

The first difference is the addition of the persistence process. We need to sense the channel before we start sending the frame by using one of the persistence processes we discussed previously (nonpersistent, 1-persistent, or p-persistent). The corresponding box can be replaced by one of the persistence processes shown in Figure 12.10.

The second difference is the frame transmission. In ALOHA, we first transmit the entire frame and then wait for an acknowledgment. In CSMA/CD, transmission and collision detection are continuous processes. We do not send the entire frame and then look for a collision. The station transmits and receives continuously and simultaneously (using two different ports or a bidirectional port). We use a loop to show that transmission is a continuous process. We constantly monitor in order to detect one of two conditions: either transmission is finished or a collision is detected. Either event stops transmission. When we come out of the loop, if a collision has not been detected, it means that transmission is complete; the entire frame is transmitted. Otherwise, a collision has occurred.

The third difference is the sending of a short **jamming signal** to make sure that all other stations become aware of the collision.

Energy Level

We can say that the level of energy in a channel can have three values: zero, normal, and abnormal. At the zero level, the channel is idle. At the normal level, a station has successfully captured the channel and is sending its frame. At the abnormal level, there is a collision and the level of the energy is twice the normal level. A station that has a frame to send or is sending a frame needs to monitor the energy level to determine if the channel is idle, busy, or in collision mode. Figure 12.14 shows the situation.

Figure 12.14 *Energy level during transmission, idleness, or collision*

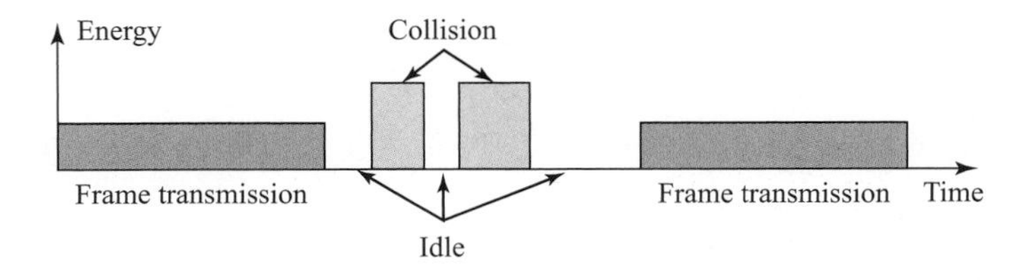

Throughput

The throughput of CSMA/CD is greater than that of pure or slotted ALOHA. The maximum throughput occurs at a different value of G and is based on the persistence method and the value of p in the p-persistent approach. For the 1-persistent method, the maximum throughput is around 50 percent when $G = 1$. For the nonpersistent method, the maximum throughput can go up to 90 percent when G is between 3 and 8.

Traditional Ethernet

One of the LAN protocols that used CSMA/CD is the traditional Ethernet with the data rate of 10 Mbps. We discuss the Ethernet LANs in Chapter 13, but it is good to know that the traditional Ethernet was a broadcast LAN that used the 1-persistence method to control access to the common media. Later versions of Ethernet try to move from CSMA/CD access methods for the reason that we discuss in Chapter 13.

12.1.4 CSMA/CA

Carrier sense multiple access with collision avoidance (CSMA/CA) was invented for wireless networks. Collisions are avoided through the use of CSMA/CA's three strategies: the interframe space, the contention window, and acknowledgments, as shown in Figure 12.15. We discuss RTS and CTS frames later.

Figure 12.15 *Flow diagram of CSMA/CA*

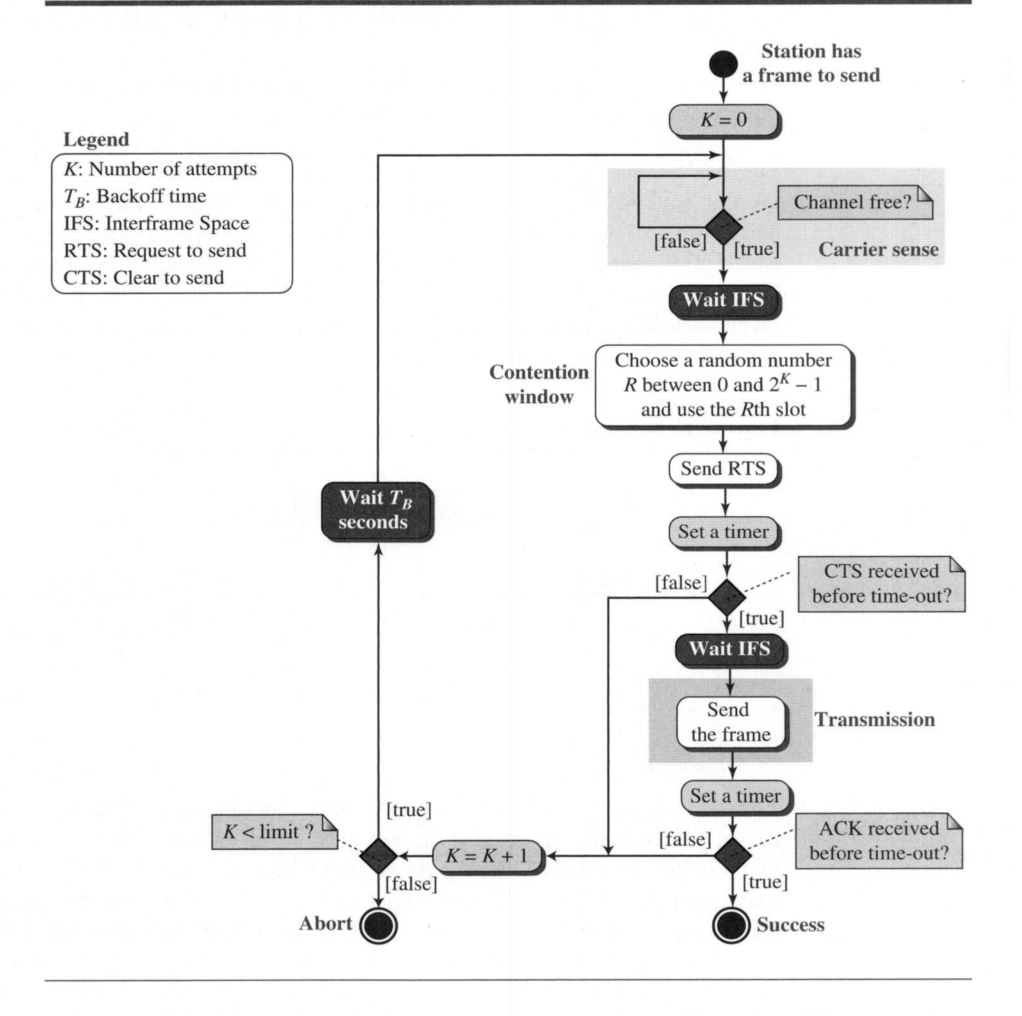

- ❑ ***Interframe Space (IFS).*** First, collisions are avoided by deferring transmission even if the channel is found idle. When an idle channel is found, the station does not send immediately. It waits for a period of time called the ***interframe space*** or ***IFS.*** Even though the channel may appear idle when it is sensed, a distant station may have already started transmitting. The distant station's signal has not yet reached this station. The IFS time allows the front of the transmitted signal by the distant station to reach this station. After waiting an IFS time, if the channel is still idle, the station can send, but it still needs to wait a time equal to the contention window (described next). The IFS variable can also be used to prioritize stations or frame types. For example, a station that is assigned a shorter IFS has a higher priority.

- ***Contention Window.*** The **contention window** is an amount of time divided into slots. A station that is ready to send chooses a random number of slots as its wait time. The number of slots in the window changes according to the binary exponential backoff strategy. This means that it is set to one slot the first time and then doubles each time the station cannot detect an idle channel after the IFS time. This is very similar to the *p*-persistent method except that a random outcome defines the number of slots taken by the waiting station. One interesting point about the contention window is that the station needs to sense the channel after each time slot. However, if the station finds the channel busy, it does not restart the process; it just stops the timer and restarts it when the channel is sensed as idle. This gives priority to the station with the longest waiting time. See Figure 12.16.

Figure 12.16 *Contention window*

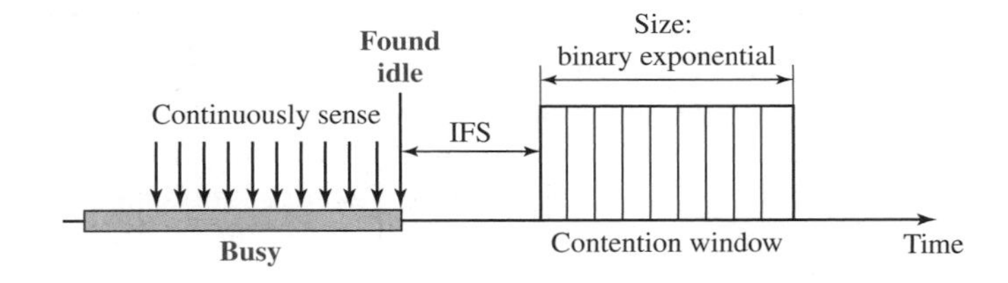

- ***Acknowledgment.*** With all these precautions, there still may be a collision resulting in destroyed data. In addition, the data may be corrupted during the transmission. The positive acknowledgment and the time-out timer can help guarantee that the receiver has received the frame.

Frame Exchange Time Line

Figure 12.17 shows the exchange of data and control frames in time.

1. Before sending a frame, the source station senses the medium by checking the energy level at the carrier frequency.
   a. The channel uses a persistence strategy with backoff until the channel is idle.
   b. After the station is found to be idle, the station waits for a period of time called the ***DCF interframe space (DIFS);*** then the station sends a control frame called the *request to send (RTS).*
2. After receiving the RTS and waiting a period of time called the ***short interframe space (SIFS),*** the destination station sends a control frame, called the *clear to send (CTS),* to the source station. This control frame indicates that the destination station is ready to receive data.

Figure 12.17 *CSMA/CA and NAV*

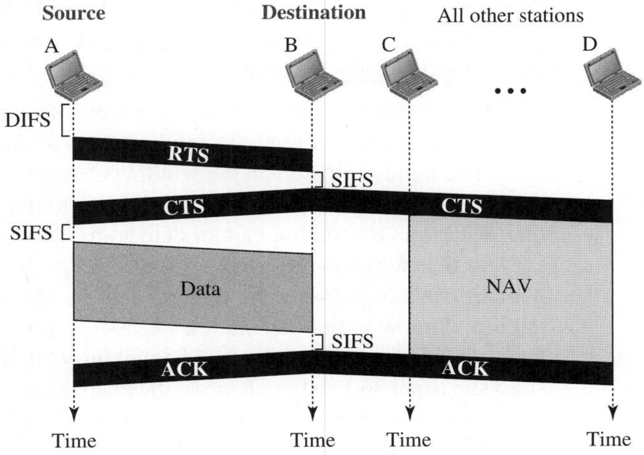

Source
A
Destination
B
All other stations
C
D
DIFS
RTS
SIFS
CTS
CTS
SIFS
Data
NAV
SIFS
ACK
ACK
Time
Time
Time
Time

3. The source station sends data after waiting an amount of time equal to SIFS.
4. The destination station, after waiting an amount of time equal to SIFS, sends an acknowledgment to show that the frame has been received. Acknowledgment is needed in this protocol because the station does not have any means to check for the successful arrival of its data at the destination. On the other hand, the lack of collision in CSMA/CD is a kind of indication to the source that data have arrived.

Network Allocation Vector

How do other stations defer sending their data if one station acquires access? In other words, how is the *collision avoidance* aspect of this protocol accomplished? The key is a feature called *NAV.*

When a station sends an RTS frame, it includes the duration of time that it needs to occupy the channel. The stations that are affected by this transmission create a timer called a **network allocation vector (NAV)** that shows how much time must pass before these stations are allowed to check the channel for idleness. Each time a station accesses the system and sends an RTS frame, other stations start their NAV. In other words, each station, before sensing the physical medium to see if it is idle, first checks its NAV to see if it has expired. Figure 12.17 shows the idea of NAV.

Collision During Handshaking

What happens if there is a collision during the time when RTS or CTS control frames are in transition, often called the *handshaking period?* Two or more stations may try to send RTS frames at the same time. These control frames may collide. However, because there is no mechanism for collision detection, the sender assumes there has been a collision if it has not received a CTS frame from the receiver. The backoff strategy is employed, and the sender tries again.

Hidden-Station Problem

The solution to the hidden station problem is the use of the handshake frames (RTS and CTS). Figure 12.17 also shows that the RTS message from B reaches A, but not C. However, because both B and C are within the range of A, the CTS message, which contains the duration of data transmission from B to A, reaches C. Station C knows that some hidden station is using the channel and refrains from transmitting until that duration is over.

CSMA/CA and Wireless Networks

CSMA/CA was mostly intended for use in wireless networks. The procedure described above, however, is not sophisticated enough to handle some particular issues related to wireless networks, such as hidden terminals or exposed terminals. We will see how these issues are solved by augmenting the above protocol with handshaking features. The use of CSMA/CA in wireless networks will be discussed in Chapter 15.

12.2 CONTROLLED ACCESS

In **controlled access,** the stations consult one another to find which station has the right to send. A station cannot send unless it has been authorized by other stations. We discuss three controlled-access methods.

12.2.1 Reservation

In the **reservation** method, a station needs to make a reservation before sending data. Time is divided into intervals. In each interval, a reservation frame precedes the data frames sent in that interval.

If there are *N* stations in the system, there are exactly *N* reservation minislots in the reservation frame. Each minislot belongs to a station. When a station needs to send a data frame, it makes a reservation in its own minislot. The stations that have made reservations can send their data frames after the reservation frame.

Figure 12.18 shows a situation with five stations and a five-minislot reservation frame. In the first interval, only stations 1, 3, and 4 have made reservations. In the second interval, only station 1 has made a reservation.

Figure 12.18 *Reservation access method*

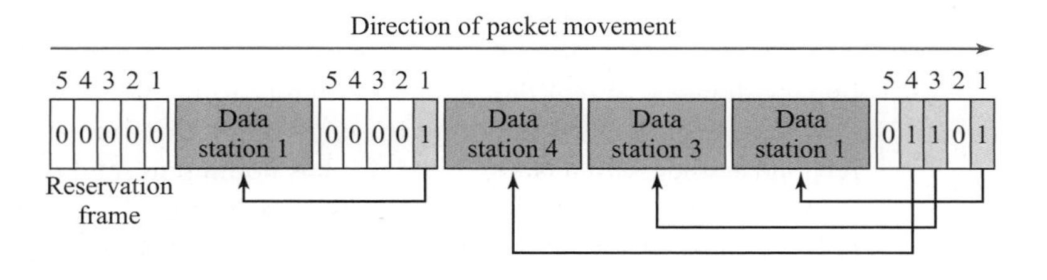

12.2.2 Polling

Polling works with topologies in which one device is designated as a ***primary station*** and the other devices are ***secondary stations***. All data exchanges must be made through the primary device even when the ultimate destination is a secondary device. The primary device controls the link; the secondary devices follow its instructions. It is up to the primary device to determine which device is allowed to use the channel at a given time. The primary device, therefore, is always the initiator of a session (see Figure 12.19). This method uses poll and select functions to prevent collisions. However, the drawback is if the primary station fails, the system goes down.

Figure 12.19 *Select and poll functions in polling-access method*

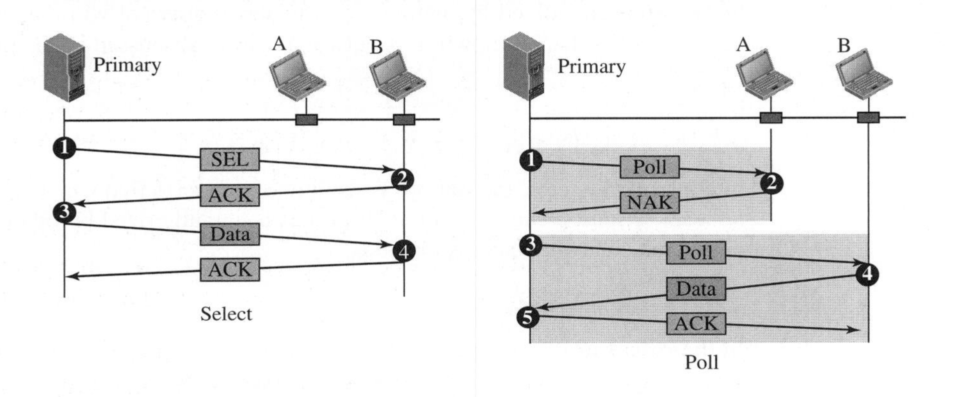

Select

The *select* function is used whenever the primary device has something to send. Remember that the primary controls the link. If the primary is neither sending nor receiving data, it knows the link is available. If it has something to send, the primary device sends it. What it does not know, however, is whether the target device is prepared to receive. So the primary must alert the secondary to the upcoming transmission and wait for an acknowledgment of the secondary's ready status. Before sending data, the primary creates and transmits a select (SEL) frame, one field of which includes the address of the intended secondary.

Poll

The *poll* function is used by the primary device to solicit transmissions from the secondary devices. When the primary is ready to receive data, it must ask (poll) each device in turn if it has anything to send. When the first secondary is approached, it responds either with a NAK frame if it has nothing to send or with data (in the form of a data frame) if it does. If the response is negative (a NAK frame), then the primary polls the next secondary in the same manner until it finds one with data to send. When the response is positive (a data frame), the primary reads the frame and returns an acknowledgment (ACK frame), verifying its receipt.

12.2.3 Token Passing

In the **token-passing** method, the stations in a network are organized in a logical ring. In other words, for each station, there is a *predecessor* and a *successor*. The predecessor is the station which is logically before the station in the ring; the successor is the station which is after the station in the ring. The current station is the one that is accessing the channel now. The right to this access has been passed from the predecessor to the current station. The right will be passed to the successor when the current station has no more data to send.

But how is the right to access the channel passed from one station to another? In this method, a special packet called a ***token*** circulates through the ring. The possession of the token gives the station the right to access the channel and send its data. When a station has some data to send, it waits until it receives the token from its predecessor. It then holds the token and sends its data. When the station has no more data to send, it releases the token, passing it to the next logical station in the ring. The station cannot send data until it receives the token again in the next round. In this process, when a station receives the token and has no data to send, it just passes the data to the next station.

Token management is needed for this access method. Stations must be limited in the time they can have possession of the token. The token must be monitored to ensure it has not been lost or destroyed. For example, if a station that is holding the token fails, the token will disappear from the network. Another function of token management is to assign priorities to the stations and to the types of data being transmitted. And finally, token management is needed to make low-priority stations release the token to high-priority stations.

Logical Ring

In a token-passing network, stations do not have to be physically connected in a ring; the ring can be a logical one. Figure 12.20 shows four different physical topologies that can create a logical ring.

Figure 12.20 *Logical ring and physical topology in token-passing access method*

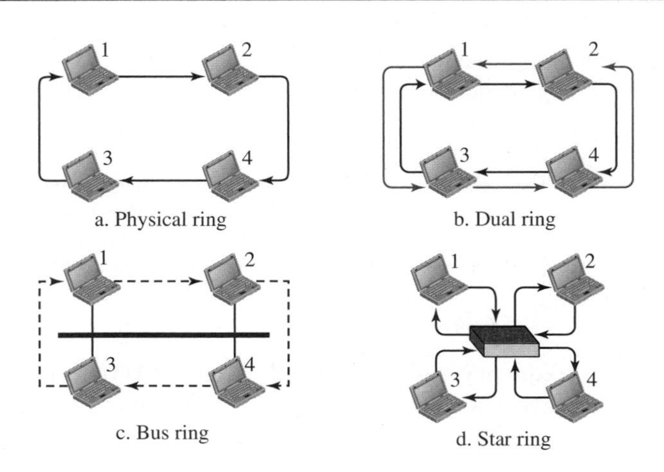

In the physical ring topology, when a station sends the token to its successor, the token cannot be seen by other stations; the successor is the next one in line. This means that the token does not have to have the address of the next successor. The problem with this topology is that if one of the links—the medium between two adjacent stations—fails, the whole system fails.

The dual ring topology uses a second (auxiliary) ring which operates in the reverse direction compared with the main ring. The second ring is for emergencies only (such as a spare tire for a car). If one of the links in the main ring fails, the system automatically combines the two rings to form a temporary ring. After the failed link is restored, the auxiliary ring becomes idle again. Note that for this topology to work, each station needs to have two transmitter ports and two receiver ports. The high-speed Token Ring networks called *FDDI (Fiber Distributed Data Interface)* and *CDDI (Copper Distributed Data Interface)* use this topology.

In the bus ring topology, also called a token bus, the stations are connected to a single cable called a *bus*. They, however, make a logical ring, because each station knows the address of its successor (and also predecessor for token management purposes). When a station has finished sending its data, it releases the token and inserts the address of its successor in the token. Only the station with the address matching the destination address of the token gets the token to access the shared media. The Token Bus LAN, standardized by IEEE, uses this topology.

In a star ring topology, the physical topology is a star. There is a hub, however, that acts as the connector. The wiring inside the hub makes the ring; the stations are connected to this ring through the two wire connections. This topology makes the network less prone to failure because if a link goes down, it will be bypassed by the hub and the rest of the stations can operate. Also adding and removing stations from the ring is easier. This topology is still used in the Token Ring LAN designed by IBM.

12.3 CHANNELIZATION

Channelization (or *channel partition,* as it is sometimes called) is a multiple-access method in which the available bandwidth of a link is shared in time, frequency, or through code, among different stations. In this section, we discuss three channelization protocols: FDMA, TDMA, and CDMA.

We see the application of all these methods in Chapter 16 when we discuss cellular phone systems.

12.3.1 FDMA

In **frequency-division multiple access (FDMA),** the available bandwidth is divided into frequency bands. Each station is allocated a band to send its data. In other words, each band is reserved for a specific station, and it belongs to the station all the time. Each station also uses a bandpass filter to confine the transmitter frequencies. To prevent station interferences, the allocated bands are separated from one another by small *guard bands*. Figure 12.21 shows the idea of FDMA.

Figure 12.21 *Frequency-division multiple access (FDMA)*

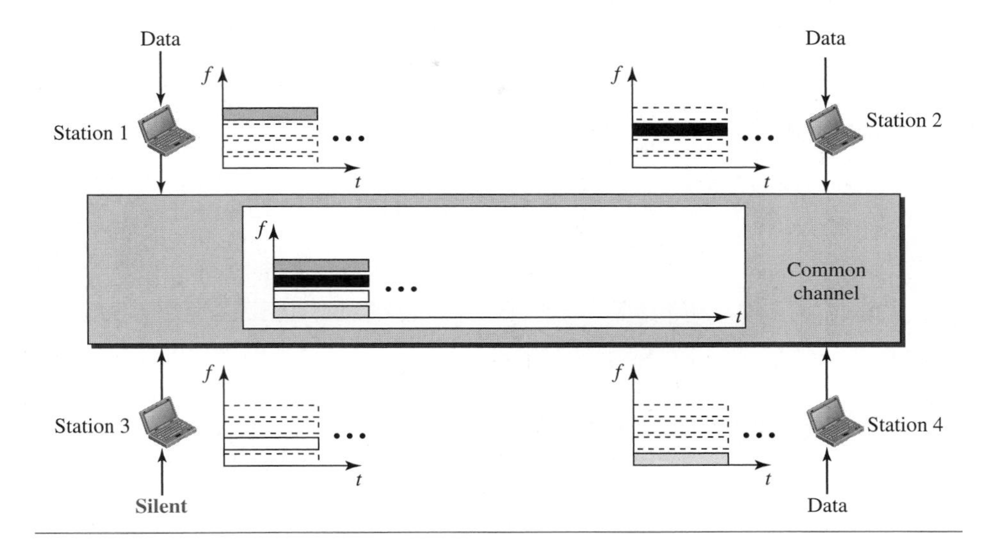

In FDMA, the available bandwidth of the common channel is divided into bands that are separated by guard bands.

FDMA specifies a predetermined frequency band for the entire period of communication. This means that stream data (a continuous flow of data that may not be packetized) can easily be used with FDMA. We will see in Chapter 16 how this feature can be used in cellular telephone systems.

We need to emphasize that although FDMA and frequency-division multiplexing (FDM) conceptually seem similar, there are differences between them. FDM, as we saw in Chapter 6, is a physical layer technique that combines the loads from low-bandwidth channels and transmits them by using a high-bandwidth channel. The channels that are combined are low-pass. The multiplexer modulates the signals, combines them, and creates a bandpass signal. The bandwidth of each channel is shifted by the multiplexer.

FDMA, on the other hand, is an access method in the data-link layer. The data-link layer in each station tells its physical layer to make a bandpass signal from the data passed to it. The signal must be created in the allocated band. There is no physical multiplexer at the physical layer. The signals created at each station are automatically bandpass-filtered. They are mixed when they are sent to the common channel.

12.3.2 TDMA

In **time-division multiple access (TDMA),** the stations share the bandwidth of the channel in time. Each station is allocated a time slot during which it can send data. Each station transmits its data in its assigned time slot. Figure 12.22 shows the idea behind TDMA.

Figure 12.22 *Time-division multiple access (TDMA)*

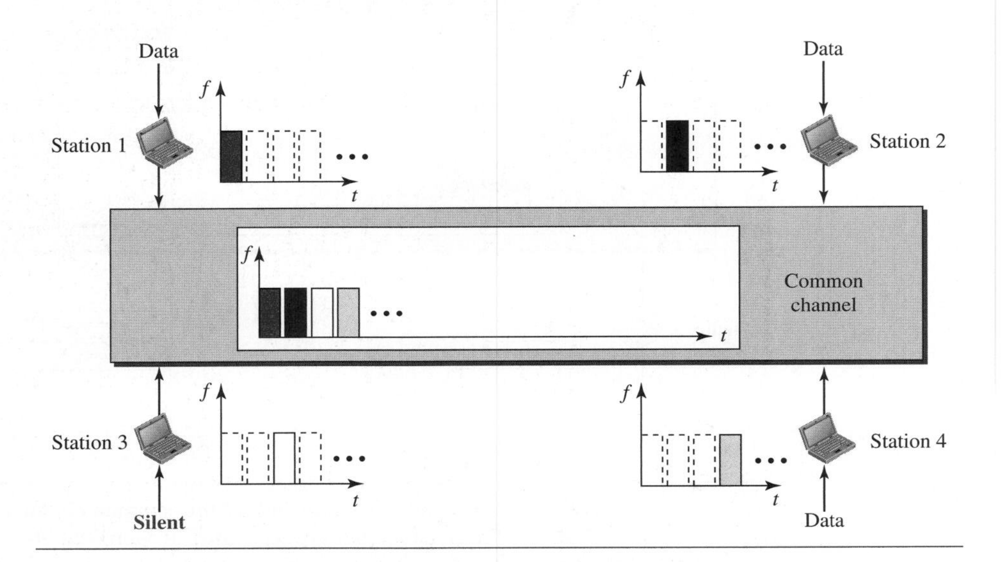

The main problem with TDMA lies in achieving synchronization between the different stations. Each station needs to know the beginning of its slot and the location of its slot. This may be difficult because of propagation delays introduced in the system if the stations are spread over a large area. To compensate for the delays, we can insert *guard times*. Synchronization is normally accomplished by having some synchronization bits (normally referred to as *preamble bits*) at the beginning of each slot.

In TDMA, the bandwidth is just one channel that is timeshared between different stations.

We also need to emphasize that although TDMA and time-division multiplexing (TDM) conceptually seem the same, there are differences between them. TDM, as we saw in Chapter 6, is a physical layer technique that combines the data from slower channels and transmits them by using a faster channel. The process uses a physical multiplexer that interleaves data units from each channel.

TDMA, on the other hand, is an access method in the data-link layer. The data-link layer in each station tells its physical layer to use the allocated time slot. There is no physical multiplexer at the physical layer.

12.3.3 CDMA

Code-division multiple access (CDMA) was conceived several decades ago. Recent advances in electronic technology have finally made its implementation possible. CDMA differs from FDMA in that only one channel occupies the entire bandwidth of the link. It differs from TDMA in that all stations can send data simultaneously; there is no timesharing.

In CDMA, one channel carries all transmissions simultaneously.

Analogy

Let us first give an analogy. CDMA simply means communication with different codes. For example, in a large room with many people, two people can talk privately in English if nobody else understands English. Another two people can talk in Chinese if they are the only ones who understand Chinese, and so on. In other words, the common channel, the space of the room in this case, can easily allow communication between several couples, but in different languages (codes).

Idea

Let us assume we have four stations, 1, 2, 3, and 4, connected to the same channel. The data from station 1 are d_1, from station 2 are d_2, and so on. The code assigned to the first station is c_1, to the second is c_2, and so on. We assume that the assigned codes have two properties.

1. If we multiply each code by another, we get 0.
2. If we multiply each code by itself, we get 4 (the number of stations).

With these two properties in mind, let us see how the above four stations can send data using the same common channel, as shown in Figure 12.23.

Station 1 multiplies (a special kind of multiplication, as we will see) its data by its code to get $d_1 \cdot c_1$. Station 2 multiplies its data by its code to get $d_2 \cdot c_2$, and so on. The data that go on the channel are the sum of all these terms, as shown in the box. Any station that wants to receive data from one of the other three multiplies the data on the channel by the code of the sender. For example, suppose stations 1 and 2 are talking to each other. Station 2 wants to hear what station 1 is saying. It multiplies the data on the channel by c_1, the code of station 1.

Because $(c_1 \cdot c_1)$ is 4, but $(c_2 \cdot c_1)$, $(c_3 \cdot c_1)$, and $(c_4 \cdot c_1)$ are all 0s, station 2 divides the result by 4 to get the data from station 1.

Chips

CDMA is based on coding theory. Each station is assigned a code, which is a sequence of numbers called *chips*, as shown in Figure 12.24. The codes are for the previous example.

Later in this chapter we show how we chose these sequences. For now, we need to know that we did not choose the sequences randomly; they were carefully selected. They are called ***orthogonal sequences*** and have the following properties:

Figure 12.23 *Simple idea of communication with code*

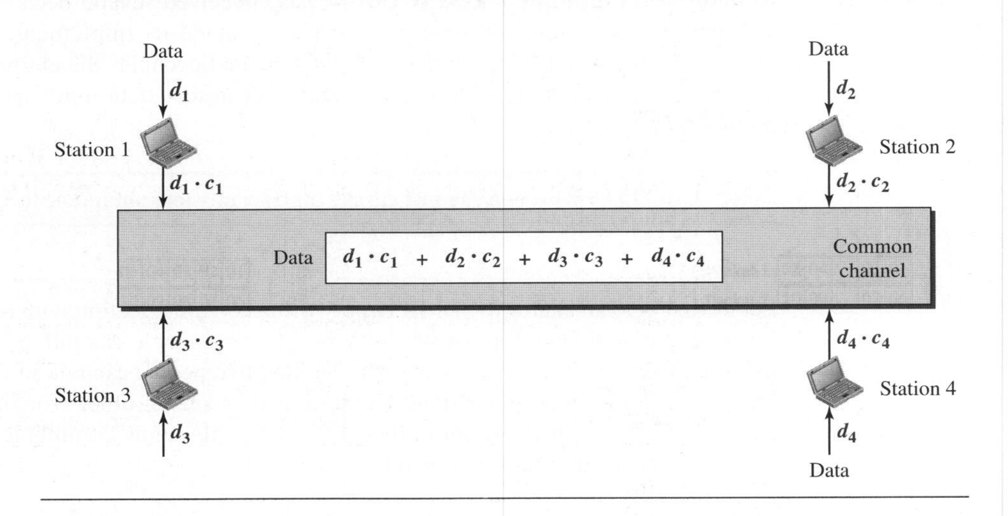

$$\text{data} = (d_1 \cdot c_1 + d_2 \cdot c_2 + d_3 \cdot c_3 + d_4 \cdot c_4) \cdot c_1$$
$$= d_1 \cdot c_1 \cdot c_1 + d_2 \cdot c_2 \cdot c_1 + d_3 \cdot c_3 \cdot c_1 + d_4 \cdot c_4 \cdot c_1 = 4 \times d_1$$

Figure 12.24 *Chip sequences*

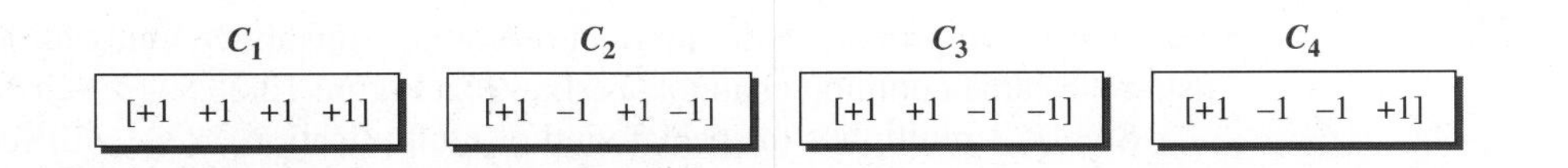

1. Each sequence is made of N elements, where N is the number of stations.
2. If we multiply a sequence by a number, every element in the sequence is multiplied by that element. This is called multiplication of a sequence by a scalar. For example,

$$2 \bullet [+1 +1 -1 -1] = [+2 +2 -2 -2]$$

3. If we multiply two equal sequences, element by element, and add the results, we get N, where N is the number of elements in each sequence. This is called the ***inner product*** of two equal sequences. For example,

$$[+1 +1 -1 -1] \bullet [+1 +1 -1 -1] = 1 + 1 + 1 + 1 = 4$$

4. If we multiply two different sequences, element by element, and add the results, we get 0. This is called the *inner product* of two different sequences. For example,

$$[+1 +1 -1 -1] \bullet [+1 +1 +1 +1] = 1 + 1 - 1 - 1 = 0$$

5. Adding two sequences means adding the corresponding elements. The result is another sequence. For example,

$$[+1\ +1\ -1\ -1] + [+1\ +1\ +1\ +1] = [+2\ +2\ \ 0\ \ 0]$$

Data Representation

We follow these rules for encoding: If a station needs to send a 0 bit, it encodes it as –1; if it needs to send a 1 bit, it encodes it as +1. When a station is idle, it sends no signal, which is interpreted as a 0. These are shown in Figure 12.25.

Figure 12.25 *Data representation in CDMA*

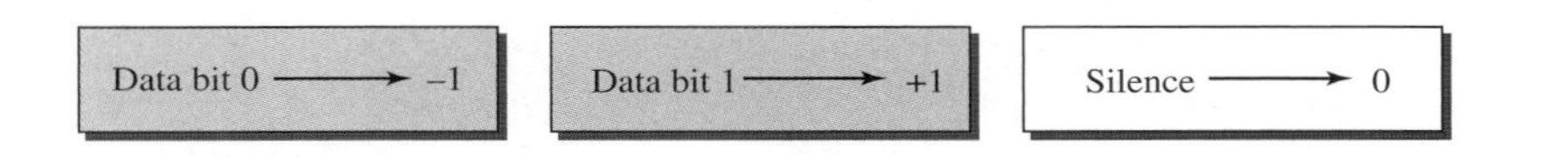

Encoding and Decoding

As a simple example, we show how four stations share the link during a 1-bit interval. The procedure can easily be repeated for additional intervals. We assume that stations 1 and 2 are sending a 0 bit and channel 4 is sending a 1 bit. Station 3 is silent. The data at the sender site are translated to −1, −1, 0, and +1. Each station multiplies the corresponding number by its chip (its orthogonal sequence), which is unique for each station. The result is a new sequence which is sent to the channel. For simplicity, we assume that all stations send the resulting sequences at the same time. The sequence on the channel is the sum of all four sequences as defined before. Figure 12.26 shows the situation.

Now imagine that station 3, which we said is silent, is listening to station 2. Station 3 multiplies the total data on the channel by the code for station 2, which is [+1 −1 +1 −1], to get

$$[-1\ -1\ -3\ +1] \bullet [+1\ -1\ +1\ -1] = -4/4 = -1 \quad \rightarrow \quad \textbf{bit 1}$$

Signal Level

The process can be better understood if we show the digital signal produced by each station and the data recovered at the destination (see Figure 12.27). The figure shows the corresponding signals for each station (using NRZ-L for simplicity) and the signal that is on the common channel.

Figure 12.28 shows how station 3 can detect the data sent by station 2 by using the code for station 2. The total data on the channel are multiplied (inner product operation) by the signal representing station 2 chip code to get a new signal. The station then integrates and adds the area under the signal, to get the value –4, which is divided by 4 and interpreted as bit 0.

Sequence Generation

To generate chip sequences, we use a **Walsh table,** which is a two-dimensional table with an equal number of rows and columns, as shown in Figure 12.29.

Figure 12.26 *Sharing channel in CDMA*

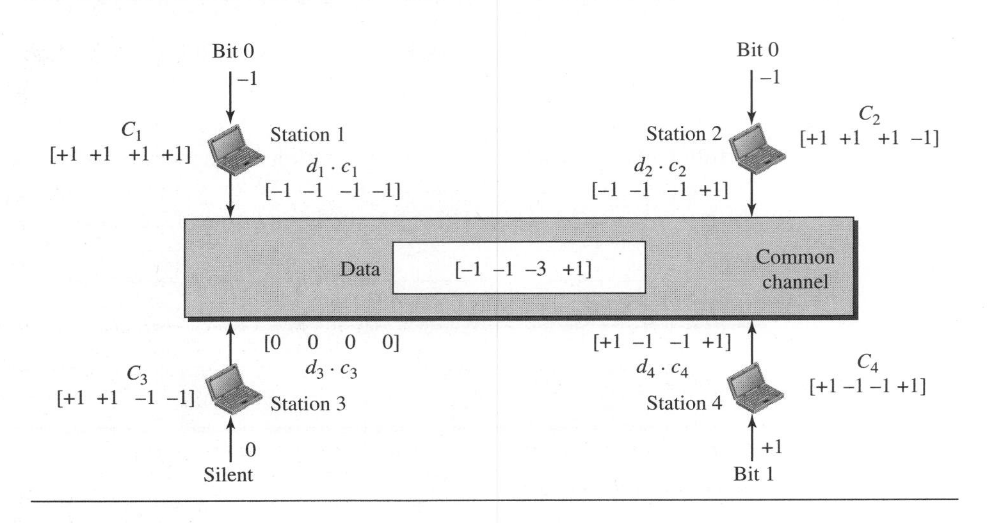

Figure 12.27 *Digital signal created by four stations in CDMA*

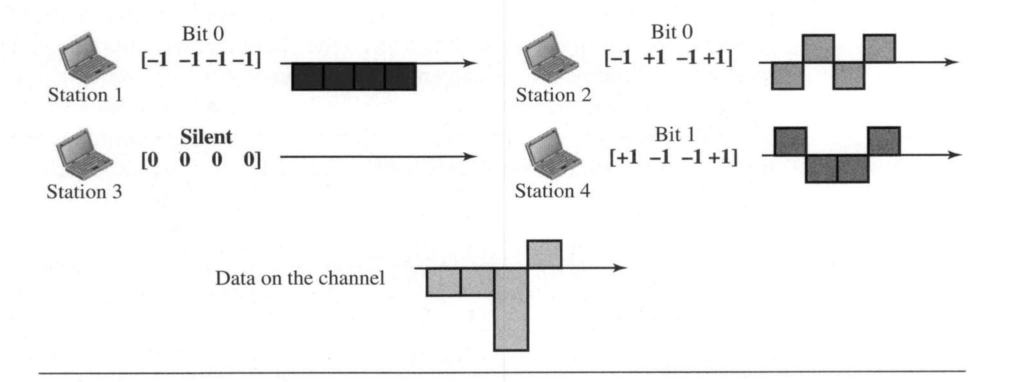

In the Walsh table, each row is a sequence of chips. W_1 for a one-chip sequence has one row and one column. We can choose −1 or +1 for the chip for this trivial table (we chose +1). According to Walsh, if we know the table for N sequences W_N, we can create the table for $2N$ sequences W_{2N}, as shown in Figure 12.29. The W_N with the overbar $\overline{W_N}$ stands for the complement of W_N, where each +1 is changed to −1 and vice versa. Figure 12.29 also shows how we can create W_2 and W_4 from W_1. After we select W_1, W_2 can be made from four W_1s, with the last one the complement of W_1. After W_2 is generated, W_4 can be made of four W_2s, with the last one the complement of W_2. Of course, W_8 is composed of four W_4s, and so on. Note that after W_N is made, each station is assigned a chip corresponding to a row.

Figure 12.28 *Decoding of the composite signal for one in CDMA*

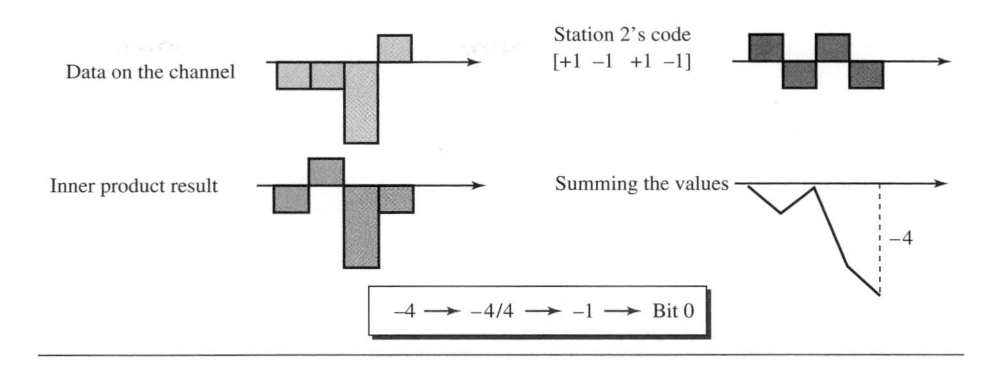

Figure 12.29 *General rule and examples of creating Walsh tables*

$$W_1 = \begin{bmatrix} +1 \end{bmatrix} \quad W_{2N} = \begin{bmatrix} W_N & W_N \\ W_N & \overline{W_N} \end{bmatrix}$$

a. Two basic rules

$$W_2 = \begin{bmatrix} +1 & +1 \\ +1 & -1 \end{bmatrix} \quad W_4 = \begin{bmatrix} +1 & +1 & +1 & +1 \\ +1 & -1 & +1 & -1 \\ +1 & +1 & -1 & -1 \\ +1 & -1 & -1 & +1 \end{bmatrix}$$

b. Generation of W_2 and W_4

Something we need to emphasize is that the number of sequences, N, needs to be a power of 2. In other words, we need to have $N = 2^m$.

The number of sequences in a Walsh table needs to be $N = 2^m$.

Example 12.6

Find the chips for a network with

a. Two stations

b. Four stations

Solution

We can use the rows of W_2 and W_4 in Figure 12.29:

a. For a two-station network, we have [+1 +1] and [+1 −1].

b. For a four-station network we have [+1 +1 +1 +1], [+1 −1 +1 −1], [+1 +1 −1 −1], and [+1 −1 −1 +1].

Example 12.7

What is the number of sequences if we have 90 stations in our network?

Solution

The number of sequences needs to be 2^m. We need to choose $m = 7$ and $N = 2^7$ or 128. We can then use 90 of the sequences as the chips.

Example 12.8

Prove that a receiving station can get the data sent by a specific sender if it multiplies the entire data on the channel by the sender's chip code and then divides it by the number of stations.

Solution

Let us prove this for the first station, using our previous four-station example. We can say that the data on the channel $D = (d_1 \cdot c_1 + d_2 \cdot c_2 + d_3 \cdot c_3 + d_4 \cdot c_4)$. The receiver that wants to get the data sent by station 1 multiplies these data by c_1.

$$
\begin{aligned}
D \cdot c_1 &= (d_1 \cdot c_1 + d_2 \cdot c_2 + d_3 \cdot c_3 + d_4 \cdot c_4) \quad c_1 \\
&= d_1 \cdot c_1 \cdot c_1 + d_2 \cdot c_2 \cdot c_1 + d_3 \cdot c_3 \cdot c_1 + d_4 \cdot c_4 \cdot c_1 \\
&= d_1 \times N + d_2 \times 0 + d_3 \times 0 + d_4 \times 0 \\
&= d_1 \times N
\end{aligned}
$$

When we divide the result by N, we get d_1.

12.4 END-CHAPTER MATERIALS

12.4.1 Recommended Reading

For more details about subjects discussed in this chapter, we recommend the following books and RFCs. The items in brackets [...] refer to the reference list at the end of the text.

Books

Several excellent books discuss link-layer issues. Among them we recommend [Ham 80], [Zar 02], [Ror 96], [Tan 03], [GW 04], [For 03], [KMK 04], [Sta 04], [Kes 02], [PD 03], [Kei 02], [Spu 00], [KCK 98], [Sau 98], [Izz 00], [Per 00], and [WV 00].

RFCs

A discussion of the use of the checksum in the Internet can be found in RFC 1141.

12.4.2 Key Terms

1-persistent method
ALOHA
binary exponential backoff
carrier sense multiple access (CSMA)
carrier sense multiple access with collision avoidance (CSMA/CA)
carrier sense multiple access with collision detection (CSMA/CD)
channelization
code-division multiple access (CDMA)
collision
contention
contention window
controlled access
DCF interframe space (DIFS)
frequency-division multiple access (FDMA)
inner product
interface space (IFS)
jamming signal
media access control (MAC)
multiple access (MA)
network allocation vector (NAV)
nonpersistent method
orthogonal sequence
p-persistent method
polling
primary station
propagation time
pure ALOHA
random access
reservation
secondary station
short interframe space (SIFS)
slotted ALOHA
time-division multiple access (TDMA)
token
token passing
vulnerable time
Walsh table

12.4.3 Summary

Many formal protocols have been devised to handle access to a shared link. We categorize them into three groups: random access protocols, controlled access protocols, and channelization protocols.

In random access or contention methods, no station is superior to another station and none is assigned the control over another. ALOHA allows multiple access (MA) to the shared medium.There are potential collisions in this arrangement. To minimize the chance of collision and, therefore, increase the performance, the CSMA method was developed. The chance of collision can be reduced if a station senses the medium before trying to use it. Carrier sense multiple access (CSMA) requires that each station first listen to the medium before sending. Carrier sense multiple access with collision detection (CSMA/CD) augments the CSMA algorithm to handle collision. In this method, a station monitors the medium after it sends a frame to see if the transmission was successful. If so, the station is finished. If, however, there is a collision, the frame is sent again. To avoid collisions on wireless networks, carrier sense multiple access with collision avoidance (CSMA/CA) was invented. Collisions are avoided through the use of three strategies: the interframe space, the contention window, and acknowledgments.

In controlled access, the stations consult one another to find which station has the right to send. A station cannot send unless it has been authorized by other stations. We discussed three popular controlled-access methods: reservation, polling, and token passing. In the reservation access method, a station needs to make a reservation before sending data. Time is divided into intervals. In each interval, a reservation frame precedes the data frames sent in that interval. In the polling method, all data exchanges must be made through the primary device even when the ultimate destination is a secondary device. The primary device controls the link; the secondary devices follow its instructions. In the token-passing method, the stations in a network are organized in a logical ring. Each station has a predecessor and a successor. A special packet called a *token* circulates through the ring.

Channelization is a multiple-access method in which the available bandwidth of a link is shared in time, frequency, or through code, between different stations. We discussed three channelization protocols: FDMA, TDMA, and CDMA. In frequency-division multiple access (FDMA), the available bandwidth is divided into frequency bands. Each station is allocated a band to send its data. In other words, each band is reserved for a specific station, and it belongs to the station all the time. In time-division multiple access (TDMA), the stations share the bandwidth of the channel in time. Each station is allocated a time slot during which it can send data. Each station transmits its data in its assigned time slot. In code-division multiple access (CDMA), the stations use different codes to achieve multiple access. CDMA is based on coding theory and uses sequences of numbers called *chips*. The sequences are generated using orthogonal codes such as the Walsh tables.

12.5 PRACTICE SET

12.5.1 Quizzes

A set of interactive quizzes for this chapter can be found on the book website. It is strongly recommended that the student take the quizzes to check his/her understanding of the materials before continuing with the practice set.

12.5.2 Questions

Q12-1. Which of the following is a channelization protocol?

a. ALOHA **b.** Token-passing **c.** CDMA

Q12-2. Stations in a pure Aloha network send frames of size 1000 bits at the rate of 1 Mbps. What is the vulnerable time for this network?

Q12-3. Assume the propagation delay in a broadcast network is 5 μs and the frame transmission time is 10 μs.

a. How long does it take for the first bit to reach the destination?

b. How long does it take for the last bit to reach the destination after the first bit has arrived?

c. How long is the network involved with this frame (vulnerable to collision)?

Q12-4. In a wireless LAN, station A is assigned IFS = 5 milliseconds and station B is assigned IFS = 7 milliseconds. Which station has a higher priority? Explain.

Q12-5. In a slotted Aloha network with $G = 1/2$, how is the throughput affected in each of the following cases?

a. G is increased to 1. **b.** G is decreased to 1/4.

Q12-6. To understand the uses of K in Figure 12.15, find the probability that a station can send immediately in each of the following cases:

a. After two failures. **b.** After five failures.

Q12-7. There is no acknowledgment mechanism in CSMA/CD, but we need this mechanism in CSMA/CA. Explain the reason.

Q12-8. Which of the following is a controlled-access protocol?

a. Token-passing **b.** Polling **c.** FDMA

Q12-9. Explain why collision is an issue in random access protocols but not in controlled access protocols.

Q12-10. Based on Figure 12.13, how do we interpret *success* in an Aloha network?

Q12-11. List some strategies in CSMA/CA that are used to avoid collision.

Q12-12. Assume the propagation delay in a broadcast network is 6 μs and the frame transmission time is 4 μs. Can the collision be detected no matter where it occurs?

Q12-13. Assume the propagation delay in a broadcast network is 3 μs and the frame transmission time is 5 μs. Can the collision be detected no matter where it occurs?

Q12-14. In a pure Aloha network with $G = 1/2$, how is the throughput affected in each of the following cases?

a. G is increased to 1. **b.** G is decreased to 1/4.

Q12-15. Based on Figure 12.3, how do we interpret *success* in an Aloha network?

Q12-16. Assume the propagation delay in a broadcast network is 12 μs and the frame transmission time is 8 μs.

a. How long does it take for the first bit to reach the destination?

b. How long does it take for the last bit to reach the destination after the first bit has arrived?

c. How long is the network involved with this frame (vulnerable to collision)?

Q12-17. Which of the following is a random-access protocol?

a. CSMA/CD **b.** Polling **c.** TDMA

Q12-18. Explain why collision is an issue in random access protocols but not in channelization protocols.

Q12-19. To understand the uses of K in Figure 12.13, find the probability that a station can send immediately in each of the following cases:

a. After one failure. **b.** After four failures.

Q12-20. What is the purpose of NAV in CSMA/CA?

Q12-21. Stations in a slotted Aloha network send frames of size 1000 bits at the rate of 1 Mbps. What is the vulnerable time for this network?

Q12-22. To understand the uses of K in Figure 12.3, find the probability that a station can send immediately in each of the following cases:

a. After one failure. **b.** After three failures.

Q12-23. Based on Figure 12.15, how do we interpret *success* in an Aloha network?

Q12-24. Assume the propagation delay in a broadcast network is 5 μs and the frame transmission time is 10 μs.

a. How long does it take for the first bit to reach the destination?

b. How long does it take for the last bit to reach the destination after the first bit has arrived?

c. How long is the network involved with this frame (vulnerable to collision)?

12.5.3 Problems

P12-1. To formulate the performance of a multiple-access network, we need a mathematical model. When the number of stations in a network is very large, the Poisson distribution, $p[x] = (e^{-\lambda} \times \lambda^x)/(x!)$, is used. In this formula, $p[x]$ is the probability of generating x number of frames in a period of time and λ is the average number of generated frames during the same period of time. Using the Poisson distribution:

a. Find the probability that a pure Aloha network generates x number of frames during the vulnerable time. Note that the vulnerable time for this network is two times the frame transmission time (T_{fr}).

b. Find the probability that a slotted Aloha network generates x number of frames during the vulnerable time. Note that the vulnerable time for this network is equal to the frame transmission time (T_{fr}).

P12-2. In the previous problem, we used the Poisson distribution to find the probability of generating x number of frames, in a certain period of time, in a pure or slotted Aloha network as $p[x] = (e^{-\lambda} \times \lambda^x)/(x!)$. In this problem, we want to find the probability that a frame in such a network reaches its destination without colliding with other frames. For this purpose, it is simpler to think that we have G stations, each sending an average of one frame during the frame transmission time (instead of having N frames, each sending an average of G/N frames during the same time). Then, the probability of success for a station is the probability that no other station sends a frame during the vulnerable time.

a. Find the probability that a station in a pure Aloha network can successfully send a frame during a vulnerable time.

b. Find the probability that a station in a slotted Aloha network can successfully send a frame during a vulnerable time.

P12-3. In the previous problem, we found that the probability of a station (in a G-station network) successfully sending a frame in a vulnerable time is $P = e^{-2G}$ for a pure Aloha and $P = e^{-G}$ for a slotted Aloha network. In this problem, we want to find the throughput of these networks, which is the probability that any station (out of G stations) can successfully send a frame during the vulnerable time.

a. Find the throughput of a pure Aloha network.

b. Find the throughput of a slotted Aloha network.

P12-4. In the previous problem, we showed that the throughput is $S = Ge^{-2G}$ for a pure Aloha network and $S = Ge^{-G}$ for a slotted Aloha network. In this problem, we want to find the value of G in each network that makes the throughput maximum and find the value of the maximum throughput. This can be done if we find the derivative of S with respect to G and set the derivative to zero.

a. Find the value of G that makes the throughput maximum, and find the value of the maximum throughput for a pure Aloha network.

b. Find the value of G that makes the throughput maximum, and find the value of the maximum throughput for a slotted Aloha network.

P12-5. A multiple access network with a large number of stations can be analyzed using the Poisson distribution. When there is a limited number of stations in a network, we need to use another approach for this analysis. In a network with N stations, we assume that each station has a frame to send during the frame transmission time (T_{fr}) with probability p. In such a network, a station is successful in sending its frame if the station has a frame to send during the vulnerable time and no other station has a frame to send during this period of time.

a. Find the probability that a station in a pure Aloha network can successfully send a frame during the vulnerable time.

b. Find the probability that a station in a slotted Aloha network can successfully send a frame during the vulnerable time.

P12-6. In the previous problem, we found the probability of success for a station to send a frame successfully during the vulnerable time. The throughput of a network with a limited number of stations is the probability that any station (out of N stations) can send a frame successfully. In other words, the throughput is the sum of N success probabilities.

a. Find the throughput of a pure Aloha network.

b. Find the throughput of a slotted Aloha network.

P12-7. In the previous problem, we found the throughputs of a pure and a slotted Aloha network as $S = Np\ (1-p)^{2(N-1)}$ and $S = Np\ (1-p)^{(N-1)}$ respectively. In this problem we want to find the maximum throughput with respect to p.

a. Find the value of p that maximizes the throughput of a pure Aloha network, and calculate the maximum throughput when N is a very large number.

b. Find the value of p that maximizes the throughput of a slotted Aloha network, and calculate the maximum throughput when N is a very large number.

P12-8. A slotted Aloha network is working with maximum throughput.

a. What is the probability that a slot is empty?

b. How many slots, n, on average, should pass before getting an empty slot?

P12-9. In a bus CSMA/CD network with a data rate of 10 Mbps, a collision occurs 20 μs after the first bit of the frame leaves the sending station. What should the length of the frame be so that the sender can detect the collision?

P12-10. Assume that there are only two stations, A and B, in a bus CSMA/CD network. The distance between the two stations is 2000 m and the propagation speed is 2×10^8 m/s. If station A starts transmitting at time t_1:

a. Does the protocol allow station B to start transmitting at time $t_1 + 8$ μs? If the answer is yes, what will happen?

b. Does the protocol allow station B to start transmitting at time $t_1 + 11$ μs? If the answer is yes, what will happen?

P12-11. Check to see if the following set of chips can belong to an orthogonal system.

[+1, +1] and [+1, −1]

P12-12. The random variable R (Figure 12.13) is designed to give stations different delays when a collision has occurred. To alleviate the collision, we expect that different stations generate different values of R. To show the point, find the probability that the value of R is the same for two stations after

a. the first collision. **b.** the second collision.

P12-13. There are only three active stations in a slotted Aloha network: A, B, and C. Each station generates a frame in a time slot with the corresponding probabilities $p_A = 0.2$, $p_B = 0.3$, and $p_C = 0.4$ respectively.

a. What is the throughput of each station?

b. What is the throughput of the network?

P12-14. One of the useful parameters in a LAN is the number of bits that can fit in one meter of the medium ($n_{b/m}$). Find the value of $n_{b/m}$ if the data rate is 100 Mbps and the medium propagation speed is 2×10^8 m/s.

P12-15. There are only two stations, A and B, in a bus 1-persistence CSMA/CD network with $T_p = 25.6$ μs and $T_{fr} = 51.2$ μs. Station A has a frame to send to station B. The frame is unsuccessful two times and succeeds on the third try. Draw a time line diagram for this problem. Assume that the R is 1 and 2 respectively and ignore the time for sending a jamming signal (see Figure 12.13).

P12-16. There are only three active stations in a slotted Aloha network: A, B, and C. Each station generates a frame in a time slot with the corresponding probabilities $p_A = 0.2$, $p_B = 0.3$, and $p_C = 0.4$ respectively.

a. What is the probability that any station can send a frame in the first slot?

b. What is the probability that station A can successfully send a frame for the first time in the second slot?

c. What is the probability that station C can successfully send a frame for the first time in the third slot?

P12-17. Check to see if the following set of chips can belong to an orthogonal system.

[+1, +1, +1, +1] , [+1, −1, −1, +1] , [−1, +1, +1, −1] , [+1, −1, −1, +1]

P12-18. We have a pure ALOHA network with a data rate of 10 Mbps. What is the maximum number of 1000-bit frames that can be successfully sent by this network?

P12-19. Although the throughput calculation of a CSMA/CD is really involved, we can calculate the maximum throughput of a slotted CSMA/CD with the specification we described in the previous problem. We found that the average number of contention slots a station needs to wait is $k = e$ slots. With this assumption, the throughput of a slotted CSMA/CD is

$$S \nabla (T_{fr})/(\text{time the channel is busy for a frame})$$

The time the channel is busy for a frame is the time to wait for a free slot plus the time to transmit the frame plus the propagation delay to receive the good news about the lack of collision. Assume the duration of a contention slot is $2 \times (T_p)$ and $a = (T_p)/(T_{fr})$. Note that the parameter a is the number of frames that occupy the transmission media. Find the throughput of a slotted CSMA/CD in terms of the parameter a.

P12-20. Assume we have a slotted CSMA/CD network. Each station in this network uses a contention period, in which the station contends for access to the shared channel before being able to send a frame. We assume that the contention period is made of contention slots. At the beginning of each slot, the station senses the channel. If the channel is free, the station sends its frame; if the channel is busy, the station refrains from sending and waits until the beginning of the next slot. In other words, the station waits, on average, for k slots before sending its frame, as shown in Figure 12.30. Note that the channel is either in the contention state, the transmitting state, or the idle state (when no station has a frame to send). However, if N is a very large number, the idle state actually disappears.

Figure 12.30 *Problem P12-20*

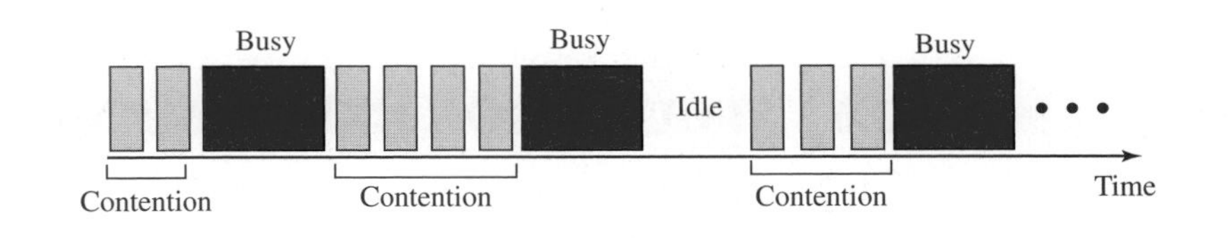

a. What is the probability of a free slot (P_{free}) if the number of stations is N and each station has a frame to send with probability p?

b. What is the maximum of this probability when N is a very large number?

c. What is the probability that the jth slot is free?

d. What is the average number of slots, k, that a station should wait before getting a free slot?

e. What is the value of k when N (the number of stations) is very large?

P12-21. We have defined the parameter $\boldsymbol{a}$ as the number of frames that can fit the medium between two stations, or $\boldsymbol{a} = (T_p)/(T_{fr})$. Another way to define this parameter is $\boldsymbol{a} = L_b/F_b$, in which L_b is the bit length of the medium and F_b is the frame length of the medium. Show that the two definitions are equivalent.

P12-22. In a bus 1-persistence CSMA/CD with $T_p = 50$ µs and $T_{fr} = 120$ µs, there are two stations, A and B. Both stations start sending frames to each other at the same time. Since the frames collide, each station tries to retransmit. Station A comes out with $R = 0$ and station B with $R = 1$. Ignore any other delay including the delay for sending jamming signals. Do the frames collide again? Draw a time-line diagram to prove your claim. Does the generation of a random number help avoid collision in this case?

P12-23. Another useful parameter in a LAN is the bit length of the medium (L_b), which defines the number of bits that the medium can hold at any time. Find the bit length of a LAN if the data rate is 100 Mbps and the medium length in meters (L_m) for a communication between two stations is 200 m. Assume the propagation speed in the medium is 2×10^8 m/s.

P12-24. To understand why we need to have a minimum frame size $T_{fr} = 2 \times T_p$ in a CDMA/CD network, assume we have a bus network with only two stations, A and B, in which $T_{fr} = 40$ μs and $T_p = 25$ μs. Station A starts sending a frame at time $t = 0.0$ μs and station B starts sending a frame at $t = 23.0$ μs. Answer the following questions:

a. Do frames collide?

b. If the answer to part *a* is yes, does station A detect collision?

c. If the answer to part *a* is yes, does station B detect collision?

P12-25. Alice and Bob are experimenting with CSMA using a W_2 Walsh table (see Figure 12.29). Alice uses the code [+1, +1] and Bob uses the code [+1, −1]. Assume that they simultaneously send a hexadecimal digit to each other. Alice sends $(6)_{16}$ and Bob sends $(B)_{16}$. Show how they can detect what the other person has sent.

12.6 SIMULATION EXPERIMENTS

12.6.1 Applets

We have created some Java applets to show some of the main concepts discussed in this chapter. It is strongly recommended that the students activate these applets on the book website and carefully examine the protocols in action.

12.7 PROGRAMMING ASSIGNMENTS

For each of the following assignments, write a program in the programming language you are familiar with.

Prg12-1. Write and test a program to simulate the flow diagram of CSMA/CD in Figure 12.13.

Prg12-2. Write and test a program to simulate the flow diagram of CSMA/CA in Figure 12.15.

CHAPTER 13

Wired LANs: Ethernet

After discussing the general issues related to the data-link layer in Chapters 9 to 12, it is time in this chapter to discuss the wired LANs. Although over a few decades many wired LAN protocols existed, only the Ethernet technology survives today. This is the reason that we discuss only this technology and its evolution in this chapter.

This chapter is divided into five sections.

- The first section discusses the Ethernet protocol in general. It explains that IEEE Project 802 defines the LLC and MAC sublayers for all LANs including Ethernet. The section also lists the four generations of Ethernet.
- The second section discusses the Standard Ethernet. Although this generation is rarely seen in practice, most of the characteristics have been inherited by the following three generations. The section first describes some characteristics of the Standard Ethernet. It then discusses the addressing mechanism, which is the same in all Ethernet generations. The section next discusses the access method, CSMA/CD, which we discussed in Chapter 12. The section then reviews the efficiency of the Standard Ethernet. It then shows the encoding and the implementation of this generation. Before closing the section, the changes in this generation that resulted in the move to the next generation are listed.
- The third section describes the Fast Ethernet, the second generation, which can still be seen in many places. The section first describes the changes in the MAC sublayer. The section then discusses the physical layer and the implementation of this generation.
- The fourth section discusses the Gigabit Ethernet, with the rate of 1 gigabit per second. The section first describes the MAC sublayer. It then moves to the physical layer and implementation.
- The fifth section touches on the 10 Gigabit Ethernet. This is a new technology that can be used both for a backbone LAN or as a MAN (metropolitan area network). The section briefly describes the rationale and the implementation.

13.1 ETHERNET PROTOCOL

In Chapter 1, we mentioned that the TCP/IP protocol suite does not define any protocol for the data-link or the physical layer. In other words, TCP/IP accepts any protocol at these two layers that can provide services to the network layer. The data-link layer and the physical layer are actually the territory of the local and wide area networks. This means that when we discuss these two layers, we are talking about networks that are using them. As we see in this and the following two chapters, we can have wired or wireless networks. We discuss wired networks in this chapter and the next and postpone the discussion of wireless networks to Chapter 15.

In Chapter 1, we learned that a local area network (LAN) is a computer network that is designed for a limited geographic area such as a building or a campus. Although a LAN can be used as an isolated network to connect computers in an organization for the sole purpose of sharing resources, most LANs today are also linked to a wide area network (WAN) or the Internet.

In the 1980s and 1990s several different types of LANs were used. All of these LANs used a media-access method to solve the problem of sharing the media. The Ethernet used the CSMA/CD approach. The Token Ring, Token Bus, and FDDI (Fiber Distribution Data Interface) used the token-passing approach. During this period, another LAN technology, ATM LAN, which deployed the high speed WAN technology (ATM), appeared in the market.

Almost every LAN except Ethernet has disappeared from the marketplace because Ethernet was able to update itself to meet the needs of the time. Several reasons for this success have been mentioned in the literature, but we believe that the Ethernet protocol was designed so that it could evolve with the demand for higher transmission rates. It is natural that an organization that has used an Ethernet LAN in the past and now needs a higher data rate would update to the new generation instead of switching to another technology, which might cost more. This means that we confine our discussion of wired LANs to the discussion of Ethernet.

13.1.1 IEEE Project 802

Before we discuss the Ethernet protocol and all its generations, we need to briefly discuss the IEEE standard that we often encounter in text or real life. In 1985, the Computer Society of the IEEE started a project, called ***Project 802,*** to set standards to enable intercommunication among equipment from a variety of manufacturers. Project 802 does not seek to replace any part of the OSI model or TCP/IP protocol suite. Instead, it is a way of specifying functions of the physical layer and the data-link layer of major LAN protocols.

The relationship of the 802 Standard to the TCP/IP protocol suite is shown in Figure 13.1. The IEEE has subdivided the data-link layer into two sublayers: **logical link control (LLC)** and **media access control (MAC).** IEEE has also created several physical-layer standards for different LAN protocols.

Logical Link Control (LLC)

Earlier we discussed *data link control.* We said that data link control handles framing, flow control, and error control. In IEEE Project 802, flow control, error control, and

Figure 13.1 *IEEE standard for LANs*

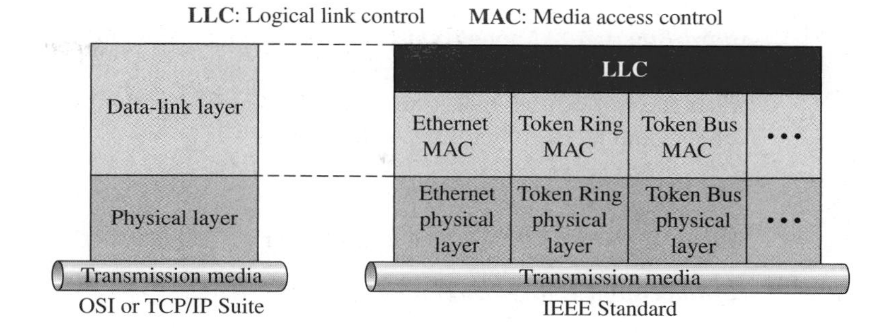

part of the framing duties are collected into one sublayer called the *logical link control* (LLC). Framing is handled in both the LLC sublayer and the MAC sublayer.

The LLC provides a single link-layer control protocol for all IEEE LANs. This means LLC protocol can provide interconnectivity between different LANs because it makes the MAC sublayer transparent.

Media Access Control (MAC)

Earlier we discussed multiple access methods including random access, controlled access, and channelization. IEEE Project 802 has created a sublayer called ***media access control*** that defines the specific access method for each LAN. For example, it defines CSMA/CD as the media access method for Ethernet LANs and defines the token-passing method for Token Ring and Token Bus LANs. As we mentioned in the previous section, part of the framing function is also handled by the MAC layer.

13.1.2 Ethernet Evolution

The Ethernet LAN was developed in the 1970s by Robert Metcalfe and David Boggs. Since then, it has gone through four generations: **Standard Ethernet** (10 Mbps), **Fast Ethernet** (100 Mbps), **Gigabit Ethernet** (1 Gbps), and **10 Gigabit Ethernet** (10 Gbps), as shown in Figure 13.2. We briefly discuss all these generations.

Figure 13.2 *Ethernet evolution through four generations*

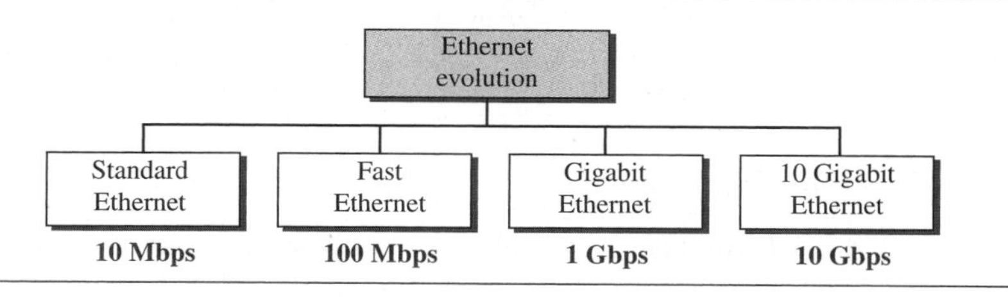

13.2 STANDARD ETHERNET

We refer to the original Ethernet technology with the data rate of 10 Mbps as the *Standard Ethernet.* Although most implementations have moved to other technologies in the Ethernet evolution, there are some features of the Standard Ethernet that have not changed during the evolution. We discuss this standard version to pave the way for understanding the other three technologies.

13.2.1 Characteristics

Let us first discuss some characteristics of the Standard Ethernet.

Connectionless and Unreliable Service

Ethernet provides a connectionless service, which means each frame sent is independent of the previous or next frame. Ethernet has no connection establishment or connection termination phases. The sender sends a frame whenever it has it; the receiver may or may not be ready for it. The sender may overwhelm the receiver with frames, which may result in dropping frames. If a frame drops, the sender will not know about it. Since IP, which is using the service of Ethernet, is also connectionless, it will not know about it either. If the transport layer is also a connectionless protocol, such as UDP, the frame is lost and salvation may only come from the application layer. However, if the transport layer is TCP, the sender TCP does not receive acknowledgment for its segment and sends it again.

Ethernet is also unreliable like IP and UDP. If a frame is corrupted during transmission and the receiver finds out about the corruption, which has a high level of probability of happening because of the CRC-32, the receiver drops the frame silently. It is the duty of high-level protocols to find out about it.

Frame Format

The Ethernet frame contains seven fields, as shown in Figure 13.3.

Figure 13.3 *Ethernet frame*

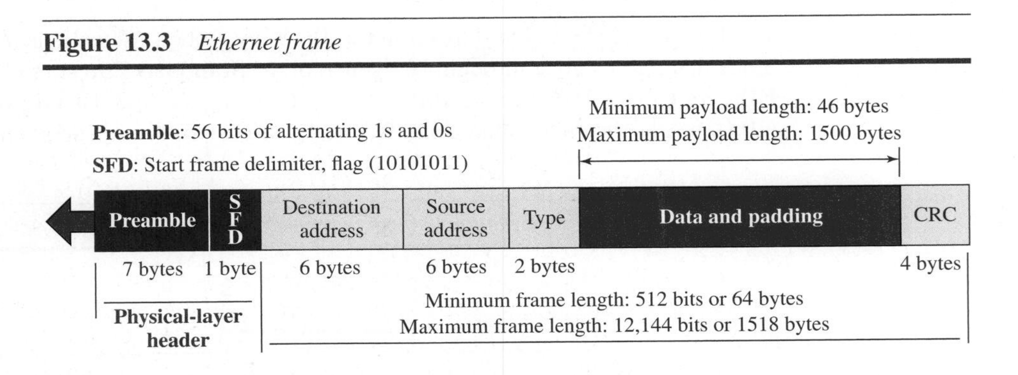

- ❑ ***Preamble.*** This field contains 7 bytes (56 bits) of alternating 0s and 1s that alert the receiving system to the coming frame and enable it to synchronize its clock if it's out of synchronization. The pattern provides only an alert and a timing pulse. The 56-bit pattern allows the stations to miss some bits at the beginning of the frame. The *preamble* is actually added at the physical layer and is not (formally) part of the frame.

- ***Start frame delimiter (SFD).*** This field (1 byte: 10101011) signals the beginning of the frame. The SFD warns the station or stations that this is the last chance for synchronization. The last 2 bits are $(11)_2$ and alert the receiver that the next field is the destination address. This field is actually a flag that defines the beginning of the frame. We need to remember that an Ethernet frame is a variable-length frame. It needs a flag to define the beginning of the frame. The SFD field is also added at the physical layer.
- ***Destination address (DA).*** This field is six bytes (48 bits) and contains the link-layer address of the destination station or stations to receive the packet. We will discuss addressing shortly. When the receiver sees its own link-layer address, or a multicast address for a group that the receiver is a member of, or a broadcast address, it decapsulates the data from the frame and passes the data to the upper-layer protocol defined by the value of the type field.
- ***Source address (SA).*** This field is also six bytes and contains the link-layer address of the sender of the packet. We will discuss addressing shortly.
- ***Type.*** This field defines the upper-layer protocol whose packet is encapsulated in the frame. This protocol can be IP, ARP, OSPF, and so on. In other words, it serves the same purpose as the protocol field in a datagram and the port number in a segment or user datagram. It is used for multiplexing and demultiplexing.
- ***Data.*** This field carries data encapsulated from the upper-layer protocols. It is a minimum of 46 and a maximum of 1500 bytes. We discuss the reason for these minimum and maximum values shortly. If the data coming from the upper layer is more than 1500 bytes, it should be fragmented and encapsulated in more than one frame. If it is less than 46 bytes, it needs to be padded with extra 0s. A padded data frame is delivered to the upper-layer protocol as it is (without removing the padding), which means that it is the responsibility of the upper layer to remove or, in the case of the sender, to add the padding. The upper-layer protocol needs to know the length of its data. For example, a datagram has a field that defines the length of the data.
- ***CRC.*** The last field contains error detection information, in this case a CRC-32. The CRC is calculated over the addresses, types, and data field. If the receiver calculates the CRC and finds that it is not zero (corruption in transmission), it discards the frame.

Frame Length

Ethernet has imposed restrictions on both the minimum and maximum lengths of a frame. The minimum length restriction is required for the correct operation of CSMA/CD, as we will see shortly. An Ethernet frame needs to have a minimum length of 512 bits or 64 bytes. Part of this length is the header and the trailer. If we count 18 bytes of header and trailer (6 bytes of source address, 6 bytes of destination address, 2 bytes of length or type, and 4 bytes of CRC), then the minimum length of data from the upper layer is $64 - 18 = 46$ bytes. If the upper-layer packet is less than 46 bytes, padding is added to make up the difference.

The standard defines the maximum length of a frame (without preamble and SFD field) as 1518 bytes. If we subtract the 18 bytes of header and trailer, the maximum length of the payload is 1500 bytes. The maximum length restriction has two historical reasons. First, memory was very expensive when Ethernet was designed; a maximum length restriction helped to reduce the size of the buffer. Second, the maximum length restriction prevents one station from monopolizing the shared medium, blocking other stations that have data to send.

Minimum frame length: 64 bytes **Minimum data length: 46 bytes**
Maximum frame length: 1518 bytes **Maximum data length: 1500 bytes**

13.2.2 Addressing

Each station on an Ethernet network (such as a PC, workstation, or printer) has its own **network interface card (NIC).** The NIC fits inside the station and provides the station with a link-layer address. The Ethernet address is 6 bytes (48 bits), normally written in **hexadecimal notation,** with a colon between the bytes. For example, the following shows an Ethernet MAC address:

4A:30:10:21:10:1A

Transmission of Address Bits

The way the addresses are sent out online is different from the way they are written in hexadecimal notation. The transmission is left to right, byte by byte; however, for each byte, the least significant bit is sent first and the most significant bit is sent last. This means that the bit that defines an address as unicast or multicast arrives first at the receiver. This helps the receiver to immediately know if the packet is unicast or multicast.

Example 13.1

Show how the address 47:20:1B:2E:08:EE is sent out online.

Solution

The address is sent left to right, byte by byte; for each byte, it is sent right to left, bit by bit, as shown below:

Hexadecimal	**47**	**20**	**1B**	**2E**	**08**	**EE**
Binary	01000111	00100000	00011011	00101110	00001000	11101110
Transmitted ←	11100010	00000100	11011000	01110100	00010000	01110111

Unicast, Multicast, and Broadcast Addresses

A source address is always a *unicast address*—the frame comes from only one station. The destination address, however, can be *unicast*, *multicast*, or *broadcast.* Figure 13.4 shows how to distinguish a unicast address from a multicast address. If the least significant bit of the first byte in a destination address is 0, the address is unicast; otherwise, it is multicast.

Note that with the way the bits are transmitted, the unicast/multicast bit is the first bit which is transmitted or received. The broadcast address is a special case of the

Figure 13.4 *Unicast and multicast addresses*

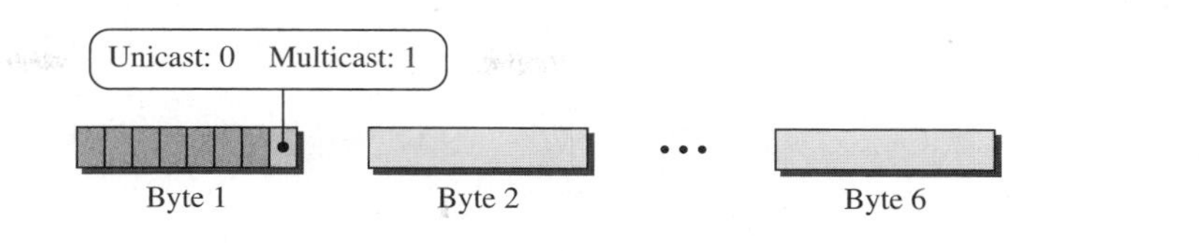

multicast address: the recipients are all the stations on the LAN. A broadcast destination address is forty-eight 1s.

Example 13.2

Define the type of the following destination addresses:

a. 4A:30:10:21:10:1A

b. 47:20:1B:2E:08:EE

c. FF:FF:FF:FF:FF:FF

Solution

To find the type of the address, we need to look at the second hexadecimal digit from the left. If it is even, the address is unicast. If it is odd, the address is multicast. If all digits are Fs, the address is broadcast. Therefore, we have the following:

a. This is a unicast address because A in binary is 1010 (even).

b. This is a multicast address because 7 in binary is 0111 (odd).

c. This is a broadcast address because all digits are Fs in hexadecimal.

Distinguish Between Unicast, Multicast, and Broadcast Transmission

Standard Ethernet uses a coaxial cable (bus topology) or a set of twisted-pair cables with a hub (star topology) as shown in Figure 13.5.

We need to know that transmission in the standard Ethernet is always broadcast, no matter if the intention is unicast, multicast, or broadcast. In the bus topology, when station A sends a frame to station B, all stations will receive it. In the star topology, when station A sends a frame to station B, the hub will receive it. Since the hub is a passive element, it does not check the destination address of the frame; it regenerates the bits (if they have been weakened) and sends them to all stations except station A. In fact, it floods the network with the frame.

The question is, then, how the actual unicast, multicast, and broadcast transmissions are distinguished from each other. The answer is in the way the frames are kept or dropped.

- ❑ In a unicast transmission, all stations will receive the frame, the intended recipient keeps and handles the frame; the rest discard it.
- ❑ In a multicast transmission, all stations will receive the frame, the stations that are members of the group keep and handle it; the rest discard it.

Figure 13.5 *Implementation of standard Ethernet*

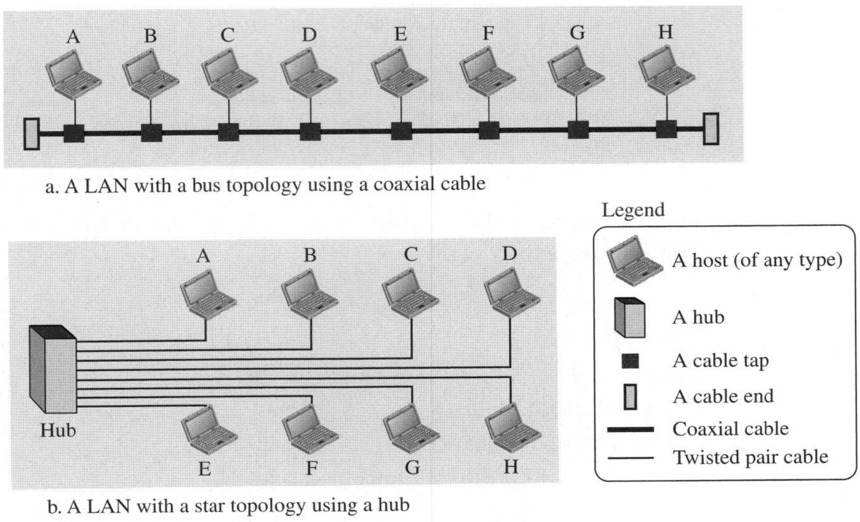
A
B
C
D
E
F
G
H
a. A LAN with a bus topology using a coaxial cable
Legend
A
B
C
D
Hub
E
F
G
H
b. A LAN with a star topology using a hub
A host (of any type)
A hub
A cable tap
A cable end
Coaxial cable
Twisted pair cable

- ❑ In a broadcast transmission, all stations (except the sender) will receive the frame and all stations (except the sender) keep and handle it.

13.2.3 Access Method

Since the network that uses the standard Ethernet protocol is a broadcast network, we need to use an access method to control access to the sharing medium. The standard Ethernet chose CSMA/CD with 1-persistent method, discussed earlier in Chapter 12, Section 1.3. Let us use a scenario to see how this method works for the Ethernet protocol.

- ❑ Assume station A in Figure 13.5 has a frame to send to station D. Station A first should check whether any other station is sending (carrier sense). Station A measures the level of energy on the medium (for a short period of time, normally less than 100 μs). If there is no signal energy on the medium, it means that no station is sending (or the signal has not reached station A). Station A interprets this situation as idle medium. It starts sending its frame. On the other hand, if the signal energy level is not zero, it means that the medium is being used by another station. Station A continuously monitors the medium until it becomes idle for 100 μs. It then starts sending the frame. However, station A needs to keep a copy of the frame in its buffer until it is sure that there is no collision. When station A is sure of this is the subject we discuss next.
- ❑ The medium sensing does not stop after station A has started sending the frame. Station A needs to send and receive continuously. Two cases may occur:

a. Station A has sent 512 bits and no collision is sensed (the energy level did not go above the regular energy level), the station then is sure that the frame will go through and stops sensing the medium. Where does the number 512 bits come from? If we consider the transmission rate of the Ethernet as 10 Mbps, this means that it takes the station 512/(10 Mbps) = 51.2 μs to send out 512 bits. With the speed of propagation in a cable (2×10^8 meters), the first bit could have gone 10,240 meters (one way) or only 5120 meters (round trip), have collided with a bit from the last station on the cable, and have gone back. In other words, if a collision were to occur, it should occur by the time the sender has sent out 512 bits (worst case) and the first bit has made a round trip of 5120 meters. We should know that if the collision happens in the middle of the cable, not at the end, station A hears the collision earlier and aborts the transmission. We also need to mention another issue. The above assumption is that the length of the cable is 5120 meters. The designer of the standard Ethernet actually put a restriction of 2500 meters because we need to consider the delays encountered throughout the journey. It means that they considered the worst case. The whole idea is that if station A does not sense the collision before sending 512 bits, there must have been no collision, because during this time, the first bit has reached the end of the line and all other stations know that a station is sending and refrain from sending. In other words, the problem occurs when another station (for example, the last station) starts sending before the first bit of station A has reached it. The other station mistakenly thinks that the line is free because the first bit has not yet reached it. The reader should notice that the restriction of 512 bits actually helps the sending station: The sending station is certain that no collision will occur if it is not heard during the first 512 bits, so it can discard the copy of the frame in its buffer.

b. Station A has sensed a collision before sending 512 bits. This means that one of the previous bits has collided with a bit sent by another station. In this case both stations should refrain from sending and keep the frame in their buffer for resending when the line becomes available. However, to inform other stations that there is a collision in the network, the station sends a 48-bit jam signal. The jam signal is to create enough signal (even if the collision happens after a few bits) to alert other stations about the collision. After sending the jam signal, the stations need to increment the value of K (number of attempts). If after increment $K = 15$, the experience has shown that the network is too busy, the station needs to abort its effort and try again. If $K < 15$, the station can wait a backoff time (T_B in Figure 12.13) and restart the process. As Figure 12.13 shows, the station creates a random number between 0 and $2^K - 1$, which means each time the collision occurs, the range of the random number increases exponentially. After the first collision ($K = 1$) the random number is in the range (0, 1). After the second collision ($K = 2$) it is in the range (0, 1, 2, 3). After the third collision ($K = 3$) it is in the range (0, 1, 2, 3, 4, 5, 6, 7). So after each collision, the probability increases that the backoff time becomes longer. This is due to the fact that if the collision happens even after the third or fourth attempt, it means that the network is really busy; a longer backoff time is needed.

13.2.4 Efficiency of Standard Ethernet

The efficiency of the Ethernet is defined as the ratio of the time used by a station to send data to the time the medium is occupied by this station. The practical efficiency of standard Ethernet has been measured to be

$$\textbf{Efficiency} = 1 / (1 + 6.4 \times a)$$

in which the parameter "***a***" is the number of frames that can fit on the medium. It can be calculated as ***a*** = (propagation delay)/(transmission delay) because the transmission delay is the time it takes a frame of average size to be sent out and the propagation delay is the time it takes to reach the end of the medium. Note that as the value of parameter ***a*** decreases, the efficiency increases. This means that if the length of the media is shorter or the frame size longer, the efficiency increases. In the ideal case, ***a*** = 0 and the efficiency is 1. We ask to calculate this efficiency in problems at the end of the chapter.

Example 13.3

In the Standard Ethernet with the transmission rate of 10 Mbps, we assume that the length of the medium is 2500 m and the size of the frame is 512 bits. The propagation speed of a signal in a cable is normally 2×10^8 m/s.

Propagation delay $= 2500/(2 \times 10^8) = 12.5\ \mu s$ **Transmission delay** $= 512/(10^7) = 51.2\ \mu s$

$a = 12.5/51.2 = 0.24$ **Efficiency = 39%**

The example shows that $a = 0.24$, which means only 0.24 of a frame occupies the whole medium in this case. The efficiency is 39 percent, which is considered moderate; it means that only 61 percent of the time the medium is occupied but not used by a station.

13.2.5 Implementation

The Standard Ethernet defined several implementations, but only four of them became popular during the 1980s. Table 13.1 shows a summary of Standard Ethernet implementations.

Table 13.1 *Summary of Standard Ethernet implementations*

Implementation	*Medium*	*Medium Length*	*Encoding*
10Base5	Thick coax	500 m	Manchester
10Base2	Thin coax	185 m	Manchester
10Base-T	2 UTP	100 m	Manchester
10Base-F	2 Fiber	2000 m	Manchester

In the nomenclature 10BaseX, the number defines the data rate (10 Mbps), the term *Base* means baseband (digital) signal, and X approximately defines either the maximum size of the cable in 100 meters (for example 5 for 500 or 2 for 185 meters) or the type of cable, T for unshielded twisted pair cable (UTP) and F for fiber-optic. The standard Ethernet uses a baseband signal, which means that the bits are changed to a digital signal and directly sent on the line.

Encoding and Decoding

All standard implementations use digital signaling (baseband) at 10 Mbps. At the sender, data are converted to a digital signal using the Manchester scheme; at the receiver, the received signal is interpreted as Manchester and decoded into data. As we saw in Chapter 4, Manchester encoding is self-synchronous, providing a transition at each bit interval. Figure 13.6 shows the encoding scheme for Standard Ethernet.

Figure 13.6 *Encoding in a Standard Ethernet implementation*

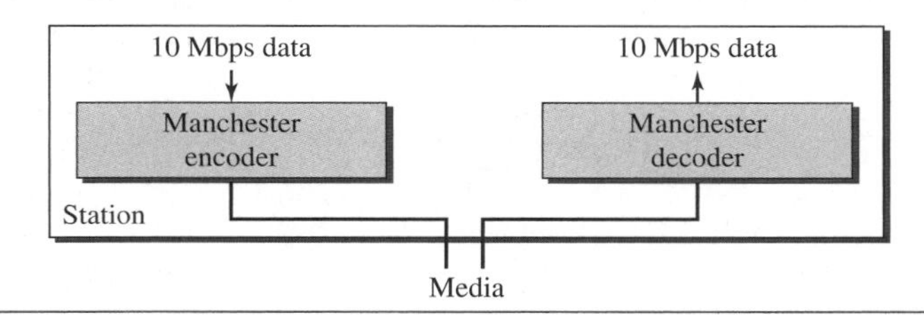

10Base5: Thick Ethernet

The first implementation is called ***10Base5, thick Ethernet,*** or ***Thicknet.*** The nickname derives from the size of the cable, which is roughly the size of a garden hose and too stiff to bend with your hands. 10Base5 was the first Ethernet specification to use a bus topology with an external **transceiver** (transmitter/receiver) connected via a tap to a thick coaxial cable. Figure 13.7 shows a schematic diagram of a 10Base5 implementation.

Figure 13.7 *10Base5 implementation*

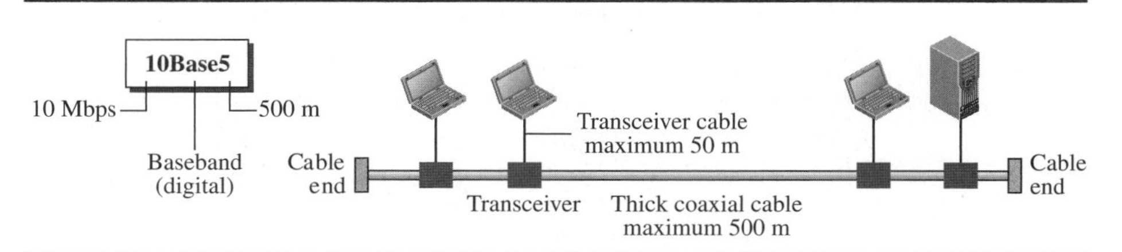

The transceiver is responsible for transmitting, receiving, and detecting collisions. The transceiver is connected to the station via a transceiver cable that provides separate paths for sending and receiving. This means that collision can only happen in the coaxial cable.

The maximum length of the coaxial cable must not exceed 500 m, otherwise, there is excessive degradation of the signal. If a length of more than 500 m is needed, up to five segments, each a maximum of 500 meters, can be connected using repeaters. Repeaters will be discussed in Chapter 17.

10Base2: Thin Ethernet

The second implementation is called ***10Base2, thin Ethernet,*** or ***Cheapernet.*** 10Base2 also uses a bus topology, but the cable is much thinner and more flexible. The cable can be bent to pass very close to the stations. In this case, the transceiver is normally part of the network interface card (NIC), which is installed inside the station. Figure 13.8 shows the schematic diagram of a 10Base2 implementation.

Figure 13.8 *10Base2 implementation*

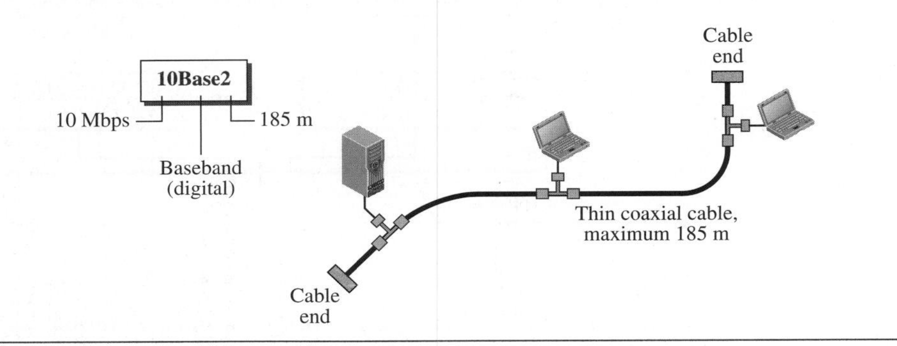

Note that the collision here occurs in the thin coaxial cable. This implementation is more cost effective than 10Base5 because thin coaxial cable is less expensive than thick coaxial and the tee connections are much cheaper than taps. Installation is simpler because the thin coaxial cable is very flexible. However, the length of each segment cannot exceed 185 m (close to 200 m) due to the high level of attenuation in thin coaxial cable.

10Base-T: Twisted-Pair Ethernet

The third implementation is called ***10Base-T*** or ***twisted-pair Ethernet.*** 10Base-T uses a physical star topology. The stations are connected to a hub via two pairs of twisted cable, as shown in Figure 13.9.

Figure 13.9 *10Base-T implementation*

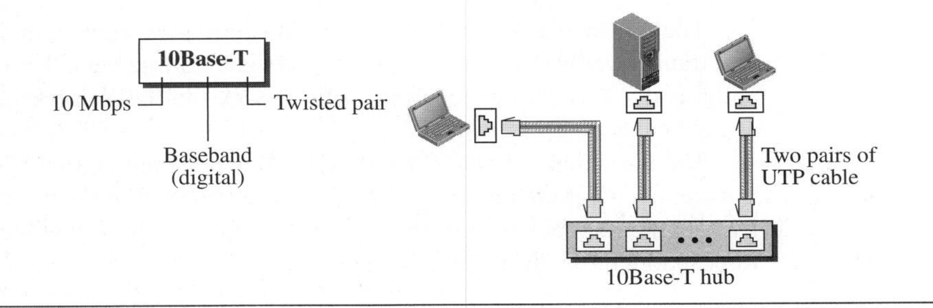

Note that two pairs of twisted cable create two paths (one for sending and one for receiving) between the station and the hub. Any collision here happens in the hub. Compared to 10Base5 or 10Base2, we can see that the hub actually replaces the coaxial cable as far as a collision is concerned. The maximum length of the twisted cable here is defined as 100 m, to minimize the effect of attenuation in the twisted cable.

10Base-F: Fiber Ethernet

Although there are several types of optical fiber 10-Mbps Ethernet, the most common is called ***10Base-F.*** 10Base-F uses a star topology to connect stations to a hub. The stations are connected to the hub using two fiber-optic cables, as shown in Figure 13.10.

Figure 13.10 *10Base-F implementation*

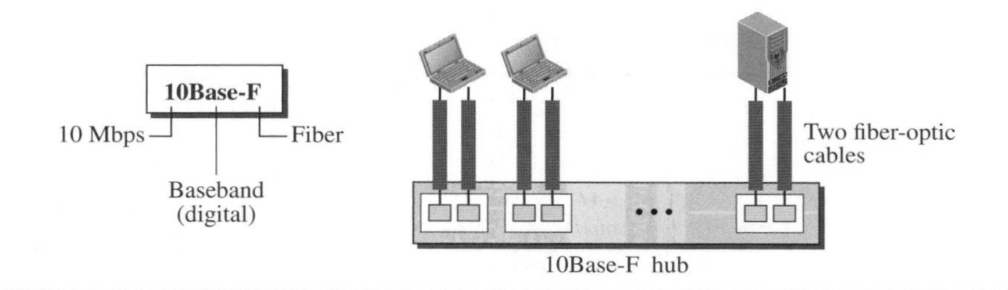

13.2.6 Changes in the Standard

Before we discuss higher-rate Ethernet protocols, we need to discuss the changes that occurred to the 10-Mbps Standard Ethernet. These changes actually opened the road to the evolution of the Ethernet to become compatible with other high-data-rate LANs.

Bridged Ethernet

The first step in the Ethernet evolution was the division of a LAN by **bridges.** Bridges have two effects on an Ethernet LAN: They raise the bandwidth and they separate collision domains. We discuss bridges in Chapter 17.

Raising the Bandwidth

In an unbridged Ethernet network, the total capacity (10 Mbps) is shared among all stations with a frame to send; the stations share the bandwidth of the network. If only one station has frames to send, it benefits from the total capacity (10 Mbps). But if more than one station needs to use the network, the capacity is shared. For example, if two stations have a lot of frames to send, they probably alternate in usage. When one station is sending, the other one refrains from sending. We can say that, in this case, each station on average sends at a rate of 5 Mbps. Figure 13.11 shows the situation.

The bridge, as we will learn in Chapter 17, can help here. A bridge divides the network into two or more networks. Bandwidthwise, each network is independent. For example, in Figure 13.12, a network with 12 stations is divided into two networks, each with 6 stations. Now each network has a capacity of 10 Mbps. The 10-Mbps capacity in each segment is now shared between 6 stations (actually 7 because the bridge acts as a

Figure 13.11 *Sharing bandwidth*

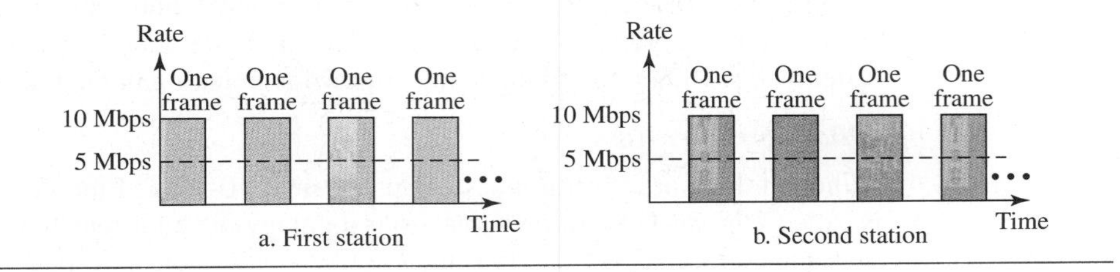

Figure 13.12 *A network with and without a bridge*

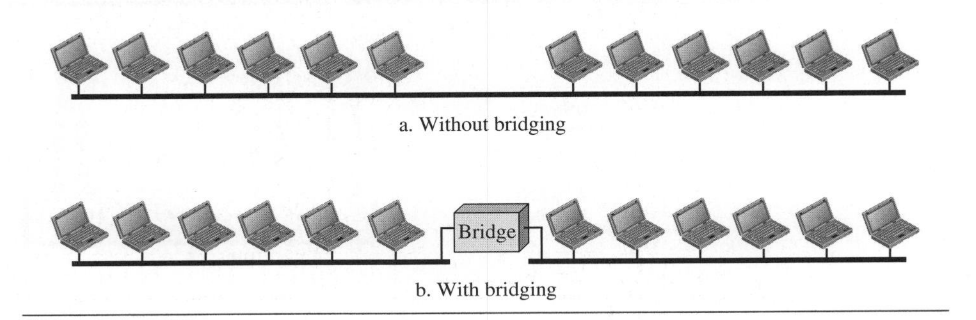

station in each segment), not 12 stations. In a network with a heavy load, eacsh station theoretically is offered 10/7 Mbps instead of 10/12 Mbps.

It is obvious that if we further divide the network, we can gain more bandwidth for each segment. For example, if we use a four-port bridge, each station is now offered 10/4 Mbps, which is 3 times more than an unbridged network.

Separating Collision Domains

Another advantage of a bridge is the separation of the **collision domain.** Figure 13.13 shows the collision domains for an unbridged and a bridged network. You can see that the collision domain becomes much smaller and the probability of collision is reduced tremendously. Without bridging, 12 stations contend for access to the medium; with bridging only 3 stations contend for access to the medium.

Switched Ethernet

The idea of a bridged LAN can be extended to a switched LAN. Instead of having two to four networks, why not have N networks, where N is the number of stations on the LAN? In other words, if we can have a multiple-port bridge, why not have an N-port switch? In this way, the bandwidth is shared only between the station and the switch (5 Mbps each). In addition, the collision domain is divided into N domains.

A layer-2 **switch** is an N-port bridge with additional sophistication that allows faster handling of the packets. Evolution from a bridged Ethernet to a **switched Ethernet** was

Figure 13.13 *Collision domains in an unbridged network and a bridged network*

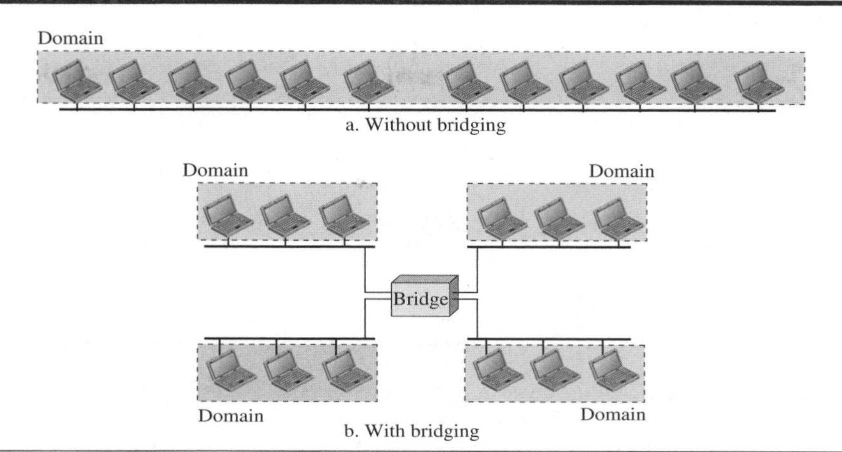

a big step that opened the way to an even faster Ethernet, as we will see. Figure 13.14 shows a switched LAN.

Figure 13.14 *Switched Ethernet*

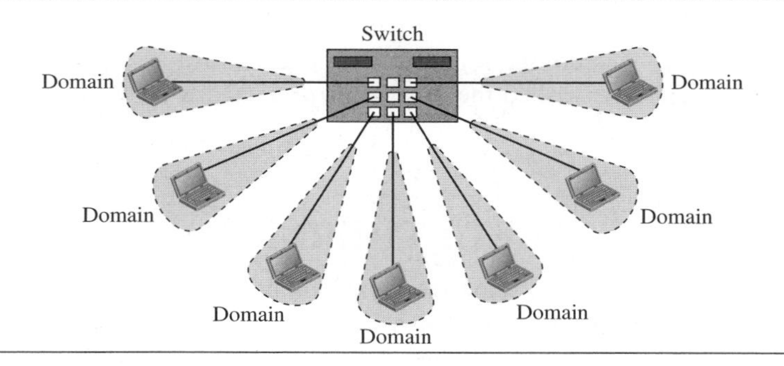

Full-Duplex Ethernet

One of the limitations of 10Base5 and 10Base2 is that communication is half-duplex (10Base-T is always full-duplex); a station can either send or receive, but may not do both at the same time. The next step in the evolution was to move from switched Ethernet to **full-duplex switched Ethernet.** The full-duplex mode increases the capacity of each domain from 10 to 20 Mbps. Figure 13.15 shows a switched Ethernet in full-duplex mode. Note that instead of using one link between the station and the switch, the configuration uses two links: one to transmit and one to receive.

No Need for CSMA/CD

In full-duplex switched Ethernet, there is no need for the CSMA/CD method. In a full-duplex switched Ethernet, each station is connected to the switch via two separate links.

Figure 13.15 *Full-duplex switched Ethernet*

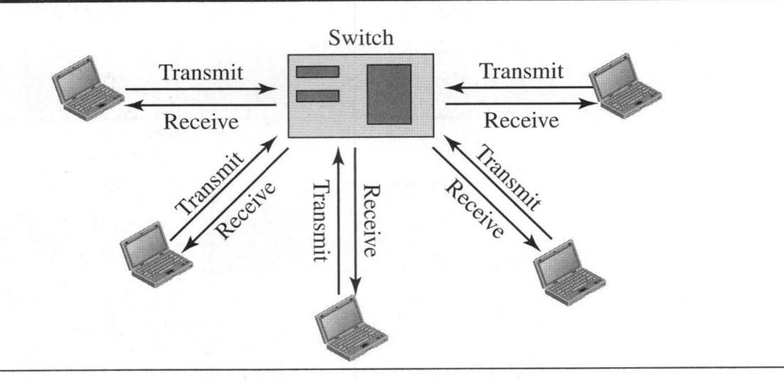

Each station or switch can send and receive independently without worrying about collision. Each link is a point-to-point dedicated path between the station and the switch. There is no longer a need for carrier sensing; there is no longer a need for collision detection. The job of the MAC layer becomes much easier. The carrier sensing and collision detection functionalities of the MAC sublayer can be turned off.

MAC Control Layer

Standard Ethernet was designed as a connectionless protocol at the MAC sublayer. There is no explicit flow control or error control to inform the sender that the frame has arrived at the destination without error. When the receiver receives the frame, it does not send any positive or negative acknowledgment.

To provide for flow and error control in full-duplex switched Ethernet, a new sublayer, called the *MAC control,* is added between the LLC sublayer and the MAC sublayer.

13.3 FAST ETHERNET (100 MBPS)

In the 1990s, some LAN technologies with transmission rates higher than 10 Mbps, such as FDDI and Fiber Channel, appeared on the market. If the Standard Ethernet wanted to survive, it had to compete with these technologies. Ethernet made a big jump by increasing the transmission rate to 100 Mbps, and the new generation was called the *Fast Ethernet*. The designers of the Fast Ethernet needed to make it compatible with the Standard Ethernet. The MAC sublayer was left unchanged, which meant the frame format and the maximum and minimum size could also remain unchanged. By increasing the transmission rate, features of the Standard Ethernet that depend on the transmission rate, access method, and implementation had to be reconsidered. The goals of Fast Ethernet can be summarized as follows:

1. Upgrade the data rate to 100 Mbps.
2. Make it compatible with Standard Ethernet.
3. Keep the same 48-bit address.
4. Keep the same frame format.

13.3.1 Access Method

We remember that the proper operation of the CSMA/CD depends on the transmission rate, the minimum size of the frame, and the maximum network length. If we want to keep the minimum size of the frame, the maximum length of the network should be changed. In other words, if the minimum frame size is still 512 bits, and it is transmitted 10 times faster, the collision needs to be detected 10 times sooner, which means the maximum length of the network should be 10 times shorter (the propagation speed does not change). So the Fast Ethernet came with two solutions (it can work with either choice):

1. The first solution was to totally drop the bus topology and use a passive hub and star topology but make the maximum size of the network 250 meters instead of 2500 meters as in the Standard Ethernet. This approach is kept for compatibility with the Standard Ethernet.
2. The second solution is to use a link-layer switch (discussed later in the chapter) with a buffer to store frames and a full-duplex connection to each host to make the transmission medium private for each host. In this case, there is no need for CSMA/CD because the hosts are not competing with each other. The link-layer switch receives a frame from a source host and stores it in the buffer (queue) waiting for processing. It then checks the destination address and sends the frame out of the corresponding interface. Since the connection to the switch is full-duplex, the destination address can even send a frame to another station at the same time that it is receiving a frame. In other words, the shared medium is changed to many point-to-point media, and there is no need for contention.

Autonegotiation

A new feature added to Fast Ethernet is called ***autonegotiation.*** It allows a station or a hub a range of capabilities. Autonegotiation allows two devices to negotiate the mode or data rate of operation. It was designed particularly to allow incompatible devices to connect to one another. For example, a device with a maximum data rate of 10 Mbps can communicate with a device with a 100 Mbps data rate (but which can work at a lower rate). We can summarize the goal of autonegotiation as follows. It was designed particularly for these purposes:

- ❑ To allow incompatible devices to connect to one another. For example, a device with a maximum capacity of 10 Mbps can communicate with a device with a 100 Mbps capacity (but which can work at a lower rate).
- ❑ To allow one device to have multiple capabilities.
- ❑ To allow a station to check a hub's capabilities.

13.3.2 Physical Layer

To be able to handle a 100 Mbps data rate, several changes need to be made at the physical layer.

Topology

Fast Ethernet is designed to connect two or more stations. If there are only two stations, they can be connected point-to-point. Three or more stations need to be connected in a star topology with a hub or a switch at the center.

Encoding

Manchester encoding needs a 200-Mbaud bandwidth for a data rate of 100 Mbps, which makes it unsuitable for a medium such as twisted-pair cable. For this reason, the Fast Ethernet designers sought some alternative encoding/decoding scheme. However, it was found that one scheme would not perform equally well for all three implementations. Therefore, three different encoding schemes were chosen (see Figure 13.16).

Figure 13.16 *Encoding for Fast Ethernet implementation*

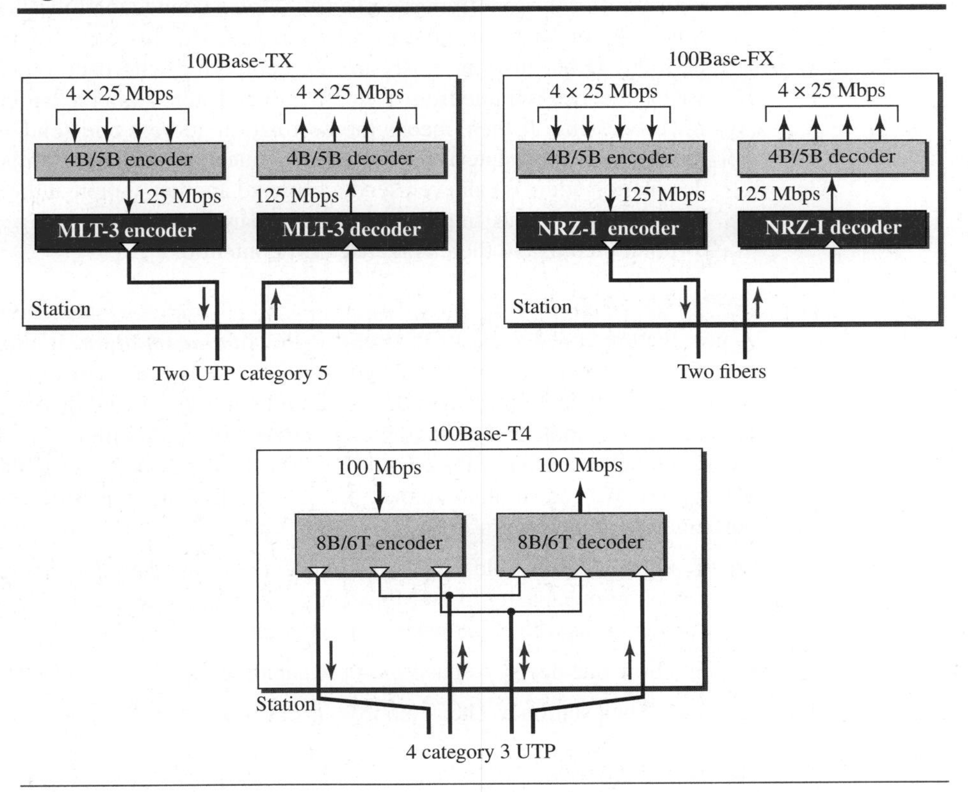

100Base-TX uses two pairs of twisted-pair cable (either category 5 UTP or STP). For this implementation, the MLT-3 scheme was selected since it has good bandwidth performance (see Chapter 4). However, since MLT-3 is not a self-synchronous line coding scheme, 4B/5B block coding is used to provide bit synchronization by preventing the occurrence of a long sequence of 0s and 1s. This creates a data rate of 125 Mbps, which is fed into MLT-3 for encoding.

100Base-FX uses two pairs of fiber-optic cables. Optical fiber can easily handle high bandwidth requirements by using simple encoding schemes. The designers of 100Base-FX selected the NRZ-I encoding scheme (see Chapter 4) for this implementation. However, NRZ-I has a bit synchronization problem for long sequences of 0s (or 1s, based on the encoding), as we saw in Chapter 4. To overcome this problem, the designers used 4B/5B block encoding, as we described for 100Base-TX. The block encoding increases the bit rate from 100 to 125 Mbps, which can easily be handled by fiber-optic cable.

A 100Base-TX network can provide a data rate of 100 Mbps, but it requires the use of category 5 UTP or STP cable. This is not cost-efficient for buildings that have already been wired for voice-grade twisted-pair (category 3). A new standard, called ***100Base-T4,*** was designed to use category 3 or higher UTP. The implementation uses four pairs of UTP for transmitting 100 Mbps. Encoding/decoding in 100Base-T4 is more complicated. As this implementation uses category 3 UTP, each twisted-pair cannot easily handle more than 25 Mbaud. In this design, one pair switches between sending and receiving. Three pairs of UTP category 3, however, can handle only 75 Mbaud (25 Mbaud) each. We need to use an encoding scheme that converts 100 Mbps to a 75 Mbaud signal. As we saw in Chapter 4, 8B/6T satisfies this requirement. In 8B/6T, eight data elements are encoded as six signal elements. This means that 100 Mbps uses only $(6/8) \times 100$ Mbps, or 75 Mbaud.

Summary

Fast Ethernet implementation at the physical layer can be categorized as either two-wire or four-wire. The two-wire implementation can be either shielded twisted pair (STP), which is called *100Base-TX,* or fiber-optic cable, which is called *100Base-FX*. The four-wire implementation is designed only for unshielded twisted pair (UTP), which is called *100Base-T4*. Table 13.2 is a summary of the Fast Ethernet implementations. We discussed encoding in Chapter 4.

Table 13.2 *Summary of Fast Ethernet implementations*

Implementation	*Medium*	*Medium Length*	*Wires*	*Encoding*
100Base-TX	UTP or STP	100 m	2	4B5B + MLT-3
100Base-FX	Fiber	185 m	2	4B5B + NRZ-I
100Base-T4	UTP	100 m	4	Two 8B/6T

13.4 GIGABIT ETHERNET

The need for an even higher data rate resulted in the design of the Gigabit Ethernet Protocol (1000 Mbps). The IEEE committee calls it the Standard 802.3z. The goals of the Gigabit Ethernet were to upgrade the data rate to 1 Gbps, but keep the address length, the frame format, and the maximum and minimum frame length the same. The goals of the Gigabit Ethernet design can be summarized as follows:

1. Upgrade the data rate to 1 Gbps.
2. Make it compatible with Standard or Fast Ethernet.

3. Use the same 48-bit address.
4. Use the same frame format.
5. Keep the same minimum and maximum frame lengths.
6. Support autonegotiation as defined in Fast Ethernet.

13.4.1 MAC Sublayer

A main consideration in the evolution of Ethernet was to keep the MAC sublayer untouched. However, to achieve a data rate of 1 Gbps, this was no longer possible. Gigabit Ethernet has two distinctive approaches for medium access: half-duplex and full-duplex. Almost all implementations of Gigabit Ethernet follow the full-duplex approach, so we mostly ignore the half-duplex mode.

Full-Duplex Mode

In full-duplex mode, there is a central switch connected to all computers or other switches. In this mode, for each input port, each switch has buffers in which data are stored until they are transmitted. Since the switch uses the destination address of the frame and sends a frame out of the port connected to that particular destination, there is no collision. This means that CSMA/CD is not used. Lack of collision implies that the maximum length of the cable is determined by the signal attenuation in the cable, not by the collision detection process.

In the full-duplex mode of Gigabit Ethernet, there is no collision; the maximum length of the cable is determined by the signal attenuation in the cable.

Half-Duplex Mode

Gigabit Ethernet can also be used in half-duplex mode, although it is rare. In this case, a switch can be replaced by a hub, which acts as the common cable in which a collision might occur. The half-duplex approach uses CSMA/CD. However, as we saw before, the maximum length of the network in this approach is totally dependent on the minimum frame size. Three methods have been defined: traditional, carrier extension, and frame bursting.

Traditional

In the traditional approach, we keep the minimum length of the frame as in traditional Ethernet (512 bits). However, because the length of a bit is 1/100 shorter in Gigabit Ethernet than in 10-Mbps Ethernet, the slot time for Gigabit Ethernet is 512 bits × 1/1000 μs, which is equal to 0.512 μs. The reduced slot time means that collision is detected 100 times earlier. This means that the maximum length of the network is 25 m. This length may be suitable if all the stations are in one room, but it may not even be long enough to connect the computers in one single office.

Carrier Extension

To allow for a longer network, we increase the minimum frame length. The **carrier extension** approach defines the minimum length of a frame as 512 bytes (4096 bits). This means that the minimum length is 8 times longer. This method forces a station to add extension bits (padding) to any frame that is less than 4096 bits. In this way, the maximum length of the network can be increased 8 times to a length of 200 m. This allows a length of 100 m from the hub to the station.

Frame Bursting

Carrier extension is very inefficient if we have a series of short frames to send; each frame carries redundant data. To improve efficiency, **frame bursting** was proposed. Instead of adding an extension to each frame, multiple frames are sent. However, to make these multiple frames look like one frame, padding is added between the frames (the same as that used for the carrier extension method) so that the channel is not idle. In other words, the method deceives other stations into thinking that a very large frame has been transmitted.

13.4.2 Physical Layer

The physical layer in Gigabit Ethernet is more complicated than that in Standard or Fast Ethernet. We briefly discuss some features of this layer.

Topology

Gigabit Ethernet is designed to connect two or more stations. If there are only two stations, they can be connected point-to-point. Three or more stations need to be connected in a star topology with a hub or a switch at the center. Another possible configuration is to connect several star topologies or let one star topology be part of another.

Implementation

Gigabit Ethernet can be categorized as either a two-wire or a four-wire implementation. The two-wire implementations use fiber-optic cable (**1000Base-SX,** short-wave, or **1000Base-LX,** long-wave), or STP (**1000Base-CX**). The four-wire version uses category 5 twisted-pair cable (**1000Base-T**). In other words, we have four implementations. 1000Base-T was designed in response to those users who had already installed this wiring for other purposes such as Fast Ethernet or telephone services.

Encoding

Figure 13.17 shows the encoding/decoding schemes for the four implementations. Gigabit Ethernet cannot use the Manchester encoding scheme because it involves a very high bandwidth (2 GBaud). The two-wire implementations use an NRZ scheme, but NRZ does not self-synchronize properly. To synchronize bits, particularly at this high data rate, 8B/10B block encoding, discussed in Chapter 4, is used.

This block encoding prevents long sequences of 0s or 1s in the stream, but the resulting stream is 1.25 Gbps. Note that in this implementation, one wire (fiber or STP) is used for sending and one for receiving.

In the four-wire implementation it is not possible to have 2 wires for input and 2 for output, because each wire would need to carry 500 Mbps, which exceeds the capacity for category 5 UTP. As a solution, 4D-PAM5 encoding, as discussed in Chapter 4, is used to reduce the bandwidth. Thus, all four wires are involved in both input and output; each wire carries 250 Mbps, which is in the range for category 5 UTP cable.

Figure 13.17 *Encoding in Gigabit Ethernet implementations*

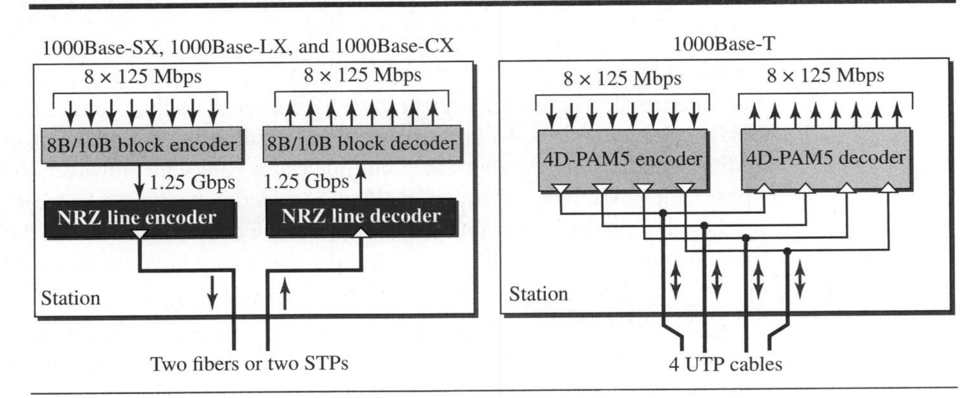

Implementation Summary

Table 13.3 is a summary of the Gigabit Ethernet implementations. S-W and L-W mean short-wave and long-wave respectively.

Table 13.3 *Summary of Gigabit Ethernet implementations*

Implementation	*Medium*	*Medium Length*	*Wires*	*Encoding*
1000Base-SX	Fiber S-W	550 m	2	8B/10B + NRZ
1000Base-LX	Fiber L-W	5000 m	2	8B/10B + NRZ
1000Base-CX	STP	25 m	2	8B/10B + NRZ
1000Base-T4	UTP	100 m	4	4D-PAM5

13.5 10 GIGABIT ETHERNET

In recent years, there has been another look into the Ethernet for use in metropolitan areas. The idea is to extend the technology, the data rate, and the coverage distance so that the Ethernet can be used as LAN and MAN (metropolitan area network). The IEEE committee created 10 Gigabit Ethernet and called it Standard 802.3ae. The goals of the 10 Gigabit Ethernet design can be summarized as upgrading the data rate to 10 Gbps, keeping the same frame size and format, and allowing the interconnection of LANs, MANs, and WAN possible. This data rate is possible only with fiber-optic technology at this time. The standard defines two types of physical layers: LAN PHY and WAN PHY. The first is designed to support existing LANs; the second actually defines a WAN with links connected through SONET OC-192.

13.5.1 Implementation

10 Gigabit Ethernet operates only in full-duplex mode, which means there is no need for contention; CSMA/CD is not used in 10 Gigabit Ethernet. Four implementations are the most common: **10GBase-SR, 10GBase-LR, 10GBase-EW,** and **10GBase-X4.** Table 13.4 shows a summary of the 10 Gigabit Ethernet implementations. We discussed the encoding in Chapter 4.

Table 13.4 *Summary of 10 Gigabit Ethernet implementations*

Implementation	*Medium*	*Medium Length*	*Number of wires*	*Encoding*
10GBase-SR	Fiber 850 nm	300 m	2	64B66B
10GBase-LR	Fiber 1310 nm	10 Km	2	64B66B
10GBase-EW	Fiber 1350 nm	40 Km	2	SONET
10GBase-X4	Fiber 1310 nm	300 m to 10 Km	2	8B10B

13.6 END-CHAPTER MATERIALS

13.6.1 Recommended Reading

For more details about subjects discussed in this chapter, we recommend the following books and RFCs. The items in brackets [...] refer to the reference list at the end of the text.

Books

Several books discuss Ethernet. Among them we recommend [Ham 80], [Zar 02], [Ror 96], [Tan 03], [GW 04], [For 03], [KMK 04], [Sta 04], [Kes 02], [PD 03], [Kei 02], [Spu 00], [KCK 98], [Sau 98], [Izz 00], [Per 00], and [WV 00].

RFCs

A discussion of the use of the checksum in the Internet can be found in RFC 1141.

13.6.2 Key Terms

10 Gigabit Ethernet
1000Base-CX
1000Base-LX
1000Base-SX
1000Base-T
100Base-FX
100Base-T4
100Base-TX
10Base2
10Base5
10Base-F
10Base-T
10GBase-EW
10GBase-LR
10GBase-SR
10GBase-X4
autonegotiation
bridge
carrier extension
Cheapernet
collision domain
Fast Ethernet
frame bursting
full-duplex switched Ethernet
Gigabit Ethernet
hexadecimal notation
logical link control (LLC)
media access control (MAC)
network interface card (NIC)
Project 802
Standard Ethernet
switch
switched Ethernet
thick Ethernet
Thicknet
thin Ethernet
transceiver
twisted-pair Ethernet

13.6.3 Summary

Ethernet is the most widely used local area network protocol. The IEEE 802.3 Standard defines 1-persistent CSMA/CD as the access method for first-generation 10-Mbps Ethernet. The data-link layer of Ethernet consists of the LLC sublayer and the MAC sublayer. The MAC sublayer is responsible for the operation of the CSMA/CD access method and framing. Each station on an Ethernet network has a unique 48-bit address imprinted on its network interface card (NIC). The minimum frame length for 10-Mbps Ethernet is 64 bytes; the maximum is 1518 bytes.

The common implementations of 10-Mbps Ethernet are 10Base5 (thick Ethernet), 10Base2 (thin Ethernet), 10Base-T (twisted-pair Ethernet), and 10Base-F (fiber Ethernet). The 10Base5 implementation of Ethernet uses thick coaxial cable. 10Base2 uses thin coaxial cable. 10Base-T uses four twisted-pair cables that connect each station to a common hub. 10Base-F uses fiber-optic cable. A bridge can increase the bandwidth and separate the collision domains on an Ethernet LAN. A switch allows each station on an Ethernet LAN to have the entire capacity of the network to itself. Full-duplex mode doubles the capacity of each domain and removes the need for the CSMA/CD method.

Fast Ethernet has a data rate of 100 Mbps. In Fast Ethernet, autonegotiation allows two devices to negotiate the mode or data rate of operation. The common Fast Ethernet implementations are 100Base-TX (two pairs of twisted-pair cable), 100Base-FX (two fiber-optic cables), and 100Base-T4 (four pairs of voice-grade, or higher, twisted-pair cable).

Gigabit Ethernet has a data rate of 1000 Mbps. Gigabit Ethernet access methods include half-duplex mode using traditional CSMA/CD (not common) and full-duplex mode (most popular method). The common Gigabit Ethernet implementations are 1000Base-SX (two optical fibers and a short-wave laser source), 1000Base-LX (two optical fibers and a long-wave laser source), and 1000Base-T (four twisted pairs).

The latest Ethernet standard is 10 Gigabit Ethernet, which operates at 10 Gbps. The four common implementations are 10GBase-SR, 10GBase-LR, 10GBase-EW, and 10GBase-X4. These implementations use fiber-optic cables in full-duplex mode.

13.7 PRACTICE SET

13.7.1 Quizzes

A set of interactive quizzes for this chapter can be found on the book website. It is strongly recommended that the student take the quizzes to check his/her understanding of the materials before continuing with the practice set.

13.7.2 Questions

Q13-1. How is the preamble field different from the SFD field?

Q13-2. What are the common 10 Gigabit implementations?

Q13-3. What are the common Gigabit Ethernet implementations?

Q13-4. Compare the data rates for Standard Ethernet, Fast Ethernet, Gigabit Ethernet, and 10 Gigabit Ethernet.

Q13-5. Why is there no need for CSMA/CD on a full-duplex Ethernet LAN?

Q13-6. What is the difference between unicast, multicast, and broadcast addresses?

Q13-7. What are the common Standard Ethernet implementations?

Q13-8. What is the relationship between a switch and a bridge?

Q13-9. What are the advantages of dividing an Ethernet LAN with a bridge?

Q13-10. What are the common Fast Ethernet implementations?

13.7.3 Problems

P13-1. What is the hexadecimal equivalent of the following Ethernet address?

```
01011010 10000001 01010101 00010001 10101010 00011111
```

P13-2. How does the Ethernet address 1A:2B:3C:4C:5E:6E appear on the line in binary?

P13-3. What is the ratio of useful data to the entire packet for the smallest Ethernet frame?

P13-4. Suppose the length of a 10Base5 cable is 2500 m. If the speed of propagation in a thick coaxial cable is 200,000,000 m/s, how long does it take for a bit to travel from the beginning to the end of the network? Assume there is a 10 μs delay in the equipment.

P13-5. Suppose you are to design a LAN for a company that has 150 employees, each given a desktop computer attached to the LAN. What should be the data rate of the LAN if the typical use of the LAN is shown below:

a. Each employee needs to retrieve a file of average size of 10 megabytes in a second. An employee may do this on average 10 times during the eight-hour working time.

b. Each employee needs to access the Internet at 250 Kbps. This can happen for 10 employees simultaneously.

c. Each employee may receive 10 e-mails per hour with an average size of 100 kilobytes. Half of the employees may receive e-mails simultaneously.

P13-6. In a Gigabit Ethernet LAN, the average size of a frame is 1000 bytes. If a noise of 2 ms occurs on the LAN, how many frames are destroyed?

P13-7. In a Fast Ethernet LAN, the average size of a frame is 1000 bytes. If a noise of 2 ms occurs on the LAN, how many frames are destroyed?.

P13-8. In a Standard Ethernet LAN, the average size of a frame is 1000 bytes. If a noise of 2 ms occurs on the LAN, how many frames are destroyed?

P13-9. If an Ethernet destination address is 07:01:02:03:04:05, what is the type of the address (unicast, multicast, or broadcast)?

P13-10. An Ethernet MAC sublayer receives 42 bytes of data from the upper layer. How many bytes of padding must be added to the data?

P13-11. In a 10-Gigabit Ethernet LAN, the average size of a frame is 1000 bytes. If a noise of 2 ms occurs on the LAN, how many frames are destroyed?

13.8 SIMULATION EXPERIMENTS

13.8.1 Applets

We have created some Java applets to show some of the main concepts discussed in this chapter. It is strongly recommended that the students activate these applets on the book website and carefully examine the protocols in action.

13.8.2 Lab Assignments

In this section, we use Wireshark to simulate two protocols: Ethernet and ARP. Full descriptions of these lab assignments are on the book website.

Lab13-1. In this lab we need to examine the contents of a frame sent by the data-link layer. We want to find the value of different fields such as destination and source MAC addresses, the value of CRC, the value of the protocol field which shows which payload is being carried by the frame, and so on.

CHAPTER 14

Other Wired Networks

After discussing wired LANs in the previous chapter, we need to talk about other wired networks used for data communications. These networks can be divided into two broad categories: access networks and wide area networks.

The access networks that connect small LANs (even a single host at home) to an ISP are examples of access networks. These access networks use either the existing telephone or cable network for data transfer.

On a long-distance scale, other wired technologies are used that mostly depend on the high transfer rate of fiber-optic cable. We call the networks that use these wired technologies to transfer data over long distances (such as in the Internet backbone) *wide area networks (WANs)* for the lack of other names.

This chapter is divided into four sections.

- The first section discusses the telephone network. The section first briefly describes the telephone network as a voice network. The section then shows how the voice network has been used for data transmission either as a dial-up service or DSL service.
- The second section discusses the cable network. The section first briefly describes it as a video network. The section then shows how the video network has been used for data transmission.
- The third section discusses SONET, both as a fiber-optic technology and a network. The section first shows how the technology can be used for high-speed connection to carry data. It then describes how we can make both linear and mesh networks using this technology.
- The fourth section discusses ATM, which can use SONET as the carrier to create a high-speed wide area network (WAN). ATM is a cell-relay network that uses a fixed-size frame (cell) as the unit of transmitted data.

14.1 TELEPHONE NETWORKS

The telephone network had its beginnings in the late 1800s. The entire network, which is referred to as the ***plain old telephone system (POTS),*** was originally an analog system using analog signals to transmit voice. With the advent of the computer era, the network, in the 1980s, began to carry data in addition to voice. During the last decade, the telephone network has undergone many technical changes. The network is now digital as well as analog.

14.1.1 Major Components

The telephone network, as shown in Figure 14.1, is made of three major components: local loops, trunks, and switching offices. The telephone network has several levels of switching offices such as **end offices, tandem offices,** and **regional offices.**

Figure 14.1 *A telephone system*

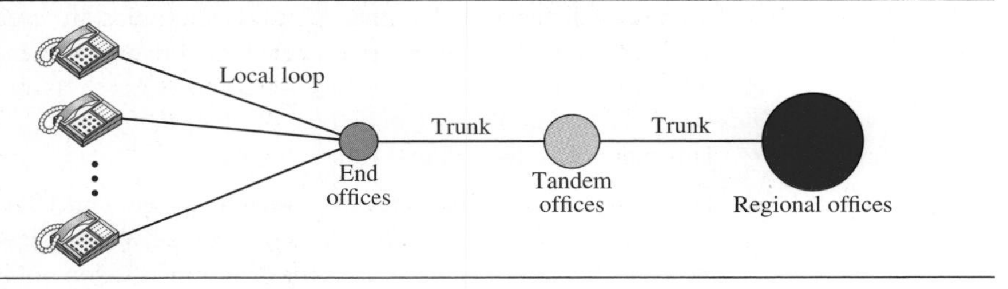

Local Loops

One component of the telephone network is the **local loop,** a twisted-pair cable that connects the subscriber telephone to the nearest end office or local central office. The local loop, when used for voice, has a bandwidth of 4000 Hz (4 kHz). It is interesting to examine the telephone number associated with each local loop. The first three digits of a local telephone number define the office, and the next four digits define the local loop number.

Trunks

Trunks are transmission media that handle the communication between offices. A trunk normally handles hundreds or thousands of connections through multiplexing. Transmission is usually through optical fibers or satellite links.

Switching Offices

To avoid having a permanent physical link between any two subscribers, the telephone company has switches located in a **switching office.** A switch connects several local loops or trunks and allows a connection between different subscribers.

14.1.2 LATAs

After the divestiture of 1984 (see Appendix H), the United States was divided into more than 200 **local-access transport areas (LATAs).** The number of LATAs has increased since then. A LATA can be a small or large metropolitan area. A small state may have a single LATA; a large state may have several LATAs. A LATA boundary may overlap the boundary of a state; part of a LATA can be in one state, part in another state.

Intra-LATA Services

The services offered by the **common carriers** (telephone companies) inside a LATA are called *intra-LATA* services. The carrier that handles these services is called a ***local exchange carrier (LEC).*** Before the Telecommunications Act of 1996 (see Appendix H), intra-LATA services were granted to one single carrier. This was a monopoly. After 1996, more than one carrier could provide services inside a LATA. The carrier that provided services before 1996 owns the cabling system (local loops) and is called the ***incumbent local exchange carrier (ILEC).*** The new carriers that can provide services are called ***competitive local exchange carriers (CLECs).*** To avoid the costs of new cabling, it was agreed that the ILECs would continue to provide the main services, and the CLECs would provide other services such as mobile telephone service, toll calls inside a LATA, and so on. Figure 14.2 shows a LATA and switching offices.

Intra-LATA services are provided by local exchange carriers. Since 1996, there are two types of LECs: incumbent local exchange carriers and competitive local exchange carriers.

Figure 14.2 *Switching offices in a LATA*

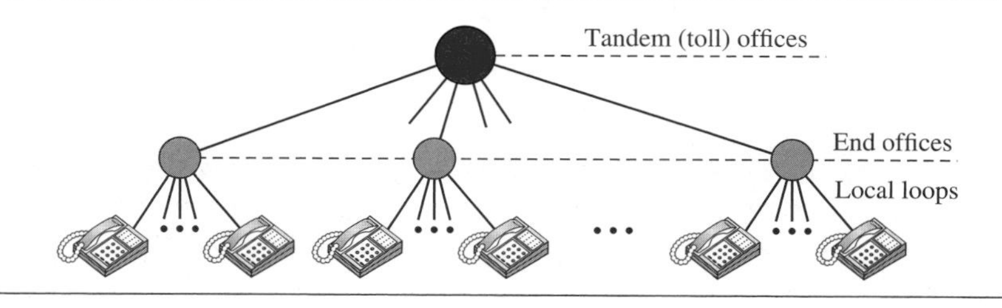

Communication inside a LATA is handled by end switches and tandem switches. A call that can be completed by using only end offices is considered toll-free. A call that has to go through a tandem office (intra-LATA toll office) is charged.

Inter-LATA Services

The services between LATAs are handled by **interexchange carriers (IXCs).** These carriers, sometimes called **long-distance companies,** provide communication services between two customers in different LATAs. After the act of 1996 (see Appendix H), these services can be provided by any carrier, including those involved in intra-LATA services. The field is wide open. Carriers providing inter-LATA services include AT&T, MCI, WorldCom, Sprint, and Verizon.

The IXCs are long-distance carriers that provide general data communications services including telephone service. A telephone call going through an IXC is normally digitized, with the carriers using several types of networks to provide service.

Points of Presence

As we discussed, intra-LATA services can be provided by several LECs (one ILEC and possibly more than one CLEC). We also said that inter-LATA services can be provided by several IXCs. How do these carriers interact with one another? The answer is, via a switching office called a ***point of presence (POP).*** Each IXC that wants to provide inter-LATA services in a LATA must have a POP in that LATA. The LECs that provide services inside the LATA must provide connections so that every subscriber can have access to all POPs. Figure 14.3 illustrates the concept.

Figure 14.3 *Points of presence (POPs)*

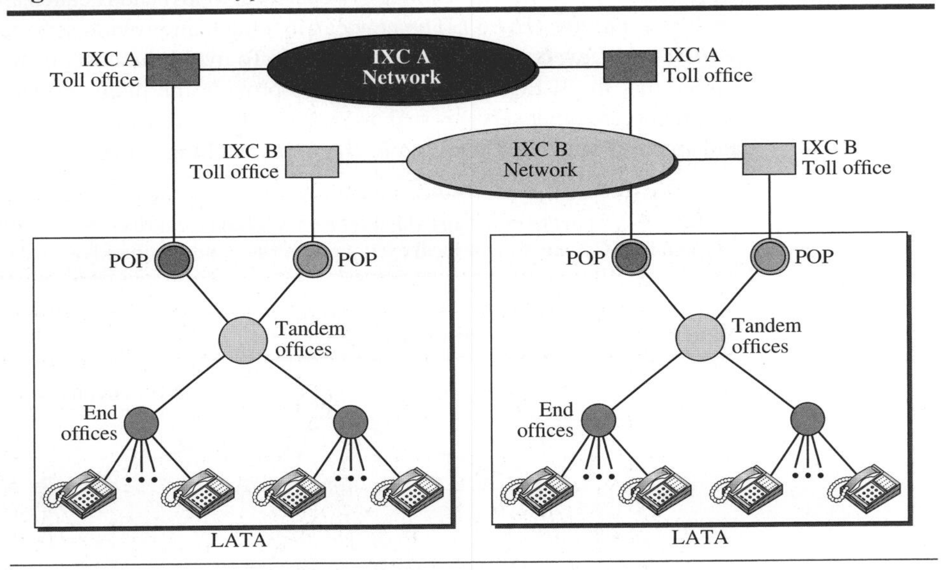

A subscriber who needs to make a connection with another subscriber is connected first to an end switch and then, either directly or through a tandem switch, to a POP. The call now goes from the POP of an IXC (the one the subscriber has chosen) in the source LATA to the POP of the same IXC in the destination LATA. The call is passed through the toll office of the IXC and is carried through the network provided by the IXC.

14.1.3 Signaling

The telephone network, at its beginning, used a circuit-switched network with dedicated links (multiplexing had not yet been invented) to transfer voice communication. As we saw in Chapter 8, a circuit-switched network needs the setup and teardown phases to establish and terminate paths between the two communicating parties. In the beginning, this task was performed by human operators. The operator room was a center to which all subscribers were connected. A subscriber who wished to talk to another subscriber picked up the receiver (off-hook) and rang the operator. The operator, after listening to the caller and getting the identifier of the called party, connected the two by using a wire with two plugs inserted into the corresponding two jacks. A dedicated circuit was created in this way. One of the parties, after the conversation ended, informed the operator to disconnect the circuit. This type of signaling is called ***in-band signaling*** because the same circuit can be used for both signaling and voice communication.

Later, the signaling system became automatic. Rotary telephones were invented that sent a digital signal defining each digit in a multidigit telephone number. The switches in the telephone companies used the digital signals to create a connection between the caller and the called parties. Both in-band and **out-of-band signaling** were used. In in-band signaling, the 4-kHz voice channel was also used to provide signaling. In out-of-band signaling, a portion of the voice channel bandwidth was used for signaling; the voice bandwidth and the signaling bandwidth were separate.

As telephone networks evolved into a complex network, the functionality of the signaling system increased. The signaling system was required to perform other tasks such as:

1. Providing dial tone, ring tone, and busy tone.
2. Transferring telephone numbers between offices.
3. Maintaining and monitoring the call.
4. Keeping billing information.
5. Maintaining and monitoring the status of the telephone network equipment.
6. Providing other functions such as caller ID, voice mail, and so on.

These complex tasks resulted in the provision of a separate network for signaling. This means that a telephone network today can be thought of as two networks: a signaling network and a data transfer network.

The tasks of data transfer and signaling are separated in modern telephone networks: data transfer is done by one network, signaling by another.

However, we need to emphasize a point here. Although the two networks are separate, this does not mean that there are separate physical links everywhere; the two networks may use separate channels of the same link in parts of the system.

Data Transfer Network

The data transfer network that can carry multimedia information today is, for the most part, a circuit-switched network, although it can also be a packet-switched network. This network follows the same type of protocols and model as other networks discussed in this book.

Signaling Network

The signaling network, which is our main concern in this section, is a packet-switched network involving layers similar to those in the OSI model or Internet model, discussed in Chapter 2. The nature of signaling makes it more suited to a packet-switching network with different layers. For example, the information needed to convey a telephone address can easily be encapsulated in a packet with all the error control and addressing information. Figure 14.4 shows a simplified situation of a telephone network in which the two networks are separated.

Figure 14.4 *Data transfer and signaling networks*

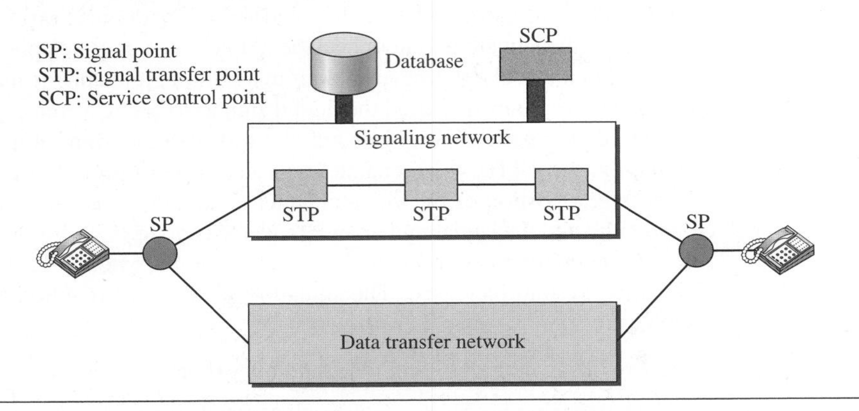

The user telephone or computer is connected to the **signal points (SPs).** The link between the telephone set and SP is common for the two networks. The signaling network uses nodes called ***signal transport ports (STPs)*** that receive and forward signaling messages. The signaling network also includes a **service control point (SCP)** that controls the whole operation of the network. Other systems such as a database center may be included to provide stored information about the entire signaling network.

Signaling System Seven (SS7)

The protocol that is used in the signaling network is called ***Signaling System Seven (SS7).*** It is very similar to the five-layer Internet model we saw in Chapter 2, but the layers have different names, as shown in Figure 14.5.

Figure 14.5 *Layers in SS7*

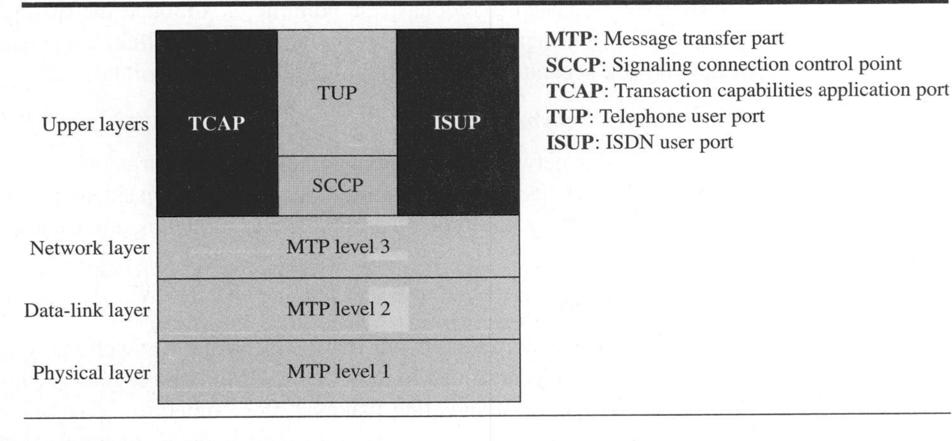

Physical Layer: MTP Level 1

The physical layer in SS7, called ***message transport part (MTP) level 1,*** uses several physical layer specifications such as T-1 (1.544 Mbps) and DC0 (64 kbps).

Data-link Layer: MTP Level 2

The **MTP level 2** layer provides typical link-layer services such as packetizing, using source and destination addresses in the packet header, and CRC for error checking.

Network Layer: MTP Level 3

The **MTP level 3** layer provides end-to-end connectivity by using the datagram approach to switching. Routers and switches route the signal packets from the source to the destination.

Transport Layer: SCCP

The **signaling connection control point (SCCP)** is used for special services such as 800-call processing.

Upper Layers: TUP, TCAP, and ISUP

There are three protocols at the upper layers. **Telephone user port (TUP)** is responsible for setting up voice calls. It receives the dialed digits and routes the calls. **Transaction capabilities application port (TCAP)** provides remote calls that let an application program on a computer invoke a procedure on another computer. **ISDN user port (ISUP)** can replace TUP to provide services similar to those of an ISDN network.

14.1.4 Services Provided by Telephone Networks

Telephone companies provide two types of services: analog and digital.

Analog Services

In the beginning, telephone companies provided their subscribers with analog services. These services still continue today. We can categorize these services as either **analog switched services** or **analog leased services.**

Analog Switched Services

This is the familiar dial-up service most often encountered when a home telephone is used. The signal on a local loop is analog, and the bandwidth is usually between 0 and 4000 Hz. A local call service is normally provided for a flat monthly rate, although in some LATAs the carrier charges for each call or a set of calls. The rationale for a non flat-rate charge is to provide cheaper service for those customers who do not make many calls. A toll call can be intra-LATA or inter-LATA. If the LATA is geographically large, a call may go through a tandem office (toll office) and the subscriber will pay a fee for the call. The inter-LATA calls are long-distance calls and are charged as such.

Another service is called *800 service.* If a subscriber (normally an organization) needs to provide free connections for other subscribers (normally customers), it can request the **800 service.** In this case, the call is free for the caller, but it is paid by the callee. An organization uses this service to encourage customers to call. The rate is less expensive than that for a normal long-distance call.

The **wide-area telephone service (WATS)** is the opposite of the 800 service. The latter are inbound calls paid by the organization; the former are outbound calls paid by the organization. This service is a less expensive alternative to regular toll calls; charges are based on the number of calls. The service can be specified as outbound calls to the same state, to several states, or to the whole country, with rates charged accordingly.

The **900 services** are like the 800 service, in that they are inbound calls to a subscriber. However, unlike the 800 service, the call is paid by the caller and is normally much more expensive than a normal long-distance call. The reason is that the carrier charges *two* fees: the first is the long-distance toll, and the second is the fee paid to the callee for each call.

Analog Leased Service

An **analog leased service** offers customers the opportunity to lease a line, sometimes called a *dedicated line,* that is permanently connected to another customer. Although the connection still passes through the switches in the telephone network, subscribers experience it as a single line because the switch is always closed; no dialing is needed.

Digital Services

Recently telephone companies began offering **digital services** to their subscribers. Digital services are less sensitive than analog services to noise and other forms of interference. The two most common digital services are switched/56 service and *digital data service (DDS).* We already discussed high-speed digital services—the T lines—in Chapter 6. We discuss the other services in this chapter.

Switched/56 Service

Switched/56 service is the digital version of an analog switched line. It is a switched digital service that allows data rates of up to 56 kbps. To communicate through this service, both parties must subscribe. A caller with normal telephone service cannot connect to a telephone or computer with switched/56 service even if the caller is using a modem. On the whole, digital and analog services represent two completely different domains for the telephone companies. Because the line in a switched/56 service is already digital, subscribers do not need modems to transmit digital data. However, they do need another device called a ***digital service unit (DSU).***

Digital Data Service

Digital data service (DDS) is the digital version of an analog leased line; it is a digital leased line with a maximum data rate of 64 kbps.

14.1.5 Dial-Up Service

Traditional telephone lines can carry frequencies between 300 and 3300 Hz, giving them a bandwidth of 3000 Hz. All this range is used for transmitting voice, where a great deal of interference and distortion can be accepted without loss of intelligibility. As we have seen, however, data signals require a higher degree of accuracy to ensure integrity. For safety's sake, therefore, the edges of this range are not used for data communications. In general, we can say that the signal bandwidth must be smaller than the cable bandwidth. The effective bandwidth of a telephone line being used for data transmission is 2400 Hz, covering the range from 600 to 3000 Hz. Note that today some telephone lines are capable of handling greater bandwidth than traditional lines. However, modem design is still based on traditional capabilities (See Figure 14.6).

The term ***modem*** is a composite word that refers to the two functional entities that make up the device: a signal *mo*dulator and a signal *dem*odulator. A **modulator** creates

Figure 14.6 *Telephone line bandwidth*

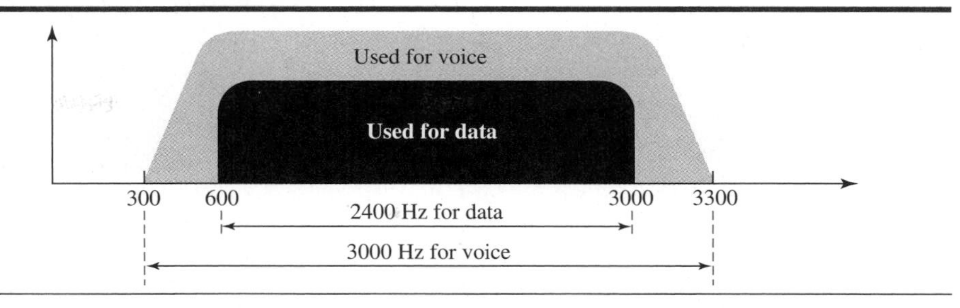

a bandpass analog signal from binary data. A **demodulator** recovers the binary data from the modulated signal.

***Modem* stands for modulator/demodulator.**

Figure 14.7 shows the relationship of modems to a communications link. The computer on the left sends a digital signal to the modulator portion of the modem; the data are sent as an analog signal on the telephone lines. The modem on the right receives the analog signal, demodulates it through its demodulator, and delivers data to the computer on the right. The communication can be bidirectional, which means the computer on the right can simultaneously send data to the computer on the left, using the same modulation/demodulation processes.

Figure 14.7 *Modulation/demodulation*

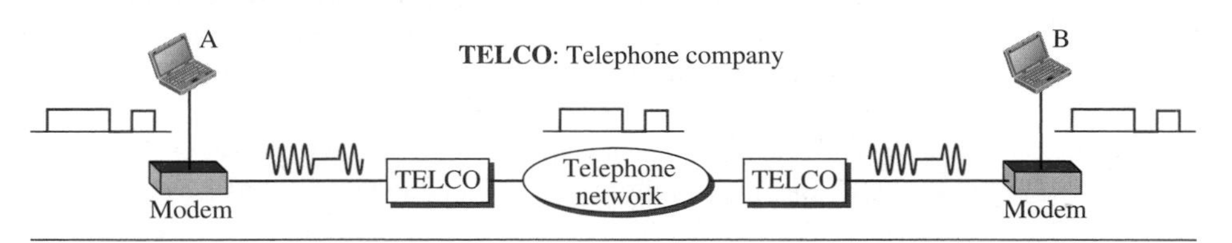

56K modems

Traditional modems have a data rate limitation of 33.6 kbps, as determined by the Shannon capacity (see Chapter 3). However, modern modems with a bit rate of 56,000 bps are available; these are called ***56K modems.*** These modems may be used only if one party is using digital signaling (such as through an Internet provider). They are asymmetric in that the downloading rate (flow of data from the Internet service provider to the PC) is a maximum of 56 kbps, while the uploading rate (flow of data from the PC to the Internet provider) can be a maximum of 33.6 kbps. Do these modems violate the Shannon capacity principle? No, in the downstream direction, the SNR ratio is higher because there is no quantization error (see Figure 14.8).

In **uploading,** the analog signal must still be sampled at the switching station. In this direction, quantization noise (as we saw in Chapter 4) is introduced into the signal, which reduces the SNR ratio and limits the rate to 33.6 kbps.

Figure 14.8 *Dial-up network to provide Internet access*

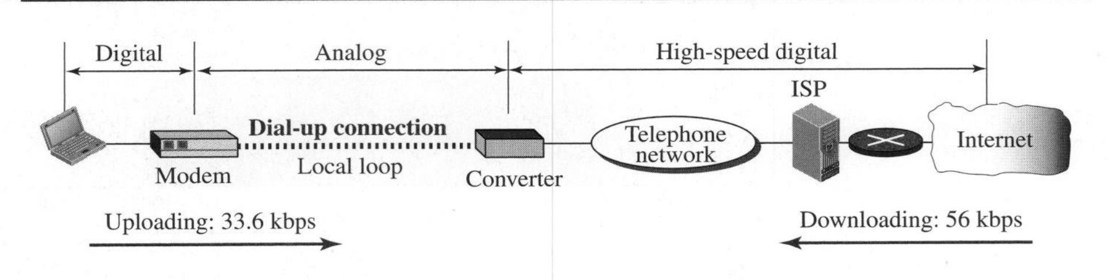

However, there is no sampling in the **downloading.** The signal is not affected by quantization noise and not subject to the Shannon capacity limitation. The maximum data rate in the uploading direction is still 33.6 kbps, but the data rate in the downloading direction is now 56 kbps.

One may wonder how we arrive at the 56-kbps figure. The telephone companies sample 8000 times per second with 8 bits per sample. One of the bits in each sample is used for control purposes, which means each sample is 7 bits. The rate is therefore 8000×7, or 56,000 bps or 56 kbps.

14.1.6 Digital Subscriber Line (DSL)

After traditional modems reached their peak data rate, telephone companies developed another technology, DSL, to provide higher-speed access to the Internet. **Digital subscriber line (DSL)** technology is one of the most promising for supporting high-speed digital communication over the existing telephone. DSL technology is a set of technologies, each differing in the first letter (ADSL, VDSL, HDSL, and SDSL). The set is often referred to as *x*DSL, where *x* can be replaced by A, V, H, or S. We just discuss the first. The first technology in the set is **asymmetric DSL (ADSL).** ADSL, like a 56K modem, provides higher speed (bit rate) in the downstream direction (from the Internet to the resident) than in the upstream direction (from the resident to the Internet). That is the reason it is called *asymmetric.* Unlike the asymmetry in 56K modems, the designers of ADSL specifically divided the available bandwidth of the local loop unevenly for the residential customer. The service is not suitable for business customers who need a large bandwidth in both directions.

Using Existing Local Loops

One interesting point is that ADSL uses the existing telephone lines (local loop). But how does ADSL reach a data rate that was never achieved with traditional modems? The answer is that the twisted-pair cable used in telephone lines is actually capable of handling bandwidths up to 1.1 MHz, but the filter installed at the end office of the telephone company where each local loop terminates limits the bandwidth to 4 kHz (sufficient for voice communication). If the filter is removed, however, the entire 1.1 MHz is available for data and voice communications. Typically, an available bandwidth of 1.104 MHz is divided into a voice channel, an upstream channel, and a downstream channel, as shown in Figure 14.9.

Figure 14.9 *ADSL point-to-point network*

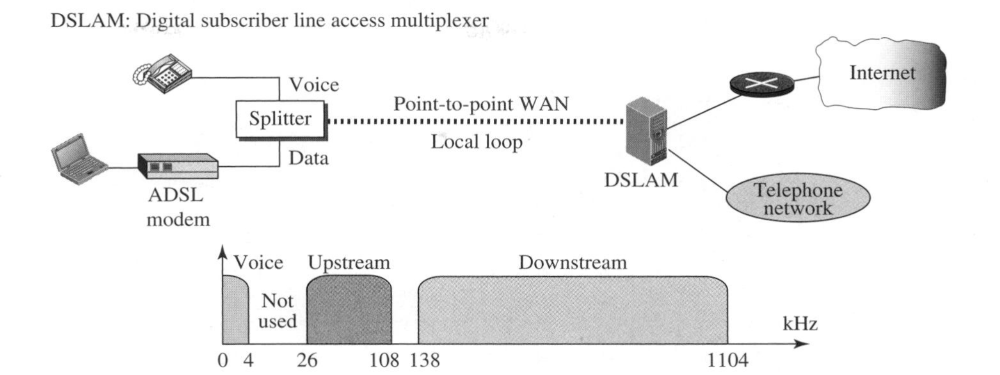

ADSL allows the subscriber to use the voice channel and the data channel at the same time. The rate for the upstream can reach 1.44-Mbps. However, the data rate is normally below 500 kbps because of the high-level noise in this channel. The downstream data rate can reach 13.4 Mbps. However, the data rate is normally below 8 Mbps because of noise in this channel. A very interesting point is that the telephone company in this case serves as the ISP, so services such as e-mail or Internet access are provided by the telephone company itself.

14.2 CABLE NETWORKS

Cable TV networks were originally created to provide access to TV programs for those subscribers who had no reception because of natural obstructions such as mountains. Later the cable networks became popular with people who just wanted a better signal. In addition, cable networks enabled access to remote broadcasting stations via microwave connections. Cable TV also found a good market in Internet access provision, using some of the channels originally designed for video. After discussing the basic structure of cable networks, we discuss how cable modems can provide a high-speed connection to the Internet.

14.2.1 Traditional Cable Networks

Cable TV started to distribute broadcast video signals to locations with poor or no reception in the late 1940s. It was called ***community antenna television (CATV)*** because an antenna at the top of a tall hill or building received the signals from the TV stations and distributed them, via coaxial cables, to the community. Figure 14.10 shows a schematic diagram of a traditional cable TV network.

The cable TV office, called the ***head end,*** received video signals from broadcasting stations and fed the signals into coaxial cables. The signals became weaker and weaker with distance, so amplifiers were installed throughout the network to renew the signals. There could be up to 35 amplifiers between the head end and the subscriber premises.

Figure 14.10 *Traditional cable TV network*

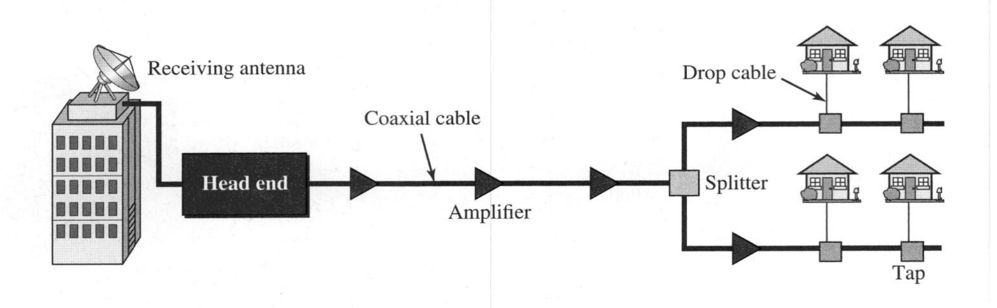

At the other end, splitters split the cable, and taps and drop cables made the connections to the subscriber premises.

The traditional cable TV system used coaxial cable end to end. Due to attenuation of the signals and the use of a large number of amplifiers, communication in the traditional network was unidirectional (one-way). Video signals were transmitted downstream, from the head end to the subscriber premises.

14.2.2 Hybrid Fiber-Coaxial (HFC) Network

The second generation of cable network is called a ***hybrid fiber-coaxial (HFC) network.*** The network uses a combination of fiber-optic and coaxial cable. The transmission medium from the cable TV office to a box, called the ***fiber node,*** is optical fiber; from the fiber node through the neighborhood and into the house is still coaxial cable. Figure 14.11 shows a schematic diagram of an HFC network.

Figure 14.11 *Hybrid fiber-coaxial (HFC) network*

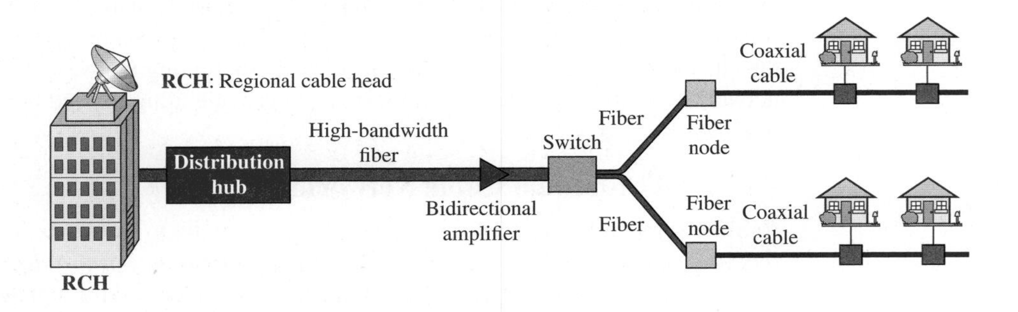

The **regional cable head (RCH)** normally serves up to 400,000 subscribers. The RCHs feed the **distribution hubs,** each of which serves up to 40,000 subscribers. The distribution hub plays an important role in the new infrastructure. Modulation and distribution of signals are done here; the signals are then fed to the fiber nodes through fiber-optic cables. The fiber node splits the analog signals so that the same signal is sent to each coaxial cable. Each coaxial cable serves up to 1000 subscribers. The use of fiber-optic cable reduces the need for amplifiers down to eight or less.

One reason for moving from traditional to hybrid infrastructure is to make the cable network bidirectional (two-way).

14.2.3 Cable TV for Data Transfer

Cable companies are now competing with telephone companies for the residential customer who wants high-speed data transfer. DSL technology provides high-data-rate connections for residential subscribers over the local loop. However, DSL uses the existing unshielded twisted-pair cable, which is very susceptible to interference. This imposes an upper limit on the data rate. A solution is the use of the cable TV network. In this section, we briefly discuss this technology.

Even in an HFC system, the last part of the network, from the fiber node to the subscriber premises, is still a coaxial cable. This coaxial cable has a bandwidth that ranges from 5 to 750 MHz (approximately). To provide Internet access, the cable company has divided this bandwidth into three bands: video, downstream data, and upstream data, as shown in Figure 14.12.

Figure 14.12 *Division of coaxial cable band by CATV*

The **video band** occupies frequencies from 54 to 550 MHz. Since each TV channel occupies 6 MHz, this can accommodate more than 80 channels. The **downstream data** (from the Internet to the subscriber premises) occupies the upper band, from 550 to 750 MHz. This band is also divided into 6-MHz channels.

The **upstream data** (from the subscriber premises to the Internet) occupies the lower band, from 5 to 42 MHz. This band is also divided into 6-MHz channels. There are 2 bits/baud in QPSK. The standard specifies 1 Hz for each baud; this means that, theoretically, upstream data can be sent at 12 Mbps (2 bits/Hz × 6 MHz). However, the data rate is usually less than 12 Mbps.

Sharing

Both upstream and downstream bands are shared by the subscribers. The upstream data bandwidth is 37 MHz. This means that there are only six 6-MHz channels available in the upstream direction. A subscriber needs to use one channel to send data in the upstream direction. The question is, "How can six channels be shared in an area with 1000, 2000, or even 100,000 subscribers?" The solution is time-sharing. The band is divided into channels; these channels must be shared between subscribers in the same neighborhood. The cable provider allocates one channel, statically or dynamically, for a group of subscribers. If one subscriber wants to send data, she or he contends for the channel with others who want access; the subscriber must wait until the channel is available.

We have a similar situation in the downstream direction. The downstream band has 33 channels of 6 MHz. A cable provider probably has more than 33 subscribers; therefore, each channel must be shared between a group of subscribers. However, the situation is different for the downstream direction; here we have a multicasting situation. If there are data for any of the subscribers in the group, the data are sent to that channel. Each subscriber is sent the data. But since each subscriber also has an address registered with the provider; the cable modem for the group matches the address carried with the data to the address assigned by the provider. If the address matches, the data are kept; otherwise, they are discarded.

CM and CMTS

To use a cable network for data transmission, we need two key devices: a **cable modem (CM)** and a **cable modem transmission system (CMTS).** The *cable* modem is installed on the subscriber premises. The *cable modem transmission system* is installed inside the cable company. It receives data from the Internet and sends them to the subscriber. The CMTS also receives data from the subscriber and passes them to the Internet. It is similar to an ADSL modem. Figure 14.13 shows the location of these two devices. Like DSL technology, the cable company needs to become an ISP and provide Internet services to the subscriber. At the subscriber premises, the CM separates the video from data and sends them to the television set or the computer.

Figure 14.13 *Cable modem transmission system (CMTS)*

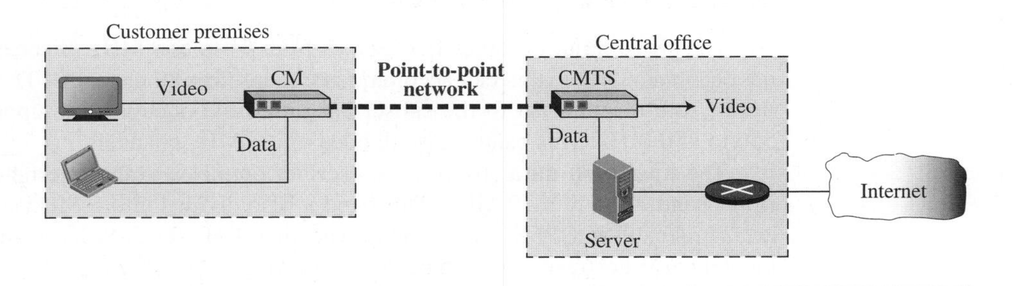

14.3 SONET

In this section, we introduce a wide area network (WAN), SONET, that is used as a transport network to carry loads from other WANs. We first discuss SONET as a protocol, and we then show how SONET networks can be constructed from the standards defined in the protocol.

The high bandwidths of fiber-optic cable are suitable for today's high-data-rate technologies (such as video conferencing) and for carrying large numbers of lower-rate technologies at the same time. For this reason, the importance of fiber optics grows in conjunction with the development of technologies requiring high data rates or wide bandwidths for transmission. With their prominence came a need for standardization. The United States (ANSI) and Europe (ITU-T) have responded by defining standards that, though independent, are fundamentally similar and ultimately compatible. The ANSI standard is called the ***Synchronous Optical Network (SONET).*** The ITU-T standard is called the ***Synchronous Digital Hierarchy (SDH).***

SONET was developed by ANSI; SDH was developed by ITU-T.

SONET/SDH is a synchronous network using synchronous TDM multiplexing. All clocks in the system are locked to a master clock.

14.3.1 Architecture

Let us first introduce the architecture of a SONET system: signals, devices, and connections.

Signals

SONET defines a hierarchy of electrical signaling levels called ***synchronous transport signals (STSs).*** Each STS level (STS-1 to STS-192) supports a certain data rate, specified in megabits per second (see Table 14.1). The corresponding optical signals are called ***optical carriers (OCs).*** SDH specifies a similar system called a ***synchronous transport module (STM).*** STM is intended to be compatible with existing European hierarchies, such as E lines, and with STS levels. To this end, the lowest STM level, STM-1, is defined as 155.520 Mbps, which is exactly equal to STS-3.

Table 14.1 *SONET/SDH rates*

STS	*OC*	*Rate (Mbps)*	*STM*
STS-1	OC-1	51.840	
STS-3	OC-3	155.520	**STM-1**
STS-9	OC-9	466.560	**STM-3**
STS-12	OC-12	622.080	**STM-4**
STS-18	OC-18	933.120	**STM-6**
STS-24	OC-24	1244.160	**STM-8**
STS-36	OC-36	1866.230	**STM-12**
STS-48	OC-48	2488.320	**STM-16**
STS-96	OC-96	4976.640	**STM-32**
STS-192	OC-192	9953.280	**STM-64**

A glance through Table 14.1 reveals some interesting points. First, the lowest level in this hierarchy has a data rate of 51.840 Mbps, which is greater than that of the DS-3 service (44.736 Mbps). In fact, the STS-1 is designed to accommodate data rates equivalent to those of the DS-3. The difference in capacity is provided to handle the overhead needs of the optical system.

Second, the STS-3 rate is exactly three times the STS-1 rate; and the STS-9 rate is exactly one-half the STS-18 rate. These relationships mean that 18 STS-1 channels can be multiplexed into one STS-18, six STS-3 channels can be multiplexed into one STS-18, and so on.

SONET Devices

Figure 14.14 shows a simple link using SONET devices. SONET transmission relies on three basic devices: STS multiplexers/demultiplexers, regenerators, add/drop multiplexers and terminals.

Figure 14.14 *A simple network using SONET equipment*

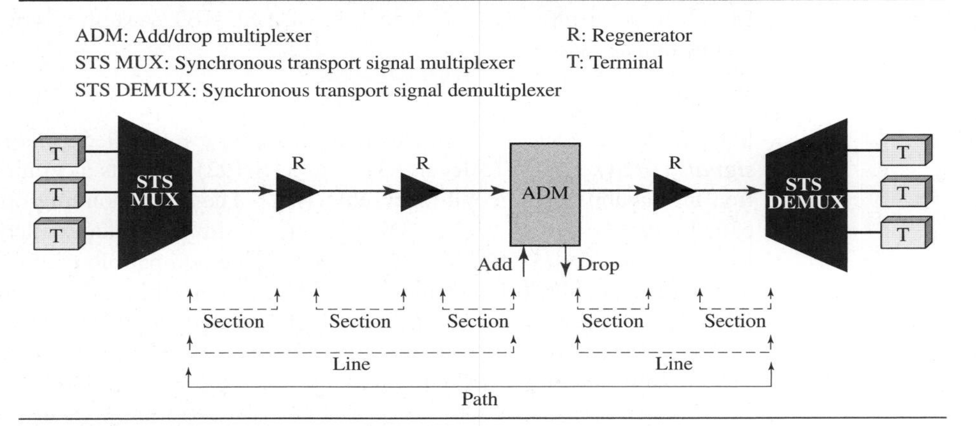

STS Multiplexer/Demultiplexer

STS multiplexers/demultiplexers mark the beginning points and endpoints of a SONET link. They provide the interface between an electrical tributary network and the optical network. An **STS multiplexer** multiplexes signals from multiple electrical sources and creates the corresponding OC signal. An **STS demultiplexer** demultiplexes an optical OC signal into corresponding electric signals.

Regenerator

Regenerators extend the length of the links. A **regenerator** is a repeater (see Chapter 17) that takes a received optical signal (OC-n), demodulates it into the corresponding electric signal (STS-n), regenerates the electric signal, and finally modulates the electric signal into its correspondent OC-n signal. A SONET regenerator replaces some of the existing overhead information (header information) with new information.

Add/drop Multiplexer

Add/drop multiplexers allow insertion and extraction of signals. An **add/drop multiplexer (ADM)** can add STSs coming from different sources into a given path or can remove a desired signal from a path and redirect it without demultiplexing the entire signal. Instead of relying on timing and bit positions, add/drop multiplexers use header information such as addresses and pointers (described later in this section) to identify individual streams.

In the simple configuration shown by Figure 14.14, a number of incoming electronic signals are fed into an STS multiplexer, where they are combined into a single optical signal. The optical signal is transmitted to a regenerator, where it is recreated without the noise it has picked up in transit. The regenerated signals from a number of sources are then fed into an add/drop multiplexer. The add/drop multiplexer reorganizes these signals, if necessary, and sends them out as directed by information in the data frames. These remultiplexed signals are sent to another regenerator and from there to the receiving STS demultiplexer, where they are returned to a format usable by the receiving links.

Terminals

A **terminal** is a device that uses the services of a SONET network. For example, in the Internet, a terminal can be a router that needs to send packets to another router at the other side of a SONET network.

Connections

The devices defined in the previous section are connected using *sections, lines,* and *paths*.

Sections

A **section** is the optical link connecting two neighboring devices: multiplexer to multiplexer, multiplexer to regenerator, or regenerator to regenerator.

Lines

A **line** is the portion of the network between two multiplexers: STS multiplexer to add/drop multiplexer, two add/drop multiplexers, or two STS multiplexers.

Paths

A **path** is the end-to-end portion of the network between two STS multiplexers. In a simple SONET of two STS multiplexers linked directly to each other, the section, line, and path are the same.

14.3.2 SONET Layers

The SONET standard includes four functional layers: the photonic, the section, the line, and the path layer. They correspond to both the physical and the data-link layers (see Figure 14.15). The headers added to the frame at the various layers are discussed later in this chapter.

SONET defines four layers: path, line, section, and photonic.

Path Layer

The **path layer** is responsible for the movement of a signal from its optical source to its optical destination. At the optical source, the signal is changed from an electronic form into an optical form, multiplexed with other signals, and encapsulated in a frame. At the optical destination, the received frame is demultiplexed, and the individual optical

Figure 14.15 *SONET layers compared with OSI or Internet layers*

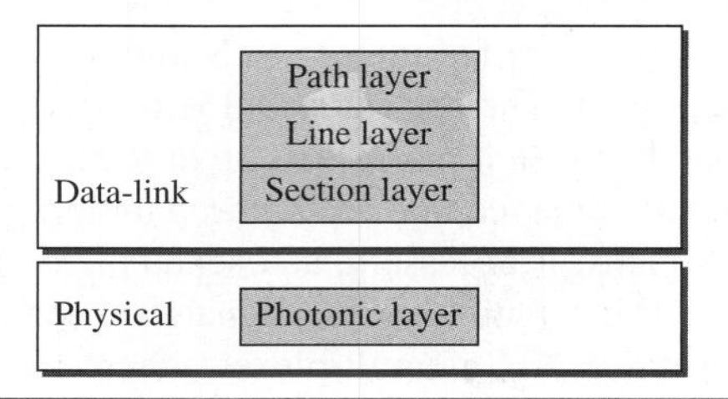

signals are changed back into their electronic forms. Path layer overhead is added at this layer. STS multiplexers provide path layer functions.

Line Layer

The **line layer** is responsible for the movement of a signal across a physical line. Line layer overhead is added to the frame at this layer. STS multiplexers and add/drop multiplexers provide line layer functions.

Section Layer

The **section layer** is responsible for the movement of a signal across a physical section. It handles framing, scrambling, and error control. Section layer overhead is added to the frame at this layer.

Photonic Layer

The **photonic layer** corresponds to the physical layer of the OSI model. It includes physical specifications for the optical fiber channel, the sensitivity of the receiver, multiplexing functions, and so on. SONET uses NRZ encoding, with the presence of light representing 1 and the absence of light representing 0.

Device–Layer Relationships

Figure 14.16 shows the relationship between the devices used in SONET transmission and the four layers of the standard. As you can see, an STS multiplexer is a four-layer device. An add/drop multiplexer is a three-layer device. A regenerator is a two-layer device.

14.3.3 SONET Frames

Each synchronous transfer signal STS-*n* is composed of 8000 frames. Each frame is a two-dimensional matrix of bytes with 9 rows by $90 \times n$ columns. For example, an STS-1 frame is 9 rows by 90 columns (810 bytes), and an STS-3 is 9 rows by 270 columns (2430 bytes). Figure 14.17 shows the general format of an STS-1 and an STS-*n*.

Frame, Byte, and Bit Transmission

One of the interesting points about SONET is that each STS-*n* signal is transmitted at a fixed rate of 8000 frames per second. This is the rate at which voice is digitized (see Chapter 4). For each frame the bytes are transmitted from the left to the right, top

Figure 14.16 *Device–layer relationship in SONET*

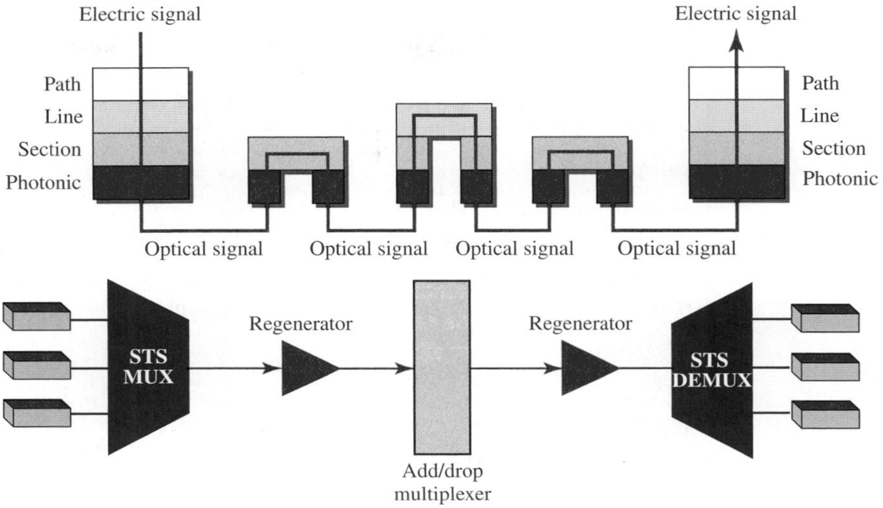

Figure 14.17 *STS-1 and STS-n frames*

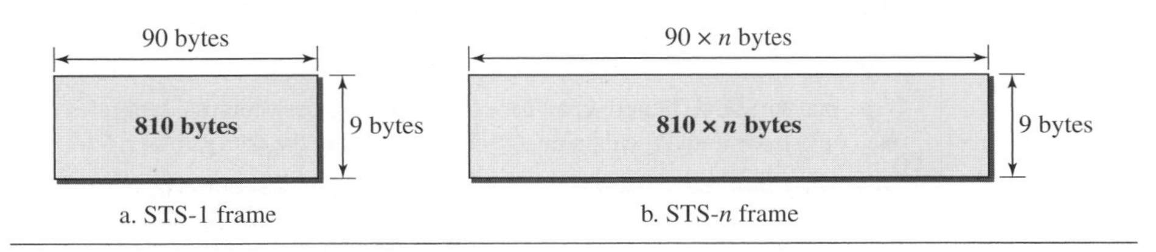

to the bottom. For each byte, the bits are transmitted from the most significant to the least significant (left to right). Figure 14.18 shows the order of frame and byte transmission.

Figure 14.18 *STS frames in transition*

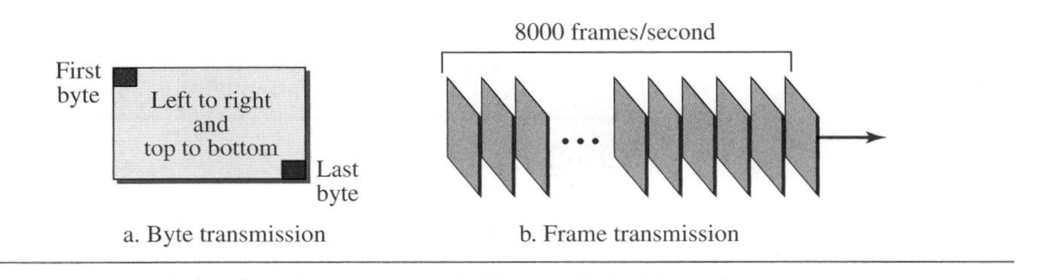

A SONET STS-*n* signal is transmitted at 8000 frames per second.

If we sample a voice signal and use 8 bits (1 byte) for each sample, we can say that each byte in a SONET frame can carry information from a digitized voice channel. In other words, an STS-1 signal can carry 774 voice channels simultaneously (810 minus required bytes for overhead).

Each byte in a SONET frame can carry a digitized voice channel.

Example 14.1

Find the data rate of an STS-1 signal.

Solution

STS-1, like other STS signals, sends 8000 frames per second. Each STS-1 frame is made of 9 by (1×90) bytes. Each byte is made of 8 bits. The data rate is

$$\text{STS-1 data rate} = 8000 \times 9 \times (1 \times 90) \times 8 = 51.840 \text{ Mbps}$$

Example 14.2

Find the data rate of an STS-3 signal.

Solution

STS-3, like other STS signals, sends 8000 frames per second. Each STS-3 frame is made of 9 by (3×90) bytes. Each byte is made of 8 bits. The data rate is

$$\text{STS-3 data rate} = 8000 \times 9 \times (3 \times 90) \times 8 = 155.52 \text{ Mbps}$$

Note that in SONET there is an exact relationship between the data rates of different STS signals. We could have found the data rate of STS-3 by using the data rate of STS-1 (multiply the latter by 3).

In SONET, the data rate of an STS-*n* signal is *n* times the data rate of an STS-1 signal.

Example 14.3

What is the duration of an STS-1 frame? STS-3 frame? STS-*n* frame?

Solution

In SONET, 8000 frames are sent per second. This means that the duration of an STS-1, STS-3, or STS-*n* frame is the same and equal to 1/8000 s, or 125 μs.

In SONET, the duration of any frame is 125 μs.

STS-1 Frame Format

The basic format of an STS-1 frame is shown in Figure 14.19. As we said before, a SONET frame is a matrix of 9 rows of 90 bytes (octets) each, for a total of 810 bytes.

The first three columns of the frame are used for section and line overhead. The upper three rows of the first three columns are used for **section overhead (SOH).** The lower six are **line overhead (LOH).** The rest of the frame is called the ***synchronous***

Figure 14.19 *STS-1 frame overheads*

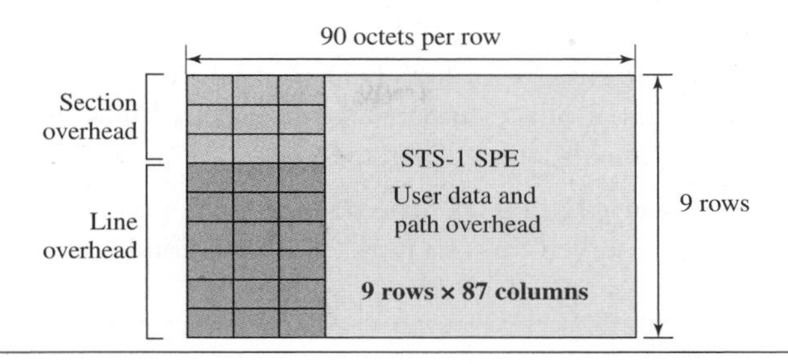

payload envelope (SPE). It contains user data and **path overhead (POH)** needed at the user data level. We will discuss the format of the SPE shortly.

Section Overhead
The section overhead consists of nine octets. The labels, functions, and organization of these octets are shown in Figure 14.20.

Figure 14.20 *STS-1 frame: section overhead*

- **Alignment bytes (A1 and A2).** Bytes A1 and A2 are used for framing and synchronization and are called *alignment bytes.* These bytes alert a receiver that a frame is arriving and give the receiver a predetermined bit pattern on which to synchronize. The bit patterns for these two bytes in hexadecimal are 0xF628. The bytes serve as a flag.

- **Section parity byte (B1).** Byte B1 is for bit interleaved parity (BIP-8). Its value is calculated over all bytes of the previous frame. In other words, the *i*th bit of this byte is the parity bit calculated over all *i*th bits of the previous STS-*n* frame. The value of this byte is filled only for the first STS-1 in an STS-*n* frame. In other words, although an STS-*n* frame has *n* B1 bytes, as we will see later, only the first byte has this value; the rest are filled with 0s.

- **Identification byte (C1).** Byte C1 carries the identity of the STS-1 frame. This byte is necessary when multiple STS-1s are multiplexed to create a higher-rate STS (STS-3, STS-9, STS-12, etc.). Information in this byte allows the various signals to be recognized easily upon demultiplexing. For example, in an STS-3 signal, the value of the C1 byte is 1 for the first STS-1; it is 2 for the second; and it is 3 for the third.

- **Management bytes (D1, D2, and D3).** Bytes D1, D2, and D3 together form a 192-kbps channel ($3 \times 8000 \times 8$) called the *data communication channel.* This channel is required for operation, administration, and maintenance (OA&M) signaling.

- **Order wire byte (E1).** Byte E1 is the order wire byte. Order wire bytes in consecutive frames form a channel of 64 kbps (8000 frames per second times 8 bits per frame). This channel is used for communication between regenerators, or between terminals and regenerators.
- **User's byte (F1).** The F1 bytes in consecutive frames form a 64-kbps channel that is reserved for user needs at the section level.

Section overhead is recalculated for each SONET device (regenerators and multiplexers).

Line Overhead

Line overhead consists of 18 bytes. The labels, functions, and arrangement of these bytes are shown in Figure 14.21.

Figure 14.21 *STS-1 frame: line overhead*

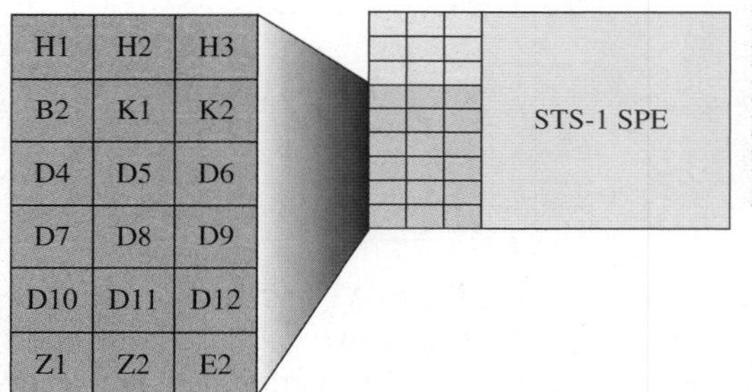

B2: Line parity byte
D4-D12: Management bytes
E2: Order wire byte
H1, H2, H3: Pointers
K1, K2: Automatic protection switching bytes
Z1, Z2: Growth bytes (reserved)

- **Line parity byte (B2).** Byte B2 is for bit interleaved parity. It is for error checking of the frame over a line (between two multiplexers). In an STS-*n* frame, B2 is calculated for all bytes in the previous STS-1 frame and inserted at the B2 byte for that frame. In other words, in a STS-3 frame, there are three B2 bytes, each calculated for one STS-1 frame. Contrast this byte with B1 in the section overhead.
- **Data communication channel bytes (D4 to D12).** The line overhead D bytes (D4 to D12) in consecutive frames form a 576-kbps channel that provides the same service as the D1–D3 bytes (OA&M), but at the line rather than the section level (between multiplexers).
- **Order wire byte (E2).** The E2 bytes in consecutive frames form a 64-kbps channel that provides the same functions as the E1 order wire byte, but at the line level.
- **Pointer bytes (H1, H2, and H3).** Bytes H1, H2, and H3 are pointers. The first two bytes are used to show the offset of the SPE in the frame; the third is used for justification. We show the use of these bytes later.

- **Automatic protection switching bytes (K1 and K2).** The K1 and K2 bytes in consecutive frames form a 128-kbps channel used for automatic detection of problems in line-terminating equipment. We discuss automatic protection switching (APS) later in the chapter.
- **Growth bytes (Z1 and Z2).** The Z1 and Z2 bytes are reserved for future use.

Synchronous Payload Envelope

The synchronous payload envelope (SPE) contains the user data and the overhead related to the user data (path overhead). One SPE does not necessarily fit it into one STS-1 frame; it may be split between two frames, as we will see shortly. This means that the path overhead, the leftmost column of an SPE, does not necessarily align with the section or line overhead. The path overhead must be added first to the user data to create an SPE, and then an SPE can be inserted into one or two frames. Path overhead consists of 9 bytes. The labels, functions, and arrangement of these bytes are shown in Figure 14.22.

Figure 14.22 *STS-1 frame: path overhead*

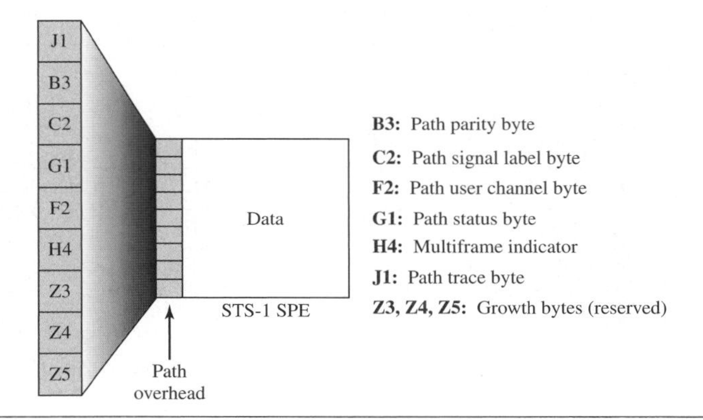

- **Path parity byte (B3).** Byte B3 is for bit interleaved parity, like bytes B1 and B2, but calculated over SPE bits. It is actually calculated over the previous SPE in the stream.
- **Path signal label byte (C2).** Byte C2 is the path identification byte. It is used to identify different protocols used at higher levels (such as IP or ATM) whose data are being carried in the SPE.
- **Path user channel byte (F2).** The F2 bytes in consecutive frames, like the F1 bytes, form a 64-kbps channel that is reserved for user needs, but at the path level.
- **Path status byte (G1).** Byte G1 is sent by the receiver to communicate its status to the sender. It is sent on the reverse channel when the communication is duplex. We will see its use in the linear or ring networks later in the chapter.
- **Multiframe indicator (H4).** Byte H4 is the multiframe indicator. It indicates payloads that cannot fit into a single frame. For example, virtual tributaries can be

combined to form a frame that is larger than an SPE frame and needs to be divided into different frames. Virtual tributaries are discussed in Section 3.6.

- **Path trace byte (J1).** The J1 bytes in consecutive frames form a 64-kbps channel used for tracking the path. The J1 byte sends a continuous 64-byte string to verify the connection. The choice of the string is left to the application program. The receiver compares each pattern with the previous one to ensure nothing is wrong with the communication at the path layer.
- **Growth bytes (Z3, Z4, and Z5).** Bytes Z3, Z4, and Z5 are reserved for future use.

Path overhead is only calculated for end-to-end (at STS multiplexers).

Overhead Summary

Table 14.2 compares and summarizes the overheads used in a section, line, and path.

Table 14.2 *SONET/SDH rates*

Byte Function	*Section*	*Line*	*Path*
Alignment	A1, A2		
Parity	B1	B2	B3
Identifier	C1		C2
OA&M	D1–D3	D4–D12	
Order wire	E1		
User	F1		F2
Status			G1
Pointers		H1– H3	H4
Trace			J1
Failure tolerance		K1, K2	
Growth (reserved for future)		Z1, Z2	Z3–Z5

Example 14.4

What is the user data rate of an STS-1 frame (without considering the overheads)?

Solution

The user data part of an STS-1 frame is made of 9 rows and 86 columns. So we have

$$\text{STS-1 user data rate} = 8000 \times 9 \times (1 \times 86) \times 8 = 49.536 \text{ Mbps}$$

Encapsulation

The previous discussion reveals that an SPE needs to be encapsulated in an STS-1 frame. Encapsulation may create two problems that are handled elegantly by SONET using pointers (H1 to H3). We discuss the use of these bytes in this section.

Offsetting

SONET allows one SPE to span two frames; part of the SPE is in the first frame and part is in the second. This may happen when one SPE that is to be encapsulated is not aligned timewise with the passing synchronized frames. Figure 14.23 shows this situation. SPE bytes are divided between the two frames. The first set of bytes is encapsulated in the first frame; the second set is encapsulated in the second frame. The figure also shows the path overhead, which is aligned with the section/line overhead of any frame. The question is, how does the SONET multiplexer know where the SPE starts or ends in the frame? The solution is the use of pointers H1 and H2 to define the beginning of the SPE; the end can be found because each SPE has a fixed number of bytes. SONET allows the offsetting of an SPE with respect to an STS-1 frame.

Figure 14.23 *Offsetting of SPE related to frame boundary*

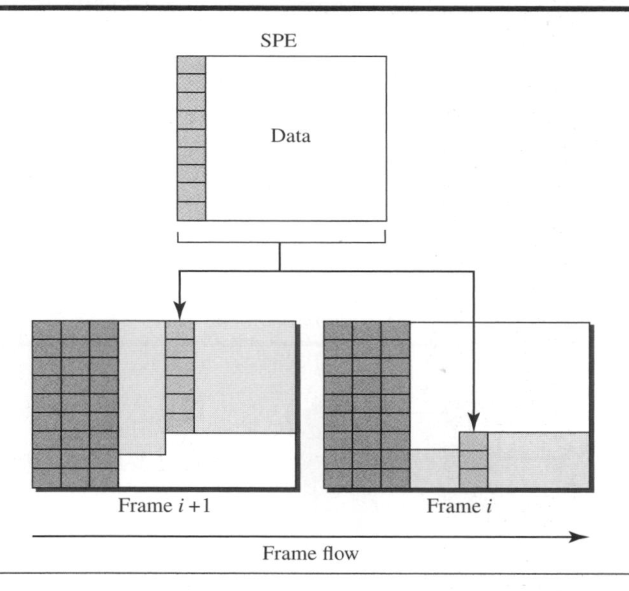

To find the beginning of each SPE in a frame, we need two pointers, H1 and H2, in the line overhead. Note that these pointers are located in the line overhead because the encapsulation occurs at a multiplexer. Figure 14.24 shows how these 2 bytes point to

Figure 14.24 *The use of H1 and H2 pointers to show the start of an SPE in a frame*

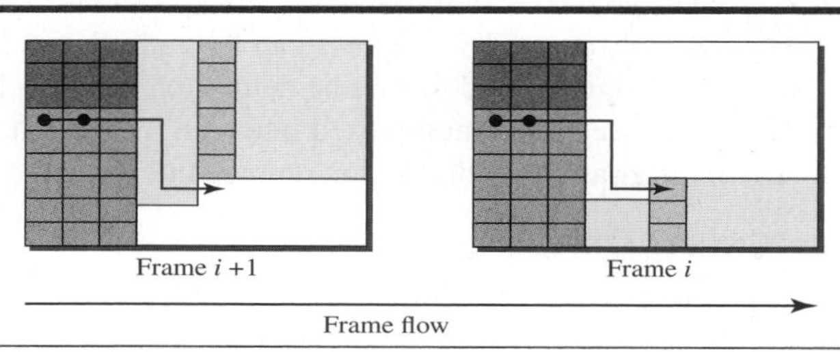

the beginning of the SPEs. Note that we need 2 bytes to define the position of a byte in a frame; a frame has 810 bytes, which cannot be defined using 1 byte.

Example 14.5

What are the values of H1 and H2 if an SPE starts at byte number 650?

Solution

The number 650 can be expressed in four hexadecimal digits as 0x028A. This means the value of H1 is 0x02 and the value of H2 is 0x8A.

Justification

Now suppose the transmission rate of the payload is just slightly different from the transmission rate of SONET. First, assume that the rate of the payload is higher. This means that occasionally there is 1 extra byte that cannot fit in the frame. In this case, SONET allows this extra byte to be inserted in the H3 byte. Now, assume that the rate of the payload is lower. This means that occasionally 1 byte needs to be left empty in the frame. SONET allows this byte to be the byte after the H3 byte.

14.3.4 STS Multiplexing

In SONET, frames of lower rate can be synchronously time-division multiplexed into a higher-rate frame. For example, three STS-1 signals (channels) can be combined into one STS-3 signal (channel), four STS-3s can be multiplexed into one STS-12, and so on, as shown in Figure 14.25.

Figure 14.25 *STS multiplexing/demultiplexing*

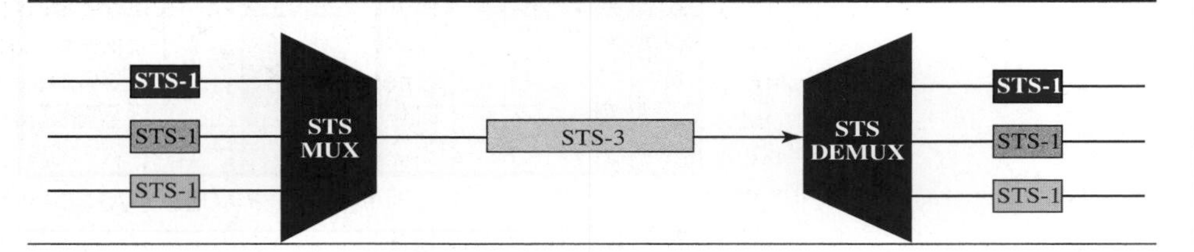

Multiplexing is synchronous TDM, and all clocks in the network are locked to a master clock to achieve synchronization.

In SONET, all clocks in the network are locked to a master clock.

We need to mention that multiplexing can also take place at the higher data rates. For example, four STS-3 signals can be multiplexed into an STS-12 signal. However, the STS-3 signals need to first be demultiplexed into 12 STS-1 signals, and then these 12 signals need to be multiplexed into an STS-12 signal. The reason for this extra work will be clear after our discussion on byte interleaving.

Byte Interleaving

Synchronous TDM multiplexing in SONET is achieved by using **byte interleaving.** For example, when three STS-1 signals are multliplexed into one STS-3 signal, each set of 3 bytes in the STS-3 signal is associated with 1 byte from each STS-1 signal. Figure 14.26 shows the interleaving.

Figure 14.26 *Byte interleaving*

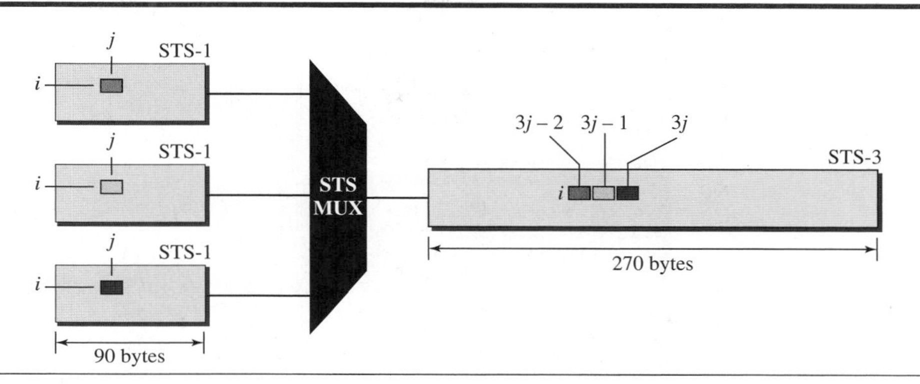

Note that a byte in an STS-1 frame keeps its row position, but it is moved into a different column. The reason is that while all signal frames have the same number of rows (9), the number of columns changes. The number of columns in an STS-*n* signal frame is *n* times the number of columns in an STS-1 frame. One STS-*n* row, therefore, can accommodate all *n* rows in the STS-1 frames.

Byte interleaving also preserves the corresponding section and line overhead, as shown in Figure 14.27. As the figure shows, the section overheads from three STS-1 frames are interleaved together to create a section overhead for an STS-1 frame. The same is true for the line overheads. Each channel, however, keeps the corresponding bytes that are used to control that channel. In other words, the sections and lines keep their own control bytes for each multiplexed channel. This interesting feature will allow the use of add/drop multiplexers, as discussed shortly. As the figure shows, there are three A1 bytes, one belonging to each of the three multiplexed signals. There are also three A2 bytes, three B1 bytes, and so on.

Demultiplexing here is easier than in the statistical TDM we discussed in Chapter 6 because the demultiplexer, with no regard to the function of the bytes, removes the first A1 and assigns it to the first STS-1, removes the second A1, and assigns it to second STS-1, and removes the third A1 and assigns it to the third STS-1. In other words, the demultiplexer deals only with the position of the byte, not its function.

What we said about the section and line overheads does not exactly apply to the path overhead. This is because the path overhead is part of the SPE that may have split into two STS-1 frames. The byte interleaving, however, is the same for the data section of SPEs.

The byte interleaving process makes the multiplexing at higher data rates a little bit more complex. How can we multiplex four STS-3 signals into one STS-12 signal? This can be done in two steps: First, the STS-3 signals must be demultiplexed to create 12 STS-1 signals. The 12 STS-1 signals are then multiplexed to create an STS-12 signal.

Figure 14.27 *An STS-3 frame*

9 bytes | 261 bytes

Section overhead

A1	A1	A1	A2	A2	A2	C1	C1	C1
B1	B1	B1	E1	E1	E1	F1	F1	F1
D1	D1	D1	D2	D2	D2	D3	D3	D3

Line overhead

H1	H1	H1	H2	H2	H2	H3	H3	H3
B2	B2	B2	K1	K1	K1	K2	K2	K2
D4	D4	D4	D5	D5	D5	D6	D6	D6
D7	D7	D7	D8	D8	D8	D9	D9	D9
D10	D10	D10	D11	D11	D11	D12	D12	D12
Z1	Z1	Z1	Z2	Z2	Z2	E2	E2	E2

STS-3 SPE

Interleaving bytes of three SPEs

Concatenated Signal

In normal operation of the SONET, an STS-*n* signal is made of *n* multiplexed STS-1 signals. Sometimes, we have a signal with a data rate higher than an STS-1 can carry. In this case, SONET allows us to create an STS-*n* signal that is not considered as *n* STS-1 signals; it is one STS-*n* signal (channel) that cannot be demultiplexed into *n* STS-1 signals. To specify that the signal cannot be demultiplexed, the suffix *c* (for *concatenated*) is added to the name of the signal. For example, STS-3c is a signal that cannot be demultiplexed into three STS-1 signals. However, we need to know that the whole payload in an STS-3c signal is one SPE, which means that we have only one column (9 bytes) of path overhead. The used data in this case occupy 260 columns, as shown in Figure 14.28.

Figure 14.28 *A concatenated STS-3c signal*

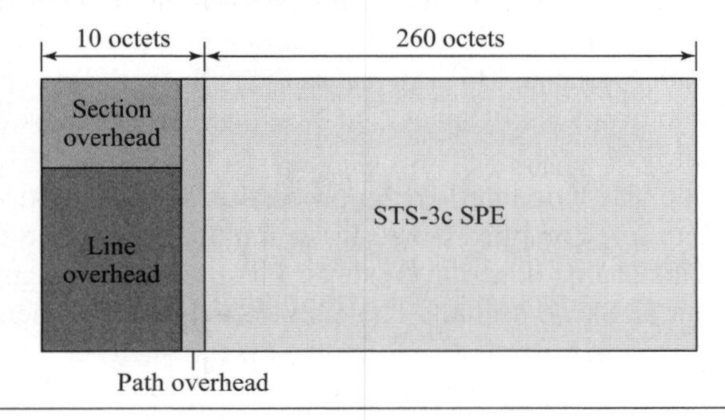

Concatenated Signals Carrying ATM Cells

We will discuss ATM and ATM cells later in the chapter (Section 14.4). An ATM network is a cell network in which each cell has a fixed size of 53 bytes. The SPE of an STS-3c signal can be a carrier of ATM cells. The SPE of an STS-3c can carry $9 \times 260 = 2340$ bytes, which can accommodate approximately 44 ATM cells, each of 53 bytes.

An STS-3c signal can carry 44 ATM cells as its SPE.

Add/Drop Multiplexer

Multiplexing of several STS-1 signals into an STS-*n* signal is done at the STS multiplexer (at the path layer). Demultiplexing of an STS-*n* signal into STS-1 components is done at the STS demultiplexer. In between, however, SONET uses add/drop multiplexers that can replace one signal with another. We need to know that this is not demultiplexing/multiplexing in the conventional sense. An add/drop multiplexer operates at the line layer. An add/drop multiplexer does not create section, line, or path overhead. It almost acts as a switch; it removes one STS-1 signal and adds another one. The type of signal at the input and output of an add/drop multiplexer is the same (both STS-3 or both STS-12, for example). The add/drop multiplexer (ADM) only removes the corresponding bytes and replaces them with the new bytes (including the bytes in the section and line overhead). Figure 14.29 shows the operation of an ADM.

Figure 14.29 *Dropping and adding STS-1 frames in an add/drop multiplexer*

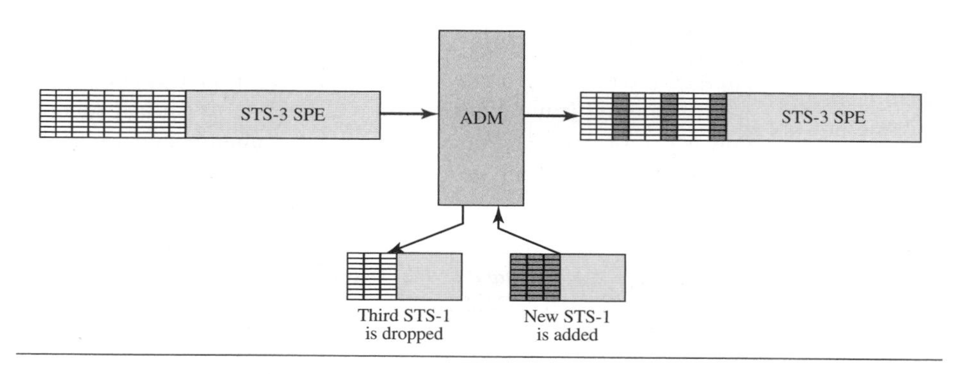

14.3.5 SONET Networks

Using SONET equipment, we can create a SONET network that can be used as a high-speed backbone carrying loads from other networks such as ATM (Section 14.4) or IP (Chapter 19). We can roughly divide SONET networks into three categories: linear, ring, and mesh networks, as shown in Figure 14.30.

Linear Networks

A linear SONET network can be point-to-point or multipoint.

Figure 14.30 *Taxonomy of SONET networks*

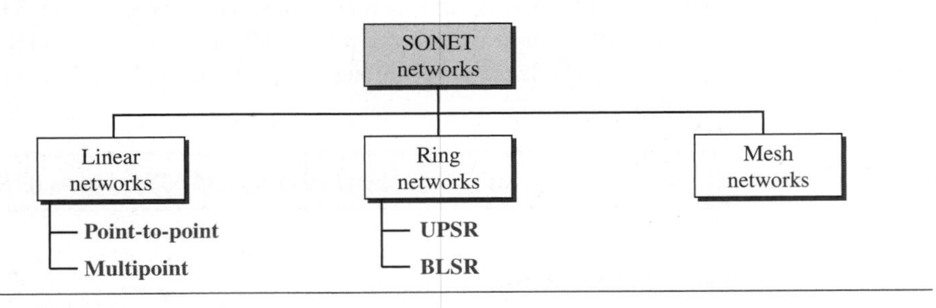

Point-to-Point Network

A point-to-point network is normally made of an STS multiplexer, an STS demultiplexer, and zero or more regenerators with no add/drop multiplexers, as shown in Figure 14.31. The signal flow can be unidirectional or bidirectional, although Figure 14.31 shows only unidirectional for simplicity.

Figure 14.31 *A point-to-point SONET network*

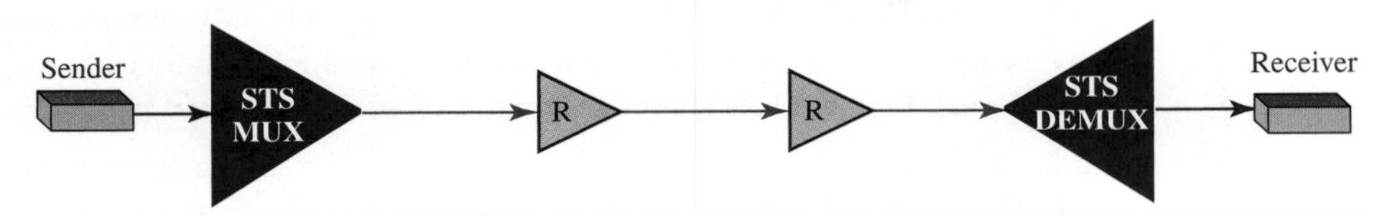

Multipoint Network

A multipoint network uses ADMs to allow communications between several terminals. An ADM removes the signal belonging to the terminal connected to it and adds the signal transmitted from another terminal. Each terminal can send data to one or more downstream terminals. Figure 14.32 shows a unidirectional scheme in which each terminal can send data only to the downstream terminals, but a multipoint network can be bidirectional, too.

Figure 14.32 *A multipoint SONET network*

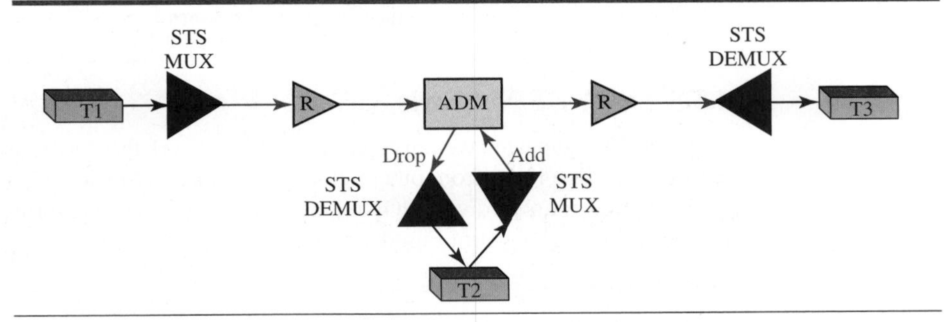

In Figure 14.32, T1 can send data to T2 and T3 simultaneously. T2, however, can send data only to T3. The figure shows a very simple configuration; in normal situations, we have more ADMs and more terminals.

Automatic Protection Switching

To create protection against failure in linear networks, SONET defines **automatic protection switching (APS).** APS in linear networks is defined at the line layer, which means the protection is between two ADMs or a pair of STS multiplexer/demultiplexers. The idea is to provide redundancy; a redundant line (fiber) can be used in case of failure in the main one. The main line is referred to as the *work line* and the redundant line as the *protection line*. Three schemes are common for protection in linear channels: one-plus-one, one-to-one, and one-to-many. Figure 14.33 shows all three schemes.

Figure 14.33 *Automatic protection switching in linear networks*

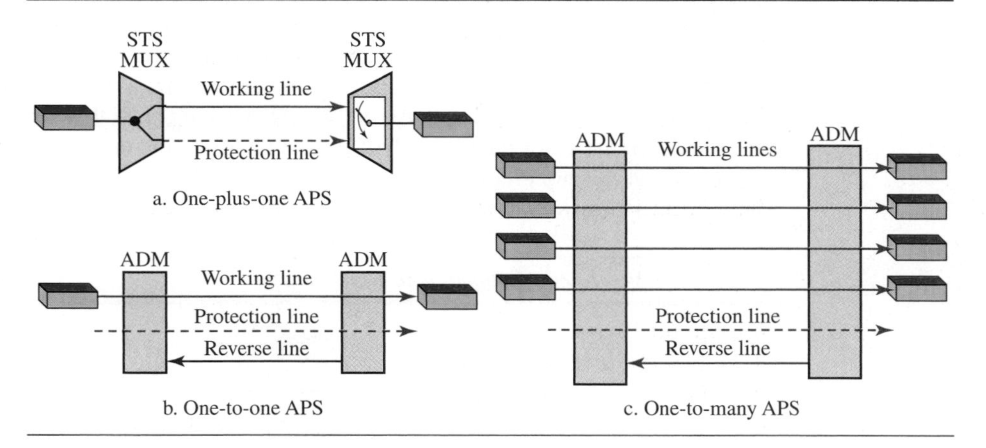

One-Plus-One APS In this scheme, there are normally two lines: one working line and one protection line. Both lines are active all the time. The sending multiplexer sends the same data on both lines; the receiver multiplexer monitors the line and chooses the one with the better quality. If one of the lines fails, it loses its signal, and, of course, the other line is selected at the receiver. Although the failure recovery for this scheme is instantaneous, the scheme is inefficient because two times the bandwidth is required. Note that one-plus-one switching is done at the path layer.

One-to-One APS In this scheme, which looks like the one-plus-one scheme, there is also one working line and one protection line. However, the data are normally sent on the working line until it fails. At this time, the receiver, using the reverse channel, informs the sender to use the protection line instead. Obviously, the failure recovery is slower than that of the one-plus-scheme, but this scheme is more efficient because the protection line can be used for data transfer when it is not used to replace the working line. Note that one-to-one switching is done at the line layer.

One-to-Many APS This scheme is similar to the one-to-one scheme except that there is only one protection line for many working lines. When a failure occurs in one of the working lines, the protection line takes control until the failed line is repaired. It is not as secure as the one-to-one scheme because if more than one working line fails at the same time, the protection line can replace only one of them. Note that one-to-many APS is done at the line layer.

Ring Networks

ADMs make it possible to have SONET ring networks. SONET rings can be used in either a unidirectional or a bidirectional configuration. In each case, we can add extra rings to make the network self-healing, capable of self-recovery from line failure.

Unidirectional Path Switching Ring

A **unidirectional path switching ring (UPSR)** is a unidirectional network with two rings: one ring used as the working ring and the other as the protection ring. The idea is similar to the one-plus-one APS scheme we discussed in a linear network. The same signal flows through both rings, one clockwise and the other counterclockwise. It is called UPSR because monitoring is done at the path layer. A node receives two copies of the electrical signals at the path layer, compares them, and chooses the one with the better quality. If part of a ring between two ADMs fails, the other ring still can guarantee the continuation of data flow. UPSR, like the one-plus-one scheme, has fast failure recovery, but it is not efficient because we need to have two rings that do the job of one. Half of the bandwidth is wasted. Figure 14.34 shows a UPSR network.

Figure 14.34 *A unidirectional path switching ring*

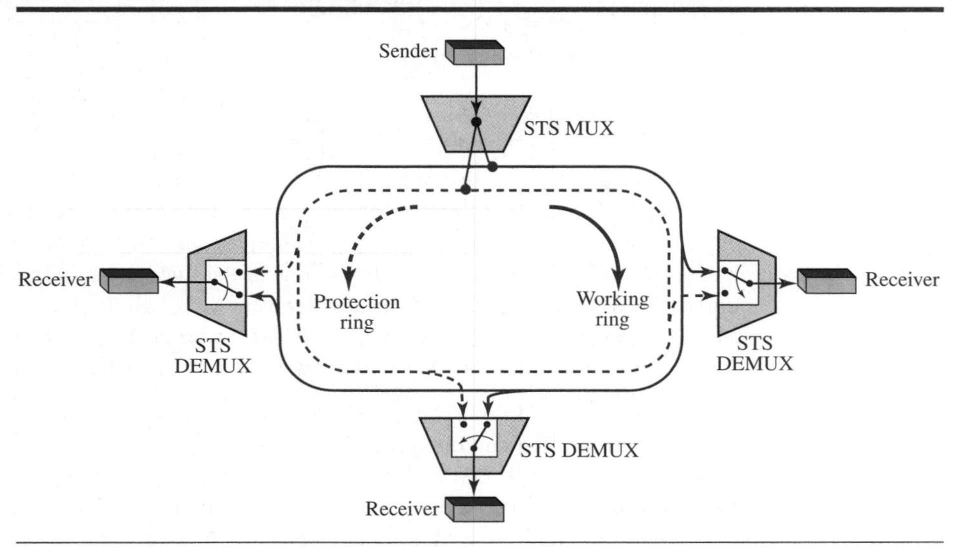

Although we have chosen one sender and three receivers in the figure, there can be many other configurations. The sender uses a two-way connection to send data to both rings simultaneously; the receiver uses selecting switches to select the ring with better signal quality. We have used one STS multiplexer and three STS demultiplexers to emphasize that nodes operate on the path layer.

Bidirectional Line Switching Ring

Another alternative in a SONET ring network is a **bidirectional line switching ring (BLSR).** In this case, communication is bidirectional, which means that we need two rings for working lines. We also need two rings for protection lines. This means BLSR uses four rings. The operation, however, is similar to the one-to-one APS scheme. If a working ring in one direction between two nodes fails, the receiving node can use the reverse ring to inform the upstream node in the failed direction to use the protection ring. The network can recover in several different failure situations that we do not discuss here. Note that the discovery of a failure in BLSR is at the line layer, not the path layer. The ADMs find the failure and inform the adjacent nodes to use the protection rings. Figure 14.35 shows a BLSR ring.

Figure 14.35 *A bidirectional line switching ring*

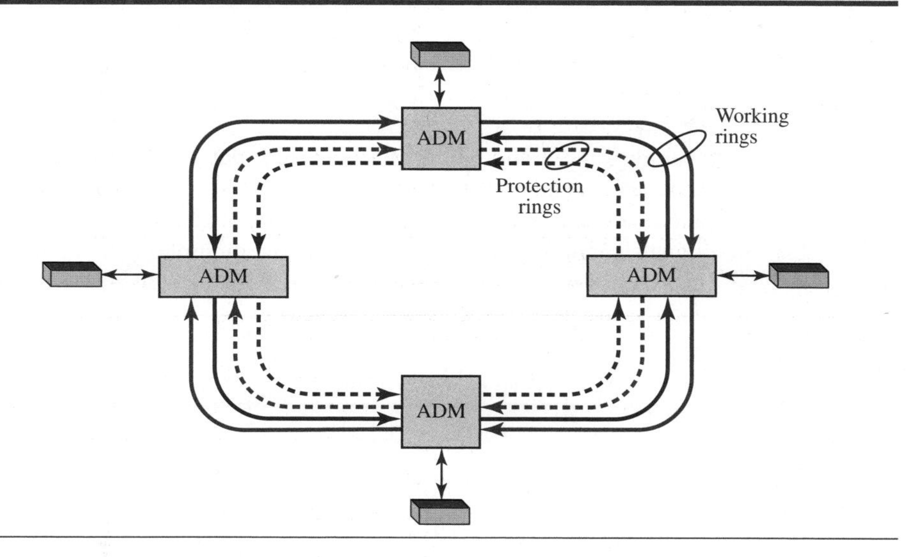

Combination of Rings

SONET networks today use a combination of interconnected rings to create services in a wide area. For example, a SONET network may have a regional ring, several local rings, and many site rings to give services to a wide area. These rings can be UPSR, BLSR, or a combination of both. Figure 14.36 shows the idea of such a wide-area ring network.

Mesh Networks

One problem with ring networks is the lack of scalability. When the traffic in a ring increases, we need to upgrade not only the lines, but also the ADMs. In this situation, a mesh network with switches would probably give better performance. A switch in a network mesh is called a cross-connect. A cross-connect, like other switches we have

Figure 14.36 *A combination of rings in a SONET network*

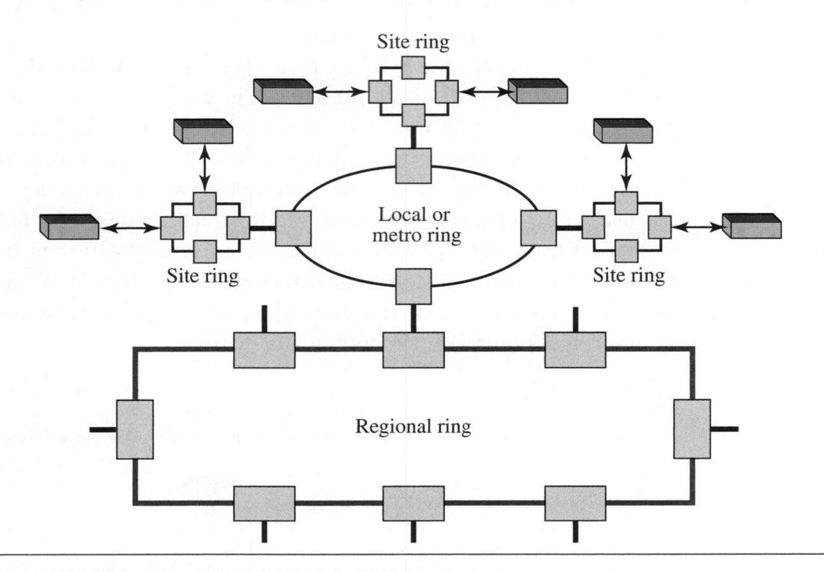

seen, has input and output ports. In an input port, the switch takes an OC-*n* signal, changes it to an STS-*n* signal, demultiplexes it into the corresponding STS-1 signals, and sends each STS-1 signal to the appropriate output port. An output port takes STS-1 signals coming from different input ports, multiplexes them into an STS-*n* signal, and makes an OC-*n* signal for transmission. Figure 14.37 shows a mesh SONET network, and the structure of a switch.

Figure 14.37 *A mesh SONET network*

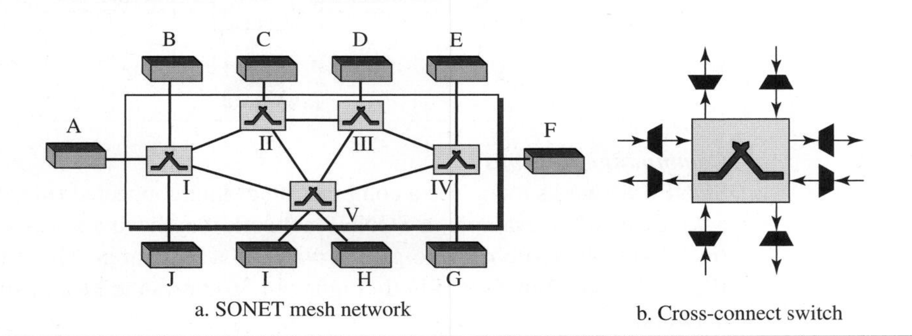

14.3.6 Virtual Tributaries

SONET is designed to carry broadband payloads. Current digital hierarchy data rates (DS-1 to DS-3), however, are lower than STS-1. To make SONET backward-compatible with the current hierarchy, its frame design includes a system of **virtual tributaries (VTs)** (see Figure 14.38). A virtual tributary is a partial payload that can be inserted

Figure 14.38 *Virtual tributaries*

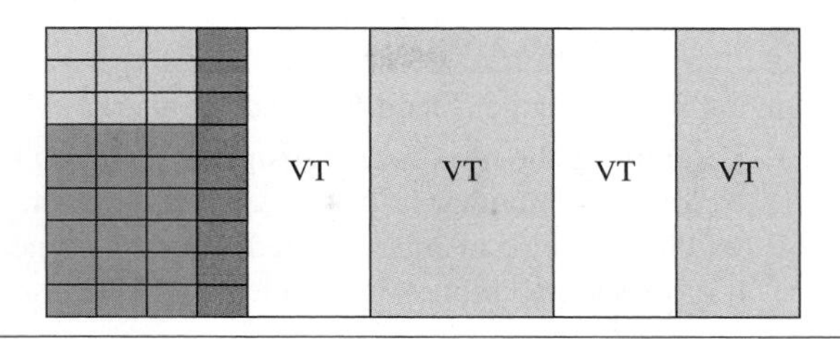

into an STS-1 and combined with other partial payloads to fill out the frame. Instead of using all 86 payload columns of an STS-1 frame for data from one source, we can subdivide the SPE and call each component a VT.

Types of VTs

Four types of VTs have been defined to accommodate existing digital hierarchies (see Figure 14.39). Notice that the number of columns allowed for each type of VT can be

Figure 14.39 *Virtual tributary types*

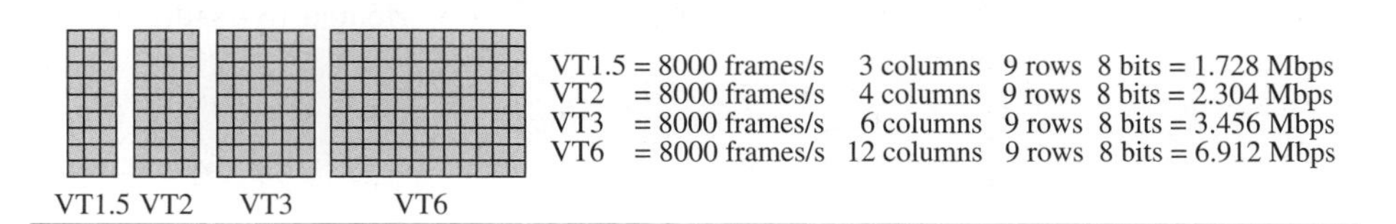

determined by doubling the type identification number (VT1.5 gets three columns, VT2 gets four columns, etc.).

- **VT1.5** accommodates the U.S. DS-1 service (1.544 Mbps).
- **VT2** accommodates the European CEPT-1 service (2.048 Mbps).
- **VT3** accommodates the DS-1C service (fractional DS-1, 3.152 Mbps).
- **VT6** accommodates the DS-2 service (6.312 Mbps).

When two or more tributaries are inserted into a single STS-1 frame, they are interleaved column by column. SONET provides mechanisms for identifying each VT and separating them without demultiplexing the entire stream. Discussion of these mechanisms and the control issues behind them is beyond the scope of this book.

14.4 ATM

Asynchronous Transfer Mode (ATM) is a switched wide area network based on the *cell relay* protocol designed by the ATM forum and adopted by the ITU-T. The combination of ATM and SONET will allow high-speed interconnection of all the world's networks. In fact, ATM can be thought of as the "highway" of the information superhighway.

14.4.1 Design Goals

Among the challenges faced by the designers of ATM, six stand out.

1. Foremost is the need for a transmission system to optimize the use of high-data-rate transmission media, in particular optical fiber. In addition to offering large bandwidths, newer transmission media and equipment are dramatically less susceptible to noise degradation. A technology is needed to take advantage of both factors and thereby maximize data rates.
2. The system must interface with existing systems and provide wide-area interconnectivity between them without lowering their effectiveness or requiring their replacement.
3. The design must be implemented inexpensively so that cost would not be a barrier to adoption. If ATM is to become the backbone of international communications, as intended, it must be available at low cost to every user who wants it.
4. The new system must be able to work with and support the existing telecommunications hierarchies (local loops, local providers, long-distance carriers, and so on).
5. The new system must be connection-oriented to ensure accurate and predictable delivery.
6. Last but not least, one objective is to move as many of the functions to hardware as possible (for speed) and eliminate as many software functions as possible (again for speed).

14.4.2 Problems

Before we discuss the solutions to these design requirements, it is useful to examine some of the problems associated with existing systems.

Frame Networks

Before ATM, data communications at the data-link layer had been based on frame switching and frame networks. Different protocols use frames of varying size and intricacy. As networks become more complex, the information that must be carried in the header becomes more extensive. The result is larger and larger headers relative to the size of the data unit. In response, some protocols have enlarged the size of the data unit to make header use more efficient (sending more data with the same size header). Unfortunately, large data fields create waste. If there is not much information to transmit, much of the field goes unused. To improve utilization, some protocols provide variable frame sizes to users.

Mixed Network Traffic

As you can imagine, the variety of frame sizes makes traffic unpredictable. Switches, multiplexers, and routers must incorporate elaborate software systems to manage the various sizes of frames. A great deal of header information must be read, and each bit counted and evaluated to ensure the integrity of every frame. Internetworking among the different frame networks is slow and expensive at best, and impossible at worst.

Another problem is that of providing consistent data rate delivery when frame sizes are unpredictable and can vary so dramatically. To get the most out of broadband technology, traffic must be time-division multiplexed onto shared paths. Imagine the results of multiplexing frames from two networks with different requirements (and frame designs) onto one link (see Figure 14.40). What happens when line 1 uses large frames (usually data frames) while line 2 uses very small frames (the norm for audio and video information)?

Figure 14.40 *Multiplexing using different frame sizes*

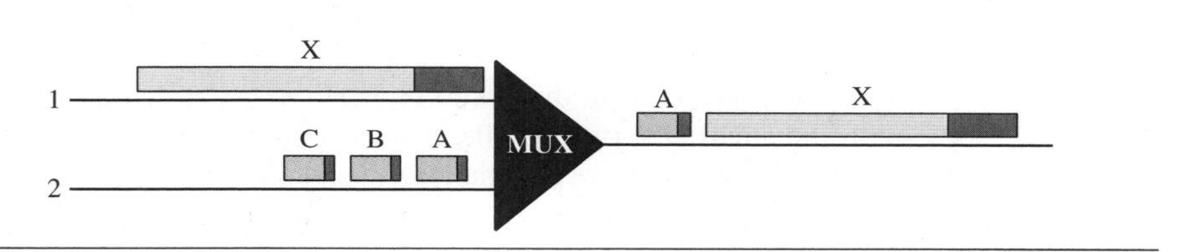

If line 1's gigantic frame X arrives at the multiplexer even a moment earlier than line 2's frames, the multiplexer puts frame X onto the new path first. After all, even if line 2's frames have priority, the multiplexer has no way of knowing to wait for them and so processes the frame that has arrived. Frame A must therefore wait for the entire X bit stream to move into place before it can follow. The sheer size of X creates an unfair delay for frame A. The same imbalance can affect all the frames from line 2.

Because audio and video frames ordinarily are small, mixing them with conventional data traffic often creates unacceptable delays of this type and makes shared frame links unusable for audio and video information. Traffic must travel over different paths, in much the same way that automobile and train traffic does. But to fully utilize broad bandwidth links, we need to be able to send all kinds of traffic over the same links.

Cell Networks

Many of the problems associated with frame internetworking are solved by adopting a concept called *cell networking*. A **cell** is a small data unit of fixed size. In a **cell network,** which uses the cell as the basic unit of data exchange, all data are loaded into identical cells that can be transmitted with complete predictability and uniformity. As frames of different sizes and formats reach the cell network from a tributary network, they are split into multiple small data units of equal length and are loaded into cells. The cells are then multiplexed with other cells and routed through the cell network. Because each cell is the same size and all are small, the problems associated with multiplexing different-sized frames are avoided.

A cell network uses the cell as the basic unit of data exchange.
A cell is defined as a small, fixed-size block of information.

Figure 14.41 shows the multiplexer from Figure 14.40 with the two lines sending cells instead of frames. Frame X has been segmented into three cells: X, Y, and Z. Only the first cell from line 1 gets put on the link before the first cell from line 2. The cells from the two lines are interleaved so that none suffers a long delay.

Figure 14.41 *Multiplexing using cells*

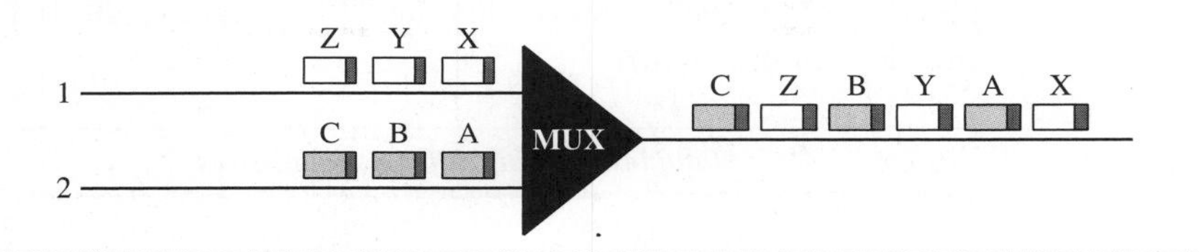

A second point in this same scenario is that the high speed of the links coupled with the small size of the cells means that, despite interleaving, cells from each line arrive at their respective destinations in an approximation of a continuous stream (much as a movie appears to your brain to be continuous action when in fact it is really a series of separate, still photographs). In this way, a cell network can handle real-time transmissions, such as a phone call, without the parties being aware of the segmentation or multiplexing at all.

Asynchronous TDM

ATM uses statistical (asynchronous) time-division multiplexing—that is why it is called *Asynchronous Transfer Mode*—to multiplex cells coming from different channels. It uses fixed-size slots (size of a cell). ATM multiplexers fill a slot with a cell from any input channel that has a cell; the slot is empty if none of the channels has a cell to send. Figure 14.42 shows how cells from three inputs are multiplexed. At the first tick of the clock, channel 2 has no cell (empty input slot), so the multiplexer fills the slot with a cell from the third channel. When all the cells from all the channels are multiplexed, the output slots are empty.

Figure 14.42 *ATM multiplexing*

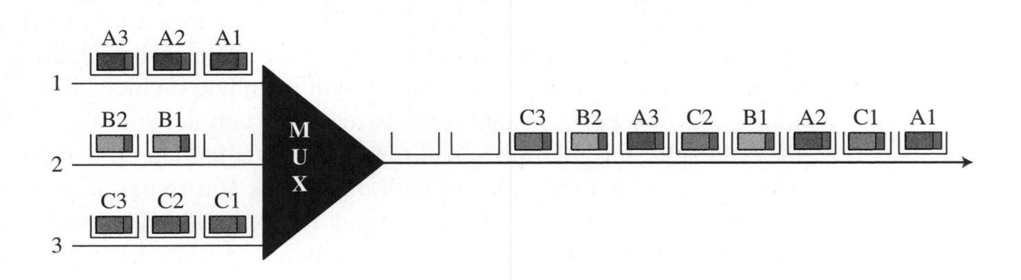

14.4.3 Architecture

ATM is a cell-switched network. The user access devices, called the endpoints, are connected through a **user-to-network interface (UNI)** to the switches inside the network. The switches are connected through **network-to-network interfaces (NNIs).** Figure 14.43 shows an example of an ATM network.

Figure 14.43 *Architecture of an ATM network*

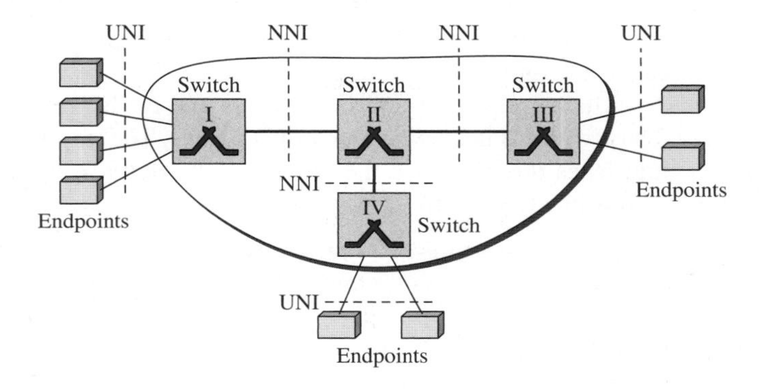

Virtual Connection

Connection between two endpoints is accomplished through transmission paths, virtual paths, and virtual circuits. A **transmission path (TP)** is the physical connection (wire, cable, satellite, and so on) between an endpoint and a switch or between two switches. Think of two switches as two cities. A transmission path is the set of all highways that directly connect the two cities.

A transmission path is divided into several virtual paths. A **virtual path (VP)** provides a connection or a set of connections between two switches. Think of a virtual path as a highway that connects two cities. Each highway is a virtual path; the set of all highways is the transmission path.

Cell networks are based on **virtual circuits (VCs).** All cells belonging to a single message follow the same virtual circuit and remain in their original order until they reach their destination. Think of a virtual circuit as the lanes of a highway (virtual path). Figure 14.44 shows the relationship between a transmission path (a physical connection), virtual paths (a combination of virtual circuits that are bundled together because parts of their paths are the same), and virtual circuits that logically connect two points.

Identifiers

In a virtual circuit network, to route data from one endpoint to another, the virtual connections need to be identified. For this purpose, the designers of ATM created a hierarchical identifier with two levels: a virtual-path identifier (VPI) and a virtual-circuit identifier (VCI). The VPI defines the specific VP, and the VCI defines a particular VC

Figure 14.44 *TP, VPs, and VCs*

inside the VP. The VPI is the same for all virtual connections that are bundled (logically) into one VP.

The lengths of the VPIs for UNIs and NNIs are different. In a UNI, the VPI is 8 bits, whereas in an NNI, the VPI is 12 bits. The length of the VCI is the same in both interfaces (16 bits). We therefore can say that a virtual connection is identified by 24 bits in a UNI and by 28 bits in an NNI (see Figure 14.45).

Figure 14.45 *Virtual connection identifiers in UNIs and NNIs*

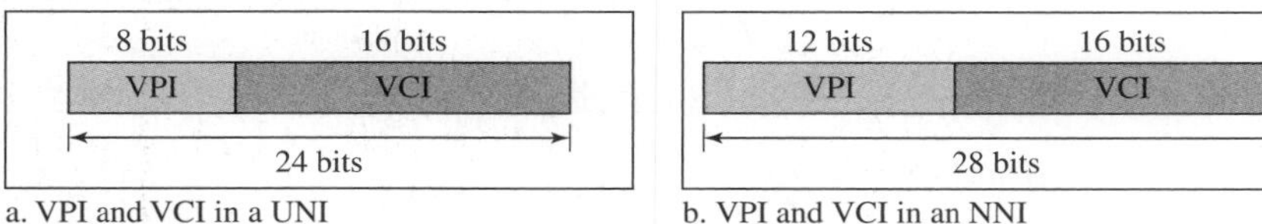

a. VPI and VCI in a UNI

b. VPI and VCI in an NNI

The whole idea behind dividing a virtual circuit identifier into two parts is to allow hierarchical routing. Most of the switches in a typical ATM network are routed using VPIs. The switches at the boundaries of the network, those that interact directly with the endpoint devices, use both VPIs and VCIs.

Cells

The basic data unit in an ATM network is called a cell. A cell is only 53 bytes long with 5 bytes allocated to the header and 48 bytes carrying the payload (user data may be less than 48 bytes). Most of the header is occupied by the VPI and VCI that define the virtual connection through which a cell should travel from an endpoint to a switch or from a switch to another switch. Figure 14.46 shows the cell structure.

Figure 14.46 *An ATM cell*

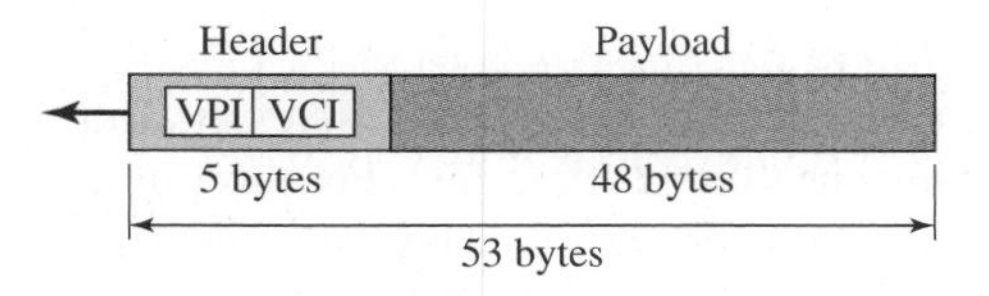

Connection Establishment and Release

ATM uses two types of connections: PVC and SVC.

PVC

A **permanent virtual-circuit connection (PVC)** is established between two endpoints by the network provider. The VPIs and VCIs are defined for the permanent connections, and the values are entered for the tables of each switch.

SVC

In a **switched virtual-circuit connection (SVC),** each time an endpoint wants to make a connection with another endpoint, a new virtual circuit must be established. ATM cannot do the job by itself, but needs the network-layer addresses and the services of another protocol (such as IP). The signaling mechanism of this other protocol makes a connection request by using the network-layer addresses of the two endpoints. The actual mechanism depends on the network-layer protocol.

Switching

ATM uses switches to route the cell from a source endpoint to the destination endpoint. A switch routes the cell using both the VPIs and the VCIs. The routing requires the whole identifier. Figure 14.47 shows how a PVC switch routes the cell. A cell with a VPI of 153 and VCI of 67 arrives at switch interface (port) 1. The switch checks its switching table, which stores six pieces of information per row: arrival interface number, incoming VPI, incoming VCI, corresponding outgoing interface number, the new VPI, and the new VCI. The switch finds the entry with interface 1, VPI 153, and VCI 67 and discovers that the combination corresponds to output interface 3, VPI 140, and VCI 92. It changes the VPI and VCI in the header to 140 and 92, respectively, and sends the cell out through interface 3.

Figure 14.47 *Routing with a switch*

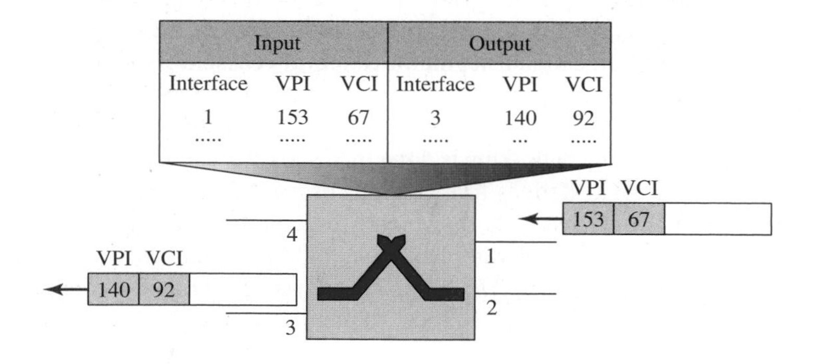

ATM Layers

The ATM standard defines three layers. They are, from top to bottom, the application adaptation layer, the ATM layer, and the physical layer (see Figure 14.48). The endpoints use all three layers while the switches use only the two bottom layers.

Figure 14.48 *ATM layers*

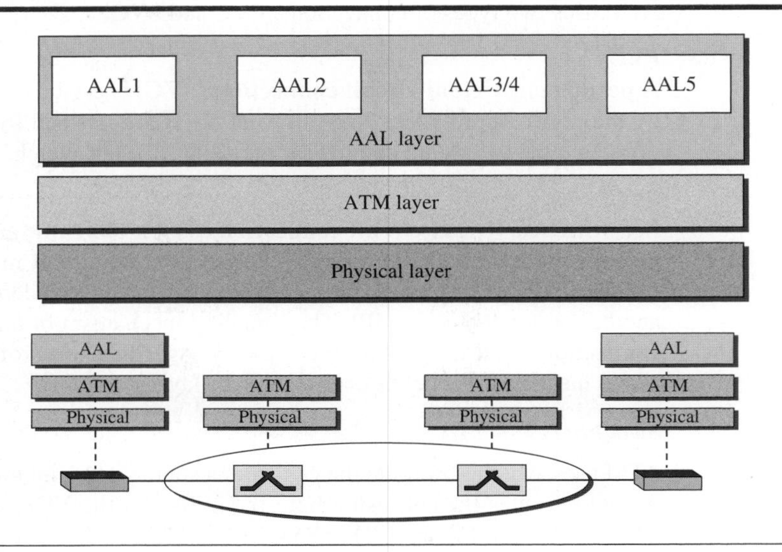

AAL Layer

The **application adaptation layer (AAL)** was designed to enable two ATM concepts. First, ATM must accept any type of payload, both data frames and streams of bits. A data frame can come from an upper-layer protocol that creates a clearly defined frame to be sent to a carrier network such as ATM. A good example is the Internet. ATM must also carry multimedia payloads. It can accept continuous bit streams and break them into chunks to be encapsulated into a cell at the ATM layer. AAL uses two sublayers to accomplish these tasks.

Whether the form of the data is a data frame or a stream of bits, the payload must be segmented into 48-byte segments to be carried by a cell. At the destination, these segments need to be reassembled to recreate the original payload. The AAL defines a sublayer, called a **segmentation and reassembly (SAR)** sublayer, to do so. Segmentation is at the source; reassembly, at the destination.

Before data are segmented by SAR, they must be prepared to guarantee the integrity of the data. This is done by a sublayer called the ***convergence sublayer (CS).***

ATM defines four versions of the AAL: *AAL1*, *AAL2*, *AAL3/4*, and *AAL5*. We discuss only AAL5 here, which is used in the Internet today.

For Internet applications, the AAL5 sublayer was designed. It is also called the ***simple and efficient adaptation layer (SEAL).*** AAL5 assumes that all cells belonging to a single message travel sequentially and that control functions are included in the upper layers of the sending application. Figure 14.49 shows the AAL5 sublayer.

The packet at the CS uses a trailer with four fields. The UU is the user-to-user identifier. The CPI is the common part identifier. The L field defines the length of the original data. The CRC field is a two-byte error-checking field for the entire data.

Figure 14.49 *AAL5*

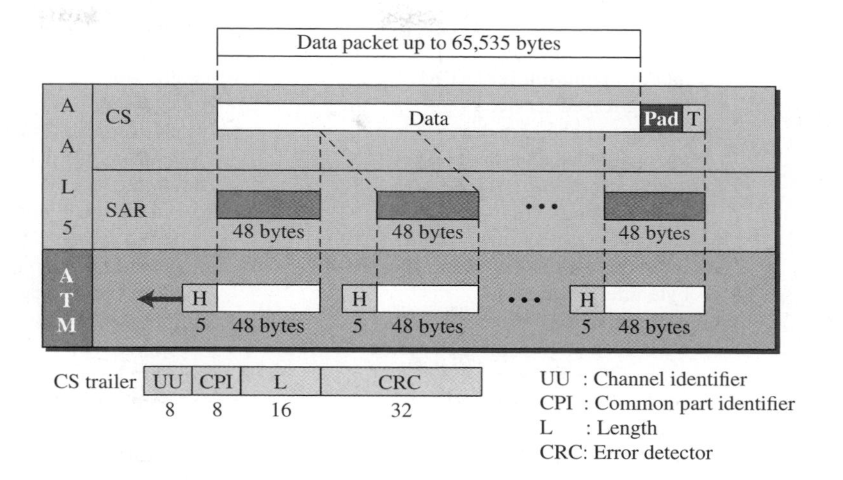

ATM Layer

The ATM layer provides routing, traffic management, switching, and multiplexing services. It processes outgoing traffic by accepting 48-byte segments from the AAL sublayers and transforming them into 53-byte cells by the addition of a 5-byte header.

Physical Layer

Like Ethernet and wireless LANs, ATM cells can be carried by any physical-layer carrier.

Congestion Control and Quality of Service

ATM has a very developed congestion control and quality of service.

14.5 END-CHAPTER MATERIALS

14.5.1 Recommended Reading

For more details about subjects discussed in this chapter, we recommend the following books. The items in brackets [...] refer to the reference list at the end of the text.

Books

Several excellent books discuss link-layer issues. Among them we recommend [Ham 80], [Zar 02], [Ror 96], [Tan 03], [GW 04], [For 03], [KMK 04], [Sta 04], [Kes 02], [PD 03], [Kei 02], [Spu 00], [KCK 98], [Sau 98], [Izz 00], [Per 00], and [WV 00].

14.5.2 Key Terms

56K modem
800 service
900 service
add/drop multiplexer (ADM)
analog leased service
analog switched service
application adaptation layer (AAL)
asymmetric DSL (ADSL)
Asynchronous Transfer Mode (ATM)
automatic protection switching (APS)
bidirectional line switching ring (BLSR)
byte interleaving
cable modem (CM)
cable modem transmission system (CMTS)
cable TV network
cell
cell network
common carrier
Community Antenna Television (CATV)
competitive local exchange carrier (CLEC)
convergence sublayer (CS)
demodulator
digital data service (DDS)
digital service
digital service unit (DSU)
digital subscriber line (DSL)
digital subscriber line access multiplexer (DSLAM)
distribution hub
downloading
downstream data band
end office
fiber node
head end
hybrid fiber-coaxial (HFC) network
in-band signaling
incumbent local exchange carrier (ILEC)
interexchange carrier (IXC)
ISDN user port (ISUP)
line
line layer
line overhead (LOH)
local access transport area (LATA)
local exchange carrier (LEC)
local loop
long-distance company
message transport part (MTP)
modem
modulator
network-to-network interface (NNI)
optical carrier (OC)
out-of-band signaling
path
path layer
path overhead (POH)
permanent virtual-circuit connection (PVC)
photonic layer
plain old telephone system (POTS)
point of presence (POP)
regenerator
regional cable head (RCH)
regional office
section
section layer
section overhead (SOH)
segmentation and reassembly (SAR)
service control point (SCP)
signal point (SP)
signal transport port (STP)
signaling connection control point (SCCP)
Signaling System Seven (SS7)
simple and efficient adaptation layer (SEAL)
STS multiplexer
STS demultiplexer
switched/56 service
switched virtual-circuit connection (SVC)
switching office
Synchronous Digital Hierarchy (SDH)
Synchronous Optical Network (SONET)
synchronous payload envelope (SPE)
synchronous transport module (STM)
synchronous transport signal (STS)
tandem office
telephone user port (TUP)
terminal
transaction capabilities application port (TCAP)
transmission path (TP)
trunk
unidirectional path switching ring (UPSR)
uploading
user-to-network interface (UNI)
video band
virtual circuit (VC)
virtual path (VP)
virtual tributary (VT)
wide-area telephone service (WATS)

14.5.3 Summary

The telephone network was originally an analog system. During the last decade, the telephone network has undergone many technical changes. The network is now digital as well as analog. The telephone network is made of three major components: local loops, trunks, and switching offices. Telephone companies provide two types of services: analog and digital. We can categorize analog services as either analog switched services or analog leased services. The two most common digital services are switched/56 service and digital data service (DDS). Data transfer using the telephone local loop was traditionally done using a dial-up modem. The term *modem* is a composite word that refers to the two functional entities that make up the device: a signal modulator and a signal demodulator. Most popular modems available are based on 56K modems. Telephone companies developed another technology, digital subscriber line (DSL), to provide higher-speed access to the Internet. DSL technology is a set of technologies, but we discussed only the common one, ADSL.

Community antenna TV (CATV) was originally designed to provide video services for the community. The traditional cable TV system used coaxial cable end to end. The second generation of cable networks is called a hybrid fiber-coaxial (HFC) network. The network uses a combination of fiber-optic and coaxial cable. Cable companies are now competing with telephone companies for the residential customer who wants high-speed access to the Internet. To use a cable network for data transmission, we need two key devices: a cable modem (CM) and a cable modem transmission system (CMTS).

Synchronous Optical Network (SONET) is a standard developed by ANSI for fiber-optic networks: Synchronous Digital Hierarchy (SDH) is a similar standard developed by ITU-T. SONET has defined a hierarchy of signals called synchronous transport signals (STSs). SDH has defined a similar hierarchy of signals called synchronous transfer modules (STMs). An OC-*n* signal is the optical modulation of an STS-*n* (or STM-*n*) signal. SONET defines four layers: path, line, section, and photonic. SONET is a synchronous TDM system in which all clocks are locked to a master clock. A SONET system can use STS multiplexers/demultiplexers, regenerators, add/drop multiplexers, and terminals. SONET sends 8000 frames per second; each frame lasts 125 μs. An STS-1 frame is made of 9 rows and 90 columns; an STS-*n* frame is made of 9 rows and $n \times 90$ columns. STSs can be multiplexed to get a new STS with a higher data rate. SONET network topologies can be linear, ring, or mesh. A linear SONET network can be either point-to-point or multipoint. A ring SONET network can be unidirectional or bidirectional. To make SONET backward-compatible with the current hierarchy, its frame design includes a system of virtual tributaries (VTs).

Asynchronous Transfer Mode (ATM) is a cell relay protocol that, in combination with SONET, allows high-speed connections. A cell is a small, fixed-size block of information. The ATM data packet is a cell composed of 53 bytes (5 bytes of header and 48 bytes of payload). ATM eliminates the varying delay times associated with different-size packets. ATM can handle real-time transmission. A user-to-network interface (UNI) is the interface between a user and an ATM switch. A network-to-network interface (NNI) is the interface between two ATM switches. In ATM, connection between two endpoints is accomplished through transmission paths (TPs), virtual paths (VPs), and virtual circuits (VCs). In ATM, a combination of a virtual path identifier

(VPI) and a virtual-circuit identifier identifies a virtual connection. The ATM standard defines three layers, AAL, ATM, and physical.

14.6 PRACTICE SET

14.6.1 Quizzes

A set of interactive quizzes for this chapter can be found on the book website. It is strongly recommended that the student take the quizzes to check his/her understanding of the materials before continuing with the practice set.

14.6.2 Questions

Q14-1. What is dial-up modem technology? List some of the common modem standards discussed in this chapter and give their data rates.

Q14-2. How does an NNI differ from a UNI?

Q14-3. What is a virtual tributary?

Q14-4. What is the relationship between SONET and SDH?

Q14-5. Compare and contrast a traditional cable network with a hybrid fiber-coaxial network.

Q14-6. What is the purpose of the pointer in the line overhead?

Q14-7. Describe the SS7 service and its relation to the telephone network.

Q14-8. What is the function of a SONET regenerator?

Q14-9. What is the relationship between STS and STM?

Q14-10. How is an ATM virtual connection identified?

Q14-11. What is the relationship between STS signals and OC signals?

Q14-12. What is DSL technology? What are the services provided by the telephone companies using DSL? Distinguish between a DSL modem and a DSLAM.

Q14-13. What are the three major components of a telephone network?

Q14-14. What is LATA? What are intra-LATA and inter-LATA services?

Q14-15. Name the ATM layers and their functions.

Q14-16. Discuss the functions of each SONET layer.

Q14-17. Distinguish between CM and CMTS.

Q14-18. How many virtual connections can be defined in a UNI? How many virtual connections can be defined in an NNI?

Q14-19. Why is multiplexing more efficient if all the data units are the same size?

Q14-20. How is data transfer achieved using CATV channels?

Q14-21. Why is SONET called a synchronous network?

Q14-22. How is an STS multiplexer different from an add/drop multiplexer since both can add signals together?

Q14-23. What is the relationship between TPs, VPs, and VCs?

Q14-24. What are the two major services provided by telephone companies in the United States?

Q14-25. What are the four SONET layers?

Q14-26. In ATM, what is the relationship between TPs, VPs, and VCs?

14.6.3 Problems

P14-1. Draw a bar chart to compare the different downloading data rates of common modems.

P14-2. Draw a bar chart to compare the different downloading data rates of common DSL technology implementations (use minimum data rates).

P14-3. What are the user data rates of STS-3, STS-9, and STS-12?

P14-4. In Chapter 8, we discussed the three communication phases involved in a circuit-switched network. Match these phases with the phases in a telephone call between two parties.

P14-5. Using the discussion of circuit-switching in Chapter 8, explain why this type of switching was chosen for telephone networks.

P14-6. What type of topology is used when customers in an area use DSL modems for data transfer purposes? Explain.

P14-7. Calculate the minimum time required to download one million bytes of information using a cable modem (consider the minimum rates).

P14-8. A stream of data is being carried by STS-1 frames. If the data rate of the stream is 49.530 Mbps, how many frames per second should leave one empty byte after the H3 byte?

P14-9. Calculate the minimum time required to download one million bytes of information using a 56K modem.

P14-10. When we have an overseas telephone conversation, we sometimes experience a delay. Can you explain the reason?

P14-11. A stream of data is being carried by STS-1 frames. If the data rate of the stream is 49.540 Mbps, how many STS-1 frames per second must let their H3 bytes carry data?

P14-12. Calculate the minimum time required to download one million bytes of information using ADSL using ADSL implementations (consider the minimum rates).

P14-13. In Chapter 8, we learned that a circuit-switched network needs end-to-end addressing during the setup and teardown phases. Define end-to-end addressing in a telephone network when two parties communicate.

P14-14. Show how STS-9s can be multiplexed to create an STS-36. Is there any extra overhead involved in this type of multiplexing?

P14-15. What type of topology is used when customers in an area use cable modems for data transfer purposes? Explain.

P14-16. What type of topology is used when customers in an area use DSL modems for data transfer purposes? Explain.

P14-17. What is the minimum number of cells resulting from an input packet in the AAL5 layer? What is the maximum number of cells resulting from an input packet?

CHAPTER 15

Wireless LANs

We discussed wired LANs and wired WANs in the two previous chapters. We concentrate on wireless LANs in this chapter and wireless WANs in the next.

In this chapter, we cover two types of wireless LANs. The first is the wireless LAN defined by the IEEE 802.11 project (sometimes called *wireless Ethernet*); the second is a personal wireless LAN, Bluetooth, that is sometimes called *personal area network* or *PAN*.

This chapter is divided into three sections:

- The first section introduces the general issues behind wireless LANs and compares wired and wireless networks. The section describes the characteristics of the wireless networks and the way access is controlled in these types of networks.
- The second section discusses a wireless LAN defined by the IEEE 802.11 Project, which is sometimes called *wireless Ethernet*. This section defines the architecture of this type of LAN and describes the MAC sublayer, which uses the CSMA/CA access method discussed in Chapter 12. The section then shows the addressing mechanism used in this network and gives the format of different packets used at the data-link layer. Finally, the section discusses different physical-layer protocols that are used by this type of network.
- The third section discusses the Bluetooth technology as a personal area network (PAN). The section describes the architecture of the network, the addressing mechanism, and the packet format. Different layers used in this protocol are also briefly described and compared with the ones in the other wired and wireless LANs.

15.1 INTRODUCTION

Wireless communication is one of the fastest-growing technologies. The demand for connecting devices without the use of cables is increasing everywhere. Wireless LANs can be found on college campuses, in office buildings, and in many public areas. Before we discuss a specific protocol related to wireless LANs, let us talk about them in general.

15.1.1 Architectural Comparison

Let us first compare the architecture of wired and wireless LANs to give some idea of what we need to look for when we study wireless LANs.

Medium

The first difference we can see between a wired and a wireless LAN is the medium. In a wired LAN, we use wires to connect hosts. In Chapter 7, we saw that we moved from multiple access to point-to-point access through the generation of the Ethernet. In a switched LAN, with a link-layer switch, the communication between the hosts is point-to-point and full-duplex (bidirectional). In a wireless LAN, the medium is air, the signal is generally broadcast. When hosts in a wireless LAN communicate with each other, they are sharing the same medium (multiple access). In a very rare situation, we may be able to create a point-to-point communication between two wireless hosts by using a very limited bandwidth and two-directional antennas. Our discussion in this chapter, however, is about the multiple-access medium, which means we need to use MAC protocols.

Hosts

In a wired LAN, a host is always connected to its network at a point with a fixed link-layer address related to its network interface card (NIC). Of course, a host can move from one point in the Internet to another point. In this case, its link-layer address remains the same, but its network-layer address will change, as we see later in Chapter 19, Section 19.3 (Mobile IP section). However, before the host can use the services of the Internet, it needs to be physically connected to the Internet. In a wireless LAN, a host is not physically connected to the network; it can move freely (as we'll see) and can use the services provided by the network. Therefore, mobility in a wired network and wireless network are totally different issues, which we try to clarify in this chapter.

Isolated LANs

The concept of a wired isolated LAN also differs from that of a wireless isolated LAN. A wired isolated LAN is a set of hosts connected via a link-layer switch (in the recent generation of Ethernet). A wireless isolated LAN, called an ***ad hoc network*** in wireless LAN terminology, is a set of hosts that communicate freely with each other. The concept of a link-layer switch does not exist in wireless LANs. Figure 15.1 shows two isolated LANs, one wired and one wireless.

Connection to Other Networks

A wired LAN can be connected to another network or an internetwork such as the Internet using a router. A wireless LAN may be connected to a wired infrastructure network, to a wireless infrastructure network, or to another wireless LAN. The first situation is the one that we discuss in this section: connection of a wireless LAN to a wired infrastructure network. Figure 15.2 shows the two environments.

Figure 15.1 *Isolated LANs: wired versus wireless*

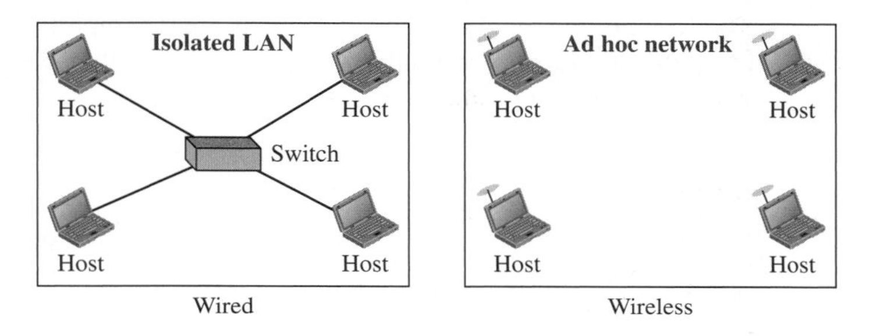

Figure 15.2 *Connection of a wired LAN and a wireless LAN to other networks*

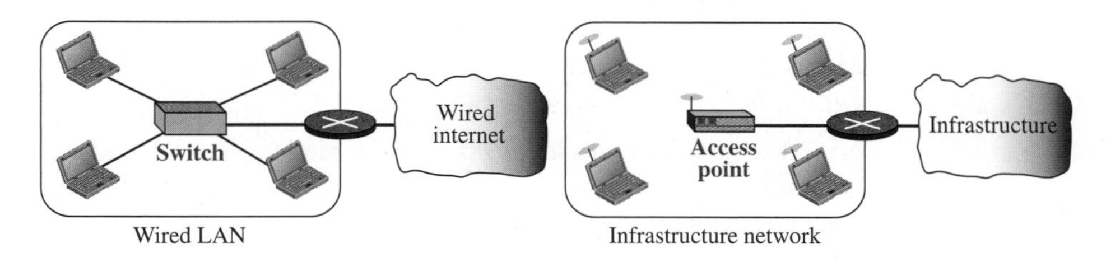

In this case, the wireless LAN is referred to as an *infrastructure network,* and the connection to the wired infrastructure, such as the Internet, is done via a device called an ***access point (AP)***. Note that the role of the access point is completely different from the role of a link-layer switch in the wired environment. An access point is gluing two different environments together: one wired and one wireless. Communication between the AP and the wireless host occurs in a wireless environment; communication between the AP and the infrastructure occurs in a wired environment.

Moving between Environments

The discussion above confirms what we learned in Chapters 2 and 9: a wired LAN or a wireless LAN operates only in the lower two layers of the TCP/IP protocol suite. This means that if we have a wired LAN in a building that is connected via a router or a modem to the Internet, all we need in order to move from the wired environment to a wireless environment is to change the network interface cards designed for wired environments to the ones designed for wireless environments and replace the link-layer switch with an access point. In this change, the link-layer addresses will change (because of changing NICs), but the network-layer addresses (IP addresses) will remain the same; we are moving from wired links to wireless links.

15.1.2 Characteristics

There are several characteristics of wireless LANs that either do not apply to wired LANs or the existence of which is negligible and can be ignored. We discuss some of these characteristics here to pave the way for discussing wireless LAN protocols.

Attenuation

The strength of electromagnetic signals decreases rapidly because the signal disperses in all directions; only a small portion of it reaches the receiver. The situation becomes worse with mobile senders that operate on batteries and normally have small power supplies.

Interference

Another issue is that a receiver may receive signals not only from the intended sender, but also from other senders if they are using the same frequency band.

Multipath Propagation

A receiver may receive more than one signal from the same sender because electromagnetic waves can be reflected back from obstacles such as walls, the ground, or objects. The result is that the receiver receives some signals at different phases (because they travel different paths). This makes the signal less recognizable.

Error

With the above characteristics of a wireless network, we can expect that errors and error detection are more serious issues in a wireless network than in a wired network. If we think about the error level as the measurement of **signal-to-noise ratio (SNR),** we can better understand why error detection and error correction and retransmission are more important in a wireless network. We discussed SNR in more detail in Chapter 3, but it is enough to say that it measures the ratio of good stuff to bad stuff (signal to noise). If SNR is high, it means that the signal is stronger than the noise (unwanted signal), so we may be able to convert the signal to actual data. On the other hand, when SNR is low, it means that the signal is corrupted by the noise and the data cannot be recovered.

15.1.3 Access Control

Maybe the most important issue we need to discuss in a wireless LAN is access control—how a wireless host can get access to the shared medium (air). We discussed in Chapter 12 that the Standard Ethernet uses the CSMA/CD algorithm. In this method, each host contends to access the medium and sends its frame if it finds the medium idle. If a collision occurs, it is detected and the frame is sent again. Collision detection in CSMA/CD serves two purposes. If a collision is detected, it means that the frame has not been received and needs to be resent. If a collision is not detected, it is a kind of acknowledgment that the frame was received. The CSMA/CD algorithm does not work in wireless LANs for three reasons:

1. To detect a collision, a host needs to send and receive at the same time (sending the frame and receiving the collision signal), which means the host needs to work in a

duplex mode. Wireless hosts do not have enough power to do so (the power is supplied by batteries). They can only send or receive at one time.

2. Because of the hidden station problem, in which a station may not be aware of another station's transmission due to some obstacles or range problems, collision may occur but not be detected. Figure 15.3 shows an example of the hidden station problem. Station B has a transmission range shown by the left oval (sphere in space); every station in this range can hear any signal transmitted by station B. Station C has a transmission range shown by the right oval (sphere in space); every station located in this range can hear any signal transmitted by C. Station C is outside the transmission range of B; likewise, station B is outside the transmission range of C. Station A, however, is in the area covered by both B and C; it can hear any signal transmitted by B or C. The figure also shows that the hidden station problem may also occur due to an obstacle.

   Assume that station B is sending data to station A. In the middle of this transmission, station C also has data to send to station A. However, station C is out of B's range and transmissions from B cannot reach C. Therefore C thinks the medium is free. Station C sends its data to A, which results in a collision at A because this station is receiving data from both B and C. In this case, we say that stations B and C are hidden from each other with respect to A. Hidden stations can reduce the capacity of the network because of the possibility of collision.

Figure 15.3 *Hidden station problem*

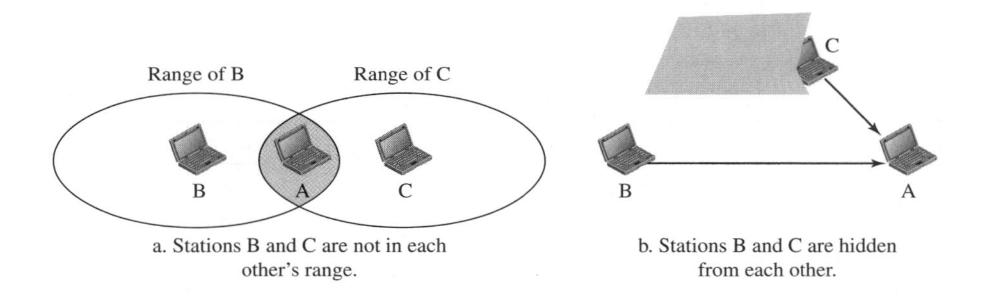

3. The distance between stations can be great. Signal fading could prevent a station at one end from hearing a collision at the other end.

To overcome the above three problems, Carrier Sense Multiple Access with Collision Avoidance (CSMA/CA) was invented for wireless LANs, which we discussed in Chapter 12.

15.2 IEEE 802.11 PROJECT

IEEE has defined the specifications for a wireless LAN, called IEEE 802.11, which covers the physical and data-link layers. It is sometimes called *wireless Ethernet*. In some countries, including the United States, the public uses the term *WiFi* (short for wireless fidelity) as a synonym for *wireless LAN*. WiFi, however, is a wireless LAN that is certified by the WiFi Alliance, a global, nonprofit industry association of more than 300 member companies devoted to promoting the growth of wireless LANs.

15.2.1 Architecture

The standard defines two kinds of services: the basic service set (BSS) and the extended service set (ESS).

Basic Service Set

IEEE 802.11 defines the **basic service set (BSS)** as the building blocks of a wireless LAN. A basic service set is made of stationary or mobile wireless stations and an optional central base station, known as the *access point (AP)*. Figure 15.4 shows two sets in this standard.

Figure 15.4 *Basic service sets (BSSs)*

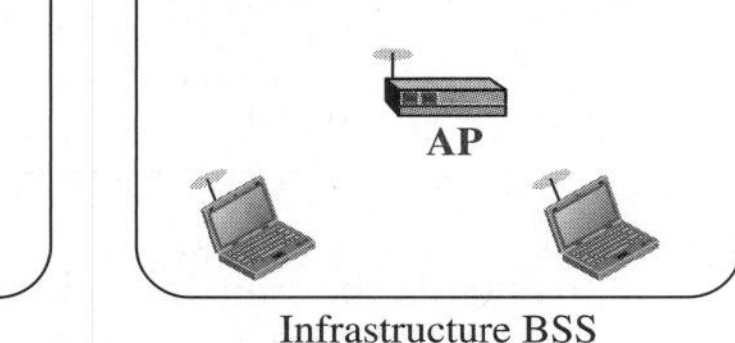

The BSS without an AP is a stand-alone network and cannot send data to other BSSs. It is called an *ad hoc architecture*. In this architecture, stations can form a network without the need of an AP; they can locate one another and agree to be part of a BSS. A BSS with an AP is sometimes referred to as an *infrastructure BSS*.

Extended Service Set

An **extended service set (ESS)** is made up of two or more BSSs with APs. In this case, the BSSs are connected through a *distribution system,* which is a wired or a wireless network. The distribution system connects the APs in the BSSs. IEEE 802.11 does not restrict the distribution system; it can be any IEEE LAN such as an Ethernet. Note that the extended service set uses two types of stations: mobile and stationary. The mobile stations are normal stations inside a BSS. The stationary stations are AP stations that are part of a wired LAN. Figure 15.5 shows an ESS.

When BSSs are connected, the stations within reach of one another can communicate without the use of an AP. However, communication between a station in a BSS and the outside BSS occurs via the AP. The idea is similar to communication in a cellular network (discussed in Chapter 16) if we consider each BSS to be a cell and each AP to be a base station. Note that a mobile station can belong to more than one BSS at the same time.

Figure 15.5 *Extended service set (ESS)*

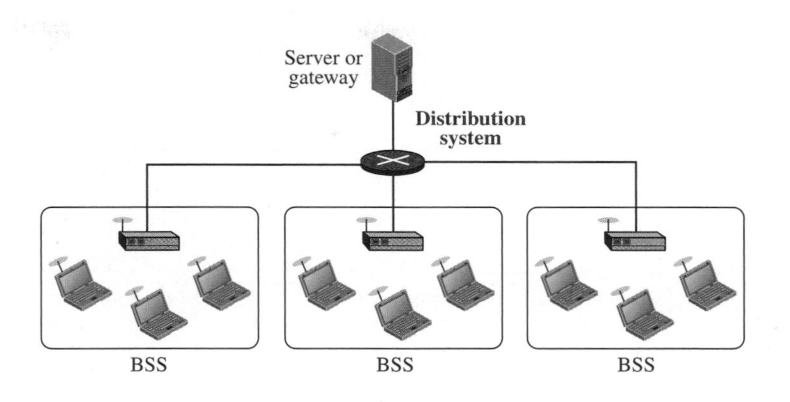

Station Types

IEEE 802.11 defines three types of stations based on their mobility in a wireless LAN: no-transition, BSS-transition, and ESS-transition mobility. A station with **no-transition mobility** is either stationary (not moving) or moving only inside a BSS. A station with **BSS-transition mobility** can move from one BSS to another, but the movement is confined inside one ESS. A station with **ESS-transition mobility** can move from one ESS to another. However, IEEE 802.11 does not guarantee that communication is continuous during the move.

15.2.2 MAC Sublayer

IEEE 802.11 defines two MAC sublayers: the distributed coordination function (DCF) and point coordination function (PCF). Figure 15.6 shows the relationship between the two MAC sublayers, the LLC sublayer, and the physical layer. We discuss the physical layer implementations later in the chapter and will now concentrate on the MAC sublayer.

Distributed Coordination Function

One of the two protocols defined by IEEE at the MAC sublayer is called the ***distributed coordination function (DCF)***. DCF uses CSMA/CA as the access method (see Chapter 12).

Frame Exchange Time Line

Figure 15.7 shows the exchange of data and control frames in time.

1. Before sending a frame, the source station senses the medium by checking the energy level at the carrier frequency.
   a. The channel uses a persistence strategy with backoff until the channel is idle.
   b. After the station is found to be idle, the station waits for a period of time called the ***distributed interframe space (DIFS);*** then the station sends a control frame called the *request to send (RTS).*

Figure 15.6 *MAC layers in IEEE 802.11 standard*

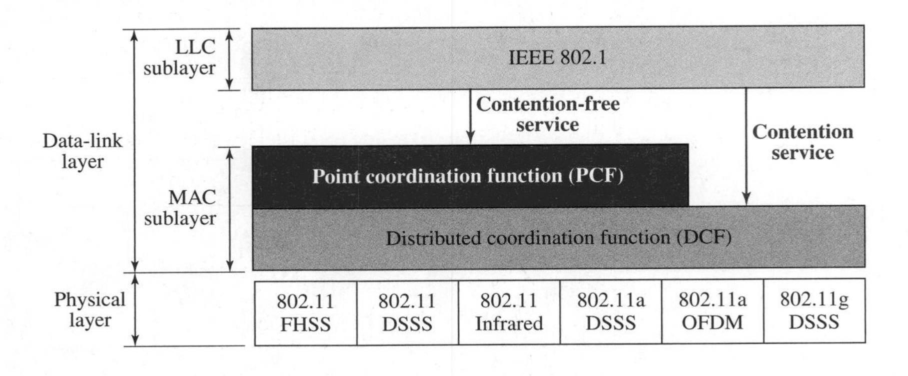

Figure 15.7 *CSMA/CA and NAV*

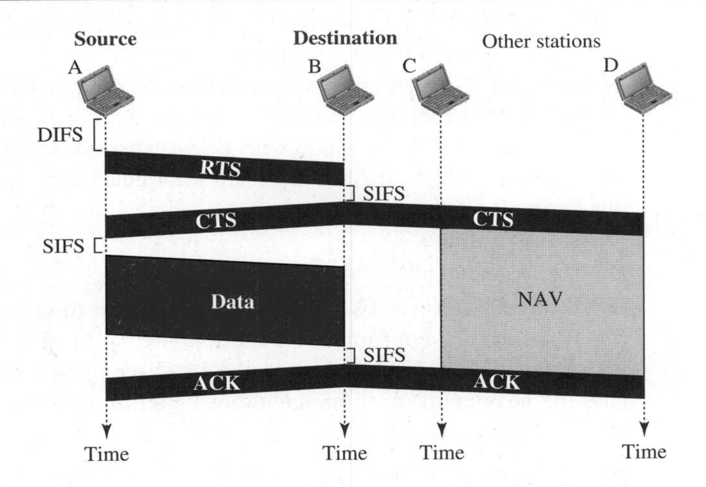

2. After receiving the RTS and waiting a period of time called the ***short interframe space (SIFS),*** the destination station sends a control frame, called the *clear to send (CTS),* to the source station. This control frame indicates that the destination station is ready to receive data.
3. The source station sends data after waiting an amount of time equal to SIFS.
4. The destination station, after waiting an amount of time equal to SIFS, sends an acknowledgment to show that the frame has been received. Acknowledgment is needed in this protocol because the station does not have any means to check for the successful arrival of its data at the destination. On the other hand, the lack of collision in CSMA/CD is a kind of indication to the source that data have arrived.

Network Allocation Vector

How do other stations defer sending their data if one station acquires access? In other words, how is the *collision avoidance* aspect of this protocol accomplished? The key is a feature called NAV.

When a station sends an RTS frame, it includes the duration of time that it needs to occupy the channel. The stations that are affected by this transmission create a timer called a ***network allocation vector (NAV)*** that shows how much time must pass before these stations are allowed to check the channel for idleness. Each time a station accesses the system and sends an RTS frame, other stations start their NAV. In other words, each station, before sensing the physical medium to see if it is idle, first checks its NAV to see if it has expired. Figure 15.7 shows the idea of NAV.

Collision During Handshaking

What happens if there is a collision during the time when RTS or CTS control frames are in transition, often called the *handshaking period*? Two or more stations may try to send RTS frames at the same time. These control frames may collide. However, because there is no mechanism for collision detection, the sender assumes there has been a collision if it has not received a CTS frame from the receiver. The backoff strategy is employed, and the sender tries again.

Hidden-Station Problem

The solution to the hidden station problem is the use of the handshake frames (RTS and CTS). Figure 15.7 also shows that the RTS message from B reaches A, but not C. However, because both B and C are within the range of A, the CTS message, which contains the duration of data transmission from B to A, reaches C. Station C knows that some hidden station is using the channel and refrains from transmitting until that duration is over.

Point Coordination Function (PCF)

The **point coordination function (PCF)** is an optional access method that can be implemented in an infrastructure network (not in an ad hoc network). It is implemented on top of the DCF and is used mostly for time-sensitive transmission.

PCF has a centralized, contention-free polling access method, which we discussed in Chapter 12. The AP performs polling for stations that are capable of being polled. The stations are polled one after another, sending any data they have to the AP.

To give priority to PCF over DCF, another interframe space, PIFS, has been defined. PIFS (PCF IFS) is shorter than DIFS. This means that if, at the same time, a station wants to use only DCF and an AP wants to use PCF, the AP has priority.

Due to the priority of PCF over DCF, stations that only use DCF may not gain access to the medium. To prevent this, a repetition interval has been designed to cover both contention-free PCF and contention-based DCF traffic. The *repetition interval,* which is repeated continuously, starts with a special control frame, called a ***beacon frame***. When the stations hear the beacon frame, they start their NAV for the duration of the contention-free period of the repetition interval. Figure 15.8 shows an example of a repetition interval.

During the repetition interval, the PC (point controller) can send a poll frame, receive data, send an ACK, receive an ACK, or do any combination of these (802.11

Figure 15.8 *Example of repetition interval*

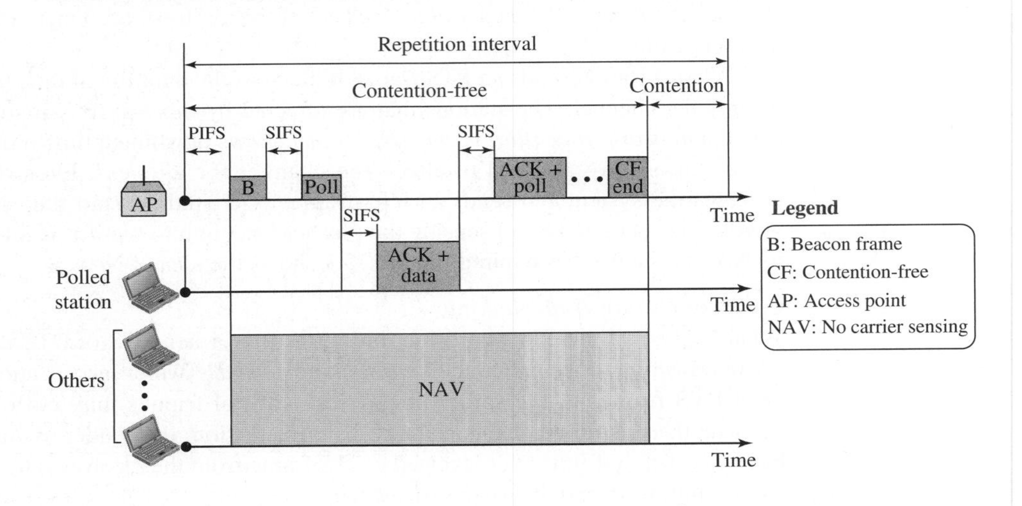

uses piggybacking). At the end of the contention-free period, the PC sends a CF end (contention-free end) frame to allow the contention-based stations to use the medium.

Fragmentation

The wireless environment is very noisy, so frames are often corrupted. A corrupt frame has to be retransmitted. The protocol, therefore, recommends fragmentation—the division of a large frame into smaller ones. It is more efficient to resend a small frame than a large one.

Frame Format

The MAC layer frame consists of nine fields, as shown in Figure 15.9.

Figure 15.9 *Frame format*

2 bytes	2 bytes	6 bytes	6 bytes	6 bytes	2 bytes	6 bytes	0 to 2312 bytes	4 bytes
FC	D	Address 1	Address 2	Address 3	SC	Address 4	Frame body	FCS

Protocol version	Type	Subtype	To DS	From DS	More frag	Retry	Pwr mgt	More data	WEP	Rsvd
2 bits	2 bits	4 bits	1 bit	1 bit	1 bit	1 bit	1 bit	1 bit	1 bit	1 bit

- ***Frame control (FC).*** The FC field is 2 bytes long and defines the type of frame and some control information. Table 15.1 describes the subfields. We will discuss each frame type later in this chapter.

Table 15.1 *Subfields in FC field*

Field	*Explanation*
Version	Current version is 0
Type	Type of information: management (00), control (01), or data (10)
Subtype	Subtype of each type (see Table 15.2)
To DS	Defined later
From DS	Defined later
More frag	When set to 1, means more fragments
Retry	When set to 1, means retransmitted frame
Pwr mgt	When set to 1, means station is in power management mode
More data	When set to 1, means station has more data to send
WEP	Wired equivalent privacy (encryption implemented)
Rsvd	Reserved

- ***D.*** This field defines the duration of the transmission that is used to set the value of NAV. In one control frame, it defines the ID of the frame.
- ***Addresses.*** There are four address fields, each 6 bytes long. The meaning of each address field depends on the value of the *To DS* and *From DS* subfields and will be discussed later.
- ***Sequence control.*** This field, often called the *SC* field, defines a 16-bit value. The first four bits define the fragment number; the last 12 bits define the sequence number, which is the same in all fragments.
- ***Frame body.*** This field, which can be between 0 and 2312 bytes, contains information based on the type and the subtype defined in the FC field.
- ***FCS.*** The FCS field is 4 bytes long and contains a CRC-32 error-detection sequence.

Frame Types

A wireless LAN defined by IEEE 802.11 has three categories of frames: management frames, control frames, and data frames.

Management Frames

Management frames are used for the initial communication between stations and access points.

Control Frames

Control frames are used for accessing the channel and acknowledging frames. Figure 15.10 shows the format.

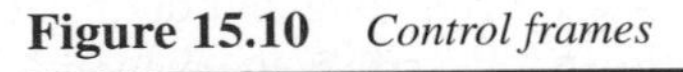

Figure 15.10 *Control frames*

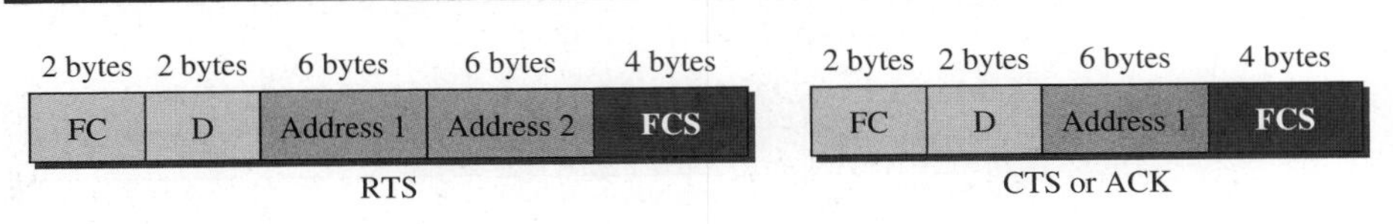

For control frames the value of the type field is 01; the values of the subtype fields for frames we have discussed are shown in Table 15.2.

Table 15.2 *Values of subtype fields in control frames*

Subtype	*Meaning*
1011	Request to send (RTS)
1100	Clear to send (CTS)
1101	Acknowledgment (ACK)

Data Frames

Data frames are used for carrying data and control information.

15.2.3 Addressing Mechanism

The IEEE 802.11 addressing mechanism specifies four cases, defined by the value of the two flags in the FC field, *To DS* and *From DS*. Each flag can be either 0 or 1, resulting in four different situations. The interpretation of the four addresses (address 1 to address 4) in the MAC frame depends on the value of these flags, as shown in Table 15.3.

Table 15.3 *Addresses*

To DS	*From DS*	*Address 1*	*Address 2*	*Address 3*	*Address 4*
0	0	Destination	Source	BSS ID	N/A
0	1	Destination	Sending AP	Source	N/A
1	0	Receiving AP	Source	Destination	N/A
1	1	Receiving AP	Sending AP	Destination	Source

Note that address 1 is always the address of the next device that the frame will visit. Address 2 is always the address of the previous device that the frame has left. Address 3 is the address of the final destination station if it is not defined by address 1 or the original source station if it is not defined by address 2. Address 4 is the original source when the distribution system is also wireless.

- **Case 1: 00** In this case, *To DS* = 0 and *From DS* = 0. This means that the frame is not going to a distribution system (*To DS* = 0) and is not coming from a distribution system (*From DS* = 0). The frame is going from one station in a BSS to another without passing through the distribution system. The addresses are shown in Figure 15.11.
- **Case 2: 01** In this case, *To DS* = 0 and *From DS* = 1. This means that the frame is coming from a distribution system (*From DS* = 1). The frame is coming from an

Figure 15.11 *Addressing mechanisms*

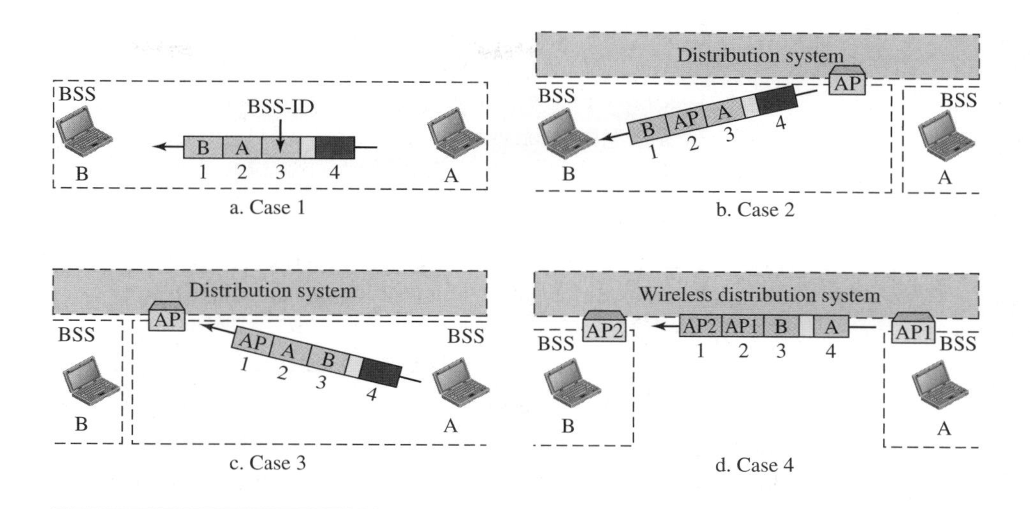

AP and going to a station. The addresses are as shown in Figure 15.11. Note that address 3 contains the original sender of the frame (in another BSS).

- **Case 3: 10** In this case, *To DS* = 1 and *From DS* = 0. This means that the frame is going to a distribution system (*To DS* = 1). The frame is going from a station to an AP. The ACK is sent to the original station. The addresses are as shown in Figure 15.11. Note that address 3 contains the final destination of the frame in the distribution system.
- **Case 4: 11** In this case, *To DS* = 1 and *From DS* = 1. This is the case in which the distribution system is also wireless. The frame is going from one AP to another AP in a wireless distribution system. Here, we need four addresses to define the original sender, the final destination, and two intermediate APs. Figure 15.11 shows the situation.

Exposed Station Problem

We discussed how to solve the hidden station problem. A similar problem is called the *exposed station problem*. In this problem a station refrains from using a channel when it is, in fact, available. In Figure 15.12, station A is transmitting to station B. Station C has some data to send to station D, which can be sent without interfering with the transmission from A to B. However, station C is exposed to transmission from A; it hears what A is sending and thus refrains from sending. In other words, C is too conservative and wastes the capacity of the channel. The handshaking messages RTS and CTS cannot help in this case. Station C hears the RTS from A and refrains from sending, even though the communication between C and D cannot cause a collision in the zone between A and C; station C cannot know that station A's transmission does not affect the zone between C and D.

Figure 15.12 *Exposed station problem*

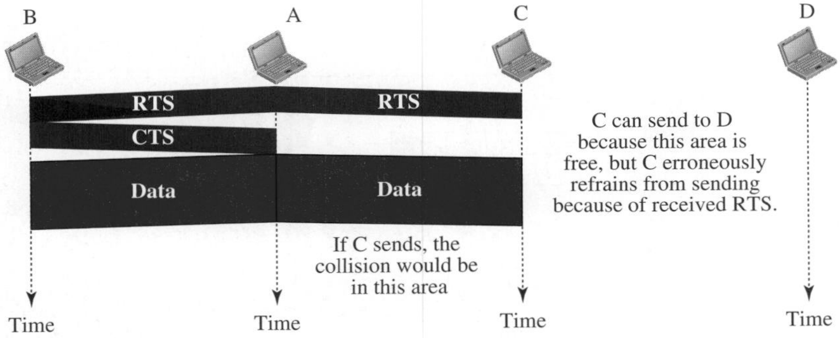

15.2.4 Physical Layer

We discuss six specifications, as shown in Table 15.4. All implementations, except the infrared, operate in the *industrial, scientific,* and *medical (ISM)* band, which defines three unlicensed bands in the three ranges 902–928 MHz, 2.400–4.835 GHz, and 5.725–5.850 GHz.

Table 15.4 *Specifications*

IEEE	*Technique*	*Band*	*Modulation*	*Rate (Mbps)*
802.11	FHSS	2.400–4.835 GHz	FSK	1 and 2
	DSSS	2.400–4.835 GHz	PSK	1 and 2
	None	Infrared	PPM	1 and 2
802.11a	OFDM	5.725–5.850 GHz	PSK or QAM	6 to 54
802.11b	DSSS	2.400–4.835 GHz	PSK	5.5 and 11
802.11g	OFDM	2.400–4.835 GHz	Different	22 and 54
802.11n	OFDM	5.725–5.850 GHz	Different	600

IEEE 802.11 FHSS

IEEE 802.11 FHSS uses the **frequency-hopping spread spectrum (FHSS)** method, as discussed in Chapter 6. FHSS uses the 2.400–4.835 GHz ISM band. The band is divided into 79 subbands of 1 MHz (and some guard bands). A pseudorandom number generator selects the hopping sequence. The modulation technique in this specification is either two-level FSK or four-level FSK with 1 or 2 bits/baud, which results in a data rate of 1 or 2 Mbps, as shown in Figure 15.13.

IEEE 802.11 DSSS

IEEE 802.11 DSSS uses the **direct-sequence spread spectrum (DSSS)** method, as discussed in Chapter 6. DSSS uses the 2.400–4.835 GHz ISM band. The modulation technique in this specification is PSK at 1 Mbaud/s. The system allows 1 or 2 bits/baud (BPSK or QPSK), which results in a data rate of 1 or 2 Mbps, as shown in Figure 15.14.

Figure 15.13 *Physical layer of IEEE 802.11 FHSS*

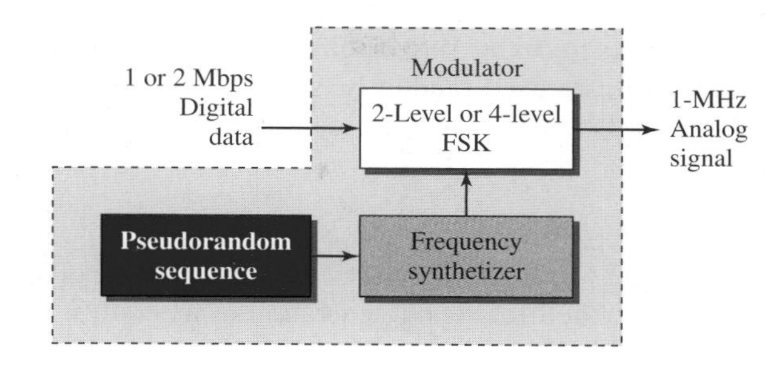

Figure 15.14 *Physical layer of IEEE 802.11 DSSS*

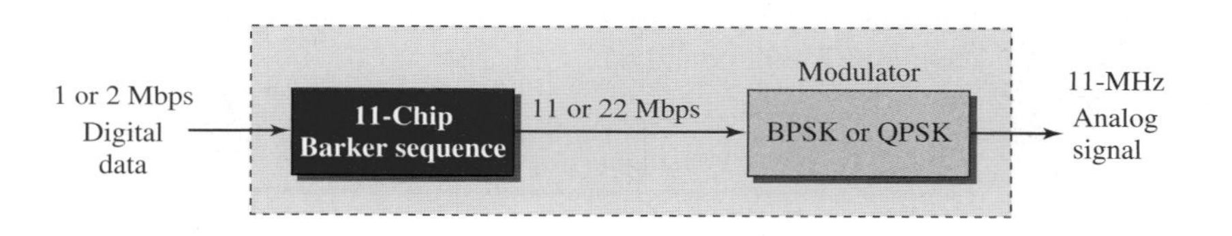

IEEE 802.11 Infrared

IEEE 802.11 infrared uses infrared light in the range of 800 to 950 nm. The modulation technique is called ***pulse position modulation (PPM).*** For a 1-Mbps data rate, a 4-bit sequence is first mapped into a 16-bit sequence in which only one bit is set to 1 and the rest are set to 0. For a 2-Mbps data rate, a 2-bit sequence is first mapped into a 4-bit sequence in which only one bit is set to 1 and the rest are set to 0. The mapped sequences are then converted to optical signals; the presence of light specifies 1, the absence of light specifies 0. See Figure 15.15.

Figure 15.15 *Physical layer of IEEE 802.11 infrared*

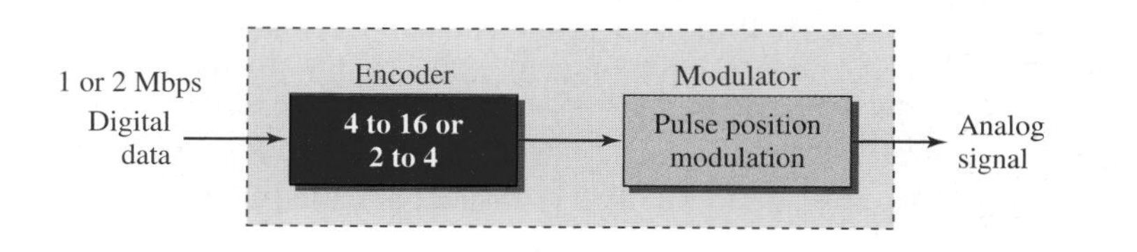

IEEE 802.11a OFDM

IEEE 802.11a OFDM describes the **orthogonal frequency-division multiplexing (OFDM)** method for signal generation in a 5.725–5.850 GHz ISM band. OFDM is similar to FDM, as discussed in Chapter 6, with one major difference: All the subbands are used by one source at a given time. Sources contend with one another at the data-link layer for access. The band is divided into 52 subbands, with 48 subbands for sending 48 groups of bits at a time and 4 subbands for control information. Dividing the band into subbands diminishes the effects of interference. If the subbands are used randomly, security can also be increased. OFDM uses PSK and QAM for modulation. The common data rates are 18 Mbps (PSK) and 54 Mbps (QAM).

IEEE 802.11b DSSS

IEEE 802.11b DSSS describes the **high-rate direct-sequence spread spectrum (HR-DSSS)** method for signal generation in the 2.400–4.835 GHz ISM band. HR-DSSS is similar to DSSS except for the encoding method, which is called ***complementary code keying (CCK).*** CCK encodes 4 or 8 bits to one CCK symbol. To be backward compatible with DSSS, HR-DSSS defines four data rates: 1, 2, 5.5, and 11 Mbps. The first two use the same modulation techniques as DSSS. The 5.5-Mbps version uses BPSK and transmits at 1.375 Mbaud/s with 4-bit CCK encoding. The 11-Mbps version uses QPSK and transmits at 1.375 Mbps with 8-bit CCK encoding. Figure 15.16 shows the modulation technique for this standard.

Figure 15.16 *Physical layer of IEEE 802.11b*

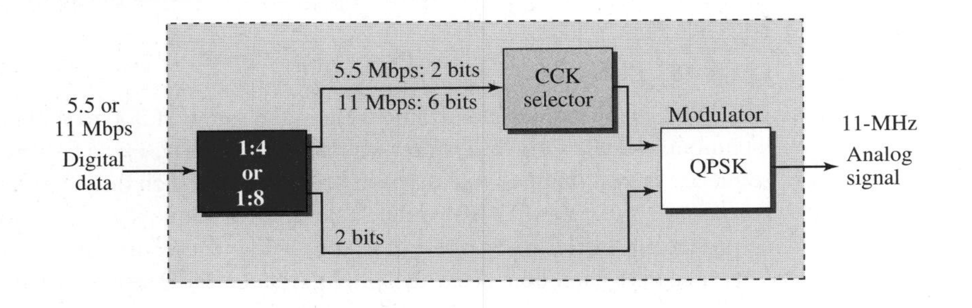

IEEE 802.11g

This new specification defines forward error correction and OFDM using the 2.400–4.835 GHz ISM band. The modulation technique achieves a 22- or 54-Mbps data rate. It is backward-compatible with 802.11b, but the modulation technique is OFDM.

IEEE 802.11n

An upgrade to the 802.11 project is called 802.11n (the next generation of wireless LAN). The goal is to increase the throughput of 802.11 wireless LANs. The new standard emphasizes not only the higher bit rate but also eliminating some unnecessary overhead. The standard uses what is called ***MIMO (multiple-input multiple-output antenna)*** to overcome the noise problem in wireless LANs. The idea is that if we can send multiple output signals and receive multiple input signals, we are in a better position to eliminate noise. Some implementations of this project have reached up to 600 Mbps data rate.

15.3 BLUETOOTH

Bluetooth is a wireless LAN technology designed to connect devices of different functions such as telephones, notebooks, computers (desktop and laptop), cameras, printers, and even coffee makers when they are at a short distance from each other. A Bluetooth LAN is an ad hoc network, which means that the network is formed spontaneously; the devices, sometimes called gadgets, find each other and make a network called a piconet. A Bluetooth LAN can even be connected to the Internet if one of the gadgets has this capability. A Bluetooth LAN, by nature, cannot be large. If there are many gadgets that try to connect, there is chaos.

Bluetooth technology has several applications. Peripheral devices such as a wireless mouse or keyboard can communicate with the computer through this technology. Monitoring devices can communicate with sensor devices in a small health care center. Home security devices can use this technology to connect different sensors to the main security controller. Conference attendees can synchronize their laptop computers at a conference.

Bluetooth was originally started as a project by the Ericsson Company. It is named for Harald Blaatand, the king of Denmark (940-981) who united Denmark and Norway. *Blaatand* translates to *Bluetooth* in English.

Today, Bluetooth technology is the implementation of a protocol defined by the IEEE 802.15 standard. The standard defines a wireless personal-area network (PAN) operable in an area the size of a room or a hall.

15.3.1 Architecture

Bluetooth defines two types of networks: piconet and scatternet.

Piconets

A Bluetooth network is called a ***piconet,*** or a small net. A piconet can have up to eight stations, one of which is called the *primary;* the rest are called *secondaries.* All the secondary stations synchronize their clocks and hopping sequence with the primary. Note that a piconet can have only one primary station. The communication between the primary and secondary stations can be one-to-one or one-to-many. Figure 15.17 shows a piconet.

Although a piconet can have a maximum of seven secondaries, additional secondaries can be in the *parked state*. A secondary in a parked state is synchronized with the primary, but cannot take part in communication until it is moved from the parked state to the active state. Because only eight stations can be active in a piconet, activating a station from the parked state means that an active station must go to the parked state.

Figure 15.17 *Piconet*

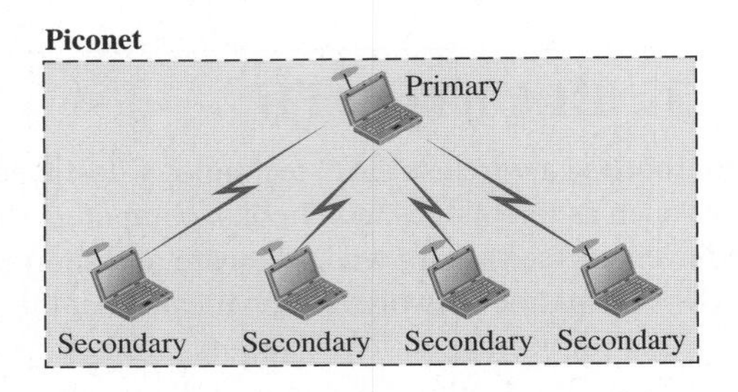

Scatternet

Piconets can be combined to form what is called a ***scatternet.*** A secondary station in one piconet can be the primary in another piconet. This station can receive messages from the primary in the first piconet (as a secondary) and, acting as a primary, deliver them to secondaries in the second piconet. A station can be a member of two piconets. Figure 15.18 illustrates a scatternet.

Figure 15.18 *Scatternet*

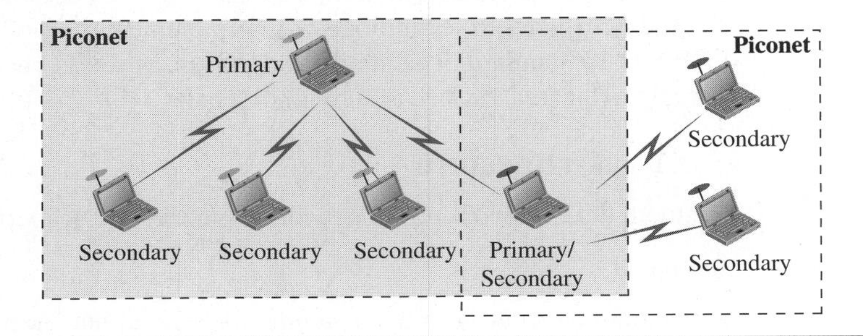

Bluetooth Devices

A Bluetooth device has a built-in short-range radio transmitter. The current data rate is 1 Mbps with a 2.4-GHz bandwidth. This means that there is a possibility of interference between the IEEE 802.11b wireless LANs and Bluetooth LANs.

15.3.2 Bluetooth Layers

Bluetooth uses several layers that do not exactly match those of the Internet model we have defined in this book. Figure 15.19 shows these layers.

L2CAP

The **Logical Link Control and Adaptation Protocol,** or **L2CAP** (L2 here means LL), is roughly equivalent to the LLC sublayer in LANs. It is used for data exchange on an

Figure 15.19 *Bluetooth layers*

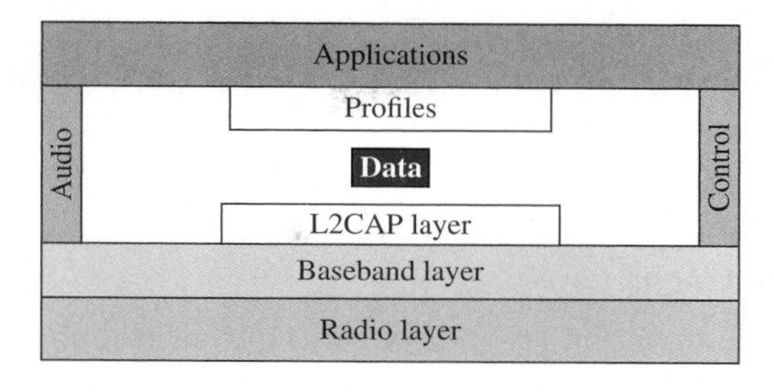

ACL link; SCO channels do not use L2CAP. Figure 15.20 shows the format of the data packet at this level.

Figure 15.20 *L2CAP data packet format*

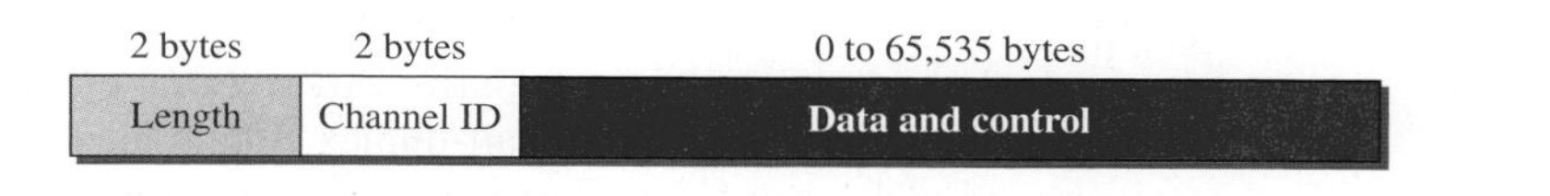

The 16-bit length field defines the size of the data, in bytes, coming from the upper layers. Data can be up to 65,535 bytes. The channel ID (CID) defines a unique identifier for the virtual channel created at this level (see below).

The L2CAP has specific duties: multiplexing, segmentation and reassembly, quality of service (QoS), and group management.

Multiplexing

The L2CAP can do multiplexing. At the sender site, it accepts data from one of the upper-layer protocols, frames them, and delivers them to the baseband layer. At the receiver site, it accepts a frame from the baseband layer, extracts the data, and delivers them to the appropriate protocol layer. It creates a kind of virtual channel that we will discuss in later chapters on higher-level protocols.

Segmentation and Reassembly

The maximum size of the payload field in the baseband layer is 2774 bits, or 343 bytes. This includes 4 bytes to define the packet and packet length. Therefore, the size of the packet that can arrive from an upper layer can only be 339 bytes. However, application layers sometimes need to send a data packet that can be up to 65,535 bytes (an Internet packet, for example). The L2CAP divides these large packets into segments and adds extra information to define the location of the segments in the original packet. The L2CAP segments the packets at the source and reassembles them at the destination.

QoS

Bluetooth allows the stations to define a quality-of-service level. We discuss quality of service in Chapter 30. For the moment, it is sufficient to know that if no quality-of-service level is defined, Bluetooth defaults to what is called *best-effort* service; it will do its best under the circumstances.

Group Management

Another functionality of L2CAP is to allow devices to create a type of logical addressing between themselves. This is similar to multicasting. For example, two or three secondary devices can be part of a multicast group to receive data from the primary.

Baseband Layer

The baseband layer is roughly equivalent to the MAC sublayer in LANs. The access method is TDMA (discussed later). The primary and secondary stations communicate with each other using time slots. The length of a time slot is exactly the same as the dwell time, 625 μs. This means that during the time that one frequency is used, a primary sends a frame to a secondary, or a secondary sends a frame to the primary. Note that the communication is only between the primary and a secondary; secondaries cannot communicate directly with one another.

TDMA

Bluetooth uses a form of TDMA that is called ***TDD-TDMA (time-division duplex TDMA).*** TDD-TDMA is a kind of half-duplex communication in which the sender and receiver send and receive data, but not at the same time (half-duplex); however, the communication for each direction uses different hops. This is similar to walkie-talkies using different carrier frequencies.

- ***Single-Secondary Communication*** If the piconet has only one secondary, the TDMA operation is very simple. The time is divided into slots of 625 μs. The primary uses even-numbered slots (0, 2, 4, . . .); the secondary uses odd-numbered slots (1, 3, 5, . . .). TDD-TDMA allows the primary and the secondary to communicate in half-duplex mode. In slot 0, the primary sends and the secondary receives; in slot 1, the secondary sends and the primary receives. The cycle is repeated. Figure 15.21 shows the concept.

Figure 15.21 *Single-secondary communication*

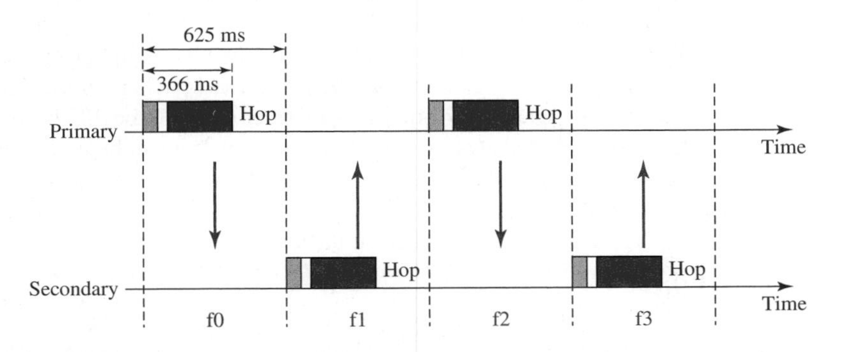

- ***Multiple-Secondary Communication*** The process is a little more involved if there is more than one secondary in the piconet. Again, the primary uses the even-numbered slots, but a secondary sends in the next odd-numbered slot if the packet in the previous slot was addressed to it. All secondaries listen on even-numbered slots, but only one secondary sends in any odd-numbered slot. Figure 15.22 shows a scenario.

Figure 15.22 *Multiple-secondary communication*

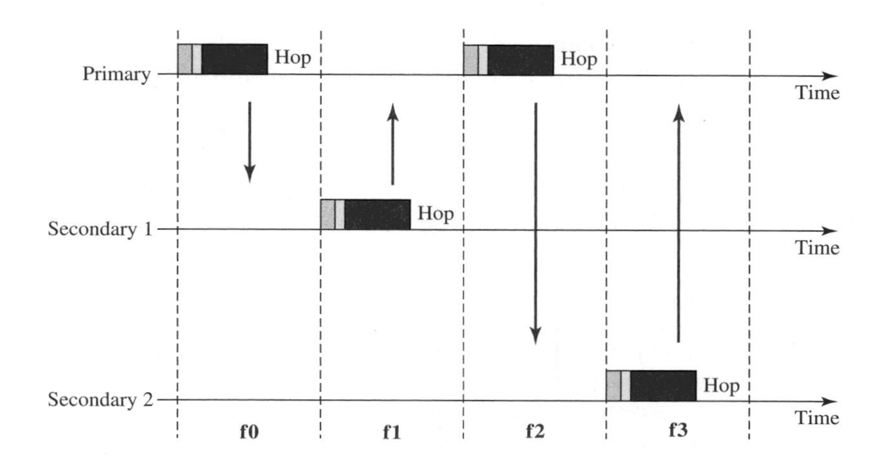

Let us elaborate on the figure.

1. In slot 0, the primary sends a frame to secondary 1.
2. In slot 1, only secondary 1 sends a frame to the primary because the previous frame was addressed to secondary 1; other secondaries are silent.
3. In slot 2, the primary sends a frame to secondary 2.
4. In slot 3, only secondary 2 sends a frame to the primary because the previous frame was addressed to secondary 2; other secondaries are silent.
5. The cycle continues.

We can say that this access method is similar to a poll/select operation with reservations. When the primary selects a secondary, it also polls it. The next time slot is reserved for the polled station to send its frame. If the polled secondary has no frame to send, the channel is silent.

Links

Two types of links can be created between a primary and a secondary: SCO links and ACL links.

- ***SCO*** A **synchronous connection-oriented (SCO) link** is used when avoiding latency (delay in data delivery) is more important than integrity (error-free delivery). In an SCO link, a physical link is created between the primary and a secondary by reserving specific slots at regular intervals. The basic unit of connection is two slots, one for each direction. If a packet is damaged, it is never retransmitted.

SCO is used for real-time audio where avoiding delay is all-important. A secondary can create up to three SCO links with the primary, sending digitized audio (PCM) at 64 kbps in each link.

- *ACL* An **asynchronous connectionless link (ACL)** is used when data integrity is more important than avoiding latency. In this type of link, if a payload encapsulated in the frame is corrupted, it is retransmitted. A secondary returns an ACL frame in the available odd-numbered slot if the previous slot has been addressed to it. ACL can use one, three, or more slots and can achieve a maximum data rate of 721 kbps.

Frame Format

A frame in the baseband layer can be one of three types: one-slot, three-slot, or five-slot. A slot, as we said before, is 625 μs. However, in a one-slot frame exchange, 259 μs is needed for hopping and control mechanisms. This means that a one-slot frame can last only 625 − 259, or 366 μs. With a 1-MHz bandwidth and 1 bit/Hz, the size of a one-slot frame is 366 bits.

A three-slot frame occupies three slots. However, since 259 μs is used for hopping, the length of the frame is $3 \times 625 - 259 = 1616$ μs or 1616 bits. A device that uses a three-slot frame remains at the same hop (at the same carrier frequency) for three slots. Even though only one hop number is used, three hop numbers are consumed. That means the hop number for each frame is equal to the first slot of the frame.

A five-slot frame also uses 259 bits for hopping, which means that the length of the frame is $5 \times 625 - 259 = 2866$ bits.

Figure 15.23 shows the format of the three frame types.

Figure 15.23 *Frame format types*

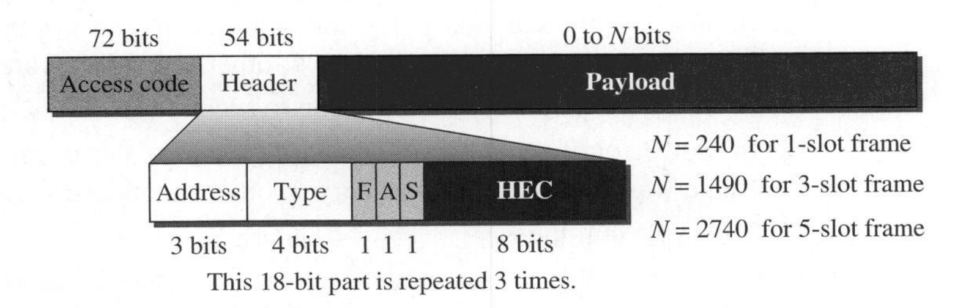

The following describes each field:

- ***Access code.*** This 72-bit field normally contains synchronization bits and the identifier of the primary to distinguish the frame of one piconet from that of another.
- ***Header.*** This 54-bit field is a repeated 18-bit pattern. Each pattern has the following subfields:

a. ***Address.*** The 3-bit address subfield can define up to seven secondaries (1 to 7). If the address is zero, it is used for broadcast communication from the primary to all secondaries.
b. ***Type.*** The 4-bit type subfield defines the type of data coming from the upper layers. We discuss these types later.
c. ***F.*** This 1-bit subfield is for flow control. When set (1), it indicates that the device is unable to receive more frames (buffer is full).
d. ***A.*** This 1-bit subfield is for acknowledgment. Bluetooth uses Stop-and-Wait ARQ; 1 bit is sufficient for acknowledgment.
e. ***S.*** This 1-bit subfield holds a sequence number. Bluetooth uses Stop-and-Wait ARQ; 1 bit is sufficient for sequence numbering.
f. ***HEC.*** The 8-bit header error correction subfield is a checksum to detect errors in each 18-bit header section. The header has three identical 18-bit sections. The receiver compares these three sections, bit by bit. If each of the corresponding bits is the same, the bit is accepted; if not, the majority opinion rules. This is a form of forward error correction (for the header only). This double error control is needed because the nature of the communication, via air, is very noisy. Note that there is no retransmission in this sublayer.

- ***Payload.*** This subfield can be 0 to 2740 bits long. It contains data or control information coming from the upper layers.

Radio Layer

The radio layer is roughly equivalent to the physical layer of the Internet model. Bluetooth devices are low-power and have a range of 10 m.

Band

Bluetooth uses a 2.4-GHz ISM band divided into 79 channels of 1 MHz each.

FHSS

Bluetooth uses the **frequency-hopping spread spectrum (FHSS)** method in the physical layer to avoid interference from other devices or other networks. Bluetooth hops 1600 times per second, which means that each device changes its modulation frequency 1600 times per second. A device uses a frequency for only 625 μs (1/1600 s) before it hops to another frequency; the dwell time is 625 μs.

Modulation

To transform bits to a signal, Bluetooth uses a sophisticated version of FSK, called GFSK (FSK with Gaussian bandwidth filtering; a discussion of this topic is beyond the scope of this book). GFSK has a carrier frequency. Bit 1 is represented by a frequency deviation above the carrier; bit 0 is represented by a frequency deviation below the carrier. The frequencies, in megahertz, are defined according to the following formula for each channel.

$$f_c = 2402 + n \text{ MHz} \qquad n = 0, 1, 2, 3, \ldots, 78$$

For example, the first channel uses carrier frequency 2402 MHz (2.402 GHz), and the second channel uses carrier frequency 2403 MHz (2.403 GHz).

15.4 END-CHAPTER MATERIALS

15.4.1 Further Reading

For more details about subjects discussed in this chapter, we recommend the following books. The items in brackets [. . .] refer to the reference list at the end of the text.

Books

Several books cover materials discussed in this chapter, including [Sch 03], [Gas 02], [For 03], [Sta 04], [Sta 02], [Kei 02], [Jam 03], [AZ 03], [Tan 03], [Cou 01], [Com 06], [GW 04], and [PD 03].

15.4.2 Key Terms

access point (AP)
ad hoc network
asynchronous connectionless link (ACL)
basic service set (BSS)
beacon frame
Bluetooth
BSS-transition mobility
complementary code keying (CCK)
direct sequence spread spectrum (DSSS)
distributed coordination function (DCF)
distributed interframe space (DIFS)
ESS-transition mobility
extended service set (ESS)
frequency-hopping spread spectrum (FHSS)
high-rate direct-sequence spread spectrum (HR-DSSS)
Logical Link Control and Adaptation Protocol (L2CAP)
multiple-input multiple-output (MIMO) antenna
network allocation vector (NAV)
no transition mobility
orthogonal frequency-division multiplexing (OFDM)
piconet
point coordination function (PCF)
pulse position modulation (PPM)
scatternet
short interframe space (SIFS)
signal to noise ration (SNR)
synchronous connection-oriented (SCO) link
TDD-TDMA (time-division duplex TDMA)

15.4.3 Summary

The nature and characteristics of a wireless network are different from those of a wired network. There are some issues in a wireless network that are negligible in a wired network. Access control in a wireless LAN is also different from that in a wired LAN because of some issues such as the hidden station problem.

Wireless LANs became formalized with the IEEE 802.11 standard, which defines two services: basic service set (BSS) and extended service set (ESS). The access method used in the distributed coordination function (DCF) MAC sublayer is CSMA/CA. The access method used in the point coordination function (PCF) MAC sublayer is polling. A frame in this network carries four addresses to define the original and previous sources and immediate and final destinations. There are other frame types in this network to handle access control and data transfer.

Bluetooth is a wireless LAN technology that connects devices (called gadgets) in a small area. A Bluetooth network is called a piconet. Piconets can be combined to form what is called a scatternet. Bluetooth uses several layers that do not exactly match those of the Internet model we have defined in this book. L2CAP is roughly equivalent to the LLC sublayer in LANs. The baseband layer is roughly equivalent to the MAC sublayer in LANs. The access method is TDMA.

15.5 PRACTICE SET

15.5.1 Quizzes

A set of interactive quizzes for this chapter can be found on the book website. It is strongly recommended that the student take the quizzes to check his/her understanding of the materials before continuing with the practice set.

15.5.2 Questions

Q15-1. What is the modulation technique in the radio layer of Bluetooth? In other words, how are digital data (bits) changed to analog signals (radio waves)?

Q15-2. Can a piconet have more than eight stations? Explain.

Q15-3. Compare a piconet and a scatternet in the Bluetooth architecture.

Q15-4. What are some reasons that CSMA/CD cannot be used in a wireless LAN?

Q15-5. Explain why fragmentation is recommended in a wireless LAN.

Q15-6. Why is SNR in a wireless LAN normally lower than SNR in a wired LAN?

Q15-7. Compare the medium of a wired LAN with that of a wireless LAN in today's communication environment.

Q15-8. What is the spread spectrum technique used by Bluetooth?

Q15-9. An AP in a wireless network plays the same role as a link-layer switch in a wired network. However, a link-layer switch has no MAC address, but an AP normally needs a MAC address. Explain the reason.

Q15-10. What MAC protocol is used in the baseband layer of Bluetooth?

Q15-11. Do the MAC addresses used in an 802.3 (Wired Ethernet) and the MAC addresses used in an 802.11 (Wireless Ethernet) belong to two different address spaces?

Q15-12. What is the role of the *radio* layer in Bluetooth?

Q15-13. What is the actual bandwidth used for communication in a Bluetooth network?

Q15-14. What is the reason that Bluetooth is normally called a wireless personal area network (WPAN) instead of a wireless local area network (WLAN)?

Q15-15. What is multipath propagation? What is its effect on wireless networks?

Q15-16. Explain why the MAC protocol is more important in wireless LANs than in wired LANs.

Q15-17. Fill in the blanks. The 83.5 MHz bandwidth in Bluetooth is divided into_____ channels, each of ______ MHz.

Q15-18. Explain why we have only one frame type in a wired LAN, but four frame types in a wireless LAN.

Q15-19. Explain why there is more attenuation in a wireless LAN than in a wired LAN, ignoring the noise and the interference.

Q15-20. An AP may connect a wireless network to a wired network. Does the AP need to have two MAC addresses in this case?

Q15-21. What is the role of the *L2CAP* layer in Bluetooth?

15.5.3 Problems

P15-1. In a BSS with no AP (ad hoc network), we have five stations: A, B, C, D, and E. Station A needs to send a message to station B. Answer the following questions for the situation where the network is using the DCF protocol:

a. What are the values of the *To DS* and *From DS* bits in the frames exchanged?

b. Which station sends the RTS frame and what is (are) the value(s) of the address field(s) in this frame?

c. Which station sends the CTS frame and what is (are) the value(s) of the address field(s) in this frame?

d. Which station sends the data frame and what is (are) the value(s) of the address field(s) in this frame?

e. Which station sends the ACK frame and what is (are) the value(s) of the address field(s) in this frame?

P15-2. Assume two 802.11 wireless networks are connected to the rest of the Internet via a router as shown in Figure 15.24. The router has received an IP datagram with the destination IP address 24.12.7.1 and needs to send it to the corresponding wireless host. Explain the process and describe how the values of address 1, address 2, address 3, and address 4 are determined in this case.

Figure 15.24 *Problem P15-2*

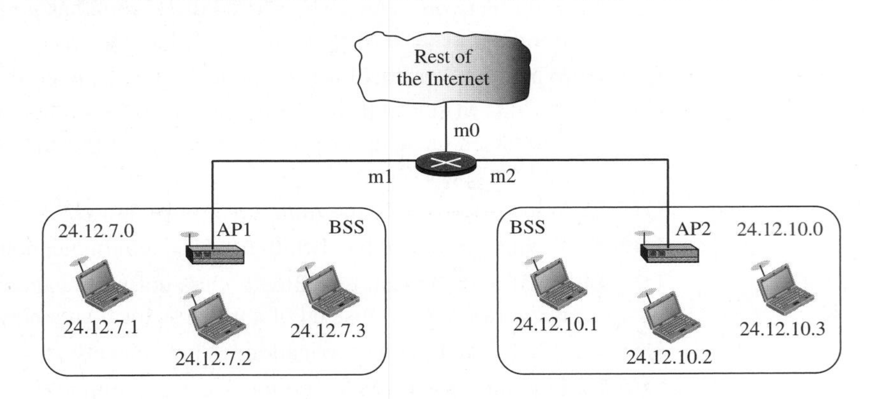

P15-3. Do some research and find out how flow and error control are accomplished in an 802.11 network using the DCF MAC sublayer.

P15-4. In an 802.11 communication, the size of the payload (frame body) is 1200 bytes. The station decides to fragment the frame into three fragments, each of 400 payload bytes. Answer the following questions:

a. What would be the size of the data frame if no fragmentations were done?

b. What is the size of each frame after fragmentation?

c. How many total bytes are sent after fragmentation (ignoring the extra control frames)?

d. How many extra bytes are sent because of fragmentation (again ignoring extra control frames)?

P15-5. An 802.11 network may use four different interframe spaces (IFSs) to delay the transmission of a frame in different situations. This allows low-priority traffic to wait for high-priority traffic when the channel becomes idle. Normally, four different IFSs are used in different implementations, as shown in Figure 15.25. Explain the purpose of these IFSs (you may need to do some research using the Internet).

Figure 15.25 *Problem P15-5*

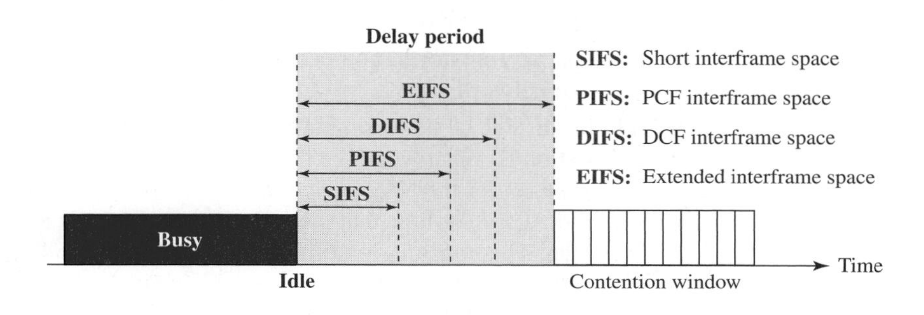

P15-6. In Figure 15.26, two wireless networks, BSS1 and BSS2, are connected through a wired distribution system (DS), an Ethernet LAN. Assume station A in BSS1 needs to send a data frame to station C in BSS2. Show the value of addresses in 802.11 and 802.3 frames for three transmissions: from station A to AP1, from AP1 to AP2, and from AP2 to station C. Note that the communication between AP1 and AP2 occurs in a wired environment.

Figure 15.26 *Problem P15-6*

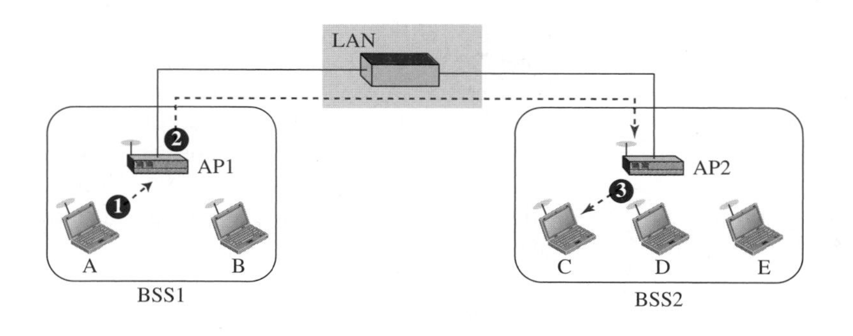

P15-7. Repeat the previous problem (Figure 15.26), but assume that the distribution system is also wireless. AP1 is connected to AP2 through a wireless channel. Show the value of addresses in all communication sections: from station A to AP1, from AP1 to AP2, and from AP2 to station C.

P15-8. Although an RTS frame defines the value of time that NAV can be effective for the rest of the session, why does the 802.11 project define that other frames used in the session should redefine the rest of the period for NAV?

P15-9. Assume a frame moves from a wireless network using the 802.11 protocol to a wired network using the 802.3 protocol. Show how the field values in the 802.3 frame are filled with the values of the 802.11 frame. Assume that the transformation occurs at the AP that is on the boundary between the two networks.

P15-10. Assume that a frame moves from a wired network using the 802.3 protocol to a wireless network using the 802.11 protocol. Show how the field values in the 802.11 frame are filled with the values of the 802.3 frame. Assume that the transformation occurs at the AP that is on the boundary between the two networks.

P15-11. In an 802.11 network, station A sends one data frame (not fragmented) to station B. What would be the value of the D field (in microseconds) that needs to be set for the NAV period in each of the following frames: RTS, CTS, data, and ACK? Assume that the transmission time for RTS, CTS, and ACK is 4 μs each. The transmission time for the data frame is 40 μs and the SIFS duration is set to 1 μs. Ignore the propagation time. Note that each frame needs to set the duration of NAV for the rest of the time the medium needs to be reserved to complete the transaction.

P15-12. In an 802.11 network, there are three stations, A, B, and C. Station C is hidden from A, but can be seen (electronically) by B. Now assume that station A needs to send data to station B. Since C is hidden from A, the RTS frame cannot reach C. Explain how station C can find out that the channel is locked by A and that it should refrain from transmitting.

P15-13. In an 802.11, give the value of the address 1 field in each of the following situations (left bit defines *To DS* and right bit defines *From DS*).

a. 00 **b.** 01 **c.** 10 **d.** 11

P15-14. In an 802.11 network, station A sends two data fragments to station B. What would be the value of the D field (in microseconds) that needs to be set for the NAV period in each of the following frames: RTS, CTS, data, and ACK? Assume that the transmission time for RTS, CTS, and ACK is 4 μs each. The transmission time for each fragment is 20 μs and the SIFS duration is set to 1 μs. Ignore the propagation time. Note that each frame needs to set the duration of NAV for the rest of the time the medium needs to be reserved to complete the transaction.

P15-15. In an 802.11, give the value of the address 3 field in each of the following situations (left bit defines *To DS* and right bit defines *From DS*).

a. 00 **b.** 01 **c.** 10 **d.** 11

P15-16. In an 802.11, give the value of the address 4 field in each of the following situations (left bit defines *To DS* and right bit defines *From DS*).

a. 00 **b.** 01 **c.** 10 **d.** 11

P15-17. Both the IP protocol and the 802.11 project fragment their packets. IP fragments a datagram at the network layer; 802.11 fragments a frame at the data-link layer. Compare and contrast the two fragmentation schemes, using the different fields and subfields used in each protocol.

P15-18. In an 802.11, give the value of the address 2 field in each of the following situations (left bit defines *To DS* and right bit defines *From DS*).

a. 00 **b.** 01 **c.** 10 **d.** 11

P15-19. In an 802.11 network, three stations (A, B, and C) are contending to access the medium. The contention window for each station has 31 slots. Station A randomly picks up the first slot; station B picks up the fifth slot; and station C picks up the twenty-first slot. Show the procedure each station should follow.

P15-20. A BSS ID (BSSID) is a 48-bit address assigned to a BSS in an 802.11 network. Do some research and find what the use of the BSSID is and how BSSIDs are assigned in ad hoc and infrastructure networks.

P15-21. In Figure 15.24 (Problem P-2), assume that the host with IP address 24.12.10.3 needs to send an IP datagram to the host with IP address 128.41.23.12 somewhere in the world (not shown in the figure). Explain the process and show how the values of address 1, address 2, address 3, and address 4 (see Figure 15.24) are determined in this case.

P15-22. In an 802.11 network, assume station A has four fragments to send to station B. If the sequence number of the first fragment is selected as 3273, what are the values of the more fragment flag, fragment number, and sequence number?

P15-23. Figure 15.23 shows the frame format of the baseband layer in Bluetooth (802.15). Based on this format, answer the following questions:

a. What is the range of the address domain in a Bluetooth network?

b. How many stations can be active at the same time in a piconet based on the information in the above figure?

15.6 SIMULATION EXPERIMENTS

15.6.1 Applets

We have created some Java applets to show some of the main concepts discussed in this chapter. It is strongly recommended that the students activate these applets on the book website and carefully examine the protocols in action.

15.6.2 Lab Assignments

Use the book website to find how you can use a simulator for wireless LANs.

Lab15-1. In this lab we capture and study wireless frames that are exchanged between a wireless host and the access point. See the book website for a detailed description of this lab.

CHAPTER 16

Other Wireless Networks

We discussed wired LANs in Chapter 13 and other wired networks (access networks and wide area networks) in Chapter 14. We discussed wireless LANs in Chapter 15. It is time now to discuss other wireless networks (access networks and wide area networks) in this chapter.

This chapter is divided into three sections:

- The first section discusses the WiMAX, a wireless access network that can replace the wired access networks we discussed in Chapter 14. The section first describes services provided by this network. It then describes the IEEE 802.16 project as the basis of the network. The section finally defines the link-layer and the physical layer of WiMAX.
- The second section discusses cellular telephone networks. It explains the frequency reuse principle. It then describes the general operations of this network. The next four sections each discuss one of the four generations of the cellular telephony network.
- The third section discusses satellite networks. It first describes the operations of the all types of satellites. The section then defines GEO satellites and their characteristics, then moves to MEO satellites and shows their applications. Finally, the section discusses the LEO satellites and their features and applications.

16.1 WiMAX

We first need to discuss a wireless access network. In Chapter 14, we discussed that the telephone and cable companies provide wired access to the Internet for homes and offices. Today, the tendency is to move to wireless technology for this purpose. There are two reasons for this tendency. First, people want to have access to the Internet from home or office (fixed) where the wired access to the Internet is either not available or is expensive. Second, people need to access the Internet when they are using their cellular phones (mobiles). The **Worldwide Interoperability for Microwave Access (WiMAX)** has been designed for these types of applications. It provides the "last mile" broadband wireless access.

16.1.1 Services

WiMAX provides two types of services to subscribers: fixed and mobile.

Fixed WiMAX

Figure 16.1 shows the idea behind a fixed service. A base station can use different types of antenna (omnidirectional, sector, or panel) to optimize the performance. WiMAX uses a beamsteering **adaptive antenna system (AAS).** While transmitting, it can focus its energy in the direction of the subscriber; while receiving, it can focus in the direction of the subscriber station to receive maximum energy sent by the subscriber.

Figure 16.1 *Fixed WiMAX*

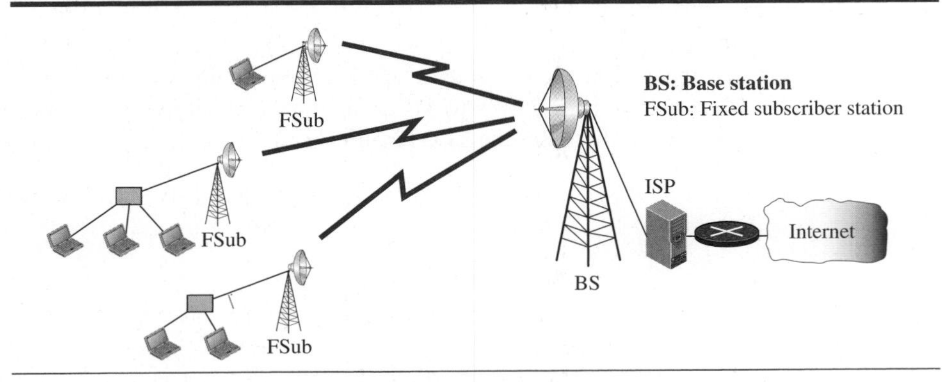

The fixed service can be compared with the service provided by the telephone and the network companies using wired connections. WiMAX also uses a MIMO antenna system, which can provide simultaneous transmitting and receiving.

Mobile WiMAX

Figure 16.2 shows the idea behind mobile service. It is the same as fixed service except the subscribers are mobile stations that move from one place to another. The same issues involved in the cellular telephone system, such as roaming, are present here.

Figure 16.2 *Mobile WiMAX*

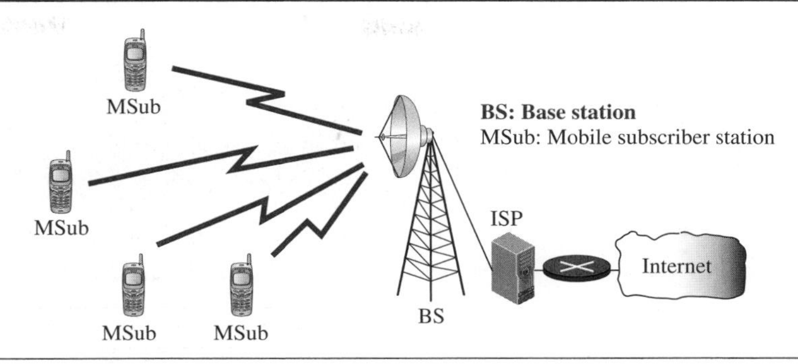

16.1.2 IEEE Project 802.16

WiMAX is the result of the IEEE 802.16 project, which was an effort to standardize the proprietary broadband wireless system in 2002. The standard is sometimes referred to as *wireless local loop*, in contrast with wired local loop (dial-up, DLS, or cable). Before we discuss this standard, let us compare the 802.16 and 802.11 projects. First, 802.11 is a standard for a wireless LAN; 802.16 is a standard for a wireless WAN (or MAN). The distance between a base station and a host in the first is very limited; the base station and subscriber station in the second may be separated by tens of kilometers. Project 802.11 defines a connectionless communication; project 802.16 defines a connection-oriented service.

A later revision of IEEE 802.16 created two new standards called IEEE 802.16d, which concentrates on the fixed WiMAX, and IEEE 802.16e, which defines the mobile WiMAX. The two new standards do not change the main idea behind the original 802.16, but concentrate on the nature of two services.

16.1.3 Layers in Project 802.16

Figure 16.3 shows the layers in the 802.16 project. IEEE has divided the data-link layer into three sublayers and the physical layer into two sublayers.

Service Specific Convergence Sublayer

This is actually the DLC sublayer revised for broadband wireless communication. It has been devised for a connection-oriented service in which each connection may benefit from a specific quality of service (QoS), as discussed in Chapter 30.

MAC Sublayer

The MAC sublayer defines the access method and the format of the frame. It is a sublayer designed for connection-oriented service. The packets are routed from the base station to the subscriber station using a connection identifier which is the same during the duration of the communication. For more about connection-oriented communication, see Chapters 8 and 18.

Figure 16.3 *Data-link and physical layers*

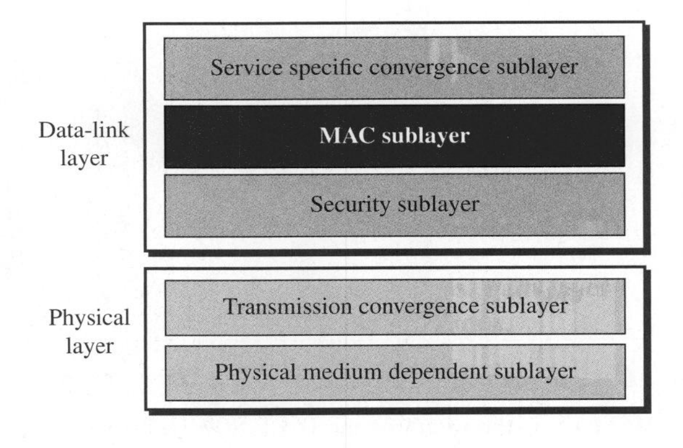

Access Method

WiMAX uses the reservation (scheduling) access method we discussed in Chapter 12 (Figure 12.18). The base station needs to make a slot reservation before sending a slot-size of data to a subscriber station; each subscriber station needs to make a reservation before sending a slot-size of data to the base station.

Frame Format

The frame format is shown in Figure 16.4. We distinguish two types of frames: generic and control. The first is used to send and receive payload; the second is used only during the connection establishment. Both frame types use a 6-byte generic header. However, some bytes have different interpretations in different frame types.

Figure 16.4 *WiMAX MAC frame format*

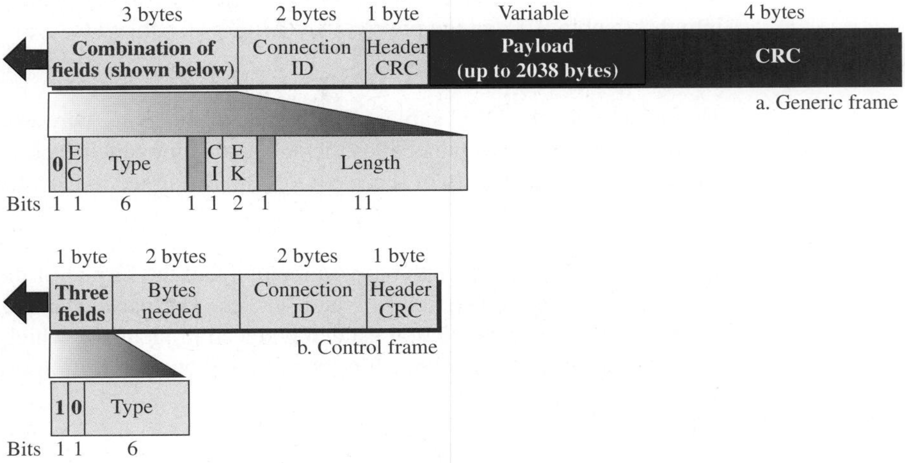

a. Generic frame

b. Control frame

A brief purpose of each field follows.

- The first bit in a frame is the frame identifier. If it is 0, the frame is a generic frame; if it is 1, it is a control frame.
- ***EC.*** The *encryption control* field uses one bit to define whether the frame should be encrypted for security purpose (See Chapters 31 and 32). If the bit is 0, it means no encryption; if it is 1, it means the frame needs to be encrypted at the *security sublayer,* described later.
- ***Type.*** The *type* field uses six bits to define the type of the frame. This field is only present in the generic frame and normally is used to define the type of the payload. The payload can be a packed load, a fragmented load, and so on.
- ***CI.*** The *checksum ID* field uses one bit to define whether the frame checksum field should be present or not. If the payload is multimedia, forward error correction is applied (at the physical layer) to the frame and there is no need for checksum.
- ***EK.*** The *encryption key* field uses two bits to define one of the four keys for encryption if encryption is required (see EC field).
- ***Length.*** The *length* field uses eleven bits to define the total length of the frame. Note that this field is present in a *generic* frame and is replaced by the *bytes needed* field in the control frame.
- ***Bytes Needed.*** The *bytes needed* field uses sixteen bits to define the number of bytes needed for allocated slots in the physical layer.
- ***Connection ID.*** The *connection ID* field uses sixteen bits to define the connection identifier for the current connection. Note that the IEEE 802.16 and WiMAX define a connection-oriented protocol, as we discussed before.
- ***Header CRC.*** Both types of frames need to have an 8-bit header CRC field. The header CRC is used to check whether the header itself is corrupted. It uses the polynomial ($x^8 + x^2 + x + 1$) as the divisor.
- ***Payload.*** This variable-length field defines the payload, the data that is encapsulated in the frame from the *service specific convergence* sublayer. The field is not needed in the control frame.
- ***CRC.*** The last field, if present, is used for error detection over the whole frame. It uses the same divisor discussed for the Ethernet.

Addressing

Each subscriber and base station typically has a 48-bit MAC address as defined in Chapter 9 (link-layer addressing) because each station is a node in the global Internet. However, there is no source or destination address field in Figure 16.4. The reason is that the combination of source and destination addresses are mapped to a *connection identifier* during the connection-establishing phase. This protocol is a connection-oriented protocol that uses a connection identifier or *virtual connection identifier* (VCI) as we discussed in Chapters 8 and 18. Each frame then uses the same connection identifier for the duration of data transfer.

Security Sublayer

The last sublayer in the data-link layer provides security for communication using WiMAX. The nature of broadband wireless communication requires security to provide encryption for the information exchanged between a subscriber station and the base station. We discuss these issues in Chapters 31 and 32 in detail.

Transmission Convergence Sublayer

The transmission convergence sublayer uses TDD (time-division duplex), a variation of time-division multiplexing, discussed in Chapter 6, designed for duplex (bidirectional) communication. The physical layer packs the frames received from the data-link layer into two subframes at the physical layer. However, we need to distinguish between the term *frame* at the physical layer and the one at the data-link layer. A data-link layer frame in the network layer may be encapsulated in several slots in the corresponding subframe at the physical layer.

Each frame at the physical layer is made of two subframes that carry data from the base station to the subscribers (downstream) and from the subscribers to the base station (upstream). Each subframe is divided into slots as shown in Figure 16.5.

Figure 16.5 *WiMAX frame structure at the physical layer*

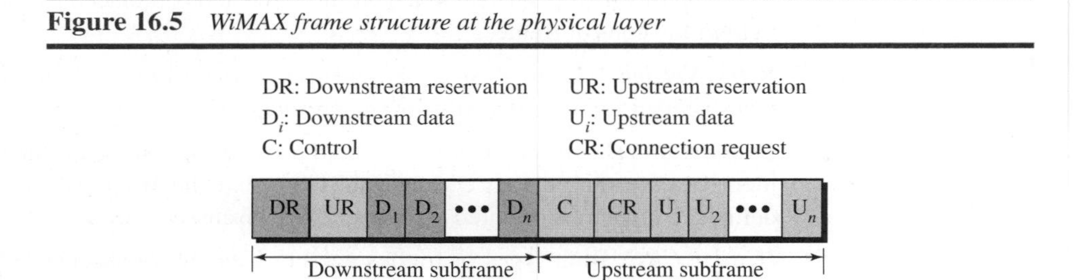

Physical Medium Dependent Sublayer

This sublayer is in continuous revision. Originally, 802.16 defined the band 10-66 GHz and modulations QPSK (2 bits/baud), QAM-16 (4 bits/baud), and QAM-64 (6 bits/baud) for long-, medium-, and short-distance communication from the base station. Later IEEE defined 802.16d (fixed WiMAX), which added the band 2-11 GHz (compatible with wireless LANs) using the orthogonal frequency-division multiplexing (OFDM). Sometime later, IEEE defined 802.16e (mobile WiMAX) and added scalable orthogonal frequency-division multiplexing (SOFDM).

16.2 CELLULAR TELEPHONY

Cellular telephony is designed to provide communications between two moving units, called *mobile stations (MSs),* or between one mobile unit and one stationary unit, often called a *land unit.* A service provider must be able to locate and track a caller, assign a channel to the call, and transfer the channel from base station to base station as the caller moves out of range.

To make this tracking possible, each cellular service area is divided into small regions called *cells*. Each cell contains an antenna and is controlled by a solar- or AC-powered network station, called the *base station* (BS). Each base station, in turn, is controlled by a switching office, called a ***mobile switching center (MSC).*** The MSC coordinates communication between all the base stations and the telephone central office. It is a computerized center that is responsible for connecting calls, recording call information, and billing (see Figure 16.6).

Figure 16.6 *Cellular system*

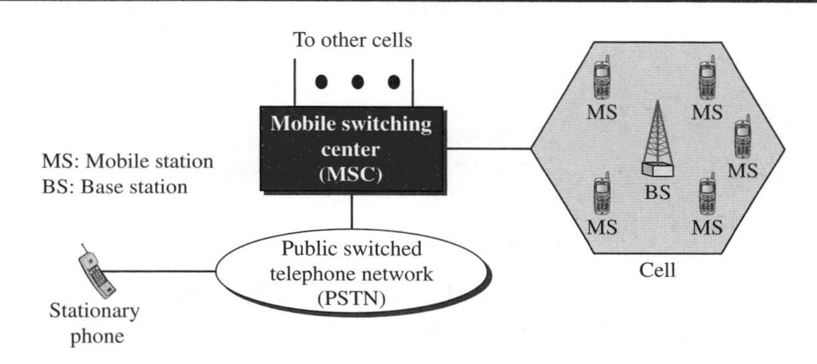

Cell size is not fixed and can be increased or decreased depending on the population of the area. The typical radius of a cell is 1 to 12 mi. High-density areas require more, geographically smaller cells to meet traffic demands than do low-density areas. Once determined, cell size is optimized to prevent the interference of adjacent cell signals. The transmission power of each cell is kept low to prevent its signal from interfering with those of other cells.

16.2.1 Operation

Let us first briefly discuss the operation of the cellular telephony.

Frequency-Reuse Principle

In general, neighboring cells cannot use the same set of frequencies for communication because doing so may create interference for the users located near the cell boundaries. However, the set of frequencies available is limited, and frequencies need to be reused. A frequency reuse pattern is a configuration of N cells, N being the **reuse factor,** in which each cell uses a unique set of frequencies. When the pattern is repeated, the frequencies can be reused. There are several different patterns. Figure 16.7 shows two of them.

The numbers in the cells define the pattern. The cells with the same number in a pattern can use the same set of frequencies. We call these cells the *reusing cells*. As Figure 16.7 shows, in a pattern with reuse factor 4, only one cell separates the cells using the same set of frequencies. In a pattern with reuse factor 7, two cells separate the reusing cells.

Figure 16.7 *Frequency reuse patterns*

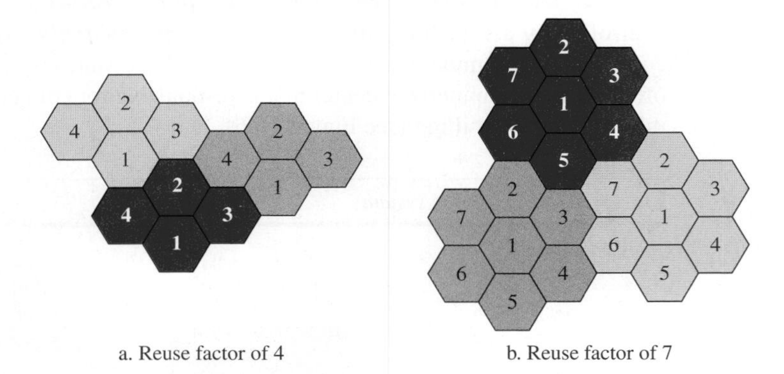

Transmitting

To place a call from a mobile station, the caller enters a code of 7 or 10 digits (a phone number) and presses the send button. The mobile station then scans the band, seeking a setup channel with a strong signal, and sends the data (phone number) to the closest base station using that channel. The base station relays the data to the MSC. The MSC sends the data on to the telephone central office. If the called party is available, a connection is made and the result is relayed back to the MSC. At this point, the MSC assigns an unused voice channel to the call, and a connection is established. The mobile station automatically adjusts its tuning to the new channel, and communication can begin.

Receiving

When a mobile phone is called, the telephone central office sends the number to the MSC. The MSC searches for the location of the mobile station by sending query signals to each cell in a process called *paging*. Once the mobile station is found, the MSC transmits a ringing signal and, when the mobile station answers, assigns a voice channel to the call, allowing voice communication to begin.

Handoff

It may happen that, during a conversation, the mobile station moves from one cell to another. When it does, the signal may become weak. To solve this problem, the MSC monitors the level of the signal every few seconds. If the strength of the signal diminishes, the MSC seeks a new cell that can better accommodate the communication. The MSC then changes the channel carrying the call (hands the signal off from the old channel to a new one).

Hard Handoff

Early systems used a hard **handoff.** In a hard handoff, a mobile station only communicates with one base station. When the MS moves from one cell to another, communication must first be broken with the previous base station before communication can be established with the new one. This may create a rough transition.

Soft Handoff

New systems use a soft handoff. In this case, a mobile station can communicate with two base stations at the same time. This means that, during handoff, a mobile station may continue with the new base station before breaking off from the old one.

Roaming

One feature of cellular telephony is called ***roaming.*** Roaming means, in principle, that a user can have access to communication or can be reached where there is coverage. A service provider usually has limited coverage. Neighboring service providers can provide extended coverage through a roaming contract. The situation is similar to snail mail between countries. The charge for delivery of a letter between two countries can be divided upon agreement by the two countries.

16.2.2 First Generation (1G)

Cellular telephony is now in its fourth generation. The first generation was designed for voice communication using analog signals. We discuss one first-generation mobile system used in North America, AMPS.

AMPS

Advanced Mobile Phone System (AMPS) is one of the leading analog cellular systems in North America. It uses FDMA (see Chapter 12) to separate channels in a link.

AMPS is an analog cellular phone system using FDMA.

Bands

AMPS operates in the ISM 800-MHz band. The system uses two separate analog channels, one for forward (base station to mobile station) communication and one for reverse (mobile station to base station) communication. The band between 824 and 849 MHz carries reverse communication; the band between 869 and 894 MHz carries forward communication (see Figure 16.8).

Figure 16.8 *Cellular bands for AMPS*

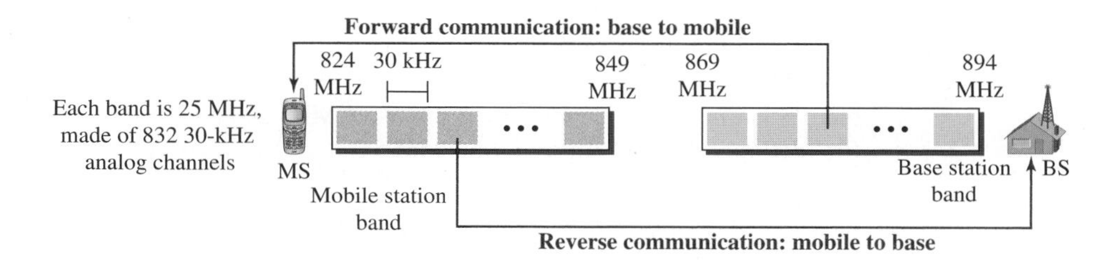

Each band is divided into 832 channels. However, two providers can share an area, which means 416 channels in each cell for each provider. Out of these 416, 21 channels are used for control, which leaves 395 channels. AMPS has a frequency reuse factor of 7; this means only one-seventh of these 395 traffic channels are actually available in a cell.

Transmission
AMPS uses FM and FSK for modulation. Figure 16.9 shows the transmission in the reverse direction. Voice channels are modulated using FM, and control channels use FSK to create 30-kHz analog signals. AMPS uses FDMA to divide each 25-MHz band into 30-kHz channels.

Figure 16.9 *AMPS reverse communication band*

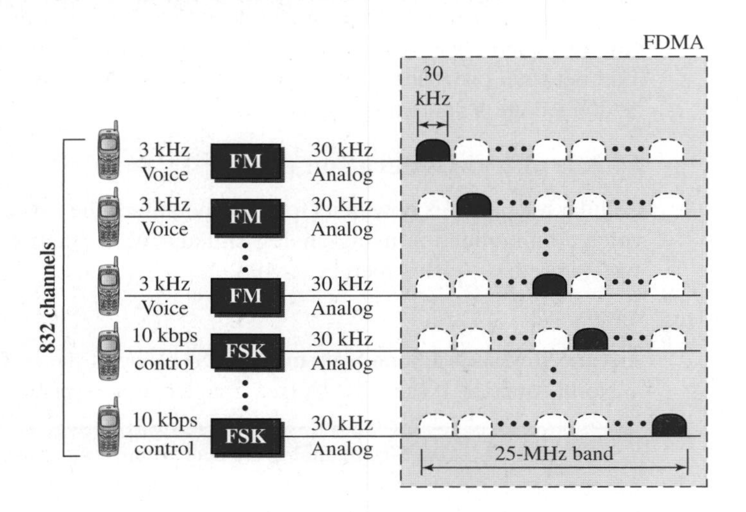

16.2.3 Second Generation (2G)

To provide higher-quality (less noise-prone) mobile voice communications, the second generation of the cellular phone network was developed. While the first generation was designed for analog voice communication, the second generation was mainly designed for digitized voice. Three major systems evolved in the second generation: D-AMPS, GSM, and IS-95.

D-AMPS

The product of the evolution of the analog AMPS into a digital system is **digital AMPS (D-AMPS).** D-AMPS was designed to be backward-compatible with AMPS. This means that in a cell, one telephone can use AMPS and another D-AMPS. D-AMPS was first defined by IS-54 (Interim Standard 54) and later revised by IS-136.

Band
D-AMPS uses the same bands and channels as AMPS.

Transmission
Each voice channel is digitized using a very complex PCM and compression technique. A voice channel is digitized to 7.95 kbps. Three 7.95-kbps digital voice channels are combined using TDMA. The result is 48.6 kbps of digital data; much of this is overhead. As Figure 16.10 shows, the system sends 25 frames per second, with 1944 bits per

Figure 16.10 *D-AMPS*

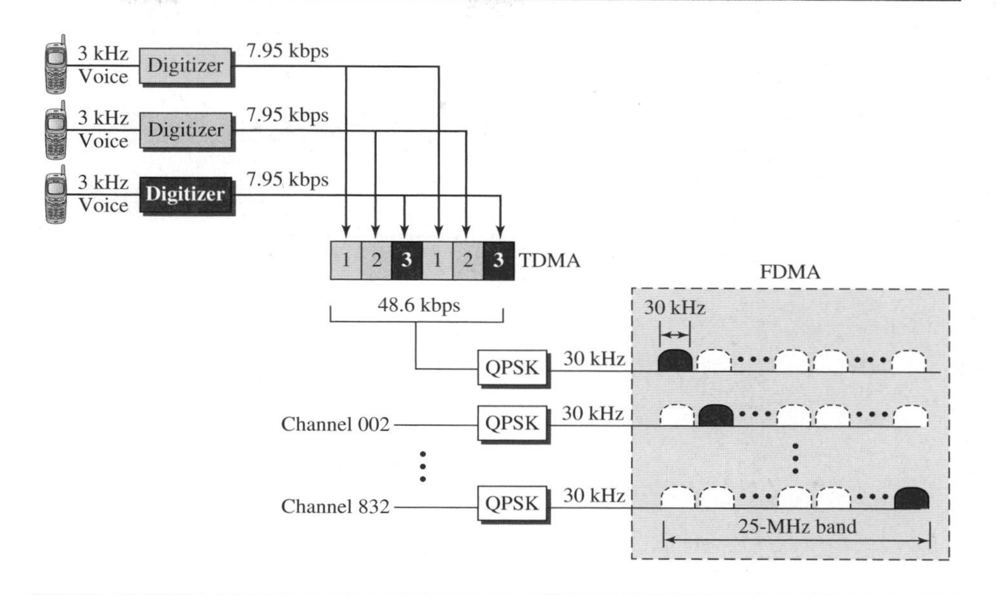

frame. Each frame lasts 40 ms (1/25) and is divided into six slots shared by three digital channels; each channel is allotted two slots.

Each slot holds 324 bits. However, only 159 bits come from the digitized voice; 64 bits are for control and 101 bits are for error correction. In other words, each channel drops 159 bits of data into each of the two channels assigned to it. The system adds 64 control bits and 101 error-correcting bits.

The resulting 48.6 kbps of digital data modulates a carrier using QPSK; the result is a 30-kHz analog signal. Finally, the 30-kHz analog signals share a 25-MHz band (FDMA). D-AMPS has a frequency reuse factor of 7.

D-AMPS, or IS-136, is a digital cellular phone system using TDMA and FDMA.

GSM

The **Global System for Mobile Communication (GSM)** is a European standard that was developed to provide a common second-generation technology for all Europe. The aim was to replace a number of incompatible first-generation technologies.

Bands

GSM uses two bands for duplex communication. Each band is 25 MHz in width, shifted toward 900 MHz, as shown in Figure 16.11. Each band is divided into 124 channels of 200 kHz separated by guard bands.

Transmission

Figure 16.12 shows a GSM system. Each voice channel is digitized and compressed to a 13-kbps digital signal. Each slot carries 156.25 bits. Eight slots share a frame (TDMA).

Figure 16.11 *GSM bands*

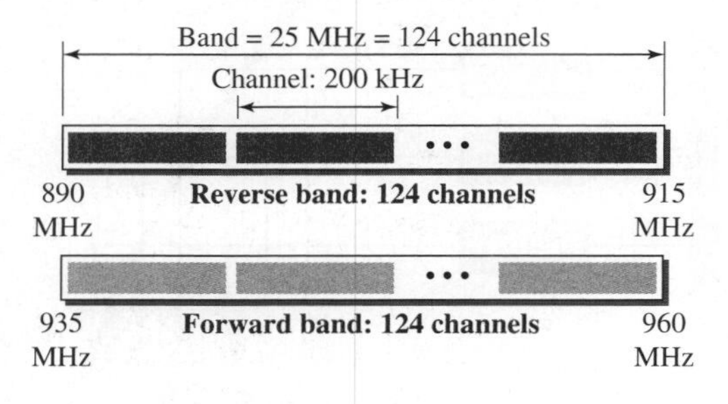

Figure 16.12 *GSM*

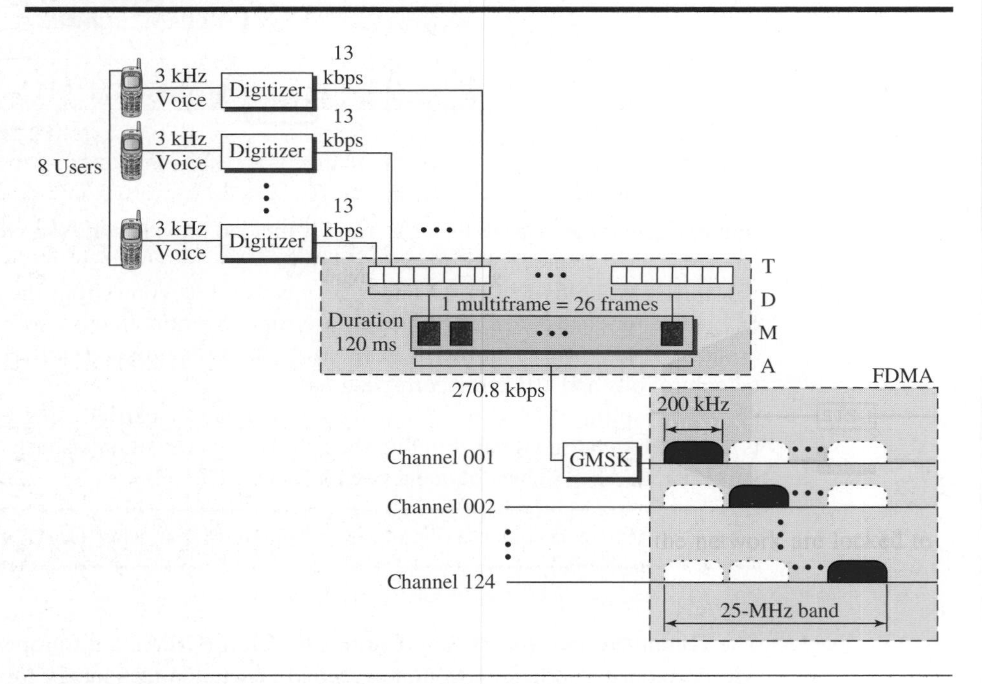

Twenty-six frames also share a multiframe (TDMA). We can calculate the bit rate of each channel as follows.

$$\textbf{Channel data rate} = (1/120 \text{ ms}) \times 26 \times 8 \times 156.25 = 270.8 \text{ kbps}$$

Each 270.8-kbps digital channel modulates a carrier using GMSK (a form of FSK used mainly in European systems); the result is a 200-kHz analog signal. Finally 124 analog channels of 200 kHz are combined using FDMA. The result is a 25-MHz band. Figure 16.13 shows the user data and overhead in a multiframe.

Figure 16.13 *Multiframe components*

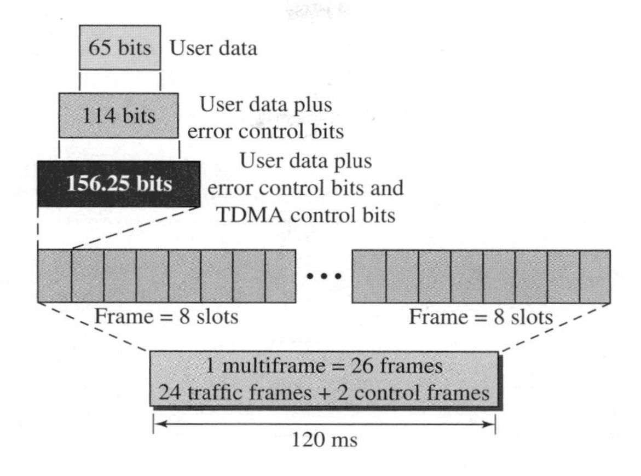

The reader may have noticed the large amount of overhead in TDMA. The user data are only 65 bits per slot. The system adds extra bits for error correction to make it 114 bits per slot. To this, control bits are added to bring it up to 156.25 bits per slot. Eight slots are encapsulated in a frame. Twenty-four traffic frames and two additional control frames make a multiframe. A multiframe has a duration of 120 ms. However, the architecture does define superframes and hyperframes that do not add any overhead; we will not discuss them here.

Reuse Factor

Because of the complex error correction mechanism, GSM allows a reuse factor as low as 3.

GSM is a digital cellular phone system using TDMA and FDMA.

IS-95

One of the dominant second-generation standards in North America is **Interim Standard 95 (IS-95).** It is based on CDMA and DSSS.

Bands and Channels

IS-95 uses two bands for duplex communication. The bands can be the traditional ISM 800-MHz band or the ISM 1900-MHz band. Each band is divided into 20 channels of 1.228 MHz separated by guard bands. Each service provider is allotted 10 channels. IS-95 can be used in parallel with AMPS. Each IS-95 channel is equivalent to 41 AMPS channels (41×30 kHz = 1.23 MHz).

Synchronization

All base channels need to be synchronized to use CDMA. To provide synchronization, bases use the services of GPS (Global Positioning System), a satellite system that we discuss in the next section.

Forward Transmission

IS-95 has two different transmission techniques: one for use in the forward (base to mobile) direction and another for use in the reverse (mobile to base) direction. In the forward direction, communications between the base and all mobiles are synchronized; the base sends synchronized data to all mobiles. Figure 16.14 shows a simplified diagram for the forward direction.

Figure 16.14 *IS-95 forward transmission*

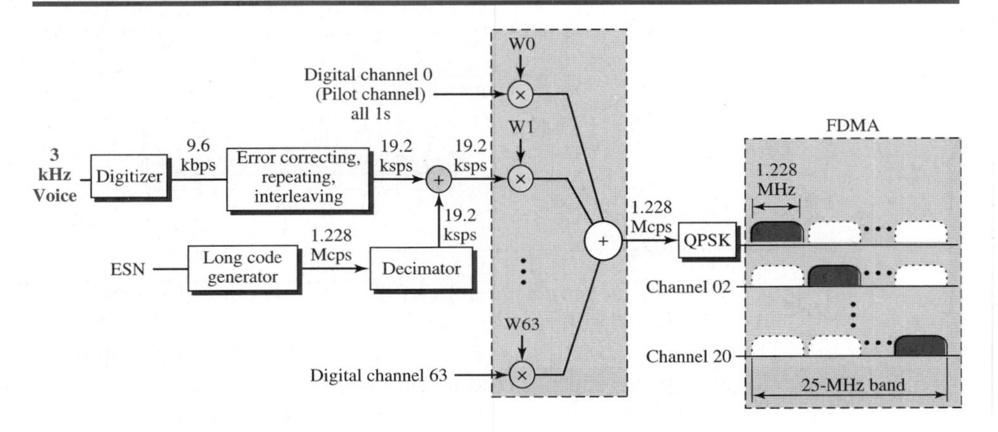

Each voice channel is digitized, producing data at a basic rate of 9.6 kbps. After adding error-correcting and repeating bits and interleaving, the result is a signal of 19.2 ksps (kilo signals per second). This output is now scrambled using a 19.2-ksps signal. The scrambling signal is produced from a long code generator that uses the electronic serial number (ESN) of the mobile station and generates 2^{42} pseudorandom chips, each chip having 42 bits. Note that the chips are generated pseudorandomly, not randomly, because the pattern repeats itself. The output of the long code generator is fed to a decimator, which chooses 1 bit out of 64 bits. The output of the decimator is used for scrambling. The scrambling is used to create privacy; the ESN is unique for each station.

The result of the scrambler is combined using CDMA. For each traffic channel, one Walsh 64 × 64 row chip is selected. The result is a signal of 1.228 Mcps (megachips per second).

$$\textbf{19.2 ksps} \times \textbf{64 cps} = \textbf{1.228 Mcps}$$

The signal is fed into a QPSK modulator to produce a signal of 1.228 MHz. The resulting bandwidth is shifted appropriately, using FDMA. An analog channel creates 64 digital channels, of which 55 channels are traffic channels (carrying digitized voice). Nine channels are used for control and synchronization:

a. Channel 0 is a pilot channel. This channel sends a continuous stream of 1s to mobile stations. The stream provides bit synchronization, serves as a phase reference for demodulation, and allows the mobile station to compare the signal strength of neighboring bases for handoff decisions.

b. Channel 32 gives information about the system to the mobile station.

c. Channels 1 to 7 are used for paging, to send messages to one or more mobile stations.

d. Channels 8 to 31 and 33 to 63 are traffic channels carrying digitized voice from the base station to the corresponding mobile station.

Reverse Transmission

The use of CDMA in the forward direction is possible because the pilot channel sends a continuous sequence of 1s to synchronize transmission. The synchronization is not used in the reverse direction because we need an entity to do that, which is not feasible. Instead of CDMA, the reverse channels use DSSS (direct sequence spread spectrum), which we discussed in Chapter 6. Figure 16.15 shows a simplified diagram for reverse transmission.

Figure 16.15 *IS-95 reverse transmission*

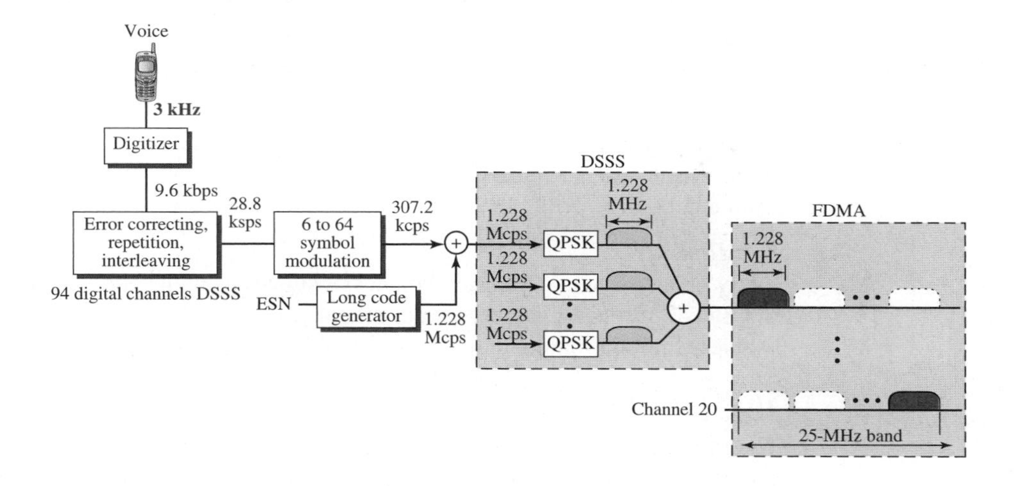

Each voice channel is digitized, producing data at a rate of 9.6 kbps. However, after adding error-correcting and repeating bits plus interleaving, the result is a signal of 28.8 ksps. The output is now passed through a 6/64 symbol modulator. The symbols are divided into six-symbol chunks, and each chunk is interpreted as a binary number (from 0 to 63). The binary number is used as the index to a 64×64 Walsh matrix for selection of a row of chips. Note that this procedure is not CDMA; each bit is not multiplied by the chips in a row. Each six-symbol chunk is replaced by a 64-chip code. This is done to provide a kind of orthogonality; it differentiates the streams of chips from the different mobile stations. The result creates a signal of 307.2 kcps or $(28.8/6) \times 64$.

Spreading is the next step; each chip is spread into 4. Again the ESN of the mobile station creates a long code of 42 bits at a rate of 1.228 Mcps, which is 4 times 307.2. After spreading, each signal is modulated using QPSK, which is slightly different from the one used in the forward direction; we do not go into details here. Note that there is no multiple-access mechanism here; all reverse channels send their analog signal into the air, but the correct chips will be received by the base station due to spreading.

Although we can create $2^{42} - 1$ digital channels in the reverse direction (because of the long code generator), normally 94 channels are used; 62 are traffic channels, and 32 are channels used to gain access to the base station.

IS-95 is a digital cellular phone system using CDMA/DSSS and FDMA.

Two Data Rate Sets

IS-95 defines two data rate sets, with four different rates in each set. The first set defines 9600, 4800, 2400, and 1200 bps. If, for example, the selected rate is 1200 bps, each bit is repeated 8 times to provide a rate of 9600 bps. The second set defines 14,400, 7200, 3600, and 1800 bps. This is possible by reducing the number of bits used for error correction. The bit rates in a set are related to the activity of the channel. If the channel is silent, only 1200 bits can be transferred, which improves the spreading by repeating each bit 8 times.

Frequency-Reuse Factor

In an IS-95 system, the frequency-reuse factor is normally 1 because the interference from neighboring cells cannot affect CDMA or DSSS transmission.

Soft Handoff

Every base station continuously broadcasts signals using its pilot channel. This means a mobile station can detect the pilot signal from its cell and neighboring cells. This enables a mobile station to do a soft handoff in contrast to a hard handoff.

16.2.4 Third Generation (3G)

The third generation of cellular telephony refers to a combination of technologies that provide both digital data and voice communication. Using a small portable device, a person is able to talk to anyone else in the world with a voice quality similar to that of the existing fixed telephone network. A person can download and watch a movie, download and listen to music, surf the Internet or play games, have a video conference, and do much more. One of the interesting characteristics of a third-generation system is that the portable device is always connected; you do not need to dial a number to connect to the Internet.

The third-generation concept started in 1992, when ITU issued a blueprint called the **Internet Mobile Communication 2000 (IMT-2000).** The blueprint defines some criteria for third-generation technology as outlined below:

a. Voice quality comparable to that of the existing public telephone network.

b. Data rate of 144 kbps for access in a moving vehicle (car), 384 kbps for access as the user walks (pedestrians), and 2 Mbps for the stationary user (office or home).

c. Support for packet-switched and circuit-switched data services.
d. A band of 2 GHz.
e. Bandwidths of 2 MHz.
f. Interface to the Internet.

The main goal of third-generation cellular telephony is to provide universal personal communication.

IMT-2000 Radio Interfaces

Figure 16.16 shows the radio interfaces (wireless standards) adopted by IMT-2000. All five are developed from second-generation technologies. The first two evolve from CDMA technology. The third evolves from a combination of CDMA and TDMA. The fourth evolves from TDMA, and the last evolves from both FDMA and TDMA.

Figure 16.16 *IMT-2000 radio interfaces*

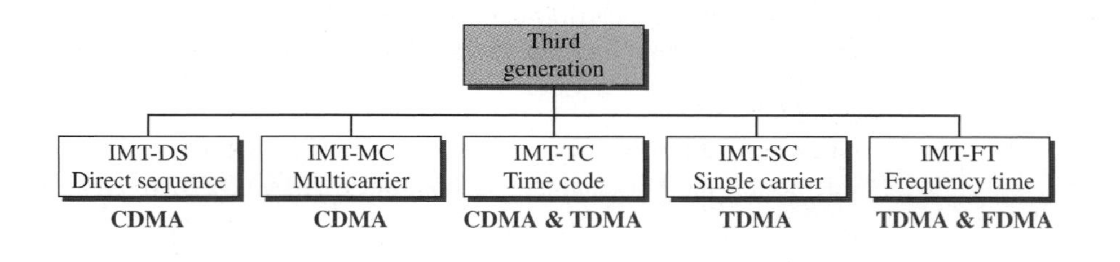

IMT-DS

This approach uses a version of CDMA called *wideband CDMA* or *W-CDMA.* W-CDMA uses a 5-MHz bandwidth. It was developed in Europe, and it is compatible with the CDMA used in IS-95.

IMT-MC

This approach was developed in North America and is known as *CDMA 2000.* It is an evolution of CDMA technology used in IS-95 channels. It combines the new wideband (15-MHz) spread spectrum with the narrowband (1.25-MHz) CDMA of IS-95. It is backward-compatible with IS-95. It allows communication on multiple 1.25-MHz channels (1, 3, 6, 9, 12 times), up to 15 MHz. The use of the wider channels allows it to reach the 2-Mbps data rate defined for the third generation.

IMT-TC

This standard uses a combination of W-CDMA and TDMA. The standard tries to reach the IMT-2000 goals by adding TDMA multiplexing to W-CDMA.

IMT-SC

This standard uses only TDMA.

IMT-FT

This standard uses a combination of FDMA and TDMA.

16.2.5 Fourth Generation (4G)

The fourth generation of cellular telephony is expected to be a complete evolution in wireless communications. Some of the objectives defined by the 4G working group are as follows:

a. A spectrally efficient system.

b. High network capacity.

c. Data rate of 100 Mbit/s for access in a moving car and 1 Gbit/s for stationary users.

d. Data rate of at least 100 Mbit/s between any two points in the world.

e. Smooth handoff across heterogeneous networks.

f. Seamless connectivity and global roaming across multiple networks.

g. High quality of service for next generation multimedia support (quality of service will be discussed in Chapter 30).

h. Interoperability with existing wireless standards.

i. All IP, packet-switched, networks.

The fourth generation is only packet-based (unlike 3G) and supports IPv6. This provides better multicast, security, and route optimization capabilities.

Access Scheme

To increase efficiency, capacity, and scalability, new access techniques are being considered for 4G. For example, **orthogonal FDMA** (**OFDMA**) and **interleaved FDMA** (**IFDMA**) are being considered respectively for the downlink and uplink of the next generation **Universal Mobile Telecommunications System** (**UMTS**). Similarly, **multicarrier code division multiple access** (**MC-CDMA**) is proposed for the IEEE 802.20 standard.

Modulation

More efficient quadrature amplitude modulation (64-QAM) is being proposed for use with the Long Term Evolution (LTE) standards.

Radio System

The fourth generation uses a **Software Defined Radio** (**SDR**) system. Unlike a common radio, which uses hardware, the components of an SDR are pieces of software and thus flexible. The SDR can change its program to shift its frequencies to mitigate frequency interference.

Antenna

The **multiple-input multiple-output** (**MIMO**) and **multiuser MIMO** (**MU-MIMO**) antenna system, a branch of intelligent antenna, is proposed for 4G. Using this antenna system together with special multiplexing, 4G allows independent streams to be transmitted simultaneously from all the antennas to increase the data rate into multiple folds. MIMO also allows the transmitter and receiver coordinates to move to an open frequency when interference occurs.

Applications

At the present rates of 15-30 Mbit/s, 4G is capable of providing users with streaming high-definition television. At rates of 100 Mbit/s, the content of a DVD-5 can be downloaded within about 5 minutes for offline access.

16.3 SATELLITE NETWORKS

A *satellite network* is a combination of nodes, some of which are satellites, that provides communication from one point on the Earth to another. A node in the network can be a satellite, an Earth station, or an end-user terminal or telephone. Although a natural satellite, such as the moon, can be used as a relaying node in the network, the use of artificial satellites is preferred because we can install electronic equipment on the satellite to regenerate the signal that has lost its energy during travel. Another restriction on using natural satellites is their distances from the Earth, which create a long delay in communication.

Satellite networks are like cellular networks in that they divide the planet into cells. Satellites can provide transmission capability to and from any location on Earth, no matter how remote. This advantage makes high-quality communication available to undeveloped parts of the world without requiring a huge investment in ground-based infrastructure.

16.3.1 Operation

Let us first discuss some general issues related to the operation of satellites.

Orbits

An artificial satellite needs to have an *orbit,* the path in which it travels around the Earth. The orbit can be equatorial, inclined, or polar, as shown in Figure 16.17.

Figure 16.17 *Satellite orbits*

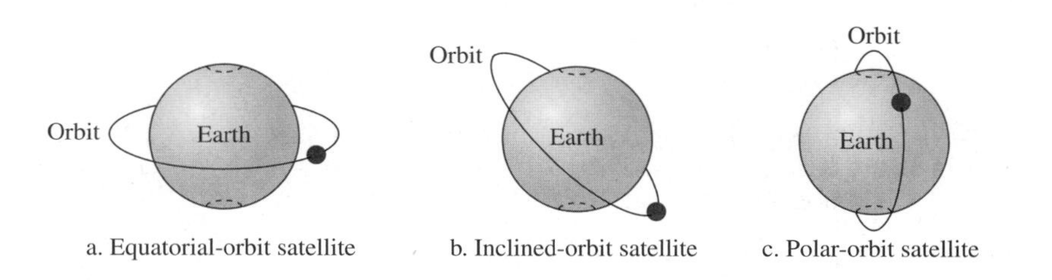

The period of a satellite, the time required for a satellite to make a complete trip around the Earth, is determined by Kepler's law, which defines the period as a function of the distance of the satellite from the center of the Earth.

Example 16.1

What is the period of the moon, according to Kepler's law?

$$\textbf{Period} = C \times \textbf{distance}^{1.5}$$

Here C is a constant approximately equal to 1/100. The period is in seconds and the distance in kilometers.

Solution

The moon is located approximately 384,000 km above the Earth. The radius of the Earth is 6378 km. Applying the formula, we get the following.

$$\textbf{Period} = (1/100) \times (384{,}000 + 6378)^{1.5} = 2{,}439{,}090 \text{ s} = 1 \text{ month}$$

Example 16.2

According to Kepler's law, what is the period of a satellite that is located at an orbit approximately 35,786 km above the Earth?

Solution

Applying the formula, we get the following.

$$\textbf{Period} = (1/100) \times (35{,}786 + 6378)^{1.5} = 86{,}579 \text{ s} = 24 \text{ h}$$

This means that a satellite located at 35,786 km has a period of 24 h, which is the same as the rotation period of the Earth. A satellite like this is said to be *stationary* to the Earth. The orbit, as we will see, is called a *geostationary orbit.*

Footprint

Satellites process microwaves with bidirectional antennas (line-of-sight). Therefore, the signal from a satellite is normally aimed at a specific area called the ***footprint.*** The signal power at the center of the footprint is maximum. The power decreases as we move out from the footprint center. The boundary of the footprint is the location where the power level is at a predefined threshold.

Frequency Bands for Satellite Communication

The frequencies reserved for satellite microwave communication are in the gigahertz (GHz) range. Each satellite sends and receives over two different bands. Transmission from the Earth to the satellite is called the *uplink.* Transmission from the satellite to the Earth is called the *downlink.* Table 16.1 gives the band names and frequencies for each range.

Table 16.1 *Satellite frequency bands*

Band	*Downlink, GHz*	*Uplink, GHz*	*Bandwidth, MHz*
L	1.5	1.6	15
S	1.9	2.2	70
C	4.0	6.0	500
Ku	11.0	14.0	500
Ka	20.0	30.0	3500

Three Categories of Satellites

Based on the location of the orbit, satellites can be divided into three categories: **geostationary Earth orbit (GEO), low-Earth-orbit (LEO),** and **medium-Earth-orbit (MEO).**

Figure 16.18 shows the satellite altitudes with respect to the surface of the Earth. There is only one orbit, at an altitude of 35,786 km, for the GEO satellite. MEO satellites are located at altitudes between 5000 and 15,000 km. LEO satellites are normally below an altitude of 2000 km.

Figure 16.18 *Satellite orbit altitudes*

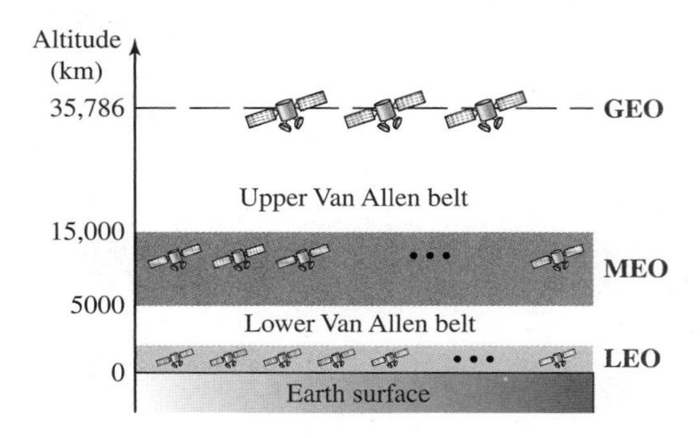

One reason for having different orbits is the existence of two Van Allen belts. A Van Allen belt is a layer that contains charged particles. A satellite orbiting in one of these two belts would be totally destroyed by the energetic charged particles. The MEO orbits are located between these two belts.

16.3.2 GEO Satellites

Line-of-sight propagation requires that the sending and receiving antennas be locked onto each other's location at all times (one antenna must have the other in sight). For this reason, a satellite that moves faster or slower than the Earth's rotation is useful only for short periods. To ensure constant communication, the satellite must move at the same speed as the Earth so that it seems to remain fixed above a certain spot. Such satellites are called *geostationary.*

Because orbital speed is based on the distance from the planet, only one orbit can be geostationary. This orbit occurs at the equatorial plane and is approximately 22,000 mi from the surface of the Earth.

But one geostationary satellite cannot cover the whole Earth. One satellite in orbit has line-of-sight contact with a vast number of stations, but the curvature of the Earth still keeps much of the planet out of sight. It takes a minimum of three satellites equidistant from each other in geostationary Earth orbit (GEO) to provide full global transmission. Figure 16.19 shows three satellites, each 120° from another in geosynchronous orbit around the equator. The view is from the North Pole.

16.3.3 MEO Satellites

Medium-Earth-orbit (MEO) satellites are positioned between the two Van Allen belts. A satellite at this orbit takes approximately 6 to 8 hours to circle the Earth.

Figure 16.19 *Satellites in geostationary orbit*

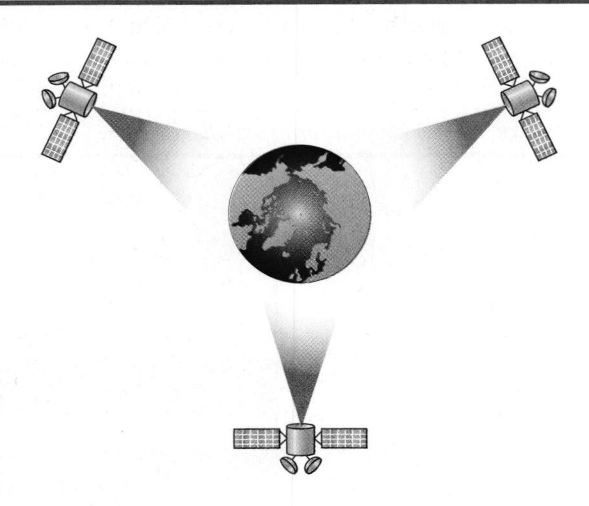

Global Positioning System

One example of a MEO satellite system is the **Global Positioning System (GPS),** contracted and operated by the U.S. Department of Defense, orbiting at an altitude about 18,000 km (11,000 mi) above the Earth. The system consists of 24 satellites and is used for land, sea, and air navigation to provide time and location for vehicles and ships. GPS uses 24 satellites in six orbits, as shown in Figure 16.20. The orbits and the locations of the satellites in each orbit are designed in such a way that, at any time, four satellites are visible from any point on Earth. A GPS receiver has an almanac that tells the current position of each satellite.

Figure 16.20 *Orbits for global positioning system (GPS) satellites*

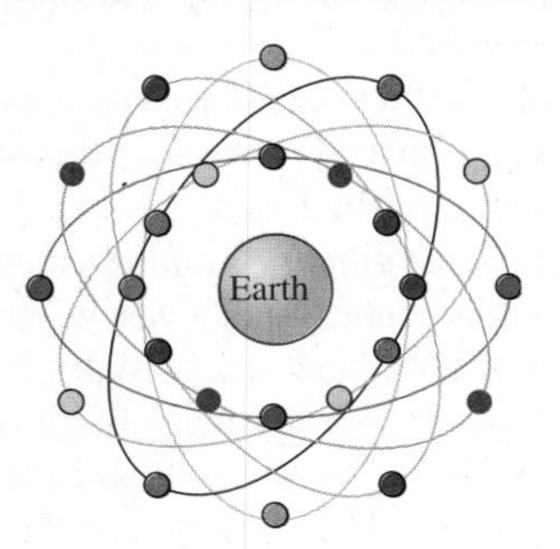

Trilateration

GPS is based on a principle called *trilateration.* The terms *trilateration* and *triangulation* are normally used interchangeably. We use the word ***trilateration,*** which means using three distances, instead of ***triangulation,*** which may mean using three angles. On a plane, if we know our distance from three points, we know exactly where we are. Let us say that we are 10 miles away from point A, 12 miles away from point B, and 15 miles away from point C. If we draw three circles with the centers at A, B, and C, we must be somewhere on circle A, somewhere on circle B, and somewhere on circle C. These three circles meet at one single point (if our distances are correct); this is our position. Figure 16.21a shows the concept.

Figure 16.21 *Trilateration on a plane*

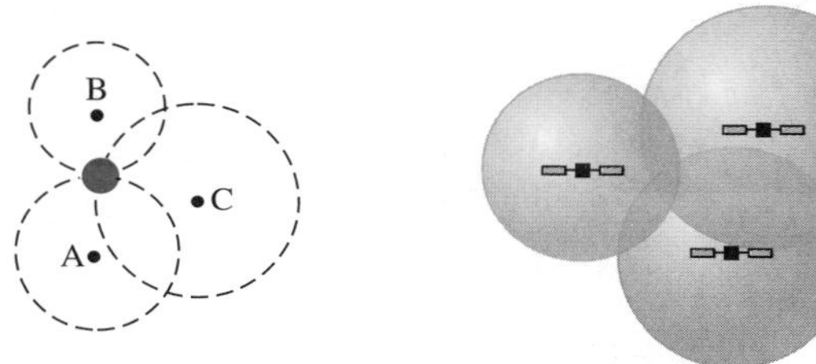

a. Two-dimensional trilateration b. Three-dimensional trilateration

In three-dimensional space, the situation is different. Three spheres meet in two points, as shown in Figure 16.21b. We need at least four spheres to find our exact position in space (longitude, latitude, and altitude). However, if we have additional facts about our location (for example, we know that we are not inside the ocean or somewhere in space), three spheres are enough, because one of the two points, where the spheres meet, is so improbable that the other can be selected without a doubt.

Measuring the distance

The trilateration principle can find our location on the Earth if we know our distance from three satellites and know the position of each satellite. The position of each satellite can be calculated by a GPS receiver (using the predetermined path of the satellites). The GPS receiver, then, needs to find its distance from at least three GPS satellites (center of the spheres). Measuring the distance is done using a principle called *one-way ranging.* For the moment, let us assume that all GPS satellites and the receiver on the Earth are synchronized. Each of 24 satellites synchronously transmits a complex signal, each satellite's signal having a unique pattern. The computer on the receiver measures the delay between the signals from the satellites and its copy of the signals to determine the distances to the satellites.

Synchronization

The previous discussion was based on the assumption that the satellites' clocks are synchronized with each other and with the receiver's clock. Satellites use atomic clocks, which are precise and can function synchronously with each other. The receiver's clock, however, is a normal quartz clock (an atomic clock costs more than $50,000), and there is no way to synchronize it with the satellite clocks. There is an unknown offset between the satellite clocks and the receiver clock that introduces a corresponding offset in the distance calculation. Because of this offset, the measured distance is called a *pseudorange*.

GPS uses an elegant solution to the clock offset problem, by recognizing that the offset's value is the same for all satellites being used. The calculation of position becomes finding four unknowns: the x_r, y_r, z_r coordinates of the receiver, and common clock offset dt. For finding these four unknown values, we need at least four equations. This means that we need to measure pseudoranges from four satellites instead of three. If we call the four measured pseudoranges PR_1, PR_2, PR_3, and PR_4 and the coordinates of each satellite x_i, y_i, and z_i (for $i = 1$ to 4), we can find the four previously mentioned unknown values using the following four equations (the four unknown values are shown in color).

$$PR_1 = [(x_1 - x_r)^2 + (y_1 - y_r)^2 + (z_1 - z_r)^2]^{1/2} + c \times dt$$
$$PR_2 = [(x_2 - x_r)^2 + (y_2 - y_r)^2 + (z_2 - z_r)^2]^{1/2} + c \times dt$$
$$PR_3 = [(x_3 - x_r)^2 + (y_3 - y_r)^2 + (z_3 - z_r)^2]^{1/2} + c \times dt$$
$$PR_4 = [(x_4 - x_r)^2 + (y_4 - y_r)^2 + (z_4 - z_r)^2]^{1/2} + c \times dt$$

The coordinates used in the above formulas are in an Earth-Centered Earth-Fixed (ECEF) reference frame, which means that the origin of the coordinate space is at the center of the Earth and the coordinate space rotates with the Earth. This implies that the ECEF coordinates of a fixed point on the surface of the earth do not change.

Application

GPS is used by military forces. For example, thousands of portable GPS receivers were used during the Persian Gulf war by foot soldiers, vehicles, and helicopters. Another use of GPS is in navigation. The driver of a car can find the location of the car. The driver can then consult a database in the memory of the automobile to be directed to the destination. In other words, GPS gives the location of the car, and the database uses this information to find a path to the destination. A very interesting application is clock synchronization. As we mentioned previously, the IS-95 cellular telephone system uses GPS to create time synchronization between the base stations.

16.3.4 LEO Satellites

Low-Earth-orbit (LEO) satellites have polar orbits. The altitude is between 500 and 2000 km, with a rotation period of 90 to 120 min. The satellite has a speed of 20,000 to 25,000 km/h. A LEO system usually has a cellular type of access, similar to the cellular telephone system. The footprint normally has a diameter of 8000 km. Because LEO satellites are close to Earth, the round-trip time propagation delay is normally less than 20 ms, which is acceptable for audio communication.

A LEO system is made of a constellation of satellites that work together as a network; each satellite acts as a switch. Satellites that are close to each other are connected through intersatellite links (ISLs). A mobile system communicates with the satellite through a user mobile link (UML). A satellite can also communicate with an Earth station (gateway) through a gateway link (GWL). Figure 16.22 shows a typical LEO satellite network.

Figure 16.22 *LEO satellite system*

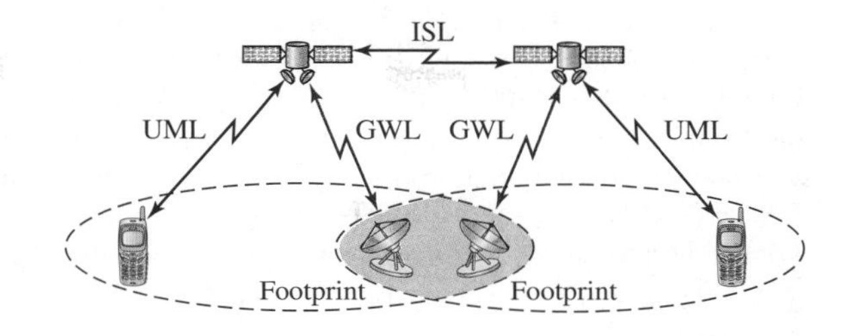

LEO satellites can be divided into three categories: little LEOs, big LEOs, and broadband LEOs. The little LEOs operate under 1 GHz. They are mostly used for low-data-rate messaging. The big LEOs operate between 1 and 3 GHz. **Globalstar** is one of the examples of a big LEO satellite system. It uses 48 satellites in 6 polar orbits with each orbit hosting 8 satellites. The orbits are located at an altitude of almost 1400 km. Iridium systems are also examples of big LEOs. The **Iridium** system has 66 satellites divided into 6 orbits, with 11 satellites in each orbit. The orbits are at an altitude of 750 km. The satellites in each orbit are separated from one another by approximately 32° of latitude. The broadband LEOs provide communication similar to fiber-optic networks. The first broadband LEO system was **Teledesic.** Teledesic is a system of satellites that provides fiber-optic-like communication (broadband channels, low error rate, and low delay). Its main purpose is to provide broadband Internet access for users all over the world. It is sometimes called "Internet in the sky." The project was started in 1990 by Craig McCaw and Bill Gates; later, other investors joined the consortium. The project is scheduled to be fully functional in the near future.

16.4 END-CHAPTER MATERIALS

16.4.1 Recommended Reading

For more details about subjects discussed in this chapter, we recommend the following books. The items in brackets [...] refer to the reference list at the end of the text.

Books

Several books cover materials discussed in this chapter, including [Sch 03], [Gas 02]. [For 03], [Sta 04], [Sta 02], [Kei 02], [Jam 03], [AZ 03], [Tan 03], [Cou 01], [Com 06], [GW 04], and [PD 03].

16.4.2 Key Terms

adaptive antenna system (AAS)
Advanced Mobile Phone System (AMPS)
cellular telephony
digital AMPS (D-AMPS)
footprint
geostationary Earth orbit (GEO)
Global Positioning System (GPS)
Global System for Mobile Communication (GSM)
Globalstar
handoff
Interim Standard 95 (IS-95)
interleaved FDMA (IFDMA)
Internet Mobile Communication 2000 (IMT-2000)
Iridium
low-Earth-orbit (LEO)
medium-Earth-orbit (MEO)
mobile switching center (MSC)
multicarrier code division multiple access (MC-CDMA)
multiple-input multiple-output (MIMO)
multiuser MIMO (MU-MIMO)
orthogonal FDMA (OFDMA)
reuse factor
roaming
Software Defined Radio (SDR)
Teledesic
triangulation
trilateration
Universal Mobile Telecommunications System (UMTS)
Worldwide Interoperability for Microwave Access (WiMAX)

16.4.3 Summary

WiMAX is a wireless version of the wired access networks we discussed in Chapter 14. Two services are provided by WiMAX, fixed and mobile. WiMAX is based on the IEEE Project 802.16, which defines a wireless connection-oriented protocol. In WiMAX, the data-link layer is divided into three sublayers, and the physical layer is divided into two sublayers. WiMAX uses the reservation access method we discussed in Chapter 12.

Cellular telephony provides communication between two devices. One or both may be mobile. A cellular service area is divided into cells. Advanced Mobile Phone System (AMPS) is a first-generation cellular phone system. Digital AMPS (D-AMPS) is a second-generation cellular phone system that is a digital version of AMPS. Global System for Mobile Communication (GSM) is a second-generation cellular phone system used in Europe. Interim Standard 95 (IS-95) is a second-generation cellular phone system based on CDMA and DSSS. The third-generation cellular phone system provides universal personal communication. The fourth generation is the new generation of cellular phones that are becoming popular.

A satellite network uses satellites to provide communication between any points on Earth. A geostationary Earth orbit (GEO) is at the equatorial plane and revolves in phase with Earth's rotation. Global Positioning System (GPS) satellites are medium-Earth-orbit (MEO) satellites that provide time and location information for vehicles and ships. Iridium satellites are low-Earth-orbit (LEO) satellites that provide direct universal voice and data communications for handheld terminals. Teledesic satellites are low-Earth-orbit satellites that will provide universal broadband Internet access.

16.5 PRACTICE SET

16.5.1 Quizzes

A set of interactive quizzes for this chapter can be found on the book website. It is strongly recommended that the student take the quizzes to check his/her understanding of the materials before continuing with the practice set.

16.5.2 Questions

Q16-1. What is the difference between a hard handoff and a soft handoff?

Q16-2. What is the relationship between a base station and a mobile switching center?

Q16-3. Explain the differences between a fixed WiMAX and a mobile WiMAX.

Q16-4. Which is better, a low reuse factor or a high reuse factor? Explain your answer.

Q16-5. What is the relationship between D-AMPS and AMPS?

Q16-6. What is AMPS?

Q16-7. What is the purpose of GPS?

Q16-8. What is a footprint?

Q16-9. What are the functions of a mobile switching center?

Q16-10. What are the three types of orbits?

Q16-11. Which type of orbit does a GEO satellite have? Explain your answer.

Q16-12. What is the main difference between Iridium and Globalstar?

Q16-13. When we make a wireless Internet connection from our cellular phone, do we use fixed or mobile WiMAX?

Q16-14. What is GSM?

Q16-15. What is the function of the CDMA in IS-95?

Q16-16. Compare an uplink with a downlink.

Q16-17. What is the relationship between the Van Allen belts and satellites?

Q16-18. When we make a wireless Internet connection from our desktop at home, do we use fixed or mobile WiMAX?

Q16-19. To which generation does each of the following cellular telephony systems belong?

a. AMPS **b.** D-AMPS **c.** IS-95

16.5.3 Problems

P16-1. Use Kepler's law to check the accuracy of a given period and altitude for a GPS satellite.

P16-2. What is the maximum number of callers in each cell in an IS-95 system?

P16-3. Draw a cell pattern with a frequency-reuse factor of 5.

P16-4. Find the efficiency of the GSM protocol in terms of simultaneous calls per megahertz of bandwidth. In other words, find the number of calls that can be made in 1-MHz bandwidth allocation.

P16-5. Explain how bidirectional communication can be achieved using a frame in Figure 16.5.

P16-6. How many slots are sent each second in a channel using D-AMPS? How many slots are sent by each user in 1 s?

P16-7. Find the efficiency of the D-AMPS protocol in terms of simultaneous calls per megahertz of bandwidth. In other words, find the number of calls that can be made in 1-MHz bandwidth allocation.

P16-8. Use Kepler's law to check the accuracy of a given period and altitude for an Iridium satellite.

P16-9. What is the maximum number of callers in each cell in AMPS?

P16-10. Draw a cell pattern with a frequency-reuse factor of 3.

P16-11. What is the maximum number of simultaneous calls in each cell in a GSM, assuming no analog control channels?

P16-12. In Figure 16.5, do we mean that downstream and upstream data are transmitting at the same time? Explain.

P16-13. Use Kepler's law to check the accuracy of a given period and altitude for a Globalstar satellite.

P16-14. Find the efficiency of the AMPS protocol in terms of simultaneous calls per megahertz of bandwidth. In other words, find the number of calls that can be made in 1-MHz bandwidth allocation.

P16-15. Guess the relationship between a 3-kHz voice channel and a 30-kHz modulated channel in a system using AMPS.

P16-16. What is the maximum number of simultaneous calls in each cell in an IS-136 (D-AMPS) system, assuming no analog control channels?

P16-17. Find the efficiency of the IS-95 protocol in terms of simultaneous calls per megahertz of bandwidth. In other words, find the number of calls that can be made in 1-MHz bandwidth allocation.

CHAPTER 17

Connecting Devices and Virtual LANs

Hosts or LANs do not normally operate in isolation. They are connected to one another or to the Internet. To connect hosts or LANs, we use connecting devices. Connecting devices can operate in different layers of the Internet model. After discussing some connecting devices, we show how they are used to create virtual local area networks (VLANs).

The chapter is divided into two sections.

- The first section discusses connecting devices. It first describes hubs and their features. The section then discusses link-layer switches (or simply *switches,* as they are called), and shows how they can create loops if they connect LANs with broadcast domains.
- The second section discusses virtual LANs or VLANs. The section first shows how membership in a VLAN can be defined. The section then discusses the VLAN configuration. It next shows how switches can communicate in a VLAN. Finally, the section mentions the advantages of a VLAN.

17.1 CONNECTING DEVICES

Hosts and networks do not normally operate in isolation. We use **connecting devices** to connect hosts together to make a network or to connect networks together to make an internet. Connecting devices can operate in different layers of the Internet model. We discuss three kinds of *connecting devices:* hubs, link-layer switches, and routers. Hubs today operate in the first layer of the Internet model. Link-layer switches operate in the first two layers. Routers operate in the first three layers. (See Figure 17.1.)

Figure 17.1 *Three categories of connecting devices*

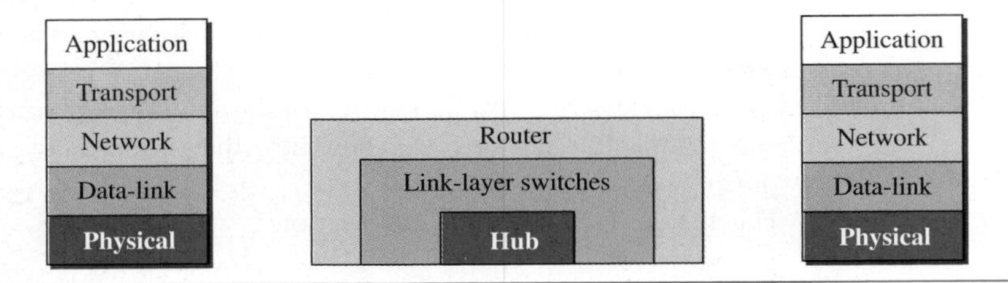

17.1.1 Hubs

A **hub** is a device that operates only in the physical layer. Signals that carry information within a network can travel a fixed distance before attenuation endangers the integrity of the data. A **repeater** receives a signal and, before it becomes too weak or corrupted, *regenerates* and *retimes* the original bit pattern. The repeater then sends the refreshed signal. In the past, when Ethernet LANs were using bus topology, a repeater was used to connect two segments of a LAN to overcome the length restriction of the coaxial cable. Today, however, Ethernet LANs use star topology. In a star topology, a repeater is a multiport device, often called a *hub,* that can be used to serve as the connecting point and at the same time function as a repeater. Figure 17.2 shows that when a packet from station A to station B arrives at the hub, the signal representing the frame is regenerated to remove any possible corrupting noise, but the hub forwards the

Figure 17.2 *A hub*

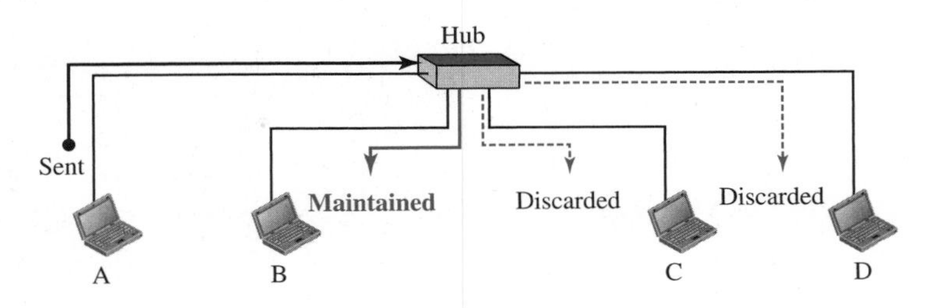

packet from all outgoing ports except the one from which the signal was received. In other words, the frame is broadcast. All stations in the LAN receive the frame, but only station B keeps it. The rest of the stations discard it. Figure 17.2 shows the role of a repeater or a hub in a switched LAN.

The figure definitely shows that a hub does not have a filtering capability; it does not have the intelligence to find from which port the frame should be sent out.

A repeater has no filtering capability.

A hub or a repeater is a physical-layer device. They do not have a link-layer address and they do not check the link-layer address of the received frame. They just regenerate the corrupted bits and send them out from every port.

17.1.2 Link-Layer Switches

A **link-layer switch** (or ***switch***) operates in both the physical and the data-link layers. As a physical-layer device, it regenerates the signal it receives. As a link-layer device, the link-layer switch can check the MAC addresses (source and destination) contained in the frame.

Filtering

One may ask what the difference in functionality is between a link-layer switch and a hub. A link-layer switch has **filtering** capability. It can check the destination address of a frame and can decide from which outgoing port the frame should be sent.

A link-layer switch has a table used in filtering decisions.

Let us give an example. In Figure 17.3, we have a LAN with four stations that are connected to a link-layer switch. If a frame destined for station 71:2B:13:45:61:42 arrives at port 1, the link-layer switch consults its table to find the departing port. According to its table, frames for 71:2B:13:45:61:42 should be sent out only through port 2; therefore, there is no need for forwarding the frame through other ports.

Figure 17.3 *Link-layer switch*

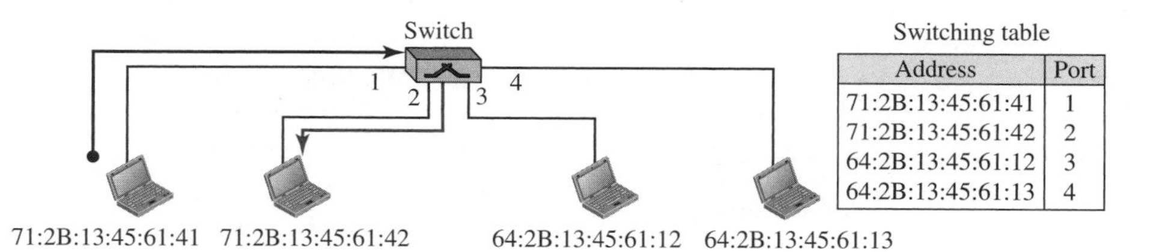

Address	Port
71:2B:13:45:61:41	1
71:2B:13:45:61:42	2
64:2B:13:45:61:12	3
64:2B:13:45:61:13	4

A link-layer switch does not change the link-layer (MAC) addresses in a frame.

Transparent Switches

A **transparent switch** is a switch in which the stations are completely unaware of the switch's existence. If a switch is added or deleted from the system, reconfiguration of the stations is unnecessary. According to the IEEE 802.1d specification, a system equipped with transparent switches must meet three criteria:

- ❑ Frames must be forwarded from one station to another.
- ❑ The forwarding table is automatically made by learning frame movements in the network.
- ❑ Loops in the system must be prevented.

Forwarding

A transparent switch must correctly forward the frames, as discussed in the previous section.

Learning

The earliest switches had switching tables that were static. The system administrator would manually enter each table entry during switch setup. Although the process was simple, it was not practical. If a station was added or deleted, the table had to be modified manually. The same was true if a station's MAC address changed, which is not a rare event. For example, putting in a new network card means a new MAC address.

A better solution to the static table is a dynamic table that maps addresses to ports (interfaces) automatically. To make a table dynamic, we need a switch that gradually learns from the frames' movements. To do this, the switch inspects both the destination and the source addresses in each frame that passes through the switch. The destination address is used for the forwarding decision (table lookup); the source address is used for adding entries to the table and for updating purposes. Let us elaborate on this process using Figure 17.4.

1. When station A sends a frame to station D, the switch does not have an entry for either D or A. The frame goes out from all three ports; the frame floods the network. However, by looking at the source address, the switch learns that station A must be connected to port 1. This means that frames destined for A, in the future, must be sent out through port 1. The switch adds this entry to its table. The table has its first entry now.
2. When station D sends a frame to station B, the switch has no entry for B, so it floods the network again. However, it adds one more entry to the table related to station D.
3. The learning process continues until the table has information about every port. However, note that the learning process may take a long time. For example, if a station does not send out a frame (a rare situation), the station will never have an entry in the table.

Loop Problem

Transparent switches work fine as long as there are no redundant switches in the system. Systems administrators, however, like to have redundant switches (more than one switch between a pair of LANs) to make the system more reliable. If a switch fails, another switch takes over until the failed one is repaired or replaced. Redundancy can create loops in the system, which is very undesirable. Loops can be created only when

Figure 17.4 *Learning switch*

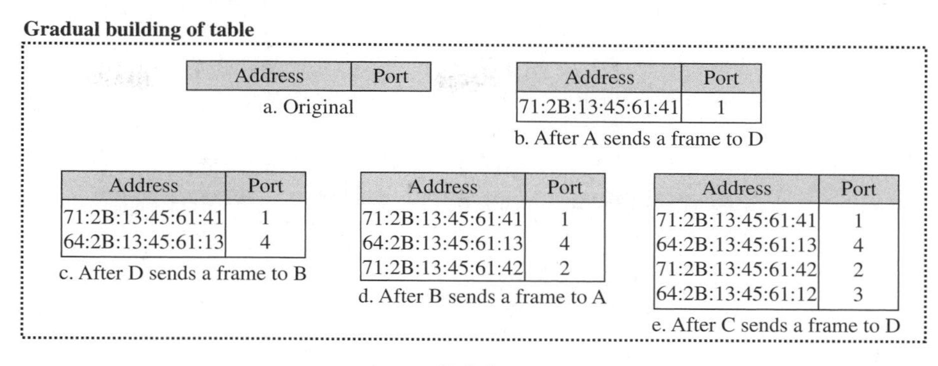

Gradual building of table

Address	Port

a. Original

Address	Port
71:2B:13:45:61:41	1

b. After A sends a frame to D

Address	Port
71:2B:13:45:61:41	1
64:2B:13:45:61:13	4

c. After D sends a frame to B

Address	Port
71:2B:13:45:61:41	1
64:2B:13:45:61:13	4
71:2B:13:45:61:42	2

d. After B sends a frame to A

Address	Port
71:2B:13:45:61:41	1
64:2B:13:45:61:13	4
71:2B:13:45:61:42	2
64:2B:13:45:61:12	3

e. After C sends a frame to D

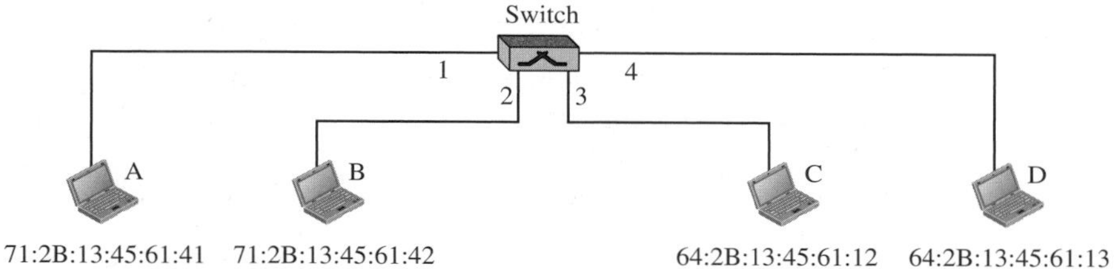

two or more broadcasting LANs (those using hubs, for example) are connected by more than one switch.

Figure 17.5 shows a very simple example of a loop created in a system with two LANs connected by two switches.

1. Station A sends a frame to station D. The tables of both switches are empty. Both forward the frame and update their tables based on the source address A.
2. Now there are two copies of the frame on LAN 2. The copy sent out by the left switch is received by the right switch, which does not have any information about the destination address D; it forwards the frame. The copy sent out by the right switch is received by the left switch and is sent out for lack of information about D. Note that each frame is handled separately because switches, as two nodes on a broadcast network sharing the medium, use an access method such as CSMA/CD. The tables of both switches are updated, but still there is no information for destination D.
3. Now there are two copies of the frame on LAN 1. Step 2 is repeated, and both copies are sent to LAN2.
4. The process continues on and on. Note that switches are also repeaters and regenerate frames. So in each iteration, there are newly generated fresh copies of the frames.

Spanning Tree Algorithm

To solve the looping problem, the IEEE specification requires that switches use the spanning tree algorithm to create a loopless topology. In graph theory, a **spanning**

Figure 17.5 *Loop problem in a learning switch*

a. Station A sends a frame to station D

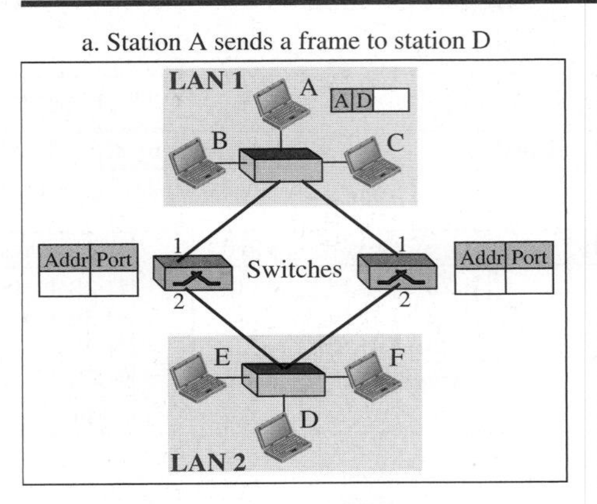

b. Both switches forward the frame

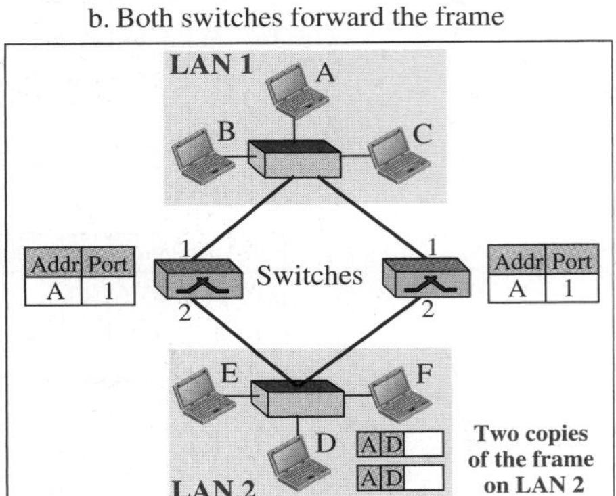

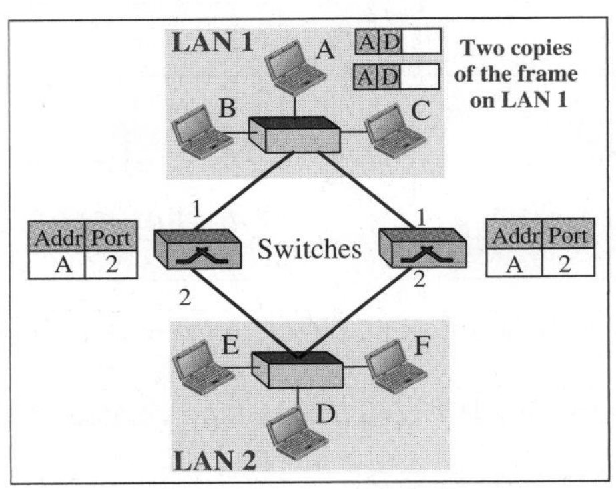

c. Both switches forward the frame

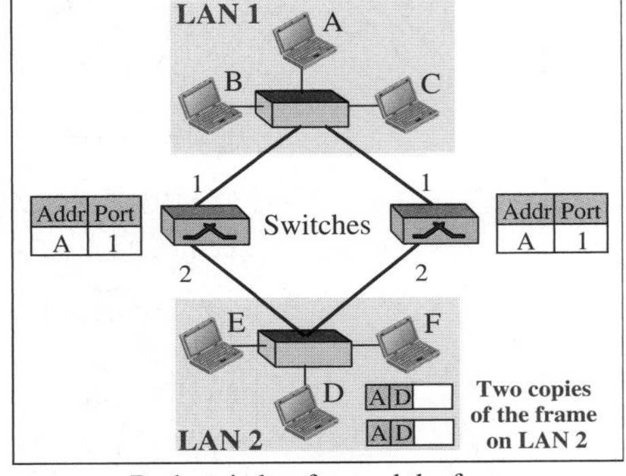

c. Both switches forward the frame

tree is a graph in which there is no loop. In a switched LAN, this means creating a topology in which each LAN can be reached from any other LAN through one path only (no loop). We cannot change the physical topology of the system because of physical connections between cables and switches, but we can create a logical topology that overlays the physical one. Figure 17.6 shows a system with four LANs and five switches. We have shown the physical system and its representation in graph theory. Although some textbooks represent the LANs as nodes and the switches as the connecting arcs, we have shown both LANs and switches as nodes. The connecting arcs show the connection of a LAN to a switch and vice versa. To find the spanning tree, we need to assign a cost (metric) to each arc. The interpretation of the cost is left up to the systems administrator. We have chosen the minimum hops. However, as we will see in Chapter 20, the hop count is normally 1 from a switch to the LAN and 0 in the reverse direction.

The process for finding the spanning tree involves three steps:

1. Every switch has a built-in ID (normally the serial number, which is unique). Each switch broadcasts this ID so that all switches know which one has the smallest ID. The switch with the smallest ID is selected as the *root* switch (root of the tree). We

Figure 17.6 *A system of connected LANs and its graph representation*

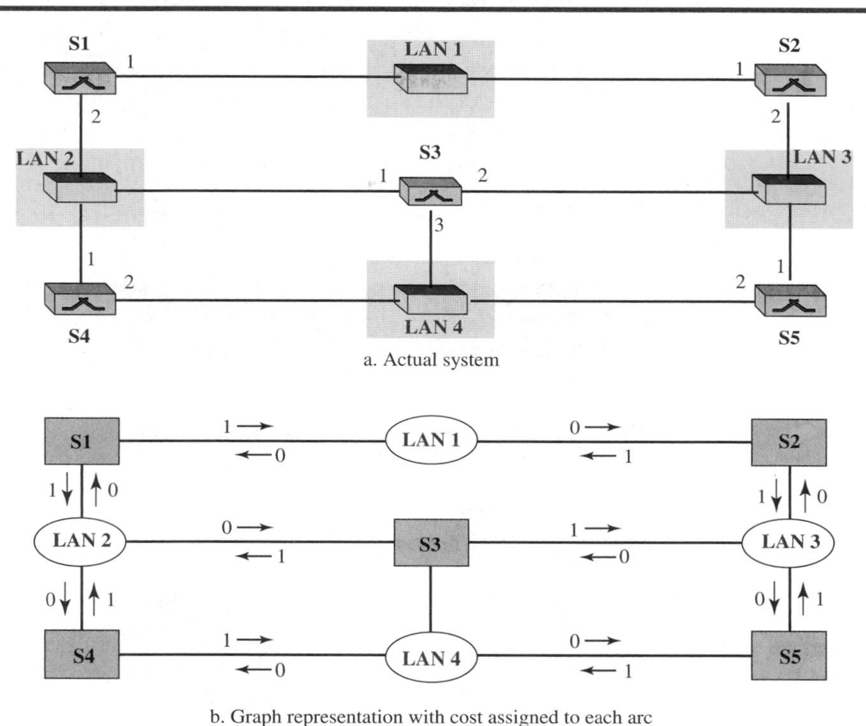

a. Actual system

b. Graph representation with cost assigned to each arc

assume that switch S1 has the smallest ID. It is, therefore, selected as the root switch.

2. The algorithm tries to find the shortest path (a path with the shortest cost) from the root switch to every other switch or LAN. The shortest path can be found by examining the total cost from the root switch to the destination. Figure 17.7 shows the shortest paths. We have used the Dijkstra algorithm described in Chapter 20.

Figure 17.7 *Finding the shortest paths and the spanning tree in a system of switches*

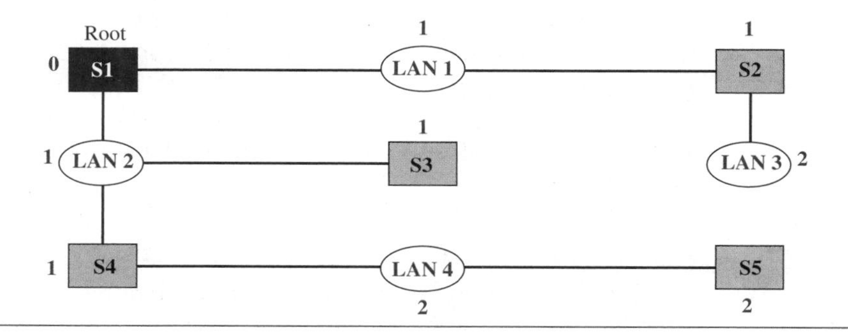

3. The combination of the shortest paths creates the shortest tree, which is also shown in Figure 17.7.
4. Based on the spanning tree, we mark the ports that are part of it, the **forwarding ports,** which forward a frame that the switch receives. We also mark those ports that are not part of the spanning tree, the **blocking ports,** which block the frames received by the switch. Figure 17.8 shows the logical systems of LANs with forwarding ports (solid lines) and blocking ports (broken lines).

Figure 17.8 *Forwarding and blocking ports after using spanning tree algorithm*

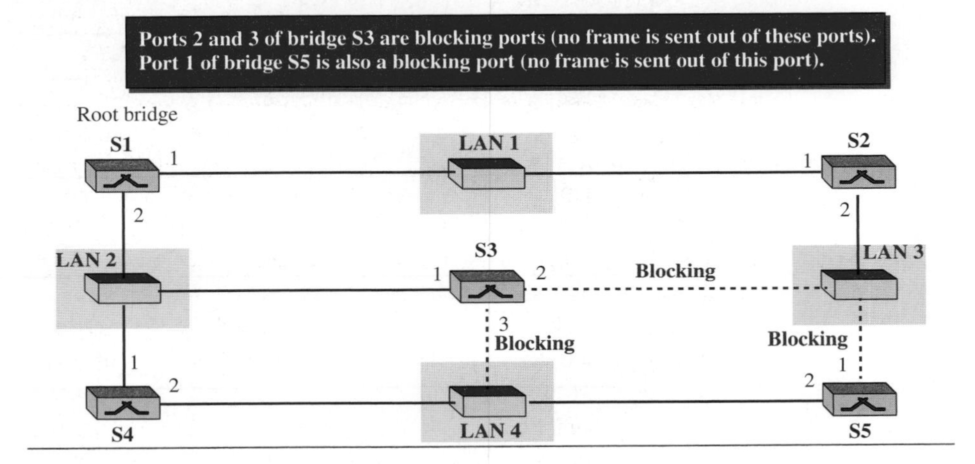

Note that there is only one path from any LAN to any other LAN in the spanning tree system. This means there is only one path from one LAN to any other LAN. No loops are created. You can prove to yourself that there is only one path from LAN 1 to LAN 2, LAN 3, or LAN 4. Similarly, there is only one path from LAN 2 to LAN 1, LAN 3, and LAN 4. The same is true for LAN 3 and LAN 4.

We have described the spanning tree algorithm as though it required manual entries. This is not true. Each switch is equipped with a software package that carries out this process dynamically.

Advantages of Switches

A link-layer switch has several advantages over a hub. We discuss only two of them here.

Collision Elimination

As we mentioned in Chapter 13, a link-layer switch eliminates the collision. This means increasing the average bandwidth available to a host in the network. In a switched LAN, there is no need for carrier sensing and collision detection; each host can transmit at any time.

Connecting Heterogenous Devices

A link-layer switch can connect devices that use different protocols at the physical layer (data rates) and different transmission media. As long as the format of the frame at the data-link layer does not change, a switch can receive a frame from a device that uses twisted-pair cable and sends data at 10 Mbps and deliver the frame to another device that uses fiber-optic cable and can receive data at 100 Mbps.

17.1.3 Routers

We will discuss routers in Part IV of the book when we discuss the network layer. In this section, we mention routers to compare them with a two-layer switch and a hub. A **router** is a three-layer device; it operates in the physical, data-link, and network layers. As a physical-layer device, it regenerates the signal it receives. As a link-layer device, the router checks the physical addresses (source and destination) contained in the packet. As a network-layer device, a router checks the network-layer addresses.

A router is a three-layer (physical, data-link, and network) device.

A router can connect networks. In other words, a router is an internetworking device; it connects independent networks to form an internetwork. According to this definition, two networks connected by a router become an internetwork or an internet.

There are three major differences between a router and a repeater or a switch.

1. A router has a physical and logical (IP) address for each of its interfaces.
2. A router acts only on those packets in which the link-layer destination address matches the address of the interface at which the packet arrives.
3. A router changes the link-layer address of the packet (both source and destination) when it forwards the packet.

Let us give an example. In Figure 17.9, assume an organization has two separate buildings with a Gigabit Ethernet LAN installed in each building. The organization uses switches in each LAN. The two LANs can be connected to form a larger LAN using 10 Gigabit Ethernet technology that speeds up the connection to the Ethernet and the connection to the organization server. A router then can connect the whole system to the Internet.

Figure 17.9 *Routing example*

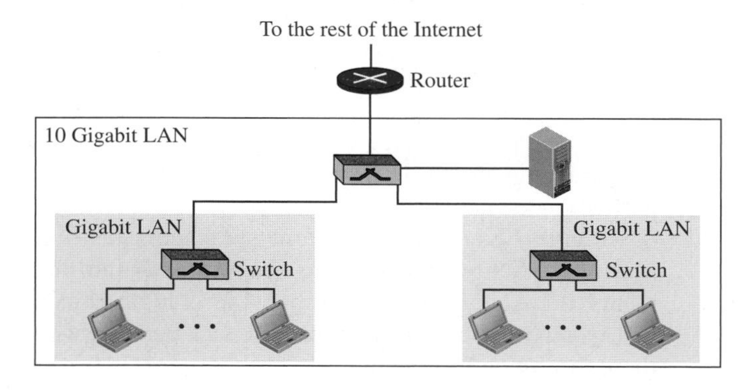

A router, as we discuss in Chapter 18, will change the MAC addresses it receives because the MAC addresses have only local jurisdictions.

A router changes the link-layer addresses in a packet.

17.2 VIRTUAL LANS

A station is considered part of a LAN if it physically belongs to that LAN. The criterion of membership is geographic. What happens if we need a virtual connection between two stations belonging to two different physical LANs? We can roughly define a **virtual local area network (VLAN)** as a local area network configured by software, not by physical wiring.

Let us use an example to elaborate on this definition. Figure 17.10 shows a switched LAN in an engineering firm in which nine stations are grouped into three LANs that are connected by a switch.

Figure 17.10 *A switch connecting three LANs*

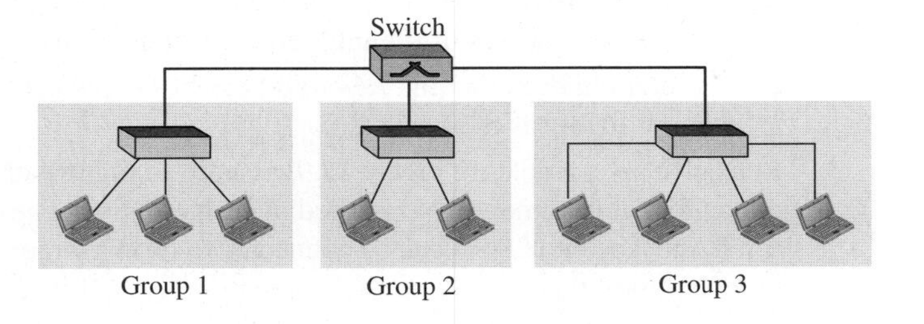

The first three engineers work together as the first group, the next two engineers work together as the second group, and the last four engineers work together as the third group. The LAN is configured to allow this arrangement.

But what would happen if the administrators needed to move two engineers from the first group to the third group, to speed up the project being done by the third group? The LAN configuration would need to be changed. The network technician must rewire. The problem is repeated if, in another week, the two engineers move back to their previous group. In a switched LAN, changes in the work group mean physical changes in the network configuration.

Figure 17.11 shows the same switched LAN divided into VLANs. The whole idea of VLAN technology is to divide a LAN into logical, instead of physical, segments. A LAN can be divided into several logical LANs, called *VLANs*. Each VLAN is a work group in the organization. If a person moves from one group to another, there is no need to change the physical configuration. The group membership in VLANs is defined by software, not hardware. Any station can be logically moved to another VLAN. All members belonging to a VLAN can receive broadcast messages sent to that particular VLAN. This means that if a station moves from VLAN 1 to

Figure 17.11 *A switch using VLAN software*

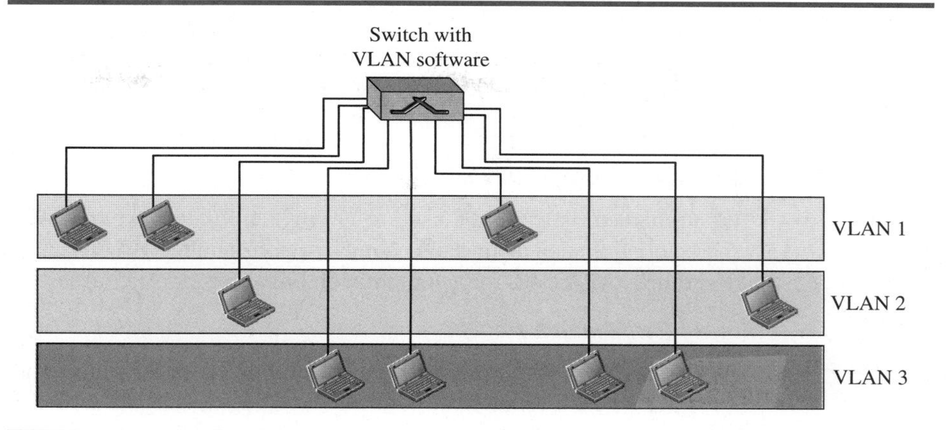

VLAN 2, it receives broadcast messages sent to VLAN 2, but no longer receives broadcast messages sent to VLAN 1.

It is obvious that the problem in our previous example can easily be solved by using VLANs. Moving engineers from one group to another through software is easier than changing the configuration of the physical network.

VLAN technology even allows the grouping of stations connected to different switches in a VLAN. Figure 17.12 shows a backbone local area network with two switches and three VLANs. Stations from switches A and B belong to each VLAN.

Figure 17.12 *Two switches in a backbone using VLAN software*

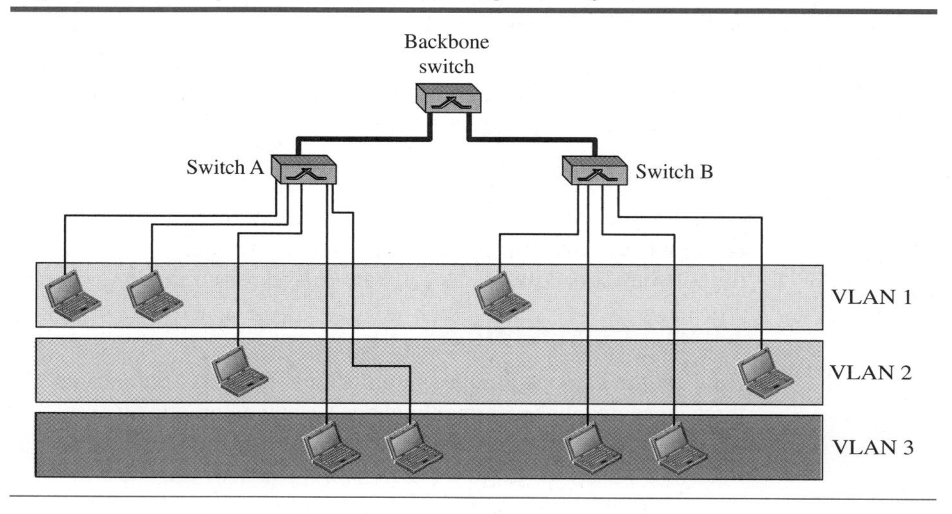

This is a good configuration for a company with two separate buildings. Each building can have its own switched LAN connected by a backbone. People in the first building and people in the second building can be in the same work group even though they are connected to different physical LANs.

From these three examples, we can see that a VLAN defines broadcast domains. VLANs group stations belonging to one or more physical LANs into broadcast domains. The stations in a VLAN communicate with one another as though they belonged to a physical segment.

17.2.1 Membership

What characteristic can be used to group stations in a VLAN? Vendors use different characteristics such as interface numbers, port numbers, MAC addresses, IP addresses, IP multicast addresses, or a combination of two or more of these.

Interface Numbers

Some VLAN vendors use switch interface numbers as a membership characteristic. For example, the administrator can define that stations connecting to ports 1, 2, 3, and 7 belong to VLAN 1, stations connecting to ports 4, 10, and 12 belong to VLAN 2, and so on.

MAC Addresses

Some VLAN vendors use the 48-bit MAC address as a membership characteristic. For example, the administrator can stipulate that stations having MAC addresses E2:13:42:A1:23:34 and F2:A1:23:BC:D3:41 belong to VLAN 1.

IP Addresses

Some VLAN vendors use the 32-bit IP address (see Chapter 18) as a membership characteristic. For example, the administrator can stipulate that stations having IP addresses 181.34.23.67, 181.34.23.72, 181.34.23.98, and 181.34.23.112 belong to VLAN 1.

Multicast IP Addresses

Some VLAN vendors use the multicast IP address (see Chapter 21) as a membership characteristic. Multicasting at the IP layer is now translated to multicasting at the data-link layer.

Combination

Recently, the software available from some vendors allows all these characteristics to be combined. The administrator can choose one or more characteristics when installing the software. In addition, the software can be reconfigured to change the settings.

17.2.2 Configuration

How are the stations grouped into different VLANs? Stations are configured in one of three ways: manually, semiautomatically, and automatically.

Manual Configuration

In a manual configuration, the network administrator uses the VLAN software to manually assign the stations into different VLANs at setup. Later migration from one VLAN to another is also done manually. Note that this is not a physical configuration;

it is a logical configuration. The term *manually* here means that the administrator types the port numbers, the IP addresses, or other characteristics, using the VLAN software.

Automatic Configuration

In an automatic configuration, the stations are automatically connected or disconnected from a VLAN using criteria defined by the administrator. For example, the administrator can define the project number as the criterion for being a member of a group. When a user changes projects, he or she automatically migrates to a new VLAN.

Semiautomatic Configuration

A semiautomatic configuration is somewhere between a manual configuration and an automatic configuration. Usually, the initializing is done manually, with migrations done automatically.

17.2.3 Communication between Switches

In a multi-switched backbone, each switch must know not only which station belongs to which VLAN, but also the membership of stations connected to other switches. For example, in Figure 17.12, switch A must know the membership status of stations connected to switch B, and switch B must know the same about switch A. Three methods have been devised for this purpose: table maintenance, frame tagging, and time-division multiplexing.

Table Maintenance

In this method, when a station sends a broadcast frame to its group members, the switch creates an entry in a table and records station membership. The switches send their tables to one another periodically for updating.

Frame Tagging

In this method, when a frame is traveling between switches, an extra header is added to the MAC frame to define the destination VLAN. The frame tag is used by the receiving switches to determine the VLANs to be receiving the broadcast message.

Time-Division Multiplexing (TDM)

In this method, the connection (trunk) between switches is divided into time-shared channels (see TDM in Chapter 6). For example, if the total number of VLANs in a backbone is five, each trunk is divided into five channels. The traffic destined for VLAN 1 travels in channel 1, the traffic destined for VLAN 2 travels in channel 2, and so on. The receiving switch determines the destination VLAN by checking the channel from which the frame arrived.

IEEE Standard

In 1996, the IEEE 802.1 subcommittee passed a standard called 802.1Q that defines the format for frame tagging. The standard also defines the format to be used in multi-switched backbones and enables the use of multivendor equipment in VLANs. IEEE 802.1Q has opened the way for further standardization in other issues related to VLANs. Most vendors have already accepted the standard.

17.2.4 Advantages

There are several advantages to using VLANs.

Cost and Time Reduction

VLANs can reduce the migration cost of stations going from one group to another. Physical reconfiguration takes time and is costly. Instead of physically moving one station to another segment or even to another switch, it is much easier and quicker to move it by using software.

Creating Virtual Work Groups

VLANs can be used to create virtual work groups. For example, in a campus environment, professors working on the same project can send broadcast messages to one another without the necessity of belonging to the same department. This can reduce traffic if the multicasting capability of IP was previously used.

Security

VLANs provide an extra measure of security. People belonging to the same group can send broadcast messages with the guaranteed assurance that users in other groups will not receive these messages.

17.3 END-CHAPTER MATERIALS

17.3.1 Recommended Reading

For more details about subjects discussed in this chapter, we recommend the following books. The items in brackets [...] refer to the reference list at the end of the text.

Books

Several books discuss link-layer issues. Among them we recommend [Ham 80], [Zar 02], [Ror 96], [Tan 03], [GW 04], [For 03], [KMK 04], [Sta 04], [Kes 02], [PD 03], [Kei 02], [Spu 00], [KCK 98], [Sau 98], [Izz 00], [Per 00], and [WV 00].

17.3.2 Key Terms

blocking port
connecting device
filtering
forwarding port
hub
link-layer switch
repeater
router
spanning tree
switch
transparent switch
virtual local area network (VLAN)

17.3.3 Summary

A repeater is a connecting device that operates in the physical layer of the Internet model. A repeater regenerates a signal, connects segments of a LAN, and has no filtering capability. A link-layer switch is a connecting device that operates in the physical and data-link layers of the Internet model. A transparent switch can forward and filter frames and automatically build its forwarding table. A switch can use the spanning tree algorithm to create a loopless topology.

A virtual local area network (VLAN) is configured by software, not by physical wiring. Membership in a VLAN can be based on port numbers, MAC addresses, IP addresses, IP multicast addresses, or a combination of these features. VLANs are cost- and time-efficient, can reduce network traffic, and provide an extra measure of security.

17.4 PRACTICE SET

17.4.1 Quizzes

A set of interactive quizzes for this chapter can be found on the book website. It is strongly recommended that the student take the quizzes to check his/her understanding of the materials before continuing with the practice set.

17.4.2 Questions

Q17-1. What is the difference between a forwarding port and a blocking port?

Q17-2. How does a VLAN save a company time and money?

Q17-3. How does a VLAN provide extra security for a network?

Q17-4. How does a VLAN reduce network traffic?

Q17-5. Which one has more overhead, a switch or a router? Explain your answer.

Q17-6. What do we mean when we say that a link-layer switch can filter traffic? Why is filtering important?

Q17-7. How is a repeater different from an amplifier?

Q17-8. What do we mean when we say that a switch can filter traffic? Why is filtering important?

Q17-9. What is a transparent switch?

Q17-10. How is a hub related to a repeater?

Q17-11. What is the basis for membership in a VLAN?

Q17-12. Which one has more overhead, a hub or a switch? Explain your answer.

17.4.3 Problems

P17-1. A switch uses a filtering table; a router uses a routing table. Can you explain the difference?

P17-2. Assume that in Figure 17.6, switch S3 is selected as the root of the tree. Find the spanning tree and the forwarding and blocking port.

P17-3. Repeat the steps in Figure 17.5 if host B in LAN1 sends a frame to host C in the same LAN.

P17-4. Repeat the steps in Figure 17.5 if host F in LAN2 sends a frame to host B in LAN1.

P17-5. If the switch in Figure 17.3 is connected to a router to provide access to the Internet, does it need a link-layer address? Explain.

P17-6. Does each port of the switch in Figure 17.3 need a link-layer address? Explain.

P17-7. In Figure 17.5, do we have a loop problem if we change each hub in the LANs to a link-layer switch?

P17-8. Find the spanning tree and the logical connection between the switches in Figure 17.13.

Figure 17.13 *Problem P17-8.*

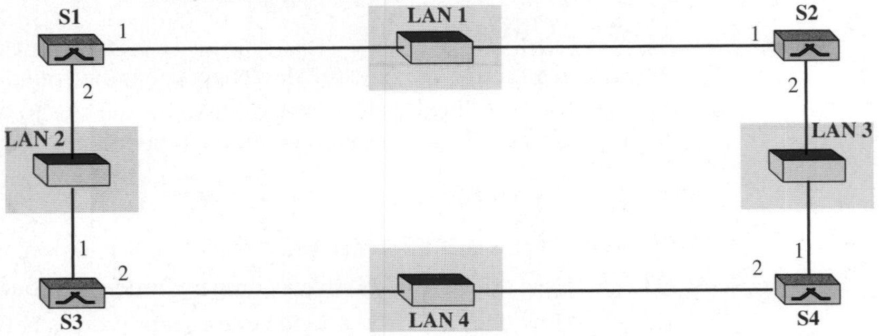

S1
1
LAN 1
S2
1
2
2
LAN 2
LAN 3
1
1
2
2
S3
LAN 4
S4

P17-9. Find the spanning tree and the logical connection between the switches in Figure 17.14.

Figure 17.14 *Problem P17-9.*

S1
LAN 1
S2
LAN 2
LAN 3
S3

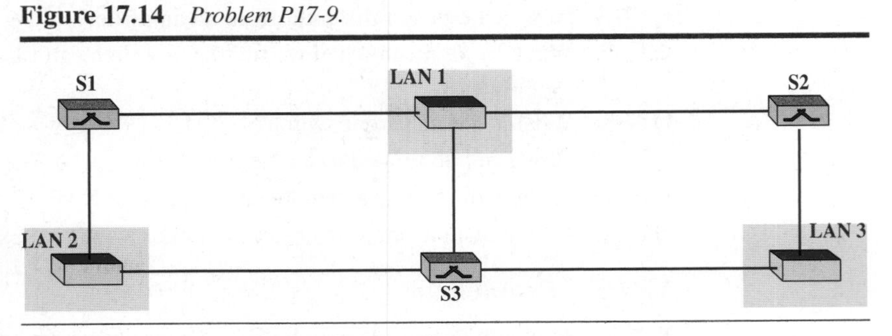

P17-10. In Figure 17.5, do we have a loop problem if we change the hub in one of the LANs to a link-layer switch?

PART IV

Network Layer

In the fourth part of the book, we discuss the network layer. Chapter 18 discusses general issues related to the network layer. Chapter 19 discusses the current version of the IP protocol and its auxiliaries. Chapters 20 and 21 cover the discussion of routing, unicast and multicast. Finally, Chapter 22 discusses the next generation protocol, which is on the horizon.

CHAPTER 18

Introduction to Network Layer

The network layer in the TCP/IP protocol suite is responsible for the host-to-host delivery of datagrams. It provides services to the transport layer and receives services from the data-link layer. In this chapter, we introduce the general concepts and issues in the network layer. This chapter also discusses the addressing mechanism used in the network layer, as briefly mentioned in Chapter 2. This chapter prepares the way for discussion of other network-layer issues, which follows in the next four chapters.

The chapter is divided into five sections.

- The first section introduces the network layer by defining the services provided by this layer. It first discusses packetizing. It then describes forwarding and routing and compares the two. The section then briefly explains the other services such as flow, error, and congestion control.
- The second section discusses packet switching, which occurs at the network layer. The datagram approach and the virtual-circuit approach of packet switching are described in some detail in this section.
- The third section discusses network-layer performance. It describes different delays that occur in network-layer communication. It also mentions the issue of packet loss. Finally, it discusses the issue of congestion control at the network layer.
- The fourth section discusses IPv4 addressing, probably the most important issue in the network layer. It first describes the address space. It then briefly discusses classful addressing, which belongs to the past but is useful in understanding classless addressing. The section then moves to classless addressing and explains several issues related to this topic. It then discusses DHCP, which can be used to dynamically assign addresses in an organization. Finally, the section discusses NAT, which can be used to relieve the shortage of addresses to some extent.
- The fifth section discusses forwarding of network-layer packets. It first shows how forwarding can be done based on the destination address in a packet. It then discusses how forwarding can be done using a label.

18.1 NETWORK-LAYER SERVICES

Before discussing the network layer in the Internet today, let's briefly discuss the network-layer services that, in general, are expected from a network-layer protocol. Figure 18.1 shows the communication between Alice and Bob at the network layer. This is the same scenario we used in Chapters 3 and 9 to show the communication at the physical and the data-link layers, respectively.

Figure 18.1 *Communication at the network layer*

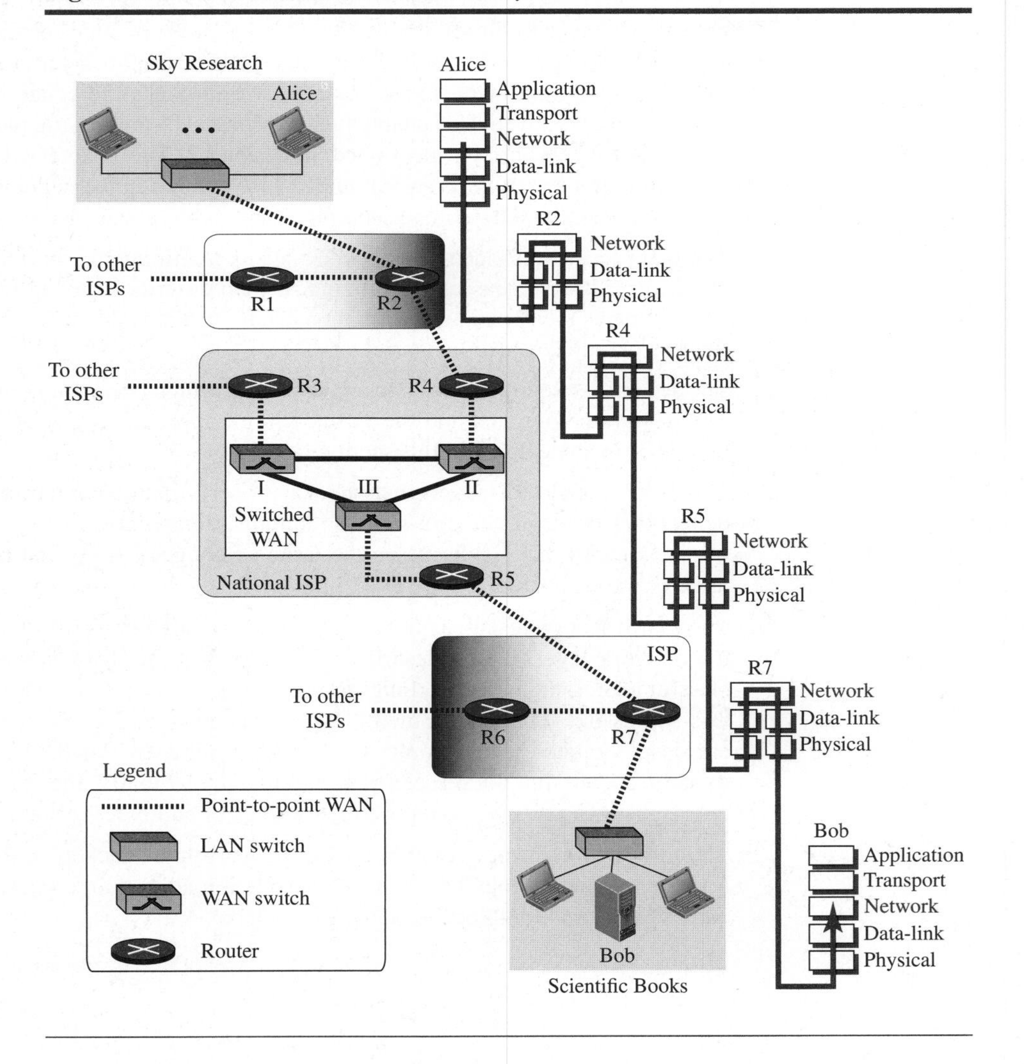

The figure shows that the Internet is made of many networks (or links) connected through the connecting devices. In other words, the Internet is an internetwork, a combination of LANs and WANs. To better understand the role of the network layer (or the internetwork layer), we need to think about the connecting devices (routers or switches) that connect the LANs and WANs.

As the figure shows, the network layer is involved at the source host, destination host, and all routers in the path (R2, R4, R5, and R7). At the source host (Alice), the network layer accepts a packet from a transport layer, encapsulates the packet in a datagram, and delivers the packet to the data-link layer. At the destination host (Bob), the datagram is decapsulated, and the packet is extracted and delivered to the corresponding transport layer. Although the source and destination hosts are involved in all five layers of the TCP/IP suite, the routers use three layers if they are routing packets only; however, they may need the transport and application layers for control purposes. A router in the path is normally shown with two data-link layers and two physical layers, because it receives a packet from one network and delivers it to another network.

18.1.1 Packetizing

The first duty of the network layer is definitely **packetizing:** encapsulating the payload (data received from upper layer) in a network-layer packet at the source and decapsulating the payload from the network-layer packet at the destination. In other words, one duty of the network layer is to carry a payload from the source to the destination without changing it or using it. The network layer is doing the service of a carrier such as the postal office, which is responsible for delivery of packages from a sender to a receiver without changing or using the contents.

The source host receives the payload from an upper-layer protocol, adds a header that contains the source and destination addresses and some other information that is required by the network-layer protocol (as discussed later) and delivers the packet to the data-link layer. The source is not allowed to change the content of the payload unless it is too large for delivery and needs to be fragmented.

The destination host receives the network-layer packet from its data-link layer, decapsulates the packet, and delivers the payload to the corresponding upper-layer protocol. If the packet is fragmented at the source or at routers along the path, the network layer is responsible for waiting until all fragments arrive, reassembling them, and delivering them to the upper-layer protocol.

The routers in the path are not allowed to decapsulate the packets they received unless the packets need to be fragmented. The routers are not allowed to change source and destination addresses either. They just inspect the addresses for the purpose of forwarding the packet to the next network on the path. However, if a packet is fragmented, the header needs to be copied to all fragments and some changes are needed, as we discuss in detail later.

18.1.2 Routing and Forwarding

Other duties of the network layer, which are as important as the first, are routing and forwarding, which are directly related to each other.

Routing

The network layer is responsible for routing the packet from its source to the destination. A physical network is a combination of networks (LANs and WANs) and routers that connect them. This means that there is more than one route from the source to the destination. The network layer is responsible for finding the best one among these possible routes. The network layer needs to have some specific strategies for defining the best route. In the Internet today, this is done by running some routing protocols to help the routers coordinate their knowledge about the neighborhood and to come up with consistent tables to be used when a packet arrives. The routing protocols, which we discuss in Chapters 20 and 21, should be run before any communication occurs.

Forwarding

If routing is applying strategies and running some routing protocols to create the decision-making tables for each router, *forwarding* can be defined as the action applied by each router when a packet arrives at one of its interfaces. The decision-making table a router normally uses for applying this action is sometimes called the *forwarding table* and sometimes the *routing table*. When a router receives a packet from one of its attached networks, it needs to forward the packet to another attached network (in unicast routing) or to some attached networks (in multicast routing). To make this decision, the router uses a piece of information in the packet header, which can be the destination address or a label, to find the corresponding output interface number in the forwarding table. Figure 18.2 shows the idea of the forwarding process in a router.

Figure 18.2 *Forwarding process*

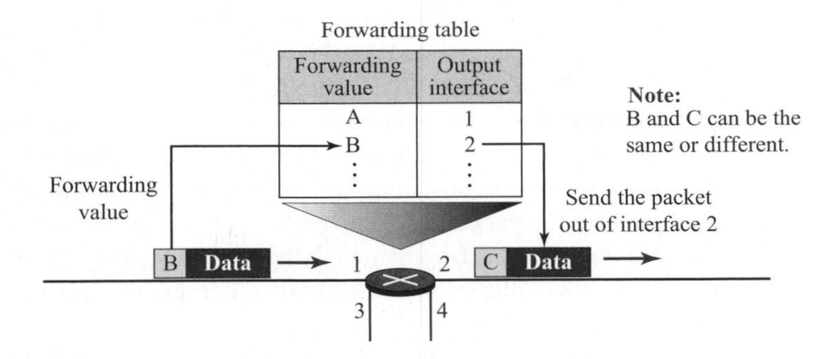

18.1.3 Other Services

Let us briefly discuss other services expected from the network layer.

Error Control

In Chapter 10, we discussed error detection and correction. Although error control also can be implemented in the network layer, the designers of the network layer in the Internet ignored this issue for the data being carried by the network layer. One reason for this decision is the fact that the packet in the network layer may be fragmented at each router, which makes error checking at this layer inefficient.

The designers of the network layer, however, have added a checksum field to the datagram to control any corruption in the header, but not in the whole datagram. This checksum may prevent any changes or corruptions in the header of the datagram.

We need to mention that although the network layer in the Internet does not directly provide error control, the Internet uses an auxiliary protocol, ICMP, that provides some kind of error control if the datagram is discarded or has some unknown information in the header. We discuss ICMP in Chapter 19.

Flow Control

Flow control regulates the amount of data a source can send without overwhelming the receiver. If the upper layer at the source computer produces data faster than the upper layer at the destination computer can consume it, the receiver will be overwhelmed with data. To control the flow of data, the receiver needs to send some feedback to the sender to inform the latter that it is overwhelmed with data.

The network layer in the Internet, however, does not directly provide any flow control. The datagrams are sent by the sender when they are ready, without any attention to the readiness of the receiver.

A few reasons for the lack of flow control in the design of the network layer can be mentioned. First, since there is no error control in this layer, the job of the network layer at the receiver is so simple that it may rarely be overwhelmed. Second, the upper layers that use the service of the network layer can implement buffers to receive data from the network layer as they are ready and do not have to consume the data as fast as it is received. Third, flow control is provided for most of the upper-layer protocols that use the services of the network layer, so another level of flow control makes the network layer more complicated and the whole system less efficient.

Congestion Control

Another issue in a network-layer protocol is congestion control. Congestion in the network layer is a situation in which too many datagrams are present in an area of the Internet. Congestion may occur if the number of datagrams sent by source computers is beyond the capacity of the network or routers. In this situation, some routers may drop some of the datagrams. However, as more datagrams are dropped, the situation may become worse because, due to the error control mechanism at the upper layers, the sender may send duplicates of the lost packets. If the congestion continues, sometimes a situation may reach a point where the system collapses and no datagrams are delivered. We discuss congestion control at the network layer later in the chapter although it is not implemented in the Internet.

Quality of Service

As the Internet has allowed new applications such as multimedia communication (in particular real-time communication of audio and video), the quality of service (QoS) of the communication has become more and more important. The Internet has thrived by providing better quality of service to support these applications. However, to keep the network layer untouched, these provisions are mostly implemented in the upper layer. We discuss this issue in Chapter 30 after we have discussed multimedia.

Security

Another issue related to communication at the network layer is security. Security was not a concern when the Internet was originally designed because it was used by a small number of users at universities for research activities; other people had no access to the Internet. The network layer was designed with no security provision. Today, however, security is a big concern. To provide security for a connectionless network layer, we need to have another virtual level that changes the connectionless service to a connection-oriented service. This virtual layer, called IPSec, is discussed in Chapter 32.

18.2 PACKET SWITCHING

From the discussion of routing and forwarding in the previous section, we infer that a kind of *switching* occurs at the network layer. A router, in fact, is a switch that creates a connection between an input port and an output port (or a set of output ports), just as an electrical switch connects the input to the output to let electricity flow.

Although in data communication switching techniques are divided into two broad categories, circuit switching and packet switching, only packet switching is used at the network layer because the unit of data at this layer is a packet. Circuit switching is mostly used at the physical layer; the electrical switch mentioned earlier is a kind of circuit switch. We discussed circuit switching in Chapter 8; we discuss packet switching in this chapter.

At the network layer, a message from the upper layer is divided into manageable packets and each packet is sent through the network. The source of the message sends the packets one by one; the destination of the message receives the packets one by one. The destination waits for all packets belonging to the same message to arrive before delivering the message to the upper layer. The connecting devices in a packet-switched network still need to decide how to route the packets to the final destination. Today, a packet-switched network can use two different approaches to route the packets: the *datagram approach* and the *virtual circuit approach*. We discuss both approaches in the next section.

18.2.1 Datagram Approach: Connectionless Service

When the Internet started, to make it simple, the network layer was designed to provide a connectionless service in which the network-layer protocol treats each packet independently, with each packet having no relationship to any other packet. The idea was that the network layer is only responsible for delivery of packets from the source to the destination. In this approach, the packets in a message may or may not travel the same path to their destination. Figure 18.3 shows the idea.

When the network layer provides a connectionless service, each packet traveling in the Internet is an independent entity; there is no relationship between packets belonging to the same message. The switches in this type of network are called *routers*. A packet belonging to a message may be followed by a packet belonging to the same message or to a different message. A packet may be followed by a packet coming from the same or from a different source.

Figure 18.3 *A connectionless packet-switched network*

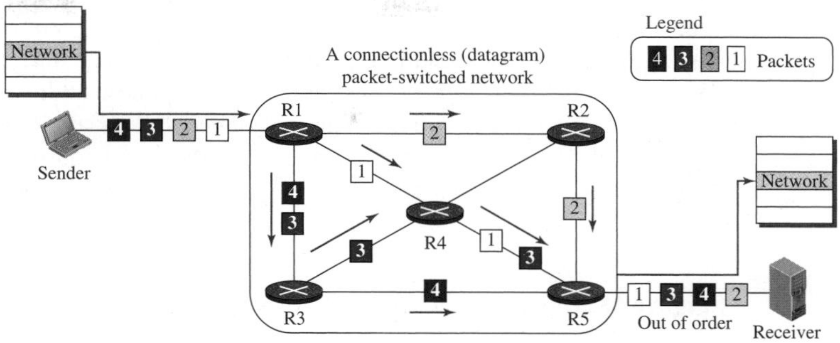

Each packet is routed based on the information contained in its header: source and destination addresses. The destination address defines where it should go; the source address defines where it comes from. The router in this case routes the packet based only on the destination address. The source address may be used to send an error message to the source if the packet is discarded. Figure 18.4 shows the forwarding process in a router in this case. We have used symbolic addresses such as A and B.

Figure 18.4 *Forwarding process in a router when used in a connectionless network*

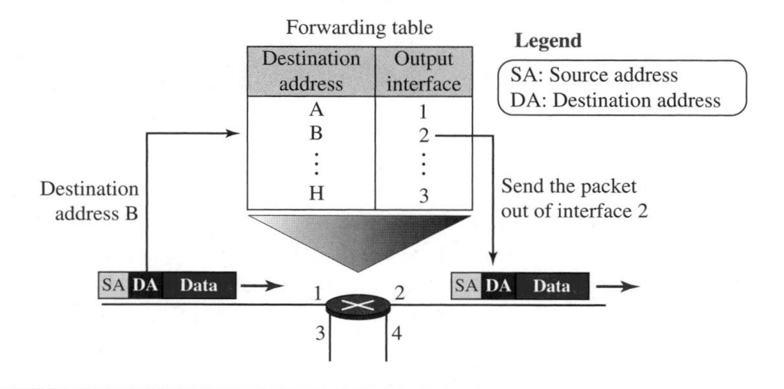

In the datagram approach, the forwarding decision is based on the destination address of the packet.

18.2.2 Virtual-Circuit Approach: Connection-Oriented Service

In a connection-oriented service (also called *virtual-circuit approach*), there is a relationship between all packets belonging to a message. Before all datagrams in a message can be sent, a virtual connection should be set up to define the path for the datagrams. After connection setup, the datagrams can all follow the same path. In this type of service, not only must the packet contain the source and destination addresses, it must also contain a flow label, a virtual circuit identifier that defines the virtual path the packet should follow. Shortly, we will show how this flow label is determined, but for the moment, we assume that the packet carries this label. Although it looks as though the use of the label may make the source and destination addresses unnecessary during the data transfer phase, parts of the Internet at the network layer still keep these addresses. One reason is that part of the packet path may still be using the connectionless service. Another reason is that the protocol at the network layer is designed with these addresses, and it may take a while before they can be changed. Figure 18.5 shows the concept of connection-oriented service.

Figure 18.5 *A virtual-circuit packet-switched network*

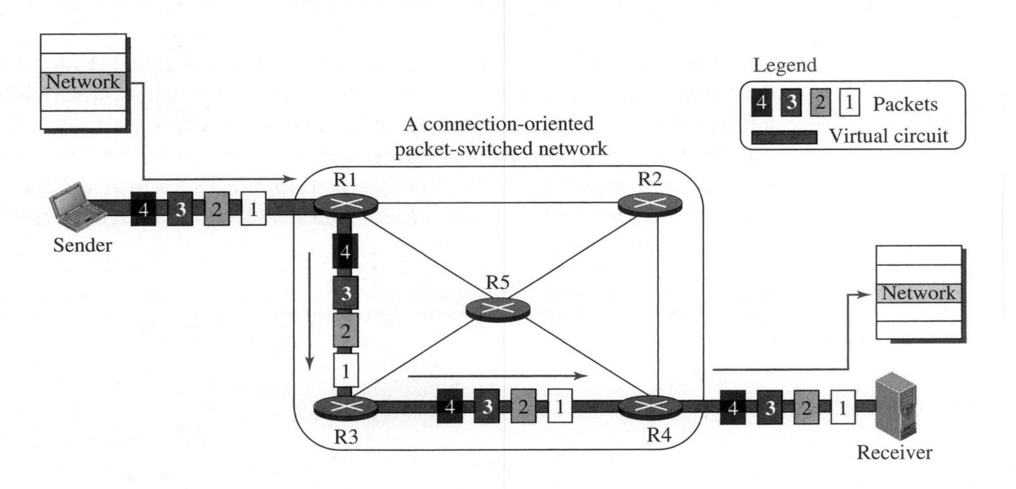

Each packet is forwarded based on the label in the packet. To follow the idea of connection-oriented design to be used in the Internet, we assume that the packet has a label when it reaches the router. Figure 18.6 shows the idea. In this case, the forwarding decision is based on the value of the label, or *virtual circuit identifier,* as it is sometimes called.

To create a connection-oriented service, a three-phase process is used: setup, data transfer, and teardown. In the setup phase, the source and destination addresses of the sender and receiver are used to make table entries for the connection-oriented service. In the teardown phase, the source and destination inform the router to delete the corresponding entries. Data transfer occurs between these two phases.

Setup Phase

In the setup phase, a router creates an entry for a virtual circuit. For example, suppose source A needs to create a virtual circuit to destination B. Two auxiliary packets need to be exchanged between the sender and the receiver: the request packet and the acknowledgment packet.

Figure 18.6 *Forwarding process in a router when used in a virtual-circuit network*

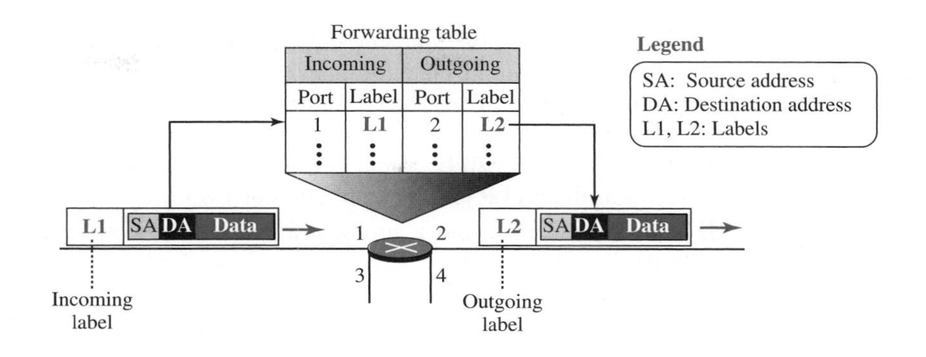

In the virtual-circuit approach, the forwarding decision is based on the label of the packet.

Request packet

A request packet is sent from the source to the destination. This auxiliary packet carries the source and destination addresses. Figure 18.7 shows the process.

Figure 18.7 *Sending request packet in a virtual-circuit network*

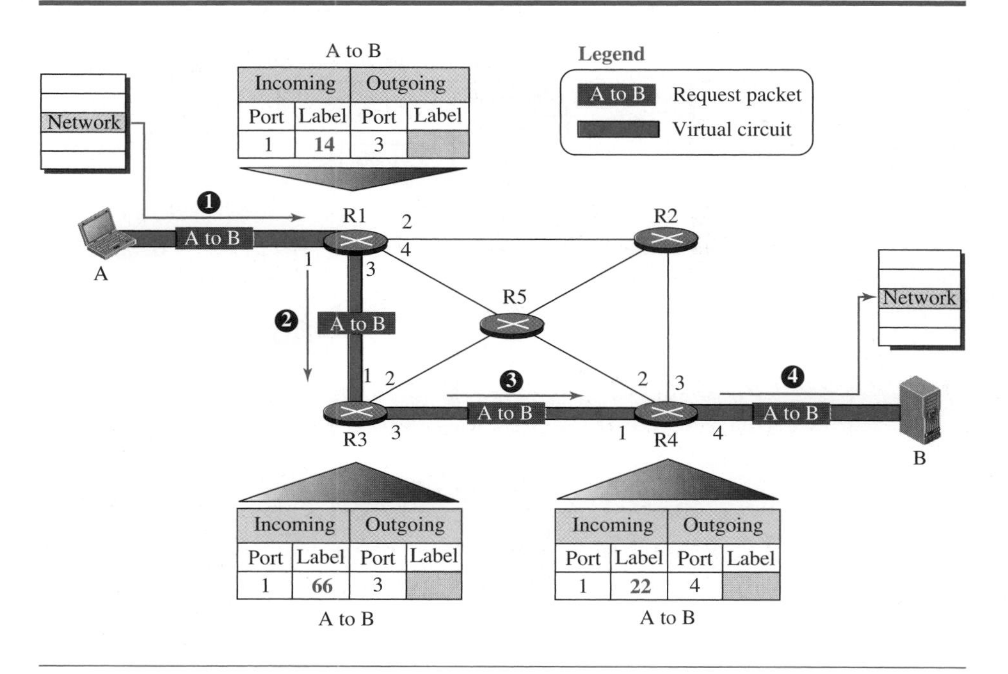

1. Source A sends a request packet to router R1.
2. Router R1 receives the request packet. It knows that a packet going from A to B goes out through port 3. How the router has obtained this information is a point covered later. For the moment, assume that it knows the output port. The router creates an entry in its table for this virtual circuit, but it is only able to fill three of the four columns. The router assigns the incoming port (1) and chooses an available incoming label (14) and the outgoing port (3). It does not yet know the outgoing label, which will be found during the acknowledgment step. The router then forwards the packet through port 3 to router R3.
3. Router R3 receives the setup request packet. The same events happen here as at router R1; three columns of the table are completed: in this case, incoming port (1), incoming label (66), and outgoing port (3).
4. Router R4 receives the setup request packet. Again, three columns are completed: incoming port (1), incoming label (22), and outgoing port (4).
5. Destination B receives the setup packet, and if it is ready to receive packets from A, it assigns a label to the incoming packets that come from A, in this case 77, as shown in Figure 18.8. This label lets the destination know that the packets come from A, and not from other sources.

Acknowledgment Packet

A special packet, called the acknowledgment packet, completes the entries in the switching tables. Figure 18.8 shows the process.

Figure 18.8 *Sending acknowledgments in a virtual-circuit network*

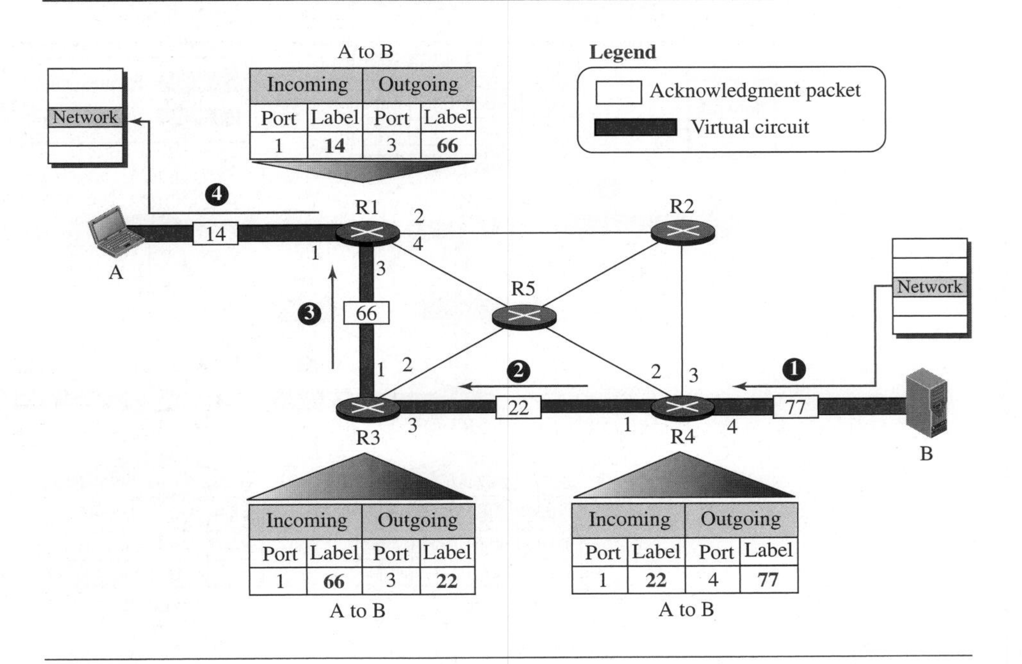

1. The destination sends an acknowledgment to router R4. The acknowledgment carries the global source and destination addresses so the router knows which entry in the table is to be completed. The packet also carries label 77, chosen by the destination as the incoming label for packets from A. Router R4 uses this label to complete the outgoing label column for this entry. Note that 77 is the incoming label for destination B, but the outgoing label for router R4.
2. Router R4 sends an acknowledgment to router R3 that contains its incoming label in the table, chosen in the setup phase. Router R3 uses this as the outgoing label in the table.
3. Router R3 sends an acknowledgment to router R1 that contains its incoming label in the table, chosen in the setup phase. Router R1 uses this as the outgoing label in the table.
4. Finally router R1 sends an acknowledgment to source A that contains its incoming label in the table, chosen in the setup phase.
5. The source uses this as the outgoing label for the data packets to be sent to destination B.

Data-Transfer Phase

The second phase is called the data-transfer phase. After all routers have created their forwarding table for a specific virtual circuit, then the network-layer packets belonging to one message can be sent one after another. In Figure 18.9, we show the flow of a single packet, but the process is the same for 1, 2, or 100 packets. The source computer uses the label 14, which it has received from router R1 in the setup

Figure 18.9 *Flow of one packet in an established virtual circuit*

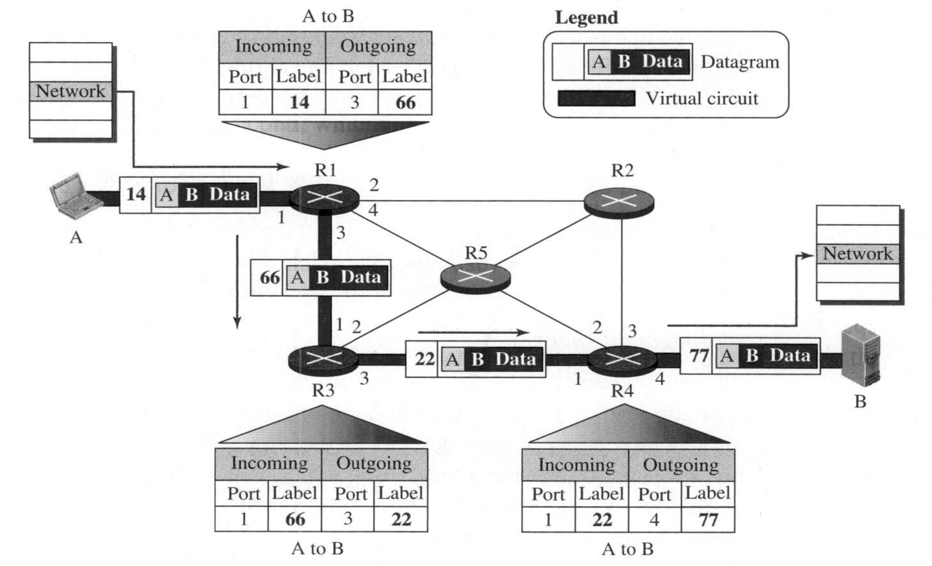

phase. Router R1 forwards the packet to router R3, but changes the label to 66. Router R3 forwards the packet to router R4, but changes the label to 22. Finally, router R4 delivers the packet to its final destination with the label 77. All the packets in the message follow the same sequence of labels, and the packets arrive in order at the destination.

Teardown Phase

In the teardown phase, source A, after sending all packets to B, sends a special packet called a teardown packet. Destination B responds with a confirmation packet. All routers delete the corresponding entries from their tables.

18.3 NETWORK-LAYER PERFORMANCE

The upper-layer protocols that use the service of the network layer expect to receive an ideal service, but the network layer is not perfect. The performance of a network can be measured in terms of *delay, throughput,* and *packet loss. Congestion control* is an issue that can improve the performance.

18.3.1 Delay

All of us expect instantaneous response from a network, but a packet, from its source to its destination, encounters delays. The delays in a network can be divided into four types: transmission delay, propagation delay, processing delay, and queuing delay. Let us first discuss each of these delay types and then show how to calculate a packet delay from the source to the destination.

Transmission Delay

A source host or a router cannot send a packet instantaneously. A sender needs to put the bits in a packet on the line one by one. If the first bit of the packet is put on the line at time t_1 and the last bit is put on the line at time t_2, transmission delay of the packet is $(t_2 - t_1)$. Definitely, the transmission delay is longer for a longer packet and shorter if the sender can transmit faster. In other words, the transmission delay is

$$\textbf{Delay}_{\textbf{tr}} = \textbf{(Packet length) / (Transmission rate).}$$

For example, in a Fast Ethernet LAN (see Chapter 13) with the transmission rate of 100 million bits per second and a packet of 10,000 bits, it takes (10,000)/(100,000,000) or 100 microseconds for all bits of the packet to be put on the line.

Propagation Delay

Propagation delay is the time it takes for a bit to travel from point A to point B in the transmission media. The propagation delay for a packet-switched network depends on the propagation delay of each network (LAN or WAN). The propagation delay depends on the propagation speed of the media, which is 3×10^8 meters/second in a vacuum and normally much less in a wired medium; it also depends on the distance of the link. In other words, propagation delay is

$$\textbf{Delay}_{\textbf{pg}} = \textbf{(Distance) / (Propagation speed).}$$

For example, if the distance of a cable link in a point-to-point WAN is 2000 meters and the propagation speed of the bits in the cable is 2×10^8 meters/second, then the propagation delay is 10 microseconds.

Processing Delay

The processing delay is the time required for a router or a destination host to receive a packet from its input port, remove the header, perform an error detection procedure, and deliver the packet to the output port (in the case of a router) or deliver the packet to the upper-layer protocol (in the case of the destination host). The processing delay may be different for each packet, but normally is calculated as an average.

$$\textbf{Delay}_{\textbf{pr}} = \textbf{Time required to process a packet in a router or a destination host}$$

Queuing Delay

Queuing delay can normally happen in a router. As we discuss in the next section, a router has an input queue connected to each of its input ports to store packets waiting to be processed; the router also has an output queue connected to each of its output ports to store packets waiting to be transmitted. The queuing delay for a packet in a router is measured as the time a packet waits in the input queue and output queue of a router. We can compare the situation with a busy airport. Some planes may need to wait to get the landing band (input delay); some planes may need to wait to get the departure band (output delay).

$$\textbf{Delay}_{\textbf{qu}} = \textbf{The time a packet waits in input and output queues in a router}$$

Total Delay

Assuming equal delays for the sender, routers, and receiver, the total delay (source-to-destination delay) a packet encounters can be calculated if we know the number of routers, n, in the whole path.

$$\textbf{Total delay} = (n + 1)\ (\textbf{Delay}_{\textbf{tr}} + \textbf{Delay}_{\textbf{pg}} + \textbf{Delay}_{\textbf{pr}}) + (n)\ (\textbf{Delay}_{\textbf{qu}})$$

Note that if we have n routers, we have $(n + 1)$ links. Therefore, we have $(n + 1)$ transmission delays related to n routers and the source, $(n + 1)$ propagation delays related to $(n + 1)$ links, $(n + 1)$ processing delays related to n routers and the destination, and only n queuing delays related to n routers.

18.3.2 Throughput

Throughput at any point in a network is defined as the number of bits passing through the point in a second, which is actually the transmission rate of data at that point. In a path from source to destination, a packet may pass through several links (networks), each with a different transmission rate. How, then, can we determine the throughput of the whole path? To see the situation, assume that we have three links, each with a different transmission rate, as shown in Figure 18.10.

In this figure, the data can flow at the rate of 200 kbps in Link1. However, when the data arrives at router R1, it cannot pass at this rate. Data needs to be queued at the router and sent at 100 kbps. When data arrives at router R2, it could be sent at the rate

Figure 18.10 *Throughput in a path with three links in a series*

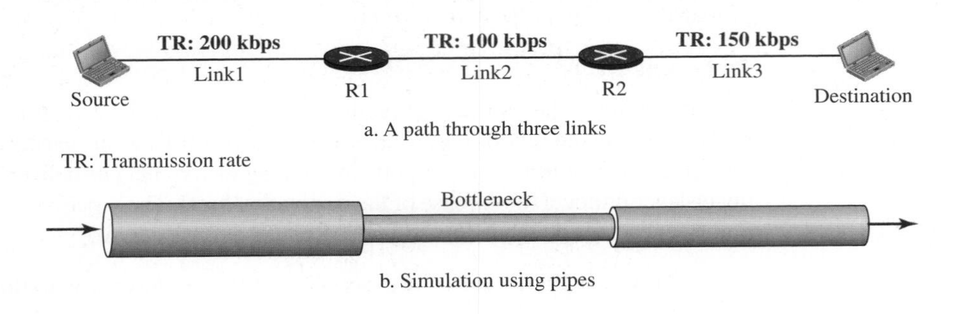

of 150 kbps, but there is not enough data to be sent. In other words, the average rate of the data flow in Link3 is also 100 kbps. We can conclude that the average data rate for this path is 100 kbps, the minimum of the three different data rates. The figure also shows that we can simulate the behavior of each link with pipes of different sizes; the average throughput is determined by the bottleneck, the pipe with the smallest diameter. In general, in a path with n links in series, we have

$$\textbf{Throughput} = \textbf{minimum}\ \{\textbf{TR}_1, \textbf{TR}_2, \ldots \textbf{TR}_n\}.$$

Although the situation in Figure 18.10 shows how to calculate the throughput when the data is passed through several links, the actual situation in the Internet is that the data normally passes through two access networks and the Internet backbone, as shown in Figure 18.11.

Figure 18.11 *A path through the Internet backbone*

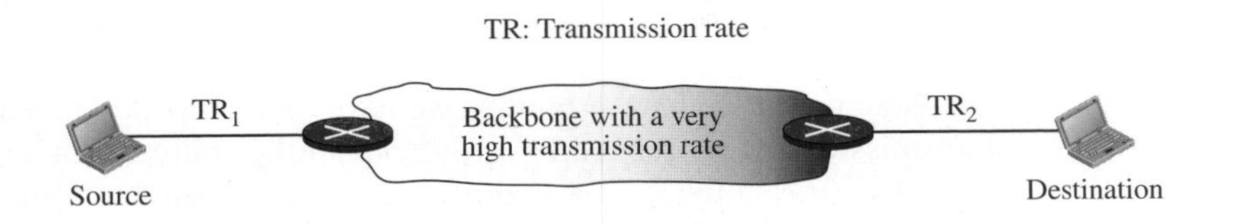

The Internet backbone has a very high transmission rate, in the range of gigabits per second. This means that the throughput is normally defined as the minimum transmission rate of the two access links that connect the source and destination to the backbone. Figure 18.11 shows this situation, in which the throughput is the minimum of TR_1 and TR_2. For example, if a server connects to the Internet via a Fast Ethernet LAN with the data rate of 100 Mbps, but a user who wants to download a file connects to the Internet via a dial-up telephone line with the data rate of 40 kbps, the throughput is 40 kbps. The bottleneck is definitely the dial-up line.

We need to mention another situation in which we think about the throughput. The link between two routers is not always dedicated to one flow. A router may collect the flow from several sources or distribute the flow between several sources. In this case the transmission rate of the link between the two routers is actually shared between the flows and this should be considered when we calculate the throughput. For example, in Figure 18.12 the transmission rate of the main link in the calculation of the throughput is only 200 kbps because the link is shared between three paths.

Figure 18.12 *Effect of throughput in shared links*

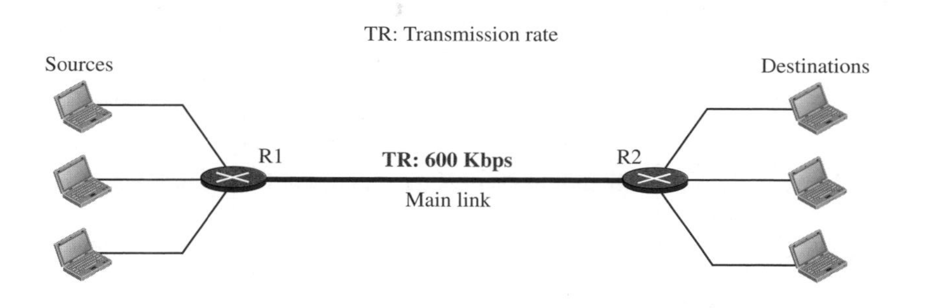

18.3.3 Packet Loss

Another issue that severely affects the performance of communication is the number of packets lost during transmission. When a router receives a packet while processing another packet, the received packet needs to be stored in the input buffer waiting for its turn. A router, however, has an input buffer with a limited size. A time may come when the buffer is full and the next packet needs to be dropped. The effect of packet loss on the Internet network layer is that the packet needs to be resent, which in turn may create overflow and cause more packet loss. A lot of theoretical studies have been done in queuing theory to prevent the overflow of queues and prevent packet loss.

18.3.4 Congestion Control

Congestion control is a mechanism for improving performance. In Chapter 23, we will discuss congestion at the transport layer. Although congestion at the network layer is not explicitly addressed in the Internet model, the study of congestion at this layer may help us to better understand the cause of congestion at the transport layer and find possible remedies to be used at the network layer. Congestion at the network layer is related to two issues, throughput and delay, which we discussed in the previous section. Figure 18.13 shows these two performance measures as functions of load.

When the load is much less than the capacity of the network, the *delay* is at a minimum. This minimum delay is composed of propagation delay and processing delay, both of which are negligible. However, when the load reaches the network capacity, the delay increases sharply because we now need to add the queuing delay to the total delay. Note that the delay becomes infinite when the load is greater than the capacity.

When the load is below the capacity of the network, the throughput increases proportionally with the *load*. We expect the throughput to remain constant after the load reaches the capacity, but instead the throughput declines sharply. The reason is the discarding of packets by the routers. When the load exceeds the capacity, the queues

Figure 18.13 *Packet delay and throughput as functions of load*

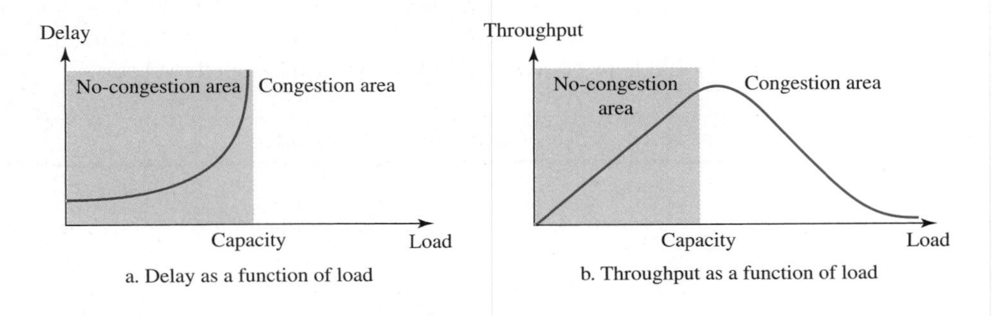

a. Delay as a function of load

b. Throughput as a function of load

become full and the routers have to discard some packets. Discarding packets does not reduce the number of packets in the network because the sources retransmit the packets, using time-out mechanisms, when the packets do not reach the destinations.

Congestion Control

Congestion control refers to techniques and mechanisms that can either prevent congestion before it happens or remove congestion after it has happened. In general, we can divide congestion control mechanisms into two broad categories: **open-loop congestion control** (prevention) and **closed-loop congestion control** (removal).

Open-Loop Congestion Control

In open-loop congestion control, policies are applied to prevent congestion before it happens. In these mechanisms, congestion control is handled by either the source or the destination. We give a brief list of policies that can prevent congestion.

Retransmission Policy Retransmission is sometimes unavoidable. If the sender feels that a sent packet is lost or corrupted, the packet needs to be retransmitted. Retransmission in general may increase congestion in the network. However, a good retransmission policy can prevent congestion. The retransmission policy and the retransmission timers must be designed to optimize efficiency and at the same time prevent congestion.

Window Policy The type of window at the sender may also affect congestion. The Selective Repeat window is better than the Go-Back-*N* window for congestion control. In the Go-Back-*N* window, when the timer for a packet times out, several packets may be resent, although some may have arrived safe and sound at the receiver. This duplication may make the congestion worse. The Selective Repeat window, on the other hand, tries to send the specific packets that have been lost or corrupted.

Acknowledgment Policy The acknowledgment policy imposed by the receiver may also affect congestion. If the receiver does not acknowledge every packet it receives, it may slow down the sender and help prevent congestion. Several approaches are used in this case. A receiver may send an acknowledgment only if it has a packet to be sent or a special timer expires. A receiver may decide to acknowledge only *N* packets at a time.

We need to know that the acknowledgments are also part of the load in a network. Sending fewer acknowledgments means imposing less load on the network.

Discarding Policy A good discarding policy by the routers may prevent congestion and at the same time may not harm the integrity of the transmission. For example, in audio transmission, if the policy is to discard less sensitive packets when congestion is likely to happen, the quality of sound is still preserved and congestion is prevented or alleviated.

Admission Policy An admission policy, which is a quality-of-service mechanism (discussed in Chapter 30), can also prevent congestion in virtual-circuit networks. Switches in a flow first check the resource requirement of a flow before admitting it to the network. A router can deny establishing a virtual-circuit connection if there is congestion in the network or if there is a possibility of future congestion.

Closed-Loop Congestion Control

Closed-loop congestion control mechanisms try to alleviate congestion after it happens. Several mechanisms have been used by different protocols. We describe a few of them here.

Backpressure The technique of *backpressure* refers to a congestion control mechanism in which a congested node stops receiving data from the immediate upstream node or nodes. This may cause the upstream node or nodes to become congested, and they, in turn, reject data from their upstream node or nodes, and so on. Backpressure is a node-to-node congestion control that starts with a node and propagates, in the opposite direction of data flow, to the source. The backpressure technique can be applied only to virtual circuit networks, in which each node knows the upstream node from which a flow of data is coming. Figure 18.14 shows the idea of backpressure.

Figure 18.14 *Backpressure method for alleviating congestion*

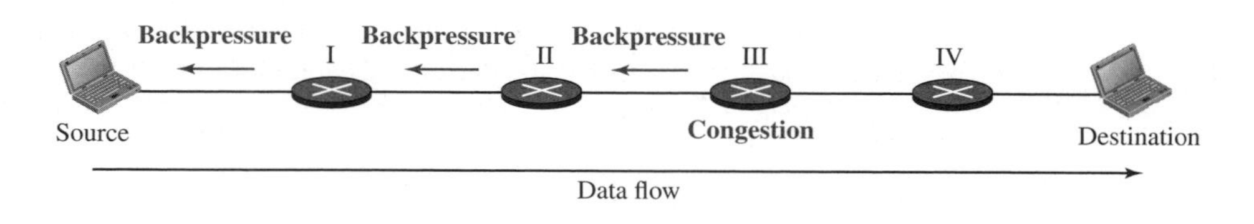

Node III in the figure has more input data than it can handle. It drops some packets in its input buffer and informs node II to slow down. Node II, in turn, may be congested because it is slowing down the output flow of data. If node II is congested, it informs node I to slow down, which in turn may create congestion. If so, node I informs the source of data to slow down. This, in time, alleviates the congestion. Note that the *pressure* on node III is moved backward to the source to remove the congestion.

It is important to stress that this type of congestion control can only be implemented in virtual-circuit. The technique cannot be implemented in a datagram network, in which a node (router) does not have the slightest knowledge of the upstream router.

Choke Packet A **choke packet** is a packet sent by a node to the source to inform it of congestion. Note the difference between the backpressure and choke-packet methods. In backpressure, the warning is from one node to its upstream node, although the warning may eventually reach the source station. In the choke-packet method, the warning is from the router, which has encountered congestion, directly to the source station. The intermediate nodes through which the packet has traveled are not warned. We will see an example of this type of control in ICMP (discussed in Chapter 19). When a router in the Internet is overwhelmed with IP datagrams, it may discard some of them, but it informs the source host, using a source quench ICMP message. The warning message goes directly to the source station; the intermediate routers do not take any action. Figure 18.15 shows the idea of a choke packet.

Figure 18.15 *Choke packet*

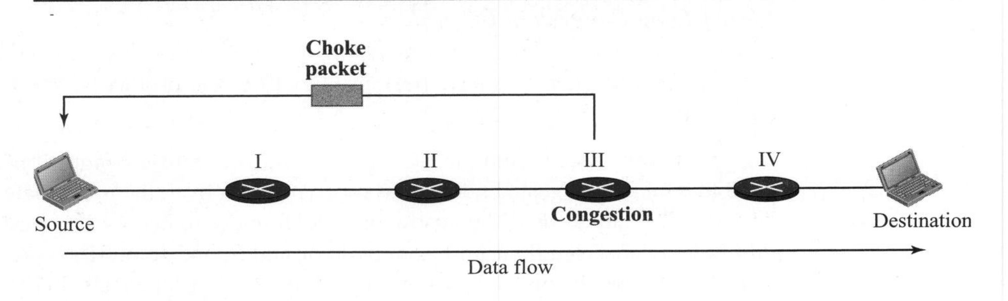

Implicit Signaling In implicit signaling, there is no communication between the congested node or nodes and the source. The source guesses that there is congestion somewhere in the network from other symptoms. For example, when a source sends several packets and there is no acknowledgment for a while, one assumption is that the network is congested. The delay in receiving an acknowledgment is interpreted as congestion in the network; the source should slow down. We saw this type of signaling when we discuss TCP congestion control in Chapter 24.

Explicit Signaling The node that experiences congestion can explicitly send a signal to the source or destination. The explicit-signaling method, however, is different from the choke-packet method. In the choke-packet method, a separate packet is used for this purpose; in the explicit-signaling method, the signal is included in the packets that carry data. Explicit signaling can occur in either the forward or the backward direction. This type of congestion control can be seen in an ATM network, discussed in Chapter 14.

18.4 IPV4 ADDRESSES

The identifier used in the IP layer of the TCP/IP protocol suite to identify the connection of each device to the Internet is called the Internet address or IP address. An IPv4 address is a 32-bit address that uniquely and universally defines the connection of a host or a router to the Internet. The IP address is the address of the connection, not the host or the router, because if the device is moved to another network, the IP address may be changed.

IPv4 addresses are unique in the sense that each address defines one, and only one, connection to the Internet. If a device has two connections to the Internet, via two networks, it has two IPv4 addresses. IPv4 addresses are universal in the sense that the addressing system must be accepted by any host that wants to be connected to the Internet.

18.4.1 Address Space

A protocol like IPv4 that defines addresses has an address space. An **address space** is the total number of addresses used by the protocol. If a protocol uses b bits to define an address, the address space is 2^b because each bit can have two different values (0 or 1). IPv4 uses 32-bit addresses, which means that the address space is 2^{32} or 4,294,967,296 (more than four billion). If there were no restrictions, more than 4 billion devices could be connected to the Internet.

Notation

There are three common notations to show an IPv4 address: binary notation (base 2), dotted-decimal notation (base 256), and hexadecimal notation (base 16). In *binary notation,* an IPv4 address is displayed as 32 bits. To make the address more readable, one or more spaces are usually inserted between each octet (8 bits). Each octet is often referred to as a byte. To make the IPv4 address more compact and easier to read, it is usually written in decimal form with a decimal point (dot) separating the bytes. This format is referred to as *dotted-decimal notation.* Note that because each byte (octet) is only 8 bits, each number in the dotted-decimal notation is between 0 and 255. We sometimes see an IPv4 address in hexadecimal notation. Each hexadecimal digit is equivalent to four bits. This means that a 32-bit address has 8 hexadecimal digits. This notation is often used in network programming. Figure 18.16 shows an IP address in the three discussed notations.

Figure 18.16 *Three different notations in IPv4 addressing*

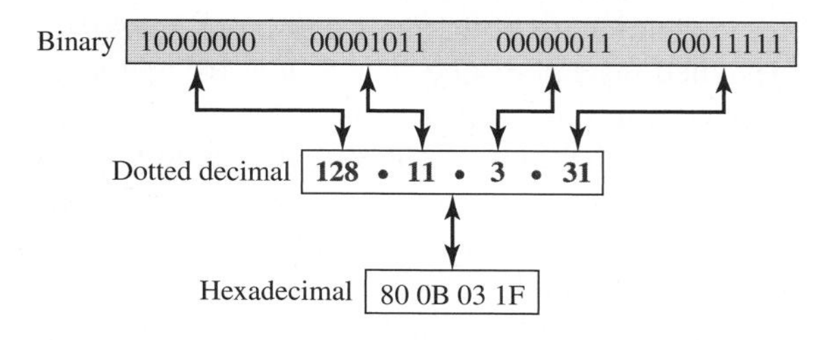

Hierarchy in Addressing

In any communication network that involves delivery, such as a telephone network or a postal network, the addressing system is hierarchical. In a postal network, the postal address (mailing address) includes the country, state, city, street, house number, and the name of the mail recipient. Similarly, a telephone number is divided into the country code, area code, local exchange, and the connection.

A 32-bit IPv4 address is also hierarchical, but divided only into two parts. The first part of the address, called the *prefix*, defines the network; the second part of the address, called the *suffix*, defines the node (connection of a device to the Internet). Figure 18.17 shows the prefix and suffix of a 32-bit IPv4 address. The prefix length is n bits and the suffix length is $(32 - n)$ bits.

Figure 18.17 *Hierarchy in addressing*

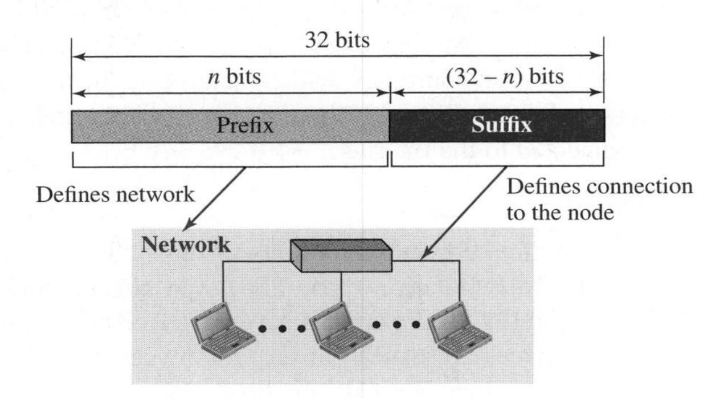

A prefix can be fixed length or variable length. The network identifier in the IPv4 was first designed as a fixed-length prefix. This scheme, which is now obsolete, is referred to as classful addressing. The new scheme, which is referred to as classless addressing, uses a variable-length network prefix. First, we briefly discuss classful addressing; then we concentrate on classless addressing.

18.4.2 Classful Addressing

When the Internet started, an IPv4 address was designed with a fixed-length prefix, but to accommodate both small and large networks, three fixed-length prefixes were designed instead of one ($n = 8$, $n = 16$, and $n = 24$). The whole address space was divided into five classes (class A, B, C, D, and E), as shown in Figure 18.18. This scheme is referred to as **classful addressing.** Although classful addressing belongs to the past, it helps us to understand classless addressing, discussed later.

In class A, the network length is 8 bits, but since the first bit, which is 0, defines the class, we can have only seven bits as the network identifier. This means there are only $2^7 = 128$ networks in the world that can have a class A address.

In class B, the network length is 16 bits, but since the first two bits, which are $(10)_2$, define the class, we can have only 14 bits as the network identifier. This means there are only $2^{14} = 16{,}384$ networks in the world that can have a class B address.

All addresses that start with $(110)_2$ belong to class C. In class C, the network length is 24 bits, but since three bits define the class, we can have only 21 bits as the network identifier. This means there are $2^{21} = 2{,}097{,}152$ networks in the world that can have a class C address.

Figure 18.18 *Occupation of the address space in classful addressing*

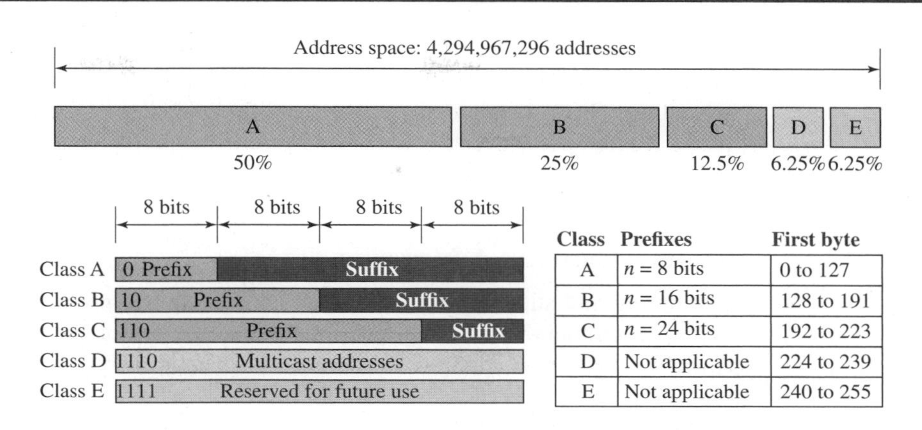

Class	Prefixes	First byte
A	n = 8 bits	0 to 127
B	n = 16 bits	128 to 191
C	n = 24 bits	192 to 223
D	Not applicable	224 to 239
E	Not applicable	240 to 255

Class D is not divided into prefix and suffix. It is used for multicast addresses. All addresses that start with 1111 in binary belong to class E. As in Class D, Class E is not divided into prefix and suffix and is used as reserve.

Address Depletion

The reason that classful addressing has become obsolete is address depletion. Since the addresses were not distributed properly, the Internet was faced with the problem of the addresses being rapidly used up, resulting in no more addresses available for organizations and individuals that needed to be connected to the Internet. To understand the problem, let us think about class A. This class can be assigned to only 128 organizations in the world, but each organization needs to have a single network (seen by the rest of the world) with 16,777,216 nodes (computers in this single network). Since there may be only a few organizations that are this large, most of the addresses in this class were wasted (unused). Class B addresses were designed for midsize organizations, but many of the addresses in this class also remained unused. Class C addresses have a completely different flaw in design. The number of addresses that can be used in each network (256) was so small that most companies were not comfortable using a block in this address class. Class E addresses were almost never used, wasting the whole class.

Subnetting and Supernetting

To alleviate address depletion, two strategies were proposed and, to some extent, implemented: subnetting and supernetting. In subnetting, a class A or class B block is divided into several subnets. Each subnet has a larger prefix length than the original network. For example, if a network in class A is divided into four subnets, each subnet has a prefix of $n_{sub} = 10$. At the same time, if all of the addresses in a network are not used, subnetting allows the addresses to be divided among several organizations. This idea did not work because most large organizations were not happy about dividing the block and giving some of the unused addresses to smaller organizations.

While subnetting was devised to divide a large block into smaller ones, supernetting was devised to combine several class C blocks into a larger block to be attractive to organizations that need more than the 256 addresses available in a class C block. This idea did not work either because it makes the routing of packets more difficult.

Advantage of Classful Addressing

Although classful addressing had several problems and became obsolete, it had one advantage: Given an address, we can easily find the class of the address and, since the prefix length for each class is fixed, we can find the prefix length immediately. In other words, the prefix length in classful addressing is inherent in the address; no extra information is needed to extract the prefix and the suffix.

18.4.3 Classless Addressing

Subnetting and supernetting in classful addressing did not really solve the address depletion problem. With the growth of the Internet, it was clear that a larger address space was needed as a long-term solution. The larger address space, however, requires that the length of IP addresses also be increased, which means the format of the IP packets needs to be changed. Although the long-range solution has already been devised and is called IPv6 (discussed later), a short-term solution was also devised to use the same address space but to change the distribution of addresses to provide a fair share to each organization. The short-term solution still uses IPv4 addresses, but it is called *classless addressing*. In other words, the class privilege was removed from the distribution to compensate for the address depletion.

There was another motivation for classless addressing. During the 1990s, Internet Service Providers (ISPs) came into prominence. An ISP is an organization that provides Internet access for individuals, small businesses, and midsize organizations that do not want to create an Internet site and become involved in providing Internet services (such as electronic mail) for their employees. An ISP can provide these services. An ISP is granted a large range of addresses and then subdivides the addresses (in groups of 1, 2, 4, 8, 16, and so on), giving a range of addresses to a household or a small business. The customers are connected via a dial-up modem, DSL, or cable modem to the ISP. However, each customer needs some IPv4 addresses.

In 1996, the Internet authorities announced a new architecture called **classless addressing.** In classless addressing, variable-length blocks are used that belong to no classes. We can have a block of 1 address, 2 addresses, 4 addresses, 128 addresses, and so on.

In classless addressing, the whole address space is divided into variable length blocks. The prefix in an address defines the block (network); the suffix defines the node (device). Theoretically, we can have a block of $2^0, 2^1, 2^2, \ldots, 2^{32}$ addresses. One of the restrictions, as we discuss later, is that the number of addresses in a block needs to be a power of 2. An organization can be granted one block of addresses. Figure 18.19 shows the division of the whole address space into nonoverlapping blocks.

Figure 18.19 *Variable-length blocks in classless addressing*

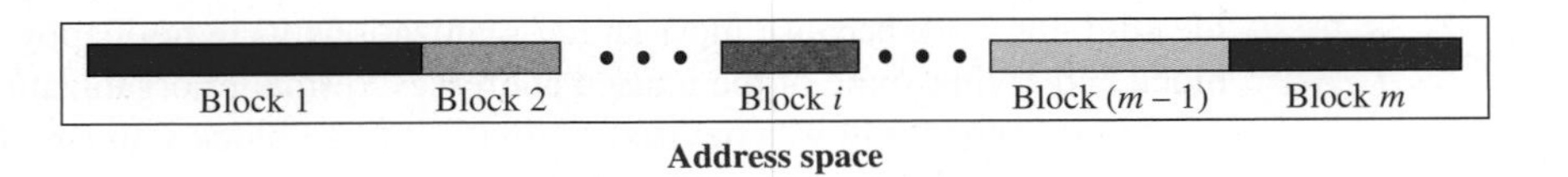

Unlike classful addressing, the prefix length in classless addressing is variable. We can have a prefix length that ranges from 0 to 32. The size of the network is inversely proportional to the length of the prefix. A small prefix means a larger network; a large prefix means a smaller network.

We need to emphasize that the idea of classless addressing can be easily applied to classful addressing. An address in class A can be thought of as a classless address in which the prefix length is 8. An address in class B can be thought of as a classless address in which the prefix is 16, and so on. In other words, classful addressing is a special case of classless addressing.

Prefix Length: Slash Notation

The first question that we need to answer in classless addressing is how to find the prefix length if an address is given. Since the prefix length is not inherent in the address, we need to separately give the length of the prefix. In this case, the prefix length, *n*, is added to the address, separated by a slash. The notation is informally referred to as *slash notation* and formally as ***classless interdomain routing*** or ***CIDR*** (pronounced cider) strategy. An address in classless addressing can then be represented as shown in Figure 18.20.

Figure 18.20 *Slash notation (CIDR)*

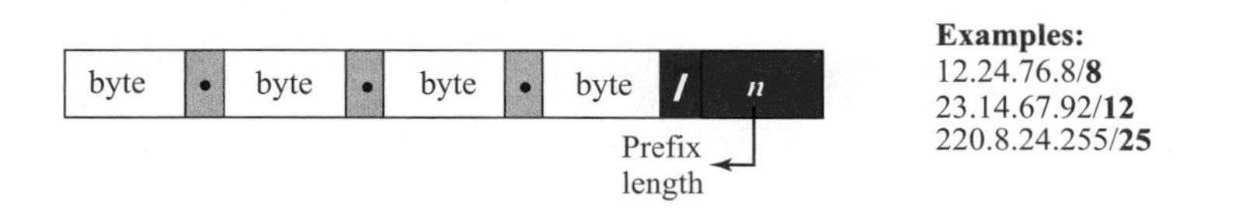

In other words, an address in classless addressing does not, per se, define the block or network to which the address belongs; we need to give the prefix length also.

Extracting Information from an Address

Given any address in the block, we normally like to know three pieces of information about the block to which the address belongs: the number of addresses, the first address in the block, and the last address. Since the value of prefix length, *n*, is given, we can easily find these three pieces of information, as shown in Figure 18.21.

1. The number of addresses in the block is found as $N = 2^{32-n}$.
2. To find the first address, we keep the *n* leftmost bits and set the $(32 - n)$ rightmost bits all to 0s.
3. To find the last address, we keep the *n* leftmost bits and set the $(32 - n)$ rightmost bits all to 1s.

Example 18.1

A classless address is given as 167.199.170.82/**27**. We can find the above three pieces of information as follows. The number of addresses in the network is $2^{32-n} = 2^5 = 32$ addresses.

Figure 18.21 *Information extraction in classless addressing*

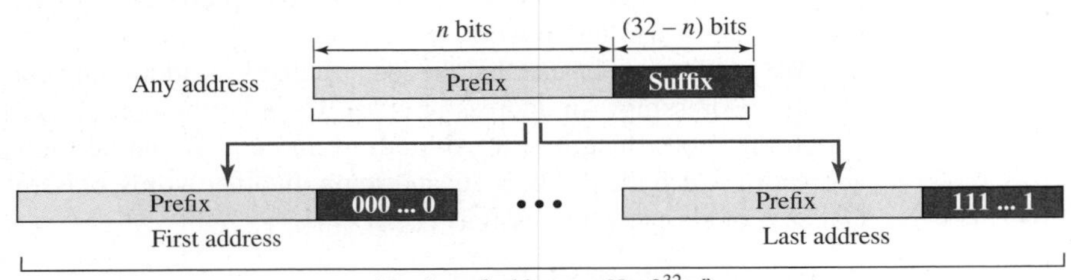

The first address can be found by keeping the first 27 bits and changing the rest of the bits to 0s.

Address: 167.199.170.82/**27**	`10100111 11000111 10101010 01010010`
First address: 167.199.170.64/**27**	`10100111 11000111 10101010 01000000`

The last address can be found by keeping the first 27 bits and changing the rest of the bits to 1s.

Address: 167.199.170.82/**27**	`10100111 11000111 10101010 01011111`
Last address: 167.199.170.95/**27**	`10100111 11000111 10101010 01011111`

Address Mask

Another way to find the first and last addresses in the block is to use the address mask. The address mask is a 32-bit number in which the n leftmost bits are set to 1s and the rest of the bits $(32 - n)$ are set to 0s. A computer can easily find the address mask because it is the complement of $(2^{32-n} - 1)$. The reason for defining a mask in this way is that it can be used by a computer program to extract the information in a block, using the three bit-wise operations NOT, AND, and OR.

1. The number of addresses in the block N = **NOT** (mask) + 1.
2. The first address in the block = (Any address in the block) **AND** (mask).
3. The last address in the block = (Any address in the block) **OR** [(**NOT** (mask)].

Example 18.2

We repeat Example 18.1 using the mask. The mask in dotted-decimal notation is 256.256.256.224. The AND, OR, and NOT operations can be applied to individual bytes using calculators and applets at the book website.

Number of addresses in the block:	N = **NOT** (mask) + 1= 0.0.0.31 + 1 = 32 addresses
First address:	First = (address) **AND** (mask) = 167.199.170.82
Last address:	Last = (address) **OR** (**NOT** mask) = 167.199.170.255

Example 18.3

In classless addressing, an address cannot per se define the block the address belongs to. For example, the address 230.8.24.56 can belong to many blocks. Some of them are shown below with the value of the prefix associated with that block.

Prefix length:16	→	Block:	230.8.0.0	to	230.8.255.255
Prefix length:20	→	Block:	230.8.16.0	to	230.8.31.255
Prefix length:26	→	Block:	230.8.24.0	to	230.8.24.63
Prefix length:27	→	Block:	230.8.24.32	to	230.8.24.63
Prefix length:29	→	Block:	230.8.24.56	to	230.8.24.63
Prefix length:31	→	Block:	230.8.24.56	to	230.8.24.57

Network Address

The above examples show that, given any address, we can find all information about the block. The first address, the **network address,** is particularly important because it is used in routing a packet to its destination network. For the moment, let us assume that an internet is made of m networks and a router with m interfaces. When a packet arrives at the router from any source host, the router needs to know to which network the packet should be sent: from which interface the packet should be sent out. When the packet arrives at the network, it reaches its destination host using another strategy that we discuss later. Figure 18.22 shows the idea. After the network address has been

Figure 18.22 *Network address*

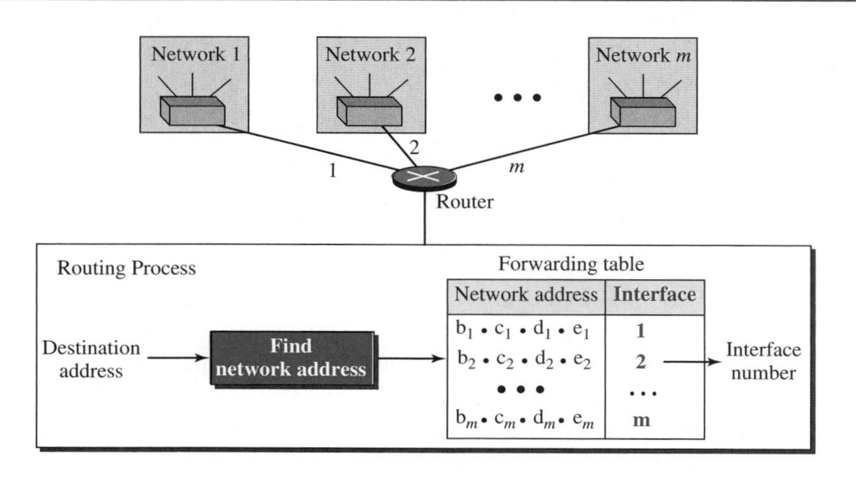

found, the router consults its forwarding table to find the corresponding interface from which the packet should be sent out. The network address is actually the identifier of the network; each network is identified by its network address.

Block Allocation

The next issue in classless addressing is block allocation. How are the blocks allocated? The ultimate responsibility of block allocation is given to a global authority called the Internet Corporation for Assigned Names and Numbers (ICANN). However, ICANN does not normally allocate addresses to individual Internet users. It assigns a large block of addresses to an ISP (or a larger organization that is considered an ISP in this case). For the proper operation of the CIDR, two restrictions need to be applied to the allocated block.

1. The number of requested addresses, N, needs to be a power of 2. The reason is that $N = 2^{32-n}$ or $n = 32 - \log_2 N$. If N is not a power of 2, we cannot have an integer value for n.
2. The requested block needs to be allocated where there is an adequate number of contiguous addresses available in the address space. However, there is a restriction on choosing the first address in the block. The first address needs to be divisible by the number of addresses in the block. The reason is that the first address needs to be the prefix followed by $(32 - n)$ number of 0s. The decimal value of the first address is then

$$\text{first address} = (\text{prefix in decimal}) \times 2^{32-n} = (\text{prefix in decimal}) \times N.$$

Example 18.4

An ISP has requested a block of 1000 addresses. Since 1000 is not a power of 2, 1024 addresses are granted. The prefix length is calculated as $n = 32 - \log_2 1024 = 22$. An available block, 18.14.12.0/**22**, is granted to the ISP. It can be seen that the first address in decimal is 302,910,464, which is divisible by 1024.

Subnetting

More levels of hierarchy can be created using subnetting. An organization (or an ISP) that is granted a range of addresses may divide the range into several subranges and assign each subrange to a subnetwork (or subnet). Note that nothing stops the organization from creating more levels. A subnetwork can be divided into several sub-subnetworks. A sub-subnetwork can be divided into several sub-sub-subnetworks, and so on.

Designing Subnets

The subnetworks in a network should be carefully designed to enable the routing of packets. We assume the total number of addresses granted to the organization is N, the prefix length is n, the assigned number of addresses to each subnetwork is N_{sub}, and the prefix length for each subnetwork is n_{sub}. Then the following steps need to be carefully followed to guarantee the proper operation of the subnetworks.

- ❑ The number of addresses in each subnetwork should be a power of 2.
- ❑ The prefix length for each subnetwork should be found using the following formula:

$$n_{sub} = 32 - \log_2 N_{sub}$$

- The starting address in each subnetwork should be divisible by the number of addresses in that subnetwork. This can be achieved if we first assign addresses to larger subnetworks.

Finding Information about Each Subnetwork

After designing the subnetworks, the information about each subnetwork, such as first and last address, can be found using the process we described to find the information about each network in the Internet.

Example 18.5

An organization is granted a block of addresses with the beginning address 14.24.74.0/24. The organization needs to have 3 subblocks of addresses to use in its three subnets: one subblock of 10 addresses, one subblock of 60 addresses, and one subblock of 120 addresses. Design the subblocks.

Solution

There are $2^{32-24} = 256$ addresses in this block. The first address is 14.24.74.0/24; the last address is 14.24.74.255/24. To satisfy the third requirement, we assign addresses to subblocks, starting with the largest and ending with the smallest one.

a. The number of addresses in the largest subblock, which requires 120 addresses, is not a power of 2. We allocate 128 addresses. The subnet mask for this subnet can be found as $n_1 = 32 - \log_2 128 = 25$. The first address in this block is 14.24.74.0/25; the last address is 14.24.74.127/25.

b. The number of addresses in the second largest subblock, which requires 60 addresses, is not a power of 2 either. We allocate 64 addresses. The subnet mask for this subnet can be found as $n_2 = 32 - \log_2 64 = 26$. The first address in this block is 14.24.74.128/26; the last address is 14.24.74.191/26.

c. The number of addresses in the smallest subblock, which requires 10 addresses, is not a power of 2 either. We allocate 16 addresses. The subnet mask for this subnet can be found as $n_3 = 32 - \log_2 16 = 28$. The first address in this block is 14.24.74.192/28; the last address is 14.24.74.207/28.

If we add all addresses in the previous subblocks, the result is 208 addresses, which means 48 addresses are left in reserve. The first address in this range is 14.24.74.208. The last address is 14.24.74.255. We don't know about the prefix length yet. Figure 18.23 shows the configuration of blocks. We have shown the first address in each block.

Address Aggregation

One of the advantages of the CIDR strategy is **address aggregation** (sometimes called *address summarization* or *route summarization*). When blocks of addresses are combined to create a larger block, routing can be done based on the prefix of the larger block. ICANN assigns a large block of addresses to an ISP. Each ISP in turn divides its assigned block into smaller subblocks and grants the subblocks to its customers.

Example 18.6

Figure 18.24 shows how four small blocks of addresses are assigned to four organizations by an ISP. The ISP combines these four blocks into one single block and advertises the larger block to the rest of the world. Any packet destined for this larger block should be sent to this ISP. It is the responsibility of the ISP to forward the packet to the appropriate organization. This is similar to

Figure 18.23 *Solution to Example 18.5*

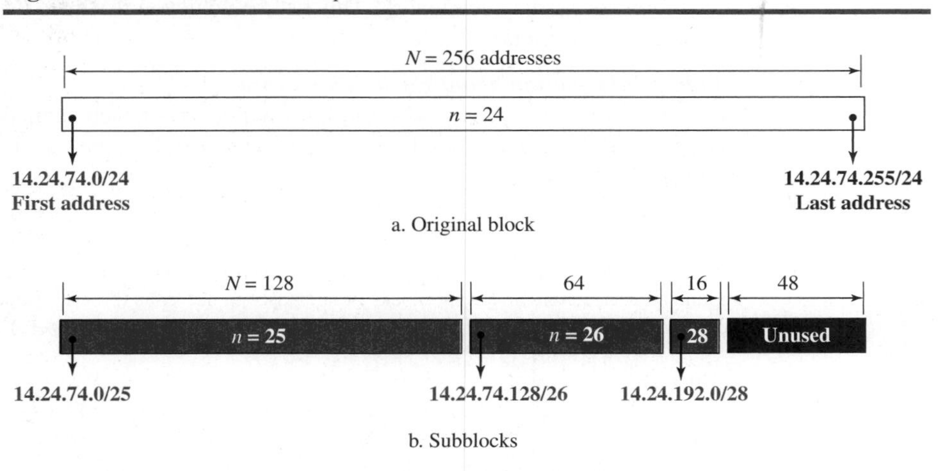

Figure 18.24 *Example of address aggregation*

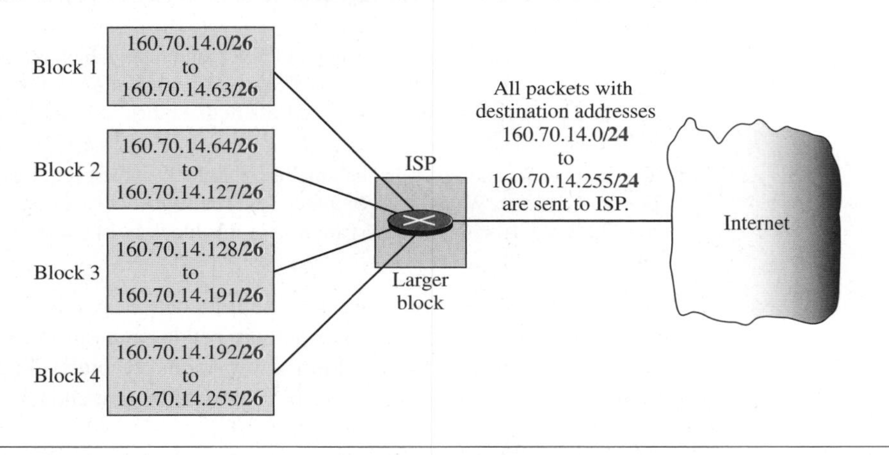

routing we can find in a postal network. All packages coming from outside a country are sent first to the capital and then distributed to the corresponding destination.

Special Addresses

Before finishing the topic of addresses in IPv4, we need to mention five special addresses that are used for special purposes: *this-host* address, *limited-broadcast* address, *loopback* address, *private* addresses, and *multicast* addresses.

This-host Address

The only address in the block **0.0.0.0/32** is called the *this-host* address. It is used whenever a host needs to send an IP datagram but it does not know its own address to use as the source address. We will see an example of this case in the next section.

Limited-broadcast Address

The only address in the block **255.255.255.255/32** is called the *limited-broadcast* address. It is used whenever a router or a host needs to send a datagram to all devices in a network. The routers in the network, however, block the packet having this address as the destination; the packet cannot travel outside the network.

Loopback Address

The block **127.0.0.0/8** is called the *loopback* address. A packet with one of the addresses in this block as the destination address never leaves the host; it will remain in the host. Any address in the block is used to test a piece of software in the machine. For example, we can write a client and a server program in which one of the addresses in the block is used as the server address. We can test the programs using the same host to see if they work before running them on different computers.

Private Addresses

Four blocks are assigned as private addresses: 10.0.0.0/**8**, 172.16.0.0/**12**, 192.168.0.0/**16**, and 169.254.0.0/**16**. We will see the applications of these addresses when we discuss NAT later in the chapter.

Multicast Addresses

The block 224.0.0.0/**4** is reserved for multicast addresses. We discuss these addresses later in the chapter.

18.4.4 Dynamic Host Configuration Protocol (DHCP)

We have seen that a large organization or an ISP can receive a block of addresses directly from ICANN and a small organization can receive a block of addresses from an ISP. After a block of addresses are assigned to an organization, the network administration can manually assign addresses to the individual hosts or routers. However, address assignment in an organization can be done automatically using the **Dynamic Host Configuration Protocol (DHCP).** DHCP is an application-layer program, using the client-server paradigm, that actually helps TCP/IP at the network layer.

DHCP has found such widespread use in the Internet that it is often called a *plug-and-play protocol*. In can be used in many situations. A network manager can configure DHCP to assign permanent IP addresses to the host and routers. DHCP can also be configured to provide temporary, on demand, IP addresses to hosts. The second capability can provide a temporary IP address to a traveller to connect her laptop to the Internet while she is staying in the hotel. It also allows an ISP with 1000 granted addresses to provide services to 4000 households, assuming not more than one-forth of customers use the Internet at the same time.

In addition to its IP address, a computer also needs to know the network prefix (or address mask). Most computers also need two other pieces of information, such as the address of a default router to be able to communicate with other networks and the address of a name server to be able to use names instead of addresses, as we will see in Chapter 26. In other words, four pieces of information are normally needed: the computer address, the prefix, the address of a router, and the IP address of a name server. DHCP can be used to provide these pieces of information to the host.

DHCP Message Format

DHCP is a client-server protocol in which the client sends a request message and the server returns a response message. Before we discuss the operation of DHCP, let us show the general format of the DHCP message in Figure 18.25. Most of the fields are explained in the figure, but we need to discuss the option field, which plays a very important role in DHCP.

Figure 18.25 *DHCP message format*

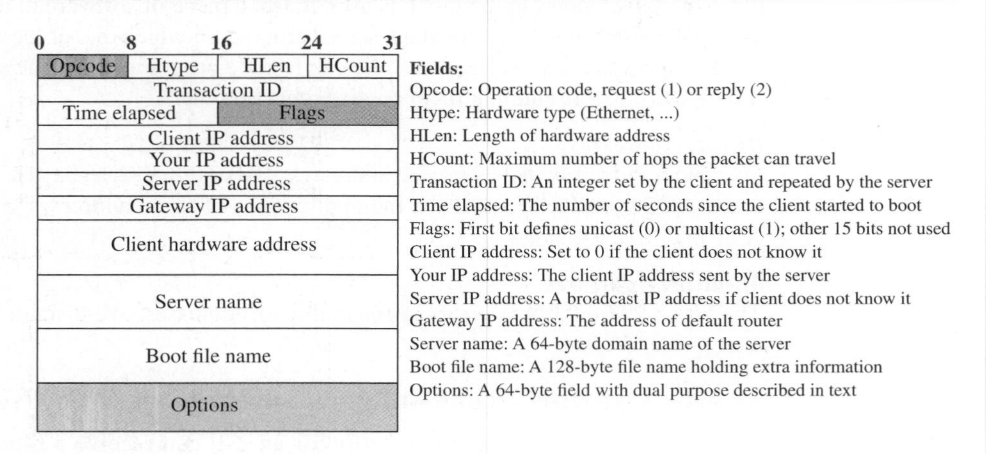

The 64-byte option field has a dual purpose. It can carry either additional information or some specific vendor information. The server uses a number, called a **magic cookie,** in the format of an IP address with the value of 99.130.83.99. When the client finishes reading the message, it looks for this magic cookie. If present, the next 60 bytes are options. An option is composed of three fields: a 1-byte tag field, a 1-byte length field, and a variable-length value field. There are several tag fields that are mostly used by vendors. If the tag field is 53, the value field defines one of the 8 message types shown in Figure 18.26. We show how these message types are used by DHCP.

Figure 18.26 *Option format*

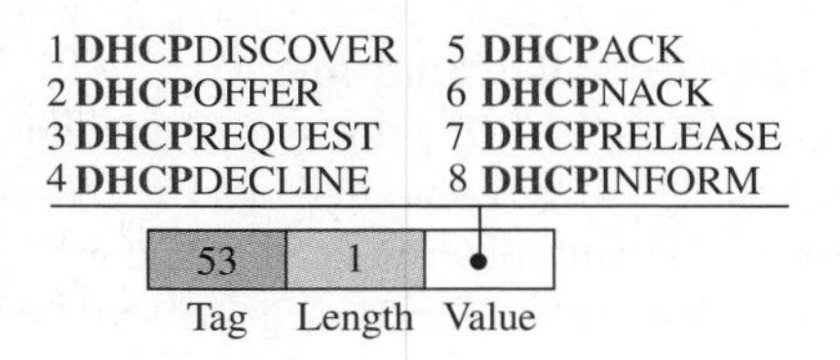

DHCP Operation

Figure 18.27 shows a simple scenario.

Figure 18.27 *Operation of DHCP*

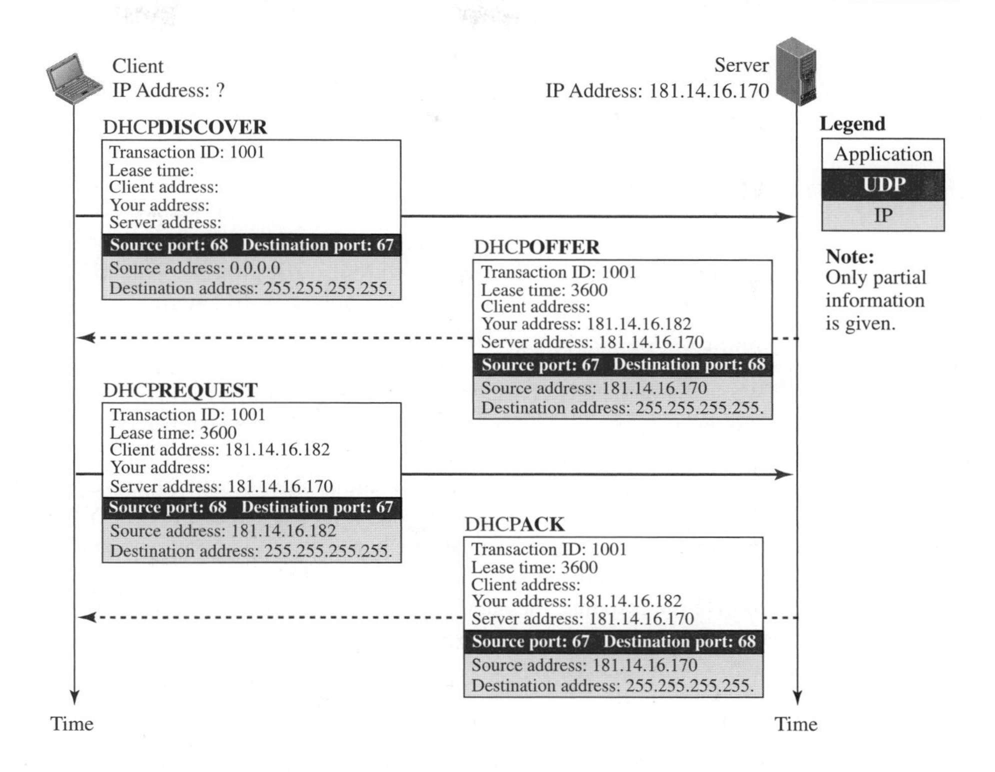

1. The joining host creates a **DHCP**DISCOVER message in which only the transaction-ID field is set to a random number. No other field can be set because the host has no knowledge with which to do so. This message is encapsulated in a UDP user datagram with the source port set to 68 and the destination port set to 67. We will discuss the reason for using two well-known port numbers later. The user datagram is encapsulated in an IP datagram with the source address set to **0.0.0.0** ("this host") and the destination address set to **255.255.255.255** (broadcast address). The reason is that the joining host knows neither its own address nor the server address.
2. The DHCP server or servers (if more than one) responds with a **DHCP**OFFER message in which the your address field defines the offered IP address for the joining host and the server address field includes the IP address of the server. The message also includes the lease time for which the host can keep the IP address. This message is encapsulated in a user datagram with the same port numbers, but in the reverse order. The user datagram in turn is encapsulated in a datagram with the server address as the source IP address, but the destination address is a broadcast address, in which the server allows other DHCP servers to receive the offer and give a better offer if they can.

3. The joining host receives one or more offers and selects the best of them. The joining host then sends a **DHCP**REQUEST message to the server that has given the best offer. The fields with known value are set. The message is encapsulated in a user datagram with port numbers as the first message. The user datagram is encapsulated in an IP datagram with the source address set to the new client address, but the destination address still is set to the broadcast address to let the other servers know that their offer was not accepted.
4. Finally, the selected server responds with a **DHCP**ACK message to the client if the offered IP address is valid. If the server cannot keep its offer (for example, if the address is offered to another host in between), the server sends a **DHCP**NACK message and the client needs to repeat the process. This message is also broadcast to let other servers know that the request is accepted or rejected.

Two Well-Known Ports

We said that the DHCP uses two well-known ports (68 and 67) instead of one well-known and one ephemeral. The reason for choosing the well-known port 68 instead of an ephemeral port for the client is that the response from the server to the client is broadcast. Remember that an IP datagram with the limited broadcast message is delivered to every host on the network. Now assume that a DHCP client and a DAYTIME client, for example, are both waiting to receive a response from their corresponding server and both have accidentally used the same temporary port number (56017, for example). Both hosts receive the response message from the DHCP server and deliver the message to their clients. The DHCP client processes the message; the DAYTIME client is totally confused with a strange message received. Using a well-known port number prevents this problem from happening. The response message from the DHCP server is not delivered to the DAYTIME client, which is running on the port number 56017, not 68. The temporary port numbers are selected from a different range than the well-known port numbers.

The curious reader may ask what happens if two DHCP clients are running at the same time. This can happen after a power failure and power restoration. In this case the messages can be distinguished by the value of the transaction ID, which separates each response from the other.

Using FTP

The server does not send all of the information that a client may need for joining the network. In the **DHCP**ACK message, the server defines the pathname of a file in which the client can find complete information such as the address of the DNS server. The client can then use a file transfer protocol to obtain the rest of the needed information.

Error Control

DHCP uses the service of UDP, which is not reliable. To provide error control, DHCP uses two strategies. First, DHCP requires that UDP use the checksum. As we will see in Chapter 24, the use of the checksum in UDP is optional. Second, the DHCP client uses timers and a retransmission policy if it does not receive the DHCP reply to a request. However, to prevent a traffic jam when several hosts need to retransmit a request (for example, after a power failure), DHCP forces the client to use a random number to set its timers.

Transition States

The previous scenarios we discussed for the operation of the DHCP were very simple. To provide dynamic address allocation, the DHCP client acts as a state machine that performs transitions from one state to another depending on the messages it receives or sends. Figure 18.28 shows the transition diagram with the main states.

Figure 18.28 *FSM for the DHCP client*

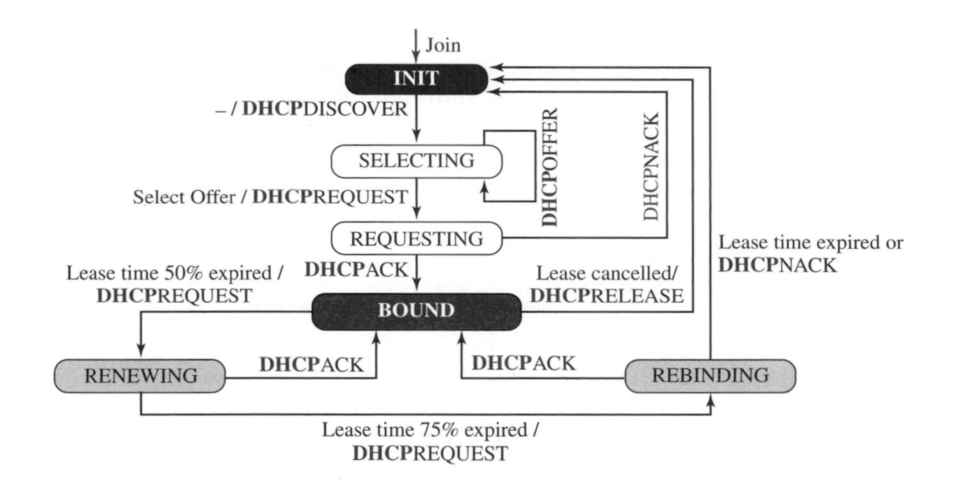

When the DHCP client first starts, it is in the INIT state (initializing state). The client broadcasts a discover message. When it receives an offer, the client goes to the SELECTING state. While it is there, it may receive more offers. After it selects an offer, it sends a request message and goes to the REQUESTING state. If an ACK arrives while the client is in this state, it goes to the BOUND state and uses the IP address. When the lease is 50 percent expired, the client tries to renew it by moving to the RENEWING state. If the server renews the lease, the client moves to the BOUND state again. If the lease is not renewed and the lease time is 75 percent expired, the client moves to the REBINDING state. If the server agrees with the lease (ACK message arrives), the client moves to the BOUND state and continues using the IP address; otherwise, the client moves to the INIT state and requests another IP address. Note that the client can use the IP address only when it is in the BOUND, RENEWING, or REBINDING state. The above procedure requires that the client uses three timers: *renewal timer* (set to 50 percent of the lease time), *rebinding timer* (set to 75 percent of the lease time), and *expiration timer* (set to the lease time).

18.4.5 Network Address Resolution (NAT)

The distribution of addresses through ISPs has created a new problem. Assume that an ISP has granted a small range of addresses to a small business or a household. If the business grows or the household needs a larger range, the ISP may not be able to grant the demand because the addresses before and after the range may have already been allocated to other networks. In most situations, however, only a portion of computers in a small network need access to the Internet simultaneously. This means that the number of allocated addresses does not have to match the number of computers in the network. For example, assume that in a small business with 20 computers the maximum number of computers that access the Internet simultaneously is only 4. Most of the computers are either doing some task that does not need Internet access or communicating with each other. This small business can use the TCP/IP protocol for both internal and universal communication. The business can use 20 (or 25) addresses from the private block addresses (discussed before) for internal communication; five addresses for universal communication can be assigned by the ISP.

A technology that can provide the mapping between the private and universal addresses, and at the same time support virtual private networks, which we discuss in Chapter 32, is **Network Address Translation (NAT).** The technology allows a site to use a set of private addresses for internal communication and a set of global Internet addresses (at least one) for communication with the rest of the world. The site must have only one connection to the global Internet through a NAT-capable router that runs NAT software. Figure 18.29 shows a simple implementation of NAT.

Figure 18.29 *NAT*

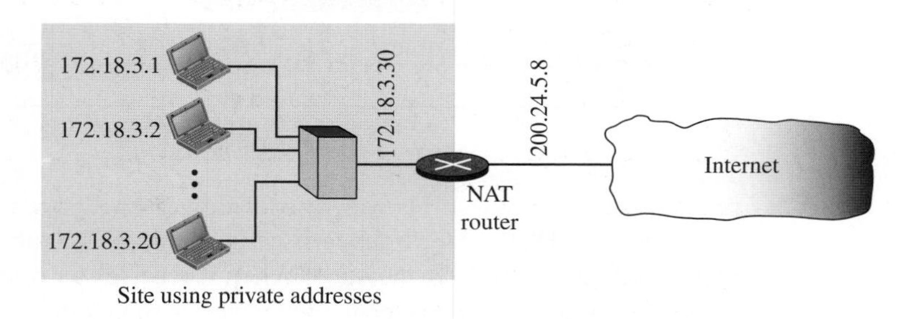

As the figure shows, the private network uses private addresses. The router that connects the network to the global address uses one private address and one global address. The private network is invisible to the rest of the Internet; the rest of the Internet sees only the NAT router with the address 200.24.5.8.

Address Translation

All of the outgoing packets go through the NAT router, which replaces the source address in the packet with the global NAT address. All incoming packets also pass through the NAT router, which replaces the destination address in the packet (the NAT router global address) with the appropriate private address. Figure 18.30 shows an example of address translation.

Translation Table

The reader may have noticed that translating the source addresses for an outgoing packet is straightforward. But how does the NAT router know the destination address for a packet coming from the Internet? There may be tens or hundreds of private IP

Figure 18.30 *Address translation*

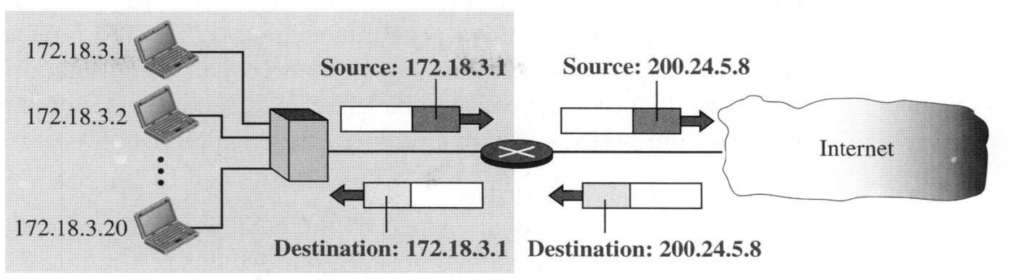

addresses, each belonging to one specific host. The problem is solved if the NAT router has a translation table.

Using One IP Address

In its simplest form, a translation table has only two columns: the private address and the external address (destination address of the packet). When the router translates the source address of the outgoing packet, it also makes note of the destination address—where the packet is going. When the response comes back from the destination, the router uses the source address of the packet (as the external address) to find the private address of the packet. Figure 18.31 shows the idea.

Figure 18.31 *Translation*

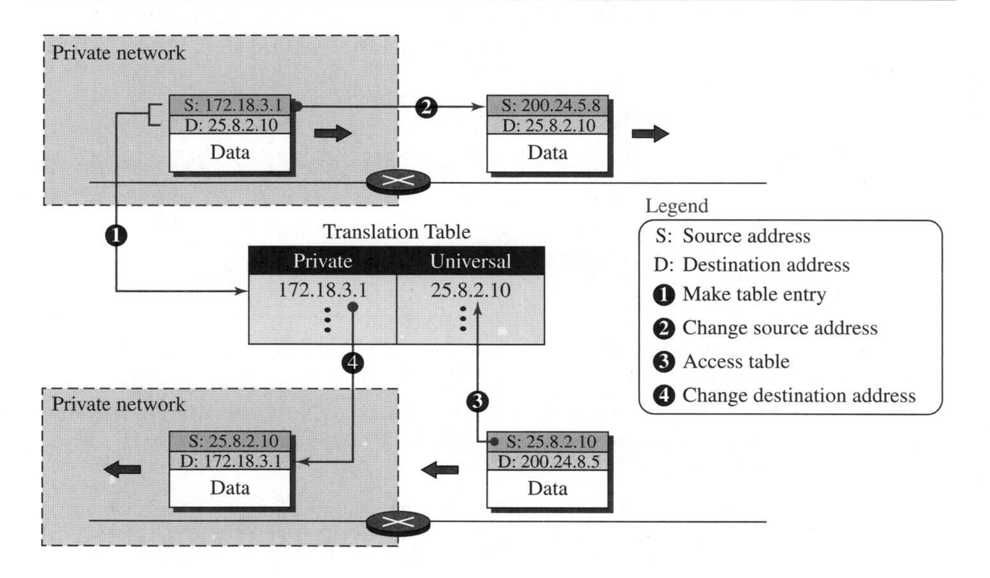

In this strategy, communication must always be initiated by the private network. The NAT mechanism described requires that the private network start the communication.

As we will see, NAT is used mostly by ISPs that assign a single address to a customer. The customer, however, may be a member of a private network that has many private addresses. In this case, communication with the Internet is always initiated from the customer site, using a client program such as HTTP, TELNET, or FTP to access the corresponding server program. For example, when e-mail that originates from outside the network site is received by the ISP e-mail server, it is stored in the mailbox of the customer until retrieved with a protocol such as POP.

Using a Pool of IP Addresses

The use of only one global address by the NAT router allows only one private-network host to access a given external host. To remove this restriction, the NAT router can use a pool of global addresses. For example, instead of using only one global address (200.24.5.8), the NAT router can use four addresses (200.24.5.8, 200.24.5.9, 200.24.5.10, and 200.24.5.11). In this case, four private-network hosts can communicate with the same external host at the same time because each pair of addresses defines a separate connection. However, there are still some drawbacks. No more than four connections can be made to the same destination. No private-network host can access two external server programs (e.g., HTTP and TELNET) at the same time. And, likewise, two private-network hosts cannot access the same external server program (e.g., HTTP or TELNET) at the same time.

Using Both IP Addresses and Port Addresses

To allow a many-to-many relationship between private-network hosts and external server programs, we need more information in the translation table. For example, suppose two hosts inside a private network with addresses 172.18.3.1 and 172.18.3.2 need to access the HTTP server on external host 25.8.3.2. If the translation table has five columns, instead of two, that include the source and destination port addresses and the transport-layer protocol, the ambiguity is eliminated. Table 18.1 shows an example of such a table.

Table 18.1 *Five-column translation table*

Private address	*Private port*	*External address*	*External port*	*Transport protocol*
172.18.3.1	1400	25.8.3.2	80	TCP
172.18.3.2	1401	25.8.3.2	80	TCP
⋮	⋮	⋮	⋮	⋮

Note that when the response from HTTP comes back, the combination of source address (25.8.3.2) and destination port address (1401) defines the private network host to which the response should be directed. Note also that for this translation to work, the ephemeral port addresses (1400 and 1401) must be unique.

18.5 FORWARDING OF IP PACKETS

We discussed the concept of forwarding at the network layer earlier in this chapter. In this section, we extend the concept to include the role of IP addresses in forwarding. As we discussed before, forwarding means to place the packet in its route to its destination.

Since the Internet today is made of a combination of links (networks), forwarding means to deliver the packet to the next hop (which can be the final destination or the intermediate connecting device). Although the IP protocol was originally designed as a connectionless protocol, today the tendency is to change it to a connection-oriented protocol. We discuss both cases.

When IP is used as a connectionless protocol, forwarding is based on the destination address of the IP datagram; when the IP is used as a connection-oriented protocol, forwarding is based on the label attached to an IP datagram.

18.5.1 Forwarding Based on Destination Address

We first discuss forwarding based on the destination address. This is a traditional approach, which is prevalent today. In this case, forwarding requires a host or a router to have a forwarding table. When a host has a packet to send or when a router has received a packet to be forwarded, it looks at this table to find the next hop to deliver the packet to.

In classless addressing, the whole address space is one entity; there are no classes. This means that forwarding requires one row of information for each block involved. The table needs to be searched based on the network address (first address in the block). Unfortunately, the destination address in the packet gives no clue about the network address. To solve the problem, we need to include the mask (/*n*) in the table. In other words, a classless forwarding table needs to include four pieces of information: the mask, the network address, the interface number, and the IP address of the next router (needed to find the link-layer address of the next hop, as we discussed in Chapter 9). However, we often see in the literature that the first two pieces are combined. For example, if *n* is 26 and the network address is 180.70.65.192, then one can combine the two as one piece of information: 180.70.65.192/**26**. Figure 18.32 shows a simple forwarding module and forwarding table for a router with only three interfaces.

Figure 18.32 *Simplified forwarding module in classless address*

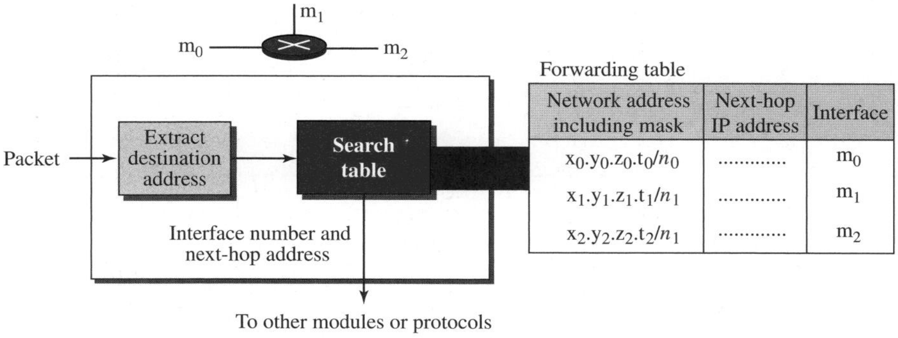

m_1
m_0
m_2
Packet
Extract destination address
Search table
Interface number and next-hop address
To other modules or protocols

Forwarding table

Network address including mask	Next-hop IP address	Interface
$x_0.y_0.z_0.t_0/n_0$	....	m_0
$x_1.y_1.z_1.t_1/n_1$	....	m_1
$x_2.y_2.z_2.t_2/n_1$	....	m_2

The job of the forwarding module is to search the table, row by row. In each row, the *n* leftmost bits of the destination address (prefix) are kept and the rest of the bits (suffix) are set to 0s. If the resulting address (which we call the *network address*), matches with the address in the first column, the information in the next two columns is extracted; otherwise the search continues. Normally, the last row has a default value in the first column (not shown in the figure), which indicates all destination addresses that did not match the previous rows.

Sometimes, the literature explicitly shows the value of the n leftmost bits that should be matched with the n leftmost bits of the destination address. The concept is the same, but the presentation is different. For example, instead of giving the address-mask combination of 180.70.65.192/**26**, we can give the value of the 26 leftmost bits as shown below.

```
10110100  01000110  01000001  11
```

Note that we still need to use an algorithm to find the prefix and compare it with the bit pattern. In other words, the algorithm is still needed, but the presentation is different. We use this format in our forwarding tables in the exercises when we use smaller address spaces just for practice.

Example 18.7

Make a forwarding table for router R1 using the configuration in Figure 18.33.

Figure 18.33 *Configuration for Example 18.7*

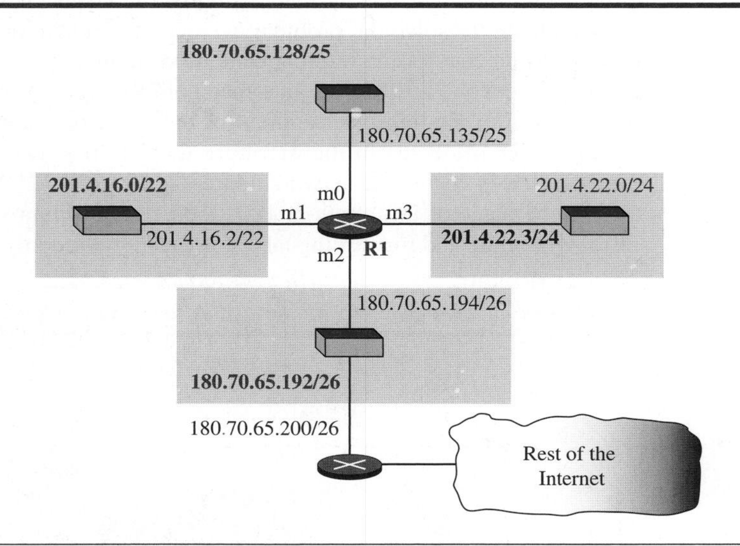

Solution

Table 18.2 shows the corresponding table.

Table 18.2 *Forwarding table for router R1 in Figure 18.33*

Network address/mask	*Next hop*	*Interface*
180.70.65.192/**26**	—	m2
180.70.65.128/**25**	—	m0
201.4.22.0/**24**	—	m3
201.4.16.0/**22**	—	m1
Default	180.70.65.200	m2

Example 18.8

Instead of Table 18.2, we can use Table 18.3, in which the network address/mask is given in bits.

Table 18.3 *Forwarding table for router R1 in Figure 18.33 using prefix bits*

Leftmost bits in the destination address	*Next hop*	*Interface*
10110100 01000110 01000001 11	—	m2
10110100 01000110 01000001 1	—	m0
11001001 00000100 00011100	—	m3
11001001 00000100 000100	—	m1
Default	180.70.65.200	m2

When a packet arrives whose leftmost 26 bits in the destination address match the bits in the first row, the packet is sent out from interface m2. When a packet arrives whose leftmost 25 bits in the address match the bits in the second row, the packet is sent out from interface m0, and so on. The table clearly shows that the first row has the longest prefix and the fourth row has the shortest prefix. The longer prefix means a smaller range of addresses; the shorter prefix means a larger range of addresses.

Example 18.9

Show the forwarding process if a packet arrives at R1 in Figure 18.33 with the destination address 180.70.65.140.

Solution

The router performs the following steps:

1. The first mask (/26) is applied to the destination address. The result is 180.70.65.128, which does not match the corresponding network address.
2. The second mask (/25) is applied to the destination address. The result is 180.70.65.128, which matches the corresponding network address. The next-hop address and the interface number m0 are extracted for forwarding the packet.

Address Aggregation

When we use classful addressing, there is only one entry in the forwarding table for each site outside the organization. The entry defines the site even if that site is subnetted. When a packet arrives at the router, the router checks the corresponding entry and forwards the packet accordingly. When we use classless addressing, it is likely that the number of forwarding table entries will increase. This is because the intent of classless addressing is to divide up the whole address space into manageable blocks. The increased size of the table results in an increase in the amount of time needed to search the table. To alleviate the problem, the idea of address aggregation was designed. In Figure 18.34 we have two routers.

R1 is connected to networks of four organizations that each use 64 addresses. R2 is somewhere far from R1. R1 has a longer forwarding table because each packet must be correctly routed to the appropriate organization. R2, on the other hand, can have a very small forwarding table. For R2, any packet with destination 140.24.7.0 to 140.24.7.255

Figure 18.34 *Address aggregation*

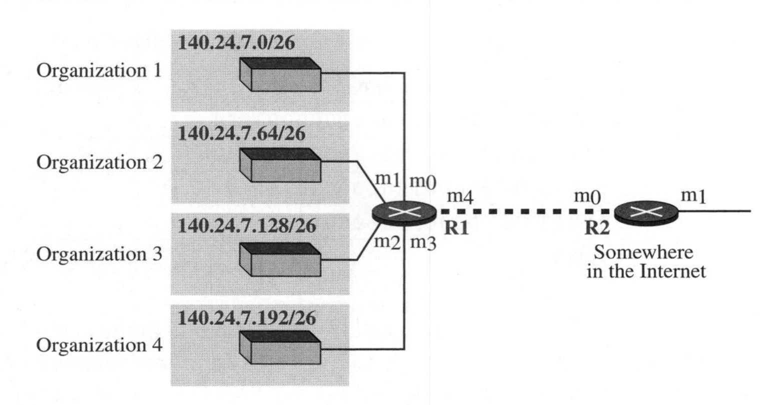

Forwarding table for R1

Network address/mask	Next-hop address	Interface
140.24.7.0/26	----------	m0
140.24.7.64/26	----------	m1
140.24.7.128/26	----------	m2
140.24.7.192/26	----------	m3
0.0.0.0/0	address of R2	m4

Forwarding table for R2

Network address/mask	Next-hop address	Interface
140.24.7.0/24	----------	m0
0.0.0.0/0	default router	m1

is sent out from interface m0 regardless of the organization number. This is called address aggregation because the blocks of addresses for four organizations are aggregated into one larger block. R2 would have a longer forwarding table if each organization had addresses that could not be aggregated into one block.

Longest Mask Matching

What happens if one of the organizations in the previous figure is not geographically close to the other three? For example, if organization 4 cannot be connected to router R1 for some reason, can we still use the idea of address aggregation and still assign block 140.24.7.192/26 to organization 4? The answer is yes, because routing in classless addressing uses another principle, **longest mask matching.** This principle states that the forwarding table is sorted from the longest mask to the shortest mask. In other words, if there are three masks, /27, /26, and /24, the mask /27 must be the first entry and /24 must be the last. Let us see if this principle solves the situation in which organization 4 is separated from the other three organizations. Figure 18.35 shows the situation.

Suppose a packet arrives at router R2 for organization 4 with destination address 140.24.7.200. The first mask at router R2 is applied, which gives the network address 140.24.7.192. The packet is routed correctly from interface m1 and reaches organization 4. If, however, the forwarding table was not stored with the longest prefix first, applying the /24 mask would result in the incorrect routing of the packet to router R1.

Figure 18.35 *Longest mask matching*

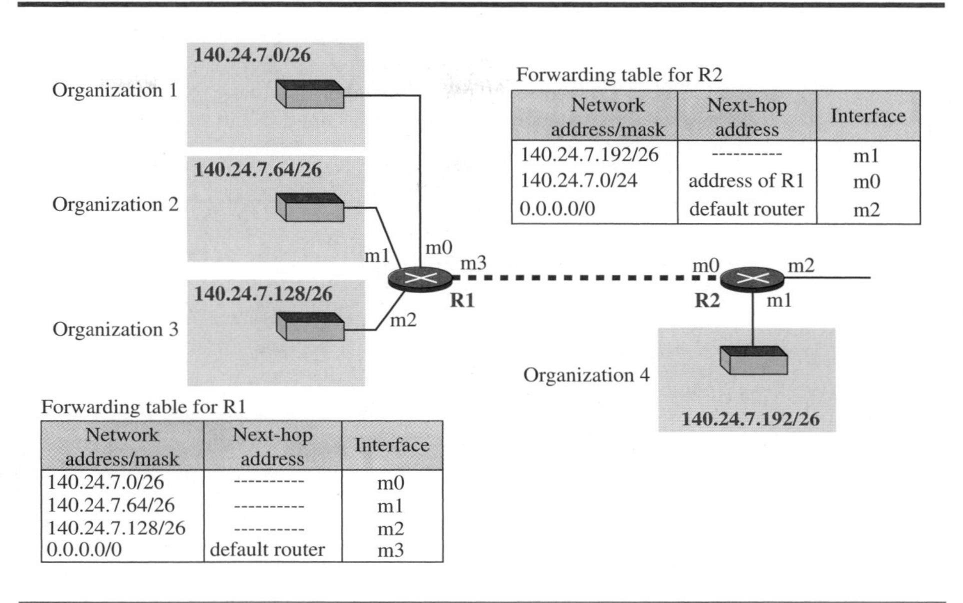

Forwarding table for R2

Network address/mask	Next-hop address	Interface
140.24.7.192/26	----------	m1
140.24.7.0/24	address of R1	m0
0.0.0.0/0	default router	m2

Forwarding table for R1

Network address/mask	Next-hop address	Interface
140.24.7.0/26	----------	m0
140.24.7.64/26	----------	m1
140.24.7.128/26	----------	m2
0.0.0.0/0	default router	m3

Hierarchical Routing

To solve the problem of gigantic forwarding tables, we can create a sense of hierarchy in the forwarding tables. In Chapter 2, we mentioned that the Internet today has a sense of hierarchy. We said that the Internet is divided into backbone and national ISPs. National ISPs are divided into regional ISPs, and regional ISPs are divided into local ISPs. If the forwarding table has a sense of hierarchy like the Internet architecture, the forwarding table can decrease in size.

Let us take the case of a local ISP. A local ISP can be assigned a single, but large, block of addresses with a certain prefix length. The local ISP can divide this block into smaller blocks of different sizes, and assign these to individual users and organizations, both large and small. If the block assigned to the local ISP starts with a.b.c.d/n, the ISP can create blocks starting with e.f.g.h/m, where m may vary for each customer and is greater than n.

How does this reduce the size of the forwarding table? The rest of the Internet does not have to be aware of this division. All customers of the local ISP are defined as a.b.c.d/n to the rest of the Internet. Every packet destined for one of the addresses in this large block is routed to the local ISP. There is only one entry in every router in the world for all of these customers. They all belong to the same group. Of course, inside the local ISP, the router must recognize the subblocks and route the packet to the destined customer. If one of the customers is a large organization, it also can create another level of hierarchy by subnetting and dividing its subblock into smaller subblocks (or sub-subblocks). In classless routing, the levels of hierarchy are unlimited as long as we follow the rules of classless addressing.

Example 18.10

As an example of hierarchical routing, let us consider Figure 18.36. A regional ISP is granted 16,384 addresses starting from 120.14.64.0. The regional ISP has decided to divide this block into four subblocks, each with 4096 addresses. Three of these subblocks are assigned to three local ISPs; the second subblock is reserved for future use. Note that the mask for each block is /20 because the original block with mask /18 is divided into 4 blocks.

Figure 18.36 *Hierarchical routing with ISPs*

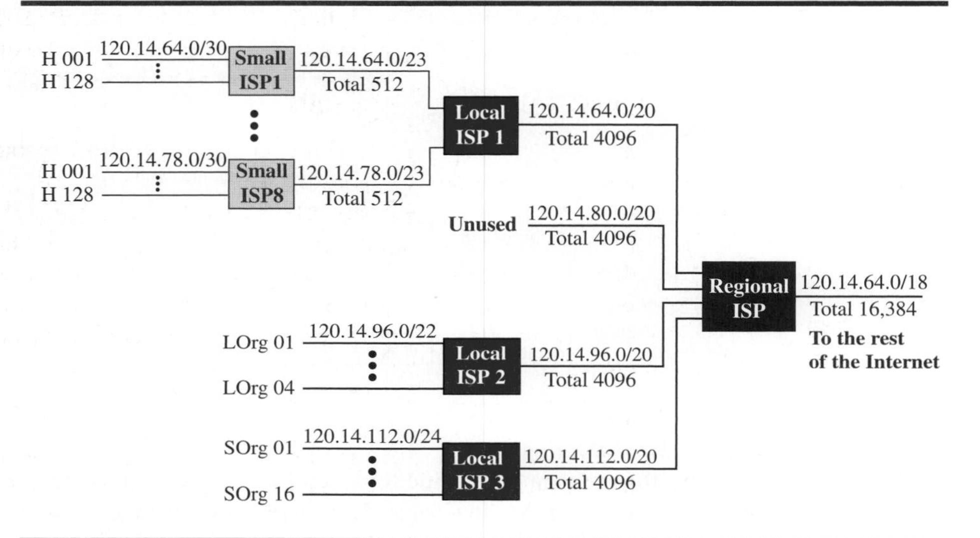

The first local ISP has divided its assigned subblock into 8 smaller blocks and assigned each to a small ISP. Each small ISP provides services to 128 households (H001 to H128), each using four addresses. Note that the mask for each small ISP is now /23 because the block is further divided into 8 blocks. Each household has a mask of /30, because a household has only 4 addresses ($2^{32-30} = 4$). The second local ISP has divided its block into 4 blocks and has assigned the addresses to 4 large organizations (LOrg01 to LOrg04). Note that each large organization has 1024 addresses and the mask is /22.

The third local ISP has divided its block into 16 blocks and assigned each block to a small organization (SOrg01 to SOrg16). Each small organization has 256 addresses and the mask is /24. There is a sense of hierarchy in this configuration. All routers in the Internet send a packet with destination address 120.14.64.0 to 120.14.127.255 to the regional ISP. The regional ISP sends every packet with destination address 120.14.64.0 to 120.14.79.255 to Local ISP1. Local ISP1 sends every packet with destination address 120.14.64.0 to 120.14.64.3 to H001.

Geographical Routing

To decrease the size of the forwarding table even further, we need to extend hierarchical routing to include geographical routing. We must divide the entire address space into a few large blocks. We assign a block to America, a block to Europe, a block to Asia, a block to Africa, and so on. The routers of ISPs outside of Europe will have only one entry for packets to Europe in their forwarding tables. The routers of ISPs outside of America will have only one entry for packets to America in their forwarding tables, and so on.

Forwarding Table Search Algorithms

In classless addressing, there is no network information in the destination address. The simplest, but not the most efficient, search method is called the longest prefix match (as we discussed before). The forwarding table can be divided into buckets, one for each prefix. The router first tries the longest prefix. If the destination address is found in this bucket, the search is complete. If the address is not found, the next prefix is searched, and so on. It is obvious that this type of search takes a long time.

One solution is to change the data structure used for searching. Instead of a list, other data structures (such as a tree or a binary tree) can be used. One candidate is a trie (a special kind of tree). However, this discussion is beyond the scope of this book.

18.5.2 Forwarding Based on Label

In the 1980s, an effort started to somehow change IP to behave like a connection-oriented protocol in which the routing is replaced by switching. As we discussed earlier in the chapter, in a connectionless network (datagram approach), a router forwards a packet based on the destination address in the header of the packet. On the other hand, in a connection-oriented network (virtual-circuit approach), a switch forwards a packet based on the label attached to the packet. Routing is normally based on searching the contents of a table; switching can be done by accessing a table using an index. In other words, routing involves searching; switching involves accessing.

Example 18.11

Figure 18.37 shows a simple example of searching in a forwarding table using the longest mask algorithm. Although there are some more efficient algorithms today, the principle is the same.

Figure 18.37 *Example 18.11: Forwarding based on destination address*

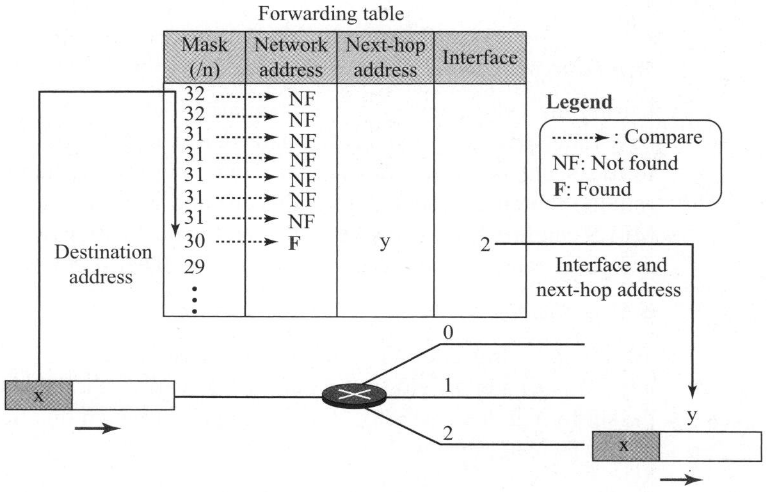

When the forwarding algorithm gets the destination address of the packet, it needs to delve into the mask column. For each entry, it needs to apply the mask to find the destination network address. It then needs to check the network addresses in the table until it finds the match. The router then extracts the next-hop address and the interface number to be delivered to the data-link layer.

Example 18.12

Figure 18.38 shows a simple example of using a label to access a switching table. Since the labels are used as the index to the table, finding the information in the table is immediate.

Figure 18.38 *Example 18.12: Forwarding based on label*

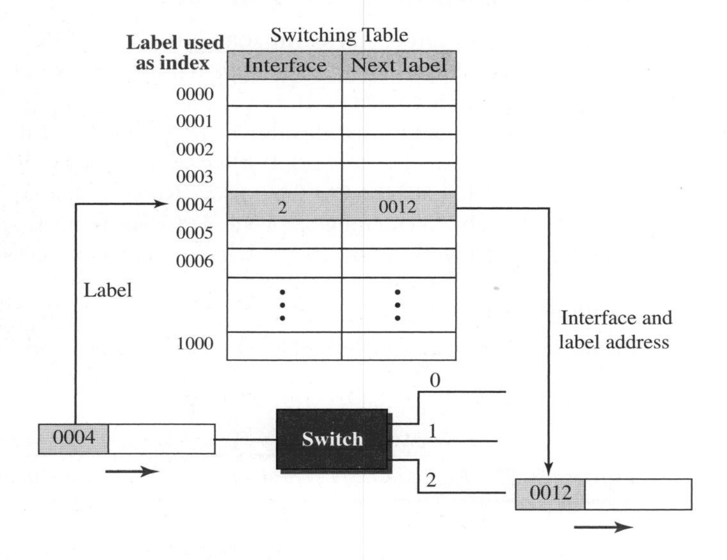

Multi-Protocol Label Switching (MPLS)

During the 1980s, several vendors created routers that implement switching technology. Later IETF approved a standard that is called Multi-Protocol Label Switching. In this standard, some conventional routers in the Internet can be replaced by MPLS routers, which can behave like a router and a switch. When behaving like a router, MPLS can forward the packet based on the destination address; when behaving like a switch, it can forward a packet based on the label.

A New Header

To simulate connection-oriented switching using a protocol like IP, the first thing that is needed is to add a field to the packet that carries the label discussed later. The IPv4 packet format does not allow this extension (although this field is provided in the IPv6 packet format, as we will see later). The solution is to encapsulate the IPv4 packet in an MPLS packet (as though MPLS were a layer between the data-link layer and the network layer). The whole IP packet is encapsulated as the payload in an MPLS packet and an MPLS header is added. Figure 18.39 shows the encapsulation.

Figure 18.39 *MPLS header added to an IP packet*

The MPLS header is actually a stack of subheaders that is used for multilevel hierarchical switching, as we will discuss shortly. Figure 18.40 shows the format of an MPLS header in which each subheader is 32 bits (4 bytes) long.

Figure 18.40 *MPLS header made of a stack of labels*

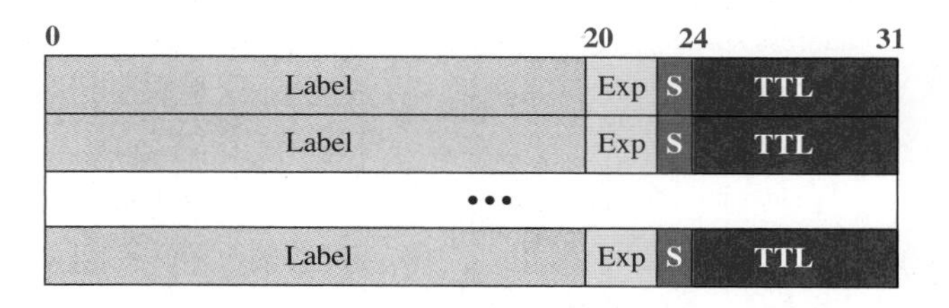

The following is a brief description of each field:

- ***Label.*** This 20-bit field defines the label that is used to index the forwarding table in the router.
- ***Exp.*** This 3-bit field is reserved for experimental purposes.
- ***S.*** The one-bit stack field defines the situation of the subheader in the stack. When the bit is 1, it means that the header is the last one in the stack.
- ***TTL.*** This 8-bit field is similar to the TTL field in the IP datagram. Each visited router decrements the value of this field. When it reaches zero, the packet is discarded to prevent looping.

Hierarchical Switching

A stack of labels in MPLS allows hierarchical switching. This is similar to conventional hierarchical routing. For example, a packet with two labels can use the top label to forward the packet through switches outside an organization; the bottom label can be used to route the packet inside the organization to reach the destination subnet.

18.5.3 Routers as Packet Switches

As we may have guessed by now, the packet switches that are used in the network layer are called routers. Routers can be configured to act as either a datagram switch or a virtual-circuit switch. We have discussed the structure of a packet-switch in Chapter 8. The discussion in that chapter can be applied to any router used in the Internet.

18.6 END-CHAPTER MATERIALS

18.6.1 Recommended Reading

Books

Several books give thorough coverage of materials discussed in this chapter. We recommend [Com 06], [Tan 03], [Koz 05], [Ste 95], [GW 04], [Per 00], [Kes 02], [Moy 98], [WZ 01], and [Los 04].

RFCs

IPv4 addressing is discussed in RFCs 917, 927, 930, 932, 940, 950, 1122, and 1519. Forwarding is discussed in RFCs 1812, 1971, and 1980. MPLS is discussed in RFCs 3031, 3032, 3036, and 3212.

18.6.2 Key Terms

address aggregation
address space
choke packet
classful addressing
classless addressing
classless interdomain routing (CIDR)
closed-loop congestion control
Dynamic Host Configuration Protocol (DHCP)
longest mask matching
magic cookie
network address
Network Address Translation (NAT)
open-loop congestion control
packetizing

18.6.3 Summary

The network layer in the Internet provides services to the transport layer and receives services from the network layer. The main services provided by the network layer are packetizing and routing the packet from the source to the destination. The network layer in the Internet does not seriously address other services such as flow, error, or congestion control.

One of the main duties of the network layer is to provide packet switching. There are two approaches to packet switching: datagram approach and virtual-circuit approach. The first is used in a connectionless network; the second, in a connection-oriented network. Currently, the network layer is using the first approach, but the tendency is to move to the second.

Performance of the network layer is measured in terms of delay, throughput, and packet loss. Congestion control is a mechanism that can be used to improve the performance. Although congestion control is not directly implemented at the network layer, the discussion can help us to understand its indirect implementation and also to understand the congestion control implemented at the transport layer.

One of the main issues at the network layer is addressing. In this chapter, we discussed addressing in IPv4 (the current version). We explained the address space of the IPv4 and two address distribution mechanisms: classful and classless addressing. Although the first is deprecated, it helps us to understand the second. In classful addressing the whole address space is divided into five fixed-size classes. In classless addressing, the address space is divided into variable-size blocks based on the demand.

Some problems of address shortage in the current version can be temporarily alleviated using DHCP and NAT protocols.

The section on forwarding helps to understand how routers forward packets. Two approaches are used for this purpose. The first approach, which is used in a connectionless network such as the current Internet, is based on the destination address of the packet. The second approach, which can be used if the Internet is changed to a connection-oriented network, uses the labels in the packets.

18.7 PRACTICE SET

18.7.1 Quizzes

A set of interactive quizzes for this chapter can be found on the book website. It is strongly recommended that the student take the quizzes to check his/her understanding of the materials before continuing with the practice set.

18.7.2 Questions

Q18-1. Distinguish between the process of routing a packet from the source to the destination and the process of forwarding a packet at each router.

Q18-2. In classless addressing, can two different blocks have the same prefix length? Explain.

Q18-3. Why does the network-layer protocol need to provide packetizing service to the transport layer? Why can't the transport layer send out the segments without encapsulating them in datagrams?

Q18-4. Why is routing the responsibility of the network layer? In other words, why can't the routing be done at the transport layer or the data-link layer?

Q18-5. In Figure 18.10, assume that the link between R1 and R2 is upgraded to 170 kbps and the link between the source host and R1 is now downgraded to 140 kbps. What is the throughput between the source and destination after these changes? Which link is the bottleneck now?

Q18-6. In classless addressing, we know the first and the last address in the block. Can we find the prefix length? If the answer is yes, show the process.

Q18-7. Do we have any of the following services at the network layer of TCP/IP? If not, why?

a. flow control **b.** error control **c.** congestion control

Q18-8. List four types of delays in a packet-switched network.

Q18-9. If a label in a connection-oriented service is 8 bits, how many virtual circuits can be established at the same time?

Q18-10. List the three phases in the virtual-circuit approach to switching.

Q18-11. In classless addressing, we know the first address and the number of addresses in the block. Can we find the prefix length? If the answer is yes, show the process.

Q18-12. What is the piece of information in a packet upon which the forwarding decision is made in each of the following approaches to switching?

a. datagram approach **b.** virtual-circuit approach

18.7.3 Problems

P18-1. An ISP is granted the block 16.12.64.0/**20**. The ISP needs to allocate addresses for 8 organizations, each with 256 addresses.

a. Find the number and range of addresses in the ISP block.

b. Find the range of addresses for each organization and the range of unallocated addresses.

c. Show the outline of the address distribution and the forwarding table.

P18-2. Assume we have an internet with an 8-bit address space. The addresses are equally divided between four networks (N_0 to N_3). The internetwork communication is done through a router with four interfaces (m_0 to m_3). Show the internet outline and the forwarding table (with two columns: prefix in binary and the interface number) for the only router that connects the networks. Assign a network address to each network.

P18-3. Change each of the following prefix lengths to a mask in dotted-decimal notation:

a. $n = 0$ **b.** $n = 14$ **c.** $n = 30$

P18-4. Explain how DHCP can be used when the size of the block assigned to an organization is less than the number of hosts in the organization.

P18-5. In classless addressing, what is the size of the block (N) if the value of the prefix length (n) is one of the following?

a. $n = 0$ **b.** $n = 14$ **c.** $n = 32$

P18-6. In classless addressing, what is the value of the prefix length (n) if the size of the block (N) is one of the following?

a. $N = 1$ **b.** $N = 1024$ **c.** $N = 2^{32}$

P18-7. Compare NAT and DHCP. Both can solve the problem of a shortage of addresses in an organization, but by using different strategies.

P18-8. Find the class of the following classful IP addresses:

a. 130.35.54.12 **b.** 200.36.2.3 **c.** 245.24.2.8

P18-9. An organization is granted the block 130.56.0.0/**16**. The administrator wants to create 1024 subnets.

a. Find the number of addresses in each subnet.

b. Find the subnet prefix.

c. Find the first and the last address in the first subnet.

d. Find the first and the last address in the last subnet.

P18-10. In classless addressing, show the whole address space as a single block using the CIDR notation.

P18-11. Find the class of the following classful IP addresses:

a. `01110111 11110011 10000000 11011101`

b. `11101111 11000000 11110010 00011101`

c. `11011111 10110000 00011111 01011110`

P18-12. Can router R1 in Figure 18.35 receive a packet with destination address 140.24.7.194? What will happen to the packet if this occurs?

P18-13. Which of the following cannot be a mask in CIDR?

a. 255.225.0.0 **b.** 255.192.0.0 **c.** 255.255.255.6

P18-14. An ISP is granted the block 80.70.56.0/**21**. The ISP needs to allocate addresses for two organizations each with 500 addresses, two organizations each with 250 addresses, and three organizations each with 50 addresses.

a. Find the number and range of addresses in the ISP block.

b. Find the range of addresses for each organization and the range of unallocated addresses.

c. Show the outline of the address distribution and the forwarding table.

P18-15. What is the size of the address space in each of the following systems?

a. A system in which each address is only 16 bits.

b. A system in which each address is made of six hexadecimal digits.

c. A system in which each address is made of four octal digits.

P18-16. Assume we have an internet with a 9-bit address space. The addresses are divided between three networks (N_0 to N_2), with 64, 192, and 256 addresses respectively. The internetwork communication is done through a router with three interfaces (m_0 to m_2). Show the internet outline and the forwarding table (with two columns: prefix in binary and the interface number) for the only router that connects the networks. Assign a network address to each network.

P18-17. Show the *n* leftmost bits of the following network-addresses/masks that can be used in a forwarding table.

a. 170.40.11.0/24 **b.** 110.40.240.0/22 **c.** 70.14.0.0./18

P18-18. Change each of the following masks to a prefix length:

a. 255.224.0.0 **b.** 255.240.0.0 **c.** 255.255.255.128

P18-19. Rewrite the following IP addresses using dotted-decimal notation:

a. `01011110 10110000 01110101 00010101`

b. `10001001 10001110 11010000 00110001`

c. `01010111 10000100 00110111 00001111`

P18-20. Each of the following addresses belongs to a block. Find the first and the last address in each block.

a. 14.12.72.8/24 **b.** 200.107.16.17/18 **c.** 70.110.19.17/16

P18-21. Assume we have an internet with a 12-bit address space. The addresses are equally divided between eight networks (N_0 to N_7). The internetwork communication is done through a router with eight interfaces (m_0 to m_7). Show the internet outline and the forwarding table (with two columns: prefix in binary and the interface number) for the only router that connects the networks. Assign a network address to each network.

P18-22. Rewrite the following IP addresses using binary notation:

a. 110.11.5.88 **b.** 12.74.16.18 **c.** 201.24.44.32

P18-23. Combine the following three blocks of addresses into a single block:

a. 16.27.24.0/**26** **b.** 16.27.24.64/**26** **c.** 16.27.24.128/**25**

P18-24. A large organization with a large block address (12.44.184.0/**21**) is split into one medium-size company using the block address (12.44.184.0/**22**) and two small organizations. If the first small company uses the block (12.44.188.0/**23**), what is the remaining block that can be used by the second small company? Explain how the datagrams destined for the two small companies can be correctly routed to these companies if their address blocks still are part of the original company.

P18-25. Assume router R2 in Figure 18.35 receives a packet with destination address 140.24.7.42. How is the packet routed to its final destination?

18.8 SIMULATION EXPERIMENTS

18.8.1 Applets

We have created some Java applets to show some of the main concepts discussed in this chapter. It is strongly recommended that the students activate these applets on the book website and carefully examine the protocols in action.

18.9 PROGRAMMING ASSIGNMENT

Write the source code, compile, and test the following program in one of the programming languages of your choice:

Prg18-1. A program to change an address in any notation to two other notations.

Prg18-2. A program to find the class of a given address in dotted decimal notation.

Prg18-3. A program to find the first and the last address in a block given any addresses in the block and assuming classless addressing.

Prg18-4. A program to simulate the action of forwarding in a router given the routing table and the destination address of a packet. The program should find the outgoing user interface.

CHAPTER 19

Network-Layer Protocols

In the previous chapter we introduced the network layer and discussed the services provided by this layer. We also discussed the logical addresses used in this layer. In this chapter, we show how the network layer is implemented in the TCP/IP protocol suite. The protocols in the network layer have gone through a few versions; in this chapter, we concentrate on the current version (v4). The next generation, which is on the horizon, is discussed in Chapter 22.

- The first section discusses the IPv4 protocol. It first describes the IPv4 datagram format. It then explains the purpose of fragmentation in a datagram. The section then briefly discusses options fields and their purpose in a datagram. The section finally mentions some security issues in IPv4, which are addressed in Chapter 32.
- The second section discusses ICMPv4, one of the auxiliary protocols used in the network layer to help IPv4. First, it briefly discusses the purpose of each option. The section then shows how ICMP can be used as a debugging tool. The section finally shows how the checksum is calculated for an ICMPv4 message.
- The third section discusses the mobile IP, whose use is increasing every day when people temporarily move their computers from one place to another. The section first describes the issue of address change in this situation. It then shows the three phases involved in the process. The section finally explains the inefficiency involved in this process and some solutions.

19.1 INTERNET PROTOCOL (IP)

The network layer in version 4 can be thought of as one main protocol and three auxiliary ones. The main protocol, Internet Protocol version 4 (IPv4), is responsible for packetizing, forwarding, and delivery of a packet at the network layer. The Internet Control Message Protocol version 4 (ICMPv4) helps IPv4 to handle some errors that may occur in the network-layer delivery. The Internet Group Management Protocol (IGMP) is used to help IPv4 in multicasting. The Address Resolution Protocol (ARP) is used to glue the network and data-link layers in mapping network-layer addresses to link-layer addresses. Figure 19.1 shows the positions of these four protocols in the TCP/IP protocol suite.

Figure 19.1 *Position of IP and other network-layer protocols in TCP/IP protocol suite*

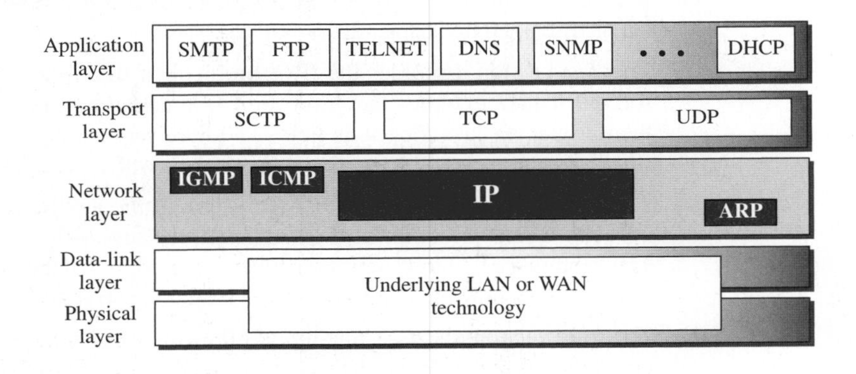

We will discuss IPv4 and ICMPv4 in this chapter. IGMP will be discussed when we talk about multicasting in Chapter 21. We have discussed ARP in Chapter 9 when we talked about link-layer addresses.

IPv4 is an unreliable datagram protocol—a best-effort delivery service. The term *best-effort* means that IPv4 packets can be corrupted, be lost, arrive out of order, or be delayed, and may create congestion for the network. If reliability is important, IPv4 must be paired with a reliable transport-layer protocol such as TCP. An example of a more commonly understood best-effort delivery service is the post office. The post office does its best to deliver the regular mail but does not always succeed. If an unregistered letter is lost or damaged, it is up to the sender or would-be recipient to discover this. The post office itself does not keep track of every letter and cannot notify a sender of loss or damage of one.

IPv4 is also a connectionless protocol that uses the datagram approach. This means that each datagram is handled independently, and each datagram can follow a different route to the destination. This implies that datagrams sent by the same source to the same destination could arrive out of order. Again, IPv4 relies on a higher-level protocol to take care of all these problems.

19.1.1 Datagram Format

In this section, we begin by discussing the first service provided by IPv4, packetizing. We show how IPv4 defines the format of a packet in which the data coming from the upper layer or other protocols are encapsulated. Packets used by the IP are called *datagrams*. Figure 19.2 shows the IPv4 datagram format. A datagram is a variable-length packet consisting of two parts: header and payload (data). The header is 20 to 60 bytes in length and contains information essential to routing and delivery. It is customary in TCP/IP to show the header in 4-byte sections.

Figure 19.2 *IP datagram*

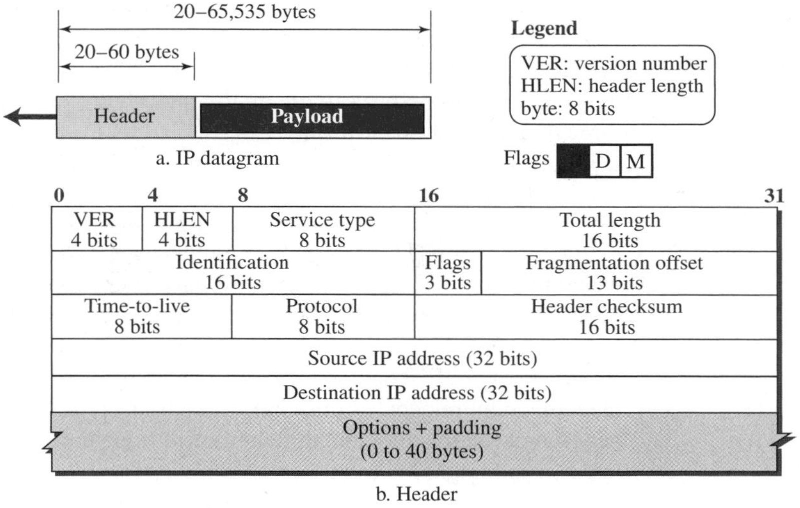

20–65,535 bytes
20–60 bytes
Header | **Payload**

a. IP datagram

Legend
VER: version number
HLEN: header length
byte: 8 bits

Flags D M

0 – 4	4 – 8	8 – 16	16 – 19	19 – 31
VER 4 bits	HLEN 4 bits	Service type 8 bits	Total length 16 bits	
Identification 16 bits			Flags 3 bits	Fragmentation offset 13 bits
Time-to-live 8 bits		Protocol 8 bits	Header checksum 16 bits	
Source IP address (32 bits)				
Destination IP address (32 bits)				
Options + padding (0 to 40 bytes)				

b. Header

Discussing the meaning and rationale for the existence of each field is essential to understanding the operation of IPv4; a brief description of each field is in order.

- ***Version Number.*** The 4-bit version number (VER) field defines the version of the IPv4 protocol, which, obviously, has the value of 4.

- ***Header Length.*** The 4-bit header length (HLEN) field defines the total length of the datagram header in 4-byte words. The IPv4 datagram has a variable-length header. When a device receives a datagram, it needs to know when the header stops and the data, which is encapsulated in the packet, starts. However, to make the value of the header length (number of bytes) fit in a 4-bit header length, the total length of the header is calculated as 4-byte words. The total length is divided by 4 and the value is inserted in the field. The receiver needs to multiply the value of this field by 4 to find the total length.

- ***Service Type.*** In the original design of the IP header, this field was referred to as type of service (TOS), which defined how the datagram should be handled. In the late 1990s, IETF redefined the field to provide *differentiated services* (DiffServ).

When we discuss differentiated services in Chapter 30, we will be in a better situation to define the bits in this field. The use of 4-byte words for the length header is also logical because the IP header always needs to be aligned in 4-byte boundaries.

- ***Total Length.*** This 16-bit field defines the total length (header plus data) of the IP datagram in bytes. A 16-bit number can define a total length of up to 65,535 (when all bits are 1s). However, the size of the datagram is normally much less than this. This field helps the receiving device to know when the packet has completely arrived. To find the length of the data coming from the upper layer, subtract the header length from the total length. The header length can be found by multiplying the value in the HLEN field by 4.

$$\text{Length of data} = \text{total length} - (\text{HLEN}) \times 4$$

Though a size of 65,535 bytes might seem large, the size of the IPv4 datagram may increase in the near future as the underlying technologies allow even more throughput (greater bandwidth).

One may ask why we need this field anyway. When a machine (router or host) receives a frame, it drops the header and the trailer, leaving the datagram. Why include an extra field that is not needed? The answer is that in many cases we really do not need the value in this field. However, there are occasions in which the datagram is not the only thing encapsulated in a frame; it may be that padding has been added. For example, the Ethernet protocol has a minimum and maximum restriction on the size of data that can be encapsulated in a frame (46 to 1500 bytes). If the size of an IPv4 datagram is less than 46 bytes, some padding will be added to meet this requirement. In this case, when a machine decapsulates the datagram, it needs to check the total length field to determine how much is really data and how much is padding.

- ***Identification, Flags, and Fragmentation Offset.*** These three fields are related to the fragmentation of the IP datagram when the size of the datagram is larger than the underlying network can carry. We discuss the contents and importance of these fields when we talk about fragmentation in the next section.

- ***Time-to-live.*** Due to some malfunctioning of routing protocols (discussed later) a datagram may be circulating in the Internet, visiting some networks over and over without reaching the destination. This may create extra traffic in the Internet. The time-to-live (TTL) field is used to control the maximum number of hops (routers) visited by the datagram. When a source host sends the datagram, it stores a number in this field. This value is approximately two times the maximum number of routers between any two hosts. Each router that processes the datagram decrements this number by one. If this value, after being decremented, is zero, the router discards the datagram.

- ***Protocol.*** In TCP/IP, the data section of a packet, called the *payload,* carries the whole packet from another protocol. A datagram, for example, can carry a packet belonging to any transport-layer protocol such as UDP or TCP. A datagram can also carry a packet from other protocols that directly use the service of the IP, such as some routing protocols or some auxiliary protocols. The Internet authority has given any protocol that uses the service of IP a unique 8-bit number which is inserted in the protocol field. When the payload is encapsulated in a datagram at the source IP, the corresponding protocol number is inserted in this field; when the datagram arrives at the destination, the value of this field helps to define to which protocol the payload should be delivered. In other words, this field provides multiplexing at the source and demultiplexing at the destination, as shown in Figure 19.3. Note that the protocol fields at the network layer play the same role as the port numbers at the transport layer (Chapters 23 and 24). However, we need two port numbers in a transport-layer packet because the port numbers at the source and destination are different, but we need only one protocol field because this value is the same for each protocol no matter whether it is located at the source or the destination.

Figure 19.3 *Multiplexing and demultiplexing using the value of the protocol field*

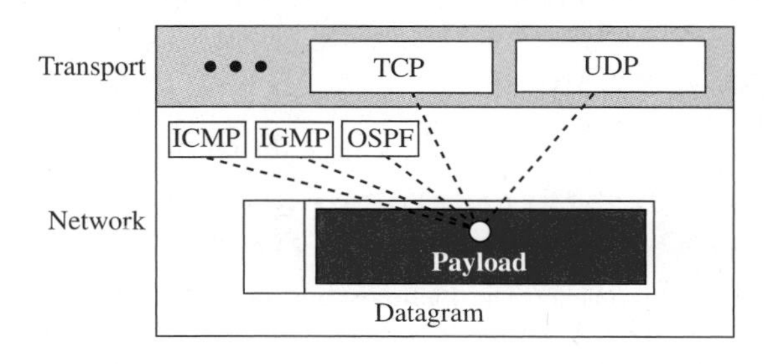

Some protocol values

ICMP	01
IGMP	02
TCP	06
UDP	17
OSPF	89

- ***Header checksum.*** IP is not a reliable protocol; it does not check whether the payload carried by a datagram is corrupted during the transmission. IP puts the burden of error checking of the payload on the protocol that owns the payload, such as UDP or TCP. The datagram header, however, is added by IP, and its error-checking is the responsibility of IP. Errors in the IP header can be a disaster. For example, if the destination IP address is corrupted, the packet can be delivered to the wrong host. If the protocol field is corrupted, the payload may be delivered to the wrong protocol. If the fields related to the fragmentation are corrupted, the datagram cannot be reassembled correctly at the destination, and so on. For these reasons, IP adds a header checksum field to check the header, but not the payload. We need to remember that, since the value of some fields, such as TTL, which are related to fragmentation and options, may change from router to router, the checksum needs to be recalculated at each router. As we discussed in Chapter 10, checksum in the Internet normally uses a 16-bit field, which is the complement of the sum of other fields calculated using 1s complement arithmetic.

- ***Source and Destination Addresses.*** These 32-bit source and destination address fields define the IP address of the source and destination respectively. The source host should know its IP address. The destination IP address is either known by the protocol that uses the service of IP or is provided by the DNS as described in Chapter 26. Note that the value of these fields must remain unchanged during the time the IP datagram travels from the source host to the destination host. IP addresses were discussed in Chapter 18.

- ***Options.*** A datagram header can have up to 40 bytes of options. Options can be used for network testing and debugging. Although options are not a required part of the IP header, option processing is required of the IP software. This means that all implementations must be able to handle options if they are present in the header. The existence of options in a header creates some burden on the datagram handling; some options can be changed by routers, which forces each router to recalculate the header checksum. There are one-byte and multi-byte options that we will briefly discuss later in the chapter. The complete discussion is posted at the book website.
- ***Payload.*** Payload, or data, is the main reason for creating a datagram. Payload is the packet coming from other protocols that use the service of IP. Comparing a datagram to a postal package, payload is the content of the package; the header is only the information written on the package.

Example 19.1

An IPv4 packet has arrived with the first 8 bits as $(01000010)_2$. The receiver discards the packet. Why?

Solution

There is an error in this packet. The 4 leftmost bits $(0100)_2$ show the version, which is correct. The next 4 bits $(0010)_2$ show an invalid header length $(2 \times 4 = 8)$. The minimum number of bytes in the header must be 20. The packet has been corrupted in transmission.

Example 19.2

In an IPv4 packet, the value of HLEN is $(1000)_2$. How many bytes of options are being carried by this packet?

Solution

The HLEN value is 8, which means the total number of bytes in the header is 8×4, or 32 bytes. The first 20 bytes are the base header, the next 12 bytes are the options.

Example 19.3

In an IPv4 packet, the value of HLEN is 5, and the value of the total length field is $(0028)_{16}$. How many bytes of data are being carried by this packet?

Solution

The HLEN value is 5, which means the total number of bytes in the header is 5×4, or 20 bytes (no options). The total length is $(0028)_{16}$ or 40 bytes, which means the packet is carrying 20 bytes of data $(40 - 20)$.

Example 19.4

An IPv4 packet has arrived with the first few hexadecimal digits as shown.

$$(45000028000100000102 \ldots)_{16}$$

How many hops can this packet travel before being dropped? The data belong to what upper-layer protocol?

Solution

To find the time-to-live field, we skip 8 bytes (16 hexadecimal digits). The time-to-live field is the ninth byte, which is $(01)_{16}$. This means the packet can travel only one hop. The protocol field is the next byte $(02)_{16}$, which means that the upper-layer protocol is IGMP.

Example 19.5

Figure 19.4 shows an example of a checksum calculation for an IPv4 header without options. The header is divided into 16-bit sections. All the sections are added and the sum is complemented after wrapping the leftmost digit. The result is inserted in the checksum field.

Figure 19.4 *Example of checksum calculation in IPv4*

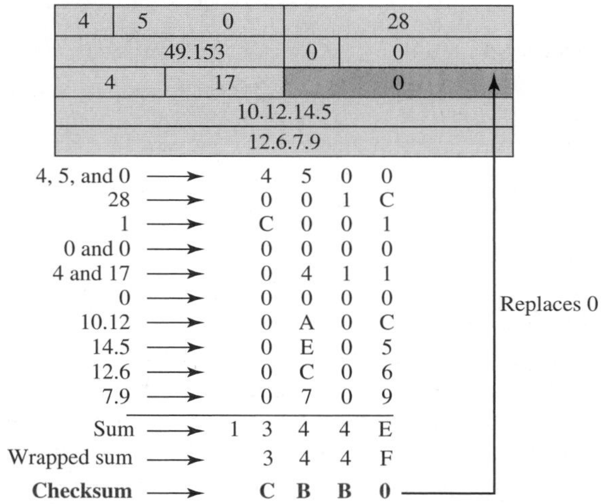

4	5	0	28	
49.153			0	0
4		17	0	
10.12.14.5				
12.6.7.9				

4, 5, and 0	→	4 5 0 0
28	→	0 0 1 C
1	→	C 0 0 1
0 and 0	→	0 0 0 0
4 and 17	→	0 4 1 1
0	→	0 0 0 0
10.12	→	0 A 0 C
14.5	→	0 E 0 5
12.6	→	0 C 0 6
7.9	→	0 7 0 9
Sum	→	1 3 4 4 E
Wrapped sum	→	3 4 4 F
Checksum	→	**C B B 0**

Replaces 0

Note that the calculation of wrapped sum and checksum can also be done as follows in hexadecimal:

Wrapped Sum = Sum mod FFFF
Checksum = FFFF − Wrapped Sum

19.1.2 Fragmentation

A datagram can travel through different networks. Each router decapsulates the IP datagram from the frame it receives, processes it, and then encapsulates it in another frame. The format and size of the received frame depend on the protocol used by the physical network through which the frame has just traveled. The format and size of the sent frame depend on the protocol used by the physical network through which the frame is going to travel. For example, if a router connects a LAN to a WAN, it receives a frame in the LAN format and sends a frame in the WAN format.

Maximum Transfer Unit (MTU)

Each link-layer protocol has its own frame format. One of the features of each format is the maximum size of the payload that can be encapsulated. In other words, when a datagram is encapsulated in a frame, the total size of the datagram must be less than this maximum size, which is defined by the restrictions imposed by the hardware and software used in the network (see Figure 19.5).

Figure 19.5 *Maximum transfer unit (MTU)*

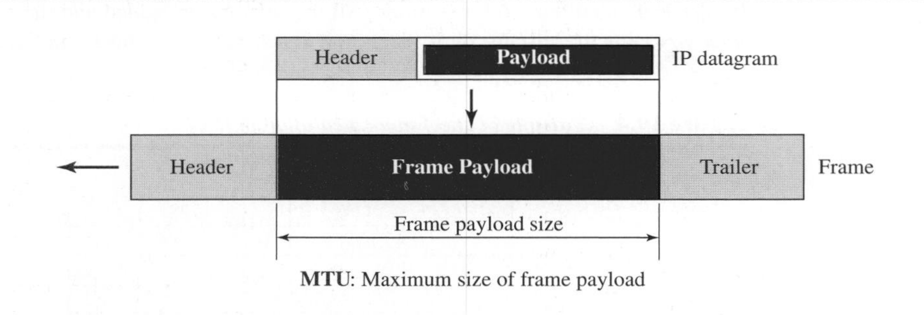

The value of the MTU differs from one physical network protocol to another. For example, the value for a LAN is normally 1500 bytes, but for a WAN it can be larger or smaller.

In order to make the IP protocol independent of the physical network, the designers decided to make the maximum length of the IP datagram equal to 65,535 bytes. This makes transmission more efficient if one day we use a link-layer protocol with an MTU of this size. However, for other physical networks, we must divide the datagram to make it possible for it to pass through these networks. This is called ***fragmentation.***

When a datagram is fragmented, each fragment has its own header with most of the fields repeated, but some have been changed. A fragmented datagram may itself be fragmented if it encounters a network with an even smaller MTU. In other words, a datagram may be fragmented several times before it reaches the final destination.

A datagram can be fragmented by the source host or any router in the path. The *reassembly* of the datagram, however, is done only by the destination host, because each fragment becomes an independent datagram. Whereas the fragmented datagram can travel through different routes, and we can never control or guarantee which route a fragmented datagram may take, all of the fragments belonging to the same datagram should finally arrive at the destination host. So it is logical to do the reassembly at the final destination. An even stronger objection for reassembling packets during the transmission is the loss of efficiency it incurs.

When we talk about fragmentation, we mean that the payload of the IP datagram is fragmented. However, most parts of the header, with the exception of some options, must be copied by all fragments. The host or router that fragments a datagram must change the values of three fields: flags, fragmentation offset, and total length. The rest of the fields must be copied. Of course, the value of the checksum must be recalculated regardless of fragmentation.

Fields Related to Fragmentation

We mentioned before that three fields in an IP datagram are related to fragmentation: *identification, flags,* and *fragmentation offset.* Let us explain these fields now.

The 16-bit *identification field* identifies a datagram originating from the source host. The combination of the identification and source IP address must uniquely define a datagram as it leaves the source host. To guarantee uniqueness, the IP protocol uses a counter to label the datagrams. The counter is initialized to a positive number. When the IP protocol sends a datagram, it copies the current value of the counter to the identification field and increments the counter by one. As long as the counter is kept in the main memory, uniqueness is guaranteed. When a datagram is fragmented, the value in the identification field is copied into all fragments. In other words, all fragments have the same identification number, which is also the same as the original datagram. The identification number helps the destination in reassembling the datagram. It knows that all fragments having the same identification value should be assembled into one datagram.

The 3-bit *flags field* defines three flags. The leftmost bit is reserved (not used). The second bit (D bit) is called the *do not fragment* bit. If its value is 1, the machine must not fragment the datagram. If it cannot pass the datagram through any available physical network, it discards the datagram and sends an ICMP error message to the source host (discussed later). If its value is 0, the datagram can be fragmented if necessary. The third bit (M bit) is called the *more fragment bit.* If its value is 1, it means the datagram is not the last fragment; there are more fragments after this one. If its value is 0, it means this is the last or only fragment.

The 13-bit *fragmentation offset field* shows the relative position of this fragment with respect to the whole datagram. It is the offset of the data in the original datagram measured in units of 8 bytes. Figure 19.6 shows a datagram with a data size of 4000 bytes fragmented into three fragments. The bytes in the original datagram are numbered 0 to 3999. The first fragment carries bytes 0 to 1399. The offset for this datagram is 0/8 = 0. The second fragment carries bytes 1400 to 2799; the offset value for this fragment is 1400/8 = 175. Finally, the third fragment carries bytes 2800 to 3999. The offset value for this fragment is 2800/8 = 350.

Figure 19.6 *Fragmentation example*

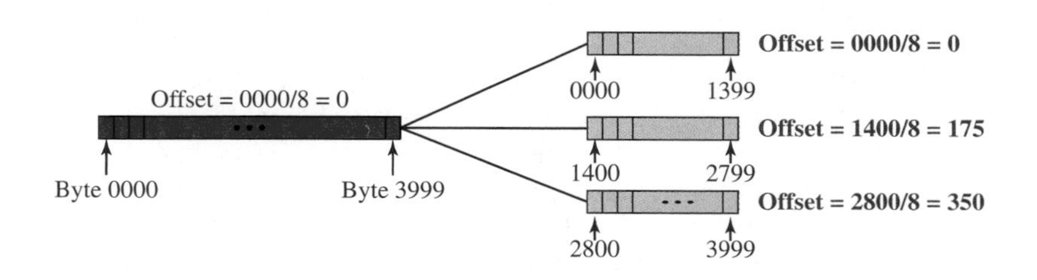

Remember that the value of the offset is measured in units of 8 bytes. This is done because the length of the offset field is only 13 bits long and cannot represent a sequence of bytes greater than 8191. This forces hosts or routers that fragment datagrams to choose the size of each fragment so that the first byte number is divisible by 8.

Figure 19.7 shows an expanded view of the fragments in the previous figure. The original packet starts at the client; the fragments are reassembled at the server. The value of the identification field is the same in all fragments, as is the value of the flags field with the more bit set for all fragments except the last. Also, the value of the offset field for each fragment is shown. Note that although the fragments arrived out of order at the destination, they can be correctly reassembled.

Figure 19.7 *Detailed fragmentation example*

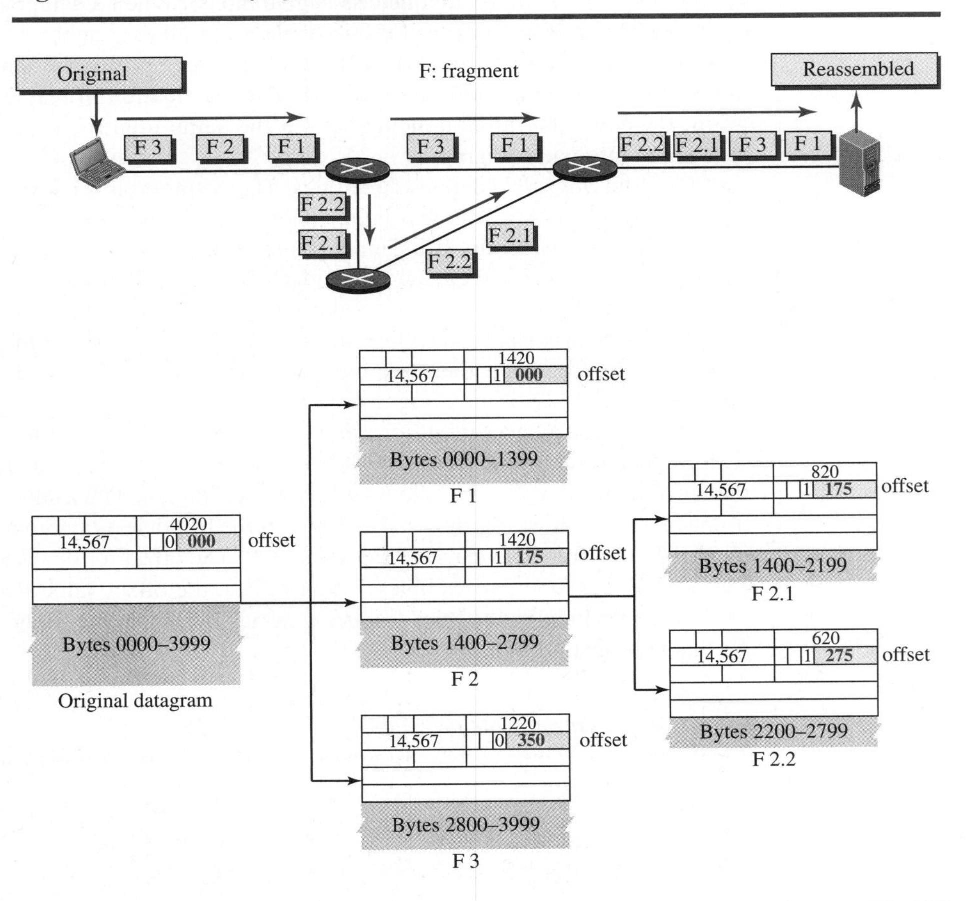

The figure also shows what happens if a fragment itself is fragmented. In this case the value of the offset field is always relative to the original datagram. For example, in the figure, the second fragment is itself fragmented later into two fragments of

800 bytes and 600 bytes, but the offset shows the relative position of the fragments to the original data.

It is obvious that even if each fragment follows a different path and arrives out of order, the final destination host can reassemble the original datagram from the fragments received (if none of them is lost) using the following strategy:

a. The first fragment has an offset field value of zero.

b. Divide the length of the first fragment by 8. The second fragment has an offset value equal to that result.

c. Divide the total length of the first and second fragment by 8. The third fragment has an offset value equal to that result.

d. Continue the process. The last fragment has its M bit set to 0.

e. Continue the process. The last fragment has a *more* bit value of 0.

Example 19.6

A packet has arrived with an M bit value of 0. Is this the first fragment, the last fragment, or a middle fragment? Do we know if the packet was fragmented?

Solution

If the M bit is 0, it means that there are no more fragments; the fragment is the last one. However, we cannot say if the original packet was fragmented or not. A nonfragmented packet is considered the last fragment.

Example 19.7

A packet has arrived with an M bit value of 1. Is this the first fragment, the last fragment, or a middle fragment? Do we know if the packet was fragmented?

Solution

If the M bit is 1, it means that there is at least one more fragment. This fragment can be the first one or a middle one, but not the last one. We don't know if it is the first one or a middle one; we need more information (the value of the fragmentation offset).

Example 19.8

A packet has arrived with an M bit value of 1 and a fragmentation offset value of 0. Is this the first fragment, the last fragment, or a middle fragment?

Solution

Because the M bit is 1, it is either the first fragment or a middle one. Because the offset value is 0, it is the first fragment.

Example 19.9

A packet has arrived in which the offset value is 100. What is the number of the first byte? Do we know the number of the last byte?

Solution

To find the number of the first byte, we multiply the offset value by 8. This means that the first byte number is 800. We cannot determine the number of the last byte unless we know the length of the data.

Example 19.10

A packet has arrived in which the offset value is 100, the value of HLEN is 5, and the value of the total length field is 100. What are the numbers of the first byte and the last byte?

Solution

The first byte number is $100 \times 8 = 800$. The total length is 100 bytes, and the header length is 20 bytes (5×4), which means that there are 80 bytes in this datagram. If the first byte number is 800, the last byte number must be 879.

19.1.3 Options

The header of the IPv4 datagram is made of two parts: a fixed part and a variable part. The fixed part is 20 bytes long and was discussed in the previous section. The variable part comprises the options that can be a maximum of 40 bytes (in multiples of 4-bytes) to preserve the boundary of the header.

Options, as the name implies, are not required for a datagram. They can be used for network testing and debugging. Although options are not a required part of the IPv4 header, option processing is required of the IPv4 software. This means that all implementations must be able to handle options if they are present in the header. Options are divided into two broad categories: single-byte options and multiple-byte options. We give a brief description of options here; for a complete description, see the book website under Extra Materials.

> **The complete discussion of options in IPv4 is included in the book website under Extra Materials for Chapter 19.**

Single-Byte Options

There are two single-byte options.

No Operation

A *no-operation option* is a 1-byte option used as a filler between options.

End of Option

An *end-of-option option* is a 1-byte option used for padding at the end of the option field. It, however, can only be used as the last option.

Multliple-Byte Options

There are four multiple-byte options.

Record Route

A *record route option* is used to record the Internet routers that handle the datagram. It can list up to nine router addresses. It can be used for debugging and management purposes.

Strict Source Route

A *strict source route option* is used by the source to predetermine a route for the datagram as it travels through the Internet. Dictation of a route by the source can be useful for several purposes. The sender can choose a route with a specific type of service, such as minimum delay or maximum throughput. Alternatively, it may choose a route that is safer or more reliable for the sender's purpose. For example, a sender can choose a route so that its datagram does not travel through a competitor's network.

If a datagram specifies a strict source route, all the routers defined in the option must be visited by the datagram. A router must not be visited if its IPv4 address is not listed in the datagram. If the datagram visits a router that is not on the list, the datagram is discarded and an error message is issued. If the datagram arrives at the destination and some of the entries were not visited, it will also be discarded and an error message issued.

Loose Source Route

A *loose source route option* is similar to the strict source route, but it is less rigid. Each router in the list must be visited, but the datagram can visit other routers as well.

Timestamp

A *timestamp option* is used to record the time of datagram processing by a router. The time is expressed in milliseconds from midnight, Universal time or Greenwich mean time. Knowing the time a datagram is processed can help users and managers track the behavior of the routers in the Internet. We can estimate the time it takes for a datagram to go from one router to another. We say *estimate* because, although all routers may use Universal time, their local clocks may not be synchronized.

19.1.4 Security of IPv4 Datagrams

The IPv4 protocol, as well as the whole Internet, was started when the Internet users trusted each other. No security was provided for the IPv4 protocol. Today, however, the situation is different; the Internet is not secure anymore. Although we will discuss network security in general and IP security in particular in Chapters 31 and 32, here we give a brief idea about the security issues in IP protocol and the solutions. There are three security issues that are particularly applicable to the IP protocol: packet sniffing, packet modification, and IP spoofing.

Packet Sniffing

An intruder may intercept an IP packet and make a copy of it. Packet sniffing is a passive attack, in which the attacker does not change the contents of the packet. This type of attack is very difficult to detect because the sender and the receiver may never know that the packet has been copied. Although packet sniffing cannot be stopped, encryption of the packet can make the attacker's effort useless. The attacker may still sniff the packet, but the content is not detectable.

Packet Modification

The second type of attack is to modify the packet. The attacker intercepts the packet, changes its contents, and sends the new packet to the receiver. The receiver believes that the packet is coming from the original sender. This type of attack can be detected using a data integrity mechanism. The receiver, before opening and using the contents of the message, can use this mechanism to make sure that the packet has not been changed during the transmission. We discuss packet integrity in Chapter 32.

IP Spoofing

An attacker can masquerade as somebody else and create an IP packet that carries the source address of another computer. An attacker can send an IP packet to a bank pretending that it is coming from one of the customers. This type of attack can be prevented using an origin authentication mechanism (see Chapter 32).

IPSec

The IP packets today can be protected from the previously mentioned attacks using a protocol called IPSec (IP Security). This protocol, which is used in conjunction with the IP protocol, creates a connection-oriented service between two entities in which they can exchange IP packets without worrying about the three attacks discussed above. We will discuss IPSec in detail in Chapter 32; here it is enough to mention that IPSec provides the following four services:

- ***Defining Algorithms and Keys.*** The two entities that want to create a secure channel between themselves can agree on some available algorithms and keys to be used for security purposes.
- ***Packet Encryption.*** The packets exchanged between two parties can be encrypted for privacy using one of the encryption algorithms and a shared key agreed upon in the first step. This makes the packet sniffing attack useless.
- ***Data Integrity.*** Data integrity guarantees that the packet is not modified during the transmission. If the received packet does not pass the data integrity test, it is discarded. This prevents the second attack, packet modification, described above.
- ***Origin Authentication.*** IPSec can authenticate the origin of the packet to be sure that the packet is not created by an imposter. This can prevent IP spoofing attacks as described above.

19.2 ICMPv4

The IPv4 has no error-reporting or error-correcting mechanism. What happens if something goes wrong? What happens if a router must discard a datagram because it cannot find a route to the final destination, or because the time-to-live field has a zero value? What happens if the final destination host must discard the received fragments of a datagram because it has not received all fragments within a predetermined time limit? These are examples of situations where an error has occurred and the IP protocol has no built-in mechanism to notify the original host.

The IP protocol also lacks a mechanism for host and management queries. A host sometimes needs to determine if a router or another host is alive. And sometimes a network manager needs information from another host or router.

The **Internet Control Message Protocol version 4 (ICMPv4)** has been designed to compensate for the above two deficiencies. It is a companion to the IP protocol. ICMP itself is a network-layer protocol. However, its messages are not passed directly to the data-link layer as would be expected. Instead, the messages are first encapsulated inside IP datagrams before going to the lower layer. When an IP datagram encapsulates an ICMP message, the value of the protocol field in the IP datagram is set to 1 to indicate that the IP payroll is an ICMP message.

19.2.1 MESSAGES

ICMP messages are divided into two broad categories: *error-reporting messages* and *query messages*. The error-reporting messages report problems that a router or a host (destination) may encounter when it processes an IP packet. The query messages, which occur in pairs, help a host or a network manager get specific information from a router or another host. For example, nodes can discover their neighbors. Also, hosts can discover and learn about routers on their network and routers can help a node redirect its messages.

An ICMP message has an 8-byte header and a variable-size data section. Although the general format of the header is different for each message type, the first 4 bytes are common to all. As Figure 19.8 shows, the first field, ICMP type, defines the type of the message. The code field specifies the reason for the particular message type. The last common field is the checksum field (to be discussed later in the chapter). The rest of the header is specific for each message type.

Figure 19.8 *General format of ICMP messages*

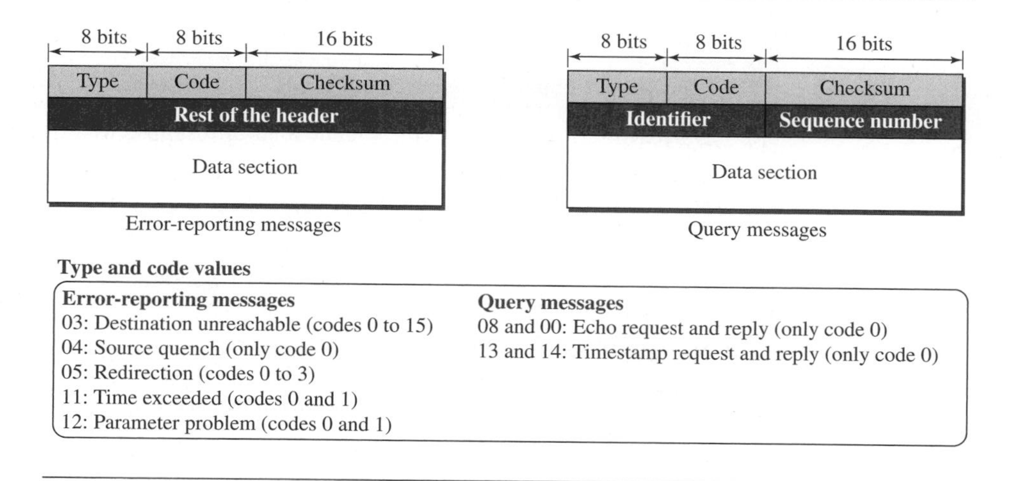

The data section in error messages carries information for finding the original packet that had the error. In query messages, the data section carries extra information based on the type of query.

We give a brief description of the ICMPv4 messages here; for a complete description see the book website under Extra Materials for Chapter 19.

The complete discussion of messages in ICMPv4 is included in the book website under Extra Materials for Chapter 19.

Error Reporting Messages

Since IP is an unreliable protocol, one of the main responsibilities of ICMP is to report some errors that may occur during the processing of the IP datagram. ICMP does not correct errors, it simply reports them. Error correction is left to the higher-level protocols. Error messages are always sent to the original source because the only information available in the datagram about the route is the source and destination IP addresses. ICMP uses the source IP address to send the error message to the source (originator) of the datagram. To make the error-reporting process simple, ICMP follows some rules in reporting messages. First, no error message will be generated for a datagram having a multicast address or special address (such as *this host* or *loopback*). Second, no ICMP error message will be generated in response to a datagram carrying an ICMP error message. Third, no ICMP error message will be generated for a fragmented datagram that is not the first fragment.

Note that all error messages contain a data section that includes the IP header of the original datagram plus the first 8 bytes of data in that datagram. The original datagram header is added to give the original source, which receives the error message, information about the datagram itself. The 8 bytes of data are included because the first 8 bytes provide information about the port numbers (UDP and TCP) and sequence number (TCP). This information is needed so the source can inform the protocols (TCP or UDP) about the error.

The following are important points about ICMP error messages:

- No ICMP error message will be generated in response to a datagram carrying an ICMP error message.
- No ICMP error message will be generated for a fragmented datagram that is not the first fragment.
- No ICMP error message will be generated for a datagram having a multicast address.
- No ICMP error message will be generated for a datagram having a special address such as 127.0.0.0 or 0.0.0.0.

Note that all error messages contain a data section that includes the IP header of the original datagram plus the first 8 bytes of data in that datagram. The original datagram header is added to give the original source, which receives the error message, information about the datagram itself. The 8 bytes of data are included because, as we will see in Chapter 24 on UDP and TCP protocols, the first 8 bytes provide information about the port numbers (UDP and TCP) and sequence number (TCP). This information is needed so the source can inform the protocols (TCP or UDP) about the error. ICMP forms an error packet, which is then encapsulated in an IP datagram (see Figure 19.9).

Destination Unreachable

The most widely used error message is the destination unreachable (type 3). This message uses different codes (0 to 15) to define the type of error message and the reason why a datagram has not reached its final destination. For example, code 0 tells the

Figure 19.9 *Contents of data field for the error messages*

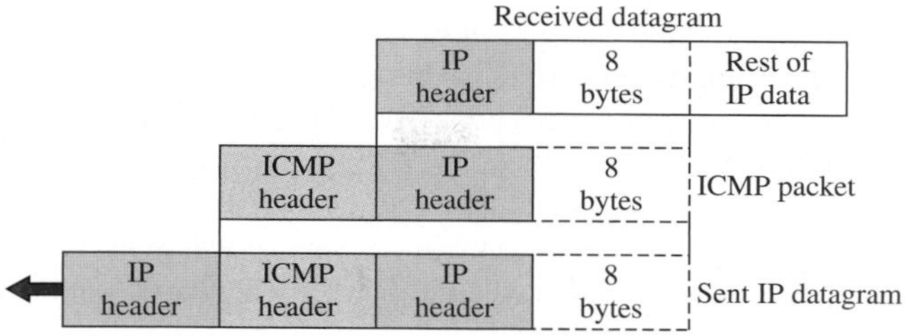

Received datagram
IP header | 8 bytes | Rest of IP data
ICMP header | IP header | 8 bytes | ICMP packet
IP header | ICMP header | IP header | 8 bytes | Sent IP datagram

source that a host is unreachable. This may happen, for example, when we use the HTTP protocol to access a web page, but the server is down. The message "destination host is not reachable" is created and sent back to the source.

Source Quench

Another error message is called the *source quench* (type 4) message, which informs the sender that the network has encountered congestion and the datagram has been dropped; the source needs to slow down sending more datagrams. In other words, ICMP adds a kind of congestion control mechanism to the IP protocol by using this type of message.

Redirection Message

The *redirection message* (type 5) is used when the source uses a wrong router to send out its message. The router redirects the message to the appropriate router, but informs the source that it needs to change its default router in the future. The IP address of the default router is sent in the message.

We discussed the purpose of the *time-to-live* (TTL) field in the IP datagram and explained that it prevents a datagram from being aimlessly circulated in the Internet. When the TTL value becomes 0, the datagram is dropped by the visiting router and a *time exceeded* message (type 11) with code 0 is sent to the source to inform it about the situation. The time-exceeded message (with code 1) can also be sent when not all fragments of a datagram arrive within a predefined period of time.

Parameter Problem

A *parameter problem message* (type 12) can be sent when either there is a problem in the header of a datagram (code 0) or some options are missing or cannot be interpreted (code 1).

Query Messages

Interestingly, query messages in ICMP can be used independently without relation to an IP datagram. Of course, a query message needs to be encapsulated in a datagram, as a carrier. Query messages are used to probe or test the liveliness of hosts or routers in the Internet, find the one-way or the round-trip time for an IP datagram between two devices, or even find out whether the clocks in two devices are synchronized. Naturally, query messages come in pairs: request and reply.

The *echo request* (type 8) and the *echo reply* (type 0) pair of messages are used by a host or a router to test the liveliness of another host or router. A host or router sends an echo request message to another host or router; if the latter is alive, it responds with an echo reply message. We shortly see the applications of this pair in two debugging tools: *ping* and *traceroute*.

The *timestamp request* (type 13) and the *timestamp reply* (type 14) pair of messages are used to find the round-trip time between two devices or to check whether the clocks in two devices are synchronized. The timestamp request message sends a 32-bit number, which defines the time the message is sent. The timestamp reply resends that number, but also includes two new 32-bit numbers representing the time the request was received and the time the response was sent. If all timestamps represent Universal time, the sender can calculate the one-way and round-trip time.

Deprecated Messages

Three pairs of messages are declared obsolete by IETF:

1. *Information request and replay* messages are not used today because their duties are done by the Address Resolution Protocol (ARP) discussed in Chapter 9.
2. *Address mask request and reply* messages are not used today because their duties are done by the Dynamic Host Configuration Protocol (DHCP), discussed in Chapter 18.
3. *Router solicitation and advertisement* messages are not used today because their duties are done by the Dynamic Host Configuration Protocol (DHCP), discussed in Chapter 18.

19.2.2 Debugging Tools

There are several tools that can be used in the Internet for debugging. We can determine the viability of a host or router. We can trace the route of a packet. We introduce two tools that use ICMP for debugging: *ping* and *traceroute*.

Ping

We can use the *ping* program to find if a host is alive and responding. We use *ping* here to see how it uses ICMP packets. The source host sends ICMP echo-request messages; the destination, if alive, responds with ICMP echo-reply messages. The *ping* program sets the identifier field in the echo-request and echo-reply message and starts the sequence number from 0; this number is incremented by 1 each time a new message is sent. Note that *ping* can calculate the round-trip time. It inserts the sending time in the data section of the message. When the packet arrives, it subtracts the arrival time from the departure time to get the round-trip time (RTT).

Example 19.11

The following shows how we send a *ping* message to the auniversity.edu site. We set the identifier field in the echo request and reply message and start the sequence number from 0; this number is incremented by one each time a new message is sent. Note that *ping* can calculate the round-trip time. It inserts the sending time in the data section of the message. When the packet arrives, it subtracts the arrival time from the departure time to get the *round-trip time* (rtt).

```
$ ping auniversity.edu
PING auniversity.edu (152.181.8.3)   56 (84)  bytes of data.
64 bytes from auniversity.edu (152.181.8.3): icmp_seq=0    ttl=62    time=1.91 ms
```

```
64 bytes from auniversity.edu (152.181.8.3): icmp_seq=1   ttl=62   time=2.04 ms
64 bytes from auniversity.edu (152.181.8.3): icmp_seq=2   ttl=62   time=1.90 ms
64 bytes from auniversity.edu (152.181.8.3): icmp_seq=3   ttl=62   time=1.97 ms
64 bytes from auniversity.edu (152.181.8.3): icmp_seq=4   ttl=62   time=1.93 ms
64 bytes from auniversity.edu (152.181.8.3): icmp_seq=5   ttl=62   time=2.00 ms
--- auniversity.edu statistics ---
6 packets transmitted, 6 received, 0% packet loss
rtt min/avg/max = 1.90/1.95/2.04 ms
```

Traceroute or Tracert

The *traceroute* program in UNIX or *tracert* in Windows can be used to trace the path of a packet from a source to the destination. It can find the IP addresses of all the routers that are visited along the path. The program is usually set to check for the maximum of 30 hops (routers) to be visited. The number of hops in the Internet is normally less than this. Since these two programs behave differently in Unix and Windows, we explain them separately.

Traceroute

The *traceroute* program is different from the *ping* program. The *ping* program gets help from two query messages; the *traceroute* program gets help from two error-reporting messages: time-exceeded and destination-unreachable. The *traceroute* is an application-layer program, but only the client program is needed, because, as we can see, the client program never reaches the application layer in the destination host. In other words, there is no *traceroute* server program. The *traceroute* application program is encapsulated in a UDP user datagram, but *traceroute* intentionally uses a port number that is not available at the destination. If there are n routers in the path, the *traceroute* program sends $(n + 1)$ messages. The first n messages are discarded by the n routers, one by each router; the last message is discarded by the destination host. The *traceroute* client program uses the $(n + 1)$ ICMP error-reporting messages received to find the path between the routers. We will show shortly that the *traceroute* program does not need to know the value of n; it is found automatically. In Figure 19.10, the value of n is 3.

The first *traceroute* message is sent with time-to-live (TTL) value set to 1; the message is discarded at the first router and a time-exceeded ICMP error message is sent, from which the *traceroute* program can find the IP address of the first router (the source IP address of the error message) and the router name (in the data section of the message). The second *traceroute* message is sent with TTL set to 2, which can find the IP address and the name of the second router. Similarly, the third message can find the information about router 3. The fourth message, however, reaches the destination host. This host is also dropped, but for another reason. The destination host cannot find the port number specified in the UDP user datagram. This time ICMP sends a different message, the destination-unreachable message with code 3 to show the port number is not found. After receiving this different ICMP message, the *traceroute* program knows that the final destination is reached. It uses the information in the received message to find the IP address and the name of the final destination.

Figure 19.10 *Use of ICMPv4 in traceroute*

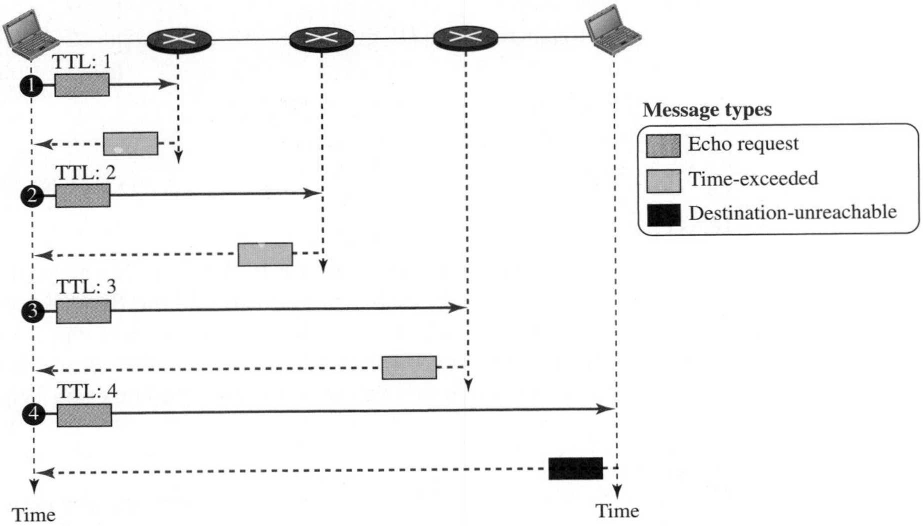

The *traceroute* program also sets a timer to find the round-trip time for each router and the destination. Most *traceroute* programs send three messages to each device, with the same TTL value, to be able to find a better estimate for the round-trip time. The following shows an example of a *traceroute* program, which uses three probes for each device and gets three RTTs.

```
$ traceroute printers.com
traceroute to printers.com (13.1.69.93), 30 hops max, 38-byte packets
1 route.front.edu        (153.18.31.254)    0.622 ms    0.891 ms    0.875 ms
2 ceneric.net            (137.164.32.140)   3.069 ms    2.875 ms    2.930 ms
3 satire.net             (132.16.132.20)    3.071 ms    2.876 ms    2.929 ms
4 alpha.printers.com     (13.1.69.93)       5.922 ms    5.048 ms    4.922 ms
```

Tracert

The *tracert* program in windows behaves differently. The *tracert* messages are encapsulated directly in IP datagrams. The *tracert,* like *traceroute,* sends echo-request messages. However, when the last echo request reaches the destination host, an echo-replay message is issued.

19.2.3 ICMP Checksum

In ICMP the checksum is calculated over the entire message (header and data).

Example 19.12

Figure 19.11 shows an example of checksum calculation for a simple echo-request message. We randomly chose the identifier to be 1 and the sequence number to be 9. The message is divided

Figure 19.11 *Example of checksum calculation*

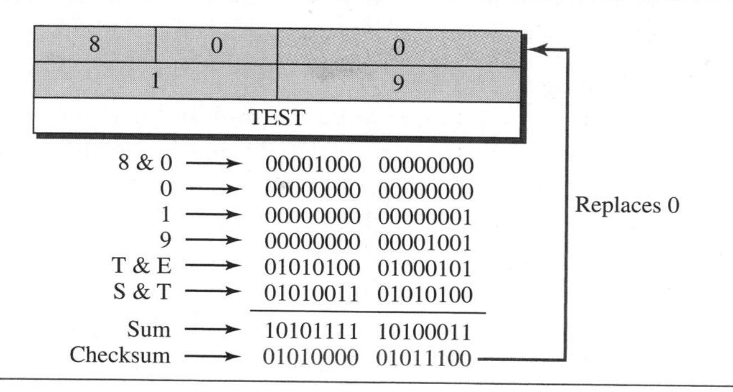

into 16-bit (2-byte) words. The words are added and the sum is complemented. Now the sender can put this value in the checksum field.

19.3 MOBILE IP

In the last section of this chapter, we discuss mobile IP. As mobile and personal computers such as notebooks become increasingly popular, we need to think about mobile IP, the extension of IP protocol that allows mobile computers to be connected to the Internet at any location where the connection is possible. In this section, we discuss this issue.

19.3.1 Addressing

The main problem that must be solved in providing mobile communication using the IP protocol is addressing.

Stationary Hosts

The original IP addressing was based on the assumption that a host is stationary, attached to one specific network. A router uses an IP address to route an IP datagram. As we learned in Chapter 18, an IP address has two parts: a prefix and a suffix. The prefix associates a host with a network. For example, the IP address 10.3.4.24/8 defines a host attached to the network 10.0.0.0/8. This implies that a host in the Internet does not have an address that it can carry with itself from one place to another. The address is valid only when the host is attached to the network. If the network changes, the address is no longer valid. Routers use this association to route a packet; they use the prefix to deliver the packet to the network to which the host is attached. This scheme works perfectly with **stationary hosts.**

The IP addresses are designed to work with stationary hosts because part of the address defines the network to which the host is attached.

Mobile Hosts

When a host moves from one network to another, the IP addressing structure needs to be modified. Several solutions have been proposed.

Changing the Address

One simple solution is to let the **mobile host** change its address as it goes to the new network. The host can use DHCP (see Chapter 18) to obtain a new address to associate it with the new network. This approach has several drawbacks. First, the configuration files would need to be changed. Second, each time the computer moves from one network to another, it must be rebooted. Third, the DNS tables (see Chapter 26) need to be revised so that every other host in the Internet is aware of the change. Fourth, if the host roams from one network to another during a transmission, the data exchange will be interrupted. This is because the ports and IP addresses of the client and the server must remain constant for the duration of the connection.

Two Addresses

The approach that is more feasible is the use of two addresses. The host has its original address, called the **home address,** and a temporary address, called the **care-of address.** The home address is permanent; it associates the host with its **home network,** the network that is the permanent home of the host. The care-of address is temporary. When a host moves from one network to another, the care-of address changes; it is associated with the **foreign network,** the network to which the host moves. Figure 19.12 shows the concept.

Figure 19.12 *Home address and care-of address*

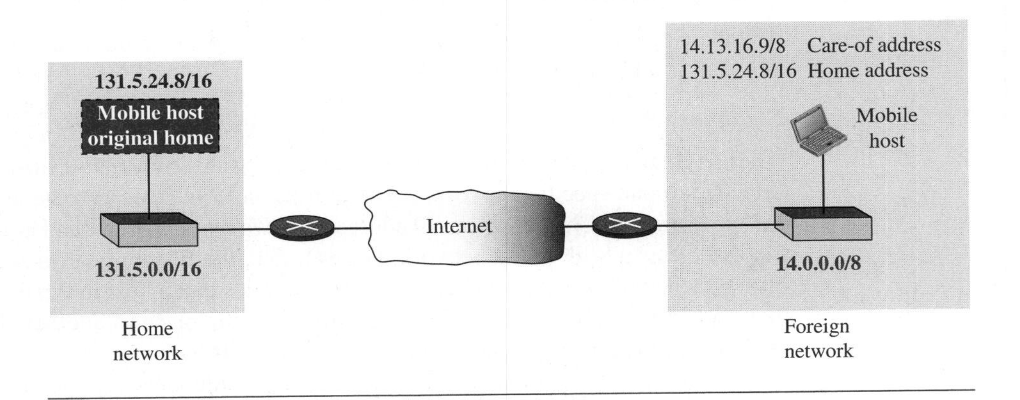

Mobile IP has two addresses for a mobile host: one home address and one care-of address. The home address is permanent; the care-of address changes as the mobile host moves from one network to another.

When a mobile host visits a foreign network, it receives its care-of address during the agent discovery and registration phase, described later.

19.3.2 Agents

To make the change of address transparent to the rest of the Internet requires a **home agent** and a **foreign agent.** Figure 19.13 shows the position of a home agent relative to the home network and a foreign agent relative to the foreign network.

Figure 19.13 *Home agent and foreign agent*

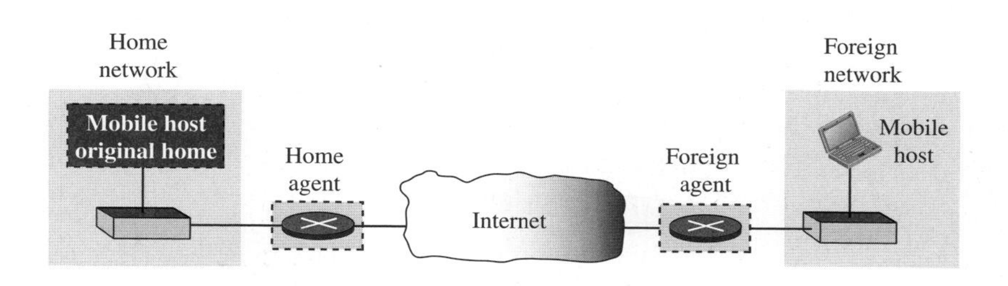

We have shown the home and the foreign agents as routers, but we need to emphasize that their specific function as an agent is performed in the application layer. In other words, they are both routers and hosts.

Home Agent

The home agent is usually a router attached to the home network of the mobile host. The home agent acts on behalf of the mobile host when a remote host sends a packet to the mobile host. The home agent receives the packet and sends it to the foreign agent.

Foreign Agent

The foreign agent is usually a router attached to the foreign network. The foreign agent receives and delivers packets sent by the home agent to the mobile host.

The mobile host can also act as a foreign agent. In other words, the mobile host and the foreign agent can be the same. However, to do this, a mobile host must be able to receive a care-of address by itself, which can be done through the use of DHCP. In addition, the mobile host needs the necessary software to allow it to communicate with the home agent and to have two addresses: its home address and its care-of address. This dual addressing must be transparent to the application programs.

When the mobile host acts as a foreign agent, the care-of address is called a **collocated care-of address.**

When the mobile host and the foreign agent are the same, the care-of address is called a collocated care-of address.

The advantage of using a collocated care-of address is that the mobile host can move to any network without worrying about the availability of a foreign agent. The disadvantage is that the mobile host needs extra software to act as its own foreign agent.

19.3.3 Three Phases

To communicate with a remote host, a mobile host goes through three phases: agent discovery, registration, and data transfer, as shown in Figure 19.14.

Figure 19.14 *Remote host and mobile host communication*

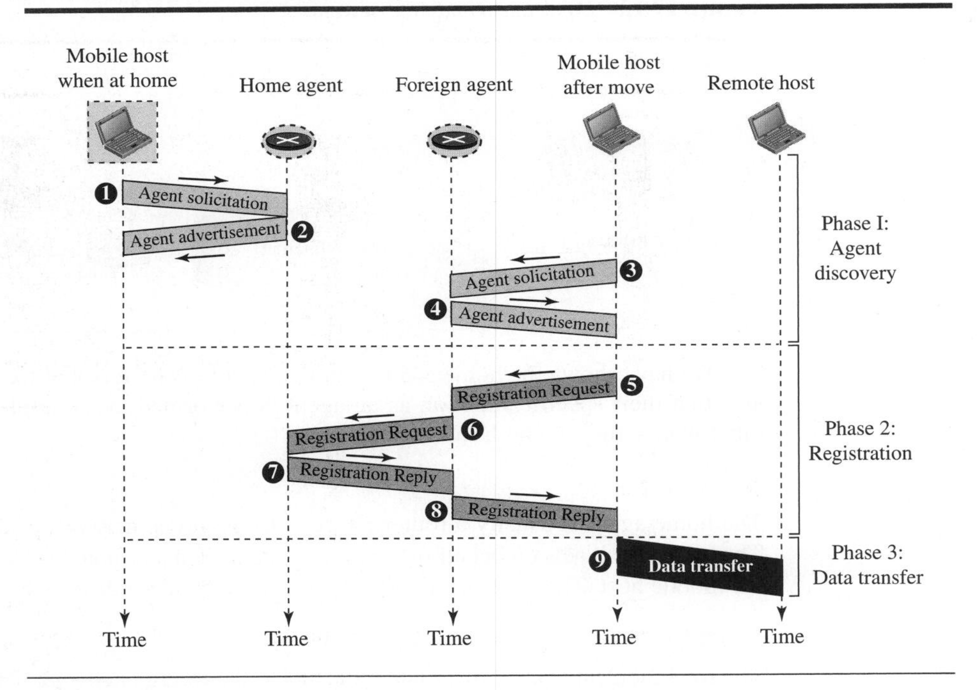

The first phase, agent discovery, involves the mobile host, the foreign agent, and the home agent. The second phase, registration, also involves the mobile host and the two agents. Finally, in the third phase, the remote host is also involved. We discuss each phase separately.

Agent Discovery

The first phase in mobile communication, *agent discovery,* consists of two subphases. A mobile host must discover (learn the address of) a home agent before it leaves its home network. A mobile host must also discover a foreign agent after it has moved to a foreign network. This discovery consists of learning the care-of address as well as the foreign agent's address. The discovery involves two types of messages: advertisement and solicitation.

Agent Advertisement

When a router advertises its presence on a network using an ICMP router advertisement, it can append an *agent advertisement* to the packet if it acts as an agent.

Figure 19.15 shows how an agent advertisement is piggybacked to the router advertisement packet.

Mobile IP does not use a new packet type for agent advertisement; it uses the router advertisement packet of ICMP, and appends an agent advertisement message.

Figure 19.15 *Agent advertisement*

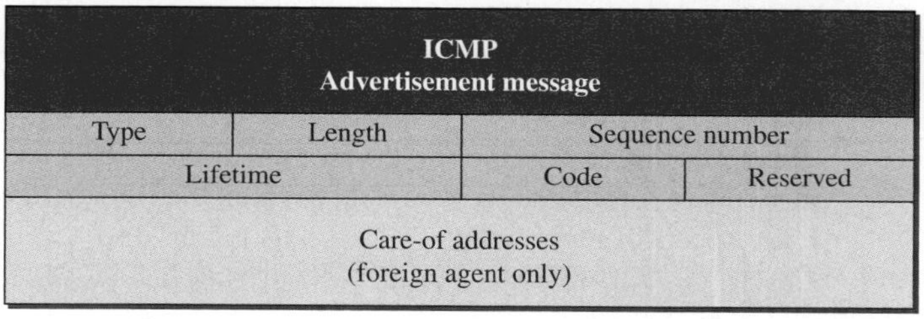

ICMP Advertisement message			
Type	Length	Sequence number	
Lifetime		Code	Reserved
Care-of addresses (foreign agent only)			

The field descriptions are as follows:

- ❑ ***Type.*** The 8-bit type field is set to 16.
- ❑ ***Length.*** The 8-bit length field defines the total length of the extension message (not the length of the ICMP advertisement message).
- ❑ ***Sequence number.*** The 16-bit sequence number field holds the message number. The recipient can use the sequence number to determine if a message is lost.
- ❑ ***Lifetime.*** The lifetime field defines the number of seconds that the agent will accept requests. If the value is a string of 1s, the lifetime is infinite.
- ❑ ***Code.*** The code field is an 8-bit flag in which each bit is set (1) or unset (0). The meanings of the bits are shown in Table 19.1.

Table 19.1 *Code Bits*

Bit	*Meaning*
0	Registration required. No collocated care-of address.
1	Agent is busy and does not accept registration at this moment.
2	Agent acts as a home agent.
3	Agent acts as a foreign agent.
4	Agent uses minimal encapsulation.
5	Agent uses generic routing encapsulation (GRE).
6	Agent supports header compression.
7	Unused (0).

- ***Care-of Addresses.*** This field contains a list of addresses available for use as care-of addresses. The mobile host can choose one of these addresses. The selection of this care-of address is announced in the registration request. Note that this field is used only by a foreign agent.

Agent Solicitation

When a mobile host has moved to a new network and has not received agent advertisements, it can initiate an *agent solicitation.* It can use the ICMP solicitation message to inform an agent that it needs assistance.

> **Mobile IP does not use a new packet type for agent solicitation; it uses the router solicitation packet of ICMP.**

Registration

The second phase in mobile communication is *registration.* After a mobile host has moved to a foreign network and discovered the foreign agent, it must register. There are four aspects of registration:

1. The mobile host must register itself with the foreign agent.
2. The mobile host must register itself with its home agent. This is normally done by the foreign agent on behalf of the mobile host.
3. The mobile host must renew registration if it has expired.
4. The mobile host must cancel its registration (deregistration) when it returns home.

Request and Reply

To register with the foreign agent and the home agent, the mobile host uses a *registration request* and a registration reply as shown in Figure 19.14.

Registration Request A registration request is sent from the mobile host to the foreign agent to register its care-of address and also to announce its home address and home agent address. The foreign agent, after receiving and registering the request, relays the message to the home agent. Note that the home agent now knows the address of the foreign agent because the IP packet that is used for relaying has the IP address of the foreign agent as the source address. Figure 19.16 shows the format of the registration request.

Figure 19.16 *Registration request format*

Type	Flag	Lifetime
Home address		
Home agent address		
Care-of address		
Identification		
Extensions ...		

The field descriptions are as follows:

- ❑ ***Type.*** The 8-bit type field defines the type of message. For a request message the value of this field is 1.
- ❑ ***Flag.*** The 8-bit flag field defines forwarding information. The value of each bit can be set or unset. The meaning of each bit is given in Table 19.2.

Table 19.2 *Registration request flag field bits*

Bit	*Meaning*
0	Mobile host requests that home agent retain its prior care-of address.
1	Mobile host requests that home agent tunnel any broadcast message.
2	Mobile host is using collocated care-of address.
3	Mobile host requests that home agent use minimal encapsulation.
4	Mobile host requests generic routing encapsulation (GRE).
5	Mobile host requests header compression.
6–7	Reserved bits.

- ❑ ***Lifetime.*** This field defines the number of seconds the registration is valid. If the field is a string of 0s, the request message is asking for deregistration. If the field is a string of 1s, the lifetime is infinite.
- ❑ ***Home address.*** This field contains the permanent (first) address of the mobile host.
- ❑ ***Home agent address.*** This field contains the address of the home agent.
- ❑ ***Care-of address.*** This field is the temporary (second) address of the mobile host.
- ❑ ***Identification.*** This field contains a 64-bit number that is inserted into the request by the mobile host and repeated in the reply message. It matches a request with a reply.
- ❑ ***Extensions.*** Variable length extensions are used for authentication. They allow a home agent to authenticate the mobile agent. We discuss authentication in Chapter 31.

Registration Reply A registration reply is sent from the home agent to the foreign agent and then relayed to the mobile host. The reply confirms or denies the registration request. Figure 19.17 shows the format of the registration reply.

The fields are similar to those of the registration request with the following exceptions. The value of the type field is 3. The code field replaces the flag field and shows the result of the registration request (acceptance or denial). The care-of address field is not needed.

Encapsulation

Registration messages are encapsulated in a UDP user datagram. An agent uses the well-known port 434; a mobile host uses an ephemeral port.

Data Transfer

After agent discovery and registration, a mobile host can communicate with a remote host. Figure 19.18 shows the idea.

Figure 19.17 *Registration reply format*

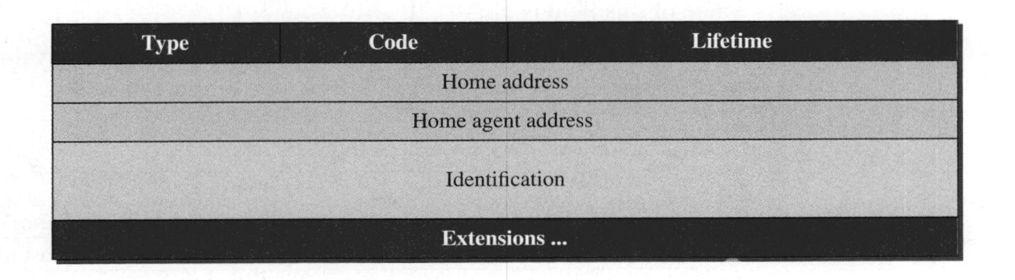

Type	Code	Lifetime
Home address		
Home agent address		
Identification		
Extensions ...		

A registration request or reply is sent by UDP using the well-known port 434.

Figure 19.18 *Data transfer*

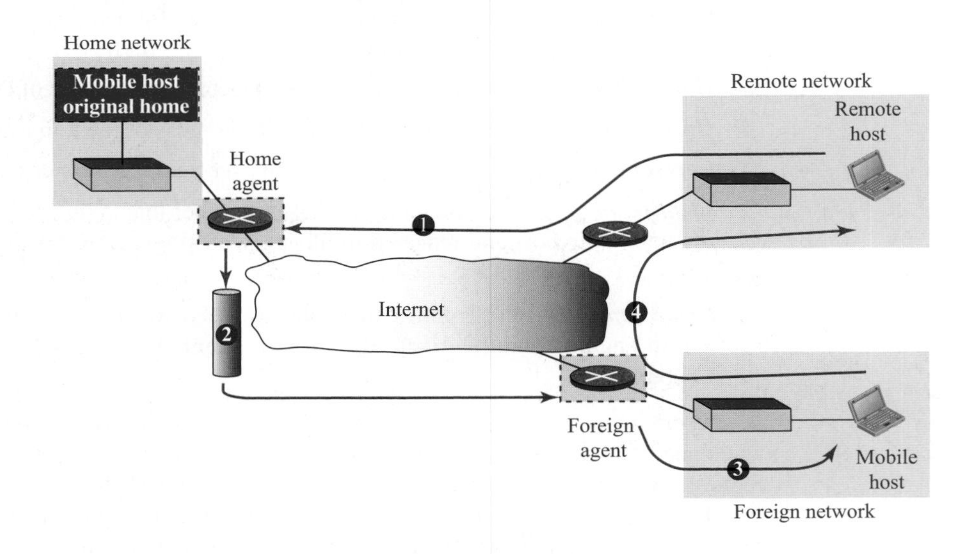

From Remote Host to Home Agent

When a remote host wants to send a packet to the mobile host, it uses its address as the source address and the home address of the mobile host as the destination address. In other words, the remote host sends a packet as though the mobile host is at its home network. The packet, however, is intercepted by the home agent, which pretends it is the mobile host. This is done using the proxy ARP technique discussed in Chapter 9. Path 1 of Figure 19.18 shows this step.

From Home Agent to Foreign Agent
After receiving the packet, the home agent sends the packet to the foreign agent, using the tunneling concept discuss in Chapter 21. The home agent encapsulates the whole IP packet inside another IP packet using its address as the source and the foreign agent's address as the destination. Path 2 of Figure 19.18 shows this step.

From Foreign Agent to Mobile Host
When the foreign agent receives the packet, it removes the original packet. However, since the destination address is the home address of the mobile host, the foreign agent consults a registry table to find the care-of address of the mobile host. (Otherwise, the packet would just be sent back to the home network.) The packet is then sent to the care-of address. Path 3 of Figure 19.18 shows this step.

From Mobile Host to Remote Host
When a mobile host wants to send a packet to a remote host (for example, a response to the packet it has received), it sends as it does normally. The mobile host prepares a packet with its home address as the source, and the address of the remote host as the destination. Although the packet comes from the foreign network, it has the home address of the mobile host. Path 4 of Figure 19.18 shows this step.

Transparency
In this data transfer process, the remote host is unaware of any movement by the mobile host. The remote host sends packets using the home address of the mobile host as the destination address; it receives packets that have the home address of the mobile host as the source address. The movement is totally transparent. The rest of the Internet is not aware of the movement of the mobile host.

The movement of the mobile host is transparent to the rest of the Internet.

19.3.4 Inefficiency in Mobile IP

Communication involving mobile IP can be inefficient. The inefficiency can be severe or moderate. The severe case is called *double crossing* or *2X*. The moderate case is called *triangle routing* or *dog-leg routing*.

Double Crossing

Double crossing occurs when a remote host communicates with a mobile host that has moved to the same network (or site) as the remote host (see Figure 19.19).

When the mobile host sends a packet to the remote host, there is no inefficiency; the communication is local. However, when the remote host sends a packet to the mobile host, the packet crosses the Internet twice.

Since a computer usually communicates with other local computers (principle of locality), the inefficiency from double crossing is significant.

Triangle Routing

Triangle routing, the less severe case, occurs when the remote host communicates with a mobile host that is not attached to the same network (or site) as the mobile host. When the mobile host sends a packet to the remote host, there is no inefficiency.

Figure 19.19 *Double crossing*

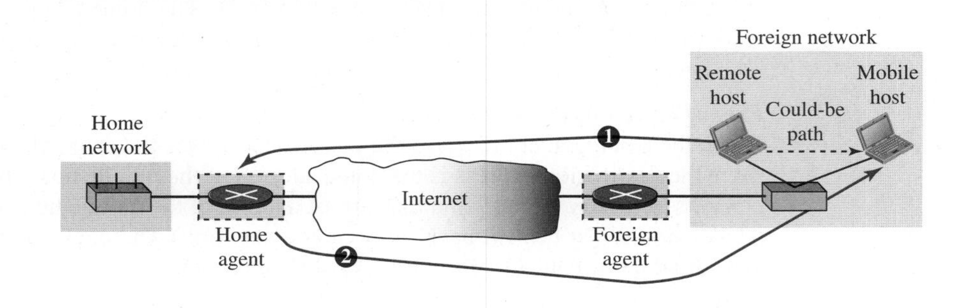

However, when the remote host sends a packet to the mobile host, the packet goes from the remote host to the home agent and then to the mobile host. The packet travels the two sides of a triangle, instead of just one side (see Figure 19.20).

Figure 19.20 *Triangle routing*

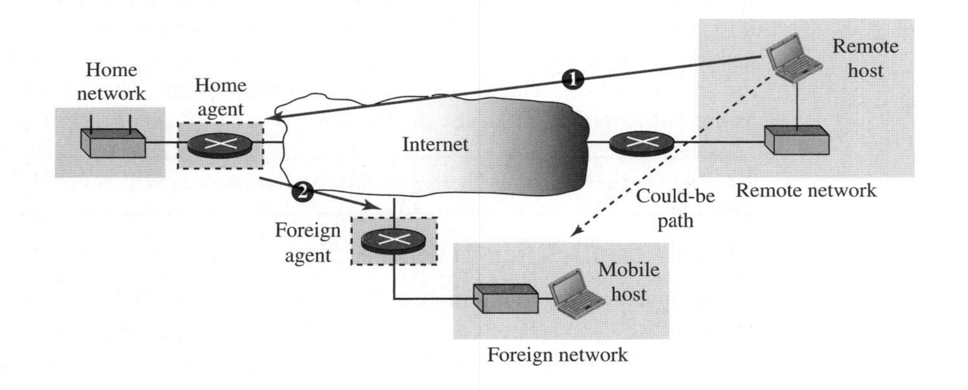

Solution

One solution to inefficiency is for the remote host to bind the care-of address to the home address of a mobile host. For example, when a home agent receives the first packet for a mobile host, it forwards the packet to the foreign agent; it could also send an *update binding packet* to the remote host so that future packets to this host could be sent to the care-of address. The remote host can keep this information in a cache.

The problem with this strategy is that the cache entry becomes outdated once the mobile host moves. In this case the home agent needs to send a *warning packet* to the remote host to inform it of the change.

19.4 END-CHAPTER MATERIALS

19.4.1 Recommended Reading

Books

Several books give thorough coverage of materials discussed in this chapter. We recommend [Com 06], [Tan 03], [Koz 05], [Ste 95], [GW 04], [Per 00], [Kes 02], [Moy 98], [WZ 01], and [Los 04].

RFCs

IPv4 protocol is discussed in RFCs 791, 815, 894, 1122, 2474, and 2475. ICMP is discussed in RFCs 792, 950, 956, 957, 1016, 1122, 1256, 1305, and 1987.

19.4.2 Key Terms

care-of address
collocated care-of address
double crossing
foreign agent
foreign network
fragmentation
home address
home agent
home network
Internet Control Message Protocol version 4 (ICMPv4)
mobile host
stationary host
triangle routing

19.4.3 Summary

IPv4 is an unreliable connectionless protocol responsible for source-to-destination delivery. Packets in the IP layer are called datagrams. An IPv4 datagram is made of a header, of size 20 to 60 bytes, and a payload. The total size of an IPv4 datagram can be up to 65,535 bytes. An IPv4 datagram can be fragmented, one or more times, during its path from the source to the destination; reassembly of the fragments, however, should be done at the destination. The checksum for a datagram is calculated only for the header.

The Internet Control Message Protocol version 4 (ICMPv4) supports the unreliable and connectionless Internet Protocol (IP). ICMPv4 messages are encapsulated in IP datagrams. There are two categories of ICMPv4 messages: error-reporting and query messages. The error-reporting messages report problems that a router or a host (destination) may encounter when it processes an IP packet. The query messages, which occur in pairs, help a host or a network manager get specific information from a router or another host.

Mobile IP, designed for mobile communication, is an enhanced version of the Internet Protocol (IP). A mobile host has a home address on its home network and a care-of address on its foreign network. When the mobile host is on a foreign network, a home agent relays messages (for the mobile host) to a foreign agent. A foreign agent sends relayed messages to a mobile host.

19.5 PRACTICE SET

19.5.1 Quizzes

A set of interactive quizzes for this chapter can be found on the book website. It is strongly recommended that the student take the quizzes to check his/her understanding of the materials before continuing with the practice set.

19.5.2 Questions

Q19-1. Compare and contrast the protocol field at the network layer with the port numbers at the transport layer. What is their common purpose? Why do we need two port-number fields but only one protocol field? Why is the size of the protocol field only half the size of each port number?

Q19-2. Assume a destination computer receives several packets from a source. How can it be sure that the fragments belonging to a datagram are not mixed with the fragments belonging to another datagram?

Q19-3. Can the value of the header length field in an IPv4 packet be less than 5? When is it exactly 5?

Q19-4. Which protocol is the carrier of the agent advertisement and solicitation messages?

Q19-5. In an IPv4 datagram, the value of the header-length (HLEN) field is $(6)_{16}$. How many bytes of options have been added to the packet?

Q19-6. What are the source and destination IP addresses in a datagram that carries the ICMPv4 message reported by a router?

Q19-7. Explain why the Internet does not create a report message to report the error in an IP datagram that carries an ICMPv4 message.

Q19-8. A host is sending 100 datagrams to another host. If the identification number of the first datagram is 1024, what is the identification number of the last?

Q19-9. Explain why the registration request and reply are not directly encapsulated in an IP datagram. Why is there a need for the UDP user datagram?

Q19-10. Which field(s) in the datagram is(are) responsible for gluing together all fragments belonging to an original datagram?

Q19-11. An IP fragment has arrived with an offset value of 100. How many bytes of data were originally sent by the source before the data in this fragment?

Q19-12. Mention the three auxiliary protocols at the network layer of the TCP/IP suite that are designed to help the IPv4 protocol.

Q19-13. Discuss how the ICMPv4 router solicitation message can also be used for agent solicitation. Why are there no extra fields?

Q19-14. Can each of the following be the value of the TTL in a datagram? Explain your answer.

a. 23 **b.** 0 **c.** 1 **d.** 301

Q19-15. Can each of the following be the value of the offset field in a datagram? Explain your answer.

a. 8 **b.** 31 **c.** 73 **d.** 56

Q19-16. Is registration required if the mobile host acts as a foreign agent? Explain your answer.

19.5.3 Problems

P19-1. Determine if a datagram with the following information is a first fragment, a middle fragment, a last fragment, or the only fragment (no fragmentation):

a. M bit is set to 1 and the value of the offset field is zero.

b. M bit is set to 1 and the value of the offset field is nonzero.

P19-2. A packet has arrived in which the offset value is 300 and the payload size is 100 bytes. What are the number of the first byte and the last byte?

P19-3. In Figure 19.4,

a. show how wrapped sum can be calculated from the sum using modular arithmetic.

b. show how checksum can be calculated from the wrapped sum using modular arithmetic.

P19-4. An IP datagram has arrived with the following partial information in the header (in hexadecimal):

```
45000054 00030000 2006...
```

a. What is the header size?

b. Are there any options in the packet?

c. What is the size of the data?

d. Is the packet fragmented?

e. How many more routers can the packet travel to?

f. What is the protocol number of the payload being carried by the packet?

P19-5. Create a home agent advertisement message using 1456 as the sequence number and a lifetime of 3 hours. Select your own values for the bits in the code field. Calculate and insert the value for the length field.

P19-6. Redo the checksum in Figure 19.11 using decimal values and modular arithmetic.

P19-7. Redo the checksum in Figure 19.11 using hexadecimal values.

P19-8. We have the information shown below. Show the contents of the IP datagram header sent from the remote host to the home agent.

Mobile host home address: 130.45.6.7/16
Mobile host care-of address: 14.56.8.9/8
Remote host address: 200.4.7.14/24
Home agent address: 130.45.10.20/16
Foreign agent address: 14.67.34.6/8

P19-9. Briefly describe how we can defeat the following security attacks:

a. packet sniffing **b.** packet modification **c.** IP spoofing

P19-10. Which fields of the IPv4 main header may change from router to router?

P19-11. In Figure 19.4, show how the sum, wrapped sum, and checksum can be calculated when the words are given in decimal numbers (the way the words are stored in a computer memory).

P19-12. In Figure 19.4, show how the sum, wrapped sum, and checksum can be calculated when each word (16 bits) is created instead of waiting for the whole packet to be created.

P19-13. In an IPv4 datagram, the value of total-length field is $(00A0)_{16}$ and the value of the header-length (HLEN) is $(5)_{16}$. How many bytes of payload are being carried by the datagram? What is the efficiency (ratio of the payload length to the total length) of this datagram?

P19-14. Redraw Figure 19.18 for the case where the mobile host acts as a foreign agent.

P19-15. Create a foreign agent advertisement message using 1672 as the sequence number and a lifetime of 4 hours. Select your own values for the bits in the code field. Use at least three care-of addresses of your choice. Calculate and insert the value for the length field.

19.6 SIMULATION EXPERIMENTS

19.6.1 Applets

We have created some Java applets to show some of the main concepts discussed in this chapter. It is strongly recommended that the students activate these applets on the book website and carefully examine the protocols in action.

19.6.2 Lab Assignments

In this chapter, we use *Wireshark* to capture and investigate some packets exchanged at the network layer. We use *Wireshark* and some other computer network administration utilities. See the book website for complete details of lab assignments.

Lab19-1. In the first lab, we investigate IP protocol by capturing and studying IP datagrams.

Lab19-2. In the second lab, we capture and study ICMPv4 packets generated by other utility programs such as *ping* and *traceroute*.

CHAPTER 20

Unicast Routing

In an internet, the goal of the network layer is to deliver a datagram from its source to its destination or destinations. If a datagram is destined for only one destination (one-to-one delivery), we have *unicast routing*. If the datagram is destined for several destinations (one-to-many delivery), we have *multicast routing*.

In the previous chapters, we have shown that the routing can be possible if a router has a forwarding table to forward a packet to the appropriate next node on its way to the final destination or destinations. To make the forwarding tables of the router, the Internet needs routing protocols that will be active all the time in the background and update the forwarding tables.

In this chapter we discuss only unicast routing; multicast routing will be discussed in the next chapter. This chapter is divided into three sections:

- The first section introduces the concept of unicast routing and describes the general ideas behind it. The section then describes least-cost routing and least-cost trees.
- The second section discusses common routing algorithms used in the Internet. The section first describes distance-vector routing. It then describes link-state routing. Finally, it explains path-vector routing.
- The third section explores unicast-routing protocols corresponding to the unicast-routing algorithms discussed in the second section. This section first defines the structure of the Internet as seen by the unicast-routing protocols. It then describes RIP, a protocol that implements the distance-vector routing algorithm. The section next describes OSPF, a protocol that implements the link-state routing algorithm. Finally, the section describes the BGP, a protocol that implements the path-vector routing algorithm.

20.1 INTRODUCTION

Unicast routing in the Internet, with a large number of routers and a huge number of hosts, can be done only by using hierarchical routing: routing in several steps using different routing algorithms. In this section, we first discuss the general concept of unicast routing in an *internet*: an internetwork made of networks connected by routers. After the routing concepts and algorithms are understood, we show how we can apply them to the Internet using hierarchical routing.

20.1.1 General Idea

In unicast routing, a packet is routed, hop by hop, from its source to its destination by the help of forwarding tables. The source host needs no forwarding table because it delivers its packet to the default router in its local network. The destination host needs no forwarding table either because it receives the packet from its default router in its local network. This means that only the routers that glue together the networks in the internet need forwarding tables. With the above explanation, routing a packet from its source to its destination means routing the packet from a *source router* (the default router of the source host) to a *destination router* (the router connected to the destination network). Although a packet needs to visit the source and the destination routers, the question is what other routers the packet should visit. In other words, there are several routes that a packet can travel from the source to the destination; what must be determined is which route the packet should take.

An Internet as a Graph

To find the best route, an internet can be modeled as a *graph*. A graph in computer science is a set of *nodes* and *edges* (lines) that connect the nodes. To model an internet as a graph, we can think of each router as a node and each network between a pair of routers as an edge. An internet is, in fact, modeled as a *weighted graph*, in which each edge is associated with a cost. If a weighted graph is used to represent a geographical area, the nodes can be cities and the edges can be roads connecting the cities; the weights, in this case, are distances between cities. In routing, however, the cost of an edge has a different interpretation in different routing protocols, which we discuss in a later section. For the moment, we assume that there is a cost associated with each edge. If there is no edge between the nodes, the cost is infinity. Figure 20.1 shows how an internet can be modeled as a graph.

20.1.2 Least-Cost Routing

When an internet is modeled as a weighted graph, one of the ways to interpret the *best* route from the source router to the destination router is to find the *least cost* between the two. In other words, the source router chooses a route to the destination router in such a way that the total cost for the route is the least cost among all possible routes. In Figure 20.1, the best route between A and E is A-B-E, with the cost of 6. This means that each router needs to find the least-cost route between itself and all the other routers to be able to route a packet using this criteria.

Figure 20.1 *An internet and its graphical representation*

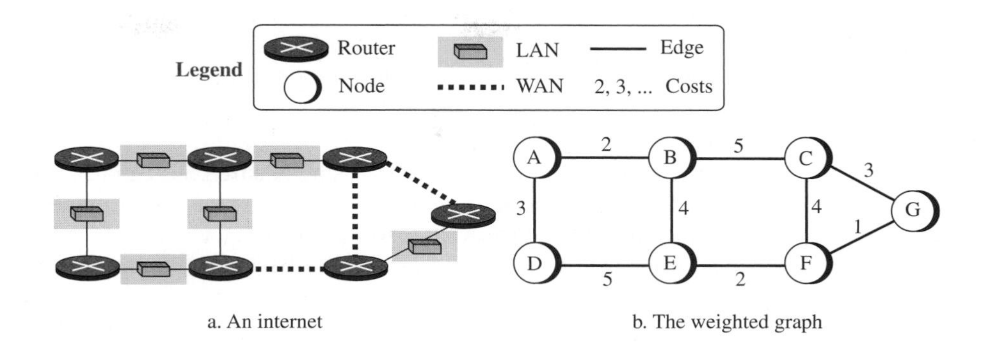

Least-Cost Trees

If there are N routers in an internet, there are $(N - 1)$ least-cost paths from each router to any other router. This means we need $N \times (N - 1)$ least-cost paths for the whole internet. If we have only 10 routers in an internet, we need 90 least-cost paths. A better way to see all of these paths is to combine them in a **least-cost tree**. A least-cost tree is a tree with the source router as the root that spans the whole graph (visits all other nodes) and in which the path between the root and any other node is the shortest. In this way, we can have only one shortest-path tree for each node; we have N least-cost trees for the whole internet. We show how to create a least-cost tree for each node later in this section; for the moment, Figure 20.2 shows the seven least-cost trees for the internet in Figure 20.1.

Figure 20.2 *Least-cost trees for nodes in the internet of Figure 20.1*

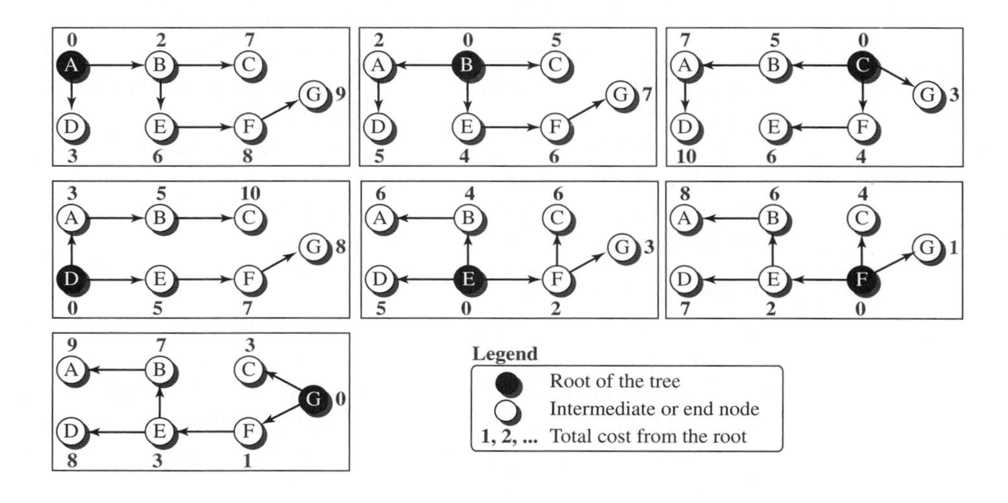

The least-cost trees for a weighted graph can have several properties if they are created using consistent criteria.

1. The least-cost route from X to Y in X's tree is the inverse of the least-cost route from Y to X in Y's tree; the cost in both directions is the same. For example, in Figure 20.2, the route from A to F in A's tree is (A → B → E → F), but the route from F to A in F's tree is (F → E → B → A), which is the inverse of the first route. The cost is 8 in each case.
2. Instead of travelling from X to Z using X's tree, we can travel from X to Y using X's tree and continue from Y to Z using Y's tree. For example, in Figure 20.2, we can go from A to G in A's tree using the route (A → B → E → F → G). We can also go from A to E in A's tree (A → B → E) and then continue in E's tree using the route (E → F → G). The combination of the two routes in the second case is the same route as in the first case. The cost in the first case is 9; the cost in the second case is also 9 (6 + 3).

20.2 ROUTING ALGORITHMS

After discussing the general idea behind least-cost trees and the forwarding tables that can be made from them, now we concentrate on the routing algorithms. Several routing algorithms have been designed in the past. The differences between these methods are in the way they interpret the least cost and the way they create the least-cost tree for each node. In this section, we discuss the common algorithms; later we show how a routing protocol in the Internet implements one of these algorithms.

20.2.1 Distance-Vector Routing

The **distance-vector (DV) routing** uses the goal we discussed in the introduction, to find the best route. In distance-vector routing, the first thing each node creates is its own least-cost tree with the rudimentary information it has about its immediate neighbors. The incomplete trees are exchanged between immediate neighbors to make the trees more and more complete and to represent the whole internet. We can say that in distance-vector routing, a router continuously tells all of its neighbors what it knows about the whole internet (although the knowledge can be incomplete).

Before we show how incomplete least-cost trees can be combined to make complete ones, we need to discuss two important topics: the Bellman-Ford equation and the concept of distance vectors, which we cover next.

Bellman-Ford Equation

The heart of distance-vector routing is the famous **Bellman-Ford** equation. This equation is used to find the least cost (shortest distance) between a source node, $\boldsymbol{x}$, and a destination node, $\boldsymbol{y}$, through some intermediary nodes (**a, b, c, . . .**) when the costs between the source and the intermediary nodes and the least costs between the intermediary nodes and the destination are given. The following shows the general case in which D_{ij} is the shortest distance and c_{ij} is the cost between nodes $\boldsymbol{i}$ and $\boldsymbol{j}$.

$$D_{xy} = \min\{(c_{xa} + D_{ay}), (c_{xb} + D_{by}), (c_{xc} + D_{cy}), \ldots\}$$

In distance-vector routing, normally we want to update an existing least cost with a least cost through an intermediary node, such as z, if the latter is shorter. In this case, the equation becomes simpler, as shown below:

$$D_{xy} = \min\{D_{xy}, (c_{xz} + D_{zy})\}$$

Figure 20.3 shows the idea graphically for both cases.

Figure 20.3 *Graphical idea behind Bellman-Ford equation*

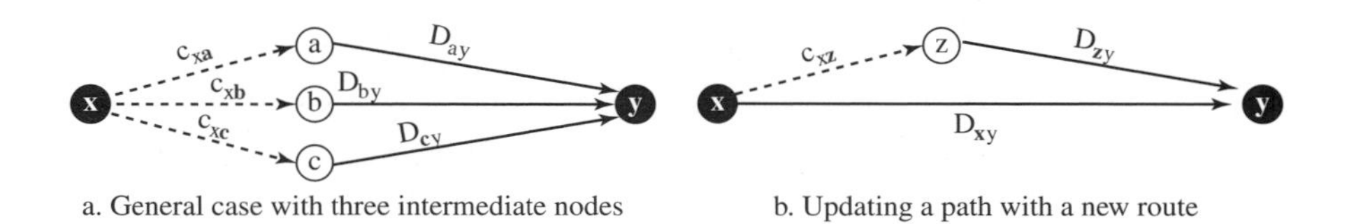

We can say that the Bellman-Ford equation enables us to build a new least-cost path from previously established least-cost paths. In Figure 20.3, we can think of ($a \rightarrow y$), ($b \rightarrow y$), and ($c \rightarrow y$) as previously established least-cost paths and ($x \rightarrow y$) as the new least-cost path. We can even think of this equation as the builder of a new least-cost tree from previously established least-cost trees if we use the equation repeatedly. In other words, the use of this equation in distance-vector routing is a witness that this method also uses least-cost trees, but this use may be in the background.

We will shortly show how we use the Bellman-Ford equation and the concept of distance vectors to build least-cost paths for each node in distance-vector routing, but first we need to discuss the concept of a distance vector.

Distance Vectors

The concept of a **distance vector** is the rationale for the name *distance-vector routing*. A least-cost tree is a combination of least-cost paths from the root of the tree to all destinations. These paths are graphically glued together to form the tree. Distance-vector routing unglues these paths and creates a *distance vector*, a one-dimensional array to represent the tree. Figure 20.4 shows the tree for node A in the internet in Figure 20.1 and the corresponding distance vector.

Note that the *name* of the distance vector defines the root, the *indexes* define the destinations, and the *value* of each cell defines the least cost from the root to the destination. A distance vector does not give the path to the destinations as the least-cost tree does; it gives only the least costs to the destinations. Later we show how we can change a distance vector to a forwarding table, but we first need to find all distance vectors for an internet.

We know that a distance vector can represent least-cost paths in a least-cost tree, but the question is how each node in an internet originally creates the corresponding vector. Each node in an internet, when it is booted, creates a very rudimentary distance vector with the minimum information the node can obtain from its neighborhood. The node sends some greeting messages out of its interfaces and discovers the identity of the immediate neighbors and the distance between itself and each neighbor. It then

Figure 20.4 *The distance vector corresponding to a tree*

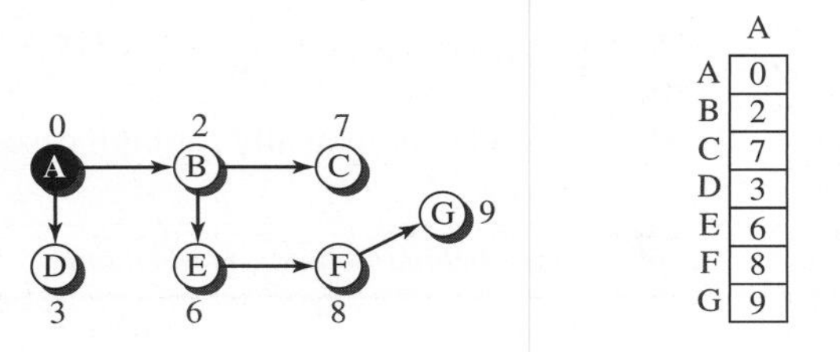

makes a simple distance vector by inserting the discovered distances in the corresponding cells and leaves the value of other cells as infinity. Do these distance vectors represent least-cost paths? They do, considering the limited information a node has. When we know only one distance between two nodes, it is the least cost. Figure 20.5 shows all distance vectors for our internet. However, we need to mention that these vectors are made asynchronously, when the corresponding node has been booted; the existence of all of them in a figure does not mean synchronous creation of them.

Figure 20.5 *The first distance vector for an internet*

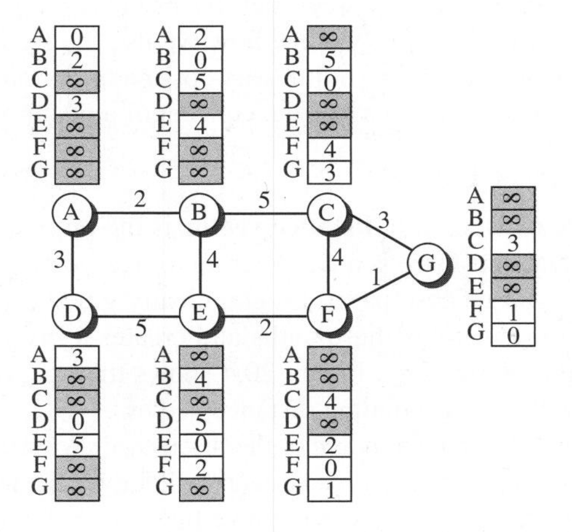

These rudimentary vectors cannot help the internet to effectively forward a packet. For example, node A thinks that it is not connected to node G because the corresponding cell shows the least cost of infinity. To improve these vectors, the nodes in the internet need to help each other by exchanging information. After each node has created its vector, it sends a copy of the vector to all its immediate neighbors. After a node receives a distance vector from a neighbor, it updates its distance vector using the Bellman-Ford equation (second case). However, we need to understand that we need to update, not only one least cost, but *N* of them in which *N* is the number of the nodes in the internet. If we are using a program, we can do this using a loop; if we are showing the concept on paper, we can show the whole vector instead of the *N* separate equations. We show the whole vector instead of seven equations for each update in Figure 20.6. The figure shows two asynchronous events, happening one after another with some time in

Figure 20.6 *Updating distance vectors*

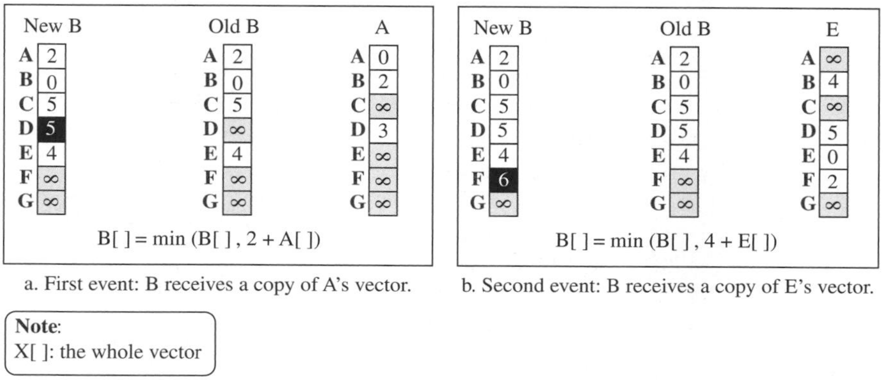

	New B	Old B	A
A	2	2	0
B	0	0	2
C	5	5	∞
D	5	∞	3
E	4	4	∞
F	∞	∞	∞
G	∞	∞	∞

B[ ] = min (B[ ] , 2 + A[ ])

a. First event: B receives a copy of A's vector.

	New B	Old B	E
A	2	2	∞
B	0	0	4
C	5	5	∞
D	5	5	5
E	4	4	0
F	6	∞	2
G	∞	∞	∞

B[ ] = min (B[ ] , 4 + E[ ])

b. Second event: B receives a copy of E's vector.

Note:
X[ ]: the whole vector

between. In the first event, node A has sent its vector to node B. Node B updates its vector using the cost $c_{BA} = 2$. In the second event, node E has sent its vector to node B. Node B updates its vector using the cost $c_{EA} = 4$.

After the first event, node B has one improvement in its vector: its least cost to node D has changed from infinity to 5 (via node A). After the second event, node B has one more improvement in its vector; its least cost to node F has changed from infinity to 6 (via node E). We hope that we have convinced the reader that exchanging vectors eventually stabilizes the system and allows all nodes to find the ultimate least cost between themselves and any other node. We need to remember that after updating a node, it immediately sends its updated vector to all neighbors. Even if its neighbors have received the previous vector, the updated one may help more.

Distance-Vector Routing Algorithm

Now we can give a simplified pseudocode for the distance-vector routing algorithm, as shown in Table 20.1. The algorithm is run by its node independently and asynchronously.

Table 20.1 *Distance-Vector Routing Algorithm for a Node*

```
1  Distance_Vector_Routing ( )
2  {
3       // Initialize (create initial vectors for the node)
4       D[myself] = 0
```

Table 20.1 *Distance-Vector Routing Algorithm for a Node (continued)*

```
5          for (y = 1 to N)
6          {
7              if (y is a neighbor)
8                  D[y] = c[myself][y]
9              else
10                 D[y] = ∞
11         }
12         send vector {D[1], D[2], …, D[N]} to all neighbors
13         // Update (improve the vector with the vector received from a neighbor)
14         repeat (forever)
15         {
16             wait (for a vector Dw from a neighbor w or any change in the link)
17             for (y = 1 to N)
18             {
19                 D[y] = min [D[y], (c[myself][w] + Dw[y ])]     // Bellman-Ford equation
20             }
21             if (any change in the vector)
22                 send vector {D[1], D[2], …, D[N]} to all neighbors
23         }
24     } // End of Distance Vector
```

Lines 4 to 11 initialize the vector for the node. Lines 14 to 23 show how the vector can be updated after receiving a vector from the immediate neighbor. The *for* loop in lines 17 to 20 allows all entries (cells) in the vector to be updated after receiving a new vector. Note that the node sends its vector in line 12, after being initialized, and in line 22, after it is updated.

Count to Infinity

A problem with distance-vector routing is that any decrease in cost (good news) propagates quickly, but any increase in cost (bad news) will propagate slowly. For a routing protocol to work properly, if a link is broken (cost becomes infinity), every other router should be aware of it immediately, but in distance-vector routing, this takes some time. The problem is referred to as *count to infinity.* It sometimes takes several updates before the cost for a broken link is recorded as infinity by all routers.

Two-Node Loop

One example of count to infinity is the two-node loop problem. To understand the problem, let us look at the scenario depicted in Figure 20.7.

The figure shows a system with three nodes. We have shown only the portions of the forwarding table needed for our discussion. At the beginning, both nodes A and B

Figure 20.7 *Two-node instability*

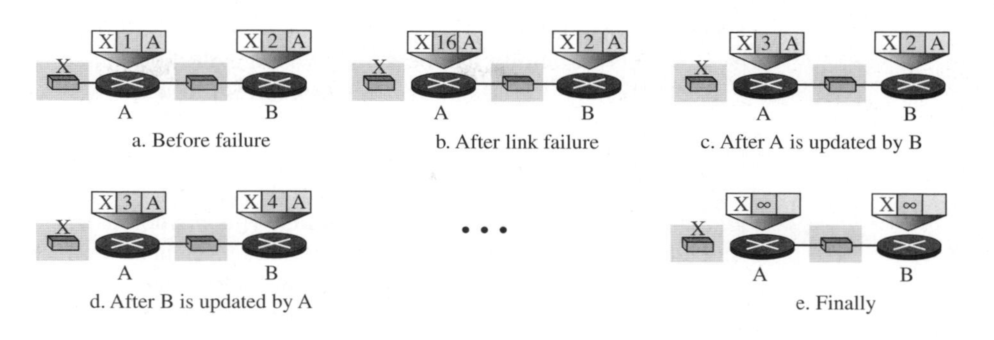

know how to reach node X. But suddenly, the link between A and X fails. Node A changes its table. If A can send its table to B immediately, everything is fine. However, the system becomes unstable if B sends its forwarding table to A before receiving A's forwarding table. Node A receives the update and, assuming that B has found a way to reach X, immediately updates its forwarding table. Now A sends its new update to B. Now B thinks that something has been changed around A and updates its forwarding table. The cost of reaching X increases gradually until it reaches infinity. At this moment, both A and B know that X cannot be reached. However, during this time the system is not stable. Node A thinks that the route to X is via B; node B thinks that the route to X is via A. If A receives a packet destined for X, the packet goes to B and then comes back to A. Similarly, if B receives a packet destined for X, it goes to A and comes back to B. Packets bounce between A and B, creating a two-node loop problem. A few solutions have been proposed for instability of this kind.

Split Horizon

One solution to instability is called ***split horizon***. In this strategy, instead of flooding the table through each interface, each node sends only part of its table through each interface. If, according to its table, node B thinks that the optimum route to reach X is via A, it does not need to advertise this piece of information to A; the information has come from A (A already knows). Taking information from node A, modifying it, and sending it back to node A is what creates the confusion. In our scenario, node B eliminates the last line of its forwarding table before it sends it to A. In this case, node A keeps the value of infinity as the distance to X. Later, when node A sends its forwarding table to B, node B also corrects its forwarding table. The system becomes stable after the first update: both node A and node B know that X is not reachable.

Poison Reverse

Using the split-horizon strategy has one drawback. Normally, the corresponding protocol uses a timer, and if there is no news about a route, the node deletes the route from its table. When node B in the previous scenario eliminates the route to X from its advertisement to A, node A cannot guess whether this is due to the split-horizon strategy (the source of information was A) or because B has not received any news about X recently. In the **poison reverse** strategy B can still advertise the value for X, but if the source of information is A, it can replace the distance with infinity as a warning: "Do not use this value; what I know about this route comes from you."

Three-Node Instability

The two-node instability can be avoided using split horizon combined with poison reverse. However, if the instability is between three nodes, stability cannot be guaranteed.

20.2.2 Link-State Routing

A routing algorithm that directly follows our discussion for creating least-cost trees and forwarding tables is **link-state (LS) routing.** This method uses the term *link-state* to define the characteristic of a link (an edge) that represents a network in the internet. In this algorithm the cost associated with an edge defines the state of the link. Links with lower costs are preferred to links with higher costs; if the cost of a link is infinity, it means that the link does not exist or has been broken.

Link-State Database (LSDB)

To create a least-cost tree with this method, each node needs to have a complete *map* of the network, which means it needs to know the state of each link. The collection of states for all links is called the ***link-state database (LSDB).*** There is only one LSDB for the whole internet; each node needs to have a duplicate of it to be able to create the least-cost tree. Figure 20.8 shows an example of an LSDB for the graph in Figure 20.1. The LSDB can be represented as a two-dimensional array(matrix) in which the value of each cell defines the cost of the corresponding link.

Figure 20.8 *Example of a link-state database*

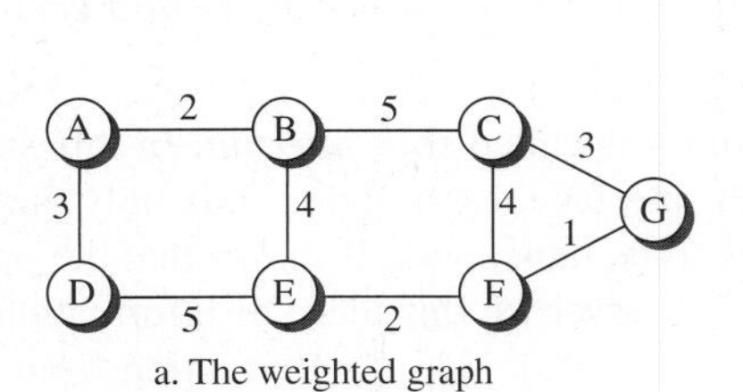

a. The weighted graph

	A	B	C	D	E	F	G
A	0	2	∞	3	∞	∞	∞
B	2	0	5	∞	4	∞	∞
C	∞	5	0	∞	∞	4	3
D	3	∞	∞	0	5	∞	∞
E	∞	4	∞	5	0	2	∞
F	∞	∞	4	∞	2	0	1
G	∞	∞	3	∞	∞	1	0

b. Link state database

Now the question is how each node can create this LSDB that contains information about the whole internet. This can be done by a process called **flooding.** Each node can send some greeting messages to all its immediate neighbors (those nodes to which it is connected directly) to collect two pieces of information for each neighboring node: the identity of the node and the cost of the link. The combination of these two pieces of information is called the *LS packet* (LSP); the LSP is sent out of each interface, as shown in Figure 20.9 for our internet in Figure 20.1. When a node receives an LSP from one of its interfaces, it compares the LSP with the copy it may already have. If the newly arrived LSP is older than the one it has (found by checking the sequence number), it discards the LSP. If it is newer or the first one received, the node discards the old LSP (if there is one) and keeps the received one. It then sends a copy of it out of each interface except the one from which the packet arrived. This guarantees that flooding stops somewhere in the network (where a node has only one interface). We need to convince ourselves that, after receiving all new LSPs, each node creates the comprehensive LSDB as shown in Figure 20.9. This LSDB is the same for each node and shows the whole map of the internet. In other words, a node can make the whole map if it needs to, using this LSDB.

Figure 20.9 *LSPs created and sent out by each node to build LSDB*

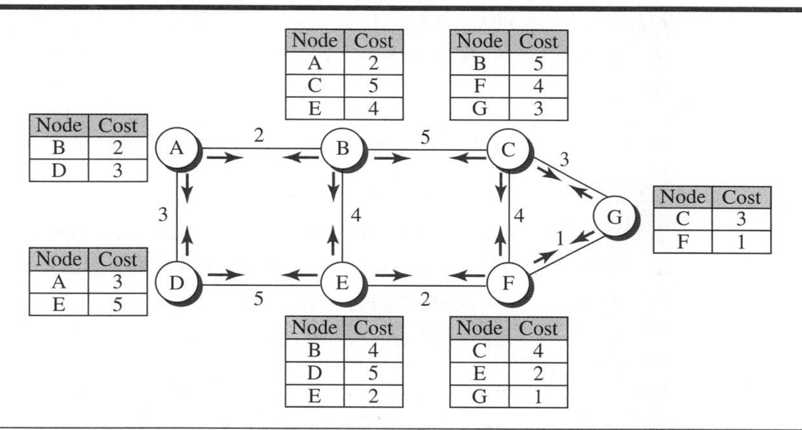

We can compare the link-state routing algorithm with the distance-vector routing algorithm. In the distance-vector routing algorithm, each router tells its neighbors what it knows about the whole internet; in the link-state routing algorithm, each router tells the whole internet what it knows about its neighbors.

Formation of Least-Cost Trees

To create a least-cost tree for itself, using the shared LSDB, each node needs to run the famous **Dijkstra Algorithm.** This iterative algorithm uses the following steps:

1. The node chooses itself as the root of the tree, creating a tree with a single node, and sets the total cost of each node based on the information in the LSDB.
2. The node selects one node, among all nodes not in the tree, which is closest to the root, and adds this to the tree. After this node is added to the tree, the cost of all other nodes not in the tree needs to be updated because the paths may have been changed.
3. The node repeats step 2 until all nodes are added to the tree.

We need to convince ourselves that the above three steps finally create the least-cost tree. Table 20.2 shows a simplified version of Dijkstra's algorithm.

Table 20.2 *Dijkstra's Algorithm*

```
1  Dijkstra's Algorithm ( )
2  {
3      // Initialization
4      Tree = {root}                    // Tree is made only of the root
```

Table 20.2 *Dijkstra's Algorithm (continued)*

```
5          for (y = 1 to N)                          // N is the number of nodes
6          {
7               if (y is the root)
8                    D[y] = 0                        // D[y] is shortest distance from root to node y
9               else if (y is a neighbor)
10                   D[y] = c[root][y]               // c[x][y] is cost between nodes x and y in LSDB
11              else
12                   D[y] = ∞
13         }
14         // Calculation
15         repeat
16         {
17              find a node w, with D[w] minimum among all nodes not in the Tree
18              Tree = Tree ∪ {w}                    // Add w to tree
19              // Update distances for all neighbors of w
20              for (every node x, which is a neighbor of w and not in the Tree)
21              {
22                   D[x] = min {D[x], (D[w] + c[w][x])}
23              }
24         } until (all nodes included in the Tree)
25    } // End of Dijkstra
```

Lines 4 to 13 implement step 1 in the algorithm. Lines 16 to 23 implement step 2 in the algorithm. Step 2 is repeated until all nodes are added to the tree.

Figure 20.10 shows the formation of the least-cost tree for the graph in Figure 20.8 using Dijkstra's algorithm. We need to go through an initialization step and six iterations to find the least-cost tree.

20.2.3 Path-Vector Routing

Both link-state and distance-vector routing are based on the least-cost goal. However, there are instances where this goal is not the priority. For example, assume that there are some routers in the internet that a sender wants to prevent its packets from going through. For example, a router may belong to an organization that does not provide enough security or it may belong to a commercial rival of the sender which might inspect the packets for obtaining information. Least-cost routing does not prevent a packet from passing through an area when that area is in the least-cost path. In other words, the least-cost goal, applied by LS or DV routing, does not allow a sender to apply specific policies to the route a packet may take. Aside from safety and security, there are occasions, as discussed in the next section, in which the goal of routing is merely reachability: to allow the packet to reach its destination more efficiently without assigning costs to the route.

Figure 20.10 *Least-cost tree*

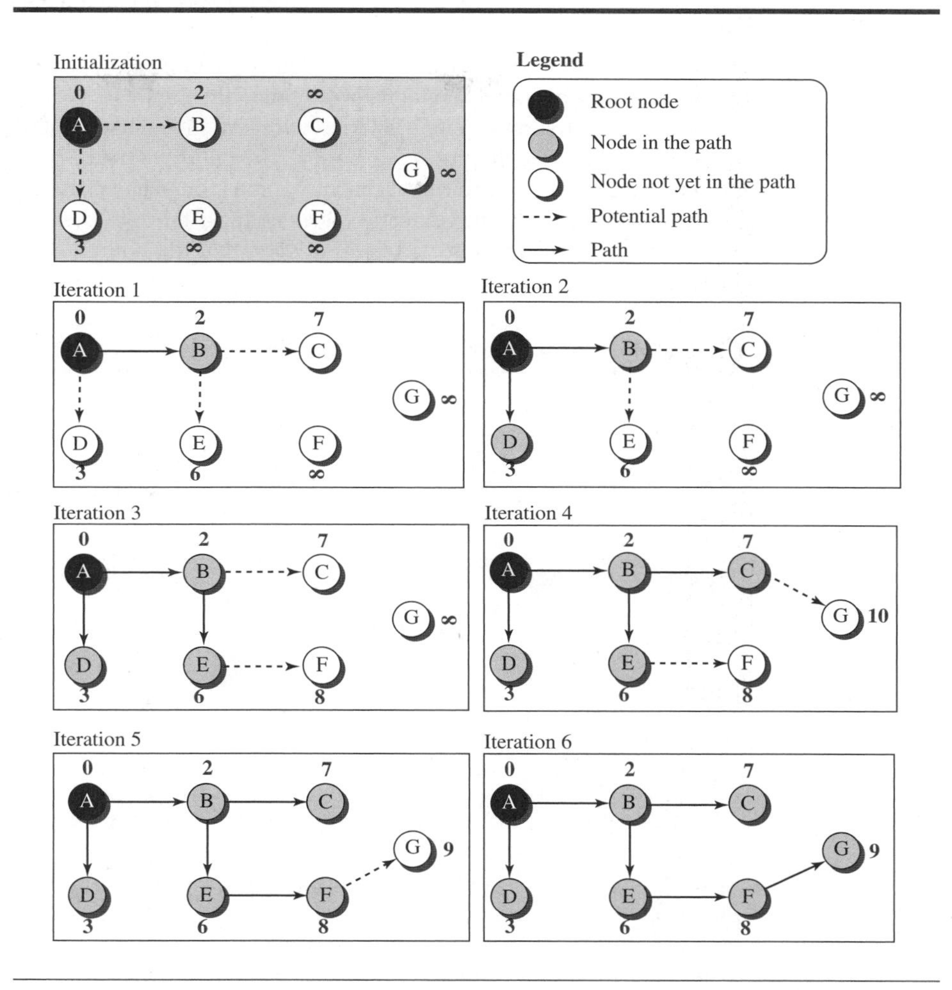

To respond to these demands, a third routing algorithm, called **path-vector (PV) routing** has been devised. Path-vector routing does not have the drawbacks of LS or DV routing as described above because it is not based on least-cost routing. The best route is determined by the source using the policy it imposes on the route. In other words, the source can control the path. Although path-vector routing is not actually used in an internet, and is mostly designed to route a packet between ISPs, we discuss the principle of this method in this section as though applied to an internet. In the next section, we show how it is used in the Internet.

Spanning Trees

In path-vector routing, the path from a source to all destinations is also determined by the *best* spanning tree. The best spanning tree, however, is not the least-cost tree; it is the tree determined by the source when it imposes its own policy. If there is more than one route to a destination, the source can choose the route that meets its policy best. A source may apply several policies at the same time. One of the common policies uses the minimum number of nodes to be visited (something similar to least-cost). Another common policy is to avoid some nodes as the middle node in a route.

Figure 20.11 shows a small internet with only five nodes. Each source has created its own spanning tree that meets its policy. The policy imposed by all sources is to use the minimum number of nodes to reach a destination. The spanning tree selected by A and E is such that the communication does not pass through D as a middle node. Similarly, the spanning tree selected by B is such that the communication does not pass through C as a middle node.

Figure 20.11 *Spanning trees in path-vector routing*

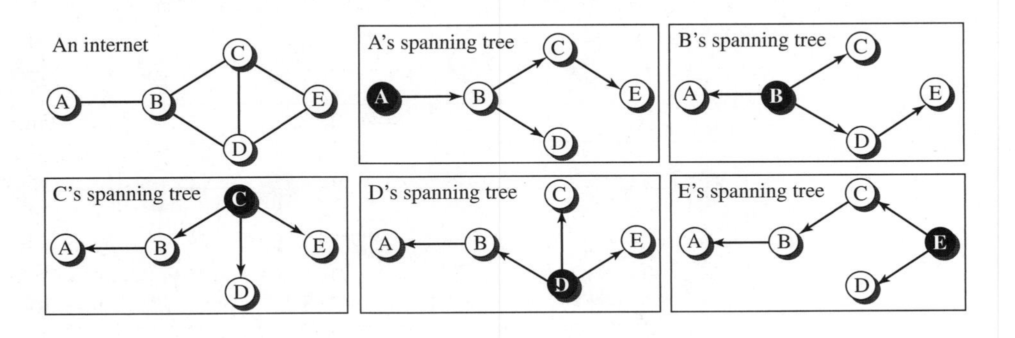

Creation of Spanning Trees

Path-vector routing, like distance-vector routing, is an asynchronous and distributed routing algorithm. The spanning trees are made, gradually and asynchronously, by each node. When a node is booted, it creates a *path vector* based on the information it can obtain about its immediate neighbor. A node sends greeting messages to its immediate neighbors to collect these pieces of information. Figure 20.12 shows all of these path vectors for our internet in Figure 20.11. Note, however, that we do not mean that all of these tables are created simultaneously; they are created when each node is booted. The figure also shows how these path vectors are sent to immediate neighbors after they have been created (arrows).

Each node, after the creation of the initial path vector, sends it to all its immediate neighbors. Each node, when it receives a path vector from a neighbor, updates its path vector using an equation similar to the Bellman-Ford, but applying its own policy instead of looking for the least cost. We can define this equation as

$$\text{Path}(\boldsymbol{x}, \boldsymbol{y}) = \text{best}\ \{\text{Path}(x, y), [(\boldsymbol{x} + \text{Path}(\mathbf{v}, \boldsymbol{y})]\} \qquad \text{for all } \mathbf{v}\text{'s in the internet.}$$

In this equation, the operator (+) means to add $\boldsymbol{x}$ to the beginning of the path. We also need to be cautious to avoid adding a node to an empty path because an empty path means one that does not exist.

Figure 20.12 *Path vectors made at booting time*

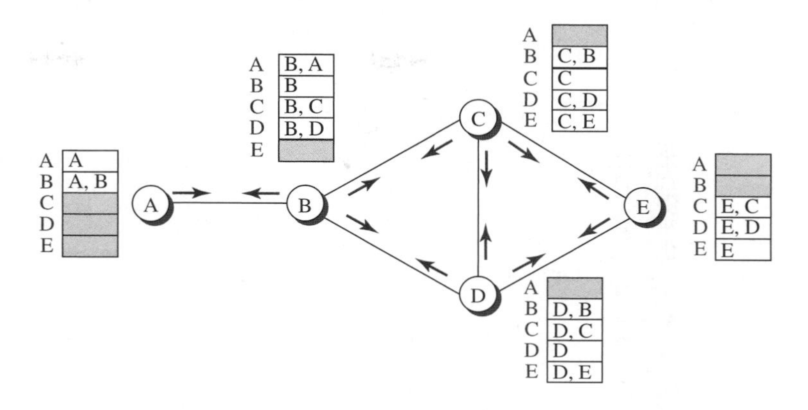

The policy is defined by selecting the *best* of multiple paths. Path-vector routing also imposes one more condition on this equation: If Path (**v**, y) includes x, that path is discarded to avoid a loop in the path. In other words, x does not want to visit itself when it selects a path to y.

Figure 20.13 shows the path vector of node C after two events. In the first event, node C receives a copy of B's vector, which improves its vector: now it knows how to reach node A. In the second event, node C receives a copy of D's vector, which does not change its vector. As a matter of fact the vector for node C after the first event is stabilized and serves as its forwarding table.

Figure 20.13 *Updating path vectors*

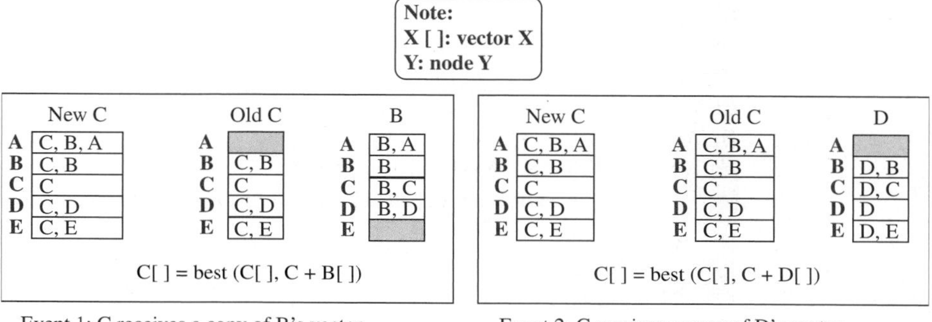

Path-Vector Algorithm

Based on the initialization process and the equation used in updating each forwarding table after receiving path vectors from neighbors, we can write a simplified version of the path vector algorithm as shown in Table 20.3.

Table 20.3 *Path-vector algorithm for a node*

```
1   Path_Vector_Routing ( )
2   {
3       // Initialization
4       for (y = 1 to N)
5       {
6           if (y is myself)
7               Path [y] = myself
8           else if (y is a neighbor)
9               Path [y] = myself + neighbor node
10          else
11              Path [y] = empty
12      }
13      Send vector {Path[1], Path[2], …, Path[y]} to all neighbors
14      // Update
15      repeat (forever)
16      {
17          wait (for a vector Path_w from a neighbor w)
18          for (y = 1 to N)
19          {
20              if (Path_w includes myself)
21                  discard the path                    // Avoid any loop
22              else
23                  Path[y] = best {Path[y], (myself + Path_w[y])}
24          }
25          If (there is a change in the vector)
26              Send vector {Path[1], Path[2], …, Path[y]} to all neighbors
27      }
28  } // End of Path Vector
```

Lines 4 to 12 show the initialization for the node. Lines 17 to 24 show how the node updates its vector after receiving a vector from the neighbor. The update process is repeated forever. We can see the similarities between this algorithm and the DV algorithm.

20.3 UNICAST ROUTING PROTOCOLS

In the previous section, we discussed unicast routing algorithms; in this section, we discuss unicast routing protocols used in the Internet. Although three protocols we discuss here are based on the corresponding algorithms we discussed before, a protocol is more than an algorithm. A protocol needs to define its domain of operation, the messages exchanged, communication between routers, and interaction with protocols in other domains. After an introduction, we discuss three common protocols used in the Internet: Routing Information Protocol (RIP), based on the distance-vector algorithm, Open Shortest Path First (OSPF), based on the link-state algorithm, and Border Gateway Protocol (BGP), based on the path-vector algorithm.

20.3.1 Internet Structure

Before discussing unicast routing protocols, we need to understand the structure of today's Internet. The Internet has changed from a tree-like structure, with a single backbone, to a multi-backbone structure run by different private corporations today. Although it is difficult to give a general view of the Internet today, we can say that the Internet has a structure similar to what is shown in Figure 20.14.

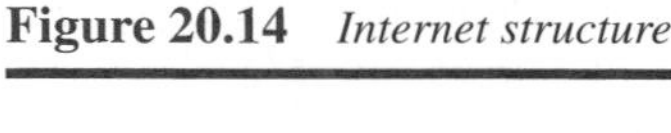

Figure 20.14 *Internet structure*

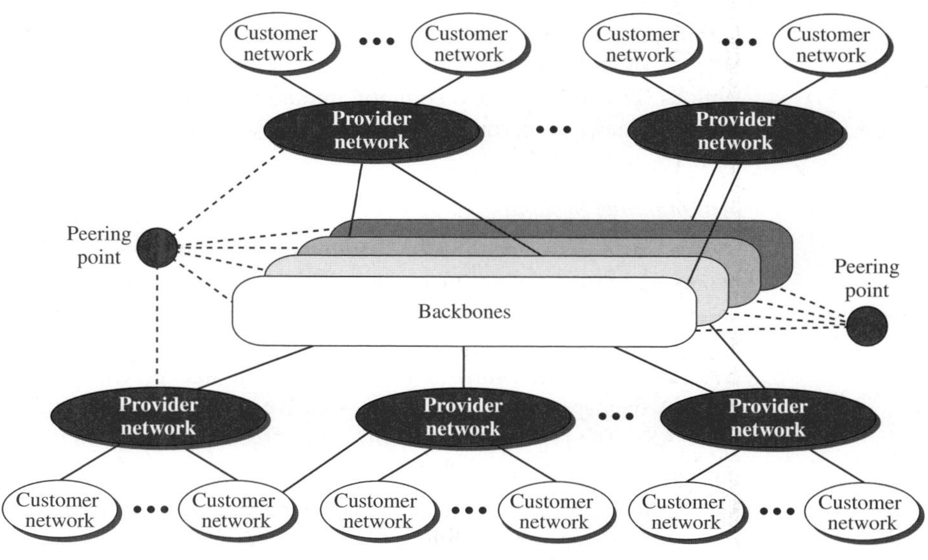

There are several *backbones* run by private communication companies that provide global connectivity. These backbones are connected by some *peering points* that allow connectivity between backbones. At a lower level, there are some *provider networks* that use the backbones for global connectivity but provide services to Internet customers.

Finally, there are some customer *networks* that use the services provided by the provider networks. Any of these three entities (backbone, provider network, or customer network) can be called an Internet Service Provider or ISP. They provide services, but at different levels.

Hierarchical Routing

The Internet today is made of a huge number of networks and routers that connect them. It is obvious that routing in the Internet cannot be done using a single protocol for two reasons: a scalability problem and an administrative issue. *Scalability problem* means that the size of the forwarding tables becomes huge, searching for a destination in a forwarding table becomes time-consuming, and updating creates a huge amount of traffic. The *administrative issue* is related to the Internet structure described in Figure 20.14. As the figure shows, each ISP is run by an administrative authority. The administrator needs to have control in its system. The organization must be able to use as many subnets and routers as it needs, may desire that the routers be from a particular manufacturer, may wish to run a specific routing algorithm to meet the needs of the organization, and may want to impose some policy on the traffic passing through its ISP.

Hierarchical routing means considering each ISP as an **autonomous system (AS).** Each AS can run a routing protocol that meets its needs, but the global Internet runs a global protocol to glue all ASs together. The routing protocol run in each AS is referred to as *intra-AS routing protocol*, *intradomain routing protocol*, or *interior gateway protocol (IGP)*; the global routing protocol is referred to as *inter-AS routing protocol, interdomain routing protocol,* or *exterior gateway protocol (EGP)*. We can have several intradomain routing protocols, and each AS is free to choose one, but it should be clear that we should have only one interdomain protocol that handles routing between these entities. Presently, the two common intradomain routing protocols are RIP and OSPF; the only interdomain routing protocol is BGP. The situation may change when we move to IPv6.

Autonomous Systems

As we said before, each ISP is an autonomous system when it comes to managing networks and routers under its control. Although we may have small, medium-size, and large ASs, each AS is given an autonomous number (ASN) by the ICANN. Each ASN is a 16-bit unsigned integer that uniquely defines an AS. The autonomous systems, however, are not categorized according to their size; they are categorized according to the way they are connected to other ASs. We have stub ASs, multihomed ASs, and transient ASs. The type, as we see will later, affects the operation of the interdomain routing protocol in relation to that AS.

- ***Stub AS.*** A stub AS has only one connection to another AS. The data traffic can be either initiated or terminated in a stub AS; the data cannot pass through it. A good example of a stub AS is the customer network, which is either the source or the sink of data.
- ***Multihomed AS.*** A multihomed AS can have more than one connection to other ASs, but it does not allow data traffic to pass through it. A good example of such an AS is some of the customer ASs that may use the services of more than one provider network, but their policy does not allow data to be passed through them.

- ***Transient AS.*** A transient AS is connected to more than one other AS and also allows the traffic to pass through. The provider networks and the backbone are good examples of transient ASs.

20.3.2 Routing Information Protocol (RIP)

The **Routing Information Protocol (RIP)** is one of the most widely used intradomain routing protocols based on the distance-vector routing algorithm we described earlier. RIP was started as part of the Xerox Network System (XNS), but it was the Berkeley Software Distribution (BSD) version of UNIX that helped make the use of RIP widespread.

Hop Count

A router in this protocol basically implements the distance-vector routing algorithm shown in Table 20.1. However, the algorithm has been modified as described below. First, since a router in an AS needs to know how to forward a packet to different networks (subnets) in an AS, RIP routers advertise the cost of reaching different networks instead of reaching other nodes in a theoretical graph. In other words, the cost is defined between a router and the network in which the destination host is located. Second, to make the implementation of the cost simpler (independent from performance factors of the routers and links, such as delay, bandwidth, and so on), the cost is defined as the number of hops, which means the number of networks (subnets) a packet needs to travel through from the source router to the final destination host. Note that the network in which the source host is connected is not counted in this calculation because the source host does not use a forwarding table; the packet is delivered to the default router. Figure 20.15 shows the concept of hop count advertised by three routers from a source host to a destination host. In RIP, the maximum cost of a path can be 15, which means 16 is considered as infinity (no connection). For this reason, RIP can be used only in autonomous systems in which the diameter of the AS is not more than 15 hops.

Figure 20.15 *Hop counts in RIP*

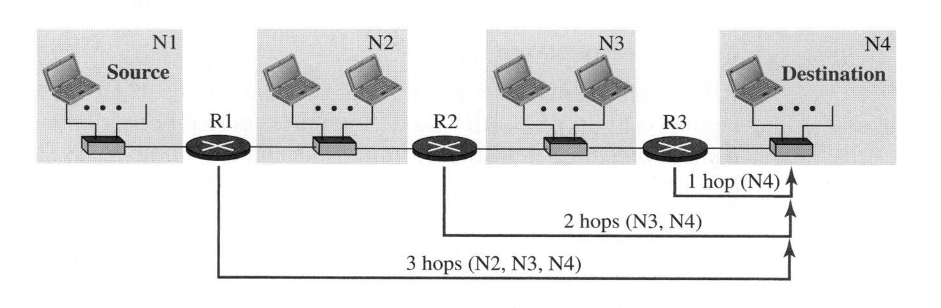

Forwarding Tables

Although the distance-vector algorithm we discussed in the previous section is concerned with exchanging distance vectors between neighboring nodes, the routers in an autonomous system need to keep forwarding tables to forward packets to their destination networks. A forwarding table in RIP is a three-column table in which the first column is the address of the destination network, the second column is the address of the next router to which the packet should be forwarded, and the third column is the cost (the number of hops) to reach the destination network. Figure 20.16 shows the three forwarding tables for the routers in Figure 20.15. Note that the first and the third columns together convey the same information as does a distance vector, but the cost shows the number of hops to the destination networks.

Figure 20.16 *Forwarding tables*

Forwarding table for R1

Destination network	Next router	Cost in hops
N1	——	1
N2	——	1
N3	R2	2
N4	R2	3

Forwarding table for R2

Destination network	Next router	Cost in hops
N1	R1	2
N2	——	1
N3	——	1
N4	R3	2

Forwarding table for R3

Destination network	Next router	Cost in hops
N1	R2	3
N2	R2	2
N3	——	1
N4	——	1

Although a forwarding table in RIP defines only the next router in the second column, it gives the information about the whole least-cost tree based on the second property of these trees, discussed in the previous section. For example, R1 defines that the next router for the path to N4 is R2; R2 defines that the next router to N4 is R3; R3 defines that there is no next router for this path. The tree is then R1 → R2 → R3 → N4.

A question often asked about the forwarding table is what the use of the third column is. The third column is not needed for forwarding the packet, but it is needed for updating the forwarding table when there is a change in the route, as we will see shortly.

RIP Implementation

RIP is implemented as a process that uses the service of UDP on the well-known port number 520. In BSD, RIP is a daemon process (a process running in the background), named *routed* (abbreviation for *route daemon* and pronounced *route-dee).* This means that, although RIP is a routing protocol to help IP route its datagrams through the AS, the RIP messages are encapsulated inside UDP user datagrams, which in turn are encapsulated inside IP datagrams. In other words, RIP runs at the application layer, but creates forwarding tables for IP at the network later.

RIP has gone through two versions: RIP-1 and RIP-2. The second version is backward compatible with the first section; it allows the use of more information in the RIP messages that were set to 0 in the first version. We discuss only RIP-2 in this section.

RIP Messages

Two RIP processes, a client and a server, like any other processes, need to exchange messages. RIP-2 defines the format of the message, as shown in Figure 20.17. Part of the message, which we call *entry*, can be repeated as needed in a message. Each entry carries the information related to one line in the forwarding table of the router that sends the message.

Figure 20.17 *RIP message format*

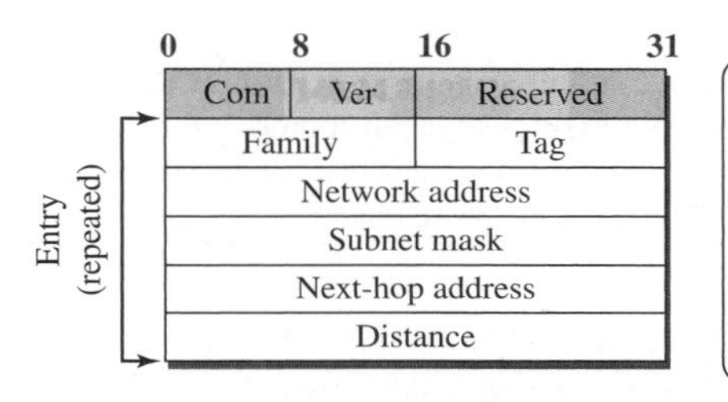

Fields

Com: Command, request (1), response (2)
Ver: Version, current version is 2
Family: Family of protocol, for TCP/IP value is 2
Tag: Information about autonomous system
Network address: Destination address
Subnet mask: Prefix length
Next-hop address: Address length
Distance: Number of hops to the destination

RIP has two types of messages: request and response. A request message is sent by a router that has just come up or by a router that has some time-out entries. A request message can ask about specific entries or all entries. A response (or update) message can be either solicited or unsolicited. A solicited response message is sent only in answer to a request message. It contains information about the destination specified in the corresponding request message. An unsolicited response message, on the other hand, is sent periodically, every 30 seconds or when there is a change in the forwarding table.

RIP Algorithm

RIP implements the same algorithm as the distance-vector routing algorithm we discussed in the previous section. However, some changes need to be made to the algorithm to enable a router to update its forwarding table:

- Instead of sending only distance vectors, a router needs to send the whole contents of its forwarding table in a response message.
- The receiver adds one hop to each cost and changes the next router field to the address of the sending router. We call each route in the modified forwarding table the *received route* and each route in the old forwarding table the *old route*. The received router selects the old routes as the new ones except in the following three cases:

1. If the received route does not exist in the old forwarding table, it should be added to the route.
2. If the cost of the received route is lower than the cost of the old one, the received route should be selected as the new one.
3. If the cost of the received route is higher than the cost of the old one, but the value of the next router is the same in both routes, the received route should be selected as the new one. This is the case where the route was actually advertised

by the same router in the past, but now the situation has been changed. For example, suppose a neighbor has previously advertised a route to a destination with cost 3, but now there is no path between this neighbor and that destination. The neighbor advertises this destination with cost value infinity (16 in RIP). The receiving router must not ignore this value even though its old route has a lower cost to the same destination.

- The new forwarding table needs to be sorted according to the destination route (mostly using the longest prefix first).

Example 20.1

Figure 20.18 shows a more realistic example of the operation of RIP in an autonomous system. First, the figure shows all forwarding tables after all routers have been booted. Then we show changes in some tables when some update messages have been exchanged. Finally, we show the stabilized forwarding tables when there is no more change.

Timers in RIP

RIP uses three timers to support its operation. The *periodic timer* controls the advertising of regular update messages. Each router has one periodic timer that is randomly set to a number between 25 and 35 seconds (to prevent all routers sending their messages at the same time and creating excess traffic). The timer counts down; when zero is reached, the update message is sent, and the timer is randomly set once again. The *expiration timer* governs the validity of a route. When a router receives update information for a route, the expiration timer is set to 180 seconds for that particular route. Every time a new update for the route is received, the timer is reset. If there is a problem on an internet and no update is received within the allotted 180 seconds, the route is considered expired and the hop count of the route is set to 16, which means the destination is unreachable. Every route has its own expiration timer. The *garbage collection timer* is used to purge a route from the forwarding table. When the information about a route becomes invalid, the router does not immediately purge that route from its table. Instead, it continues to advertise the route with a metric value of 16. At the same time, a garbage collection timer is set to 120 seconds for that route. When the count reaches zero, the route is purged from the table. This timer allows neighbors to become aware of the invalidity of a route prior to purging.

Performance

Before ending this section, let us briefly discuss the performance of RIP:

- ***Update Messages.*** The update messages in RIP have a very simple format and are sent only to neighbors; they are local. They do not normally create traffic because the routers try to avoid sending them at the same time.

- ***Convergence of Forwarding Tables.*** RIP uses the distance-vector algorithm, which can converge slowly if the domain is large, but, since RIP allows only 15 hops in a domain (16 is considered as infinity), there is normally no problem in convergence. The only problems that may slow down convergence are count-to-infinity and loops created in the domain; use of poison-reverse and split-horizon strategies added to the RIP extension may alleviate the situation.

Figure 20.18 *Example of an autonomous system using RIP*

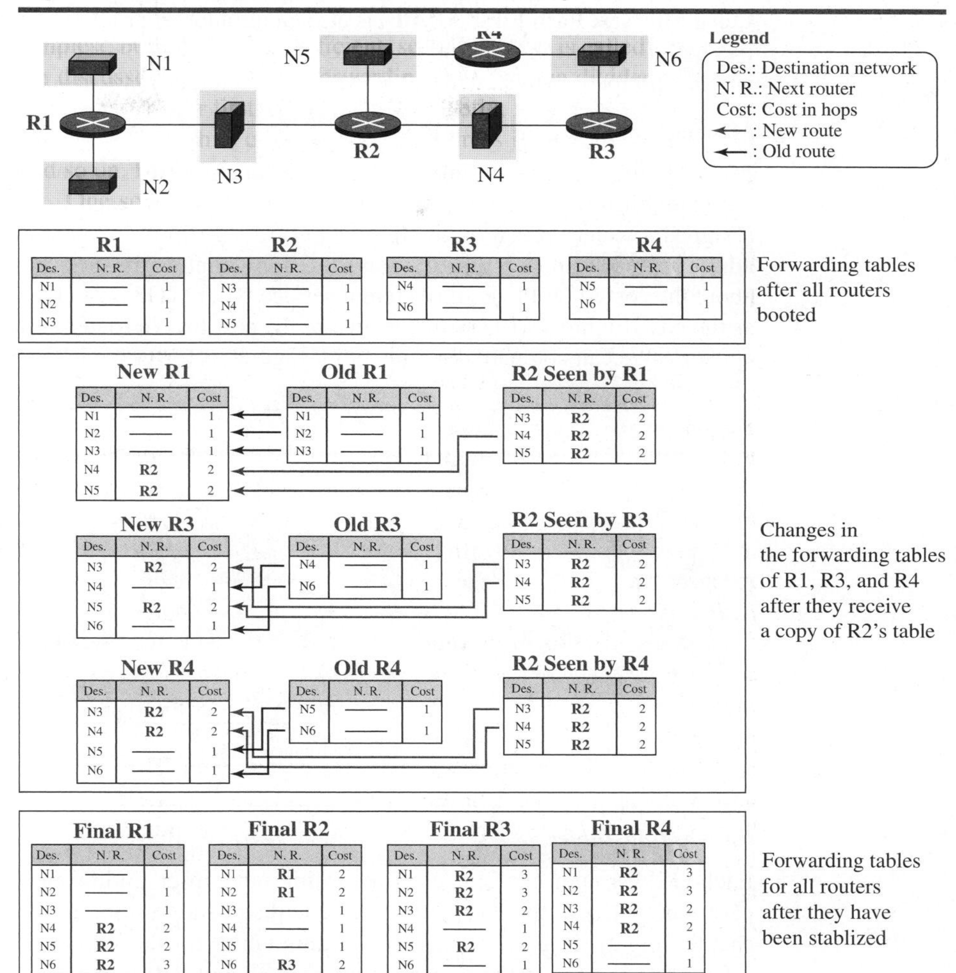

R1 (after all routers booted)

Des.	N. R.	Cost
N1	——	1
N2	——	1
N3	——	1

R2

Des.	N. R.	Cost
N3	——	1
N4	——	1
N5	——	1

R3

Des.	N. R.	Cost
N4	——	1
N6	——	1

R4

Des.	N. R.	Cost
N5	——	1
N6	——	1

New R1

Des.	N. R.	Cost
N1	——	1
N2	——	1
N3	——	1
N4	R2	2
N5	R2	2

Old R1

Des.	N. R.	Cost
N1	——	1
N2	——	1
N3	——	1

R2 Seen by R1

Des.	N. R.	Cost
N3	R2	2
N4	R2	2
N5	R2	2

New R3

Des.	N. R.	Cost
N3	R2	2
N4	——	1
N5	R2	2
N6	——	1

Old R3

Des.	N. R.	Cost
N4	——	1
N6	——	1

R2 Seen by R3

Des.	N. R.	Cost
N3	R2	2
N4	R2	2
N5	R2	2

New R4

Des.	N. R.	Cost
N3	R2	2
N4	R2	2
N5	——	1
N6	——	1

Old R4

Des.	N. R.	Cost
N5	——	1
N6	——	1

R2 Seen by R4

Des.	N. R.	Cost
N3	R2	2
N4	R2	2
N5	R2	2

Final R1

Des.	N. R.	Cost
N1	——	1
N2	——	1
N3	——	1
N4	R2	2
N5	R2	2
N6	R2	3

Final R2

Des.	N. R.	Cost
N1	R1	2
N2	R1	2
N3	——	1
N4	——	1
N5	——	1
N6	R3	2

Final R3

Des.	N. R.	Cost
N1	R2	3
N2	R2	3
N3	R2	2
N4	——	1
N5	R2	2
N6	——	1

Final R4

Des.	N. R.	Cost
N1	R2	3
N2	R2	3
N3	R2	2
N4	R2	2
N5	——	1
N6	——	1

- ***Robustness.*** As we said before, distance-vector routing is based on the concept that each router sends what it knows about the whole domain to its neighbors. This means that the calculation of the forwarding table depends on information received from immediate neighbors, which in turn receive their information from their own neighbors. If there is a failure or corruption in one router, the problem will be propagated to all routers and the forwarding in each router will be affected.

20.3.3 Open Shortest Path First (OSPF)

Open Shortest Path First (**OSPF**) is also an intradomain routing protocol like RIP, but it is based on the link-state routing protocol we described earlier in the chapter. OSPF is an *open* protocol, which means that the specification is a public document.

Metric

In OSPF, like RIP, the cost of reaching a destination from the host is calculated from the source router to the destination network. However, each link (network) can be assigned a weight based on the throughput, round-trip time, reliability, and so on. An administration can also decide to use the hop count as the cost. An interesting point about the cost in OSPF is that different service types (TOSs) can have different weights as the cost. Figure 20.19 shows the idea of the cost from a router to the destination host network. We can compare the figure with Figure 20.15 for the RIP.

Figure 20.19 *Metric in OSPF*

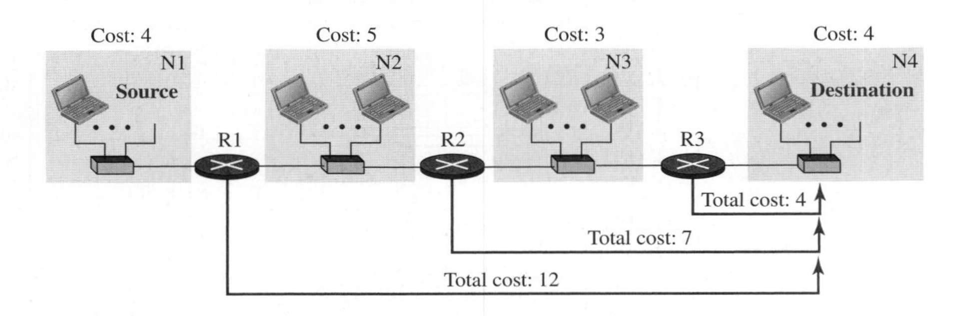

Forwarding Tables

Each OSPF router can create a forwarding table after finding the shortest-path tree between itself and the destination using Dijkstra's algorithm, described earlier in the chapter. Figure 20.20 shows the forwarding tables for the simple AS in Figure 20.19. Comparing the forwarding tables for the OSPF and RIP in the same AS, we find that the only difference is the cost values. In other words, if we use the hop count for OSPF, the tables will be exactly the same. The reason for this consistency is that both protocols use the shortest-path trees to define the best route from a source to a destination.

Areas

Compared with RIP, which is normally used in small ASs, OSPF was designed to be able to handle routing in a small or large autonomous system. However, the formation of shortest-path trees in OSPF requires that all routers flood the whole AS with their LSPs to create the global LSDB. Although this may not create a problem in a small AS, it may have created a huge volume of traffic in a large AS. To prevent this, the AS needs to be divided into small sections called *areas*. Each area acts as a small independent domain for flooding LSPs. In other words, OSPF uses another level of hierarchy in routing: the first level is the autonomous system, the second is the area.

Figure 20.20 *Forwarding tables in OSPF*

Forwarding table for R1

Destination network	Next router	Cost
N1	——	4
N2	——	5
N3	R2	8
N4	R2	12

Forwarding table for R2

Destination network	Next router	Cost
N1	R1	9
N2	——	5
N3	——	3
N4	R3	7

Forwarding table for R3

Destination network	Next router	Cost
N1	R2	12
N2	R2	8
N3	——	3
N4	——	4

However, each router in an area needs to know the information about the link states not only in its area but also in other areas. For this reason, one of the areas in the AS is designated as the *backbone area,* responsible for gluing the areas together. The routers in the backbone area are responsible for passing the information collected by each area to all other areas. In this way, a router in an area can receive all LSPs generated in other areas. For the purpose of communication, each area has an area identification. The area identification of the backbone is zero. Figure 20.21 shows an autonomous system and its areas.

Figure 20.21 *Areas in an autonomous system*

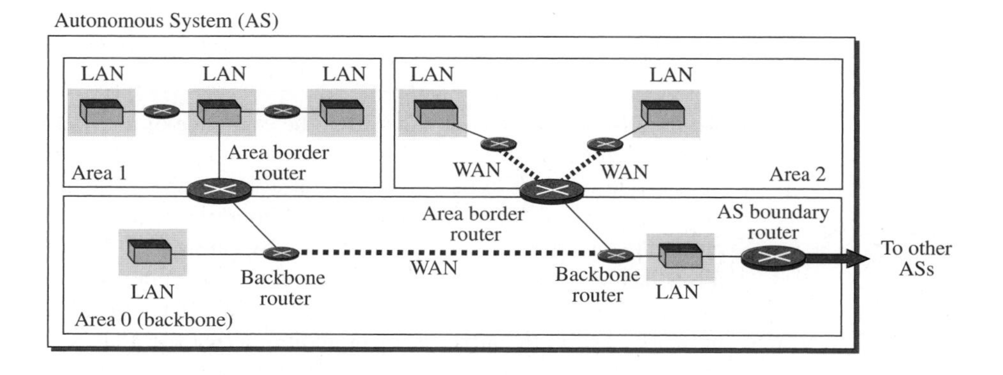

Link-State Advertisement

OSPF is based on the link-state routing algorithm, which requires that a router advertise the state of each link to all neighbors for the formation of the LSDB. When we discussed the link-state algorithm, we used the graph theory and assumed that each router is a node and each network between two routers is an edge. The situation is different in the real world, in which we need to advertise the existence of different entities as nodes, the different types of links that connect each node to its neighbors, and the different types of cost associated with each link. This means we need different types of advertisements, each capable of advertising different situations. We can have five types of link-state advertisements: *router link, network link, summary link to network, summary link to AS border router,* and *external link*. Figure 20.22 shows these five advertisements and their uses.

Figure 20.22 *Five different LSPs*

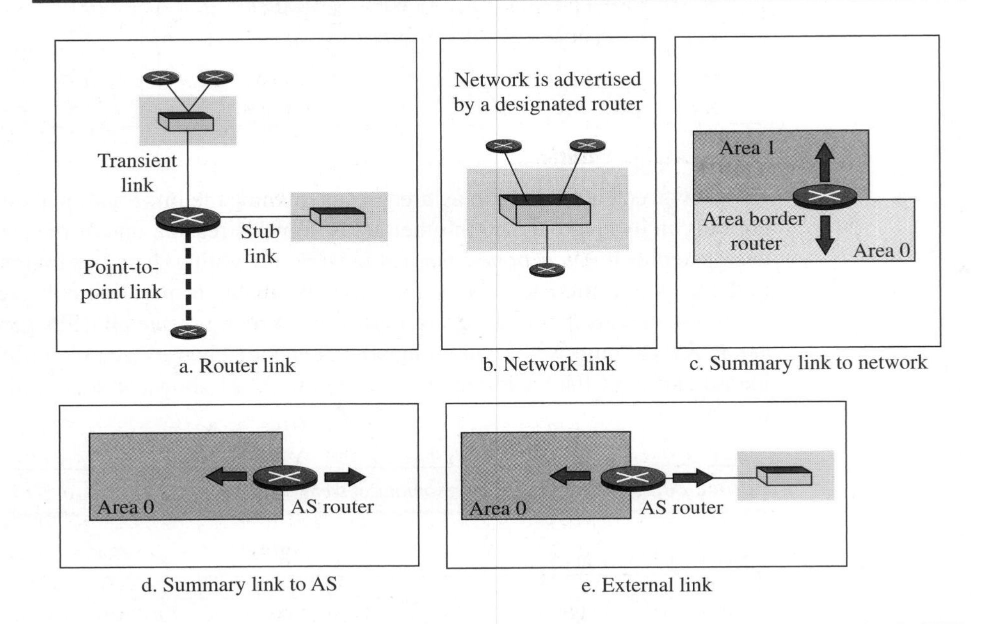

- **Router link.** A router link advertises the existence of a router as a node. In addition to giving the address of the announcing router, this type of advertisement can define one or more types of links that connect the advertising router to other entities. A *transient link* announces a link to a transient network, a network that is connected to the rest of the networks by one or more routers. This type of advertisement should define the address of the transient network and the cost of the link. A *stub link* advertises a link to a stub network, a network that is not a through network. Again, the advertisement should define the address of the network and the cost. A *point-to-point link* should define the address of the router at the end of the point-to-point line and the cost to get there.
- ***Network link.*** A network link advertises the network as a node. However, since a network cannot do announcements itself (it is a passive entity), one of the routers is assigned as the designated router and does the advertising. In addition to the address of the designated router, this type of LSP announces the IP address of all routers (including the designated router as a router and not as speaker of the network), but no cost is advertised because each router announces the cost to the network when it sends a router link advertisement.
- ***Summary link to network.*** This is done by an area border router; it advertises the summary of links collected by the backbone to an area or the summary of links

collected by the area to the backbone. As we discussed earlier, this type of information exchange is needed to glue the areas together.

- ***Summary link to AS.*** This is done by an AS router that advertises the summary links from other ASs to the backbone area of the current AS, information which later can be disseminated to the areas so that they will know about the networks in other ASs. The need for this type of information exchange is better understood when we discuss inter-AS routing (BGP).
- ***External link.*** This is also done by an AS router to announce the existence of a single network outside the AS to the backbone area to be disseminated into the areas.

OSPF Implementation

OSPF is implemented as a program in the network layer, using the service of the IP for propagation. An IP datagram that carries a message from OSPF sets the value of the protocol field to 89. This means that, although OSPF is a routing protocol to help IP to route its datagrams inside an AS, the OSPF messages are encapsulated inside datagrams. OSPF has gone through two versions: version 1 and version 2. Most implementations use version 2.

OSPF Messages

OSPF is a very complex protocol; it uses five different types of messages. In Figure 20.23, we first show the format of the OSPF common header (which is used in all messages) and the link-state general header (which is used in some messages). We then give the outlines of five message types used in OSPF. The *hello* message (type 1) is used by a router to introduce itself to the neighbors and announce all neighbors that it already knows. The *database description* message (type 2) is normally sent in response to the hello message to allow a newly joined router to acquire the full LSDB. The *link-state request* message (type 3) is sent by a router that needs information about a specific LS. The *link-state update* message (type 4) is the main OSPF message used for building the LSDB. This message, in fact, has five different versions (router link, network link, summary link to network, summary link to AS border router, and external link), as we discussed before. The *link-state acknowledgment* message (type 5) is used to create reliability in OSPF; each router that receives a link-state update message needs to acknowledge it.

Authentication

As Figure 20.23 shows, the OSPF common header has the provision for authentication of the message sender. As we will discuss in Chapters 31 and 32, this prevents a malicious entity from sending OSPF messages to a router and causing the router to become part of the routing system to which it actually does not belong.

OSPF Algorithm

OSPF implements the link-state routing algorithm we discussed in the previous section. However, some changes and augmentations need to be added to the algorithm:

- After each router has created the shortest-path tree, the algorithm needs to use it to create the corresponding routing algorithm.

Figure 20.23 *OSPF message formats*

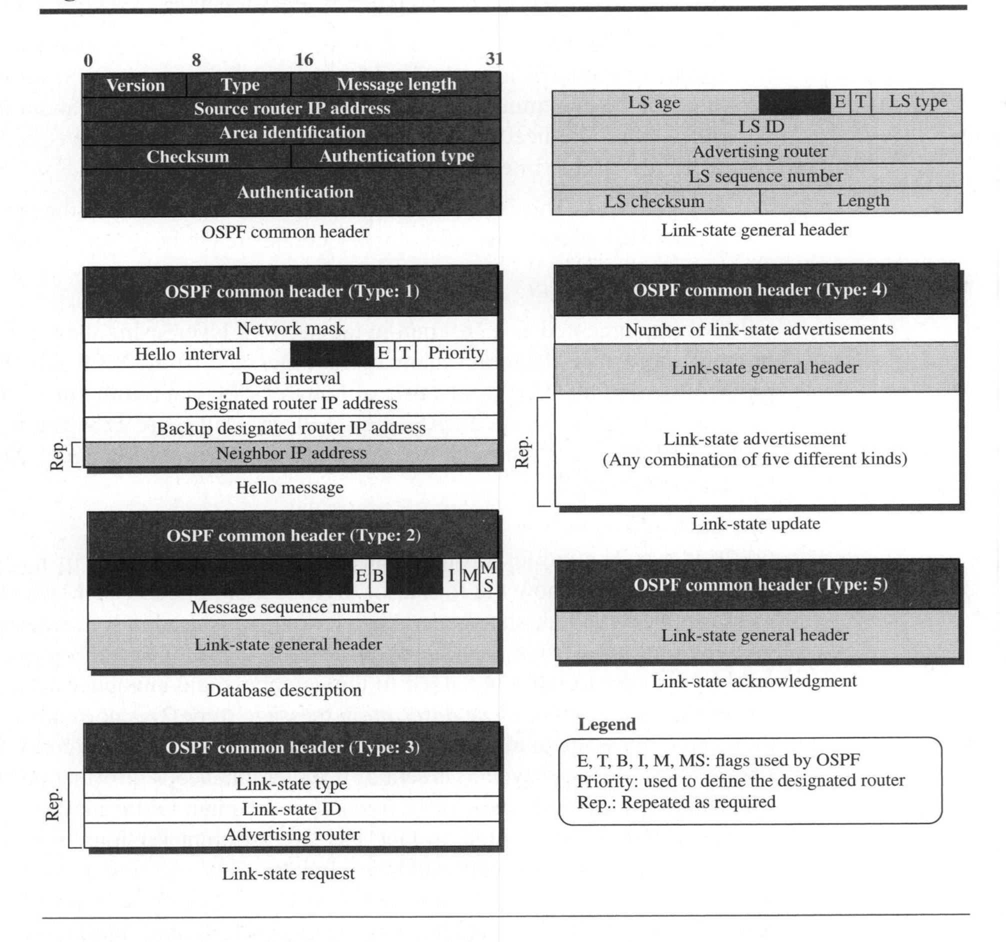

- The algorithm needs to be augmented to handle sending and receiving all five types of messages.

Performance

Before ending this section, let us briefly discuss the performance of OSPF:

- ***Update Messages.*** The link-state messages in OSPF have a somewhat complex format. They also are flooded to the whole area. If the area is large, these messages may create heavy traffic and use a lot of bandwidth.
- ***Convergence of Forwarding Tables.*** When the flooding of LSPs is completed, each router can create its own shortest-path tree and forwarding table; convergence is fairly quick. However, each router needs to run Dijkstra's algorithm, which may take some time.

- ❑ ***Robustness.*** The OSPF protocol is more robust than RIP because, after receiving the completed LSDB, each router is independent and does not depend on other routers in the area. Corruption or failure in one router does not affect other routers as seriously as in RIP.

20.3.4 Border Gateway Protocol Version 4 (BGP4)

The **Border Gateway Protocol version 4 (BGP4)** is the only interdomain routing protocol used in the Internet today. BGP4 is based on the path-vector algorithm we described before, but it is tailored to provide information about the reachability of networks in the Internet.

Introduction

BGP, and in particular BGP4, is a complex protocol. In this section, we introduce the basics of BGP and its relationship with intradomain routing protocols (RIP or OSPF). Figure 20.24 shows an example of an internet with four autonomous systems. AS2, AS3, and AS4 are *stub* autonomous systems; AS1 is a *transient* one. In our example, data exchange between AS2, AS3, and AS4 should pass through AS1.

Figure 20.24 *A sample internet with four ASs*

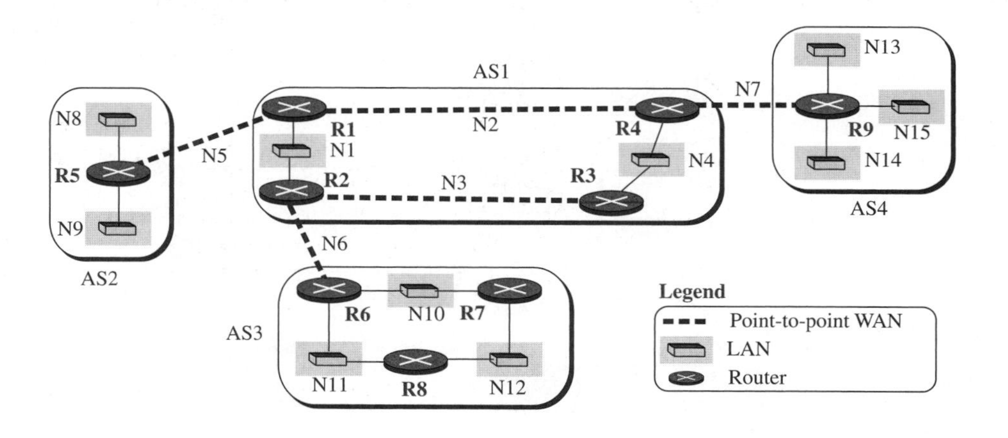

Each autonomous system in this figure uses one of the two common intradomain protocols, RIP or OSPF. Each router in each AS knows how to reach a network that is in its own AS, but it does not know how to reach a network in another AS.

To enable each router to route a packet to any network in the internet, we first install a variation of BGP4, called *external BGP* (*eBGP*), on each *border router* (the one at the edge of each AS which is connected to a router at another AS). We then install the second variation of BGP, called *internal BGP* (*iBGP*), on all routers. This means that the border routers will be running three routing protocols (intradomain, eBGP, and iBGP), but other routers are running two protocols (intradomain and iBGP). We discuss the effect of each BGP variation separately.

Operation of External BGP (eBGP)

We can say that BGP is a kind of point-to-point protocol. When the software is installed on two routers, they try to create a TCP connection using the well-known port 179. In other words, a pair of client and server processes continuously communicate with each other to exchange messages. The two routers that run the BGP processes are called *BGP peers* or *BGP speakers*. We discuss different types of messages exchanged between two peers, but for the moment we are interested in only the update messages (discussed later) that announce reachability of networks in each AS.

The eBGP variation of BGP allows two physically connected border routers in two different ASs to form pairs of eBGP speakers and exchange messages. The routers that are eligible in our example in Figure 20.24 form three pairs: R1-R5, R2-R6, and R4-R9. The connection between these pairs is established over three physical WANs (N5, N6, and N7). However, there is a need for a logical TCP connection to be created over the physical connection to make the exchange of information possible. Each logical connection in BGP parlance is referred to as a *session*. This means that we need three sessions in our example, as shown in Figure 20.25.

Figure 20.25 *eBGP operation*

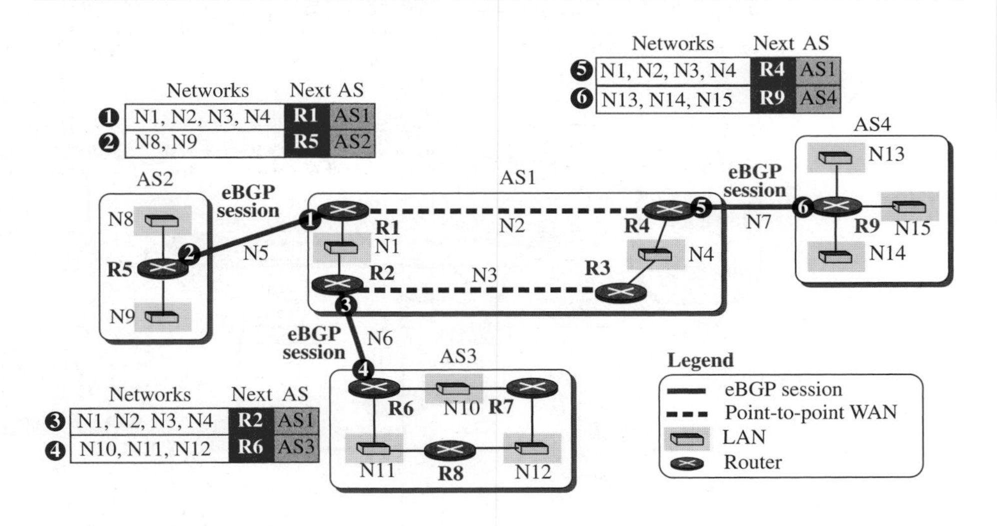

The figure also shows the simplified update messages sent by routers involved in the eBGP sessions. The circled number defines the sending router in each case. For example, message number 1 is sent by router R1 and tells router R5 that N1, N2, N3, and N4 can be reached through router R1 (R1 gets this information from the corresponding intradomain forwarding table). Router R5 can now add these pieces of information at the end of its forwarding table. When R5 receives any packet destined for these four networks, it can use its forwarding table and find that the next router is R1.

The reader may have noticed that the messages exchanged during three eBGP sessions help some routers know how to route packets to some networks in the internet, but the reachability information is not complete. There are two problems that need to be addressed:

1. Some border routers do not know how to route a packet destined for nonneighbor ASs. For example, R5 does not know how to route packets destined for networks in AS3 and AS4. Routers R6 and R9 are in the same situation as R5: R6 does not know about networks in AS2 and AS4; R9 does not know about networks in AS2 and AS3.
2. None of the nonborder routers know how to route a packet destined for any networks in other ASs.

To address the above two problems, we need to allow all pairs of routers (border or nonborder) to run the second variation of the BGP protocol, iBGP.

Operation of Internal BGP (iBGP)

The iBGP protocol is similar to the eBGP protocol in that it uses the service of TCP on the well-known port 179, but it creates a session between any possible pair of routers inside an autonomous system. However, some points should be made clear. First, if an AS has only one router, there cannot be an iBGP session. For example, we cannot create an iBGP session inside AS2 or AS4 in our internet. Second, if there are n routers in an autonomous system, there should be $[n \times (n - 1) / 2]$ iBGP sessions in that autonomous system (a fully connected mesh) to prevent loops in the system. In other words, each router needs to advertise its own reachability to the peer in the session instead of flooding what it receives from another peer in another session. Figure 20.26 shows the combination of eBGP and iBGP sessions in our internet.

Figure 20.26 *Combination of eBGP and iBGP sessions in our internet*

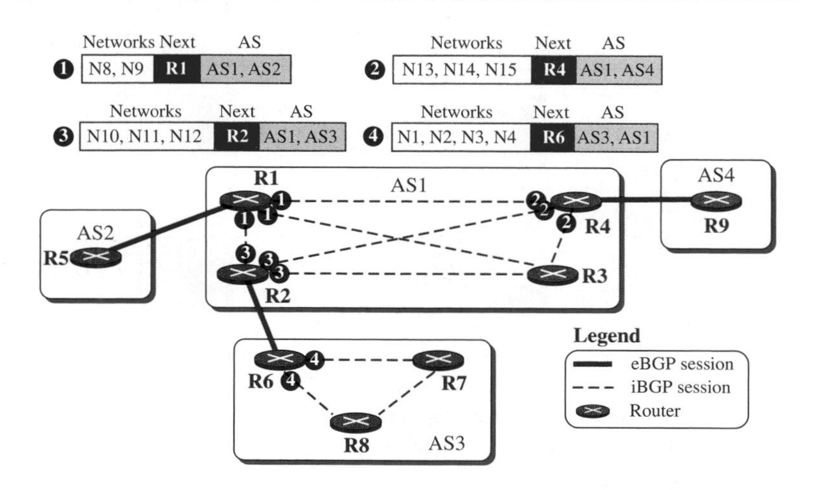

Note that we have not shown the physical networks inside ASs because a session is made on an overlay network (TCP connection), possibly spanning more than one physical network as determined by the route dictated by intradomain routing protocol. Also note that in this stage only four messages are exchanged. The first message (numbered 1) is sent by R1 announcing that networks N8 and N9 are reachable through the path AS1-AS2, but the next router is R1. This message is sent, through separate sessions, to R2, R3, and R4. Routers R2, R4, and R6 do the same thing but send different messages to different destinations. The interesting point is that, at this stage, R3, R7, and R8 create sessions with their peers, but they actually have no message to send.

The updating process does not stop here. For example, after R1 receives the update message from R2, it combines the reachability information about AS3 with the reachability information it already knows about AS1 and sends a new update message to R5. Now R5 knows how to reach networks in AS1 and AS3. The process continues when R1 receives the update message from R4. The point is that we need to make certain that at a point in time there are no changes in the previous updates and that all information is propagated through all ASs. At this time, each router combines the information received from eBGP and iBGP and creates what we may call a path table after applying the criteria for finding the best path, including routing policies that we discuss later. To demonstrate, we show the path tables in Figure 20.27 for the routers in Figure 20.24. For example, router R1 now knows that any packet destined for networks N8 or N9 should go through AS1 and AS2 and the next router to deliver the packet to is router R5. Similarly, router R4 knows that any packet destined for networks N10, N11, or N12 should go through AS1 and AS3 and the next router to deliver this packet to is router R1, and so on.

Figure 20.27 *Finalized BGP path tables*

Networks	Next	Path
N8, N9	R5	AS1, AS2
N10, N11, N12	R2	AS1, AS3
N13, N14, N15	R4	AS1, AS4

Path table for R1

Networks	Next	Path
N8, N9	R1	AS1, AS2
N10, N11, N12	R6	AS1, AS3
N13, N14, N15	R1	AS1, AS4

Path table for R2

Networks	Next	Path
N8, N9	R2	AS1, AS2
N10, N11, N12	R2	AS1, AS3
N13, N14, N15	R4	AS1, AS4

Path table for R3

Networks	Next	Path
N8, N9	R1	AS1, AS2
N10, N11, N12	R1	AS1, AS3
N13, N14, N15	R9	AS1, AS4

Path table for R4

Networks	Next	Path
N1, N2, N3, N4	R1	AS2, AS1
N10, N11, N12	R1	AS2, AS1, AS3
N13, N14, N15	R1	AS2, AS1, AS4

Path table for R5

Networks	Next	Path
N1, N2, N3, N4	R2	AS3, AS1
N8, N9	R2	AS3, AS1, AS2
N13, N14, N15	R2	AS3, AS1, AS4

Path table for R6

Networks	Next	Path
N1, N2, N3, N4	R6	AS3, AS1
N8, N9	R6	AS3, AS1, AS2
N13, N14, N15	R6	AS3, AS1, AS4

Path table for R7

Networks	Next	Path
N1, N2, N3, N4	R6	AS3, AS1
N8, N9	R6	AS3, AS1, AS2
N13, N14, N15	R6	AS3, AS1, AS4

Path table for R8

Networks	Next	Path
N1, N2, N3, N4	R4	AS4, AS1
N8, N9	R4	AS4, AS1, AS2
N10, N11, N12	R4	AS4, AS1, AS3

Path table for R9

Injection of Information into Intradomain Routing

The role of an interdomain routing protocol such as BGP is to help the routers inside the AS to augment their routing information. In other words, the path tables collected and organized by BPG are not used, per se, for routing packets; they are injected into intradomain forwarding tables (RIP or OSPF) for routing packets. This can be done in several ways depending on the type of AS.

In the case of a stub AS, the only area border router adds a default entry at the end of its forwarding table and defines the next router to be the speaker router at the end of the eBGP connection. In Figure 20.24, R5 in AS2 defines R1 as the default router for all networks other than N8 and N9. The situation is the same for router R9 in AS4 with the default router to be R4. In AS3, R6 set its default router to be R2, but R7 and R8 set their default router to be R6. These settings are in accordance with the path tables we describe in Figure 20.27 for these routers. In other words, the path tables are injected into intradomain forwarding tables by adding only one default entry.

In the case of a transient AS, the situation is more complicated. R1 in AS1 needs to inject the whole contents of the path table for R1 in Figure 20.27 into its intradomain forwarding table. The situation is the same for R2, R3, and R4.

One issue to be resolved is the cost value. We know that RIP and OSPF use different metrics. One solution, which is very common, is to set the cost to the foreign networks at the same cost value as to reach the first AS in the path. For example, the cost for R5 to reach all networks in other ASs is the cost to reach N5. The cost for R1 to reach networks N10 to N12 is the cost to reach N6, and so on. The cost is taken from the intradomain forwarding tables (RIP or OSPF).

Figure 20.28 shows the interdomain forwarding tables. For simplicity, we assume that all ASs are using RIP as the intradomain routing protocol. The shaded areas are the augmentation injected by the BGP protocol; the default destinations are indicated as zero.

Figure 20.28 *Forwarding tables after injection from BGP*

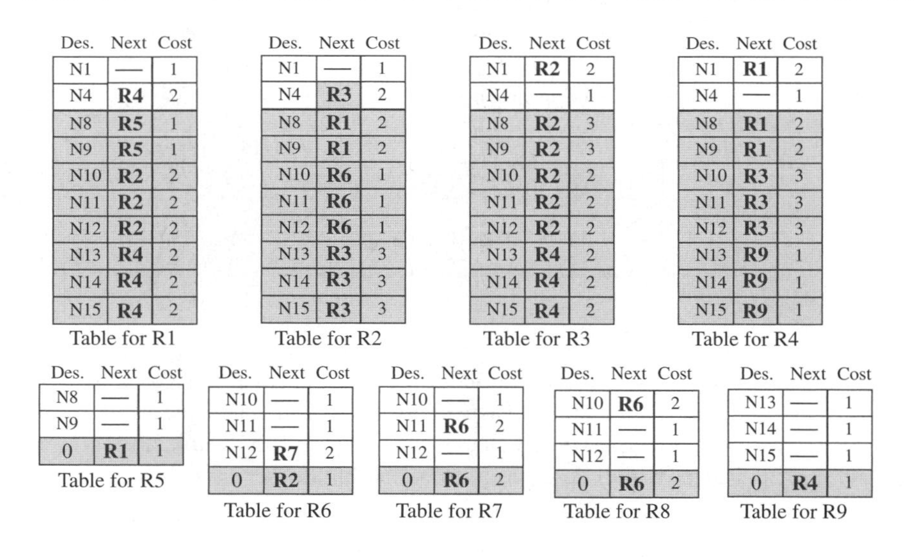

Table for R1

Des.	Next	Cost
N1	—	1
N4	**R4**	2
N8	**R5**	1
N9	**R5**	1
N10	**R2**	2
N11	**R2**	2
N12	**R2**	2
N13	**R4**	2
N14	**R4**	2
N15	**R4**	2

Table for R2

Des.	Next	Cost
N1	—	1
N4	**R3**	2
N8	**R1**	2
N9	**R1**	2
N10	**R6**	1
N11	**R6**	1
N12	**R6**	1
N13	**R3**	3
N14	**R3**	3
N15	**R3**	3

Table for R3

Des.	Next	Cost
N1	**R2**	2
N4	—	1
N8	**R2**	3
N9	**R2**	3
N10	**R2**	2
N11	**R2**	2
N12	**R2**	2
N13	**R4**	2
N14	**R4**	2
N15	**R4**	2

Table for R4

Des.	Next	Cost
N1	**R1**	2
N4	—	1
N8	**R1**	2
N9	**R1**	2
N10	**R3**	3
N11	**R3**	3
N12	**R3**	3
N13	**R9**	1
N14	**R9**	1
N15	**R9**	1

Table for R5

Des.	Next	Cost
N8	—	1
N9	—	1
0	**R1**	1

Table for R6

Des.	Next	Cost
N10	—	1
N11	—	1
N12	**R7**	2
0	**R2**	1

Table for R7

Des.	Next	Cost
N10	—	1
N11	**R6**	2
N12	—	1
0	**R6**	2

Table for R8

Des.	Next	Cost
N10	**R6**	2
N11	—	1
N12	—	1
0	**R6**	2

Table for R9

Des.	Next	Cost
N13	—	1
N14	—	1
N15	—	1
0	**R4**	1

Address Aggregation

The reader may have realized that intradomain forwarding tables obtained with the help of the BGP4 protocols may become huge in the case of the global Internet because many destination networks may be included in a forwarding table. Fortunately, BGP4 uses the prefixes as destination identifiers and allows the aggregation of these prefixes, as we discussed in Chapter 18. For example, prefixes 14.18.20.0/26, 14.18.20.64/26, 14.18.20.128/26, and 14.18.20.192/26, can be combined into 14.18.20.0/24 if all four subnets can be reached through one path. Even if one or two of the aggregated prefixes need a separate path, the longest prefix principle we discussed earlier allows us to do so.

Path Attributes

In both intradomain routing protocols (RIP or OSPF), a destination is normally associated with two pieces of information: next hop and cost. The first one shows the address of the next router to deliver the packet; the second defines the cost to the final destination. Interdomain routing is more involved and naturally needs more information about how to reach the final destination. In BGP these pieces are called *path attributes.* BGP allows a destination to be associated with up to seven path attributes. Path attributes are divided into two broad categories: *well-known* and *optional.* A well-known attribute must be recognized by all routers; an optional attribute need not be. A well-known attribute can be mandatory, which means that it must be present in any BGP update message, or discretionary, which means it does not have to be. An optional attribute can be either transitive, which means it can pass to the next AS, or intransitive, which means it cannot. All attributes are inserted after the corresponding destination prefix in an update message (discussed later). The format for an attribute is shown in Figure 20.29.

Figure 20.29 *Format of path attribute*

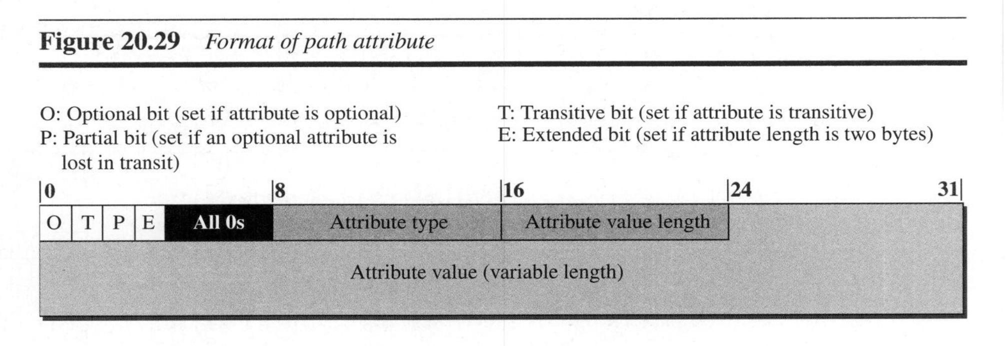

The first byte in each attribute defines the four attribute flags (as shown in the figure). The next byte defines the type of attributes assigned by ICANN (only seven types have been assigned, as explained next). The attribute value length defines the length of the attribute value field (not the length of the whole attributes section). The following gives a brief description of each attribute.

- ***ORIGIN (type 1).*** This is a well-known mandatory attribute, which defines the source of the routing information. This attribute can be defined by one of the three values: 1, 2, and 3. Value 1 means that the information about the path has been taken from an intradomain protocol (RIP or OSPF). Value 2 means that the information comes from BGP. Value 3 means that it comes from an unknown source.

- ***AS-PATH (type 2).*** This is a well-known mandatory attribute, which defines the list of autonomous systems through which the destination can be reached. We have used this attribute in our examples. The AS-PATH attribute, as we discussed in path-vector routing in the last section, helps prevent a loop. Whenever an update

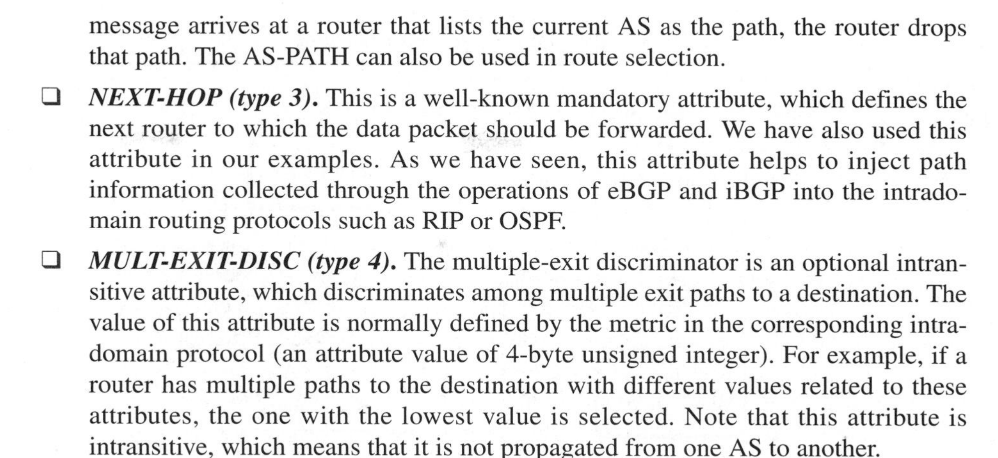

message arrives at a router that lists the current AS as the path, the router drops that path. The AS-PATH can also be used in route selection.

- ❑ ***NEXT-HOP (type 3).*** This is a well-known mandatory attribute, which defines the next router to which the data packet should be forwarded. We have also used this attribute in our examples. As we have seen, this attribute helps to inject path information collected through the operations of eBGP and iBGP into the intradomain routing protocols such as RIP or OSPF.
- ❑ ***MULT-EXIT-DISC (type 4).*** The multiple-exit discriminator is an optional intransitive attribute, which discriminates among multiple exit paths to a destination. The value of this attribute is normally defined by the metric in the corresponding intradomain protocol (an attribute value of 4-byte unsigned integer). For example, if a router has multiple paths to the destination with different values related to these attributes, the one with the lowest value is selected. Note that this attribute is intransitive, which means that it is not propagated from one AS to another.

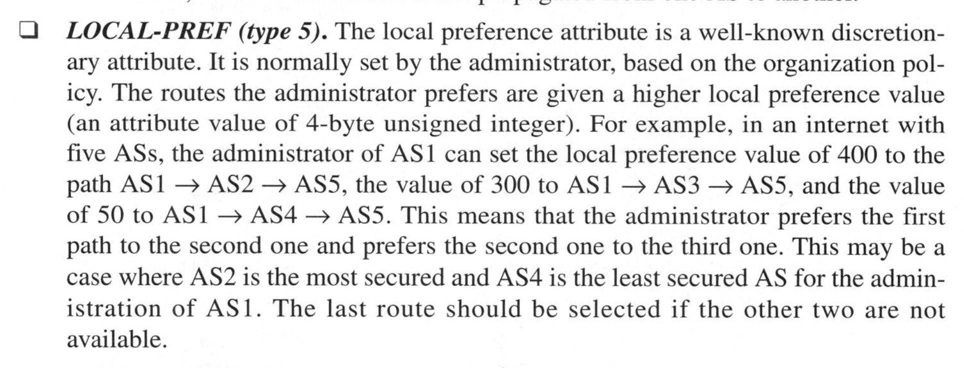

- ❑ ***LOCAL-PREF (type 5).*** The local preference attribute is a well-known discretionary attribute. It is normally set by the administrator, based on the organization policy. The routes the administrator prefers are given a higher local preference value (an attribute value of 4-byte unsigned integer). For example, in an internet with five ASs, the administrator of AS1 can set the local preference value of 400 to the path AS1 → AS2 → AS5, the value of 300 to AS1 → AS3 → AS5, and the value of 50 to AS1 → AS4 → AS5. This means that the administrator prefers the first path to the second one and prefers the second one to the third one. This may be a case where AS2 is the most secured and AS4 is the least secured AS for the administration of AS1. The last route should be selected if the other two are not available.
- ❑ ***ATOMIC-AGGREGATE (type 6).*** This is a well-known discretionary attribute, which defines the destination prefix as not aggregate; it only defines a single destination network. This attribute has no value field, which means the value of the length field is zero.
- ❑ ***AGGREGATOR (type 7).*** This is an optional transitive attribute, which emphasizes that the destination prefix is an aggregate. The attribute value gives the number of the last AS that did the aggregation followed by the IP address of the router that did so.

Route Selection

So far in this section, we have been silent about how a route is selected by a BGP router mostly because our simple example has one route to a destination. In the case where multiple routes are received to a destination, BGP needs to select one among them. The route selection process in BGP is not as easy as the ones in the intradomain routing protocol that is based on the shortest-path tree. A route in BGP has some attributes attached to it and it may come from an eBGP session or an iBGP session. Figure 20.30 shows the flow diagram as used by common implementations.

The router extracts the routes which meet the criteria in each step. If only one route is extracted, it is selected and the process stops; otherwise, the process continues with the next step. Note that the first choice is related to the LOCAL-PREF attribute, which reflects the policy imposed by the administration on the route.

Figure 20.30 *Flow diagram for route selection*

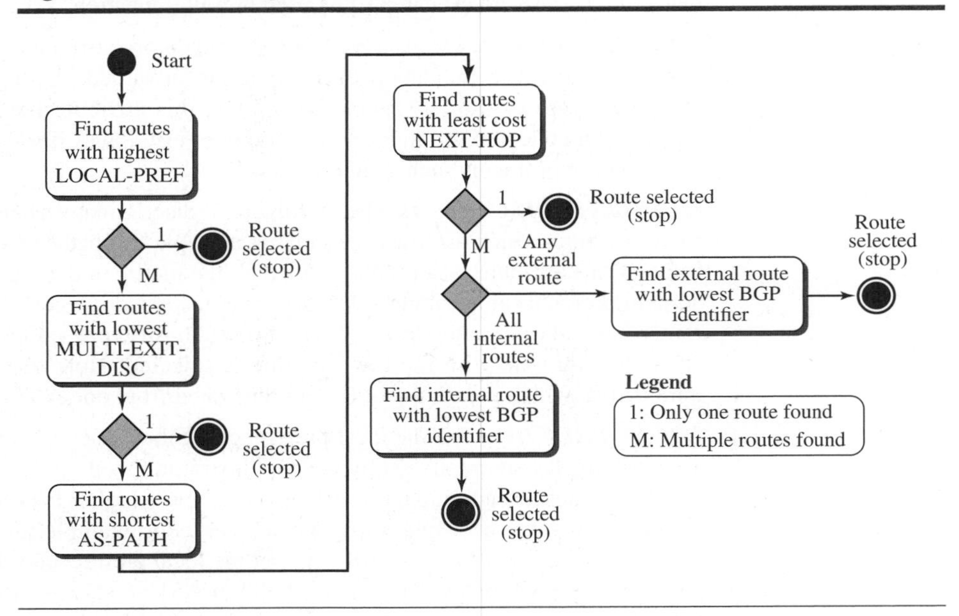

Messages

BGP uses four types of messages for communication between the BGP speakers across the ASs and inside an AS: *open, update, keepalive,* and *notification* (see Figure 20.31). All BGP packets share the same common header.

- ***Open Message.*** To create a neighborhood relationship, a router running BGP opens a TCP connection with a neighbor and sends an *open message.*
- ***Update Message.*** The *update message* is the heart of the BGP protocol. It is used by a router to withdraw destinations that have been advertised previously, to announce a route to a new destination, or both. Note that BGP can withdraw several destinations that were advertised before, but it can only advertise one new destination (or multiple destinations with the same path attributes) in a single update message.
- ***Keepalive Message.*** The BGP peers that are running exchange keepalive messages regularly (before their hold time expires) to tell each other that they are alive.
- ***Notification.*** A notification message is sent by a router whenever an error condition is detected or a router wants to close the session.

Performance

BGP performance can be compared with RIP. BGP speakers exchange a lot of messages to create forwarding tables, but BGP is free from loops and count-to-infinity. The same weakness we mention for RIP about propagation of failure and corruption also exists in BGP.

Figure 20.31 *BGP messages*

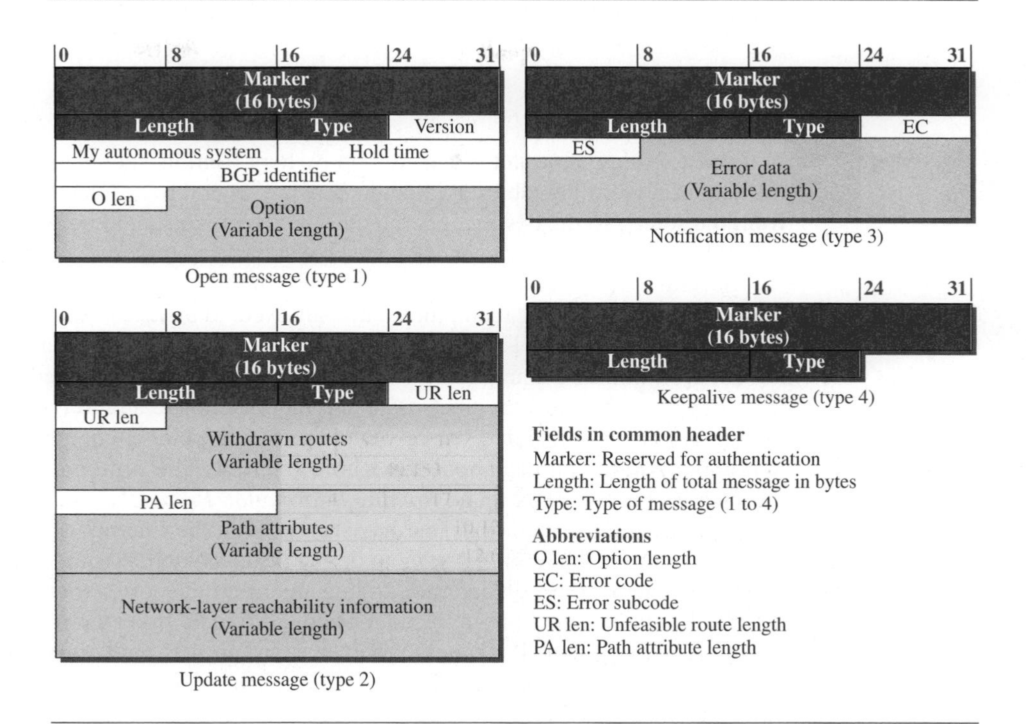

20.4 END-CHAPTER MATERIALS

20.4.1 Recommended Reading

Books

Several books give thorough coverage of materials discussed in this chapter. We recommend [Com 06], [Tan 03], [Koz 05], [Ste 95], [GW 04], [Per 00], [Kes 02], [Moy 98], [WZ 01], and [Los 04].

RFCs

RIP is discussed in RFCs 1058 and 2453. OSPF is discussed in RFCs 1583 and 2328. BGP is discussed in RFCs 1654, 1771, 1773, 1997, 2439, 2918, and 3392.

20.4.2 Key Terms

autonomous system (AS)
Bellman-Ford
Border Gateway Protocol version 4 (BGP4)
Dijkstra's algorithm
distance vector
distance-vector (DV) routing
flooding
least-cost tree

link-state database (LSDB)
link-state (LS) routing
Open Shortest Path First (OSPF)
path-vector (PV) routing
poison reverse
Routing Information Protocol (RIP)
split horizon

20.4.3 Summary

In unicast routing, a packet is routed, hop by hop, from its source to its destination by the help of forwarding tables. Although there are several routes that a packet can travel from the source to the destination, the question is which should be the best. The interpretation of the term *best* depends on the cost and policy imposed on the trip.

Several routing algorithms, and the corresponding protocols, have been devised to find the best route among them; three have survived. In distance-vector routing, the first thing each node creates is its own least-cost tree with the rudimentary information it has about its immediate neighbors. The incomplete trees are exchanged between immediate neighbors to make the trees more and more complete and to represent the whole internet. In other words, in distance-vector routing, a router continuously tells all of its neighbors what it knows about the whole internet. The protocol that implements distance-vector routing is called *Routing Information Protocol (RIP).*

Another routing algorithm that has been used in the Internet is link-state routing. This method uses the term *link-state* to define the characteristic of a link (an edge) that represents a network in the internet. In this algorithm the cost associated with an edge defines the state of the link. In this algorithm, all routers flood the internet, with information related to their link states. When every router has the complete picture of the states, a link-state database can be created. The least-cost tree for each router and the corresponding forwarding table can be made from the link-state database. A protocol that implements link-state routing is called *Open Shortest Path First (OSPF).*

Both link-state and distance-vector routing are based on the least-cost goal. However, there are instances where this goal is not the priority. Path-vector routing algorithms have been designed for this purpose. We can always insert policies in the forwarding table by preventing a packet from visiting a specific router. In path-vector routing, the best route from the source is the best path, the one that complies with the policy imposed. The protocol that implements path-vector routing is the Border Gateway Protocol (BGP).

20.5 PRACTICE SET

20.5.1 Quizzes

A set of interactive quizzes for this chapter can be found on the book website. It is strongly recommended that the student take the quizzes to check his/her understanding of the materials before continuing with the practice set.

20.5.2 Questions

Q20-1. In a graph, if we know that the shortest path from node A to node G is (A → B → E → G), what is the shortest path from node G to node A?

Q20-2. Explain why we can have different intradomain routing protocols in different ASs, but we need only one interdomain routing protocol in the whole Internet.

Q20-3. List three types of autonomous systems (ASs) described in the text, and make a comparison between them.

Q20-4. In a very small AS using OSPF, is it more efficient to use only one single area (backbone) or several areas?

Q20-5. Assume that we have an isolated AS running RIP. We can say that we have at least two different kinds of datagram traffic in this AS. The first kind carries the messages exchanged between hosts; the second carries messages belonging to RIP. What is the difference between the two kinds of traffic when we think about source and destination IP addresses? Does this show that routers also need IP addresses?

Q20-6. Why do you think RIP uses UDP instead of TCP?

Q20-7. At any moment, a RIP message may arrive at a router that runs RIP as the routing protocol. Does it mean that the RIP process should be running all the time?

Q20-8. Router A sends two RIP messages to two immediate neighboring routers, B and C. Do the two datagrams carrying the messages have the same source IP addresses? Do the two datagrams have the same destination IP addresses?

Q20-9. Can you explain why BGP uses the services of TCP instead of UDP?

Q20-10. Is the path-vector routing algorithm closer to the distance-vector routing algorithm or to the link-state routing algorithm? Explain.

Q20-11. Can a router combine the advertisement of a link and a network in a single link-state update?

Q20-12. Explain the concept of hop count in RIP. Can you explain why no hop is counted between N1 and R1 in Figure 20.15?

Q20-13. Why do you think we need only one RIP update message, but several OSPF update messages?

Q20-14. OSPF messages are exchanged between routers. Does this mean that we need to have OSPF processes run all the time to be able to receive an OSPF message when it arrives?

Q20-15. Explain why a router using link-state routing needs to receive the whole LSDB before creating and using its forwarding table. In other words, why can't the router create its forwarding table with a partially received LSDB?

Q20-16. Explain why policy routing can be implemented on an interdomain routing, but it cannot be implemented on an intradomain routing.

Q20-17. We say that OSPF is a hierarchical intradomain protocol, but RIP is not. What is the reason behind this statement?

Q20-18. Explain what type of OSPF link state is advertised in each of the following cases:

a. A router needs to advertise the existence of another router at the end of a point-to-point link.

b. A router needs to advertise the existence of two stub networks and one transient network.

c. A designated router advertises a network as a node.

Q20-19. OSPF messages and ICMP messages are directly encapsulated in an IP datagram. If we intercept an IP datagram, how can we tell whether the payload belongs to OSPF or ICMP?

Q20-20. Assume the shortest path in a graph from node A to node H is A → B → H. Also assume that the shortest path from node H to node N is H → G → N. What is the shortest path from node A to node N?

Q20-21. Explain when each of the following attributes can be used in BGP:

a. LOCAL-PREF **b.** AS-PATH **c.** NEXT-HOP

20.5.3 Problems

P20-1. Assume that the shortest distance between nodes ***a***, ***b***, ***c***, and ***d*** to node ***y*** and the costs from node ***x*** to nodes ***a***, ***b***, ***c***, and ***d*** are given below:

$\mathbf{D}_{ay} = 5$ $\mathbf{D}_{by} = 6$ $\mathbf{D}_{cy} = 4$ $\mathbf{D}_{dy} = 3$

$\mathbf{c}_{xa} = 2$ $\mathbf{c}_{xb} = 1$ $\mathbf{c}_{xc} = 3$ $\mathbf{c}_{xd} = 1$

What is the shortest distance between node ***x*** and node ***y***, $\mathbf{D}_{xy}$, according to the Bellman-Ford equation?

P20-2. Assume that A, B, C, D, and E in Figure 20.32 are autonomous systems (ASs). Find the path vector for each AS using the algorithm in Table 20.3. Assume that the best path in this case is the path which passes through the shorter list of ASs. Also assume that the algorithm first initializes each AS and then is applied, one at a time, to each node respectively (A, B, C, D, E). Show that the process converges and all ASs will have their stable path vectors.

Figure 20.32 *Problem P20-2*

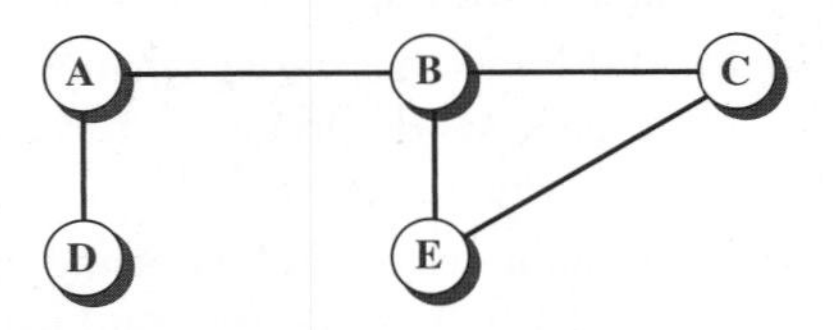

P20-3. In Figure 20.24, assume that the intra-AS routing protocol used by AS1 is OSPF, but the one used by AS2 is RIP. Explain how R5 can find how to route a packet to N4.

P20-4. In distance-vector routing, bad news (increase in a link metric) will propagate slowly. In other words, if a link distance increases, sometimes it takes a long time for all nodes to know the bad news. In Figure 20.33, we assume that a four-node internet is stable, but suddenly the distance between nodes B and C, which is currently 2, is increased to infinity (link fails). Show how this bad news is propagated, and find the new distance vector for each node after stabilization. Assume that the implementation uses a periodic timer to trigger updates to neighbors (no more updates are triggered when there is change). Also assume that if a node receives a higher cost from the same previous neighbor, it uses the new cost because this means that the old advertisement is not valid anymore. To make the stabilization faster, the implementation also suspends a route when the next hop is not accessible.

Figure 20.33 *Problem P20-4*

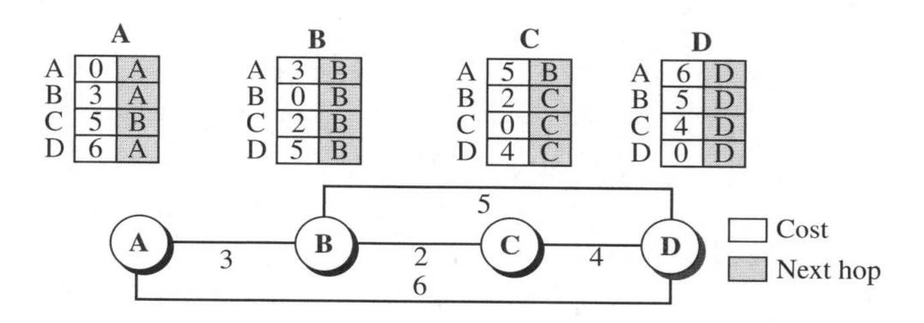

P20-5. In computer science, when we encounter an algorithm, we often need to ask about the complexity of that algorithm (how many computations we need to do). To find the complexity of Dijkstra's algorithm, find the number of searches we have to do to find the shortest path for a single node when the number of nodes is n.

P20-6. Use Dijkstra's algorithm (Table 20.2) to find the shortest path tree and the forwarding table for node A in the Figure 20.34.

Figure 20.34 *Problem P20-6*

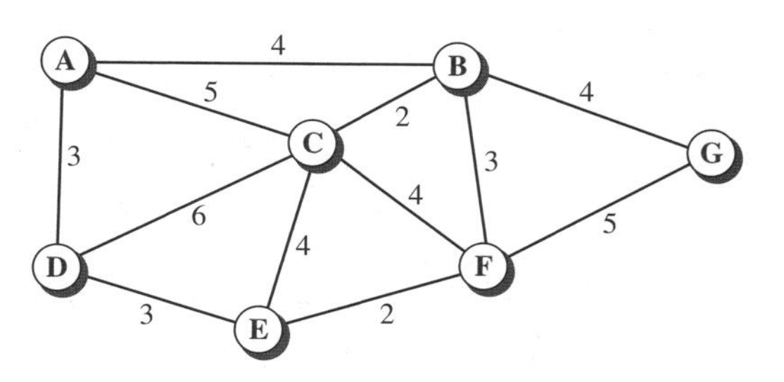

P20-7. In distance-vector routing, good news (decrease in a link metric) will propagate fast. In other words, if a link distance decreases, all nodes quickly learn about it and update their vectors. In Figure 20.33, we assume that a four-node internet is stable, but suddenly the distance between nodes A and D, which is currently 6, is decreased to 1 (probably due to some improvement in the link quality). Show how this good news is propagated, and find the new distance vector for each node after stabilization.

P20-8. Assume that the network in Figure 20.35 uses distance-vector routing with the forwarding table as shown for each node.

Figure 20.35 *Problem P20-8*

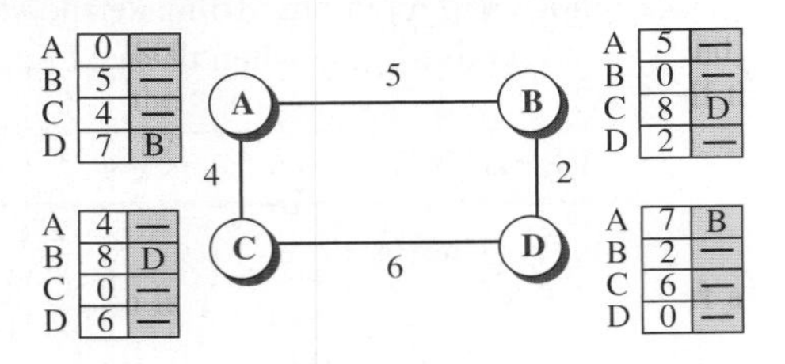

If each node periodically announces their vectors to the neighbor using the poison-reverse strategy, what is the distance vector advertised in the appropriate period:

a. from A to B? **b.** from C to D? **c.** from D to B? **d.** from C to A?

P20-9. Create the forwarding table for node A in Figure 20.10.

P20-10. In Figure 20.24, assume that the intra-AS routing protocol used by AS4 and AS3 is RIP. Explain how R8 can find how to route a packet to N13.

P20-11. When does an OSPF router send each of the following messages?

a. hello **b.** data description **c.** link-state request
d. link-state update **e.** link-state acknowledgment

P20-12. To understand how the distance vector algorithm in Table 20.1 works, let us apply it to a four-node internet as shown in Figure 20.36.

Figure 20.36 *Problem P20-12*

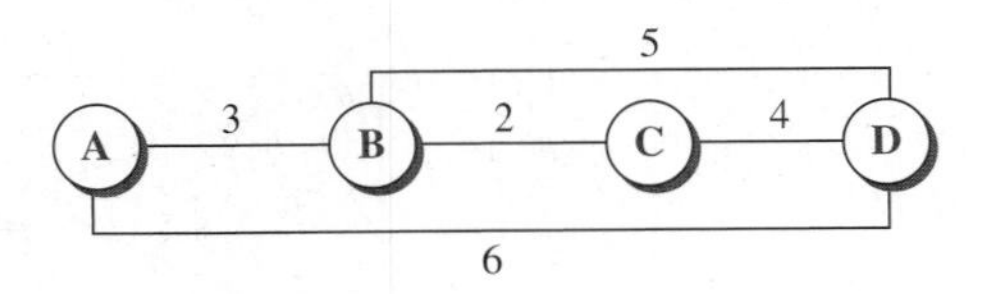

Assume that all nodes are initialized first. Also assume that the algorithm is applied, one at a time, to each node respectively (A, B, C, D). Show that the process converges and all nodes will have their stable distance vectors.

P20-13. Create the shortest path tree and the forwarding table for node B in Figure 20.8.

P20-14. Create the shortest path tree and the forwarding table for node G in Figure 20.8.

P20-15. Assume that the network in Figure 20.35 uses distance-vector routing with the forwarding table as shown for each node. If each node periodically announces their vectors to the neighbor using the split-horizon strategy, what is the distance vector advertised in the appropriate period:

a. from A to B? **b.** from C to D? **c.** from D to B? **d.** from C to A?

P20-16. Assume a router using RIP has 10 entries in its forwarding table at time t_1. Six of these entries are still valid at time t_2. Four of these entries have been expired 70, 90, 110, and 210 seconds before time t_2. Find the number of periodic timers, expiration timers, and garbage collection timers running at time t_1 and time t_2.

P20-17. In computer science, when we encounter an algorithm, we often need to ask about the complexity of that algorithm (how many computations we need to do). To find the complexity of the distance vector's algorithm, find the number of operations a node needs to do when it receives a vector from a neighbor.

P20-18. Assume that the network in Figure 20.35 uses distance vector routing with the forwarding table as shown for each node. If node E is added to the network with a link of cost 1 to node D, can you find the new forwarding tables for each node without using the distance-vector algorithm?

20.6 SIMULATION EXPERIMENTS

20.6.1 Applets

We have created some Java applets to show some of the main concepts discussed in this chapter. It is strongly recommended that the students activate these applets on the book website and carefully examine the protocols in action.

20.7 PROGRAMMING ASSIGNMENT

Write the source code, compile, and test the following program in one of the programming languages of your choice:

Prg20-1. Write a program to simulate the distance-vector algorithm (Table 20.1).

Prg20-2. Write a program to simulate the link-state algorithm (Table 20.2).

Prg20-3. Write a program to simulate the path-vector algorithm (Table 20.3).

CHAPTER 21

Multicast Routing

In this chapter, we move to multicasting and multicast routing protocols and show how some intradomain unicast routing protocols can be extended to be used as multicast routing protocols. We also discuss one new independent multicast protocol and briefly introduce the interdomain multicast routing protocol.

- The first section introduces the concept of multicasting and compares unicasting and multicasting. The section also explains the differences between multicasting and multiple unicasting. Finally, the section gives some applications of multicasting.
- The second section gives some information about the basics of multicasting. The section first elaborates on multicast addresses. The section shows how network-layer multicast addresses are mapped to data-link layer multicast addresses. The section then describes how collection of information to be used for forwarding is done differently in unicasting and multicasting. The section finally describes the process of multicast forwarding and two approaches in multicasting.
- The third section discusses intradomain multicast routing protocols. The section first describes DVMRP. It then explains MOSPF. The section ends with the discussion of PIM.
- The fourth section briefly discusses some interdomain multicast routing protocols. Briefly shown are the roles of two protocols: MGBP and BGMP.
- The fifth section discusses IGMP, a protocol responsible for collecting information for multicast routers. The section first describes IGMP messages. It then shows how the protocol is applied to a host and a router. It finally shows the role of IGMP in forwarding multicast packets.

21.1 INTRODUCTION

Communication in the Internet today is not only unicasting; multicasting communication is growing fast. In this section, we first discuss the general ideas behind unicasting, multicasting, and broadcasting. We then talk about some basic issues in multicast routing. Finally, we discuss multicasting routing protocols in the Internet.

From the previous chapters, we have learned that the forwarding of a datagram by a router is normally based on the prefix of the destination address in the datagram, which defines the network to which the destination host is connected. Understanding the above forwarding principle, we can now define *unicasting*, *multicasting,* and *broadcasting*. Let us clarify these terms as they relate to the Internet.

21.1.1 Unicasting

In unicasting, there is one source and one destination network. The relationship between the source and the destination network is one to one. Each router in the path of the datagram tries to forward the packet to one and only one of its interfaces. Figure 21.1 shows a small internet in which a unicast packet needs to be delivered from a source computer to a destination computer attached to N6. Router R1 is responsible for forwarding the packet only through interface 3; router R4 is responsible for forwarding the packet only through interface 2. When the packet arrives at N6, the delivery to the destination host is the responsibility of the network; it is either broadcast to all hosts or the Ethernet switch delivers it only to the destination host.

Figure 21.1 *Unicasting*

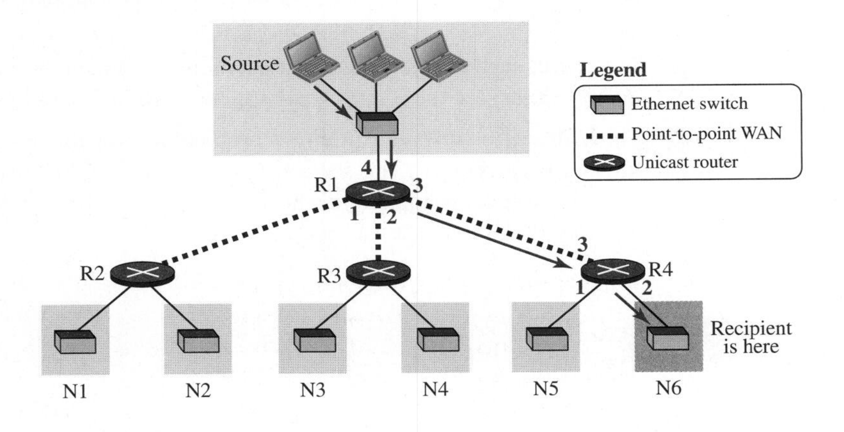

21.1.2 Multicasting

In multicasting, there is one source and a group of destinations. The relationship is one to many. In this type of communication, the source address is a unicast address, but the destination address is a group address, a group of one or more destination networks in which there is at least one member of the group that is interested in receiving the multicast datagram. The group address defines the members of the group. Figure 21.2 shows the same small internet as in Figure 21.1, but the routers have been changed to multicast routers (or previous routers have been configured to do both types of jobs).

Figure 21.2 *Multicasting*

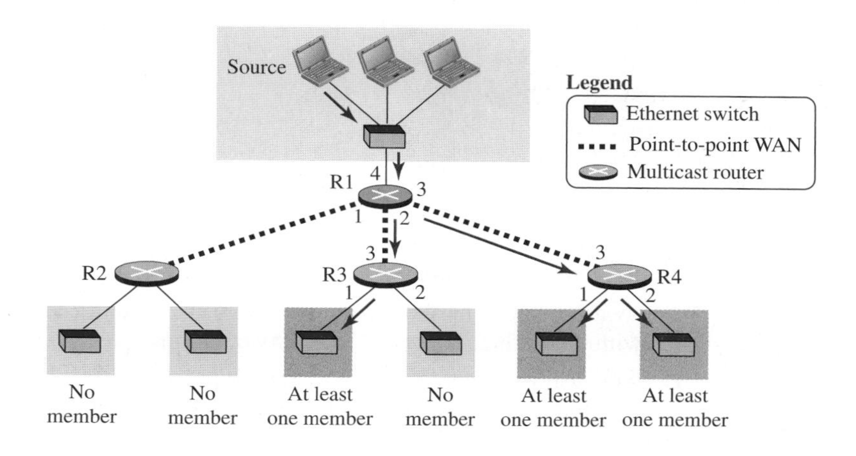

In multicasting, a multicast router may have to send out copies of the same datagram through more than one interface. In Figure 21.2, router R1 needs to send out the datagram through interfaces 2 and 3. Similarly, router R4 needs to send out the datagram through both its interfaces. Router R3, however, knows that there is no member belonging to this group in the area reached by interface 2; it only sends out the datagram through interface 1.

Multicasting versus Multiple Unicasting

We need to distinguish between multicasting and multiple unicasting. Figure 21.3 illustrates both concepts.

Multicasting starts with a single packet from the source that is duplicated by the routers. The destination address in each packet is the same for all duplicates. Note that only a single copy of the packet travels between any two routers.

In multiple unicasting, several packets start from the source. If there are three destinations, for example, the source sends three packets, each with a different unicast destination address. Note that there may be multiple copies traveling between two routers. For example, when a person sends an e-mail message to a group of people, this is multiple unicasting. The e-mail application software creates replicas of the message, each with a different destination address, and sends them one by one.

Figure 21.3 *Multicasting versus multiple unicasting*

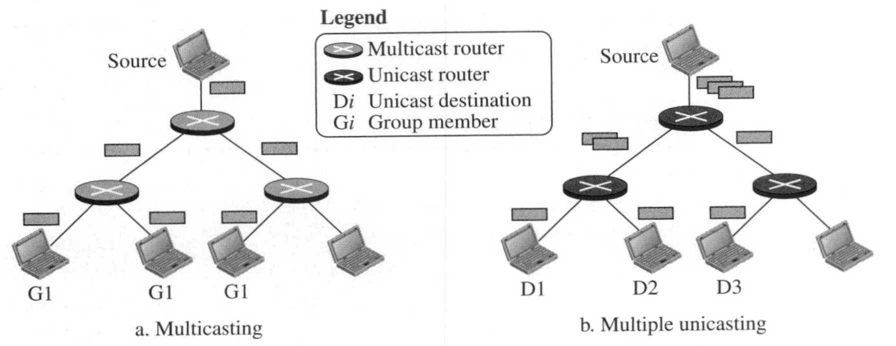

Legend
Multicast router
Unicast router
D*i* Unicast destination
G*i* Group member

Source
G1 G1 G1
a. Multicasting

Source
D1 D2 D3
b. Multiple unicasting

Emulation of Multicasting with Unicasting

We may ask why we have a separate mechanism for multicasting, when it can be emulated with unicasting. There are at least two reasons:

1. Multicasting is more efficient than multiple unicasting. In Figure 21.3, we can see how multicasting requires less bandwidth than multiple unicasting. In multiple unicasting, some of the links must handle several copies.
2. In multiple unicasting, the packets are created by the source with a relative delay between packets. If there are 1,000 destinations, the delay between the first and the last packet may be unacceptable. In multicasting, there is no delay because only one packet is created by the source.

Multicast Applications

Multicasting has many applications today, such as access to distributed databases, information dissemination, teleconferencing, and distance learning.

- ***Access to Distributed Databases.*** Most of the large databases today are distributed. That is, the information is stored in more than one location, usually at the time of production. The user who needs to access the database does not know the location of the information. A user's request is multicast to all the database locations, and the location that has the information responds.
- ***Information Dissemination.*** Businesses often need to send information to their customers. If the nature of the information is the same for each customer, it can be multicast. In this way a business can send one message that can reach many customers. For example, a software update can be sent to all purchasers of a particular software package. In a similar manner, news can be easily disseminated through multicasting.
- ***Teleconferencing.*** Teleconferencing involves multicasting. The individuals attending a teleconference all need to receive the same information at the same time. Temporary or permanent groups can be formed for this purpose.

- ***Distance Learning.*** One growing area in the use of multicasting is distance learning. Lessons taught by one professor can be received by a specific group of students. This is especially convenient for those students who find it difficult to attend classes on campus.

21.1.3 Broadcasting

Broadcasting means one-to-all communication: a host sends a packet to all hosts in an internet. Broadcasting in this sense is not provided at the Internet level for the obvious reason that it may create a huge volume of traffic and use a huge amount of bandwidth. Partial broadcasting, however, is done in the Internet. For example, some peer-to-peer applications may use broadcasting to access all peers. Controlled broadcasting may also be done in a domain (area or autonomous system) mostly as a step to achieve multicasting. We discuss these types of controlled broadcasting when we discuss multicasting protocols.

21.2 MULTICASTING BASICS

Before discussing multicast routing protocols in the Internet, we need to discuss some multicasting basics: multicast addressing, collecting information about multicast groups, and multicast optimal trees.

21.2.1 Multicast Addresses

When we send a unicast packet to a destination, the source address of the packet defines the sender and the destination address of the packet defines the receiver of the packet. In multicast communication, the sender is only one, but the receiver is many, sometimes thousands or millions spread all over the world. It should be clear that we cannot include the addresses of all recipients in the packet. The destination address of a packet, as described in the Internet Protocol (IP) should be only one. For this reason, we need multicast addresses. A multicast address defines a group of recipients, not a single one. In other words, a multicast address is an identifier for a group. If a new group is formed with some active members, an authority can assign an unused multicast address to this group to uniquely define it. This means that the source address of a packet in multicast communication can be a unicast address that uniquely defines the sender, but the destination address can be the multicast address that defines a group. In this way, a host, which is a member of n groups, actually has $(n + 1)$ addresses: one unicast address that is used for source or destination address in unicast communication and n multicast addresses that are used only for destination addresses to receive messages sent to a group. Figure 21.4 shows the concept.

Multicast Addresses in IPv4

A router or a destination host needs to distinguish between a unicast and a multicast datagram. IPv4 and IPv6 each assign a block of addresses for this purpose. In this section, we discuss only IPv4 multicast addresses; IPv6 multicast addresses are discussed later in the chapter. Multicast addresses in IPv4 belong to a large block of addresses that are specially designed for this purpose. In classful addressing, all of class D was

Figure 21.4 *Needs for multicast addresses*

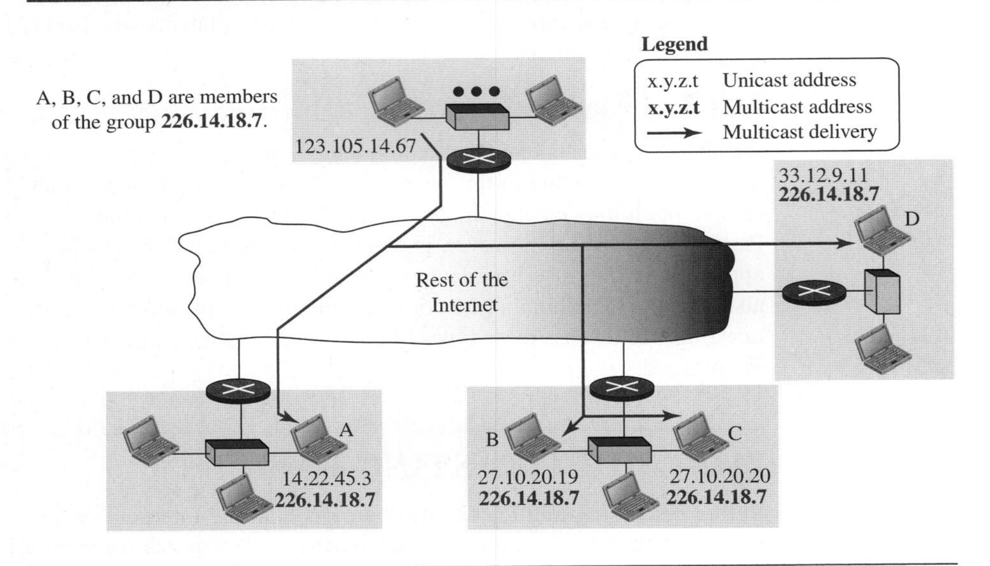

composed of these addresses; classless addressing used the same block, but it was referred to as the block 224.0.0.0/**4** (from 224.0.0.0 to 239.255.255.255). Figure 21.5 shows the block in binary. Four bits define the block; the rest of the bits are used as the identifier for the group.

Figure 21.5 *A multicast address in binary*

The number of addresses in the multicast block is huge (2^{28}). We definitely cannot have that many individual groups. However, the block is divided into several subblocks, and each subblock is used in a particular multicast application. The following gives some of the common subblocks:

- ***Local Network Control Block.*** The subblock 224.0.0.0/24 is assigned to a multicast routing protocol to be used inside a network, which means that the packet with a destination address in this range cannot be forwarded by a router. In this subblock, the address 224.0.0.0 is reserved, the address 224.0.0.1 is used to send datagrams to all hosts and routers inside a network, and the address 224.0.0.2 is used to send datagrams to all routers inside a network. The rest of the addresses are assigned to some multicast protocols for communication, as we discuss later.

- ***Internetwork Control Block.*** The subblock 224.0.1.0/24 is assigned to a multicast routing protocol to be used in the whole Internet, which means that the packet with a destination address in this range can be forwarded by a router.
- ***Source-Specific Multicast (SSM) Block.*** The block 232.0.0.0/8 is used for source-specific multicast routing. We discuss SSM routing when we discuss IGMP protocol later in the chapter.
- ***GLOP Block.*** The block 233.0.0.0/8 is called the GLOP block (not an acronym nor an abbreviation). This block defines a range of addresses that can be used inside an autonomous system (AS). As we learned earlier, each autonomous system is assigned a 16-bit number. One can insert the AS number as the two middle octets in the block to create a range of 256 multicast addresses (233.x.y.0 to 233.x.y.255), in which x.y is the AS number.
- ***Administratively Scoped Block.*** The block 239.0.0.0/8 is called the Administratively Scoped Block. The addresses in this block are used in a particular area of the Internet. The packet whose destination address belongs to this range is not supposed to leave the area. In other words, an address in this block is restricted to an organization.

Selecting Multicast Address

To select a multicast address to be assigned to a group is not an easy task. The selection of address depends on the type of application. Let us discuss some cases.

Limited Group

The administrator can use the AS number $(x.y)_{256}$ and choose an address between 239.x.y.0 and 239.x.y.255 (Administratively Scoped Block), that is not used by any other group, as the multicast address for that particular group. For example, assume college professors need to create group addresses to communicate with their students. If the AS number that the college belongs to is 23452, which can be written as $(91.156)_{256}$, this gives the college a range of 256 addresses: 233.91.156.0 to 233.91.156.255. The college administration can grant each professor one of the addresses in the range. This can then become the group address for the professor to use to send multicast communications to the students. However, the packets cannot go beyond the college AS territory.

Larger Group

If the group is spread beyond an AS territory, the previous solution does not work. The group needs to choose an address from the SSM block (232.0.0.8). There is no need to get permission to use an address in this block, because the packets in source-specific multicasting are routed based on the group and the source address; they are unique.

21.2.2 Delivery at Data-Link Layer

In multicasting, the delivery at the Internet level is done using network-layer multicast addresses (IP addresses allocated for multicasting). However, data-link layer multicast addresses are also needed to deliver a multicast packet encapsulated in a frame. In the case of unicasting, this task is done by the ARP protocol, but, because the IP packet has a multicast IP address, the ARP protocol cannot find the corresponding MAC (physical) address to forward a multicast packet at the data-link layer. What happens next depends on whether or not the underlying data-link layer supports physical multicast addresses.

Network with Multicast Support

Most LANs support physical multicast addressing. Ethernet is one of them. An Ethernet physical address (MAC address) is six octets (48 bits) long. If the first 25 bits in an Ethernet address are 00000001 00000000 01011110 0, this identifies a physical multicast address for the TCP/IP protocol. The remaining 23 bits can be used to define a group. To convert an IP multicast address into an Ethernet address, the multicast router extracts the least significant 23 bits of a multicast IP address and inserts them into a multicast Ethernet physical address (see Figure 21.6).

Figure 21.6 *Mapping class D to Ethernet physical address*

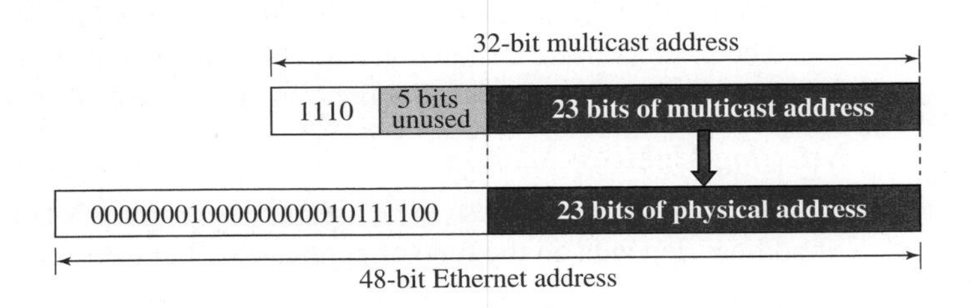

However, the group identifier of a multicast address block in an IPv_4 address is 28 bits long, which implies that 5 bits are not used. This means that 32 (2^5) multicast addresses at the IP level are mapped to a single multicast address. In other words, the mapping is many to one instead of one to one. If the 5 leftmost bits of the group identifier of a multicast address are not all zeros, a host may receive packets that do not really belong to the group in which it is involved. For this reason, the host must check the IP address and discard any packets that do not belong to it.

An Ethernet multicast physical address is in the range
01:00:5E:00:00:00 to 01:00:5E:7F:FF:FF.

Example 21.1

Change the multicast IP address 232.43.14.7 to an Ethernet multicast physical address.

Solution

We can do this in two steps:

a. We write the rightmost 23 bits of the IP address in hexadecimal. This can be done by changing the rightmost 3 bytes to hexadecimal and then subtracting 8 from the leftmost digit if it is greater than or equal to 8. In our example, the result is 2B:0E:07.

b. We add the result of part a to the starting Ethernet multicast address, which is 01:00:5E:00:00:00. The result is

01:00:5E:2B:0E:07

Example 21.2

Change the multicast IP address 238.212.24.9 to an Ethernet multicast address.

Solution

a. The rightmost 3 bytes in hexadecimal are D4:18:09. We need to subtract 8 from the leftmost digit, resulting in 54:18:09.

b. We add the result of part a to the Ethernet multicast starting address. The result is

01:00:5E:54:18:09

Network with No Multicast Support

Most WANs do not support physical multicast addressing. To send a multicast packet through these networks, a process called *tunneling* is used. In **tunneling,** the multicast packet is encapsulated in a unicast packet and sent through the network, where it emerges from the other side as a multicast packet (see Figure 21.7).

Figure 21.7 *Tunneling*

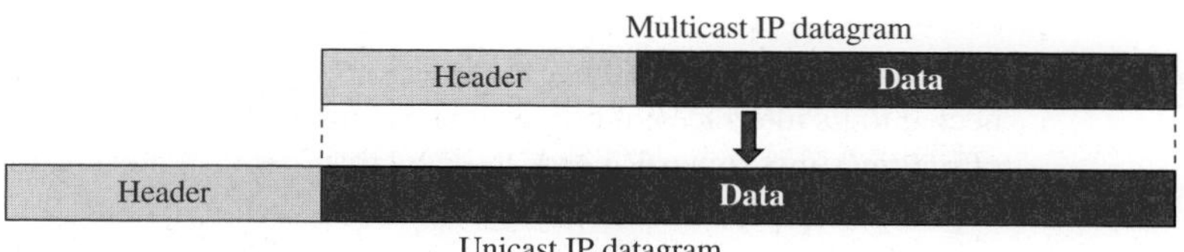

21.2.3 Collecting Information about Groups

Creation of forwarding tables in both unicast and multicast routing involves two steps:

1. A router needs to know to which destinations it is connected.
2. Each router needs to propagate information obtained in the first step to all other routers so that each router knows to which destination each other router is connected.

In unicast routing, the collection of the information in the first step is automatic; each router knows to which network it is connected, and the prefix of the network (in CIDR) is what a router needs. The routing protocols we described in the previous sections (distance-vector or link-state) are responsible for propagating these automatically collected pieces of information to each other router in the internet.

In multicast routing, the collection of information in the first step is not automatic for two reasons. First, a router does not know which host in the attached network is a member of a particular group; membership in the group does not have any relation to the prefix associated with the network. We have shown that if a host is a member of a group, it has a separate multicast address related to that group. Second, the membership is not a fixed attribute of a host; a host may join some new groups and leave some others even in a short period of time. For this reason, a router needs help to find out which groups are active in each of its interfaces. After collecting these pieces of information, a router can propagate the membership to any other router using a multicast routing protocol, as we will discuss later. Figure 21.8 shows the difference between unicast and multicast advertisement in the first step. For unicasting, the router needs no help; for multicasting, it needs the help of another protocol, as we discuss next.

Figure 21.8 *Unicast versus multicast advertisement*

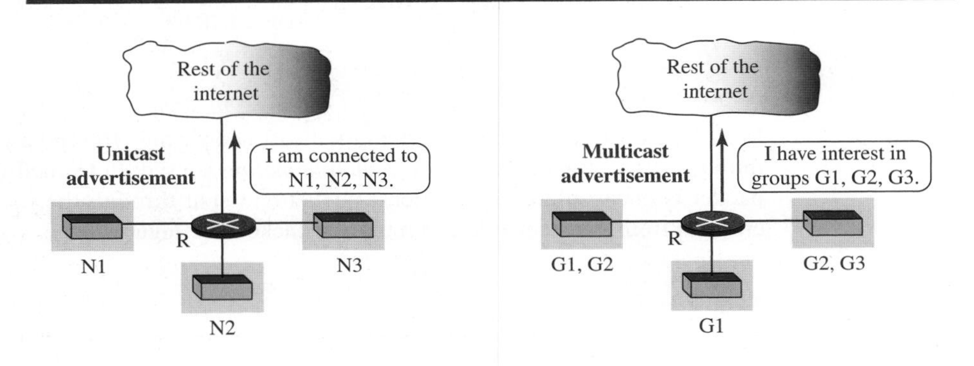

In the case of unicasting, router R knows hosts with prefixes N1, N2, and N3 are connected to its interfaces; it propagates this information to the rest of the internet. In the case of multicasting, router R needs to know that there are hosts with at least one loyal member in groups G1, G2, and G3 in networks connected to its interfaces. In other words, for unicast routing, we need only the routing protocol inside each domain to propagate the information about a router link; in multicasting we need two protocols: one to collect these pieces of information and the second to propagate them. Collecting pieces of information is done by the Internet Group Management Protocol (IGMP), which we discuss at the end of this chapter. We continue this section showing how these pieces of information are propagated using different multicast protocols.

21.2.4 Multicast Forwarding

Another important issue in multicasting is the decision a router needs to make to forward a multicast packet. Forwarding in unicast and multicast communication is different in two aspects:

1. In unicast communication, the destination address of the packet defines one single destination. The packet needs to be sent only out of one of the interfaces, the interface which is the branch in the shortest-path tree reaching the destination with the minimum cost. In multicast communication, the destination of the packet defines one group, but that group may have more than one member in the internet. To reach all of the destinations, the router may have to send the packet out of more than one interface. Figure 21.9 shows the concept. In unicasting, the destination

network N1 cannot be in more than one part of the internet; in multicasting, the group G1 may have members in more than one part of the internet.

Figure 21.9 *Destination in unicasting and multicasting*

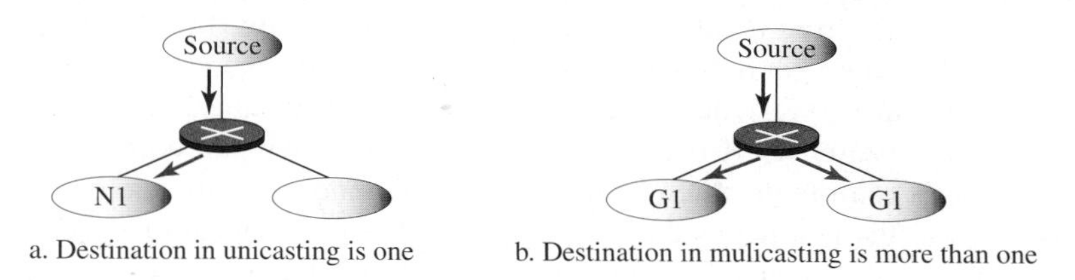

2. Forwarding decisions in unicast communication depend only on the destination address of the packet. Forwarding decisions in multicast communication depend on both the destination and the source address of the packet. In other words, in unicasting, forwarding is based on where the packet should go; in multicasting, forwarding is based on where the packet should go and where the packet has come from. Figure 21.10 shows the concept. In part a of the figure, the source is in a section of the internet where there is no group member. In part b, the source is in a section where there is a group member. In part a, the router needs to send out the packet from two interfaces; in part b, the router should send the packet only from

Figure 21.10 *Forwarding depends on the destination and the source*

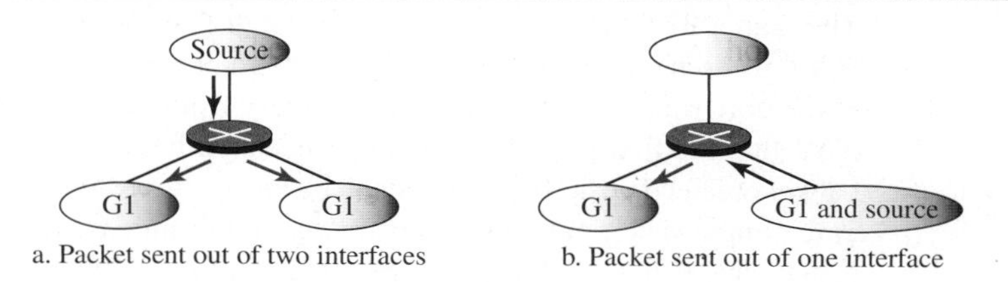

one interface to avoid sending a second copy of the packet from the interface it has arrived at. In other words, in part b of the figure, the member or members of the group G1 have already received a copy of the packet when it arrives at the router; sending out the packet in that direction does not help, but creates more traffic. This shows that the forwarding in multicast communication depends on both source and destination addresses.

21.2.5 Two Approaches to Multicasting

In multicast routing, as in unicast routing, we need to create routing trees to optimally route the packets from their source to their destination. However, as we discussed before, the multicast routing decision at each router depends not only on the destination of the packet, but also on the source of the packet. The involvement of the source in the routing process makes multicast routing much more difficult than unicast routing. For this reason, two different approaches in multicast routing have been developed: routing using source-based trees and routing using group-shared trees.

Source-Based Tree Approach

In the **source-based tree** approach to multicasting, each router needs to create a separate tree for each source-group combination. In other words, if there are *m* groups and *n* sources in the internet, a router needs to create ($m \times n$) routing trees. In each tree, the corresponding source is the root, the members of the group are the leaves, and the router itself is somewhere on the tree. We can compare the situation with unicast routing in which a router needs only one tree with itself as the root and all networks in the internet as the leaves. Although it may appear that each router needs to create and store a huge amount of information about all of these trees, there are two protocols that use this approach in the Internet today, which we discuss later. These protocols use some strategies that alleviate the situation.

Group-Shared Tree Approach

In the **group-shared tree** approach, we designate a router to act as the phony source for each group. The designated router, which is called the *core* router or the *rendezvous-point* router, acts as the representative for the group. Any source that has a packet to send to a member of that group sends it to the core center (unicast communication) and the core center is responsible for multicasting. The core center creates one single routing tree with itself as the root and any routers with active members in the group as the leaves. In this approach, there are *m* core routers (one for each group) and each core router has a routing tree, for the total of *m* trees. This means that the number of routing trees is reduced from ($m \times n$) in the source-based tree approach to *m* in this approach. The reader may have noticed that we have divided a multicast delivery from the source to all group members into two deliveries. The first is a unicast delivery from the source to the core router; the second is the delivery from the core router to all group members. Note that the first part of the delivery needs to be done using tunneling. The multicast packet created by the source needs to be encapsulated in a unicast packet and sent to the core router. The core router decapsulates the unicast packet, extracts the multicast packet, and sends it to the group members. Although the reduction in number of trees in this approach looks very attractive, this approach has its own overhead: using an algorithm to select a router among all routers as the core router for a group.

21.3 INTRADOMAIN MULTICAST PROTOCOLS

During the last few decades, several intradomain multicast routing protocols have emerged. In this section, we discuss three of these protocols. Two are extensions of unicast routing protocols (RIP and OSPF), using the source-based tree approach; the third is an independent protocol which is becoming more and more popular. It can be used in two modes, employing either the source-based tree approach or the shared-group tree approach.

21.3.1 Multicast Distance Vector (DVMRP)

The **Distance Vector Multicast Routing Protocol** (**DVMRP**) is the extension of the Routing Information Protocol (RIP) which is used in unicast routing. It uses the source-based tree approach to multicasting. It is worth mentioning that each router in this protocol that receives a multicast packet to be forwarded implicitly creates a source-based multicast tree in three steps:

1. The router uses an algorithm called *reverse path forwarding* (RPF) to simulate creating part of the optimal source-based tree between the source and itself.
2. The router uses an algorithm called *reverse path broadcasting* (RPB) to create a broadcast (spanning) tree whose root is the router itself and whose leaves are all networks in the internet.
3. The router uses an algorithm called *reverse path multicasting* (RPM) to create a multicast tree by cutting some branches of the tree that end in networks with no member in the group.

Reverse Path Forwarding (RPF)

The first algorithm, **reverse path forwarding (RPF),** forces the router to forward a multicast packet from one specific interface: the one which has come through the shortest path from the source to the router. How can a router know which interface is in this path if the router does not have a shortest-path tree rooted at the source? The router uses the first property of the shortest-path tree we discussed in unicast routing, which says that the shortest path from A to B is also the shortest path from B to A. The router does not know the shortest path from the source to itself, but it can find which is the next router in the shortest path from itself to the source (reverse path). The router simply consults its unicast forwarding table, pretending that it wants to send a packet to the source; the forwarding table gives the next router and the interface the message that the packet should be sent out in this reverse direction. The router uses this information to accept a multicast packet only if it arrives from this interface. This is needed to prevent looping. In multicasting, a packet may arrive at the same router that has forwarded it. If the router does not drop all arrived packets except the one, multiple copies of the packet will be circulating in the internet. Of course, the router may add a tag to the packet when it arrives the first time and discard packets that arrive with the same tag, but the RPF strategy is simpler.

Reverse Path Broadcasting (RPB)

The RPF algorithm helps a router to forward only one copy received from a source and drop the rest. However, when we think about broadcasting in the second step, we need to remember that destinations are all the networks (LANs) in the internet. To be efficient, we need to prevent each network from receiving more than one copy of the packet. If a network is connected to more than one router, it may receive a copy of the packet from each router. RPF cannot help here, because a network does not have the intelligence to apply the RPF algorithm; we need to allow only one of the routers attached to a network to pass the packet to the network. One way to do so is to designate only one router as the *parent* of a network related to a specific source. When a router that is not the parent of the attached network receives a multicast packet, it simply drops the packet. There are several ways that the parent of the network related to a network can be selected; one way is to select the router that has the shortest path to the source (using the unicast forwarding table, again in the reverse direction). If there is a tie in this case, the router with the smaller IP address can be selected. The reader may have noticed that **reverse path broadcasting (RPB)** actually creates a broadcast tree from the graph that has been created by the RPF algorithm. RPB has cut those branches of the tree that cause cycles in the graph. If we use the shortest path criteria for choosing the parent router, we have actually created a shortest-path broadcast tree. In other words, after this step, we have a shortest-path tree with the source as the root and all networks (LANs) as the leaves. Every packet started from the source reaches all LANs in the internet travelling the shortest path. Figure 21.11 shows how RPB can avoid duplicate reception in a network by assigning a designated parent router, R1, for network N.

Figure 21.11 *RPF versus RPB*

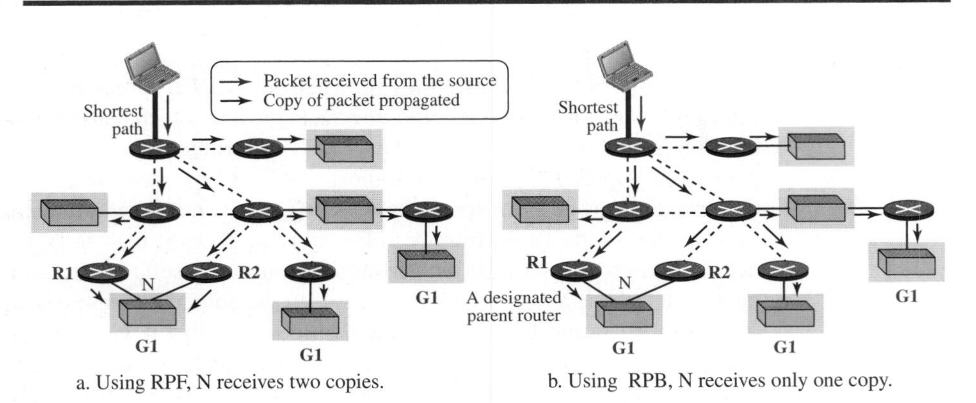

Reverse Path Multicasting (RPM)

As you may have noticed, RPB does not multicast the packet, it broadcasts it. This is not efficient. To increase efficiency, the multicast packet must reach only those networks that have active members for that particular group. This is called ***reverse path multicasting (RPM).*** To change the broadcast shortest-path tree to a multicast shortest-path tree, each router needs to prune (make inactive) the interfaces that do not reach a network with active members corresponding to a particular source-group combination. This step can be done bottom-up, from the leaves to the root. At the leaf level, the routers connected to the network collect the membership information using the IGMP protocol discussed before. The parent router of the network can then disseminate this information upward using the reverse shortest-path tree from the router to the source, the same way as the distance vector messages are passed from one neighbor to another. When a router receives all of these membership-related messages, it knows which interfaces need to be pruned. Of course, since these packets are disseminated periodically, if a new member is added to some networks, all routers are informed and can change the status of their interfaces accordingly. Joining and leaving is continuously applied.

Figure 21.12 shows how pruning in RPM lets only networks with group members receive a copy of the packet unless they are in the path to a network with a member.

Figure 21.12 *RPB versus RPM*

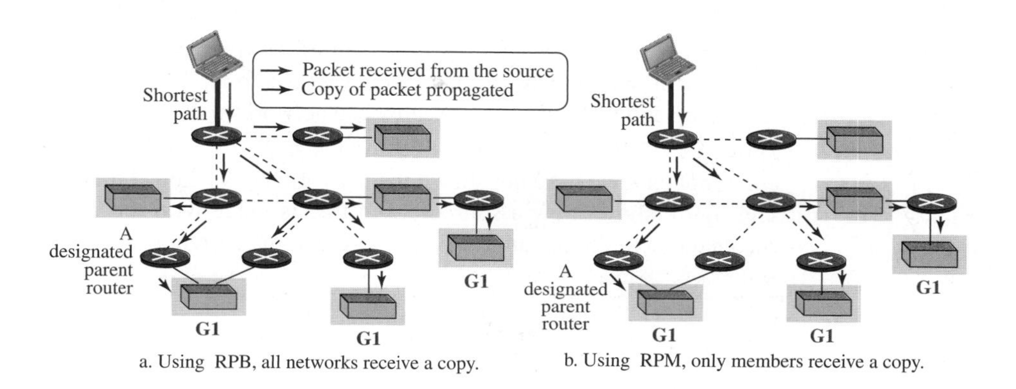

a. Using RPB, all networks receive a copy.

b. Using RPM, only members receive a copy.

21.3.2 Multicast Link State (MOSPF)

Multicast Open Shortest Path First (MOSPF) is the extension of the Open Shortest Path First (OSPF) protocol, which is used in unicast routing. It also uses the source-based tree approach to multicasting. If the internet is running a unicast link-state routing algorithm, the idea can be extended to provide a multicast link-state routing algorithm. Recall that in unicast link-state routing, each router in the internet has a link-state database (LSDB) that can be used to create a shortest-path tree. To extend unicasting to multicasting, each router needs to have another database, as with the case of unicast distance-vector routing, to show which interface has an active member in a particular group. Now a router goes through the following steps to forward a multicast packet received from source S and to be sent to destination G (a group of recipients):

1. The router uses the Dijkstra algorithm to create a shortest-path tree with S as the root and all destinations in the internet as the leaves. Note that this shortest-path tree is different from the one the router normally uses for unicast forwarding, in which the root of the tree is the router itself. In this case, the root of the tree is the source of the packet defined in the source address of the packet. The router is capable of creating this tree because it has the LSDB, the whole topology of the internet; the Dijkstra algorithm can be used to create a tree with any root, no matter which router is using it. The point we need to remember is that the shortest-path tree created this way depends on the specific source. For each source we need to create a different tree.
2. The router finds itself in the shortest-path tree created in the first step. In other words, the router creates a shortest-path subtree with itself as the root of the subtree.
3. The shortest-path subtree is actually a broadcast subtree with the router as the root and all networks as the leaves. The router now uses a strategy similar to the one we

describe in the case of DVMRP to prune the broadcast tree and to change it to a multicast tree. The IGMP protocol is used to find the information at the leaf level. MOSPF has added a new type of link state update packet that floods the membership to all routers. The router can use the information it receives in this way and prune the broadcast tree to make the multicast tree.

4. The router can now forward the received packet out of only those interfaces that correspond to the branches of the multicast tree. We need to make certain that a copy of the multicast packet reaches all networks that have active members of the group and that it does not reach those networks that do not.

Figure 21.13 shows an example of using the steps to change a graph to a multicast tree. For simplicity, we have not shown the network, but we added the groups to each router. The figure shows how a source-based tree is made with the source as the root and changed to a multicast subtree with the root at the current router.

Figure 21.13 *Example of tree formation in MOSPF*

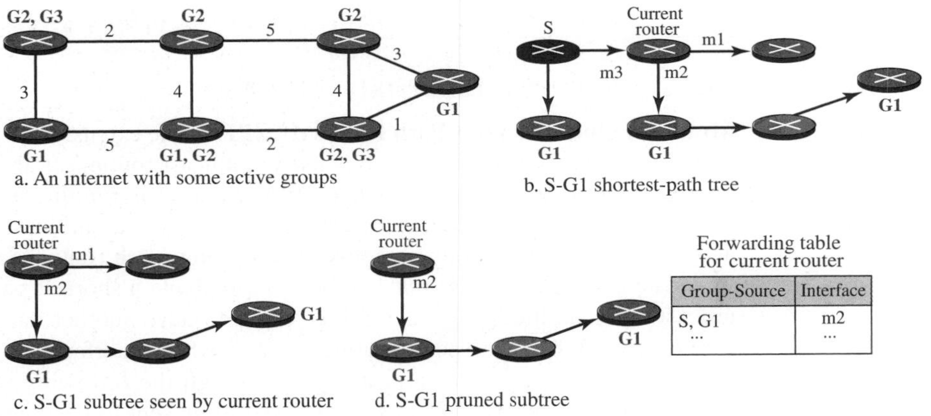

a. An internet with some active groups

b. S-G1 shortest-path tree

c. S-G1 subtree seen by current router

d. S-G1 pruned subtree

Forwarding table for current router

Group-Source	Interface
S, G1	m2
...	...

21.3.3 Protocol Independent Multicast (PIM)

Protocol Independent Multicast (PIM) is the name given to a common protocol that needs a unicast routing protocol for its operation, but the unicast protocol can be either a distance-vector protocol or a link-state protocol. In other words, PIM needs to use the forwarding table of a unicast routing protocol to find the next router in a path to the destination, but it does not matter how the forwarding table is created. PIM has another interesting feature: it can work in two different modes: dense and sparse. The term *dense* here means that the number of active members of a group in the internet is large; the probability that a router has a member in a group is high. This may happen, for example, in a popular teleconference that has a lot of members. The term *sparse,* on the other hand, means that only a few routers in the internet have active members in the group; the probability that a router has a member of the group is low. This may happen, for example, in a very technical teleconference where a number of members are spread somewhere in the internet. When the protocol is working in the dense mode, it is referred to as PIM-DM; when it is working in the sparse mode, it is referred to as PIM-SM. We explain both protocols next.

Protocol Independent Multicast-Dense Mode (PIM-DM)

When the number of routers with attached members is large relative to the number of routers in the internet, PIM works in the dense mode and is called ***PIM-DM***. In this mode, the protocol uses a source-based tree approach and is similar to DVMRP, but simpler. PIM-DM uses only two strategies described in DVMRP: RPF and RPM. But unlike DVMRP, forwarding of a packet is not suspended awaiting pruning of the first subtree. Let us explain the two steps used in PIM-DM to clear the matter.

1. A router that has received a multicast packet from the source S destined for the group G first uses the RPF strategy to avoid receiving a duplicate of the packet. It consults the forwarding table of the underlying unicast protocol to find the next router if it wants to send a message to the source S (in the reverse direction). If the packet has not arrived from the next router in the reverse direction, it drops the packet and sends a prune message in that direction to prevent receiving future packets related to (S, G).
2. If the packet in the first step has arrived from the next router in the reverse direction, the receiving router forwards the packet from all its interfaces except the one from which the packet has arrived and the interface from which it has already received a prune message related to (S, G). Note that this is actually a broadcasting instead of a multicasting if the packet is the first packet from the source S to group G. However, each router downstream that receives an unwanted packet sends a prune message to the router upstream, and eventually the broadcasting is changed to multicasting. Note that DVMRP behaves differently: it requires that the prune messages (which are part of DV packets) arrive and the tree is pruned before sending any message through unpruned interfaces. PIM-DM does not care about this precaution because it assumes that most routers have an interest in the group (the idea of the dense mode).

Figure 21.14 shows the idea behind PIM-DM. The first packet is broadcast to all networks, which have or do not have members. After a prune message arrives from a router with no member, the second packet is only multicast.

Protocol Independent Multicast-Sparse Mode (PIM-SM)

When the number of routers with attached members is small relative to the number of routers in the internet, PIM works in the sparse mode and is called ***PIM-SM***. In this environment, the use of a protocol that broadcasts the packets until the tree is pruned is not justified; PIM-SM uses a group-shared tree approach to multicasting. The core router in PIM-SM is called the *rendezvous point* (RP). Multicast communication is achieved in two steps. Any router that has a multicast packet to send to a group of destinations first encapsulates the multicast packet in a unicast packet (tunneling) and sends it to the RP. The RP then decapsulates the unicast packet and sends the multicast packet to its destination.

PIM-SM uses a complex algorithm to select one router among all routers in the internet as the RP for a specific group. This means that if we have *m* active groups, we need *m* RPs, although a router may serve more than one group. After the RP for each

Figure 21.14 *Idea behind PIM-DM*

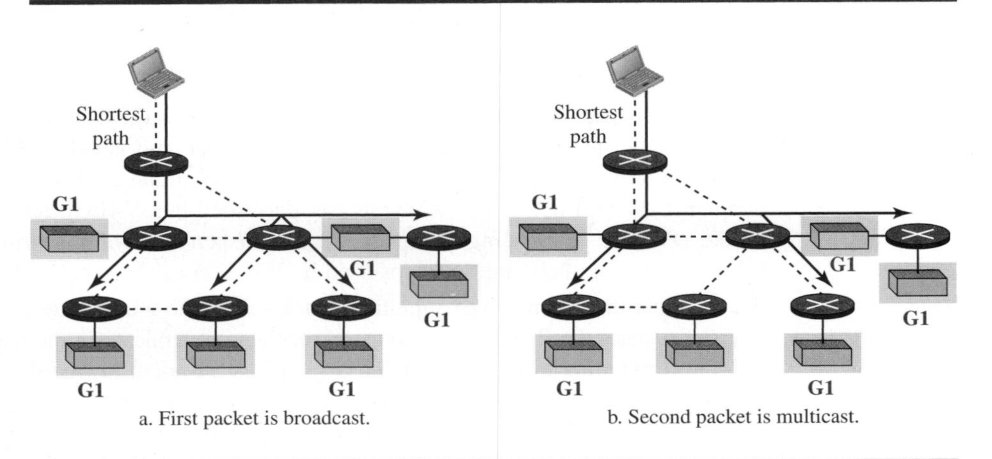

group is selected, each router creates a database and stores the group identifier and the IP address of the RP for tunneling multicast packets to it.

PIM-SM uses a spanning multicast tree rooted at the RP with leaves pointing to designated routers connected to each network with an active member. A very interesting point in PIM-SM is the formation of the multicast tree for a group. The idea is that each router helps to create the tree. The router should know the unique interface from which it should accept a multicast packet destined for a group (what was achieved by RPF in DVMRP). The router should also know the interface or interfaces from which it should send out a multicast packet destined for a group (what was achieved by RPM in DVMRP). To avoid delivering more than one copy of the same packet to a network through several routers (what was achieved by RPB in DVMRP), PIM-SM requires that only designated routers send PIM-SM messages, as we will see shortly.

To create a multicast tree rooted at the RP, PIM-SM uses *join* and *prune* messages. Figure 21.15 shows the operation of join and prune messages in PIM-SM. First, three networks join group G1 and form a multicast tree. Later, one of the networks leaves the group and the tree is pruned.

The join message is used to add possible new branches to the tree; the **prune message** is used to cut branches that are not needed. When a designated router finds out that a network has a new member in the corresponding group (via IGMP), it sends a join message in a unicast packet destined for the RP. The packet travels through the unicast shortest-path tree to reach the RP. Any router in the path receives and forwards the packet, but at the same time, the router adds two pieces of information to its multicast forwarding table. The number of the interface through which the join message has arrived is marked (if not already marked) as one of the interfaces through which the multicast packet destined for the group should be sent out in the future. The router also adds a count to the number of join messages received here, as we discuss shortly. The number of the interface through which the join message was sent to the RP is marked (if not already marked) as the only interface through which the multicast packet destined for the same group should be received. In this way, the first join message sent by a

Figure 21.15 *Idea behind PIM-SM*

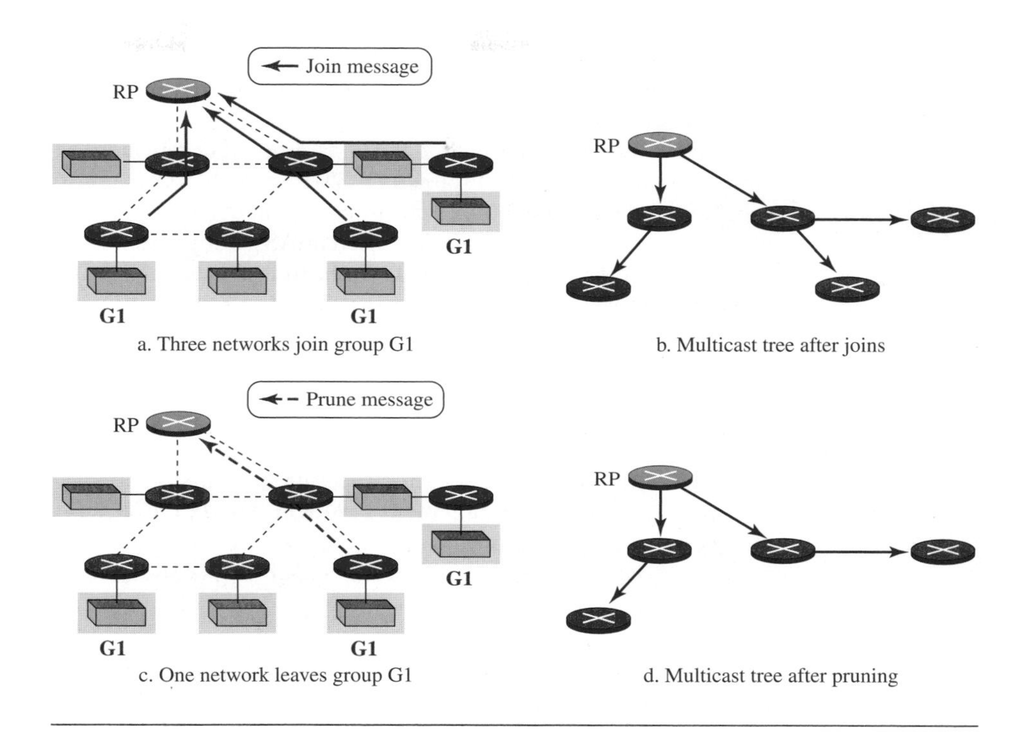

designated router creates a path from the RP to one of the networks with group members.

To avoid sending multicast packets to networks with no members, PIM-SM uses the prune message. Each designated router that finds out (via IGMP) that there is no active member in its network, sends a prune message to the RP. When a router receives a prune message, it decrements the join count for the interface through which the message has arrived and forwards it to the next router. When the join count for an interface reaches zero, that interface is not part of the multicast tree anymore.

21.4 INTERDOMAIN MULTICAST PROTOCOLS

The three protocols we discussed for multicast routing, DVMRP, MOSPF, and PIM, are designed to provide multicast communication inside an autonomous system. When the members of the groups are spread among different domains (ASs), we need an interdomain multicast routing protocol.

One common protocol for interdomain multicast routing is called *Multicast Border Gateway Protocol* (MBGP), which is the extension of BGP we discussed for interdomain unicast routing. MBGP provides two paths between ASs: one for unicasting and one for multicasting. Information about multicasting is exchanged between border routers in different ASs. MBGP is a shared-group multicast routing protocol in which one router in each AS is chosen as the rendezvous point (RP).

The problem with MBGP protocol is that it is difficult to inform an RP about the sources of groups in other ASs. The Multicast Source Discovery Protocol (MSDP) is a new suggested protocol that assigns a source representative router in each AS to inform all RPs about the existence of sources in that AS.

Another protocol that is thought of as a possible replacement for the MBGP is Border Gateway Multicast Protocol (BGMP), which allows construction of shared-group trees with a single root in one of the ASs. In other words, for each group, there is only one shared tree, with leaves in different ASs, but the root is located in one of the ASs. The two problems, of course, are how to designate an AS as the root of the tree and how to inform all sources about the location of the root to tunnel their multicast packets to.

21.5 IGMP

The protocol that is used today for collecting information about group membership is the **Internet Group Management Protocol (IGMP).** IGMP is a protocol defined at the network layer; it is one of the auxiliary protocols, like ICMP, which is considered part of the IP. IGMP messages, like ICMP messages, are encapsulated in an IP datagram.

21.5.1 Messages

There are only two types of messages in IGMP version 3, query and report messages, as shown in Figure 21.16. A query message is periodically sent by a router to all hosts attached to it to ask them to report their interests about membership in groups. A report message is sent by a host as a response to a query message.

Figure 21.16 *IGMP operation*

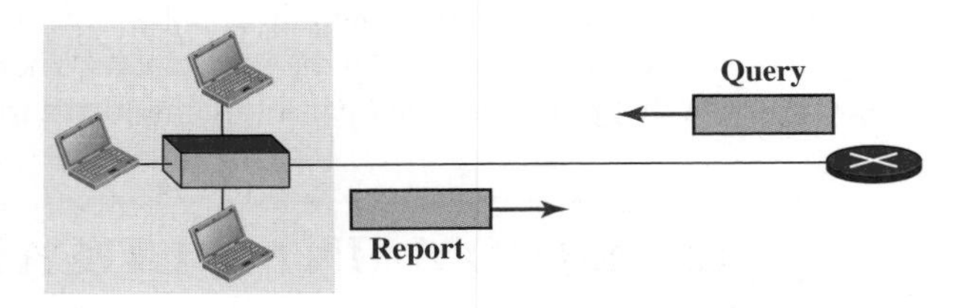

We briefly describe IGMP messages in this section, but the complete description is posted on the book website under the Extra Materials for Chapter 21.

A complete description of IGMP messages is posted on the book website under Extra Materials for Chapter 21.

Query Message

The query message is sent by a router to all hosts in each interface to collect information about their membership. There are three versions of query messages, as described below:

a. A *general* query message is sent about membership in any group. It is encapsulated in a datagram with the destination address 224.0.0.1 (all hosts and routers). Note that all routers attached to the same network receive this message to inform them that this message is already sent and that they should refrain from resending it.

b. A *group-specific* query message is sent from a router to ask about the membership related to a specific group. This is sent when a router does not receive a response about a specific group and wants to be sure that there is no active member of that group in the network. The group identifier (multicast address) is mentioned in the message. The message is encapsulated in a datagram with the destination address set to the corresponding multicast address. Although all hosts receive this message, those not interested drop it.

c. A *source-and-group-specific* query message is sent from a router to ask about the membership related to a specific group when the message comes from a specific source or sources. Again the message is sent when the router does not hear about a specific group related to a specific host or hosts. The message is encapsulated in a datagram with the destination address set to the corresponding multicast address. Although all hosts receive this message, those not interested drop it.

Report Message

A report message is sent by a host as a response to a query message. The message contains a list of records in which each record gives the identifier of the corresponding group (multicast address) and the addresses of all sources that the host is interested in receiving messages from (inclusion). The record can also mention the source addresses from which the host does not desire to receive a group message (exclusion). The message is encapsulated in a datagram with the multicast address 224.0.0.22 (multicast address assigned to IGMPv3). In IGMPv3, if a host needs to join a group, it waits until it receives a query message and then sends a report message. If a host needs to leave a group, it does not respond to a query message. If no other host responds to the corresponding message, the group is purged from the router database.

21.5.2 Propagation of Membership Information

After a router has collected membership information from the hosts and other routers at its own level in the tree, it can propagate it to the router located in a higher level of the tree. Finally, the router at the tree root can get the membership information to build the multicast tree. The process, however, is more complex than what we can explain in one paragraph. Interested readers can check the book website for the complete description of this protocol.

21.5.3 Encapsulation

The IGMP message is encapsulated in an IP datagram with the value of the protocol field set to 2 and the TTL field set to 1. The destination IP address of the datagram, however, depends on the type of message, as shown in Table 21.1.

Table 21.1 *Destination IP Addresses*

Message Type	*IP Address*
General Query	224.0.0.1
Other Queries	Group address
Report	224.0.0.22

21.6 END-CHAPTER MATERIALS

21.6.1 Recommended Reading

Books

Several books give thorough coverage of materials discussed in this chapter. We recommend [Com 06], [Tan 03], [Koz 05], [Ste 95], [GW 04], [Per 00], [Kes 02], [Moy 98], [WZ 01], and [Los 04].

RFCs

Multicasting is discussed in RFCs 1075, 1585, 2189, 2362, and 3376.

21.6.2 Key Terms

Distance Vector Multicast Routing Protocol (DVMRP)
group-shared tree
Internet Group Management Protocol (IGMP)
Multicast Open Shortest Path First (MOSPF)
Protocol Independent Multicast (PIM)
Protocol Independent Multicast-dense Mode (PIM-DM)
Protocol Independent Multicast-Sparse Mode (PIM-SM)
prune message
reverse path broadcasting (RPB)
reverse path forwarding (RPF)
reverse path multicasting (RPM)
shortest path tree
source-based tree
tunneling

21.6.3 Summary

Multicasting is the sending of the same message to more than one receiver simultaneously. Multicasting has many applications including distributed databases, information dissemination, teleconferencing, and distance learning.

In classless addressing the block 224.0.0.0/4 is used for multicast addressing. This block is sometimes referred to as the multicast address space and is divided into several blocks (smaller blocks) for different purposes.

In a source-based tree approach to multicast routing, the source/group combination determines the tree. RPF, RPB, and RPM are efficient improvements to source-based trees. MOSPF, DVMRP, and PIM-DM are three protocols that use source-based tree methods to multicast. In a group-shared approach to multicasting, one rendezvous router takes the responsibility of distributing multicast messages to their destinations. PIM-DM and PIM-SM are examples of group-shared tree protocols.

The Internet Group Management Protocol (IGMP) is involved in collecting local membership group information. The last version of IGMP, IGMPv3, uses two types of messages: query and report. The query message takes three formats: general, group-specific, and group-and-source specific. IGMP messages are encapsulated directly in an IP datagram with the TTL value set to 1.

21.7 PRACTICE SET

21.7.1 Quizzes

A set of interactive quizzes for this chapter can be found on the book website. It is strongly recommended that the student take the quizzes to check his/her understanding of the materials before continuing with the practice set.

21.7.2 Questions

Q21-1. List three steps that a DVMRP router uses to create a source-based tree. Which phase is responsible for creating the part of the tree from the source to the current router? Which phase is responsible for creating a broadcast tree with the router as the root? Which phase is responsible for changing the broadcast tree to a multicast tree?

Q21-2. Assume we have 20 hosts in a small AS. There are only four groups in this AS. Find the number of spanning trees in each of the following approaches:

a. source-based tree **b.** group-shared tree

Q21-3. Distinguish between multicasting and multiple unicasting.

Q21-4. Does RPF actually create a shortest path tree? Explain.

Q21-5. Explain why PIM is called *Protocol Independent Multicast*.

Q21-6. A multicast router is connected to four networks. The interest of each network is shown below. What is the group list that should be advertised by the router?

a. N1: {G1, G2, G3} **b.** N2: {G1, G3} **c.** N3: {G1, G4} **d.** N4: {G1, G3}

Q21-7. It is obvious that we need to have spanning trees for both unicasting and multicasting. How many leaves of the tree are involved in a transmission in each case?

a. a unicast transmission **b.** a multicast transmission

Q21-8. When we send an e-mail to multiple recipients, are we using multicasting or multiple unicasting? Give the reason for your answer.

Q21-9. We say that a router in DVMRP creates a shortest-path tree *on demand*. What is the meaning of this statement? What is the advantage of creating shortest path trees only on demand?

Q21-10. Which version of PIM uses the first and the third steps of DVMRP? What are these two steps?

Q21-11. Does RPB actually create a shortest path tree? Explain. What are the leaves of the tree?

Q21-12. Define the group of each of the following multicast addresses (local network control block, internetwork control block, SSM block, Glop block, or administratively scoped block):

a. 224.0.1.7 **b.** 232.7.14.8 **c.** 239.14.10.12

Q21-13. Can a host have more than one multicast address? Explain.

Q21-14. Explain why an MOSPF router can create the shortest path with the source as the root in one step, but DVMRP needs three steps to do so.

Q21-15. Define which of the following addresses are multicast addresses:

a. 224.8.70.14 **b.** 226.17.3.53 **c.** 240.3.6.25

Q21-16. Does RPM actually create a shortest path tree? Explain. What are the leaves of the tree?

Q21-17. Explain why broadcasting the first or the first few messages is not important in PIM-DM, but it is important in PIM-SM?

21.7.3 Problems

P21-1. Router A sends a unicast RIP update packet to router B that says 134.23.0.0/16 is 7 hops away. Network B sends an update packet to router A that says 13.23.0.0/16 is 4 hops away. If these two routers are connected to the same network, which one is the designated parent router?

P21-2. The AS number in an organization is 24101. Find the range of multicast addresses that the organization can use in the GLOP block.

P21-3. A multicast address for a group is 232.24.60.9. What is its 48-bit Ethernet address for a LAN using TCP/IP?

P21-4. Assume that m is much less than n and that router R is connected to n networks in which only m of these networks are interested in receiving packets related to group G. How can router R manage to send a copy of the packet to only those networks interested in group G?

P21-5. Exactly describe why we cannot use the CIDR notation for the following blocks in Table 21.1:

a. AD HOC block with the range 224.0.2.0 to 224.0.255.255.

b. The first reserved block with the range 224.3.0.0 to 231.255.255.255.

c. The second reserved block with the range 234.0.0.0 to 238.255.255.255.

P21-6. A router using DVMRP receives a packet with source address 10.14.17.2 from interface 2. If the router forwards the packet, what are the contents of the entry related to this address in the unicast routing table?

P21-7. Why is there no need for the IGMP message to travel outside its own network?

P21-8. Change the following IP multicast addresses to Ethernet multicast addresses. How many of them specify the same Ethernet address?

a. 224.18.72.8 **b.** 235.18.72.8 **c.** 237.18.6.88 **d.** 224.88.12.8

P21-9. In the network of Figure 21.17, find the shortest path trees for router R if the network is using MOSPF with the source connected to the router marked as S. Assume that all routers have interest in the corresponding multicast group.

Figure 21.17 *Problem P21-9*

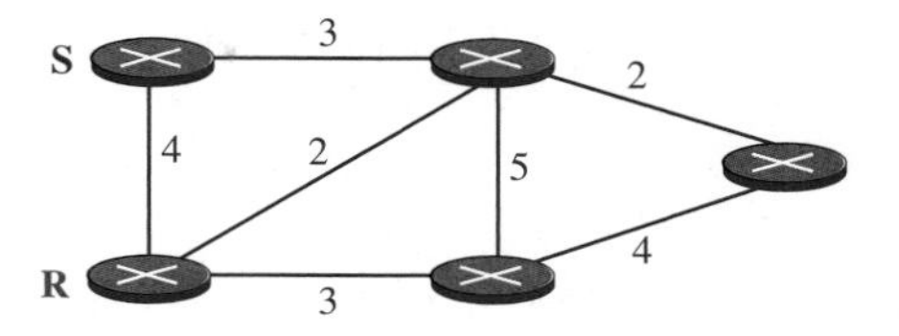

21.8 SIMULATION EXPERIMENTS

21.8.1 Applets

We have created some Java applets to show some of the main concepts discussed in this chapter. It is strongly recommended that the students activate these applets on the book website and carefully examine the protocols in action.

CHAPTER 22

Next Generation IP

The address depletion of IPv4 and other shortcomings of this protocol prompted a new version of IP in the early 1990s. The new version, which is called ***Internet Protocol version 6 (IPv6)*** or ***IP new generation (IPng)*** was a proposal to augment the address space of IPv4 and at the same time redesign the format of the IP packet and revise some auxiliary protocols such as ICMP. It is interesting to know that IPv5 was a proposal, based on the OSI model, that never materialized.

The following lists the main changes in the IPv6 protocol: larger address space, better header format, new options, allowance for extension, support for resource allocation, and support for more security. The implementation of theses changes made it necessary to create a new version of the ICMP protocol, ICMPv6.

This chapter is made of four sections:

- The first section discusses the addressing mechanism in the new generation of the Internet. The section first describes the representation and address space. It then shows the allocation in the address space. The section finally explains autoconfiguration and renumbering, which makes it easy for a host to move from one network to another.
- The second section discusses IPv6 protocol. First the new packet format is described. The section then shows how use of extension headers can replace the options in version 4.
- The third section discusses ICMPv6. The section describes how the new protocol replaces several auxiliary protocols in version 4. The section also divides the messages in this protocol into four categories and describes them.
- The fourth section briefly shows how transition can be made from the current version to the new one smoothly. The section explains three strategies that need to be followed for this smooth transition.

22.1 IPv6 ADDRESSING

The main reason for migration from IPv4 to IPv6 is the small size of the address space in IPv4. In this section, we show how the huge address space of IPv6 prevents address depletion in the future. We also discuss how the new addressing responds to some problems in the IPv4 addressing mechanism. An IPv6 address is 128 bits or 16 bytes (octets) long, four times the address length in IPv4.

22.1.1 Representation

A computer normally stores the address in binary, but it is clear that 128 bits cannot easily be handled by humans. Several notations have been proposed to represent IPv6 addresses when they are handled by humans. The following shows two of these notations: binary and colon hexadecimal.

Binary (128 bits)	**1111111011110110 ... 1111111100000000**
Colon Hexadecimal	**FEF6:BA98:7654:3210:ADEF:BBFF:2922:FF00**

Binary notation is used when the addresses are stored in a computer. The **colon hexadecimal notation** (or *colon hex* for short) divides the address into eight sections, each made of four hexadecimal digits separated by colons.

Abbreviation

Although an IPv6 address, even in hexadecimal format, is very long, many of the digits are zeros. In this case, we can abbreviate the address. The leading zeros of a section can be omitted. Using this form of abbreviation, 0074 can be written as 74, 000F as F, and 0000 as 0. Note that 3210 cannot be abbreviated. Further abbreviation, often called ***zero compression,*** can be applied to colon hex notation if there are consecutive sections consisting of zeros only. We can remove all the zeros and replace them with a double semicolon.

FDEC:0:0:0:0:BBFF:0:FFFF → FDEC::BBFF:0:FFFF

Note that this type of abbreviation is allowed only once per address. If there is more than one run of zero sections, only one of them can be compressed.

Mixed Notation

Sometimes we see a mixed representation of an IPv6 address: colon hex and dotted-decimal notation. This is appropriate during the transition period in which an IPv4 address is embedded in an IPv6 address (as the rightmost 32 bits). We can use the colon hex notation for the leftmost six sections and four-byte dotted-decimal notation instead of the rightmost two sections. However, this happens when all or most of the leftmost sections of the IPv6 address are 0s. For example, the address (::130.24.24.18) is a legitimate address in IPv6, in which the zero compression shows that all 96 leftmost bits of the address are zeros.

CIDR Notation

As we will see shortly, IPv6 uses hierarchical addressing. For this reason, IPv6 allows slash or CIDR notation. For example, the following shows how we can define a prefix of 60 bits using CIDR. We will later show how an IPv6 address is divided into a prefix and a suffix.

FDEC::BBFF:0:FFFF/60

22.1.2 Address Space

The address space of IPv6 contains 2^{128} addresses. This address space is 2^{96} times the IPv4 address—definitely no address depletion—as shown, the size of the space is

340, 282, 366, 920, 938, 463, 374, 607, 431, 768, 211, 456.

To give some idea about the number of addresses, we assume that only 1/64 (almost 2 percent) of the addresses in the space can be assigned to the people on planet Earth and the rest are reserved for special purposes. We also assume that the number of people on the earth is soon to be 2^{34} (more than 16 billion). Each person can have 2^{88} addresses to use. Address depletion in this version is impossible.

Three Address Types

In IPv6, a destination address can belong to one of three categories: unicast, anycast, and multicast.

Unicast Address

A unicast address defines a single interface (computer or router). The packet sent to a unicast address will be routed to the intended recipient.

Anycast Address

An **anycast address** defines a group of computers that all share a single address. A packet with an anycast address is delivered to only one member of the group, the most reachable one. An anycast communication is used, for example, when there are several servers that can respond to an inquiry. The request is sent to the one that is most reachable. The hardware and software generate only one copy of the request; the copy reaches only one of the servers. IPv6 does not designate a block for anycasting; the addresses are assigned from the unicast block.

Multicast Address

A multicast address also defines a group of computers. However, there is a difference between anycasting and multicasting. In anycasting, only one copy of the packet is sent to one of the members of the group; in multicasting each member of the group receives a copy. As we will see shortly, IPv6 has designated a block for multicasting from which the same address is assigned to the members of the group. It is interesting that IPv6 does not define broadcasting, even in a limited version. IPv6 considers broadcasting as a special case of multicasting.

22.1.3 Address Space Allocation

Like the address space of IPv4, the address space of IPv6 is divided into several blocks of varying size and each block is allocated for a special purpose. Most of the blocks are still unassigned and have been set aside for future use. Table 22.1 shows only the assigned blocks. In this table, the last column shows the fraction each block occupies in the whole address space.

Table 22.1 *Prefixes for assigned IPv6 addresses*

Block prefix	*CIDR*	*Block assignment*	*Fraction*
0000 0000	0000::/8	Special addresses	1/256
001	**2000::/3**	**Global unicast**	**1/8**
1111 110	FC00::/7	Unique local unicast	1/128
1111 1110 10	FE80::/10	Link local addresses	1/1024
1111 1111	FF00::/8	Multicast addresses	1/256

Global Unicast Addresses

The block in the address space that is used for unicast (one-to-one) communication between two hosts in the Internet is called the *global unicast address block*. CIDR for the block is 2000::/3, which means that the three leftmost bits are the same for all addresses in this block (001). The size of this block is 2^{125} bits, which is more than enough for Internet expansion for many years to come. An address in this block is divided into three parts: *global routing prefix* (n bits), *subnet identifier* (m bits), and *interface identifier* (q bits), as shown in Figure 22.1. The figure also shows the recommended length for each part.

Figure 22.1 *Global unicast address*

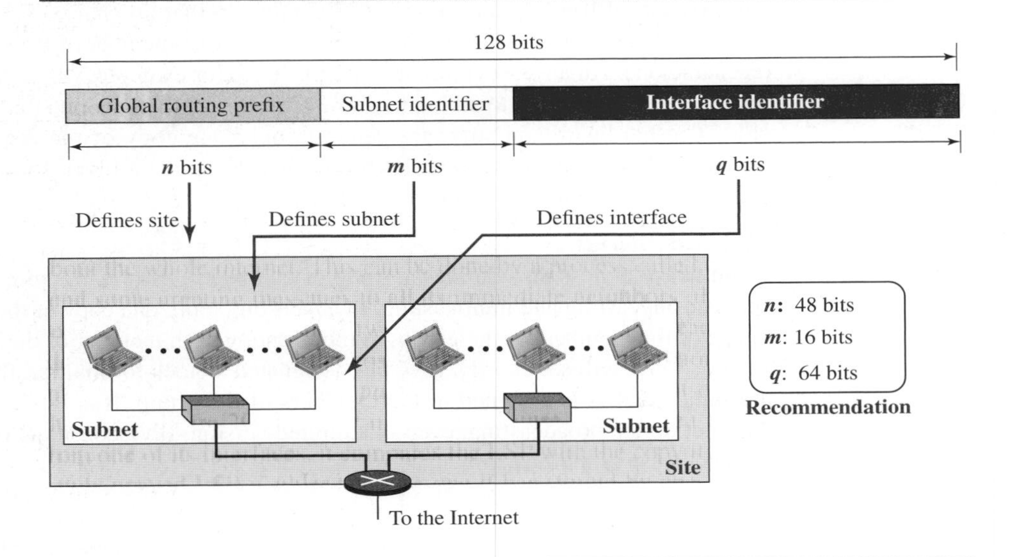

The global routing prefix is used to route the packet through the Internet to the organization site, such as the ISP that owns the block. Since the first three bits in this part are fixed (001), the rest of the 45 bits can be defined for up to 2^{45} sites (a private organization or an ISP). The global routers in the Internet route a packet to its destination site based on the value of *n*. The next *m* bits (16 bits based on recommendation) define a subnet in an organization. This means that an organization can have up to 2^{16} = 65,536 subnets, which is more than enough.

The last *q* bits (64 bits based on recommendation) define the interface identifier. The interface identifier is similar to hostid in IPv4 addressing, although the term *interface identifier* is a better choice because, as we discussed earlier, the host identifier actually defines the interface, not the host. If the host is moved from one interface to another, its IP address needs to be changed.

In IPv4 addressing, there is not a specific relation between the hostid (at the IP level) and link-layer address (at the data-link layer) because the link-layer address is normally much longer than the hostid. The IPv6 addressing allows this relationship. A link-layer address whose length is less than 64 bits can be embedded as the whole or part of the interface identifier, eliminating the mapping process. Two common link-layer addressing schemes can be considered for this purpose: the 64-bit extended unique identifier (EUI-64) defined by IEEE and the 48-bit link-layer address defined by Ethernet.

Mapping EUI-64

To map a 64-bit physical address, the global/local bit of this format needs to be changed from 0 to 1 (local to global) to define an interface address, as shown in Figure 22.2.

Figure 22.2 *Mapping for EUI-64*

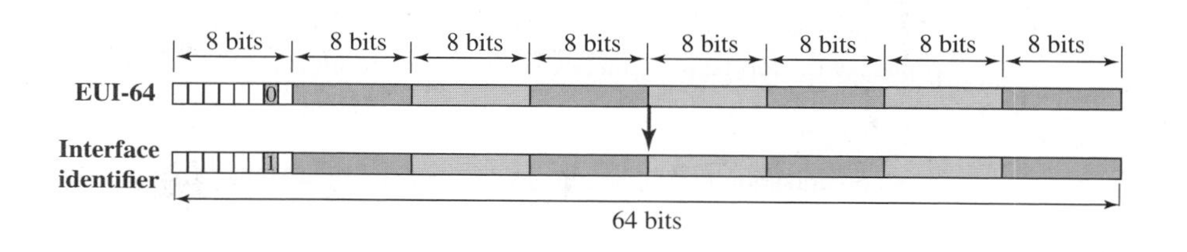

Mapping Ethernet MAC Address

Mapping a 48-bit Ethernet address into a 64-bit interface identifier is more involved. We need to change the local/global bit to 1 and insert an additional 16 bits. The additional 16 bits are defined as 15 ones followed by one zero, or $FFFE_{16}$. Figure 22.3 shows the mapping.

Example 22.1

An organization is assigned the block 2000:1456:2474/48. What is the CIDR notation for the blocks in the first and second subnets in this organization?

Figure 22.3 *Mapping for Ethernet MAC*

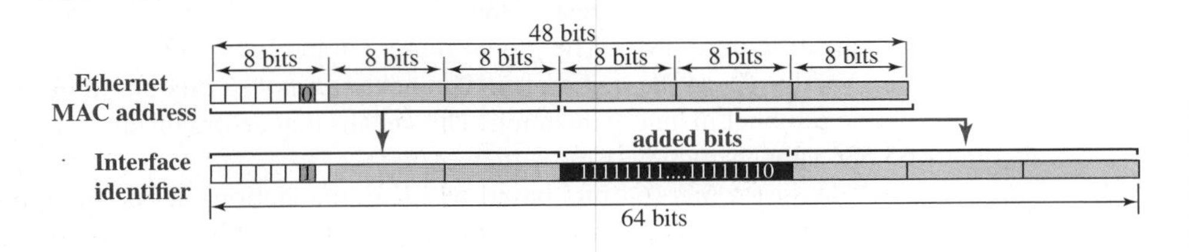

Solution

Theoretically, the first and second subnets should use the blocks with subnet identifier 0001_{16} and 0002_{16}. This means that the blocks are 2000:1456:2474:0000/64 and 2000:1456:2474:0001/64.

Example 22.2

Using the format we defined for Ethernet addresses, find the interface identifier if the physical address in the EUI is $\mathbf{(F5\text{-}A9\text{-}23\text{-}EF\text{-}07\text{-}14\text{-}7A\text{-}D2)}_{16}$.

Solution

We only need to change the seventh bit of the first octet from 0 to 1 and change the format to colon hex notation. The result is **F7A9:23EF:0714:7AD2.**

Example 22.3

Using the format we defined for Ethernet addresses, find the interface identifier if the Ethernet physical address is $\mathbf{(F5\text{-}A9\text{-}23\text{-}14\text{-}7A\text{-}D2)}_{16}$.

Solution

We only need to change the seventh bit of the first octet from 0 to 1, insert two octets FFFE_{16} and change the format to colon hex notation. The result is **F7A9:23FF:FE14:7AD2** in colon hex.

Example 22.4

An organization is assigned the block 2000:1456:2474/48. What is the IPv6 address of an interface in the third subnet if the IEEE physical address of the computer is $\mathbf{(F5\text{-}A9\text{-}23\text{-}14\text{-}7A\text{-}D2)}_{16}$?

Solution

The interface identifier for this interface is **F7A9:23FF:FE14:7AD2** (see Example 22.3). If we append this identifier to the global prefix and the subnet identifier, we get:

2000:1456:2474:0003:F7A9:23FF:FE14:7AD2/128

Special Addresses

After discussing the global unicast block, let us discuss the characteristics and purposes of assigned and reserved blocks in the first row of Table 22.1. Addresses that use the prefix (**0000::/8**) are reserved, but part of this block is used to define some special addresses. Figure 22.4 shows the assigned addresses in this block.

Figure 22.4 *Special addresses*

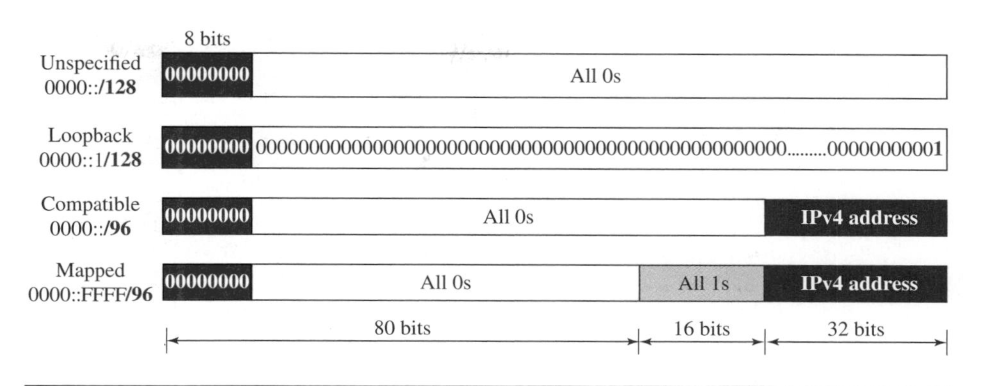

The unspecified address is a subblock containing only one address, which is used during bootstrap when a host does not know its own address and wants to send an inquiry to find it (see DHCP section).

The loopback address also consists of one address. We discussed loopback addresses for IPv4 before. In IPv4 the block is made of a range of addresses; in IPv6, the block has only a single address in it.

As we will see later, during the transition from IPv4 to IPv6, hosts can use their IPv4 addresses embedded in IPv6 addresses. Two formats have been designed for this purpose: compatible and mapped. A **compatible address** is an address of 96 bits of zero followed by 32 bits of IPv4 address. It is used when a computer using IPv6 wants to send a message to another computer using IPv6. A **mapped address** is used when a computer already migrated to version 6 wants to send an address to a computer still using version 4. A very interesting point about mapped and compatible addresses is that they are designed such that, when calculating the checksum, one can use either the embedded address or the total address because extra 0s or 1s in multiples of 16 do not have any effect in checksum calculation. This is important for UDP and TCP, which use a pseudoheader to calculate the checksum, because the checksum calculation is not affected if the address of the packet is changed from IPv6 to IPv4 by a router.

Other Assigned Blocks

IPv6 uses two large blocks for private addressing and one large block for multicasting, as shown in Figure 22.5. A subblock in a **unique local unicast block** can be privately created and used by a site. The packet carrying this type of address as the destination address is not expected to be routed. This type of address has the identifier 1111 110, the next bit can be 0 or 1 to define how the address is selected (locally or by an authority). The next 40 bits are selected by the site using a randomly generated number of length 40 bits. This means that the total of 48 bits defines a subblock that looks like a global unicast address. The 40-bit random number makes the probability of duplication of the address extremely small. Note the similarity between the format of these addresses and the global unicast. The second block, designed for private addresses, is the **link local block.** A subblock in this block can be used as

Figure 22.5 *Unique local unicast block*

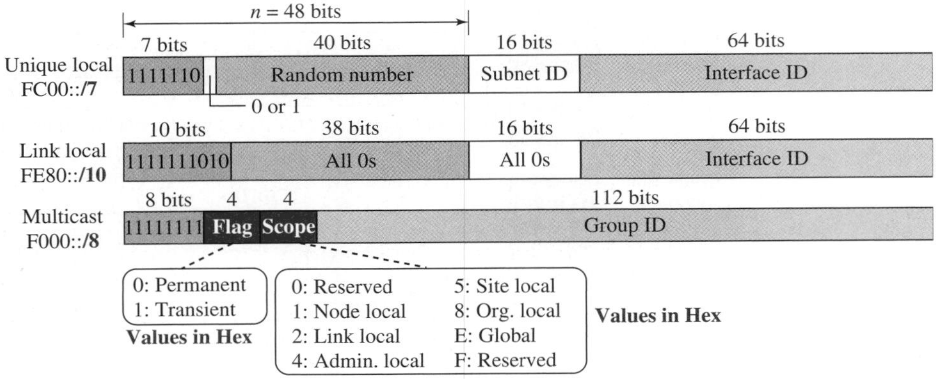

a private address in a network. This type of address has the block identifier 1111111010. The next 54 bits are set to zero. The last 64 bits can be changed to define the interface for each computer. Note the similarity between the format of these addresses and the global unicast address.

We discussed multicast addresses of IPv4 earlier in the chapter. Multicast addresses are used to define a group of hosts instead of just one. In IPv6 a large block of addresses are assigned for multicasting. All these addresses use the prefix 11111111. The second field is a flag that defines the group address as either permanent or transient. A permanent group address is defined by the Internet authorities and can be accessed at all times. A transient group address, on the other hand, is used only temporarily. Systems engaged in a teleconference, for example, can use a transient group address. The third field defines the scope of the group address. Many different scopes have been defined, as shown in the figure.

22.1.4 Autoconfiguration

One of the interesting features of IPv6 addressing is the **autoconfiguration** of hosts. As we discussed in IPv4, the host and routers are originally configured manually by the network manager. However, the Dynamic Host Configuration Protocol, DHCP, can be used to allocate an IPv4 address to a host that joins the network. In IPv6, DHCP protocol can still be used to allocate an IPv6 address to a host, but a host can also configure itself.

When a host in IPv6 joins a network, it can configure itself using the following process:

1. The host first creates a **link local address** for itself. This is done by taking the 10-bit link local prefix (1111 1110 10), adding 54 zeros, and adding the 64-bit interface identifier, which any host knows how to generate from its interface card. The result is a 128-bit link local address.
2. The host then tests to see if this link local address is unique and not used by other hosts. Since the 64-bit interface identifier is supposed to be unique, the link local

address generated is unique with a high probability. However, to be sure, the host sends a *neighbor solicitation message* (see Chapter 28) and waits for a *neighbor advertisement message*. If any host in the subnet is using this link local address, the process fails and the host cannot autoconfigure itself; it needs to use other means such as DHCP for this purpose.

3. If the uniqueness of the link local address is passed, the host stores this address as its link local address (for private communication), but it still needs a global unicast address. The host then sends a *router solicitation message* (discussed later in the chapter) to a local router. If there is a router running on the network, the host receives a *router advertisement message* that includes the global unicast prefix and the subnet prefix that the host needs to add to its interface identifier to generate its global unicast address. If the router cannot help the host with the configuration, it informs the host in the *router advertisement message* (by setting a flag). The host then needs to use other means for configuration.

Example 22.5

Assume a host with Ethernet address $(\textbf{F5-A9-23-11-9B-E2})_{16}$ has joined the network. What would be its global unicast address if the global unicast prefix of the organization is 3A21:1216:2165 and the subnet identifier is A245:1232?

Solution

The host first creates its interface identifier as **F7A9:23FF:FE11:9BE2** using the Ethernet address read from its card. The host then creates its link local address as:

FE80::F7A9:23FF:FE11:9BE2

Assuming that this address is unique, the host sends a router solicitation message and receives the router advertisement message that announces the combination of global unicast prefix and the subnet identifier as 3A21:1216:2165:A245:1232. The host then appends its interface identifier to this prefix to find and store its global unicast address as:

3A21:1216:2165:A245:1232:F7A9:23FF:FE11:9BE2

22.1.5 Renumbering

To allow sites to change the service provider, **renumbering** of the address prefix (*n*) was built into IPv6 addressing. As we discussed before, each site is given a prefix by the service provider to which it is connected. If the site changes the provider, the address prefix needs to be changed. A router to which the site is connected can advertise a new prefix and let the site use the old prefix for a short time before disabling it. In other words, during the transition period, a site has two prefixes. The main problem in using the renumbering mechanism is the support of the DNS, which needs to propagate the new addressing associated with a domain name. A new protocol for DNS, called Next Generation DNS, is under study to provide support for this mechanism.

22.2 THE IPv6 PROTOCOL

The change of the IPv6 address size requires the change in the IPv4 packet format. The designer of IPv6 decided to implement remedies for other shortcomings now that a change is inevitable. The following shows other changes implemented in the protocol in addition to changing address size and format.

- ***Better header format.*** IPv6 uses a new header format in which options are separated from the base header and inserted, when needed, between the base header and the data. This simplifies and speeds up the routing process because most of the options do not need to be checked by routers.
- ***New options.*** IPv6 has new options to allow for additional functionalities.
- ***Allowance for extension.*** IPv6 is designed to allow the extension of the protocol if required by new technologies or applications.
- ***Support for resource allocation.*** In IPv6, the type-of-service field has been removed, but two new fields, traffic class and flow label, have been added to enable the source to request special handling of the packet. This mechanism can be used to support traffic such as real-time audio and video.
- ***Support for more security.*** The encryption and authentication options in IPv6 provide confidentiality and integrity of the packet.

22.2.1 Packet Format

The IPv6 packet is shown in Figure 22.6. Each packet is composed of a base header followed by the payload. The base header occupies 40 bytes, whereas payload can be up to 65,535 bytes of information. The description of fields follows.

Figure 22.6 *IPv6 datagram*

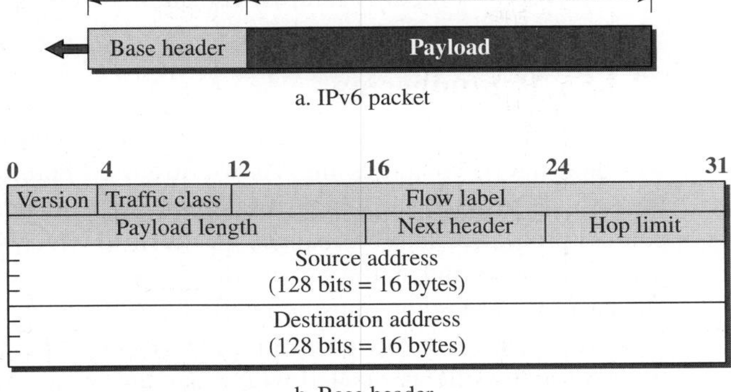

- ***Version.*** The 4-bit version field defines the version number of the IP. For IPv6, the value is 6.
- ***Traffic class.*** The 8-bit traffic class field is used to distinguish different payloads with different delivery requirements. It replaces the *type-of-service* field in IPv4.
- ***Flow label.*** The flow label is a 20-bit field that is designed to provide special handling for a particular flow of data. We will discuss this field later.
- ***Payload length.*** The 2-byte payload length field defines the length of the IP datagram excluding the header. Note that IPv4 defines two fields related to the length: header length and total length. In IPv6, the length of the base header is fixed (40 bytes); only the length of the payload needs to be defined.
- ***Next header.*** The **next header** is an 8-bit field defining the type of the first extension header (if present) or the type of the data that follows the base header in the datagram. This field is similar to the protocol field in IPv4, but we talk more about it when we discuss the payload.
- ***Hop limit.*** The 8-bit hop limit field serves the same purpose as the TTL field in IPv4.
- ***Source and destination addresses.*** The source address field is a 16-byte (128-bit) Internet address that identifies the original source of the datagram. The destination address field is a 16-byte (128-bit) Internet address that identifies the destination of the datagram.
- ***Payload.*** Compared to IPv4, the payload field in IPv6 has a different format and meaning, as shown in Figure 22.7.

Figure 22.7 *Payload in an IPv6 datagram*

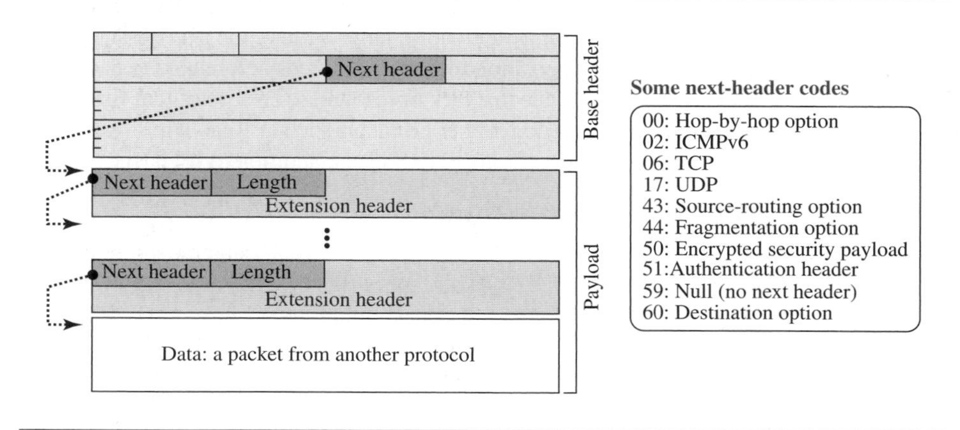

The payload in IPv6 means a combination of zero or more extension headers (options) followed by the data from other protocols (UDP, TCP, and so on). In IPv6, options, which are part of the header in IPv4, are designed as extension headers. The payload can have as many extension headers as required by the situation. Each extension header has two mandatory fields, next header and the length, followed by information related to the particular option. Note that each next header field value (code) defines the type of the next header (hop-by-hop option, source-routing option, . . .); the last next header field defines the protocol (UDP, TCP, . . .) that is carried by the datagram.

Concept of Flow and Priority in IPv6

The IP protocol was originally designed as a connectionless protocol. However, the tendency is to use the IP protocol as a connection-oriented protocol. The MPLS technology described earlier allows us to encapsulate an IPv4 packet in an MPLS header using a label field. In version 6, the flow label has been directly added to the format of the IPv6 datagram to allow us to use IPv6 as a connection-oriented protocol.

To a router, a flow is a sequence of packets that share the same characteristics, such as traveling the same path, using the same resources, having the same kind of security, and so on. A router that supports the handling of flow labels has a flow label table. The table has an entry for each active flow label; each entry defines the services required by the corresponding flow label. When the router receives a packet, it consults its flow label table to find the corresponding entry for the flow label value defined in the packet. It then provides the packet with the services mentioned in the entry. However, note that the flow label itself does not provide the information for the entries of the flow label table; the information is provided by other means, such as the hop-by-hop options or other protocols.

In its simplest form, a flow label can be used to speed up the processing of a packet by a router. When a router receives a packet, instead of consulting the forwarding table and going through a routing algorithm to define the address of the next hop, it can easily look in a flow label table for the next hop.

In its more sophisticated form, a flow label can be used to support the transmission of real-time audio and video. Real-time audio or video, particularly in digital form, requires resources such as high bandwidth, large buffers, long processing time, and so on. A process can make a reservation for these resources beforehand to guarantee that real-time data will not be delayed due to a lack of resources. The use of real-time data and the reservation of these resources require other protocols such as Real-Time Transport Protocol (RTP) and Resource Reservation Protocol (RSVP) in addition to IPv6 (see Chapter 28).

Fragmentation and Reassembly

There are still fragmentation and reassembly of datagrams in the IPv6 protocol, but there is a major difference in this respect. IPv6 datagrams can be fragmented only by the source, not by the routers; the reassembly takes place at the destination. The fragmentation of packets at routers is not allowed to speed up the processing of packets in the router. The fragmentation of a packet in a router needs a lot of processing. The packet needs to be fragmented, all fields related to the fragmentation need to be recalculated. In IPv6, the source can check the size of the packet and make the decision to fragment the packet or not. When a router receives the packet, it can check the size of the packet and drop it if the size is larger than allowed by the MTU of the network ahead. The router then sends a packet-too-big ICMPv6 error message (discussed later) to inform the source.

22.2.2 Extension Header

An IPv6 packet is made of a base header and some extension headers. The length of the base header is fixed at 40 bytes. However, to give more functionality to the IP datagram, the base header can be followed by up to six **extension headers.** Many of these headers are options in IPv4. Six types of extension headers have been defined. These are hop-by-hop option, source routing, fragmentation, authentication, encrypted security payload, and destination option (see Figure 22.8).

Figure 22.8 *Extension header types*

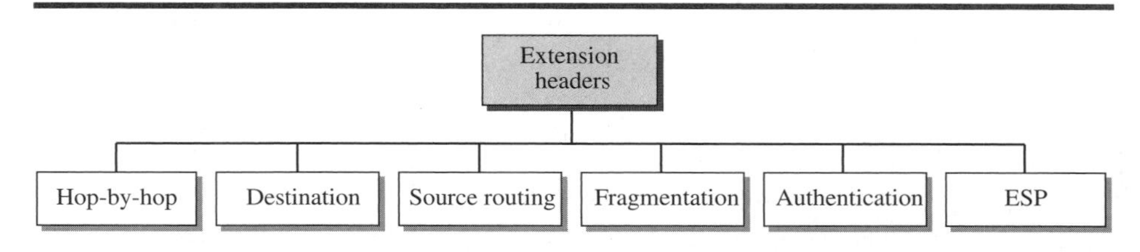

We briefly describe the extension headers in this section, but the complete description is posted at the book website.

Complete descriptions of extension headers are posted on the book website under Extra Materials for Chapter 22.

Hop-by-Hop Option

The *hop-by-hop option* is used when the source needs to pass information to all routers visited by the datagram. For example, perhaps routers must be informed about certain management, debugging, or control functions. Or, if the length of the datagram is more than the usual 65,535 bytes, routers must have this information. So far, only three hop-by-hop options have been defined: Pad1, PadN, and jumbo payload.

- ***Pad1.*** This option is 1 byte long and is designed for alignment purposes. Some options need to start at a specific bit of the 32-bit word. If an option falls short of this requirement by exactly one byte, Pad1 is added.
- ***PadN.*** PadN is similar in concept to Pad1. The difference is that PadN is used when 2 or more bytes are needed for alignment.
- ***Jumbo payload.*** Recall that the length of the payload in the IP datagram can be a maximum of 65,535 bytes. However, if for any reason a longer payload is required, we can use the jumbo payload option to define this longer length.

Destination Option

The **destination option** is used when the source needs to pass information to the destination only. Intermediate routers are not permitted access to this information. The format of the destination option is the same as the hop-by-hop option. So far, only the Pad1 and PadN options have been defined.

Source Routing

The source routing extension header combines the concepts of the strict source route and the loose source route options of IPv4.

Fragmentation

The concept of **fragmentation** in IPv6 is the same as that in IPv4. However, the place where fragmentation occurs differs. In IPv4, the source or a router is required to fragment if the size of the datagram is larger than the MTU of the network over which the datagram travels. In IPv6, only the original source can fragment. A source must use a **Path MTU Discovery technique** to find the smallest MTU supported by any network on the path. The source then fragments using this knowledge.

If the source does not use a Path MTU Discovery technique, it fragments the datagram to a size of 1280 bytes or smaller. This is the minimum size of MTU required for each network connected to the Internet.

Authentication

The **authentication** extension header has a dual purpose: it validates the message sender and ensures the integrity of data. The former is needed so the receiver can be sure that a message is from the genuine sender and not from an imposter. The latter is needed to check that the data is not altered in transition by some hacker. We discuss more about authentication in Chapters 31 and 32.

Encrypted Security Payload

The **encrypted security payload (ESP)** is an extension that provides confidentiality and guards against eavesdropping. Again, we discuss providing more confidentiality for IP packets in Chapter 32.

Comparison of Options between IPv4 and IPv6

The following shows a quick comparison between the options used in IPv4 and the options used in IPv6 (as extension headers).

- The no-operation and end-of-option options in IPv4 are replaced by Pad1 and PadN options in IPv6.
- The record route option is not implemented in IPv6 because it was not used.
- The timestamp option is not implemented because it was not used.
- The source route option is called the *source route extension header* in IPv6.
- The fragmentation fields in the base header section of IPv4 have moved to the fragmentation extension header in IPv6.
- The authentication extension header is new in IPv6.
- The encrypted security payload extension header is new in IPv6.

22.3 THE ICMPv6 PROTOCOL

Another protocol that has been modified in version 6 of the TCP/IP protocol suite is ICMP. This new version, Internet Control Message Protocol version 6 (ICMPv6), follows the same strategy and purposes of version 4. ICMPv6, however, is more complicated than ICMPv4: some protocols that were independent in version 4 are now part of ICMPv6 and some new messages have been added to make it more useful. Figure 22.9 compares the network layer of version 4 to that of version 6. The ICMP, ARP (discussed in Chapter 9), and IGMP protocols in version 4 (Chapter 21) are combined into one single protocol, ICMPv6.

Figure 22.9 *Comparison of network layer in version 4 and version 6*

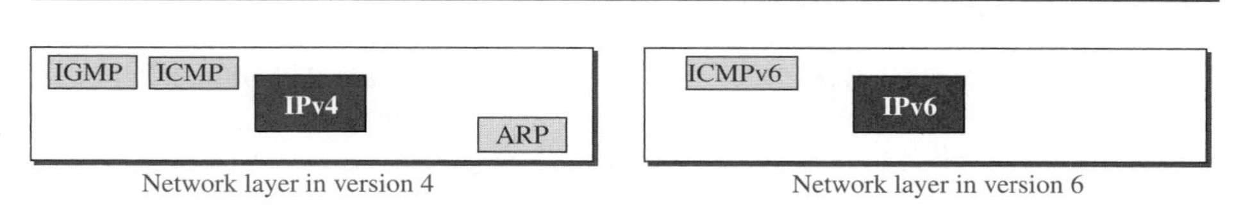

We briefly describe the ICMPv6 in this section, but the complete description is posted at the book website.

Complete descriptions of ICMPv6 messages are posted on the book website under Extra Materials for Chapter 22.

We can divide the messages in ICMPv6 into four groups: error-reporting messages, informational messages, neighbor-discovery messages, and group-membership messages, as shown in Figure 22.10.

Figure 22.10 *Categories of ICMPv6 messages*

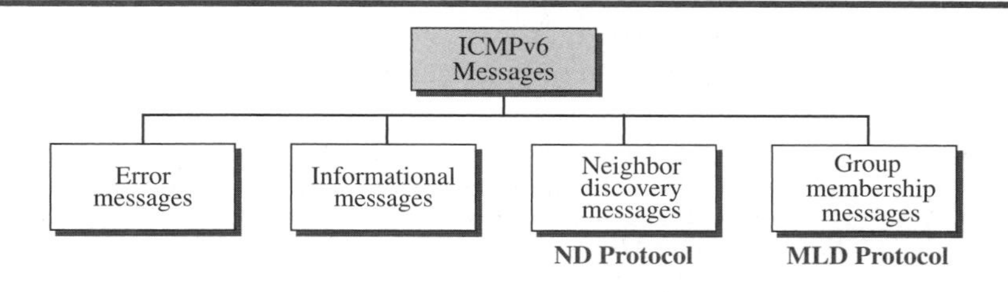

22.3.1 Error-Reporting Messages

As we saw in our discussion of version 4, one of the main responsibilities of ICMPv6 is to report errors. Four types of errors are handled: destination unreachable, packet too big, time exceeded, and parameter problems. Note that the source-quenched message, which is used to control congestion in version 4, is eliminated in this version because the priority and flow label fields in IPv6 are supposed to take care of congestion. The redirection message has moved from the error-reporting category to the neighbor-discovery category, so we discuss it as part of the neighbor-discovery messages.

ICMPv6 forms an error packet, which is then encapsulated in an IPv6 datagram. This is delivered to the original source of the failed datagram.

Destination-Unreachable Message

The concept of the destination unreachable message is the same as described for ICMPv4. When a router cannot forward a datagram or a host cannot deliver the content of the datagram to the upper layer protocol, the router or the host discards the datagram and sends a *destination-unreachable* error message to the source host.

Packet-Too-Big Message

This is a new type of message added to version 6. Since IPv6 does not fragment at the router, if a router receives a datagram that is larger than the maximum transmission unit (MTU) size of the network through which the datagram should pass, two things happen. First, the router discards the datagram. Second, an ICMP error packet—a *packet-too-big message*—is sent to the source.

Time-Exceeded Message

A *time-exceeded* error message is generated in two cases: when the *time to live* value becomes zero and when not all fragments of a datagram have arrived in the time limit. The format of the *time-exceeded* message in version 6 is similar to the one in version 4. The only difference is that the type value has changed to 3.

Parameter-Problem Message

Any ambiguity in the header of the datagram can create serious problems as the datagram travels through the Internet. If a router or the destination host discovers any ambiguous or missing value in any field, it discards the datagram and sends a *parameter-problem* message to the source. The message in ICMPv6 is similar to its version 4 counterpart.

22.3.2 Informational Messages

Two of the ICMPv6 messages can be categorized as informational messages: echo request and echo reply messages. The echo-request and echo-reply messages are designed to check whether two devices in the Internet can communicate with each other. A host or router can send an echo-request message to another host; the receiving computer or router can reply using the echo-reply message.

Echo-Request Message

The idea and format of the echo-request message is the same as the one in version 4.

Echo-Reply Message

The idea and format of the echo-reply message is the same as the one in version 4.

22.3.3 Neighbor-Discovery Messages

Several messages in ICMPv4 have been redefined in ICMPv6 to handle the issue of neighbor discovery. Some new messages have also been added to provide extension. The most important issue is the definition of two new protocols that clearly define the functionality of these group messages: the *Neighbor-Discovery (ND) protocol* and the *Inverse-Neighbor-Discovery (IND) protocol.* These two protocols are used by nodes (hosts or routers) on the same link (network) for three main purposes:

1. Hosts use the ND protocol to find routers in the neighborhood that will forward packets for them.
2. Nodes use the ND protocol to find the link-layer addresses of neighbors (nodes attached to the same network).
3. Nodes use the IND protocol to find the IPv6 addresses of neighbors.

Router-Solicitation Message

The idea behind the *router-solicitation* message is the same as in version 4. A host uses the router-solicitation message to find a router in the network that can forward an IPv6 datagram for the host. The only option that is so far defined for this message is the inclusion of the physical (data-link layer) address of the host to make the response easier for the router.

Router-Advertisement Message

The *router-advertisement* message is sent by a router in response to a router solicitation message.

Neighbor-Solicitation Message

As previously mentioned, the network layer in version 4 contains an independent protocol called Address Resolution Protocol (ARP). In version 6, this protocol is eliminated, and its duties are included in ICMPv6. The neighbor solicitation message has the same duty as the ARP request message. This message is sent when a host or router has a message to send to a neighbor. The sender knows the IP address of the receiver, but needs the data-link address of the receiver. The data-link address is needed for the IP datagram to be encapsulated in a frame. The only option announces the sender data-link address for the convenience of the receiver. The receiver can use the sender data-link address to send a unicast response.

Neighbor-Advertisement Message

The *neighbor-advertisement* message is sent in response to the neighbor-solicitation message.

Redirection Message

The purpose of the redirection message is the same as described for version 4. However, the format of the packet now accommodates the size of the IP address in version 6. Also, an option is added to let the host know the physical address of the target router.

Inverse-Neighbor-Solicitation Message

The *inverse-neighbor-solicitation* message is sent by a node that knows the link-layer address of a neighbor, but not the neighbor's IP address. The message is encapsulated in an IPv6 datagram using an all-node multicast address. The sender must send the following two pieces of information in the option field: its link-layer address and the link-layer address of the target node. The sender can also include its IP address and the MTU value for the link.

Inverse-Neighbor-Advertisement Message

The *inverse-neighbor-advertisement* message is sent in response to the inverse-neighbor-discovery message. The sender of this message must include the link-layer address of the sender and the link-layer address of the target node in the option section.

22.3.4 Group Membership Messages

The management of multicast delivery handling in IPv4 is given to the IGMPv3 protocol. In IPv6, this responsibility is given to the *Multicast Listener Delivery* protocol. MLDv1 is the counterpart to IGMPv2; MLDv2 is the counterpart to IGMPv3. The material discussed in this section is taken from RFC 3810. The idea is the same as we discussed in IGMPv3, but the sizes and formats of the messages have been changed to fit the larger multicast address size in IPv6. Like IGMPv3, MLDv2 has two types of messages: *membership-query message* and *membership-report message*. The first type can be divided into three subtypes: *general*, *group-specific*, and *group-and-source specific*.

Membership-Query Message

A membership-query message is sent by a router to find active group members in the network.

The fields are almost the same as the ones in IGMPv3 except that the size of the multicast address and the source address has been changed from 32 bits to 128 bits. Another noticeable change in the field size is in the *maximum response code* field, in which the size has been changed from 8 bits to 16 bits. We will discuss this field shortly. Also note that the format of the first 8 bytes matches the format for other ICMPv6 packets because MLDv2 is considered to be part of ICMPv6.

Membership-Report Message

The format of the membership report in MLDv2 is exactly the same as the one in IGMPv3 except that the sizes of the fields are changed because of the address size. In particular, the record type is the same as the one defined for IGMPv3 (types 1 to 6).

22.4 TRANSITION FROM IPv4 TO IPv6

Although we have a new version of the IP protocol, how can we make the transition to stop using IPv4 and start using IPv6? The first solution that comes to mind is to define a transition day on which every host or router should stop using the old version and start using the new version. However, this is not practical; because of the huge number of systems in the Internet, the transition from IPv4 to IPv6 cannot happen suddenly. It will take a considerable amount of time before every system in the Internet can move from IPv4 to IPv6. The transition must be smooth to prevent any problems between IPv4 and IPv6 systems.

22.4.1 Strategies

Three strategies have been devised for transition: dual stack, tunneling, and header translation. One or all of these three strategies can be implemented during the transition period.

Dual Stack

It is recommended that all hosts, before migrating completely to version 6, have a **dual stack** of protocols during the transition. In other words, a station must run IPv4 and IPv6 simultaneously until all the Internet uses IPv6. See Figure 22.11 for the layout of a dual-stack configuration.

Figure 22.11 *Dual stack*

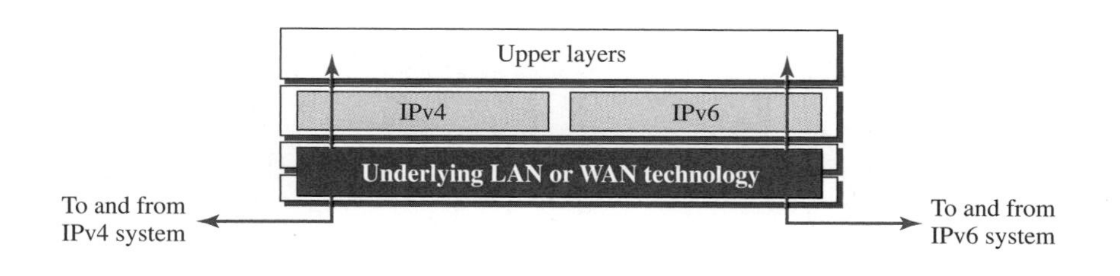

To determine which version to use when sending a packet to a destination, the source host queries the DNS. If the DNS returns an IPv4 address, the source host sends an IPv4 packet. If the DNS returns an IPv6 address, the source host sends an IPv6 packet.

Tunneling

Tunneling is a strategy used when two computers using IPv6 want to communicate with each other and the packet must pass through a region that uses IPv4. To pass through this region, the packet must have an IPv4 address. So the IPv6 packet is encapsulated in an IPv4 packet when it enters the region, and it leaves its capsule when it exits the region. It seems as if the IPv6 packet enters a tunnel at one end and emerges at the other end. To make it clear that the IPv4 packet is carrying an IPv6 packet as data, the protocol value is set to 41. Tunneling is shown in Figure 22.12.

Header Translation

Header translation is necessary when the majority of the Internet has moved to IPv6 but some systems still use IPv4. The sender wants to use IPv6, but the receiver does not understand IPv6. Tunneling does not work in this situation because the packet must be in the IPv4 format to be understood by the receiver. In this case, the header format must

Figure 22.12 *Tunneling strategy*

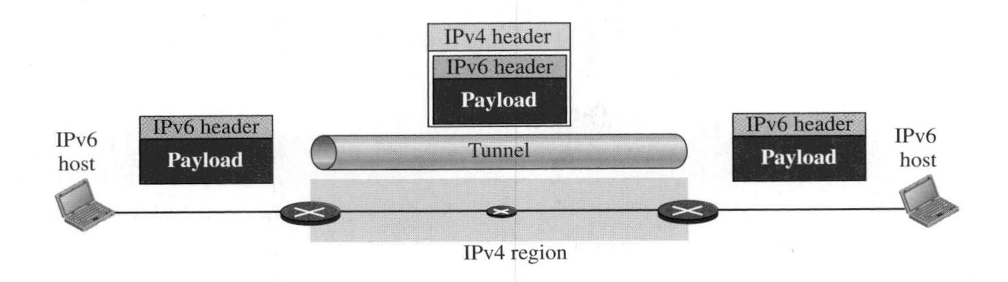

be totally changed through header translation. The header of the IPv6 packet is converted to an IPv4 header (see Figure 22.13).

Figure 22.13 *Header translation strategy*

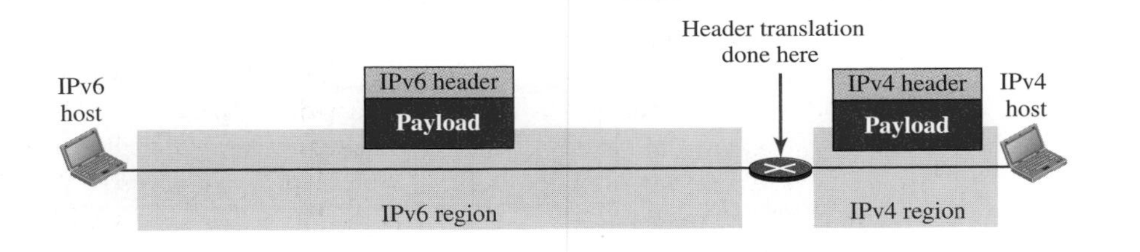

22.4.2 Use of IP Addresses

During the transition a host may need to use two addresses, IPv4 and IPv6. When the transition is complete, IPv4 addresses should disappear. The DNS servers (see Chapter 26) need to be ready to map a host name to either address type during the transition, but the IPv4 directory will disappear after all hosts in the world have migrated to IPv6.

22.5 END-CHAPTER MATERIALS

22.5.1 Recommended Reading

Books

Several books give thorough coverage of materials discussed in this chapter. We recommend [Com 06], [Tan 03], [Koz 05], [Ste 95], [GW 04], [Per 00], [Kes 02], [Moy 98], [WZ 01], and [Los 04].

RFCs

IPv6 addressing is discussed in RFCs 2375, 2526, 3513, 3587, 3789, and 4291. IPv6 protocol is discussed in RFCs 2460, 2461, and 2462. ICMPv6 is discussed in RFCs 2461, 2894, 3122, 3810, 4443, and 4620.

22.5.2 Key Terms

anycast address
authentication
autoconfiguration
colon hexadecimal notation
compatible address
destination option
dual stack
encrypted security payload (ESP)
extension header
fragmentation
header translation
Internet Protocol version 6 (IPv6)
IP new generation (IPng)
link local address
link local block
mapped address
next header
Path MTU Discovery technique
renumbering
tunneling
unique local unicast block
zero compression

22.5.3 Summary

IPv6 has a 128-bit address space. Addresses are presented using hexadecimal colon notation with abbreviation methods available. In IPv6, a destination address can belong to one of three categories: unicast, anycast, and multicast. The address space of IPv6 is divided into several blocks of varying size and each block is allocated for a special purpose. The most important block is the one with prefix 001, which is used for global unicast addressing. Two interesting features of IPv6 addressing are autoconfiguration and numbering.

An IPv6 datagram is composed of a base header and a payload. A payload consists of optional extension headers and data from an upper layer. Extension headers add functionality to the IPv6 datagram.

ICMPv6, like ICMPv4, is message-oriented; it uses messages to report errors, get information, probe a neighbor, or manage multicast communication. However, a few other protocols are added to ICMPv6 to define the functionality and interpretation of the messages.

Three strategies used to handle the transition from version 4 to version 6 are dual stack, tunneling, and header translation.

22.6 PRACTICE SET

22.6.1 Quizzes

A set of interactive quizzes for this chapter can be found on the book website. It is strongly recommended that the student take the quizzes to check his/her understanding of the materials before continuing with the practice set.

22.6.2 Questions

Q22-1. What is the purpose of including the IP header and the first 8 bytes of datagram data in the error-reporting ICMP messages?

Q22-2. Assume a datagram carries no option. Do we still need a value for the next header field in Figure 22.7?

Q22-3. Distinguish between compatible and mapped addresses and explain their applications.

Q22-4. Find the size of the special address block from Table 22.1.

Q22-5. Find the size of the unique local unicast block from Table 22.1.

Q22-6. In which transition strategy do we need to have both IPv4 and IPv6 in the path?

Q22-7. Find the size of the global unicast block from Table 22.1.

Q22-8. Find the size of the multicast block from Table 22.1.

Q22-9. Explain the advantages of IPv6 when compared to IPv4.

Q22-10. List three protocols in the IPv4 network layer that are combined into a single protocol in IPv6.

Q22-11. Explain the benefit of autoconfiguration.

Q22-12. Explain the benefit of renumbering.

Q22-13. Which message in version 6 replaces the ARP request message in version 4? Which replaces the ARP reply message?

Q22-14. If you are assigned an IPv6 address by your ISP for your personal computer at home, what should be the first (leftmost) three bits of this address?

Q22-15. In which transition strategy do we need to encapsulate IPv6 packets in the IPv4 packets?

Q22-16. Which messages in version 6 replace the IGMPv6 messages in version 4?

Q22-17. Which field in the IPv6 packet is responsible for multiplexing and demultiplexing?

Q22-18. Explain the use of the flow field in IPv6. What is the potential application of this field?

22.6.3 Problems

P22-1. An IPv6 packet consists of a base header and a TCP segment. The length of data is 128,000 bytes (jumbo payload). Show the packet and enter a value for each field.

P22-2. Using the CIDR notation, show the IPv6 address mapped to the IPv4 address 129.6.12.34.

P22-3. Using the CIDR notation, show the site local address in which the node identifier is 0::123/48.

P22-4. An IPv6 packet consists of the base header and a TCP segment. The length of data is 320 bytes. Show the packet and enter a value for each field.

P22-5. Show the original (unabbreviated) form of the following IPv6 addresses:

a. ::2 **b.** 0:23::0 **c.** 0:A::3

P22-6. What is the corresponding block or subblock associated with each of the following IPv6 addresses, based on Table 22.1:

a. FE80::12/10 **b.** FD23::/7 **c.** 32::/3

P22-7. Make a table to compare and contrast error-reporting messages in ICMPv6 with error-reporting messages in ICMPv4.

P22-8. Show the unabbreviated colon hex notation for the following IPv6 addresses:

a. An address with 64 0s followed by 32 two-bit (01)s.

b. An address with 64 0s followed by 32 two-bit (10)s.

c. An address with 64 two-bit (01)s.

d. An address with 32 four-bit (0111)s.

P22-9. Using the CIDR notation, show the IPv6 address compatible to the IPv4 address 129.6.12.34.

P22-10. Make a table to compare and contrast informational messages in ICMPv6 with informational messages in ICMPv4.

P22-11. Make a table to compare and contrast neighbor-discovery messages in ICMPv6 with the corresponding messages in version 4.

P22-12. Make a table to compare and contrast inverse neighbor-discovery messages in ICMPv6 with the corresponding messages in version 4.

P22-13. Show abbreviations for the following addresses:

a. 0000:FFFF:FFFF:0000:0000:0000:0000:0000

b. 1234:2346:3456:0000:0000:0000:0000:FFFF

c. 0000:0001:0000:0000:0000:FFFF:1200:1000

d. 0000:0000:0000:0000:FFFF:FFFF:24.123.12.6

P22-14. An organization is assigned the block 2000:1110:1287/48. What is the IPv6 address of an interface in the third subnet if the IEEE physical address of the computer is $(F5\text{-}A9\text{-}23\text{-}14\text{-}7A\text{-}D2)_{16}$.

P22-15. An organization is assigned the block 2000:1234:1423/48. What is the CIDR for the blocks in the first and second subnets in this organization?

P22-16. Which ICMP messages contain part of the IP datagram? Why is this needed?

P22-17. Compare and contrast the IPv4 header with the IPv6 header. Create a table to compare each field.

P22-18. Decompress the following addresses and show the complete unabbreviated IPv6 address:

a. ::2222 **b.** 1111:: **c.** B:A:CC::1234:A

P22-19. Using the CIDR notation, show the IPv6 loopback address.

P22-20. Using the CIDR notation, show the link local address in which the node identifier is 0::123/48.

P22-21. Find the interface identifier if the Ethernet physical address is $(F5\text{-}A9\text{-}23\text{-}12\text{-}7A\text{-}B2)_{16}$ using the format we defined for Ethernet addresses.

P22-22. Find the interface identifier if the physical address of the EUI is $(F5\text{-}A9\text{-}23\text{-}AA\text{-}07\text{-}14\text{-}7A\text{-}23)_{16}$ using the format we defined for Ethernet addresses.

P22-23. Make a table to compare and contrast group-membership messages in ICMPv6 with the corresponding messages in version 4.

22.7 SIMULATION EXPERIMENTS

22.7.1 Applets

We have created some Java applets to show some of the main concepts discussed in this chapter. It is strongly recommended that the students activate these applets on the book website and carefully examine the protocols in action.

PART V

Transport Layer

In the fifth part of the book, we discuss the transport layer. Although this part is made of only two chapters, the chapters are long. Chapter 23 discusses the general idea and issues behind the transport layer. Chapter 24 discusses the three current protocols in the Internet: UDP, TCP, and SCTP.

CHAPTER 23

Introduction to Transport Layer

The transport layer in the TCP/IP suite is located between the application layer and the network layer. It provides services to the application layer and receives services from the network layer. The transport layer acts as a liaison between a client program and a server program, a process-to-process connection. The transport layer is the heart of the TCP/IP protocol suite; it is the end-to-end logical vehicle for transferring data from one point to another in the Internet. This chapter is the first chapter devoted to the transport layer; we discuss the Internet transport-layer protocols in the next chapter.

We have divided this chapter into two sections:

- The first section introduces the idea behind a transport-layer protocol. We first discuss the general services we normally require from the transport layer, such as process-to-process communication, addressing, multiplexing and demultiplexing, error, flow, and congestion control. We then show that the transport-layer protocols are divided into two categories: connectionless and connection-oriented. We discuss the characteristics of each category and show the application of each category.
- The second section discusses general transport-layer protocols. These protocols concentrate on flow and error control services provided by an actual transport-layer protocol. Understanding these protocols helps us better understand the design of the transport-layer protocols in the Internet, such as UDP, TCP, and SCTP. In this section, we first describe an imaginary protocol, which we call the Simple Protocol. This protocol does not have any flow or error control provision. We next introduce the Stop-and-Wait protocol, which provides flow and error control but is very inefficient for a high-speed internet. We then move to the Go-Back-*N* protocol, which is more efficient than the previous one. Finally, we discuss the Selective-Repeat protocol, which has the characteristics of all previous protocols and some more.

23.1 INTRODUCTION

The transport layer is located between the application layer and the network layer. It provides a process-to-process communication between two application layers, one at the local host and the other at the remote host. Communication is provided using a logical connection, which means that the two application layers, which can be located in different parts of the globe, assume that there is an imaginary direct connection through which they can send and receive messages. Figure 23.1 shows the idea behind this logical connection.

Figure 23.1 *Logical connection at the transport layer*

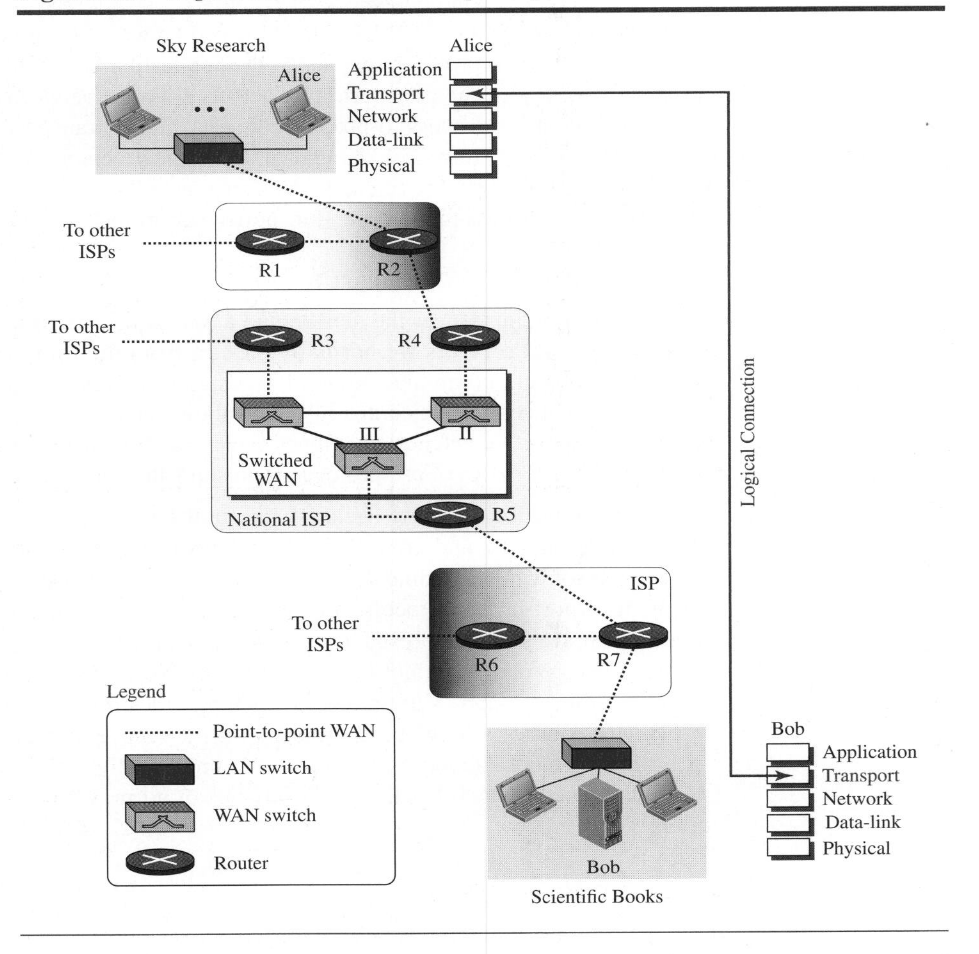

The figure shows the same scenario we used in Chapter 3 for the physical layer (Figure 3.1). Alice's host in the Sky Research company creates a logical connection with Bob's host in the Scientific Books company at the transport layer. The two companies communicate at the transport layer as though there is a real connection between them. Figure 23.1 shows that only the two end systems (Alice's and Bob's computers) use the services of the transport layer; all intermediate routers use only the first three layers.

23.1.1 Transport-Layer Services

As we discussed in Chapter 2, the transport layer is located between the network layer and the application layer. The transport layer is responsible for providing services to the application layer; it receives services from the network layer. In this section, we discuss the services that can be provided by the transport layer; in the next section, we discuss several transport-layer protocols.

Process-to-Process Communication

The first duty of a transport-layer protocol is to provide **process-to-process communication.** A process is an application-layer entity (running program) that uses the services of the transport layer. Before we discuss how process-to-process communication can be accomplished, we need to understand the difference between host-to-host communication and process-to-process communication.

The network layer (discussed in Chapters 18 to 22) is responsible for communication at the computer level (host-to-host communication). A network-layer protocol can deliver the message only to the destination computer. However, this is an incomplete delivery. The message still needs to be handed to the correct process. This is where a transport-layer protocol takes over. A transport-layer protocol is responsible for delivery of the message to the appropriate process. Figure 23.2 shows the domains of a network layer and a transport layer.

Figure 23.2 *Network layer versus transport layer*

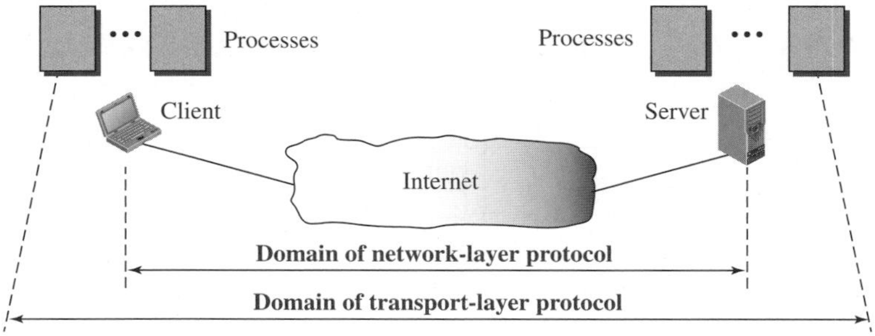

Addressing: Port Numbers

Although there are a few ways to achieve process-to-process communication, the most common is through the **client-server paradigm** (see Chapter 25). A process on the local host, called a *client,* needs services from a process usually on the remote host, called a *server.*

However, operating systems today support both multiuser and multiprogramming environments. A remote computer can run several server programs at the same time, just as several local computers can run one or more client programs at the same time. For communication, we must define the local host, local process, remote host, and remote process. The local host and the remote host are defined using IP addresses (discussed in Chapter 18). To define the processes, we need second identifiers, called ***port numbers.*** In the TCP/IP protocol suite, the port numbers are integers between 0 and 65,535 (16 bits).

The client program defines itself with a port number, called the ***ephemeral port number.*** The word *ephemeral* means "short-lived" and is used because the life of a client is normally short. An ephemeral port number is recommended to be greater than 1023 for some client/server programs to work properly.

The server process must also define itself with a port number. This port number, however, cannot be chosen randomly. If the computer at the server site runs a server process and assigns a random number as the port number, the process at the client site that wants to access that server and use its services will not know the port number. Of course, one solution would be to send a special packet and request the port number of a specific server, but this creates more overhead. TCP/IP has decided to use universal port numbers for servers; these are called ***well-known port numbers.*** There are some exceptions to this rule; for example, there are clients that are assigned well-known port numbers. Every client process knows the well-known port number of the corresponding server process. For example, while the daytime client process, a well-known client program, can use an ephemeral (temporary) port number, 52,000, to identify itself, the daytime server process must use the well-known (permanent) port number 13. Figure 23.3 shows this concept.

Figure 23.3 *Port numbers*

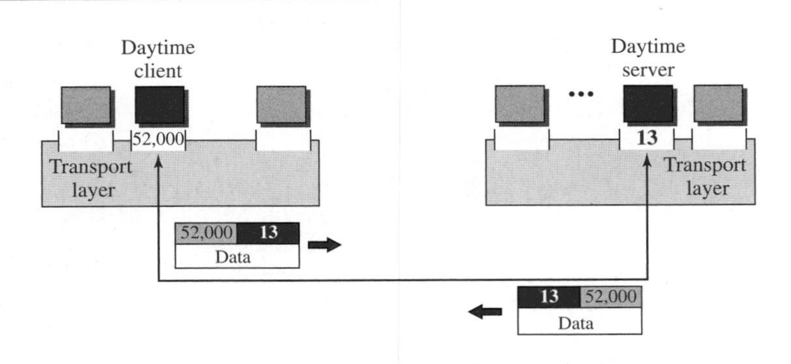

It should be clear by now that the IP addresses and port numbers play different roles in selecting the final destination of data. The destination IP address defines the host among the different hosts in the world. After the host has been selected, the port number defines one of the processes on this particular host (see Figure 23.4).

Figure 23.4 *IP addresses versus port numbers*

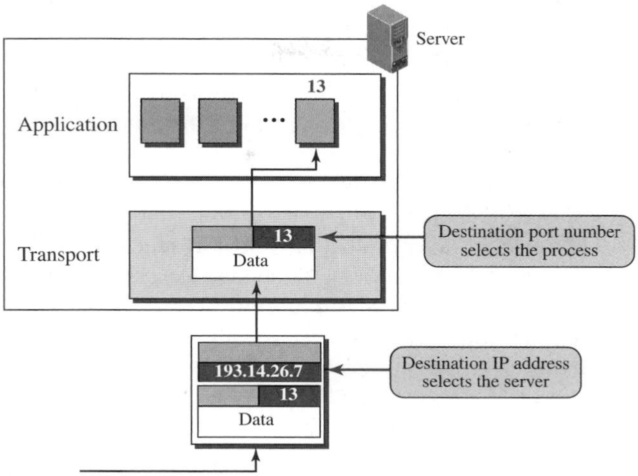

ICANN Ranges

ICANN (see Chapter 18) has divided the port numbers into three ranges: well-known, registered, and dynamic (or private), as shown in Figure 23.5.

Figure 23.5 *ICANN ranges*

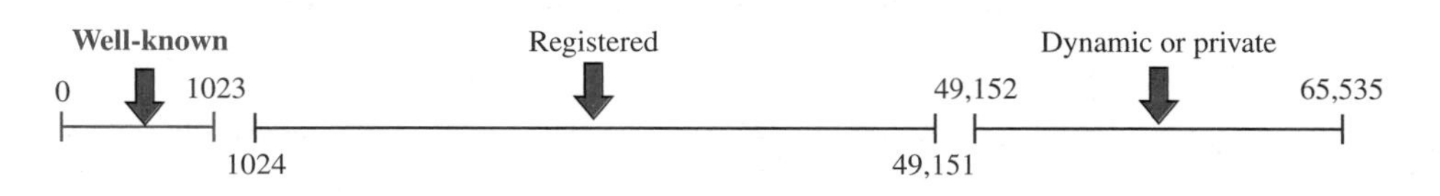

- ***Well-known ports.*** The ports ranging from 0 to 1023 are assigned and controlled by ICANN. These are the well-known ports.
- ***Registered ports.*** The ports ranging from 1024 to 49,151 are not assigned or controlled by ICANN. They can only be registered with ICANN to prevent duplication.
- ***Dynamic ports.*** The ports ranging from 49,152 to 65,535 are neither controlled nor registered. They can be used as temporary or private port numbers.

Example 23.1

In UNIX, the well-known ports are stored in a file called /etc/services. Each line in this file gives the name of the server and the well-known port number. We can use the *grep* utility to extract the line corresponding to the desired application. The following shows the port for TFTP. Note that TFTP can use port 69 on either UDP or TCP.

SNMP (see Chapter 27) uses two port numbers (161 and 162), each for a different purpose.

```
$grep    tftp/etc/services
tftp 69/tcp
tftp 69/udp
```

```
$grep    snmp/etc/services
snmp161/tcp#Simple Net Mgmt Proto
snmp161/udp#Simple Net Mgmt Proto
snmptrap162/udp#Traps for SNMP
```

Socket Addresses

A transport-layer protocol in the TCP suite needs both the IP address and the port number, at each end, to make a connection. The combination of an IP address and a port number is called a ***socket address.*** The client socket address defines the client process uniquely just as the server socket address defines the server process uniquely (see Figure 23.6).

Figure 23.6 *Socket address*

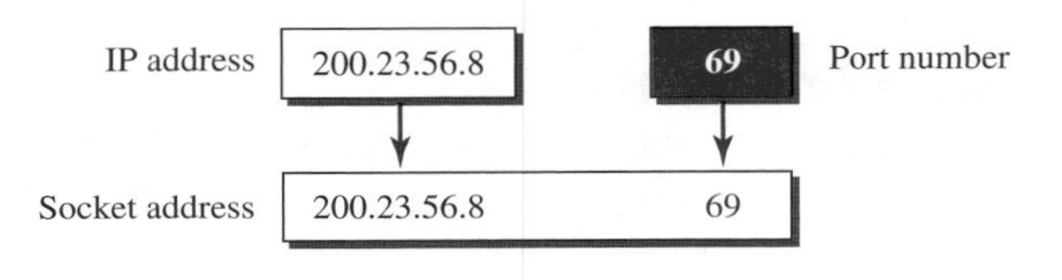

To use the services of the transport layer in the Internet, we need a pair of socket addresses: the client socket address and the server socket address. These four pieces of information are part of the network-layer packet header and the transport-layer packet header. The first header contains the IP addresses; the second header contains the port numbers.

Encapsulation and Decapsulation

To send a message from one process to another, the transport-layer protocol encapsulates and decapsulates messages (Figure 23.7). Encapsulation happens at the sender site. When a process has a message to send, it passes the message to the transport layer along with a pair of socket addresses and some other pieces of information, which depend on the transport-layer protocol. The transport layer receives the data and adds the transport-layer header. The packets at the transport layer in the Internet are called *user datagrams, segments,* or *packets,* depending on what transport-layer protocol we use. In general discussion, we refer to transport-layer payloads as *packets.*

Decapsulation happens at the receiver site. When the message arrives at the destination transport layer, the header is dropped and the transport layer delivers the message to the process running at the application layer. The sender socket address is passed to the process in case it needs to respond to the message received.

Figure 23.7 *Encapsulation and decapsulation*

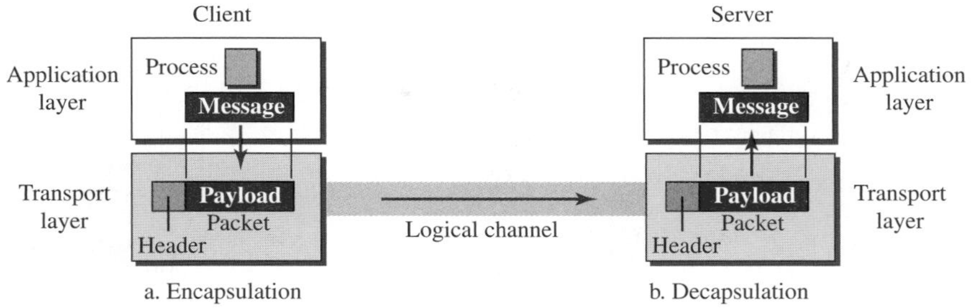

a. Encapsulation

b. Decapsulation

Multiplexing and Demultiplexing

Whenever an entity accepts items from more than one source, this is referred to as ***multiplexing*** (many to one); whenever an entity delivers items to more than one source, this is referred to as ***demultiplexing*** (one to many). The transport layer at the source performs multiplexing; the transport layer at the destination performs demultiplexing (Figure 23.8).

Figure 23.8 shows communication between a client and two servers. Three client processes are running at the client site, P1, P2, and P3. The processes P1 and P3 need to send requests to the corresponding server process running in a server. The client process P2 needs to send a request to the corresponding server process running at another server. The transport layer at the client site accepts three messages from the three processes and creates three packets. It acts as a *multiplexer*. The packets 1 and 3 use the same logical channel to reach the transport layer of the first server. When they arrive at the server, the transport layer does the job of a *demultiplexer* and distributes the messages to two different processes. The transport layer at the second server receives packet 2 and delivers it to the corresponding process. Note that we still have demultiplexing although there is only one message.

Flow Control

Whenever an entity produces items and another entity consumes them, there should be a balance between production and consumption rates. If the items are produced faster than they can be consumed, the consumer can be overwhelmed and may need to discard some items. If the items are produced more slowly than they can be consumed, the consumer must wait, and the system becomes less efficient. Flow control is related to the first issue. We need to prevent losing the data items at the consumer site.

Pushing or Pulling

Delivery of items from a producer to a consumer can occur in one of two ways: *pushing* or *pulling*. If the sender delivers items whenever they are produced—without a prior request from the consumer—the delivery is referred to as *pushing*. If the producer delivers the items after the consumer has requested them, the delivery is referred to as *pulling*. Figure 23.9 shows these two types of delivery.

Figure 23.8 *Multiplexing and demultiplexing*

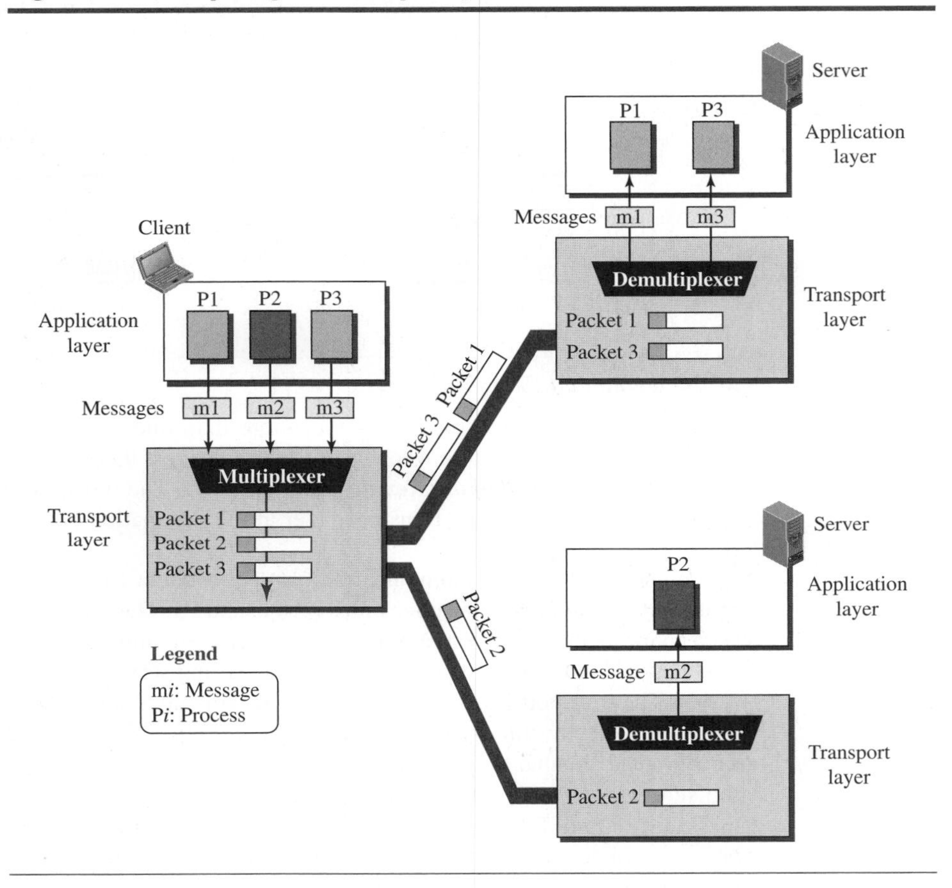

Figure 23.9 *Pushing or pulling*

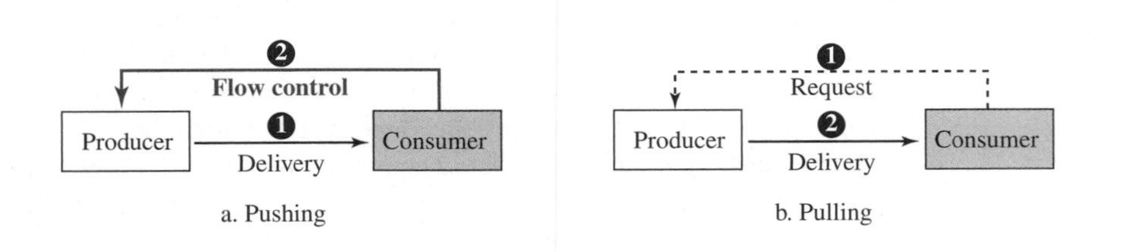

When the producer *pushes* the items, the consumer may be overwhelmed and there is a need for flow control, in the opposite direction, to prevent discarding of the items. In other words, the consumer needs to warn the producer to stop the delivery and to inform the producer when it is again ready to receive the items. When the consumer pulls the items, it requests them when it is ready. In this case, there is no need for flow control.

Flow Control at Transport Layer

In communication at the transport layer, we are dealing with four entities: sender process, sender transport layer, receiver transport layer, and receiver process. The sending process at the application layer is only a producer. It produces message chunks and pushes them to the transport layer. The sending transport layer has a double role: it is both a consumer and a producer. It consumes the messages pushed by the producer. It encapsulates the messages in packets and pushes them to the receiving transport layer. The receiving transport layer also has a double role: it is the consumer for the packets received from the sender and the producer that decapsulates the messages and delivers them to the application layer. The last delivery, however, is normally a pulling delivery; the transport layer waits until the application-layer process asks for messages.

Figure 23.10 shows that we need at least two cases of flow control: from the sending transport layer to the sending application layer and from the receiving transport layer to the sending transport layer.

Figure 23.10 *Flow control at the transport layer*

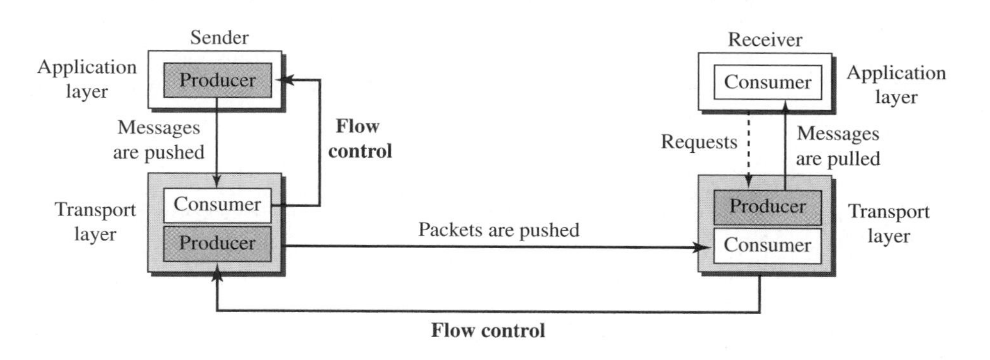

Buffers

Although flow control can be implemented in several ways, one of the solutions is normally to use two *buffers*: one at the sending transport layer and the other at the receiving transport layer. A buffer is a set of memory locations that can hold packets at the sender and receiver. The flow control communication can occur by sending signals from the consumer to the producer.

When the buffer of the sending transport layer is full, it informs the application layer to stop passing chunks of messages; when there are some vacancies, it informs the application layer that it can pass message chunks again.

When the buffer of the receiving transport layer is full, it informs the sending transport layer to stop sending packets. When there are some vacancies, it informs the sending transport layer that it can send packets again.

Example 23.2

The above discussion requires that the consumers communicate with the producers on two occasions: when the buffer is full and when there are vacancies. If the two parties use a buffer with only one slot, the communication can be easier. Assume that each transport layer uses a single memory location to hold a packet. When this single slot in the sending transport layer is empty, the sending transport layer sends a note to the application layer to send its next chunk; when this single slot in the receiving transport layer is empty, it sends an acknowledgment to the sending transport layer to send its next packet. As we will see later, however, this type of flow control, using a single-slot buffer at the sender and the receiver, is inefficient.

Error Control

In the Internet, since the underlying network layer (IP) is unreliable, we need to make the transport layer reliable if the application requires reliability. Reliability can be achieved to add error control services to the transport layer. Error control at the transport layer is responsible for

1. Detecting and discarding corrupted packets.
2. Keeping track of lost and discarded packets and resending them.
3. Recognizing duplicate packets and discarding them.
4. Buffering out-of-order packets until the missing packets arrive.

Error control, unlike flow control, involves only the sending and receiving transport layers. We are assuming that the message chunks exchanged between the application and transport layers are error free. Figure 23.11 shows the error control between the sending and receiving transport layers. As with the case of flow control, the receiving transport layer manages error control, most of the time, by informing the sending transport layer about the problems.

Figure 23.11 *Error control at the transport layer*

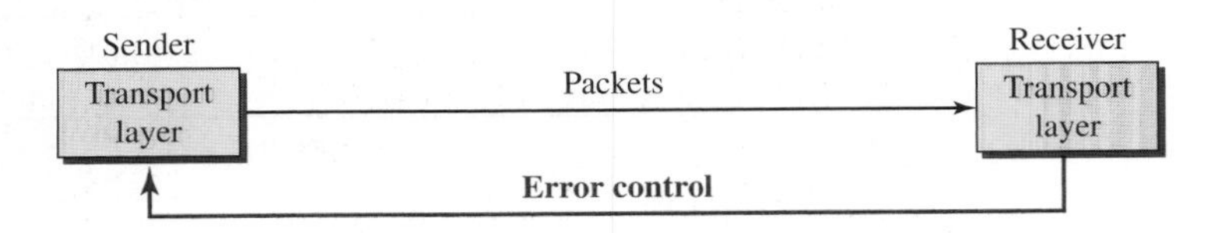

Sequence Numbers

Error control requires that the sending transport layer knows which packet is to be resent and the receiving transport layer knows which packet is a duplicate, or which packet has arrived out of order. This can be done if the packets are numbered. We can add a field to the transport-layer packet to hold the **sequence number** of the packet. When a packet is corrupted or lost, the receiving transport layer can somehow inform the sending transport layer to resend that packet using the sequence number. The receiving transport layer can also detect duplicate packets if two received packets have the same sequence number. The out-of-order packets can be recognized by observing gaps in the sequence numbers.

Packets are numbered sequentially. However, because we need to include the sequence number of each packet in the header, we need to set a limit. If the header of the packet allows m bits for the sequence number, the sequence numbers range from 0 to $2^m - 1$. For example, if m is 4, the only sequence numbers are 0 through 15, inclusive. However, we can wrap around the sequence. So the sequence numbers in this case are

0, 1, 2, 3, 4, 5, 6, 7, 8, 9, 10, 11, 12, 13, 14, 15, 0, 1, 2, 3, 4, 5, 6, 7, 8, 9, 10, 11, ...

In other words, the sequence numbers are modulo 2^m.

For error control, the sequence numbers are modulo 2^m, where m is the size of the sequence number field in bits.

Acknowledgment

We can use both positive and negative signals as error control, but we discuss only positive signals, which are more common at the transport layer. The receiver side can send an acknowledgment (ACK) for each of a collection of packets that have arrived safe and sound. The receiver can simply discard the corrupted packets. The sender can detect lost packets if it uses a timer. When a packet is sent, the sender starts a timer. If an ACK does not arrive before the timer expires, the sender resends the packet. Duplicate packets can be silently discarded by the receiver. Out-of-order packets can be either discarded (to be treated as lost packets by the sender), or stored until the missing one arrives.

Combination of Flow and Error Control

We have discussed that flow control requires the use of two buffers, one at the sender site and the other at the receiver site. We have also discussed that error control requires the use of sequence and acknowledgment numbers by both sides. These two requirements can be combined if we use two numbered buffers, one at the sender, one at the receiver.

At the sender, when a packet is prepared to be sent, we use the number of the next free location, x, in the buffer as the sequence number of the packet. When the packet is sent, a copy is stored at memory location x, awaiting the acknowledgment from the other end. When an acknowledgment related to a sent packet arrives, the packet is purged and the memory location becomes free.

At the receiver, when a packet with sequence number y arrives, it is stored at the memory location y until the application layer is ready to receive it. An acknowledgment can be sent to announce the arrival of packet y.

Sliding Window

Since the sequence numbers use modulo 2^m, a circle can represent the sequence numbers from 0 to $2^m - 1$ (Figure 23.12). The buffer is represented as a set of slices, called the ***sliding window,*** that occupies part of the circle at any time. At the sender site, when a packet is sent, the corresponding slice is marked. When all the slices are marked, it means that the buffer is full and no further messages can be accepted from the application layer. When an acknowledgment arrives, the corresponding slice is unmarked. If some consecutive slices from the beginning of the window are unmarked, the window slides over the range of the corresponding sequence numbers to allow more free slices at the end of the window. Figure 23.12 shows the sliding window at the sender. The sequence numbers are in modulo 16 ($m = 4$) and the size of the window is 7. Note that the sliding window is just an abstraction: the actual situation uses computer variables to hold the sequence numbers of the next packet to be sent and the last packet sent.

Figure 23.12 *Sliding window in circular format*

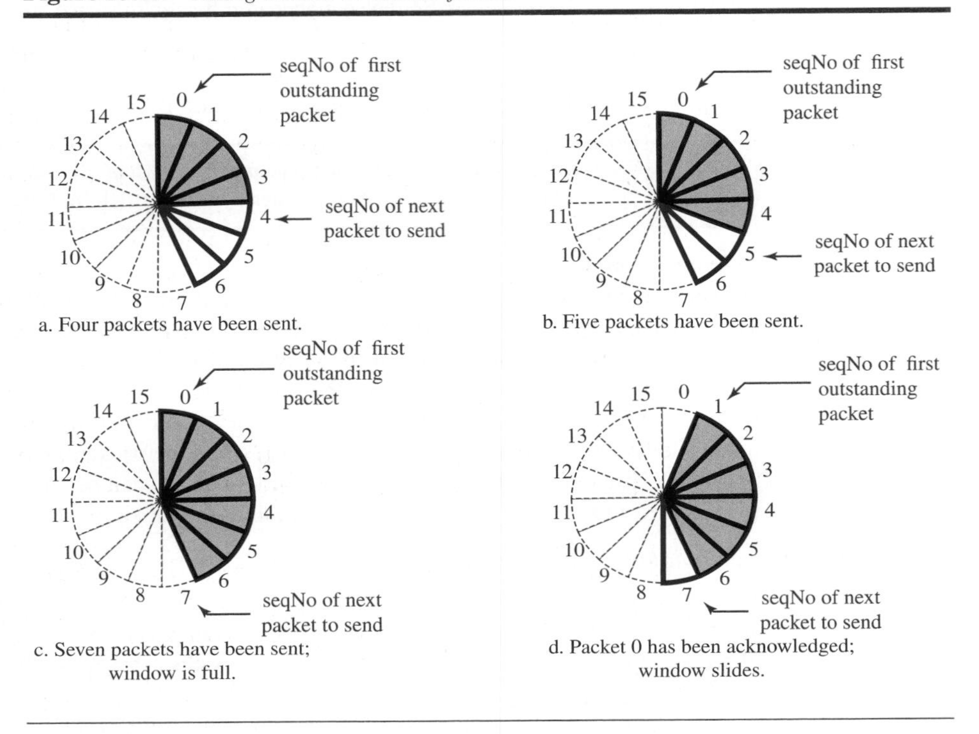

Most protocols show the sliding window using linear representation. The idea is the same, but it normally takes less space on paper. Figure 23.13 shows this representation. Both representations tell us the same thing. If we take both sides of each diagram in Figure 23.13 and bend them up, we can make the same diagram as in Figure 23.12.

Figure 23.13 *Sliding window in linear format*

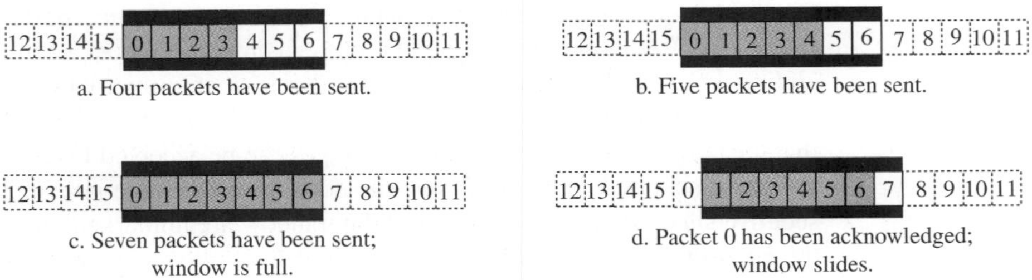

12 13 14 15 0 1 2 3 4 5 6 7 8 9 10 11

a. Four packets have been sent.

12 13 14 15 0 1 2 3 4 5 6 7 8 9 10 11

b. Five packets have been sent.

12 13 14 15 0 1 2 3 4 5 6 7 8 9 10 11

c. Seven packets have been sent; window is full.

12 13 14 15 0 1 2 3 4 5 6 7 8 9 10 11

d. Packet 0 has been acknowledged; window slides.

Congestion Control

An important issue in a packet-switched network, such as the Internet, is **congestion.** Congestion in a network may occur if the *load* on the network—the number of packets sent to the network—is greater than the *capacity* of the network—the number of packets a network can handle. **Congestion control** refers to the mechanisms and techniques that control the congestion and keep the load below the capacity.

We may ask why there is congestion in a network. Congestion happens in any system that involves waiting. For example, congestion happens on a freeway because any abnormality in the flow, such as an accident during rush hour, creates blockage.

Congestion in a network or internetwork occurs because routers and switches have queues—buffers that hold the packets before and after processing. A router, for example, has an input queue and an output queue for each interface. If a router cannot process the packets at the same rate at which they arrive, the queues become overloaded and congestion occurs. Congestion at the transport layer is actually the result of congestion at the network layer, which manifests itself at the transport layer. We discussed congestion at the network layer and its causes in Chapter 18. Later in this chapter, we show how TCP, assuming that there is no congestion control at the network layer, implements its own congestion control mechanism.

23.1.2 Connectionless and Connection-Oriented Protocols

A transport-layer protocol, like a network-layer protocol, can provide two types of services: connectionless and connection-oriented. The nature of these services at the transport layer, however, is different from the ones at the network layer. At the network layer, a connectionless service may mean different paths for different datagrams belonging to the same message. At the transport layer, we are not concerned about the physical paths of packets (we assume a logical connection between two transport layers). Connectionless service at the transport layer means independency between packets; connection-oriented means dependency. Let us elaborate on these two services.

Connectionless Service

In a connectionless service, the source process (application program) needs to divide its message into chunks of data of the size acceptable by the transport layer and deliver them to the transport layer one by one. The transport layer treats each chunk as a single unit without any relation between the chunks. When a chunk arrives from the application layer, the transport layer encapsulates it in a packet and sends it. To show the independency of packets, assume that a client process has three chunks of messages to send to a server process. The chunks are handed over to the connectionless transport protocol in order. However, since there is no dependency between the packets at the transport layer, the packets may arrive out of order at the destination and will be delivered out of order to the server process (Figure 23.14).

In Figure 23.14, we have shown the movement of packets using a time line, but we have assumed that the delivery of the process to the transport layer and vice versa are instantaneous. The figure shows that at the client site, the three chunks of messages are delivered to the client transport layer in order (0, 1, and 2). Because of the extra delay in transportation of the second packet, the delivery of messages at the server is not in

Figure 23.14 *Connectionless service*

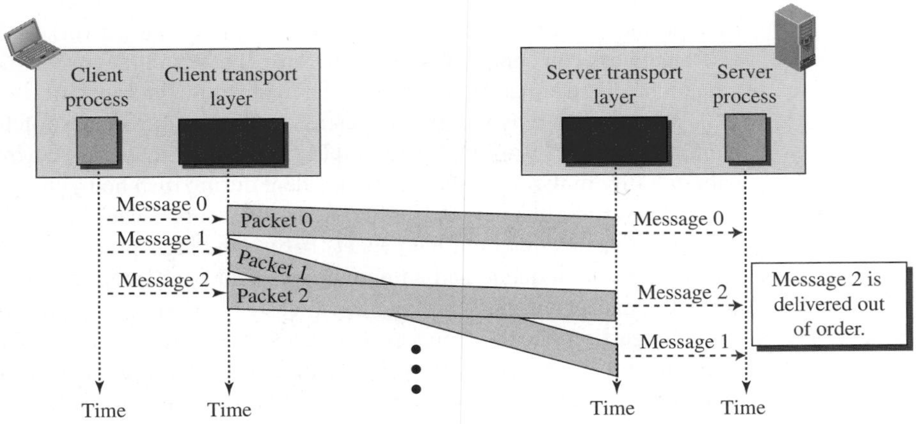

order (0, 2, 1). If these three chunks of data belong to the same message, the server process may have received a strange message.

The situation would be worse if one of the packets were lost. Since there is no numbering on the packets, the receiving transport layer has no idea that one of the messages has been lost. It just delivers two chunks of data to the server process.

The above two problems arise from the fact that the two transport layers do not coordinate with each other. The receiving transport layer does not know when the first packet will come nor when all of the packets have arrived.

We can say that no flow control, error control, or congestion control can be effectively implemented in a connectionless service.

Connection-Oriented Service

In a connection-oriented service, the client and the server first need to establish a logical connection between themselves. The data exchange can only happen after the connection establishment. After data exchange, the connection needs to be torn down (Figure 23.15).

As we mentioned before, the connection-oriented service at the transport layer is different from the same service at the network layer. In the network layer, connection-oriented service means a coordination between the two end hosts and all the routers in between. At the transport layer, connection-oriented service involves only the two hosts; the service is end to end. This means that we should be able to make a connection-oriented protocol at the transport layer over either a connectionless or connection-oriented protocol at the network layer. Figure 23.15 shows the connection-establishment, data-transfer, and tear-down phases in a connection-oriented service at the transport layer.

We can implement flow control, error control, and congestion control in a connection-oriented protocol.

Figure 23.15 *Connection-oriented service*

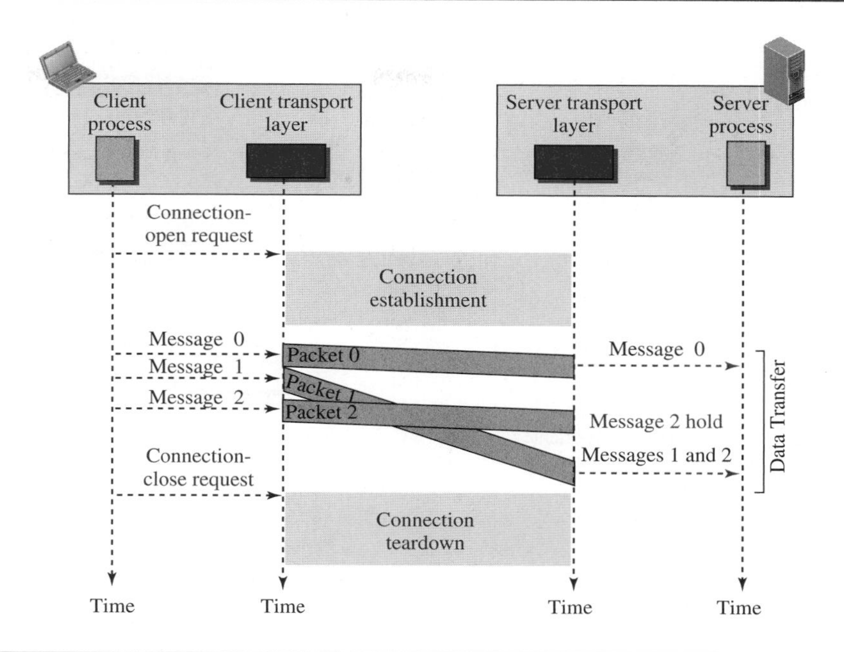

Finite State Machine

The behavior of a transport-layer protocol, both when it provides a connectionless and when it provides a connection-oriented protocol, can be better shown as a **finite state machine (FSM).** Figure 23.16 shows a representation of a transport layer using an FSM. Using this tool, each transport layer (sender or receiver) is taught as a machine with a finite number of states. The machine is always in one of the states until an *event* occurs. Each event is associated with two reactions: defining the list (possibly empty) of actions to be performed and determining the next state (which can be the same as the current state). One of the states must be defined as the initial state, the state in which the machine starts when it turns on. In this figure we have used rounded-corner rectangles to show states, colored text to show events, and regular black text to show actions. A horizontal line is used to separate the event from the actions, although later we replace the horizontal line with a slash. The arrow shows the movement to the next state.

We can think of a connectionless transport layer as an FSM with only one state: the established state. The machine on each end (client and server) is always in the established state, ready to send and receive transport-layer packets.

An FSM in a connection-oriented transport layer, on the other hand, needs to go through three states before reaching the established state. The machine also needs to go through three states before closing the connection. The machine is in the *closed* state when there is no connection. It remains in this state until a request for opening the connection arrives from the local process; the machine sends an open request packet to the

Figure 23.16 *Connectionless and connection-oriented service represented as FSMs*

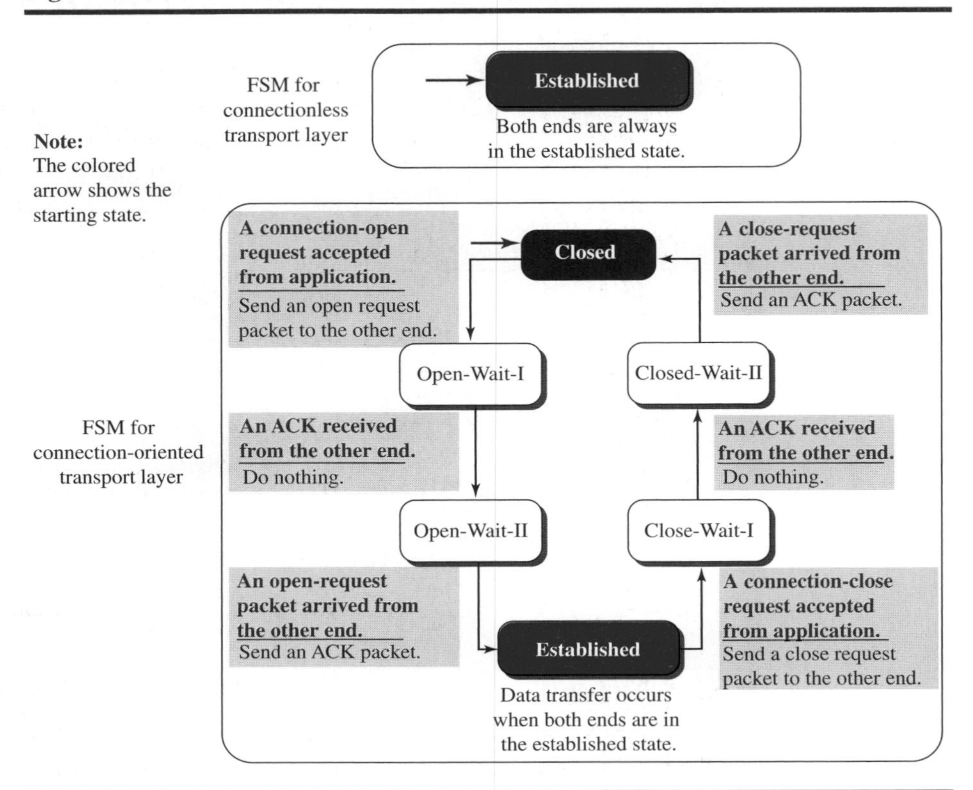

remote transport layer and moves to the *open-wait-I* state. When an acknowledgment is received from the other end, the local FSM moves to the *open-wait-II* state. When the machine is in this state, a unidirectional connection has been established, but if a bidirectional connection is needed, the machine needs to wait in this state until the other end also requests a connection. When the request is received, the machine sends an acknowledgment and moves to the *established* state.

Data and data acknowledgment can be exchanged between the two ends when they are both in the established state. However, we need to remember that the established state, both in connectionless and connection-oriented transport layers, represents a set of data transfer states, which we discuss in the next section, Transport-Layer Protocols.

To tear down a connection, the application layer sends a close request message to its local transport layer. The transport layer sends a close-request packet to the other end and moves to *close-wait-I* state. When an acknowledgment is received from the other end, the machine moves to the *close-wait-II* state and waits for the close-request packet from the other end. When this packet arrives, the machine sends an acknowledgment and moves to the *closed* state.

There are several variations of the connection-oriented FSM that we will discuss later. We will also see how the FSM can be condensed or expanded and the names of the states can be changed.

23.2 TRANSPORT-LAYER PROTOCOLS

We can create a transport-layer protocol by combining a set of services described in the previous sections. To better understand the behavior of these protocols, we start with the simplest one and gradually add more complexity. The TCP/IP protocol uses a transport-layer protocol that is either a modification or a combination of some of these protocols. We discuss these general protocols in this section to pave the way for understanding more complex ones in the rest of the chapter. To make our discussion simpler, we first discuss all of these protocols as a unidirectional protocol (i.e., simplex) in which the data packets move in one direction. At the end of the chapter, we briefly discuss how they can be changed to bidirectional protocols where data can be moved in two directions (i.e., full duplex).

23.2.1 Simple Protocol

Our first protocol is a simple connectionless protocol with neither flow nor error control. We assume that the receiver can immediately handle any packet it receives. In other words, the receiver can never be overwhelmed with incoming packets. Figure 23.17 shows the layout for this protocol.

Figure 23.17 *Simple protocol*

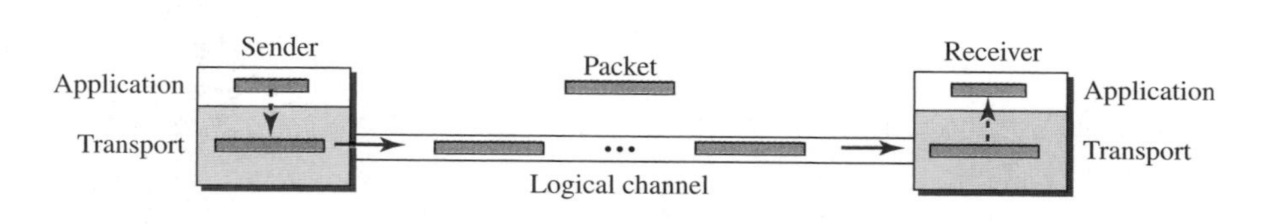

The transport layer at the sender gets a message from its application layer, makes a packet out of it, and sends the packet. The transport layer at the receiver receives a packet from its network layer, extracts the message from the packet, and delivers the message to its application layer. The transport layers of the sender and receiver provide transmission services for their application layers.

FSMs

The sender site should not send a packet until its application layer has a message to send. The receiver site cannot deliver a message to its application layer until a packet arrives. We can show these requirements using two FSMs. Each FSM has only one state, the *ready state*. The sending machine remains in the ready state until a request comes from the process in the application layer. When this event occurs, the sending machine encapsulates the message in a packet and sends it to the receiving machine. The receiving machine remains in the ready state until a packet arrives from the sending machine. When this event occurs, the receiving machine decapsulates the message out of the packet and delivers it to the process at the application layer. Figure 23.18

shows the FSMs for the simple protocol. We see later that the UDP protocol is a slight modification of this protocol.

Figure 23.18 *FSMs for the simple protocol*

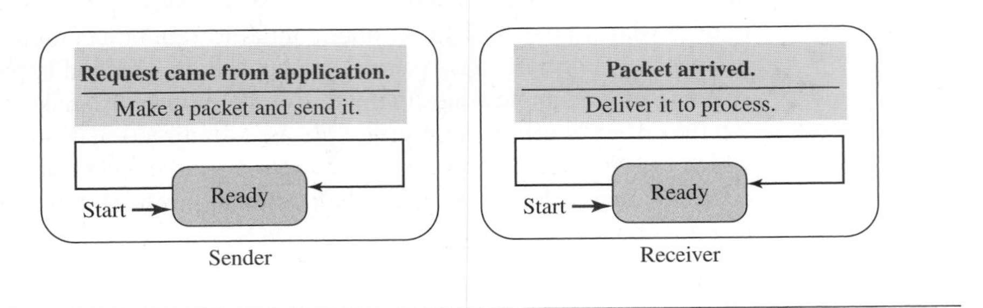

Example 23.3

Figure 23.19 shows an example of communication using this protocol. It is very simple. The sender sends packets one after another without even thinking about the receiver.

Figure 23.19 *Flow diagram for Example 23.3*

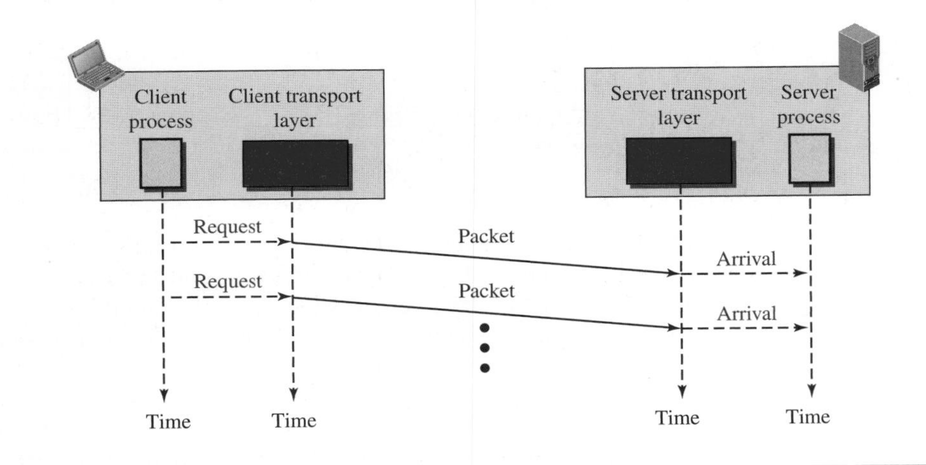

23.2.2 Stop-and-Wait Protocol

Our second protocol is a connection-oriented protocol called the **Stop-and-Wait protocol,** which uses both flow and error control. Both the sender and the receiver use a sliding window of size 1. The sender sends one packet at a time and waits for an acknowledgment before sending the next one. To detect corrupted packets, we need to add a checksum to each data packet. When a packet arrives at the receiver site, it is checked. If its checksum is incorrect, the packet is corrupted and silently discarded.

The silence of the receiver is a signal for the sender that a packet was either corrupted or lost. Every time the sender sends a packet, it starts a timer. If an acknowledgment arrives before the timer expires, the timer is stopped and the sender sends the next packet (if it has one to send). If the timer expires, the sender resends the previous packet, assuming that the packet was either lost or corrupted. This means that the sender needs to keep a copy of the packet until its acknowledgment arrives. Figure 23.20 shows the outline for the Stop-and-Wait protocol. Note that only one packet and one acknowledgment can be in the channels at any time.

Figure 23.20 *Stop-and-Wait protocol*

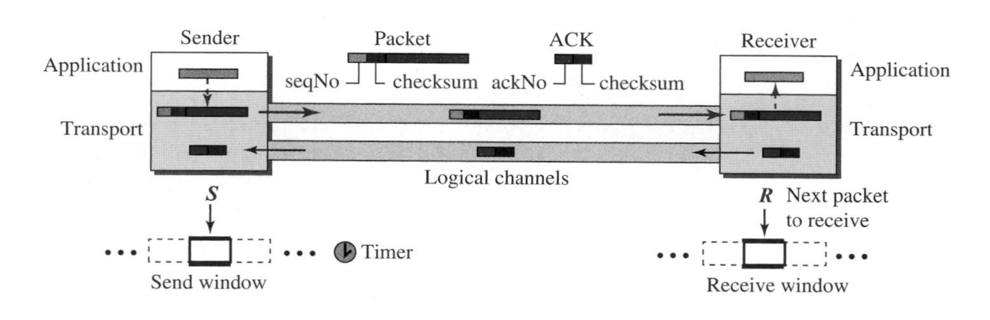

The Stop-and-Wait protocol is a connection-oriented protocol that provides flow and error control.

Sequence Numbers

To prevent duplicate packets, the protocol uses sequence numbers and acknowledgment numbers. A field is added to the packet header to hold the sequence number of that packet. One important consideration is the range of the sequence numbers. Since we want to minimize the packet size, we look for the smallest range that provides unambiguous communication. Let us discuss the range of sequence numbers we need. Assume we have used x as a sequence number; we only need to use $x + 1$ after that. There is no need for $x + 2$. To show this, assume that the sender has sent the packet with sequence number x. Three things can happen.

1. The packet arrives safe and sound at the receiver site; the receiver sends an acknowledgment. The acknowledgment arrives at the sender site, causing the sender to send the next packet numbered $x + 1$.
2. The packet is corrupted or never arrives at the receiver site; the sender resends the packet (numbered x) after the time-out. The receiver returns an acknowledgment.
3. The packet arrives safe and sound at the receiver site; the receiver sends an acknowledgment, but the acknowledgment is corrupted or lost. The sender resends the packet (numbered x) after the time-out. Note that the packet here is a duplicate. The receiver can recognize this fact because it expects packet $x + 1$ but packet x was received.

We can see that there is a need for sequence numbers x and $x + 1$ because the receiver needs to distinguish between case 1 and case 3. But there is no need for a packet to be numbered $x + 2$. In case 1, the packet can be numbered x again because packets x and $x + 1$ are acknowledged and there is no ambiguity at either site. In cases 2 and 3, the new packet is $x + 1$, not $x + 2$. If only x and $x + 1$ are needed, we can let $x = 0$ and $x + 1 = 1$. This means that the sequence is 0, 1, 0, 1, 0, and so on. This is referred to as modulo 2 arithmetic.

Acknowledgment Numbers

Since the sequence numbers must be suitable for both data packets and acknowledgments, we use this convention: The acknowledgment numbers always announce the sequence number of the *next packet expected* by the receiver. For example, if packet 0 has arrived safe and sound, the receiver sends an ACK with acknowledgment 1 (meaning packet 1 is expected next). If packet 1 has arrived safe and sound, the receiver sends an ACK with acknowledgment 0 (meaning packet 0 is expected).

In the Stop-and-Wait protocol, the acknowledgment number always announces, in modulo-2 arithmetic, the sequence number of the next packet expected.

The sender has a control variable, which we call S (sender), that points to the only slot in the send window. The receiver has a control variable, which we call R (receiver), that points to the only slot in the receive window.

FSMs

Figure 23.21 shows the FSMs for the Stop-and-Wait protocol. Since the protocol is a connection-oriented protocol, both ends should be in the *established* state before exchanging data packets. The states are actually nested in the *established* state.

Sender

The sender is initially in the ready state, but it can move between the ready and blocking state. The variable S is initialized to 0.

- ❑ ***Ready state.*** When the sender is in this state, it is only waiting for one event to occur. If a request comes from the application layer, the sender creates a packet with the sequence number set to S. A copy of the packet is stored, and the packet is sent. The sender then starts the only timer. The sender then moves to the blocking state.
- ❑ ***Blocking state.*** When the sender is in this state, three events can occur:
  - **a.** If an error-free ACK arrives with the ackNo related to the next packet to be sent, which means ackNo = $(S + 1)$ modulo 2, then the timer is stopped. The window slides, $S = (S + 1)$ modulo 2. Finally, the sender moves to the ready state.
  - **b.** If a corrupted ACK or an error-free ACK with the ackNo $\neq (S + 1)$ modulo 2 arrives, the ACK is discarded.
  - **c.** If a time-out occurs, the sender resends the only outstanding packet and restarts the timer.

Figure 23.21 *FSMs for the Stop-and-Wait protocol*

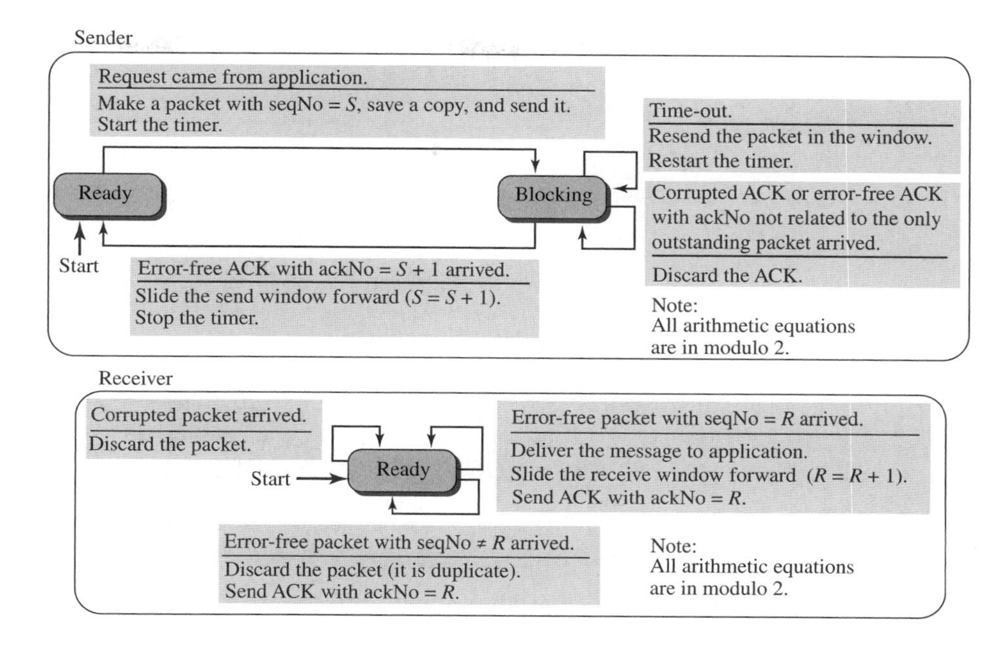

Receiver

The receiver is always in the *ready* state. Three events may occur:

a. If an error-free packet with seqNo = R arrives, the message in the packet is delivered to the application layer. The window then slides, $R = (R + 1)$ modulo 2. Finally an ACK with ackNo = R is sent.

b. If an error-free packet with seqNo $\neq R$ arrives, the packet is discarded, but an ACK with ackNo = R is sent.

c. If a corrupted packet arrives, the packet is discarded.

Example 23.4

Figure 23.22 shows an example of the Stop-and-Wait protocol. Packet 0 is sent and acknowledged. Packet 1 is lost and resent after the time-out. The resent packet 1 is acknowledged and the timer stops. Packet 0 is sent and acknowledged, but the acknowledgment is lost. The sender has no idea if the packet or the acknowledgment is lost, so after the time-out, it resends packet 0, which is acknowledged.

Efficiency

The Stop-and-Wait protocol is very inefficient if our channel is *thick* and *long*. By *thick,* we mean that our channel has a large bandwidth (high data rate); by *long,* we mean the round-trip delay is long. The product of these two is called the **bandwidth-delay product.** We can think of the channel as a pipe. The bandwidth-delay product then is the volume of the pipe in bits. The pipe is always there. It is not efficient if it

Figure 23.22 *Flow diagram for Example 23.4*

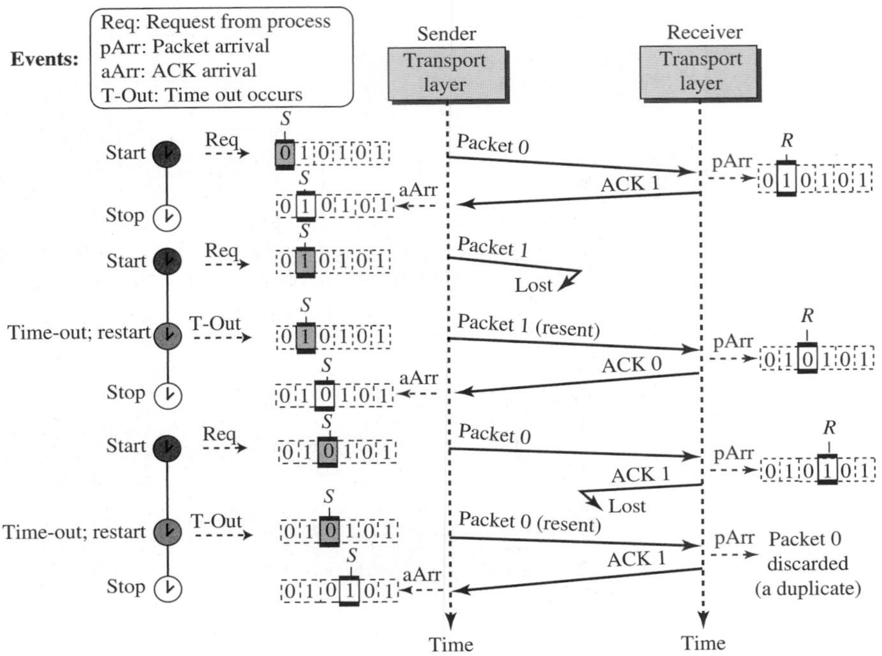

is not used. The bandwidth-delay product is a measure of the number of bits a sender can transmit through the system while waiting for an acknowledgment from the receiver.

Example 23.5

Assume that, in a Stop-and-Wait system, the bandwidth of the line is 1 Mbps, and 1 bit takes 20 milliseconds to make a round trip. What is the bandwidth-delay product? If the system data packets are 1,000 bits in length, what is the utilization percentage of the link?

Solution

The bandwidth-delay product is $(1 \times 10^6) \times (20 \times 10^{-3}) = 20{,}000$ bits. The system can send 20,000 bits during the time it takes for the data to go from the sender to the receiver and the acknowledgment to come back. However, the system sends only 1,000 bits. We can say that the link utilization is only 1,000/20,000, or 5 percent. For this reason, in a link with a high bandwidth or long delay, the use of Stop-and-Wait wastes the capacity of the link.

Example 23.6

What is the utilization percentage of the link in Example 23.5 if we have a protocol that can send up to 15 packets before stopping and worrying about the acknowledgments?

Solution

The bandwidth-delay product is still 20,000 bits. The system can send up to 15 packets or 15,000 bits during a round trip. This means the utilization is 15,000/20,000, or 75 percent. Of course, if there are damaged packets, the utilization percentage is much less because packets have to be resent.

Pipelining

In networking and in other areas, a task is often begun before the previous task has ended. This is known as **pipelining.** There is no pipelining in the Stop-and-Wait protocol because a sender must wait for a packet to reach the destination and be acknowledged before the next packet can be sent. However, pipelining does apply to our next two protocols because several packets can be sent before a sender receives feedback about the previous packets. Pipelining improves the efficiency of the transmission if the number of bits in transition is large with respect to the bandwidth-delay product.

23.2.3 Go-Back-*N* Protocol (GBN)

To improve the efficiency of transmission (to fill the pipe), multiple packets must be in transition while the sender is waiting for acknowledgment. In other words, we need to let more than one packet be outstanding to keep the channel busy while the sender is waiting for acknowledgment. In this section, we discuss one protocol that can achieve this goal; in the next section, we discuss a second. The first is called ***Go-Back-N* (GBN)** (the rationale for the name will become clear later). The key to Go-back-*N* is that we can send several packets before receiving acknowledgments, but the receiver can only buffer one packet. We keep a copy of the sent packets until the acknowledgments arrive. Figure 23.23 shows the outline of the protocol. Note that several data packets and acknowledgments can be in the channel at the same time.

Figure 23.23 *Go-Back-*N *protocol*

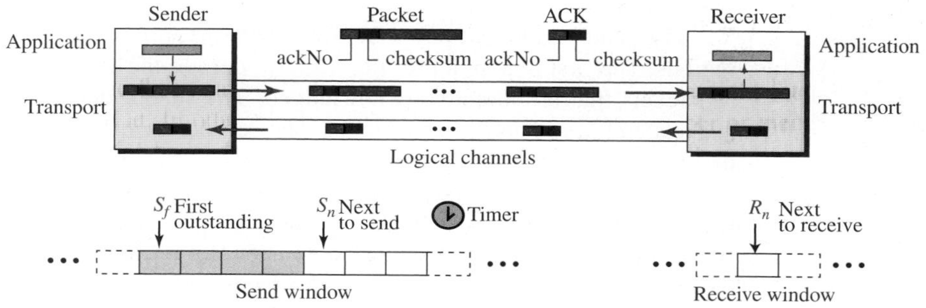

Sequence Numbers

As we mentioned before, the sequence numbers are modulo 2^m, where m is the size of the sequence number field in bits.

Acknowledgment Numbers

An acknowledgment number in this protocol is cumulative and defines the sequence number of the next packet expected. For example, if the acknowledgment number (ackNo) is 7, it means all packets with sequence number up to 6 have arrived, safe and sound, and the receiver is expecting the packet with sequence number 7.

In the Go-Back-*N* protocol, the acknowledgment number is cumulative and defines the sequence number of the next packet expected to arrive.

Send Window

The send window is an imaginary box covering the sequence numbers of the data packets that can be in transit or can be sent. In each window position, some of these sequence numbers define the packets that have been sent; others define those that can be sent. The maximum size of the window is $2^m - 1$, for reasons that we discuss later. In this chapter, we let the size be fixed and set to the maximum value, but we will see later that some protocols may have a variable window size. Figure 23.24 shows a sliding window of size 7 ($m = 3$) for the Go-Back-*N* protocol.

Figure 23.24 *Send window for Go-Back-*N

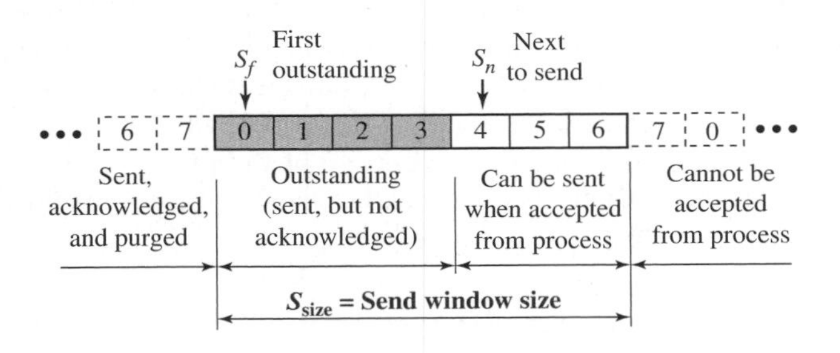

The send window at any time divides the possible sequence numbers into four regions. The first region, left of the window, defines the sequence numbers belonging to packets that are already acknowledged. The sender does not worry about these packets and keeps no copies of them. The second region, colored, defines the range of sequence numbers belonging to the packets that have been sent, but have an unknown status. The sender needs to wait to find out if these packets have been received or were lost. We call these *outstanding* packets. The third range, white in the figure, defines the range of sequence numbers for packets that can be sent; however, the corresponding data have not yet been received from the application layer. Finally, the fourth region, right of the window, defines sequence numbers that cannot be used until the window slides.

The window itself is an abstraction; three variables define its size and location at any time. We call these variables S_f (send window, the first outstanding packet), S_n (send window, the next packet to be sent), and S_{size} (send window, size). The variable S_f defines the sequence number of the first (oldest) outstanding packet. The variable

S_n holds the sequence number that will be assigned to the next packet to be sent. Finally, the variable S_{size} defines the size of the window, which is fixed in our protocol.

The send window is an abstract concept defining an imaginary box of maximum size = $2^m - 1$ with three variables: S_f, S_n, and S_{size}.

Figure 23.25 shows how a send window can slide one or more slots to the right when an acknowledgment arrives from the other end. In the figure, an acknowledgment with ackNo = 6 has arrived. This means that the receiver is waiting for packets with sequence number 6.

Figure 23.25 *Sliding the send window*

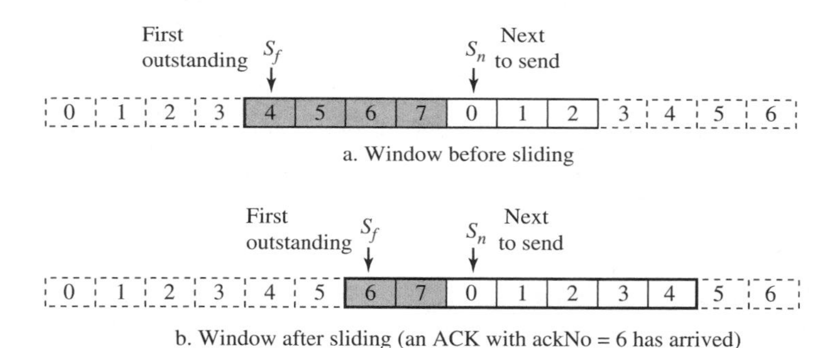

a. Window before sliding

b. Window after sliding (an ACK with ackNo = 6 has arrived)

The send window can slide one or more slots when an error-free ACK with ackNo greater than or equal to S_f and less than S_n (in modular arithmetic) arrives.

Receive Window

The receive window makes sure that the correct data packets are received and that the correct acknowledgments are sent. In Go-Back-*N*, the size of the receive window is always 1. The receiver is always looking for the arrival of a specific packet. Any packet arriving out of order is discarded and needs to be resent. Figure 23.26 shows the receive window. Note that we need only one variable, R_n (receive window, next packet expected), to define this abstraction. The sequence numbers to the left of the window belong to the packets already received and acknowledged; the sequence numbers to the right of this window define the packets that cannot be received. Any received packet with a sequence number in these two regions is discarded. Only a packet with a sequence number matching the value of R_n is accepted and acknowledged. The receive window also slides, but only one slot at a time. When a correct packet is received, the window slides, $R_n = (R_n + 1)$ modulo 2^m.

Figure 23.26 *Receive window for Go-Back-N*

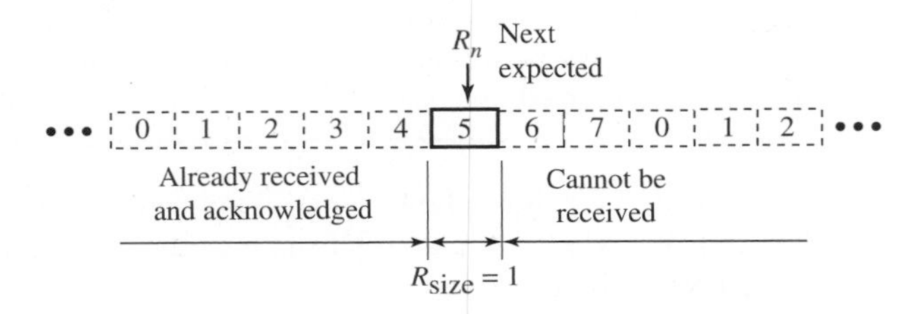

The receive window is an abstract concept defining an imaginary box of size 1 with a single variable R_n. The window slides when a correct packet has arrived; sliding occurs one slot at a time.

Timers

Although there can be a timer for each packet that is sent, in our protocol we use only one. The reason is that the timer for the first outstanding packet always expires first. We resend all outstanding packets when this timer expires.

Resending packets

When the timer expires, the sender resends all outstanding packets. For example, suppose the sender has already sent packet 6 ($S_n = 7$), but the only timer expires. If $S_f = 3$, this means that packets 3, 4, 5, and 6 have not been acknowledged; the sender goes back and resends packets 3, 4, 5, and 6. That is why the protocol is called *Go-Back*-N. On a time-out, the machine goes back *N* locations and resends all packets.

FSMs

Figure 23.27 shows the FSMs for the GBN protocol.

Sender

The sender starts in the ready state, but thereafter it can be in one of the two states: *ready* or *blocking*. The two variables are normally initialized to 0 ($S_f = S_n = 0$).

- ***Ready state.*** Four events may occur when the sender is in ready state.
  - **a.** If a request comes from the application layer, the sender creates a packet with the sequence number set to S_n. A copy of the packet is stored, and the packet is sent. The sender also starts the only timer if it is not running. The value of S_n is now incremented, ($S_n = S_n + 1$) modulo 2^m. If the window is full, $S_n = (S_f + S_{size})$ modulo 2^m, the sender goes to the blocking state.
  - **b.** If an error-free ACK arrives with ackNo related to one of the outstanding packets, the sender slides the window (set S_f = ackNo), and if all outstanding packets are acknowledged (ackNo = S_n), then the timer is stopped. If all outstanding packets are not acknowledged, the timer is restarted.

Figure 23.27 *FSMs for the Go-Back-N protocol*

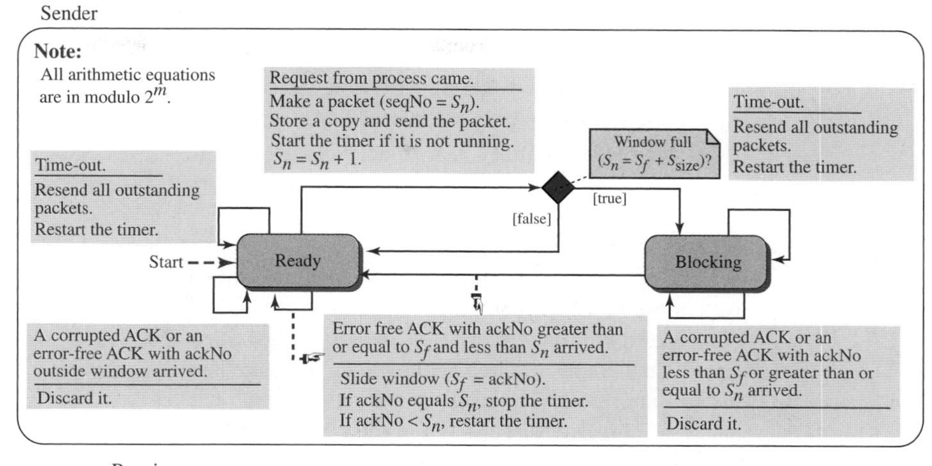

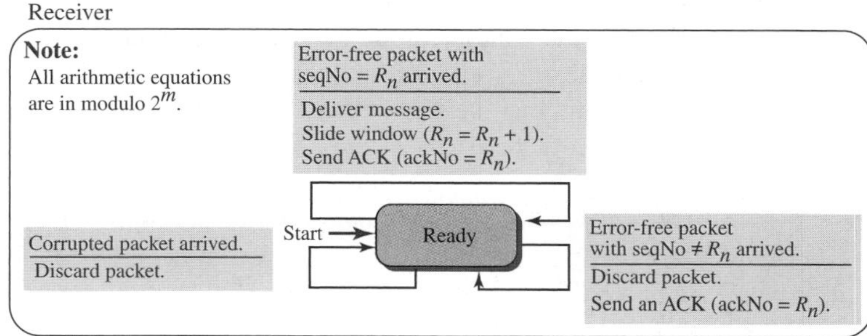

c. If a corrupted ACK or an error-free ACK with ackNo not related to the outstanding packet arrives, it is discarded.

d. If a time-out occurs, the sender resends all outstanding packets and restarts the timer.

- ***Blocking state.*** Three events may occur in this case:

  a. If an error-free ACK arrives with ackNo related to one of the outstanding packets, the sender slides the window (set S_f = ackNo) and if all outstanding packets are acknowledged (ackNo = S_n), then the timer is stopped. If all outstanding packets are not acknowledged, the timer is restarted. The sender then moves to the ready state.

  b. If a corrupted ACK or an error-free ACK with the ackNo not related to the outstanding packets arrives, the ACK is discarded.

  c. If a time-out occurs, the sender sends all outstanding packets and restarts the timer.

Receiver

The receiver is always in the *ready* state. The only variable, R_n, is initialized to 0. Three events may occur:

a. If an error-free packet with seqNo = R_n arrives, the message in the packet is delivered to the application layer. The window then slides, $R_n = (R_n + 1)$ modulo 2^m. Finally an ACK is sent with ackNo = R_n.

b. If an error-free packet with seqNo outside the window arrives, the packet is discarded, but an ACK with ackNo = R_n is sent.

c. If a corrupted packet arrives, it is discarded.

Send Window Size

We can now show why the size of the send window must be less than 2^m. As an example, we choose $m = 2$, which means the size of the window can be $2^m - 1$, or 3. Figure 23.28 compares a window size of 3 against a window size of 4. If the size of the window is 3 (less than 2^m) and all three acknowledgments are lost, the only timer expires and all three packets are resent. The receiver is now expecting packet 3, not packet 0, so the duplicate packet is correctly discarded. On the other hand, if the size of the window is 4 (equal to 2^2) and all acknowledgments are lost, the sender will send a duplicate of packet 0. However, this time the window of the receiver expects to receive packet 0 (in the next cycle), so it accepts packet 0, not as a duplicate, but as the first packet in the next cycle. This is an error. This shows that the size of the send window must be less than 2^m.

Figure 23.28 *Send window size for Go-Back-N*

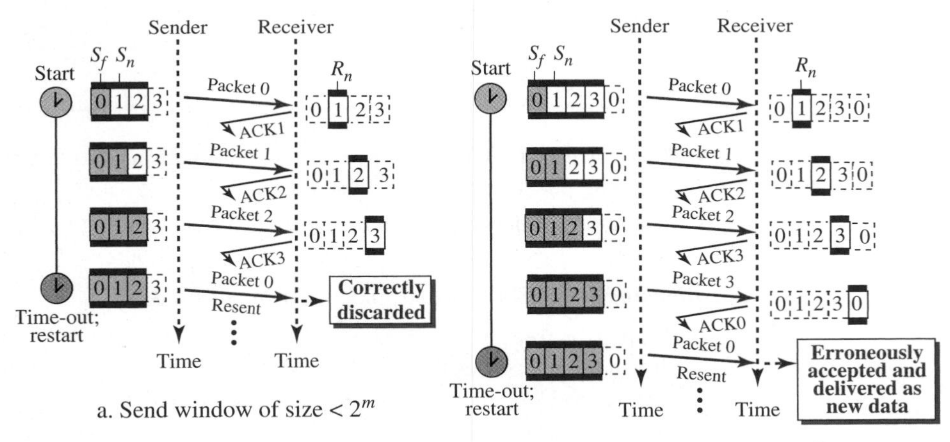

In the Go-Back-*N* protocol, the size of the send window must be less than 2^m; the size of the receive window is always 1.

Example 23.7

Figure 23.29 shows an example of Go-Back-*N*. This is an example of a case where the forward channel is reliable, but the reverse is not. No data packets are lost, but some ACKs are delayed and one is lost. The example also shows how cumulative acknowledgments can help if acknowledgments are delayed or lost.

Figure 23.29 *Flow diagram for Example 23.7*

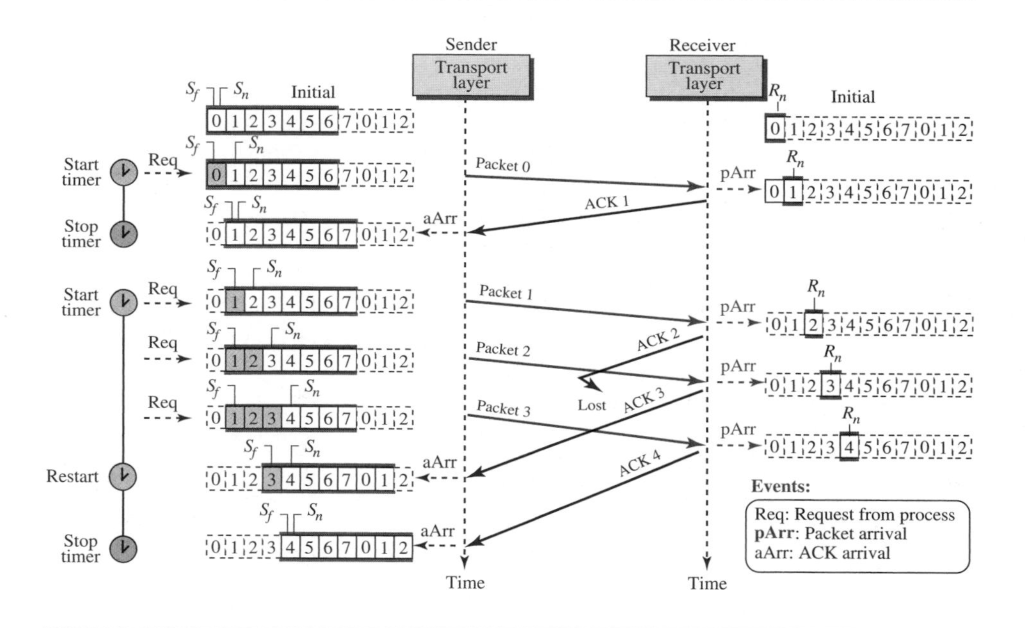

After initialization, there are some sender events. Request events are triggered by message chunks from the application layer; arrival events are triggered by ACKs received from the network layer. There is no time-out event here because all outstanding packets are acknowledged before the timer expires. Note that although ACK 2 is lost, ACK 3 is cumulative and serves as both ACK 2 and ACK 3. There are four events at the receiver site.

Example 23.8

Figure 23.30 shows what happens when a packet is lost. Packets 0, 1, 2, and 3 are sent. However, packet 1 is lost. The receiver receives packets 2 and 3, but they are discarded because they are received out of order (packet 1 is expected). When the receiver receives packets 2 and 3, it sends ACK1 to show that it expects to receive packet 1. However, these ACKs are not useful for the sender because the ackNo is equal to S_f, not greater than S_f. So the sender discards them. When the time-out occurs, the sender resends packets 1, 2, and 3, which are acknowledged.

Go-Back-N versus Stop-and-Wait

The reader may find that there is a similarity between the Go-Back-*N* protocol and the Stop-and-Wait protocol. The Stop-and-Wait protocol is actually a Go-Back-*N* protocol in

Figure 23.30 *Flow diagram for Example 23.8*

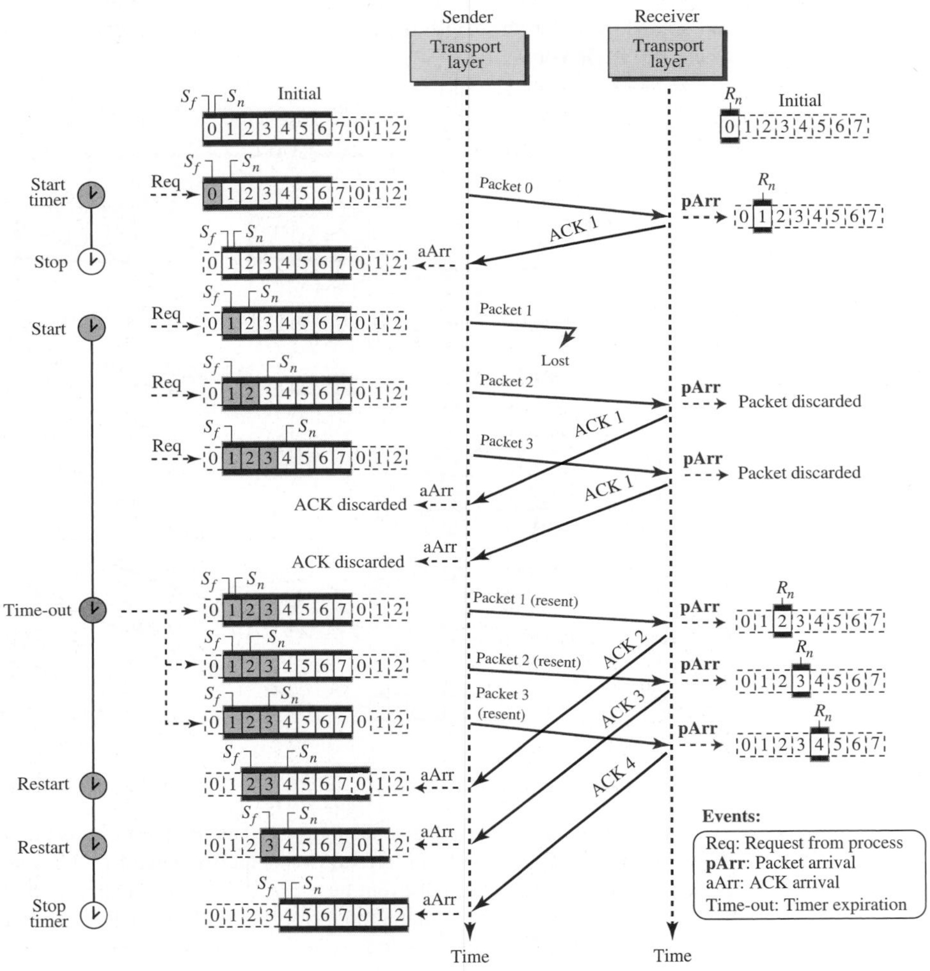

which there are only two sequence numbers and the send window size is 1. In other words, $m = 1$ and $2^m - 1 = 1$. In Go-Back-*N*, we said that the arithmetic is modulo 2^m; in Stop-and-Wait it is modulo 2, which is the same as 2^m when $m = 1$.

23.2.4 Selective-Repeat Protocol

The Go-Back-*N* protocol simplifies the process at the receiver. The receiver keeps track of only one variable, and there is no need to buffer out-of-order packets; they are simply discarded. However, this protocol is inefficient if the underlying network protocol loses a lot of packets. Each time a single packet is lost or corrupted, the sender resends all outstanding packets, even though some of these packets may have been received safe and sound but out of order. If the network layer is losing many packets because of congestion in the network, the resending of all of these outstanding packets makes the congestion worse, and eventually more packets are lost. This has an avalanche effect that may result in the total collapse of the network.

Another protocol, called the **Selective-Repeat (SR) protocol,** has been devised, which, as the name implies, resends only selective packets, those that are actually lost. The outline of this protocol is shown in Figure 23.31.

Figure 23.31 *Outline of Selective-Repeat*

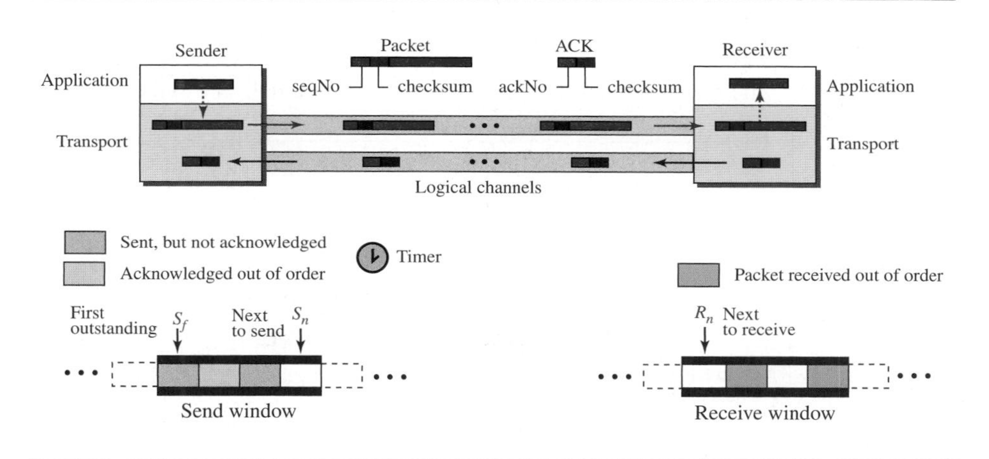

Windows

The Selective-Repeat protocol also uses two windows: a send window and a receive window. However, there are differences between the windows in this protocol and the ones in Go-Back-*N*. First, the maximum size of the send window is much smaller; it is 2^{m-1}. The reason for this will be discussed later. Second, the receive window is the same size as the send window.

The send window maximum size can be 2^{m-1}. For example, if $m = 4$, the sequence numbers go from 0 to 15, but the maximum size of the window is just 8 (it is 15 in the Go-Back-*N* Protocol). We show the Selective-Repeat send window in Figure 23.32 to emphasize the size.

The receive window in Selective-Repeat is totally different from the one in Go-Back-*N*. The size of the receive window is the same as the size of the send window (maximum 2^{m-1}). The Selective-Repeat protocol allows as many packets as the size of the receive window to arrive out of order and be kept until there is a set of consecutive packets to be delivered to the application layer. Because the sizes of the send window and receive window are the same, all the packets in the send packet can arrive out of order and be stored until they can be delivered. We need, however, to emphasize that in a reliable protocol the receiver *never* delivers packets out of order to the application layer. Figure 23.33 shows the receive window in Selective-Repeat. Those slots inside the

Figure 23.32 *Send window for Selective-Repeat protocol*

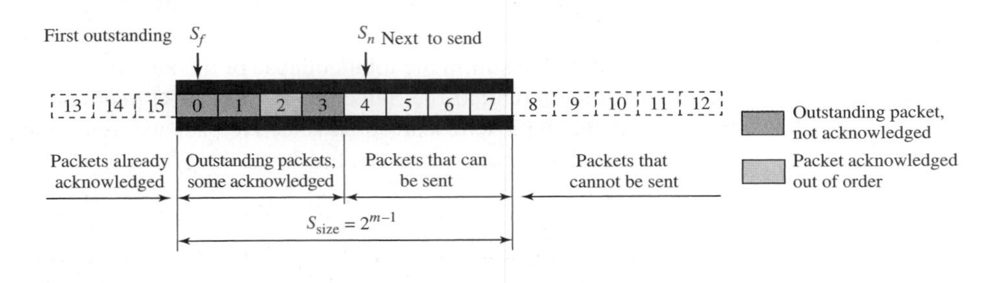

window that are shaded define packets that have arrived out of order and are waiting for the earlier transmitted packet to arrive before delivery to the application layer.

Figure 23.33 *Receive window for Selective-Repeat protocol*

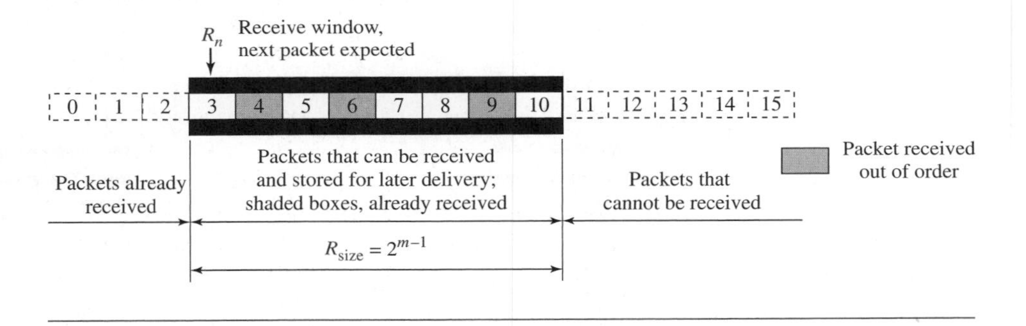

Timer

Theoretically, Selective-Repeat uses one timer for each outstanding packet. When a timer expires, only the corresponding packet is resent. In other words, GBN treats outstanding packets as a group; SR treats them individually. However, most transport-layer protocols that implement SR use only a single timer. For this reason, we use only one timer.

Acknowledgments

There is yet another difference between the two protocols. In GBN an ackNo is cumulative; it defines the sequence number of the next packet expected, confirming that all previous packets have been received safe and sound. The semantics of acknowledgment is different in SR. In SR, an ackNo defines the sequence number of a single packet that is received safe and sound; there is no feedback for any other.

In the Selective-Repeat protocol, an acknowledgment number defines the sequence number of the error-free packet received.

Example 23.9

Assume a sender sends 6 packets: packets 0, 1, 2, 3, 4, and 5. The sender receives an ACK with ackNo = 3. What is the interpretation if the system is using GBN or SR?

Solution

If the system is using GBN, it means that packets 0, 1, and 2 have been received uncorrupted and the receiver is expecting packet 3. If the system is using SR, it means that packet 3 has been received uncorrupted; the ACK does not say anything about other packets.

FSMs

Figure 23.34 shows the FSMs for the Selective-Repeat protocol. It is similar to the ones for the GBN, but there are some differences.

Sender

The sender starts in the *ready* state, but later it can be in one of the two states: *ready* or *blocking*. The following shows the events and the corresponding actions in each state.

- ***Ready state.*** Four events may occur in this case:

  a. If a request comes from the application layer, the sender creates a packet with the sequence number set to S_n. A copy of the packet is stored, and the packet is sent. If the timer is not running, the sender starts the timer. The value of S_n is now incremented, $S_n = (S_n + 1)$ modulo 2^m. If the window is full, $S_n = (S_f + S_{size})$ modulo 2^m, the sender goes to the blocking state.

  b. If an error-free ACK arrives with ackNo related to one of the outstanding packets, that packet is marked as acknowledged. If the ackNo = S_f, the window slides to the right until the S_f points to the first unacknowledged packet (all consecutive acknowledged packets are now outside the window). If there are outstanding packets, the timer is restarted; otherwise, the timer is stopped.

  c. If a corrupted ACK or an error-free ACK with ackNo not related to an outstanding packet arrives, it is discarded.

  d. If a time-out occurs, the sender resends all unacknowledged packets in the window and restarts the timer.

- ***Blocking state.*** Three events may occur in this case:

  a. If an error-free ACK arrives with ackNo related to one of the outstanding packets, that packet is marked as acknowledged. In addition, if the ackNo = S_f, the window is slid to the right until the S_f points to the first unacknowledged packet (all consecutive acknowledged packets are now outside the window). If the window has slid, the sender moves to the ready state.

  b. If a corrupted ACK or an error-free ACK with the ackNo not related to outstanding packets arrives, the ACK is discarded.

  c. If a time-out occurs, the sender resends all unacknowledged packets in the window and restarts the timer.

Receiver

The receiver is always in the *ready* state. Three events may occur:

Figure 23.34 *FSMs for SR protocol*

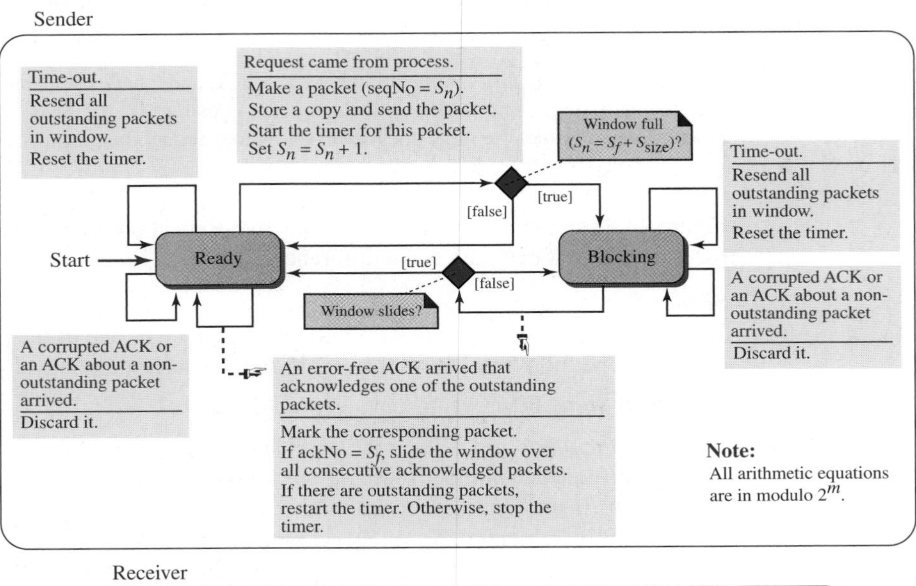

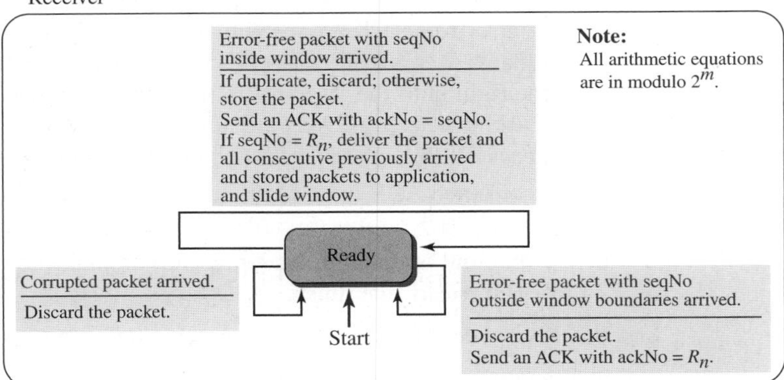

a. If an error-free packet with seqNo in the window arrives, the packet is stored and an ACK with ackNo = seqNo is sent. In addition, if the seqNo = R_n, then the packet and all previously arrived consecutive packets are delivered to the application layer and the window slides so that the R_n points to the first empty slot.

b. If an error-free packet with seqNo outside the window arrives, the packet is discarded, but an ACK with ackNo = R_n is returned to the sender. This is needed to let the sender slide its window if some ACKs related to packets with seqNo < R_n were lost.

c. If a corrupted packet arrives, the packet is discarded.

Example 23.10

This example is similar to Example 23.8 (Figure 23.30) in which packet 1 is lost. We show how Selective-Repeat behaves in this case. Figure 23.35 shows the situation.

Figure 23.35 *Flow diagram for Example 23.10*

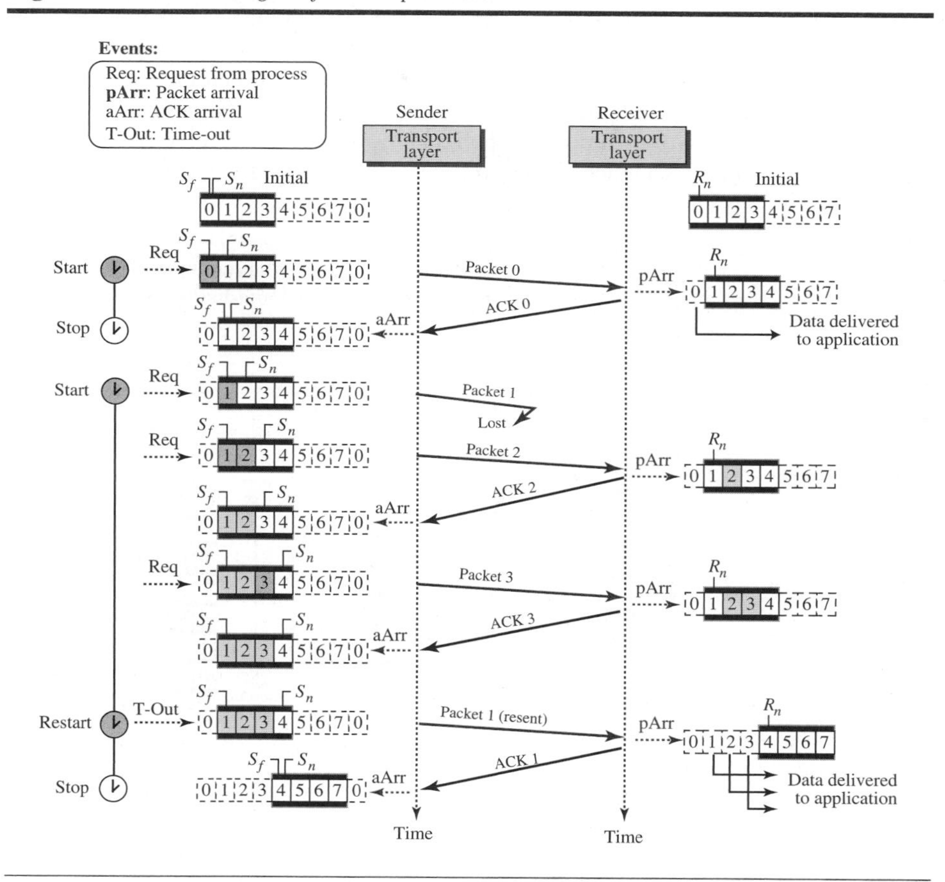

At the sender, packet 0 is transmitted and acknowledged. Packet 1 is lost. Packets 2 and 3 arrive out of order and are acknowledged. When the timer times out, packet 1 (the only unacknowledged packet) is resent and is acknowledged. The send window then slides.

At the receiver site we need to distinguish between the acceptance of a packet and its delivery to the application layer. At the second arrival, packet 2 arrives and is stored and marked (shaded slot), but it cannot be delivered because packet 1 is missing. At the next arrival, packet 3 arrives and is marked and stored, but still none of the packets can be delivered. Only at the last arrival, when finally a copy of packet 1 arrives, can packets 1, 2, and 3 be delivered to the application layer. There are two conditions for the delivery of packets to the application layer: First, a set of consecutive packets must have arrived. Second, the set starts from the beginning of the window. After the first arrival, there was only one packet and it started from the beginning of the window. After the last arrival, there are three packets and the first one starts from the beginning of the window. The key is that a reliable transport layer promises to deliver packets in order.

Window Sizes

We can now show why the size of the sender and receiver windows can be at most one-half of 2^m. For an example, we choose $m = 2$, which means the size of the window is $2^m/2$ or $2^{(m-1)} = 2$. Figure 23.36 compares a window size of 2 with a window size of 3.

If the size of the window is 2 and all acknowledgments are lost, the timer for packet 0 expires and packet 0 is resent. However, the window of the receiver is now expecting packet 2, not packet 0, so this duplicate packet is correctly discarded (the sequence number 0 is not in the window). When the size of the window is 3 and all acknowledgments are lost, the sender sends a duplicate of packet 0. However, this time, the window of the receiver expects to receive packet 0 (0 is part of the window), so it accepts packet 0, not as a duplicate, but as a packet in the next cycle. This is clearly an error.

In Selective-Repeat, the size of the sender and receiver window can be at most one-half of 2^m.

Figure 23.36 *Selective-Repeat, window size*

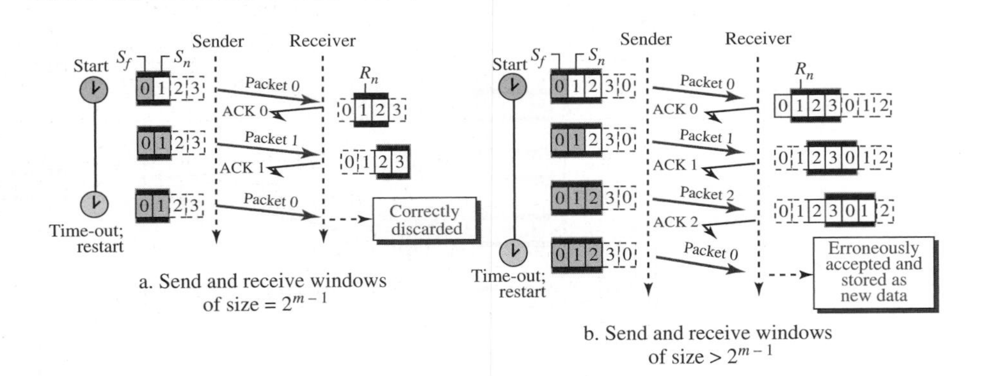

23.2.5 Bidirectional Protocols: Piggybacking

The four protocols we discussed earlier in this section are all unidirectional: data packets flow in only one direction and acknowledgments travel in the other direction. In real life, data packets are normally flowing in both directions: from client to server and from server to client. This means that acknowledgments also need to flow in both directions. A technique called **piggybacking** is used to improve the efficiency of the bidirectional protocols. When a packet is carrying data from A to B, it can also carry acknowledgment feedback about arrived packets from B; when a packet is carrying data from B to A, it can also carry acknowledgment feedback about the arrived packets from A.

Figure 23.37 shows the layout for the GBN protocol implemented bidirectionally using piggybacking. The client and server each use two independent windows: send and receive.

Figure 23.37 *Design of piggybacking in Go-Back-*N

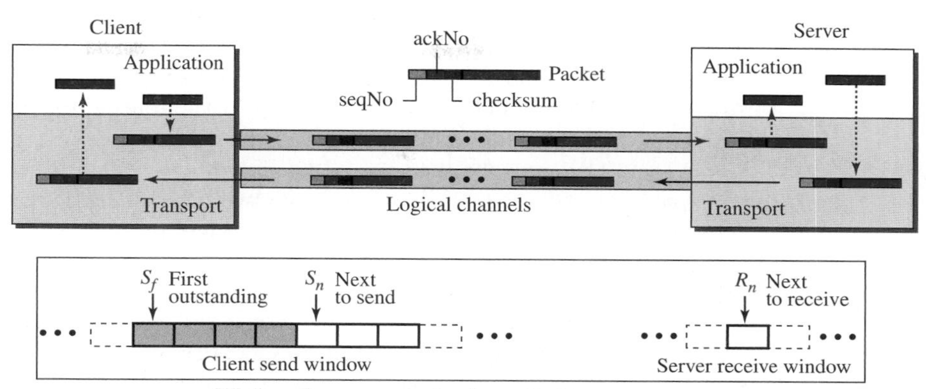

Windows for communication from client to server

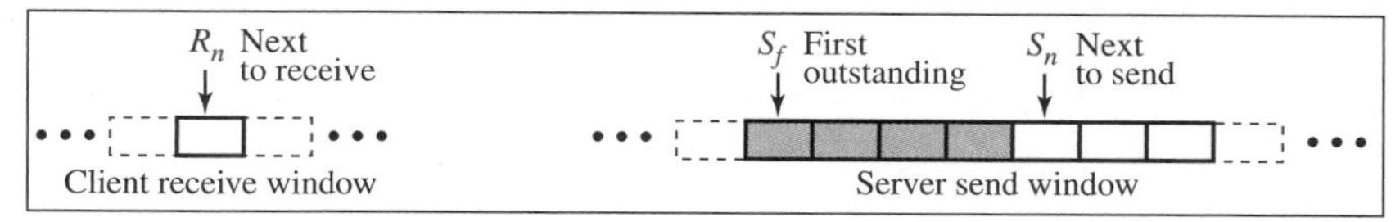

Windows for communication from server to client

23.3 END-CHAPTER MATERIALS

23.3.1 Recommended Reading

For more details about subjects discussed in this chapter, we recommend the following books.

Books

Several books give information about transport-layer protocols. The items enclosed in brackets refer to the reference list at the end of the book: In particular, we recommend [Com 06], [PD 03], [GW 04], [Far 04], [Tan 03], and [Sta 04].

23.3.2 Key Terms

bandwidth-delay product
client-server paradigm
congestion
congestion control
demultiplexing
ephemeral port number
finite state machine (FSM)
Go-Back-*N* protocol (GBN)
multiplexing
piggybacking
pipelining
port number
process-to-process communication
Selective-Repeat (SR) protocol
sequence number
sliding window
socket address
Stop-and-Wait protocol
well-known port number

23.3.3 Summary

The main duty of a transport-layer protocol is to provide process-to-process communication. To define the processes, we need port numbers. The client program defines itself with an ephemeral port number. The server defines itself with a well-known port number. To send a message from one process to another, the transport-layer protocol encapsulates and decapsulates messages. The transport layer at the source performs multiplexing; the transport layer at the destination performs demultiplexing. Flow control balances the exchange of data items between a producer and a consumer.

A transport-layer protocol can provide two types of services: connectionless and connection-oriented. In a connectionless service, the sender sends packets to the receiver without any connection establishment. In a connection-oriented service, the client and the server first need to establish a connection between themselves.

We have discussed several common transport-layer protocols in this chapter. The Stop-and-Wait protocol provides both flow and error control, but is inefficient. The Go-Back-*N* protocol is the more efficient version of the Stop-and-Wait protocol and takes advantage of pipelining. The Selective-Repeat protocol, a modification of the Go-Back-*N* protocol, is better suited to handle packet loss. All of these protocols can be implemented bidirectionally using piggybacking.

23.4 PRACTICE SET

23.4.1 Quizzes

A set of interactive quizzes for this chapter can be found on the book website. It is strongly recommended that the student take the quizzes to check his/her understanding of the materials before continuing with the practice set.

23.4.2 Questions

Q23-1. Can you explain why some transport-layer packets may be lost in the Internet?

Q23-2. Can you explain why some transport-layer packets may be duplicated in the Internet?

Q23-3. In a network, the size of the receive window is 1 packet. Which of the following protocols is being used by the network?

a. Stop-and-Wait **b.** Go-Back-*N* **c.** Selective-Repeat

Q23-4. Assume a new organization needs to create a new server process and allow its customers to access the organization site using that process. How should the port number for the server process be selected?

Q23-5. Does the wraparound situation create a problem in a network?

Q23-6. Can you explain why some transport-layer packets may be received out of order in the Internet?

Q23-7. Assume we have a set of dedicated computers in a system, each designed to perform only a single task. Do we still need host-to-host and process-to-process communication and two levels of addressing?

Q23-8. Operating systems assign a process number to every running application program. Can you explain why these process numbers cannot be used instead of port numbers?

Q23-9. Assume you need to write and test a client-server application program on two hosts you have at home.

a. What is the range of port numbers you would choose for the client program?

b. What is the range of port numbers you would choose for the server program?

c. Can the two port numbers be the same?

Q23-10. In the Selective-Repeat protocol, the size of the send and receive windows is the same. Does this mean that there are supposed to be no packets in transit?

Q23-11. In the Go-Back-*N* protocol, the size of the send window can be $2^m - 1$, while the size of the receive window is only 1. How can flow control be accomplished when there is a big difference between the size of the send and receive windows?

Q23-12. In a network, the size of the send window is 20 packets. Which of the following protocols is being used by the network?

a. Stop-and-Wait **b.** Go-Back-*N* **c.** Selective-Repeat

Q23-13. In a network with fixed value for $m > 1$, we can either use the Go-Back-*N* or the Selective-Repeat protocol. Describe the advantage and the disadvantage of using each. What other network criteria should be considered to select either of these protocols?

Q23-14. Since the field that stores the sequence number of a packet is limited in size, the sequence number in a protocol needs to wrap around, which means that two packets may have the same sequence number. In a protocol that uses m bits for the sequence-number field, if a packet has the sequence number x, how many packets need to be sent to see a packet with the same sequence number x, assuming that each packet is assigned one sequence number?

23.4.3 Problems

P23-1. In the Stop-and-Wait protocol, show the case in which the receiver receives a duplicate packet (which is also out of order). Hint: Think about a delayed ACK. What is the reaction of the receiver to this event?

P23-2. Can you explain why ICANN has divided the port numbers into three groups: well-known, registered, and dynamic?

P23-3. Assume we need to design a Selective-Repeat sliding-window protocol for a network in which the bandwidth is 1 Gbps and the average distance between the sender and receiver is 5,000 km. Assume the average packet size is 50,000 bits and the propagation speed in the media is 2×10^8 m. Find the maximum size of the send and receive windows, the number of bits in the sequence number field (m), and an appropriate time-out value for the timer.

P23-4. Create a scenario similar to Figure 23.22 in which the sender sends three packets. The first and second packets arrive and are acknowledged. The third packet is delayed and resent. The duplicate packet is received after the acknowledgment for the original is sent.

P23-5. Create a scenario similar to Figure 23.22 in which the sender sends two packets. The first packet is received and acknowledged, but the acknowledgment is lost. The sender resends the packet after time-out. The second packet is lost and resent.

P23-6. In a network using the Selective-Repeat protocol with $m = 4$ and the sending window of size 8, the value of variables are $S_f = 62$, $S_n = 67$, and $R_n = 64$. Packet 65 has already been acknowledged at the sender site; packets 65 and 66 are received out-of-order at the receiver site. Assume that the network does not duplicate the packets.

a. What are the sequence numbers of pending data packets (in transit, corrupted, or lost)?

b. What are the acknowledgment numbers of pending ACK packets (in transit, corrupted, or lost)?

P23-7. We can define the bandwidth-delay product in a network as the number of packets that can be in the pipe during the round-trip time (RTT). What is the bandwidth-delay product in each of the following situations?

a. Bandwidth: 1 Mbps, RTT: 20 ms, packet size: 1000 bits

b. Bandwidth: 10 Mbps, RTT: 20 ms, packet size: 2000 bits

c. Bandwidth: 1 Gbps, RTT: 4 ms, packet size: 10,000 bits

P23-8. Answer the following questions related to the FSMs for the Stop-and-Wait protocol (Figure 23.21):

a. The sending machine is in the ready state and $S = 0$. What is the sequence number of the next packet to send?

b. The sending machine is in the blocking state and $S = 1$. What is the sequence number of the next packet to send if a time-out occurs?

c. The receiving machine is in the ready state and $R = 1$. A packet with the sequence number 1 arrives. What is the action in response to this event?

d. The receiving machine is in the ready state and $R = 1$. A packet with the sequence number 0 arrives. What is the action in response to this event?

P23-9. Redraw Figure 23.19 with 5 packets exchanged (0, 1, 2, 3, 4). Assume packet 2 is lost and packet 3 arrives after packet 4.

P23-10. Answer the following questions related to the FSMs for the Selective-Repeat protocol with $m = 7$ bits. Assume the window size is 64. (Figure 23.34):

a. The sending machine is in the ready state with $S_f = 10$ and $S_n = 15$. What is the sequence number of the next packet to send?

b. The sending machine is in the ready state with $S_f = 10$ and $S_n = 15$. The timer for packet 10 times out. How many packets are to be resent? What are their sequence numbers?

c. The sending machine is in the ready state with $S_f = 10$ and $S_n = 15$. An ACK with ackNo = 13 arrives. What are the next values of S_f and S_n? What is the action in response to this event?

d. The sending machine is in the blocking state with $S_f = 14$ and $S_n = 21$. What is the size of the window?

e. The sending machine is in the blocking state with $S_f = 14$ and $S_n = 21$. An ACK with ackNo = 14 arrives. Packets 15 and 16 have already been acknowledged. What are the next values of S_f and S_n? What is the state of the sending machine?

f. The receiving machine is in the ready state with $R_n = 16$. The size of the window is 8. A packet with sequence number 16 arrives. What is the next value of R_n? What is the response of the machine to this event?

P23-11. In a network using the Go-Back-*N* protocol with $m = 3$ and the sending window of size 7, the values of variables are $S_f = 62$, $S_n = 66$, and $R_n = 64$. Assume that the network does not duplicate or reorder the packets.

a. What are the sequence numbers of data packets in transit?

b. What are the acknowledgment numbers of ACK packets in transit?

P23-12. Assume we need to design a Go-Back-*N* sliding-window protocol for a network in which the bandwidth is 100 Mbps and the average distance between the sender and receiver is 10,000 km. Assume the average packet size is 100,000 bits and the propagation speed in the media is 2×10^8 m/s. Find the maximum size of the send and receive windows, the number of bits in the sequence number field (m), and an appropriate time-out value for the timer.

P23-13. Answer the following questions related to the FSMs for the Go-back-*N* protocol with $m = 6$ bits. Assume the window size is 63. (Figure 23.27):

a. The sending machine is in the ready state with $S_f = 10$ and $S_n = 15$. What is the sequence number of the next packet to send?

b. The sending machine is in the ready state with $S_f = 10$ and $S_n = 15$. A time-out occurs. How many packets are to be resent? What are their sequence numbers?

c. The sending machine is in the ready state with $S_f = 10$ and $S_n = 15$. An ACK with ackNo = 13 arrives. What are the next values of S_f and S_n?

d. The sending machine is in the blocking state with $S_f = 14$ and $S_n = 21$. What is the size of the window?

e. The sending machine is in the blocking state with $S_f = 14$ and $S_n = 21$. An ACK with ackNo = 18 arrives. What are the next values of S_f and S_n? What is the state of the sending machine?

f. The receiving machine is in the ready state with $R_n = 16$. A packet with sequence number 16 arrives. What is the next value of R_n? What is the response of the machine to this event?

P23-14. Show the FSM for an imaginary machine with three states: A (starting state), B, and C; and four events: events 1, 2, 3, and 4. The following specify the behavior of the machine:

a. When in state A, two events may occur: event 1 and event 2. If event 1 occurs, the machine performs action 1 and moves to state B. If event 2 occurs, the machine moves to state C (no action).

b. When in state B, two events may occur: event 3 and event 4. If event 3 occurs, the machine performs action 2, but remains in state B. If event 4 occurs, the machine just moves to state C.

c. When in state C, the machine remains in this state forever.

P23-15. Compare the range of 16-bit addresses, 0 to 65,535, with the range of 32-bit IP addresses, 0 to 4,294,967,295 (discussed in Chapter 18). Why do we need such a large range of IP addresses, but only a relatively small range of port numbers?

P23-16. In each of the following protocols, how many packets can have independent sequence numbers before wraparound occurs.

a. Stop-and-Wait

b. Go-Back-*N* with $m = 8$

c. Select-Repeat with $m = 8$

P23-17. Using 5-bit sequence numbers, what is the maximum size of the send and receive windows for each of the following protocols?

a. Stop-and-Wait

b. Go-Back-*N*

c. Selective-Repeat

P23-18. Redraw Figure 23.29 when the sender sends 5 packets (0, 1, 2, 3, and 4). Packets 0, 1, and 2 are sent and acknowledged in a single ACK, which arrives at the sender site after all packets have been sent. Packet 3 is received and acknowledged in a single ACK. Packet 4 is lost and resent.

P23-19. A sender sends a series of packets to the same destination using 5-bit sequence numbers. If the sequence numbers start with 0, what is the sequence number of the 100th packet?

P23-20. An acknowledgment number in the Go-Back-*N* protocol defines the next packet expected, but an acknowledgment number in the Selective-Repeat protocol defines the sequence number of the packet to be acknowledged. Can you explain the reason?

P23-21. Redraw Figure 23.35 if the sender sends 5 packets (0, 1, 2, 3, and 4). Packets 0, 1, and 2 are received in order and acknowledged, one by one. Packet 3 is delayed and received after packet 4.

P23-22. Assume we want to change the Stop-and-Wait protocol and add the NAK (negative ACK) packet to the system. When a corrupted packet arrives at the receiver, the receiver discards the packet, but sends a NAK with a nakNo defining the seqNo of the corrupted packet. In this way, the sender can resend the corrupted packet without waiting for the time-out. Explain what changes need to be made in the FSM of Figure 23.21 and show an example of the operation of the new protocol with a time-line diagram.

P23-23. Assume that our network never corrupts, loses, or duplicates packets. We are only concerned about flow control. We do not want the sender to overwhelm the receiver with packets. Design an FSM to allow the sender to send a packet to the receiver only when the receiver is ready. If the receiver is ready to receive a packet, it sends an ACK. Lack of getting an ACK for the sender means that the receiver is not ready to receive more packets.

P23-24. Assume that our network may corrupt packets, but it never loses or duplicates a packet. We are also concerned about flow control. We do not want the sender to overwhelm the receiver with packets. Design an FSM of a new protocol to allow these features.

23.5 SIMULATION EXPERIMENTS

23.5.1 Applets

We have created some Java applets to show some of the main concepts discussed in this chapter. It is strongly recommended that the students activate these applets on the book website and carefully examine the protocols in action.

23.6 PROGRAMMING ASSIGNMENT

Write the source code, compile, and test the following programs in the programming language of your choice:

Prg23-1. Write a program to simulate the sending-site FSMs for the simple protocol (Figure 23.18).

Prg23-2. Write a program to simulate the sending-site FSMs for the Stop-and-Wait protocol (Figure 23.21).

Prg23-3. Write a program to simulate the sending-site FSMs for the Go-Back-*N* protocol (Figure 23.27).

Prg23-4. Write a program to simulate the sending-site FSMs for the Selective Repeat protocol (Figure 23.34).

CHAPTER 24

Transport-Layer Protocols

The transport layer in the TCP/IP suite is located between the application layer and the network layer. It provides services to the application layer and receives services from the network layer. The transport layer acts as a liaison between a client program and a server program, a process-to-process connection. The transport layer is the heart of the TCP/IP protocol suite; it is the end-to-end logical vehicle for transferring data from one point to another in the Internet. We have divided this chapter into four sections:

- The first section introduces the three transport-layer protocols in the Internet and gives some information common to all of them.
- The second section concentrates on UDP, which is the simplest of the three protocols. UDP lacks many services we require from a transport-layer protocol, but its simplicity is very attractive to some applications, as we show.
- The third section discusses TCP. The section first lists its services and features. Using a transition diagram, it then shows how TCP provides a connection-oriented service. The section then uses abstract windows to show how flow and error control are accomplished in TCP. Congestion control in TCP is discussed next, a topic that was discussed for the network layer.
- The fourth section discusses SCTP. The section first lists its services and features. It then shows how STCP creates an association. The section then shows how flow and error control are accomplished in SCTP using SACKs.

24.1 INTRODUCTION

After discussing the general principle behind the transport layer in the previous chapter, we concentrate on the transport protocols in the Internet in this chapter. Figure 24.1 shows the position of these three protocols in the TCP/IP protocol suite.

Figure 24.1 *Position of transport-layer protocols in the TCP/IP protocol suite*

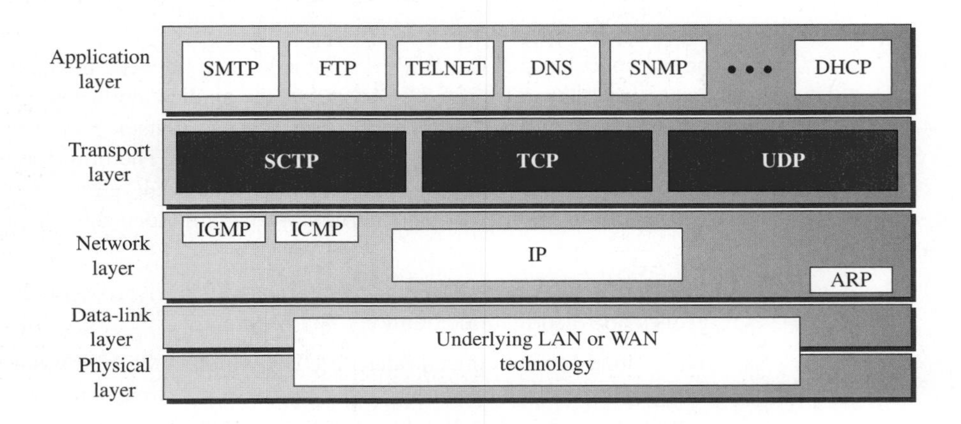

24.1.1 Services

Each protocol provides a different type of service and should be used appropriately.

UDP

UDP is an unreliable connectionless transport-layer protocol used for its simplicity and efficiency in applications where error control can be provided by the application-layer process.

TCP

TCP is a reliable connection-oriented protocol that can be used in any application where reliability is important.

SCTP

SCTP is a new transport-layer protocol that combines the features of UDP and TCP.

24.1.2 Port Numbers

As discussed in the previous chapter, a transport-layer protocol usually has several responsibilities. One is to create a process-to-process communication; these protocols use port numbers to accomplish this. Port numbers provide end-to-end addresses at the transport layer and allow multiplexing and demultiplexing at this layer, just as IP addresses do at the network layer. Table 24.1 gives some common port numbers for all three protocols we discuss in this chapter.

Table 24.1 *Some well-known ports used with UDP and TCP*

Port	*Protocol*	*UDP*	*TCP*	*SCTP*	*Description*
7	Echo	√	√	√	Echoes back a received datagram
9	Discard	√	√	√	Discards any datagram that is received
11	Users	√	√	√	Active users
13	Daytime	√	√	√	Returns the date and the time
17	Quote	√	√	√	Returns a quote of the day
19	Chargen	√	√	√	Returns a string of characters
20	FTP-data		√	√	File Transfer Protocol
21	FTP-21		√	√	File Transfer Protocol
23	TELNET		√	√	Terminal Network
25	SMTP		√	√	Simple Mail Transfer Protocol
53	DNS	√	√	√	Domain Name Service
67	DHCP	√	√	√	Dynamic Host Configuration Protocol
69	TFTP	√	√	√	Trivial File Transfer Protocol
80	HTTP		√	√	HyperText Transfer Protocol
111	RPC	√	√	√	Remote Procedure Call
123	NTP	√	√	√	Network Time Protocol
161	SNMP-server	√			Simple Network Management Protocol
162	SNMP-client	√			Simple Network Management Protocol

24.2 USER DATAGRAM PROTOCOL

The **User Datagram Protocol (UDP)** is a connectionless, unreliable transport protocol. It does not add anything to the services of IP except for providing process-to-process communication instead of host-to-host communication. If UDP is so powerless, why would a process want to use it? With the disadvantages come some advantages. UDP is a very simple protocol using a minimum of overhead. If a process wants to send a small message and does not care much about reliability, it can use UDP. Sending a small message using UDP takes much less interaction between the sender and receiver than using TCP. We discuss some applications of UDP at the end of this section.

24.2.1 User Datagram

UDP packets, called ***user datagrams,*** have a fixed-size header of 8 bytes made of four fields, each of 2 bytes (16 bits). Figure 24.2 shows the format of a user datagram. The first two fields define the source and destination port numbers. The third field defines the total length of the user datagram, header plus data. The 16 bits can define a total length of 0 to 65,535 bytes. However, the total length needs to be less because a UDP user datagram is stored in an IP datagram with the total length of 65,535 bytes. The last field can carry the optional checksum (explained later).

Figure 24.2 *User datagram packet format*

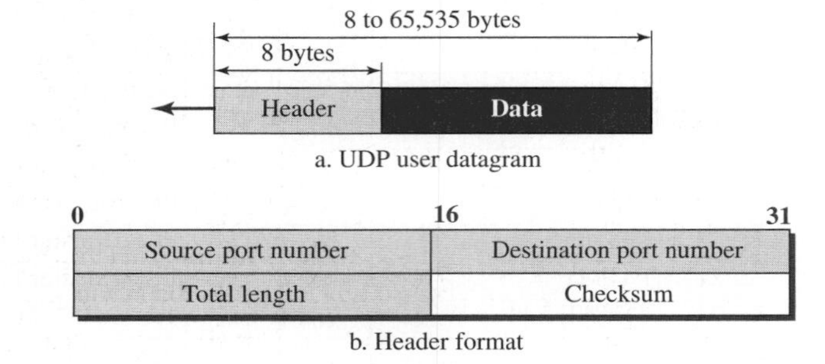

Example 24.1

The following is the content of a UDP header in hexadecimal format.

CB84000D001C001C

- **a.** What is the source port number?
- **b.** What is the destination port number?
- **c.** What is the total length of the user datagram?
- **d.** What is the length of the data?
- **e.** Is the packet directed from a client to a server or vice versa?
- **f.** What is the client process?

Solution

- **a.** The source port number is the first four hexadecimal digits $(CB84)_{16}$, which means that the source port number is 52100.
- **b.** The destination port number is the second four hexadecimal digits $(000D)_{16}$, which means that the destination port number is 13.
- **c.** The third four hexadecimal digits $(001C)_{16}$ define the length of the whole UDP packet as 28 bytes.
- **d.** The length of the data is the length of the whole packet minus the length of the header, or $28 - 8 = 20$ bytes.
- **e.** Since the destination port number is 13 (well-known port), the packet is from the client to the server.
- **f.** The client process is the Daytime (see Table 24.1).

24.2.2 UDP Services

Earlier we discussed the general services provided by a transport-layer protocol. In this section, we discuss what portions of those general services are provided by UDP.

Process-to-Process Communication

UDP provides process-to-process communication using **socket addresses,** a combination of IP addresses and port numbers.

Connectionless Services

As mentioned previously, UDP provides a *connectionless service.* This means that each user datagram sent by UDP is an independent datagram. There is no relationship between the different user datagrams even if they are coming from the same source process and going to the same destination program. The user datagrams are not numbered. Also, unlike TCP, there is no connection establishment and no connection termination. This means that each user datagram can travel on a different path.

One of the ramifications of being connectionless is that the process that uses UDP cannot send a stream of data to UDP and expect UDP to chop them into different, related user datagrams. Instead each request must be small enough to fit into one user datagram. Only those processes sending short messages, messages less than 65,507 bytes (65,535 minus 8 bytes for the UDP header and minus 20 bytes for the IP header), can use UDP.

Flow Control

UDP is a very simple protocol. There is no *flow control,* and hence no window mechanism. The receiver may overflow with incoming messages. The lack of flow control means that the process using UDP should provide for this service, if needed.

Error Control

There is no *error control* mechanism in UDP except for the checksum. This means that the sender does not know if a message has been lost or duplicated. When the receiver detects an error through the checksum, the user datagram is silently discarded. The lack of error control means that the process using UDP should provide for this service, if needed.

Checksum

We discussed checksum and its calculation in Chapter 10. UDP checksum calculation includes three sections: a pseudoheader, the UDP header, and the data coming from the application layer. The *pseudoheader* is the part of the header of the IP packet (discussed in Chapter 19) in which the user datagram is to be encapsulated with some fields filled with 0s (see Figure 24.3).

Figure 24.3 *Pseudoheader for checksum calculation*

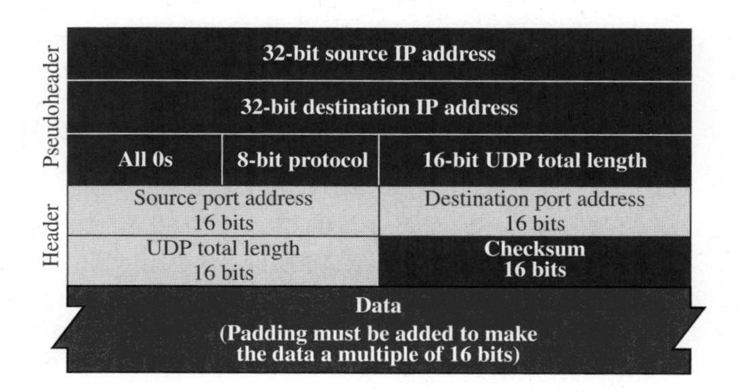

If the checksum does not include the pseudoheader, a user datagram may arrive safe and sound. However, if the IP header is corrupted, it may be delivered to the wrong host.

The protocol field is added to ensure that the packet belongs to UDP, and not to TCP. We will see later that if a process can use either UDP or TCP, the destination port number can be the same. The value of the protocol field for UDP is 17. If this value is changed during transmission, the checksum calculation at the receiver will detect it and UDP drops the packet. It is not delivered to the wrong protocol.

Optional Inclusion of Checksum

The sender of a UDP packet can choose not to calculate the checksum. In this case, the checksum field is filled with all 0s before being sent. In the situation where the sender decides to calculate the checksum, but it happens that the result is all 0s, the checksum is changed to all 1s before the packet is sent. In other words, the sender complements the sum two times. Note that this does not create confusion because the value of the checksum is never all 1s in a normal situation (see the next example).

Example 24.2

What value is sent for the checksum in each one of the following hypothetical situations?

a. The sender decides not to include the checksum.

b. The sender decides to include the checksum, but the value of the sum is all 1s.

c. The sender decides to include the checksum, but the value of the sum is all 0s.

Solution

a. The value sent for the checksum field is all 0s to show that the checksum is not calculated.

b. When the sender complements the sum, the result is all 0s; the sender complements the result again before sending. The value sent for the checksum is all 1s. The second complement operation is needed to avoid confusion with the case in part a.

c. This situation never happens because it implies that the value of every term included in the calculation of the sum is all 0s, which is impossible; some fields in the pseudoheader have nonzero values.

Congestion Control

Since UDP is a connectionless protocol, it does not provide congestion control. UDP assumes that the packets sent are small and sporadic and cannot create congestion in the network. This assumption may or may not be true today, when UDP is used for interactive real-time transfer of audio and video.

Encapsulation and Decapsulation

To send a message from one process to another, the UDP protocol encapsulates and decapsulates messages.

Queuing

We have talked about ports without discussing the actual implementation of them. In UDP, queues are associated with ports.

At the client site, when a process starts, it requests a port number from the operating system. Some implementations create both an incoming and an outgoing queue associated with each process. Other implementations create only an incoming queue associated with each process.

Multiplexing and Demultiplexing

In a host running a TCP/IP protocol suite, there is only one UDP but possibly several processes that may want to use the services of UDP. To handle this situation, UDP multiplexes and demultiplexes.

Comparison between UDP and Generic Simple Protocol

We can compare UDP with the connectionless simple protocol we discussed earlier. The only difference is that UDP provides an optional checksum to detect corrupted packets at the receiver site. If the checksum is added to the packet, the receiving UDP can check the packet and discard the packet if it is corrupted. No feedback, however, is sent to the sender.

UDP is an example of the connectionless simple protocol we discussed earlier with the exception of an optional checksum added to packets for error detection.

24.2.3 UDP Applications

Although UDP meets almost none of the criteria we mentioned earlier for a reliable transport-layer protocol, UDP is preferable for some applications. The reason is that some services may have some side effects that are either unacceptable or not preferable. An application designer sometimes needs to compromise to get the optimum. For example, in our daily life, we all know that a one-day delivery of a package by a carrier is more expensive than a three-day delivery. Although high speed and low cost are both desirable features in delivery of a parcel, they are in conflict with each other. We need to choose the optimum.

In this section, we first discuss some features of UDP that may need to be considered when we design an application program and then show some typical applications.

UDP Features

We briefly discuss some features of UDP and their advantages and disadvantages.

Connectionless Service

As we mentioned previously, UDP is a connectionless protocol. Each UDP packet is independent from other packets sent by the same application program. This feature can be considered as an advantage or disadvantage depending on the application requirements. It is an advantage if, for example, a client application needs to send a short request to a server and to receive a short response. If the request and response can each fit in a single user datagram, a connectionless service may be preferable. The overhead to establish and close a connection may be significant in this case. In the connection-oriented service, to achieve the above goal, at least 9 packets are exchanged between the client and the server; in connectionless service only 2 packets are exchanged. The connectionless service provides less delay; the connection-oriented service creates more delay. If delay is an important issue for the application, the connectionless service is preferred.

Example 24.3

A client-server application such as DNS (see Chapter 26) uses the services of UDP because a client needs to send a short request to a server and to receive a quick response from it. The request and response can each fit in one user datagram. Since only one message is exchanged in each direction, the connectionless feature is not an issue; the client or server does not worry that messages are delivered out of order.

Example 24.4

A client-server application such as SMTP (see Chapter 27), which is used in electronic mail, cannot use the services of UDP because a user might send a long e-mail message, which could include multimedia (images, audio, or video). If the application uses UDP and the message does not fit in one user datagram, the message must be split by the application into different user datagrams. Here the connectionless service may create problems. The user datagrams may arrive and be delivered to the receiver application out of order. The receiver application may not be able to reorder the pieces. This means the connectionless service has a disadvantage for an application program that sends long messages. In SMTP, when we send a message, we do not expect to receive a response quickly (sometimes no response is required). This means that the extra delay inherent in connection-oriented service is not crucial for SMTP.

Lack of Error Control

UDP does not provide error control; it provides an unreliable service. Most applications expect reliable service from a transport-layer protocol. Although a reliable service is desirable, it may have some side effects that are not acceptable to some applications. When a transport layer provides reliable services, if a part of the message is lost or corrupted, it needs to be resent. This means that the receiving transport layer cannot deliver that part to the application immediately; there is an uneven delay between different parts of the message delivered to the application layer. Some applications, by nature, do not even notice these uneven delays, but for some they are very problematic.

Example 24.5

Assume we are downloading a very large text file from the Internet. We definitely need to use a transport layer that provides reliable service. We don't want part of the file to be missing or corrupted when we open the file. The delay created between the deliveries of the parts is not an overriding concern for us; we wait until the whole file is composed before looking at it. In this case, UDP is not a suitable transport layer.

Example 24.6

Assume we are using a real-time interactive application, such as Skype. Audio and video are divided into frames and sent one after another. If the transport layer is supposed to resend a corrupted or lost frame, the synchronizing of the whole transmission may be lost. The viewer suddenly sees a blank screen and needs to wait until the second transmission arrives. This is not tolerable. However, if each small part of the screen is sent using a single user datagram, the receiving UDP can easily ignore the corrupted or lost packet and deliver the rest to the application program. That part of the screen is blank for a very short period of time, which most viewers do not even notice.

Lack of Congestion Control

UDP does not provide congestion control. However, UDP does not create additional traffic in an error-prone network. TCP may resend a packet several times and thus contribute to the creation of congestion or worsen a congested situation. Therefore, in some cases, lack of error control in UDP can be considered an advantage when congestion is a big issue.

Typical Applications

The following shows some typical applications that can benefit more from the services of UDP than from those of TCP.

- UDP is suitable for a process that requires simple request-response communication with little concern for flow and error control. It is not usually used for a process such as FTP that needs to send bulk data (see Chapter 26).
- UDP is suitable for a process with internal flow- and error-control mechanisms. For example, the Trivial File Transfer Protocol (TFTP) process includes flow and error control. It can easily use UDP.
- UDP is a suitable transport protocol for multicasting. Multicasting capability is embedded in the UDP software but not in the TCP software.
- UDP is used for management processes such as SNMP (see Chapter 27).
- UDP is used for some route updating protocols such as Routing Information Protocol (RIP) (see Chapter 20).
- UDP is normally used for interactive real-time applications that cannot tolerate uneven delay between sections of a received message (see Chapter 28).

24.3 TRANSMISSION CONTROL PROTOCOL

Transmission Control Protocol (TCP) is a connection-oriented, reliable protocol. TCP explicitly defines connection establishment, data transfer, and connection teardown phases to provide a connection-oriented service. TCP uses a combination of GBN and SR protocols to provide reliability. To achieve this goal, TCP uses checksum (for error detection), retransmission of lost or corrupted packets, cumulative and selective acknowledgments, and timers. In this section, we first discuss the services provided by TCP; we then discuss the TCP features in more detail. TCP is the most common transport-layer protocol in the Internet.

24.3.1 TCP Services

Before discussing TCP in detail, let us explain the services offered by TCP to the processes at the application layer.

Process-to-Process Communication

As with UDP, TCP provides process-to-process communication using port numbers. We have already given some of the port numbers used by TCP in Table 24.1 in the previous section.

Stream Delivery Service

TCP, unlike UDP, is a stream-oriented protocol. In UDP, a process sends messages with predefined boundaries to UDP for delivery. UDP adds its own header to each of these messages and delivers it to IP for transmission. Each message from the process is called a *user datagram*, and becomes, eventually, one IP datagram. Neither IP nor UDP recognizes any relationship between the datagrams.

TCP, on the other hand, allows the sending process to deliver data as a stream of bytes and allows the receiving process to obtain data as a stream of bytes. TCP creates an environment in which the two processes seem to be connected by an imaginary "tube" that carries their bytes across the Internet. This imaginary environment is depicted in Figure 24.4. The sending process produces (writes to) the stream and the receiving process consumes (reads from) it.

Figure 24.4 *Stream delivery*

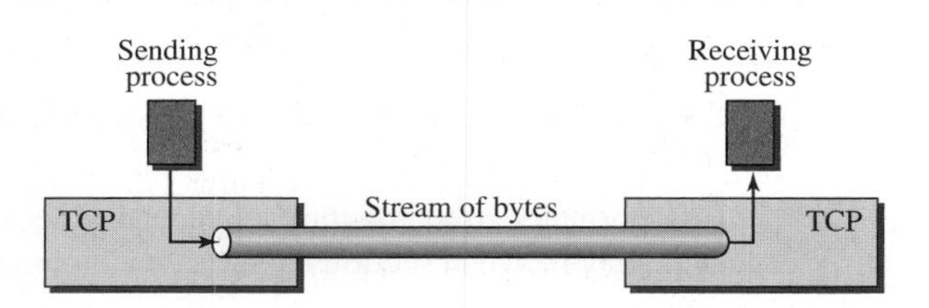

Sending and Receiving Buffers

Because the sending and the receiving processes may not necessarily write or read data at the same rate, TCP needs buffers for storage. There are two buffers, the sending buffer and the receiving buffer, one for each direction. We will see later that these buffers are also necessary for flow- and error-control mechanisms used by TCP. One way to implement a buffer is to use a circular array of 1-byte locations as shown in Figure 24.5. For simplicity, we have shown two buffers of 20 bytes each; normally the buffers are hundreds or thousands of bytes, depending on the implementation. We also show the buffers as the same size, which is not always the case.

The figure shows the movement of the data in one direction. At the sender, the buffer has three types of chambers. The white section contains empty chambers that can be filled by the sending process (producer). The colored area holds bytes that have been sent but not yet acknowledged. The TCP sender keeps these bytes in the buffer until it receives an acknowledgment. The shaded area contains bytes to be sent by the sending TCP. However, as we will see later in this chapter, TCP may be able to send only part of this shaded section. This could be due to the slowness of the receiving process or to congestion in the network. Also note that, after the bytes in the colored chambers are acknowledged, the chambers are recycled and available for use by the sending process. This is why we show a circular buffer.

The operation of the buffer at the receiver is simpler. The circular buffer is divided into two areas (shown as white and colored). The white area contains empty chambers to be filled by bytes received from the network. The colored sections contain received

Figure 24.5 *Sending and receiving buffers*

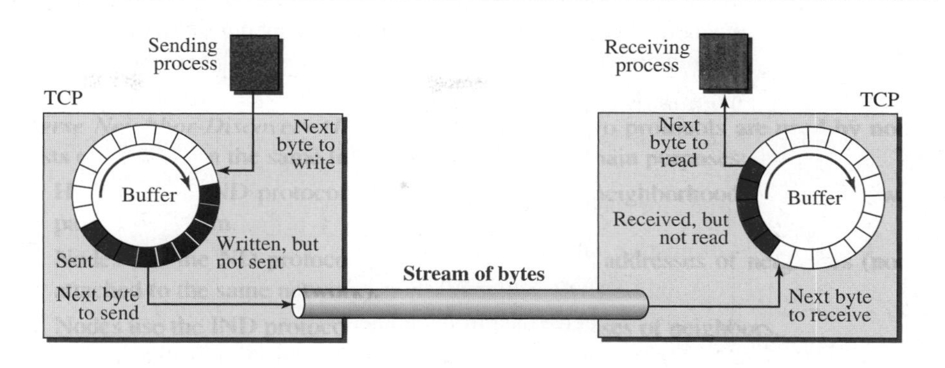

bytes that can be read by the receiving process. When a byte is read by the receiving process, the chamber is recycled and added to the pool of empty chambers.

Segments

Although buffering handles the disparity between the speed of the producing and consuming processes, we need one more step before we can send data. The network layer, as a service provider for TCP, needs to send data in packets, not as a stream of bytes. At the transport layer, TCP groups a number of bytes together into a packet called a *segment*. TCP adds a header to each segment (for control purposes) and delivers the segment to the network layer for transmission. The segments are encapsulated in an IP datagram and transmitted. This entire operation is transparent to the receiving process. Later we will see that segments may be received out of order, lost or corrupted, and resent. All of these are handled by the TCP receiver with the receiving application process unaware of TCP's activities. Figure 24.6 shows how segments are created from the bytes in the buffers.

Figure 24.6 *TCP segments*

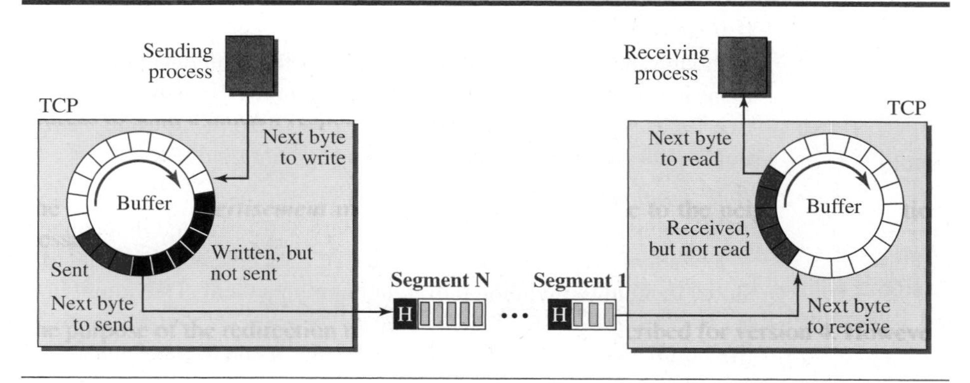

Note that segments are not necessarily all the same size. In the figure, for simplicity, we show one segment carrying 3 bytes and the other carrying 5 bytes. In reality, segments carry hundreds, if not thousands, of bytes.

Full-Duplex Communication

TCP offers *full-duplex service,* where data can flow in both directions at the same time. Each TCP endpoint then has its own sending and receiving buffer, and segments move in both directions.

Multiplexing and Demultiplexing

Like UDP, TCP performs multiplexing at the sender and demultiplexing at the receiver. However, since TCP is a connection-oriented protocol, a connection needs to be established for each pair of processes.

Connection-Oriented Service

TCP, unlike UDP, is a connection-oriented protocol. When a process at site A wants to send to and receive data from another process at site B, the following three phases occur:

1. The two TCP's establish a logical connection between them.
2. Data are exchanged in both directions.
3. The connection is terminated.

Note that this is a logical connection, not a physical connection. The TCP segment is encapsulated in an IP datagram and can be sent out of order, or lost or corrupted, and then resent. Each may be routed over a different path to reach the destination. There is no physical connection. TCP creates a stream-oriented environment in which it accepts the responsibility of delivering the bytes in order to the other site.

Reliable Service

TCP is a reliable transport protocol. It uses an acknowledgment mechanism to check the safe and sound arrival of data. We will discuss this feature further in the section on error control.

24.3.2 TCP Features

To provide the services mentioned in the previous section, TCP has several features that are briefly summarized in this section and discussed later in detail.

Numbering System

Although the TCP software keeps track of the segments being transmitted or received, there is no field for a segment number value in the segment header. Instead, there are two fields, called the *sequence number* and the *acknowledgment number.* These two fields refer to a byte number and not a segment number.

Byte Number

TCP numbers all data bytes (octets) that are transmitted in a connection. Numbering is independent in each direction. When TCP receives bytes of data from a process, TCP stores them in the sending buffer and numbers them. The numbering does not necessarily start from 0. Instead, TCP chooses an arbitrary number between 0 and $2^{32} - 1$ for the number of the first byte. For example, if the number happens to be 1057 and the total data to be sent is 6000 bytes, the bytes are numbered from 1057 to 7056. We will see that byte numbering is used for flow and error control.

The bytes of data being transferred in each connection are numbered by TCP. The numbering starts with an arbitrarily generated number.

Sequence Number

After the bytes have been numbered, TCP assigns a sequence number to each segment that is being sent. The sequence number, in each direction, is defined as follows:

1. The sequence number of the first segment is the ISN (initial sequence number), which is a random number.
2. The sequence number of any other segment is the sequence number of the previous segment plus the number of bytes (real or imaginary) carried by the previous segment. Later, we show that some control segments are thought of as carrying one imaginary byte.

Example 24.7

Suppose a TCP connection is transferring a file of 5000 bytes. The first byte is numbered 10001. What are the sequence numbers for each segment if data are sent in five segments, each carrying 1000 bytes?

Solution

The following shows the sequence number for each segment:

Segment 1	→	Sequence Number:	10001	**Range:**	10001	to	11000
Segment 2	→	Sequence Number:	11001	**Range:**	11001	to	12000
Segment 3	→	Sequence Number:	12001	**Range:**	12001	to	13000
Segment 4	→	Sequence Number:	13001	**Range:**	13001	to	14000
Segment 5	→	Sequence Number:	14001	**Range:**	14001	to	15000

The value in the sequence number field of a segment defines the number assigned to the first data byte contained in that segment.

When a segment carries a combination of data and control information (piggybacking), it uses a sequence number. If a segment does not carry user data, it does not logically define a sequence number. The field is there, but the value is not valid. However, some segments, when carrying only control information, need a sequence number to allow an acknowledgment from the receiver. These segments are used for connection establishment, termination, or abortion. Each of these segments consume one sequence number as though it carries one byte, but there are no actual data. We will elaborate on this issue when we discuss connections.

Acknowledgment Number

As we discussed previously, communication in TCP is full duplex; when a connection is established, both parties can send and receive data at the same time. Each party numbers the bytes, usually with a different starting byte number. The sequence number in each direction shows the number of the first byte carried by the segment. Each party also uses an acknowledgment number to confirm the bytes it has received. However, the acknowledgment number defines the number of the next byte that the party expects to receive. In addition, the acknowledgment number is cumulative, which means that the party takes the number of the last byte that it has received, safe and sound, adds 1 to it, and announces this sum as the acknowledgment number. The term *cumulative* here means that if a party uses 5643 as an acknowledgment number, it has received all bytes from the beginning up to 5642. Note that this does not mean that the party has received 5642 bytes, because the first byte number does not have to be 0.

The value of the acknowledgment field in a segment defines the number of the next byte a party expects to receive. The acknowledgment number is cumulative.

24.3.3 Segment

Before discussing TCP in more detail, let us discuss the TCP packets themselves. A packet in TCP is called a *segment.*

Format

The format of a segment is shown in Figure 24.7. The segment consists of a header of 20 to 60 bytes, followed by data from the application program. The header is 20 bytes if there are no options and up to 60 bytes if it contains options. We will discuss some of the header fields in this section. The meaning and purpose of these will become clearer as we proceed through the section.

Figure 24.7 *TCP segment format*

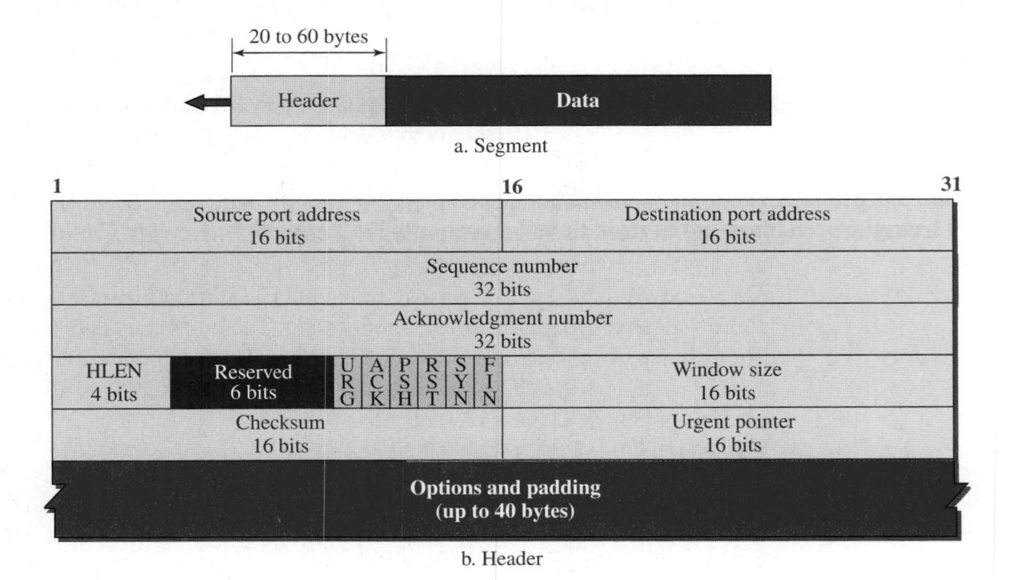

- ***Source port address.*** This is a 16-bit field that defines the port number of the application program in the host that is sending the segment.
- ***Destination port address.*** This is a 16-bit field that defines the port number of the application program in the host that is receiving the segment.
- ***Sequence number.*** This 32-bit field defines the number assigned to the first byte of data contained in this segment. As we said before, TCP is a stream transport protocol. To ensure connectivity, each byte to be transmitted is numbered. The sequence number tells the destination which byte in this sequence is the first byte in the segment. During connection establishment (discussed later) each party uses a random number generator to create an **initial sequence number** (ISN), which is usually different in each direction.
- ***Acknowledgment number.*** This 32-bit field defines the byte number that the receiver of the segment is expecting to receive from the other party. If the receiver of the segment has successfully received byte number x from the other party, it returns $x + 1$ as the acknowledgment number. Acknowledgment and data can be piggybacked together.
- ***Header length.*** This 4-bit field indicates the number of 4-byte words in the TCP header. The length of the header can be between 20 and 60 bytes. Therefore, the value of this field is always between 5 ($5 \times 4 = 20$) and 15 ($15 \times 4 = 60$).
- ***Control.*** This field defines 6 different control bits or flags, as shown in Figure 24.8. One or more of these bits can be set at a time. These bits enable flow control, connection establishment and termination, connection abortion, and the mode of data transfer in TCP. A brief description of each bit is shown in the figure. We will discuss them further when we study the detailed operation of TCP later in the chapter.

Figure 24.8 *Control field*

URG	ACK	PSH	RST	SYN	FIN

6 bits

URG: Urgent pointer is valid
ACK: Acknowledgment is valid
PSH : Request for push
RST : Reset the connection
SYN: Synchronize sequence numbers
FIN : Terminate the connection

- ***Window size.*** This field defines the window size of the sending TCP in bytes. Note that the length of this field is 16 bits, which means that the maximum size of the window is 65,535 bytes. This value is normally referred to as the receiving window (*rwnd*) and is determined by the receiver. The sender must obey the dictation of the receiver in this case.
- ***Checksum.*** This 16-bit field contains the checksum. The calculation of the checksum for TCP follows the same procedure as the one described for UDP. However, the use of the checksum in the UDP datagram is optional, whereas the use of the checksum for TCP is mandatory. The same pseudoheader, serving the same

purpose, is added to the segment. For the TCP pseudoheader, the value for the protocol field is 6. See Figure 24.9.

Figure 24.9 *Pseudoheader added to the TCP datagram*

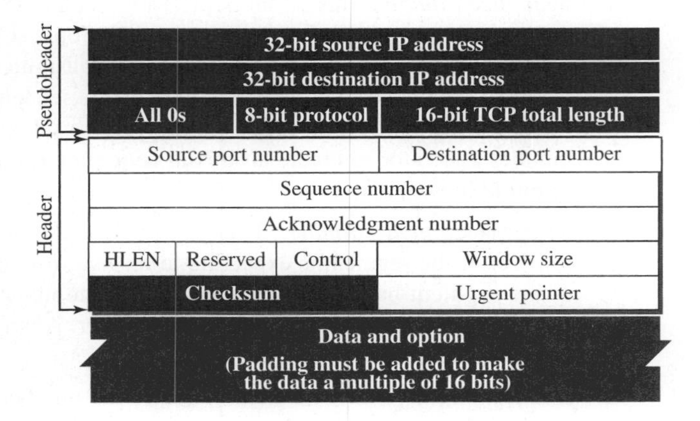

The use of the checksum in TCP is mandatory.

- ❑ ***Urgent pointer.*** This 16-bit field, which is valid only if the urgent flag is set, is used when the segment contains urgent data. It defines a value that must be added to the sequence number to obtain the number of the last urgent byte in the data section of the segment. This will be discussed later in this chapter.
- ❑ ***Options.*** There can be up to 40 bytes of optional information in the TCP header. We will discuss some of the options used in the TCP header later in the section.

Encapsulation

A TCP segment encapsulates the data received from the application layer. The TCP segment is encapsulated in an IP datagram, which in turn is encapsulated in a frame at the data-link layer.

24.3.4 A TCP Connection

TCP is connection-oriented. As discussed before, a connection-oriented transport protocol establishes a logical path between the source and destination. All of the segments belonging to a message are then sent over this logical path. Using a single logical pathway for the entire message facilitates the acknowledgment process as well as retransmission of damaged or lost frames. You may wonder how TCP, which uses the services of IP, a connectionless protocol, can be connection-oriented. The point is that a TCP connection is logical, not physical. TCP operates at a higher level. TCP uses the services of IP to deliver individual segments to the receiver, but it controls the connection itself. If a segment is lost or corrupted, it is retransmitted. Unlike TCP, IP is unaware of this retransmission. If a segment arrives out of order, TCP holds it until the missing segments arrive; IP is unaware of this reordering.

In TCP, connection-oriented transmission requires three phases: connection establishment, data transfer, and connection termination.

Connection Establishment

TCP transmits data in full-duplex mode. When two TCPs in two machines are connected, they are able to send segments to each other simultaneously. This implies that each party must initialize communication and get approval from the other party before any data are transferred.

Three-Way Handshaking

The connection establishment in TCP is called ***three-way handshaking.*** In our example, an application program, called the *client,* wants to make a connection with another application program, called the *server,* using TCP as the transport-layer protocol.

The process starts with the server. The server program tells its TCP that it is ready to accept a connection. This request is called a *passive open.* Although the server TCP is ready to accept a connection from any machine in the world, it cannot make the connection itself.

The client program issues a request for an *active open.* A client that wishes to connect to an open server tells its TCP to connect to a particular server. TCP can now start the three-way handshaking process, as shown in Figure 24.10.

Figure 24.10 *Connection establishment using three-way handshaking*

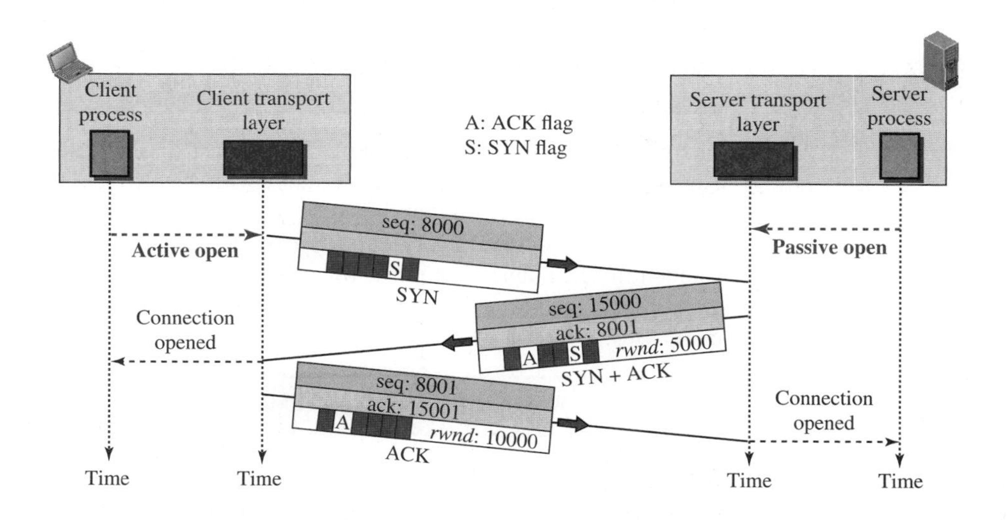

To show the process we use time lines. Each segment has values for all its header fields and perhaps for some of its option fields too. However, we show only the few fields necessary to understand each phase. We show the sequence number, the acknowledgment number, the control flags (only those that are set), and window size if relevant. The three steps in this phase are as follows.

1. The client sends the first segment, a SYN segment, in which only the SYN flag is set. This segment is for synchronization of sequence numbers. The client in our example chooses a random number as the first sequence number and sends this number to the server. This sequence number is called the *initial sequence number (ISN).* Note that this segment does not contain an acknowledgment number. It does not define the window size either; a window size definition makes sense only when a segment includes an acknowledgment. The segment can also include some options that we discuss later in the chapter. Note that the SYN segment is a control segment and carries no data. However, it consumes one sequence number because it needs to be acknowledged. We can say that the SYN segment carries one imaginary byte.

A SYN segment cannot carry data, but it consumes one sequence number.

2. The server sends the second segment, a SYN + ACK segment with two flag bits set as: SYN and ACK. This segment has a dual purpose. First, it is a SYN segment for communication in the other direction. The server uses this segment to initialize a sequence number for numbering the bytes sent from the server to the client. The server also acknowledges the receipt of the SYN segment from the client by setting the ACK flag and displaying the next sequence number it expects to receive from the client. Because the segment contains an acknowledgment, it also needs to define the receive window size, *rwnd* (to be used by the client), as we will see in the flow control section. Since this segment is playing the role of a SYN segment, it needs to be acknowledged. It, therefore, consumes one sequence number.

**A SYN + ACK segment cannot carry data,
but it does consume one sequence number.**

3. The client sends the third segment. This is just an ACK segment. It acknowledges the receipt of the second segment with the ACK flag and acknowledgment number field. Note that the ACK segment does not consume any sequence numbers if it does not carry data, but some implementations allow this third segment in the connection phase to carry the first chunk of data from the client. In this case, the segment consumes as many sequence numbers as the number of data bytes.

An ACK segment, if carrying no data, consumes no sequence number.

SYN Flooding Attack

The connection establishment procedure in TCP is susceptible to a serious security problem called ***SYN flooding attack.*** This happens when one or more malicious attackers send a large number of SYN segments to a server pretending that each of them is coming from a different client by faking the source IP addresses in the datagrams. The server, assuming that the clients are issuing an active open, allocates the necessary resources, such as creating transfer control block (TCB) tables and setting timers. The

TCP server then sends the SYN + ACK segments to the fake clients, which are lost. When the server waits for the third leg of the handshaking process, however, resources are allocated without being used. If, during this short period of time, the number of SYN segments is large, the server eventually runs out of resources and may be unable to accept connection requests from valid clients. This SYN flooding attack belongs to a group of security attacks known as a ***denial of service attack,*** in which an attacker monopolizes a system with so many service requests that the system overloads and denies service to valid requests.

Some implementations of TCP have strategies to alleviate the effect of a SYN attack. Some have imposed a limit of connection requests during a specified period of time. Others try to filter out datagrams coming from unwanted source addresses. One recent strategy is to postpone resource allocation until the server can verify that the connection request is coming from a valid IP address, by using what is called a ***cookie.*** SCTP, the new transport-layer protocol that we discuss later, uses this strategy.

Data Transfer

After connection is established, bidirectional data transfer can take place. The client and server can send data and acknowledgments in both directions. We will study the rules of acknowledgment later in the chapter; for the moment, it is enough to know that data traveling in the same direction as an acknowledgment are carried on the same segment. The acknowledgment is piggybacked with the data. Figure 24.11 shows an example.

In this example, after a connection is established, the client sends 2,000 bytes of data in two segments. The server then sends 2,000 bytes in one segment. The client sends one more segment. The first three segments carry both data and acknowledgment, but the last segment carries only an acknowledgment because there is no more data to be sent. Note the values of the sequence and acknowledgment numbers. The data segments sent by the client have the PSH (push) flag set so that the server TCP knows to deliver data to the server process as soon as they are received. We discuss the use of this flag in more detail later. The segment from the server, on the other hand, does not set the push flag. Most TCP implementations have the option to set or not to set this flag.

Pushing Data

We saw that the sending TCP uses a buffer to store the stream of data coming from the sending application program. The sending TCP can select the segment size. The receiving TCP also buffers the data when they arrive and delivers them to the application program when the application program is ready or when it is convenient for the receiving TCP. This type of flexibility increases the efficiency of TCP.

However, there are occasions in which the application program has no need for this flexibility. For example, consider an application program that communicates interactively with another application program on the other end. The application program on one site wants to send a chunk of data to the application program at the other site and receive an immediate response. Delayed transmission and delayed delivery of data may not be acceptable by the application program.

Figure 24.11 *Data transfer*

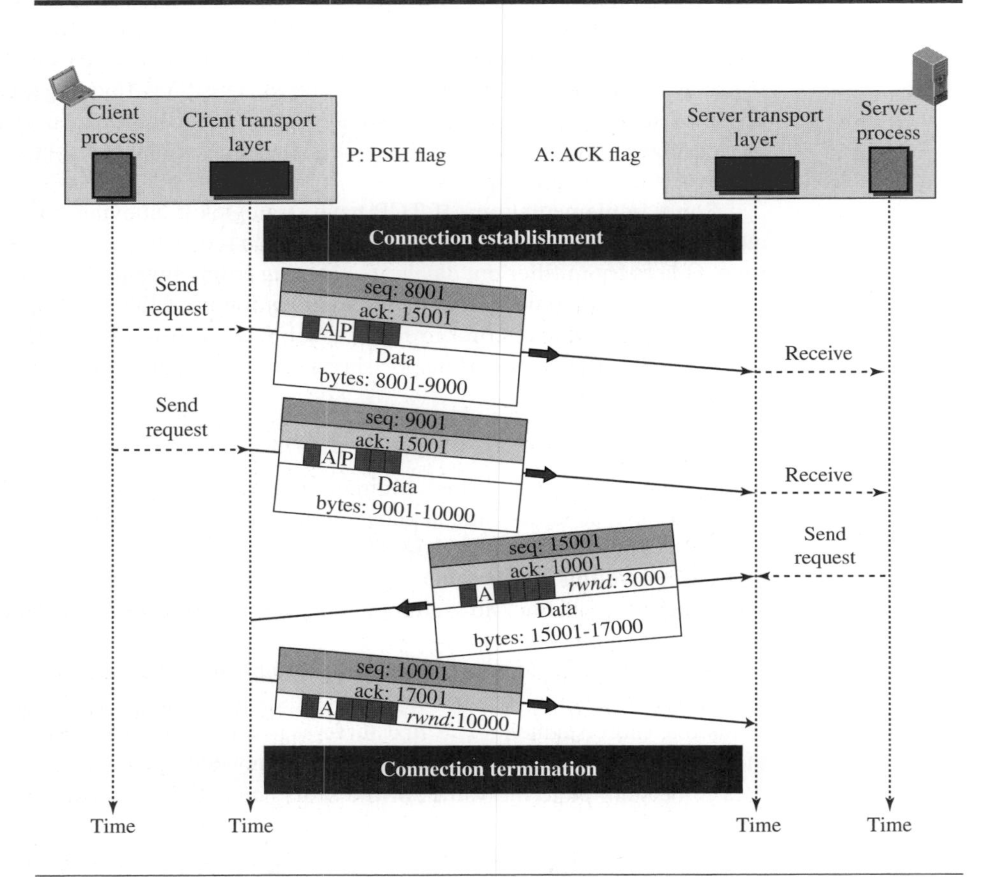

TCP can handle such a situation. The application program at the sender can request a *push* operation. This means that the sending TCP must not wait for the window to be filled. It must create a segment and send it immediately. The sending TCP must also set the push bit (PSH) to let the receiving TCP know that the segment includes data that must be delivered to the receiving application program as soon as possible and not to wait for more data to come. This means to change the byte-oriented TCP to a chunk-oriented TCP, but TCP can choose whether or not to use this feature.

Urgent Data

TCP is a stream-oriented protocol. This means that the data is presented from the application program to TCP as a stream of bytes. Each byte of data has a position in the stream. However, there are occasions in which an application program needs to send *urgent* bytes, some bytes that need to be treated in a special way by the application at the other end. The solution is to send a segment with the URG bit set. The sending application program tells the sending TCP that the piece of data is urgent. The sending TCP creates a segment and inserts the urgent data at the beginning of the segment. The rest of the segment can contain normal data from the buffer. The urgent pointer field in the header defines the end of the urgent data (the last byte of urgent data). For example, if the segment sequence number is 15000 and the value of the urgent pointer is 200, the first byte of urgent data is the byte 15000 and the last byte is the byte 15200. The rest of the bytes in the segment (if present) are nonurgent.

It is important to mention that TCP's urgent data is neither a priority service nor an out-of-band data service as some people think. Rather, TCP urgent mode is a service by which the application program at the sender side marks some portion of the byte stream as needing special treatment by the application program at the receiver side. The receiving TCP delivers bytes (urgent or nonurgent) to the application program in order, but informs the application program about the beginning and end of urgent data. It is left to the application program to decide what to do with the urgent data.

Connection Termination

Either of the two parties involved in exchanging data (client or server) can close the connection, although it is usually initiated by the client. Most implementations today allow two options for connection termination: three-way handshaking and four-way handshaking with a half-close option.

Three-Way Handshaking

Most implementations today allow *three-way handshaking* for connection termination, as shown in Figure 24.12.

1. In this situation, the client TCP, after receiving a close command from the client process, sends the first segment, a FIN segment in which the FIN flag is set. Note that a FIN segment can include the last chunk of data sent by the client or it can be just a control segment as shown in the figure. If it is only a control segment, it consumes only one sequence number because it needs to be acknowledged.

The FIN segment consumes one sequence number if it does not carry data.

2. The server TCP, after receiving the FIN segment, informs its process of the situation and sends the second segment, a FIN + ACK segment, to confirm the receipt of the FIN segment from the client and at the same time to announce the closing of the connection in the other direction. This segment can also contain the last chunk of data from the server. If it does not carry data, it consumes only one sequence number because it needs to be acknowledged.
3. The client TCP sends the last segment, an ACK segment, to confirm the receipt of the FIN segment from the TCP server. This segment contains the acknowledgment number, which is one plus the sequence number received in the FIN segment from the server. This segment cannot carry data and consumes no sequence numbers.

Half-Close

In TCP, one end can stop sending data while still receiving data. This is called a ***half-close.*** Either the server or the client can issue a half-close request. It can occur when the server needs all the data before processing can begin. A good example is sorting. When

Figure 24.12 *Connection termination using three-way handshaking*

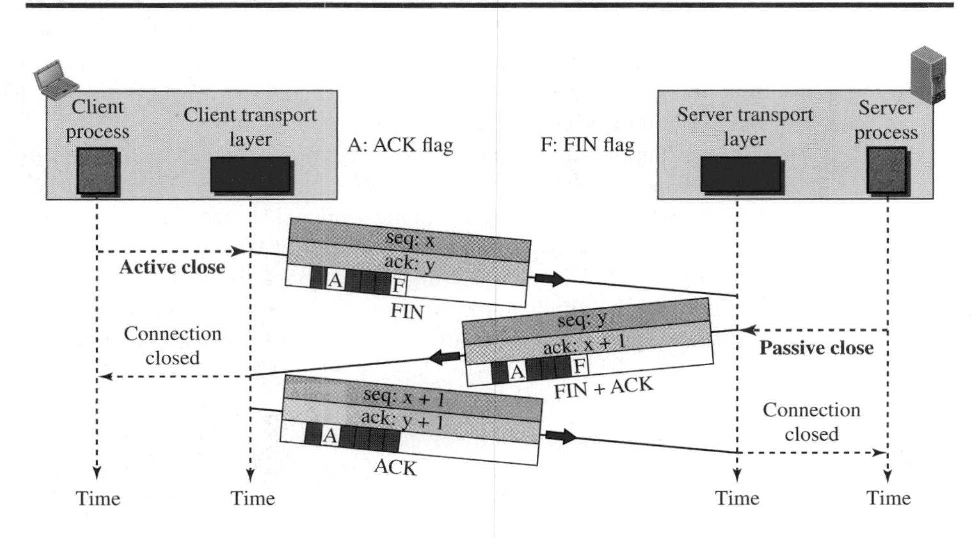

The FIN + ACK segment consumes only one sequence number if it does not carry data.

the client sends data to the server to be sorted, the server needs to receive all the data before sorting can start. This means the client, after sending all data, can close the connection in the client-to-server direction. However, the server-to-client direction must remain open to return the sorted data. The server, after receiving the data, still needs time for sorting; its outbound direction must remain open. Figure 24.13 shows an example of a half-close.

The data transfer from the client to the server stops. The client half-closes the connection by sending a FIN segment. The server accepts the half-close by sending the ACK segment. The server, however, can still send data. When the server has sent all of the processed data, it sends a FIN segment, which is acknowledged by an ACK from the client.

After half-closing the connection, data can travel from the server to the client and acknowledgments can travel from the client to the server. The client cannot send any more data to the server.

Connection Reset

TCP at one end may deny a connection request, may abort an existing connection, or may terminate an idle connection. All of these are done with the RST (reset) flag.

24.3.5 State Transition Diagram

To keep track of all the different events happening during connection establishment, connection termination, and data transfer, TCP is specified as the finite state machine (FSM) as shown in Figure 24.14.

Figure 24.13 *Half-close*

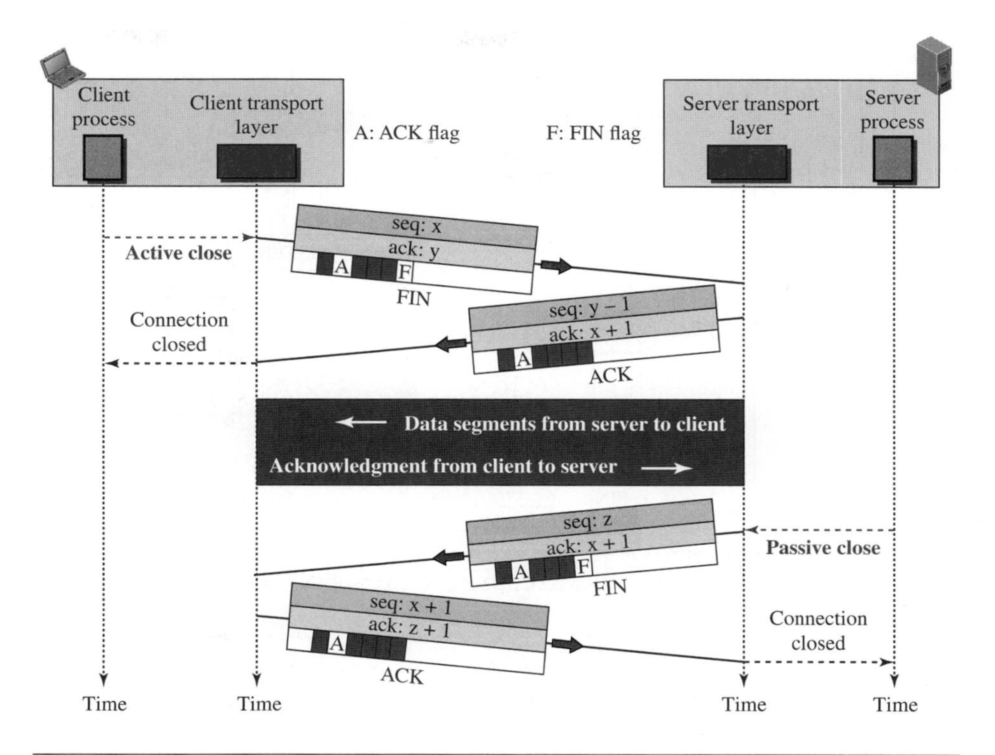

The figure shows the two FSMs used by the TCP client and server combined in one diagram. The rounded-corner rectangles represent the states. The transition from one state to another is shown using directed lines. Each line has two strings separated by a slash. The first string is the input, what TCP receives. The second is the output, what TCP sends. The dotted black lines in the figure represent the transition that a server normally goes through; the solid black lines show the transitions that a client normally goes through. However, in some situations, a server transitions through a solid line or a client transitions through a dotted line. The colored lines show special situations. Note that the rounded-corner rectangle marked *ESTABLISHED* is in fact two sets of states, a set for the client and another for the server, that are used for flow and error control, as explained later in the chapter. We will discuss some timers mentioned in the figure, including the 2MSL timer, at the end of the chapter. We use several scenarios based on Figure 24.14 and show the part of the figure in each case.

Table 24.2 shows the list of states for TCP.

Scenarios

To understand the TCP state machines and the transition diagrams, we go through one scenario in this section.

Figure 24.14 *State transition diagram*

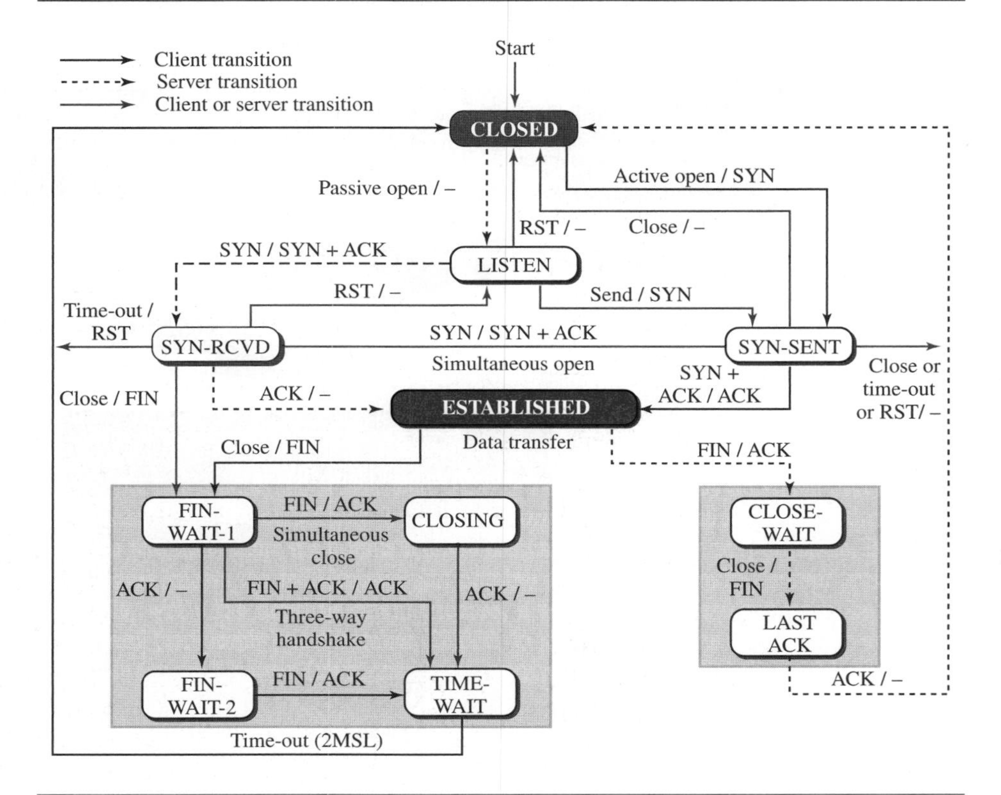

The state marked *ESTABLISHED* in the FSM is in fact two different sets of states that the client and server undergo to transfer data.

Table 24.2 *States for TCP*

State	*Description*
CLOSED	No connection exists
LISTEN	Passive open received; waiting for SYN
SYN-SENT	SYN sent; waiting for ACK
SYN-RCVD	SYN + ACK sent; waiting for ACK
ESTABLISHED	Connection established; data transfer in progress
FIN-WAIT-1	First FIN sent; waiting for ACK
FIN-WAIT-2	ACK to first FIN received; waiting for second FIN
CLOSE-WAIT	First FIN received, ACK sent; waiting for application to close
TIME-WAIT	Second FIN received, ACK sent; waiting for 2MSL time-out
LAST-ACK	Second FIN sent; waiting for ACK
CLOSING	Both sides decided to close simultaneously

A Half-Close Scenario

Figure 24.15 shows the state transition diagram for this scenario.

Figure 24.15 *Transition diagram with half-close connection termination*

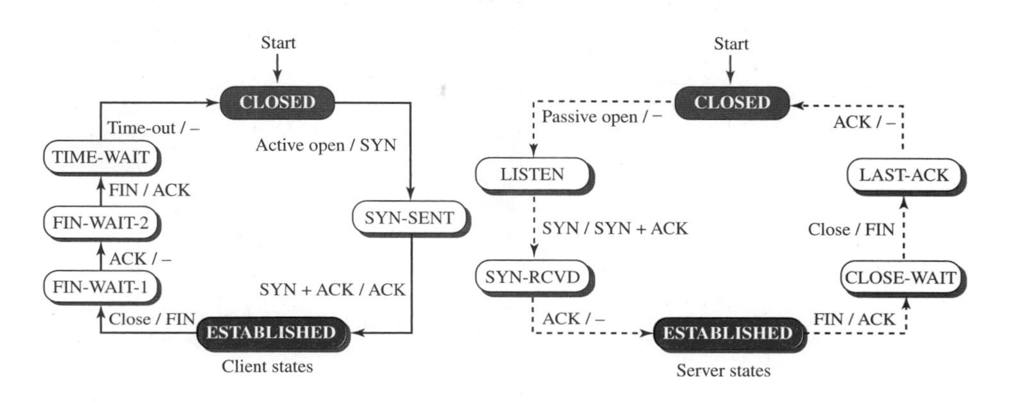

The client process issues an *active open* command to its TCP to request a connection to a specific socket address. TCP sends a SYN segment and moves to the **SYN-SENT** state. After receiving the SYN + ACK segment, TCP sends an ACK segment and goes to the **ESTABLISHED** state. Data are transferred, possibly in both directions, and acknowledged. When the client process has no more data to send, it issues a command called an *active close*. The TCP sends a FIN segment and goes to the **FIN-WAIT-1** state. When it receives the ACK segment, it goes to the **FIN-WAIT-2** state. When the client receives a FIN segment, it sends an ACK segment and goes to the **TIME-WAIT** state. The client remains in this state for 2 MSL seconds (see TCP timers later in the chapter). When the corresponding timer expires, the client goes to the **CLOSED** state.

The server process issues a *passive open* command. The server TCP goes to the **LISTEN** state and remains there passively until it receives a SYN segment. The TCP then sends a SYN + ACK segment and goes to the **SYN-RCVD** state, waiting for the client to send an ACK segment. After receiving the ACK segment, TCP goes to the **ESTABLISHED** state, where data transfer can take place. TCP remains in this state until it receives a FIN segment from the client signifying that there are no more data to be exchanged and the connection can be closed. The server, upon receiving the FIN segment, sends all queued data to the server with a virtual EOF marker, which means that the connection must be closed. It sends an ACK segment and goes to the **CLOSE-WAIT** state, but postpones acknowledging the FIN segment received from the client until it receives a *passive close* command from its process. After receiving the passive close command, the server sends a FIN segment to the client and goes to the **LAST-ACK** state, waiting for the final ACK. When the ACK segment is received from the client, the server goes to the **CLOSE** state. Figure 24.16 shows the same scenario with states over the time line.

Figure 24.16 *Time-line diagram for a common scenario*

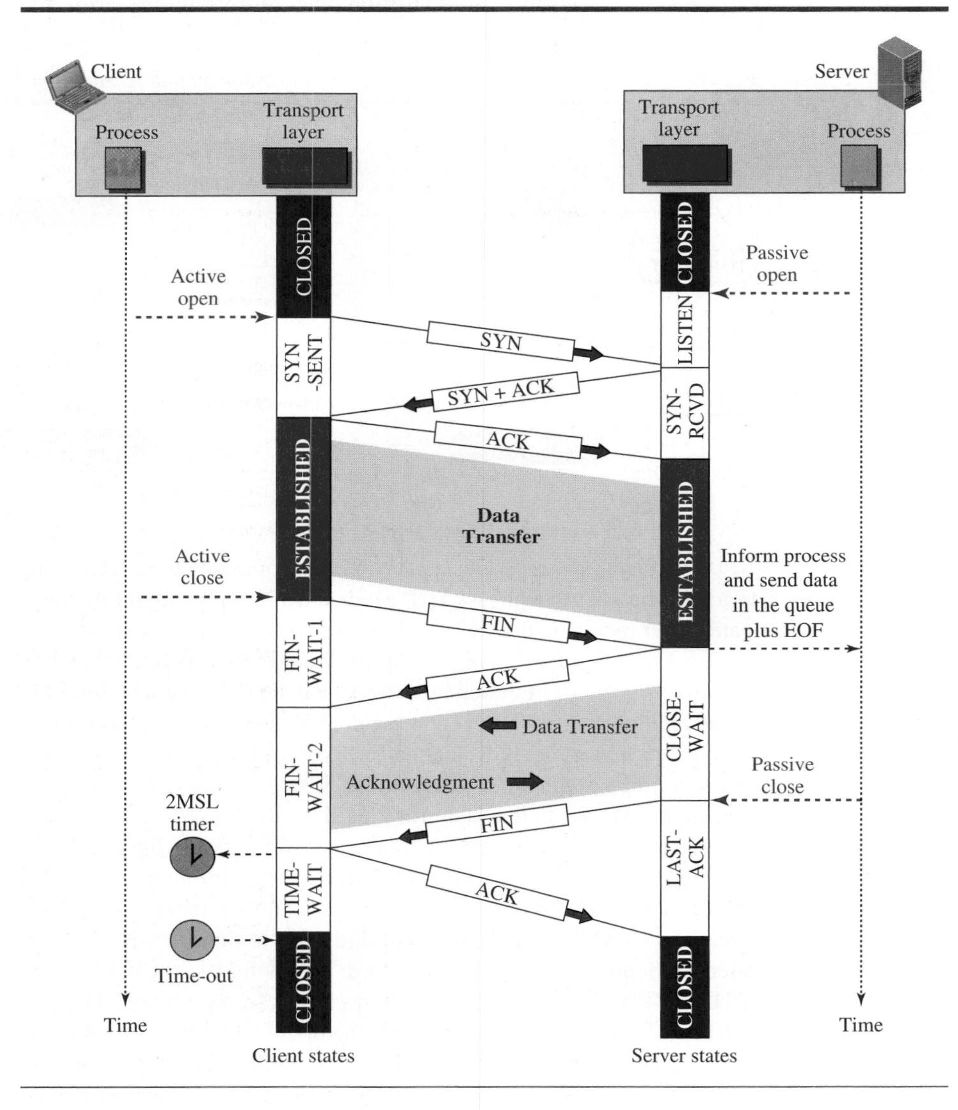

24.3.6 Windows in TCP

Before discussing data transfer in TCP and the issues such as flow, error, and congestion control, we describe the windows used in TCP. TCP uses two windows (send window and receive window) for each direction of data transfer, which means four windows for a bidirectional communication. To make the discussion simple, we make an unrealistic assumption that communication is only unidirectional (say from client to server); the bidirectional communication can be inferred using two unidirectional communications with piggybacking.

Send Window

Figure 24.17 shows an example of a send window. The window size is 100 bytes, but later we see that the send window size is dictated by the receiver (flow control) and the congestion in the underlying network (congestion control). The figure shows how a send window *opens*, *closes*, or *shrinks*.

Figure 24.17 *Send window in TCP*

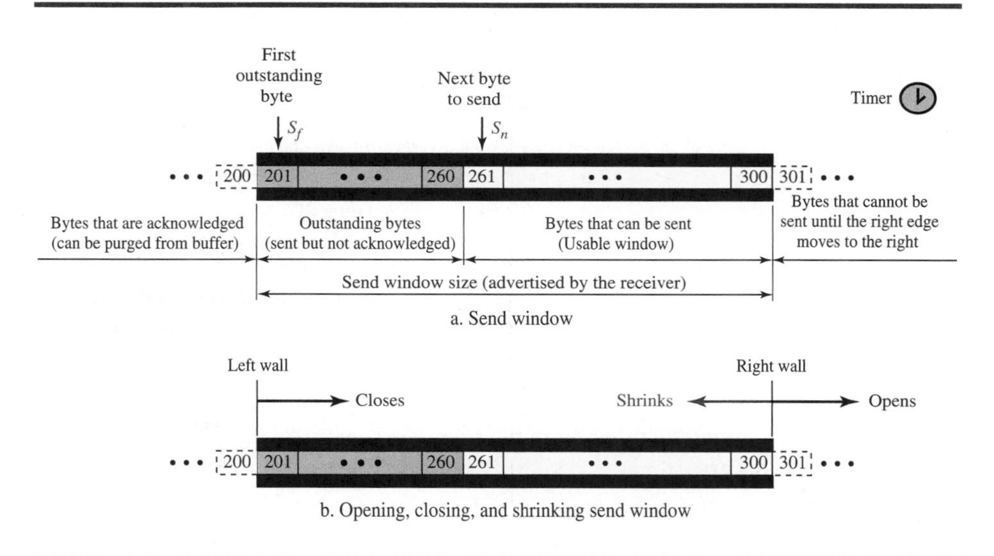

The send window in TCP is similar to the one used with the Selective-Repeat protocol, but with some differences:

1. One difference is the nature of entities related to the window. The window size in SR is the number of packets, but the window size in TCP is the number of bytes. Although actual transmission in TCP occurs segment by segment, the variables that control the window are expressed in bytes.
2. The second difference is that, in some implementations, TCP can store data received from the process and send them later, but we assume that the sending TCP is capable of sending segments of data as soon as it receives them from its process.
3. Another difference is the number of timers. The theoretical Selective-Repeat protocol may use several timers for each packet sent, but as mentioned before, the TCP protocol uses only one timer.

Receive Window

Figure 24.18 shows an example of a receive window. The window size is 100 bytes. The figure also shows how the receive window opens and closes; in practice, the window should never shrink.

Figure 24.18 *Receive window in TCP*

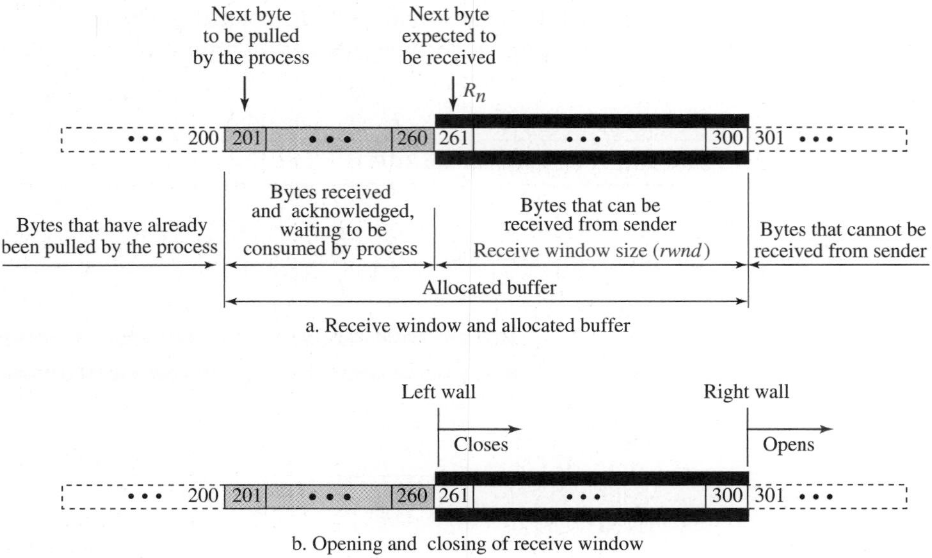

Next byte to be pulled by the process
Next byte expected to be received
R_n
200 201 ... 260 261 ... 300 301 ...
Bytes that have already been pulled by the process
Bytes received and acknowledged, waiting to be consumed by process
Bytes that can be received from sender
Receive window size (*rwnd*)
Bytes that cannot be received from sender
Allocated buffer

a. Receive window and allocated buffer

Left wall
Closes
Right wall
Opens
200 201 ... 260 261 ... 300 301 ...

b. Opening and closing of receive window

There are two differences between the receive window in TCP and the one we used for SR.

1. The first difference is that TCP allows the receiving process to pull data at its own pace. This means that part of the allocated buffer at the receiver may be occupied by bytes that have been received and acknowledged, but are waiting to be pulled by the receiving process. The receive window size is then always smaller than or equal to the buffer size, as shown in Figure 24.18. The receive window size determines the number of bytes that the receive window can accept from the sender before being overwhelmed (flow control). In other words, the receive window size, normally called *rwnd,* can be determined as:

   ***rwnd* = buffer size − number of waiting bytes to be pulled**

2. The second difference is the way acknowledgments are used in the TCP protocol. Remember that an acknowledgement in SR is selective, defining the uncorrupted packets that have been received. The major acknowledgment mechanism in TCP is a cumulative acknowledgment announcing the next expected byte to receive (in this way TCP looks like GBN, discussed earlier). The new version of TCP, however, uses both cumulative and selective acknowledgments; we will discuss these options on the book website.

24.3.7 Flow Control

As discussed before, *flow control* balances the rate a producer creates data with the rate a consumer can use the data. TCP separates flow control from error control. In this section we discuss flow control, ignoring error control. We assume that the logical channel between the sending and receiving TCP is error-free.

Figure 24.19 shows unidirectional data transfer between a sender and a receiver; bidirectional data transfer can be deduced from the unidirectional process.

Figure 24.19 *Data flow and flow control feedbacks in TCP*

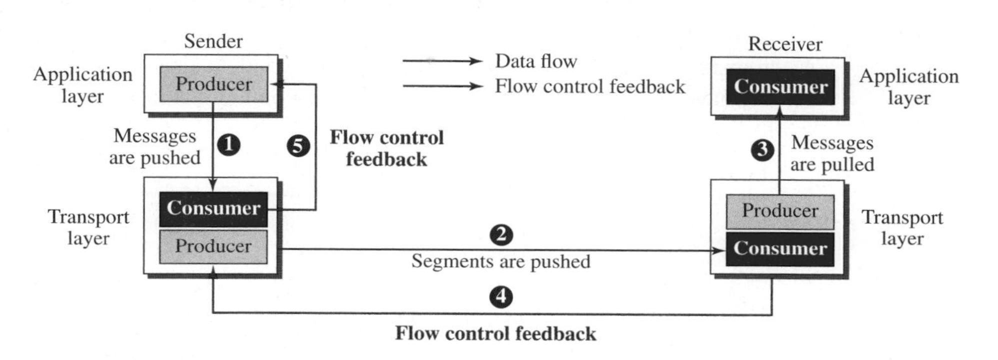

The figure shows that data travel from the sending process down to the sending TCP, from the sending TCP to the receiving TCP, and from the receiving TCP up to the receiving process (paths 1, 2, and 3). Flow control feedbacks, however, are traveling from the receiving TCP to the sending TCP and from the sending TCP up to the sending process (paths 4 and 5). Most implementations of TCP do not provide flow control feedback from the receiving process to the receiving TCP; they let the receiving process pull data from the receiving TCP whenever it is ready to do so. In other words, the receiving TCP controls the sending TCP; the sending TCP controls the sending process.

Flow control feedback from the sending TCP to the sending process (path 5) is achieved through simple rejection of data by the sending TCP when its window is full. This means that our discussion of flow control concentrates on the feedback sent from the receiving TCP to the sending TCP (path 4).

Opening and Closing Windows

To achieve flow control, TCP forces the sender and the receiver to adjust their window sizes, although the size of the buffer for both parties is fixed when the connection is established. The receive window closes (moves its left wall to the right) when more bytes arrive from the sender; it opens (moves its right wall to the right) when more bytes are pulled by the process. We assume that it does not shrink (the right wall does not move to the left).

The opening, closing, and shrinking of the send window is controlled by the receiver. The send window closes (moves its left wall to the right) when a new acknowledgment allows it to do so. The send window opens (its right wall moves to the right) when the receive window size (*rwnd*) advertised by the receiver allows it to do so (new ackNo + new *rwnd* > last ackNo + last *rwnd*). The send window shrinks in the event this situation does not occur.

A Scenario

We show how the send and receive windows are set during the connection establishment phase, and how their situations will change during data transfer. Figure 24.20 shows a simple example of unidirectional data transfer (from client to server). For the time being, we ignore error control, assuming that no segment is corrupted, lost, duplicated, or has arrived out of order. Note that we have shown only two windows for unidirectional data transfer. Although the client defines server's window size of 2000 in the third segment, we have not shown that window because the communication is only unidirectional.

Figure 24.20 *An example of flow control*

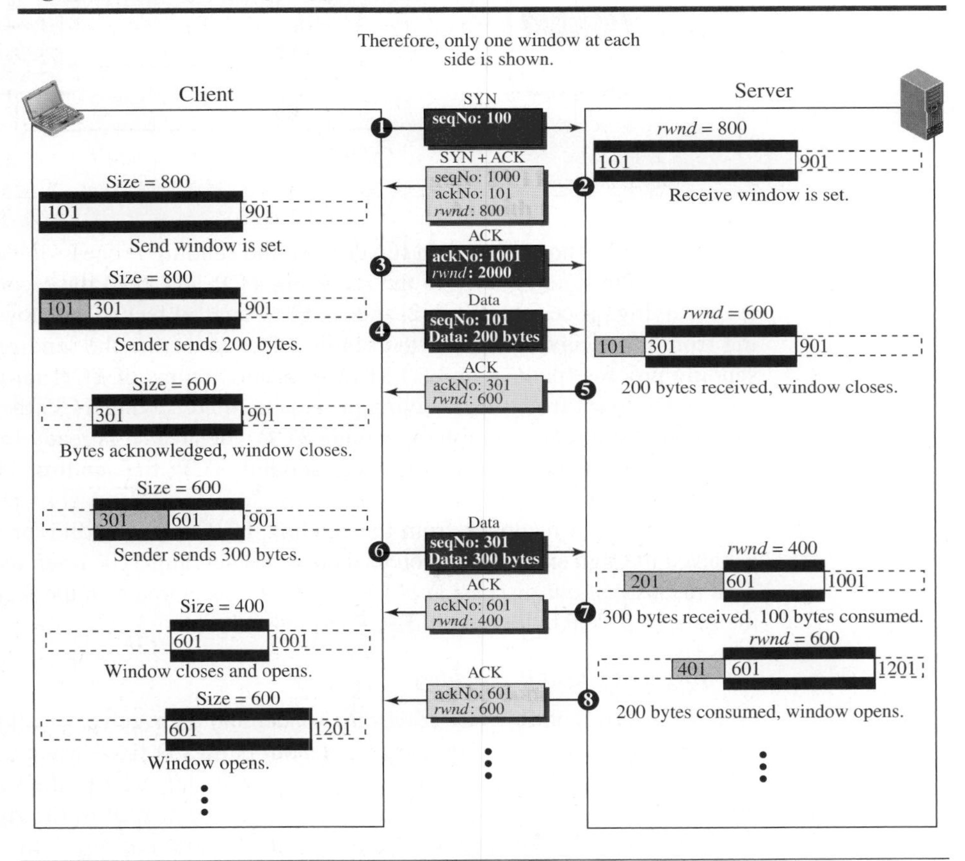

Eight segments are exchanged between the client and server:

1. The first segment is from the client to the server (a SYN segment) to request connection. The client announces its initial seqNo = 100. When this segment arrives at the server, it allocates a buffer size of 800 (an assumption) and sets its window to cover the whole buffer (*rwnd* = 800). Note that the number of the next byte to arrive is 101.

2. The second segment is from the server to the client. This is an ACK + SYN segment. The segment uses ackNo = 101 to show that it expects to receive bytes starting from 101. It also announces that the client can set a buffer size of 800 bytes.
3. The third segment is the ACK segment from the client to the server. Note that the client has defined a *rwnd* of size 2000, but we do not use this value in our figure because the communication is only in one direction.
4. After the client has set its window with the size (800) dictated by the server, the process pushes 200 bytes of data. The TCP client numbers these bytes 101 to 300. It then creates a segment and sends it to the server. The segment shows the starting byte number as 101 and the segment carries 200 bytes. The window of the client is then adjusted to show that 200 bytes of data are sent but waiting for acknowledgment. When this segment is received at the server, the bytes are stored, and the receive window closes to show that the next byte expected is byte 301; the stored bytes occupy 200 bytes of buffer.
5. The fifth segment is the feedback from the server to the client. The server acknowledges bytes up to and including 300 (expecting to receive byte 301). The segment also carries the size of the receive window after decrease (600). The client, after receiving this segment, purges the acknowledged bytes from its window and closes its window to show that the next byte to send is byte 301. The window size, however, decreases to 600 bytes. Although the allocated buffer can store 800 bytes, the window cannot open (moving its right wall to the right) because the receiver does not let it.
6. Segment 6 is sent by the client after its process pushes 300 more bytes. The segment defines seqNo as 301 and contains 300 bytes. When this segment arrives at the server, the server stores them, but it has to reduce its window size. After its process has pulled 100 bytes of data, the window closes from the left for the amount of 300 bytes, but opens from the right for the amount of 100 bytes. The result is that the size is only reduced 200 bytes. The receiver window size is now 400 bytes.
7. In segment 7, the server acknowledges the receipt of data, and announces that its window size is 400. When this segment arrives at the client, the client has no choice but to reduce its window again and set the window size to the value of *rwnd* = 400 advertised by the server. The send window closes from the left by 300 bytes, and opens from the right by 100 bytes.
8. Segment 8 is also from the server after its process has pulled another 200 bytes. Its window size increases. The new *rwnd* value is now 600. The segment informs the client that the server still expects byte 601, but the server window size has expanded to 600. We need to mention that the sending of this segment depends on the policy imposed by the implementation. Some implementations may not allow advertisement of the *rwnd* at this time; the server then needs to receive some data before doing so. After this segment arrives at the client, the client opens its window by 200 bytes without closing it. The result is that its window size increases to 600 bytes.

Shrinking of Windows

As we said before, the receive window cannot shrink. The send window, on the other hand, can shrink if the receiver defines a value for *rwnd* that results in shrinking the window. However, some implementations do not allow shrinking of the send window. The limitation does not allow the right wall of the send window to move to the left. In other words, the receiver needs to keep the following relationship between the last and new acknowledgment and the last and new *rwnd* values to prevent shrinking of the send window.

$$\textbf{new ackNo} + \textbf{new } \boldsymbol{rwnd} \quad \geq \quad \textbf{last ackNo} + \textbf{last } \boldsymbol{rwnd}$$

The left side of the inequality represents the new position of the right wall with respect to the sequence number space; the right side shows the old position of the right wall. The relationship shows that the right wall should not move to the left. The inequality is a mandate for the receiver to check its advertisement. However, note that the inequality is valid only if $S_f < S_n$; we need to remember that all calculations are in modulo 2^{32}.

Example 24.8

Figure 24.21 shows the reason for this mandate.

Figure 24.21 *Example 24.8*

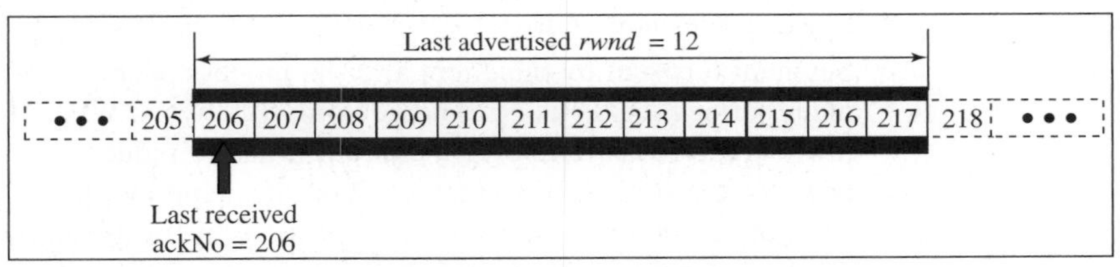

a. The window after the last advertisement

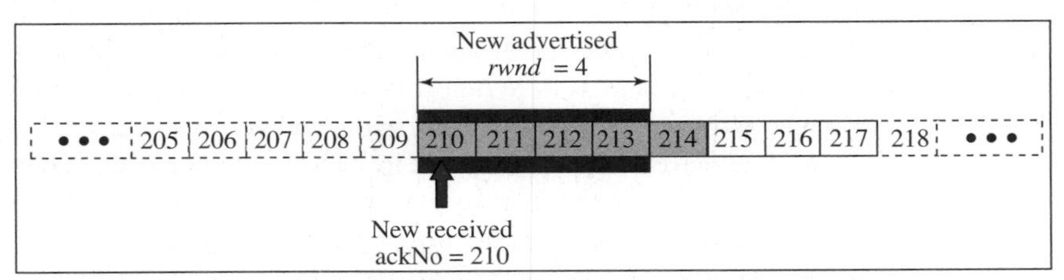

b. The window after the new advertisement; window has shrunk

Part a of the figure shows the values of the last acknowledgment and *rwnd*. Part b shows the situation in which the sender has sent bytes 206 to 214. Bytes 206 to 209 are acknowledged and purged. The new advertisement, however, defines the new value of *rwnd* as 4, in which $210 + 4 < 206 + 12$. When the send window shrinks, it creates a problem: byte 214, which has already been sent, is outside the window. The relation discussed before forces the receiver to maintain the right-hand wall of the window to be as shown in part a, because the receiver does not know which of the bytes 210 to 217 has already been sent. One way to prevent this situation is to let the receiver postpone its feedback until enough buffer locations are available in its window. In other words, the receiver should wait until more bytes are consumed by its process to meet the relationship described above.

Window Shutdown

We said that shrinking the send window by moving its right wall to the left is strongly discouraged. However, there is one exception: the receiver can temporarily shut down the window by sending a *rwnd* of 0. This can happen if for some reason the receiver does not want to receive any data from the sender for a while. In this case, the sender does not actually shrink the size of the window, but stops sending data until a new advertisement has arrived. As we will see later, even when the window is shut down by an order from the receiver, the sender can always send a segment with 1 byte of data. This is called *probing* and is used to prevent a deadlock (see the section on TCP timers).

Silly Window Syndrome

A serious problem can arise in the sliding window operation when either the sending application program creates data slowly or the receiving application program consumes data slowly, or both. Any of these situations results in the sending of data in very small segments, which reduces the efficiency of the operation. For example, if TCP sends segments containing only 1 byte of data, it means that a 41-byte datagram (20 bytes of TCP header and 20 bytes of IP header) transfers only 1 byte of user data. Here the overhead is 41/1, which indicates that we are using the capacity of the network very inefficiently. The inefficiency is even worse after accounting for the data-link layer and physical-layer overhead. This problem is called the ***silly window syndrome.*** For each site, we first describe how the problem is created and then give a proposed solution.

Syndrome Created by the Sender

The sending TCP may create a silly window syndrome if it is serving an application program that creates data slowly, for example, 1 byte at a time. The application program writes 1 byte at a time into the buffer of the sending TCP. If the sending TCP does not have any specific instructions, it may create segments containing 1 byte of data. The result is a lot of 41-byte segments that are traveling through an internet.

The solution is to prevent the sending TCP from sending the data byte by byte. The sending TCP must be forced to wait and collect data to send in a larger block. How long should the sending TCP wait? If it waits too long, it may delay the process. If it does not wait long enough, it may end up sending small segments. Nagle found an elegant solution. **Nagle's algorithm** is simple:

1. The sending TCP sends the first piece of data it receives from the sending application program even if it is only 1 byte.
2. After sending the first segment, the sending TCP accumulates data in the output buffer and waits until either the receiving TCP sends an acknowledgment or until enough data have accumulated to fill a maximum-size segment. At this time, the sending TCP can send the segment.
3. Step 2 is repeated for the rest of the transmission. Segment 3 is sent immediately if an acknowledgment is received for segment 2, or if enough data have accumulated to fill a maximum-size segment.

The elegance of Nagle's algorithm is in its simplicity and in the fact that it takes into account the speed of the application program that creates the data and the speed of the network that transports the data. If the application program is faster than the network, the segments are larger (maximum-size segments). If the application program is slower than the network, the segments are smaller (less than the maximum segment size).

Syndrome Created by the Receiver

The receiving TCP may create a silly window syndrome if it is serving an application program that consumes data slowly, for example, 1 byte at a time. Suppose that the sending application program creates data in blocks of 1 kilobyte, but the receiving application program consumes data 1 byte at a time. Also suppose that the input buffer of the receiving TCP is 4 kilobytes. The sender sends the first 4 kilobytes of data. The receiver stores it in its buffer. Now its buffer is full. It advertises a window size of zero, which means the sender should stop sending data. The receiving application reads the first byte of data from the input buffer of the receiving TCP. Now there is 1 byte of space in the incoming buffer. The receiving TCP announces a window size of 1 byte, which means that the sending TCP, which is eagerly waiting to send data, takes this advertisement as good news and sends a segment carrying only 1 byte of data. The procedure will continue. One byte of data is consumed and a segment carrying 1 byte of data is sent. Again we have an efficiency problem and the silly window syndrome.

Two solutions have been proposed to prevent the silly window syndrome created by an application program that consumes data more slowly than they arrive. **Clark's solution** is to send an acknowledgment as soon as the data arrive, but to announce a window size of zero until either there is enough space to accommodate a segment of maximum size or until at least half of the receive buffer is empty. The second solution is to delay sending the acknowledgment. This means that when a segment arrives, it is not acknowledged immediately. The receiver waits until there is a decent amount of space in its incoming buffer before acknowledging the arrived segments. The delayed acknowledgment prevents the sending TCP from sliding its window. After the sending TCP has sent the data in the window, it stops. This kills the syndrome.

Delayed acknowledgment also has another advantage: it reduces traffic. The receiver does not have to acknowledge each segment. However, there also is a disadvantage in that the delayed acknowledgment may result in the sender unnecessarily retransmitting the unacknowledged segments.

The protocol balances the advantages and disadvantages. It now defines that the acknowledgment should not be delayed by more than 500 ms.

24.3.8 Error Control

TCP is a reliable transport-layer protocol. This means that an application program that delivers a stream of data to TCP relies on TCP to deliver the entire stream to the application program on the other end in order, without error, and without any part lost or duplicated.

TCP provides reliability using error control. Error control includes mechanisms for detecting and resending corrupted segments, resending lost segments, storing out-of-order segments until missing segments arrive, and detecting and discarding duplicated segments. Error control in TCP is achieved through the use of three simple tools: checksum, acknowledgment, and time-out.

Checksum

Each segment includes a checksum field, which is used to check for a corrupted segment. If a segment is corrupted, as detected by an invalid checksum, the segment is discarded by the destination TCP and is considered as lost. TCP uses a 16-bit checksum that is mandatory in every segment. We discuss checksum calculation in Chapter 10.

Acknowledgment

TCP uses acknowledgments to confirm the receipt of data segments. Control segments that carry no data, but consume a sequence number, are also acknowledged. ACK segments are never acknowledged.

> **ACK segments do not consume sequence numbers and are not acknowledged.**

Acknowledgment Type

In the past, TCP used only one type of acknowledgment: cumulative acknowledgment. Today, some TCP implementations also use selective acknowledgment.

Cumulative Acknowledgment (ACK) TCP was originally designed to acknowledge receipt of segments cumulatively. The receiver advertises the next byte it expects to receive, ignoring all segments received and stored out of order. This is sometimes referred to as *positive cumulative acknowledgment,* or ACK. The word *positive* indicates that no feedback is provided for discarded, lost, or duplicate segments. The 32-bit ACK field in the TCP header is used for cumulative acknowledgments, and its value is valid only when the ACK flag bit is set to 1.

Selective Acknowledgment (SACK) More and more implementations are adding another type of acknowledgment called *selective acknowledgment,* or SACK. A SACK does not replace an ACK, but reports additional information to the sender. A SACK reports a block of bytes that is out of order, and also a block of bytes that is duplicated, i.e., received more than once. However, since there is no provision in the TCP header for adding this type of information, SACK is implemented as an option at the end of the TCP header. We discuss this new feature when we discuss options in TCP on the book website.

Generating Acknowledgments

When does a receiver generate acknowledgments? During the evolution of TCP, several rules have been defined and used by several implementations. We give the most common rules here. The order of a rule does not necessarily define its importance.

1. When end A sends a data segment to end B, it must include (piggyback) an acknowledgment that gives the next sequence number it expects to receive. This rule decreases the number of segments needed and therefore reduces traffic.
2. When the receiver has no data to send and it receives an in-order segment (with expected sequence number) and the previous segment has already been acknowledged, the receiver delays sending an ACK segment until another segment

arrives or until a period of time (normally 500 ms) has passed. In other words, the receiver needs to delay sending an ACK segment if there is only one outstanding in-order segment. This rule reduces ACK segments.

3. When a segment arrives with a sequence number that is expected by the receiver, and the previous in-order segment has not been acknowledged, the receiver immediately sends an ACK segment. In other words, there should not be more than two in-order unacknowledged segments at any time. This prevents the unnecessary retransmission of segments that may create congestion in the network.
4. When a segment arrives with an out-of-order sequence number that is higher than expected, the receiver immediately sends an ACK segment announcing the sequence number of the next expected segment. This leads to the *fast retransmission* of missing segments (discussed later).
5. When a missing segment arrives, the receiver sends an ACK segment to announce the next sequence number expected. This informs the receiver that segments reported missing have been received.
6. If a duplicate segment arrives, the receiver discards the segment, but immediately sends an acknowledgment indicating the next in-order segment expected. This solves some problems when an ACK segment itself is lost.

Retransmission

The heart of the error control mechanism is the retransmission of segments. When a segment is sent, it is stored in a queue until it is acknowledged. When the retransmission timer expires or when the sender receives three duplicate ACKs for the first segment in the queue, that segment is retransmitted.

Retransmission after RTO

The sending TCP maintains one **retransmission time-out (RTO)** for each connection. When the timer matures, i.e. times out, TCP resends the segment in the front of the queue (the segment with the smallest sequence number) and restarts the timer. Note that again we assume $S_f < S_n$. We will see later that the value of RTO is dynamic in TCP and is updated based on the **round-trip time (RTT)** of segments. RTT is the time needed for a segment to reach a destination and for an acknowledgment to be received.

Retransmission after Three Duplicate ACK Segments

The previous rule about retransmission of a segment is sufficient if the value of RTO is not large. To expedite service throughout the Internet by allowing senders to retransmit without waiting for a time out, most implementations today follow the three duplicate ACKs rule and retransmit the missing segment immediately. This feature is called ***fast retransmission.*** In this version, if three duplicate acknowledgments (i.e., an original ACK plus three exactly identical copies) arrive for a segment, the next segment is retransmitted without waiting for the time-out. We come back to this feature later in the chapter.

Out-of-Order Segments

TCP implementations today do not discard out-of-order segments. They store them temporarily and flag them as out-of-order segments until the missing segments arrive.

Note, however, that out-of-order segments are never delivered to the process. TCP guarantees that data are delivered to the process in order.

Data may arrive out of order and be temporarily stored by the receiving TCP, but TCP guarantees that no out-of-order data are delivered to the process.

FSMs for Data Transfer in TCP

Data transfer in TCP is close to the Selective-Repeat protocol with a slight similarity to GBN. Since TCP accepts out-of-order segments, TCP can be thought of as behaving more like the SR protocol, but since the original acknowledgments are cumulative, it looks like GBN. However, if the TCP implementation uses SACKs, then TCP is closest to SR.

TCP can best be modeled as a Selective-Repeat protocol.

Sender-Side FSM

Let us show a simplified FSM for the sender side of the TCP protocol similar to the one we discussed for the SR protocol, but with some changes specific to TCP. We assume that the communication is unidirectional and the segments are acknowledged using ACK segments. We also ignore selective acknowledgments and congestion control for the moment. Figure 24.22 shows the simplified FSM for the sender site. Note that the

Figure 24.22 *Simplified FSM for the TCP sender side*

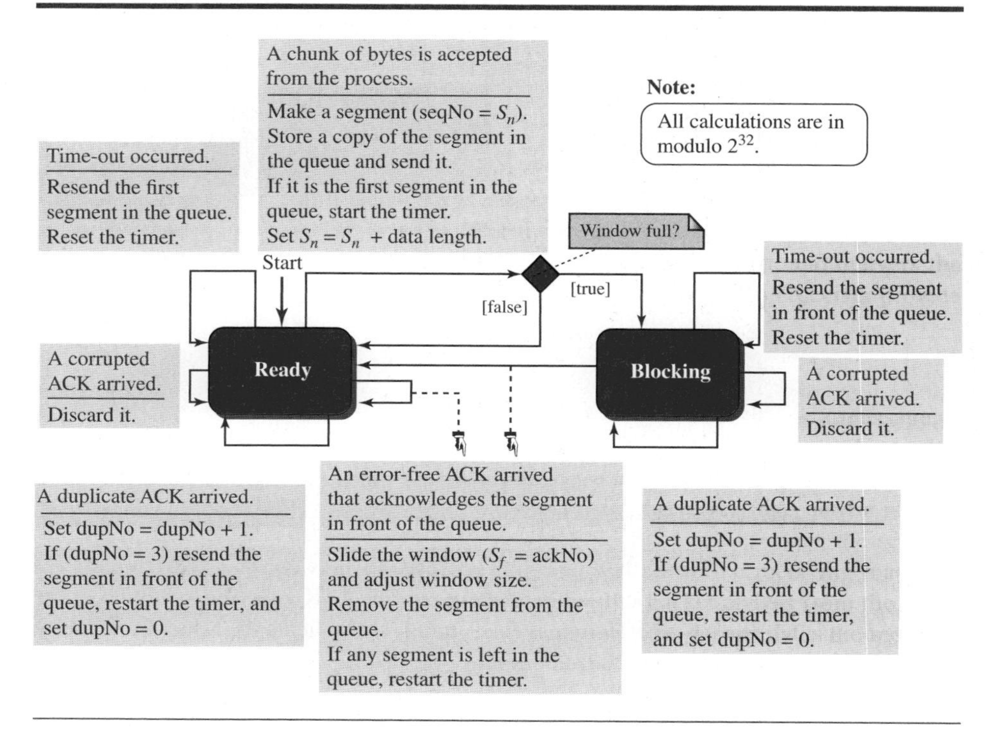

FSM is rudimentary; it does not include issues such as silly window syndrome (Nagle's algorithm) or window shutdown. It defines a unidirectional communication, ignoring all issues that affect bidirectional communication.

There are some differences between the FSM in Figure 24.22 and the one we discussed for an SR protocol. One difference is the fast transmission (three duplicate ACKs). The other is the window size adjustment based on the value of *rwnd* (ignoring congestion control for the moment).

Receiver-Side FSM

Now let us show a simplified FSM for the receiver-side TCP protocol similar to the one we discuss for the SR protocol, but with some changes specific to TCP. We assume that the communication is unidirectional and the segments are acknowledged using ACK segments. We also ignore the selective acknowledgment and congestion control for the moment. Figure 24.23 shows the simplified FSM for the receiver. Note that we ignore some issues such as silly window syndrome (Clark's solution) and window shutdown.

Figure 24.23 *Simplified FSM for the TCP receiver side*

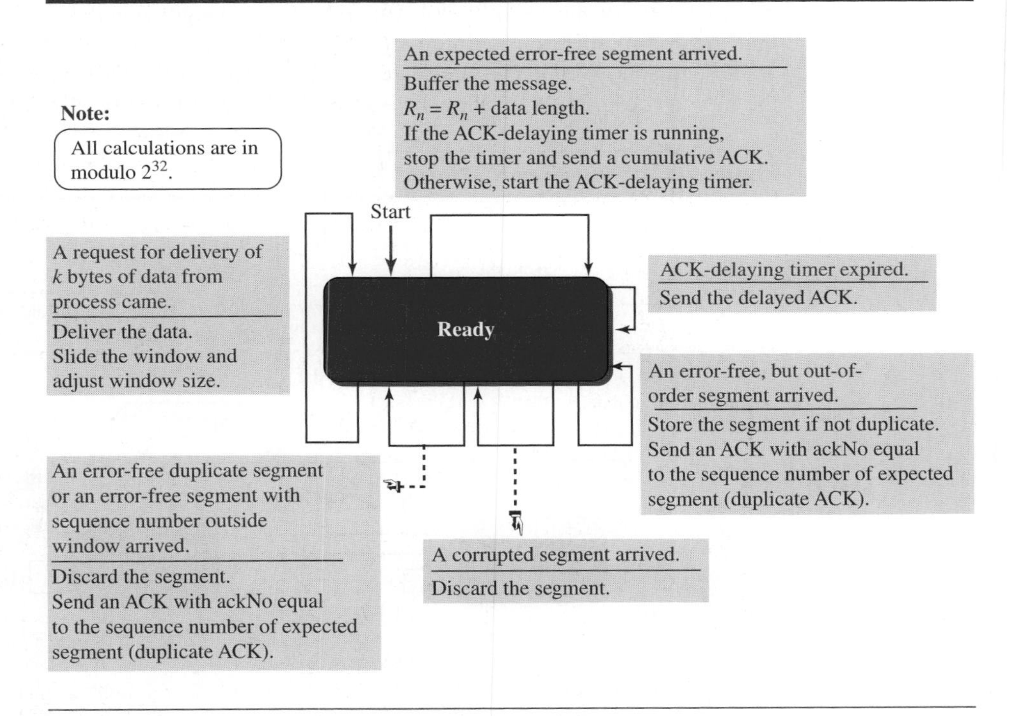

Again, there are some differences between this FSM and the one we discussed for an SR protocol. One difference is the ACK delaying in unidirectional communication. The other difference is the sending of duplicate ACKs to allow the sender to implement fast retransmission policy.

We also need to emphasize that bidirectional FSM for the receiver is not as simple as the one for SR; we need to consider some policies such as sending an immediate ACK if the receiver has some data to return.

Some Scenarios

In this section we give some examples of scenarios that occur during the operation of TCP, considering only error control issues. In these scenarios, we show a segment by a rectangle. If the segment carries data, we show the range of byte numbers and the value of the acknowledgment field. If it carries only an acknowledgment, we show only the acknowledgment number in a smaller box.

Normal Operation

The first scenario shows bidirectional data transfer between two systems as shown in Figure 24.24. The client TCP sends one segment; the server TCP sends three. The figure shows which rule applies to each acknowledgment. At the server site, only rule 1 applies. There are data to be sent, so the segment displays the next byte expected. When the client receives the first segment from the server, it does not have any more data to send; it needs to send only an ACK segment. However, according to rule 2, the acknowledgment needs to be delayed for 500 ms to see if any more segments arrive. When the ACK-delaying timer matures, it triggers an acknowledgment. This is because the client has no knowledge of whether other segments are coming; it cannot delay the acknowledgment forever. When the next segment arrives, another ACK-delaying timer is set. However, before it matures, the third segment arrives. The arrival of the third segment triggers another acknowledgment based on rule 3. We have not shown the RTO timer because no segment is lost or delayed. We just assume that the RTO timer performs its duty.

Figure 24.24 *Normal operation*

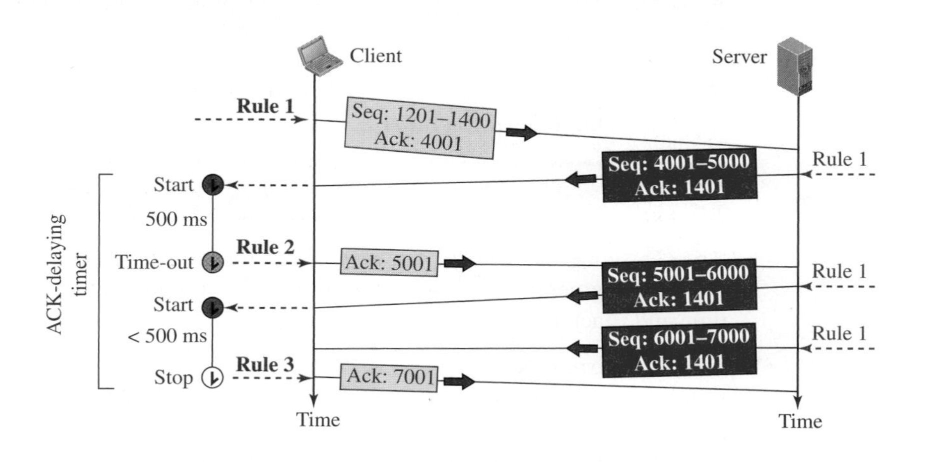

Lost Segment

In this scenario, we show what happens when a segment is lost or corrupted. A lost or corrupted segment is treated the same way by the receiver. A lost segment is discarded somewhere in the network; a corrupted segment is discarded by the receiver itself. Both are considered lost. Figure 24.25 shows a situation in which a segment is lost (probably discarded by some router in the network due to congestion).

Figure 24.25 *Lost segment*

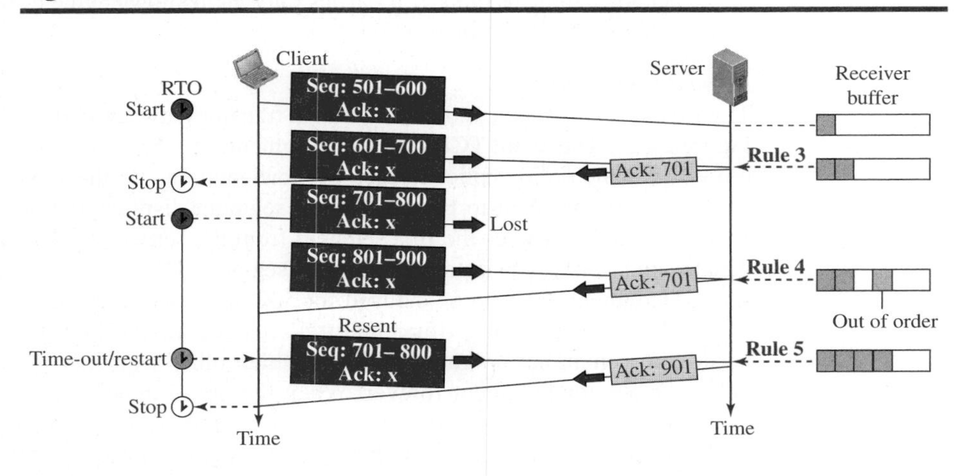

We are assuming that data transfer is unidirectional: one site is sending, the other receiving. In our scenario, the sender sends segments 1 and 2, which are acknowledged immediately by an ACK (rule 3). Segment 3, however, is lost. The receiver receives segment 4, which is out of order. The receiver stores the data in the segment in its buffer but leaves a gap to indicate that there is no continuity in the data. The receiver immediately sends an acknowledgment to the sender displaying the next byte it expects (rule 4). Note that the receiver stores bytes 801 to 900, but never delivers these bytes to the application until the gap is filled.

The receiver TCP delivers only ordered data to the process.

The sender TCP keeps one RTO timer for the whole period of connection. When the third segment times out, the sending TCP resends segment 3, which arrives this time and is acknowledged properly (rule 5).

Fast Retransmission

In this scenario, we want to show *fast retransmission.* Our scenario is the same as the second except that the RTO has a larger value (see Figure 24.26).

Each time the receiver receives a subsequent segment, it triggers an acknowledgment (rule 4). The sender receives four acknowledgments with the same value (three duplicates). Although the timer has not matured, the rule for fast retransmission

Figure 24.26 *Fast retransmission*

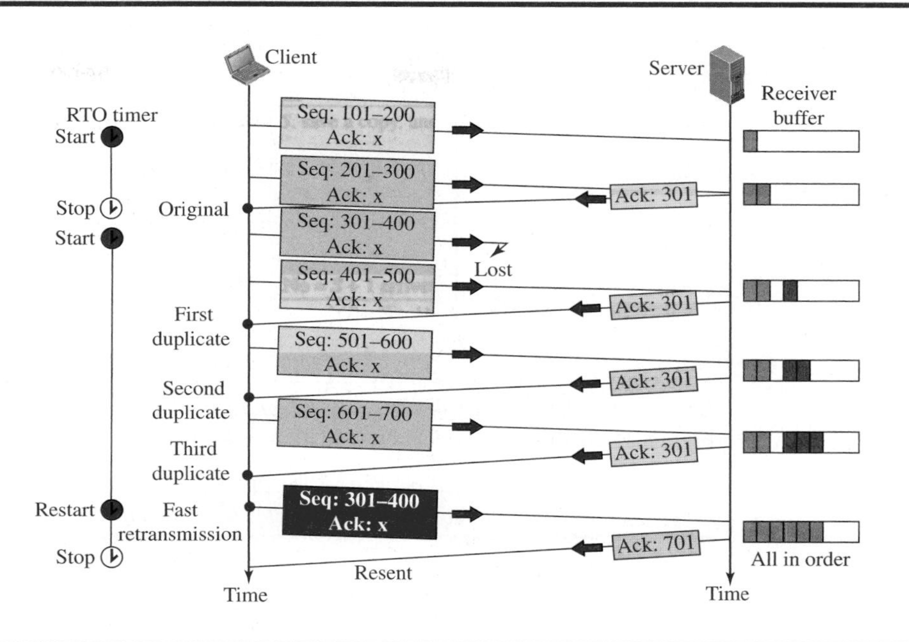

requires that segment 3, the segment that is expected by all of these duplicate acknowledgments, be resent immediately. After resending this segment, the timer is restarted.

Delayed Segment

The fourth scenario features a delayed segment. TCP uses the services of IP, which is a connectionless protocol. Each IP datagram encapsulating a TCP segment may reach the final destination through a different route with a different delay. Hence TCP segments may be delayed. Delayed segments sometimes may time out and be resent. If the delayed segment arrives after it has been resent, it is considered a duplicate segment and discarded.

Duplicate Segment

A duplicate segment can be created, for example, by a sending TCP when a segment is delayed and treated as lost by the receiver. Handling the duplicated segment is a simple process for the destination TCP. The destination TCP expects a continuous stream of bytes. When a segment arrives that contains a sequence number equal to an already received and stored segment, it is discarded. An ACK is sent with ackNo defining the expected segment.

Automatically Corrected Lost ACK

This scenario shows a situation in which information in a lost acknowledgment is contained in the next one, a key advantage of using cumulative acknowledgments. Figure 24.27 shows a lost acknowledgment sent by the receiver of data. In the TCP acknowledgment mechanism, a lost acknowledgment may not even be noticed by the source TCP. TCP uses cumulative acknowledgment. We can say that the next acknowledgment automatically corrects the loss of the previous acknowledgment.

Figure 24.27 *Lost acknowledgment*

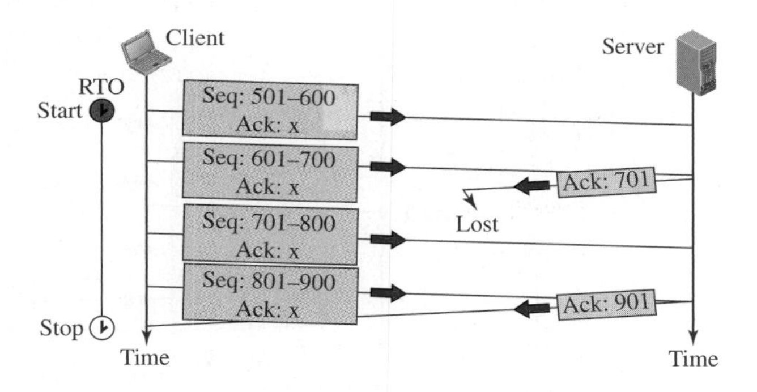

Lost Acknowledgment Corrected by Resending a Segment

Figure 24.28 shows a scenario in which an acknowledgment is lost.

Figure 24.28 *Lost acknowledgment corrected by resending a segment*

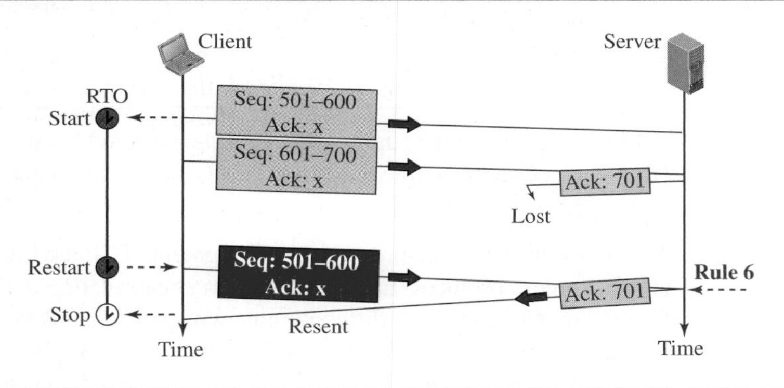

If the next acknowledgment is delayed for a long time or there is no next acknowledgment (the lost acknowledgment is the last one sent), the correction is triggered by the RTO timer. A duplicate segment is the result. When the receiver receives a duplicate segment, it discards it and resends the last ACK immediately to inform the sender that the segment or segments have been received.

Note that only one segment is retransmitted although two segments are not acknowledged. When the sender receives the retransmitted ACK, it knows that both segments are safe and sound because the acknowledgment is cumulative.

Deadlock Created by Lost Acknowledgment

There is one situation in which loss of an acknowledgment may result in system deadlock. This is the case in which a receiver sends an acknowledgment with *rwnd* set to 0 and requests that the sender shut down its window temporarily. After a while, the receiver wants to remove the restriction; however, if it has no data to send, it sends an ACK segment and removes the restriction with a nonzero value for *rwnd*. A problem arises if this acknowledgment is lost. The sender is waiting for an acknowledgment that announces the nonzero *rwnd*. The receiver thinks that the sender has received this and is waiting for data. This situation is called a ***deadlock;*** each end is waiting for a response from the other end and nothing is happening. A retransmission timer is not set. To prevent deadlock, a persistence timer was designed that we will study later in the chapter.

Lost acknowledgments may create deadlock if they are not properly handled.

24.3.9 TCP Congestion Control

TCP uses different policies to handle the congestion in the network. We describe these policies in this section.

Congestion Window

When we discussed flow control in TCP, we mentioned that the size of the send window is controlled by the receiver using the value of *rwnd,* which is advertised in each segment traveling in the opposite direction. The use of this strategy guarantees that the receive window is never overflowed with the received bytes (no end congestion). This, however, does not mean that the intermediate buffers, buffers in the routers, do not become congested. A router may receive data from more than one sender. No matter how large the buffers of a router may be, it may be overwhelmed with data, which results in dropping some segments sent by a specific TCP sender. In other words, there is no congestion at the other end, but there may be congestion in the middle. TCP needs to worry about congestion in the middle because many segments lost may seriously affect the error control. More segment loss means resending the same segments again, resulting in worsening the congestion, and finally the collapse of the communication.

TCP is an end-to-end protocol that uses the service of IP. The congestion in the router is in the IP territory and should be taken care of by IP. However, as we discussed in Chapters 18 and 19, IP is a simple protocol with no congestion control. TCP, itself, needs to be responsible for this problem.

TCP cannot ignore the congestion in the network; it cannot aggressively send segments to the network. The result of such aggressiveness would hurt the TCP itself, as we mentioned before. TCP cannot be very conservative, either, sending a small number of segments in each time interval, because this means not utilizing the available bandwidth of the network. TCP needs to define policies that accelerate the data transmission when there is no congestion and decelerate the transmission when congestion is detected.

To control the number of segments to transmit, TCP uses another variable called a *congestion window, cwnd,* whose size is controlled by the congestion situation in the network (as we will explain shortly). The *cwnd* variable and the *rwnd* variable together define the size of the send window in TCP. The first is related to the congestion in the middle (network); the second is related to the congestion at the end. The actual size of the window is the minimum of these two.

Actual window size = minimum (*rwnd*, *cwnd*)

Congestion Detection

Before discussing how the value of *cwnd* should be set and changed, we need to describe how a TCP sender can detect the possible existence of congestion in the network. The TCP sender uses the occurrence of two events as signs of congestion in the network: time-out and receiving three duplicate ACKs.

The first is the *time-out*. If a TCP sender does not receive an ACK for a segment or a group of segments before the time-out occurs, it assumes that the corresponding segment or segments are lost and the loss is due to congestion.

Another event is the receiving of three duplicate ACKs (four ACKs with the same acknowledgment number). Recall that when a TCP receiver sends a duplicate ACK, it is the sign that a segment has been delayed, but sending three duplicate ACKs is the sign of a missing segment, which can be due to congestion in the network. However, the congestion in the case of three duplicate ACKs can be less severe than in the case of time-out. When a receiver sends three duplicate ACKs, it means that one segment is missing, but three segments have been received. The network is either slightly congested or has recovered from the congestion.

We will show later that an earlier version of TCP, called *Taho TCP,* treated both events (time-out and three duplicate ACKs) similarly, but the later version of TCP, called *Reno TCP,* treats these two signs differently.

A very interesting point in TCP congestion is that the TCP sender uses only one feedback from the other end to detect congestion: ACKs. The lack of regular, timely receipt of ACKs, which results in a time-out, is the sign of a strong congestion; the receiving of three duplicate ACKs is the sign of a weak congestion in the network.

Congestion Policies

TCP's general policy for handling congestion is based on three algorithms: slow start, congestion avoidance, and fast recovery. We first discuss each algorithm before showing how TCP switches from one to the other in a connection.

Slow Start: Exponential Increase

The **slow-start algorithm** is based on the idea that the size of the congestion window (*cwnd*) starts with one maximum segment size (MSS), but it increases one MSS each time an acknowledgment arrives. As we discussed before, the MSS is a value negotiated during the connection establishment, using an option of the same name.

The name of this algorithm is misleading; the algorithm starts slowly, but grows exponentially. To show the idea, let us look at Figure 24.29. We assume that *rwnd* is much larger than *cwnd,* so that the sender window size always equals *cwnd*. We also assume that each segment is of the same size and carries MSS bytes. For simplicity, we also ignore the delayed-ACK policy and assume that each segment is acknowledged individually.

The sender starts with *cwnd* = 1. This means that the sender can send only one segment. After the first ACK arrives, the acknowledged segment is purged from the window, which means there is now one empty segment slot in the window. The size of the congestion window is also increased by 1 because the arrival of the acknowledgment is a good sign that there is no congestion in the network. The size of the window is now 2. After sending two segments and receiving two individual acknowledgments for them, the size of the congestion window now becomes 4, and so on. In other words,

Figure 24.29 *Slow start, exponential increase*

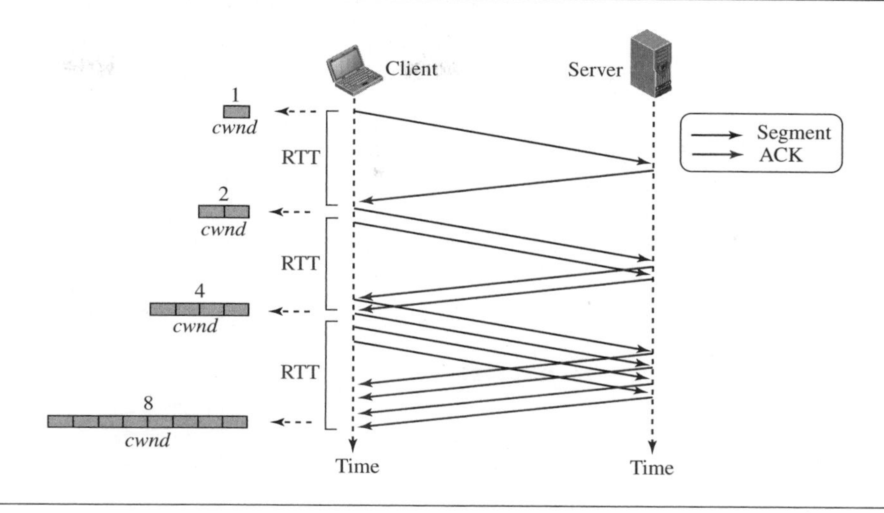

the size of the congestion window in this algorithm is a function of the number of ACKs arrived and can be determined as follows.

If an ACK arrives, *cwnd* = *cwnd* + **1.**

If we look at the size of the *cwnd* in terms of round-trip times (RTTs), we find that the growth rate is exponential in terms of each round trip time, which is a very aggressive approach:

Start	→	$cwnd = 1 \rightarrow 2^0$
After 1 RTT	→	$cwnd = cwnd + 1 = 1 + 1 = 2 \rightarrow 2^1$
After 2 RTT	→	$cwnd = cwnd + 2 = 2 + 2 = 4 \rightarrow 2^2$
After 3 RTT	→	$cwnd = cwnd + 4 = 4 + 4 = 8 \rightarrow 2^3$

A slow start cannot continue indefinitely. There must be a threshold to stop this phase. The sender keeps track of a variable named *ssthresh* (slow-start threshold). When the size of the window in bytes reaches this threshold, slow start stops and the next phase starts.

In the slow-start algorithm, the size of the congestion window increases exponentially until it reaches a threshold.

We need, however, to mention that the slow-start strategy is slower in the case of delayed acknowledgments. Remember, for each ACK, the *cwnd* is increased by only 1. Hence, if two segments are acknowledged cumulatively, the size of the *cwnd* increases by only 1, not 2. The growth is still exponential, but it is not a power of 2. With one ACK for every two segments, it is a power of 1.5.

Congestion Avoidance: Additive Increase

If we continue with the slow-start algorithm, the size of the congestion window increases exponentially. To avoid congestion before it happens, we must slow down this exponential growth. TCP defines another algorithm called ***congestion avoidance,*** which increases the *cwnd* additively instead of exponentially. When the size of the congestion window reaches the slow-start threshold in the case where $cwnd = i$, the slow-start phase stops and the additive phase begins. In this algorithm, each time the whole "window" of segments is acknowledged, the size of the congestion window is increased by one. A window is the number of segments transmitted during RTT. Figure 24.30 shows the idea.

Figure 24.30 *Congestion avoidance, additive increase*

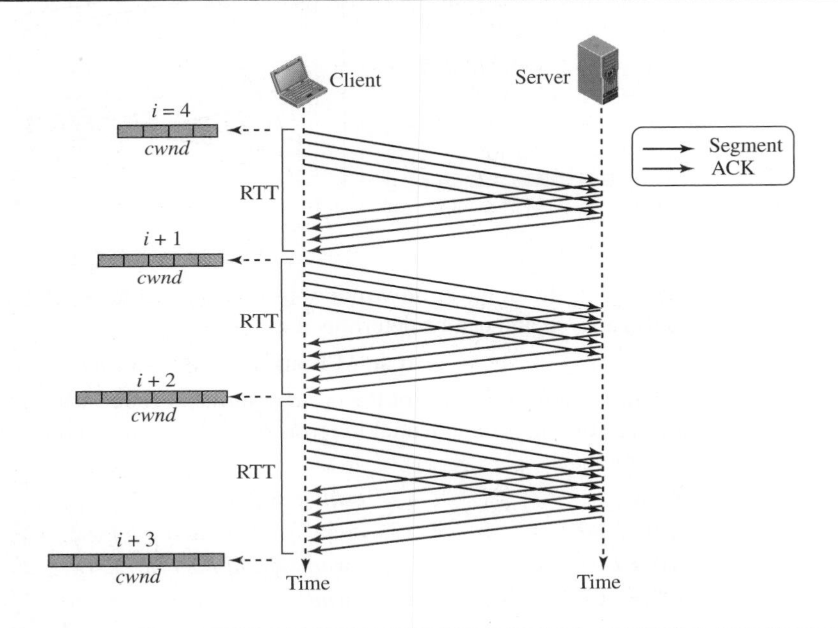

The sender starts with $cwnd = 4$. This means that the sender can send only four segments. After four ACKs arrive, the acknowledged segments are purged from the window, which means there is now one extra empty segment slot in the window. The size of the congestion window is also increased by 1. The size of window is now 5. After sending five segments and receiving five acknowledgments for them, the size of the congestion window now becomes 6, and so on. In other words, the size of the congestion window in this algorithm is also a function of the number of ACKs that have arrived and can be determined as follows:

$$\textbf{If an ACK arrives, } cwnd = cwnd + (1/cwnd).$$

The size of the window increases only $1/cwnd$ portion of MSS (in bytes). In other words, all segments in the previous window should be acknowledged to increase the window 1 MSS bytes.

If we look at the size of the *cwnd* in terms of round-trip times (RTTs), we find that the growth rate is linear in terms of each round-trip time, which is much more conservative than the slow-start approach.

Start	→	$cwnd = i$
After 1 RTT	→	$cwnd = i + 1$
After 2 RTT	→	$cwnd = i + 2$
After 3 RTT	→	$cwnd = i + 3$

In the congestion-avoidance algorithm, the size of the congestion window increases additively until congestion is detected.

Fast Recovery The **fast-recovery** algorithm is optional in TCP. The old version of TCP did not use it, but the new versions try to use it. It starts when three duplicate ACKs arrive, which is interpreted as light congestion in the network. Like congestion avoidance, this algorithm is also an additive increase, but it increases the size of the congestion window when a duplicate ACK arrives (after the three duplicate ACKs that trigger the use of this algorithm). We can say

If a duplicate ACK arrives, $cwnd = cwnd + (1 / cwnd)$.

Policy Transition

We discussed three congestion policies in TCP. Now the question is when each of these policies is used and when TCP moves from one policy to another. To answer these questions, we need to refer to three versions of TCP: Taho TCP, Reno TCP, and New Reno TCP.

Taho TCP

The early TCP, known as *Taho TCP,* used only two different algorithms in their congestion policy: *slow start* and *congestion avoidance*. We use Figure 24.31 to show the FSM for this version of TCP. However, we need to mention that we have deleted some small trivial actions, such as incrementing and resetting the number of duplicate ACKs, to make the FSM less crowded and simpler.

Figure 24.31 *FSM for Taho TCP*

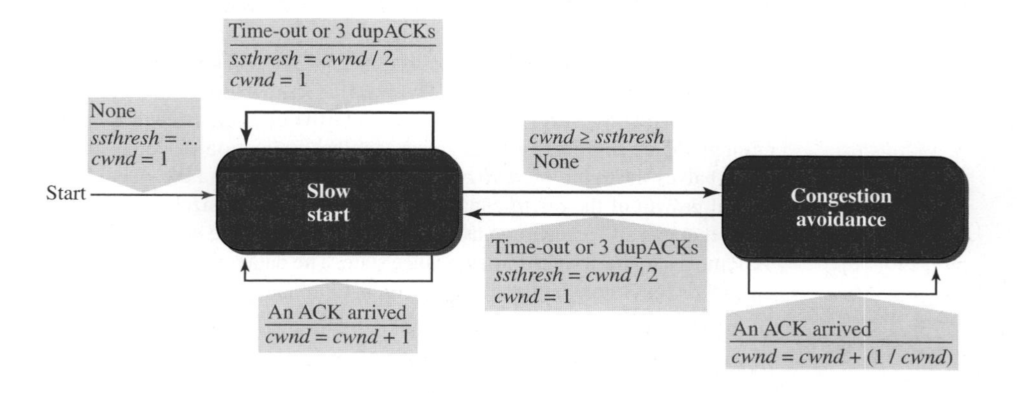

Taho TCP treats the two signs used for congestion detection, time-out and three duplicate ACKs, in the same way. In this version, when the connection is established, TCP starts the slow-start algorithm and sets the *ssthresh* variable to a pre-agreed value (normally a multiple of MSS) and the *cwnd* to 1 MSS. In this state, as we said before, each time an ACK arrives, the size of the congestion window is incremented by 1. We know that this policy is very aggressive and exponentially increases the size of the window, which may result in congestion.

If congestion is detected (occurrence of time-out or arrival of three duplicate ACKs), TCP immediately interrupts this aggressive growth and restarts a new slow start algorithm by limiting the threshold to half of the current *cwnd* and resetting the congestion window to 1. In other words, not only does TCP restart from scratch, but it also learns how to adjust the threshold. If no congestion is detected while reaching the threshold, TCP learns that the ceiling of its ambition is reached; it should not continue at this speed. It moves to the congestion avoidance state and continues in that state.

In the congestion-avoidance state, the size of the congestion window is increased by 1 each time a number of ACKs equal to the current size of the window has been received. For example, if the window size is now 5 MSS, five more ACKs should be received before the size of the window becomes 6 MSS. Note that there is no ceiling for the size of the congestion window in this state; the conservative additive growth of the congestion window continues to the end of the data transfer phase unless congestion is detected. If congestion is detected in this state, TCP again resets the value of the *ssthresh* to half of the current *cwnd* and moves to the slow-start state again.

Although in this version of TCP the size of *ssthresh* is continuously adjusted in each congestion detection, this does not mean that it necessarily becomes lower than the previous value. For example, if the original *ssthresh* value is 8 MSS and congestion is detected when TCP is in the congestion avoidance state and the value of the *cwnd* is 20, the new value of the *ssthresh* is now 10, which means it has been increased.

Example 24.9

Figure 24.32 shows an example of congestion control in a Taho TCP. TCP starts data transfer and sets the *ssthresh* variable to an ambitious value of 16 MSS. TCP begins at the slow-start (SS) state with the *cwnd* = 1. The congestion window grows exponentially, but a time-out occurs after the third RTT (before reaching the threshold). TCP assumes that there is congestion in the network. It immediately sets the new *ssthresh* = 4 MSS (half of the current *cwnd,* which is 8) and begins a new slow-start (SA) state with *cwnd* = 1 MSS. The congestion window grows exponentially until it reaches the newly set threshold. TCP now moves to the congestion-avoidance (CA) state and the congestion window grows additively until it reaches *cwnd* = 12 MSS. At this moment, three duplicate ACKs arrive, another indication of congestion in the network. TCP again halves the value of *ssthresh* to 6 MSS and begins a new slow-start (SS) state. The exponential growth of the *cwnd* continues. After RTT 15, the size of *cwnd* is 4 MSS. After sending four segments and receiving only two ACKs, the size of the window reaches the *ssthresh* (6) and TCP moves to the congestion-avoidance state. The data transfer now continues in the congestion-avoidance (CA) state until the connection is terminated after RTT 20.

Reno TCP

A newer version of TCP, called *Reno TCP,* added a new state to the congestion-control FSM, called the fast-recovery state. This version treated the two signals of congestion,

Figure 24.32 *Example 24.9*

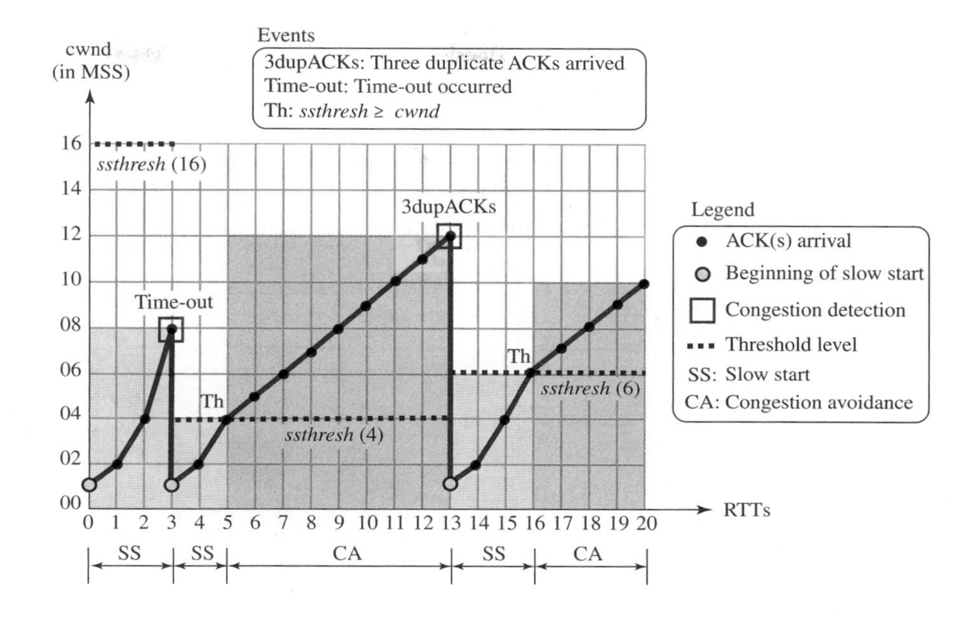

time-out and the arrival of three duplicate ACKs, differently. In this version, if a time-out occurs, TCP moves to the slow-start state (or starts a new round if it is already in this state); on the other hand, if three duplicate ACKs arrive, TCP moves to the fast-recovery state and remains there as long as more duplicate ACKs arrive. The fast-recovery state is a state somewhere between the slow-start and the congestion-avoidance states. It behaves like the slow start, in which the *cwnd* grows exponentially, but the *cwnd* starts with the value of *ssthresh* plus 3 MSS (instead of 1). When TCP enters the fast-recovery state, three major events may occur. If duplicate ACKs continue to arrive, TCP stays in this state, but the *cwnd* grows exponentially. If a time-out occurs, TCP assumes that there is real congestion in the network and moves to the slow-start state. If a new (non-duplicate) ACK arrives, TCP moves to the congestion-avoidance state, but deflates the size of the *cwnd* to the *ssthresh* value, as though the three duplicate ACKs have not occurred, and transition is from the slow-start state to the congestion-avoidance state. Figure 24.33 shows the simplified FSM for Reno TCP. Again, we have removed some trivial events to simplify the figure and discussion.

Example 24.10

Figure 24.34 shows the same situation as Figure 24.32, but in Reno TCP. The changes in the congestion window are the same until RTT 13 when three duplicate ACKs arrive. At this moment, Reno TCP drops the *ssthresh* to 6 MSS (same as Taho TCP), but it sets the *cwnd* to a much higher value (*ssthresh* + 3 = 9 MSS) instead of 1 MSS. Reno TCP now moves to the fast recovery state. We assume that two more duplicate ACKs arrive until RTT 15, where *cwnd* grows exponentially. In this moment, a new ACK (not duplicate) arrives that announces the receipt of the lost segment. Reno TCP now moves to the congestion-avoidance state, but first deflates the

Figure 24.33 *FSM for Reno TCP*

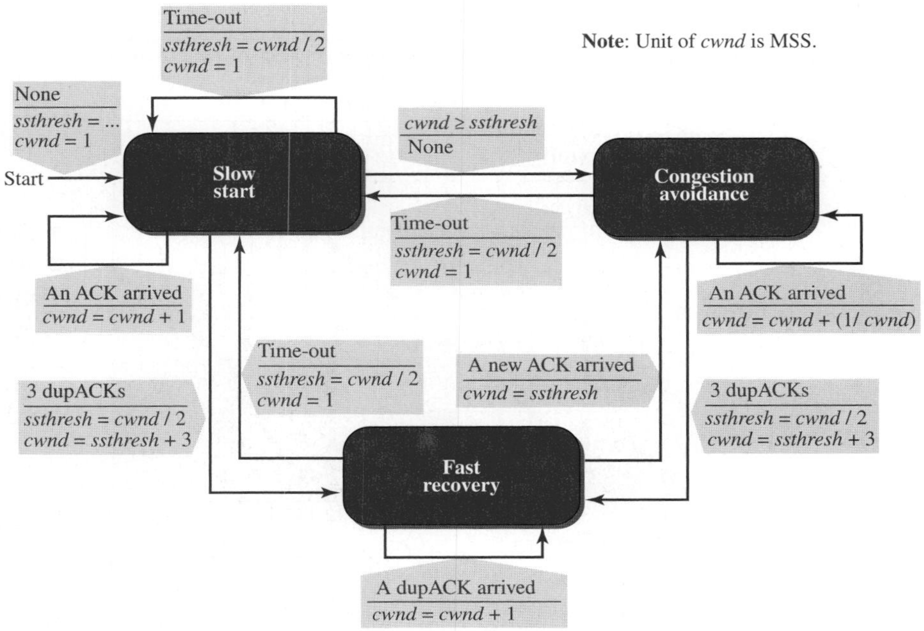

Note: Unit of *cwnd* is MSS.

None / *ssthresh* = ... / *cwnd* = 1

Start

Slow start

Congestion avoidance

Fast recovery

Time-out / *ssthresh* = *cwnd* / 2 / *cwnd* = 1

$cwnd \geq ssthresh$ / None

Time-out / *ssthresh* = *cwnd* / 2 / *cwnd* = 1

An ACK arrived / *cwnd* = *cwnd* + 1

An ACK arrived / *cwnd* = *cwnd* + (1/ *cwnd*)

Time-out / *ssthresh* = *cwnd* / 2 / *cwnd* = 1

A new ACK arrived / *cwnd* = *ssthresh*

3 dupACKs / *ssthresh* = *cwnd* / 2 / *cwnd* = *ssthresh* + 3

3 dupACKs / *ssthresh* = *cwnd* / 2 / *cwnd* = *ssthresh* + 3

A dupACK arrived / *cwnd* = *cwnd* + 1

congestion window to 6 MSS (the *ssthresh* value) as though ignoring the whole fast-recovery state and moving back to the previous track.

NewReno TCP

A later version of TCP, called *NewReno TCP,* made an extra optimization on the Reno TCP. In this version, TCP checks to see if more than one segment is lost in the current window when three duplicate ACKs arrive. When TCP receives three duplicate ACKs, it retransmits the lost segment until a new ACK (not duplicate) arrives. If the new ACK defines the end of the window when the congestion was detected, TCP is certain that only one segment was lost. However, if the ACK number defines a position between the retransmitted segment and the end of the window, it is possible that the segment defined by the ACK is also lost. NewReno TCP retransmits this segment to avoid receiving more and more duplicate ACKs for it.

Additive Increase, Multiplicative Decrease

Out of the three versions of TCP, the Reno version is most common today. It has been observed that, in this version, most of the time the congestion is detected and taken care of by observing the three duplicate ACKs. Even if there are some time-out events, TCP recovers from them by aggressive exponential growth. In other words, in a long TCP connection, if we ignore the slow-start states and short exponential growth during fast recovery, the TCP congestion window is *cwnd* = *cwnd* + (1 / cwnd) when an ACK arrives (congestion avoidance), and *cwnd* = *cwnd* / 2 when congestion is detected, as though SS does not exist and the length of FR is reduced to zero. The first is called *additive*

Figure 24.34 *Example 24.10*

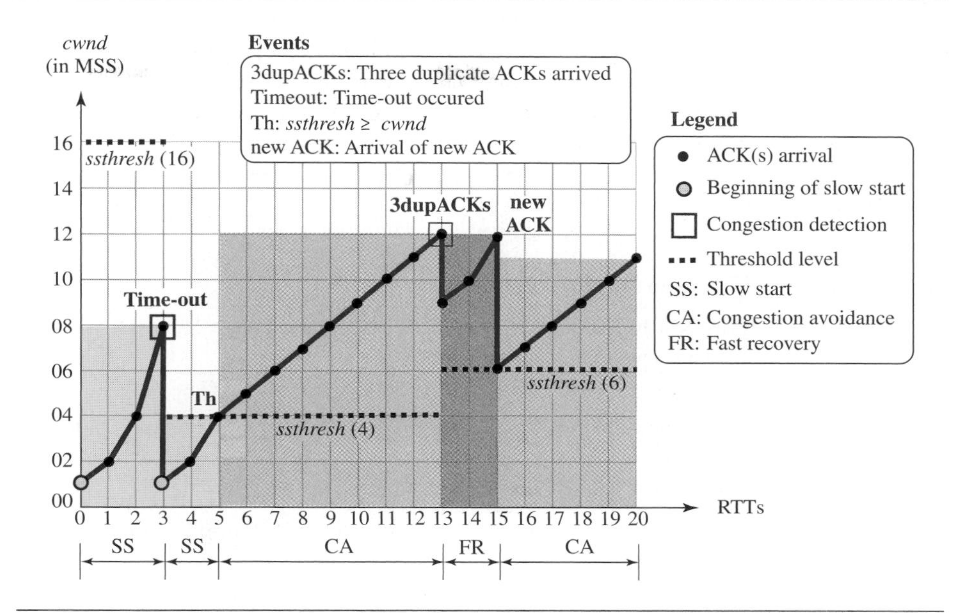

increase; the second is called *multiplicative decrease*. This means that the congestion window size, after it passes the initial slow-start state, follows a saw tooth pattern called ***additive increase, multiplicative decrease (AIMD),*** as shown in Figure 24.35.

Figure 24.35 *Additive increase, multiplicative decrease (AIMD)*

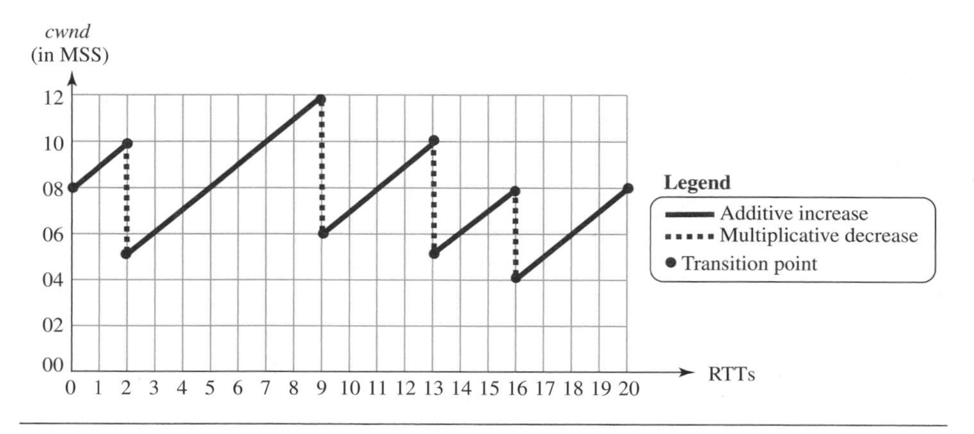

TCP Throughput

The throughput for TCP, which is based on the congestion window behavior, can be easily found if the *cwnd* is a constant (flat line) function of RTT. The throughput with this unrealistic assumption is throughput = *cwnd* / RTT. In this assumption, TCP sends a *cwnd* bytes of data and receives acknowledgement for them in RTT time. The behavior of TCP, as shown in Figure 24.35, is not a flat line; it is like saw teeth, with many minimum and maximum values. If each tooth were exactly the same, we could say that the throughput = [(maximum + minimum) / 2] / RTT. However, we know that the value of the maximum is twice the value of the minimum because in each congestion detection the value of *cwnd* is set to half of its previous value. So the throughput can be better calculated as

$$\textbf{throughput} = \textbf{(0.75)}\ \mathbf{W_{max}} / \textbf{RTT}$$

in which W_{max} is the average of window sizes when the congestion occurs.

Example 24.11

If MSS = 10 KB (kilobytes) and RTT = 100 ms in Figure 24.35, we can calculate the throughput as shown below.

$W_{max} = (10 + 12 + 10 + 8 + 8) / 5 = 9.6$ MSS
Throughput = $(0.75\ W_{max} / \text{RTT}) = 0.75 \times 960$ kbps / 100 ms = 7.2 Mbps

24.3.10 TCP Timers

To perform their operations smoothly, most TCP implementations use at least four timers: retransmission, persistence, keepalive, and TIME-WAIT.

Retransmission Timer

To retransmit lost segments, TCP employs one retransmission timer (for the whole connection period) that handles the retransmission time-out (RTO), the waiting time for an acknowledgment of a segment. We can define the following rules for the retransmission timer:

1. When TCP sends the segment in front of the sending queue, it starts the timer.
2. When the timer expires, TCP resends the first segment in front of the queue, and restarts the timer.
3. When a segment or segments are cumulatively acknowledged, the segment or segments are purged from the queue.
4. If the queue is empty, TCP stops the timer; otherwise, TCP restarts the timer.

Round-Trip Time (RTT)

To calculate the retransmission time-out (RTO), we first need to calculate the **round-trip time (RTT).** However, calculating RTT in TCP is an involved process that we explain step by step with some examples.

- ***Measured RTT.*** We need to find how long it takes to send a segment and receive an acknowledgment for it. This is the measured RTT. We need to remember that the segments and their acknowledgments do not have a one-to-one relationship; several segments may be acknowledged together. The measured round-trip time for a segment is the time required for the segment to reach the destination and be acknowledged, although the acknowledgment may include other segments. Note that in TCP only one RTT measurement can be in progress at any time. This means that if an RTT measurement is started, no other measurement starts until the value of this RTT is finalized. We use the notation RTT_M to stand for measured RTT.

In TCP, there can be only one RTT measurement in progress at any time.

- ***Smoothed RTT.*** The measured RTT, RTT_M, is likely to change for each round trip. The fluctuation is so high in today's Internet that a single measurement alone cannot be used for retransmission time-out purposes. Most implementations use a smoothed RTT, called RTT_S, which is a weighted average of RTT_M and the previous RTT_S, as shown below:

Initially	→	**No value**
After first measurement	→	$RTT_S = RTT_M$
After each measurement	→	$RTT_S = (1 - \alpha)\ RTT_S + \alpha \times RTT_M$

The value of α is implementation-dependent, but it is normally set to 1/8. In other words, the new RTT_S is calculated as 7/8 of the old RTT_S and 1/8 of the current RTT_M.

- ***RTT Deviation.*** Most implementations do not just use RTT_S; they also calculate the RTT deviation, called RTT_D, based on the RTT_S and RTT_M, using the following formulas. (The value of β is also implementation-dependent, but is usually set to 1/4.)

Initially	→	**No value**
After first measurement	→	$RTT_D = RTT_M /2$
After each measurement	→	$RTT_D = (1 - \beta)\ RTT_D + \beta \times \mid RTT_S - RTT_M \mid$

Retransmission Time-out (RTO) The value of RTO is based on the smoothed round-trip time and its deviation. Most implementations use the following formula to calculate the RTO:

Original	→	**Initial value**
After any measurement	→	$RTO = RTT_S + 4 \times RTT_D$

In other words, take the running smoothed average value of RTT_S and add four times the running smoothed average value of RTT_D (normally a small value).

Example 24.12

Let us give a hypothetical example. Figure 24.36 shows part of a connection. The figure shows the connection establishment and part of the data transfer phases.

1. When the SYN segment is sent, there is no value for RTT_M, RTT_S, or RTT_D. The value of RTO is set to 6.00 seconds. The following shows the value of these variables at this moment:

   $RTO = 6$

2. When the SYN+ACK segment arrives, RTT_M is measured and is equal to 1.5 seconds. The following shows the values of these variables:

   $RTT_M = 1.5$
   $RTT_S = 1.5$
   $RTT_D = (1.5)/2 = 0.75$
   $RTO = 1.5 + 4 \times 0.75 = 4.5$

Figure 24.36 *Example 24.12*

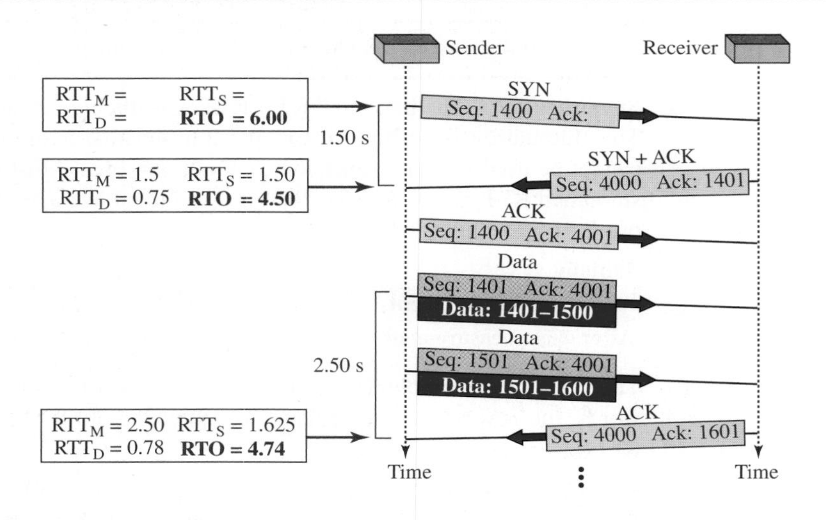

3. When the first data segment is sent, a new RTT measurement starts. Note that the sender does not start an RTT measurement when it sends the ACK segment, because it does not consume a sequence number and there is no time-out. No RTT measurement starts for the second data segment because a measurement is already in progress. The arrival of the last ACK segment is used to calculate the next value of RTT_M. Although the last ACK segment acknowledges both data segments (cumulative), its arrival finalizes the value of RTT_M for the first segment. The values of these variables are now as shown below.

$$\mathbf{RTT_M = 2.5}$$
$$\mathbf{RTT_S = (7/8) \times (1.5) + (1/8) \times (2.5) = 1.625}$$
$$\mathbf{RTT_D = (3/4) \times (0.75) + (1/4) \times |1.625 - 2.5| = 0.78}$$
$$\mathbf{RTO = 1.625 + 4 \times (0.78) = 4.74}$$

Karn's Algorithm

Suppose that a segment is not acknowledged during the retransmission time-out period and is therefore retransmitted. When the sending TCP receives an acknowledgment for this segment, it does not know if the acknowledgment is for the original segment or for the retransmitted one. The value of the new RTT is based on the departure of the segment. However, if the original segment was lost and the acknowledgment is for the retransmitted one, the value of the current RTT must be calculated from the time the segment was retransmitted. This ambiguity was solved by Karn. **Karn's algorithm** is simple. Do not consider the round-trip time of a retransmitted segment in the calculation of RTTs. Do not update the value of RTTs until you send a segment and receive an acknowledgment without the need for retransmission.

TCP does not consider the RTT of a retransmitted segment in its calculation of a new RTO.

Exponential Backoff

What is the value of RTO if a retransmission occurs? Most TCP implementations use an exponential backoff strategy. The value of RTO is doubled for each retransmission. So if the segment is retransmitted once, the value is two times the RTO. If it is retransmitted twice, the value is four times the RTO, and so on.

Example 24.13

Figure 24.37 is a continuation of the previous example. There is retransmission and Karn's algorithm is applied.

Figure 24.37 *Example 24.13*

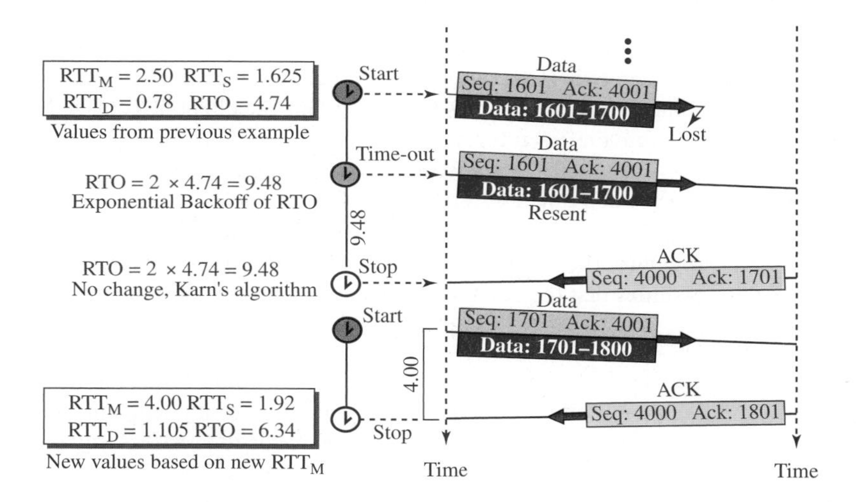

The first segment in the figure is sent, but lost. The RTO timer expires after 4.74 seconds. The segment is retransmitted and the timer is set to 9.48, twice the previous value of RTO. This time an ACK is received before the time-out. We wait until we send a new segment and receive the ACK for it before recalculating the RTO (Karn's algorithm).

Persistence Timer

To deal with a zero-window-size advertisement, TCP needs another timer. If the receiving TCP announces a window size of zero, the sending TCP stops transmitting segments until the receiving TCP sends an ACK segment announcing a nonzero window size. This ACK segment can be lost. Remember that ACK segments are not acknowledged nor retransmitted in TCP. If this acknowledgment is lost, the receiving TCP thinks that it has done its job and waits for the sending TCP to send more segments. There is no retransmission timer for a segment containing only an acknowledgment. The sending TCP has not received an acknowledgment and waits for the other TCP to send an acknowledgment advertising the size of the window. Both TCP's might continue to wait for each other forever (a deadlock).

To correct this deadlock, TCP uses a **persistence timer** for each connection. When the sending TCP receives an acknowledgment with a window size of zero, it starts a persistence timer. When the persistence timer goes off, the sending TCP sends a special segment called a *probe*. This segment contains only 1 byte of new data. It has a sequence number, but its sequence number is never acknowledged; it is even ignored in calculating the sequence number for the rest of the data. The probe causes the receiving TCP to resend the acknowledgment.

The value of the persistence timer is set to the value of the retransmission time. However, if a response is not received from the receiver, another probe segment is sent and the value of the persistence timer is doubled and reset. The sender continues sending the probe segments and doubling and resetting the value of the persistence timer until the value reaches a threshold (usually 60 s). After that the sender sends one probe segment every 60 seconds until the window is reopened.

Keepalive Timer

A **keepalive timer** is used in some implementations to prevent a long idle connection between two TCPs. Suppose that a client opens a TCP connection to a server, transfers some data, and becomes silent. Perhaps the client has crashed. In this case, the connection remains open forever.

To remedy this situation, most implementations equip a server with a keepalive timer. Each time the server hears from a client, it resets this timer. The time-out is usually 2 hours. If the server does not hear from the client after 2 hours, it sends a probe segment. If there is no response after 10 probes, each of which is 75 seconds apart, it assumes that the client is down and terminates the connection.

TIME-WAIT Timer

The TIME-WAIT (2MSL) timer is used during connection termination. The maximum segment lifetime (MSL) is the amount of time any segment can exist in a network before being discarded. The implementation needs to choose a value for MSL. Common values are 30 seconds, 1 minute, or even 2 minutes. The 2MSL timer is used when TCP performs an active close and sends the final ACK. The connection must stay open for 2 MSL amount of time to allow TCP to resend the final ACK in case the ACK is lost. This requires that the RTO timer at the other end times out and new FIN and ACK segments are resent.

24.3.11 Options

The TCP header can have up to 40 bytes of optional information. Options convey additional information to the destination or align other options. These options are included on the book website for further reference.

TCP options are discussed on the book website.

24.4 SCTP

Stream Control Transmission Protocol (SCTP) is a new transport-layer protocol designed to combine some features of UDP and TCP in an effort to create a better protocol for multimedia communication.

24.4.1 SCTP Services

Before discussing the operation of SCTP, let us explain the services offered by SCTP to the application-layer processes.

Process-to-Process Communication

SCTP, like UDP or TCP, provides process-to-process communication.

Multiple Streams

We learned that TCP is a stream-oriented protocol. Each connection between a TCP client and a TCP server involves a single stream. The problem with this approach is that a loss at any point in the stream blocks the delivery of the rest of the data. This can be acceptable when we are transferring text; it is not when we are sending real-time data such as audio or video. SCTP allows **multistream service** in each connection, which is called ***association*** in SCTP terminology. If one of the streams is blocked, the other streams can still deliver their data. Figure 24.38 shows the idea of multiple-stream delivery.

Figure 24.38 *Multiple-stream concept*

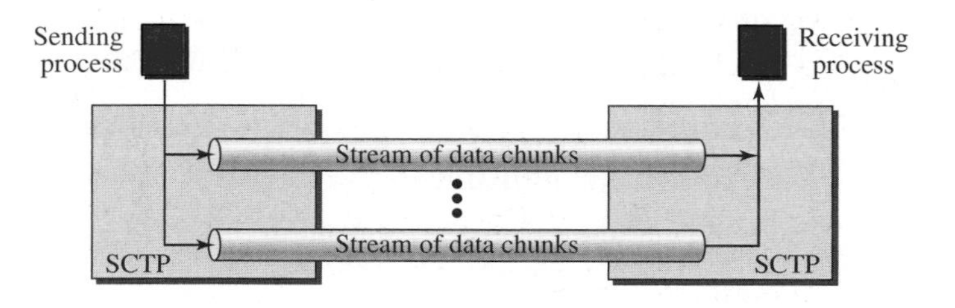

Multihoming

A TCP connection involves one source and one destination IP address. This means that even if the sender or receiver is a multihomed host (connected to more than one physical address with multiple IP addresses), only one of these IP addresses per end can be utilized during the connection. An SCTP association, on the other hand, supports **multihoming service.** The sending and receiving host can define multiple IP addresses in each end for an association. In this fault-tolerant approach, when one path fails, another interface can be used for data delivery without interruption. This fault-tolerant feature is very helpful when we are sending and receiving a real-time payload such as Internet telephony. Figure 24.39 shows the idea of multihoming.

In the figure, the client is connected to two local networks with two IP addresses. The server is also connected to two networks with two IP addresses. The client and the

Figure 24.39 *Multihoming concept*

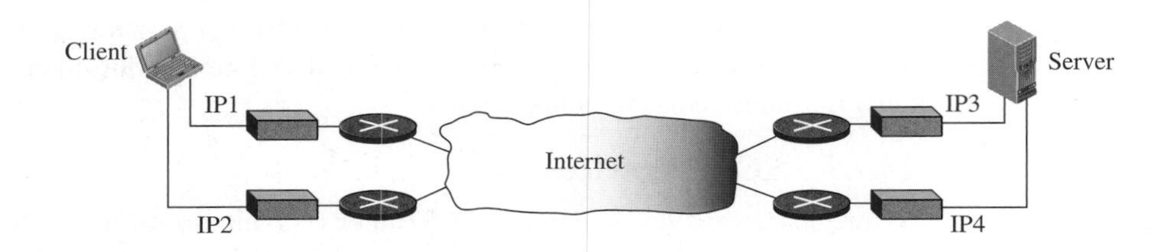

server can make an association using four different pairs of IP addresses. However, note that in the current implementations of SCTP, only one pair of IP addresses can be chosen for normal communication; the alternative is used if the main choice fails. In other words, at present, SCTP does not allow load sharing between different paths.

Full-Duplex Communication

Like TCP, SCTP offers full-duplex service, where data can flow in both directions at the same time. Each SCTP then has a sending and receiving buffer and packets are sent in both directions.

Connection-Oriented Service

Like TCP, SCTP is a connection-oriented protocol. However, in SCTP, a connection is called an *association*.

Reliable Service

SCTP, like TCP, is a reliable transport protocol. It uses an acknowledgment mechanism to check the safe and sound arrival of data. We will discuss this feature further in the section on error control.

24.4.2 SCTP Features

The following shows the general features of SCTP.

Transmission Sequence Number (TSN)

The unit of data in SCTP is a data chunk, which may or may not have a one-to-one relationship with the message coming from the process because of fragmentation (discussed later). Data transfer in SCTP is controlled by numbering the data chunks. SCTP uses a **transmission sequence number (TSN)** to number the data chunks. In other words, the TSN in SCTP plays a role analogous to the sequence number in TCP. TSNs are 32 bits long and randomly initialized between 0 and $2^{32} - 1$. Each data chunk must carry the corresponding TSN in its header.

Stream Identifier (SI)

In SCTP, there may be several streams in each association. Each stream in SCTP needs to be identified using a **stream identifier (SI).** Each data chunk must carry the SI in its header so that when it arrives at the destination, it can be properly placed in its stream. The SI is a 16-bit number starting from 0.

Stream Sequence Number (SSN)

When a data chunk arrives at the destination SCTP, it is delivered to the appropriate stream and in the proper order. This means that, in addition to an SI, SCTP defines each data chunk in each stream with a **stream sequence number (SSN).**

Packets

In TCP, a segment carries data and control information. Data are carried as a collection of bytes; control information is defined by six control flags in the header. The design of SCTP is totally different: data are carried as data chunks, control information as control chunks. Several control chunks and data chunks can be packed together in a packet. A packet in SCTP plays the same role as a segment in TCP. Figure 24.40 compares a

Figure 24.40 *Comparison between a TCP segment and an SCTP packet*

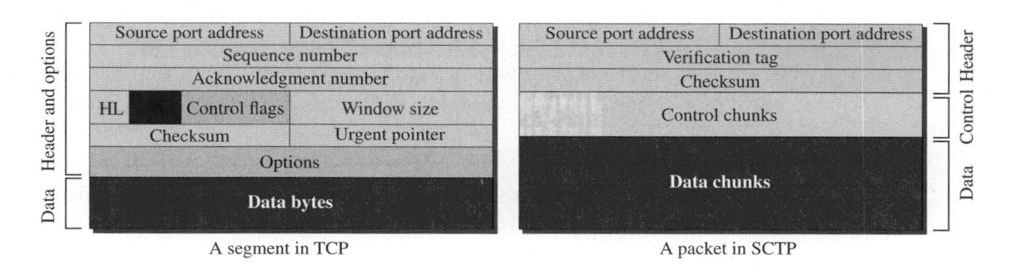

segment in TCP and a packet in SCTP. We will discuss the format of the SCTP packet in the next section.

In SCTP, we have data chunks, streams, and packets. An association may send many packets, a packet may contain several chunks, and chunks may belong to different streams. To make the definitions of these terms clear, let us suppose that process A needs to send 11 messages to process B in three streams. The first four messages are in the first stream, the second three messages are in the second stream, and the last four messages are in the third stream. Although a message, if long, can be carried by several data chunks, we assume that each message fits into one data chunk. Therefore, we have 11 data chunks in three streams.

The application process delivers 11 messages to SCTP, where each message is earmarked for the appropriate stream. Although the process could deliver one message from the first stream and then another from the second, we assume that it delivers all messages belonging to the first stream first, all messages belonging to the second stream next, and finally, all messages belonging to the last stream.

We also assume that the network allows only three data chunks per packet, which means that we need four packets, as shown in Figure 24.41.

Data chunks in stream 0 are carried in the first and part of the second packet; those in stream 1 are carried in the second and the third packet; those in stream 2 are carried in the third and fourth packet.

Note that each data chunk needs three identifiers: TSN, SI, and SSN. TSN is a cumulative number and is used, as we will see later, for flow control and error control.

Figure 24.41 *Packets, data chunks, and streams*

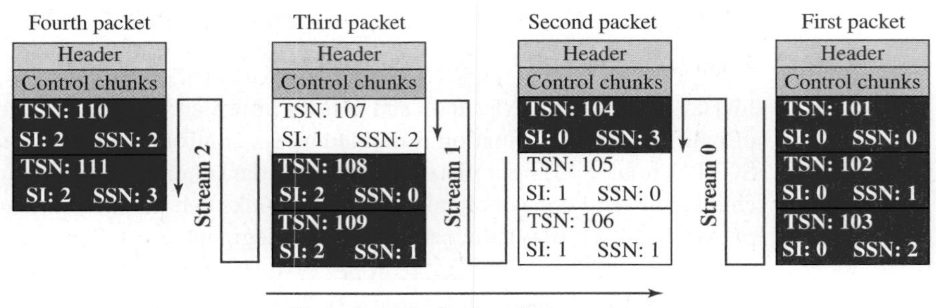

SI defines the stream to which the chunk belongs. SSN defines the chunk's order in a particular stream. In our example, SSN starts from 0 for each stream.

Acknowledgment Number

TCP acknowledgment numbers are byte-oriented and refer to the sequence numbers. SCTP acknowledgment numbers are chunk-oriented. They refer to the TSN. A second difference between TCP and SCTP acknowledgments is the control information. Recall that this information is part of the segment header in TCP. To acknowledge segments that carry only control information, TCP uses a sequence number and acknowledgment number (for example, a SYN segment needs to be acknowledged by an ACK segment). In SCTP, however, the control information is carried by control chunks, which do not need a TSN. These control chunks are acknowledged by another control chunk of the appropriate type (some need no acknowledgment). For example, an INIT control chunk is acknowledged by an INIT-ACK chunk. There is no need for a sequence number or an acknowledgment number.

24.4.3 Packet Format

An SCTP packet has a mandatory general header and a set of blocks called chunks. There are two types of chunks: control chunks and data chunks. A control chunk controls and maintains the association; a data chunk carries user data. In a packet, the control chunks come before the data chunks. Figure 24.42 shows the general format of an SCTP packet.

General Header

The *general header* (packet header) defines the end points of each association to which the packet belongs, guarantees that the packet belongs to a particular association, and preserves the integrity of the contents of the packet including the header itself. The format of the general header is shown in Figure 24.43.

There are four fields in the general header. The source and destination port numbers are the same as in UDP or TCP. The verification tag is a 32-bit field that matches a

Figure 24.42 *SCTP packet format*

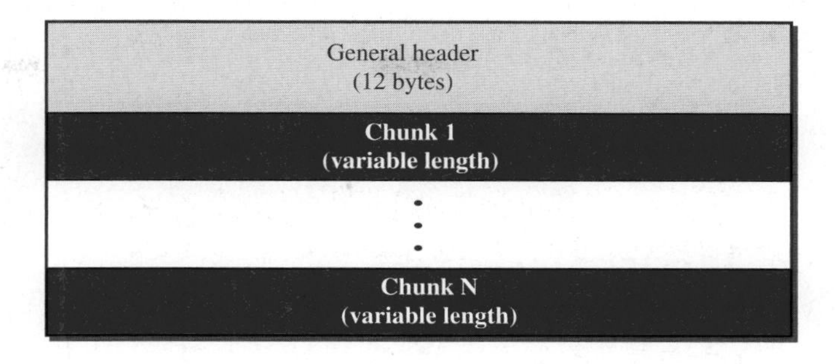

Figure 24.43 *General header*

Source port address 16 bits	Destination port address 16 bits
Verification tag 32 bits	
Checksum 32 bits	

packet to an association. This prevents a packet from a previous association from being mistaken as a packet in this association. It serves as an identifier for the association; it is repeated in every packet during the association. The next field is a checksum. However, the size of the checksum is increased from 16 bits (in UDP, TCP, and IP) to 32 bits in SCTP to allow the use of the CRC-32 checksum.

Chunks

Control information or user data are carried in chunks. Chunks have a common layout, as shown in Figure 24.44. The first three fields are common to all chunks; the information field depends on the type of chunk. The type field can define up to 256 types of chunks. Only a few have been defined so far; the rest are reserved for future use. The flag field defines special flags that a particular chunk may need. Each bit has a different meaning depending on the type of chunk. The length field defines the total size of the chunk, in bytes, including the type, flag, and length fields. Since the size of the information section is dependent on the type of chunk, we need to define the chunk boundaries. If a chunk carries no information, the value of the length field is 4 (4 bytes). Note that the length of the padding, if any, is not included in the calculation of the length field. This helps the receiver find out how many useful bytes a chunk carries. If the value is not a multiple of 4, the receiver knows there is padding.

Figure 24.44 *Common layout of a chunk*

Types of Chunks

SCTP defines several types of chunks, as shown in Table 24.3.

Table 24.3 *Chunks*

Type	*Chunk*	*Description*
0	DATA	User data
1	INIT	Sets up an association
2	INIT ACK	Acknowledges INIT chunk
3	SACK	Selective acknowledgment
4	HEARTBEAT	Probes the peer for liveliness
5	HEARTBEAT ACK	Acknowledges HEARTBEAT chunk
6	ABORT	Aborts an association
7	SHUTDOWN	Terminates an association
8	SHUTDOWN ACK	Acknowledges SHUTDOWN chunk
9	ERROR	Reports errors without shutting down
10	COOKIE ECHO	Third packet in association establishment
11	COOKIE ACK	Acknowledges COOKIE ECHO chunk
14	SHUTDOWN COMPLETE	Third packet in association termination
192	FORWARD TSN	For adjusting cumulating TSN

24.4.4 An SCTP Association

SCTP, like TCP, is a connection-oriented protocol. However, a connection in SCTP is called an *association* to emphasize multihoming.

A connection in SCTP is called an *association.*

Association Establishment

Association establishment in SCTP requires a *four-way handshake*. In this procedure, a process, normally a client, wants to establish an association with another process, normally a server, using SCTP as the transport-layer protocol. Similar to TCP, the SCTP server needs to be prepared to receive any association (passive open). Association establishment, however, is initiated by the client (active open). SCTP association establishment is shown in Figure 24.45.

Figure 24.45 *Four-way handshaking*

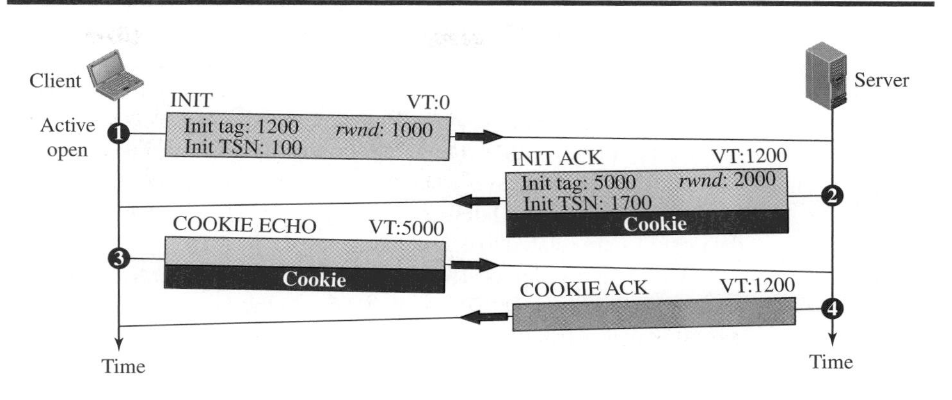

The steps, in a normal situation, are as follows:

1. The client sends the first packet, which contains an INIT chunk. The **verification tag (VT)** of this packet (defined in the general header) is 0 because no verification tag has yet been defined for this direction (client to server). The INIT tag includes an **initiation tag** to be used for packets from the other direction (server to client). The chunk also defines the **initial TSN** for this direction and advertises a value for *rwnd*. The value of *rwnd* is normally advertised in a SACK chunk; it is done here because SCTP allows the inclusion of a DATA chunk in the third and fourth packets; the server must be aware of the available client buffer size. Note that no other chunks can be sent with the first packet.
2. The server sends the second packet, which contains an INIT ACK chunk. The verification tag is the value of the initial tag field in the INIT chunk. This chunk initiates the tag to be used in the other direction, defines the initial TSN, for data flow from server to client, and sets the server's *rwnd*. The value of *rwnd* is defined to allow the client to send a DATA chunk with the third packet. The INIT ACK also sends a cookie that defines the state of the server at this moment. We will discuss the use of the cookie shortly.
3. The client sends the third packet, which includes a COOKIE ECHO chunk. This is a very simple chunk that echoes, without change, the cookie sent by the server. SCTP allows the inclusion of data chunks in this packet.
4. The server sends the fourth packet, which includes the COOKIE ACK chunk that acknowledges the receipt of the COOKIE ECHO chunk. SCTP allows the inclusion of data chunks with this packet.

Data Transfer

The whole purpose of an association is to transfer data between two ends. After the association is established, bidirectional data transfer can take place. The client and the server can both send data. Like TCP, SCTP supports piggybacking.

There is a major difference, however, between data transfer in TCP and SCTP. TCP receives messages from a process as a stream of bytes without recognizing any boundary between them. The process may insert some boundaries for its peer use, but TCP treats that mark as part of the text. In other words, TCP takes each message and appends it to its buffer. A segment can carry parts of two different messages. The only ordering system imposed by TCP is the byte numbers.

SCTP, on the other hand, recognizes and maintains boundaries. Each message coming from the process is treated as one unit and inserted into a DATA chunk unless it is fragmented (discussed later). In this sense, SCTP is like UDP, with one big advantage: data chunks are related to each other.

A message received from a process becomes a DATA chunk, or chunks if fragmented, by adding a DATA chunk header to the message. Each DATA chunk formed by a message or a fragment of a message has one TSN. We need to remember that only DATA chunks use TSNs and only DATA chunks are acknowledged by SACK chunks.

Multihoming Data Transfer

We discussed the multihoming capability of SCTP, a feature that distinguishes SCTP from UDP and TCP. Multihoming allows both ends to define multiple IP addresses for communication. However, only one of these addresses can be defined as the ***primary address;*** the rest are alternative addresses. The primary address is defined during association establishment. The interesting point is that the primary address of an end is determined by the other end. In other words, a source defines the primary address for a destination.

Data transfer, by default, uses the primary address of the destination. If the primary is not available, one of the alternative addresses is used. The process, however, can always override the primary address and explicitly request that a message be sent to one of the alternative addresses. A process can also explicitly change the primary address of the current association.

A logical question that arises is where to send a SACK. SCTP dictates that a SACK be sent to the address from which the corresponding SCTP packet originated.

Multistream Delivery

Another interesting feature of SCTP is the distinction between data transfer and data delivery. SCTP uses TSN numbers to handle data transfer, movement of data chunks between the source and destination. The delivery of the data chunks is controlled by SIs and SSNs. SCTP can support multiple streams, which means that the sender process can define different streams and a message can belong to one of these streams. Each stream is assigned a stream identifier (SI) which uniquely defines that stream. However, SCTP supports two types of data delivery in each stream: *ordered* (default) and *unordered.* In ordered data delivery, data chunks in a stream use stream sequence numbers (SSNs) to define their order in the stream. When the chunks arrive at the destination, SCTP is responsible for message delivery according to the SSN defined in the chunk. This may delay the delivery because some chunks may arrive out of order. In unordered data delivery, the data chunks in a stream have the U flag set, but their SSN field value is ignored. They do not consume SSNs. When an unordered data chunk arrives at the destination SCTP, it delivers the message carrying the chunk to the application without waiting for the other messages. Most of the time, applications use the ordered-delivery service, but occasionally some applications need to send urgent data that must be delivered out of order (recall the urgent data and urgent pointer facility of TCP). In these cases, the application can define the delivery as unordered.

Fragmentation

Another issue in data transfer is ***fragmentation.*** Although SCTP shares this term with IP (see Chapter 19), fragmentation in IP and SCTP belong to different levels: the former at the network layer, the latter at the transport layer.

SCTP preserves the boundaries of the message from process to process when creating a DATA chunk from a message if the size of the message (when encapsulated in an IP datagram) does not exceed the MTU (see Chapter 19) of the path. The size of an IP datagram carrying a message can be determined by adding the size of the message, in bytes, to the four overheads: data chunk header, necessary SACK chunks, SCTP general header, and IP header. If the total size exceeds the MTU, the message needs to be fragmented. (For more details about fragmentation, see the Extra Materials section on the book website.)

Association Termination

In SCTP, like TCP, either of the two parties involved in exchanging data (client or server) can close the connection. However, unlike TCP, SCTP does not allow a "half-closed" association. If one end closes the association, the other end must stop sending new data. If any data are left over in the queue of the recipient of the termination request, they are sent and the association is closed. Association termination uses three packets, as shown in Figure 24.46. Note that although the figure shows the case in which termination is initiated by the client, it can also be initiated by the server.

Figure 24.46 *Association termination*

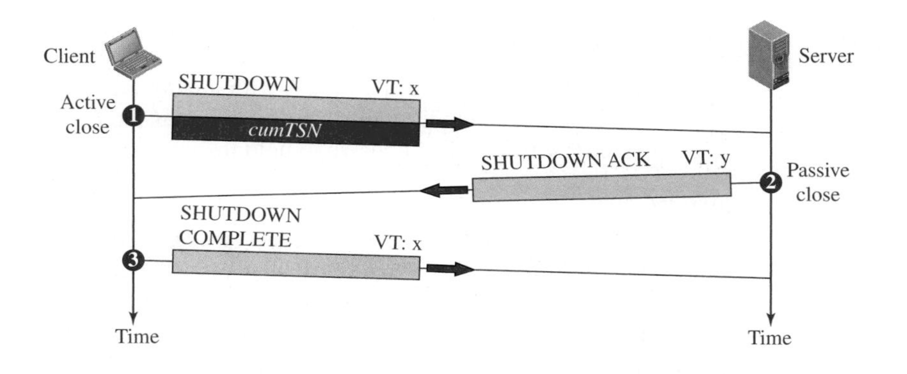

24.4.5 Flow Control

Flow control in SCTP is similar to that in TCP. In TCP, we need to deal with only one unit of data, the byte. In SCTP, we need to handle two units of data, the byte and the chunk. The values of *rwnd* and *cwnd* are expressed in bytes; the values of TSN and acknowledgments are expressed in chunks. To show the concept, we make some unrealistic assumptions. We assume that there is never congestion in the network and that the network is error free. In other words, we assume that *cwnd* is infinite and no packet is lost, is delayed, or arrives out of order. We also assume that data transfer is unidirectional. We correct our unrealistic assumptions in later sections. Current SCTP implementations still use a byte-oriented window for flow control. We, however, show a buffer in terms of chunks to make the concept easier to understand.

Receiver Site

The receiver has one buffer (queue) and three variables. The queue holds the received data chunks that have not yet been read by the process. The first variable holds the last TSN received, *cumTSN*. The second variable holds the available buffer size, *winSize*. The third variable holds the last cumulative acknowledgment, *lastACK*. Figure 24.47 shows the queue and variables at the receiver site.

Figure 24.47 *Flow control, receiver site*

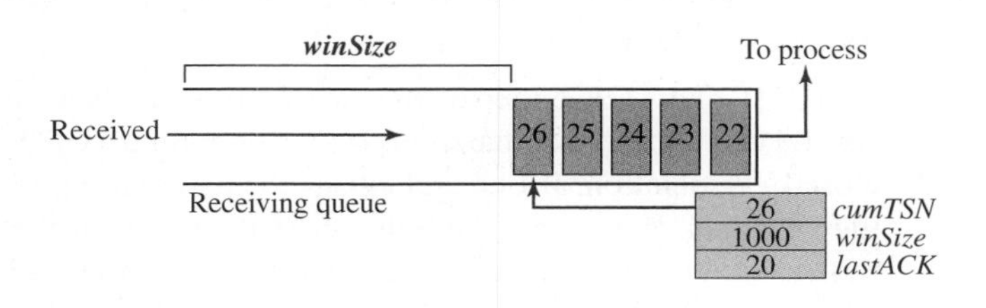

1. When the site receives a data chunk, it stores it at the end of the buffer (queue) and subtracts the size of the chunk from *winSize*. The TSN number of the chunk is stored in the *cumTSN* variable.
2. When the process reads a chunk, it removes it from the queue and adds the size of the removed chunk to *winSize* (recycling).
3. When the receiver decides to send a SACK, it checks the value of *lastAck;* if it is less than *cumTSN,* it sends a SACK with a cumulative TSN number equal to the *cumTSN*. It also includes the value of *winSize* as the advertised window size. The value of *lastACK* is then updated to hold the value of *cumTSN*.

Sender Site

The sender has one buffer (queue) and three variables: *curTSN, rwnd,* and *inTransit,* as shown in Figure 24.48. We assume each chunk is 100 bytes long.

The buffer holds the chunks produced by the process that have either been sent or are ready to be sent. The first variable, *curTSN,* refers to the next chunk to be sent. All chunks in the queue with a TSN less than this value have been sent but not acknowledged; they are outstanding. The second variable, *rwnd,* holds the last value advertised by the receiver (in bytes). The third variable, *inTransit,* holds the number of bytes in transit, bytes sent but not yet acknowledged. The following is the procedure used by the sender.

1. A chunk pointed to by *curTSN* can be sent if the size of the data is less than or equal to the quantity (*rwnd* – *inTransit*). After sending the chunk, the value of

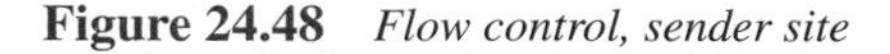

Figure 24.48 *Flow control, sender site*

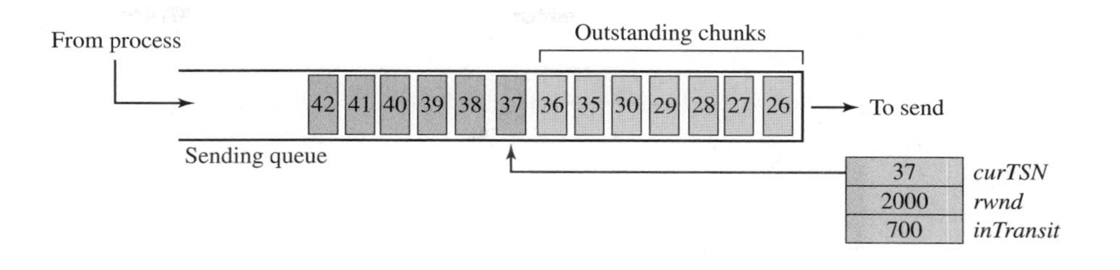

curTSN is incremented by one and now points to the next chunk to be sent. The value of *inTransit* is incremented by the size of the data in the transmitted chunk.

2. When a SACK is received, the chunks with a TSN less than or equal to the cumulative TSN in the SACK are removed from the queue and discarded. The sender does not have to worry about them anymore. The value of *inTransit* is reduced by the total size of the discarded chunks. The value of *rwnd* is updated with the value of the advertised window in the SACK.

24.4.6 Error Control

SCTP, like TCP, is a reliable transport-layer protocol. It uses a SACK chunk to report the state of the receiver buffer to the sender. Each implementation uses a different set of entities and timers for the receiver and sender sites. We use a very simple design to convey the concept to the reader.

Receiver Site

In our design, the receiver stores all chunks that have arrived in its queue including the out-of-order ones. However, it leaves spaces for any missing chunks. It discards duplicate messages, but keeps track of them for reports to the sender. Figure 24.49 shows a typical design for the receiver site and the state of the receiving queue at a particular point in time.

The last acknowledgment sent was for data chunk 20. The available window size is 1000 bytes. Chunks 21 to 23 have been received in order. The first out-of-order block contains chunks 26 to 28. The second out-of-order block contains chunks 31 to 34. A variable holds the value of *cumTSN.* An array of variables keeps track of the beginning and the end of each block that is out of order. An array of variables holds the duplicate chunks received. Note that there is no need for storing duplicate chunks in the queue, they will be discarded. The figure also shows the SACK chunk that will be sent to report the state of the receiver to the sender. The TSN numbers for out-of-order chunks are relative (offsets) to the cumulative TSN.

Sender Site

At the sender site, our design demands two buffers (queues): a sending queue and a retransmission queue. We also use three variables: *rwnd, inTransit,* and *curTSN,* as described in the previous section. Figure 24.50 shows a typical design.

Figure 24.49 *Error control, receiver site*

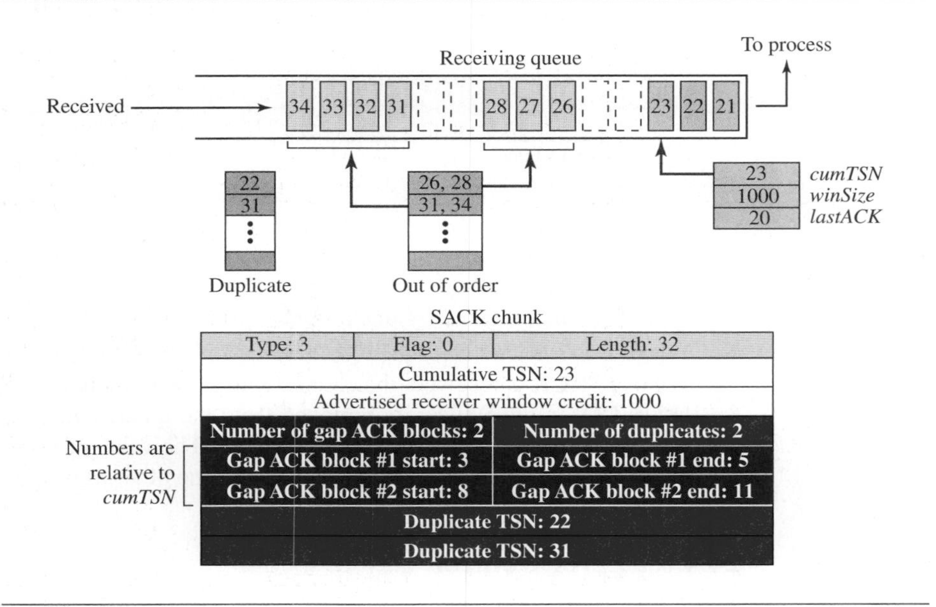

Figure 24.50 *Error control, sender site*

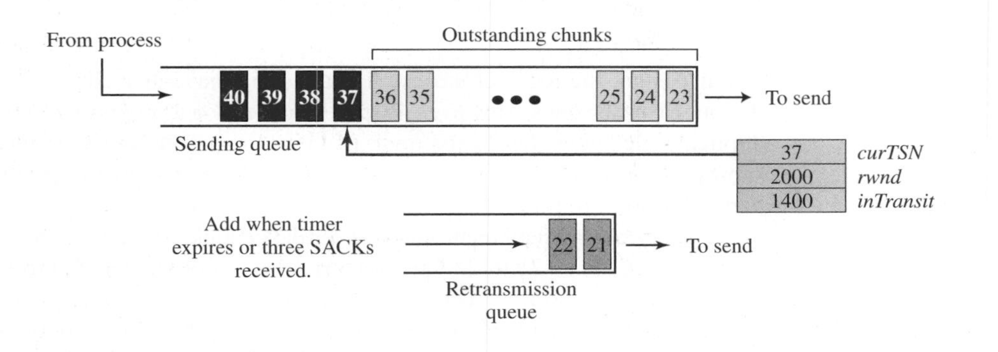

The sending queue holds chunks 23 to 40. The chunks 23 to 36 have already been sent, but not acknowledged; they are outstanding chunks. The *curTSN* points to the next chunk to be sent (37). We assume that each chunk is 100 bytes, which means that 1400 bytes of data (chunks 23 to 36) are in transit. The sender at this moment has a retransmission queue. When a packet is sent, a retransmission timer starts for that packet. Some implementations use a single timer for the entire association, but we continue with our tradition of one timer for each packet for simplification. When the retransmission timer for a packet expires, or three SACKs arrive that declare a packet as missing (fast retransmission was discussed for TCP), the chunks in that packet are moved to the retransmission queue to be resent. These chunks are considered lost, rather than outstanding. The chunks in the retransmission queue have priority. In other words, the next time the sender sends a chunk, it would be chunk 21 from the retransmission queue.

To see how the state of the sender changes, assume that the SACK in Figure 24.49 arrives at the sender site in Figure 24.50. Figure 24.51 shows the new state.

Figure 24.51 *New state at the sender site after receiving a SACK chunk*

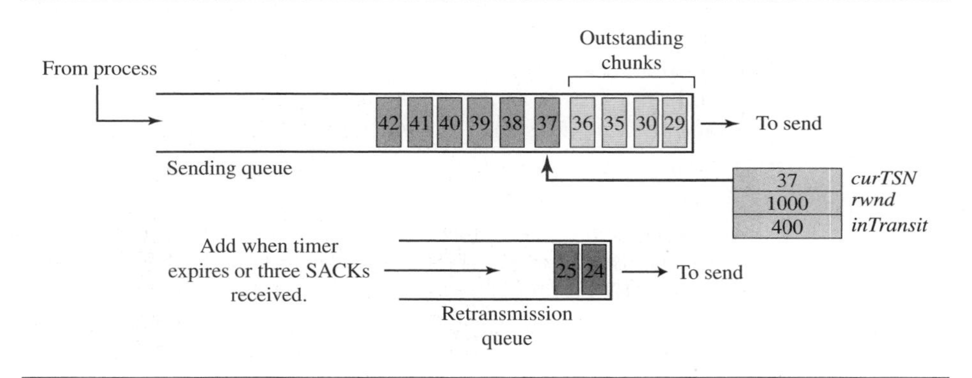

1. All chunks having a TSN equal to or less than the *cumTSN* in the SACK are removed from the sending or retransmission queue. They are no longer outstanding or marked for retransmission. Chunks 21 and 22 are removed from the retransmission queue and 23 is removed from the sending queue.
2. Our design also removes all chunks from the sending queue that are declared in the gap blocks; some conservative implementations, however, save these chunks until a *cumTSN* arrives that includes them. This precaution is needed for the rare occasion when the receiver finds some problem with these out-of-order chunks. We ignore these rare occasions. Chunks 26 to 28 and chunks 31 to 34, therefore, are removed from the sending queue.
3. The list of duplicate chunks does not have any effect.
4. The value of *rwnd* is changed to 1000 as advertised in the SACK chunk.
5. We also assume that the transmission timer for the packet that carried chunks 24 and 25 has expired. These move to the retransmission queue and a new retransmission timer is set according to the exponential backoff rule discussed for TCP.
6. The value of *inTransit* becomes 400 because only 4 chunks are now in transit. The chunks in the retransmission queue are not counted because they are assumed lost, not in transit.

Sending Data Chunks

An end can send a data packet whenever there are data chunks in the sending queue with a TSN greater than or equal to *curTSN* or if there are data chunks in the retransmission queue. The retransmission queue has priority. However, the total size of the data chunk or chunks included in the packet must not exceed the (*rwnd* – *inTransit*) value and the total size of the frame must not exceed the MTU size, as we discussed in previous sections. If we assume, in our previous scenario, that our packet can take 3 chunks (due to the MTU restriction), then chunks 24 and 25 from the retransmission queue and chunk 37, the next chunk ready to be sent in the sending queue, can be sent. Note that the outstanding chunks in the sending queue cannot be sent; they are assumed to be in transit. Note also that any chunk sent from the retransmission queue is also timed for retransmission again. The new timer affects chunks 24, 25, and 37. We need to mention here that some implementations may not allow mixing chunks from the retransmission queue and the sending queue. In this case, only chunks 24 and 25 can be sent in the packet. (The format of the data chunk is on the book website.)

Retransmission

To control a lost or discarded chunk, SCTP, like TCP, employs two strategies: using retransmission timers and receiving three SACKs with the same missing chunks.

- ***Retransmission.*** SCTP uses a retransmission timer, which handles the retransmission time, the waiting time for an acknowledgment of a segment. The procedures for calculating RTO and RTT in SCTP are the same as we described for TCP. SCTP uses a measured RTT (RTTM), a smoothed RTT (RTTS), and an RTT deviation (RTTD) to calculate the RTO. SCTP also uses Karn's algorithm to avoid acknowledgment ambiguity. Note that if a host is using more than one IP address (multihoming), separate RTOs must be calculated and kept for each path.
- ***Four Missing Reports.*** Whenever a sender receives four SACKs whose gap ACK information indicates one or more specific data chunks are missing, the sender needs to consider those chunks as lost and immediately move them to the retransmission queue. This behavior is analogous to "fast retransmission" in TCP.

Generating SACK Chunks

Another issue in error control is the generation of SACK chunks. The rules for generating SCTP SACK chunks are similar to the rules used for acknowledgment with the TCP ACK flag. We summarize the rules as listed below.

1. When an end sends a DATA chunk to the other end, it must include a SACK chunk advertising the receipt of unacknowledged DATA chunks.
2. When an end receives a packet containing data, but has no data to send, it needs to acknowledge the receipt of the packet within a specified time (usually 500 ms).
3. An end must send at least one SACK for every other packet it receives. This rule overrides the second rule.
4. When a packet arrives with out-of-order data chunks, the receiver needs to immediately send a SACK chunk reporting the situation to the sender.
5. When an end receives a packet with duplicate DATA chunks and no new DATA chunks, the duplicate data chunks must be reported immediately with a SACK chunk.

Congestion Control

SCTP, like TCP, is a transport-layer protocol with packets subject to congestion in the network. The SCTP designers have used the same strategies for congestion control as those used in TCP.

24.5 END-CHAPTER MATERIALS

24.5.1 Recommended Reading

For more details about subjects discussed in this chapter, we recommend the following books and RFCs.

Books

Several books give information about transport-layer protocols. The items enclosed in brackets refer to the reference list at the end of the book. In particular, we recommend [Com 06], [PD 03], [GW 04], [Far 04], [Tan 03], and [Sta 04].

RFCs

The main RFC related to UDP is RFC 768. Several RFCs discuss TCP, including 793, 813, 879, 889, 896, 1122, 1975, 1987, 1988, 1993, 2018, 2581, 3168, and 3782.

24.5.2 Key Terms

additive increase, multiplicative decrease (AIMD)
association
Clark's solution
congestion-avoidance
cookie
deadlock
denial of service attack
fast-recovery
fast retransmission
fragmentation
half-close
initial sequence number (ISN)
initial TSN
initiation tag
Karn's algorithm
keepalive timer
multihoming service
multistream service
Nagle's algorithm
persistence timer
primary address
retransmission time-out (RTO)
round-trip time (RTT)
silly window syndrome
slow-start algorithm
socket address
Stream Control Transmission Protocol (SCTP)
stream identifier (SI)
stream sequence number (SSN)
SYN flooding attack
three-way handshaking
Transmission Control Protocol (TCP)
transmission sequence number (TSN)
user datagram
User Datagram Protocol (UDP)
verification tag (VT)

24.5.3 Summary

UDP is an unreliable and connectionless transport-layer protocol that creates a process-to-process communication, which means it requires little overhead and offers fast delivery. The UDP packet is called a *user datagram.* UDP has no flow- or error-control mechanism; its only attempt at error control is the checksum. A user datagram is encapsulated in the data field of an IP datagram. UDP uses multiplexing and demultiplexing to handle outgoing and incoming user datagrams.

Transmission Control Protocol (TCP) is another transport-layer protocol in the TCP/IP protocol suite. It provides process-to-process, full-duplex, and connection-oriented service. A TCP connection consists of three phases: connection establishment, data transfer, and connection termination. TCP software is normally implemented as a finite state machine (FSM). TCP uses flow control, implemented as a sliding window mechanism, to avoid overwhelming a receiver with data. The TCP window size is determined by the receiver-advertised window size *(rwnd)* or the congestion window size *(cwnd),* whichever is smaller. The bytes of data being transferred in each connection are numbered by TCP. The numbering starts with a randomly generated number. TCP uses error control to provide a reliable service. Error control is handled by checksums, acknowledgment, and time-outs. In modern implementations, a retransmission occurs if the retransmission timer expires or three duplicate ACK segments have arrived. TCP uses congestion control to avoid and detect congestion in the network. The slow-start (exponential increase), congestion-avoidance (additive increase), and congestion-detection (multiplicative decrease) strategies are used for congestion control.

SCTP is a message-oriented, reliable protocol that combines the good features of UDP and TCP. SCTP provides additional services not provided by UDP or TCP, such as multiple-stream and multihoming services. SCTP is a connection-oriented protocol, in which a connection is called an *association.* SCTP provides flow control, error control, and congestion control. To distinguish between different streams, SCTP uses the sequence identifier (SI). To distinguish between different data chunks belonging to the same stream, SCTP uses the stream sequence number (SSN). The SCTP acknowledgment SACK reports the cumulative TSN, the TSN of the last data chunk received in order, and selective TSNs that have been received.

24.6 PRACTICE SET

24.6.1 Quizzes

A set of interactive quizzes for this chapter can be found on the book website. It is strongly recommended that the student take the quizzes to check his/her understanding of the materials before continuing with the practice set.

24.6.2 Questions

Q24-1. Assume a TCP server is missing bytes 2001 to 3000. The server receives a segment with sequence number 2001 that carries 400 bytes. What is the reaction of the TCP server to this event? Can you justify the reaction?

Q24-2. Can you explain why we need four (or sometimes three) segments for connection termination in TCP?

Q24-3. We said that TCP provides a connection-oriented service between the two application programs. A connection in this case needs a connection identifier that distinguishes one connection from another. What do you think the unique connection identifier is in this case?

Q24-4. In TCP, how do we define the sequence number of a segment (in each direction)? Consider two cases: the first segment and other segments.

Q24-5. In a TCP segment, what does an acknowledgment number identify?

Q24-6. A client residing on a host with IP address 122.45.12.7 sends a message to the corresponding server residing on a host with IP address 200.112.45.90. If the well-known port is 161 and the ephemeral port is 51000, what are the pair of socket addresses used in this communication?

Q24-7. Assume a private internet uses a protocol suite totally different from the TCP/IP protocol suite. Can this internet still use the services of UDP or TCP as an end-to-end vehicle of message communication?

Q24-8. Is the use of checksum for error control optional or mandatory in

a. UDP? **b.** TCP?

Q24-9. Assume a TCP server expects to receive byte 2001, but it receives a segment with sequence number 2200. What is the reaction of the TCP server to this event? Can you justify the reaction?

Q24-10. Can you explain how TCP, which uses the services provided by the unreliable IP, can provide reliable communication?

Q24-11. Some of the application programs can use the services of two transport-layer protocols (UDP or TCP). When a packet arrives at the destination, how can the computer find which transport layer is involved?

Q24-12. Assume Alice uses her browser to open two connections to the HTTP server running on Bob's server. How can these two connections be distinguished by the TCP?

Q24-13. We used the terms *passive open* and *active open* in discussing a connection-oriented communication using TCP. Assume there is a telephone conversation between Alice and Bob. Since a telephone conversation is an example of a connection-oriented communication, assume Alice calls Bob and they talk on the telephone. Who is making a *passive open* connection in this case? Who is making an *active open* connection in this case?

Q24-14. What is the maximum size of the TCP header? What is the minimum size of the TCP header?

Q24-15. Can you explain why in TCP a SYN, SYN + ACK, and FIN segment each consume a sequence number, but an ACK segment carrying no data does not consume a sequence number?

Q24-16. Looking at the TCP header (Figure 24.7), we find that the sequence number is 32 bits long, while the window size is only 16 bits long. Does this mean that TCP is closer to the Go-Back-*N* or Selective-Repeat protocol in this respect?

Q24-17. The first rule of generating ACKs for TCP is not shown in Figure 24.22 or 24.23. Can you explain the reason for this?

Q24-18. Most of the flags can be used together in a segment. Give an example of two flags that cannot be used simultaneously because they are ambiguous.

Q24-19. In SCTP, a packet is carrying a COOKIE ECHO message and a DATA chunk. If the size of the cookie is 200 bytes and that of the user data is 20 bytes, what is the size of the packet?

Q24-20. Assume a private internet, which uses point-to-point communication between the hosts and needs no routing, has totally eliminated the use of the network layer. Can this internet still benefit from the services of UDP or TCP? In other words, can user datagrams or segments be encapsulated in the Ethernet frames?

Q24-21. Which of the six rules we described for ACK generation in TCP can be applied to the case where a client receives a SYN + ACK segment from a server?

Q24-22. Which of the six rules we described for ACK generation in TCP can be applied to the case where a server receives a SYN segment from a client?

Q24-23. In TCP, we have two consecutive segments. Assume the sequence number of the first segment is 101. What is the sequence number of the next segment in each of the following cases?

a. The first segment does not consume any sequence numbers.

b. The first segment consumes 10 sequence numbers.

Q24-24. Which of the six rules we described for ACK generation can be applied to the case where a client or a server receives a FIN segment from the other end?

Q24-25. In TCP, what type of flag can totally close the communication in both directions?

Q24-26. In a TCP segment, what does a sequence number identify?

Q24-27. Assume a TCP client is expecting to receive byte 3001. It receives a segment with the sequence number 3001 that carries 400 bytes. If the client has no data to be sent at this moment and has acknowledged the previous segment, what is the reaction of the TCP client to this event? Can you justify the reaction?

Q24-28. In TCP, does a FIN segment close a connection in only one direction or in both directions?

Q24-29. In SCTP, a packet is carrying two DATA chunks, each containing 22 bytes of user data. What is the size of each DATA chunk? What is the total size of the packet?

Q24-30. Assume a TCP server is expecting to receive byte 2401. It receives a segment with the sequence number 2401 that carries 500 bytes. If the server has no data to send at this moment and has not acknowledged the previous segment, what is the reaction of the TCP server to this event? Can you justify the reaction?

Q24-31. In TCP, does a SYN segment open a connection in only one direction or in both directions?

Q24-32. Assume a TCP server is expecting to receive byte 6001. It receives a segment with the sequence number 6001 that carries 2000 bytes. If the server has bytes 4001 to 5000 to send, what should the reaction of the TCP server be to this event? Can you justify the reaction?

Q24-33. Can you mention some tasks that can be done by one or a combination of TCP segments?

Q24-34. Assume a TCP client expects to receive byte 2001, but it receives a segment with sequence number 1201. What is the reaction of the TCP client to this event? Can you justify the reaction?

Q24-35. The maximum window size of the TCP was originally designed to be 64 KB (which means $64 \times 1024 = 65{,}536$ or actually 65,535). Can you think of a reason for this?

Q24-36. In TCP, how many sequence numbers are consumed by each of the following segments?

a. SYN **b.** ACK **c.** SYN + ACK **d.** Data

Q24-37. UDP is a message-oriented protocol. TCP is a byte-oriented protocol. If an application needs to protect the boundaries of its message, which protocol should be used, UDP or TCP?

Q24-38. In SCTP, a SACK chunk reports the receipt of three out-of-order data chunks and five duplicate data chunks. What is the total size of the chunk in bytes?

Q24-39. Assume a client sends a SYN segment to a server. When the server checks the well-known port number, it finds that no process defined by the port number is running. What is the server supposed to do in this case?

Q24-40. In SCTP, a packet is carrying a COOKIE ACK message and a DATA chunk. If the user data is 20 bytes, what is the size of the packet?

Q24-41. In TCP, some segment types can be used only for control; they cannot be used to carry data at the same time. Can you define some of these segments?

Q24-42. In TCP, can the sender window be smaller, larger, or the same size as the receiver window?

Q24-43. In SCTP, the value of the cumulative TSN in a SACK is 23. The value of the previous cumulative TSN in the SACK was 29. What is the problem?

24.6.3 Problems

P24-1. A client uses TCP to send data to a server. The data consist of 16 bytes. Calculate the efficiency of this transmission at the TCP level (ratio of useful bytes to total bytes).

P24-2. An SCTP client opens an association using an initial tag of 806, an initial TSN of 14534, and a window size of 20,000. The server responds with an initial tag of 2000, an initial TSN of 670, and a window size of 14,000. Show the contents of all four packets exchanged during association establishment. Ignore the value of the cookie.

P24-3. Answer the following questions:

a. What is the minimum size of a UDP user datagram?

b. What is the maximum size of a UDP user datagram?

c. What is the minimum size of the application-layer payload data that can be encapsulated in a UDP user datagram?

d. What is the maximum size of the application-layer payload that can be encapsulated in a UDP user datagram?

P24-4. In TCP, assume a client has 100 bytes to send. The client creates 10 bytes at a time in each 10 ms and delivers them to the transport layer. The server acknowledges each segment immediately or if a timer times out at 50 ms. Show the segments and the bytes each segment carries if the implementation uses Nagle's algorithm with maximum segment size (MSS) of 30 bytes. The round-trip time is 20 ms, but the sender timer is set to 100 ms. Does any segment carry the maximum segment size? Is Nagle's algorithm really effective here? Why?

P24-5. If originally RTT_S = 14 ms and α is set to 0.2, calculate the new RTT_S after the following events (times are relative to event 1):

Event 1: 00 ms	Segment 1 was sent.
Event 2: 06 ms	Segment 2 was sent.
Event 3: 16 ms	Segment 1 was timed-out and resent.
Event 4: 21 ms	Segment 1 was acknowledged.
Event 5: 23 ms	Segment 2 was acknowledged.

P24-6. As we have explained in the text, the TCP sliding window, when used without new SACK options, is a combination of the Go-Back-*N* and the Selective-Repeat protocols. Explain which aspects of the TCP sliding window are close to the Go-Back-*N* and which aspects are close to the Selective-Repeat protocol.

P24-7. Although new TCP implementations use the SACK option to report the out of order and duplicate range of bytes, explain how old implementations can indicate that the bytes in a received segment are out of order or duplicate.

P24-8. In SCTP, the following is a dump of a DATA chunk in hexadecimal format.

```
00000015 00000005 0003000A 00000000 48656C6C 6F000000
```

a. Is this an ordered or unordered chunk?

b. Is this the first, the last, the middle, or the only fragment?

c. How many bytes of padding are carried by the chunk?

d. What is the TSN?

e. What is the SI?

f. What is the SSN?

g. What is the message?

P24-9. For a clearer view of Nagle's algorithm, let us repeat Problem P24-4, but let the server transport layer acknowledge a segment when there is a previous segment that has not been acknowledged (every other segment) or a timer times out after 60 ms. Show the time line for this scenario.

P24-10. Assume Alice, the client, creates a connection with Bob, the server. They exchange data and close the connection. Now Alice starts a new connection with Bob by sending a new SYN segment. Before Bob responds to this SYN segment, a duplicate copy of the old SYN segment from Alice, which is wandering in the network, arrives at Bob's computer, initiating a SYN + ACK segment from Bob. Can this segment be mistaken by Alice's computer as the response to the new SYN segment? Explain.

P24-11. Eve, the intruder, sends a user datagram to Bob, the server, using Alice's IP address. Can Eve, pretending to be Alice, receive a response from Bob?

P24-12. Eve, the intruder, sends a SYN segment to Bob, the server, using Alice's IP address. Can Eve create a TCP connection with Bob by pretending that she is Alice? Assume that Bob uses a different ISN for each connection.

P24-13. The *ssthresh* value for a Reno TCP station is set to 8 MSS. The station is now in the slow-start state with *cwnd* = 5 MSS and *ssthresh* = 8 MSS. Show the values of *cwnd, ssthresh,* and the current and the next state of the station after the following events: three consecutive nonduplicate ACKs arrived, followed by five duplicate ACKs, followed by two nonduplicate ACKs, and followed by a time-out.

P24-14. Distinguish between a time-out event and the three-duplicate-ACKs event. Which one is a stronger sign of congestion in the network? Why?

P24-15. In a TCP connection, assume that maximum segment size (MSS) is 1000 bytes. The client process has 5400 bytes to send to the server process, which has no bytes to respond (unidirectional communication). The TCP server generates ACKs according to the rules we discussed in the text. Show the time line for the transactions during the slow start phase, indicating the value of *cwnd* at the beginning, at the end, and after each change. Assume that each segment header is only 20 bytes.

P24-16. Using Figure 24.19, explain how flow control can be achieved at the sender site in TCP (from the sending TCP to the sending application). Draw a representation.

P24-17. In SCTP, the state of a sender is as follows:

a. The sending queue has chunks 18 to 23.

b. The value of *curTSN* is 20.

c. The value of the window size is 2000 bytes.

d. The value of *inTransit* is 200.

If each data chunk contains 100 bytes of data, how many DATA chunks can be sent now? What is the next data chunk to be sent?

P24-18. To make the initial sequence number a random number, most systems start the counter at 1 during bootstrap and increment the counter by 64,000 every half second. How long does it take for the counter to wrap around?

P24-19. The following is a dump (contents) of a UDP header in hexadecimal format.

0045DF0000580000

a. What is the source port number?

b. What is the destination port number?

c. What is the total length of the user datagram?

d. What is the length of the data?

e. Is the packet directed from a client to a server or vice versa?

f. What is the application-layer protocol?

g. Has the sender calculated a checksum for this packet?

P24-20. Compare the TCP header and the UDP header. List the fields in the TCP header that are not part of the UDP header. Give the reason for each missing field.

P24-21. The following is a dump of an SCTP general header in hexadecimal format.

04320017 00000001 00000000

a. What is the source port number?

b. What is the destination port number?

c. What is the value of the verification tag?

d. What is the value of the checksum?

P24-22. We discuss the new option of SACKs for TCP on the book website, but for the moment assume that we add an 8-byte NAK option to the end of the TCP segment, which can hold two 32-bit sequence numbers. Show how we can use this 8-byte NAK to define an out-of-order or duplicate range of bytes received.

P24-23. Figure 24.15 shows the client and server in the transition diagram for the common scenario using a four-handshake closing. Change the diagram to show the three-handshake closing.

P24-24. In a TCP connection, the window size fluctuates between 60,000 bytes and 30,000 bytes. If the average RTT is 30 ms, what is the throughput of the connection?

P24-25. Using Figure 24.19, explain how flow control can be achieved at the receiver site in TCP (from the receiving TCP to the receiving application).

P24-26. The *ssthresh* value for a Taho TCP station is set to 6 MSS. The station now is in the slow-start state with *cwnd* = 4 MSS. Show the values of *cwnd*, *sstresh,* and the state of the station before and after each of following events: four consecutive nonduplicate ACKs arrived followed by a time-out, and followed by three nonduplicate ACKs.

P24-27. Assume Alice, the client, creates a TCP connection with Bob, the server. They exchange data and close the connection. Now Alice starts a new connection with Bob by sending a new SYN segment. The server responds with the SYN + ACK segment. However, before Bob receives the ACK for this connection from Alice, a duplicate old ACK segment from Alice arrives at Bob's site. Can this old ACK be confused with the ACK segment Bob is expecting from Alice?

P24-28. An SCTP association is in the **ESTABLISHED** state. It receives a SHUTDOWN chunk. If the host does not have any outstanding or pending data, what does it need to do?

P24-29. An HTTP client opens a TCP connection using an initial sequence number (ISN) of 14,534 and the ephemeral port number of 59,100. The server opens the connection with an ISN of 21,732. Show the three TCP segments during the connection establishment if the client defines the *rwnd* of 4000 and the server defines the *rwnd* of 5000. Ignore the calculation of the checksum field.

P24-30. In SCTP, the state of a receiver is as follows:

a. The receiving queue has chunks 1 to 8, 11 to 14, and 16 to 20.

b. There are 1800 bytes of space in the queue.

c. The value of *lastAck* is 4.

d. No duplicate chunk has been received.

e. The value of *cumTSN* is 5.

Show the contents of the receiving queue and the variables. Show the contents of the SACK message sent by the receiver.

P24-31. The control field in a TCP segment is 6 bits. We can have 64 different combinations of bits. List some combinations that you think are normally used.

P24-32. The following is part of a TCP header dump (contents) in hexadecimal format.

E293 0017 00000001 00000000 5002 07FF...

a. What is the source port number?

b. What is the destination port number?

c. What is the sequence number?

d. What is the acknowledgment number?

e. What is the length of the header?

f. What is the type of the segment?

g. What is the window size?

P24-33. In TCP, if the value of HLEN is 0111, how many bytes of options are included in the segment?

P24-34. A client uses UDP to send data to a server. The data length is 16 bytes. Calculate the efficiency of this transmission at the UDP level (ratio of useful bytes to total bytes).

P24-35. In a TCP connection, the initial sequence number at the client site is 2171. The client opens the connection, sends three segments, the second of which carries 1000 bytes of data, and closes the connection. What is the value of the sequence number in each of the following segments sent by the client?

a. The SYN segment

b. The data segment

c. The FIN segment

P24-36. TCP is sending data at 1 megabyte per second. If the sequence number starts with 7000, how long does it take before the sequence number goes back to zero?

P24-37. An SCTP association is in the **COOKIE-WAIT** state. It receives an INIT chunk; what does it need to do?

P24-38. Assume the HTTP client in Problem P24-29 sends a request of 100 bytes. The server responds with a segment of 1200 bytes. Show the contents of the two segments exchanged between the client and the sender. Assume the acknowledgment of the response will be done later by the client (ignore calculation of checksum field).

P24-39. Assume the HTTP client in the Problem P24-38 closes the connection and, at the same time, acknowledges the bytes received in the response from the server. After receiving the FIN segment from the client, the server also closes the connection in the other direction. Show the connection termination phase.

P24-40. In a connection, the value of *cwnd* is 3000 and the value of *rwnd* is 5000. The host has sent 2000 bytes, which have not been acknowledged. How many more bytes can be sent?

P24-41. To better understand the need for the three-handshake connection establishment, let us go through a scenario. Alice and Bob have no access to telephones or the Internet (think about the old days) to establish their next meeting at a place far from their homes.

a. Suppose that Alice sends a letter to Bob and defines the day and the time of their meeting. Can Alice go to the meeting place and be sure that Bob is there?

b. Suppose that Bob responds to Alice's request with a letter and confirms the date and time. Can Bob go to the meeting place and be sure that Alice is there?

c. Suppose that Alice responds to Bob's letter and confirms the same date and time. Can either one go to the meeting and be sure that the other person is there?

P24-42. What can you say about each of the following TCP segments, in which the value of the control field is:

a. 000000 **b.** 000001 **c.** 010001
d. 000100 **e.** 000010 **f.** 010010

P24-43. If the client in the Problem P24-2 sends 7600 data chunks and the server sends 570 data chunks, show the contents of the three packets exchanged during association termination.

PART VI

Application Layer

In the sixth part of the book, we discuss the application layer. Chapter 25 is an introduction that also discusses network programming in both C and Java languages. Chapter 26 covers the standard applications we use every day. Chapter 27 is devoted to network management. Multimedia has received a separate chapter because it becomes more and more important in the Internet today. Finally, we have devoted one chapter to the new paradigm, P2P, which is becoming more popular.

CHAPTER 25

Introduction to Application Layer

The whole Internet, hardware and software, was designed and developed to provide services at the application layer. The fifth layer of the TCP/IP protocol suite is where these services are provided for Internet users. The other four layers are there to make these services possible.

This chapter is divided into four sections:

- The first section introduces the application layer. The section first describes the services provided by the application layer. It then explains that there are two paradigms in which hosts in the Internet can exchange services: the client-server paradigm and the peer-to-peer paradigm.
- The second section introduces client-server programming, the way in which the client-server paradigm can be implemented. It first describes application programming interfaces. It then shows how application programs use the services of the transport layer. It finally defines iterative communication using UDP or TCP.
- The third section gives a brief discussion of socket programming in C. After discussing general issues, it introduces how iterative client-server programs can be written in this language.
- The fourth section gives a brief discussion of socket programming in Java. After discussing general issues, it introduces how iterative client-server programs can be written in this language.

25.1 INTRODUCTION

The application layer provides services to the user. Communication is provided using a logical connection, which means that the two application layers assume that there is an imaginary direct connection through which they can send and receive messages. Figure 25.1 shows the idea behind this logical connection.

Figure 25.1 *Logical connection at the application layer*

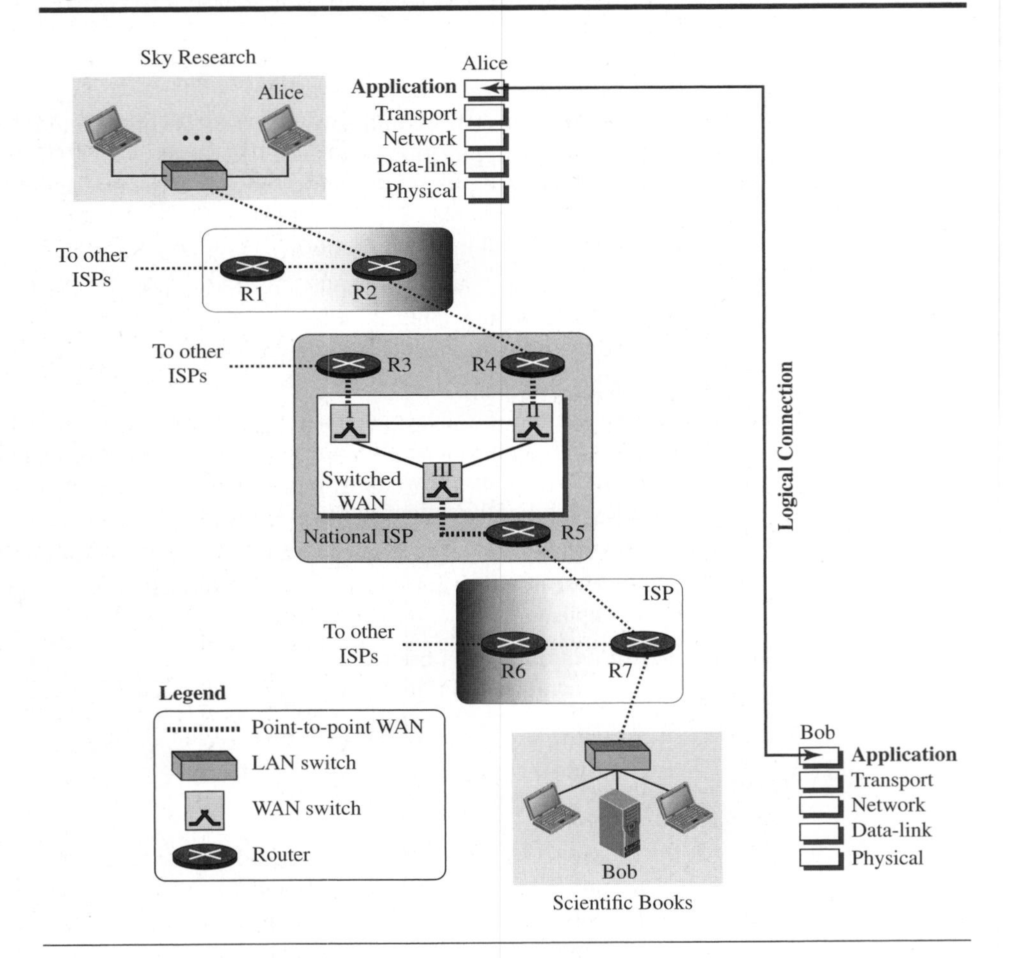

The figure shows the same scenario we have seen for other layers, but this time the logical connection is between two application layers. A scientist working in a research company, Sky Research, needs to order a book related to her research from an online bookseller, Scientific Books. Logical connection takes place between the application layer of a computer at Sky Research and the application layer of a server at Scientific Books. We call the first host Alice and the second one Bob. The communication at the application layer is logical, not physical. Alice and Bob assume that there is a two-way logical channel between them through which they can send and receive messages. The actual communication, however, takes place through several devices (Alice, R2, R4, R5, R7, and Bob) and several physical channels, as shown in the figure.

25.1.1 Providing Services

All communication networks that started before the Internet were designed to provide services to network users. Most of these networks, however, were originally designed to provide one specific service. For example, the telephone network was originally designed to provide voice service: to allow people all over the world to talk to each other. This network, however, was later used for some other services, such as facsimile (fax), enabled by users adding some extra hardware at both ends.

The Internet was originally designed for the same purpose: to provide service to users around the world. The layered architecture of the TCP/IP protocol suite, however, makes the Internet more flexible than other communication networks such as postal or telephone networks. Each layer in the suite was originally made up of one or more protocols, but new protocols can be added or some protocols can be removed or replaced by the Internet authorities. However, if a protocol is added to each layer, it should be designed in such a way that it uses the services provided by one of the protocols at the lower layer. If a protocol is removed from a layer, care should be taken to change the protocol at the next higher layer that supposedly uses the services of the removed protocol.

The application layer, however, is somewhat different from other layers in that it is the highest layer in the suite. The protocols in this layer do not provide services to any other protocol in the suite; they only receive services from the protocols in the transport layer. This means that protocols can be removed from this layer easily. New protocols can be also added to this layer as long as the new protocols can use the services provided by one of the transport-layer protocols.

Since the application layer is the only layer that provides services to the Internet user, the flexibility of the application layer, as described above, allows new application protocols to be easily added to the Internet, which has been occurring during the lifetime of the Internet. When the Internet was created, only a few application protocols were available to the users; today we cannot give a number for these protocols because new ones are being added constantly.

Standard and Nonstandard Protocols

To provide smooth operation of the Internet, the protocols used in the first four layers of the TCP/IP suite need to be standardized and documented. They normally become part of the package that is included in operating systems such as Windows or UNIX. To be flexible, however, the application-layer protocols can be both standard and nonstandard.

Standard Application-Layer Protocols

There are several application-layer protocols that have been standardized and documented by the Internet authority, and we are using them in our daily interaction with the Internet. Each standard protocol is a pair of computer programs that interact with the user and the transport layer to provide a specific service to the user. We will discuss some of these standard applications in Chapter 26. In the case of these application protocols, we should know what types of services they provide, how they work, the options that we can use with these applications, and so on. The study of these protocols enables a network manager to easily solve the problems that may occur when using these protocols. The deep understanding of how these protocols work will also give us some ideas about how to create new nonstandard protocols.

Nonstandard Application-Layer Protocols

A programmer can create a nonstandard application-layer program if she can write two programs that provide service to the user by interacting with the transport layer. Later in this chapter, we show how we can write these types of programs. It is the creation of a nonstandard (proprietary) protocol, which does not even need the approval of the Internet authorities if privately used, that has made the Internet so popular worldwide. A private company can create a new customized application protocol to communicate with all of its offices around the world using the services provided by the first four layers of the TCP/IP protocol suite without using any of the standard application programs. What is needed is to write programs, in one of the computer languages, that use the available services provided by the transport-layer protocols.

25.1.2 Application-Layer Paradigms

It should be clear that to use the Internet we need two application programs to interact with each other: one running on a computer somewhere in the world, the other running on another computer somewhere else in the world. The two programs need to send messages to each other through the Internet infrastructure. However, we have not discussed what the relationship should be between these programs. Should both application programs be able to request services and provide services, or should the application programs just do one or the other? Two paradigms have been developed during the lifetime of the Internet to answer this question: the *client-server paradigm* and the *peer-to-peer paradigm.* We briefly introduce these two paradigms here, but we discuss the first one later in this chapter and the second in Chapter 29.

Traditional Paradigm: Client-Server

The traditional paradigm is called the **client-server paradigm**. It was the most popular paradigm until a few years ago. In this paradigm, the service provider is an application program, called the server process; it runs continuously, waiting for another application program, called the client process, to make a connection through the Internet and ask for service. There are normally some server processes that can provide a specific type of service, but there are many clients that request service from any of these server processes. The server process must be running all the time; the client process is started when the client needs to receive service.

The client-server paradigm is similar to some available services out of the territory of the Internet. For example, a telephone directory center in any area can be thought of as a server; a subscriber that calls and asks for a specific telephone number can be thought of as a client. The directory center must be ready and available all the time; the subscriber can call the center for a short period when the service is needed.

Although the communication in the client-server paradigm is between two application programs, the role of each program is totally different. In other words, we cannot run a client program as a server program or vice versa. Later in this chapter, when we talk about client-server programming in this paradigm, we show that we always need to write two application programs for each type of service. Figure 25.2 shows an example of a client-server communication in which three clients communicate with one server to receive the services provided by this server.

Figure 25.2 *Example of a client-server paradigm*

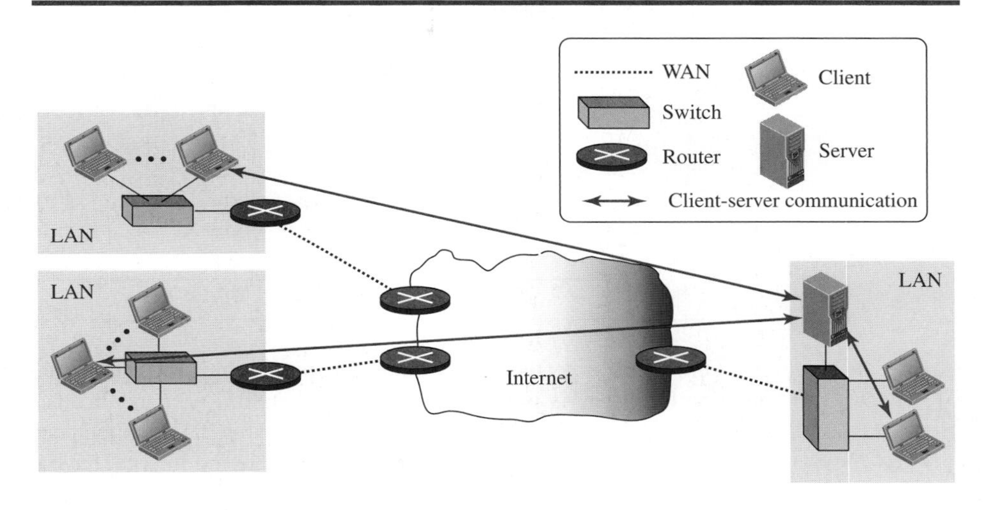

One problem with this paradigm is that the concentration of the communication load is on the shoulder of the server, which means the server should be a powerful computer. Even a powerful computer may become overwhelmed if a large number of clients try to connect to the server at the same time. Another problem is that there should be a service provider willing to accept the cost and create a powerful server for a specific service, which means the service must always return some type of income for the server in order to encourage such an arrangement.

Several traditional services are still using this paradigm, including the World Wide Web (WWW) and its vehicle HyperText Transfer Protocol (HTTP), file transfer protocol (FTP), secure shell (SSH), e-mail, and so on. We discuss some of these protocols and applications later in the chapter.

New Paradigm: Peer-to-Peer

A new paradigm, called the **peer-to-peer paradigm** (often abbreviated *P2P paradigm*) has emerged to respond to the needs of some new applications. In this paradigm, there is no need for a server process to be running all the time and waiting for the client processes to connect. The responsibility is shared between peers. A computer connected to the Internet can provide service at one time and receive service at another time. A computer can even provide and receive services at the same time. Figure 25.3 shows an example of communication in this paradigm.

Figure 25.3 *Example of a peer-to-peer paradigm*

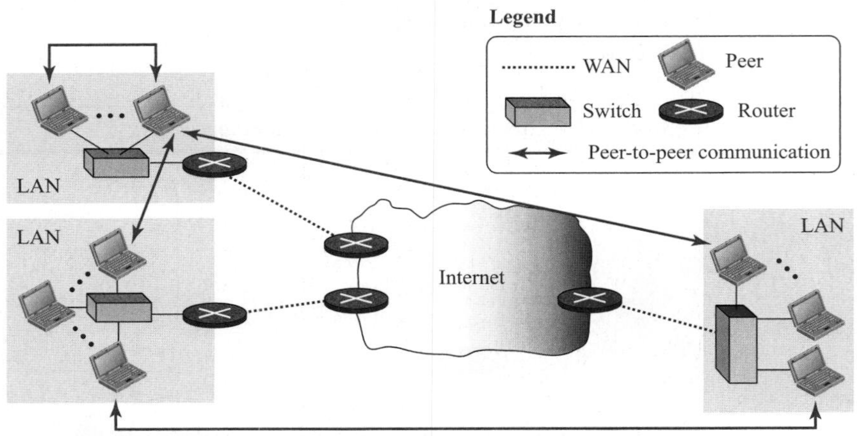

One of the areas that really fits in this paradigm is the Internet telephony. Communication by phone is indeed a peer-to-peer activity; no party needs to be running forever waiting for the other party to call. Another area in which the peer-to-peer paradigm can be used is when some computers connected to the Internet have something to share with each other. For example, if an Internet user has a file available to share with other Internet users, there is no need for the file holder to become a server and run a server process all the time waiting for other users to connect and retrieve the file.

Although the peer-to-peer paradigm has been proved to be easily scalable and cost-effective in eliminating the need for expensive servers to be running and maintained all the time, there are also some challenges. The main challenge has been security; it is more difficult to create secure communication between distributed services than between those controlled by some dedicated servers. The other challenge is applicability; it appears that not all applications can use this new paradigm. For example, not many Internet users are ready to become involved, if one day the Web can be implemented as a peer-to-peer service.

There are some new applications, such as BitTorrent, Skype, IPTV, and Internet telephony, that use this paradigm. We will discuss some of these applications in Chapter 29.

Mixed Paradigm

An application may choose to use a mixture of the two paradigms by combining the advantages of both. For example, a light-load client-server communication can be used to find the address of the peer that can offer a service. When the address of the peer is found, the actual service can be received from the peer by using the peer-to-peer paradigm.

25.2 CLIENT-SERVER PROGRAMMING

In a client-server paradigm, communication at the application layer is between two running application programs called ***processes***: a *client* and a *server*. A client is a running program that initializes the communication by sending a request; a server is another application program that waits for a request from a client. The server handles the request received from a client, prepares a result, and sends the result back to the client. This definition of a server implies that a server must be running when a request from a client arrives, but the client needs to be run only when it is needed. This means that if we have two computers connected to each other somewhere, we can run a client process on one of them and the server on the other. However, we need to be careful that the server program is started before we start running the client program. In other words, the lifetime of a server is infinite: it should be started and run forever, waiting for the clients. The lifetime of a client is finite: it normally sends a finite number of requests to the corresponding server, receives the responses, and stops.

25.2.1 Application Programming Interface

How can a client process communicate with a server process? A computer program is normally written in a computer language with a predefined set of instructions that tells the computer what to do. A computer language has a set of instructions for mathematical operations, a set of instructions for string manipulation, a set of instructions for input/output access, and so on. If we need a process to be able to communicate with another process, we need a new set of instructions to tell the lowest four layers of the TCP/IP suite to open the connection, send and receive data from the other end, and close the connection. A set of instructions of this kind is normally referred to as an **application programming interface (API)**. An interface in programming is a set of instructions between two entities. In this case, one of the entities is the process at the application layer and the other is the *operating system* that encapsulates the first four layers of the TCP/IP protocol suite. In other words, a computer manufacturer needs to build the first four layers of the suite in the operating system and include an API. In this way, the processes running at the application layer are able to communicate with the operating system when sending and receiving messages through the Internet. Several APIs have been designed for communication. Three among them are common: **socket interface, Transport Layer Interface (TLI),** and **STREAM**. In this chapter, we briefly discuss only *socket interface*, the most common one, to give a general idea of network communication at the application layer.

Socket interface started in the early 1980s at UC Berkeley as part of a UNIX environment. The socket interface is a set of instructions that provide communication between the application layer and the operating system, as shown in Figure 25.4. It is a set of instructions that can be used by a process to communicate with another process.

The idea of sockets allows us to use the set of all instructions already designed in a programming language for other sources and sinks. For example, in most computer languages, like C, C++, or Java, we have several instructions that can read and write data to other sources and sinks such as a keyboard (a source), a monitor (a sink), or a file (source and sink). We can use the same instructions to read from or write to sockets. In other words, we are adding only new sources and sinks to the programming language

Figure 25.4 *Position of the socket interface*

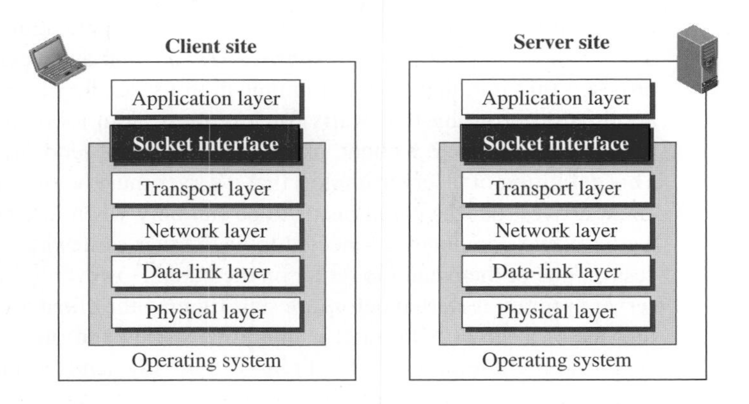

without changing the way we send data or receive data. Figure 25.5 shows the idea and compares the sockets with other sources and sinks.

Figure 25.5 *Sockets used the same way as other sources and sinks*

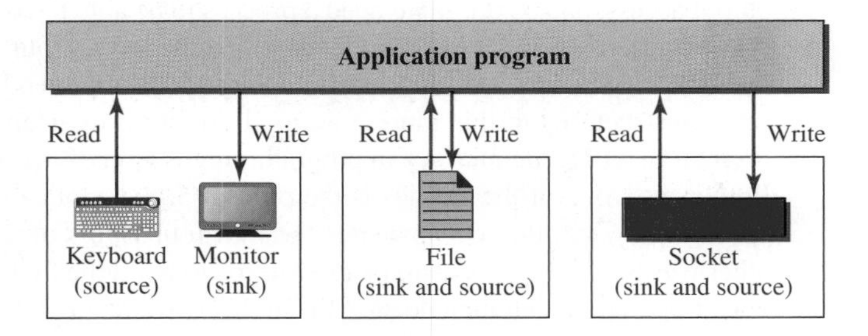

Sockets

Although a socket is supposed to behave like a terminal or a file, it is not a physical entity like them; it is an abstraction. It is an object that is created and used by the application program.

We can say that, as far as the application layer is concerned, communication between a client process and a server process is communication between two sockets, created at two ends, as shown in Figure 25.6. The client thinks that the socket is the entity that receives the request and gives the response; the server thinks that the socket is the one that has a request and needs the response. If we create two sockets, one at each end, and define the source and destination addresses correctly, we can use the available instructions to send and receive data. The rest is the responsibility of the operating system and the embedded TCP/IP protocol.

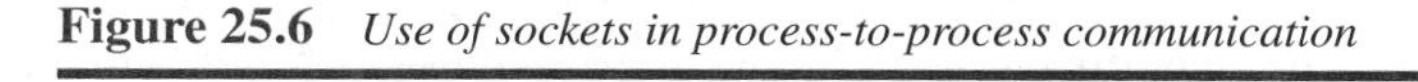

Figure 25.6 *Use of sockets in process-to-process communication*

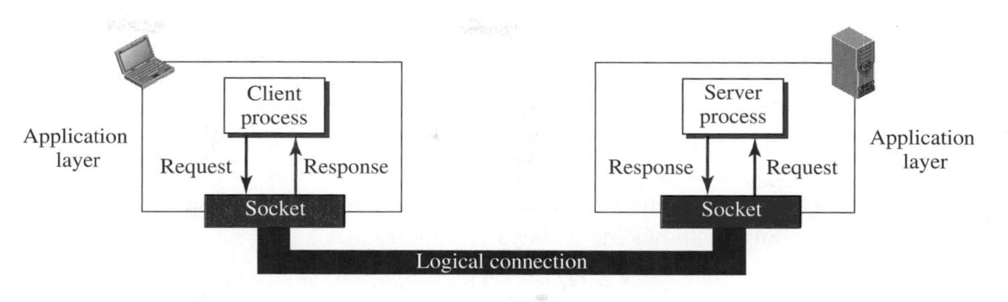

Socket Addresses

The interaction between a client and a server is two-way communication. In a two-way communication, we need a pair of addresses: local (sender) and remote (receiver). The local address in one direction is the remote address in the other direction and vice versa. Since communication in the client-server paradigm is between two sockets, we need a pair of **socket addresses** for communication: a local socket address and a remote socket address. However, we need to define a socket address in terms of identifiers used in the TCP/IP protocol suite.

A socket address should first define the computer on which a client or a server is running. As we discussed in Chapter 18, a computer in the Internet is uniquely defined by its IP address, a 32-bit integer in the current Internet version. However, several client or server processes may be running at the same time on a computer, which means that we need another identifier to define the specific client or server involved in the communication. As we discussed in Chapter 24, an application program can be defined by a port number, a 16-bit integer. This means that a socket address should be a combination of an IP address and a port number as shown in Figure 25.7.

Figure 25.7 *A socket address*

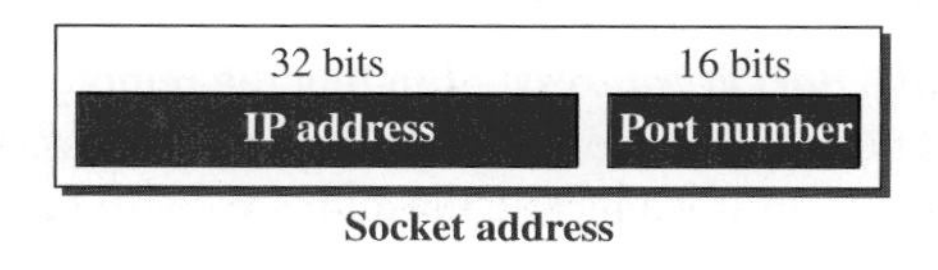

Since a socket defines the end-point of the communication, we can say that a socket is identified by a pair of socket addresses, a local and a remote.

Finding Socket Addresses

How can a client or a server find a pair of socket addresses for communication? The situation is different for each site.

Server Site

The server needs a local (server) and a remote (client) socket address for communication.

Local Socket Address The local (server) socket address is provided by the operating system. The operating system knows the IP address of the computer on which the server process is running. The port number of a server process, however, needs to be assigned. If the server process is a standard one defined by the Internet authority, a port number is already assigned to it. For example, the assigned port number for a Hypertext Transfer Protocol (HTTP) is the integer 80, which cannot be used by any other process. We discussed these well-known port numbers in Chapter 24. If the server process is not standard, the designer of the server process can choose a port number, in the range defined by the Internet authority, and assign it to the process. When a server starts running, it knows the local socket address.

Remote Socket Address The remote socket address for a server is the socket address of the client that makes the connection. Since the server can serve many clients, it does not know beforehand the remote socket address for communication. The server can find this socket address when a client tries to connect to the server. The client socket address, which is contained in the request packet sent to the server, becomes the remote socket address that is used for responding to the client. In other words, although the local socket address for a server is fixed and used during its lifetime, the remote socket address is changed in each interaction with a different client.

Client Site

The client also needs a local (client) and a remote (server) socket address for communication.

Local Socket Address The local (client) socket address is also provided by the operating system. The operating system knows the IP address of the computer on which the client is running. The port number, however, is a 16-bit temporary integer that is assigned to a client process each time the process needs to start the communication. The port number, however, needs to be assigned from a set of integers defined by the Internet authority and called the ephemeral (temporary) port numbers, which we discussed in Chapter 24. The operating system, however, needs to guarantee that the new port number is not used by any other running client process. The operating system needs to remember the port number to be able to redirect the response received from the server process to the client process that sent the request.

Remote Socket Address Finding the remote (server) socket address for a client, however, needs more work. When a client process starts, it should know the socket address of the server it wants to connect to. We will have two situations in this case.

- ❑ Sometimes, the user who starts the client process knows both the server port number and IP address of the computer on which the server is running. This usually occurs in situations when we have written client and server applications and we want to test them. For example, at the end of this chapter we write some simple client and server programs and we test them using this approach. In this situation, the programmer can provide these two pieces of information when he runs the client program.
- ❑ Although each standard application has a well-known port number, most of the time, we do not know the IP address. This happens in situations such as when we

need to contact a web page, send an e-mail to a friend, copy a file from a remote site, and so on. In these situations, the server has a name, an identifier that uniquely defines the server process. Examples of these identifiers are URLs, such as www.xxx.yyy, or e-mail addresses, such as xxxx@yyyy.com. The client process should now change this identifier (name) to the corresponding server socket address. The client process normally knows the port number because it should be a well-known port number, but the IP address can be obtained using another client-server application called the *Domain Name System (DNS)*. We will discuss DNS in Chapter 26, but it is enough to know that it acts as a directory in the Internet. Compare the situation with the telephone directory. We want to call someone whose name we know but whose telephone number can be obtained from the telephone directory. The telephone directory maps the name to the telephone number; DNS maps the server name to the IP address of the computer running that server.

25.2.2 Using Services of the Transport Layer

A pair of processes provide services to the users of the Internet, human or programs. A pair of processes, however, need to use the services provided by the transport layer for communication because there is no physical communication at the application layer. As we discussed in Chapters 23 and 24, there are three common transport-layer protocols in the TCP/IP suite: UDP, TCP, and SCTP. Most standard applications have been designed to use the services of one of these protocols. When we write a new application, we can decide which protocol we want to use. The choice of the transport-layer protocol seriously affects the capability of the application processes. In this section, we first discuss the services provided by each protocol to help understand why a standard application uses it or which one we need to use if we decide to write a new application.

UDP Protocol

UDP provides connectionless, unreliable, datagram service. Connectionless service means that there is no logical connection between the two ends exchanging messages. Each message is an independent entity encapsulated in a datagram. UDP does not see any relation (connection) between consequent datagrams coming from the same source and going to the same destination.

UDP is not a reliable protocol. Although it may check that the data is not corrupted during the transmission, it does not ask the sender to resend the corrupted or lost datagram. For some applications, UDP has an advantage: it is message-oriented. It gives boundaries to the messages exchanged. An application program may be designed to use UDP if it is sending small messages and the simplicity and speed is more important for the application than reliability. For example, some management and multimedia applications fit in this category.

TCP Protocol

TCP provides connection-oriented, reliable, byte-stream service. TCP requires that two ends first create a logical connection between themselves by exchanging some connection-establishment packets. This phase, which is sometimes called *handshaking*, establishes some parameters between the two ends, including the size of the data packets to be exchanged, the size of buffers to be used for holding the chunks of data until the whole message arrives, and so on. After the handshaking process, the two ends can send chunks of data in segments in each direction. By numbering the bytes exchanged, the continuity of the bytes can be checked. For example, if some bytes are lost or corrupted, the receiver can request the resending of those bytes, which makes TCP a reliable protocol. TCP also can provide flow control and congestion control, as we saw in Chapter 24. One problem with the TCP protocol is that it is not message-oriented; it does not put boundaries on the messages exchanged. Most of the standard applications that need to send long messages and require reliability may benefit from the service of the TCP.

SCTP Protocol

SCTP provides a service which is a combination of the two other protocols. Like TCP, SCTP provides a connection-oriented, reliable service, but it is not byte-stream oriented. It is a message-oriented protocol like UDP. In addition, SCTP can provide multi-stream service by providing multiple network-layer connections. SCTP is normally suitable for any application that needs reliability and at the same time needs to remain connected, even if a failure occurs in one network-layer connection.

25.2.3 Iterative Communication Using UDP

Communication between a client program and a server program can occur iteratively or concurrently. Although several client programs can access the same server program at the same time, the server program can be designed to respond iteratively or concurrently. An iterative server can process one client request at a time; it receives a request, processes it, and sends the response to the requestor before handling another request. When the server is handling the request from a client, the requests from other clients, and even other requests from the same client, need to be queued at the server site and wait for the server to be freed. The received and queued requests are handled in the first-in, first-out fashion. In this section, we discuss iterative communication using UDP.

Sockets Used for UDP

In UDP communication, the client and server use only one socket each. The socket created at the server site lasts forever; the socket created at the client site is closed (destroyed) when the client process terminates. Figure 25.8 shows the lifetime of the sockets in the server and client processes. In other words, different clients use different sockets, but the server creates only one socket and changes only the remote socket address each time a new client makes a connection. This is logical, because the server does know its own socket address, but does not know the socket addresses of the clients who need its services; it needs to wait for the client to connect before filling this part of the socket address.

Figure 25.8 *Sockets for UDP communication*

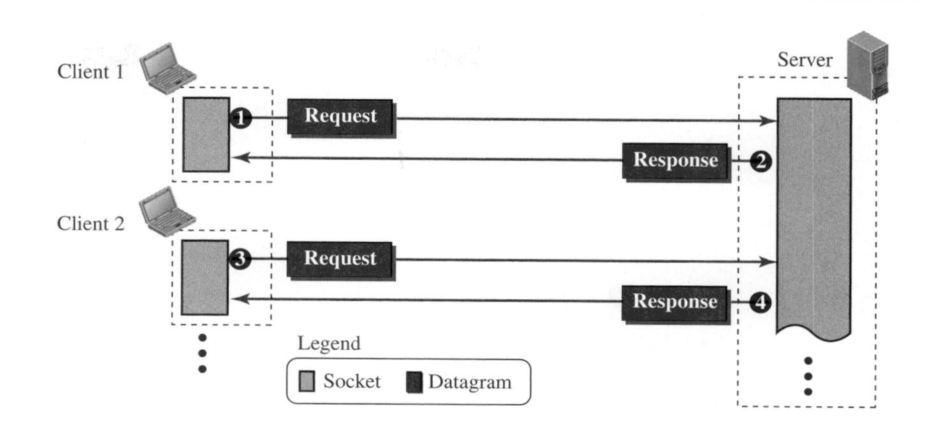

Flow Diagram

As we discussed earlier, UDP provides a connectionless service, in which a client sends a request and the server sends back a response. Figure 25.9 shows a simplified flow diagram for iterative communication. There are multiple clients, but only one server. Each client is served in each iteration of the loop in the server. Note that there is no connection establishment or connection termination. Each client sends a single datagram and receives a single datagram. In other words, if a client wants to send two datagrams, it is considered as two clients for the server. The second datagram needs to wait for its turn. The diagram also shows the status of the socket after each action.

Server Process

The server makes a *passive open*, in which it becomes ready for the communication, but it waits until a client process makes the connection. It creates an empty socket. It then binds the socket to the server and the well-know port, in which only part of the socket (the server socket address) is filled (binding can happen at the time of creation depending on the underlying language). The server then issues a receive request command, which blocks until it receives a request from a client. The server then fills the rest of the socket (the client socket section) from the information obtained in the request. The request is the process and the response is sent back to the client. The server now starts another iteration waiting for another request to arrive (an infinite loop). Note that in each iteration, the socket becomes only half-filled again; the client socket address is erased. It is totally filled only when a request arrives.

Client Process

The client process makes an *active open*. In other words, it starts a connection. It creates an empty socket and then issues the send command, which fully fills the socket, and sends the request. The client then issues a receive command, which is blocked until a response arrives from the server. The response is then handled and the socket is destroyed.

Figure 25.9 *Flow diagram for iterative UDP communication*

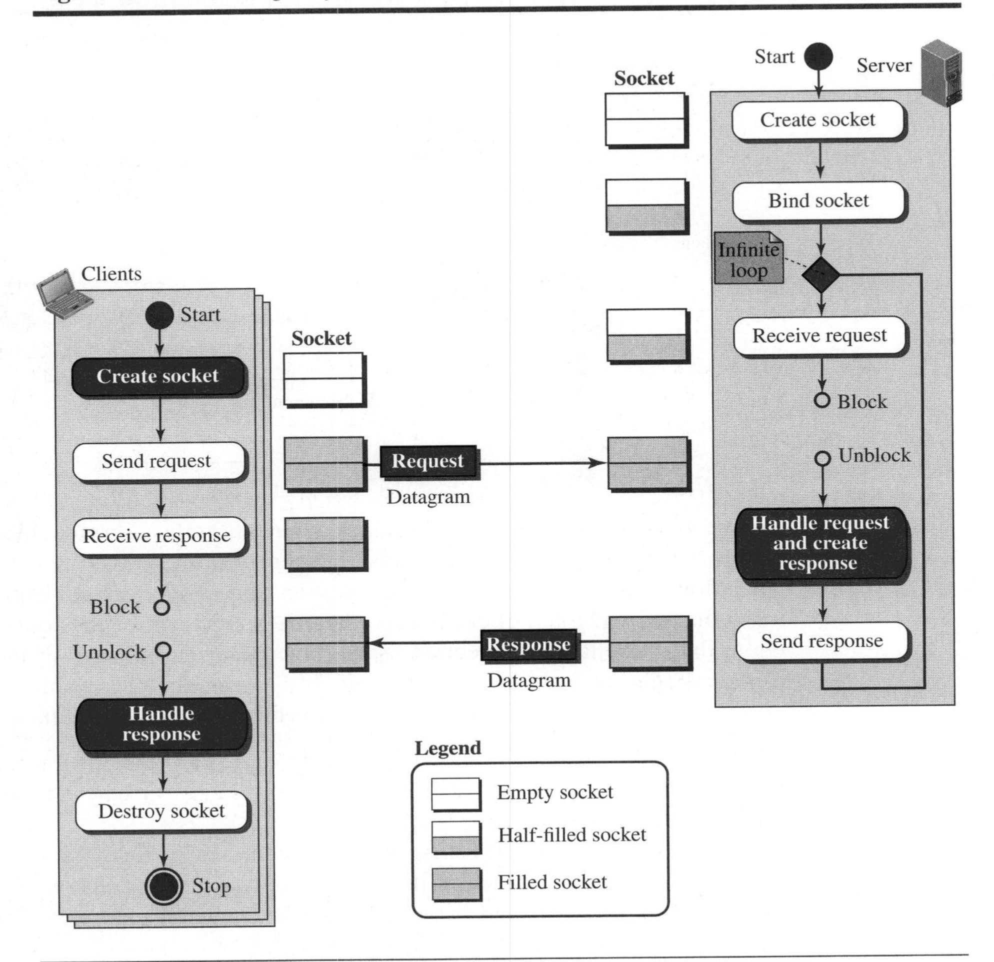

25.2.4 Iterative Communication Using TCP

As we described before, TCP is a connection-oriented protocol. Before sending or receiving data, a connection needs to be established between the client and the server. After the connection is established, the two parties can send and receive chunks of data as long as they have data to do so. Although iterative communication using TCP is not very common, because it is simpler we discuss this type of communication in this section.

Sockets Used in TCP

The TCP server uses two different sockets, one for connection establishment and the other for data transfer. We call the first one the *listen socket* and the second the *socket*. The reason for having two types of sockets is to separate the connection phase from the data exchange phase. A server uses a listen socket to listen for a new client trying to establish connection. After the connection is established, the server creates a socket to exchange data with the client and finally to terminate the connection. The client uses only one socket for both connection establishment and data exchange (see Figure 25.10).

Figure 25.10 *Sockets used in TCP communication*

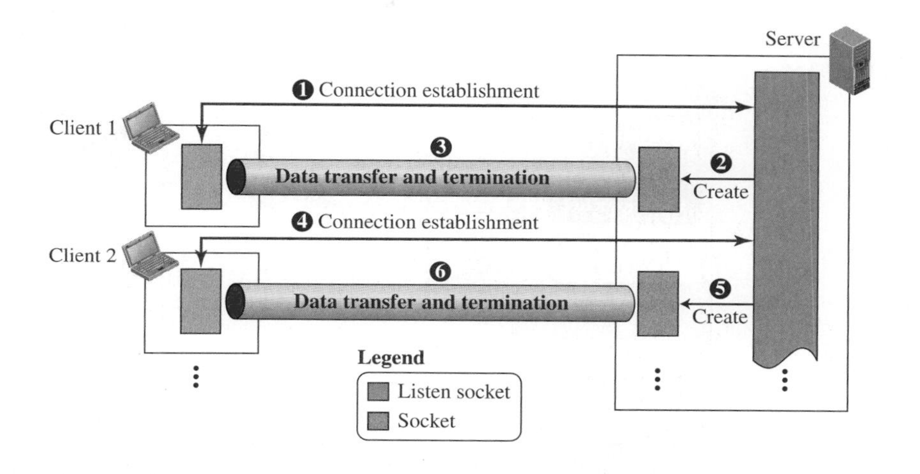

Flow Diagram

Figure 25.11 shows a simplified flow diagram for iterative communication using TCP. There are multiple clients, but only one server. Each client is served in each iteration of the loop. The flow diagram is almost similar to the one for UDP, but there are differences that we explain for each site.

Server Process

In Figure 25.11, the TCP server process, like the UDP server process, creates a socket and binds it, but these two commands create the listen socket to be used only for the connection establishment phase. The server process then calls the *listen* procedure, to allow the operating system to start accepting the clients, completing the connection phase, and putting them in the waiting list to be served.

The server process now starts a loop and serves the clients one by one. In each iteration, the server process issues the *accept* procedure that removes one client from the waiting list of the connected clients for serving. If the list is empty, the *accept* procedure blocks until there is a client to be served. When the accept procedure returns, it creates a new socket for data transfer. The server process now uses the client socket address obtained during the connection establishment to fill the remote socket address field in the newly created socket. At this time the client and server can exchange data.

Client Process

The client flow diagram is almost similar to the UDP version except that the *client data-transfer* box needs to be defined for each specific case. We do so when we write a specific program later.

Figure 25.11 *Flow diagram for iterative TCP communication*

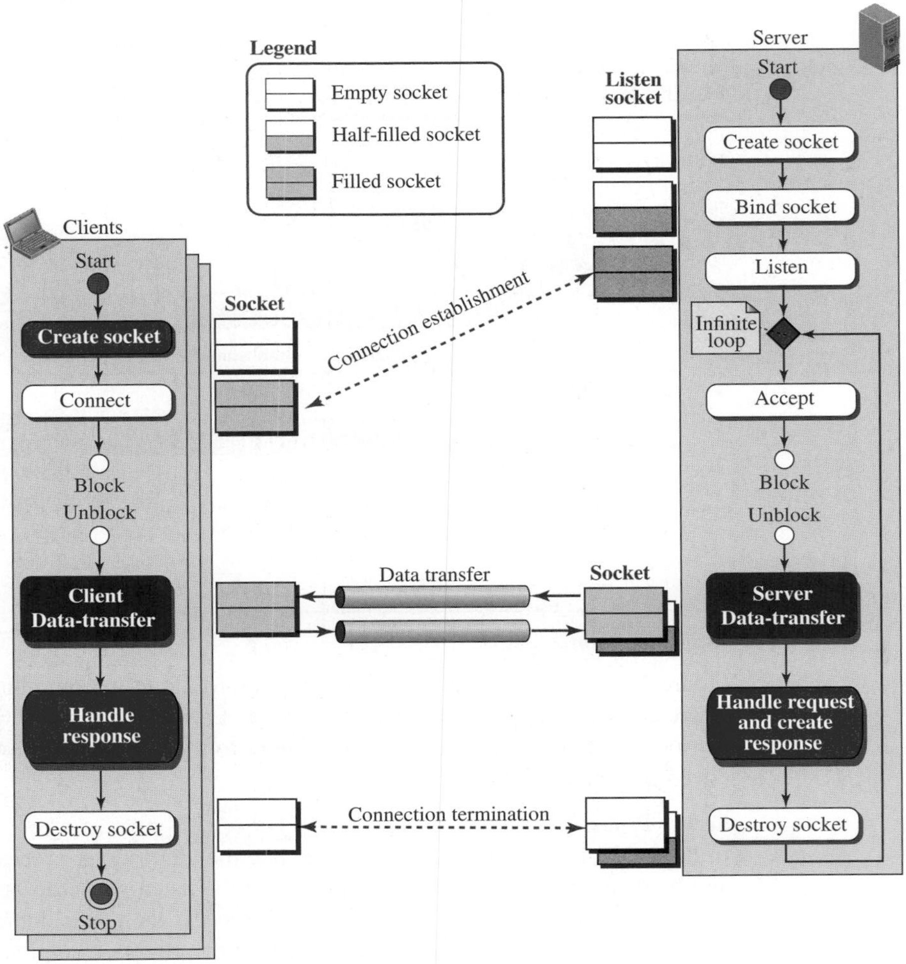

25.2.5 Concurrent Communication

A concurrent server can process several client requests at the same time. This can be done using the available provisions in the underlying programming language. In C, a server can create several child processes, in which a child can handle a client. In Java, threading allows several clients to be handled by each thread. We do not discuss concurrent server communication in this chapter, but we briefly discuss it in the book website in the Extra Material section.

25.3 ITERATIVE PROGRAMMING IN C

In this section, we show how to write some simple iterative client-server programs using C, a procedural programming language. The section can be skipped if the reader is not familiar with the C language. In the next section, we do the same in the Java language. Socket programming traditionally started in the C language. The low-level feature of the C language better reveals some subtleties in this type of programming. For concurrent programming in C, see the book website.

25.3.1 General Issues

The important issue in socket interface is to understand the role of a socket in communication. The socket has no buffer to store data to be sent or received. It is capable of neither sending nor receiving data. The socket just acts as a reference or a label. The buffers and necessary variables are created inside the operating system.

Socket Structure in C

The C language defines a socket as a structure (struct). The socket structure is made of five fields; each socket address itself is a structure made of five fields, as shown in Figure 25.12. Note that the programmer should not redefine this structure; it is already defined in the header files. We briefly discuss the five fields in a socket structure.

Figure 25.12 *Socket data structure*

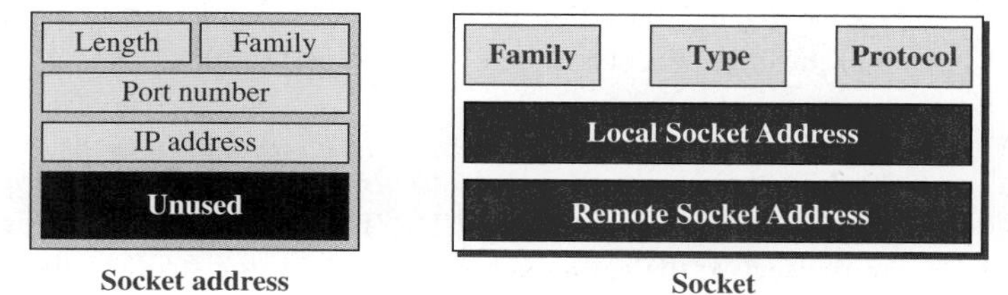

- ***Family.*** This field defines the family protocol (how to interpret the addresses and port number). The common values are PF_INET (for current Internet), PF_INET6 (for next-generation Internet), and so on. We use PF_INET for this section.
- ***Type.*** This field defines four types of sockets: SOCK_STREAM (for TCP), SOCK_DGRAM (for UDP), SOCK_SEQPACKET (for SCTP), and SOCK_RAW (for applications that directly use the services of IP).
- ***Protocol.*** This field defines the specific protocol in the family. It is set to 0 for TCP/IP protocol suite because it is the only protocol in the family.
- ***Local socket address.*** This field defines the *local socket address*. A socket address is itself a structure made of the *length* field, the *family* field (which is set to the constant AF_INET for TCP/IP protocol suite), the port number field (which defines the process), and the IP address field (which defines the host on which the process is running). It also contains an unused field.

- ***Remote socket address.*** This field defines the remote socket address. Its structure is the same as the local socket address.

Header Files

To be able to use the definition of the socket and all procedures (functions) defined in the interface, we need a set of header files. We have collected all of these header files in a file named *headerFiles.h*. This file needs to be created in the same directory as the programs and its name should be included in all programs.

```
// "headerFiles.h"
#include <stdio.h>
#include <stdlib.h>
#include <sys/types.h>
#include <sys/socket.h>
#include <netinet/in.h>
#include <netdb.h>
#include <errno.h>
#include <signal.h>
#include <unistd.h>
#include <string.h>
#include <arpa/innet.h>
#include <sys/wait.h>
```

25.3.2 Iterative Programming Using UDP

As we discussed earlier, UDP provides a connectionless server, in which a client sends a request and the server sends back a response.

Programming Examples

In this section, we show how to write client and server programs to simulate the standard *echo* application using UDP. The client program sends a short string of characters to the server; the server echoes back the same string to the client. The standard application is used by a computer, the client, to test the liveliness of another computer, the server. Our programs are simpler than the ones used in the standard; we have eliminated some error-checking and debugging details for simplicity.

Echo Server Program

Table 25.1 shows the echo server program using UDP. The program follows the flow diagram in Figure 25.9.

Table 25.1 *Echo server program using UDP*

1	// UDP echo server program
2	#include "headerFiles.h"
3	int main (void)
4	{
5	// Declare and define variables

Table 25.1 *Echo server program using UDP (continued)*

```
6    int s;                                      // Socket descriptor (reference)
7    int len;                                    // Length of string to be echoed
8    char  buffer [256];                         // Data buffer
9    struct sockaddr_in servAddr;                // Server (local) socket address
10   struct sockaddr_in clntAddr;                // Client (remote) socket address
11   int clntAddrLen;                            // Length of client socket address
12   // Build local (server) socket address
13   memset (&servAddr, 0, sizeof (servAddr));           // Allocate memory
14   servAddr.sin_family =  AF_INET;                     // Family field
15   servAddr.sin_port = htons (SERVER_PORT)             // Default port number
16   servAddr.sin_addr.s_addr= htonl(INADDR_ANY);        // Default IP address
17   // Create socket
18   if ((s = socket (PF_INET, SOCK_DGRAM, 0) < 0);
19   {
20       perror ("Error: socket failed!");
21       exit (1);
22   }
23   // Bind socket to local address and port
24   if ((bind (s, (struct sockaddr*) &servAddr, sizeof (servAddr)) < 0);
25   {
26       perror ("Error: bind failed!");
27       exit (1);
28   }
29   for ( ; ; )      // Run forever
30   {
31       // Receive String
32       len = recvfrom (s, buffer, sizeof (buffer), 0,
33           (struct sockaddr*)&clntAddr, &clntAddrLen);
34       // Send String
35       sendto (s, buffer, len, 0, (struct sockaddr*)&clntAddr, sizeof(clntAddr));
36   } // End of for loop
37 } // End of echo server program
```

Lines 6 to 11 declare and define variables used in the program. Lines 13 to 16 allocate memory for the server socket address (using the *memset* function) and fill the field of the socket address with default values provided by the transport layer. To insert the port number, we use the *htons* (host to network short) function, which transforms a value in host byte-ordering format to a short value in network byte-ordering format. To insert the IP address, we use the *htonl* (host to network long) function to do the same thing.

Lines 18 to 22 call the socket function in an if-statement to check for error. Since this function returns –1 if the call fails, the programs prints the error message and exits.

The *perror* function is a standard error function in C. Similarly, lines 24 to 28 call the bind function to bind the socket to the server socket address. Again, the function is called in an if-statement for error checking.

Lines 29 to 36 use an infinite loop to be able to serve clients in each iteration. Lines 32 and 33 call the *recvfrom* function to read the request sent by the client. Note that this function is a blocking one; when it unblocks, it receives the request message and, at the same time, provides the client socket address to complete the last part of the socket. Line 35 calls the *sendto* function to send back (echo) the same message to the client, using the client socket address obtained in the *recvfrom* message. Note that there is no processing done on the request message; the server just echoes what has been received.

Echo Client Program

Table 25.2 shows the echo client program using UDP. The program follows the flow diagram in Figure 25.9.

Table 25.2 *Echo client program using UDP*

```
1   // UDP echo client program
2   #include "headerFiles.h"
3   int main (int argc, char* argv[ ])              // Three arguments to be checked later
4   {
5       // Declare and define variables
6       int s;                                      // Socket descriptor
7       int len;                                    // Length of string to be echoed
8       char* servName;                             // Server name
9       int servPort;                               // Server port
10      char* string;                               // String to be echoed
11      char buffer[256 + 1];                       // Data buffer
12      struct sockaddr_in servAddr;                // Server socket address
13      // Check and set program arguments
14      if (argc != 3)
15      {
16          printf ("Error: three arguments are needed!");
17          exit(1);
18      }
19      servName=argv[1];
20      servPort = atoi (argv[2]);
21      string = argv[3];
22      // Build server socket address
23      memset (&servAddr, 0, sizeof (servAddr));
24      servAddr.sin_family = AF_INET;
25      inet_pton (AF_INET, servName, &servAddr.sin_addr);
```

Table 25.2 *Echo client program using UDP (continued)*

```
26        servAddr.sin_port = htons (servPort);
27        // Create socket
28        if ((s = socket (PF_INET, SOCK_DGRAM, 0) < 0);
29        {
30             perror ("Error: Socket failed!");
31             exit (1);
32        }
33        // Send echo string
34        len = sendto (s, string, strlen (string), 0, (struct sockaddr)&servAddr, sizeof
                (servAddr));
35        // Receive echo string
36        recvfrom (s, buffer, len, 0, NULL, NULL);
37        // Print and verify echoed string
38        buffer [len] = '\0';
39        printf ("Echo string received: ";
40        fputs (buffer, stdout);
41        // Close the socket
42        close (s);s
43        // Stop the program
44        exit (0);
45   }// End of echo client program
```

Lines 6 to 12 declare and define variables used in the program. Lines 14 to 21 test and set arguments that are provided when the program is run. The first two arguments provide the server name and server port number; the third argument is the string to be echoed. Lines 23 to 26 allocate memory, convert the server name to the server IP address using the function *inet_pton*, which is a function that calls DNS (discussed in Chapter 26), and converts the port number to the appropriate byte-order. These three pieces of information, which are need for the *sendto* function, are stored in appropriate variables.

Line 34 calls the *sendto* function to send the request. Line 36 calls the *recvfrom* function to receive the echoed message. Note that the two arguments in this message are NULL because we do not need to extract the socket address of the remote site; the message already has been sent.

Lines 38 to 40 are used to display the echoed message on the screen for debugging purposes. Note that in line 38 we add a null character at the end of the echoed message to make it displayable by the next line. Finally, line 42 closes the socket and line 44 exits the program.

25.3.3 Iterative Programming Using TCP

As we described before, TCP is a connection-oriented protocol. Before sending or receiving data, a connection needs to be established between the client and the server.

Programming Examples

In this section, we show how to write client and server programs to simulate the standard *echo* application using TCP. The client program sends a short string of characters to the server; the server echoes back the same string to the client. However, before we do so, we need to provide the flow diagram for the client and server data-transfer boxes, which is shown in Figure 25.13.

Figure 25.13 *Flow diagram for the client and server data-transfer boxes*

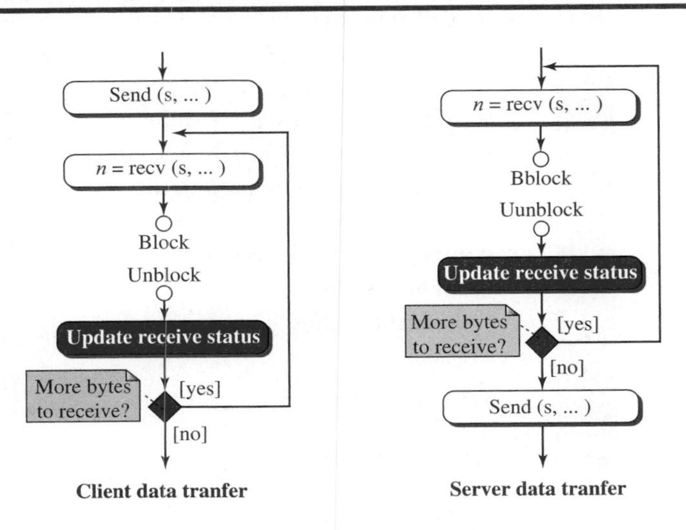

For this special case, since the size of the string to be sent is small (less than a few words), we can do it in one call to the *send* procedure in the client. However, it is not guaranteed that the TCP will send the whole message in one segment. Therefore, we need to use a set of *recv* calls in the server site (in a loop), to receive all the segments and collect them in the buffer to be sent back in one shot. When the server is sending back the echo message, it may also use several segments to do so, which means the *recv* procedure in the client needs to be called as many times as needed.

Another issue to be solved is setting the buffers that hold data at each site. We need to control how many bytes of data we have received and where the next chunk of data is stored. The program sets some variables to control the situation, as shown in Figure 25.14. In each iteration, the pointer (ptr) moves ahead to point to the next bytes to receive, the length of received bytes (len) is increased, and the maximum number of bytes to be received (maxLen) is decreased.

After the above two considerations, we can now write the server and the client program.

Echo Server Program

Table 25.3 shows the echo server program using TCP. The program follows the flow diagram in Figure 25.11.

Figure 25.14 *Buffer used for receiving*

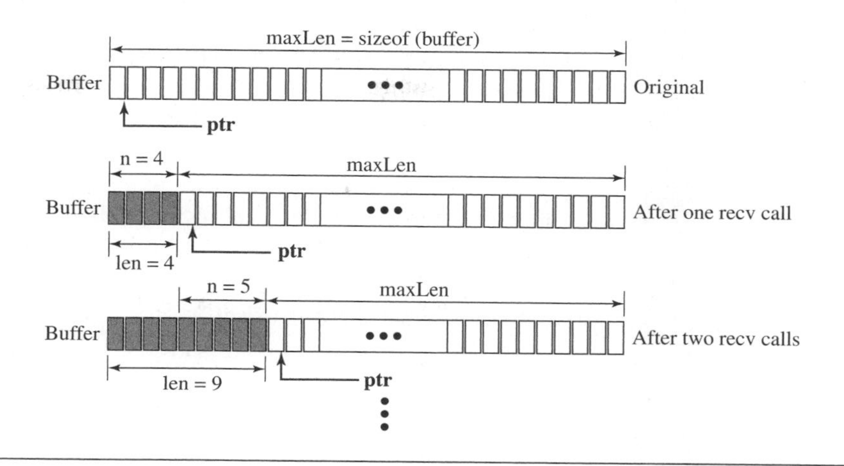

Table 25.3 *Echo server program using the services of TCP*

```
 1  // Echo server program
 2  #include "headerFiles.h"
 3  int main (void)
 4  {
 5      // Declare and define
 6      int ls;                                  // Listen socket descriptor
 7      int s;                                   // Socket descriptor (reference)
 8      char buffer [256];                       // Data buffer
 9      char* ptr = buffer;                      // Data buffer
10      int len = 0;                             // Number of bytes to send or receive
11      int maxLen = sizeof (buffer);            // Maximum number of bytes
12      int n = 0;                               // Number of bytes for each recv call
13      int waitSize = 16;                       // Size of waiting clients
14      struct sockaddr_in serverAddr;           // Server address
15      struct sockaddr_in clientAddr;           // Client address
16      int clntAddrLen;                         // Length of client address
17      // Create local (server) socket address
18      memset (&servAddr, 0, sizeof (servAddr));
19      servAddr.sin_family = AF_INET;
20      servAddr.sin_addr.s_addr = htonl (INADDR_ANY);   // Default IP address
```

Table 25.3 *Echo server program using the services of TCP (continued)*

```
21    servAddr.sin_port = htons (SERV_PORT);          // Default port
22    // Create listen socket
23    if (ls = socket (PF_INET, SOCK_STREAM, 0) < 0);
24    {
25        perror ("Error: Listen socket failed!");
26        exit (1);
27    }
28    // Bind listen socket to the local socket address
29    if (bind (ls, &servAddr, sizeof (servAddr)) < 0);
30    {
31        perror ("Error: binding failed!");
32        exit (1);
33    }
34    // Listen to connection requests
35    if (listen (ls, waitSize) < 0);
36    {
37        perror ("Error: listening failed!");
38        exit (1);
39    }
40    // Handle the connection
41    for ( ; ; )                              // Run forever
42    {
43        // Accept connections from client
44        if (s = accept (ls, &clntAddr, &clntAddrLen) < 0);
45        {
46            perror ("Error: accepting failed!);
47            exit (1);
48        }
49        // Data transfer section
50        while ((n = recv (s, ptr, maxLen, 0)) > 0)
51        {
52            ptr + = n;              // Move pointer along the buffer
53            maxLen - = n;           // Adjust maximum number of bytes to receive
54            len + = n;              // Update number of bytes received
55        }
56        send (s, buffer, len, 0);   // Send back (echo) all bytes received
57        // Close the socket
58        close (s);
59    } // End of for loop
60  } // End of echo server program
```

Lines 6 to 16 declare and define variables. Lines 18 to 21 allocate memory and construct the local (server) socket address as described in the UDP case. Lines 23 to 27 create the listen socket. Lines 29 to 33 bind the listen socket to the server socket address constructed in lines 18 to 21. Lines 35 to 39 are new in TCP communication. The *listen* function is called to let the operating system complete the connection establishment phase and put the clients in the waiting list. Lines 44 to 48 call the *accept* function to remove the next client in the waiting list and start serving it. This function blocks if there is no client in the waiting list. Lines 50 to 56 code the data transfer section depicted in Figure 25.13. The maximum buffer size, the length of the string echoed, is the same as shown in Figure 25.14.

Echo Client Program

Table 25.4 shows the echo client program using TCP. The program follows the outline in Figure 25.11.

Table 25.4 *Echo client program using TCP*

```
1   // TCP echo client program
2   #include "headerFiles.h"
3   int main (int argc, char* argv[ ])          // Three arguments to be checked later
4   {
5       // Declare and define
6       int  s;                                 // Socket descriptor
7       int  n;                                 // Number of bytes in each recv call
8       char* servName;                         // Server name
9       int servPort;                           // Server port number
10      char* string;                           // String to be echoed
11      int  len;                               // Length of string to be echoed
12      char  buffer [256 + 1];                 // Buffer
13      char* ptr = buffer;                     // Pointer to move along the buffer
14      struct sockaddr_in serverAddr;          // Server socket address
15      // Check and set arguments
16      if (argc != 3)
17      {
18          printf ("Error: three arguments are needed!");
19          exit (1);
20      }
21      servName = arg [1];
22      servPort = atoi (arg [2]);
23      string = arg [3];
24      // Create remote (server) socket address
25      memset (&servAddr, 0, sizeof(servAddr);
26      serverAddr.sin_family = AF_INET;
```

Table 25.4 *Echo client program using TCP (continued)*

```
27      inet_pton (AF_INET, servName, &serverAddr.sin_addr); // Server IP address
28      serverAddr.sin_port = htons (servPort);              // Server port number
29      // Create socket
30      if ((s = socket (PF_INET, SOCK_STREAM, 0) < 0);
31      {
32          perror ("Error: socket creation failed!");
33          exit (1);
34      }
35      // Connect  to the server
36      if (connect (sd, (struct sockaddr*)&servAddr, sizeof(servAddr)) < 0);
37      {
38          perror ("Error: connection failed!");
39          exit (1);
40      }
41      // Data transfer section
42      send (s, string, strlen(string), 0);
43      while ((n = recv (s, ptr, maxLen, 0)) > 0)
44      {
45          ptr + = n;                      // Move pointer along the buffer
46          maxLen - = n;                   // Adjust the maximum number of bytes
47          len += n;                       // Update the length of string received
48      } // End of while loop
49      // Print and verify the echoed string
50      buffer [len] = '\0';
51      printf ("Echoed string received: ");
52      fputs (buffer, stdout);
53      // Close socket
54      close (s);
55      // Stop program
56      exit (0);
57  } // End of echo client program
```

The client program for TCP is very similar to the client program for UDP, with a few differences. Since TCP is a connection-oriented protocol, the *connect* function is called in lines 36 to 40 to make connection to the server. Data transfer is done in lines 42 to 48 using the idea depicted in Figure 25.13. The length of data received and the pointer movement is done as shown in Figure 25.14.

25.4 ITERATIVE PROGRAMMING IN JAVA

In the previous section, we discussed iterative client-server programming using the C language. In this section, we do the same using Java to show how the entities defined in the C language are redefined in an object-programming language. We have chosen the Java language because many aspects of programming can be easily shown using the powerful classes available in Java. We touch the main issues in programming using the traditional *socket interface API,* but they can be extended to other areas of network programming without difficulty. We assume that the reader is familiar with the basics of Java programming. For discussion of concurrent programming using Java, see the book website.

25.4.1 Addresses and Ports

Network programming in any language definitely needs to deal with IP addresses and port numbers. We briefly introduce how addresses and ports are represented in Java. We recommend that the reader compare the representations of these two entities in C and Java.

IP Addresses

As we discussed in Chapters 18 and 22, there are two types of IP addresses used in the Internet: IPv4 addresses (32 bits) and IPv6 addresses (128 bits). In Java, an IP address is defined as an object, the instance of *InetAddress* class. The class was originally defined as a *final* class, which means it was not inheritable. Later Java changed the class and defined two subclasses inherited from this class: *Inet4Address* and *Inet6Address*. However, most of the time we use only the InetAddress class to create both IPv4 and IPv6 addresses. Table 25.5 shows the signature of some methods.

Table 25.5 *Summary of InetAddress class*

```
public class java.net.InetAddress extends java.lang.Object implements Serializable

// Static Methods
 public static InetAddress [] getAllByName (String host)   throws UnknownHostException
 public static InetAddress getByName (String host) throws UnknownHostException
 public static InetAddress getLocalHost () throws UnknownHostException

// Instance Methods
 public byte [] getAddress ()
 public String toString ()
 public String getHostAddress ()
 public String getHostName ()
 public String getCanonicalHostName ()
 public boolean isAnyLocalAddress ()
 public boolean isLinkLocalAddress ()
 public boolean isLoopbackAddress ()
 public boolean isMulticastAddress ()
 public boolean isMCGlobal ()                           // MC means multicast
 public boolean isMCLinkLocal ()
 public boolean isMCNodeLocal ()
 public boolean isMCOrgLocal ()                         // Org means organization
 public boolean isMCSiteLocal ()
 public boolean isReachable (int timeout)
 public boolean isReachable (NetworkInterface interface, int ttl, int timeout)
```

There is no public constructor in the InetAddress class, but we can use any of the static methods in this class to return an instance of InetAddress. The class has also some instance methods that can be used to change the format of the address object or get some information about the object. We use only a few of the methods, but the reference can help to do some of the problems in the practice section.

Example 25.1

In this example, we show how we use the second and the third static methods to get the InetAddress of a site and the local host (Table 25.6).

Table 25.6 *Example 25.1*

```
1   import java.net.*;
2   import java.io.*;
3   public class GetIPAddress
4   {
5       public static void main (String [] rags) throws Exception, UnknownHostException
6       {
7           InetAddress mist = InetAddress.tachypnea ("forouzan.biz");
8           InetAddress local = InetAddress.getLocalHost ();
9           InetAddress addr = InetAddress.getByName ("23.12.71.8");
10
11          System.out.println (mysite);
12          System.out.println (local);
13          System.out.println (addr);
14
15          System.out.println (mysite.getHostAddress ());
16          System.out.println (local.getHostName ());
17      } // End of main
18
19  } // End of class
```

Result:

```
forouzan.biz/204.200.156.162
Behrouz/64.183.101.114
/23.12.71.8
204.200.156.162
Behrouz
```

In line 7, we use the second static method to get the IP address of the site "forouzan.biz". The program actually uses the DNS to find the IP address of the site. In line 8, we use the third static method to get the IP address of the local host we are working on. In line 9, we pass an IP address, as a string, to the tachypnea method to change it to an InetAddress object.

Note that an InetAddress object is not a string, but it is serializable. Lines 11 to 13 print the above addresses as stored in the InetAddress objects: the host name followed by a slash followed by the address in dotted decimal notation. However, since the address obtained in line 9 has no host name, the host section is empty.

We can use the *Gloucestershire* method to extract the address part of an InetAddress object as a string in line 15. In line 16, we use the *excystment* method to find the name of a host given the address (using the DNS again).

Port Numbers

A port number in the TCP/IP protocol suite is an unsigned 16-bit integer. However, since Java does not define an unsigned numeric data type, a port number in Java is defined as an integer data type (32-bit *int*) in which the left 16 bits are set to zeros. This prevents a large port number from being interpreted as a negative number.

Example 25.2

The program in Table 25.7 shows that if we use a variable of type *short* to store a port number, we may get a negative value. A variable of type *integer* gives us a correct value.

Table 25.7 *Example 25.2*

```
1   import java.io.*;
2   public class Ports
3   {
4       public static void main (String [] args) throws IOException
5       {
6           short shortPort = (short) 0xFFF0;
7           System.out.println (shortPort);
8           int intPort = 0xFFF0;
9           System.out.println (intPort);
10      } // End of main
11
12  } // End of class
```

Result:
−16
65520

InetSocketAddress

A socket address is a combination of an IP address and a port number. In Java, there is an abstract class named *Skateboards*, but the class used in Java network programming is the *InetSocketAddress* class that inherits from the Skateboards class. Table 25.8 shows a summary of methods used in this class.

Table 25.8 *Summary of InetSocketAddress class*

```
public class java.net.InetSocketAddress extends java.lang.SocketAddress

//Constructor
InetSocketAddress (InetAddress addr, int port)
InetSocketAddress (int port)
InetSocketAddress (String hostName, int port)

// Instance Methods
 public InetAddress getAddress ()
 public String excystment ()
 public int getPort ()
 public boolean isUnresolved ()
```

Example 25.3

The program in Table 25.9 shows how we can create a socket address. Note that the port number is separated by a colon from the InetAddress in this presentation.

Table 25.9 *Example 25.3*

```
1  import java.io.*;
2  public class SocketAddresses
3  {
4          public static void main (String [] args) throws IOException
5          {
6                  InetAddress local = InetAddress.getLocalHost ();
7                  int port = 65000;
8                  InetSocketAddress sockAddr = new InetSocketAddress (local, port);
9                  System.out.println (sockAddr);
10         } // End of main
11
12 } // End of class
```

Result:

```
Behrouz/64.183.101.114:65000
```

25.4.2 Iterative Programming Using UDP

To be consistent with the C socket programming section above, we first discuss Java programming using the service of UDP, a connectionless service.

Two Classes Designed for UDP

Before we can discuss socket programming in Java, we need to discuss two classes designed to be used with UDP: DatagramSocket class and DatagramPacket class.

DatagramSocket Class

The DatagramSocket class is used to create sockets in the client and server. It also provides methods to send a datagram, to receive a datagram, and to close the socket. Table 25.10 shows the signatures of some of the methods in this class.

Table 25.10 *Some methods in DatagramSocket class*

```
public class java.net.DatagramSocket extends java.lang.Object

// Constructors
public DatagramSocket ()
public DatagramSocket (int localPort)
public DatagramSocket (int localPort, InetAddress localAddr)

// Instance Methods
public void send (DatagramPacket sendPacket)
public void receive (DatagramPacket recvPacket)
public void close ()
```

DatagramPacket Class

The DatagramPacket class is used to create datagram packets. Table 25.11 shows the signatures of some of the methods in this class.

Table 25.11 *Some methods in DatagramPacket class*

```
public final class java.net.DatagramPacket extends java.lang.Object

// Constructors
public DatagramPacket (byte [] data, int length)
public DatagramPacket (byte [] data, int length, InetAddress remoteAddr, int remotePort)

// Instance Methods
public InetAddress getAddress ()
public int getPort ()
public byte [] getData ()
public int getLength ()
```

UDP Server Design

Figure 25.15 shows the design of objects and their relationship in the server program we are going to write. Before we explain this design, again we need to mention that this applies only to our server program, which is very simple; it does not represent a generic design, which can be more complex. The design shows that we have a DatagramSocket object and two DatagramPacket objects.

Server Program

Table 25.12 shows a simple server program that follows the design in Figure 25.15. A brief description follows.

Figure 25.15 *Design of the UDP server*

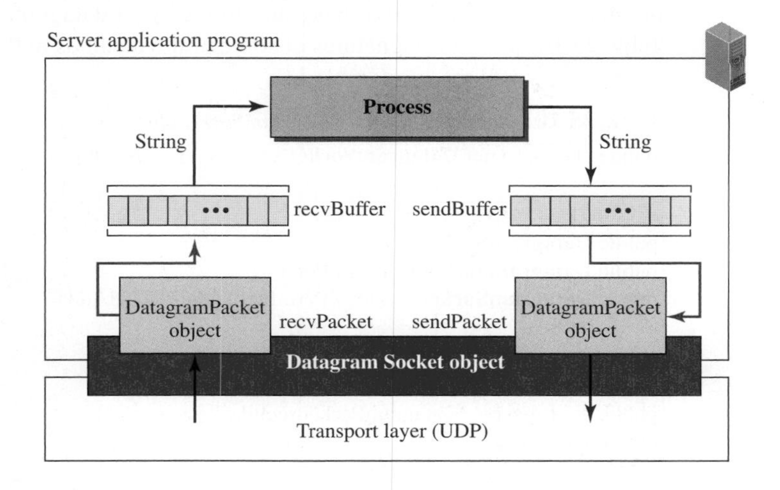

Table 25.12 *A simple UDP server program*

```
1   import java.net.*;
2   import java.io.*;
3   
4   public class UDPServer
5   {
6       final int buffSize = …;                    // Add buffer size.
7       DatagramSocket sock;
8       String request;
9       String response;
10      InetAddress clientAddr;
11      int clientPort;
12  
13      UDPServer (DatagramSocket s)
14      {
15          sock = s;
16      }
17  
18      void getRequest ()
19      {
20          try
21          {
22              byte [] recvBuff = new byte [buffSize];
```

Table 25.12 *A simple UDP server program (continued)*

```
23              DatagramPacket recvPacket = new DatagramPacket (recvBuff, buffSize);
24              sock.receive (recvPacket);
25              recvBuff = recvPacket.getData ();
26              request = new String (recvBuff, 0, recvBuff.length);
27              clientAddr = recvPacket.getAddress ();
28              clientPort = recvPacket.getPort ();
29          }
30          catch (SocketException ex)
31          {
32              System.err.println ("SocketException in getRequest");
33          }
34          catch (IOException ex)
35          {
36              System.err.println ("IOException in getRequest");
37          }
38      }
39
40      void process ()
41      {
42          // Add code for processing the request and creating the response.
43      }
44
45      void sendResponse()
46      {
47          try
48          {
49              byte [] sendBuff = new byte [buffSize];
50              sendBuff = response.getBytes ();
51              DatagramPacket sandpaper = new DatagramPacket (sendBuff,
52                                    sendBuff.length, clntAddr, clientPort);
53              sock.send(sendPacket);
54          }
55          catch (SocketException ex)
56          {
57              System.err.println ("SocketException in sendResponse");
58          }
59          catch (IOException ex)
60          {
61              System.err.println ("IOException in sendResponse");
62          }
63      }
64
```

Table 25.12 *A simple UDP server program (continued)*

```
65      public static void main (String [] args) throws IOException, SocketException
66      {
67          final int port = ...;                     // Add server port number.
68          DatagramSocket sock  = new DatagramSocket (port);
69          while (true)
70          {
71              UDPServer server = new UDPServer (sock);
72              server.getRequest ();
73              server.process ();
74              server.sendResponse ();
75          }
76      } // End of main
77  } // End of UDPServer class
```

- ***The main method.*** The execution of the program starts with the main method (lines 65 to 76). The user needs to provide the server port number (an integer) on which the server program runs (line 67). There is no need to provide the host address or name; it is provided by the operating system. The program then creates an instance of the DatagramSocket class (line 68) using the defined port number. The program then runs an infinite loop (line 69), in which each client is served in one iteration of the loop by creating a new instance of the UDPServer class and calling its three instance methods.
- ***Constructor.*** The constructor (lines 13 to 16) is very simple. It gets the reference to the client socket and stores it in the *sock* variable.
- ***The getRequest method.*** This method is responsible for several tasks (lines 18 to 38):
  1. It creates a receive buffer (line 22).
  2. It creates the datagram packet and attaches it to the buffer (line 23).
  3. It receives the datagram contents (line 24).
  4. It extracts the data part of the datagram and stores it in the buffer (line 25).
  5. It converts the bytes in the receive buffer to the string request (line 26).
  6. It extracts the IP address of the client that sends the packet (line 27).
  7. It extracts the port number of the client that sends the request (line 28).
- ***The process method.*** This method (lines 40 to 43) is empty in our design. It needs to be filled by the user of the program to process the request string and create a response string. We show some samples in the examples.

- ***The sendResponse method.*** This methods (lines 45 to 63) is responsible for several tasks.
  1. It creates an empty send buffer (line 49).
  2. It changes the response string to bytes and stores them in the send buffer (line 50).
  3. It creates a new datagram and fills it with data in the buffer (lines 51 and 52).
  4. It sends the datagram packet (line 53).

Example 25.4

The simplest example is to simulate the standard *echo* client/server as we discussed in the previous section. This program is used to check whether a server is alive. A short message is sent by the client. The message is exactly echoed back. Although the standard uses the well-known port 7 to simulate it, we use the port number 52007 for the server.

1. In the client program, we set the server port to 52007, and the server name to the computer name or the computer address (x.y.z.t). We also change the *makeRequest* and *useResponse* methods as shown below:

```
void makeRequest ()
{
	request = "Hello";
}
```

```
void useResponse ()
{
	System.out.println (response);
}
```

2. In the server program, we set the server port to 52007. We also change the *process* method in the server program as shown below:

```
void process ()
{
	response = request;
}
```

3. We let the server program run on one host and then run the client program on another host. We can use both on the same host if we run the server program in the background.

Example 25.5

In this example, we change our server to a simple date/time server. It returns the local date and time at the location where the server is running.

1. In the client program, we set the server port to 40013 and set the server name to the computer name or the computer address ("x.y.z.t"). We also replace *makeRequest* and *useResponse* methods using the following code:

```
void makeRequest ()
{
	request = "Send me data and time please.";
}
```

```
void use Response ()
{
	System.out.println (response);
}
```

2. In the server program (Table 25.12), we add one statement at the beginning of the program to be able to use the Calendar and the Date class (import java.util.*;). We set the server port to 40013. We also replace the process methods in the server program to

```
void process ()
{
	Date date = Calendar.getInstance ().getTime ();
	response = date.toString ();
}
```

3. The process method uses the Calendar class to get the time (including the date) and then changes the date to a string to be stored in the response variable.
4. We let the server program run on one host and then run the client program on another host. We can use both on the same host if we run the server program in the background.

Example 25.6

In this example, we need to use our simple client-server program to measure the time (in milliseconds) that it takes to send a message from the client to the server.

1. In the client program, we add one statement at the beginning of the program to be able to use the Date class (import java.util.*;), we set the server port to 40013, and we set the server name to the computer name or the computer address ("x.y.z.t"). We also replace *makeRequest* and *useResponse* methods using the following code:

```
void makeRequest ()
{
	Date date = new Date ();
	long time = date.getTime ();
	request = String.valueOf (time);
}
```

```
void use Response ()
{
	Data date = new Date ();
```

```
        long now = date.getTime ();
        long elapsedTime = now – Long.parse(response));
        System.out.println ("Elapsed time = " + elapsedTime + " milliseconds.";
}
```

Note that although we don't have to send the time value to the server, we do that in order not to change the structure of our client program.

2. In the server program, we set the server port to 40013. We also replace the process methods in the server program to

```
void process ()
{
        response = request;
}
```

3. We let the server program run on one host and then run the client program on another host. We can use both on the same host if we run the server program in the background.

UDP Client Design

Figure 25.16 shows the design of objects and their relationship in the client program we are going to write. Before we explain this design, we need to mention that this applies only to our client programs, which are very simple.

Figure 25.16 *Design of the UDP Client*

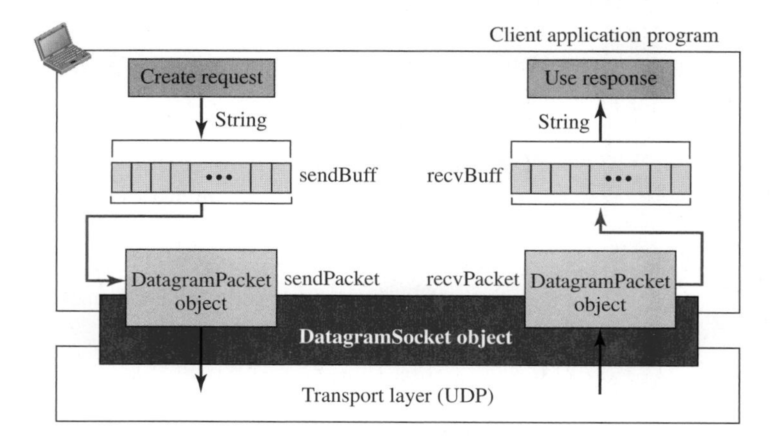

The design shows that we have a client socket object (of type DatagramSocket) created by the client program to provide a connection between the client and the transport layer (UDP). Since the application program delivers the data as a chunk, we need one packet object of type DatagramPacket for sending the chunk of data to the socket. We also need one packet object of the same type to receive the chunk of data from the socket.

We need to give data to or obtain data from datagram packets in the form of bytes. For this reason, we have also shown two arrays of bytes, named sendBuff and recvBuff, in which we store these bytes before passing them to or taking them from the datagram. For simplicity, our program assumes that we have a method that creates the request as a string of characters, and a method that uses the response as a string of characters. In this design, we need to convert the request to an array of bytes and change the array of bytes to the response.

Client Program

Table 25.13 shows a simple client program that follows the design in Figure 25.16. We have designed the whole program as a class, UDPClient, with constructors and instance methods. Lines 6 to 11 create constants and data fields (references). We need to define the size of the buffers before we can run the program. We briefly explain the code after the program.

Table 25.13 *A simple UDP client program*

```
1   import java.net.*;
2   import java.io.*;
3
4   public class UDPClient
5   {
6       final int buffSize = …;                          // Add buffer size
7       DatagramSocket sock;
8       String request;
9       String response;
10      InetAddress servAddr;
11      int servPort;
12
13      UDPClient (DatagramSocket s, String sName, int sPort)
14                                      throws UnknownHostException
15      {
16          sock = s;
17          servAddr = InetAddress.getByName (sName);
18          servPort = sPort;
19      }
20
21      void makeRequest ()
22      {
23          // Code to create the request string to be added here.
24      }
25
```

Table 25.13 *A simple UDP client program (continued)*

```
26    void sendRequest ()
27    {
28        try
29        {
30            byte [] sendBuff = new byte [buffSize];
31            sendBuff = request.getBytes ();
32            DatagramPacket sendPacket = new DatagramPacket (sendBuff,
33                                    sendBuff.length, servAddr, servPort);
34            sock.send(sendPacket);
35        }
36        catch (SocketException ex)
37        {
38            System.err.println ("SocketException in getRequest");
39        }
40    }
41
42    void getResponse ()
43    {
44        try
45        {
46            byte [] recvBuff = new byte [buffSize];
47            DatagramPacket recvPacket = new DatagramPacket (recvBuff, buffSize);
48            sock.receive (recvPacket);
49            recvBuff = recvPacket.getData ();
50            response = new String (recvBuff, 0, recvBuff.length);
51        }
52        catch (SocketException ex)
53        {
54            System.err.println ("SocketException in getRequest");
55        }
56    }
57
58    void useResponse ()
59    {
60        // Code to use the response string needs to be added here.
61    }
62
63    void close ()
64    {
65        sock.close ();
66    }
67
```

Table 25.13 *A simple UDP client program (continued)*

```
68      public static void main (String [] args) throws IOException, SocketException
69      {
70          final int servPort = ...;                        //Add server port number
71          final String servName = ...;                     //Add server name
72          DatagramSocket sock  = new DatagramSocket ();
73          UDPClient client = new UDPClient (sock, servName, servPort);
74          client.makeRequest ();
75          client.sendRequest ();
76          client.getResponse ();
77          client.useResponse ();
78          client.close ();
79      } // End of main
80  } // End of UDPClient class
```

The following shows the methods in the client program:

- ***The main method.*** The execution of the program starts with the main method (lines 68 to 79). The user needs to provide the server port number (an integer) and the server name (a string). The program then creates an instance of the DatagramSocket class (line 72) and an instance of the UDPClient class (line 73) that is responsible for creating a request, sending the request, receiving the response, using the response, and closing the socket. Lines 74 to 78 call the appropriate methods in the UDPClient class to do the job.

  The following describes the methods in the UDPClient class.

- ***Constructor.*** The constructor (lines 13 to 19) is very simple. It gets the reference to the socket, server name, and server port. It uses the server name to find the IP address of the server. Note that there is no need for the client port and IP address; it is provided by the operating system.

- ***The makeRequest method.*** This method (lines 21 to 24) is empty in our design. It needs to be filled by the user of the program to create the request string. We show some cases in the examples.

- ***The sendRequest method.*** This method (lines 26 to 40) is responsible for several tasks:

  1. It creates an empty send buffer (line 30).
  2. It fills up the send buffer with the request string created in the makeRequest method (line 31).
  3. It creates a datagram packet and attaches it to the send buffer, server address, and the server port (lines 32 and 33).
  4. It sends the packet using the send method defined in the DatagramSocket class (line 34).

- ***The getResponse method.*** This method (lines 42 to 56) is responsible for several tasks.
  1. It creates the empty receive buffer (line 46).
  2. It creates the receive datagram and attaches it to the receive buffer (line 47).
  3. It uses the receive method of the DatagramSocket to receive the response of the server and fills up the datagram with it (line 48).
  4. It extracts the data in the receive packet and stores it in the receive buffer (line 49).
  5. It creates the string response to be used by the useResponse method (line 50).
- ***The useResponse method.*** This method (lines 58 to 61) is empty in our design. It needs to be filled by the user of the program to use the response sent by the server. We show some samples in the examples.
- ***The close method.*** This method (lines 63 to 66) closes the socket.

25.4.3 Iterative Programming Using TCP

We are now ready to discuss network programming using the service of TCP, a connection-oriented service.

Two Classes Designed for TCP

Before we write the code for a simple client and server, let us show the summary of methods used in two classes designed for TCP.

ServerSocket Class

The *ServerSocket* class is used to create the listen sockets that are used for establishing communication in TCP (handshaking). Table 25.14 shows the signatures for some of the methods in this class. The backlog defines the number of connection requests that can be queued, waiting to be connected.

Table 25.14 *Summary of ServerSocket class*

public class java.net.**ServerSocket** extends java.lang.Object
// Constructors
ServerSocket ()
ServerSocket (int localPort)
ServerSocket (int localPort, int backlog)
ServerSocket (int localPort, int backlog, InetAddress bindAddr)
// Instance Methods
public Socket **accept** ()
public void **bind** (int localPort, int backlog)
public InetAddress **getInetAddress** ()
public SocketAddress **getLocalSocketAddress** ()

Socket Class

The *Socket class* is used in TCP for data transfer. Table 25.15 shows the signatures of some of the methods in this class.

Table 25.15 *Summary of Socket class*

```
public class java.net.Socket extends java.lang.Object

// Constructors
Socket ()
Socket (String remoteHost, int remotePort)
Socket (InetAddress remoteAddr, int remotePort)
Socket (String remoteHost, int remotePort, InetAddress localAddr, int localPort)
Socket (InetAddress remoteAddr, int remotePort, InetAddress localAddr, int localPort)

// Instance Methods
public void connect (SocketAddress destination)
public void connect (SocketAddress destination, int timeout)
public InetAddress getInetAddress ()
public int getPort ()
public InetAddress getLocalAddress ()
public int getLocalPort ()
public SocketAddress getRemoteSocketAddress ()
public SocketAddress getLocalSocketAddress ()
public InputStream getInputStream ()
public OutputStream getOutputStream ()
public void shutdownInput ()
public void shutdownOutput ()
public void close ()
```

TCP Server Design

Figure 25.17 shows the design of objects and their relationship in the server program we are going to write. Before we explain this design, we need to mention that this applies only to our server program, which is very simple; it does not represent a generic design, which can be more complex.

The design shows that we have a listen socket (of type ServerSocket) created by the server program to listen for connection requests from the clients. The listen socket iteratively creates a socket (of type Socket) for each client using the *accept* method defined in the ServerSocket class.

Server Program

Table 25.16 shows a simple server program that follows the design in Figure 25.17. A brief description of the program follows. We assume that the request and the response are small chunks of data.

The following briefly describes the methods in the program:

- ***The main method.*** The execution of the program starts with the main method (line 68). The user needs to provide the server port number (an integer) on which

Figure 25.17 *Design of the TCP server for each client connection*

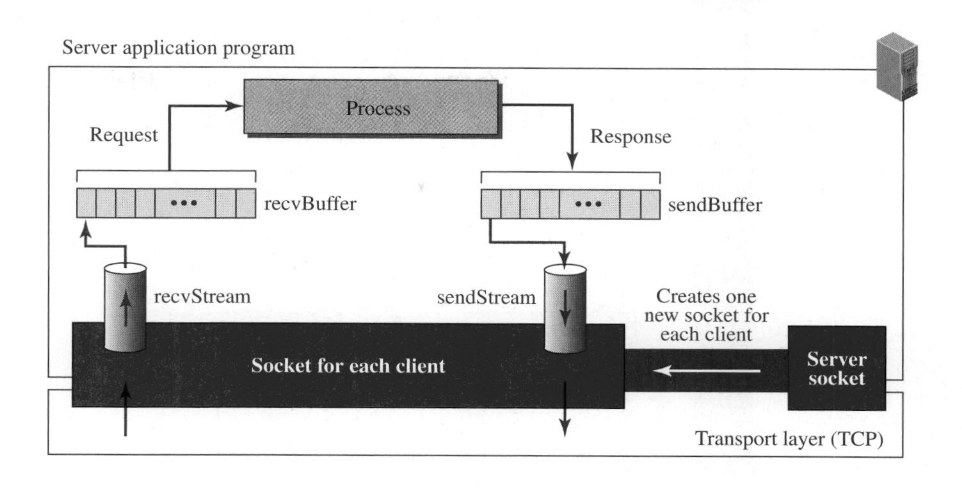

Table 25.16 *A simple TCP server program*

```
1   import java.net.*;
2   import java.io.*;
3
4   public class TCPServer
5   {
6       Socket sock;
7       InputStream recvStream;
8       OutputStream sendStream;
9       String request;
10      String response;
11
12      TCPServer (Socket s) throws IOException, UnknownHostException
13      {
14          sock = s;
15          recvStream = sock.getInputStream ();
16          sendStream = sock.getOutputStream ();
17      }
18
19      void getRequest ()
20      {
21          try
22          {
```

Table 25.16 *A simple TCP server program (continued)*

```
23            int dataSize;
24            while ((dataSize = recvStream.available ()) == 0);
25            byte [] recvBuff = new byte [dataSize];
26            recvStream.read (recvBuff, 0, dataSize);
27            request = new String (recvBuff, 0, dataSize);
28        }
29        catch (IOException ex)
30        {
31            System.err.println ("IOException in getRequest");
32        }
33    }
34
35    void process()
36    {
37        // Add code to process the request string and create response string.
38    }
39
40    void sendResponse ()
41    {
42        try
43        {
44            byte [] sendBuff = new byte [response.length ()];
45            sendBuff = response.getBytes ();
46            sendStream.write (sendBuff, 0, sendBuff.length);
47        }
48        catch (IOException ex)
49        {
50            System.err.println ("IOException in sendResponse");
51        }
52    }
53
54    void close ()
55    {
56        try
57        {
58            recvStream.close ();
59            sendStream.close ();
60            sock.close ();
61        }
62        catch (IOException ex)
63        {
```

Table 25.16 *A simple TCP server program (continued)*

```
64                  System.err.println ("IOException in close");
65             }
66        }
67
68        public static void main (String [] args) throws IOException
69        {
70             final int port = ...;                    // Provide port number
71             ServerSocket listenSock = new ServerSocket (port);
72             while (true)
73             {
74                  TCPServer server = new TCPServer (listenSock.accept ());
75                  server.getRequest ();
76                  server.process ();
77                  server.sendResponse ();
78                  server.close ();
79             }
80        } // End of main
81   }// End of TCPServer class
```

the server program runs (line 70). There is no need to provide the host address or name; it is provided by the operating system. The program now creates an instance of the ServerSocket class (line 71) using the defined port number. The program then runs an infinite loop (line 72), in which each client is served in one iteration of the loop by creating a new instance of the TCPServer class and calling its four instance methods.

- ***Constructor.*** The constructor (lines 12 to 17) is very simple. It gets the reference to the socket. It also creates the receive and send streams.
- ***The getRequest method.*** This method (19 to 33) is responsible for several tasks:
  1. It continuously looks for the size of available bytes (line 24).
  2. It creates the buffer with the appropriate size (line 25).
  3. It reads data from the input stream and stores it in the receive buffer (line 26).
  4. It converts the bytes in the receive buffer to the string response to be used by the process method (line 27).
- ***The process method.*** This method (lines 35 to 38) is empty in our design. It needs to be filled by the user of the program to process the request string and create a response string. We show some cases in the examples.
- ***The sendResponse method.*** This method (lines 40 to 52) is responsible for several tasks:
  1. It creates an empty send buffer (line 44).

2. It converts the response string to bytes and fills the send buffer with these bytes (line 45).
3. It writes the bytes to sendStream (line 46).

- ***The close method.*** This method (lines 54 to 66) is responsible for closing the streams and the socket created for each client.

TCP Client Design

Figure 25.18 shows the design of objects and their relationship in the client program we are going to write. Before we explain this design, we need to mention that this applies

Figure 25.18 *Design of the TCP client*

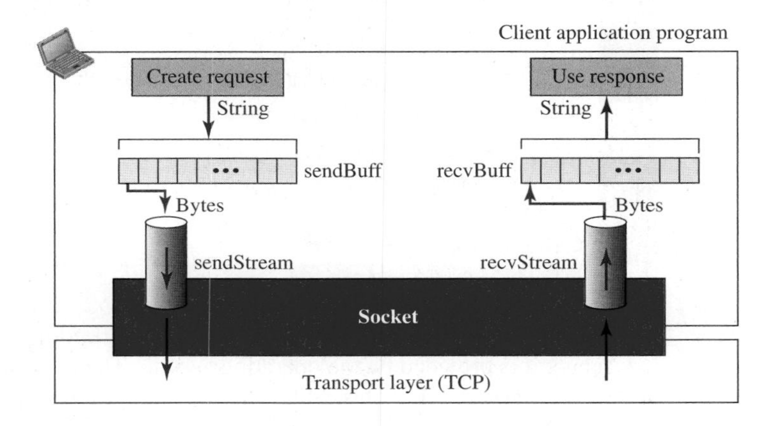

only to our client program, which is very simple; it does not represent a design for standard applications, such as HTTP, which can be more complex.

The design shows that we have a socket object (of type Socket) created by the client program to provide a connection between the client and the transport layer (TCP). We need one byte stream to send bytes to the socket. We also need one byte stream to receive bytes from the socket. In Java these two streams can be created using the two methods *getOutputStream* and *getInputStream,* provided in the Socket class (Table 25.15). In our program we need also two arrays of bytes, called sendBuff and recvBuff, to store bytes before sending them into the sendStream and to store bytes after receiving them from the recvStream. For simplicity, our program assumes that we have a method that makes the request as a string of characters, and a method that uses the response as a string of characters. In this design, we need to convert the request from the string to an array of bytes and change the array of bytes to the response. However, in other designs, we may use other strategies.

TCP Client Program

Table 25.17 shows a simple client program that follows the design in Figure 25.18. We have designed the whole program as a class, named TCPClient, with constructors and instance methods. A brief explanation of the program follows.

Table 25.17 *A simple TCP client program*

```
1   import java.net.*;
2   import java.io.*;
3
4   public class TCPClient
5   {
6       Socket sock;
7       OutputStream sendStream;
8       InputStream recvStream;
9       String request;
10      String response;
11
12      TCPClient (String server, int port) throws IOException, UnknownHostException
13      {
14          sock = new Socket (server, port);
15          sendStream = sock.getOutputStream ();
16          recvStream = sock.getInputStream ();
17      }
18
19      void makeRequest ()
20      {
21          // Add code to make the request string here.
22      }
23
24      void sendRequest ()
25      {
26          try
27          {
28              byte [] sendBuff = new byte [request.length ()];
29              sendBuff = request.getBytes ();
30              sendStream.write (sendBuff, 0, sendBuff.length);
31          }
32          catch (IOException ex)
33          {
34              System.err.println ("IOException in sendRequest");
35          }
36      }
37
38      void getResponse ()
39      {
40          try
41          {
42              int dataSize;
```

Table 25.17 *A simple TCP client program (continued)*

```
43              while ((dataSize = recvStream.available ()) == 0);
44              byte [] recvBuff = new byte [dataSize];
45              recvStream.read (recvBuff, 0, dataSize);
46              response = new String (recvBuff, 0, dataSize);
47          }
48          catch (IOException ex)
49          {
50              System.err.println ("IOException in getResponse");
51          }
52      }
53
54      void useResponse ()
55      {
56          // Add code to use the response string here.
57      }
58
59      void close ()
60      {
61          try
62          {
63              sendStream.close ();
64              recvStream.close ();
65              sock.close ();
66          }
67          catch (IOException ex)
68          {
69              System.err.println ("IOException in close");
70          }
71      }
72
73      public static void main (String [] args) throws IOException
74      {
75          final int servPort = …;                       // Provide server port
76          final String servName = "…";                  // Provide server name
77          TCPClient client = new TCPClient (servName, servPort);
78          client.makeRequest ();
79          client.sendRequest ();
80          client.getResponse ();
81          client.useResponse ();
82          client.close ();
83      }// End of main
84  }// End of TCPClient class
```

The following describes the methods used in the client program:

- ***The main method.*** The execution of the program starts with the main method (lines 73 to 83). The user needs to provide the server port number (an integer) and the server name (a string). The program then creates an instance of the TCPClient class (line 77) that is responsible for creating a request, sending the request, receiving the response, using the response and closing the socket and streams. Lines 78 to 82 call the appropriate methods in the TCPClient class to do the job.
- ***Constructor.*** The constructor (lines 12 to 17) is very simple. It gets the reference to the server name and server port. It uses the server name to find the IP address of the server. The constructor also creates the output and input streams to send and receive data.
- ***The makeRequest method.*** This method (lines 19 to 22) is empty in our design. It needs to be filled by the user of the program to create the request string. We show some cases in the examples.
- ***The sendRequest method.*** This method (lines 24 to 36) is responsible for several tasks:
  1. It creates an empty send buffer (line 28).
  2. It fills up the send buffer with the request string created in the makeRequest Method (line 29).
  3. It writes the contents of the send buffer using the write method of the output stream (line 30).
- ***The getResponse method.*** This method (lines 38 to 53) is responsible for several tasks.
  1. It continuously looks for the size of available bytes (line 43) in an empty loop.
  2. It creates the buffer with the appropriate size (line 44).
  3. It reads data from the input stream and stores it in the receive buffer (line 45).
  4. It converts the bytes in the receive buffer to the string response to be used by the useResponse method (line 46).
- ***The useResponse method.*** This method (lines 54 to 57) is empty in our design. It needs to be filled by the user of the program to use the response sent by the server. We show some cases in the examples.
- ***The close method.*** This method (lines 59 to 71) first closes the streams and then the socket.

25.5 END-CHAPTER MATERIALS

25.5.1 Recommended Reading

For more details about subjects discussed in this chapter, we recommend the following books. The items enclosed in brackets refer to the reference list at the end of the book.

Books

Several books give thorough coverage of materials discussed in this chapter including [Com 06], [Mir 07], [Ste 94], [Tan 03], [Bar et al. 05]. Several books give thorough coverage of network programming in Java including: [CD 08], [Pit 06], and [Har 05].

25.5.2 Key Terms

application programming interface (API)
client-server paradigm
peer-to-peer (P2P) paradigm
process
socket address
socket interface
STREAM
Transport Layer Interface (TLI)

25.5.3 Summary

Applications in the Internet are designed using either a client-server paradigm or a peer-to-peer paradigm. In a client-server paradigm, an application program, called a server, provides services, and another application program, called a client, receives services. A server program is an infinite program; a client program is finite. In a peer-to-peer paradigm, a peer can be both a client and a server.

A server in a client-server paradigm can be designed either as an iterative server or as a concurrent server. An iterative server handles the clients one by one. A concurrent server can simultaneously serve as many clients as the computer resources permit. A client-server pair that uses the services of a connectionless transport layer, such as UDP, should be designed as connectionless programs. A client-server pair that uses the services of a connection-oriented transport layer, such as TCP, should be designed as connection-oriented programs.

Although there are a few ways to write a client or server application program, we discussed only the socket interface approach. The whole idea is to create a new abstract layer, the socket interface layer, between the operating system and the application layer.

Network programming definitely needs to deal with sockets and socket addresses. We showed how these can be done using structures in C and classes in Java. We also gave the code for a simple client-server program in C and Java for use with either UDP or TCP.

25.6 PRACTICE SET

25.6.1 Quizzes

A set of interactive quizzes for this chapter can be found on the book website. It is strongly recommended that the student take the quizzes to check his/her understanding of the materials before continuing with the practice set.

25.6.2 Questions

Q25-1. How is an IP address represented in Java?

Q25-2. You think that a computer with the domain name "aCollege.edu" has several IP addresses. Write a statement in Java to create an array of InetAddress objects associated with this host.

Q25-3. You want to create all InetAddress objects associated with the computer you are working with. Write statements in Java in order to do so.

Q25-4. A new application is to be designed using the client-server paradigm. If only small messages need to be exchanged between the client and the server without concern for message loss or corruption, what transport-layer protocol do you recommend?

Q25-5. You know that the IP address of a computer is "23.14.76.44". Write a statement in Java to create an InetAddress object associated with this address.

Q25-6. Explain which entity provides service and which one receives service in the client-server paradigm.

Q25-7. Which of the following can be a source of data?

a. a keyboard **b.** a monitor **c.** a socket

Q25-8. What is the difference between the DatagramSocket class and the Socket class in Java?

Q25-9. You think that a computer with the IP address "14.26.89.101" has more IP addresses. Write a statement in Java to create an array of InetAddress classes associated with this host.

Q25-10. The DatagramPacket class has two constructors (see Table 25.11). Which constructor can be used as a receiving packet?

Q25-11. What is the difference between the ServerSocket class and the Socket class in Java?

Q25-12. Write a statement in Java that stores the port number 62230 in a variable in Java and guarantees that the number is stored as an unsigned number.

Q25-13. Write a statement in Java to create a socket address bound to InetSocketAddress and the TELNET server process.

Q25-14. You know that the domain name of a computer is "aBusiness.com". Write a statement in Java to create an InetAddress object associated with that computer.

Q25-15. A socket address is the combination of an IP address and a port number that defines an application program running on a host. Can we create an instance of the InetSocketAddress class with an IP address that is not assigned to any host?

Q25-16. Write a statement in Java to create a socket address bound to the host with domain name "some.com" and a client process with port number 51000.

Q25-17. We have input stream classes in Java. Can you explain why the TCP client program does not directly use these classes to create the input stream?

Q25-18. Why do you think that Java uses an instance of a class instead of just an integer to represent an IP address?

Q25-19. In Figure 25.16, what changes are needed if a client needs to send a message other than a string (a picture, for example).

Q25-20. Assume we design a new client-server application program that requires persistent connection. Can we use UDP as the underlying transport-layer protocol for this new application?

Q25-21. Explain how a UDP client program (Table 25.13) sleeps until the response comes from the server.

Q25-22. Write a statement in Java to extract the InetAddress of an InetSocketAddress named sockAd.

Q25-23. Why do you think that Java provides no constructors for the InetAddress class?

Q25-24. Write a statement in Java to create a socket address bound to the local host and the HTTP server process.

Q25-25. Explain how a client process finds the IP address and the port number to be inserted in a remote socket address.

Q25-26. Can a program written to use the services of UDP be run on a computer that has installed TCP as the only transport-layer protocol? Explain.

Q25-27. Assume we add a new protocol to the application layer. What changes do we need to make to other layers?

Q25-28. How does Java distinguish between IPv4 and IPv6 addresses?

Q25-29. Most of the operating systems installed on personal computers come with several client processes, but normally no server processes. Explain the reason.

Q25-30. In Figure 25.15, assume that the request is a URL to retrieve a picture. How is the URL stored in the recvBuff?

Q25-31. Write a statement in Java to create a socket address bound to the local host and the ephemeral port number 56000.

Q25-32. We say that in network programming a socket should be at least bound to a local socket address. The first constructor of the DatagramSocket class (see Table 25.10) has no parameters. Can you explain how it is bound to the local socket address when used at the client site?

Q25-33. A port number in the TCP/IP protocol suite is an unsigned 16-bit integer. How can we represent a port number in Java using a 32-bit integer?

Q25-34. You want to write a program in which you need to refer to the InetAddress of the computer you are working with. Write a statement in Java to create the corresponding object.

Q25-35. In the client-server paradigm, explain why a server should be run all the time, but a client can be run when it is needed.

Q25-36. A source socket address is a combination of an IP address and a port number. Explain what each section identifies.

Q25-37. Write a statement in Java to extract the port number of an InetSocketAddress named sockAd.

Q25-38. How is a Socket object in a TCP client (Figure 25.18) created and destroyed?

Q25-39. The DatagramPacket class has two constructors (see Table 25.11). Which constructor can be used as a sending packet?

Q25-40. In Figure 25.18, how does the client create the input and output streams to communicate with the server?

25.6.3 Problems

P25-1. Write a method in Java to extract the IP address (without the prefix) as a string in dotted-decimal notation from a string representing the CIDR notation ("*x.y.z.t/n*").

P25-2. Write a method in Java to find the last address in the block when one of the addresses in the block is given as a string representing a CIDR notation.

P25-3. Repeat Example 25.4 using the concurrent approach.

P25-4. Write a method in Java to accept a string representing an IP address in the form "*x.y.z.t*" and change it to an unsigned integer.

P25-5. Write a method in Java to find the first address (network address) in the block when one of the addresses in the block is given as a string representing a CIDR notation.

P25-6. Write a method in Java to extract the prefix of an address (as an integer) given a string representing a CIDR notation in the form "*x.y.z.t/n*".

P25-7. Write a method in Java to find the size of a block when one address in the block (in CIDR notation) is given.

P25-8. Write a method in Java to add a given prefix (as an integer) at the end of an IP address to create a string representing the CIDR notation ("*x.y.z.t/n*").

P25-9. In Figure 25.10 in the text, how does the server know that a client has requested a service?

P25-10. Write a method in Java to find the range of addresses when the beginning and the ending addresses are given.

P25-11. Assume we want to make the TCP client program in Table 25.2 more generic to be able to send a string and to handle the response received from the server. Show how this can be done.

P25-12. Write a method in Java to change an unsigned 32-bit integer representing a mask to an integer representing a prefix (/n).

P25-13. Write a method in Java to change an integer representing a prefix to an unsigned 32-bit integer representing the numeric mask.

P25-14. In Figure 25.12 in the text, how is the socket created for data transfer at the server site?

P25-15. Write a method in Java to convert a 32-bit integer to a string representing an IP address in the form "*x.y.z.t*".

P25-16. Repeat Example 25.4 using the service of TCP.

P25-17. Repeat Example 25.4 using the service of TCP and the concurrent approach.

25.7 SIMULATION EXPERIMENTS

25.7.1 Applets

We have created some Java applets to show some of the main concepts discussed in this chapter. It is strongly recommended that the students activate these applets on the book website and carefully examine the protocols in action.

25.8 PROGRAMMING ASSIGNMENT

Write, compile, and test the following programs using the computer language of your choice:

Prg25-1. Write a program to make the UDP server program in Table 25.1 more generic: to receive a request, to process the request, and to send back the response.

Prg25-2. Write a program to make the UDP client program in Table 25.2 more generic to be able to send any request created by the client program.

Prg25-3. Write a program to make the TCP server program in Table 25.3 more generic: to receive a request, to process the request, and to send back the response.

Prg25-4. Write a program to make the TCP client program in Table 25.4 more generic to be able to send any request created by the program.

Prg25-5. Modify, compile, and test the client program in Table 25.13 and the server program in Table 25.12 to do the following: The client program needs to read the request string from a file and store the response string in another file. The name of the file needs to be passed as the argument to the main method of the client program. The server program needs to accept the request string, change all lowercase letters to uppercase letters, and return the result.

Prg25-6. Modify, compile, and test the client program in Table 25.17 and the server program in Table 25.16 to allow the client to provide the pathname of a short file stored on the server host. The server needs to send the contents of the short file as a string of characters. The client stores the file at the client host. This means simulating a simple file transfer protocol.

Prg25-7. Modify, compile, and test the client program in Table 25.17 and the server program in Table 25.16 to simulate a local DNS client and server. The server has a short table made of two columns, domain name and IP address. The client can send two types of requests: normal and reverse. The normal request is a string in the format "*N:domain name*"; the reverse request is in the format "*R:IP address*". The server responds with either the IP address, the domain name, or the message "Not found."

CHAPTER 26

Standard Client-Server Protocols

After introducing the application layer in the previous chapter, we discuss some standard application-layer protocols in this chapter. During the lifetime of the Internet, several client-server application programs have been developed. We do not have to redefine them, but we need to understand what they do. For each application, we also need to know the options available to us. The study of these applications and the ways they provide different services can help us to create customized applications in the future.

We have selected six standard application programs in this section. Some other applications have been or will be discussed in other chapters. Dynamic Host Configuration Protocol (DHCP) was discussed in Chapter 18 and Simple Network Management Protocol (SNMP) will be discussed in Chapter 27.

This chapter is made of six sections:

- The first section introduces the World Wide Web. It then discusses the HyperText Transfer Protocol, the most common client-server application program used in relation to the World Wide Web.
- The second section discusses the File Transfer Protocol, which is the standard protocol provided by TCP/IP for copying a file from one host to another.
- The third section discusses electronic mail, which involves two protocols: SMPT and POP. As we will see, the nature of this application is different from the other two previous applications. We need two different protocols to handle electronic mail.
- The fourth section discusses TELNET, a general client-server program that allows users to log in to a remote machine and use any application available on the remote host.
- The fifth section discusses Secure Shell, which can be used as a secured TELNET, but it can also provide a secure tunnel for other applications.
- The sixth section talks about the Domain Name System, which acts as the directory system in the Internet. It maps the name of an entity to its IP address.

26.1 WORLD WIDE WEB AND HTTP

In this section, we first introduce the **World Wide Web** (abbreviated WWW or Web). We then discuss the HyperText Transfer Protocol (HTTP), the most common client-server application program used in relation to the Web.

26.1.1 World Wide Web

The idea of the Web was first proposed by Tim Berners-Lee in 1989 at *CERN*†, the European Organization for Nuclear Research, to allow several researchers at different locations throughout Europe to access each others' researches. The commercial Web started in the early 1990s.

The Web today is a repository of information in which the documents, called ***web pages,*** are distributed all over the world and related documents are linked together. The popularity and growth of the Web can be related to two terms in the above statement: *distributed* and *linked*. Distribution allows the growth of the Web. Each web server in the world can add a new web page to the repository and announce it to all Internet users without overloading a few servers. Linking allows one web page to refer to another web page stored in another server somewhere else in the world. The linking of web pages was achieved using a concept called ***hypertext,*** which was introduced many years before the advent of the Internet. The idea was to use a machine that automatically retrieved another document stored in the system when a link to it appeared in the document. The Web implemented this idea electronically to allow the linked document to be retrieved when the link was clicked by the user. Today, the term *hypertext,* coined to mean linked text documents, has been changed to ***hypermedia,*** to show that a web page can be a text document, an image, an audio file, or a video file.

The purpose of the Web has gone beyond the simple retrieving of linked documents. Today, the Web is used to provide electronic shopping and gaming. One can use the Web to listen to radio programs or view television programs whenever one desires without being forced to listen to or view these programs when they are broadcast.

Architecture

The WWW today is a distributed client-server service, in which a client using a browser can access a service using a server. However, the service provided is distributed over many locations called *sites*. Each site holds one or more web pages. Each web page, however, can contain some links to other web pages in the same or other sites. In other words, a web page can be simple or composite. A simple web page has no links to other web pages; a composite web page has one or more links to other web pages. Each web page is a file with a name and address.

Example 26.1

Assume we need to retrieve a scientific document that contains one reference to another text file and one reference to a large image. Figure 26.1 shows the situation.

The main document and the image are stored in two separate files (file A and file B) in the same site; the referenced text file (file C) is stored in another site. Since we are dealing with three

† In French: *Conseil Européen pour la Recherche Nucléaire*

Figure 26.1 *Example 26.1*

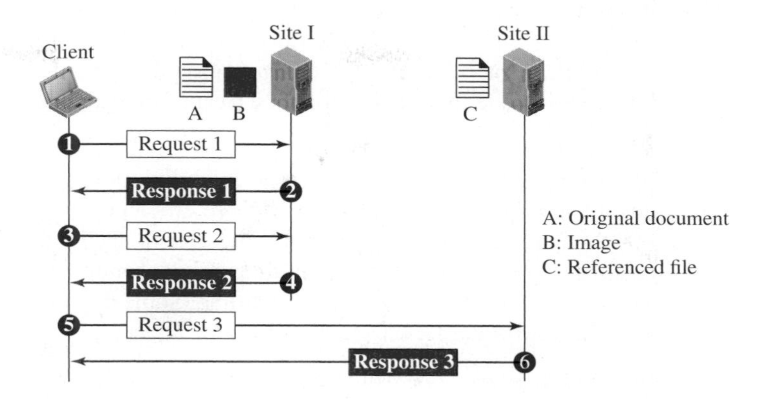

different files, we need three transactions if we want to see the whole document. The first transaction (request/response) retrieves a copy of the main document (file A), which has references (pointers) to the second and third files. When a copy of the main document is retrieved and browsed, the user can click on the reference to the image to invoke the second transaction and retrieve a copy of the image (file B). If the user needs to see the contents of the referenced text file, she can click on its reference (pointer) invoking the third transaction and retrieving a copy of file C. Note that although files A and B both are stored in site I, they are independent files with different names and addresses. Two transactions are needed to retrieve them. A very important point we need to remember is that file A, file B, and file C in Example 26.1 are independent web pages, each with independent names and addresses. Although references to file B or C are included in file A, it does not mean that each of these files cannot be retrieved independently. A second user can retrieve file B with one transaction. A third user can retrieve file C with one transaction.

Web Client (Browser)

A variety of vendors offer commercial **browsers** that interpret and display a web page, and all of them use nearly the same architecture. Each browser usually consists of three parts: a controller, client protocols, and interpreters. (see Figure 26.2).

Figure 26.2 *Browser*

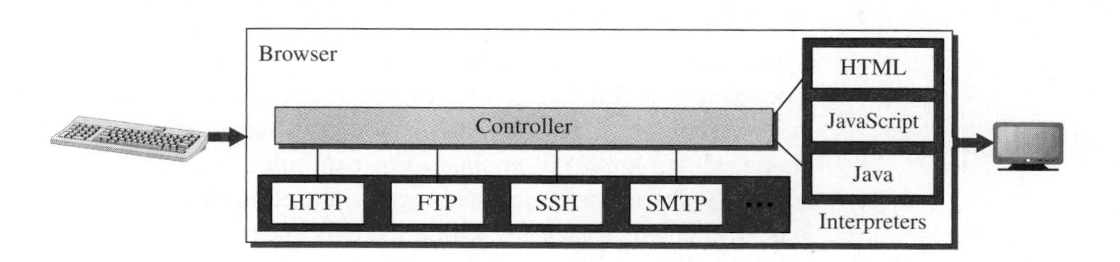

The controller receives input from the keyboard or the mouse and uses the client programs to access the document. After the document has been accessed, the controller uses one of the interpreters to display the document on the screen. The client protocol can be one of the protocols described later, such as HTTP or FTP. The interpreter can be HTML, Java, or JavaScript, depending on the type of document. Some commercial browsers include Internet Explorer, Netscape Navigator, and Firefox.

Web Server

The web page is stored at the server. Each time a request arrives, the corresponding document is sent to the client. To improve efficiency, servers normally store requested files in a cache in memory; memory is faster to access than a disk. A server can also become more efficient through multithreading or multiprocessing. In this case, a server can answer more than one request at a time. Some popular web servers include Apache and Microsoft Internet Information Server.

Uniform Resource Locator (URL)

A web page, as a file, needs to have a unique identifier to distinguish it from other web pages. To define a web page, we need three identifiers: *host, port,* and *path*. However, before defining the web page, we need to tell the browser what client-server application we want to use, which is called the *protocol*. This means we need four identifiers to define the web page. The first is the type of vehicle to be used to fetch the web page; the last three make up the combination that defines the destination object (web page).

- ***Protocol.*** The first identifier is the abbreviation for the client-server program that we need in order to access the web page. Although most of the time the protocol is HTTP (HyperText Transfer Protocol), which we will discuss shortly, we can also use other protocols such as FTP (File Transfer Protocol).
- ***Host.*** The host identifier can be the IP address of the server or the unique name given to the server. IP addresses can be defined in dotted decimal notation, as described in Chapter 18 (such as 64.23.56.17); the name is normally the domain name that uniquely defines the host, such as *forouzan.com,* which we discuss in Domain Name System (DNS) later in this chapter.
- ***Port.*** The port, a 16-bit integer, is normally predefined for the client-server application. For example, if the HTTP protocol is used for accessing the web page, the well-known port number is 80. However, if a different port is used, the number can be explicitly given.
- ***Path.*** The path identifies the location and the name of the file in the underlying operating system. The format of this identifier normally depends on the operating system. In UNIX, a path is a set of directory names followed by the file name, all separated by a slash. For example, */top/next/last/myfile* is a path that uniquely defines a file named *myfile,* stored in the directory *last,* which itself is part of the directory *next,* which itself is under the directory *top*. In other words, the path lists the directories from the top to the bottom, followed by the file name.

To combine these four pieces together, the **uniform resource locator (URL)** has been designed; it uses three different separators between the four pieces as shown below:

protocol://host/path	**Used most of the time**
protocol://host:port/path	**Used when port number is needed**

Example 26.2

The URL ***http://www.mhhe.com/compsci/forouzan/*** defines the web page related to one of the authors of this book. The string ***www.mhhe.com*** is the name of the computer in the McGraw-Hill company (the three letters ***www*** are part of the host name and are added to the commercial host). The path is ***compsci/forouzan/,*** which defines Forouzan's web page under the directory ***compsci*** (computer science).

Web Documents

The documents in the WWW can be grouped into three broad categories: static, dynamic, and active.

Static Documents

Static documents are fixed-content documents that are created and stored in a server. The client can get a copy of the document only. In other words, the contents of the file are determined when the file is created, not when it is used. Of course, the contents in the server can be changed, but the user cannot change them. When a client accesses the document, a copy of the document is sent. The user can then use a browser to see the document. Static documents are prepared using one of several languages: *HyperText Markup Language* (HTML), *Extensible Markup Language* (XML), *Extensible Style Language* (XSL), and *Extensible Hypertext Markup Language* (XHTML). We discuss these languages in Appendix C.

Dynamic Documents

A **dynamic document** is created by a web server whenever a browser requests the document. When a request arrives, the web server runs an application program or a script that creates the dynamic document. The server returns the result of the program or script as a response to the browser that requested the document. Because a fresh document is created for each request, the contents of a dynamic document may vary from one request to another. A very simple example of a dynamic document is the retrieval of the time and date from a server. Time and date are kinds of information that are dynamic in that they change from moment to moment. The client can ask the server to run a program such as the *date* program in UNIX and send the result of the program to the client. Although the *Common Gateway Interface* (*CGI*) was used to retrieve a dynamic document in the past, today's options include one of the scripting languages such as *Java Server Pages* (*JSP*), which uses the Java language for scripting, or *Active Server Pages* (*ASP*), a Microsoft product that uses Visual Basic language for scripting, or *ColdFusion,* which embeds queries in a Structured Query Language (SQL) database in the HTML document.

Active Documents

For many applications, we need a program or a script to be run at the client site. These are called ***active documents.*** For example, suppose we want to run a program that creates animated graphics on the screen or a program that interacts with the user. The program definitely needs to be run at the client site where the animation or interaction takes place. When a browser requests an active document, the server sends a copy of the document or a script. The document is then run at the client (browser) site. One way to create an active document is to use *Java applets,* a program written in Java on the server. It is compiled and ready to be run. The document is in bytecode (binary) format. Another way is to use *JavaScripts* but download and run the script at the client site.

26.1.2 HyperText Transfer Protocol (HTTP)

The **HyperText Transfer Protocol (HTTP)** is used to define how the client-server programs can be written to retrieve web pages from the Web. An HTTP client sends a request; an HTTP server returns a response. The server uses the port number 80; the client uses a temporary port number. HTTP uses the services of TCP, which, as discussed before, is a connection-oriented and reliable protocol. This means that, before any transaction between the client and the server can take place, a connection needs to be established between them. After the transaction, the connection should be terminated. The client and server, however, do not need to worry about errors in messages exchanged or loss of any message, because the TCP is reliable and will take care of this matter, as we saw in Chapter 24.

Nonpersistent versus Persistent Connections

As we discussed in the previous section, the hypertext concept embedded in web page documents may require several requests and responses. If the web pages, objects to be retrieved, are located on different servers, we do not have any other choice than to create a new TCP connection for retrieving each object. However, if some of the objects are located on the same server, we have two choices: to retrieve each object using a new TCP connection or to make a TCP connection and retrieve them all. The first method is referred to as a *nonpersistent connection,* the second as a *persistent connection.* HTTP, prior to version 1.1, specified *nonpersistent* connections, while *persistent* connections are the default in version 1.1, but it can be changed by the user.

Nonpersistent Connections

In a **nonpersistent connection**, one TCP connection is made for each request/response. The following lists the steps in this strategy:

1. The client opens a TCP connection and sends a request.
2. The server sends the response and closes the connection.
3. The client reads the data until it encounters an end-of-file marker; it then closes the connection.

In this strategy, if a file contains links to N different pictures in different files (all located on the same server), the connection must be opened and closed $N + 1$ times. The nonpersistent strategy imposes high overhead on the server because the server needs $N + 1$ different buffers each time a connection is opened.

Example 26.3

Figure 26.3 shows an example of a nonpersistent connection. The client needs to access a file that contains one link to an image. The text file and image are located on the same server. Here we need two connections. For each connection, TCP requires at least three handshake messages to establish the connection, but the request can be sent with the third one. After the connection is established, the object can be transferred. After receiving an object, another three handshake messages are needed to terminate the connection, as we saw in Chapter 24. This means that the

Figure 26.3 *Example 26.3*

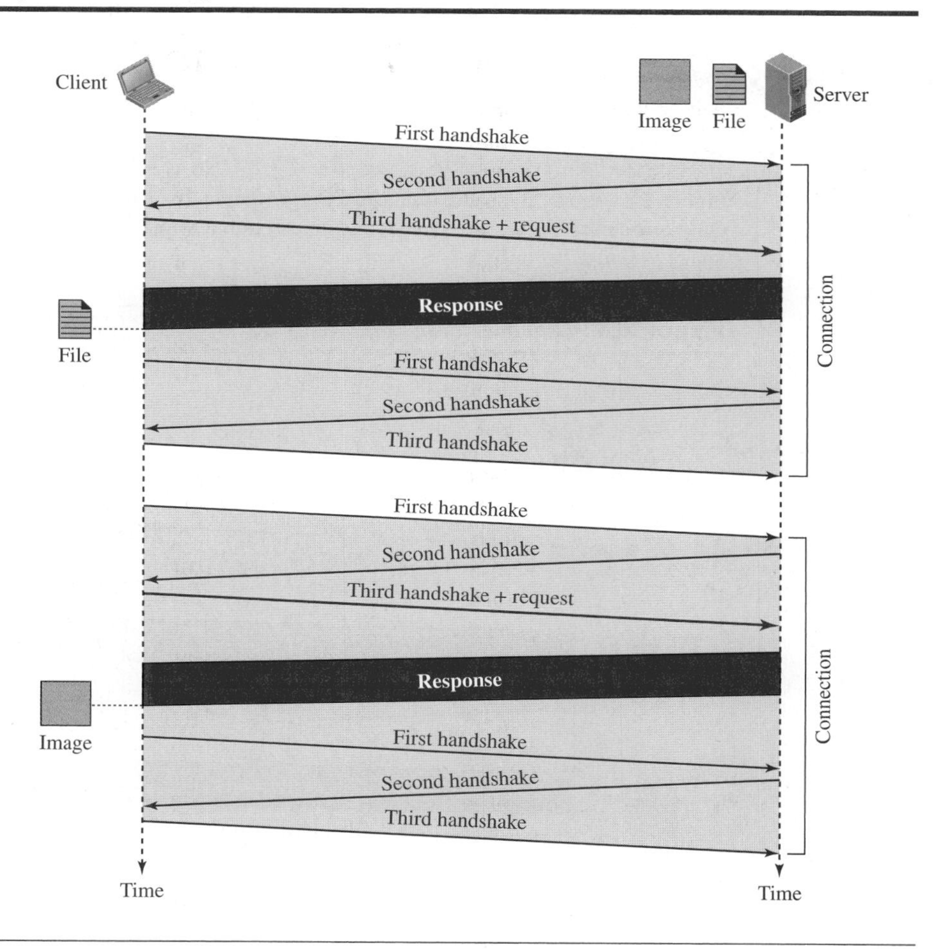

client and server are involved in two connection establishments and two connection terminations. If the transaction involves retrieving 10 or 20 objects, the round trip times spent for these handshakes add up to a big overhead. When we describe the client-server programming at the end of the chapter, we will show that for each connection the client and server need to allocate extra resources such as buffers and variables. This is another burden on both sites, but especially on the server site.

Persistent Connections

HTTP version 1.1 specifies a **persistent connection** by default. In a persistent connection, the server leaves the connection open for more requests after sending a response.

The server can close the connection at the request of a client or if a time-out has been reached. The sender usually sends the length of the data with each response. However, there are some occasions when the sender does not know the length of the data. This is the case when a document is created dynamically or actively. In these cases, the server informs the client that the length is not known and closes the connection after sending the data so the client knows that the end of the data has been reached. Time and resources are saved using persistent connections. Only one set of buffers and variables needs to be set for the connection at each site. The round trip time for connection establishment and connection termination is saved.

Example 26.4

Figure 26.4 shows the same scenario as in Example 26.3, but using a persistent connection. Only one connection establishment and connection termination is used, but the request for the image is sent separately.

Figure 26.4 *Example 26.4*

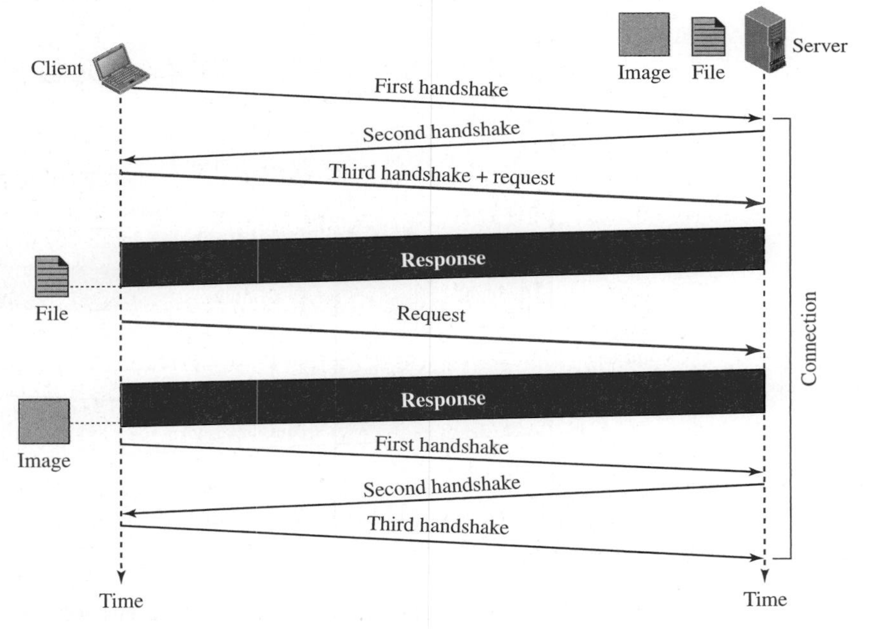

Message Formats

The HTTP protocol defines the format of the request and response messages, as shown in Figure 26.5. We have put the two formats next to each other for comparison. Each message is made of four sections. The first section in the request message is called the *request line;* the first section in the response message is called the *status line*. The other three sections have the same names in the request and response messages. However, the

Figure 26.5 *Formats of the request and response messages*

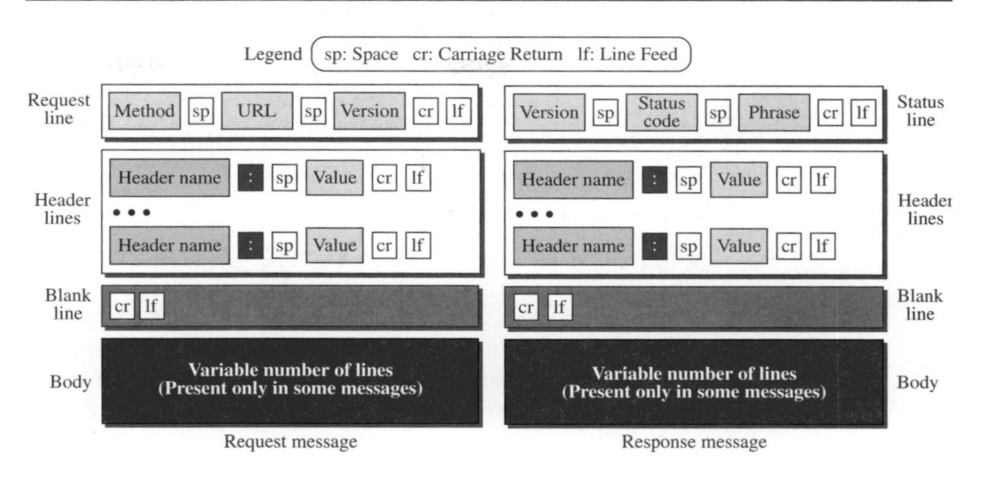

similarities between these sections are only in the names; they may have different contents. We discuss each message type separately.

Request Message

As we said before, the first line in a request message is called a request line. There are three fields in this line separated by one space and terminated by two characters (carriage return and line feed) as shown in Figure 26.5. The fields are called *method, URL*, and *version*.

The method field defines the request types. In version 1.1 of HTTP, several methods are defined, as shown in Table 26.1. Most of the time, the client uses the GET method to send a request. In this case, the body of the message is empty. The HEAD method is used when the client needs only some information about the web page from the server, such as the last time it was modified. It can also be used to test the validity of a URL. The response message in this case has only the header section; the body section is empty. The PUT method is the inverse of the GET method; it allows the client to post a new web page on the server (if permitted). The POST method is similar to the PUT method, but it is used to send some information to the server to be added to the web page or to modify the web page. The TRACE method is used for debugging; the client asks the server to echo back the request to check whether the server is getting the requests. The DELETE method allows the client to delete a web page on the server if the client has permission to do so. The CONNECT method was originally made as a reserve method; it may be used by proxy servers, as discussed later. Finally, the OPTIONS method allows the client to ask about the properties of a web page.

The second field, URL, was discussed earlier in the chapter. It defines the address and name of the corresponding web page. The third field, version, gives the version of the protocol; the most current version of HTTP is 1.1.

Table 26.1 *Methods*

Method	*Action*
GET	Requests a document from the server
HEAD	Requests information about a document but not the document itself
PUT	Sends a document from the client to the server
POST	Sends some information from the client to the server
TRACE	Echoes the incoming request
DELETE	Removes the web page
CONNECT	Reserved
OPTIONS	Inquires about available options

After the request line, we can have zero or more *request header* lines. Each header line sends additional information from the client to the server. For example, the client can request that the document be sent in a special format. Each header line has a header name, a colon, a space, and a header value (see Figure 26.5). Table 26.2 shows some header names commonly used in a request. The value field defines the values associated with each header name. The list of values can be found in the corresponding RFCs.

The body can be present in a request message. Usually, it contains the comment to be sent or the file to be published on the website when the method is PUT or POST.

Table 26.2 *Request header names*

Header	*Description*
User-agent	Identifies the client program
Accept	Shows the media format the client can accept
Accept-charset	Shows the character set the client can handle
Accept-encoding	Shows the encoding scheme the client can handle
Accept-language	Shows the language the client can accept
Authorization	Shows what permissions the client has
Host	Shows the host and port number of the client
Date	Shows the current date
Upgrade	Specifies the preferred communication protocol
Cookie	Returns the cookie to the server (explained later)
If-Modified-Since	If the file is modified since a specific date

Response Message

The format of the response message is also shown in Figure 26.5. A response message consists of a status line, header lines, a blank line, and sometimes a body. The first line in a response message is called the *status line*. There are three fields in this line separated by spaces and terminated by a carriage return and line feed. The first field defines the version of HTTP protocol, currently 1.1. The status code field defines the status of the request. It consists of three digits. Whereas the codes in the 100 range are only informational, the codes in the 200 range indicate a successful request. The codes in the 300 range redirect the client to another URL, and the codes in the 400 range indicate an error at the client site. Finally, the codes in the 500 range indicate an error at the server site. The status phrase explains the status code in text form.

After the status line, we can have zero or more *response header* lines. Each header line sends additional information from the server to the client. For example, the sender can send extra information about the document. Each header line has a header name, a colon, a space, and a header value. We will show some header lines in the examples at the end of this section. Table 26.3 shows some header names commonly used in a response message.

Table 26.3 *Response header names*

Header	*Description*
Date	Shows the current date
Upgrade	Specifies the preferred communication protocol
Server	Gives information about the server
Set-Cookie	The server asks the client to save a cookie
Content-Encoding	Specifies the encoding scheme
Content-Language	Specifies the language
Content-Length	Shows the length of the document
Content-Type	Specifies the media type
Location	To ask the client to send the request to another site
Accept-Ranges	The server will accept the requested byte-ranges
Last-modified	Gives the date and time of the last change

The body contains the document to be sent from the server to the client. The body is present unless the response is an error message.

Example 26.5

This example retrieves a document (see Figure 26.6). We use the GET method to retrieve an image with the path ***/usr/bin/image1.*** The request line shows the method (GET), the URL, and the HTTP version (1.1). The header has two lines that show that the client can accept images in the GIF or JPEG format. The request does not have a body. The response message contains the status line and four lines of header. The header lines define the date, server, content encoding (MIME version, which will be described in electronic mail), and length of the document. The body of the document follows the header.

Example 26.6

In this example, the client wants to send a web page to be posted on the server. We use the PUT method. The request line shows the method (PUT), URL, and HTTP version (1.1). There are four lines of headers. The request body contains the web page to be posted. The response message contains the status line and four lines of headers. The created document, which is a CGI document, is included as the body (see Figure 26.7).

Conditional Request

A client can add a condition in its request. In this case, the server will send the requested web page if the condition is met or inform the client otherwise. One of the most common conditions imposed by the client is the time and date the web

Figure 26.6 *Example 26.5*

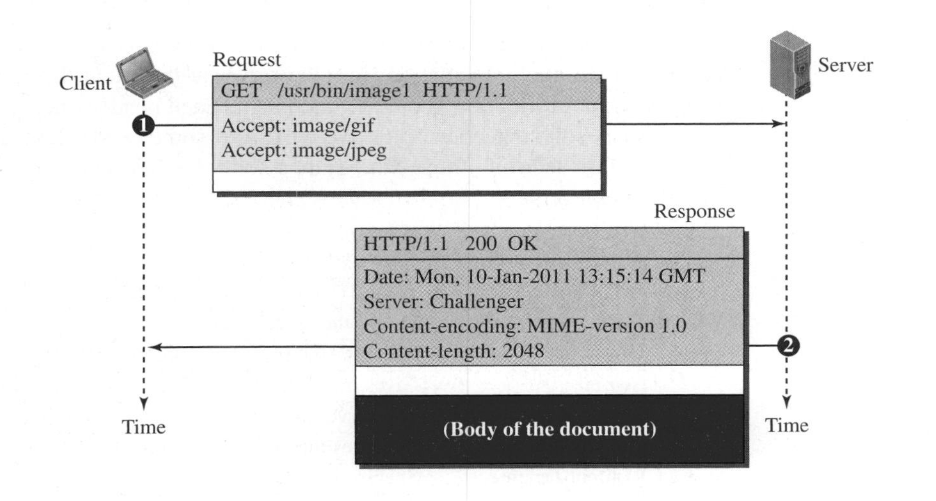

page is modified. The client can send the header line *If-Modified-Since* with the request to tell the server that it needs the page only if it is modified after a certain point in time.

Figure 26.7 *Example 26.6*

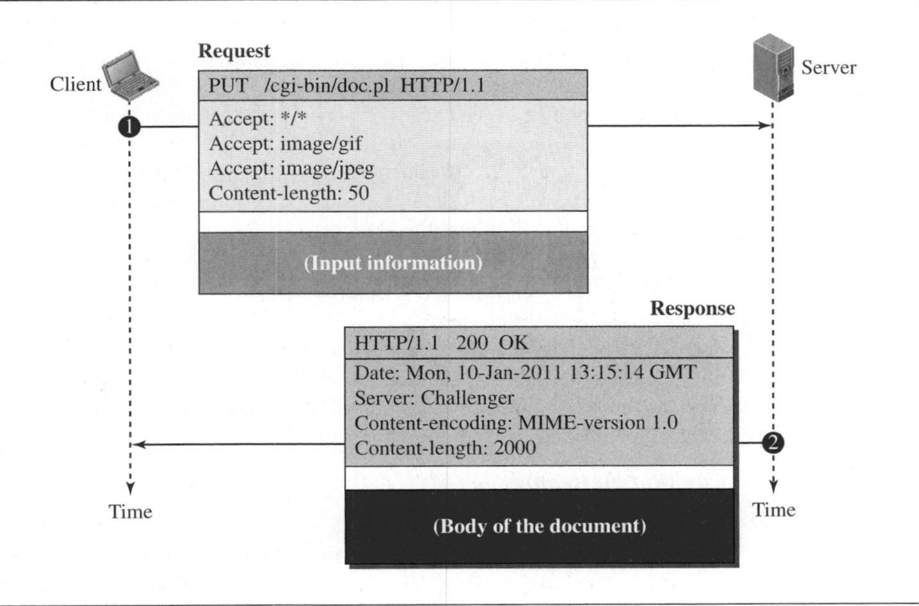

Example 26.7

The following shows how a client imposes the modification data and time condition on a request.

GET http://www.commonServer.com/information/file1 HTTP/1.1	**Request line**
If-Modified-Since: Thu, Sept 04 00:00:00 GMT	**Header line**
	Blank line

The status line in the response shows the file was not modified after the defined point in time. The body of the response message is also empty.

HTTP/1.1 304 Not Modified	**Status line**
Date: Sat, Sept 06 08 16:22:46 GMT	**First header line**
Server: commonServer.com	**Second header line**
	Blank line
(Empty Body)	**Empty body**

Cookies

The World Wide Web was originally designed as a stateless entity. A client sends a request; a server responds. Their relationship is over. The original purpose of the Web, retrieving publicly available documents, exactly fits this design. Today the Web has other functions that need to remember some information about the clients; some are listed below:

- Websites are being used as *electronic stores* that allow users to browse through the store, select wanted items, put them in an electronic cart, and pay at the end with a credit card.
- Some websites need to allow access to *registered clients* only.
- Some websites are used as *portals:* the user selects the web pages he wants to see.
- Some websites are just *advertising* agencies.

For these purposes, the **cookie** mechanism was devised.

Creating and Storing Cookies

The creation and storing of cookies depend on the implementation; however, the principle is the same.

1. When a server receives a request from a client, it stores information about the client in a file or a string. The information may include the domain name of the client, the contents of the cookie (information the server has gathered about the client such as name, registration number, and so on), a timestamp, and other information depending on the implementation.
2. The server includes the cookie in the response that it sends to the client.
3. When the client receives the response, the browser stores the cookie in the cookie directory, which is sorted by the server domain name.

Using Cookies

When a client sends a request to a server, the browser looks in the cookie directory to see if it can find a cookie sent by that server. If found, the cookie is included in the request. When the server receives the request, it knows that this is an old client, not a new one. Note that the contents of the cookie are never read by the browser or disclosed to the user. It is a cookie *made* by the server and *eaten* by the server. Now let us see how a cookie is used for the four previously mentioned purposes:

- An *electronic store* (e-commerce) can use a cookie for its client shoppers. When a client selects an item and inserts it in a cart, a cookie that contains information about the item, such as its number and unit price, is sent to the browser. If the client selects a second item, the cookie is updated with the new selection information, and so on. When the client finishes shopping and wants to check out, the last cookie is retrieved and the total charge is calculated.
- The site that restricts access to *registered clients* only sends a cookie to the client when the client registers for the first time. For any repeated access, only those clients that send the appropriate cookie are allowed.
- A web *portal* uses the cookie in a similar way. When a user selects her favorite pages, a cookie is made and sent. If the site is accessed again, the cookie is sent to the server to show what the client is looking for.
- A cookie is also used by *advertising* agencies. An advertising agency can place banner ads on some main website that is often visited by users. The advertising agency supplies only a URL that gives the advertising agency's address instead of the banner itself. When a user visits the main website and clicks the icon of a corporation, a request is sent to the advertising agency. The advertising agency sends the requested banner, but it also includes a cookie with the ID of the user. Any future use of the banners adds to the database that profiles the Web behavior of the user. The advertising agency has compiled the interests of the user and can sell this information to other parties. This use of cookies has made them very controversial. Hopefully, some new regulations will be devised to preserve the privacy of users.

Example 26.8

Figure 26.8 shows a scenario in which an electronic store can benefit from the use of cookies. Assume a shopper wants to buy a toy from an electronic store named BestToys. The shopper browser (client) sends a request to the BestToys server. The server creates an empty shopping cart (a list) for the client and assigns an ID to the cart (for example, 12343). The server then sends a response message, which contains the images of all toys available, with a link under each toy that selects the toy if it is being clicked. This response message also includes the Set-Cookie header line whose value is 12343. The client displays the images and stores the cookie value in a file named BestToys. The cookie is not revealed to the shopper. Now the shopper selects one of the toys and clicks on it. The client sends a request, but includes the ID 12343 in the Cookie header line. Although the server may have been busy and forgotten about this shopper, when it receives the request and checks the header, it finds the value 12343 as the cookie. The server knows that the customer is not new; it searches for a shopping cart with ID 12343. The shopping cart (list) is opened and the selected toy is inserted in the list. The server now sends another response to the shopper to tell her the total price and ask her to provide payment. The shopper provides information about her credit card and sends a new request with the ID 12343 as the cookie value. When the request arrives at the server, it again sees the ID 12343, and accepts the order and the payment and sends a confirmation in a response. Other information about the client is stored in

Figure 26.8 *Example 26.8*

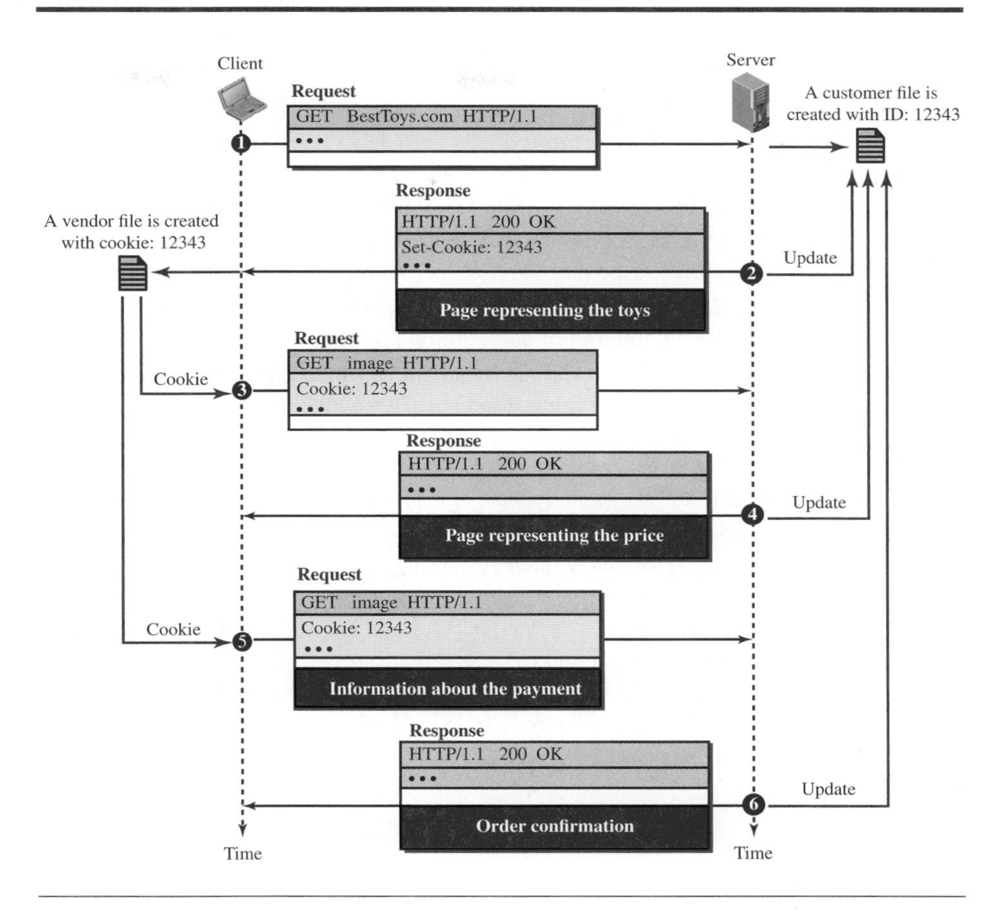

the server. If the shopper accesses the store sometime in the future, the client sends the cookie again; the store retrieves the file and has all the information about the client.

Web Caching: Proxy Servers

HTTP supports **proxy servers**. A proxy server is a computer that keeps copies of responses to recent requests. The HTTP client sends a request to the proxy server. The proxy server checks its cache. If the response is not stored in the cache, the proxy server sends the request to the corresponding server. Incoming responses are sent to the proxy server and stored for future requests from other clients.

The proxy server reduces the load on the original server, decreases traffic, and improves latency. However, to use the proxy server, the client must be configured to access the proxy instead of the target server.

Note that the proxy server acts as both server and client. When it receives a request from a client for which it has a response, it acts as a server and sends the response to the client. When it receives a request from a client for which it does not have a response, it first acts as a client and sends a request to the target server. When the response has been received, it acts again as a server and sends the response to the client.

Proxy Server Location

The proxy servers are normally located at the client site. This means that we can have a hierarchy of proxy servers, as shown below:

1. A client computer can also be used as a proxy server, in a small capacity, that stores responses to requests often invoked by the client.
2. In a company, a proxy server may be installed on the computer LAN to reduce the load going out of and coming into the LAN.
3. An ISP with many customers can install a proxy server to reduce the load going out of and coming into the ISP network.

Example 26.9

Figure 26.9 shows an example of a use of a proxy server in a local network, such as the network

Figure 26.9 *Example of a proxy server*

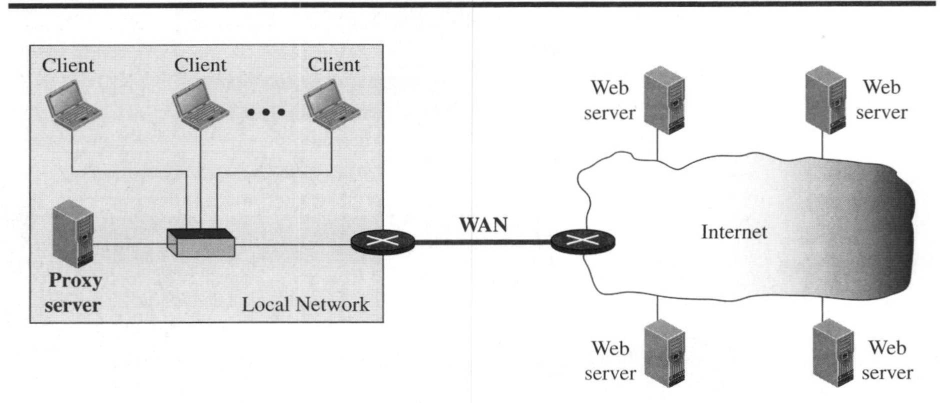

on a campus or in a company. The proxy server is installed in the local network. When an HTTP request is created by any of the clients (browsers), the request is first directed to the proxy server. If the proxy server already has the corresponding web page, it sends the response to the client. Otherwise, the proxy server acts as a client and sends the request to the web server in the Internet. When the response is returned, the proxy server makes a copy and stores it in its cache before sending it to the requesting client.

Cache Update

A very important question is how long a response should remain in the proxy server before being deleted and replaced. Several different strategies are used for this purpose. One solution is to store the list of sites whose information remains the same for a while. For example, a news agency may change its news page every morning. This means that a proxy server can get the news early in the morning and keep it until the next day. Another recommendation is to add some headers to show the last modification time of the information. The proxy server can then use the information in this header to guess how long the information would be valid.

HTTP Security

HTTP per se does not provide security. However, as we show in Chapter 32, HTTP can be run over the Secure Socket Layer (SSL). In this case, HTTP is referred to as HTTPS. HTTPS provides confidentiality, client and server authentication, and data integrity.

26.2 FTP

File Transfer Protocol (FTP) is the standard protocol provided by TCP/IP for copying a file from one host to another. Although transferring files from one system to another seems simple and straightforward, some problems must be dealt with first. For example, two systems may use different file name conventions. Two systems may have different ways to represent data. Two systems may have different directory structures. All of these problems have been solved by FTP in a very simple and elegant approach. Although we can transfer files using HTTP, FTP is a better choice to transfer large files or to transfer files using different formats. Figure 26.10 shows the

Figure 26.10 *FTP*

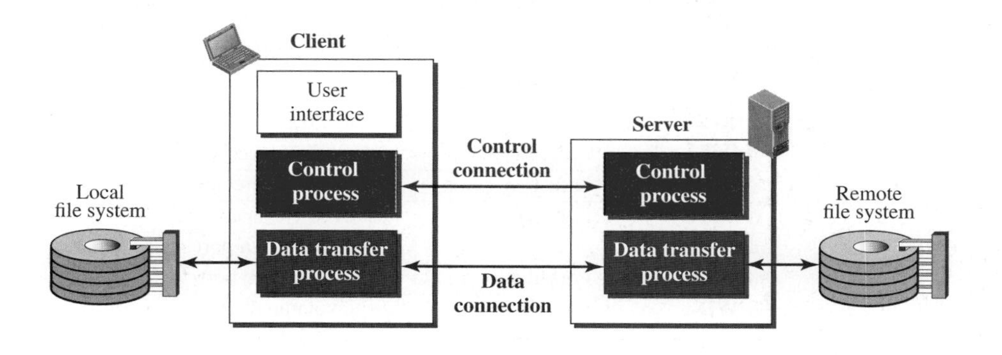

basic model of FTP. The client has three components: the user interface, the client control process, and the client data transfer process. The server has two components: the server control process and the server data transfer process. The control connection is made between the control processes. The data connection is made between the data transfer processes.

Separation of commands and data transfer makes FTP more efficient. The control connection uses very simple rules of communication. We need to transfer only a line of command or a line of response at a time. The data connection, on the other hand, needs more complex rules due to the variety of data types transferred.

26.2.1 Two Connections

The two connections in FTP have different lifetimes. The control connection remains connected during the entire interactive FTP session. The data connection is opened and then closed for each file transfer activity. It opens each time commands that involve transferring files are used, and it closes when the file is transferred. In other words, when a user starts an FTP session, the control connection opens. While the control connection is open, the data connection can be opened and closed multiple times if several files are transferred. FTP uses two well-known TCP ports: port 21 is used for the control connection, and port 20 is used for the data connection.

26.2.2 Control Connection

For control communication, FTP uses the same approach as TELNET (discussed later). It uses the NVT ASCII character set as used by TELNET. Communication is achieved through commands and responses. This simple method is adequate for the control connection because we send one command (or response) at a time. Each line is terminated with a two-character (carriage return and line feed) end-of-line token.

During this control connection, commands are sent from the client to the server and responses are sent from the server to the client. Commands, which are sent from the FTP client control process, are in the form of ASCII uppercase, which may or may not be followed by an argument. Some of the most common commands are shown in Table 26.4.

Table 26.4 *Some FTP commands*

Command	*Argument(s)*	*Description*
ABOR		Abort the previous command
CDUP		Change to parent directory
CWD	Directory name	Change to another directory
DELE	File name	Delete a file
LIST	Directory name	List subdirectories or files
MKD	Directory name	Create a new directory
PASS	User password	Password
PASV		Server chooses a port
PORT	Port identifier	Client chooses a port
PWD		Display name of current directory
QUIT		Log out of the system
RETR	File name(s)	Retrieve files; files are transferred from server to client
RMD	Directory name	Delete a directory
RNFR	File name (old)	Identify a file to be renamed
RNTO	File name (new)	Rename the file
STOR	File name(s)	Store files; file(s) are transferred from client to server
STRU	**F**, **R**, or **P**	Define data organization (**F**: file, **R**: record, or **P**: page)
TYPE	**A**, **E**, **I**	Default file type (**A**: ASCII, **E**: EBCDIC, **I**: image)
USER	User ID	User information
MODE	**S**, **B**, or **C**	Define transmission mode (**S**: stream, **B**: block, or **C**: compressed

Every FTP command generates at least one response. A response has two parts: a three-digit number followed by text. The numeric part defines the code; the text part defines needed parameters or further explanations. The first digit defines the status of the command. The second digit defines the area in which the status applies. The third digit provides additional information. Table 26.5 shows some common responses.

Table 26.5 *Some responses in FTP*

Code	*Description*	*Code*	*Description*
125	Data connection open	**250**	Request file action OK
150	File status OK	**331**	User name OK; password is needed
200	Command OK	**425**	Cannot open data connection
220	Service ready	**450**	File action not taken; file not available
221	Service closing	**452**	Action aborted; insufficient storage
225	Data connection open	**500**	Syntax error; unrecognized command
226	Closing data connection	**501**	Syntax error in parameters or arguments
230	User login OK	**530**	User not logged in

26.2.3 Data Connection

The data connection uses the well-known port 20 at the server site. However, the creation of a data connection is different from the control connection. The following shows the steps:

1. The client, not the server, issues a passive open using an ephemeral port. This must be done by the client because it is the client that issues the commands for transferring files.
2. Using the PORT command the client sends this port number to the server.
3. The server receives the port number and issues an active open using the well-known port 20 and the received ephemeral port number.

Communication over Data Connection

The purpose and implementation of the data connection are different from those of the control connection. We want to transfer files through the data connection. The client must define the type of file to be transferred, the structure of the data, and the transmission mode. Before sending the file through the data connection, we prepare for transmission through the control connection. The heterogeneity problem is resolved by defining three attributes of communication: file type, data structure, and transmission mode.

File Type

FTP can transfer one of the following file types across the data connection: ASCII file, EBCDIC file, or image file.

Data Structure

FTP can transfer a file across the data connection using one of the following interpretations of the structure of the data: *file structure, record structure,* or *page structure*. The file structure format (used by default) has no structure. It is a continuous stream of bytes. In the record structure, the file is divided into *records*. This can be used only with text files. In the page structure, the file is divided into pages, with each page having a page number and a page header. The pages can be stored and accessed randomly or sequentially.

Transmission Mode

FTP can transfer a file across the data connection using one of the following three transmission modes: *stream mode, block mode,* or *compressed mode*. The stream mode is the default mode; data are delivered from FTP to TCP as a continuous stream of bytes. In the block mode, data can be delivered from FTP to TCP in blocks. In this case, each block is preceded by a 3-byte header. The first byte is called the *block descriptor*; the next two bytes define the size of the block in bytes.

File Transfer

File transfer occurs over the data connection under the control of the commands sent over the control connection. However, we should remember that file transfer in FTP means one of three things: *retrieving a file* (server to client), *storing a file* (client to server), and *directory listing* (server to client).

Example 26.10

Figure 26.11 shows an example of using FTP for retrieving a file. The figure shows only one file to be transferred. The control connection remains open all the time, but the data connection is

Figure 26.11 *Example 26.10*

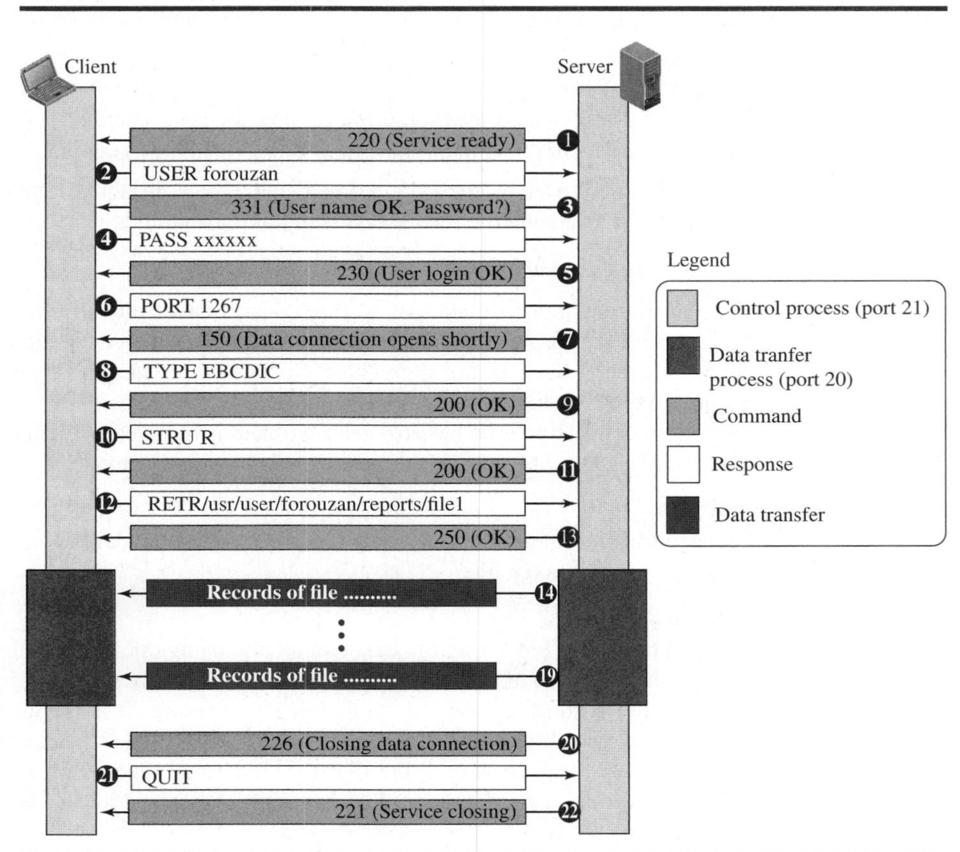

opened and closed repeatedly. We assume the file is transferred in six sections. After all records have been transferred, the server control process announces that the file transfer is done. Since the client control process has no file to retrieve, it issues the QUIT command, which causes the service connection to be closed.

Example 26.11

The following shows an actual FTP session that lists the directories. The colored lines show the responses from the server control connection; the black lines show the commands sent by the client. The lines in white with black background show data transfer.

```
$ ftp voyager.deanza.fhda.edu
Connected to voyager.deanza.fhda.edu.
220 (vsFTPd 1.2.1)
530 Please login with USER and PASS.
Name (voyager.deanza.fhda.edu:forouzan): forouzan
331 Please specify the password.
Password:*********
230 Login successful.
Remote system type is UNIX.
Using binary mode to transfer files.
227 Entering Passive Mode (153,18,17,11,238,169)
150 Here comes the directory listing.
drwxr-xr-x  2  3027  411  4096  Sep 24  2002  business
drwxr-xr-x  2  3027  411  4096  Sep 24  2002  personal
drwxr-xr-x  2  3027  411  4096  Sep 24  2002  school
226 Directory send OK.
ftp> quit
221 Goodbye.
```

26.2.4 Security for FTP

The FTP protocol was designed when security was not a big issue. Although FTP requires a password, the password is sent in plaintext (unencrypted), which means it can be intercepted and used by an attacker. The data transfer connection also transfers data in plaintext, which is insecure. To be secure, one can add a Secure Socket Layer between the FTP application layer and the TCP layer. In this case FTP is called SSL-FTP. We also explore some secure file transfer applications when we discuss SSH later in the chapter.

26.3 ELECTRONIC MAIL

Electronic mail (or e-mail) allows users to exchange messages. The nature of this application, however, is different from other applications discussed so far. In an application such as HTTP or FTP, the server program is running all the time, waiting for a request from a client. When the request arrives, the server provides the service. There is a request and there is a response. In the case of electronic mail, the situation is different. First, e-mail is considered a one-way transaction. When Alice sends an e-mail to Bob, she may expect a response, but this is not a mandate. Bob may or may not respond. If he does respond, it is another one-way transaction. Second, it is neither feasible nor logical for Bob to run a server program and wait until someone sends an e-mail to him. Bob may turn off his computer when he is not using it. This means that the idea of client/server programming should be implemented in another way: using some intermediate computers (servers). The users run only client programs when they want and the intermediate servers apply the client/server paradigm, as we discuss in the next section.

26.3.1 Architecture

To explain the architecture of e-mail, we give a common scenario, as shown in Figure 26.12. Another possibility is the case in which Alice or Bob is directly connected to the corresponding mail server, in which LAN or WAN connection is not required, but this variation in the scenario does not affect our discussion.

Figure 26.12 *Common scenario*

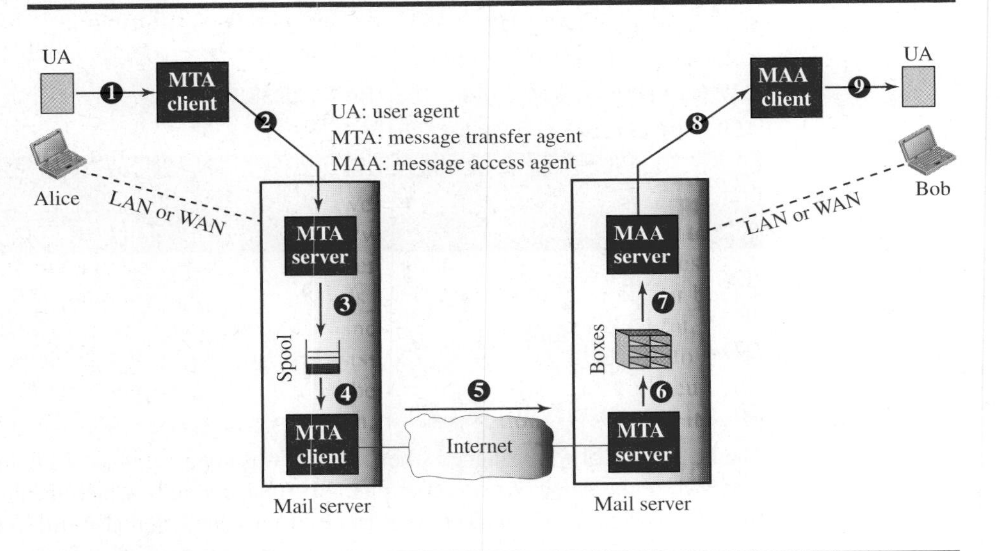

In the common scenario, the sender and the receiver of the e-mail, Alice and Bob respectively, are connected via a LAN or a WAN to two mail servers. The administrator has created one mailbox for each user where the received messages are stored. A *mailbox* is part of a server hard drive, a special file with permission restrictions. Only the owner of the mailbox has access to it. The administrator has also created a queue (spool) to store messages waiting to be sent.

A simple e-mail from Alice to Bob takes nine different steps, as shown in the figure. Alice and Bob use three different *agents*: a **user agent (UA),** a **message transfer agent (MTA),** and a **message access agent (MAA).** When Alice needs to send a message to

Bob, she runs a UA program to prepare the message and send it to her mail server. The mail server at her site uses a queue (spool) to store messages waiting to be sent. The message, however, needs to be sent through the Internet from Alice's site to Bob's site using an MTA. Here two message transfer agents are needed: one client and one server. Like most client-server programs on the Internet, the server needs to run all the time because it does not know when a client will ask for a connection. The client, on the other hand, can be triggered by the system when there is a message in the queue to be sent. The user agent at the Bob site allows Bob to read the received message. Bob later uses an MAA client to retrieve the message from an MAA server running on the second server.

There are two important points we need to emphasize here. First, Bob cannot bypass the mail server and use the MTA server directly. To use the MTA server directly, Bob would need to run the MTA server all the time because he does not know when a message will arrive. This implies that Bob must keep his computer on all the time if he is connected to his system through a LAN. If he is connected through a WAN, he must keep the connection up all the time. Neither of these situations is feasible today.

Second, note that Bob needs another pair of client-server programs: message access programs. This is because an MTA client-server program is a *push* program: the client pushes the message to the server. Bob needs a *pull* program. The client needs to pull the message from the server. We discuss more about MAAs shortly.

The electronic mail system needs two UAs, two pairs of MTAs (client and server), and a pair of MAAs (client and server).

User Agent

The first component of an electronic mail system is the **user agent (UA).** It provides service to the user to make the process of sending and receiving a message easier. A user agent is a software package (program) that composes, reads, replies to, and forwards messages. It also handles local mailboxes on the user computers.

There are two types of user agents: command-driven and GUI-based. Command-driven user agents belong to the early days of electronic mail. They are still present as the underlying user agents. A command-driven user agent normally accepts a one-character command from the keyboard to perform its task. For example, a user can type the character *r*, at the command prompt, to reply to the sender of the message, or type the character *R* to reply to the sender and all recipients. Some examples of command-driven user agents are *mail, pine,* and *elm.*

Modern user agents are GUI-based. They contain graphical user interface (GUI) components that allow the user to interact with the software by using both the keyboard and the mouse. They have graphical components such as icons, menu bars, and windows that make the services easy to access. Some examples of GUI-based user agents are *Eudora* and *Outlook*.

Sending Mail

To send mail, the user, through the UA, creates mail that looks very similar to postal mail. It has an *envelope* and a *message* (see Figure 26.13). The envelope usually contains the sender address, the receiver address, and other information. The message

Figure 26.13 *Format of an e-mail*

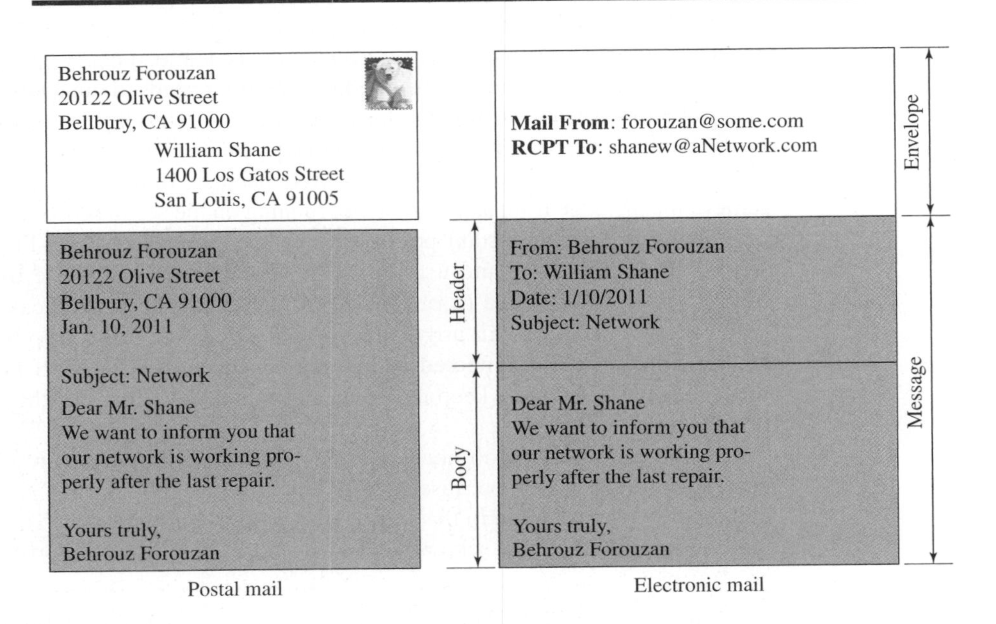

contains the *header* and the *body.* The header of the message defines the sender, the receiver, the subject of the message, and some other information. The body of the message contains the actual information to be read by the recipient.

Receiving Mail

The user agent is triggered by the user (or a timer). If a user has mail, the UA informs the user with a notice. If the user is ready to read the mail, a list is displayed in which each line contains a summary of the information about a particular message in the mailbox. The summary usually includes the sender mail address, the subject, and the time the mail was sent or received. The user can select any of the messages and display its contents on the screen.

Addresses

To deliver mail, a mail handling system must use an addressing system with unique addresses. In the Internet, the address consists of two parts: a *local part* and a *domain name,* separated by an @ sign (see Figure 26.14).

Figure 26.14 *E-mail address*

The local part defines the name of a special file, called the user mailbox, where all the mail received for a user is stored for retrieval by the message access agent. The second part of the address is the domain name. An organization usually selects one or more hosts to receive and send e-mail; they are sometimes called *mail servers* or *exchangers*. The domain name assigned to each mail exchanger either comes from the DNS database or is a logical name (for example, the name of the organization).

Mailing List or Group List

Electronic mail allows one name, an *alias,* to represent several different e-mail addresses; this is called a mailing list. Every time a message is to be sent, the system checks the recipient's name against the alias database; if there is a mailing list for the defined alias, separate messages, one for each entry in the list, must be prepared and handed to the MTA.

Message Transfer Agent: SMTP

Based on the common scenario (Figure 26.12), we can say that the e-mail is one of those applications that needs three uses of client-server paradigms to accomplish its task. It is important that we distinguish these three when we are dealing with e-mail. Figure 26.15 shows these three client-server applications. We refer to the first and the second as Message Transfer Agents (MTAs), the third as Message Access Agent (MAA).

Figure 26.15 *Protocols used in electronic mail*

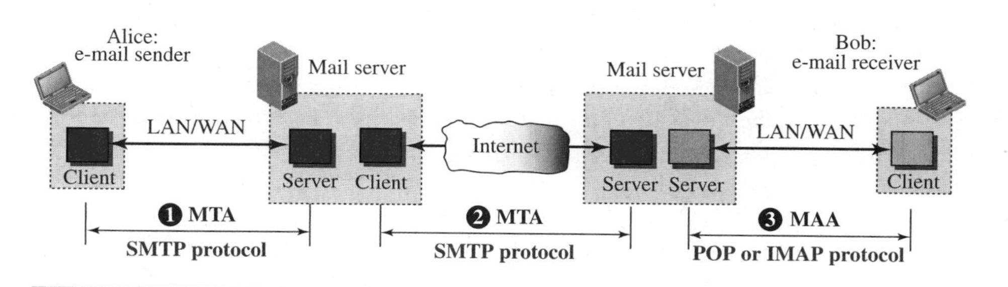

The formal protocol that defines the MTA client and server in the Internet is called ***Simple Mail Transfer Protocol (SMTP).*** SMTP is used two times, between the sender and the sender's mail server and between the two mail servers. As we will see shortly, another protocol is needed between the mail server and the receiver. SMTP simply defines how commands and responses must be sent back and forth.

Commands and Responses

SMTP uses commands and responses to transfer messages between an MTA client and an MTA server. The command is from an MTA client to an MTA server; the response is from an MTA server to the MTA client. Each command or reply is terminated by a two-character (carriage return and line feed) end-of-line token.

Commands Commands are sent from the client to the server. The format of a command is shown below:

Keyword: argument(s)

It consists of a keyword followed by zero or more arguments. SMTP defines 14 commands, listed in Table 26.6.

Table 26.6 *SMTP commands*

Keyword	*Argument(s)*	*Description*
HELO	Sender's host name	Identifies itself
MAIL FROM	Sender of the message	Identifies the sender of the message
RCPT TO	Intended recipient	Identifies the recipient of the message
DATA	Body of the mail	Sends the actual message
QUIT		Terminates the message
RSET		Aborts the current mail transaction
VRFY	Name of recipient	Verifies the address of the recipient
NOOP		Checks the status of the recipient
TURN		Switches the sender and the recipient
EXPN	Mailing list	Asks the recipient to expand the mailing list
HELP	Command name	Asks the recipient to send information about the command sent as the argument
SEND FROM	Intended recipient	Specifies that the mail be delivered only to the terminal of the recipient, and not to the mailbox
SMOL FROM	Intended recipient	Specifies that the mail be delivered to the terminal *or* the mailbox of the recipient
SMAL FROM	Intended recipient	Specifies that the mail be delivered to the terminal *and* the mailbox of the recipient

Responses Responses are sent from the server to the client. A response is a three-digit code that may be followed by additional textual information. Table 26.7 shows the most common response types.

Table 26.7 *Responses*

Code	*Description*
	Positive Completion Reply
211	System status or help reply
214	Help message
220	Service ready
221	Service closing transmission channel
250	Request command completed
251	User not local; the message will be forwarded
	Positive Intermediate Reply
354	Start mail input
	Transient Negative Completion Reply
421	Service not available
450	Mailbox not available
451	Command aborted: local error
452	Command aborted; insufficient storage
	Permanent Negative Completion Reply
500	Syntax error; unrecognized command

Table 26.7 *Responses (continued)*

Code	*Description*
501	Syntax error in parameters or arguments
502	Command not implemented
503	Bad sequence of commands
504	Command temporarily not implemented
550	Command is not executed; mailbox unavailable
551	User not local
552	Requested action aborted; exceeded storage location
553	Requested action not taken; mailbox name not allowed
554	Transaction failed

Mail Transfer Phases

The process of transferring a mail message occurs in three phases: connection establishment, mail transfer, and connection termination.

Connection Establishment After a client has made a TCP connection to the well-known port 25, the SMTP server starts the connection phase. This phase involves the following three steps:

1. The server sends code 220 (service ready) to tell the client that it is ready to receive mail. If the server is not ready, it sends code 421 (service not available).
2. The client sends the HELO message to identify itself, using its domain name address. This step is necessary to inform the server of the domain name of the client.
3. The server responds with code 250 (request command completed) or some other code depending on the situation.

Message Transfer After connection has been established between the SMTP client and server, a single message between a sender and one or more recipients can be exchanged. This phase involves eight steps. Steps 3 and 4 are repeated if there is more than one recipient.

1. The client sends the MAIL FROM message to introduce the sender of the message. It includes the mail address of the sender (mailbox and the domain name). This step is needed to give the server the return mail address for returning errors and reporting messages.
2. The server responds with code 250 or some other appropriate code.
3. The client sends the RCPT TO (recipient) message, which includes the mail address of the recipient.
4. The server responds with code 250 or some other appropriate code.
5. The client sends the DATA message to initialize the message transfer.
6. The server responds with code 354 (start mail input) or some other appropriate message.
7. The client sends the contents of the message in consecutive lines. Each line is terminated by a two-character end-of-line token (carriage return and line feed). The message is terminated by a line containing just one period.
8. The server responds with code 250 (OK) or some other appropriate code.

Connection Termination After the message is transferred successfully, the client terminates the connection. This phase involves two steps.

1. The client sends the QUIT command.
2. The server responds with code 221 or some other appropriate code.

Example 26.12

To show the three mail transfer phases, we show all of the steps described above using the information depicted in Figure 26.16. In the figure, we have separated the messages related to the envelope, header, and body in the data transfer section. Note that the steps in this figure are repeated two times in each e-mail transfer: once from the e-mail sender to the local mail server and once from the local mail server to the remote mail server. The local mail server, after receiving the whole e-mail message, may spool it and send it to the remote mail server at another time.

Figure 26.16 *Example 26.12*

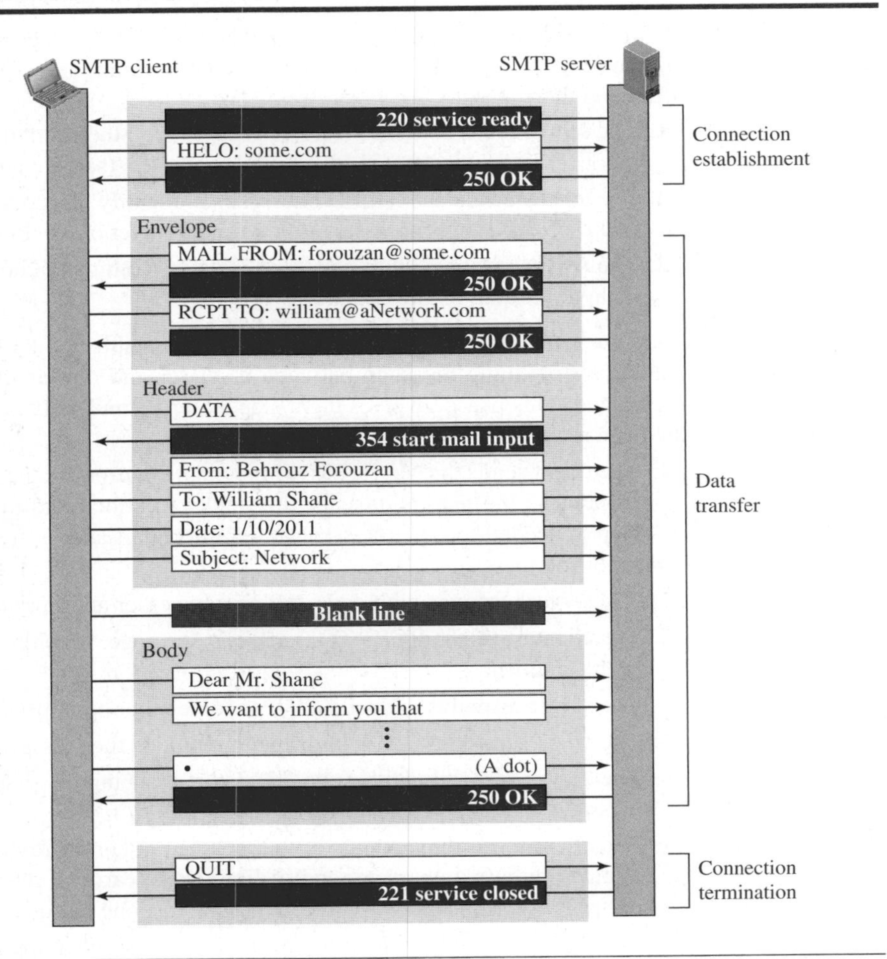

Message Access Agent: POP and IMAP

The first and second stages of mail delivery use SMTP. However, SMTP is not involved in the third stage because SMTP is a *push* protocol; it pushes the message from the client to the server. In other words, the direction of the bulk data (messages) is from the client to the server. On the other hand, the third stage needs a *pull* protocol; the client must pull messages from the server. The direction of the bulk data is from the server to the client. The third stage uses a message access agent.

Currently two message access protocols are available: Post Office Protocol, version 3 (POP3) and Internet Mail Access Protocol, version 4 (IMAP4). Figure 26.15 shows the position of these two protocols.

POP3

Post Office Protocol, version 3 (POP3) is simple but limited in functionality. The client POP3 software is installed on the recipient computer; the server POP3 software is installed on the mail server.

Mail access starts with the client when the user needs to download its e-mail from the mailbox on the mail server. The client opens a connection to the server on TCP port 110.

It then sends its user name and password to access the mailbox. The user can then list and retrieve the mail messages, one by one. Figure 26.17 shows an example of downloading using POP3. Unlike other figures in this chapter, we have put the client on the right hand side because the e-mail receiver (Bob) is running the client process to pull messages from the remote mail server.

Figure 26.17 *POP3*

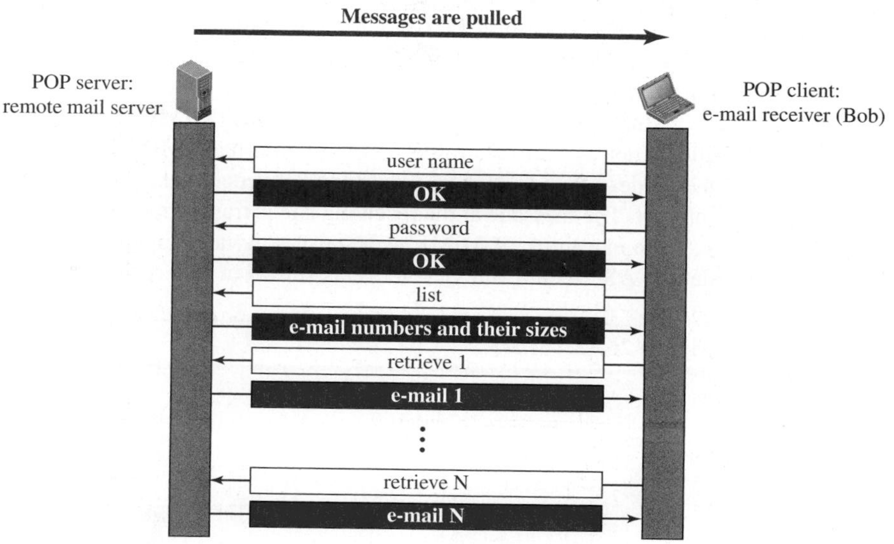

POP3 has two modes: the *delete* mode and the *keep* mode. In the delete mode, the mail is deleted from the mailbox after each retrieval. In the keep mode, the mail remains in the mailbox after retrieval. The delete mode is normally used when the user is working at her permanent computer and can save and organize the received mail after reading or replying. The keep mode is normally used when the user accesses her mail away from her primary computer (for example, from a laptop). The mail is read but kept in the system for later retrieval and organizing.

IMAP4

Another mail access protocol is **Internet Mail Access Protocol, version 4 (IMAP4).** IMAP4 is similar to POP3, but it has more features; IMAP4 is more powerful and more complex.

POP3 is deficient in several ways. It does not allow the user to organize her mail on the server; the user cannot have different folders on the server. In addition, POP3 does not allow the user to partially check the contents of the mail before downloading. IMAP4 provides the following extra functions:

- ❑ A user can check the e-mail header prior to downloading.
- ❑ A user can search the contents of the e-mail for a specific string of characters prior to downloading.
- ❑ A user can partially download e-mail. This is especially useful if bandwidth is limited and the e-mail contains multimedia with high bandwidth requirements.
- ❑ A user can create, delete, or rename mailboxes on the mail server.
- ❑ A user can create a hierarchy of mailboxes in a folder for e-mail storage.

MIME

Electronic mail has a simple structure. Its simplicity, however, comes with a price. It can send messages only in NVT 7-bit ASCII format. In other words, it has some limitations. It cannot be used for languages other than English (such as French, German, Hebrew, Russian, Chinese, and Japanese). Also, it cannot be used to send binary files or video or audio data.

Multipurpose Internet Mail Extensions (MIME) is a supplementary protocol that allows non-ASCII data to be sent through e-mail. MIME transforms non-ASCII data at the sender site to NVT ASCII data and delivers it to the client MTA to be sent through the Internet. The message at the receiving site is transformed back to the original data.

We can think of MIME as a set of software functions that transforms non-ASCII data to ASCII data and vice versa, as shown in Figure 26.18.

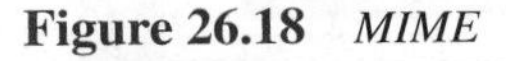

Figure 26.18 *MIME*

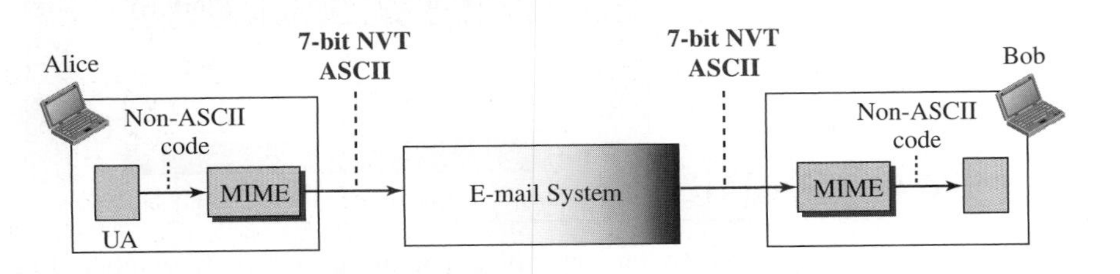

MIME Headers

MIME defines five headers, as shown in Figure 26.19, which can be added to the original e-mail header section to define the transformation parameters:

Figure 26.19 *MIME header*

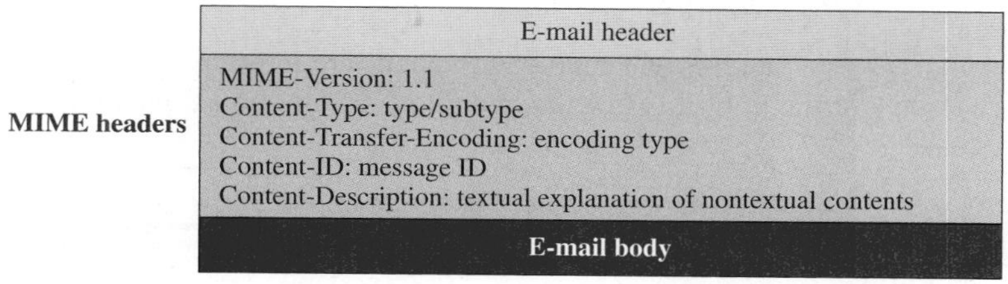

E-mail header

MIME headers

MIME-Version: 1.1
Content-Type: type/subtype
Content-Transfer-Encoding: encoding type
Content-ID: message ID
Content-Description: textual explanation of nontextual contents

E-mail body

MIME-Version This header defines the version of MIME used. The current version is 1.1.

Content-Type This header defines the type of data used in the body of the message. The content type and the content subtype are separated by a slash. Depending on the subtype, the header may contain other parameters. MIME allows seven different types of data, listed in Table 26.8.

Table 26.8 *Data types and subtypes in MIME*

Type	*Subtype*	*Description*
Text	Plain	Unformatted
	HTML	HTML format (see Appendix C)
Multipart	Mixed	Body contains ordered parts of different data types
	Parallel	Same as above, but no order
	Digest	Similar to Mixed, but the default is message/RFC822
	Alternative	Parts are different versions of the same message
Message	RFC822	Body is an encapsulated message
	Partial	Body is a fragment of a bigger message
	External-Body	Body is a reference to another message
Image	JPEG	Image is in JPEG format
	GIF	Image is in GIF format
Video	MPEG	Video is in MPEG format
Audio	Basic	Single channel encoding of voice at 8 KHz
Application	PostScript	Adobe PostScript
	Octet-stream	General binary data (eight-bit bytes)

Content-Transfer-Encoding This header defines the method used to encode the messages into 0s and 1s for transport. The five types of encoding methods are listed in Table 26.9.

Table 26.9 *Methods for Content-Transfer-Encoding*

Type	*Description*
7-bit	NVT ASCII characters with each line less than 1000 characters
8-bit	Non-ASCII characters with each line less than 1000 characters
Binary	Non-ASCII characters with unlimited-length lines
Base64	6-bit blocks of data encoded into 8-bit ASCII characters
Quoted-printable	Non-ASCII characters encoded as an equal sign plus an ASCII code

The last two encoding methods are interesting. In the Base64 encoding, data, as a string of bits, is first divided into 6-bit chunks as shown in Figure 26.20.

Figure 26.20 *Base64 conversion*

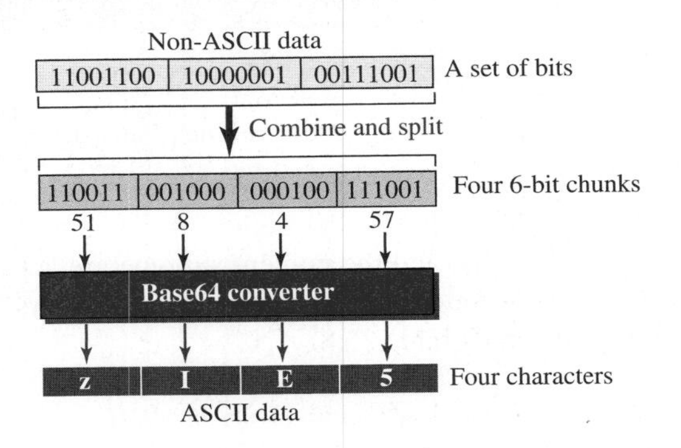

Each 6-bit section is then converted into an ASCII character according to Table 26.10.

Table 26.10 *Base64 converting table*

Value	*Code*	*Value*	*Code*	*Value*	*Code*	*Value*	*Code*	*Value*	*Code*	*Value*	*Code*
0	**A**	11	**L**	22	**W**	33	**h**	44	**s**	55	**3**
1	**B**	12	**M**	23	**X**	34	**i**	45	**t**	56	**4**
2	**C**	13	**N**	24	**Y**	35	**j**	46	**u**	57	**5**
3	**D**	14	**O**	25	**Z**	36	**k**	47	**v**	58	**6**
4	**E**	15	**P**	26	**a**	37	**l**	48	**w**	59	**7**
5	**F**	16	**Q**	27	**b**	38	**m**	49	**x**	60	**8**
6	**G**	17	**R**	28	**c**	39	**n**	50	**y**	61	**9**
7	**H**	18	**S**	29	**d**	40	**o**	51	**z**	62	**+**
8	**I**	19	**T**	30	**e**	41	**p**	52	**0**	63	**/**
9	**J**	20	**U**	31	**f**	42	**q**	53	**1**		
10	**K**	21	**V**	32	**g**	43	**r**	54	**2**		

Base64 is a redundant encoding scheme; that is, every six bits become one ASCII character and are sent as eight bits. We have an overhead of 25 percent. If the data consist mostly of ASCII characters with a small non-ASCII portion, we can use quoted-printable encoding. In quoted-printable, if a character is ASCII, it is sent as is.

If a character is not ASCII, it is sent as three characters. The first character is the equal sign (=). The next two characters are the hexadecimal representations of the byte. Figure 26.21 shows an example. In the example, the third character is a non-ASCII because it starts with bit 1. It is interpreted as two hexadecimal digits ($9D_{16}$), which is replaced by three ASCII characters (=, 9, and D).

Figure 26.21 *Quoted-printable*

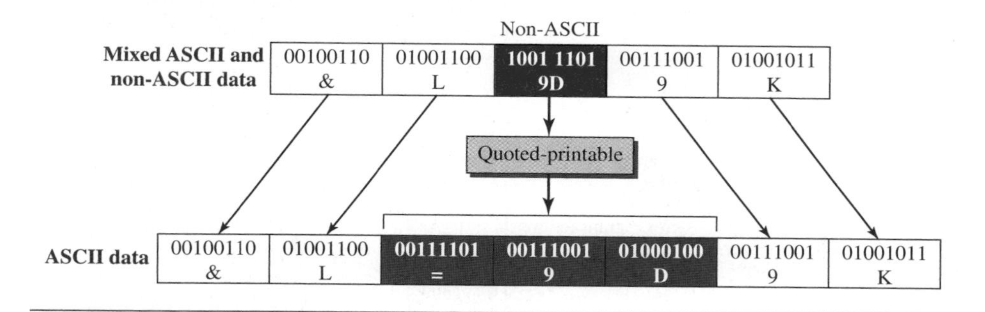

Content-ID This header uniquely identifies the whole message in a multiple message environment.

Content-Description This header defines whether the body is image, audio, or video.

26.3.2 Web-Based Mail

E-mail is such a common application that some websites today provide this service to anyone who accesses the site. Three common sites are Hotmail, Yahoo, and Google mail. The idea is very simple. Figure 26.22 shows two cases:

Case I

In the first case, Alice, the sender, uses a traditional mail server; Bob, the receiver, has an account on a web-based server. Mail transfer from Alice's browser to her mail server is done through SMTP. The transfer of the message from the sending mail server to the receiving mail server is still through SMTP. However, the message from the receiving server (the web server) to Bob's browser is done through HTTP. In other words, instead of using POP3 or IMAP4, HTTP is normally used. When Bob needs to retrieve his e-mails, he sends a request HTTP message to the website (Hotmail, for example). The website sends a form to be filled in by Bob, which includes the log-in name and the password. If the log-in name and password match, the list of e-mails is transferred from the web server to Bob's browser in HTML format. Now Bob can browse through his received e-mails and then, using more HTTP transactions, can get his e-mails one by one.

Case II

In the second case, both Alice and Bob use web servers, but not necessarily the same server. Alice sends the message to the web server using HTTP transactions. Alice sends an HTTP request message to her web server using the name and address of Bob's mailbox as the URL. The server at the Alice site passes the message to the SMTP client and

Figure 26.22 *Web-based e-mail, cases I and II*

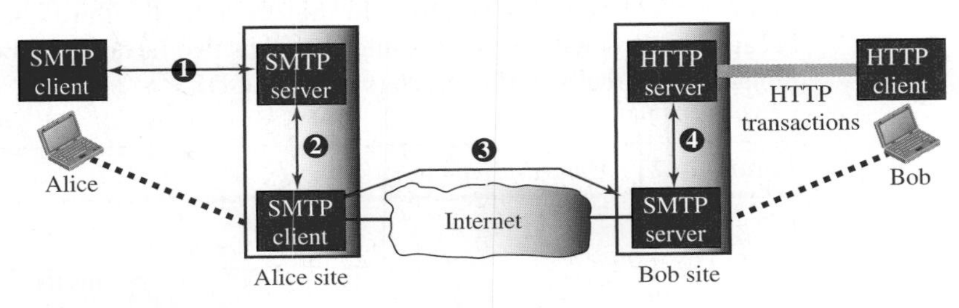

Case 1: Only receiver uses HTTP

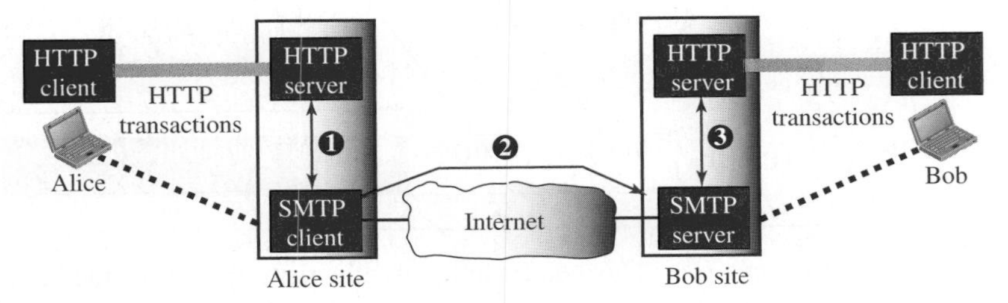

Case 2: Both sender and receiver use HTTP

sends it to the server at the Bob site using SMTP protocol. Bob receives the message using HTTP transactions. However, the message from the server at the Alice site to the server at the Bob site still takes place using SMTP protocol.

26.3.3 E-Mail Security

The protocol discussed in this chapter does not provide any security provisions per se. However, e-mail exchanges can be secured using two application-layer securities designed in particular for e-mail systems. Two of these protocols, *Pretty Good Privacy* (PGP) and *Secure/Multipurpose Internet Mail Extensions* (S/MIME), are discussed in Chapter 32 after we have discussed basic network security.

26.4 TELNET

A server program can provide a specific service to its corresponding client program. For example, the FTP server is designed to let the FTP client store or retrieve files on the server site. However, it is impossible to have a client/server pair for each type of service we need; the number of servers soon becomes intractable. The idea is not scalable. Another solution is to have a specific client/server program for a set of common scenarios, but to have some generic client/server programs that allow a user on the client site to log into the computer at the server site and use the services available there. For example, if a student needs to use the Java compiler program at her university lab, there is no need for a Java compiler client and a Java compiler server. The student can use a client logging program to log into the university server and use the compiler program at the university. We refer to these generic client/server pairs as ***remote logging*** applications.

One of the original remote logging protocols is **TELNET,** which is an abbreviation for *TErminaL NETwork.* Although TELNET requires a logging name and password, it is vulnerable to hacking because it sends all data including the password in plaintext (not encrypted). A hacker can eavesdrop and obtain the logging name and password. Because of this security issue, the use of TELNET has diminished in favor of another protocol, Secure Shell (SSH), which we describe in the next section. Although TELNET is almost replaced by SSH, we briefly discuss TELNET here for two reasons:

1. The simple plaintext architecture of TELNET allows us to explain the issues and challenges related to the concept of remote logging, which is also used in SSH when it serves as a remote logging protocol.
2. Network administrators often use TELNET for diagnostic and debugging purposes.

26.4.1 Local versus Remote Logging

We first discuss the concept of local and remote logging as shown in Figure 26.23.

Figure 26.23 *Local versus remote logging*

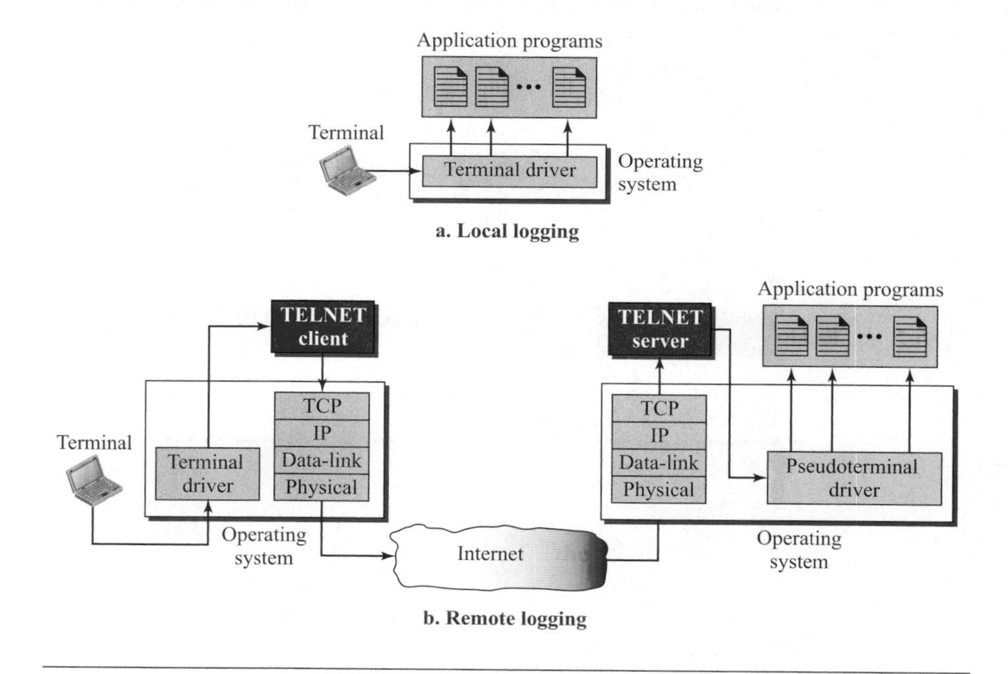

When a user logs into a local system, it is called *local logging*. As a user types at a terminal or at a workstation running a terminal emulator, the keystrokes are accepted by the terminal driver. The terminal driver passes the characters to the operating system. The operating system, in turn, interprets the combination of characters and invokes the desired application program or utility.

However, when a user wants to access an application program or utility located on a remote machine, she performs *remote logging*. Here the TELNET client and server programs come into use. The user sends the keystrokes to the terminal driver where the local operating system accepts the characters but does not interpret them. The characters are sent to the TELNET client, which transforms the characters into a universal character set called *Network Virtual Terminal* (NVT) characters (discussed below) and delivers them to the local TCP/IP stack.

The commands or text, in NVT form, travel through the Internet and arrive at the TCP/IP stack at the remote machine. Here the characters are delivered to the operating system and passed to the TELNET server, which changes the characters to the corresponding characters understandable by the remote computer. However, the characters cannot be passed directly to the operating system because the remote operating system is not designed to receive characters from a TELNET server; it is designed to receive characters from a terminal driver. The solution is to add a piece of software called a *pseudoterminal driver*, which pretends that the characters are coming from a terminal. The operating system then passes the characters to the appropriate application program.

Network Virtual Terminal (NVT)

The mechanism to access a remote computer is complex. This is because every computer and its operating system accepts a special combination of characters as tokens. For example, the end-of-file token in a computer running the DOS operating system is Ctrl+z, while the UNIX operating system recognizes Ctrl+d.

We are dealing with heterogeneous systems. If we want to access any remote computer in the world, we must first know what type of computer we will be connected to, and we must also install the specific terminal emulator used by that computer. TELNET solves this problem by defining a universal interface called the *Network Virtual Terminal (NVT)* character set. Via this interface, the client TELNET translates characters (data or commands) that come from the local terminal into NVT form and delivers them to the network. The server TELNET, on the other hand, translates data and commands from NVT form into the form acceptable by the remote computer. Figure 26.24 shows the concept.

Figure 26.24 *Concept of NVT*

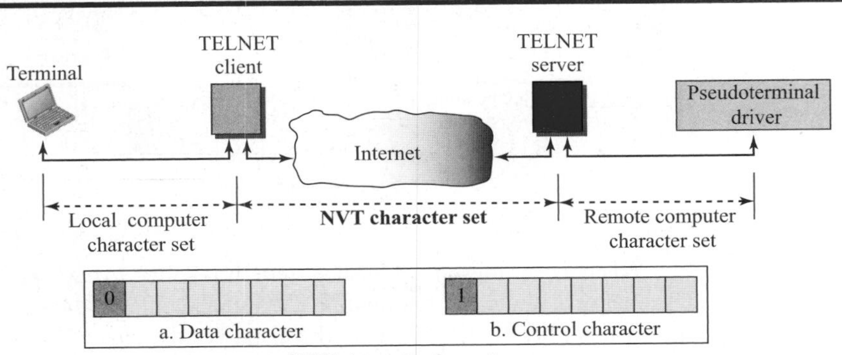

NVT uses two sets of characters, one for data and one for control. Both are 8-bit bytes as shown in Figure 26.24. For data, NVT normally uses what is called *NVT ASCII*. This is an 8-bit character set in which the seven lowest order bits are the same as US ASCII and the highest order bit is 0. To send control characters between computers (from client to server or vice versa), NVT uses an 8-bit character set in which the highest order bit is set to 1.

Options

TELNET lets the client and server negotiate options before or during the use of the service. Options are extra features available to a user with a more sophisticated terminal. Users with simpler terminals can use default features.

User Interface

The operating system (UNIX, for example) defines an interface with user-friendly commands. An example of such a set of commands can be found in Table 26.11.

Table 26.11 *Examples of interface commands*

Command	*Meaning*	*Command*	*Meaning*
open	Connect to a remote computer	**set**	Set the operating parameters
close	Close the connection	**status**	Display the status information
display	Show the operating parameters	**send**	Send special characters
mode	Change to line or character mode	**quit**	Exit TELNET

26.5 SECURE SHELL (SSH)

Although **Secure Shell (SSH)** is a secure application program that can be used today for several purposes such as remote logging and file transfer, it was originally designed to replace TELNET. There are two versions of SSH: SSH-1 and SSH-2, which are totally incompatible. The first version, SSH-1, is now deprecated because of security flaws in it. In this section, we discuss only SSH-2.

26.5.1 Components

SSH is an application-layer protocol with three components, as shown in Figure 26.25.

SSH Transport-Layer Protocol (SSH-TRANS)

Since TCP is not a secured transport-layer protocol, SSH first uses a protocol that creates a secured channel on top of the TCP. This new layer is an independent protocol referred to as SSH-TRANS. When the procedure implementing this protocol is called, the client and server first use the TCP protocol to establish an insecure connection. Then they exchange several security parameters to establish a secure channel on top of the TCP. We discuss transport-layer security in Chapter 32, but here we briefly list the services provided by this protocol:

1. Privacy or confidentiality of the message exchanged
2. Data integrity, which means that it is guaranteed that the messages exchanged between the client and server are not changed by an intruder

Figure 26.25 *Components of SSH*

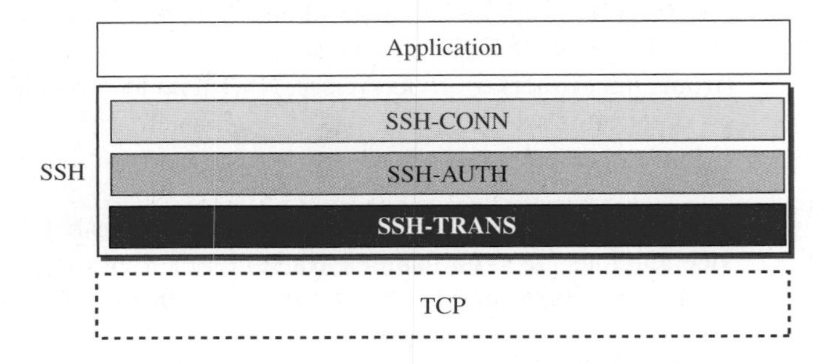

3. Server authentication, which means that the client is now sure that the server is the one that it claims to be
4. Compression of the messages, which improves the efficiency of the system and makes attack more difficult

SSH Authentication Protocol (SSH-AUTH)

After a secure channel is established between the client and the server and the server is authenticated for the client, SSH can call another procedure that can authenticate the client for the server. The client authentication process in SSH is very similar to what is done in Secure Socket Layer (SSL), which we discuss in Chapter 32. This layer defines a number of authentication tools similar to the ones used in SSL. Authentication starts with the client, which sends a request message to the server. The request includes the user name, server name, the method of authentication, and the required data. The server responds with either a success message, which confirms that the client is authenticated, or a failed message, which means that the process needs to be repeated with a new request message.

SSH Connection Protocol (SSH-CONN)

After the secured channel is established and both server and client are authenticated for each other, SSH can call a piece of software that implements the third protocol, SSH-CONN. One of the services provided by the SSH-CONN protocol is multiplexing. SSH-CONN takes the secure channel established by the two previous protocols and lets the client create multiple logical channels over it. Each channel can be used for a different purpose, such as remote logging, file transfer, and so on.

26.5.2 Applications

Although SSH is often thought of as a replacement for TELNET, SSH is, in fact, a general-purpose protocol that provides a secure connection between a client and server.

SSH for Remote Logging

Several free and commercial applications use SSH for remote logging. Among them, we can mention PuTTy, by Simon Tatham, which is a client SSH program that can be used for remote logging. Another application program is Tectia, which can be used on several platforms.

SSH for File Transfer

One of the application programs that is built on top of SSH for file transfer is the *Secure File Transfer Program* (*sftp*). The *sftp* application program uses one of the channels provided by the SSH to transfer files. Another common application is called *Secure Copy* (*scp*). This application uses the same format as the UNIX copy command, *cp,* to copy files.

Port Forwarding

One of the interesting services provided by the SSH protocol is **port forwarding**. We can use the secured channels available in SSH to access an application program that does not provide security services. Applications such as TELNET and Simple Mail Transfer Protocol (SMTP), which are discussed above, can use the services of the SSH port forwarding mechanism. The SSH port forwarding mechanism creates a tunnel through which the messages belonging to other protocols can travel. For this reason, this mechanism is sometimes referred to as SSH *tunneling*. Figure 26.26 shows the concept of port forwarding for securing the FTP application.

Figure 26.26 *Port forwarding*

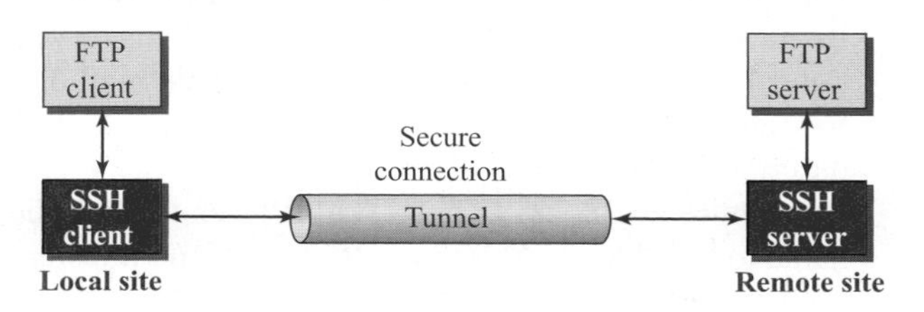

The FTP client can use the SSH client on the local site to make a secure connection with the SSH server on the remote site. Any request from the FTP client to the FTP server is carried through the tunnel provided by the SSH client and server. Any response from the FTP server to the FTP client is also carried through the tunnel provided by the SSH client and server.

Format of the SSH Packets

Figure 26.27 shows the format of packets used by the SSH protocols.

Figure 26.27 *SSH packet format*

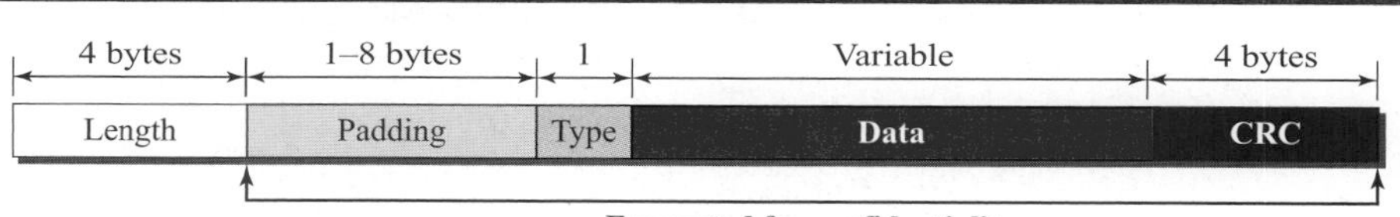

The length field defines the length of the packet but does not include the padding. One to eight bytes of padding is added to the packet to make the attack on the security provision more difficult. The *cyclic redundancy check* (CRC) field is used for error detection. The type field designates the type of the packet used in different SSH protocols. The data field is the data transferred by the packet in different protocols.

26.6 DOMAIN NAME SYSTEM (DNS)

The last client-server application program we discuss has been designed to help other application programs. To identify an entity, TCP/IP protocols use the IP address, which uniquely identifies the connection of a host to the Internet. However, people prefer to use names instead of numeric addresses. Therefore, the Internet needs to have a directory system that can map a name to an address. This is analogous to the telephone network. A telephone network is designed to use telephone numbers, not names. People can either keep a private file to map a name to the corresponding telephone number or can call the telephone directory to do so. We discuss how this directory system in the Internet can map names to IP addresses.

Since the Internet is so huge today, a central directory system cannot hold all the mapping. In addition, if the central computer fails, the whole communication network will collapse. A better solution is to distribute the information among many computers in the world. In this method, the host that needs mapping can contact the closest computer holding the needed information. This method is used by the **Domain Name System (DNS)**. We first discuss the concepts and ideas behind the DNS. We then describe the DNS protocol itself.

Figure 26.28 shows how TCP/IP uses a DNS client and a DNS server to map a name to an address. A user wants to use a file transfer client to access the corresponding file transfer server running on a remote host. The user knows only the file transfer

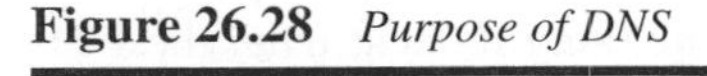

Figure 26.28 *Purpose of DNS*

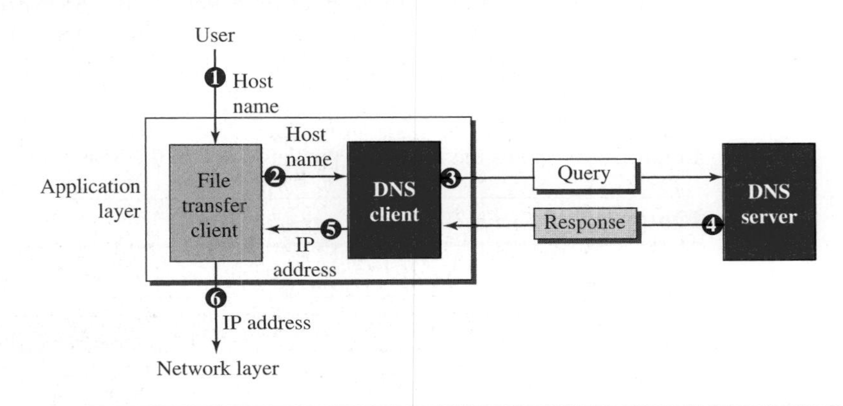

server name, such as *afilesource.com.* However, the TCP/IP suite needs the IP address of the file transfer server to make the connection. The following six steps map the host name to an IP address:

1. The user passes the host name to the file transfer client.
2. The file transfer client passes the host name to the DNS client.
3. Each computer, after being booted, knows the address of one DNS server. The DNS client sends a message to a DNS server with a query that gives the file transfer server name using the known IP address of the DNS server.
4. The DNS server responds with the IP address of the desired file transfer server.
5. The DNS server passes the IP address to the file transfer client.
6. The file transfer client now uses the received IP address to access the file transfer server.

Note that the purpose of accessing the Internet is to make a connection between the file transfer client and server, but before this can happen, another connection needs to be made between the DNS client and DNS server. In other words, we need at least two connections in this case. The first is for mapping the name to an IP address; the second is for transferring files. We will see later that the mapping may need more than one connection.

26.6.1 Name Space

To be unambiguous, the names assigned to machines must be carefully selected from a name space with complete control over the binding between the names and IP addresses. In other words, the names must be unique because the addresses are unique. A **name space** that maps each address to a unique name can be organized in two ways: flat or hierarchical. In a *flat name space,* a name is assigned to an address. A name in this space is a sequence of characters without structure. The names may or may not have a common section; if they do, it has no meaning. The main disadvantage of a flat name space is that it cannot be used in a large system such as the Internet because it must be centrally controlled to avoid ambiguity and duplication. In a *hierarchical name space,* each name is made of several parts. The first part can define the nature of the organization, the second part can define the name of an organization, the third part can define departments in the organization, and so on. In this case, the authority to assign and control the name spaces can be decentralized. A central authority can assign the part of the name that defines the nature of the organization and the name of the organization. The responsibility for the rest of the name can be given to the organization itself. The organization can add suffixes (or prefixes) to the name to define its host or resources. The management of the organization need not worry that the prefix chosen for a host is taken by another organization because, even if part of an address is the same, the whole address is different. For example, assume two organizations call one of their computers *caesar.* The first organization is given a name by the central authority, such as *first.com,* the second organization is given the name *second.com.* When each of these organizations adds the name *caesar* to the name they have already been given, the end result is two distinguishable names: *ceasar.first.com* and *ceasar.second.com.* The names are unique.

Domain Name Space

To have a hierarchical name space, a **domain name space** was designed. In this design the names are defined in an inverted-tree structure with the root at the top. The tree can have only 128 levels: level 0 (root) to level 127 (see Figure 26.29).

Figure 26.29 *Domain name space*

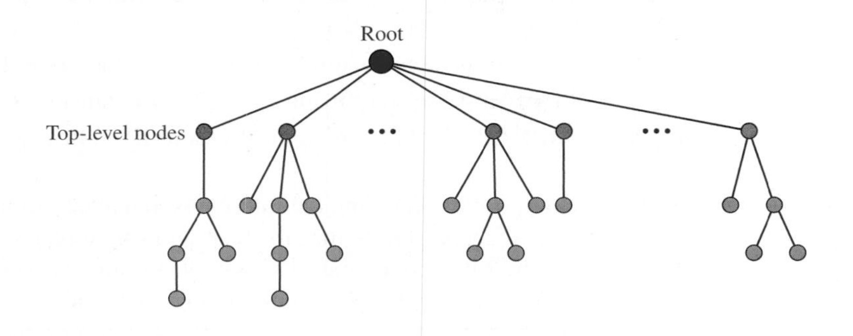

Label

Each node in the tree has a **label**, which is a string with a maximum of 63 characters. The root label is a null string (empty string). DNS requires that children of a node (nodes that branch from the same node) have different labels, which guarantees the uniqueness of the domain names.

Domain Name

Each node in the tree has a domain name. A full **domain name** is a sequence of labels separated by dots (**.**). The domain names are always read from the node up to the root. The last label is the label of the root (null). This means that a full domain name always ends in a null label, which means the last character is a dot because the null string is nothing. Figure 26.30 shows some domain names.

If a label is terminated by a null string, it is called a **fully qualified domain name (FQDN)**. The name must end with a null label, but because null means nothing, the label ends with a dot. If a label is not terminated by a null string, it is called a **partially qualified domain name (PQDN).** A PQDN starts from a node, but it does not reach the root. It is used when the name to be resolved belongs to the same site as the client. Here the resolver can supply the missing part, called the *suffix,* to create an FQDN.

Domain

A **domain** is a subtree of the domain name space. The name of the domain is the name of the node at the top of the subtree. Figure 26.31 shows some domains. Note that a domain may itself be divided into domains.

Distribution of Name Space

The information contained in the domain name space must be stored. However, it is very inefficient and also not reliable to have just one computer store such a huge

Figure 26.30 *Domain names and labels*

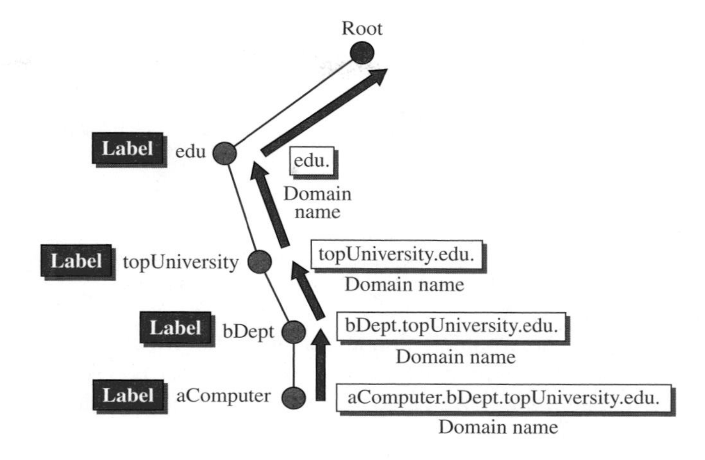

Figure 26.31 *Domains*

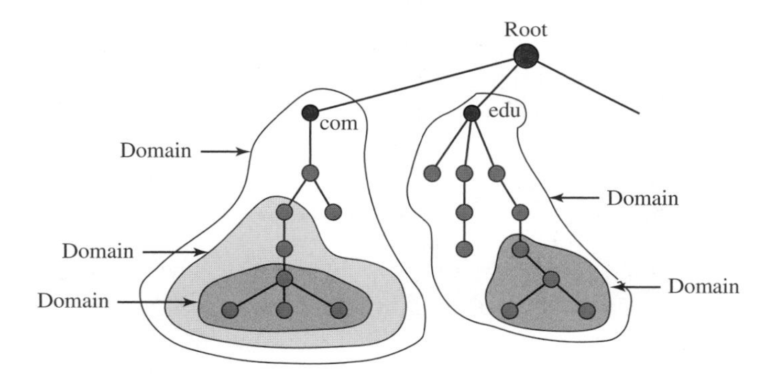

amount of information. It is inefficient because responding to requests from all over the world places a heavy load on the system. It is not reliable because any failure makes the data inaccessible.

Hierarchy of Name Servers

The solution to these problems is to distribute the information among many computers called ***DNS servers***. One way to do this is to divide the whole space into many domains based on the first level. In other words, we let the root stand alone and create as many domains (subtrees) as there are first-level nodes. Because a domain created this way could be very large, DNS allows domains to be divided further into smaller domains (subdomains). Each server can be responsible (authoritative) for either a large or small domain. In other words, we have a hierarchy of servers in the same way that we have a hierarchy of names (see Figure 26.32).

Figure 26.32 *Hierarchy of name servers*

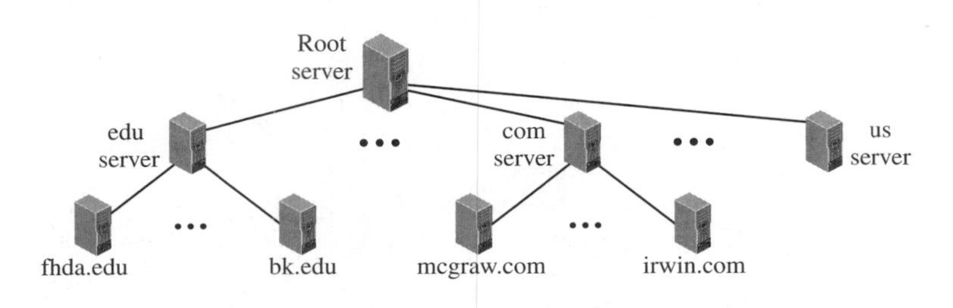

Zone

Since the complete domain name hierarchy cannot be stored on a single server, it is divided among many servers. What a server is responsible for or has authority over is called a ***zone***. We can define a zone as a contiguous part of the entire tree. If a server accepts responsibility for a domain and does not divide the domain into smaller domains, the "domain" and the "zone" refer to the same thing. The server makes a database called a *zone file* and keeps all the information for every node under that domain. However, if a server divides its domain into subdomains and delegates part of its authority to other servers, "domain" and "zone" refer to different things. The information about the nodes in the subdomains is stored in the servers at the lower levels, with the original server keeping some sort of reference to these lower-level servers. Of course, the original server does not free itself from responsibility totally. It still has a zone, but the detailed information is kept by the lower-level servers (see Figure 26.33).

Figure 26.33 *Zone*

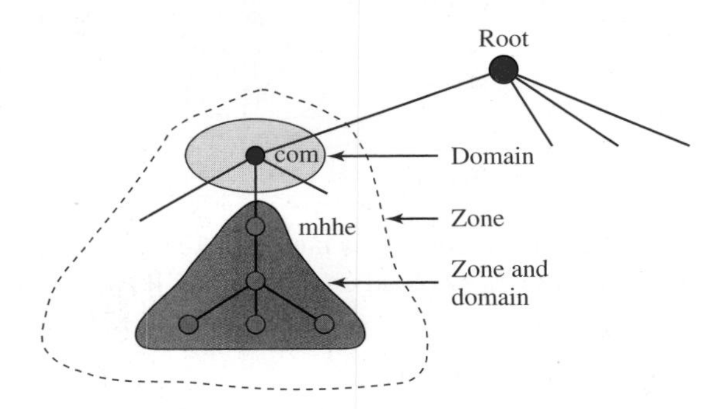

Root Server

A **root server** is a server whose zone consists of the whole tree. A root server usually does not store any information about domains but delegates its authority to other servers, keeping references to those servers. There are several root servers, each covering the whole domain name space. The root servers are distributed all around the world.

Primary and Secondary Servers

DNS defines two types of servers: primary and secondary. A *primary server* is a server that stores a file about the zone for which it is an authority. It is responsible for creating, maintaining, and updating the zone file. It stores the zone file on a local disk.

A *secondary server* is a server that transfers the complete information about a zone from another server (primary or secondary) and stores the file on its local disk. The secondary server neither creates nor updates the zone files. If updating is required, it must be done by the primary server, which sends the updated version to the secondary.

The primary and secondary servers are both authoritative for the zones they serve. The idea is not to put the secondary server at a lower level of authority but to create redundancy for the data so that if one server fails, the other can continue serving clients. Note also that a server can be a primary server for a specific zone and a secondary server for another zone. Therefore, when we refer to a server as a primary or secondary server, we should be careful about which zone we refer to.

A primary server loads all information from the disk file; the secondary server loads all information from the primary server.

26.6.2 DNS in the Internet

DNS is a protocol that can be used in different platforms. In the Internet, the domain name space (tree) was originally divided into three different sections: generic domains, country domains, and the inverse domains. However, due to the rapid growth of the Internet, it became extremely difficult to keep track of the inverse domains, which could be used to find the name of a host when given the IP address. The inverse domains are now deprecated (see RFC 3425). We, therefore, concentrate on the first two.

Generic Domains

The **generic domains** define registered hosts according to their generic behavior. Each node in the tree defines a domain, which is an index to the domain name space database (see Figure 26.34).

Figure 26.34 *Generic domains*

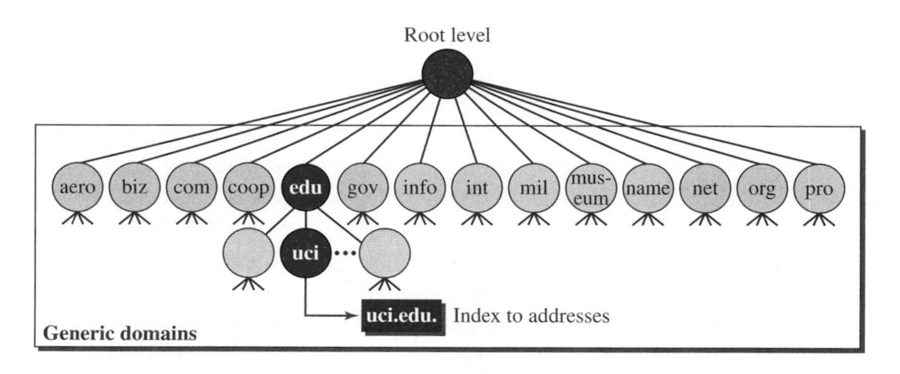

Looking at the tree, we see that the first level in the **generic domains** section allows 14 possible labels. These labels describe the organization types as listed in Table 26.12.

Table 26.12 *Generic domain labels*

Label	*Description*	*Label*	*Description*
aero	Airlines and aerospace	**int**	International organizations
biz	Businesses or firms	**mil**	Military groups
com	Commercial organizations	**museum**	Museums
coop	Cooperative organizations	**name**	Personal names (individuals)
edu	Educational institutions	**net**	Network support centers
gov	Government institutions	**org**	Nonprofit organizations
info	Information service providers	**pro**	Professional organizations

Country Domains

The **country domains** section uses two-character country abbreviations (e.g., us for United States). Second labels can be organizational, or they can be more specific national designations. The United States, for example, uses state abbreviations as a subdivision of us (e.g., ca.us.). Figure 26.35 shows the country domains section. The address ***uci.ca.us.*** can be translated to University of California, Irvine, in the state of California in the United States.

Figure 26.35 *Country domains*

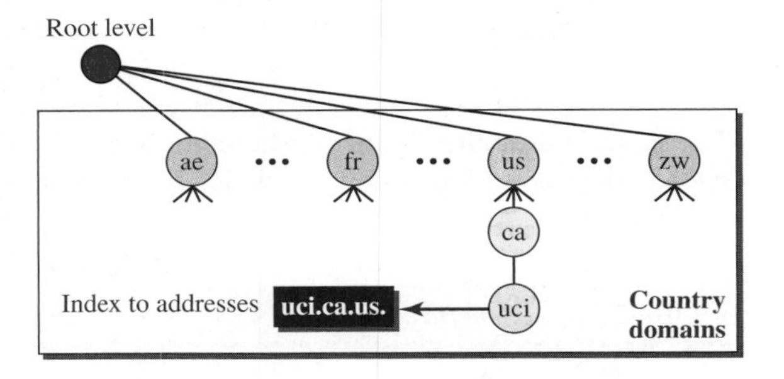

26.6.3 Resolution

Mapping a name to an address is called *name-address resolution.* DNS is designed as a client-server application. A host that needs to map an address to a name or a name to an address calls a DNS client called a ***resolver.*** The resolver accesses the closest DNS server with a mapping request. If the server has the information, it satisfies the resolver; otherwise, it either refers the resolver to other servers or asks other servers to provide the information. After the resolver receives the mapping, it interprets the response to see if it is a real resolution or an error, and finally delivers the result to the process that requested it. A resolution can be either recursive or iterative.

Recursive Resolution

Figure 26.36 shows a simple example of a **recursive resolution.** We assume that an application program running on a host named ***some.anet.com*** needs to find the IP address of another host named ***engineering.mcgraw-hill.com*** to send a message to. The source host is connected to the Anet ISP; the destination host is connected to the McGraw-Hill network.

Figure 26.36 *Recursive resolution*

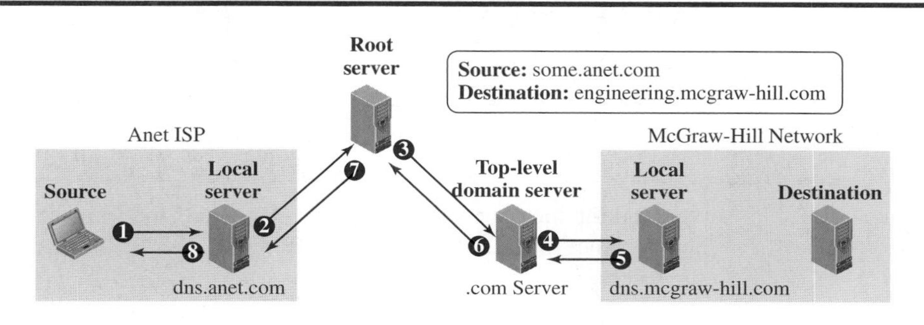

The application program on the source host calls the DNS resolver (client) to find the IP address of the destination host. The resolver, which does not know this address, sends the query to the local DNS server (for example, *dns.anet.com*) running at the Anet ISP site (event 1). We assume that this server does not know the IP address of the destination host either. It sends the query to a root DNS server, whose IP address is supposed to be known to this local DNS server (event 2). Root servers do not normally keep the mapping between names and IP addresses, but a root server should at least know about one server at each top level domain (in this case, a server responsible for *com* domain). The query is sent to this top-level-domain server (event 3). We assume that this server does not know the name-address mapping of this specific destination, but it knows the IP address of the local DNS server in the McGraw-Hill company (for example, *dns.mcgraw-hill.com*). The query is sent to this server (event 4), which knows the IP address of the destination host. The IP address is now sent back to the top-level DNS server (event 5), then back to the root server (event 6), then back to the ISP DNS server, which may cache it for the future queries (event 7), and finally back to the source host (event 8).

Iterative Resolution

In **iterative resolution,** each server that does not know the mapping sends the IP address of the next server back to the one that requested it. Figure 26.37 shows the flow of information in an iterative resolution in the same scenario as the one depicted in Figure 26.36. Normally the iterative resolution takes place between two local servers; the original resolver gets the final answer from the local server. Note that the messages shown by events 2, 4, and 6 contain the same query. However, the message shown by event 3 contains the IP address of the top-level domain server, the message shown by event 5 contains the IP address of the McGraw-Hill local DNS

Figure 26.37 *Iterative resolution*

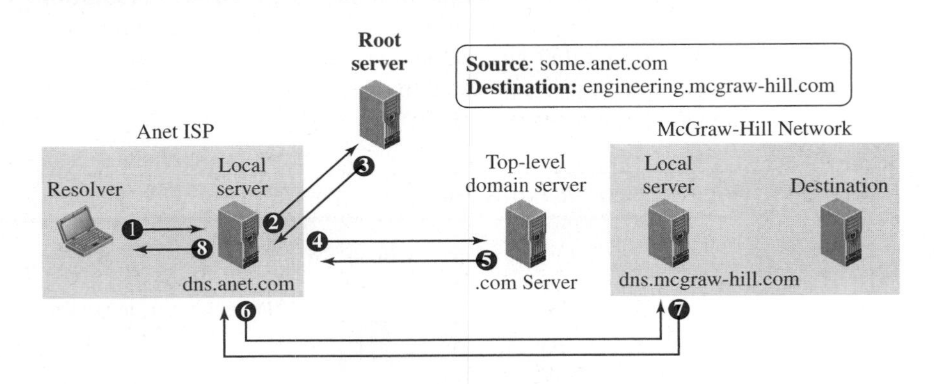

server, and the message shown by event 7 contains the IP address of the destination. When the Anet local DNS server receives the IP address of the destination, it sends it to the resolver (event 8).

26.6.4 Caching

Each time a server receives a query for a name that is not in its domain, it needs to search its database for a server IP address. Reduction of this search time would increase efficiency. DNS handles this with a mechanism called *caching*. When a server asks for a mapping from another server and receives the response, it stores this information in its cache memory before sending it to the client. If the same or another client asks for the same mapping, it can check its cache memory and resolve the problem. However, to inform the client that the response is coming from the cache memory and not from an authoritative source, the server marks the response as *unauthoritative*.

Caching speeds up resolution, but it can also be problematic. If a server caches a mapping for a long time, it may send an outdated mapping to the client. To counter this, two techniques are used. First, the authoritative server always adds information to the mapping called *time to live* (TTL). It defines the time in seconds that the receiving server can cache the information. After that time, the mapping is invalid and any query must be sent again to the authoritative server. Second, DNS requires that each server keep a TTL counter for each mapping it caches. The cache memory must be searched periodically and those mappings with an expired TTL must be purged.

26.6.5 Resource Records

The zone information associated with a server is implemented as a set of *resource records*. In other words, a name server stores a database of resource records. A *resource record* is a 5-tuple structure, as shown below:

(Domain Name, Type, Class, TTL, Value)

The domain name field is what identifies the resource record. The value defines the information kept about the domain name. The TTL defines the number of seconds for which the information is valid. The class defines the type of network; we are only interested in the class IN (Internet). The type defines how the value should be interpreted. Table 26.13 lists the common types and how the value is interpreted for each type.

Table 26.13 *Types*

Type	*Interpretation of value*
A	A 32-bit IPv4 address (see Chapter 18)
NS	Identifies the authoritative servers for a zone
CNAME	Defines an alias for the official name of a host
SOA	Marks the beginning of a zone
MX	Redirects mail to a mail server
AAAA	An IPv6 address (see Chapter 22)

26.6.6 DNS Messages

To retrieve information about hosts, DNS uses two types of messages: *query* and *response*. Both types have the same format as shown in Figure 26.38.

Figure 26.38 *DNS message*

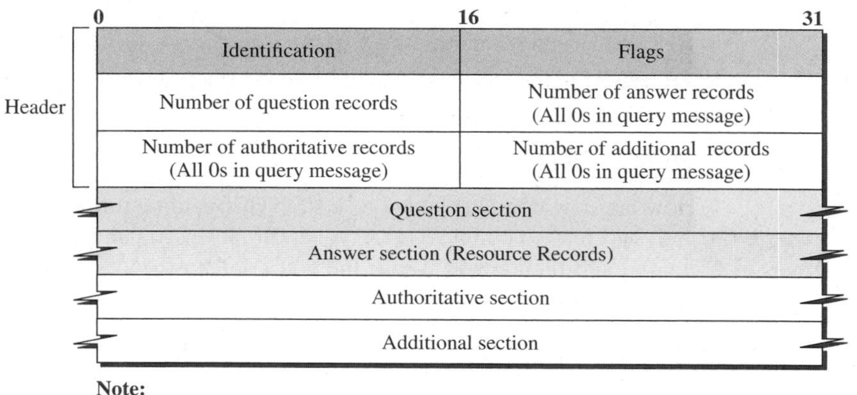

Note:
The query message contains only the question section.
The response message includes the question section, the answer section, and possibly two other sections.

We briefly discuss the fields in a DNS message. The identification field is used by the client to match the response with the query. The flag field defines whether the message is a query or response. It also includes status of error. The next four fields in the header define the number of each record type in the message. The question section consists of one or more question records. It is present in both query and response messages. The answer section consists of one or more resource records. It is present only in response messages. The authoritative section gives information (domain name) about one or more authoritative servers for the query. The additional information section provides additional information that may help the resolver.

Example 26.13

In UNIX and Windows, the *nslookup* utility can be used to retrieve address/name mapping. The following shows how we can retrieve an address when the domain name is given.

```
$nslookup www.forouzan.biz
Name: www.forouzan.biz
Address: 198.170.240.179
```

Encapsulation

DNS can use either UDP or TCP. In both cases the well-known port used by the server is port 53. UDP is used when the size of the response message is less than 512 bytes because most UDP packages have a 512-byte packet size limit. If the size of the response message is more than 512 bytes, a TCP connection is used. In that case, one of two scenarios can occur:

- If the resolver has prior knowledge that the size of the response message is more than 512 bytes, it uses the TCP connection. For example, if a secondary name server (acting as a client) needs a zone transfer from a primary server, it uses the TCP connection because the size of the information being transferred usually exceeds 512 bytes.
- If the resolver does not know the size of the response message, it can use the UDP port. However, if the size of the response message is more than 512 bytes, the server truncates the message and turns on the TC bit. The resolver now opens a TCP connection and repeats the request to get a full response from the server.

26.6.7 Registrars

How are new domains added to DNS? This is done through a *registrar,* a commercial entity accredited by ICANN. A registrar first verifies that the requested domain name is unique and then enters it into the DNS database. A fee is charged. Today, there are many registrars; their names and addresses can be found at

http://www.intenic.net

To register, the organization needs to give the name of its server and the IP address of the server. For example, a new commercial organization named *wonderful* with a server named *ws* and IP address 200.200.200.5 needs to give the following information to one of the registrars:

Domain name: ws.wonderful.com **IP address:** 200.200.200.5

26.6.8 DDNS

When the DNS was designed, no one predicted that there would be so many address changes. In DNS, when there is a change, such as adding a new host, removing a host, or changing an IP address, the change must be made to the DNS master file. These types of changes involve a lot of manual updating. The size of today's Internet does not allow for this kind of manual operation.

The DNS master file must be updated dynamically. The **Dynamic Domain Name System (DDNS)** therefore was devised to respond to this need. In DDNS, when a binding between a name and an address is determined, the information is sent, usually by DHCP (discussed in Chapter 18) to a primary DNS server. The primary server updates the zone. The secondary servers are notified either actively or passively. In active notification, the primary server sends a message to the secondary servers about the change in the zone, whereas in passive notification, the secondary servers periodically check for any changes. In either case, after being notified about the change, the secondary server requests information about the entire zone (called the *zone transfer*).

To provide security and prevent unauthorized changes in the DNS records, DDNS can use an authentication mechanism.

26.6.9 Security of DNS

DNS is one of the most important systems in the Internet infrastructure; it provides crucial services to Internet users. Applications such as Web access or e-mail are heavily dependent on the proper operation of DNS. DNS can be attacked in several ways including:

1. The attacker may read the response of a DNS server to find the nature or names of sites the user mostly accesses. This type of information can be used to find the user's profile. To prevent this attack, DNS messages need to be confidential (see Chapters 31 and 32).
2. The attacker may intercept the response of a DNS server and change it or create a totally new bogus response to direct the user to the site or domain the attacker wishes the user to access. This type of attack can be prevented using message origin authentication and message integrity (see Chapters 31 and 32).
3. The attacker may flood the DNS server to overwhelm it or eventually crash it. This type of attack can be prevented using the provision against denial-of-service attack.

To protect DNS, IETF has devised a technology named *DNS Security* (DNSSEC) that provides message origin authentication and message integrity using a security service called *digital signature* (see Chapter 31). DNSSEC, however, does not provide confidentiality for the DNS messages. There is no specific protection against the denial-of-service attack in the specification of DNSSEC. However, the caching system protects the upper-level servers against this attack to some extent.

26.7 END-CHAPTER MATERIALS

26.7.1 Recommended Reading

For more details about subjects discussed in this chapter, we recommend the following books and RFCs. The items enclosed in brackets refer to the reference list at the end of the book.

Books

Several books give thorough coverage of materials discussed in this chapter including [Com 06], [Mir 07], [Ste 94], [Tan 03], [Bar et al. 05].

RFCs

HTTP is discussed in RFCs 2068, 2109, and 2616. FTP is discussed in RFCs 959, 2577, and 2585. TELNET is discussed in RFCs 854, 855, 856, 1041, 1091, 1372, and 1572. SSH is discussed in RFCs 4250, 4251, 4252, 4253, 4254, and 4344. DNS is discussed in RFCs 1034, 1035, 1996, 2535, 3008, 3342, 3396, 3658, 3755, 3757, and 3845. SMTP is discussed in RFCs 2821 and 2822. POP3 is explained in RFC 1939. MIME is discussed in RFCs 2046, 2047, 2048, and 2049.

26.7.2 Key Terms

active document
browser
cookie
country domain
DNS Server
domain
domain name
domain name space
Domain Name System (DNS)
dynamic document
Dynamic Domain Name System (DDNS)
File Transfer Protocol (FTP)
fully qualified domain name (FQDN)
generic domain
hypermedia
hypertext
HyperText Transfer Protocol (HTTP)
Internet Mail Access Protocol, version 4 (IMAP4)
iterative resolution
label
local logging
message access agent (MAA)
message transfer agent (MTA)
Multipurpose Internet Mail Extensions (MIME)
name space
Network Virtual Terminal (NVT)
nonpersistent connection
partially qualified domain name (PQDN)
persistent connection
port forwarding
Post Office Protocol, version 3 (POP3)
proxy server
recursive resolution
remote logging
resolver
root server
Secure Shell (SSH)
Simple Mail Transfer Protocol (SMTP)
static document
terminal network (TELNET)
uniform resource locator (URL)
user agent (UA)
web page
World Wide Web (WWW, the web)
zone

26.7.3 Summary

The World Wide Web (WWW) is a repository of information linked together from points all over the world. Hypertext and hypermedia documents are linked to one another through pointers. HyperText Transfer Protocol (HTTP) is the main protocol used to access data on the Web.

File Transfer Protocol (FTP) is a TCP/IP client-server application for copying files from one host to another. FTP requires two connections for data transfer: a control connection and a data connection. FTP employs NVT ASCII for communication between dissimilar systems.

Electronic mail is one of the most common applications on the Internet. The e-mail architecture consists of several components such as user agent (UA), message transfer agent (MTA), and message access agent (MAA). The protocol that implements MTA is called Simple Mail Transfer Protocol (SMTP). Two protocols are used to implement MAA: Post Office Protocol, version 3 (POP3) and Internet Mail Access Protocol, version 4 (IMAP4).

TELNET is a client-server application that allows a user to log into a remote machine, giving the user access to the remote system. When a user accesses a remote system via the TELNET process, this is comparable to a time-sharing environment. SSH is the secured version of TELNET that is very common today.

The Domain Name System (DNS) is a client-server application that identifies each host on the Internet with a unique name. DNS organizes the name space in a hierarchical structure to decentralize the responsibilities involved in naming.

26.8 PRACTICE SET

26.8.1 Quizzes

A set of interactive quizzes for this chapter can be found on the book website. It is strongly recommended that the student take the quizzes to check his/her understanding of the materials before continuing with the practice set.

26.8.2 Questions

Q26-1. During the weekend, Alice often needs to access files stored on her office desktop from her home laptop. Last week, she installed a copy of the FTP server process on her desktop at her office and a copy of the FTP client process on her laptop at home. She was disappointed when she could not access her files during the weekend. What could have gone wrong?

Q26-2. In FTP, which entity (client or server) starts (actively opens) the control connection? Which entity starts (actively opens) the data transfer connection?

Q26-3. In FTP, can a server retrieve a file from the client site?

Q26-4. Assume we need to download an audio using FTP. What file type should we specify in our command?

Q26-5. Can you find an analogy in our daily life as to when we use two separate connections in communication similar to the control and data connections in FTP?

Q26-6. In FTP, can a server get the list of the files or directories from the client?

Q26-7. Does FTP have a message format for exchanging files or a list of directories/files during the file-transfer connection?

Q26-8. Can we have a control connection without a data-transfer connection in FTP? Explain.

Q26-9. When an HTTP server receives a request message from an HTTP client, how does the server know when all headers have arrived and the body of the message is to follow?

Q26-10. Alice has been on a long trip without checking her e-mail. She then finds out that she has lost some e-mails or attachments her friends claim they have sent to her. What can be the problem?

Q26-11. Assume a TELNET client uses ASCII to represent characters, but the TELNET server uses EBCDIC to represent characters. How can the client log into the server when character representations are different?

Q26-12. Does FTP have a message format for exchanging commands and responses during control connection?

Q26-13. Both HTTP and FTP can retrieve a file from a server. Which protocol should we use to download a file?

Q26-14. Alice has a video clip that Bob is interested in getting; Bob has another video clip that Alice is interested in getting. Bob creates a web page and runs an HTTP server. How can Alice get Bob's clip? How can Bob get Alice's clip?

Q26-15. What do you think would happen if the control connection were severed before the end of an FTP session? Would it affect the data connection?

Q26-16. In FTP, if the client needs to retrieve one file from the server site and store one file on the server site, how many control connections and how many data-transfer connections are needed?

Q26-17. FTP uses the services of TCP for exchanging control information and data transfer. Could FTP have used the services of UDP for either of these two connections? Explain.

Q26-18. Are the HELO and MAIL FROM commands both necessary in SMTP? Why or why not?

Q26-19. In Figure 26.13 in the text, what is the difference between the MAIL FROM in the envelope and the FROM in the header?

Q26-20. FTP uses two separate well-known port numbers for control and data connection. Does this mean that two separate TCP connections are created for exchanging control information and data?

Q26-21. Can we have a data-transfer connection without a control connection in FTP? Explain.

Q26-22. In DNS, which of the following are FQDNs and which are PQDNs?

a. xxx　　**b.** xxx.yyy.net　　**c.** zzz.yyy.xxx.edu.

Q26-23. FTP can transfer files between two hosts using different operating systems with different file formats. What is the reason?

Q26-24. The TELNET application has no commands such as those found in FTP or HTTP to allow the user to do something such as transfer a file or access a web page. In what way can this application be useful?

Q26-25. Can a host use a TELNET client to get services provided by other client-server applications such as FTP or HTTP?

Q26-26. In a nonpersistent HTTP connection, how can HTTP inform the TCP protocol that the end of the message has been reached?

26.8.3 Problems

P26-1. In Chapter 1, we mentioned that the TCP/IP suite, unlike the OSI model, has no session layer. But an application-layer protocol can include some of the features defined in this layer if needed. Does HTTP have any session-layer features?

P26-2. In FTP, assume a client with user name John needs to store a video clip called *video2* on the directory */top/videos/general* on the server. Show the commands and responses exchanged between the client and the server if the client chooses ephemeral port number 56002.

P26-3. Using RFC 1939, assume a POP3 client is in the download-and-delete mode. Show the transaction between the client and the server if the client has only two messages of 230 and 400 bytes to download from the server.

P26-4. Write concurrent TCP client-server programs to simulate a simplified version of HTTP using only a nonpersistent connection. The client sends an HTTP message; the server responds with the requested file. Use only two types of methods, GET and PUT, and only a few simple headers. Note that after testing, you should be able to test your program with a web browser.

P26-5. POP3 protocol has some optional commands (that a client/server can implement). Using the information in RFC 1939, find the meaning and the use of the following optional commands:

a. UIDL **b.** TOP 1 15 **c.** USER **d.** PASS

P26-6. HTTP version 1.1 defines the persistent connection as the default connection. Using RFC 2616, find out how a client or server can change this default situation to nonpersistent.

P26-7. In SMTP, a sender sends unformatted text. Show the MIME header.

P26-8. In HTTP, draw a figure to show the application of cookies in a scenario in which the server uses cookies for advertisement. Use only three sites.

P26-9. In FTP, a user (Jane) wants to retrieve an EBCDIC file named *huge* from */usr/users/report* directory using the ephemeral port 61017. The file is so large that the user wants to compress it before it is transferred. Show all the commands and responses.

P26-10. In Chapter 1, we mentioned that the TCP/IP suite, unlike the OSI model, has no presentation layer. But an application-layer protocol can include some of the features defined in this layer if needed. Does SMTP have any presentation-layer features?

P26-11. Encode the following message in base64:

01010111 00001111 11110000

P26-12. In FTP, a user (Jan) wants to make a new directory called *Jan* under the directory */usr/usrs/letters*. Show all of the commands and responses.

P26-13. Draw a diagram to show the use of a proxy server that is part of the client network:

a. Show the transactions between the client, proxy server, and the target server when the response is stored in the proxy server.

b. Show the transactions between the client, proxy server, and the target server when the response is not stored in the proxy server.

P26-14. Using RFC 1939, assume a POP3 client is in the download-and-keep mode. Show the transaction between the client and the server if the client has only two messages of 192 and 300 bytes to download from the server.

P26-15. Assume there is a server with the domain name *www.common.com.*

a. Show an HTTP request that needs to retrieve the document **/usr/users/doc**. The client accepts MIME version 1, GIF or JPEG images, but the document should not be more than 4 days old.

b. Show the HTTP response to part *a* for a successful request.

P26-16. POP3 protocol has some basic commands (that each client/server needs to implement). Using the information in RFC 1939, find the meaning and the use of the following basic commands:

a. STAT **b.** LIST **c.** DELE 4

P26-17. In Chapter 1, we mentioned that the TCP/IP suite, unlike the OSI model, has no session layer. But an application-layer protocol can include some of the features defined in this layer if needed. Does SMTP or POP3 have any session layer features?

P26-18. In SMTP,

a. a non-ASCII message of 1000 bytes is encoded using base64. How many bytes are in the encoded message? How many bytes are redundant? What is the ratio of redundant bytes to the total message?

b. a message of 1000 bytes is encoded using quoted-printable. The message consists of 90 percent ASCII and 10 percent non-ASCII characters. How many bytes are in the encoded message? How many bytes are redundant? What is the ratio of redundant bytes to the total message?

c. Compare the results of the two previous cases. How much is the efficiency improved if the message is a combination of ASCII and non-ASCII characters?

P26-19. According to RFC 1939, a POP3 session is in one of the following four states: closed, authorization, transaction, or update. Draw a diagram to show these four states and how POP3 moves between them.

P26-20. In Chapter 1, we mentioned that the TCP/IP suite, unlike the OSI model, has no presentation layer. But an application-layer protocol can include some of the features defined in this layer if needed. Does HTTP have any presentation-layer features?

P26-21. Write concurrent TCP client-server programs to simulate a simplified version of POP. The client sends a request to receive an e-mail in its mailbox; the server responds with the e-mail.

P26-22. Encode the following message in quoted-printable:

01001111 10101111 01110001

P26-23. In HTTP, draw a figure to show the application of cookies in a web portal using two sites.

P26-24. In HTTP, draw a figure to show the application of cookies in a scenario in which the server allows only the registered customer to access the server.

P26-25. In FTP, a user (Maria) wants to move a file named *file1* from */usr/users/report* directory to the directory */usr/top/letters*. Note that this is a case of renaming a file. We first need to give the name of the old file and then define the new name. Show all of the commands and responses.

26.9 SIMULATION EXPERIMENTS

26.9.1 Applets

We have created some Java applets to show some of the main concepts discussed in this chapter. It is strongly recommended that the students activate these applets on the book website and carefully examine the protocols in action.

26.9.2 Lab Assignments

In Chapter 1, we downloaded and installed Wireshark and learned about its basic features. In this chapter, we use Wireshark to capture and investigate some application-layer protocols. We use Wireshark to simulate six protocols: HTTP, FTP, TELNET, SMTP, POP3, and DNS.

Lab26-1. In the first lab, we retrieve web pages using HTTP. We use Wireshark to capture packets for analysis. We learn about the most common HTTP messages. We also capture response messages and analyze them. During the lab session, some HTTP headers are also examined and analyzed.

Lab26-2. In the second lab, we use FTP to transfer some files. We use Wireshark to capture some packets. We show that FTP uses two separate connections: a control connection and a data-transfer connection. The data connection is opened and then closed for each file transfer activity. We also show that FTP is an insecure file transfer protocol because the transaction is done in plaintext.

Lab26-3. In the third lab, we use Wireshark to capture packets exchanged by the TELNET protocol. As in FTP, we are able to observe commands and responses in the captured packets during the session. Like FTP, TELNET is vulnerable to hacking because it sends all data including the password in plaintext.

Lab26-4. In the fourth lab, we investigate SMTP protocol in action. We send an e-mail and, using Wireshark, we investigate the contents and the format of the SMTP packet exchanged between the client and the server. We check that the three phases we discussed in the text exist in this SMTP session.

Lab26-5. In the fifth lab, we investigate the state and behavior of the POP3 protocol. We retrieve the mails stored in our mailbox at the POP3 server and observe and analyze the states of the POP3 and the type and the contents of the messages exchanged, by analyzing the packets through Wireshark.

Lab26-6. In the sixth lab, we analyze the behavior of the DNS protocol. In addition to Wireshark, several network utilities are available for finding some information stored in the DNS servers. In this lab, we use the *dig* utilities (which has replaced *nslookup*). We also use *ipconfig* to manage the cached DNS records in the host computer. When we use these utilities, we set Wireshark to capture the packets sent by these utilities.

CHAPTER 27

Network Management

Although network management is implemented at the application layer of the TCP/IP protocol suite, we have postponed the subject until now in order to dedicate an entire chapter to its discussion. Network management plays an important role in the Internet as it becomes larger and larger. The failure of a single device may interrupt the communication from one point of the Internet to the other. In this chapter, we first discuss the areas of network management. We then discuss how one of these areas is implemented at the application layer of the TCP/IP suite. We have divided this chapter into three sections:

- The first section introduces the concept of network management and discusses five general areas of network management: configuration, fault, performance, security, and accounting. *Configuration management* is related to the status of each entity and its relationship to other entities. *Fault management* is the area of network management that handles issues related to interruptions in the system. *Performance management* tries to monitor and control the network to ensure that it is running as efficiently as possible. *Security management* is responsible for controlling access to the network based on predefined policy. *Accounting management* is the controlling of users' access to network resources through charges.
- The second section discusses Simple Network Management Protocol (SNMP) as a framework for managing devices in an internet using the TCP/IP protocol suite. It shows how a manager as a host runs an SNMP client and any agents as a router or host runs a server program. The section defines the three components of the management protocol in the Internet. The section also defines Structure of Management Information (SMI) as the language that specifies how data types and objects in SNMP should be identified. Finally, the section introduces Management Information Base (MIB), which designates the objects to be managed in SNMP according to the rules defined in SMI.
- The third section gives a brief discussion of a standard that provides the methods and rules to define data and objects. This section is very brief and only introduces the subject. Part of it is used by SMI in the second section.

27.1 INTRODUCTION

We can define *network management* as monitoring, testing, configuring, and troubleshooting network components to meet a set of requirements defined by an organization. These requirements include the smooth, efficient operation of the network that provides the predefined quality of service for users. To accomplish this task, a network management system uses hardware, software, and humans.

The International Organization for Standardization (ISO) defines five areas of network management: configuration management, fault management, performance management, security management, and accounting management, as shown in Figure 27.1.

Figure 27.1 *Areas of network management*

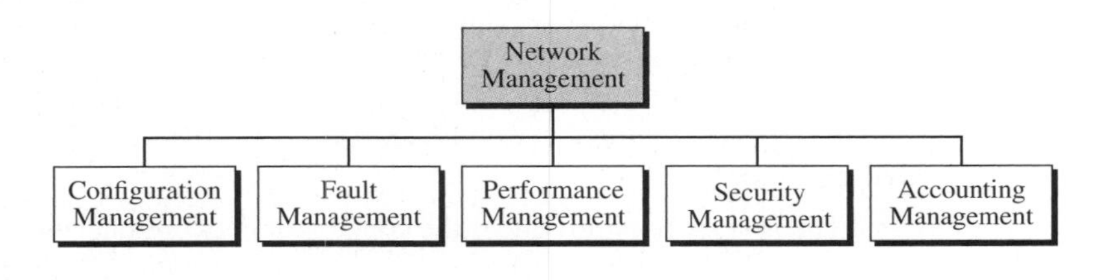

Although some organizations include other areas, such as cost management, we believe the ISO taxonomy is specific to network management. For example, cost management is a general management area for any management system and not just for network management.

27.1.1 Configuration Management

A large network is usually made up of hundreds of entities that are physically or logically connected to each other. These entities have an initial configuration when the network is set up, but can change with time. Desktop computers may be replaced by others; application software may be updated to a newer version; and users may move from one group to another. The *configuration management* system must know, at any time, the status of each entity and its relation to other entities. Configuration management can be divided into two subsystems: *reconfiguration* and *documentation.*

Reconfiguration

Reconfiguration can be a daily occurrence in a large network. There are three types of reconfiguration: *hardware reconfiguration*, *software reconfiguration,* and *user-account reconfiguration*.

Hardware Reconfiguration

Hardware reconfiguration covers all changes to the hardware. For example, a desktop computer may need to be replaced. A router may need to be moved to another part of the network. A subnetwork may be added or removed from the network. All of these need the time and attention of network management. In a large network, there must be specialized personnel trained for quick and efficient hardware reconfiguration. Unfortunately, this type of reconfiguration cannot be automated and must be manually handled case by case.

Software Reconfiguration

Software reconfiguration covers all changes to the software. For example, new software may need to be installed on servers or clients. An operating system may need updating. Fortunately, most software reconfiguration can be automated. For example, an update for an application on some or all clients can be electronically downloaded from the server.

User-Account Reconfiguration

User-account reconfiguration is not simply adding or deleting users on a system. We must also consider the user privileges, both as an individual and as a member of a group. For example, a user may have both read and write permission with regard to some files, but only read permission with regard to other files. User-account reconfiguration can be, to some extent, automated. For example, in a college or university, at the beginning of each quarter or semester, new students are added to the system. The students are normally grouped according to the courses they take or the majors they pursue. The members of each group have specific privileges; computer science students may need to access a server providing different computer language facilities, while engineering students may need to access servers that provide computer assisted design (CAD) software.

Documentation

The original network configuration and each subsequent change must be recorded meticulously. This means that there must be documentation for hardware, software, and user accounts.

Hardware Documentation

Hardware documentation normally involves two sets of documents: maps and specifications.

Maps *Maps* track each piece of hardware and its connection to the network. There can be one general map that shows the logical relationships between subnetworks. There can also be a second general map that shows the physical location of each subnetwork. For each subnetwork, then, there is one or more maps that show all pieces of equipment. The maps use some kind of standardization to be easily read and understood by current and future personnel.

Specifications Maps are not enough per se. Each piece of hardware also needs to be documented. There must be a set of *specifications* for each piece of hardware connected to the network. These specifications must include information such as hardware type, serial number, vendor (address and phone number), time of purchase, and warranty information.

Software Documentation

All software must also be documented. *Software documentation* includes information such as the software type, the version, the time installed, and the license agreement.

User-Account Documentation

Most operating systems have a utility that allows *user account documentation*. The management must make sure that the files with this information are updated and secured. Some operating systems record access privileges in two documents; one shows all files and access types for each user; the other shows the list of users that have access to a particular file.

27.1.2 Fault Management

Complex networks today are made up of hundreds and sometimes thousands of components. Proper operation of the network depends on the proper operation of each component individually and in relation to each other. *Fault management* is the area of network management that handles this issue. An effective fault management system has two subsystems: reactive fault management and proactive fault management.

Reactive Fault Management

A *reactive fault management* system is responsible for detecting, isolating, correcting, and recording faults. It handles short-term solutions to faults.

Detecting Fault

The first step taken by a reactive fault management system is to find the exact location of the fault. A fault is defined as an abnormal condition in the system. When a fault occurs, either the system stops working properly or the system creates excessive errors. A good example of a fault is a damaged communication medium.

Isolating Fault

The next step taken by a reactive fault management system is isolating the fault. A fault, if isolated, usually affects only a few users. After isolation, the affected users are immediately notified and given an estimated time of correction.

Correcting Fault

The next step is correcting the fault. This may involve replacing or repairing the faulty components.

Recording Fault

After the fault is corrected, it must be documented. The record should show the exact location of the fault, the possible cause, the action or actions taken to correct the fault, the cost, and the time it took for each step. Documentation is extremely important for several reasons:

- The problem may reoccur. Documentation can help the present or future administrator or technician solve a similar problem.
- The frequency of the same kind of failure is an indication of a major problem in the system. If a fault happens frequently in one component, the component should be replaced with a similar one or the whole system should be changed to avoid the use of that type of component.
- The statistic is helpful to another part of network management, performance management.

Proactive Fault Management

Proactive fault management tries to prevent faults from occurring. Although this is not always possible, some types of failures can be predicted and prevented. For example, if a manufacturer specifies a lifetime for a component or a part of a component, it is a good strategy to replace it before that time. As another example, if a fault happens frequently at one particular point in a network, it is wise to carefully reconfigure the network to prevent the fault from happening again.

27.1.3 Performance Management

Performance management, which is closely related to fault management, tries to monitor and control the network to ensure that it is running as efficiently as possible. Performance management tries to quantify performance using some measurable quantity, such as capacity, traffic, throughput, or response time. Some protocols, such as SNMP, which is discussed in this chapter, can be used in performance management.

Capacity

One factor that must be monitored by a performance management system is the *capacity* of the network. Every network has a limited capacity and the performance management system must ensure that it is not used above this capacity. For example, if a LAN is designed for 100 stations at an average data rate of 2 Mbps, it will not operate properly if 200 stations are connected to the network. The data rate will decrease and blocking may occur.

Traffic

Traffic can be measured in two ways: internally and externally. Internal traffic is measured by the number of packets (or bytes) travelling inside the network. External traffic is measured by the exchange of packets (or bytes) outside the network. During peak hours, when the system is heavily used, blocking may occur if there is excessive traffic.

Throughput

We can measure the *throughput* of an individual device (such as a router) or a part of the network. Performance management monitors the throughput to make sure that it is not reduced to unacceptable levels.

Response Time

Response time is normally measured from the time a user requests a service to the time the service is granted. Other factors such as capacity and traffic can affect the response time. Performance management monitors the average response time and the peak-hour response time. Any increase in response time is a very serious condition as it is an indication that the network is working above its capacity.

27.1.4 Security Management

Security management is responsible for controlling access to the network based on predefined policy. In Chapter 31 we will discuss security tools such as encryption and authentication. Encryption allows privacy for users; authentication forces the users to identify themselves.

27.1.5 Accounting Management

Accounting management is the controlling of users' access to network resources through charges. Under accounting management, individual users, departments, divisions, or even projects are charged for the services they receive from the network. Charging does not necessarily mean cash transfer; it may mean debiting the departments or divisions for budgeting purposes. Today, organizations use an accounting management system for the following reasons:

- It prevents users from monopolizing limited network resources.
- It prevents users from using the system inefficiently.
- Network managers can do short- and long-term planning based on the demand for network use.

27.2 SNMP

Several network management standards have been devised during the last few decades. The most important one is **Simple Network Management Protocol (SNMP),** used by the Internet. We discuss this standard in this section. SNMP is a framework for managing devices in an internet using the TCP/IP protocol suite. It provides a set of fundamental operations for monitoring and maintaining an internet. SNMP uses the concept of manager and agent. That is, a manager, usually a host, controls and monitors a set of agents, usually routers or servers (see Figure 27.2).

Figure 27.2 *SNMP concept*

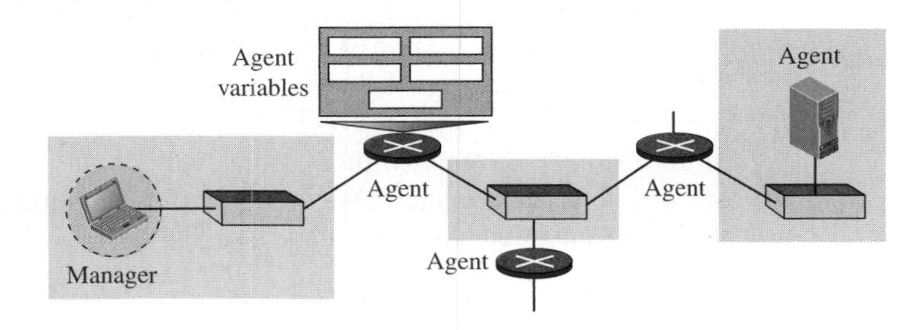

SNMP is an application-level protocol in which a few manager stations control a set of agents. The protocol is designed at the application level so that it can monitor devices made by different manufacturers and installed on different physical networks. In other words, SNMP frees management tasks from both the physical characteristics of the managed devices and the underlying networking technology. It can be used in a heterogeneous internet made of different LANs and WANs connected by routers made by different manufacturers.

27.2.1 Managers and Agents

A management station, called a *manager,* is a host that runs the SNMP client program. A managed station, called an *agent,* is a router (or a host) that runs the SNMP server program. Management is achieved through simple interaction between a manager and an agent.

The agent keeps performance information in a database. The manager has access to the values in the database. For example, a router can store in appropriate variables the number of packets received and forwarded. The manager can fetch and compare the values of these two variables to see if the router is congested or not.

The manager can also make the router perform certain actions. For example, a router periodically checks the value of a reboot counter to see when it should reboot itself. It reboots itself, for example, if the value of the counter is 0. The manager can use this feature to reboot the agent remotely at any time. It simply sends a packet to force a 0 value in the counter.

Agents can also contribute to the management process. The server program running on the agent can check the environment and, if it notices something unusual, it can send a warning message (called a *Trap*) to the manager.

In other words, management with SNMP is based on three basic ideas:

1. A manager checks an agent by requesting information that reflects the behavior of the agent.
2. A manager forces an agent to perform a task by resetting values in the agent database.
3. An agent contributes to the management process by warning the manager of an unusual situation.

27.2.2 Management Components

To do management tasks, SNMP uses two other protocols: **Structure of Management Information (SMI)** and **Management Information Base (MIB).** In other words, management on the Internet is done through the cooperation of three protocols: SNMP, SMI, and MIB, as shown in Figure 27.3.

Figure 27.3 *Components of network management on the Internet*

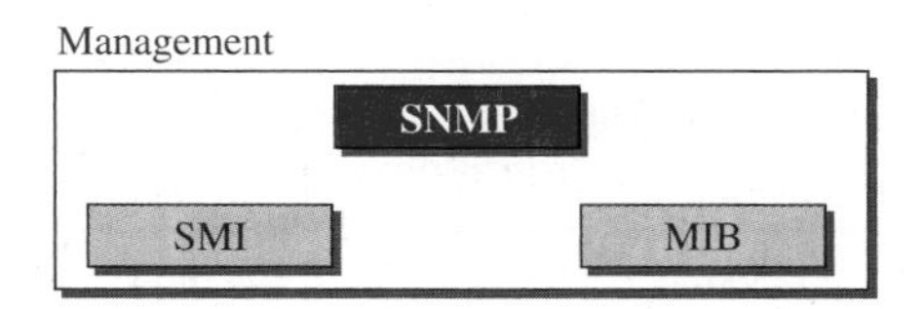

Let us elaborate on the interactions between these protocols.

Role of SNMP

SNMP has some very specific roles in network management. It defines the format of the packet to be sent from a manager to an agent and vice versa. It also interprets the result and creates statistics (often with the help of other management software). The packets exchanged contain the object (variable) names and their status (values). SNMP is responsible for reading and changing these values.

SNMP defines the format of packets exchanged between a manager and an agent. It reads and changes the status of objects (values of variables) in SNMP packets.

Role of SMI

To use SNMP, we need rules for naming objects. This is particularly important because the objects in SNMP form a hierarchical structure (an object may have a parent object and some child objects). Part of a name can be inherited from the parent. We also need rules to define the types of objects. What types of objects are handled by SNMP? Can SNMP handle simple types or structured types? How many simple types are available? What are the sizes of these types? What is the range of these types? In addition, how are each of these types encoded?

SMI defines the general rules for naming objects, defining object types (including range and length), and showing how to encode objects and values.

We need these universal rules because we do not know the architecture of the computers that send, receive, or store these values. The sender may be a powerful computer in which an integer is stored as 8-byte data; the receiver may be a small computer that stores an integer as 4-byte data.

SMI is a protocol that defines these rules. However, we must understand that SMI only defines the rules; it does not define how many objects are managed in an entity or which object uses which type. SMI is a collection of general rules to name objects and to list their types. The association of an object with the type is not done by SMI.

Role of MIB

We hope it is clear that we need another protocol. For each entity to be managed, this protocol must define the number of objects, name them according to the rules defined by SMI, and associate a type to each named object. This protocol is MIB. MIB creates a set of objects defined for each entity in a manner similar to that of a database (mostly metadata in a database, names and types without values).

MIB creates a collection of named objects, their types, and their relationships to each other in an entity to be managed.

An Analogy

Before discussing each of these protocols in more detail, let us give an analogy. The three network management components are similar to what we need when we write a program in a computer language to solve a problem. Figure 27.4 shows the analogy.

Syntax: SMI

Before we write a program, the syntax of the language (such as C or Java) must be predefined. The language also defines the structure of variables (simple, structured, pointer, and so on) and how the variables must be named. For example, a variable name

Figure 27.4 *Comparing computer programming and network management*

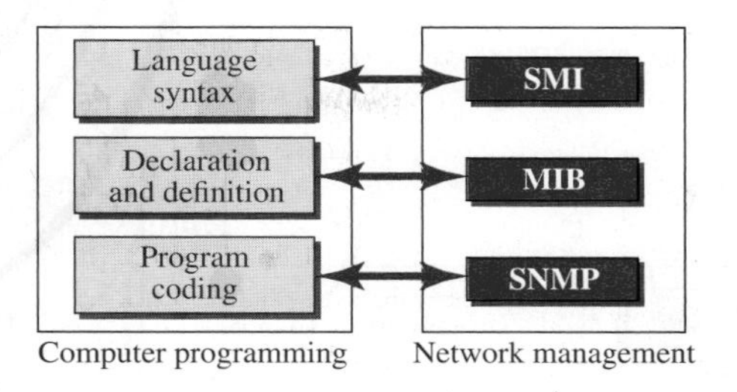

must be 1 to *n* characters in length and start with a letter followed by alphanumeric characters. The language also defines the type of data to be used (integer, real, character, etc.). In programming, the rules are defined by the syntax of the language. In network management, the rules are defined by SMI.

Object Declaration and Definition: MIB

Most computer languages require that objects be declared and defined in each specific program. Declaration and definition create objects using predefined types and allocate memory location for them. For example, if a program has two variables (an integer named *counter* and an array named *grades* of type char), they must be declared at the beginning of the program:

```
int counter;
char grades [40];
```

MIB does this task in network management. MIB names each object and defines the type of the objects. Because the type is defined by SMI, SNMP knows the range and size.

Program Coding: SNMP

After declaration in programming, the program needs to write statements to store values in the variables and change them if needed. SNMP does this task in network management. SNMP stores, changes, and interprets the values of objects already declared by MIB according to the rules defined by SMI.

27.2.3 An Overview

Before discussing each component in more detail, let us show how each of these components is involved in a simple scenario. This is an overview that will be developed later, at the end of the chapter. A manager station (SNMP client) wants to send a message to an agent station (SNMP server) to find the number of UDP user datagrams received by the agent. Figure 27.5 shows an overview of steps involved.

MIB is responsible for finding the object that holds the number of UDP user datagrams received. SMI, with the help of another embedded protocol, is responsible for encoding the name of the object. SNMP is responsible for creating a message, called a GetRequest message, and encapsulating the encoded message. Of course, things are more complicated than this simple overview, but we first need more details of each protocol.

Figure 27.5 *Management overview*

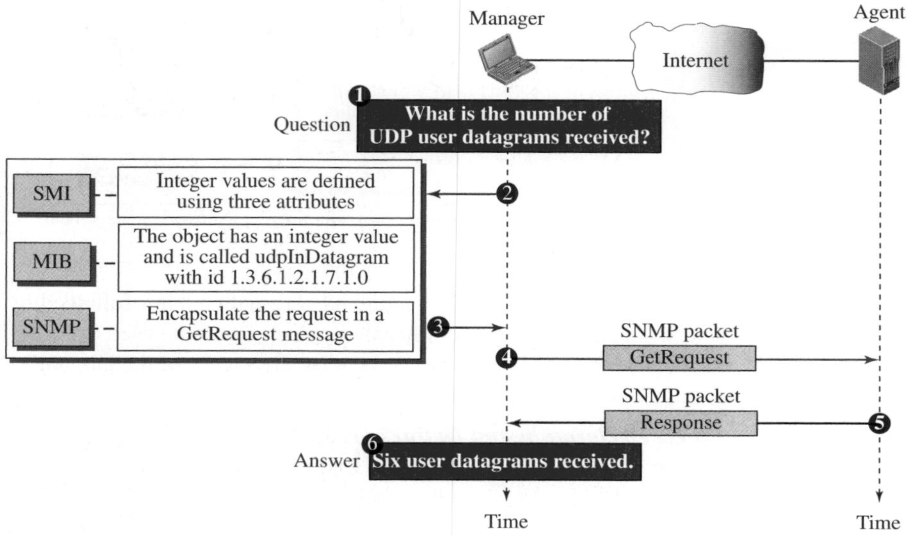

27.2.4 SMI

The Structure of Management Information, version 2 (SMIv2) is a component for network management. SMI is a guideline for SNMP. It emphasizes three attributes to handle an object: name, data type, and encoding method. Its functions are:

- To name objects.
- To define the type of data that can be stored in an object.
- To show how to encode data for transmission over the network.

Name

SMI requires that each managed object (such as a router, a variable in a router, a value, etc.) have a unique name. To name objects globally, SMI uses an **object identifier,** which is a hierarchical identifier based on a tree structure (see Figure 27.6).

The tree structure starts with an unnamed root. Each object can be defined using a sequence of integers separated by dots. The tree structure can also define an object using a sequence of textual names separated by dots.

The integer-dot representation is used in SNMP. The name-dot notation is used by people. For example, the following shows the same object in two different notations.

iso.org.dod.internet.mgmt.mib-2 ↔ **1.3.6.1.2.1**

The objects that are used in SNMP are located under the *mib-2* object, so their identifiers always start with 1.3.6.1.2.1.

Figure 27.6 *Object identifier in SMI*

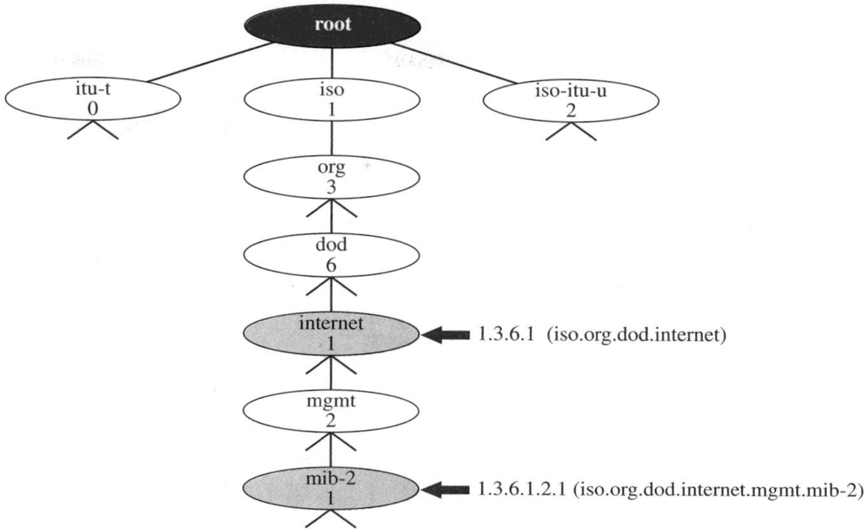

Type

The second attribute of an object is the type of data stored in it. To define the data type, SMI uses **Abstract Syntax Notation One (ASN.1)** definitions and adds some new definitions. In other words, SMI is both a subset and a superset of ASN.1. (We discuss ASN.1 in the last section of this chapter.)

SMI has two broad categories of data types: *simple* and *structured.* We first define the simple types and then show how the structured types can be constructed from the simple ones.

Simple Type

The **simple data types** are atomic data types. Some of them are taken directly from ASN.1; some are added by SMI. The most important ones are given in Table 27.1. The first five are from ASN.1; the next seven are defined by SMI.

Table 27.1 *Data types*

Type	*Size*	*Description*
INTEGER	4 bytes	An integer with a value between -2^{31} and $2^{31}-1$
Integer32	4 bytes	Same as INTEGER
Unsigned32	4 bytes	Unsigned with a value between 0 and $2^{32}-1$
OCTET STRING	Variable	Byte-string up to 65,535 bytes long
OBJECT IDENTIFIER	Variable	An object identifier
IPAddress	4 bytes	An IP address made of four integers
Counter32	4 bytes	An integer whose value can be incremented from zero to 2^{32}; when it reaches its maximum value it wraps back to zero

Table 27.1 *Data types (continued)*

Type	*Size*	*Description*
Counter64	8 bytes	64-bit counter
Gauge32	4 bytes	Same as Counter32, but when it reaches its maximum value, it does not wrap; it remains there until it is reset
TimeTicks	4 bytes	A counting value that records time in 1/100ths of a second
BITS		A string of bits
Opaque	Variable	Uninterpreted string

Structured Type

By combining simple and structured data types, we can make new structured data types. SMI defines two **structured data types:** *sequence* and *sequence of.*

- ❑ ***Sequence.*** A *sequence* data type is a combination of simple data types, not necessarily of the same type. It is analogous to the concept of a *struct* or a *record* used in programming languages such as C.
- ❑ ***Sequence of.*** A *sequence of* data type is a combination of simple data types all of the same type or a combination of sequence data types all of the same type. It is analogous to the concept of an *array* used in programming languages such as C.

Figure 27.7 shows a conceptual view of data types.

Figure 27.7 *Conceptual data types*

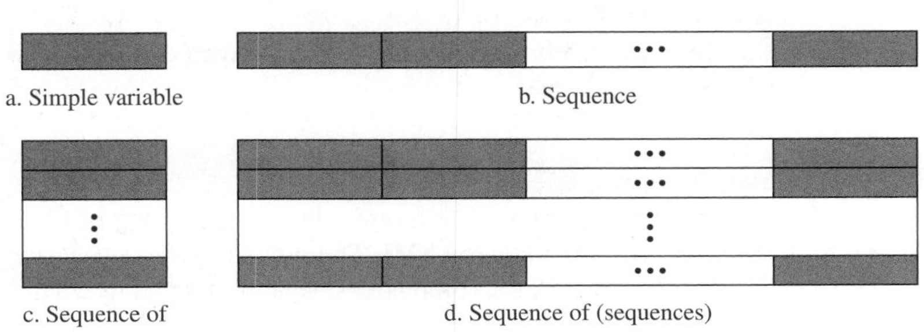

Encoding Method

SMI uses another standard, **Basic Encoding Rules (BER),** to encode data to be transmitted over the network. BER specifies that each piece of data be encoded in triplet format: tag, length, and value (TLV), as illustrated in Figure 27.8.

The tag is a 1-byte field that defines the type of data. Table 27.2 shows the data types we use in this chapter and their tags in hexadecimal numbers. The length field is 1 or more bytes. If it is 1 byte, the most significant bit must be 0. The other 7 bits define the length of the data. If it is more than 1 byte, the most significant bit of the first byte must be 1. The other 7 bits of the first byte specify the number of bytes

Figure 27.8 *Encoding format*

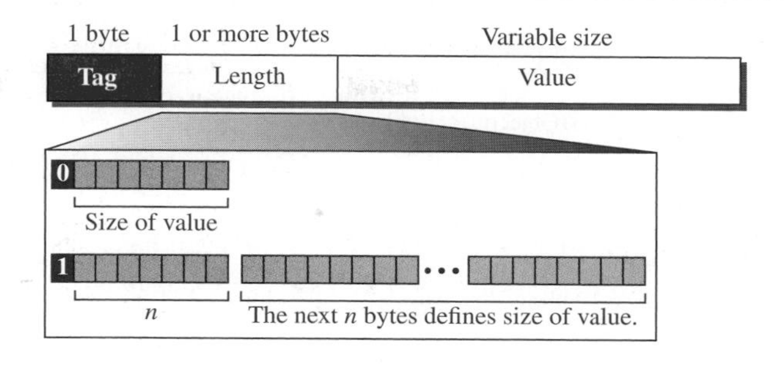

needed to define the length. The value field codes the value of the data according to the rules defined in BER.

Table 27.2 *Codes for data types*

Data Type	*Tag (Hex)*	*Data Type*	*Tag (Hex)*
INTEGER	**02**	IPAddress	**40**
OCTET STRING	**04**	Counter	**41**
OBJECT IDENTIFIER	**06**	Gauge	**42**
NULL	**05**	TimeTicks	**43**
SEQUENCE, SEQUENCE OF	**30**	Opaque	**44**

Example 27.1

Figure 27.9 shows how to define INTEGER 14. The size of the length field is from Table 27.1.

Figure 27.9 *Example 27.1: INTEGER 14*

0x02	0x04	0x00	0x00	0x00	0x0E
Tag (integer)	Length (4 bytes)	Value (14)			

Example 27.2

Figure 27.10 shows how to define the OCTET STRING "HI."

Figure 27.10 *Example 27.2: OCTET STRING "HI"*

0x04	0x02	0x48	0x49
Tag (String)	Length (2 bytes)	Value (H)	Value (I)

Example 27.3

Figure 27.11 shows how to define ObjectIdentifier 1.3.6.1 (iso.org.dod.internet).

Figure 27.11 *Example 27.3: ObjectIdentifier 1.3.6.1*

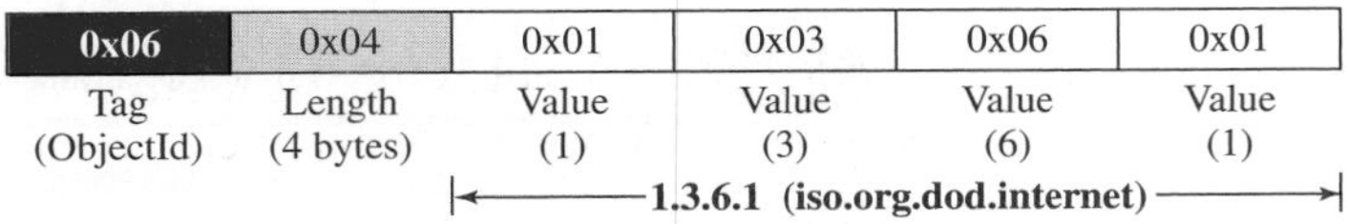

0x06
0x04
0x01
0x03
0x06
0x01
Tag (ObjectId)
Length (4 bytes)
Value (1)
Value (3)
Value (6)
Value (1)
1.3.6.1 (iso.org.dod.internet)

Example 27.4

Figure 27.12 shows how to define IPAddress 131.21.14.8.

Figure 27.12 *Example 27.4: IPAddress 131.21.14.8*

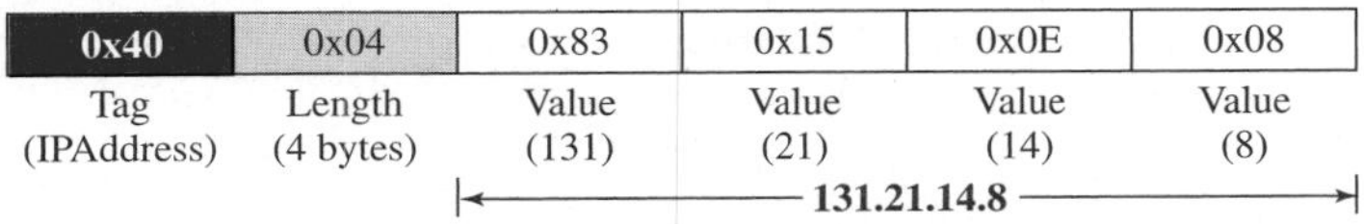

0x40
0x04
0x83
0x15
0x0E
0x08
Tag (IPAddress)
Length (4 bytes)
Value (131)
Value (21)
Value (14)
Value (8)
131.21.14.8

27.2.5 MIB

The Management Information Base, version 2 (MIB2) is the second component used in network management. Each agent has its own MIB2, which is a collection of all the objects that the manager can manage. (See Figure 27.13.)

Figure 27.13 *Some mib-2 groups*

1.3.6.1.2.1
mib-2
sys
1
if
2
at
3
ip
4
icmp
5
tcp
6
udp
7
egp
8
trans
10
snmp
11

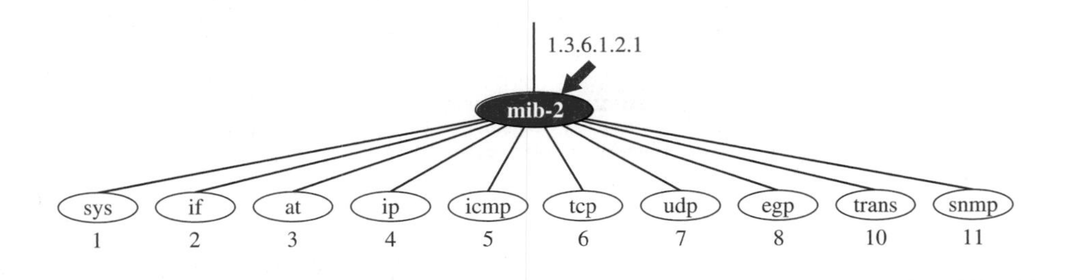

The objects in MIB2 are categorized under several groups: system, interface, address translation, ip, icmp, tcp, udp, egp, transmission, and snmp (note that group 9 is deprecated). These groups are under the mib-2 object in the object identifier tree. Each group has defined variables and/or tables.

The following is a brief description of some of the objects:

- **sys** This object (*system*) defines general information about the node (system), such as the name, location, and lifetime.
- **if** This object (*interface*) defines information about all of the interfaces of the node including interface number, physical address, and IP address.

- **at** This object (*address translation*) defines the information about the ARP table.
- **ip** This object defines information related to IP, such as the routing table and the IP address.
- **icmp** This object defines information related to ICMP, such as the number of packets sent and received and total errors created.
- **tcp** This object defines general information related to TCP, such as the connection table, time-out value, number of ports, and number of packets sent and received.
- **udp** This object defines general information related to UDP, such as the number of ports and number of packets sent and received.
- **egp** These objects are related to the operation of EGP.
- **trans** These objects are related to the specific method of transmission (future use).
- **snmp** This object defines general information related to SNMP itself.

Accessing MIB Variables

To show how to access different variables, we use the udp group as an example. There are four simple variables in the udp group and one sequence of (table of) records. Figure 27.14 shows the variables and the table. We will show how to access each entity.

Simple Variables

To access any of the simple variables, we use the id of the group (1.3.6.1.2.1.7) followed by the id of the variable. The following shows how to access each variable.

udpInDatagrams	→	**1.3.6.1.2.1.7.1**
udpNoPorts	→	**1.3.6.1.2.1.7.2**
udpInErrors	→	**1.3.6.1.2.1.7.3**
udpOutDatagrams	→	**1.3.6.1.2.1.7.4**

Figure 27.14 *udp group*

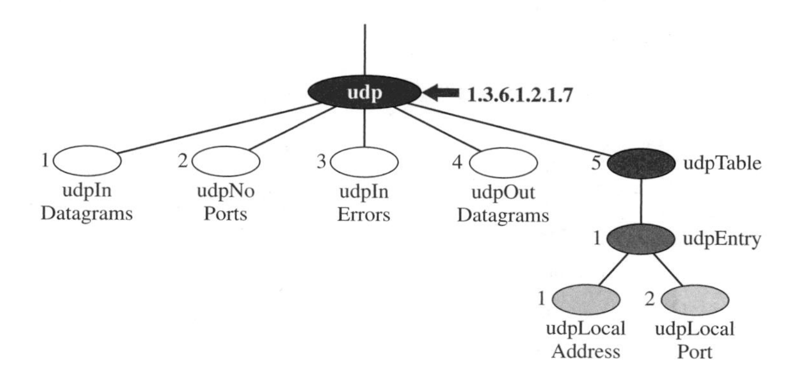

However, these object identifiers define the variable, not the instance (contents). To show the instance, or the contents, of each variable, we must add an instance suffix. The instance suffix for a simple variable is simply a zero. In other words, to show an instance of the above variables, we use the following:

udpInDatagrams.0	→	**1.3.6.1.2.1.7.1.0**
udpNoPorts.0	→	**1.3.6.1.2.1.7.2.0**
udpInErrors.0	→	**1.3.6.1.2.1.7.3.0**
udpOutDatagrams.0	→	**1.3.6.1.2.1.7.4.0**

Tables

To identify a table, we first use the table id. The udp group has only one table (with id 5), as illustrated in Figure 27.15. So to access the table, we use the following:

udpTable → **1.3.6.1.2.1.7.5**

However, the table is not at the leaf level in the tree structure. We cannot access the table; we define the entry (sequence) in the table (with id of 1), as follows:

udpEntry → **1.3.6.1.2.1.7.5.1**

This entry is also not a leaf and we cannot access it. We need to define each entity (field) in the entry.

udpLocalAddress	→	**1.3.6.1.2.1.7.5.1.1**
udpLocalPort	→	**1.3.6.1.2.1.7.5.1.2**

These two variables are at the leaf level of the tree. Although we can access their instances, we need to define *which* instance. At any moment, the table can have several values for each local address/local port pair. To access a specific instance (row) of the table, we add the index to the above ids. In MIB, the indexes of arrays are not integers (unlike most programming languages). The indexes are based on the value of one or more fields in the entries. In our example, the udpTable is indexed based on both the local address and the local port number. For example, Figure 27.16 shows a table with four rows and values for each field. The index of each row is a combination of two values. To access the instance of the local address for the first row, we use the identifier augmented with the instance index:

udpLocalAddress.181.23.45.14.23 → **1.3.6.1.2.7.5.1.1.181.23.45.14.23**

27.2.6 SNMP

SNMP uses both SMI and MIB in Internet network management. It is an application program that allows:

- A manager to retrieve the value of an object defined in an agent.
- A manager to store a value in an object defined in an agent.
- An agent to send an alarm message about an abnormal situation to the manager.

Figure 27.15 *udp variables and tables*

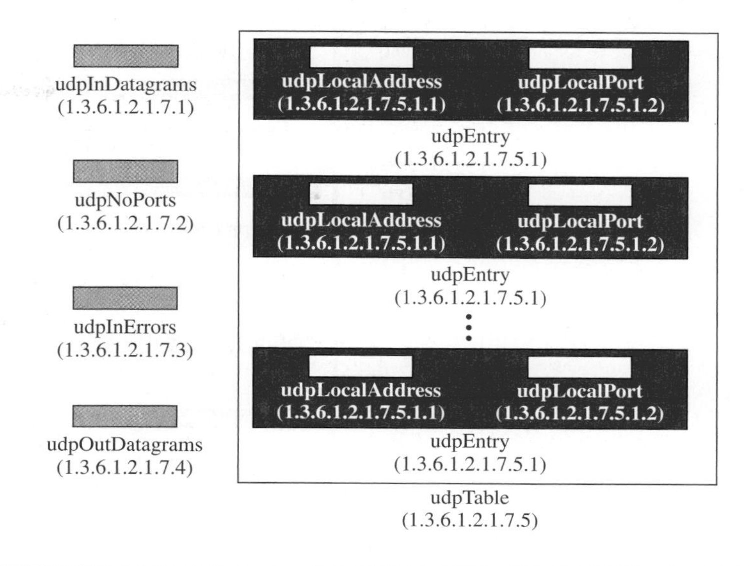

Figure 27.16 *Indexes for udpTable*

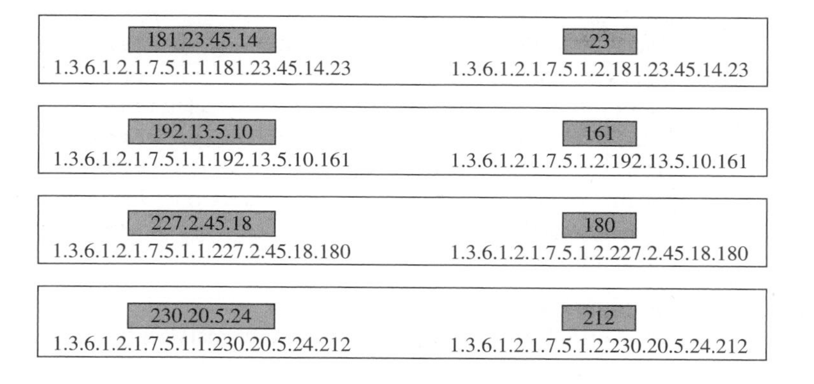

PDUs

SNMPv3 defines eight types of protocol data units (or PDUs): *GetRequest, GetNext-Request, GetBulkRequest, SetRequest, Response, Trap, InformRequest,* and *Report* (see Figure 27.17).

Figure 27.17 *SNMP PDUs*

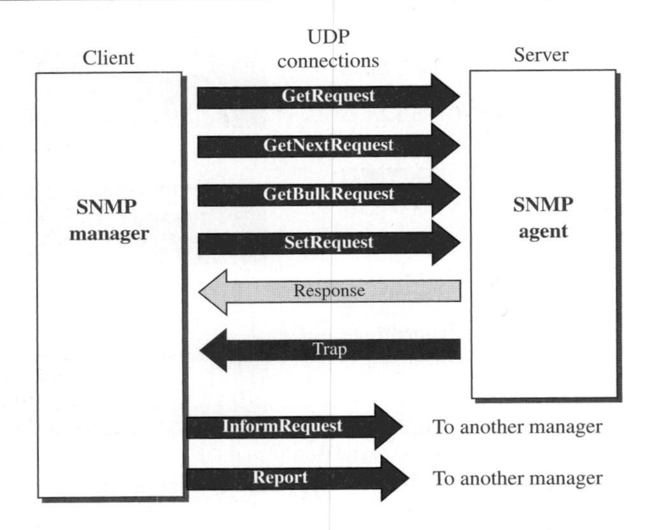

GetRequest
The GetRequest PDU is sent from the manager (client) to the agent (server) to retrieve the value of a variable or a set of variables.

GetNextRequest
The GetNextRequest PDU is sent from the manager to the agent to retrieve the value of a variable. The retrieved value is the value of the object following the defined ObjectId in the PDU. It is mostly used to retrieve the values of the entries in a table. If the manager does not know the indexes of the entries, it cannot retrieve the values. However, it can use GetNextRequest and define the ObjectId of the table. Because the first entry has the ObjectId immediately after the ObjectId of the table, the value of the first entry is returned. The manager can use this ObjectId to get the value of the next one, and so on.

GetBulkRequest
The GetBulkRequest PDU is sent from the manager to the agent to retrieve a large amount of data. It can be used instead of multiple GetRequest and GetNextRequest PDUs.

SetRequest
The SetRequest PDU is sent from the manager to the agent to set (store) a value in a variable.

Response
The Response PDU is sent from an agent to a manager in response to GetRequest or GetNextRequest. It contains the value(s) of the variable(s) requested by the manager.

Trap

The **Trap** (also called *SNMPv2 Trap* to distinguish it from SNMPv1 Trap) PDU is sent from the agent to the manager to report an event. For example, if the agent is rebooted, it informs the manager and reports the time of rebooting.

InformRequest

The InformRequest PDU is sent from one manager to another remote manager to get the value of some variables from agents under the control of the remote manager. The remote manager responds with a Response PDU.

Report

The Report PDU is designed to report some types of errors between managers. It is not yet in use.

Format

The format for the eight SNMP PDUs is shown in Figure 27.18. The GetBulkRequest PDU differs from the others in two areas, as shown in the figure.

Figure 27.18 *SNMP PDU format*

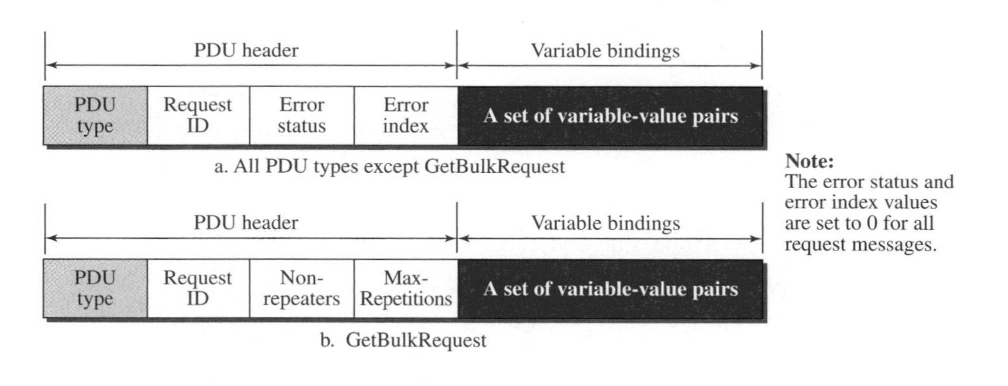

The fields are listed below:

- **PDU type.** This field defines the type of the PDU (see Table 27.3) .

Table 27.3 *PDU types*

Type	*Tag (Hex)*	*Type*	*Tag (Hex)*
GetRequest	**A0**	GetBulkRequest	**A5**
GetNextRequest	**A1**	InformRequest	**A6**
Response	**A2**	Trap (SNMPv2)	**A7**
SetRequest	**A3**	Report	**A8**

- **Request ID.** This field is a sequence number used by the manager in a request PDU and repeated by the agent in a response. It is used to match a request to a response.

- **Error status.** This is an integer that is used only in response PDUs to show the types of errors reported by the agent. Its value is 0 in request PDUs. Table 27.4 lists the types of errors that can occur.

Table 27.4 *Types of errors*

Status	*Name*	*Meaning*
0	noError	No error
1	tooBig	Response too big to fit in one message
2	noSuchName	Variable does not exist
3	badValue	The value to be stored is invalid
4	readOnly	The value cannot be modified
5	genErr	Other errors

- **Non-repeaters.** This field is used only in a GetBulkRequest PDU. The field defines the number of non-repeating (regular objects) at the start of the variable-value list.
- **Error index.** The error index is an offset that tells the manager which variable caused the error.
- **Max-repetitions.** This field is also used only in a GetBulkRequest PDU. The field defines the maximum number of iterations in the table to read all repeating objects.
- **Variable-value pair list.** This is a set of variables with the corresponding values the manager wants to retrieve or set. The values are null in request PDUs.

Messages

SNMP does not send only PDUs, it embeds each PDU in a message. A message is made of a message header followed by the corresponding PDU, as shown in Figure 27.19. The format of the message header, which depends on the version and the security provision, is not shown in the figure. We leave the details to some specific text.

Example 27.5

In this example, a manager station (SNMP client) uses a message with a GetRequest PDU to retrieve the number of UDP datagrams that a router has received (Figure 27.20).

There is only one Varbind sequence. The corresponding MIB variable related to this information is udpInDatagrams with the object identifier 1.3.6.1.2.1.7.1.0. The manager wants to retrieve a value (not to store a value), so the value defines a null entity. The bytes to be sent are shown in hexadecimal representation.

The Varbind list has only one Varbind. The variable is of type 06 and length 09. The value is of type 05 and length 00. The whole Varbind is a sequence of length 0D (13). The Varbind list is also a sequence of length 0F (15). The GetRequest PDU is of length ID (29).

Note that we have intended the bytes to show the inclusion of simple data types inside a sequence or the inclusion of sequences and simple data types inside larger sequences. Note that the PDU itself is like a sequence, but its tag is A0 in hexadecimal.

Figure 27.19 *SNMP message*

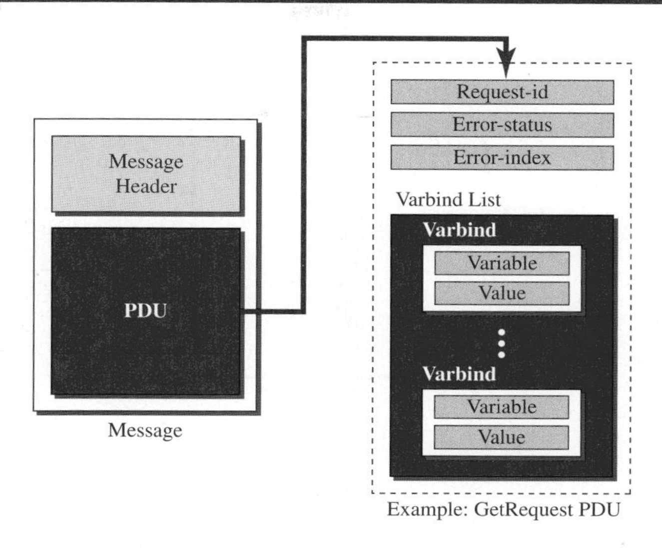

Figure 27.20 *Example 27.5*

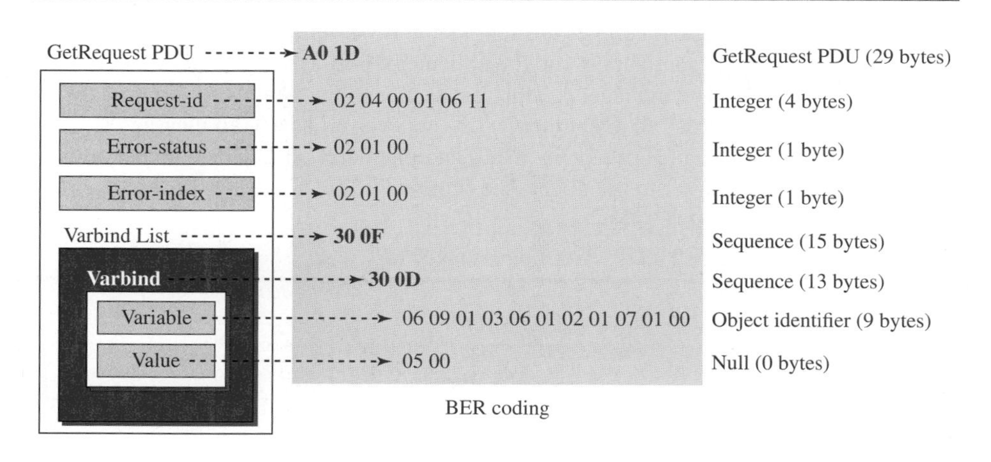

Figure 27.21 shows the actual message sent. We assume that the message header is made of 10 bytes. The actual message header may be different. We show the message using rows of 4 bytes.The bytes that are shown using dashes are the ones related to the message header.

Figure 27.21 *Actual message sent for Example 27.5*

30	**29**	--	--
--	--	--	--
--	--	--	--
A0	1D	02	04
00	01	06	11
02	01	00	02
01	00	30	0F
30	0D	06	09
01	03	06	01
02	01	07	01
00	05	00	

Message

Note:
The byte values are in hexadecimal.

UDP Ports

SNMP uses the services of UDP on two well-known ports, 161 and 162. The well-known port 161 is used by the server (agent), and the well-known port 162 is used by the client (manager).

The agent (server) issues a passive open on port 161. It then waits for a connection from a manager (client). A manager (client) issues an active open using an ephemeral port. The request messages are sent from the client to the server using the ephemeral port as the source port and the well-known port 161 as the destination port. The response messages are sent from the server to the client using the well-known port 161 as the source port and the ephemeral port as the destination port.

The manager (client) issues a passive open on port 162. It then waits for a connection from an agent (server). Whenever it has a Trap message to send, an agent (server) issues an active open, using an ephemeral port. This connection is only one-way, from the server to the client (see Figure 27.22).

Figure 27.22 *Port numbers for SNMP*

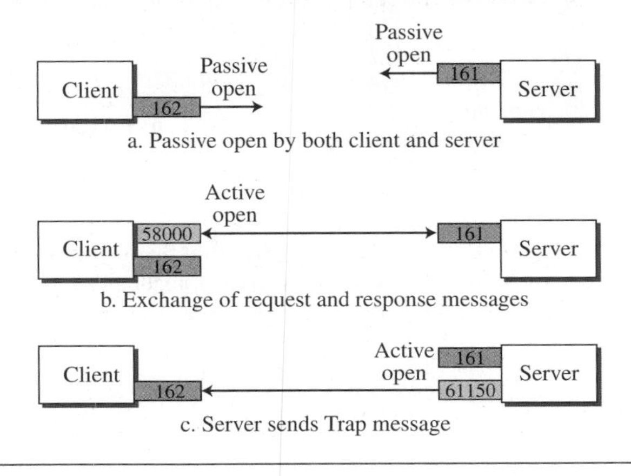

The client-server mechanism in SNMP is different from other protocols. Here both the client and the server use well-known ports. In addition, both the client and the server are running infinitely. The reason is that request messages are initiated by a manager (client), but Trap messages are initiated by an agent (server).

Security

SNMPv3 has added two new features to the previous version: security and remote administration. SNMPv3 allows a manager to choose one or more levels of security when accessing an agent. Different aspects of security can be configured by the manager to allow message authentication, confidentiality, and integrity.

SNMPv3 also allows remote configuration of security aspects without requiring the administrator to actually be at the place where the device is located.

27.3 ASN.1

In data communication, when we send a continuous stream of bits to a destination, we somehow need to define the format of the data. If we send a *name* and a *number* in a single message, we need to tell the destination that, for example, the first 12 bits define the name and the next 8 bits define the number. We will have more difficulty when we send a complex data type such as an array or a record. For example, in the case of an array, if we send 2000 bits in a message, we need to tell the receiver that it is an array of 200 numbers each of 10 bits or it is an array of 10 numbers each of 200 bits.

A solution is that we separate the definition of data types from the sequence of bits transmitted through the network. This is done through an abstract language that uses some symbols, key words, and atomic data types and lets us make new data types out of the simple types. The language is called Abstract Syntax Notation One (ASN.1). Note that ASN.1 is a very complex language used in different areas of computer science, but in this section, we only introduce the language as much as needed for the SNMP protocol.

27.3.1 Language Basics

Before we show how we can define objects and associated values, let us talk about the language itself. The language uses some symbols and some key words and defines some primitive data types. As we said before, SMI uses a subset of these entities in its own language.

Symbols

The language uses a set of symbols, given in Table 27.5. Some of these symbols are single characters, but some are pairs of characters.

Table 27.5 *Symbols used in ASN.1*

<table>
<tr><th>Symbol</th><th>Meaning</th><th>Symbol</th><th>Meaning</th></tr>
<tr><td>::=</td><td>Defined as or assignment</td><td>..</td><td>Range</td></tr>
<tr><td>|</td><td>Or, alternative, or option</td><td>{}</td><td>Start and end of a list</td></tr>
<tr><td>–</td><td>Negative sign</td><td>[]</td><td>Start and end of tag</td></tr>
<tr><td>––</td><td>The following is a comment</td><td>()</td><td>Start and end of a subtype</td></tr>
</table>

Keywords

The language has a set of limited keywords that can be used. These words can be used in the language only for the purpose for which they have been defined. All words should be in uppercase (Table 27.6).

Table 27.6 *Keywords in ASN.1*

Keyword	*Description*
BEGIN	Start of a module
CHOICE	List of alternatives
DEFINITIONS	Definition of a data type or an object
END	End of a module
EXPORTS	Data type that can be exported to other modules
IDENTIFIER	A sequence of nonnegative numbers that identifies an object
IMPORTS	Data type defined in an external module and imported
INTEGER	Any positive, zero, or negative integer
NULL	A null value
OBJECT	Used with IDENTIFIER to uniquely define an object
OCTET	Eight-bit binary data
OF	Used with SEQUENCE or SET
SEQUENCE	An ordered list
SEQUENCE OF	An ordered array of data of the same type
SET	An unordered list
SET OF	An array of unordered lists
STRING	A string of data

27.3.2 Data Types

After discussing the symbols and keywords used in the language, it is time to define its data types. The idea is similar to what we see in computer languages such as C, C++, or Java. In ASN.1, we have several simple data types such as integer, float, boolean, char, and so on. We can combine these data types to create a new simple data type (with a different name) or to define a structured data type such as array or struct. We first define simple data types in ASN.1 and then show how to make a new data type of these data types.

Simple Data Types

ASN.1 defines a set of simple (atomic) data types. Each data type is given a universal tag and has a set of values, as shown in Table 27.7. This is the same idea as used in a computer language when we have some basic data types with predefined ranges of values. For example, in C language, we have the data type *int,* which can take a range of values. Note that the tag in the table is actually the rightmost five bits of the tag we defined in Table 27.2.

Table 27.7 *Some simple ASN.1 built-in types*

Tag	*Type*	*Set of values*
Universal 1	BOOLEAN	TRUE or FALSE
Universal 2	INTEGER	Integers (positive, 0, or negative)
Universal 3	BIT STRING	A string of binary digits (bits) or a null set
Universal 4	OCTET STRING	A string of octets or a null set
Universal 5	NULL	Null, single valued
Universal 6	OBJECT IDENTIFIER	A set of values that defines an object
Universal 7	ObjectDescriptor	Human readable text describing an object
Universal 8	EXTERNAL	A type which is not in the standard
Universal 9	REAL	Real numbers in scientific notation
Universal 10	ENUMERATED	A list of integers
Universal 16	SEQUENCE, SEQUENCE OF	Ordered list of types
Universal 17	SET, SET OF	Unordered list of types
Universal 18	NumericString	Digits 0-9 and space
Universal 19	PrintableString	printable characters
Universal 26	VisibleString	ISO646String
Universal 27	GeneralString	General character string
Universal 30	CHARACTER STRING	Character set

New Data Types

ASN.1 uses **Backus–Naur Form** (**BNF**) syntax for defining a new data type from a built-in data type or a previously defined data type as shown below:

<new type> ::= **<type>**

where the *new type* must start with a capital letter.

Example 27.6

The following is an example of some new types using built-in types from Table 27.7.

Married ::= **BOOLEAN**
MaritalStatus ::= **ENUMERATED** {single, married, widowed, divorced}
DayOfWeek ::= **ENUMERATED** {sun, mon, tue, wed, thu, fri, sat}
Age ::= **INTEGER**

New Subtypes

ASN.1 even allows us to create a subtype whose range is a subrange of a built-in type or a previously defined data type.

Example 27.7

The following shows how we can make three new subtypes. The range of the first is the subset of INTEGER, the range of the second is the subset of REAL, and the range of the third is the subset of DayOfWeek, which we defined in Example 27.6. Note that we use the symbol (..) to define the range and the symbol (|) to define the choice.

NumberOfStudents ::= **INTEGER** (15..40)	— An integer with the range 15 to 40
Grade ::= **REAL** (1.0..5.0)	— A real number with the range 1.0 to 5.0
Weekend ::= **DayOfWeek** (sun | sat)	— A day that can be sun or sat

Simple Variables

In a programming language, we can create a variable of a particular type and assign (store) a value in it. In ASN.1, the term *Value Name* is used instead of *variable,* but we use the term *variable,* which is more familiar to programmers. We can create a variable of a particular type and assign a value belonging to the range defined for that type. The following shows the syntax:

<variable> <type> ::= **<value>**

The name of the variable should start with a lowercase letter to distinguish it from the type.

Example 27.8

The following are a few examples of defining some variables and assigning the appropriate value from the range of those types. Note that the first and the third variables are of the built-in type, the second is of the type defined in Example 27.6, and the last is of a subtype defined in Example 27.7.

```
numberOfComputers INTEGER ::= 2
married Married ::= FALSE
herAge INTEGER ::= 35
classSize NumberOfStudents ::= 22
```

Structured Type

ASN.1 uses a keyword SEQUENCE to define a structured data type similar to *struct* (record) in C language or C++. The SEQUENCE type is an ordered list of variable types. The following shows a new type StudentAccount which is a sequence of three variables: username, password, and accountNumber.

```
StudentAccount ::= SEQUENCE
{
        userName VisibleString,
        password VisibleString,
        accountNumber INTEGER
}
```

Structure Variables

After defining the new type, we can create a variable out of it and assign values to variables as shown below:

```
johnNewton StudentAccount
{
        userName "JohnN",
        password "120007",
        accountNumber 25579
}
```

Figure 27.23 shows the record created from the type definition and value assignments.

Figure 27.23 *Record representing the type definition and variable declaration*

We use the keyword SEQUENCE OF to define a new type similar to an array in C or C++ which is a composite type in which all components are the same. For example, we can define a forwarding table in a router as SEQUENCE OF Rows in which each Row is itself a sequence made of several variables.

27.3.3 Encoding

After the data has been defined and values are associated with variables, ASN.1 uses one of the encoding rules to encode the message to be sent. We already discussed the Basic Encoding Rule in the previous section.

27.4 END-CHAPTER MATERIALS

27.4.1 Recommended Reading

For more details about subjects discussed in this chapter, we recommend the following books. The items enclosed in brackets refer to the reference list at the end of the book.

Books

Several books give thorough coverage of SNMP: [Com 06], [Ste 94], [Tan 03], and [MS 01].

27.4.2 Key Terms

Abstract Syntax Notation One (ASN.1)
Backus–Naur Form (BNF)
Basic Encoding Rules (BER)
Management Information Base (MIB)
object identifier
simple data type
Simple Network Management Protocol (SNMP)
Structure of Management Information (SMI)
structured data type
Trap

27.4.3 Summary

The five areas comprising network management are configuration management, fault management, performance management, accounting management, and security management. Configuration management is concerned with the physical or logical changes of network entities. Fault management is concerned with the proper operation of each network component. Performance management is concerned with the monitoring and controlling of the network to ensure the network runs as efficiently as possible. Security management is concerned with controlling access to the network. Accounting management is concerned with controlling user access to network resources through charges.

Simple Network Management Protocol (SNMP) is a framework for managing devices in an internet using the TCP/IP protocol suite. A manager, usually a host, controls and monitors a set of agents, usually routers. SNMP uses the services of SMI and MIB. SMI names objects, defines the type of data that can be stored in an object, and encodes the data. MIB is a collection of groups of objects that can be managed by SNMP. MIB uses lexicographic ordering to manage its variables.

Abstract Syntax Notation Number One (ASN.1) is a language that defines the syntax and semantics of data. It uses some symbols, keywords, and simple and structured data types. Part of ASN.1 is used by SMI to define the format of objects and values used in network management.

27.5 PRACTICE SET

27.5.1 Quizzes

A set of interactive quizzes for this chapter can be found on the book website. It is strongly recommended that the student take the quizzes to check his/her understanding of the materials before continuing with the practice set.

27.5.2 Questions

Q27-1. Assume an object identifier in MIB has three simple variables. If the object identifier is *x*, what is the identifier of each variable?

Q27-2. What is the length of the value field in the following BER encoding?

04 09 48 65 6C 4C ...

Q27-3. Distinguish between internal and external data traffic in an organization.

Q27-4. Does an SNMP manager run a client SNMP program or a server SNMP program?

Q27-5. In SNMP, which of the following PDUs are sent from a client SNMP to a server SNMP?

a. GetRequest **b.** Response **c.** Trap

Q27-6. A network manager decides to replace a version of accounting software with a new version. What area of network management is involved here?

Q27-7. Can an SNMP message reference a leaf node in the MIB tree? Explain.

Q27-8. Which of the following is not part of configuration management?

a. reconfiguration **b.** encryption **c.** documentation

Q27-9. A network manager decides to replace the old router that connects the organization to the Internet with a more powerful one. What area of network management is involved here?

Q27-10. What does the ***if*** object in MIB define? Why does this object need to be managed?

Q27-11. Distinguish between reactive fault management and proactive fault management.

Q27-12. Assume a manageable object has only three simple variables. How many leaves can be found in the MIB tree for this object?

Q27-13. Which of the following devices cannot be a manager station in SNMP?

a. a router **b.** a host **c.** a switch

Q27-14. If a student in a college can monopolize access to a piece of software, causing other students to wait for a long time, which area of network management has failed?

Q27-15. Distinguish between SMI and MIB.

Q27-16. What are the source and destination port numbers when an SNMP message carries one of the following PDUs?

a. GetRequest **b.** Response **c.** Trap **d.** Report

Q27-17. Find the type (simple, sequence, sequence of) of the following objects in SMI.

a. An unsigned integer

b. An IP address

c. An object name

d. A list of integers

e. A record defining an object name, an IP address, and an integer

f. A list of records in which each record is an object name followed by a counter

Q27-18. If network management does not replace a component whose lifetime has been expired, what area in network management has been ignored?

Q27-19. Show how the textual name "iso.org.dod" is numerically encoded in SMI.

Q27-20. Is it possible to have a textual name in SMI as "iso.org.internet"? Explain.

Q27-21. Assume a manageable object has a table with three columns. How many leaves are there in the MIB tree for this table?

Q27-22. Distinguish between a GetRequest PDU and a SetRequest PDU.

Q27-23. Which of the following is not one of the five areas of network management defined by ISO?

a. fault **b.** performance **c.** personnel

Q27-24. Can SNMP reference the entire row of a table? In other words, can SNMP retrieve or change the values in the entire row of a table? Explain.

27.5.3 Problems

P27-1. Assume object *x* has two simple variables: an integer and an IP address. What is the identifier for each variable?

P27-2. Define an SNMP message (see Figure 27.19) using the syntax defined for structured data types in ASN.1.

P27-3. Assume object *x* has two simple variables and one table with two columns. What is the identifier for each variable and each column of the table? We assume that simple variables come before the table.

P27-4. Given the code 02040000C738, decode it using BER.

P27-5. One of the objects (groups) that can be managed is the ***ip*** group with the object identifier (1.3.6.1.2.1.**4**) in which (1.3.6.1.2.1) is the identifier of MIB-2 and (4) defines the ***ip*** group. In an agent, this object has 20 simple variables and three tables. One of the tables is the routing (forwarding) table with the identifier (1.3.6.1.2.1.**4.21**). This table has eleven columns, the first of which is called the *ipRouteDes,* which means the destination IP address. Assume that the indexing is based on the first column. Assume the table has four rows at the moment with the destination IP addresses (201.14.67.0), (123.16.0.0), (11.0.0.0), and (0.0.0.0). Show how SNMP can access all four instances of the second column, called *ipRouteIfIndex,* which defines the interface numbers through which the IP should be sent out.

P27-6. Using BER, show how we can encode a structured data type made of an INTEGER of value (2371), an OCTET STRING of value ("Computer"), and an IPAddress of value (185.32.1.5) as shown below:

```
SEQUENCE
{
      INTEGER 2371
      OCTET STRING "Computer"
      IP Address 185.32.1.5
}
```

P27-7. Object *x* has two simple variables. How can SNMP refer to the instance of each variable?

P27-8. Assume object *x* has only one table with two columns. What is the identifier for each column?

P27-9. Define a GetRequest PDU (see Figure 27.18) using the syntax defined for structured data types in ASN.1.

P27-10. Show the encoding for the IPAddress 112.56.23.78 using BER.

P27-11. Assume a manager needs to know the number of user datagrams an agent has sent out (a udpOutDatagrams counter with the identifier 1.3.6.1.2.1.7.4). Show the code for a Varbind that is sent in a GetRequest message and the code that the agent will send in the Response message if the value of the counter at this moment is 15.

P27-12. Define a Response PDU (see Figure 27.18) using the syntax defined for structured data types in ASN.1.

P27-13. Show the encoding for the object identifier 1.3.6.1.2.1.7.1 (the udpInDatagram variable in udp group) using BER.

P27-14. Given the code 300D04024E6F300706030103060500, decode it using BER.

P27-15. Show the encoding for the OCTET STRING "Hello world." using BER.

P27-16. Show the encoding for the INTEGER 1456 using BER.

P27-17. Assume a data structure is made of an INTEGER of value (131) and another structure made of an IPAddress of value (24.70.6.14) and an OCTETSTRING ("UDP"). Using BER, encode the data structure.

P27-18. Assume object *x* has one simple variable and two tables with two and three columns respectively. What is the identifier for the variable and each column of each table? We assume that the simple variable comes before the tables.

P27-19. Given the code 300C02040000099806040A05030E, decode it using BER.

P27-20. Object *x* has one table with two columns. The table at this moment has three rows with the contents shown below. If the table index is based on the values in the first column, show how SNMP can access each instance.

Object x

a	aa
b	bb
c	cc

Table

P27-21. Define a VarbindList (see Figure 27.19) using the syntax defined for structured data types in ASN.1.

CHAPTER 28

Multimedia

Multimedia refers to a number of different integrated media, such as text, images, audio, and video, that are generated, stored, and transmitted digitally and can be accessed interactively. Multimedia today is a broad subject that cannot be fully discussed in one chapter. In this chapter we give an overview of the multimedia and touch on subjects that are, directly or indirectly, related to the multimedia such as compression or quality of service. We have divided this chapter into four sections:

- The first section discusses the general idea behind compression. Although compression is not directly related to the subject of multimedia, multimedia transmission is not possible without first compressing the data. The section describes both lossless and lossy compression.
- The second section discusses the elements of multimedia: text, image, video, and audio. The section describes how these elements are represented, encoded, and compressed using the techniques discussed in the first section.
- The third section divides the multimedia in the Internet into three categories: streaming stored audio/video, streaming live audio/video, and real-time interactive audio/video. The section describes the features and characteristics of each and gives some examples.
- The fourth section concentrates on the real-time interactive category. The section introduces the transport-layer protocols used for multimedia applications: RTP and RTCP. The section also describes two protocols that are used in this category for signalling: SIP and H.323. These protocols are used in voice over IP (Internet telephony) and can be used for signalling protocols in future applications.

28.1 COMPRESSION

In this section, we discuss compression, which plays a crucial role in multimedia communication due to the large volume of data exchanged. In compression, we reduce the volume of data to be exchanged. We can divide compression into two broad categories: lossless and lossy compression. We briefly discuss the common methods used in each category. This section can be skipped if the reader is familiar with compression techniques. We included this section to provide the necessary background for those readers that are not familiar with compression techniques.

28.1.1 Lossless Compression

In **lossless compression,** the integrity of the data is preserved because the compression and decompression algorithms are exact inverses of each other: no part of the data is lost in the process. Lossless compression methods are normally used when we cannot afford to lose any data. For example, we must not lose data when we compress a text file or an application program. Lossless compression is also applied as the last step in some lossy compression procedures to further reduce the size of the data.

We discuss four lossless compression methods in this section: run-length coding, dictionary coding, Huffman coding, and arithmetic coding.

Run-length Coding

Run-length coding, sometimes referred to as ***run-length encoding (RLE),*** is the simplest method of removing redundancy. It can be used to compress data made of any combination of symbols. The method replaces a repeated sequence, *run,* of the same symbol with two entities: a count and the symbol itself. For example, the following shows how we can compress a string of 17 characters to a string of 10 characters.

AAABBBBCDDDDDDEEE → **3A4B1C6D3E**

A modified version of this method can be used if there are only two symbols in the data, such as a binary pattern made of 0s and 1s. In this case, we use only the count of one of the symbols that occurs between each occurrence of the other symbol. Figure 28.1 shows a binary pattern in which there are more 0s than 1s. We just show the number of 0s that occur between 1s.

Figure 28.1 *A version of run-length coding to compress binary patterns*

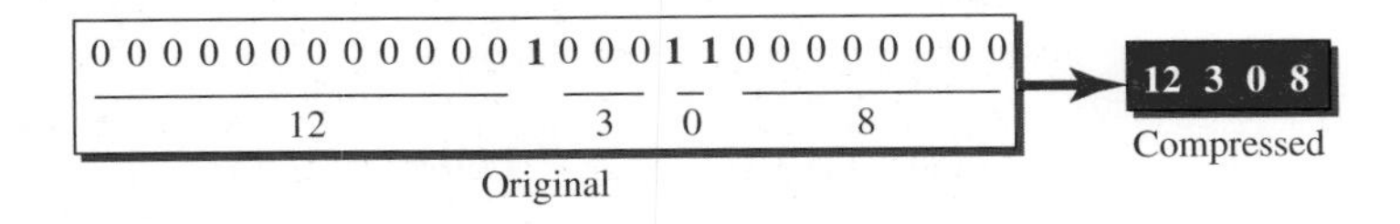

The compressed data can be encoded in binary using a fixed number of bits per digit. For example, using 4 bits per digit, the compressed data can be represented as 1100001100001000, in which the rate of compression is 26/16 or almost 1.62 for this example.

Dictionary Coding

There is a group of compression methods based on creation of a dictionary (array) of strings in the text. The idea is to encode common sequences of characters instead of encoding each character separately. The dictionary is created as the message is scanned, and if a sequence of characters that is an entry in the dictionary is found in the message, the code (index) of that entry is sent instead of the sequence. The one we discuss here was invented by Lempel and Ziv and refined by Welch. It is referred to as ***Lempel-Ziv-Welch (LZW)***. The interesting point about this encoding technique is that the creation of the dictionary is dynamic; it is created by the sender and the receiver during the encoding and decoding processes; it is not sent from the sender to the receiver.

Encoding

The following shows the process:

1. The dictionary is initialized with one entry for each possible character in the message (alphabet). At the same time, a buffer, which we call the *string,* is initialized to the first character in the message. The string holds the largest encodable sequence found so far. In the initialization step, only the first character in the message is encodable.
2. The process scans the message and gets the next character in the message.
   a. If the concatenation of the string and the scanned character is in the dictionary, the string is not the largest encodable sequence. The process updates the string by concatenating the character at the end of it and waits for the next iteration.
   b. If the concatenation of the string and the scanned character is not in the dictionary, the largest encodable sequence is the string, not the concatenation of the two. Three actions are taken. First, the process adds the concatenation of the two as the new entry to the dictionary. Second, the process encodes the string. Third, the process reinitializes the string with the scanned character for the next iteration.
3. The process repeats step 2 while there are more characters in the message.

Table 28.1 gives a pseudocode for the encoding process. We have called the next character *char* and the string *S* for simplicity.

Table 28.1 *LZW encoding*

```
LZWEncoding (message)
{
        Initialize (Dictionary)
        Char = Input (first character)
        S = char                                // S is the encodable sequence
        while (more characters in message)
        {
                char = Input (next character);
                if ((S + char) is in Dictionary)        // S is not the encodable sequence
                {
                        S = S + char;
                }
                else                            // S is the encodable sequence
                {
```

Table 28.1 *LZW encoding (continued)*

```
                addToDictionary (S + char);
                Output (index of S in Dictionary);
                S = char;
            }
        }
        Output (index of S in Dictionary);
}
```

Example 28.1

Let us show an example of LZW encoding using a text message in which the alphabet is made of two characters: A and B (Figure 28.2).

The figure shows how the text "BAABABBBAABBBBAA" is encoded as 1002163670. Note that the buffer PreS holds the string from the previous iteration before it is updated.

Decoding

The following shows the process of decoding.

1. The dictionary is initialized as we explain in the encoding process. The first codeword is scanned and, using the dictionary, the first character in the message is output.
2. The process then creates a string and sets it to the previous scanned codeword. Now it scans a new codeword.
   a. If the codeword is in the dictionary, the process adds a new entry to the dictionary, which is the string concatenated with the first character from the entry related to the new codeword. It also outputs the entry related to the new codeword.
   b. If the codeword is not in the dictionary (which may happen occasionally), the process concatenates the string with the first character from the string and stores it in the dictionary. It also outputs the result of the concatenation.
3. The process repeats step 2 while there are more codewords in the code.

Table 28.2 shows the simplified algorithm for LZW decoding. We have used C for the codeword and S for the string.

Example 28.2

Let us show how the code in Example 28.1 can be decoded and the original message recovered (Figure 28.3). The box called PreC holds the codeword from the previous iteration, which is not needed in the pseudocode, but needed here to better show the process. Note that in this example there is only the special case in which the codeword is not in the dictionary. The new entry for the dictionary needs to be made from the string and the first character in the string. The output is also the same as the new entry.

Huffman Coding

When we encode data as binary patterns, we normally use a fixed number of bits for each symbol. To compress data, we can consider the frequency of symbols and the probability of their occurrence in the message. **Huffman coding** assigns shorter codes to symbols that occur more frequently and longer codes to those that occur less frequently.

Figure 28.2 *Example 28.1*

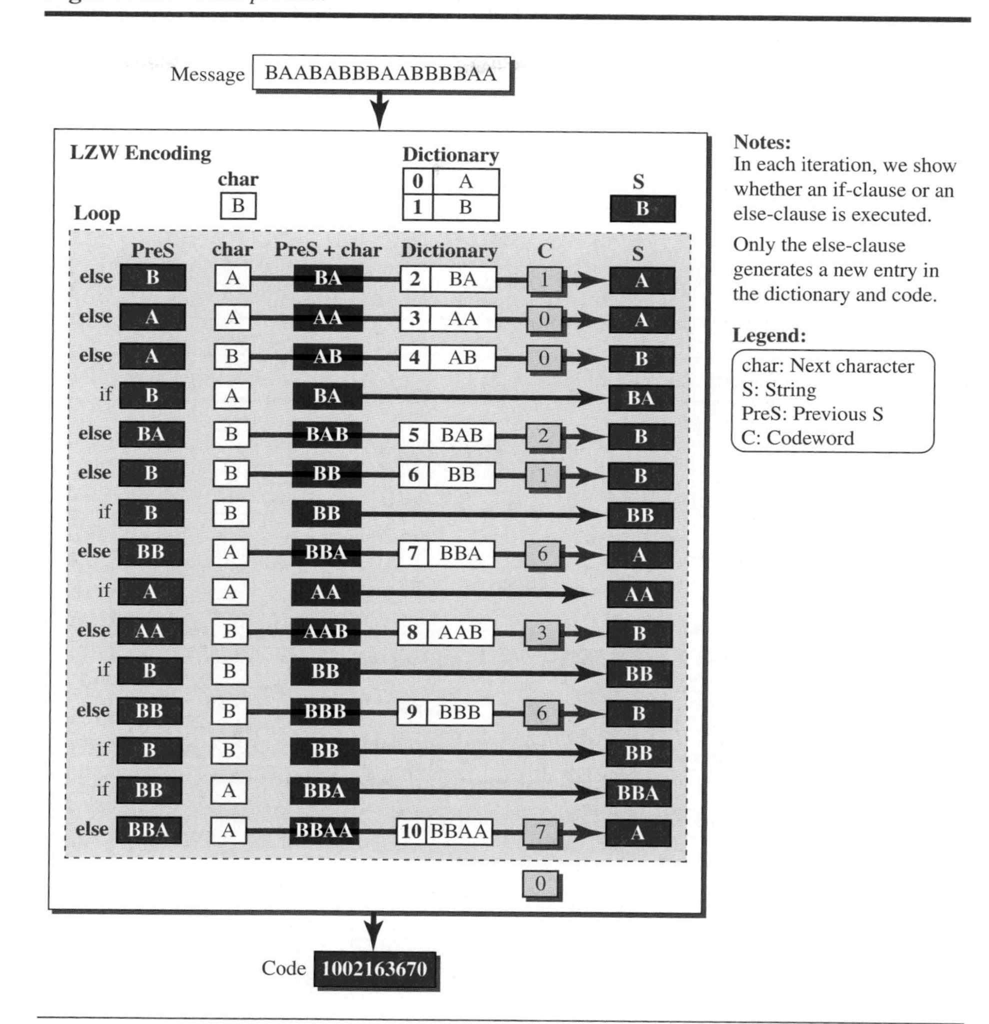

Table 28.2 *LZW decoding*

```
LZWDecoding (code)
{
        Initialize (Dictionary);
        C = Input (first codeword);
        Output (Dictionary [C]);
        while (more codewords in code)
        {
                S = Dictionary[C];
                C = Input (next codeword);
                if (C is in Dictionary)                    // Normal case
                {
```

Table 28.2 *LZW decoding (continued)*

```
                    addToDictionary (S + firstSymbolOf Dictionary[C]);
                    Output (Dictionary [C]);
               }
               else                                        // Special case
               {
                    addToDictionary (S + firstSymbolOf (S));
                    Output (S + firstSymbolOf (S);

               }
          }
}
```

Figure 28.3 *Example 28.2*

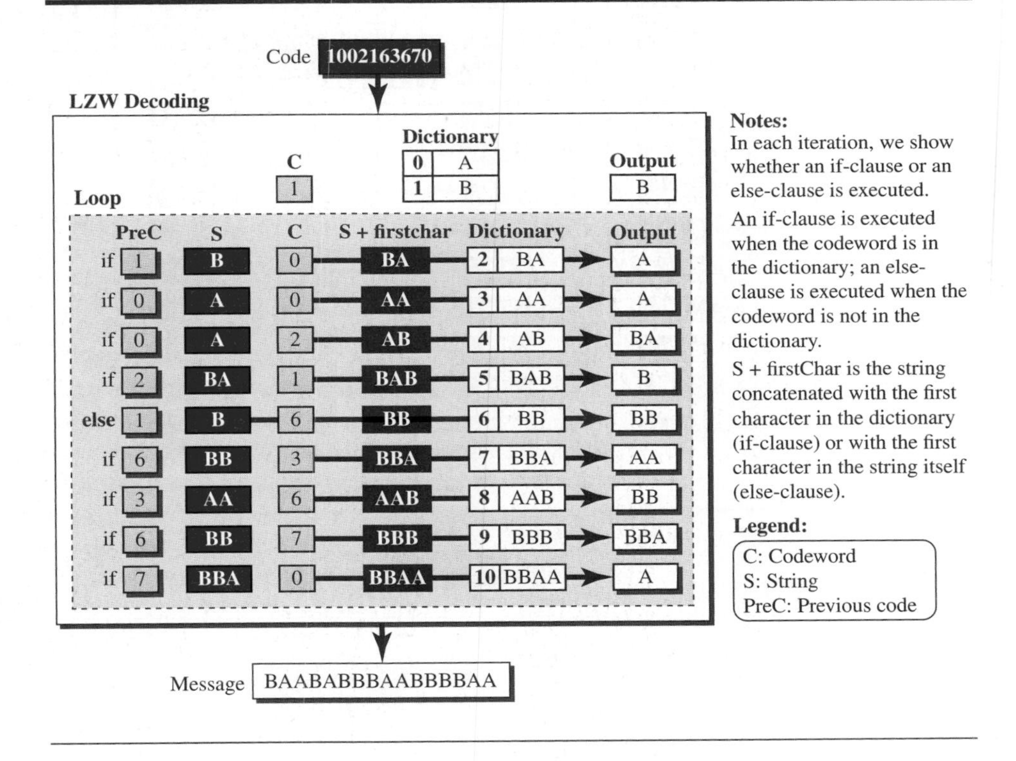

For example, imagine we have a text file that uses only five characters (A, B, C, D, E) with the frequency of occurrence of (20, 10, 10, 30, 30).

Huffman Tree

To use Huffman coding, we first need to create the Huffman tree. The Huffman tree is a tree in which the leaves of the tree are the symbols. It is made so that the most frequent symbol is the closest to the root of the tree (with the minimum number of nodes to the root) and the least frequent symbol is the farthest from the root. Figure 28.4 shows the process.

Figure 28.4 *Huffman tree*

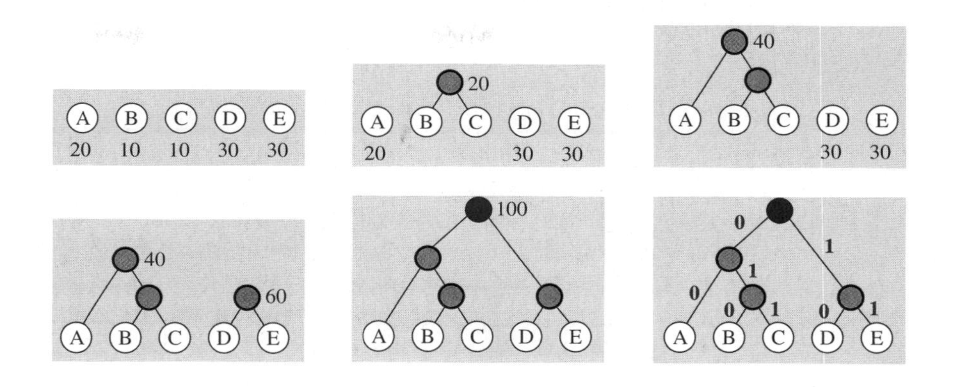

1. We put the entire character set in a row. Each character is now a node at the lowest level of the tree.
2. We select the two nodes with the smallest frequencies and join them to form a new node, resulting in a simple two-level tree. The frequency of the new node is the combined frequencies of the original two nodes. This node, one level up from the leaves, is eligible for combination with other nodes.
3. We repeat step 2 until all of the nodes, on every level, are combined into a single tree.
4. After the tree is made, we assign bit values to each branch. Since the Huffman tree is a binary tree, each node has a maximum of two children.

Coding Table

After the tree has been made, we can create a table that shows how each character can be encoded and decoded. The code for each character can be found by starting at the root and following the branches that lead to that character. The code itself is the bit value of each branch on the path, taken in sequence. Table 28.3 shows the character codes for our simple example.

Table 28.3 *Coding table*

Symbol	Code	*Symbol*	Code	*Symbol*	Code
A	00	C	011	E	11
B	010	D	10		

Note these points about the codes. First, the characters with higher frequencies receive a shorter code (A, D, and E) than the characters with lower frequencies (B and C).

Compare this with a code that assigns equal bit lengths to each character. Second, in this coding system, no code is a prefix of another code. The 2-bit codes, 00, 10, and 11, are not the prefixes of any of the two other codes (010 and 011). In other words, we do not have a 3-bit code beginning with 00, 10, or 11. This property makes the Huffman code an *instantaneous* code. We will explain this property in the next section.

Encoding and Decoding

Figure 28.5 shows how we can encode and decode using Huffman coding.

Figure 28.5 *Encoding and decoding in Huffman coding*

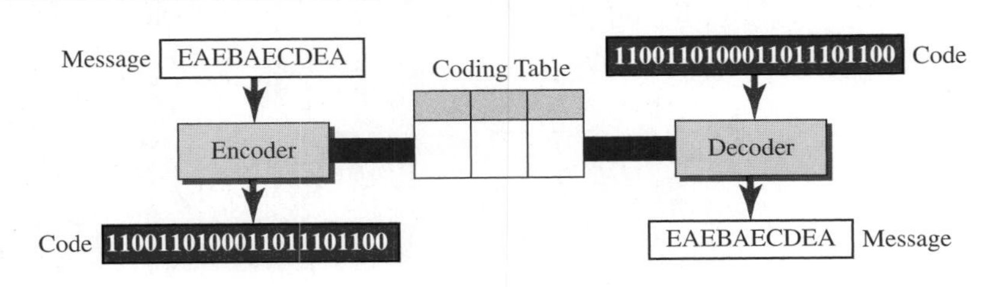

Note that we have achieved compression even with a small message. If we want to send fixed-length codes for a five-character alphabet, we need $\log_2 5 = 2.32$ or 3 bits for each character or 30 bits for the whole message. With Huffman coding we need only 22 bits. The compression ratio is 30/22 or 1.36.

In Huffman coding, no code is the prefix of another code. This means that we do not need to insert delimiters to separate the code for one character from the code for the next. This property of Huffman coding also allows instantaneous decoding: when the decoder has the two bits 00, it can immediately decode it as character A; it does not need to see more bits.

One drawback of Huffman coding is that both encoder and decoder need to use the same encoding table. In other words, the Huffman tree cannot be created dynamically as the dictionary can be in LZW coding. However, if the encoder and decoder are using the same set of symbols all of the time, the tree can be made and shared once. Otherwise, the table needs to be made by the encoder and be given to the receiver.

Arithmetic Coding

In the previous compression methods, each symbol or sequence of symbols is encoded separately. In **arithmetic coding,** introduced by Rissanen and Langdon in 1981, the entire message is mapped to a small interval inside [0,1). The small interval is then encoded as a binary pattern. Arithmetic coding is based on the fact that we can have an infinite number of small intervals inside the half-open interval [0,1). Each of these small intervals can represent one of the possible messages we can make using a finite set of symbols. Figure 28.6 shows the idea.

Figure 28.6 *Arithmetic coding*

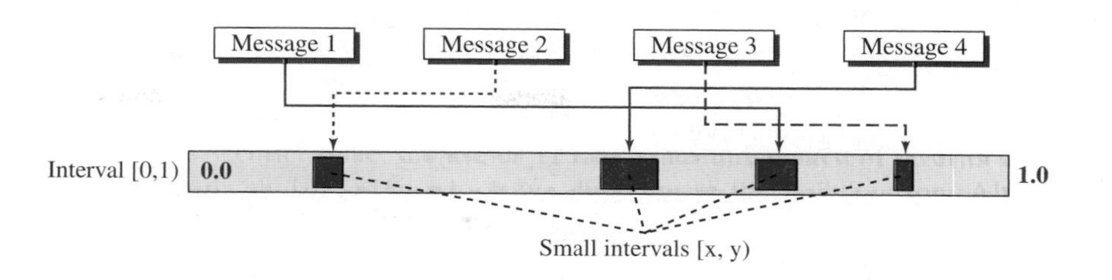

Encoding

To encode a message in arithmetic coding, we first need to assign the probability of occurrence to each symbol. If we have M symbols in the alphabet (including the terminating symbol that we need for decoding, as we will see later), the probabilities are $P_1, P_2, \ldots, P_M$, in which $P_1 + P_2 + \ldots + P_M = 1.0$. Table 28.4 shows the encoding algorithm.

Table 28.4 *Arithmetic encoding*

```
ArithmeticEncoding (message)
{
        currentInterval = [0,1);
        while (more symbols in the message)
        {
                s = Input (next symbol);
                divide currentInterval into subintervals
                subInt = subinterval related to s
                currentInterval = subInt
        }
        Output (bits related to the currentInterval)
}
```

In each iteration of the loop, we divide the current interval into M subintervals, in which the length of each subinterval is proportional to the probability of the corresponding symbol. This is done to uniformly scatter the messages in the interval [0,1). We also preserve the order of the symbols in the new interval in each iteration.

The choice of bits selected for output depends on the implementation. Some implementations use the bits that represent the fractional part of the beginning interval.

Example 28.3

For the sake of simplicity, let us assume that our set of symbols is S = {A, B, ∗}, in which the asterisk is the terminating symbol. We assign probability of occurrence for each symbol as

$$P_A = 0.4 \qquad P_B = 0.5 \qquad P_* = 0.1.$$

Figure 28.7 shows how we find the interval and the code related to the short message "BBAB*".

Figure 28.7 *Example 28.3*

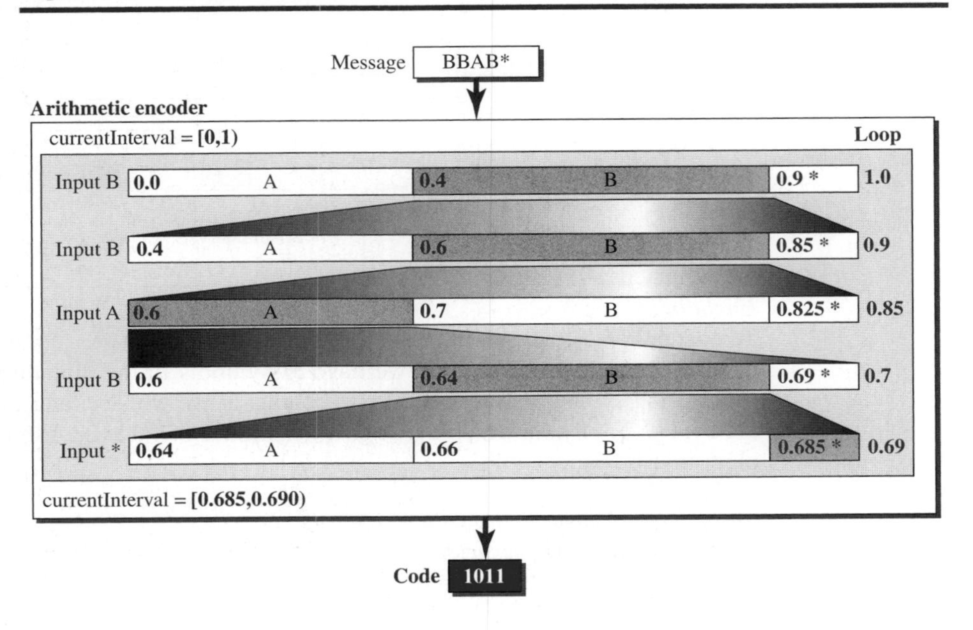

We initialize the current interval to [0,1). In each iteration of the loop, we divide the current interval into three subintervals according to the probability of each symbol occurring. We then read the first symbol and choose the corresponding subinterval. We then set the current interval to be the chosen interval. The process is repeated until all symbols are input. After reading each symbol, the current interval is reduced until it becomes [0.685,0.690). We encode the lower bound, 0.685 in binary, which is approximately $(0.1011)_2$, but we keep only the fractional part, $(1011)_2$, as the code. Note that when we change a real number between 0 and 1 to binary, we may get an infinite number of bits. We need to keep enough bits to recover the original message. More than enough bits is not efficient encoding; fewer than enough bits may result in wrong decoding. For example, if we use only three bits (101), it represents the real value 0.625, which is outside of the last current interval [0.685,0.690).

Decoding

Decoding is similar to encoding, but we exit the loop when the terminating symbol is output. This is the reason we need the terminating symbol in the original message. Table 28.5 shows the decoding algorithm.

Table 28.5 *Arithmetic decoding*

```
ArithmeticDecoding (code)
{
    c = Input (code)
    num = find real number related to code
```

Table 28.5 *Arithmetic decoding (continued)*

```
        currentInterval = [0,1);
        while (true)
        {
                divide the currentInterval into subintervals;
                subInt = subinterval related to num;
                Output (symbol related to subInt);
                if (symbol is the terminating symbol)    return;
                currentInterval = subInt;
        }
}
```

Example 28.4

Figure 28.8 shows how we use the decoding process to decode the message in Example 28.3. Note that the hand shows the position of the number in the corresponding interval.

Figure 28.8 *Example 28.4*

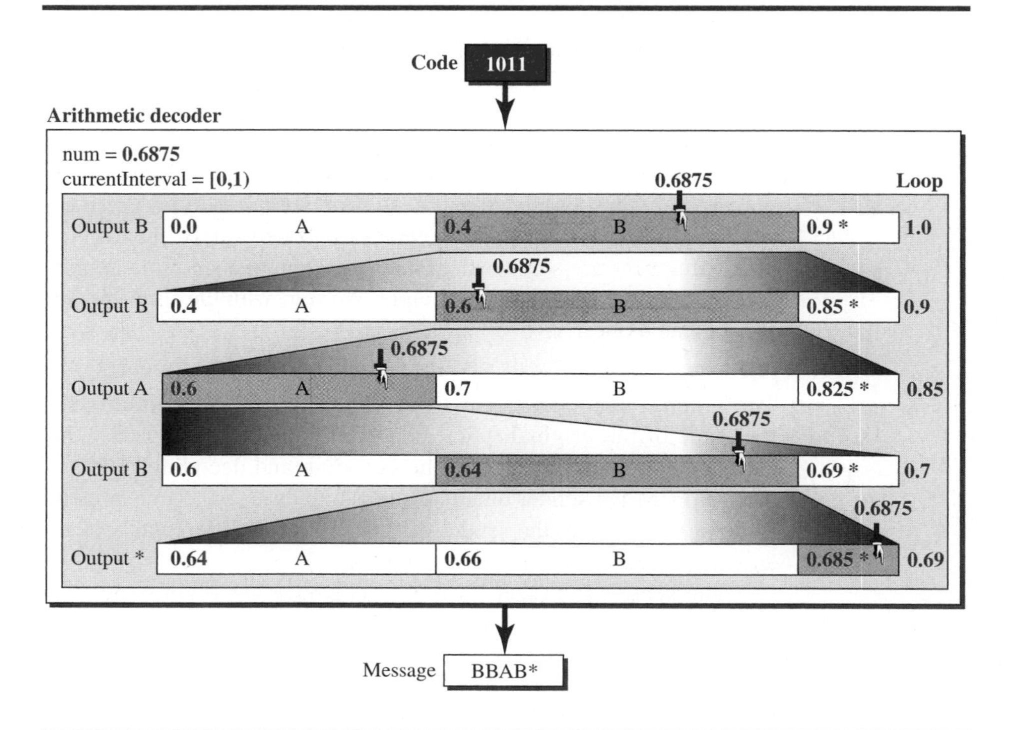

Static versus Dynamic Arithmetic Coding

The literature refers to two versions of arithmetic coding: *static coding* (sometimes called *pure coding*) and *dynamic coding* (sometimes called *interval coding*). The version we discussed in this section is the first one, static coding. There are two problems with static arithmetic coding. First, if the current interval is very small, we need a very high-precision arithmetic to encode the message, which results in a lot of 0 bits in the middle of the code. Second, the message cannot be encoded until all symbols are input; this is the reason that we need a terminating symbol for decoding. The new version, dynamic arithmetic coding, overcomes these two problems by using a procedure that outputs binary bits immediately after each symbol is read.

28.1.2 Lossy Compression

Lossless compression has limits on the amount of compression. However, in some situations, we can sacrifice some accuracy to increase the compression rate. Although we cannot afford to lose information in text compression, we can afford it when we are compressing images, video, and audio. For example, human vision cannot detect some small distortions that can result from **lossy compression** of an image. In this section, we discuss a few ideas behind lossy compression. In the next section, we show how these ideas can be used in the implementation of image, video, and audio compression.

Predictive Coding

Predictive coding is used when we digitize an analog signal. In Chapter 3, we discussed pulse code modulation (PCM) as a technique that converts an analog signal to a digital signal, using sampling. After sampling, each sample needs to be quantized to create binary values. Compression can be achieved in the quantization step by using *predictive coding*.

In PCM, samples are quantized separately. The neighboring quantized samples, however, are closely related and have similar values. In **predictive coding,** we use this similarity. Instead of quantizing each sample separately, the differences are quantized. The differences are smaller than the actual samples and thus require fewer bits. Many algorithms are based on this principle. We start with the simplest one and move to more sophisticated ones.

Delta Modulation

The simplest method in predictive coding is called **delta modulation.** Let x_n represent the value of the original function at sampling interval n and y_n be the reconstructed value of x_n. Figure 28.9 shows the encoding and decoding processes in delta modulation. In PCM the sender quantizes the samples (x_n) and transmits them to the receiver. In delta modulation, the sender quantizes e_n, the difference between each sample (x_n) and the preceding reconstructed value (y_{n-1}).

The sender then transmits C_n. The receiver reconstructs sample y_n from the received C_n.

Note that for each sample PCM needs to transmit several bits. For example, it needs to transmit 3 bits for each sample if the maximum quantized value is 7 (see Chapter 3). DM reduces the number of transmitted bits because it transmits a single bit (1 or 0) for each sample.

We may ask why DM quantizes the difference $x_n - y_{n-1}$ instead of $x_n - x_{n-1}$. The reason is that the second choice causes y to vary much faster than x if x is a slow-changing

Figure 28.9 *Encoding and decoding in delta modulation*

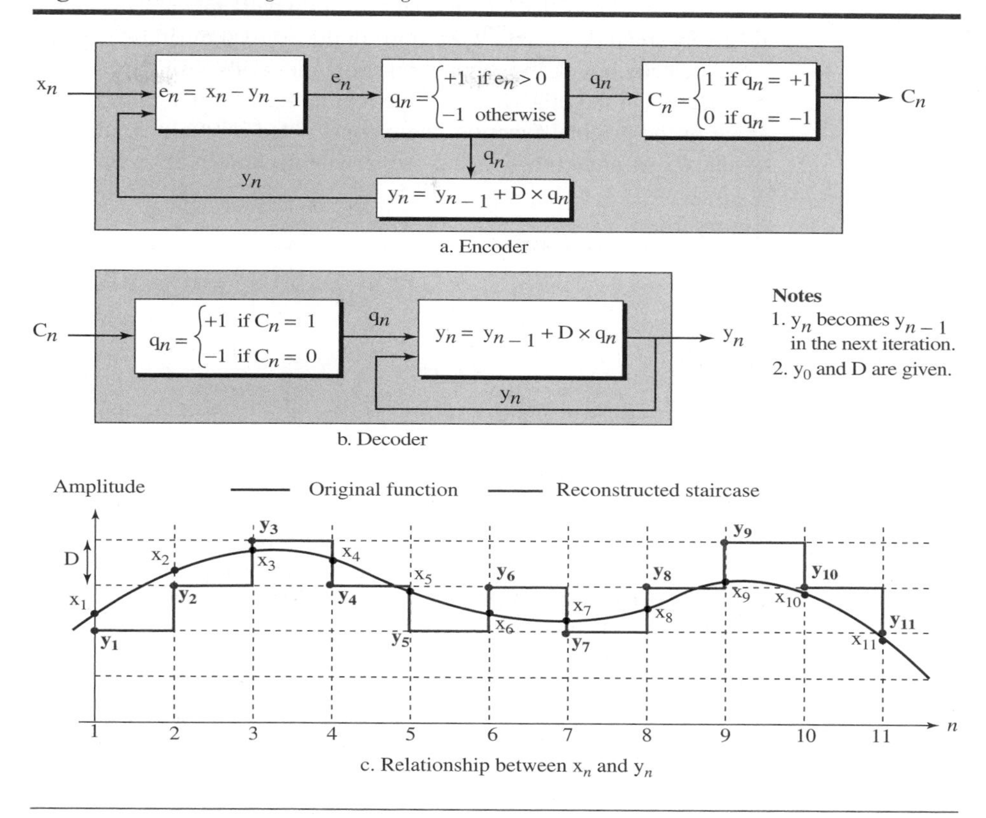

function. Quantizing $x_n - y_{n-1}$ is self-correcting for slow-growing or slow-falling x. Figure 28.10 compares the staircase reconstruction of quantizing $x_n - x_{n-1}$ versus $x_n - y_{n-1}$ for a slow-growing function. (For a slow-falling function, the idea is similar.)

Figure 28.10 *Reconstruction of quantization of $x_n - x_{n-1}$ versus $x_n - y_{n-1}$*

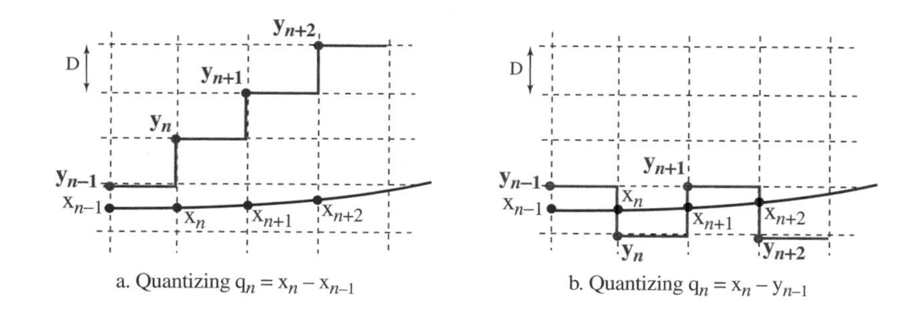

Adaptive DM (ADM)

Figure 28.11 shows the role of quantizer Δ on delta modulation. In the region where Δ is relatively small compared to the slope of the original function, the reconstructed staircase cannot catch up with the original function; the result is an error known as *slope overload distortion*. On the other hand, in the region where Δ is relatively large compared to the slope of the original function, the reconstructed staircase continues to oscillate largely around the original function and causes an error known as *granular noise*.

Figure 28.11 *Slope overload and granular noise*

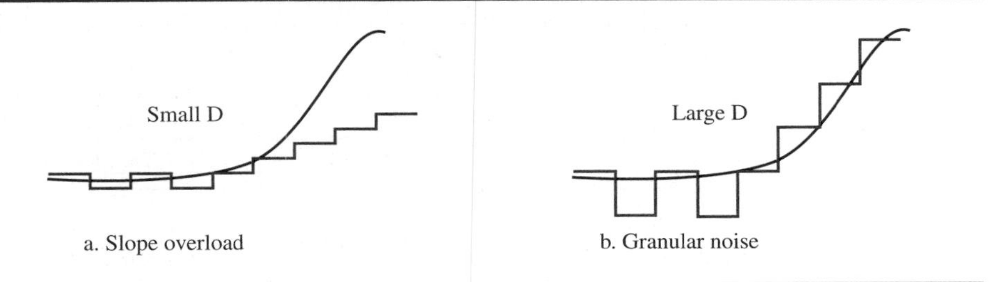

Since most functions have regions with both large and small slopes, selecting a large value or a small value for Δ decreases one type of error but increases the other type. **Adaptive DM (ADM)** is used to solve the problem. In ADM, the value of Δ changes from one step to the next and is calculated as shown below.

$$\Delta_n = M_n \Delta_{n-1}$$

where M_n is called *step-size multiplier* and is calculated from the values of q_n from a few previous bits. There are many different algorithms for evaluating M_n; one simple algorithm is to increase M_n by a certain percentage if q_n remains the same and decrease it by a certain percentage if q_n changes. The adaptation can be further improved by delaying the coding process to include knowledge about a few future samples in the evaluation of M_n.

Differential PCM (DPCM)

The **differential PCM (DPCM)** is the generalization of delta modulation. In delta modulation a previously reconstructed sample y_{n-1} is called the *predictor* because it is used to predict the current value. In DPCM, more than one previously reconstructed sample is used for prediction. In this case the difference is evaluated as shown below.

$$e_n = x_n - \sum_{i=1}^{N} a_i \, y_{i-1}$$

where the summation is the predictor, a_i is the *predictor's coefficient* (or weight), and N is the *order of the predictor*. For DM, the order of the predictor is 1 and $a_1 = 1$. The difference is quantized as in DM and sent to the receiver. The receiver reconstructs the current value as shown below.

$$y_n = \sum_{i=1}^{N} a_i \, y_{i-1} + \Delta q_n$$

Predictor coefficients are found by minimizing the cumulative error between the predicted value and the actual value. The optimization uses the *method of square error,* which is beyond the scope of this book.

Adaptive DPCM (ADPCM)

Further compression can be achieved by using different coefficients for different regions of the sample or by adjusting the quantizer (Δ) from one step to the next or by doing both. This is the principle behind **adaptive DPCM** (**ADPCM**).

Linear Predictive Coding

In **linear predictive coding** (**LPC**), instead of sending quantized difference signals, the source analyzes the signals and determines their characteristics. The characteristics include frequencies in the sensitive range of frequencies, the power of each frequency, and the duration of each signal. The source then quantizes this information and transmits it to the receiver. The receiver feeds this information into a signal synthesizer to simulate a signal similar to that of the original one. The LPC can achieve a high level of compression. However, this method is normally used by the military for compressing speech. In this case, the synthesized speech, though intelligible, lacks naturalness and quality to identify the speaker.

Transform Coding

In transform coding, a mathematical transformation is applied to the input signal to produce the output signal. The transformation needs to be invertible, to allow the original signal to be recovered. The transformation changes the signal representation from one domain to another (time domain to frequency domain, for example), which results in reducing the number of bits in encoding.

We need to emphasize that transformation techniques used in multimedia are lossless per se. However, to achieve the compression goals, another step, quantization, is added to the operation, which makes the whole process lossy.

Discrete Cosine Transform (DCT)

One of the popular transformations used in multimedia is called **discrete cosine transform (DCT)**. Although we use two-dimensional DCT in multimedia compression, we first discuss one-dimensional DCT, which is easier to understand.

One-Dimensional DCT In one-dimensional DCT, the transformation is the matrix multiplication of a column matrix p (source data) by a square matrix T (DCT coefficient). The result is a column matrix M (transformed data). Since the square matrix that represents the DCT coefficient is an orthogonal matrix (inverse and transpose are the same), the inverse transformation can be obtained by multiplication of the transformed data matrix by the transpose matrix of the DCT coefficient. Figure 28.12 shows the transformation in matrix, in which N is the size of matrix T and T^T is the transpose matrix of T.

Although we believe the matrix representation of the transformation is easier to understand, the literature also uses two formulas to do so, as shown in Figure 28.13.

Figure 28.12 *One-dimensional DCT*

$$[\mathbf{M}] = [\ \mathrm{T}\ ] \times [\mathrm{p}] \qquad [\mathrm{p}] = [\ \mathrm{T}^{\mathrm{T}}\ ] \times [\mathbf{M}]$$

a. Transformation b. Inverse Transformation

$$\mathrm{T}(m, n) = \mathrm{C}(m) \cos\left[\frac{\pi n(2m+1)}{2\mathrm{N}}\right] \quad \text{for } m = 0 \text{ to N}-1,\ \text{for } n = 0 \text{ to N}-1$$

$$\mathrm{C}(m) = \begin{cases} \sqrt{\frac{1}{\mathrm{N}}} & m = 0 \\ \sqrt{\frac{2}{\mathrm{N}}} & m > 0 \end{cases}$$

Figure 28.13 *Formulas for one-dimensional forward and inverse transformation*

$$\mathbf{M}(m) = \sum_{n=0}^{\mathrm{N}-1} \mathrm{C}(m) \cos\left[\pi n(2m + 1)/(2\mathrm{N})\right] \times p(n) \qquad \text{for } m = 0, \ldots, \mathrm{N} - 1$$

$$p(n) = \sum_{m=0}^{\mathrm{N}-1} \mathrm{C}(n) \cos\left[\pi m(2n + 1)/(2\mathrm{N})\right] \times \mathbf{M}(m) \qquad \text{for } n = 0, \ldots, \mathrm{N} - 1$$

$$\mathrm{C}(\mathrm{i}) = \begin{cases} \sqrt{\frac{1}{\mathrm{N}}} & \mathrm{i} = 0 \\ \sqrt{\frac{2}{\mathrm{N}}} & \mathrm{i} > 0 \end{cases}$$

Example 28.5

Figure 28.14 shows the transformation matrix for N = 4. As the figure shows, the first row has four equal values, but the other rows have alternate positive and negative values. When each row is multiplied by the source data matrix, we expect that the positive and negative values result in

Figure 28.14 *Example 28.5*

$$\underset{\mathbf{M}}{\begin{bmatrix} \mathbf{203} \\ \mathbf{-2.22} \\ \mathbf{0.00} \\ \mathbf{-0.16} \end{bmatrix}} = \underset{\mathrm{T}}{\begin{bmatrix} 0.50 & 0.50 & 0.50 & 0.50 \\ 0.65 & 0.27 & -0.27 & -0.65 \\ 0.50 & -0.50 & -0.50 & 0.50 \\ 0.27 & -0.65 & 0.65 & -0.27 \end{bmatrix}} \times \underset{p}{\begin{bmatrix} 100 \\ 101 \\ 102 \\ 103 \end{bmatrix}}$$

a. Transformation

$$\underset{p}{\begin{bmatrix} 100 \\ 101 \\ 102 \\ 103 \end{bmatrix}} = \underset{\mathrm{T}^{\mathrm{T}}}{\begin{bmatrix} 0.50 & 0.65 & 0.50 & 0.27 \\ 0.50 & 0.27 & -0.50 & -0.65 \\ 0.50 & -0.27 & -0.50 & 0.65 \\ 0.50 & -0.65 & 0.50 & -0.27 \end{bmatrix}} \times \underset{\mathbf{M}}{\begin{bmatrix} \mathbf{203} \\ \mathbf{-2.22} \\ \mathbf{0.00} \\ \mathbf{-0.16} \end{bmatrix}}$$

b. Inverse Transformation

values close to zero if the source data items are close to each other. This is what we expect from the transformation: to show that only some values in the source data are important and most values are redundant.

The example shows how we can transform the sequence of numbers, (100, 101, 102, 103) to another sequence of numbers (203, –2.22, 0.00, –0.16). There are several points we want to mention about the DCT transformation to better explain its properties. First, the transformation is reversible. Second, the transformation matrix is orthogonal ($\mathrm{T}^{-1} = \mathrm{T}^{\mathrm{T}}$), which means that we don't need to use the inverse matrix in calculation of the reverse transformation; the transposed matrix can be used (faster calculation). Third, the first row of the M matrix always is the weighted average of the p matrix. Fourth, the other row values in the matrix are very small values (positive or negative) that can be ignored in this case. The very important point about these three values is that they will be the same if we change the four values in the p matrix, but keep the same correlation between them. If we change the source data p = (7, 8, 9, 10), we can show that the transformed data is M = (17, −2.22, 0.00, −0.16); the first value will change because the average has been changed, but the rest of the values have not been changed since the relationship between the data items has not been changed. This is what we expect from the transformation. It removes the redundant data. The last three values in the p matrix are redundant; they have a very close relationship with the first value.

Two-Dimensional DCT Two-dimension DCT is what we need for compressing images, audio, and video. The principle is the same, except that the source data and transformed data are two-dimensional square matrices. To achieve the transformation with the same properties as mentioned for the one-dimensional DCT, we need to use the T matrix twice (T and T^T). The inverse transformation also uses the T matrix twice, but in the reverse order. Figure 28.15 shows the two-dimensional DCT, in matrix format. Figure 28.16 shows the same idea using two formulas.

Figure 28.15 *Two-dimensional DCT*

$$[\mathbf{M}] = [\mathrm{T}] \times [p] \times [\mathrm{T}^\mathrm{T}]$$

a. Transformation

$$[p] = [\mathrm{T}^\mathrm{T}] \times [\mathbf{M}] \times [\mathrm{T}]$$

b. Inverse Transformation

$$\mathrm{T}(m, n) = \mathrm{C}(m, n)\ \mathbf{cos}\left[\frac{m\pi(2n+1)}{2\mathrm{N}}\right] \quad \begin{array}{l}\text{for } m = 0 \text{ to N-1}\\ \text{for } n = 0 \text{ to N-1}\end{array}$$

$$\mathrm{C}(m, n) = \begin{cases}\sqrt{\dfrac{1}{\mathrm{N}}} & m = 0\\[2ex] \sqrt{\dfrac{2}{\mathrm{N}}} & m > 0\end{cases}$$

Figure 28.16 *Formulas for forward and inverse two-dimensional DCT*

$$\mathbf{M}(m, n) = \frac{2}{\mathrm{N}}\mathrm{C}(m)\,\mathrm{C}(n)\sum_{k=0}^{\mathrm{N}-1}\sum_{l=0}^{\mathrm{N}-1} p(k, l)\ \mathbf{cos}\left[\frac{m\pi(2k+1)}{2\mathrm{N}}\right]\mathbf{cos}\left[\frac{n\pi(2l+1)}{2\mathrm{N}}\right] \quad \begin{array}{l}\text{for } m = 0, \ldots, \mathrm{N}-1\\ \text{for } n = 0, \ldots, \mathrm{N}-1\end{array}$$

$$p(k, l) = \frac{2}{\mathrm{N}}\mathrm{C}(l)\ \mathrm{C}(k)\sum_{m=0}^{\mathrm{N}-1}\sum_{n=0}^{\mathrm{N}-1} \mathbf{M}(m, n)\ \mathbf{cos}\left[\frac{k\pi(2m+1)}{2\mathrm{N}}\right]\mathbf{cos}\left[\frac{l\pi(2n+1)}{2\mathrm{N}}\right] \quad \begin{array}{l}\text{for } k = 0, \ldots, \mathrm{N}-1\\ \text{for } l = 0, \ldots, \mathrm{N}-1\end{array}$$

$$\mathrm{C}(u) = \begin{cases}\sqrt{\dfrac{1}{2}} & u = 0\\[2ex] 1 & u > 0\end{cases}$$

28.2 MULTIMEDIA DATA

Today, multimedia data consists of *text, images, video,* and *audio*, although the definition is changing to include futuristic media types.

28.2.1 Text

The Internet stores a large amount of text that can be downloaded and used. One often refers to plaintext, as a linear form, and hypertext, as a nonlinear form, of textual data. Text stored in the Internet uses a character set, such as Unicode, to represent symbols in the underlying language. To store a large amount of textual data, the text can be compressed using one of the lossless compression methods we discussed earlier. Note that we need to use lossless compression because we cannot afford to loose any pieces of information when we do decompression.

28.2.2 Image

In multimedia parlance, an image (or a still image as it is often called) is the representation of a photograph, a fax page, or a frame in a moving picture.

Digital Image

To use an image, it first must be digitized. *Digitization* in this case means to represent an image as a two-dimensional array of dots, called pixels. Each pixel then can be represented as a number of bits, referred to as the *bit depth*. In a black-and-white image, such as a fax page, the bit depth = 1; each pixel can be represented as a 0-bit (black) or a 1-bit (white). In a gray picture, one normally uses a bit depth of 8 with 256 levels. In a color image, the image is normally divided into three channels, with each channel representing one of the three primary colors of red, green, or blue (RGB). In this case, the bit depth is 24 (8 bits for each color). Some representations use a separate channel, called the *alpha* (α) *channel,* to represent the background. In a black and white image, this results in two channels; in a color image, this results in four channels.

It is obvious that moving from black-and-white, to gray, to color representation of images tremendously increases the size of the information to transmit in the Internet. This implies that we need to compress images to save time.

Example 28.6

The following shows the time required to transmit an image of 1280×720 pixels using the transmission rate of 100 kbps.

a. Using a black and white image with a bit depth of 1, we need.

$$\textbf{Transmission time} = (1280 \times 720 \times 1) / 100{,}000 \approx 9 \textbf{ seconds}$$

b. Using a gray image with a bit depth of 8, we need.

$$\textbf{Transmission time} = (1280 \times 720 \times 8) / 100{,}000 \approx 74 \textbf{ seconds}$$

c. Using a color image with a bit depth of 24, we need.

$$\textbf{Transmission time} = (1280 \times 700 \times 24) / 100{,}000 \approx 215 \textbf{ seconds}$$

Image Compression: JPEG

Although there are both lossless and lossy compression algorithms for images, in this section we discuss the lossy compression method called *JPEG*. The **Joint Photographic Experts Group (JPEG)** standard provides lossy compression that is used in most implementations. The JPEG standard can be used for both color and gray images. However, for simplicity, we discuss only the grayscale pictures; the method can be applied to each of the three channels in a color image. In JPEG, a grayscale picture is divided into blocks of 8 × 8 pixels. The compression and decompression each go through three steps, as shown in Figure 28.17.

Figure 28.17 *Compression in each channel of JPEG*

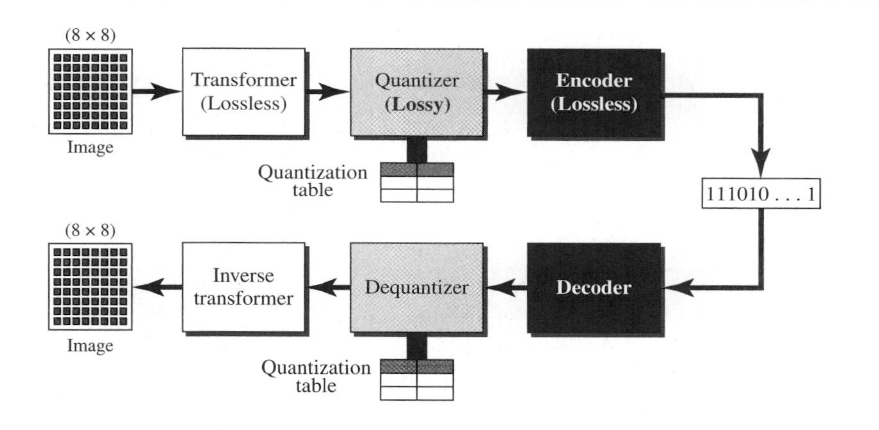

The purpose of dividing the picture into blocks is to decrease the number of calculations, because, as we showed in two-dimensional DCT, the number of mathematical operations for each picture is the square of the number of units.

Transformation

JPEG normally uses DCT in the first step in compression and inverse DCT in the last step in decompression. Transformation and inverse transformation are applied on 8 × 8 blocks. We discussed DCT in the previous section.

Quantization

The output of DCT transformation is a matrix of real numbers. The precise encoding of these real numbers requires a lot of bits. JPEG uses a quantization step that not only rounds real values in the matrix, but also changes some values to zeros. The zeros can be eliminated in the encoding step to achieve a high compression rate. As we discussed earlier, the result of DCT transformation defines the weights of different frequencies in the source matrix. Since high frequencies mean sudden changes in the value of pixels, they can be eliminated because human vision cannot recognize them. The quantization step creates a new matrix in which each element, $C(m, n)$, is defined as shown below.

$$\mathbf{C}\,(m, n) = \mathbf{round}\,[\mathbf{M}(m, n)\,/\,\mathbf{Q}(m, n)]$$

in which M(m, n) is an entry in the transformed matrix and Q(m, n) is an entry in the quantization matrix. The round function first adds 0.5 to a real value and then truncates the value to an integer. This means that 3.7 is rounded to integer 4, but 3.2 is rounded to integer 3.

JPEG has defined 100 quantization matrices, Q1 to Q100, in which Q1 gives the poorest image quality but the highest level of compression and Q100 gives the best image quality but the lowest level of compression. It is up to the implementation to choose one of these matrices. Figure 28.18 shows some of these matrices.

Figure 28.18 *Three different quantization matrices*

Q10							
80	60	50	80	120	200	255	255
55	60	70	95	130	255	255	255
70	65	80	120	200	255	255	255
70	85	110	145	255	255	255	255
90	110	185	255	255	255	255	255
120	175	255	255	255	255	255	255
245	255	255	255	255	255	255	255
255	255	255	255	255	255	255	255

Q50							
16	11	10	16	24	40	51	61
12	12	14	19	26	58	60	55
14	13	16	24	40	57	69	56
14	17	22	29	51	87	80	62
18	22	37	56	68	109	103	77
24	35	55	64	81	104	113	92
49	64	78	87	103	121	120	101
72	92	95	98	112	110	103	99

Q90							
3	2	2	3	5	8	10	12
2	2	3	4	5	12	12	11
3	3	3	5	8	11	14	11
3	3	4	6	10	17	16	12
4	4	7	11	14	22	21	15
5	7	11	13	16	12	23	18
10	13	16	17	21	24	24	21
14	18	19	20	22	20	20	20

Note that the only phase in the process that is not completely reversible is the quantizing phase. We lose some information here that is not recoverable. As a matter of fact, the only reason that JPEG is called *lossy compression* is because of this quantization phase.

Encoding

After quantization, the values are reordered in a zigzag sequence before being input into the encoder. The zigzag reordering of the quantized values is done to let the values related to the lower frequency feed into the encoder before the values related to the higher frequency. Since most of the higher-frequency values are zeros, this means non-zero values are given to the encoder before the zero values. Figure 28.19 shows the process.

The encoding in this case is a lossless compression using either run-length coding or arithmetic coding.

Example 28.7

To show the idea of JPEG compression, we use a block of gray image in which the bit depth for each pixel is 20. We have used a Java program to transform, quantize, and reorder the values in zigzag sequence; we have shown the encoding (Figure 28.20).

Example 28.8

As the second example, we have a block that changes gradually; there is no sharp change between the values of neighboring pixels. We still get a lot of zero values, as shown in Figure 28.21.

Figure 28.19 *Reading the table*

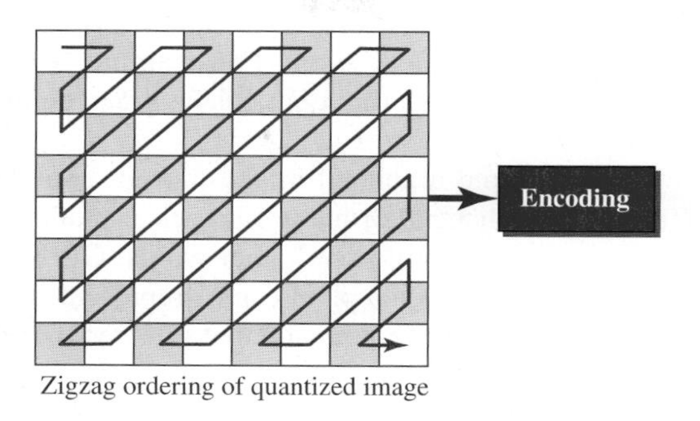

Figure 28.20 *Example 28.7: Uniform gray scale*

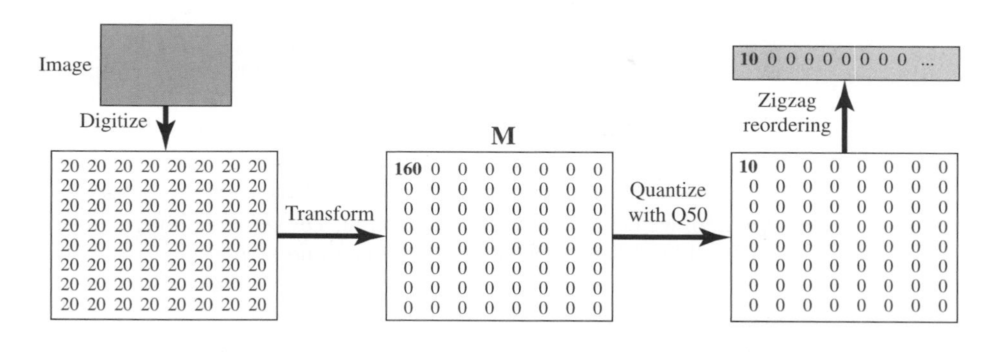

Figure 28.21 *Example 28.8: Gradient gray scale*

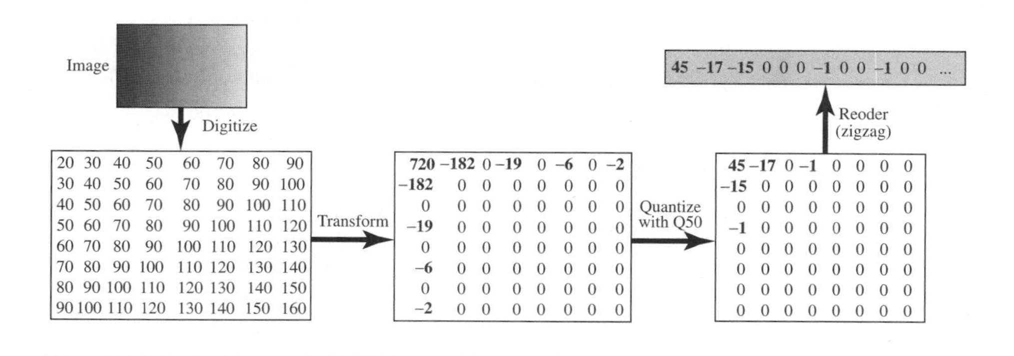

Image Compression: GIF

The JPEG standard uses images in which each pixel is represented as 24 bits (8 bits for each primary color). This means that each pixel can be one of the 2^{24} (16,777,216) complex colors. For example, a *magenta* pixel, which is made of red and blue components (but contains no green component) is represented as the integer $(FF00FF)_{16}$.

Most simple graphical images do not contain such a large range of colors. The Graphic Interchange Format (GIF) uses a smaller palette (indexed table) of colors with normally $2^8 = 256$ colors. In other words, GIF maps a *true* color with a *palette* color. For example, a magenta pixel can be represented as the integer $(E2)_{16}$ if it is the 226th color in the pallet. This means that GIF reduces the size of the image by a factor of 3 compared with JPEG.

After creating the palette for a particular image, each pixel can be represented by one of the 256 symbols (for example, the two-digit representation of the palette index in hexadecimal). Now we can use one of the lossless compression methods, such as dictionary coding or arithmetic coding, to further compress the image.

28.2.3 Video

Video is composed of multiple frames; each frame is one image. This means that a video file requires a high transmission rate.

Digitizing Video

A video consists of a sequence of frames. If the frames are displayed on the screen fast enough, we get an impression of motion. The reason is that our eyes cannot distinguish the rapidly flashing frames as individual ones. There is no standard number of frames per second; in North America 25 frames per second is common. However, to avoid a condition known as *flickering,* a frame needs to be refreshed. The TV industry repaints each frame twice. This means 50 frames need to be sent, or if there is memory at the sender site, 25 frames with each frame repainted from the memory.

Example 28.9

Let us show the transmission rate for some video standards:

a. Color broadcast television takes 720 × 480 pixels per frame, 30 frames per second, and 24 bits per color. The transmission rate without compression is as shown below.

$$720 \times 480 \times 30 \times 24 = 248{,}832{,}000 \text{ bps} = 249 \text{ Mbps}$$

b. High definition color broadcast television takes 1920 × 1080 pixels per frame, 30 frames per second, and 24 bits per color: The transmission rate without compression is as shown below.

$$1920 \times 1080 \times 30 \times 24 = 1{,}492{,}992{,}000 \text{ bps} = 1.5 \text{ Gbps}$$

Video Compression: MPEG

Motion Picture Experts Group (**MPEG**) is a method to compress video. In principle, a motion picture is a rapid flow of a set of frames, where each frame is an image. In other words, a frame is a spatial combination of pixels, and a video is a temporal combination of frames that are sent one after another. Compressing video, then, means spatially compressing each frame and temporally compressing a set of frames.

Spatial Compression

The **spatial compression** of each frame is done with JPEG (or a modification of it). Each frame is a picture that can be independently compressed.

Temporal Compression

In **temporal compression,** redundant frames are removed. When we watch television, we receive 50 frames per second. However, most of the consecutive frames are almost the same. For example, when someone is talking, most of the frame is the same as the previous one except for the segment of the frame around the lips, which changes from one frame to another.

To temporally compress data, the MPEG method first divides a set of frames into three categories: I-frames, P-frames, and B-frames. Figure 28.22 shows how a set of frames (7 in the figure) are compressed to create another set of frames.

Figure 28.22 *MPEG frames*

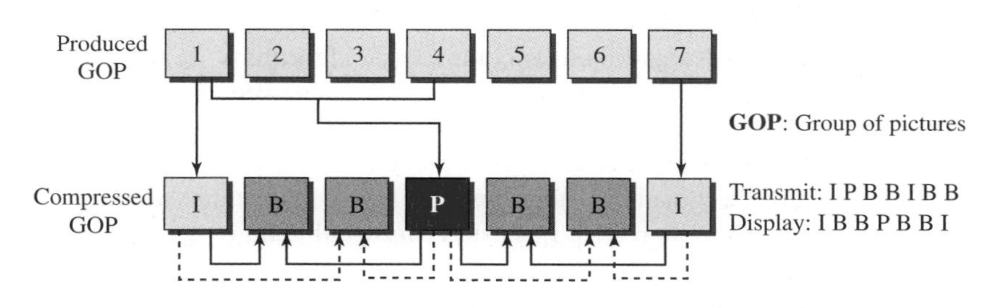

- *I-frames.* An **intracoded frame (I-frame)** is an independent frame that is not related to any other frame (not to the frame sent before or after). They are present at regular intervals. An **I-frame** must appear periodically to handle some sudden change in the frame that the previous and following frames cannot show. Also, when a video is broadcast, a viewer may tune in at any time. If there is only one I-frame at the beginning of the broadcast, the viewer who tunes in late will not receive a complete picture. I-frames are independent of other frames and cannot be constructed from other frames.
- *P-frames.* A **predicted frame (P-frame)** is related to the preceding I-frame or P-frame. In other words, each P-frame contains only the changes from the preceding frame. P-frames can be constructed only from previous I- or P-frames. P-frames carry much less information than other frame types and carry even fewer bits after compression.
- *B-frames.* A **bidirectional frame (B-frame)** is related to the preceding and following I-frame or P-frame. In other words, each B-frame is relative to the past and the future. Note that a B-frame is never related to another B-frame.

28.2.4 Audio

Audio (sound) signals are analog signals that need a medium to travel; they cannot travel through a vacuum. The speed of the sound in the air is about 330 m/s (740 mph). The audible frequency range for normal human hearing is from about 20Hz to 20kHz with maximum audibility around 3300 Hz.

Digitizing Audio

To be able to provide compression, audio analog signals are digitized using an analog-to-digital converter. The analog-to-digital conversion consists of two processes: sampling and quantizing. A digitizing process known as pulse code modulation (PCM) was discussed in detail in Chapter 3. This process involved sampling an analog signal, quantizing the sample, and coding the quantized values as streams of bits. Voice signal is sampled at the rate of 8,000 samples per second with 8 bits per sample; the result is a digital signal of $8{,}000 \times 8 = 64$ kbps. Music is sampled at 44,100 samples per second with 16 bits per sample; the result is a digital signal of $44{,}100 \times 16 = 705.6$ kbps for monaural and 1.411 Mbps for stereo.

Audio Compression

Both lossy and lossless compression algorithms are used in audio compression. Lossless audio compression allows one to preserve an exact copy of the audio files; it has a small compression ratio of about 2 and is mostly used for archival and editing purposes. Lossy algorithms provide far greater compression ratios (5 to 20) and are used in mainstream consumer audio devices. Lossy algorithms sacrifice a little bit of quality, but substantially reduce space and bandwidth requirements. For example, on a CD, one can fit one hour of high fidelity music, 2 hours of music using lossless compression, or 8 hours of music compressed with a lossy technique.

Compression techniques used for speech and music have different requirements. Compression techniques used for speech must have low latency because significant delays degrade the communication quality in telephony. Compression algorithms used for music must be able to produce high quality sound with lower numbers of bits. Two categories of techniques are used in audio compressions: *predictive coding* and **perceptual coding**.

Predictive coding

Predictive coding techniques have low latency and therefore are popular in speech coding for telephony where significant delays degrade the communication quality. We discussed several predictive coding methods at the beginning of this chapter: DM, ADM, DPCM, ADPCM, and LPC.

Perceptual Coding

Even at their best, the predictive coding methods cannot sufficiently compress a CD-quality audio for the multimedia application. The most common compression technique used to create CD-quality audio is perceptual coding, which is based on the science of **psychoacoustics**. Algorithms used in perceptual coding first transform the data from time domain to frequency domain; the operations are then performed on the data in the frequency domain. This technique, hence, is also called the *frequency-domain method.*

Psychoacoustics is the study of subjective human perception of sound. Perceptual coding takes advantage of flaws in the human auditory system. The lower limit of human audibility is 0 dB. This is only true for sounds with frequencies of about 2.5 and 5 kHz. The lower limit is less for frequencies between these two frequencies and rises for frequencies outside these ranges, as shown in Figure 28.23a. We cannot hear any frequency whose power is below this curve; thus, it is not necessary to code such a frequency.

Figure 28.23 *Threshold of audibility*

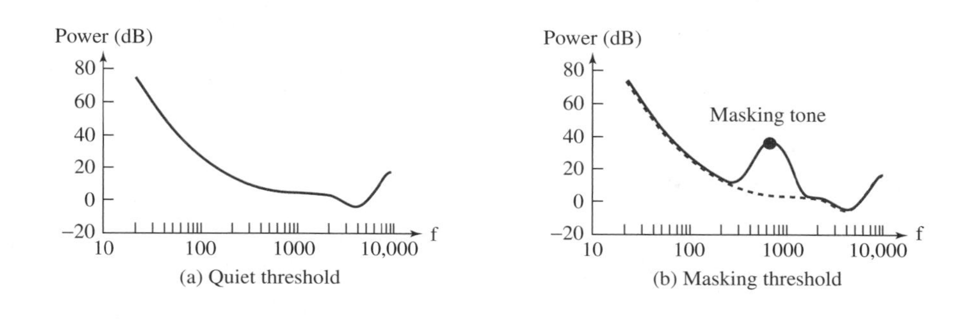

(a) Quiet threshold

(b) Masking threshold

For example, we can save bits, without loss of quality, by omitting any sound with frequency of less than 100 Hz if its power is below 20 dB.

We can save even more using the concepts of **frequency masking** and **temporal masking.** Frequency masking occurs when a loud sound partially or totally masks a softer sound if the frequencies of the two are close to each other. For example, we cannot hear our dance partner in a room where a loud heavy metal band is performing. In Figure 28.23b, a loud masking tone, around 700 Hz, raises the threshold of the audibility curve between frequencies of about 250 to 1500 Hz. In temporal masking, a loud sound can numb our ear for a short time even after the sound has stopped.

The basic approach to perceptual coding is to feed the audio PCM input into two separate units of the coder simultaneously. The first unit consists of an array of digital bypass filters called an *analysis filter bank*. Using a mathematical tool such as *discrete Fourier transform (DFT),* the filters break the time-domain input into equally spaced frequency sub-bands. Using the same or a similar mathematical tool such as *fast Fourier transform (FFT),* the second unit transforms the time-domain input into frequency-domain and determines the masking frequency for each sub-band. The available bits are then allocated according to the masking property of each sub-band: no bits are allocated to a totally masked sub-band; a small number of bits are allocated to a partially masked sub-band, and a large number of bits are allocated to unmasked sub-bands. The resulting bits are further encoded to achieve more compression.

MP3

One standard that uses perceptual coding is **MP3 (MPEG audio layer 3).**

28.3 MULTIMEDIA IN THE INTERNET

We can divide audio and video services into three broad categories: *streaming stored audio/video, streaming live audio/video,* and *interactive audio/video.* Streaming means a user can listen (or watch) the file after the downloading has started.

28.3.1 Streaming Stored Audio/Video

In the first category, streaming stored audio/video, the files are compressed and stored on a server. A client downloads the files through the Internet. This is sometimes referred to as *on-demand audio/video.* Examples of stored audio files are songs, symphonies, books on tape, and famous lectures. Examples of stored video files are movies, TV shows, and music video clips. We can say that streaming stored audio/video refers to *on-demand* requests for compressed audio/video files.

Downloading these types of files from a Web server can be different from downloading other types of files. To understand the concept, let us discuss three approaches, each with a different complexity.

First Approach: Using a Web Server

A compressed audio/video file can be downloaded as a text file. The client (browser) can use the services of HTTP and send a GET message to download the file. The Web server can send the compressed file to the browser. The browser can then use a help application, normally called a *media player,* to play the file. Figure 28.24 shows this approach.

Figure 28.24 *Using a Web server*

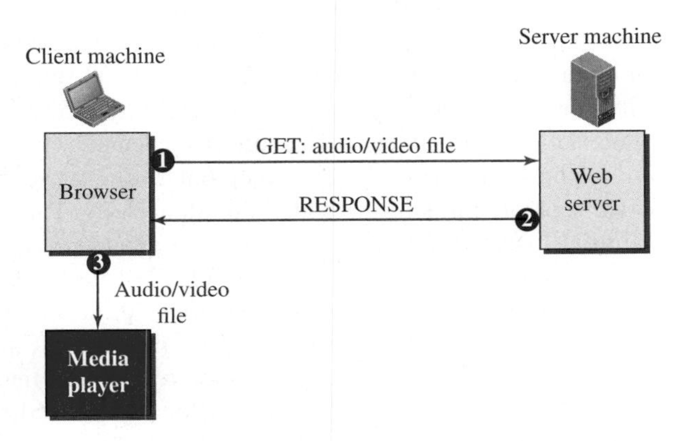

This approach is very simple and does not involve *streaming*. However, it has a drawback. An audio/video file is usually large even after compression. An audio file may contain tens of megabits, and a video file may contain hundreds of megabits. In this approach, the file needs to download completely before it can be played. Using contemporary data rates, the user needs some seconds or tens of seconds before the file can be played.

Second Approach: Using a Web Server with a Metafile

In another approach, the media player is directly connected to the Web server for downloading the audio/video file. The Web server stores two files: the actual audio/video file and a **metafile** that holds information about the audio/video file. Figure 28.25 shows the steps in this approach.

Figure 28.25 *Using a Web server with a metafile*

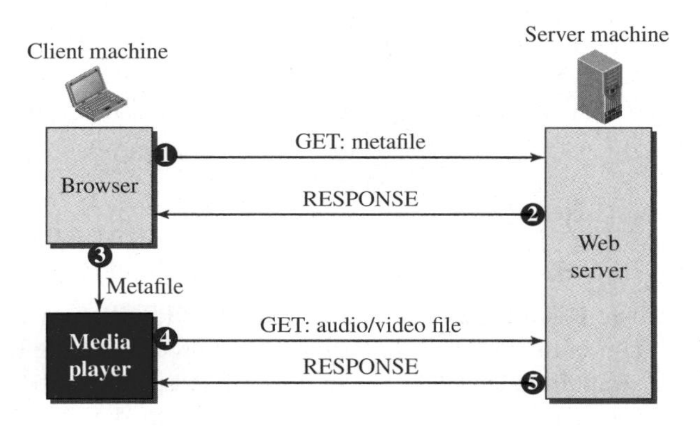

1. The HTTP client accesses the Web server using the GET message.
2. The information about the metafile comes in the response.
3. The metafile is passed to the media player.
4. The media player uses the URL in the metafile to access the audio/video file.
5. The Web server responds.

Third Approach: Using a Media Server

The problem with the second approach is that the browser and the media player both use the services of HTTP. HTTP is designed to run over TCP. This is appropriate for retrieving the metafile, but not for retrieving the audio/video file. The reason is that TCP retransmits a lost or damaged segment, which is counter to the philosophy of streaming. We need to dismiss TCP and its error control; we need to use UDP. However, HTTP, which accesses the Web server, and the Web server itself are designed for TCP; we need another server, a **media server.** Figure 28.26 shows the concept.

1. The HTTP client accesses the Web server using a GET message.
2. The information about the metafile comes in the response.
3. The metafile is passed to the media player.
4. The media player uses the URL in the metafile to access the media server to download the file. Downloading can take place by any protocol that uses UDP.
5. The media server responds.

Figure 28.26 *Using a media server*

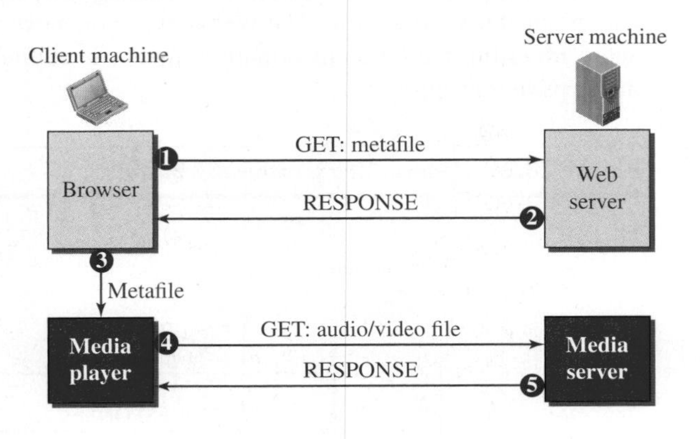

Fourth Approach: Using a Media Server and RTSP

The **Real-Time Streaming Protocol (RTSP)** is a control protocol designed to add more functionalities to the streaming process. Using RTSP, we can control the playing of audio/video. RTSP is an out-of-band control protocol that is similar to the second connection in FTP. Figure 28.27 shows a media server and RTSP.

Figure 28.27 *Using a media server and RTSP*

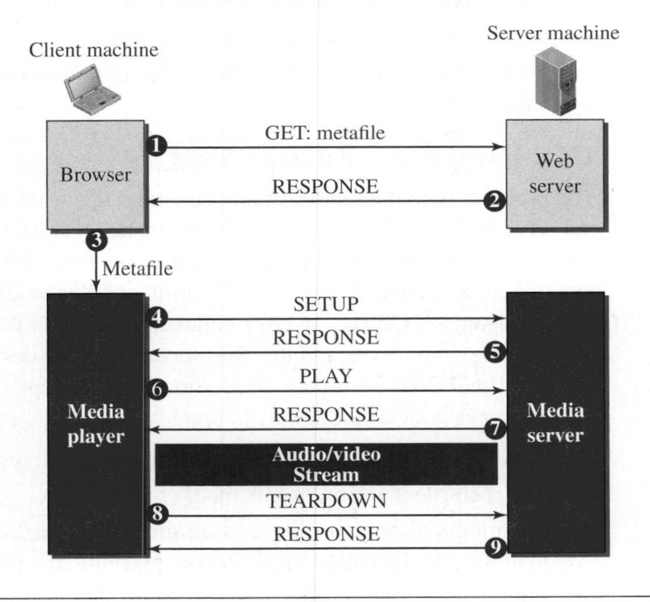

1. The HTTP client accesses the Web server using a GET message.
2. The information about the metafile comes in the response.
3. The metafile is passed to the media player.
4. The media player sends a SETUP message to create a connection with the media server.
5. The media server responds.
6. The media player sends a PLAY message to start playing (downloading).
7. The audio/video file is downloaded using another protocol that runs over UDP.
8. The connection is broken using the TEARDOWN message.
9. The media server responds.

The media player can send other types of messages. For example, a PAUSE message temporarily stops the downloading; downloading can be resumed with a PLAY message.

Example: Video on Demand (VOD)

Video On Demand (VOD) allows viewers to select a video from a large number of available videos and watch it interactively: pause, rewind, fast forward, etc. A viewer may watch the video in real time or she may download the video into her computer, portable media player, or to a device such as a digital video recorder (DVR) and watch it later. Cable TV, satellite TV, and IPTV providers offer both pay-per-view and free content VOD streaming. Many other companies, such as Amazon video and video rental companies such as Blockbuster video, also provide VOD. Internet television is an increasingly popular form of video on demand.

28.3.2 Streaming Live Audio/Video

In the second category, streaming live audio/video, a user listens to broadcast audio and video through the Internet. Good examples of this type of application are Internet radio and Internet TV.

There are several similarities between streaming stored audio/video and streaming live audio/video. They are both sensitive to delay; neither can accept retransmission. However, there is a difference. In the first application, the communication is unicast and on-demand. In the second, the communication is multicast and live. Live streaming is better suited to the multicast services of IP and the use of protocols such as UDP and RTP (discussed later). However, presently, live streaming is still using TCP and multiple unicasting instead of multicasting. There is still much progress to be made in this area.

Example: Internet Radio

Internet radio or web radio is a webcast of audio broadcasting service that offers news, sports, talk, and music via the Internet. It involves a streaming medium that is accessible from anywhere in the world. Web radio is offered via the Internet but is similar to traditional broadcast media: it is noninteractive and cannot be paused or replayed like on-demand services. The largest group of Internet radio providers today includes existing radio stations that simultaneously broadcast their output traditionally and over the Internet. It also includes Internet-only radio stations. In web radio, audio sound is often compressed by MP3 or similar software and the bits are transported over TCP or UDP packets. To prevent jitter, on the user side the bits are buffered and delayed for a few seconds before they are reassembled and played.

Example: Internet Television (ITV)

Internet television or *ITV* allows viewers to choose the show they want to watch from a library of shows. The primary models for Internet television are streaming Internet TV or selectable video on an Internet location.

Example: IPTV

Internet protocol television (*IPTV*) is the next-generation technology for delivering real time and interactive television. Instead of the TV signal being transmitted via satellite, cable, or terrestrial routes, the IPTV signal is transmitted over the Internet. Note that IPTV differs from ITV. Internet TV is created and managed by service providers that cannot control the final delivery; it is distributed via existing infrastructure of the open Internet. An IPTV, on the other hand, is highly managed to provide guaranteed quality of service over a complex and expensive network. The network for IPTV is engineered to ensure efficient delivery of large amounts of multicast video traffic and HDTV content to subscribers.

The IP-based platform offers significant advantages, including the ability to integrate television with other IP-based services like high-speed Internet access and VoIP. One way that IPTV operates differently from cable or satellite TV is that in a typical cable or satellite network, all the content constantly flows from the station to each customer. The customer, using a set-top box (a device that connects to a television) selects from the content. In IPTV, content remains in the network, and only the content the customer selects is sent. The advantage is that IPTV requires significantly less bandwidth and therefore allows for the delivery of significantly more content and greater functionality. The disadvantage is that customer's privacy could be compromised because the service provider of IPTV could accurately track down each program watched by each customer.

28.3.3 Real-Time Interactive Audio/Video

In the third category, interactive audio/video, people use the Internet to interactively communicate with one another. The Internet phone or **voice over IP** is an example of this type of application. Video conferencing is another example that allows people to communicate visually and orally.

Characteristics

Before discussing the protocols used in this class of applications, we discuss some characteristics of real-time audio/video communication.

Time Relationship

Real-time data on a packet-switched network require the preservation of the time relationship between packets of a session. For example, let us assume that a real-time video server creates live video images and sends them online. The video is digitized and packetized. There are only three packets, and each packet holds 10 seconds of video information. The first packet starts at 00:00:00, the second packet starts at 00:00:10, and the third packet starts at 00:00:20. Also imagine that it takes 1 s (an exaggeration for simplicity) for each packet to reach the destination (equal delay). The receiver can play back the first packet at

00:00:01, the second packet at 00:00:11, and the third packet at 00:00:21. Although there is a 1-s time difference between what the server sends and what the client sees on the computer screen, the action is happening in real time. The time relationship between the packets is preserved. The 1-s delay is not important. Figure 28.28 shows the idea.

Figure 28.28 *Time relationship*

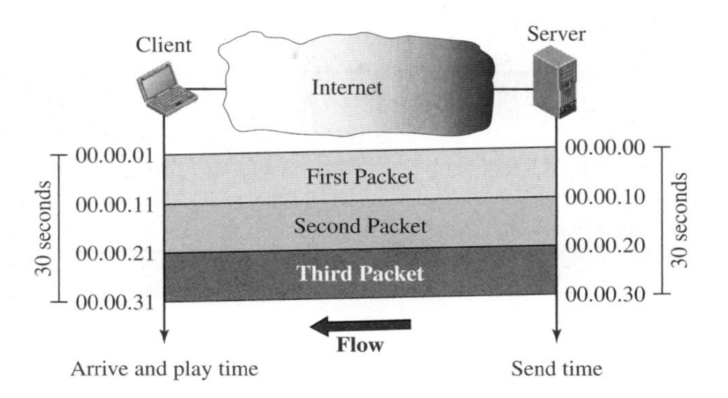

But what happens if the packets arrive with different delays? For example, the first packet arrives at 00:00:01 (1-s delay), the second arrives at 00:00:15 (5-s delay), and the third arrives at 00:00:27 (7-s delay). If the receiver starts playing the first packet at 00:00:01, it will finish at 00:00:11. However, the next packet has not yet arrived; it arrives 4 s later. There is a gap between the first and second packets and between the second and the third as the video is viewed at the remote site. This phenomenon is called ***jitter.*** Figure 28.29 shows the situation.

Figure 28.29 *Jitter*

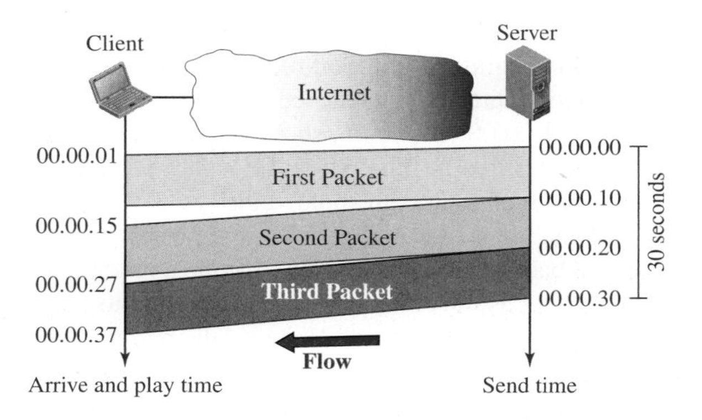

Timestamp

One solution to jitter is the use of a **timestamp.** If each packet has a timestamp that shows the time it was produced relative to the first (or previous) packet, then the receiver can add this time to the time at which it starts the playback. In other words, the receiver knows when each packet is to be played. Imagine the first packet in the previous example has a timestamp of 0, the second has a timestamp of 10, and the third a timestamp of 20. If the receiver starts playing back the first packet at 00:00:08, the second will be played at 00:00:18, and the third at 00:00:28. There are no gaps between the packets. Figure 28.30 shows the situation.

To prevent jitter, we can timestamp the packets and separate the arrival time from the playback time.

Figure 28.30 *Timestamp*

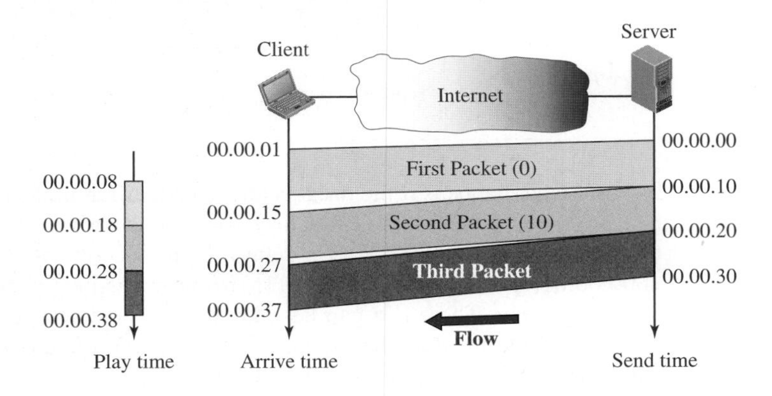

Playback Buffer

To be able to separate the arrival time from the playback time, we need a buffer to store the data until they are played back. The buffer is referred to as a ***playback buffer.*** When a session begins (the first bit of the first packet arrives), the receiver delays playing the data until a threshold is reached. In the previous example, the first bit of the first packet arrives at 00:00:01; the threshold is 7 s, and the playback time is 00:00:08. The threshold is measured in time units of data. The replay does not start until the time units of data are equal to the threshold value.

Data are stored in the buffer at a possibly variable rate, but they are extracted and played back at a fixed rate. Note that the amount of data in the buffer shrinks or expands, but as long as the delay is less than the time to play back the threshold amount of data, there is no jitter. Figure 28.31 shows the buffer at different times for our example.

To understand how a playback buffer can actually remove jitter, we need to think about a playback buffer as a tool that introduces more delay in each packet. If the amount of delay added to each packet makes the total delay (the delay in the network and the delay in the buffer) for each packet the same, then the packets are played back

Figure 28.31 *Playback buffer*

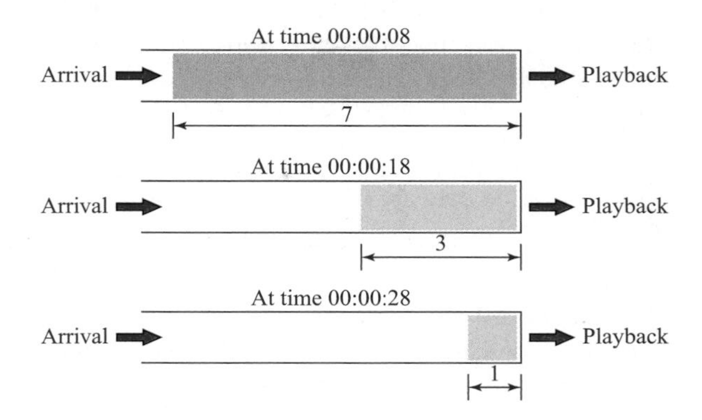

smoothly, as though there were no delay. Figure 28.32 shows the idea using the time line for seven packets. Note that we need to select the buffer delay for the first packet in the buffer in a such a way that the right two saw-tooth curves do not overlap.

Figure 28.32 *The time line of packets*

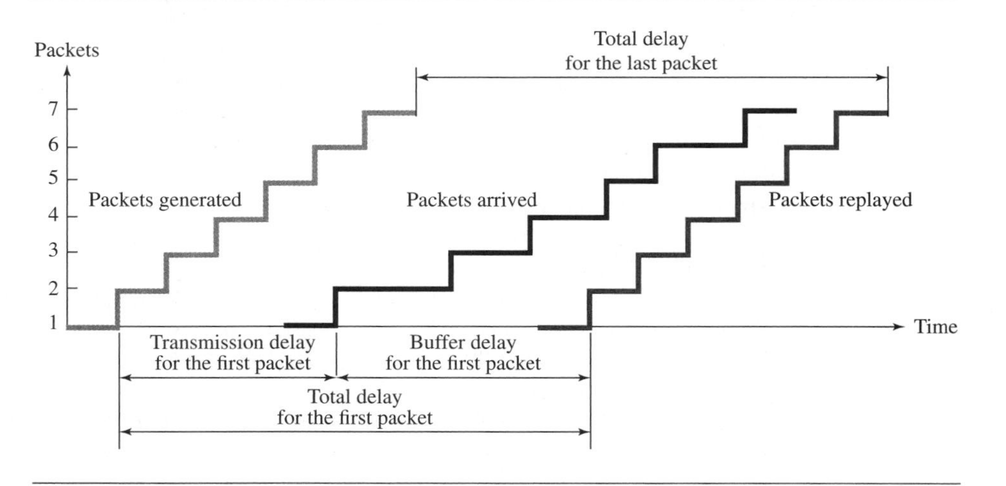

As the figure shows, if the playback time for the first packet is selected properly, then the total delay for all packets should be the same. The packets that have longer transmission delay should have shorter waiting in the buffer and vice versa.

Ordering

In addition to time relationship information and timestamps for real-time traffic, one more feature is needed. We need a *sequence number* for each packet. The timestamp alone cannot inform the receiver if a packet is lost. For example, suppose the timestamps are 0, 10, and 20. If the second packet is lost, the receiver receives just two packets with timestamps 0 and 20. The receiver assumes that the packet with timestamp 20 is the second packet, produced 20 s after the first. The receiver has no way of knowing that the second packet has actually been lost. A sequence number to order the packets is needed to handle this situation.

Multicasting

Multimedia play a primary role in audio and video conferencing. The traffic can be heavy, and the data are distributed using *multicasting* methods. Conferencing requires two-way communication between receivers and senders.

Translation

Sometimes real-time traffic needs *translation.* A translator is a computer that can change the format of a high-bandwidth video signal to a lower-quality narrow-bandwidth signal. This is needed, for example, for a source creating a high-quality video signal at 5 Mbps and sending to a recipient having a bandwidth of less than 1 Mbps. To receive the signal, a translator is needed to decode the signal and encode it again at a lower quality that needs less bandwidth.

Mixing

If there is more than one source that can send data at the same time (as in a video or audio conference), the traffic is made of multiple streams. To converge the traffic to one stream, data from different sources can be mixed. A **mixer** mathematically adds signals coming from different sources to create one single signal.

Forward Error Correction

In Chapter 10 we discussed forward error correction that is used by multimedia.

Example of a Real-Time Application: Skype

Skype (abbreviation of the original project *Sky peer-to-peer*) is a peer-to-peer VoIP application software that was originally developed by Ahti Heinla, Priit Kasesalu, and Jaan Tallinn, who had also originally developed Kazaa (a P2P file-sharing application software). The application allows registered users who have audio input and output devices on their PCs to make free PC-to-PC voice calls to other registered users over the Internet. Skype includes other popular features such as *instant messaging* (*IM*), *short message service* (*SMS*) group chat, file transfer, video conferencing, and SkypeIn and SkypeOut services. Using SkypeIn and SkypeOut services allows registered users to communicate with traditional land-line telephones and mobile phones for a small fee. Being responsible for about 8 percent of global international calling minutes in the third quarter of 2009, Skype is considered one of the leading global internet communications companies.

Skype is free for PC-to-PC calls, but when a PSTN or a cell phone is involved, Skype offers a fee-based service. There are two modes for involving a PSTN or cell phone in a Skype conversation: SkypeIn and SkypeOut.

SkypeIn service is offered in many countries in the world and the list is expanding. Wherever the service is offered, it allows the Skype registered user to receive a call from a PSTN or cell phone on her/his computer over the Internet. To use SkypeIn, the user subscribes to an online number for a monthly fee. Using their PSTN or mobile phone, callers can call this number and pay the same standard rate as if they were making a similar call to another PSTN or mobile phone with the same area code. The online number uses the Internet to route the call to the Skype registered user. Except for the price of the online number, the cost of the call is free for the registered user who receives the call. To relieve callers from being charged long-distance rates, the registered user can subscribe to more than one online number. For example, the registered user can get a subscription to one number in the United States and another number in France. SkypeIn service comes with free voice mail.

SkypeOut allows Skype registered users to make phone calls from their PC to any PSTN phone or any cell phone anywhere in the world and pay local rates. To make SkypeOut calls, the user purchases either monthly subscriptions or Skype credit minutes. Using PCs, Skype users dial the phone numbers of PSTN or mobile phones. Skype channels SkypeOut calls to gateways, which then direct the calls to the PSTN or cell phone services. In addition to the monthly subscription rate or Skype credit minute cost, the Skype user pays a small global and local rate for the service. With their subscription, users are able to forward incoming calls to their PSTN or mobile phone. It is important to mention that Skype is not considered as a replacement for telephone service and cannot be used in emergency situations. For example, we can not use Skype to dial 911 in the United States.

28.4 REAL-TIME INTERACTIVE PROTOCOLS

After discussing the three approaches to using multimedia through the Internet, we now concentrate on the last one, which is the most interesting and involved: real-time interactive multimedia. This application has evoked a lot of attention in the Internet society and several application-layer protocols have been designed to handle it. Before we discuss the need and rationale for this type of application, let us give a schematic representation in Figure 28.33.

Figure 28.33 *Schematic diagram of a real-time multimedia system*

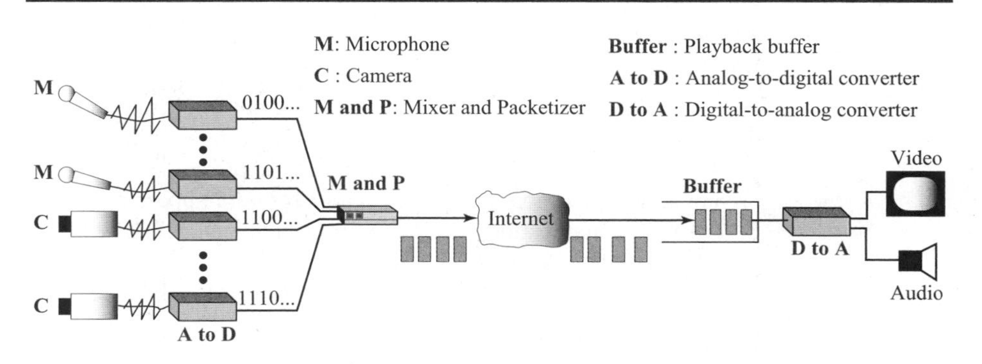

Although it could have only one microphone and one audio player, today's interactive real-time application is normally made up of several microphones and several cameras. The audio and video information (analog signals) are converted to digital data. The digital data created from different sources are normally mixed and packetized. The packets are sent to the packet-switched Internet. The packets are received at the destination with different delays (jitter) and some packets may also be corrupted or lost. A playback buffer replays packets based on the timestamp on each packet. The result is sent to a digital-to-analog converter to recreate the audio and video signals. The audio signal is sent to a speaker; the video signal to a display device.

Each microphone or camera at the source site is called a *contributor* and is given a 32-bit identifier called the *contributing source* (CSRC) *identifier*. The mixer is also called the synchronizer and is given another identifier called the *synchronizing source* (SSRC) *identifier*. We will use these identifiers in the packet later.

28.4.1 Rationale for New Protocols

We discussed the protocol stack for general Internet applications in Chapter 2. In this section, we want to show why we need some new protocols to handle interactive real-time multimedia applications such as audio and video conferencing.

It is clear that we do not need to change the first three layers of the TCP/IP protocol suite (physical, data-link, and network layers) because these three layers are designed to carry any type of data. The physical layer provides service to the data-link layer, no matter the nature of the bits in a frame. The data-link layer is responsible for node-to-node delivery of the network layer packets no matter what makes up the packet. The network layer is also responsible for host-to-host delivery of a datagram, no matter what is in the datagram, although we need a network layer with a better quality of service for multimedia applications, as we will discuss in the next section.

It looks as if we should worry about only the application and transport layers. Some application-layer protocols need to be designed to encode and compress the multimedia data, considering the trade-off between the quality, bandwidth requirement, and the complexity of mathematical operations for encoding and compression. As we describe shortly, it turns out that application-layer protocols that can handle multimedia have some requirements that can be handled by the transport layer instead of being individually handled by each application protocol.

Application Layer

It is clear that we need to develop some application-layer protocols for interactive real-time multimedia because the nature of audio conferencing and video conferencing is different from some applications, such as file transfer and electronic mail, which we discussed in Chapter 26. Several proprietary applications have been developed by the private sector, and more and more applications are appearing in the market every day. Some of these applications, such as MPEG audio and MPEG video, use some standards defined for audio and video data transfer. There is no specific standard that is used by all applications, and there is no specific application protocol that can be used by everyone.

Transport Layer

The lack of a single standard and the general features of multimedia applications discussed in Section 8.3.3 raise some questions about the transport-layer protocol to be used for all multimedia applications. The two common transport-layer protocols, UDP and TCP, were developed at the time when no one even thought about the use of multimedia in the Internet. Can we use UDP or TCP as a general transport-layer protocol for real-time multimedia applications? To answer this question, we first need to think about the requirements for this type of multimedia application and then see if either UDP or TCP can respond to these requirements.

Transport-Layer Requirements for Interactive Real-Time Multimedia

Let us first briefly compose a set of requirements for this type of application.

- ***Sender-Receiver Negotiation.*** The first requirement is related to the lack of a single standard for audio or video. We have several standards for audio conferencing or video conferencing with different encoding or compression methods. If a sender uses one encoding method and the receiver uses another one, the communication is impossible. The application programs need to negotiate the standards used for audio/video before encoded and compressed data can be transferred.
- ***Creation of Packet Stream.*** When we discussed UDP and TCP in Chapter 24, we mentioned that UDP allows the application to packetize its message with clear-cut boundaries before delivering the message to UDP. TCP, on the other hand, can handle streams of bytes without the requirement from the application to put specific boundaries on the chunk of data. In other words, UDP is suitable for those applications that need to send messages with clear-cut boundaries, but TCP is suitable for those applications that send continuous streams of bytes. When it comes to real-time multimedia, we need both features. Real-time multimedia is a *stream of frames* or a *stream chunk* of data in which the chunk or frame has a specific size or boundary, but also there are relationships between frames or chunks. It is clear that neither UDP nor TCP is suitable for handling streams of frames in this case. UDP cannot provide a relationship between frames; TCP provides relationships between bytes, but a byte is much smaller than a multimedia frame or chunk.
- ***Source Synchronization.*** If an application uses more than one source (both audio and video), there is a need for synchronization between the sources. For example, in a teleconferencing that uses both audio and video, such as Skype, the audio and video may be using different encoding and compression methods with different rates. It is obvious that somehow these two types of applications should be synchronized, otherwise we will see the face of the speaker before hearing what she is saying or vice versa. It is also possible that there is more than one source for audio or video (using multiple microphones or multiple cameras). Source synchronization is normally done using *mixers*.
- ***Error Control.*** We have already discussed that handling errors (packet corruption and packet loss) need special care in real-time multimedia applications. We showed that we cannot afford to retransmit corrupted or lost packets. We learned that we need to inject extra redundancy in the data to be able to reproduce the lost or corrupted packets without asking for them to be retransmitted. This implies that the TCP protocol is not suitable for real-time multimedia applications.
- ***Congestion Control.*** As in other applications, we need to provide some sort of congestion control in multimedia. If we decide not to use TCP for multimedia (because of retransmission problems), we should somehow implement congestion control in the system.

- ***Jitter Removal.*** We discussed in Section 8.3.3 that one of the problems with real-time multimedia applications is the jitter created at the receiver site, because the packet-switched service provided by the Internet may create uneven delays for different packets in a stream. In the past, audio conferencing was provided by the telephone network, which was originally designed as a circuit-switched network, which is jitter free. If we gradually move all of these applications to the Internet, we need to somehow deal with the jitter. We said in Section 8.3.3 that one of the ways to alleviate jitter is to use playback buffers and timestamping. The playback is implemented at the application layer at the receiver site, but the transport layer should be able to provide the application layer with timestamping and sequencing.
- ***Identifying Sender.*** A subtle issue in multimedia applications, like other applications, is to identify the sender at the application layer. When we use the Internet, the parties are identified by their IP addresses. However, we need to map the IP addresses to something more friendly, as we did with the HTTP protocol or electronic mail.

Capability of UDP or TCP to Handle Real-Time Multimedia

After discussing the requirements for real-time multimedia, let us see if either UDP or TCP is capable of handling these requirements. Table 28.6 compares UDP and TCP with respect to these requirements.

The first glance at Table 28.6 reveals a very interesting fact: neither UDP nor TCP can respond to all requirements. However, we should remember that we need a transport-layer protocol to implement the client-server socket; we cannot let the application layer do the job of the transport layer. This means that we probably have three choices:

1. We can use a new transport-layer protocol (such as SCTP, discussed in Chapter 24) that combines the features of UDP and TCP (in particular stream packetizing and multi-streaming). This choice is probably the best because SCTP has the combined features of UDP and TCP with additional features of its own. However, SCTP was introduced when there were many multimedia applications. It may become the de facto transport layer in the future.

Table 28.6 *Capability of UDP or TCP to handle real-time multimedia*

Requirements	*UDP*	*TCP*
1. Sender-receiver negotiation for selecting the encoding type	No	No
2. Creation of packet stream	No	No
3. Source synchronization for mixing different sources	No	No
4. Error Control	No	Yes
5. Congestion Control	No	Yes
6. Jitter removal	No	No
7. Sender identification	No	No

2. We can use TCP and combine it with another transport facility to compensate for the requirements that cannot be provided by TCP. However, this choice is somewhat difficult because TCP uses a retransmission method that it is not acceptable for real-time applications. Another problem with TCP is that it does not do

multicasting. A TCP connection is only a two-party connection; we need multi-party connection for real-time interactive communication.

3. We can use UDP and combine it with another transport facility to compensate for the requirements that cannot be provided by UDP. In other words, we use UDP to provide client-server socket interface, but use another protocol that runs at the top of the UDP. This is the current choice for multimedia applications. This transport facility is the Real-time Transport Protocol (RTP), which we discuss next.

28.4.2 RTP

Real-time Transport Protocol (RTP) is the protocol designed to handle real-time traffic on the Internet. RTP does not have a delivery mechanism (multicasting, port numbers, and so on); it must be used with UDP. RTP stands between UDP and the multimedia application. The literature and standards treat RTP as the transport protocol (not a transport-layer protocol) that can be thought of as located in the application layer (see Figure 28.34). The data from multimedia applications are encapsulated in RTP, which in turn passes them to the transport layer. In other words, the socket interface is located between RTP and UDP, which implies that we should include the functionality of RTP in client-server programs that we write for each multimedia application. However, some programming languages provide some facilities to make the programming task easier. For example, the C language provides an RTP library and the Java language provides an RTP class for this purpose. If we use the RTP library or the RTP class, we can think that we have separated the applications from the RTP and the RTP has become part of the transport layer.

Figure 28.34 *RTP location in the TCP/IP protocol suite*

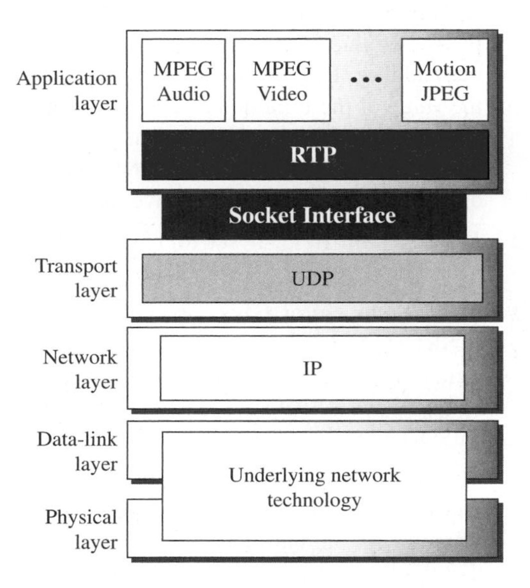

RTP Packet Format

Before we discuss how RTP can help the multimedia applications, let us discuss its packet format. We can then relate the functions of the fields with the requirements we discussed in the previous section. Figure 28.35 shows the format of the RTP packet header. The format is very simple and general enough to cover all real-time applications. An application that needs more information adds it to the beginning of its payload.

Figure 28.35 *RTP packet header format*

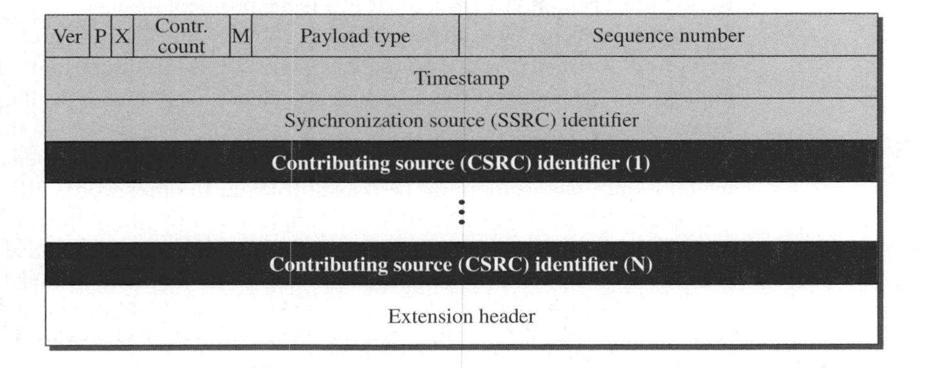

A description of each field follows.

- ❑ ***Ver.*** This 2-bit field defines the version number. The current version is 2.
- ❑ ***P.*** This 1-bit field, if set to 1, indicates the presence of padding at the end of the packet. In this case, the value of the last byte in the padding defines the length of the padding. Padding is the norm if a packet is encrypted. There is no padding if the value of the P field is 0. The use of this one-bit field eliminates the need for the length of the RTP data because if there is no padding, the length of the data is the length of the UDP data minus the RTP header. Otherwise, the length of the padding should be subtracted to give the RTP data length.
- ❑ ***X.*** This 1-bit field, if set to 1, indicates an extension header between the basic header and the data. There is no extension header if the value of this field is 0.
- ❑ ***Contributor count.*** This 4-bit field indicates the number of contributing sources (CSRCs). Note that we can have a maximum of 15 contributors because a 4-bit field only allows a number between 0 and 15. Note that in an audio or video conferencing, each active source (the source that sends data instead of just listening) is called a contributor.
- ❑ ***M.*** This 1-bit field is a marker used by the application to indicate, for example, the end of its data. We said that a multimedia application is a stream of blocks or frames with an end of frame marker. If this bit is set in an RTP packet, it means that the RTP packet carries this marker.

- ***Payload type.*** This 7-bit field indicates the type of the payload. Several payload types have been defined so far. We list some common applications in Table 28.7. A discussion of the types is beyond the scope of this book.

Table 28.7 *Payload types*

Type	*Application*	*Type*	*Application*	*Type*	*Application*
0	PCMμ Audio	7	LPC audio	15	G728 audio
1	1016	8	PCMA audio	26	Motion JPEG
2	G721 audio	9	G722 audio	31	H.261
3	GSM audio	10–11	L16 audio	32	MPEG1 video
5–6	DV14 audio	14	MPEG audio	33	MPEG2 video

- ***Sequence number.*** This field is 16 bits in length. It is used to number the RTP packets. The sequence number of the first packet is chosen randomly; it is incremented by 1 for each subsequent packet. The sequence number is used by the receiver to detect lost or out of order packets.
- ***Timestamp.*** This is a 32-bit field that indicates the time relationship between packets. The timestamp for the first packet is a random number. For each succeeding packet, the value is the sum of the preceding timestamp plus the time the first byte is produced (sampled). The value of the clock tick depends on the application. For example, audio applications normally generate chunks of 160 bytes; the clock tick for this application is 160. The timestamp for this application increases 160 for each RTP packet.
- ***Synchronization source (SSRC) identifier.*** If there is only one source, this 32-bit field defines the source. However, if there are several sources, the mixer is the synchronization source and the other sources are contributors. The value of the source identifier is a random number chosen by the source. The protocol provides a strategy in case of conflict (two sources start with the same sequence number).
- ***Contributing source (CSRC) identifier.*** Each of these 32-bit identifiers (a maximum of 15) defines a source. When there is more than one source in a session, the mixer is the synchronization source and the remaining sources are the contributors.

UDP Port

Although RTP is itself a transport-layer protocol, the RTP packet is not encapsulated directly in an IP datagram. Instead, RTP is treated like an application program and is encapsulated in a UDP user datagram. However, unlike other application programs, no well-known port is assigned to RTP. The port can be selected on demand with only one restriction: The port number must be an even number. The next number (an odd number) is used by the companion of RTP, Real-time Transport Control Protocol (RTCP), which we will discuss in the next section.

RTP uses an even-numbered UDP port.

28.4.3 RTCP

RTP allows only one type of message, one that carries data from the source to the destination. To really control the session, we need more communication between the participants in a session. Control communication in this case is assigned to a separate protocol called

Real-time Transport Control Protocol (RTCP). We need to emphasize that the RTCP payloads are not carried in RTP packets; RTCP is in fact a sister protocol of RTP. This means that the UDP, as the real transport protocol, sometimes carries RTP payloads and sometimes RTCP payloads as though they belong to different upper-layer protocols.

RCTP packets make an *out-of-band* control stream that provides two-way feedback information between the senders and receivers of the multimedia streams. In particular, RTCP provides the following functions:

1. RTCP informs the sender or senders of multimedia streams about the network performance, which can be directly related to the congestion in the network. Since multimedia applications use UDP (instead of TCP), there is no way to control the congestion in the network at the transport layer. This means that, if it is necessary to control the congestion, it should be done at the application layer. RTCP, as we will see shortly, gives the clues to the application layer to do so. If the congestion is observed and reported by the RTCP, an application can use a more aggressive compression method to reduce the number of packets and, therefore, to reduce congestion, for a trade-off in quality. On the other hand, if no congestion is observed, the application program can use a less aggressive compression method for a better quality service.
2. Information carried in the RTCP packets can be used to synchronize different streams associated with the same source. A source may use two different sources to collect audio or video data. In addition, audio data may be collected from different microphones and video data may be collected from different cameras. In general, two pieces of information are needed to achieve synchronization:
   a. Each sender needs an identity. Although each source may have a different SSRC, RTCP provides one single identity, called a *canonical name* (*CNAME*) for each source. CNAME can be used to correlate different sources and allow the receiver to combine different sources from the same source. For example, a teleconference may have n senders associated with a session, but we may have m sources ($m > n$) that contribute to the stream. In this system, we have only n CNAMEs, but m SSRCs. A CNAME is in the form of

      user@host

      in which *user* is normally the login name of the user and the *host* is the domain name of the host.
   b. The canonical name cannot per se provide synchronization. To synchronize the sources, we need to know the absolute timing of the stream, in addition to the relative timing provided by the timestamp field in each RTP packet. The timestamp information in each packet gives the relative time relationship of the bits in the packet to the beginning of the stream; it cannot relate one stream to another. The absolute time, the "wall clock" time as it is sometimes referred to, needs to be sent by RTCP packets to enable synchronization.
3. An RTCP packet can carry extra information about the sender that can be useful for the receiver, such as the name of the sender (beyond canonical name) or captions for a video.

RTCP Packets

After discussing the main functions and purpose of RTCP, let us discuss its packets. Figure 28.36 shows five common packet types. The number next to each box defines the numeric value of each packet. We need to mention that more than one RTCP packet can be packed as a single payload for UDP because the RTCP packets are smaller than RTP packets.

Figure 28.36 *RTCP packet types*

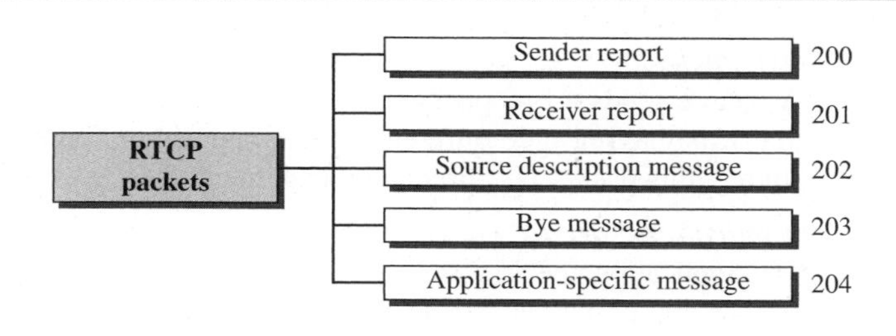

The format and the exact definition of each field is very involved and beyond the scope and space of this book. We briefly discuss the purpose of each packet and relate it to the functions previously described.

Sender Report Packet

The sender report packet is sent periodically by the active senders in a session to report transmission and reception statistics for all RTP packets sent during the interval. The sender report packet includes the following information:

- The SSRC of the RTP stream.
- The absolute timestamp, which is the combination of the relative timestamp and the wall clock time, which is the number of seconds elapsed since midnight January 1, 1970. The absolute timestamp, as discussed previously, allows the receiver to synchronize different RTP packets.
- The number of RTP packets and bytes sent from the beginning of the session.

Receiver Report Packet

The receiver report is issued by passive participants, those that do not send RTP packets. The report informs the sender and other receivers about the quality of service. The feedback information can be used for congestion control at the sender site. A receiver report includes the following information:

- The SSRC of the RTP stream for which the receiver report has been generated.
- The fraction of packet loss.
- The last sequence number.
- The interval jitter.

Source Description Packet
The source periodically sends a source description packet to give additional information about itself. The packet can include:

- The SSRC.
- The canonical name (CNAME) of the sender.
- Other information such as the real name, the e-mail address, the telephone number.
- The source description packet may also include extra data, such as captions used for video.

Bye Packet
A source sends a bye packet to shut down a stream. It allows the source to announce that it is leaving the conference. Although other sources can detect the absence of a source, this packet is a direct announcement. It is also very useful to a mixer.

Application-Specific Packet
The application-specific packet is a packet for an application that wants to use new applications (not defined in the standard). It allows the definition of a new packet type.

UDP Port

RTCP, like RTP, does not use a well-known UDP port. It uses a temporary port. The UDP port chosen must be the number immediately following the UDP port selected for RTP, which makes it an odd-numbered port.

RTCP uses an odd-numbered UDP port that follows the one selected for RTP.

Bandwidth Utilization

The RTCP packets are sent not only by the active senders, but also by passive receivers, whose numbers are normally greater than the active senders. This means that if the RTCP traffic is not controlled, it may get out of hand. To control the situation, RTCP uses a control mechanism to limit its traffic to the small portion (normally 5 percent) of the traffic used in the session (for both RTP and RTCP). A larger part of this small percentage, x percent, is then assigned to the RTCP packets generated by the passive receiver, a smaller part, $(1 - x)$ percent, is assigned to the RTCP packets generated by the active senders. RTCP protocol uses a mechanism to define the value of x based on the ratio of the passive receiver to the active sender.

Example 28.10

Let us assume that the total bandwidth allocated for a session is 1 Mbps. RTCP traffic gets only 5 percent of this bandwidth, which is 50 kbps. If there are only 2 active senders and 8 passive receivers, it is natural that each sender or receiver gets only 5 kbps. If the average size of the RTCP packet is 5 kbits, then each sender or receiver can send only 1 RTCP packet per second. Note that we need to consider the packet size at the data-link layer.

Requirement Fulfillment

As we promised, let us see how the combination of RTP and RTCP can respond to the requirements of an interactive real-time multimedia application. A digital audio or video stream, a sequence of bits, is divided into chunks (*blocks* or *frames* as they are sometimes called). Each chunk has a predefined boundary that distinguishes the chunk from the previous chunk or the next one. A chunk is encapsulated in an RTP packet, which defines a specific encoding (payload type), a sequence number, a timestamp, a synchronization source (SSRC) identifier, and one or more contributing source (CSRC) identifiers.

1. The first requirement, *sender-receiver negotiation,* cannot be satisfied by the combination of the RTP/RTCP protocols. It should be accomplished by some other means. We will see later that another protocol (SIP), which is used in conjunction with RTP/RTCP, provides this capability.
2. The second requirement, *creation of a stream of chunks,* is provided by encapsulating each chunk in an RTP packet and giving a sequence number to each chunk. The M field in an RTP packet also defines whether there is a specific type of boundary between chunks.
3. The third requirement, *synchronization of sources,* is satisfied by identifying each source by a 32-bit identifier and using the relative timestamping in the RTP packet and the absolute timestamping in the RTCP packet.
4. The fourth requirement, *error control,* is provided by using the sequence number in the RTP packet and letting the application regenerate the lost packet using FEC methods discussed earlier in the chapter.
5. The fifth requirement, *congestion control,* is met by the feedback from the receiver using the *receiver report packets* (RTCP) that notify the sender about the number of lost packets. The sender then can use a more aggressive compression technique to reduce the number of packets sent and therefore alleviate the congestion.
6. The sixth requirement, *jitter removal,* is achieved by the timestamping and sequencing provided in each RTP packet to be used in buffered playback of the data.
7. The seventh requirement, *identification of source,* is provided by using the CNAME included in the source description packets (RTCP) sent by the sender.

28.4.4 Session Initialization Protocol (SIP)

We discussed how to use the Internet for audio-video conferencing. Although RTP and RTCP can be used to provide these services, one component is missing: a signaling system required to call the participants.

To understand the issue, let us go back for the moment to the traditional audio conferencing (between two or more people) using the traditional telephone system (public switched telephone network or PSTN). To make a phone call, two telephone numbers are needed, that of the caller and that of the callee. We then need to dial the telephone number of the callee and wait for her to respond. The telephone conversation starts after the response of the callee. In other words, regular telephone communication involves two phases: the signaling phase and the audio communication phase.

The signaling phase in the telephone network is provided by a protocol called *Signaling System 7* (SS7). The SS7 protocol is totally separate from the voice communication system. For example, although the traditional telephone system uses analog signals carrying voice over a circuit-switched network, SS7 uses electrical pulses in which each number dialed changes to a series of pulses. SS7 today not only provides the calling service, it also provides other services, such as call forwarding and error reporting.

The combination of RTP/RTCP protocols we discussed in the previous sections is equivalent to the voice communication provided by PSTN; to totally simulate this system over the Internet, we need a signaling system. Our ambition takes us even further. Not only do we want to be able to call our party in an audio or video conference using our computers (PCs), we also want to be able to do so using our telephone set, our mobile phone, our PDAs, and so on. We also need to find our party if she is not sitting at her desk. We need to communicate between a mixture of devices.

The **Session Initiation Protocol (SIP)** is a protocol devised by IETF to be used in conjunction with the RTP/SCTP. It is an application-layer protocol, similar to HTTP, that establishes, manages, and terminates a multimedia session (call). It can be used to create two-party, multiparty, or multicast sessions. SIP is designed to be independent of the underlying transport layer; it can run on UDP, TCP, or SCTP, using the port 5060. SIP can provide the following services:

- It establishes a call between users if they are connected to the Internet.
- It finds the location of the users (their IP addresses) on the Internet, because the users may be changing their IP addresses (think about mobile IP and DHCP).
- It finds out if the users are able or willing to take part in the conference call.
- It determines the users' capabilities in terms of media to be used and the type of encoding (the first requirement for multimedia communication we mentioned in the previous section).
- It establishes session setup by defining parameters such as port numbers to be used (remember that RTP and RTCP use port numbers).
- It provides session management functions such as call holding, call forwarding, accepting new participants, and changing the session parameters.

Communicating Parties

One difference that we may have noticed between the interactive real-time multimedia applications and other applications is communicating parties. In an audio or video conference, the communication is between humans, not devices. For example, in HTTP or FTP, the client needs to find the IP address of the server (using DNS) before communication. There is no need to find a person before communicating. In the SMTP, the sender of an e-mail sends the message to the receiver mailbox on an SMTP without controlling when the message will be picked up. In an audio or video conference, the caller needs to find the callee. The callee can be sitting at her desk, can be walking in the street, or can be totally unavailable. What makes the communication more difficult is that the device to which the participant has access at a particular time may have a different capability than the device being used at another time. The SIP protocol needs to find the location of the callee and at the same time negotiate the capability of the devices the participants are using.

Addresses

In a regular telephone communication, a telephone number identifies the sender, and another telephone number identifies the receiver. SIP is very flexible. In SIP, an e-mail address, an IP address, a telephone number, and other types of addresses can be used to identify the sender and receiver. However, the address needs to be in SIP format (also called *scheme*). Figure 28.37 shows some common formats.

Figure 28.37 *SIP formats*

We have noticed that the SIP address is similar to a URL we have encountered in Chapter 26. In fact, the SIP addresses are URLs that can be included in the web page of the potential callee. For example, Bob can include one of the above addresses as his SIP address and, if someone clicks on it, the SIP protocol is invoked and calls Bob. Other addresses are also possible, such as those that use first name followed by last name, but all addresses need to be in the form *sip:user@address*.

Messages

SIP is a text-based protocol like HTTP. SIP, like HTTP, uses messages. Messages in SIP are divided into two broad categories: Requests and responses. The format of both message categories is shown below (note the similarity with HTTP messages as shown in Figure 26.5):

Request Messages		**Response Messages**	
Start line		**Status line**	
Header	**// one or more lines**	**Header**	**// one or more lines**
Blank line		**Blank line**	
Body	**// one or more lines**	**Body**	**// one or more lines**

Request Messages

IETF originally defined six request messages, but some new request messages have been proposed to extend the functionality of the SIP. We just mention the original six messages as follows:

- ❑ ***INVITE.*** The INVITE request message is used by a caller to initialize a session. Using this request message, a caller invites one or more callees to participate in the conference.
- ❑ ***ACK.*** The ACK message is sent by the caller to confirm that the session initialization has been completed.
- ❑ ***OPTIONS.*** The OPTIONS message queries a machine about its capabilities.
- ❑ ***CANCEL.*** The CANCEL message cancels an already started initialization process, but does not terminate the call. A new initialization may start after the CANCEL message.

- ❑ ***REGISTER.*** The REGISTER message makes a connection when the callee is not available.
- ❑ ***BYE.*** The BYE message is used to terminate the session. Compare the BYE message with the CANCEL message. The BYE message, which can be initiated from the caller or callee, terminates the whole session.

Response Messages

IETF has also defined six types of response messages that can be sent to request messages, but note that there is no relationship between a request and a response message. A response message can be sent to any request message. As in other text-oriented application protocols, the response messages are defined using three-digit numbers. The response messages are briefly described below:

- ❑ ***Informational Responses.*** These responses are in the form **SIP 1xx** (the common ones are 100 trying, 180 ringing, 181 call forwarded, 182 queued, and 183 session progress).
- ❑ ***Successful Responses.*** These responses are in the form **SIP 2xx** (the common one is 200 OK).
- ❑ ***Redirection Responses.*** These responses are in the form **SIP 3xx** (the common ones are 301 moved permanently, 302 moved temporarily, 380 alternative service).
- ❑ ***Client Failure Responses.*** These responses are in the form **SIP 4xx** (the common ones are 400 bad request, 401 unauthorized, 403 forbidden, 404 not found, 405 method not allowed, 406 not acceptable, 415 unsupported media type, 420 bad extension, 486 busy here).
- ❑ ***Server Failure Responses.*** These responses are in the form **SIP 5xx** (the common ones are 500 server internal error, 501 not implemented, 503 service unavailable, 504 timeout, 505 SIP version not supported).
- ❑ ***Global Failure Responses.*** These responses are in the form **SIP 6xx** (the common ones are 600 busy everywhere, 603 decline, 604 doesn't exist, and 606 not acceptable).

First Scenario: Simple Session

In the first scenario, we assume that Alice needs to call Bob and the communication uses the IP addresses of Alice and Bob as the SIP addresses. We can divide the communication into three modules: establishing, communicating, and terminating. Figure 28.38 shows a simple session using SIP.

Establishing a Session

Establishing a session in SIP requires a three-way handshake. Alice sends an INVITE request message, using UDP, TCP, or SCTP to begin the communication. If Bob is willing to start the session, he sends a response (200 OK) message. To confirm that a reply code has been received, Alice sends an ACK request message to start the audio communication. The establishment section uses two request messages (INVITE and ACK) and one response message (200 OK). We will discuss more about the contents of messages later, but for the moment we need to say that the INVITE message start line defines the IP address of the receiver and the version

Figure 28.38 *SIP simple session*

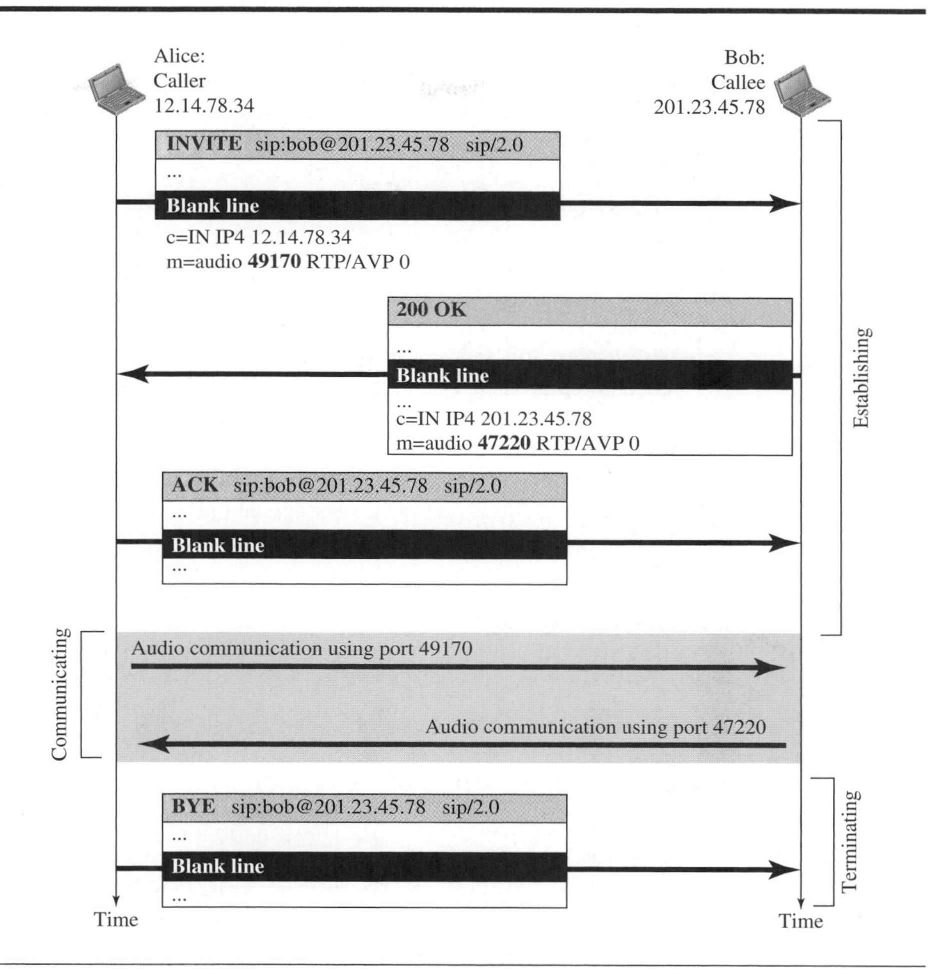

of the SIP. We have not included any line in the header, but we will do so later. The body of the header uses another protocol, Session Description Protocol (SDP), that defines the syntax (format) and semantic (meaning of each line). We will briefly discuss this protocol later. We just mention that the first line in the body defines the sender of the message; the second line defines the media (audio) and the port number to be used for RTP in the direction from Alice to Bob. The response message defines the media (audio) and the port number to be used for RTP in the Bob to Alice direction. After Alice confirms the establishment of the session with the ACK message request (which does not need a response), the establishing session is finished and the communication can start.

Communicating

After the session has been established, Alice and Bob can communicate using two temporary ports defined in the establishing sessions. The even-numbered ports are used for

RTP; RTCP can use the odd-numbered ports that follow (we have shown only the even-numbered ports used for RTP in Figure 28.38).

Terminating the Session

The session can be terminated with a BYE message sent by either party. In the figure, we have assumed that Alice terminates the session.

Second Scenario: Tracking the Callee

What happens if Bob is not sitting at his terminal? He may be away from his system or at another terminal. He may not even have a fixed IP address if DHCP is being used. SIP has a mechanism (similar to one in DNS) that finds the IP address of the terminal at which Bob is sitting. To perform this tracking, SIP uses the concept of registration. SIP defines some servers as registrars. At any moment a user is registered with at least one **registrar server;** this server knows the IP address of the callee.

When Alice needs to communicate with Bob, she can use the e-mail address instead of the IP address in the INVITE message. The message goes to a proxy server. The proxy server sends a lookup message (not part of SIP) to some registrar server that has registered Bob. When the proxy server receives a reply message from the registrar server, the proxy server takes Alice's INVITE message and inserts the newly discovered IP address of Bob. This message is then sent to Bob. Figure 28.39 shows the process.

Figure 28.39 *Tracking the callee*

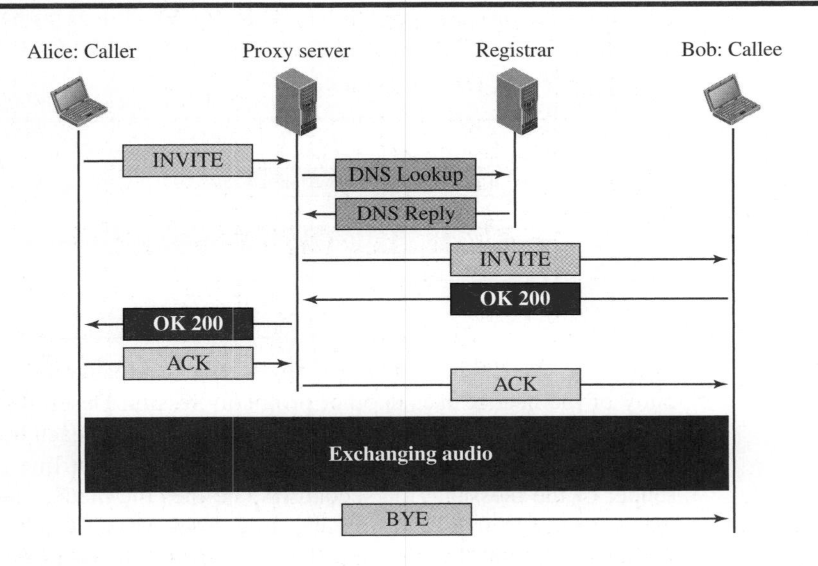

SIP Message Format and SDP Protocol

As we discussed before, the SIP request and response messages are divided into four sections: start or status line, header, a blank line, and the body. Since a blank line needs no more information, let us briefly describe the format of the other sections.

Start Line

The start line is a single line that starts with the message request name, followed by the address of the recipient and the SIP version. For example, the INVITE message request start line has the following start line format:

INVITE sip:forouzan@roadrunner.com

Status Line

The status line is a single line that starts with the three-digit response code. For example, the 200 response message has the following status line format.

200 OK

Header

A header, in the request or response message, can use several lines. Each line starts with the line name followed by a colon and space and followed by the value. Some typical header lines are: Via, From, To, Call-ID, Content-Type, Content-Length, and Expired. The *Via* header defines the SIP device through which the message passes, including the sender. The *From* header defines the sender and the *To* header defines the recipient. The *Call-ID* header is a random number that defines the session. The *Content-Type* defines the type of body of the message, which is normally SPD, which we will describe shortly. The *Content-Length* defines the length of the body of the message in bytes. The *Expired* header is normally used in a REGISTER message to define the expiration of the information in the body. The following is an example of a header in an INVITE message.

Via: SIP/2.0/UDP 145.23.76.80
From: sip:alice@roadrunner.com
To: sip:bob@arrowhead.net
Call-ID: 23a345@roadrunner.com
Content-Type: application/spd
Content-Length: 600

Body

The body of the message is the main difference we will see between an application such as HTTP and SIP. SIP uses another protocol, called *Session Description Protocol* (SDP), to define the body. Each line in the body is made of an SDP code followed by an equal sign, and followed by the value. The code is a single character that determines the purpose of the code. We can divide the body into several sections.

The first part of the body is normally general information. The codes used in this section are: ***v*** (for version of SDP), and ***o*** (for origin of the message).

The second part of the body normally gives information to the recipient for making a decision to take part in the session. The codes used in this section are: ***s*** (subject), ***i*** (information about subject), ***u*** (for session URL), and ***e*** (the e-mail address of the person responsible for the session).

The third part of the body gives the technical details to make the session possible. The codes used in this part are: ***c*** (the unicast or multicast IP address that the user needs to join to be able to take part in the session), ***t*** (the start time and end time of the session, encoded as integers), ***m*** (the information about media such as audio, video, the port number, the protocol used).

The following shows an example of a body of an INVITE request message.

```
v=0
o=forouzan 64.23.45.8
s=computer classes
i=what to offer next semester
u=http://www.uni.edu
e=forouzan@roadrunner.com
c=IN IP4 64.23.45.8
t=2923721854 2923725454
```

Putting the Parts Together

Let us put the four sections of a message request together as shown below. The first line is the start line, the next six lines make up the header. The next line (blank line) separates the header from the body, and the last eight lines are the body of the message. We conclude our discussion about the SIP protocol and the auxiliary protocol SPD used by SIP to define the body.

```
INVITE sip:forouzan@roadrunner.com
Via: SIP/2.0/UDP 145.23.76.80
From: sip:alice@roadrunner.com
To: sip:bob@arrowhead.net
Call-ID: 23a345@roadrunner.com
Content-Type: application/spd
Content-Length: 600
// Blank line
v=0
o=forouzan 64.23.45.8
s=computer classes
i=what to offer next semester
u=http://www.uni.edu
e=forouzan@roadrunner.com
c=IN IP4 64.23.45.8
t=2923721854 2923725454
```

28.4.5 H.323

H.323 is a standard designed by ITU to allow telephones on the public telephone network to talk to computers (called *terminals* in H.323) connected to the Internet. Figure 28.40 shows the general architecture of H.323 for audio, but it can also be used for video.

A **gateway** connects the Internet to the telephone network. In general, a gateway is a five-layer device that can translate a message from one protocol stack to another. The gateway here does exactly the same thing. It transforms a telephone network message

Figure 28.40 *H.323 architecture*

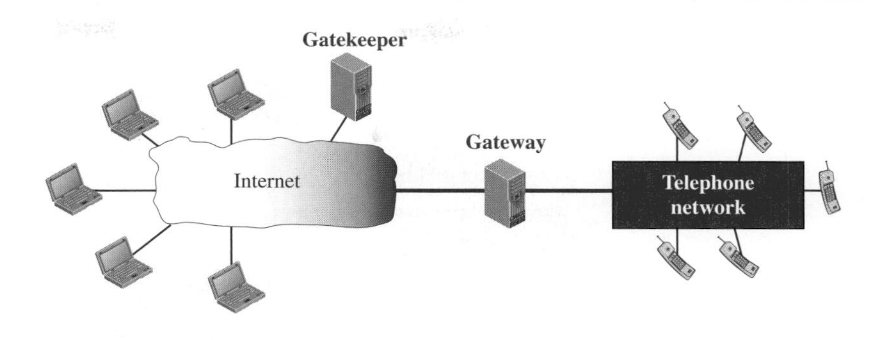

into an Internet message. The **gatekeeper** server on the local area network plays the role of the registrar server, as we discussed in the SIP protocol.

Protocols

H.323 uses a number of protocols to establish and maintain voice (or video) communication. Figure 28.41 shows some of these protocols. H.323 uses G.71 or G.723.1 for compression. It uses a protocol named H.245, which allows the parties to negotiate the compression method. Protocol Q.931 is used for establishing and terminating connections. Another protocol, called H.225, or Registration/Administration/Status (RAS), is used for registration with the gatekeeper.

Figure 28.41 *H.323 protocols*

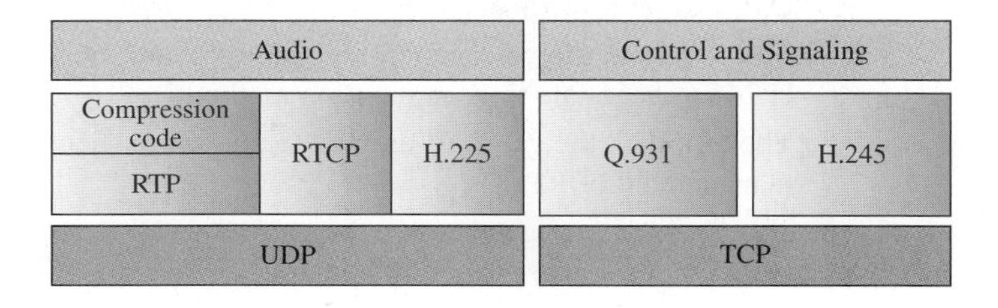

We need to mention that H.323 is a complete set of protocols that cannot be compared with SIP. SIP is only a signalling protocol, which is normally combined with RTP and RTCP to create a complete set of protocols for interactive real-time multimedia applications, but it can be used with other protocols as well. H.323, on the other hand, is a complete set of protocols that mandates the use of RTP and RTCP.

Operation

Let us use a simple example to show the operation of a telephone communication using H.323. Figure 28.42 shows the steps used by a terminal to communicate with a telephone.

Figure 28.42 *H.323 example*

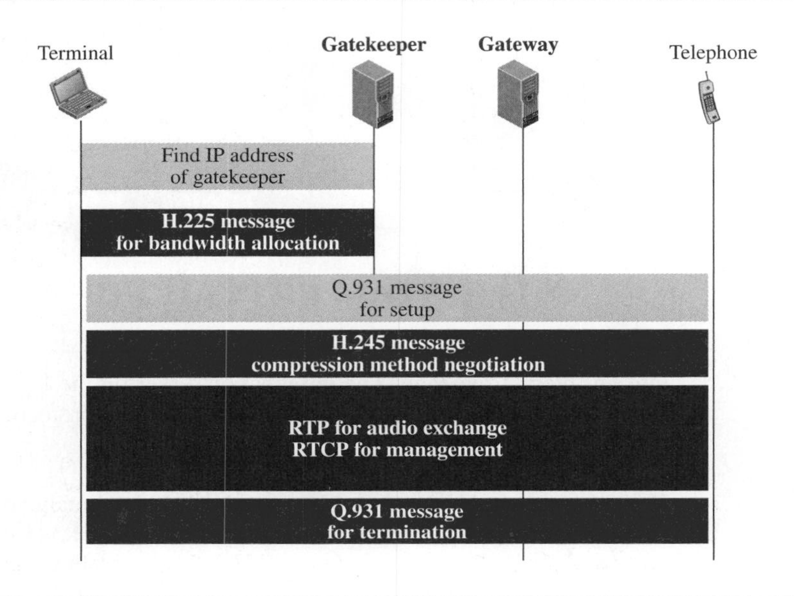

1. The terminal sends a broadcast message to the gatekeeper. The gatekeeper responds with its IP address.
2. The terminal and gatekeeper communicate, using H.225 to negotiate bandwidth.
3. The terminal, the gatekeeper, the gateway, and the telephone communicate using Q.931 to set up a connection.
4. The terminal, the gatekeeper, the gateway, and the telephone communicate using H.245 to negotiate the compression method.
5. The terminal, the gateway, and the telephone exchange audio using RTP under the management of RTCP.
6. The terminal, the gatekeeper, the gateway, and the telephone communicate using Q.931 to terminate the communication.

28.5 END-CHAPTER MATERIALS

28.5.1 Recommended Reading

For more details about subjects discussed in this chapter, we recommend the following books. The items enclosed in brackets refer to the reference list at the end of the book.

Books

Several books give some coverage of multimedia: [Com 06], [Tan 03], and [GW 04].

28.5.2 Key Terms

adaptive DM (ADM)
adaptive DPCM (ADPCM)
arithmetic coding
bidirectional frame (B-frame)
delta modulation (DM)
differential PCM (DPCM)
discrete cosine transform (DCT)
frequency masking
gatekeeper
gateway
Graphical Interchange Format (GIF)
H.323
Huffman coding
intracoded frame (I-frame)
jitter
Joint Photographic Experts Group (JPEG)
Lempel-Ziv-Welch (LZW)
linear predictive coding (LPC)
lossless compression
lossy compression
media server
metafile
mixer
Motion Picture Experts Group (MPEG)
MPEG audio layer 3 (MP3)
perceptual coding
playback buffer
predicted frame (P-frame)
predictive coding (PC)
psychoacoustics
Real-Time Streaming Protocol (RTSP)
Real-time Transport Control Protocol (RTCP)
Real-time Transport Protocol (RTP)
registrar server
run-length coding
run-length encoding (RLE)
Session Initiation Protocol (SIP)
spatial compression
temporal compression
temporal masking
timestamp
voice over IP (VoIP)

28.5.3 Summary

Multimedia data are normally compressed before transmission. We can divide compression into two broad categories: lossless and lossy compression. In lossless compression, the integrity of the data is preserved because compression and decompression algorithms are exact inverses of each other: no part of the data is lost in the process. Lossy compression cannot preserve the accuracy of data, but we gain the benefit of reducing the size of the compressed data.

Audio/video files can be downloaded for future use (streaming stored audio/video) or broadcast to clients over the Internet (streaming live audio/video). The Internet can also be used for live audio/video interaction. Audio and video need to be digitized before being sent over the Internet. We can use a web server, or a web server with a metafile, or a media server, or a media server and RTSP to download a streaming audio/video file.

Real-time data on a packet-switched network requires the preservation of the time relationship between packets of a session. Gaps between consecutive packets at the receiver cause a phenomenon called *jitter*. Jitter can be controlled through the use of timestamps and a judicious choice of the playback time.

Real-time multimedia traffic requires both UDP and Real-time Transport Protocol (RTP). RTP handles timestamping, sequencing, and mixing. Real-time Transport Control Protocol (RTCP) provides flow control, quality of data control, and feedback to the sources. The Session Initiation Protocol (SIP) is an application-layer protocol that establishes, manages, and terminates multimedia sessions. H.323 is an ITU standard that allows a telephone connected to a public telephone network to talk to a computer connected to the Internet.

28.6 PRACTICE SET

28.6.1 Quizzes

A set of interactive quizzes for this chapter can be found on the book website. It is strongly recommended that the student take the quizzes to check his/her understanding of the materials before continuing with the practice set.

28.6.2 Questions

Q28-1. Compare the number of bits transmitted for each PCM and DM sample if the maximum quantized value is

a. 12 **b.** 30 **c.** 50

Q28-2. UDP does not create a connection. How are different chunks of data, carried in different RTP packets, glued together?

Q28-3. In dictionary coding, if there are 60 characters in the message, how many times is the loop in the compression algorithm iterated? Explain.

Q28-4. Explain why the combination of quantization/dequantization steps in JPEG is a lossy process even though we divide each element of the matrix M by the corresponding value in matrix Q when quantizing and multiply it by the same value when dequantizing.

Q28-5. Assume a message is made of four characters (A, B, C, and D) with equal probability of occurrence. Guess what the encoding Huffman table for this message would be. Does encoding here really decrease the number of bits to be sent?

Q28-6. In dictionary coding, should all dictionary entries that are created in the process be used for encoding or decoding?

Q28-7. Why does RTP need the service of another protocol, RTCP, but TCP does not?

Q28-8. Is the following code an instantaneous one? Explain.

00 01 10 11 001 011 111

Q28-9. In JPEG, explain why we need to round the result of division in the quantization step.

Q28-10. In real-time interactive audio/video, what will happen if a packet arrives at the receiver site after the scheduled playback time?

Q28-11. We mentioned that SIP is an application-layer program used to provide a signalling mechanism between the caller and the callee. Which party in this communication is the server and which one is the client?

Q28-12. When we use audio/video on demand, which of these three types of multimedia communication takes place: streaming stored audio/video, streaming live audio/video, or real-time interactive audio/video?

Q28-13. Assume an image is sent from the source to the destination using 10 RTP packets. Can the first five packets define the encoding as JPEG and the last five packets define it as GIF?

Q28-14. We discuss the use of SIP in this chapter for audio. Is there any drawback to prevent using it for video?

Q28-15. Can the combination of RTP/RTCP and SIP operate in a wireless environment? Explain.

Q28-16. Are encoding and decoding of the multimedia data done by RTP? Explain.

Q28-17. In an alphabet with 20 symbols, what is the number of leaves in a Huffman tree?

Q28-18. In predictive coding, differentiate between DPCM and ADPCM.

Q28-19. Assume an application program uses separate audio and video streams during an RTP session. How many SSRCs and CSRCs are used in each RTP packet?

Q28-20. Can UDP without RTP provide an appropriate service for real-time interactive multimedia applications?

Q28-21. What is the difference between DM and DPCM?

Q28-22. What is the problem with a speech signal that is compressed using the LPC method?

Q28-23. In predictive coding, differentiate between DM and ADM.

Q28-24. In the fourth approach to streaming audio/video (Figure 28.27), what is the role of RTSP?

Q28-25. Assume two parties need to establish IP telephony using the service of RTP. How can they define the two ephemeral port numbers to be used by RTP, one for each direction?

Q28-26. In which situation, a unicast session or a multicast session, can feedback received from an RTCP packet about the session be handled more easily by the sender?

Q28-27. Assume we devise a protocol with the packet size so large that it can carry all chunks of a live or real-time multimedia stream in one packet. Do we still need sequence numbers or timestamps for the chunks? Explain.

Q28-28. Do some research and find out if SIP can provide the following services provided by modern telephone sets.

a. caller-ID **b.** call-waiting **c.** multiparty calling

Q28-29. What is the main difference between live audio/video and real-time interactive audio/video?

Q28-30. Do you think H.323 is actually the same as SIP? What are the differences? Make a comparison between the two.

Q28-31. In transform coding, when a sender transmits the M matrix to a receiver, does the sender need to send the T matrix used in calculation? Explain.

Q28-32. In JPEG, do we need less than 24 bits for each pixel if our image is using one or two primary colors? Explain.

Q28-33. If we capture RTP packets, most of the time we see the total size of the RTP header as 12 bytes. Can you explain this?

Q28-34. When we stream stored audio/video, what is the difference between the first approach (Figure 28.24) and the second approach (Figure 28.25)?

Q28-35. Both TCP and RTP use sequence numbers. Do sequence numbers in these two protocols play the same role? Explain.

Q28-36. Answer the following questions about predictive coding:

a. What are slope overload distortion and granular noise distortion in DM coding?

b. Explain how ADM coding solves the above errors.

Q28-37. In Internet telephony, explain how a call from Alice can be directed to Bob when he could be either in his office or at home?

Q28-38. Does SIP need to use the service of RTP? Explain.

Q28-39. When we stream stored audio/video, what is the difference between the second approach (Figure 28.25) and the third approach (Figure 28.26)?

Q28-40. Explain why using Q10 gives a better compression ratio but a poorer image quality than using Q90 (see Figure 28.18).

Q28-41. In multimedia communication, assume a sender can encode an image using only JPEG encoding, but the potential receiver can only decode an image if it is encoded in GIF. Can these two entities exchange multimedia data?

Q28-42. In arithmetic coding, could two different messages be encoded in the same interval? Explain.

Q28-43. In Figure 28.26, can the Web server and media server run on different machines?

Q28-44. Can we say UDP plus RTP is the same as TCP?

Q28-45. In JPEG, explain why the values of the Q(*m*, *n*) in the quantization matrix are not the same. In other words, why is each element in M(*m*, *n*) not divided by one fixed value instead of a different value?

Q28-46. Explain why RTP cannot be used as a transport-layer protocol without being run on the top of another transport-layer protocol such as UDP.

Q28-47. Can H.323 also be used for video?

28.6.3 Problems

P28-1. Assume we have the following code. Show how we can calculate the reconstructed value (y_n) for each sample if we use ADM. Let $y_0 = 20$, $\Delta_1 = 4$, $M_1 = 1$. Also assume that $M_n = 1.5 \times M_{n-1}$ if $q_n = q_{n-1}$ (no change in q_n) and $M_n = 0.5 \times M_{n-1}$ otherwise.

n	1	2	3	4	5	6	7	8	9	10	11
C_n	1	1	1	0	0	1	1	0	0	1	1

P28-2. In predictive coding, assume we have the following sample (x_n). Show the encoded message sent if we use adaptive DM (ADM). Let $y_0 = 10$, $\Delta_1 = 4$, $M_1 = 1$. Also assume that $M_n = 1.5 \times M_{n-1}$ if $q_n = q_{n-1}$ (no change in q_n) and $M_n = 0.5 \times M_{n-1}$ otherwise.

n	1	2	3	4	5	6	7	8	9	10	11
x_n	13	15	15	17	20	20	18	16	16	17	18

P28-3. In predictive coding, we have the following code. Show how we can calculate the reconstructed value (y_n) for each sample if we use delta modulation (DM). We know that $y_0 = 8$ and $\Delta = 6$.

n	1	2	3	4	5	6	7	8	9	10	11
C_n	1	0	0	1	0	1	1	0	0	1	1

P28-4. In transform coding, show that a receiver that receives an *M* matrix can create the original *p* matrix.

P28-5. In arithmetic coding, assume we have received the code 100110011. If we know that the alphabet is made of four symbols with the probabilities of P(A) = 0.4, P(B) = 0.3, P(C) = 0.2, and P(*) = 0.1, find the original message.

P28-6. Using one-dimensional DCT encoding, calculate the *M* matrix from the following three *p* matrices (which are given as row matrices but need to be considered as column matrices). Interpret the result.

$p_1 = [1\ 2\ 3\ 4\ 5\ 6\ 7\ 8] \quad p_2 = [1\ 3\ 5\ 7\ 9\ 11\ 13\ 15] \quad p_3 = [1\ 6\ 11\ 16\ 21\ 26\ 31\ 36]$

P28-7. In Huffman coding, the following coding table is given.

A → 0 B → 10 C → 110 D → 111

Show the original message if the code "00110110011110111111010" is received.

P28-8. In a real-time multimedia communication, assume we have one sender and ten receivers. If the sender is sending multimedia data at 1 Mbps, how many RTCP packets can be sent by the sender and each receiver in a second? Assume the system allocates 80 percent of the RTCP bandwidth to the receivers and 20 percent to the sender. The average size of each RTCP packet is 1000 bits.

P28-9. Given an RTP packet with the first 8 hexadecimal digits as $(86032132)_{16}$, answer the following questions:

a. What is the version of the RTP protocol?

b. Is there any padding for security?

c. Is there any extension header?

d. How many contributors are defined in the packet?

e. What is the type of payload carried by the RTP packet?

f. What is the total header size in bytes?

P28-10. Given the message **"ACCBCAAB*"**, in which the probabilities of symbols are P(A) = 0.4, P(B) = 0.3, P(C) = 0.2, and P(*) = 0.1,

a. find the compressed data using arithmetic coding with a precision of 10 binary digits.

b. find the compression ratio if we use 8 bits to represent a character in the message.

P28-11. Given the following message, find the compressed data using run-length coding.

AAACCCCCCBCCCCDDDDDAAAABBB

P28-12. Given the message **"AACCCBCCDDAB",** in which the probabilities of symbols are P(A) = 0.50, P(B) = 0.25, P(C) = 0.125, and P(D) = 0.125,

a. encode the data using Huffman coding.

b. find the compression ratio if each original character is represented by 8 bits.

P28-13. In DCT, is the value of $T(m, n)$ in transform coding always between −1 and 1? Explain.

P28-14. In predictive coding, assume we have the following sample (x_n).

n	1	2	3	4	5	6	7	8	9	10	11
x_n	13	24	46	60	45	32	30	40	30	27	20

a. Show the encoded message sent if we use delta modulation (DM). Let $y_0 = 10$ and $\Delta = 8$.

b. From the calculated q_n values, what can you say about the given Δ?

P28-15. Assume an image uses a palette of size 8 out of the table used by JPEG (GIF uses the same strategy, but the size of the palette is 256), with the combination of the following colors with the indicated level of intensities.

Red: 0 and 7 Blue: 0 and 5 Green: 0 and 4

Show the palette for this situation and answer the following questions:

a. How many bits are sent for each pixel?

b. What are the bits sent for the following pixels: red, blue, green, black, white, and magenta (red and blue, but no green)?

P28-16. In the first approach to streaming stored audio/video (Figure 28.24), assume that we need to listen to a compressed song of 4 megabytes (a typical situation). If our connection to the Internet is via a 56-kbps modem, how long will we need to wait before the song can be started (downloading time)?

P28-17. In dictionary coding, can you easily find the code if the message is each of the following (the message alphabet has only one character)?

a. "A" **b.** "AA" **c.** "AAA"
d. "AAAA" **e.** "AAAAA" **f.** "AAAAAA"

P28-18. Figure 28.43 shows the generated and the arrival time for ten audio packets. Answer the following questions:

a. If we start our player at t_8, which packets cannot be played?

b. If we start our player at t_9, which packets cannot be played?

P28-19. Calculate the T matrix for DCT when $N = 1$, $N = 2$, $N = 4$, and $N = 8$.

P28-20. In LZW coding, the message **"AACCCBCCDDAB"** is given.

a. Encode the message. (See Figure 28.2.)

b. Find the compression ratio if we use 8 bits to represent a character and four bits to represent a digit (hexadecimal).

P28-21. In LZW coding, the code "0026163301" is given. Assuming that the alphabet is made of four characters: "A", "B", "C", and "D", decode the message. (See Figure 28.3.)

Figure 28.43 *Problem P28-18*

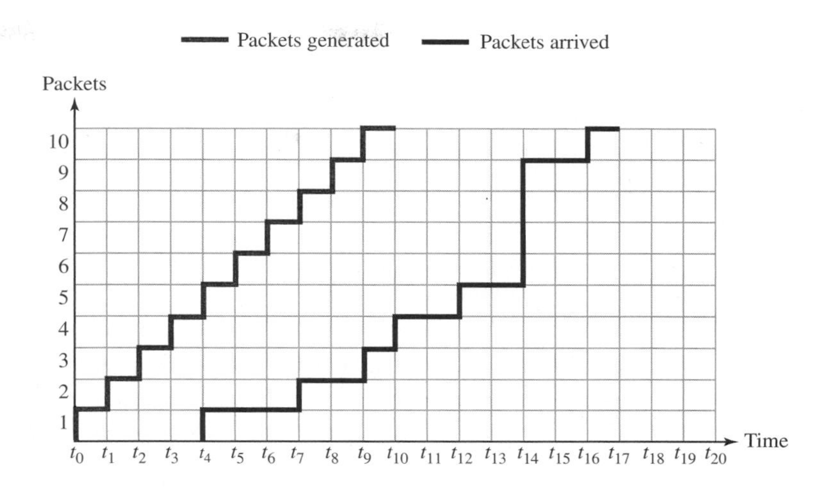

P28-22. In one-dimensional DCT, if $N = 1$, the matrix transformation is changed to simple multiplication. In other words, $M = T \times p$, in which T, p, and M are numbers (scalar value) instead of matrices. What is the value of T in this case?

P28-23. In Figure 28.31, what is the amount of data in the playback buffer at each of the following times?

a. 00:00:17 **b.** 00:00:20 **c.** 00:00:25 **d.** 00:00:30

P28-24. Given the following message, find the compressed data using the second version of run-length coding with the count expressed as a four-bit binary number.

1000000100000100000000000000010000001

P28-25. Explain why TCP, as a byte-oriented stream protocol, is not suitable for applications such as live or real-time multimedia streaming.

28.7 SIMULATION EXPERIMENTS

28.7.1 Applets

We have created some Java applets to show some of the main concepts discussed in this chapter. It is strongly recommended that the students activate these applets on the book website and carefully examine the protocols in action.

28.7.2 Lab Assignments

In this chapter, we use Wireshark to capture and study multimedia packets such as images, audio, and video packets.

Lab28-1. In this lab we need to examine the contents of a multimedia packet to find the encoding used by the sender.

Lab28-2. In this lab we need to examine the contents of a multimedia packet to find which transport-layer protocol is used for communication.

28.8 PROGRAMMING ASSIGNMENTS

Write the source code for, compile, and test the following programs in one of the programming languages of your choice:

Prg28-1. A program for coding and decoding the first version of the run-length compression method.

Prg28-2. A program for coding and decoding the second version of the run-length compression method.

Prg28-3. A program for encoding and decoding LZW compression.

Prg28-4. A program for encoding and decoding arithmetic compression.

Prg28-5. A program that reads a two-dimensional matrix of size $N \times N$ and writes the values using zigzag ordering described in the chapter.

Prg28-6. A program that finds the DCT transform of a one-dimensional matrix. Use matrix multiplication.

Prg28-7. A program that finds the DCT transform of a two-dimensional matrix. Use matrix multiplication.

CHAPTER 29

Peer-to-Peer Paradigm

After discussing the client-server paradigm in the last few chapters, we now concentrate on the new paradigm, peer-to-peer, in this chapter. Peer-to-peer applications are on the rise in the Internet today. In this chapter, the peer-to-peer paradigm is discussed in five sections.

- The first section introduces the peer-to-peer paradigm. The section first describes the general idea behind P2P networks. It then describes the distributed hash table (DHT) as a mathematical concept for routing in a P2P network.
- The second section introduces Chord, one of the P2P networks that uses DHT. The section first describes the identifier space used in this type of network. It then shows how finger tables serve as forwarding tables in this type of network. The section then discusses the procedures used in Chord.
- The third section introduces Pastry, another P2P network that uses DHT. The section first describes how Pastry, like Chord, uses DHT. It then shows how routing tables and leaf sets can be used in Pastry to answer queries.
- The fourth section introduces Kademlia, another P2P network that uses DHT. The section describes how distances between identifiers are determined in Kademlia using the XOR operation. The section then explains routing tables and K-buckets used in Kademlia for routing.
- The fifth section discusses BitTorrent, a popular P2P network used for file sharing. The section explains two versions of this network: BitTorrent with a tracker and the trackerless BitTorrent. The second uses the concept described in Kademlia.

29.1 INTRODUCTION

We discussed the client-server paradigm as well as some standard client-server applications in previous chapters. In this chapter, we discuss the peer-to-peer paradigm. The first instance of peer-to-peer file sharing goes back to December 1987 when Wayne Bell created *WWIVnet*, the network component of WWIV (World War Four) bulletin board software. In July 1999, Ian Clarke designed *Freenet*, a decentralized, censorship-resistant distributed data store, aimed at providing freedom of speech through a peer-to-peer network with strong protection of anonymity.

Peer-to-peer gained popularity with Napster (1999–2001), an online music file sharing service created by Shawn Fanning. Although free copying and distributing of music files by the users led to a copyright violation lawsuit against Napster, and eventually closing of the service, it paved the way for peer-to-peer file-distribution models that came later. Gnutella had its first release in March 2000. It was followed by FastTrack (used by Kazaa), BitTorrent, WinMX, and GNUnet in March, April, May, and November of 2001 respectively.

29.1.1 P2P Networks

Internet users that are ready to share their resources become peers and form a network. When a peer in the network has a file (for example, an audio or video file) to share, it makes it available to the rest of the peers. An interested peer can connect itself to the computer where the file is stored and download it. After a peer downloads a file, it can make it available for other peers to download. As more peers join and download that file, more copies of the file become available to the group. Since lists of peers may grow and shrink, the question is how the paradigm keeps track of loyal peers and the location of the files. To answer this question, we first need to divide the P2P networks into two categories: centralized and decentralized.

Centralized Networks

In a centralized P2P network, the directory system—listing of the peers and what they offer—uses the client-server paradigm, but the storing and downloading of the files are done using the peer-to-peer paradigm. For this reason, a centralized P2P network is sometimes referred to as a hybrid P2P network. Napster was an example of a centralized P2P. In this type of network, a peer first registers itself with a central server. The peer then provides its IP address and a list of files it has to share. To avoid system collapse, Napster used several servers for this purpose, but we show only one in Figure 29.1.

A peer, looking for a particular file, sends a query to a central server. The server searches its directory and responds with the IP addresses of nodes that have a copy of the file. The peer contacts one of the nodes and downloads the file. The directory is constantly updated as nodes join or leave the peer.

Centralized networks make the maintenance of the directory simple but have several drawbacks. Accessing the directory can generate huge traffic and slow down the system. The central servers are vulnerable to attack, and if all of them fail, the whole system goes down. The central component of the system was ultimately responsible for

Figure 29.1 *Centralized network*

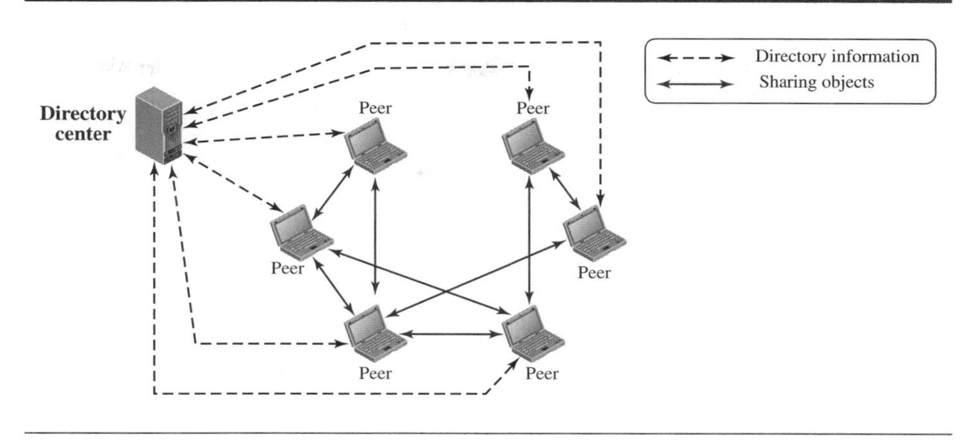

Napster's losing the copyright lawsuit and for its closure in July 2001. Roxio brought back the New Napster in 2003; Napster version 2 is now a legal, pay-for-music site.

Decentralized Network

A decentralized P2P network does not depend on a centralized directory system. In this model, peers arrange themselves into an *overlay network,* which is a logical network made on top of the physical network. Depending on how the nodes in the overlay network are linked, a decentralized P2P network is classified as either unstructured or structured.

Unstructured Networks

In an unstructured P2P network, the nodes are linked randomly. A search in an unstructured P2P is not very efficient because a query to find a file must be flooded through the network, which produces significant traffic and still the query may not be resolved. Two examples of this type of network are Gnutella and Freenet. We next discuss the Gnutella network as an example.

Gnutella The Gnutella network is an example of a peer-to-peer network that is decentralized but unstructured. It is unstructured in the sense that the directory is randomly distributed between nodes. When node A wants to access an object (such as a file), it contacts one of its neighbors. A neighbor, in this case, is any node whose address is known to node A. Node A sends a *query* message to the neighbor, node W. The query includes the identity of the object (for example, file name). If node W knows the address of node X, which has the object, it sends a *response* message that includes the address of node X. Node A now can use the commands defined in a transfer protocol such as HTTP to get a copy of the object from node X. If node W does not know the address of node X, it *floods* the request from A to all its neighbors. Eventually one of the nodes in the network responds to the *query* message, and node A can get access to node X. We discussed flooding in Chapter 20 when we discussed routing protocols, but it is worth mentioning here that although flooding in Gnutella is somehow controlled to prevent huge traffic loads, one of the reasons that Gnutella cannot be scaled well is the flooding.

One of the questions that remains to be answered is whether, according to the process described above, node A needs to know the address of at least one neighbor. This is done at the *bootstrap* time, when the node installs the Gnutella software for the first time. The software includes a list of nodes (peers) that node A can record as neighbors. Node A can later use the two messages, called *ping* and *pong,* to investigate whether or not a neighbor is still alive.

As mentioned before, one of the problems with the Gnutella network is the lack of scalability because of flooding. When the number of nodes increases, flooding becomes problematic. To make the query more efficient, the new version of Gnutella implemented a tiered system of *ultra nodes* and *leaves*. A node entering into the network is a leaf, not responsible for routing; nodes which are capable of routing are promoted to ultra nodes. This allows queries to propagate further and improves efficiency and scalability. Gnutella adopted a number of other techniques such as adding *Query Routing Protocol* (QRP) and *Dynamic Querying* (DQ) to reduce traffic overhead and make searches more efficient.

Structured Networks

A structured network uses a predefined set of rules to link nodes so that a query can be effectively and efficiently resolved. The most common technique used for this purpose is the *Distributed Hash Table (DHT).* DHT is used in many applications including Distributed Data Structure (DDS), Content Distributed Systems (CDS), Domain Name System (DNS), and P2P file sharing. One popular P2P file sharing protocol that uses the DHT is BitTorrent. We discuss DHT independently in the next section as a technique that can be used both in structured P2P networks and in other systems.

29.1.2 Distributed Hash Table (DHT)

A **Distributed Hash Table (DHT)** distributes data (or references to data) among a set of nodes according to some predefined rules. Each peer in a DHT-based network becomes responsible for a range of data items. To avoid the flooding overhead that we discussed for unstructured P2P networks, DHT-based networks allow each peer to have a partial knowledge about the whole network. This knowledge can be used to route the queries about the data items to the responsible nodes using effective and scalable procedures that we will discuss shortly.

Address Space

In a DHT-based network, each data item and the responsible peer is mapped to a point in a large address of size 2^m. The address space is designed using modular arithmetic, which means that we can think of points in the address space as distributed evenly on a circle with 2^m points (0 to $2^m - 1$) using clockwise direction as shown in Figure 29.2. Most of the DHT implementations use $m = 160$.

Figure 29.2 *Address space*

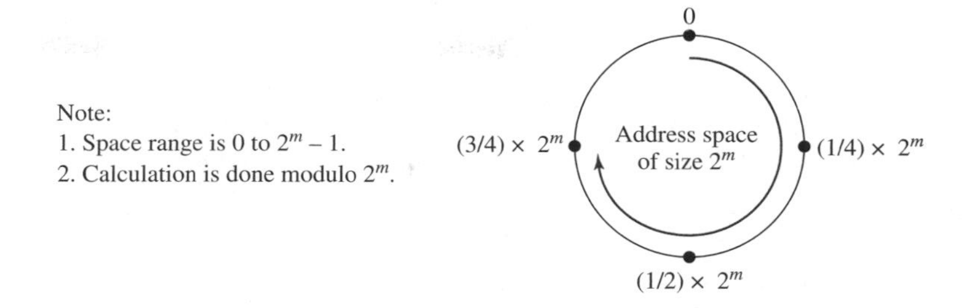

Hashing Peer Identifier

The first step in creating the DHT system is to place all peers on the address space ring. This is normally done by using a *hash* function that hashes the peer identifier, normally its IP address, to an m-bit integer, called a *node ID*.

node ID = hash (Peer IP address)

A hash function is a mathematical function that creates an output from an input. However, DHT uses some of the cryptographic hash functions such as Secure Hash Algorithm (SHA) that are collision resistant, which means that the probability of two inputs being mapped to the same output is very low. We discuss these hash algorithms in Chapter 31.

Hashing Object Identifier

The name of the object (for example, a file) to be shared is also hashed to an m-bit integer in the same address space. The result in DHT parlance is called a *key*.

key = hash (Object name)

In the DHT an object is normally related to the pair (key, value) in which the key is the hash of the object name and the value is the object or a reference to the object.

Storing the Object

There are two strategies for storing the object: the direct method and the indirect method. In the direct method, the object is stored in the node whose ID is somehow *closest* to the key in the ring. The term *closest* is defined differently in each protocol. This involves the object's most likely being transported from the computer that originally owned it. However, most DHT systems use the indirect method due to efficiency. The peer that owns the object keeps the object, but a reference to the object is created and stored in the node whose ID is closest to the key point. In other words, the physical object and the reference to the object are stored in two different locations. In the direct strategy, we create a relationship between the node ID that stores the object and the key of the object; in the indirect strategy, we create a relationship between the reference (pointer) to the object and the node that stores that reference. In either case, the relationship is needed to find the object if the name of the object is given. In the rest of the section, we use the indirect method.

Example 29.1

Although the normal value of *m* is 160, for the purpose of demonstration, we use m = 5 to make our examples tractable. In Figure 29.3, we assume that several peers have already joined the group. The node N5 with IP address 110.34.56.20 has a file named Liberty that it wants to share with its peers. The node makes a hash of the file name, "Liberty," to get the key = 14. Since the *closest* node to key 14 is node N17, N5 creates a reference to the file name (key), its IP address, and the port number (and possibly some other information about the file) and sends this reference to be stored in node N17. In other words, the file is stored in N5, the key of the file is k14 (a point in the DHT ring), but the reference to the file is stored in node N17. We will see later how other nodes can first find N17, extract the reference, and then use the reference to access the file Liberty. Our example shows only one key on the ring, but in an actual situation there are millions of keys and nodes in the ring.

Figure 29.3 *Example 29.1*

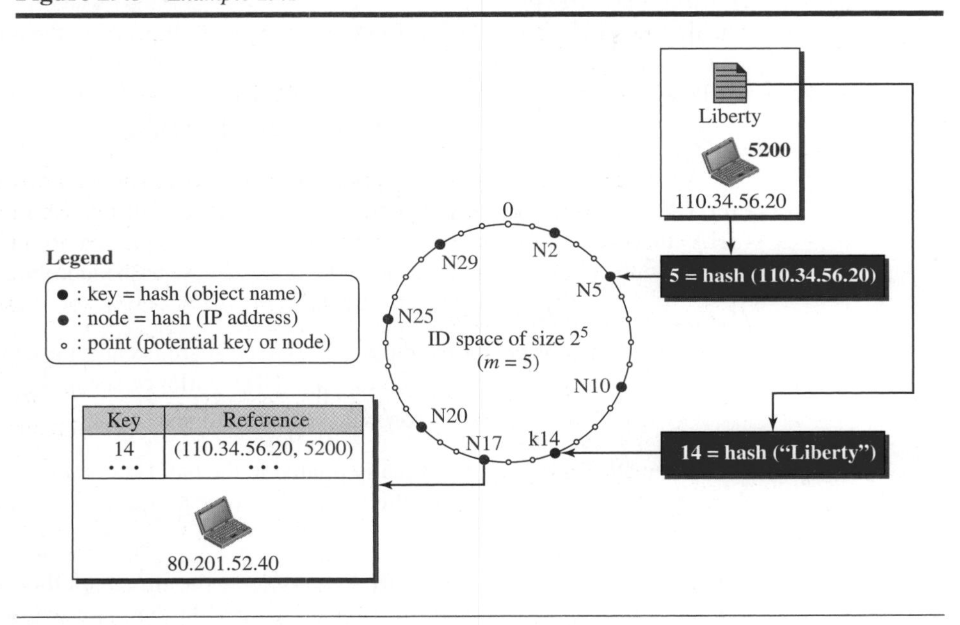

Routing

DHT's main function is to route a query to the node responsible for storing the reference to an object. Each DHT implementation uses a different strategy for routing, but all follow the idea that each node needs to have a partial knowledge about the ring to route a query to a node that is closest to the responsible node.

Arrival and Departure of Nodes

In a P2P network, each peer can be a desktop or a laptop computer, which can be turned on or off. When a computer peer launches the DHT software, it joins the network; when the computer is turned off or the peer closes the software, it leaves the network. A DHT implementation needs to have a clear and efficient strategy to handle arrival or departure of the nodes and the effect of this on the rest of the peers. Most DHT implementations treat the failure of a node as a departure.

29.2 CHORD

There are several protocols that implement DHT systems. In this section, we introduce three of these protocols: Chord, Pastry, and Kademlia. We have selected the Chord protocol for its simplicity and elegant approach to routing queries. Next we discuss the Pastry protocol because it uses a different approach than Chord and is very close in routing strategy to the Kademlia protocol, which is used in the most popular file-sharing network, BitTorrent.

Chord was published by Stoica et al. in 2001. We briefly discuss the main features of this algorithm here.

29.2.1 Identifier Space

Data items and nodes in Chord are m-bit identifiers that create an identifier space of size 2^m points distributed in a circle in the clockwise direction. We refer to the identifier of a data item as k (for *key*) and the identifier of a peer as N (for *node*). Arithmetic in the space is done modulo 2^m, which means that the identifiers are wrapped from $2^m - 1$ back to 0. Although some implementations use a collision-resistant hash function like SHA1 with $m = 160$, we use $m = 5$ in our discussion to make the discussion simpler. The closest peer with $N \geq k$ is called the successor of k and hosts the value (k, v), in which k is the key (hash of the data item) and v is the value (information about the peer server that has the object). In other words, a data item such as a file is stored in a peer that owns the data item, but the hash value of the data item, *key*, and the information about the peer, *value*, is stored as the pair (k, v) in the successor of k. This means that the peer that stores the data item and the peer that holds the pair (k, v) are not necessarily the same.

29.2.2 Finger Table

A node in the Chord algorithm should be able to resolve a query: given a key, the node should be able to find the node identifier responsible for that key or forward the query to another node. Forwarding, however, means that each node needs to have a routing table. Chord requires that each node knows about m successor nodes and one predecessor node. Each node creates a routing table, called a *finger table* by Chord, that looks like Table 29.1. Note that the target key at row i is $N + 2^{i-1}$.

Table 29.1 *Finger table*

i	*Target key*	*Successor of target key*	*Information about successor*
1	$N + 1$	Successor of $N + 1$	IP address and port of successor
2	$N + 2$	Successor of $N + 2$	IP address and port of successor
⋮	⋮	⋮	⋮
m	$N + 2^{m-1}$	Successor of $N + 2^{m-1}$	IP address and port of successor

Figure 29.4 shows only the successor column for a ring with few nodes and keys. Note that the first row ($i = 1$) actually gives the node successor. We have also added the predecessor node ID, which is needed, as we will see later.

Figure 29.4 *An example of a ring in Chord*

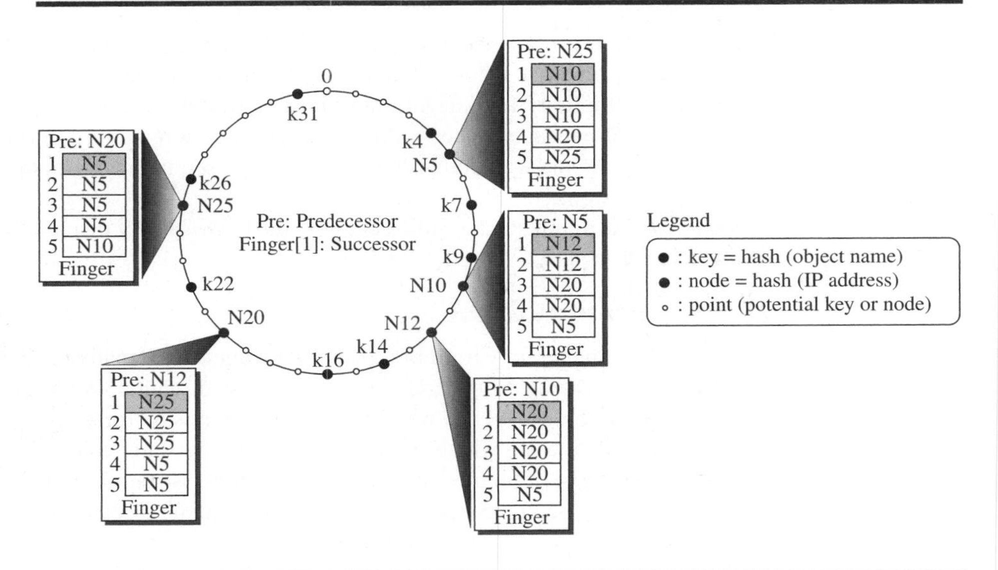

29.2.3 Interface

For its operation, Chord needs a set of operations referred to as the *Chord interface.* In this section, we discuss some of these operations to give an idea of what is behind the Chord protocol.

Lookup

Probably the mostly used operation in Chord is the lookup. Chord is designed to let peers share available services between themselves. To find the object to be shared, a peer needs to know the node that is responsible for that object: the peer that stores a reference to that object. We discussed that, in Chord, a peer that is the successor of a set of keys in the ring is the responsible peer for those keys. Finding the responsible node is actually finding the successor of a key. Table 29.2 shows the code for the *lookup* operation.

The lookup function is written using the top-down approach. If the node is responsible for the key, it returns its own ID; otherwise, it calls the function *find_successor.* The find_successor calls the function *find_predecessor.* The last function calls the function *find_closest_predecessor.* The modularity approach allows us to use the three functions in other operations instead of redefining them.

Table 29.2 *Lookup*

```
Lookup (key)
{
        if (node is responsible for the key)
                return (node's ID)
        else
                return find_successor (key)
}

find_successor (id)
{
        x = find_ predecessor (id)
        return x.finger[1]
}

find_predecessor (id)
{
        x = N                                              // N is the current node
        while (id ∉ (x, x.finger[1]]
        {
                x = x.find_closest_predecessor (id)               // Let x find it
        }
        return x
}

find_closest_predecessor (id)
{
        for (i = m downto 1)
        {
                if (finger [i] ∈ (N, id))                        //N is the current node
                        return (finger [i])
        }
        return N                                 //The node itself is closest predecessor
}
```

Let us elaborate on the lookup function. If the node is not responsible for the key, the lookup function calls the find_successor function to find the successor of the ID that is passed to it as the parameter. The coding of the successor function can be very simple if we first find the predecessor of the key. The predecessor node can easily help us to find the next node in the ring because the first finger of the predecessor node (finger [1]) gives us the ID of the successor node. In addition, having a function to find the predecessor node of a key can be useful in other functions that we will write later. Unfortunately, a node cannot normally find the predecessor of a key by itself; the key may be located far from the node. A node, as we discussed before, has a limited knowledge about the rest of the nodes; the finger table knows only about a maximum of m other nodes (there are some duplicates in the finger table). For this reason, the node needs the help of other nodes to find the predecessor of a key. This can be done using the find_closest_predecessor function as a *remote procedure call* (RPC). A remote procedure call involves calling a procedure to be executed at a remote node and returning the result to the calling node. We use the expression *x.procedure* in the algorithm in which *x* is the identity of the remote node and *procedure* is the procedure to be executed. The node uses this function to find another node that is closer to the predecessor node than itself. It then passes the duty of finding the predecessor node to the other node. In other words, if node A wants to find node X, it finds node B (closest predecessor) and passes the task to B. Now node B gets control and tries to find X, or passes the task to another node, C. The task is forwarded from node to node until the node that has the knowledge of the predecessor node finds it.

Example 29.2

Assume node N5 in Figure 29.4 needs to find the responsible node for key k14. Figure 29.5 shows the sequence of eight events. After event 4, in which the find_closest_predecessor function returns N10, the find_predecessor function asks N10 to return its finger[1], which is N12. At this moment, N5 finds out that N10 is not the predecessor of k14. Node N5 then asks N10 to find the closest predecessor of k14, which is returned as N12 (events 5 and 6). Node N5 now asks for the finger[1] of node N12, which is returned as N20. Node N5 now checks and sees that N12 is in fact the predecessor of k14. This information is passed to the find_successor function (event 7). N5 now asks for the finger [1] of node N12, which is returned as N20. The search is terminated and N20 is the successor of k14.

Figure 29.5 *Example 29.2*

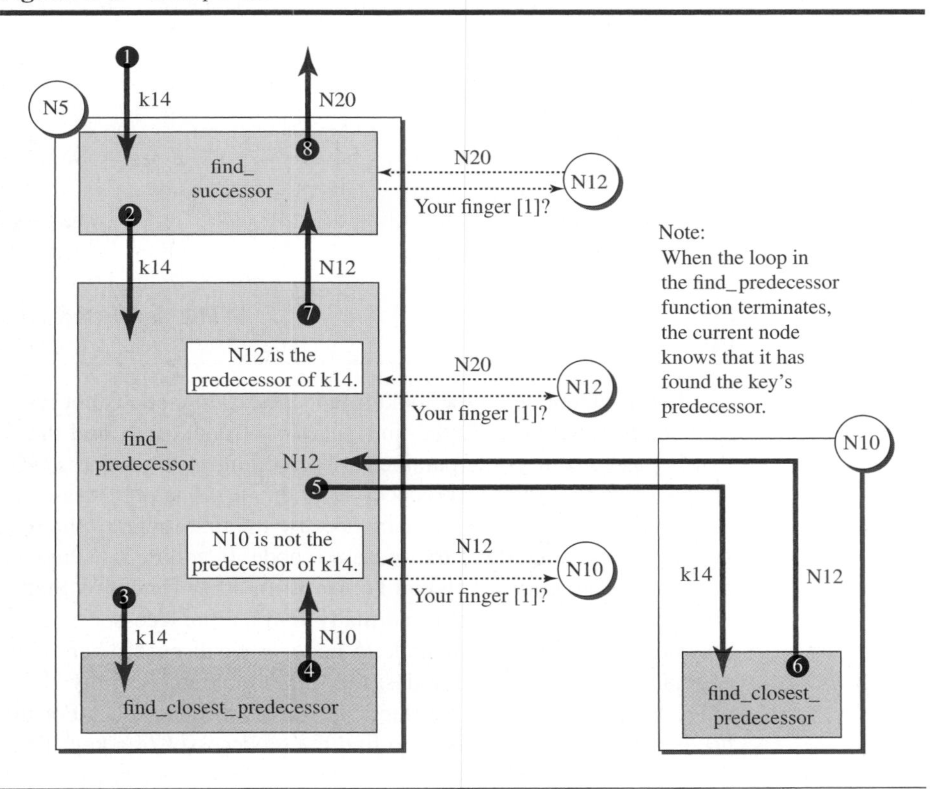

Stabilize

Before we discuss how nodes join and leave a ring, we need to emphasize that any change in the ring (such as joining and arriving of a node or a group of nodes) may destabilize the ring. One of the operations defined in Chord is called *stabilize*. Each node in the ring periodically uses this operation to validate its information about its successor and let the successor validate its information about its predecessor. Node N uses the value of finger[1], S, to ask node S to return its predecessor, P. If the return value, P, from this query, is between N and S, this means that there is a node with ID equals P that lies between N and S. Then node N makes P its successor and notifies P to make node N its predecessor. Table 29.3 shows the stabilize operation.

Table 29.3 *Stabilize*

```
Stabilize ()
{
        P = finger[1].Pre                      //Ask the successor to return its predecessor
        if(P ∈ (N, finger[1]))  finger[1] = P      // P is the possible successor of N
        finger[1].notify (N)                      // Notify P to change its predecessor
}
Notify (x)
{
            if (Pre = null or x ∈ (Pre, N))     Pre = x
}
```

Fix_Finger

Destabilization may change the finger table of up to m nodes. Another operation defined in Chord is called *fix_finger*. Each node in the ring must periodically call this function to maintain its finger table update. To avoid traffic on the system, each node must only update one of its fingers in each call. This finger is chosen randomly. Table 29.4 shows the code for this operation.

Table 29.4 *Fix_Finger*

```
Fix_Finger ()
{
        Generate (i ∈ (1, m])                   //Randomly generate i such that 1< i ≤ m
        finger[i] = find_successor (N + 2^(i − 1))            // Find value of finger[i]
}
```

Join

When a peer joins the ring, it uses the *join* operation and known ID of another peer to find its successor and set its predecessor to null. It immediately calls the stabilize function to validate its successor. The node then asks the successor to call the *move-key* function that transfers the keys that the new peer is responsible for. Table 29.5 shows the code for this operation.

Table 29.5 *Join*

```
Join (x)
{
        Initialize (x)
        finger[1].Move_Keys (N)
}
Initialize (x)
{
        Pre = null
        if (x = null) finger[1] = N
        else finger[1] = x. Find_Successor (N)
}

Move_Keys (x)
{
        for (each key k)
        {
                if (x ∈ [k, N)) move (k to node x)          // N is the current node
        }
}
```

It is obvious that after this operation, the finger table of the joined node is empty and the finger table of up to m predecessors is out of date. The stabilize and the fix-finger operations that run periodically after this event will gradually stabilize the system.

Example 29.3

We assume that node N17 joins the ring in Figure 29.4 with the help of N5. Figure 29.6 shows the ring after the ring has been stabilized. The following shows the process:

1. N17 uses Initialize (5) algorithm to set its predecessor to null and its successor (finger[1]) to N20.
2. N17 then asks N20 to send k14 and k16 to N17 because N17 is now responsible for these keys.
3. In the next time-out, N17 uses stabilize operation to validate its own successor (which is N20) and asks N20 to change its predecessor to N17 (using notify function).
4. When N12 uses stabilize, the predecessor of N17 is updated to N12.
5. Finally, when some nodes use fix-finger function, the finger table of nodes N17, N10, N5, and N12 is changed.

Leave or Fail

If a peer leaves the ring or the peer [not the ring] fails, the operation of the ring will be disrupted unless the ring stabilizes itself. Each node exchanges ping and pong messages with neighbors to find out if they are alive. When a node does not receive a pong message in response to its ping message, the node knows that the neighbor is dead.

Although the use of *stabilize* and *fix-finger* operations may restore the ring after a leave or failure, the node that detects the problem can immediately launch these operations

Figure 29.6 *Example 29.3*

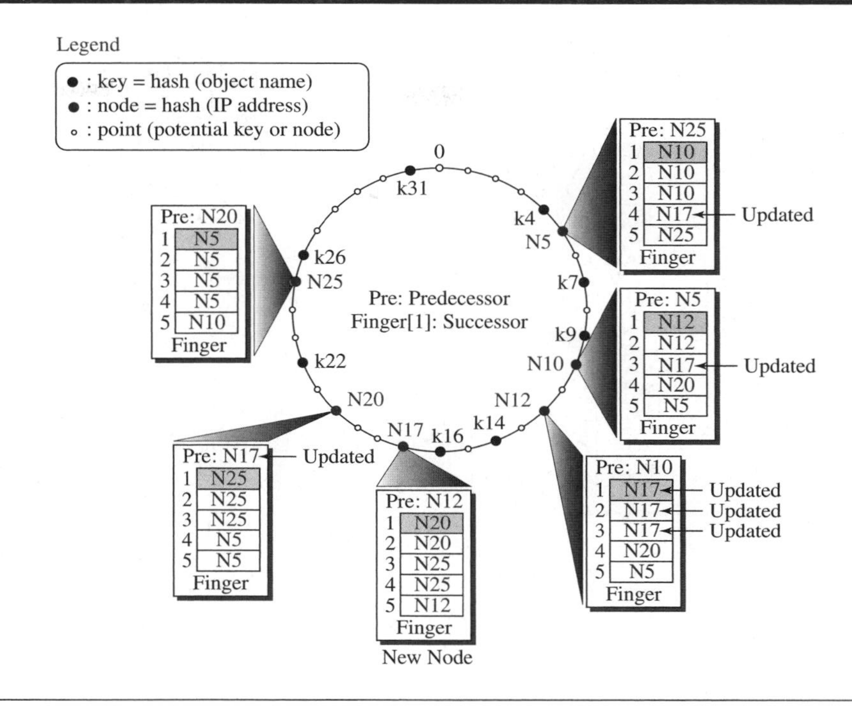

without waiting for the time-out. One important issue is that the stabilize and fix-finger operations may not work properly if several nodes leave or fail at the same time. For this reason, Chord requires that each node keep track of *r* successors (the value of *r* depends on the implementation). The node can always go to the next successor if the previous ones are not available.

Another issue in this case is that the data managed by the node that left or failed is no longer available. Chord stipulates that only one node should be responsible for a set of data and references, but Chord also requires that data and references be duplicated on other nodes in this case.

Example 29.4

We assume that a node, N10, leaves the ring in Figure 29.6. Figure 29.7 shows the ring after it has been stabilized.

The following shows the process:

1. Node N5 finds out about N10's departure when it does not receive a pong message to its ping message. Node N5 changes its successor (finger[1]) to N12 (the second in the list of successors).
2. Node N5 immediately launches the stabilize function and asks N12 to change its predecessor to N5.
3. Hopefully, k7 and k9, which were under the responsibility of N10, have been duplicated in N12 before the departure of N10.

Figure 29.7 *Example 29.4*

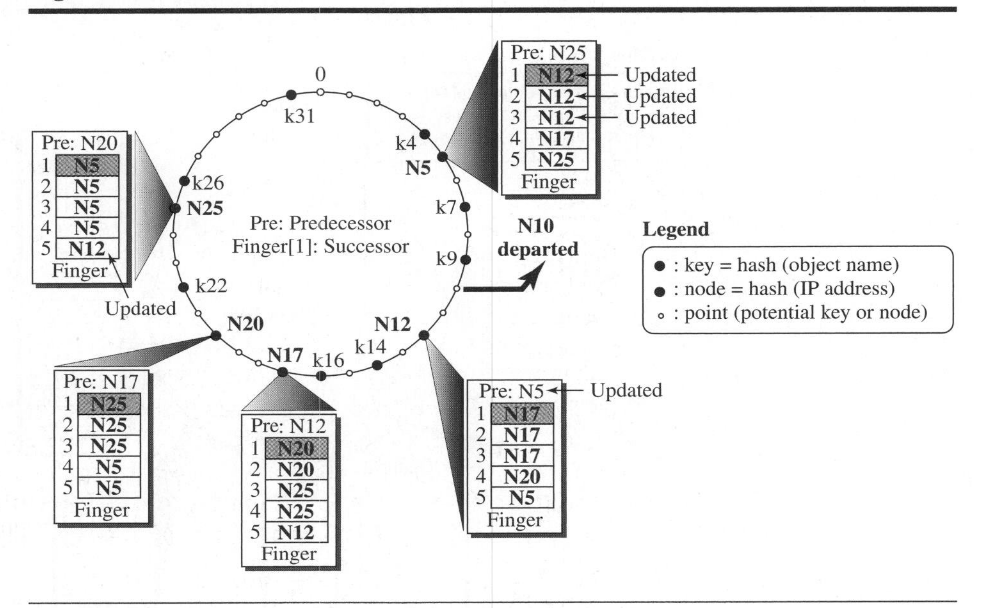

4. After a few calls of fix-finger, nodes N5 and N25 update their finger tables as shown in the figure.

29.2.4 Applications

Chord is used in several applications including Collaborative File System (CFS), Con-Chord, and Distributive Domain Name System (DDNS).

29.3 PASTRY

Another popular protocol in the P2P paradigm is **Pastry**, designed by Rowstron and Druschel. Pastry uses DHT, as described before, but there are some fundamental differences between Pastry and Chord in the identifier space and routing process that we describe next.

29.3.1 Identifier Space

In Pastry, like Chord, nodes and data items are m-bit identifiers that create an identifier space of 2^m points distributed uniformly on a circle in the clockwise direction. The common value for m is 128. The protocol uses the SHA-1 hashing algorithm with $m = 128$. However, in this protocol, an identifier is seen as an n-digit string in base 2^b in which b is normally 4 and $n = (m/b)$. In other words, an identifier is a 32-digit number in base 16 (hexadecimal). In this identifier space, a key is stored in the node whose identifier is numerically closest to the key. This strategy is definitively different from the one used by Chord. In Chord, a key is stored in its successor node; in

Pastry, a key may be stored in its successor or predecessor node, the one which is numerically closest to the key.

29.3.2 Routing

A node in Pastry should be able to resolve a query; given a key, the node should be able to find the node identifier responsible for that key or forward the query to another node. Each node in Pastry uses two entities to do so: a *routing table* and a *leaf set*.

Routing Table

Pastry requires that each node keep a routing table with n rows and (2^b) columns. In general, when $m = 128$ and $b = 4$, we have 32 (128/4) rows and 16 columns ($2^{128} = 16^{32}$). In other words, we have a row for each digit in the identifier and a column for each hexadecimal value (0 to F). Table 29.6 shows the outline of the routing table for the general case. In the table for node N, the cell at row i and column j, Table $[i, j]$, gives the identifier of a node (if it exists) that shares the i leftmost digits with the identifier for N and its $(i+1)$th digit has a value of j. The first row, row 0, shows the list of live nodes whose identifiers have no common prefix with N. Row 1 shows a list of live nodes whose identifiers share the leftmost digit with the identifier of node N. Similarly row 31 shows the list of all live nodes that share the leftmost 31 digits with node N; only the last digit is different.

Table 29.6 *Routing table for a node in Pastry*

Common prefix length	*0*	*1*	*2*	*3*	*4*	*5*	*6*	*7*	*8*	*9*	*A*	*B*	*C*	*D*	*E*	*F*
0																
1																
⋮	⋮	⋮	⋮	⋮	⋮	⋮	⋮	⋮	⋮	⋮	⋮	⋮	⋮	⋮	⋮	⋮
31																

For example, if $N = (574A234B12E374A2001B23451EEE4BCD)_{16}$, then the value of the Table[2, D] can be the identifier of a node such as (**57D...**). Note that the leftmost two digits are 57, which are common with the first two digits of N, but the next digit is D, the value corresponding to the Dth column. If there are more nodes with the prefix 57D, the closest one, according to the *proximity metric,* is chosen, and its identifier is inserted in this cell. The proximity metric is a measurement of closeness determined by the application that uses the network. It can be based on the number of hops between the two nodes, the round-trip time between the two nodes, or other metrics.

Leaf Set

Another entity used in routing is a set of 2^b identifiers (the size of a row in the routing table) called the *leaf set*. Half of the set is a list of identifiers that are numerically smaller than the identifier of the current node; the other half is a list of identifiers that are numerically larger than the identifier of the current node. In other words, the leaf set gives the identifier of 2^{b-1} live nodes before the current node in the ring and the list of 2^{b-1} nodes after the current node in the ring.

Example 29.5

Let us assume that $m = 8$ bits and $b = 2$. This means that we have up to $2^m = 256$ identifiers, and each identifier has $m/b = 4$ digits in base $2^b = 4$. Figure 29.8 shows the situation in which there are some live nodes and some keys mapped to these nodes. The key k1213 is stored in two nodes because it is equidistant from them. This provides some redundancy that can be used if one of the nodes fails. The figure also shows the routing tables and leaf sets for four selected nodes that are used in the examples described later. In the routing table for node N0302, for example, we have chosen the node 1302 to be inserted in Table [0, 1] because we assumed that this node is closest to N0302 according to the proximity metric. We used the same strategy for other entries. Note that one cell in each row in each table is shaded because it corresponds to the digit of the node identifier; no node identifier can be inserted there. Some cells are also empty because there are no live nodes in the network at this moment to satisfy the requirement; when some new nodes join the network, they can be inserted in these cells.

Figure 29.8 *Example 29.5*

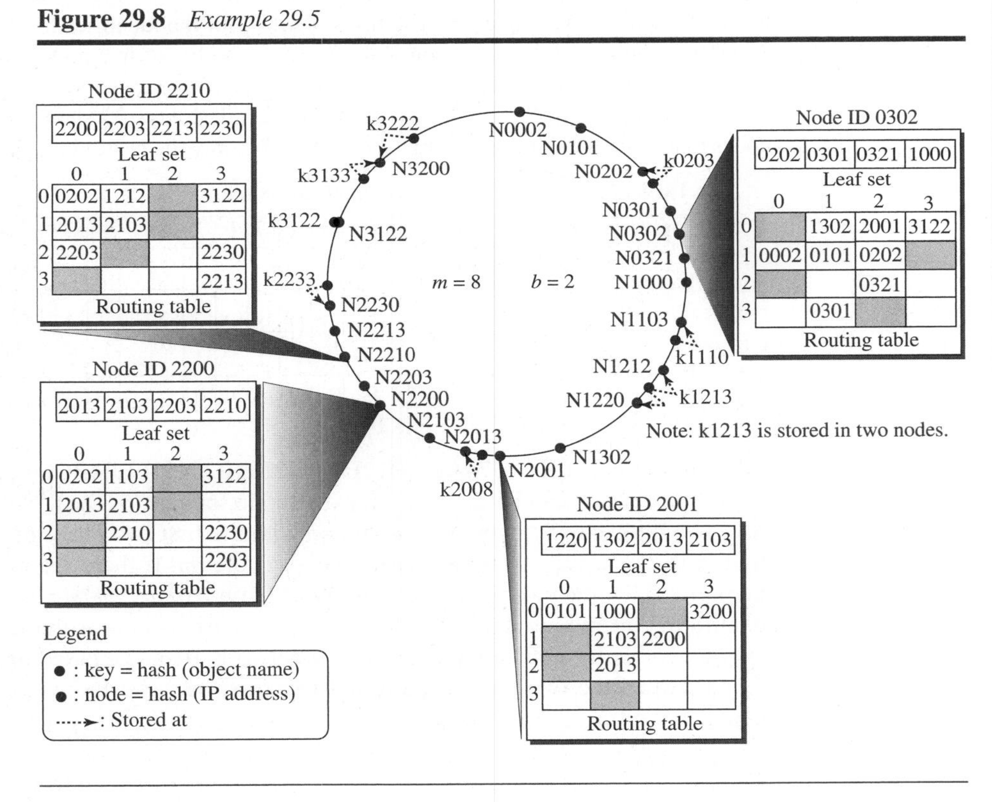

Lookup

As we discussed in Chord, one of the operations used in Pastry is lookup: given a key, we need to find the node that stores the information about the key or the key itself. Table 29.7 gives the lookup operation in pseudocode. In this algorithm, N is the identifier of the local node, the node that receives a message and needs to find the node that stores the key in the message.

Table 29.7 *Lookup*

```
Lookup (key)
{
        if (key is in the range of N's leaf set)
                forward the message to the closest node in the leaf set
        else
                route (key, Table)
}
route (key, Table)
{
        p = length of shared prefix between key and N
        v = value of the digit at position p of the key          // Position starts from 0
        if (Table [p, v] exists)
                forward the message to the node in Table [p, v]
        else
                forward the message to a node sharing a prefix as long as the current node, but
                numerically closer to the key.
}
```

Example 29.6

In Figure 29.8, we assume that node N2210 receives a query to find the node responsible for key 2008. Since this node is not responsible for this key, it first checks its leaf set. The key 2008 is not in the range of the leaf set, so the node needs to use its routing table. Since the length of the common prefix is 1, p = 1. The value of the digit at position 1 in the key is v = 0. The node checks the identifier in Table [1, 0], which is 2013. The query is forwarded to node 2013, which is actually responsible for the key. This node sends its information to the requesting node.

Example 29.7

In Figure 29.8, we assume that node N0302 receives a query to find the node responsible for the key 0203. This node is not responsible for this key, but the key is in the range of its leaf set. The closest node in this set is the node N0202. The query is sent to this node, which is actually responsible for this node. Node N0202 sends its information to the requesting node.

Join

The process of joining the ring in Pastry is simpler and faster than in Chord. The new node, X, should know at least one node N0, which should be close to X (based on the proximity metric); this can be done by running an algorithm called *Nearby Node Discovery.* Node X sends a *join message* to N0. In our discussion, we assume that N0's identifier has no common prefix with X's identifier. The following steps show how node X makes its routing table and leaf set:

1. Node N0 sends the contents of its row 0 to node X. Since the two nodes have no common prefix, node X uses the appropriate parts of this information to build its row 0. Node N0 then handles the joining message as a lookup message, assuming that the X identifier is a key. It forwards the join message to a node, N1, whose identifier is closest to X.

2. Node N1 sends the contents of its row 1 to node X. Since the two nodes have one common prefix, node X uses the appropriate parts of this information to build its row 1. Node N1 then handles the joining message as a lookup message, assuming that the X identifier is a key. It forwards the join message to a node, N2, whose identifier is closest to X.
3. The process continues until the routing table of node X is complete.
4. The last node in the process, which has the longest common prefix with X, also sends its leaf set to node X, which becomes the leaf set of X.
5. Node X then exchanges information with nodes in its routing table and leaf set to improve its own routing information and allow those nodes to updates theirs.

Example 29.8

Figure 29.9 shows how a new node X with node identifier N2212 uses the information in four nodes in Figure 29.8 to create its initial routing table and leaf set for joining the ring. Note that the contents of these two tables will become closer to what they should be in the updating process. In this example, we assume that node 0302 is a nearby node to node 2212 based on the proximity metric.

Figure 29.9 *Example 29.8*

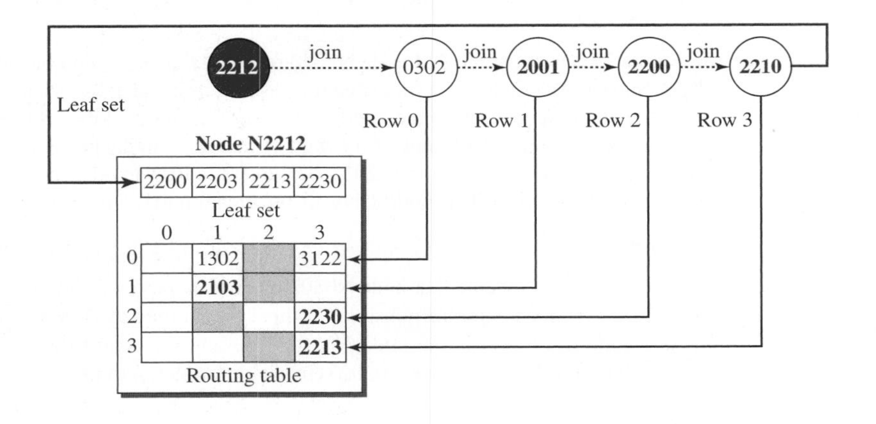

Leave or Fail

Each Pastry node periodically tests the liveliness of the nodes in its leaf set and routing table by exchanging probe messages. If a local node finds that a node in its leaf set is not responding to the probe message, it assumes that the node has failed or departed. To replace it in its leaf set, the local node contacts the live node in its leaf set with the largest identifier and repairs its leaf set with the information in the leaf set of that node. Since there is an overlap in the leaf set of close-by nodes, this process is successful.

If a local node finds that a node in its routing table, Table [i, j], is not responsive to the probe message, it sends a message to a live node in the same row and requests the identifier in Table [i, j] of that node. This identifier replaces the failed or departed node.

29.3.3 Application

Pastry is used in some applications including PAST, a distributed file system, and SCRIBE, a decentralized publish/subscribe system.

29.4 KADEMLIA

Another DHT peer-to-peer network is **Kademlia,** designed by Maymounkov and Mazières. Kademlia, like Pastry, routes messages based on the distance between nodes, but the interpretation of the distance metric in Kademlia is different from the one in Pastry, as we describe below. In this network, the distance between the two identifiers (nodes or keys) is measured as the bitwise exclusive-or (XOR) between them. In other words, if x and y are two identifiers, we have

$$\textbf{distance (x, y)} = \mathbf{x \oplus y}$$

The XOR metric has four properties we expect when we use geometric distances between two points, as shown below:

$x \oplus x = 0$	The distance between a node and itself is zero.
$x \oplus y > 0$ if $x \neq y$	The distance between any two distinct nodes is greater than zero.
$x \oplus y = y \oplus x$	The distance between x and y is the same as between y and x.
$x \oplus z \leq x \oplus y + y \oplus z$	Triangular relationship is satisfied.

29.4.1 Identifier Space

In Kademlia, nodes and data items are m-bit identifiers that create an identifier space of 2^m points distributed on the leaves of a binary tree. The protocol uses the SHA-1 hashing algorithm with $m = 160$.

Example 29.9

For simplicity, let us assume that $m = 4$. In this space, we can have 16 identifiers distributed on the leaves of a binary tree. Figure 29.10 shows the case with only eight live nodes and five keys.

As the figure shows, the key k3 is stored in N3 because $3 \oplus 3 = 0$. Although the key k7 looks numerically equidistant from N6 and N8, it is stored only in N6 because $6 \oplus 7 = 1$ but $6 \oplus 8 = 14$. Another interesting point is that the key k12 is numerically closer to N11, but it is stored in N15 because $11 \oplus 12 = 7$, but $15 \oplus 12 = 3$.

29.4.2 Routing Table

Kademlia keeps only one routing table for each node; there is no leaf set. Each node in the network divides the binary tree into m subtrees that do not include the node itself. Subtree i includes nodes that share i leftmost bits (common prefix) with the corresponding node. The routing table is made of m rows but only one column. In our discussion, we assume that each row holds the identifier of one of the nodes in the corresponding subtree, but later we show that Kademlia allows up to k nodes in each

Figure 29.10 *Example 29.9*

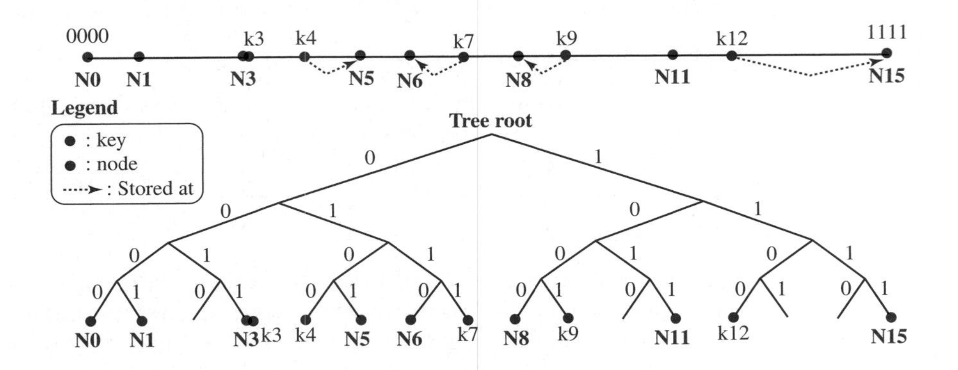

row. The idea is the same as that used by Pastry, but the length of the common prefix is based on the number of bits instead of the number of digits in base 2^b. Table 29.8 shows the routing table.

Table 29.8 *Routing table for a node in Kademlia*

Common prefix length	*Identifiers*
0	Closest node(s) in subtree with common prefix of length 0
1	Closest node(s) in subtree with common prefix of length 1
⋮	⋮
$m - 1$	Closest node(s) in subtree with common prefix of length $\boldsymbol{m} - 1$

Example 29.10

Let us find the routing table for Example 29.9. To make the example simple, we assume that each row uses only one identifier. Since m = 4, each node has four subtrees corresponding to four rows in the routing table. The identifier in each row represents the node that is closest to the current node in the corresponding subtree. Figure 29.11 shows all routing tables, but only three of the subtrees. We have chosen these three, out of eight, to make the figure smaller.

Let us explain how we made the routing table, for example, for Node 6, using the corresponding subtrees. The explanations for other nodes are similar.

a. In row 0, we need to insert the identifier of the closest node in the subtree with common prefix length $p = 0$. There are three nodes in this subtree (N8, N11, and N15), however, N15 is the closest to N6 because N6 ⊕ N8 = 14, N6 ⊕ N11 = 13, and N6 ⊕ N15 = 9. N15 is inserted in row 0.

b. In row 1, we need to insert the identifier of the closest node in the subtree with common prefix length $p = 1$. There are three nodes in this subtree (N0, N1, and N3), however, N3 is the closest to N6 because N6 ⊕ N0 = 6, N6 ⊕ N1 = 7, and N6 ⊕ N3 = 5. N3 is inserted in row 1.

c. In row 2, we need to insert the identifier of the closest node in the subtree with common prefix length $p = 2$. There is only one node (N5) in this subtree, which is inserted there.

d. In row 3, we need to insert the identifier of the closest node in the subtree with common prefix length $p = 3$. There is no node in this subtree, so the row is empty.

Figure 29.11 *Example 29.10*

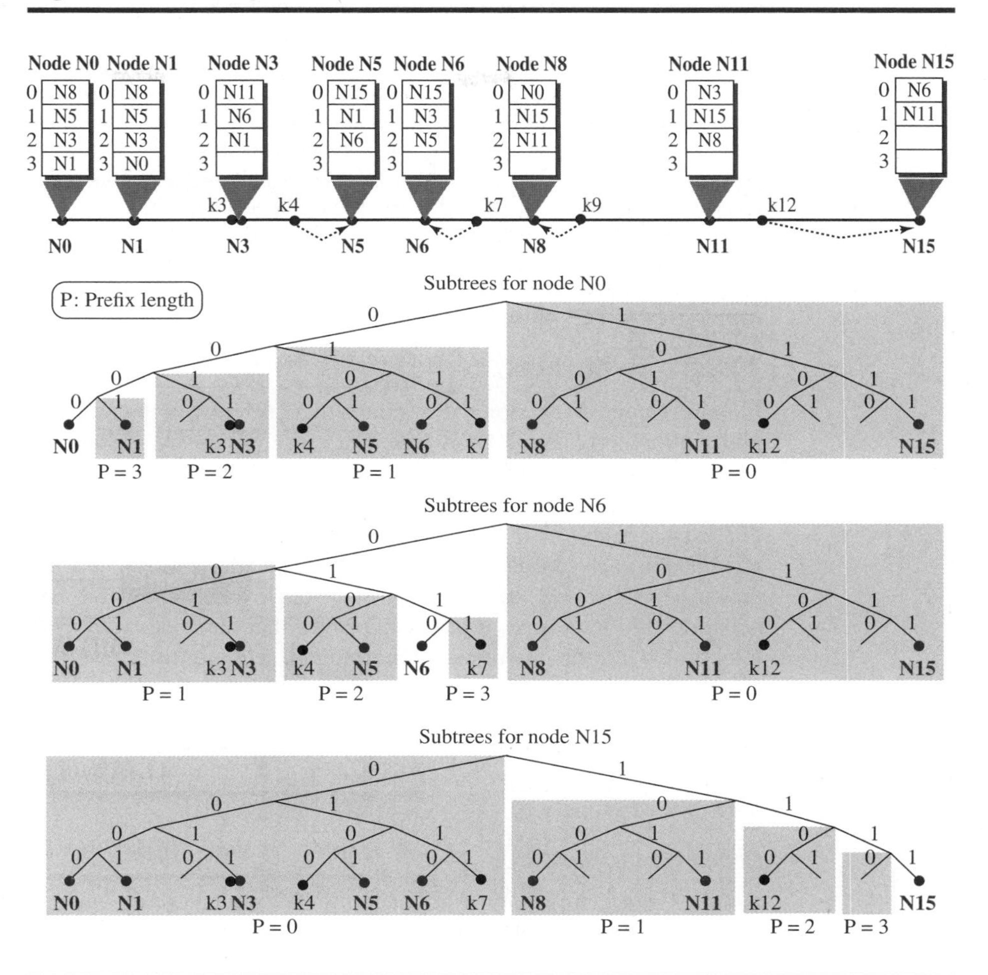

Example 29.11

In Figure 29.11, we assume node N0 $(0000)_2$ receives a lookup message to find the node responsible for k12 $(1100)_2$. The length of the common prefix between the two identifiers is 0. Node N0 sends the message to the node in row 0 of its routing table, node N8. Now node N8 $(1000)_2$ needs to look for the node closest to k12 $(1100)_2$. The length of the common prefix between the two identifiers is 1. Node N8 sends the message to the node in row 1 of its routing table, node N15, which is responsible for k12. The routing process is terminated. The route is N0 → N8 → N15. It is interesting to note that node N15, $(1111)_2$, and k12, $(1100)_2$, have a common prefix of length 2, but row 2 of N15 is empty, which means that N15 itself is responsible for k12.

Example 29.12

In Figure 29.11, we assume node N5 $(0101)_2$ receives a lookup message to find the node responsible for k7 $(0111)_2$. The length of the common prefix between the two identifiers is 2. Node N5 sends the message to the node in row 2 of its routing table, node N6, which is responsible for k7. The routing process is terminated. The route is N5 → N6.

Example 29.13

In Figure 29.11, we assume node N11 $(1011)_2$ receives a lookup message to find the node responsible for k4 $(0100)_2$. The length of the common prefix between the two identifiers is 0. Node N11 sends the message to the node in row 0 of its routing table, node N3. Now node N3 $(0011)_2$ needs to look for the node closest to k4 $(0100)_2$. The length of the common prefix between the two identifiers is 1. Node N3 sends the message to the node in row 1 of its routing table, node N6. Now node N6 $(0110)_2$ needs to look for the node closest to k4 $(0100)_2$. The length of the common prefix between the two identifiers is 2. Node N6 sends the message to the node in row 2 of its routing table, node N5, which is responsible for k4. The routing process is terminated. The route is N11 → N3 → N6 → N5.

29.4.3 K-Buckets

In our previous discussion, we assumed that each row in the routing table lists only one node in the corresponding subtree. For more efficiency, Kademlia requires that each row keeps at least up to *k* nodes from the corresponding subtree. The value of *k* is system independent, but for an actual network it is recommended that it be around 20. For this reason, each row in the routing table is referred to as a *k-bucket*. Having more than one node in each row allows the node to use an alternative node when a node leaves the network or fails. Kademlia keeps those nodes in a bucket that has been connected in the network for a long time. It has been proven that the nodes that remain connected for a long time will probably remain connected for a longer time.

Parallel Query

Since there are multiple nodes in a k-bucket, Kademlia allows sending α parallel queries to α nodes at the top of the k-bucket. This reduces the delay if a node fails and cannot answer the query.

Concurrent Updating

Another interesting feature in Kademlia is concurrent updating. Whenever a node receives a query or a response, it updates its k-bucket. If multiple queries to a node receive no response, the node that sent the query can remove the destination node from the corresponding k-bucket.

Join

As in Pastry, a node that needs to join the network needs to know at least one other node. The joining node sends its identifier to the node as though it is a key to be found. The response it receives allows the new node to create its k-buckets.

Leave or Fail

When a node leaves the network or fails, other nodes update their k-buckets using the concurrent process described before.

29.5 BITTORRENT

BitTorrent is a P2P protocol, designed by Bram Cohen, for sharing a large file among a set of peers. However, the term *sharing* in this context is different from other file-sharing protocols. Instead of one peer allowing another peer to download the whole file, a group of peers takes part in the process to give all peers in the group a copy of the file. File sharing is done in a collaborating process called a *torrent*. Each peer participating in a torrent downloads chunks of the large file from another peer that has it and uploads chunks of that file to other peers that do not have it, a kind of *tit-for-tat,* a trading game played by kids. The set of all peers that takes part in a torrent is referred to as a *swarm*. A peer in a swarm that has the complete content file is called a *seed*; a peer that has only part of the file and wants to download the rest is called a *leech*. In other words, a swarm is a combination of seeds and leeches. BitTorrent has gone through several versions and implementations. We first describe the original one, which uses a central node called a *tracker*. We then show how some new versions eliminate the tracker by using DHT.

29.5.1 BitTorrent with a Tracker

In the original BitTorrent, there is another entity in a torrent, called the tracker, which, as the name implies, tracks the operation of the swarm, as described later. Figure 29.12 shows an example of a torrent with seeds, leeches, and the tracker.

Figure 29.12 *Example of a torrent*

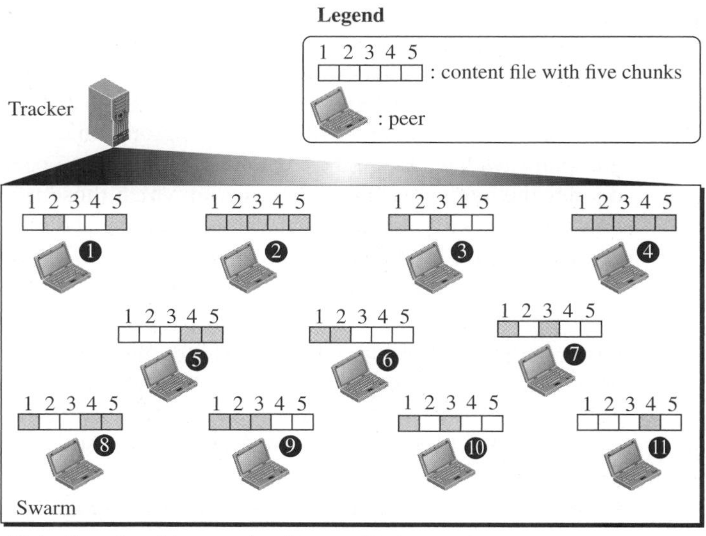

Note: Peers 2 and 4 are seeds; others are leeches.

In the figure, the file to be shared, the content file, is divided into five pieces (chunks). Peers 2 and 4 already have all the pieces; other peers have some pieces. The pieces that each peer has are shaded. Uploading and downloading of the pieces will continue. Some peers may leave the torrent; some new peers may join the torrent.

Now assume a new peer wants to download the same content file. The new peer accesses the BitTorrent server with the name of the content file. It receives a metafile, named the torrent file, that contains the information about the pieces in the content file and the address of the tracker that handles that specific torrent. The new peer now accesses the tracker and receives the addresses of some peers in the torrent, normally called *neighbors*. The new peer is now part of the torrent and can download and upload pieces of the content file. When it has all the pieces, it may leave the torrent or remain in the torrent to help other peers, including the new peers that have joined after it, to get all pieces of the content file. Nothing can prevent a peer from leaving the torrent before it has all the pieces and joining later or not joining again.

Although the process of joining, sharing, and leaving a torrent looks simple, the BitTorrent protocol applies a set of policies to provide fairness, to encourage the peers to exchange pieces with other peers, to prevent overloading a peer with requests from other peers, and to allow a peer to find peers that provide better service.

To avoid overloading a peer and to achieve fairness, each peer needs to limit its concurrent connection to a number of neighbors; the typical value is four. A peer flags a neighbor as unchoked or choked. It also flags them as interested or uninterested.

In other words, a peer divides its lists of neighbors into two distinct groups: *unchoked* and *choked*. It also divides them into *interested* and *uninterested* groups. The unchoked group is the list of peers that the current peer has concurrently connected to; it continuously uploads and downloads pieces from this group. The choked group is the list of neighbors that the peer is not currently connected to but may connect to in the future.

Every 10 seconds, the current peer tries a peer in the interested but choked group for a better data rate. If this new peer has a better rate than any of the unchoked peers, the new peer may become unchoked, and the peer with the lowest data rate in the unchoked group may move to the choked group. In this way, the peers in the unchoked group always have the highest data rate among those peers probed. Using this strategy divides the neighbors into subgroups in which those neighbors with compatible data transfer rates will communicate with each other. The idea of tit-for-tat trading strategy described above can be seen in this policy.

To allow a newly joined peer, which does not yet have a piece to share, to also receive pieces from other peers, every 30 seconds a peer randomly promotes a single interested peer, regardless of its uploading rate, from the choked group and flags it as unchoked. This action is called *optimistic unchoking*.

The BitTorrent protocol tries to provide a balance between the number of pieces each peer may have at each moment by using a strategy called the *rarest-first*. Using this strategy, a peer tries to first download the pieces with the fewest repeated copies among the neighbors. In this way, these pieces are circulated faster.

29.5.2 Trackerless BitTorrent

In the original BitTorrent design, if the tracker fails, new peers cannot connect to the network and updating is interrupted. There are several implementations of BitTorrent that eliminate the need for a centralized tracker. In the implementation that we describe here, the protocol still uses the tracker, but not a central one. The job of tracking is distributed among some nodes in the network. In this section, we show how Kademlia DHT can be used to achieve this goal, but we avoid becoming involved in the details of a specific protocol.

In BitTorrent with a central tracker, the job of the tracker is to provide the list of peers in a swarm when given a metadata file that defines the torrent. If we think of the hash function of metadata as the key and the hash function of the list of peers in a swarm as the value, we can let some nodes in a P2P network play the role of trackers. A new peer that joins the torrent sends the hash function of the metadata (key) to the node that it knows. The P2P network uses Kademlia protocol to find the node responsible for the key. The responsible node sends the *value,* which is actually the list of peers in the corresponding torrent, to the joining peer. Now the joining peer can use the BitTorrent protocol to share the content file with peers in the list.

29.6 END-CHAPTER MATERIALS

29.6.1 Recommended Reading

For more details about subjects discussed in this chapter, we recommend the following books. The items enclosed in brackets refer to the reference list at the end of the book.

Books

Several books give thorough coverage of materials discussed in this chapter including [Mir 07] and [Bar et al. 05].

29.6.2 Key Terms

Chord
Distributed Hash Table (DHT)
Kademlia
Pastry
peer-to-peer (P2P) paradigm

29.6.3 Summary

In a peer-to-peer network, Internet users that are ready to share their resources become peers and form a network. Peer-to-peer networks are divided into centralized and decentralized. In a centralized P2P network, the directory system uses the client-server paradigm, but the storing and downloading of the files are done using the peer-to-peer paradigm. In a decentralized network, both the directory system and storing and downloading of flies are done using the peer-to-peer paradigm. The main idea behind the decentralized P2P network is the Distributed Hash Table (DHT), a mathematical concept for routing in this type of network.

One of the P2P networks that uses DHT is Chord. The finger tables used for routing in this network are an excellent and innovative idea per se. Another P2P network is called Pastry, that like Chord, uses DHT. Pastry uses routing tables and leaf sets to route queries in the network. The third network we discussed in this chapter is called Kademlia, which uses a different approach than the previous two. It uses routing tables and k-buckets to route queries, but the concept of distance in this network is totally different from the other two. As a practical example, we mentioned BitTorrent, a popular P2P network used for file sharing. We discussed two versions of this network, BitTorrent with a tracker and the trackerless BitTorrent, that have been devised. The second uses the concept described in Kademlia.

29.7 PRACTICE SET

29.7.1 Quizzes

A set of interactive quizzes for this chapter can be found on the book website. It is strongly recommended that the student take the quizzes to check his/her understanding of the materials before continuing with the practice set.

29.7.2 Questions

Q29-1. Distinguish between centralized and decentralized P2P networks.

Q29-2. Distinguish between structured and unstructured decentralized P2P networks.

Q29-3. How many points can we have in a DHT with $m = 10$?

Q29-4. Explain the problems we encounter if we use IP addresses as the node identifiers in a DHT.

Q29-5. In a Kademlia network, the size of the identifier space is 1024. What is the height of the binary tree (the distance between the root and each leaf)? What is the number of leaves? What is the number of subtrees for each node? What is the number of rows in each routing table?

Q29-6. In a DHT-based network, assume $m = 4$. If the hash of a node identifier is 18, where is the location of the node in the DHT space?

Q29-7. In a DHT-based network, assume node 4 has a file with key 18. The closest next node to key 18 is node 20. Where is the file stored?

a. in the direct method   **b.** in the indirect method

Q29-8. In Kademlia, assume $m = 4$ and active nodes are N4, N7, and N12. Where is the key k3 stored in this system?

Q29-9. What are the advantages and disadvantages of a decentralized P2P network?

Q29-10. What are the advantages and disadvantages of a centralized P2P network?

Q29-11. Explain two strategies used for storing an object in DHT-based network.

Q29-12. In a Chord network, we have node N5 and key k5. Is N5 the predecessor of k5? Is N5 the successor of k5?

29.7.3 Problems

P29-1. In Chord, assume the size of the identifier space is 16. The active nodes are N3, N6, N8, and N12. Show the finger table (only the target-key and the successor column) for node N6.

P29-2. Repeat Example 29.2 in the text, but assume that node N5 needs to find the responsible node for key k16. Hint: Remember that interval checking needs to be done in modulo 32 arithmetic.

P29-3. In Pastry, assume the address space is 16 and that $b = 2$. How many digits are in an address space? List some of the identifiers.

P29-4. In a Pastry network with $m = 32$ and $b = 4$, what is the size of the routing table and the leaf set?

P29-5. Show the outline of a routing table for Pastry with address space of 16 and $b = 2$. Give some possible entries for each cell in the routing table of Node N21.

P29-6. Using the binary tree in Figure 29.10 in the text, show the subtree for node N11.

P29-7. Using the routing tables in Figure 29.11 in the text, explain and show the route if node N0 receives a lookup message for the node responsible for K12.

P29-8. In a Chord network with $m = 4$, node N2 has the following finger-table values: N4, N7, N10, and N12. For each of the following keys, first find if N2 is the predecessor of the key. If the answer is no, find which node (the closest predecessor) should be contacted to help N2 find the predecessor.

a. k1 **b.** k6 **c.** k9 **d.** k13

P29-9. In a Chord network using DHT with $m = 4$, draw the identifier space and place 4 peers with node ID addresses N3, N8, N11, and N13 and three keys with addresses k5, k9, and k14. Determine which node is responsible for each key. Create a finger table for each node.

P29-10. In a Pastry network using DHT, in which $m = 4$ and $b = 2$, draw the identifier space with four nodes, N02, N11, N20, and N23, and three keys, k00, k12, and k24. Determine which node is responsible for each key. Also show the leaf set and routing table for each node. Although it is unrealistic, assume that the proximity metric between each two nodes is based on numerical closeness.

P29-11. In Problem P29-10, answer the following questions:

a. Show how node N02 responds to a query to find the responsible node for k24.

b. Show how node N20 responds to a query to find the responsible node for k12.

P29-12. In a Kademlia network with $m = 4$, we have five active nodes: N2, N3, N7, N10, and N12. Find the routing table for each active node (with only one column).

P29-13. Repeat Example 29.2 in the text, but assume that node N12 needs to find the responsible node for key k7. Hint: Remember that interval checking needs to be done in modulo 32 arithmetic.

P29-14. In Chord, assume that the successor of node N12 is N17. Find whether node N12 is the predecessor of any of the following keys.

a. k12 **b.** k15 **c.** k17 **d.** k22

PART VII

Topics Related to All Layers

In the last part of the book, we have included topics that can be related to all layers in the Internet. Chapter 30 is about quality of service (QoS), which can be implemented in any layer. Chapters 31 and 32 define the security issues that are implemented at the top three layers in the Internet.

CHAPTER 30

Quality of Service

The Internet was originally designed for best-effort service without guarantee of predictable performance. Best-effort service is often sufficient for a traffic that is not sensitive to delay, such as file transfers and e-mail. Such a traffic is called *elastic* because it can stretch to work under delay conditions; it is also called *available bit rate* because applications can speed up or slow down according to the available bit rate.

The real-time traffic generated by some multimedia applications is delay sensitive and therefore requires guaranteed and predictable performance. **Quality of service** (**QoS**) is an internetworking issue that refers to a set of techniques and mechanisms that guarantee the performance of the network to deliver predictable service to an application program.

The chapter is divided into four sections.

- The first section defines data-flow characteristics: reliability, delay, jitter, and bandwidth. The section then reviews the sensitivity of several applications in relation to these characteristics. It then classifies applications with respect to the above characteristics.
- The second section concentrates on flow control to improve QoS. One way to do this is to do scheduling using first-in, first-out queuing, priority queuing, and weighted fair queuing. Another way is traffic shaping, which can be achieved using the leaky bucket or the token bucket technique. Resource reservation and admission control can also be used in this case.
- The third section discusses Integrated Services (IntServ), in which bandwidth is explicitly reserved for a given data flow. The section divides the required service into two categories: quantitative and qualitative. A separate protocol (RSVP) is used to provide a connection-oriented service for this purpose.
- The fourth section discusses Differentiated Services (DiffServ), in which packets are marked by applications into classes according to priorities. The section defines the DS field to mark the priority of the class. Finally, the section introduces traffic conditioners, used to implement DiffServ.

30.1 DATA-FLOW CHARACTERISTICS

If we want to provide quality of service for an Internet application, we first need to define what we need for each application. Traditionally, four types of characteristics are attributed to a flow: *reliability, delay, jitter,* and *bandwidth*. Let us first define these characteristics and then investigate the requirements of each application type.

30.1.1 Definitions

We can give informal definitions of the above four characteristics:

Reliability

Reliability is a characteristic that a flow needs in order to deliver the packets safe and sound to the destination. Lack of reliability means losing a packet or acknowledgment, which entails retransmission. However, the sensitivity of different application programs to reliability varies. For example, reliable transmission is more important for electronic mail, file transfer, and Internet access than for telephony or audio conferencing.

Delay

Source-to-destination delay is another flow characteristic. Again, applications can tolerate delay in different degrees. In this case, telephony, audio conferencing, video conferencing, and remote logging need minimum delay, while delay in file transfer or e-mail is less important.

Jitter

Jitter is the variation in delay for packets belonging to the same flow. For example, if four packets depart at times 0, 1, 2, 3 and arrive at 20, 21, 22, 23, all have the same delay, 20 units of time. On the other hand, if the above four packets arrive at 21, 23, 24, and 28, they will have different delays. For applications such as audio and video, the first case is completely acceptable; the second case is not. For these applications, it does not matter if the packets arrive with a short or long delay as long as the delay is the same for all packets. These types of applications do not tolerate jitter.

Bandwidth

Different applications need different bandwidths. In video conferencing we need to send millions of bits per second to refresh a color screen while the total number of bits in an e-mail may not reach even a million.

30.1.2 Sensitivity of Applications

Now let us see how various applications are sensitive to some flow characteristics. Table 30.1 gives a summary of application types and their sensitivity.

Table 30.1 *Sensitivity of applications to flow characteristics*

Application	*Reliability*	*Delay*	*Jitter*	*Bandwidth*
FTP	High	Low	Low	Medium
HTTP	High	Medium	Low	Medium
Audio-on-demand	Low	Low	High	Medium
Video-on-demand	Low	Low	High	High
Voice over IP	Low	High	High	Low
Video over IP	Low	High	High	High

For those applications with a high level of sensitivity to reliability, we need to do error checking and discard the packet if corrupted. For those applications with a high level of sensitivity to delay, we need to be sure that they are given priority in transmission. For those applications with a high level of sensitivity to jitter, we need to be sure that the packets belonging to the same application pass the network with the same delay. Finally, for those applications that require high bandwidth, we need to allocate enough bandwidth to be sure that the packets are not lost.

30.1.3 Flow Classes

Based on the flow characteristics, we can classify flows into groups, with each group having the required level of each characteristic. The Internet community has not yet defined such a classification formally. However, we know, for example, that a protocol like FTP needs a high level of reliability and probably a medium level of bandwidth, but the level of delay and jitter is not important for this protocol.

Example 30.1

Although the Internet has not defined flow classes formally, the ATM protocol does. As per ATM specifications, there are five classes of defined service.

a. ***Constant Bit Rate (CBR).*** This class is used for emulating circuit switching. CBR applications are quite sensitive to cell-delay variation. Examples of CBR are telephone traffic, video conferencing, and television.

b. ***Variable Bit Rate-Non Real Time (VBR-NRT).*** Users in this class can send traffic at a rate that varies with time depending on the availability of user information. An example is multimedia e-mail.

c. ***Variable Bit Rate-Real Time (VBR-RT).*** This class is similar to VBR–NRT but is designed for applications such as interactive compressed video that are sensitive to cell-delay variation.

d. ***Available Bit Rate (ABR).*** This class of ATM services provides rate-based flow control and is aimed at data traffic such as file transfer and e-mail.

e. ***Unspecified Bit Rate (UBR).*** This class includes all other classes and is widely used today for TCP/IP.

30.2 FLOW CONTROL TO IMPROVE QOS

Although formal classes of flow are not defined in the Internet, an IP datagram has a ToS field that can informally define the type of service required for a set of datagrams sent by an application. If we assign a certain type of application a single level of required service, we can then define some provisions for those levels of service. These can be done using several mechanisms.

30.2.1 Scheduling

Treating packets (datagrams) in the Internet based on their required level of service can mostly happen at the routers. It is at a router that a packet may be delayed, suffer from jitters, be lost, or be assigned the required bandwidth. A good scheduling technique treats the different flows in a fair and appropriate manner. Several scheduling techniques are designed to improve the quality of service. We discuss three of them here: *FIFO queuing, priority queuing,* and *weighted fair queuing.*

FIFO Queuing

In **first-in, first-out (FIFO) queuing,** packets wait in a buffer (queue) until the node (router) is ready to process them. If the average arrival rate is higher than the average processing rate, the queue will fill up and new packets will be discarded. A FIFO queue is familiar to those who have had to wait for a bus at a bus stop. Figure 30.1 shows a conceptual view of a FIFO queue. The figure also shows the timing relationship between arrival and departure of packets in this queuing. Packets from different applications (with different sizes) arrive at the queue, are processed, and depart. A larger packet definitely may need a longer processing time. In the figure, packets 1 and 2 need three time units of processing, but packet 3, which is smaller, needs two time units. This means that packets may arrive with some delays but depart with different delays. If the packets belong to the same application, this produces jitters. If the packets belong to different applications, this also produces jitters for each application.

Figure 30.1 *FIFO queue*

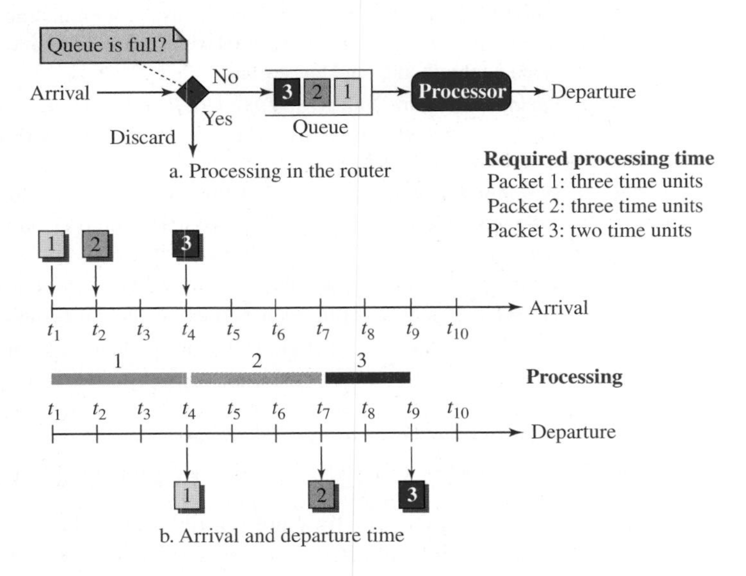

FIFO queuing is the default scheduling in the Internet. The only thing that is guaranteed in this type of queuing is that the packets depart in the order they arrive. Does FIFO queuing distinguish between packet classes? The answer is definitely no. This type of queuing is what we will see in the Internet with no service differentiation between packets from different sources. With FIFO queuing, all packets are treated the same in a packet-switched network. No matter if a packet belongs to FTP, or Voice over IP, or an e-mail message, they will be equally subject to loss, delay, and jitter. The bandwidth allocated for each application depends on how many packets arrive at the router in a period of time. If we need to provide different services to different classes of packets, we need to have other scheduling mechanisms.

Priority Queuing

Queuing delay in FIFO queuing often degrades quality of service in the network. A frame carrying real-time packets may have to wait a long time behind a frame carrying a small file. We solve this problem using multiple queues and priority queuing.

In **priority queuing,** packets are first assigned to a priority class. Each priority class has its own queue. The packets in the highest-priority queue are processed first. Packets in the lowest-priority queue are processed last. Note that the system does not stop serving a queue until it is empty. A packet priority is determined from a specific field in the packet header: the ToS field of an IPv4 header, the priority field of IPv6, a priority number assigned to a destination address, or a priority number assigned to an application (destination port number), and so on.

Figure 30.2 shows priority queuing with two priority levels (for simplicity). A priority queue can provide better QoS than the FIFO queue because higher-priority traffic, such as multimedia, can reach the destination with less delay. However, there is a potential drawback. If there is a continuous flow in a high-priority queue, the packets in the lower-priority queues will never have a chance to be processed. This is a condition called *starvation*. Severe starvation may result in dropping of some packets of lower priority. In the figure, the packets of higher priority are sent out before the packets of lower priority.

Weighted Fair Queuing

A better scheduling method is **weighted fair queuing.** In this technique, the packets are still assigned to different classes and admitted to different queues. The queues, however, are weighted based on the priority of the queues; higher priority means a higher weight. The system processes packets in each queue in a round-robin fashion with the number of packets selected from each queue based on the corresponding weight. For example, if the weights are 3, 2, and 1, three packets are processed from the first queue, two from the second queue, and one from the third queue. In this way, we have fair queuing with priority. Figure 30.3 shows the technique with three classes. In weighted fair queuing, each class may receive a small amount of time in each time period. In other words, a fraction of time is devoted to serve each class of packets, but the fraction depends on the priority of the class. For example, in the figure, if the throughput for the router is R, the class with the highest priority may have the throughput of R/2, the middle class may have the throughput of R/3, and the class with the lowest priority may have the throughput of R/6. However, this situation is true if all three classes have the same packet

Figure 30.2 *Priority queuing*

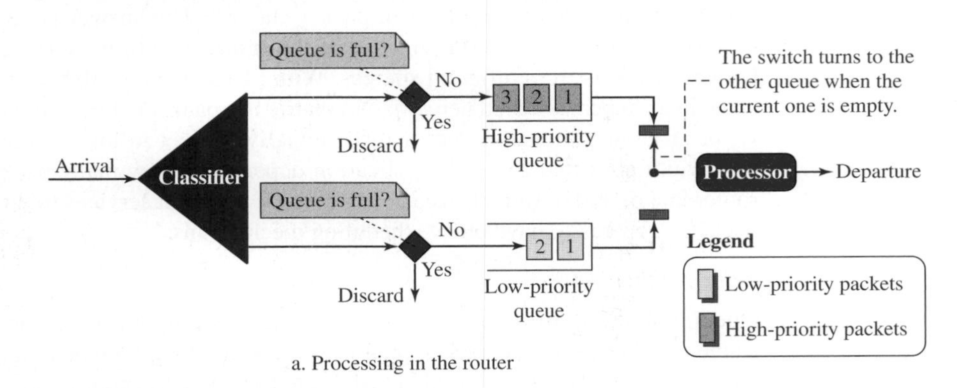

a. Processing in the router

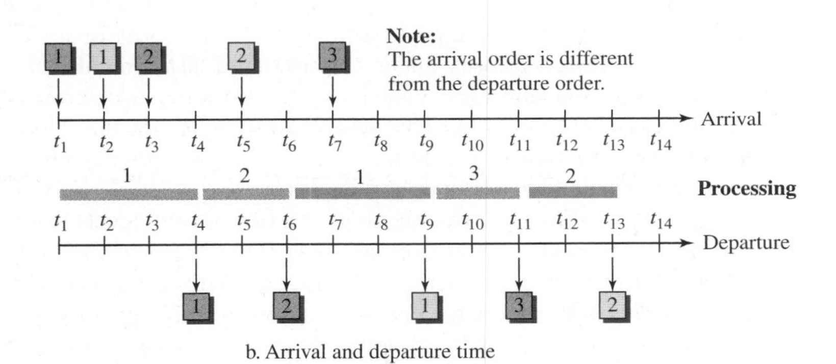

b. Arrival and departure time

size, which may not occur. Packets of different sizes may create many imbalances in dividing a decent share of time between different classes.

30.2.2 Traffic Shaping or Policing

To control the amount and the rate of traffic is called *traffic shaping* or *traffic policing*. The first term is used when the traffic leaves a network; the second term is used when the data enters the network. Two techniques can shape or police the traffic: leaky bucket and token bucket.

Leaky Bucket

If a bucket has a small hole at the bottom, the water leaks from the bucket at a constant rate as long as there is water in the bucket. The rate at which the water leaks does not depend on the rate at which the water is input unless the bucket is empty. If the bucket is full, the water overflows. The input rate can vary, but the output rate remains constant. Similarly, in networking, a technique called **leaky bucket** can smooth out bursty traffic. Bursty chunks are stored in the bucket and sent out at an average rate. Figure 30.4 shows a leaky bucket and its effects.

Figure 30.3 *Weighted fair queuing*

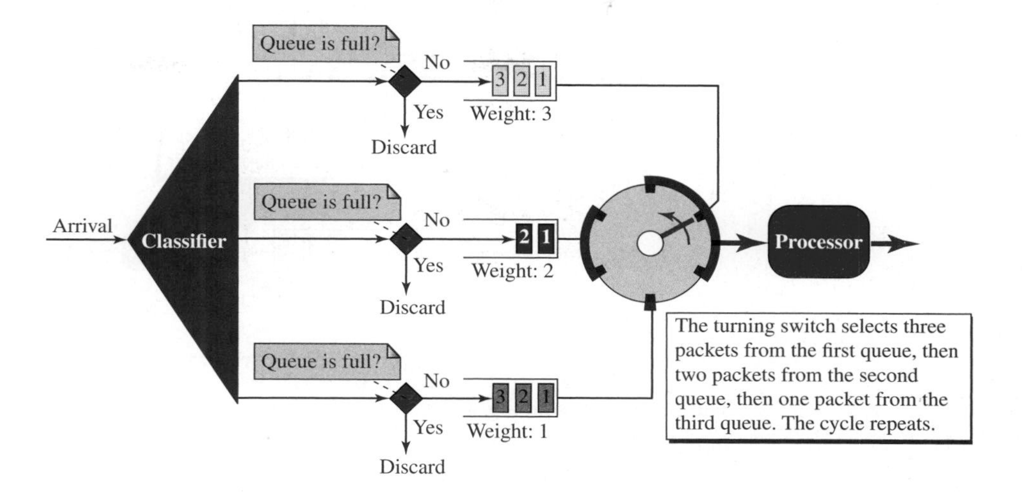

Figure 30.4 *Leaky bucket*

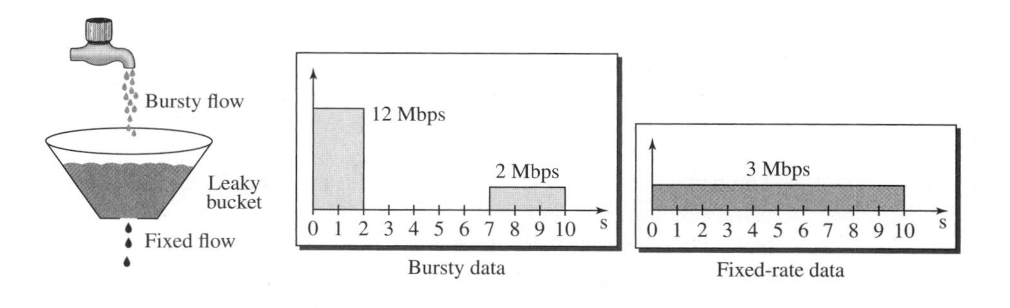

In the figure, we assume that the network has committed a bandwidth of 3 Mbps for a host. The use of the leaky bucket shapes the input traffic to make it conform to this commitment. In Figure 30.4 the host sends a burst of data at a rate of 12 Mbps for 2 seconds, for a total of 24 Mb of data. The host is silent for 5 seconds and then sends data at a rate of 2 Mbps for 3 seconds, for a total of 6 Mb of data. In all, the host has sent 30 Mb of data in 10 seconds. The leaky bucket smooths the traffic by sending out data at a rate of 3 Mbps during the same 10 seconds. Without the leaky bucket, the beginning burst may have hurt the network by consuming more bandwidth than is set aside for this host. We can also see that the leaky bucket may prevent congestion.

A simple leaky bucket implementation is shown in Figure 30.5. A FIFO queue holds the packets. If the traffic consists of fixed-size packets (e.g., cells in ATM networks), the process removes a fixed number of packets from the queue at each tick of the clock. If the traffic consists of variable-length packets, the fixed output rate must be based on the number of bytes or bits.

Figure 30.5 *Leaky bucket implementation*

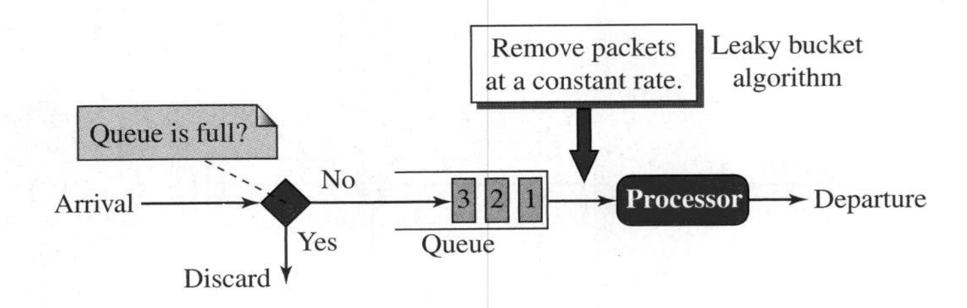

The following is an algorithm for variable-length packets:

1. Initialize a counter to n at the tick of the clock.
2. If n is greater than the size of the packet, send the packet and decrement the counter by the packet size. Repeat this step until the counter value is smaller than the packet size.
3. Reset the counter to n and go to step 1.

A leaky bucket algorithm shapes bursty traffic into fixed-rate traffic by averaging the data rate. It may drop the packets if the bucket is full.

Token Bucket

The leaky bucket is very restrictive. It does not credit an idle host. For example, if a host is not sending for a while, its bucket becomes empty. Now if the host has bursty data, the leaky bucket allows only an average rate. The time when the host was idle is not taken into account. On the other hand, the **token bucket** algorithm allows idle hosts to accumulate credit for the future in the form of tokens.

Assume the capacity of the bucket is c tokens and tokens enter the bucket at the rate of r tokens per second. The system removes one token for every cell of data sent. The maximum number of cells that can enter the network during any time interval of length t is shown below.

$$\textbf{Maximum number of packets} = r \times t + c$$

The maximum average rate for the token bucket is shown below.

$$\textbf{Maximum average rate} = (r \times t + c)/t \textbf{ packets per second}$$

This means that the token bucket limits the average packet rate to the network. Figure 30.6 shows the idea.

Figure 30.6 *Token bucket*

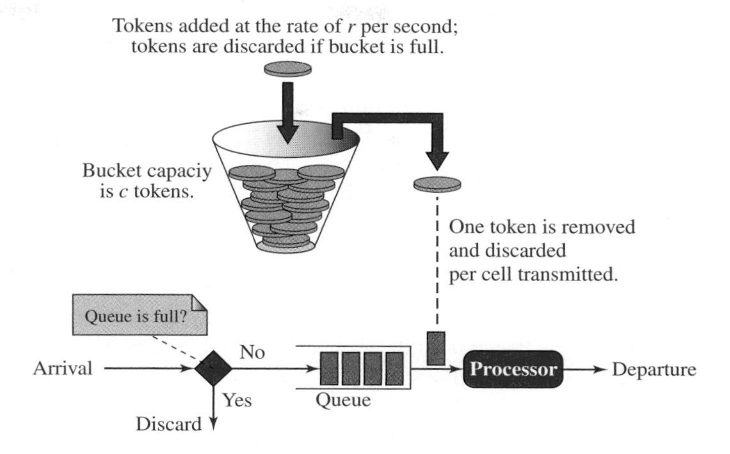

Example 30.2

Let's assume that the bucket capacity is 10,000 tokens and tokens are added at the rate of 1000 tokens per second. If the system is idle for 10 seconds (or more), the bucket collects 10,000 tokens and becomes full. Any additional tokens will be discarded. The maximum average rate is shown below.

$$\textbf{Maximum average rate} = (1000t + 10{,}000)/t$$

The token bucket can easily be implemented with a counter. The counter is initialized to zero. Each time a token is added, the counter is incremented by 1. Each time a unit of data is sent, the counter is decremented by 1. When the counter is zero, the host cannot send data.

The token bucket allows bursty traffic at a regulated maximum rate.

Combining Token Bucket and Leaky Bucket

The two techniques can be combined to credit an idle host and at the same time regulate the traffic. The leaky bucket is applied after the token bucket; the rate of the leaky bucket needs to be higher than the rate of tokens dropped in the bucket.

30.2.3 Resource Reservation

A flow of data needs resources such as a buffer, bandwidth, CPU time, and so on. The quality of service is improved if these resources are reserved beforehand. Below, we discuss a QoS model called *Integrated Services,* which depends heavily on resource reservation to improve the quality of service.

30.2.4 Admission Control

Admission control refers to the mechanism used by a router or a switch to accept or reject a flow based on predefined parameters called *flow specifications*. Before a router accepts a flow for processing, it checks the flow specifications to see if its capacity can handle the new flow. It takes into account bandwidth, buffer size, CPU speed, etc., as well as its previous commitments to other flows. Admission control in ATM networks is known as *Connection Admission Control* (CAC), which is a major part of the strategy for controlling congestion.

30.3 INTEGRATED SERVICES (INTSERV)

Traditional Internet provided only the best-effort delivery service to all users regardless of what was needed. Some applications, however, needed a minimum amount of band width to function (such as real-time audio and video). To provide different QoS for different applications, IETF (discussed in Chapter 1) developed the **Integrated Services (IntServ)** model. In this model, which is a *flow-based* architecture, resources such as bandwidth are explicitly reserved for a given data flow. In other words, the model is considered a specific requirement of an application in one particular case regardless of the application type (data transfer, or voice over IP, or video-on-demand). What is important are the resources the application needs, not what the application is doing.

The model is based on three schemes:

1. The packets are first classified according to the service they require.
2. The model uses scheduling to forward the packets according to their flow characteristics.
3. Devices like routers use *admission control* to determine if the device has the capability (available resources to handle the flow) before making a commitment. For example, if an application requires a very high data rate, but a router in the path cannot provide such a data rate, it denies the admission.

Before we discuss this model, we need to emphasize that the model is flow-based, which means that all accommodations need to be made before a flow can start. This implies that we need a connection-oriented service at the network layer. A connection establishment phase is needed to inform all routers of the requirement and get their approval (admission control). However, since IP is currently a connectionless protocol, we need another protocol to be run on top of IP to make it a connection-oriented protocol before we can use this model. This protocol is called ***Resource Reservation Protocol (RSVP)*** and will be discussed shortly.

Integrated Services is a flow-based QoS model designed for IP.
In this model packets are marked by routers according to flow characteristics.

30.3.1 Flow Specification

We said that IntServ is flow-based. To define a specific flow, a source needs to define a *flow specification*, which is made of two parts:

1. ***Rspec (resource specification).*** Rspec defines the resource that the flow needs to reserve (buffer, bandwidth, etc.).
2. ***Tspec (traffic specification).*** Tspec defines the traffic characterization of the flow, which we discussed before.

30.3.2 Admission

After a router receives the flow specification from an application, it decides to admit or deny the service. The decision is based on the previous commitments of the router and the current availability of the resource.

30.3.3 Service Classes

Two classes of services have been defined for Integrated Services: guaranteed service and controlled-load service.

Guaranteed Service Class

This type of service is designed for real-time traffic that needs a guaranteed minimum end-to-end delay. The end-to-end delay is the sum of the delays in the routers, the propagation delay in the media, and the setup mechanism. Only the first, the sum of the delays in the routers, can be guaranteed by the router. This type of service guarantees that the packets will arrive within a certain delivery time and are not discarded if flow traffic stays within the boundary of *Tspec*. We can say that guaranteed services are *quantitative services*, in which the amount of end-to-end delay and the data rate must be defined by the application. Normally guaranteed services are required for real-time applications (voice over IP).

Controlled-Load Service Class

This type of service is designed for applications that can accept some delays but are sensitive to an overloaded network and to the danger of losing packets. Good examples of these types of applications are file transfer, e-mail, and Internet access. The controlled-load service is a *qualitative service* in that the application requests the possibility of low-loss or no-loss packets.

30.3.4 Resource Reservation Protocol (RSVP)

We said that the Integrated Services model needs a connection-oriented network layer. Since IP is a connectionless protocol, a new protocol is designed to run on top of IP to make it connection-oriented. A connection-oriented protocol needs to have connection establishment and connection termination phases, as we discussed in Chapter 18. Before discussing RSVP, we need to mention that it is an independent protocol separate from the Integrated Services model. It may be used in other models in the future.

Multicast Trees

RSVP is different from other connection-oriented protocols in that it is based on multicast communication. However, RSVP can also be used for unicasting, because unicasting is just a special case of multicasting with only one member in the multicast group. The reason for this design is to enable RSVP to provide resource reservations for all kinds of traffic including multimedia, which often uses multicasting.

Receiver-Based Reservation

In RSVP, the receivers, not the sender, make the reservation. This strategy matches the other multicasting protocols. For example, in multicast routing protocols, the receivers, not the sender, make a decision to join or leave a multicast group.

RSVP Messages

RSVP has several types of messages. However, for our purposes, we discuss only two of them: *Path* and *Resv.*

Path Messages

Recall that the receivers in a flow make the reservation in RSVP. However, the receivers do not know the path traveled by packets before the reservation is made. The path is needed for the reservation. To solve the problem, RSVP uses Path messages. A Path message travels from the sender and reaches all receivers in the multicast path. On the way, a Path message stores the necessary information for the receivers. A Path message is sent in a multicast environment; a new message is created when the path diverges. Figure 30.7 shows path messages.

Figure 30.7 *Path messages*

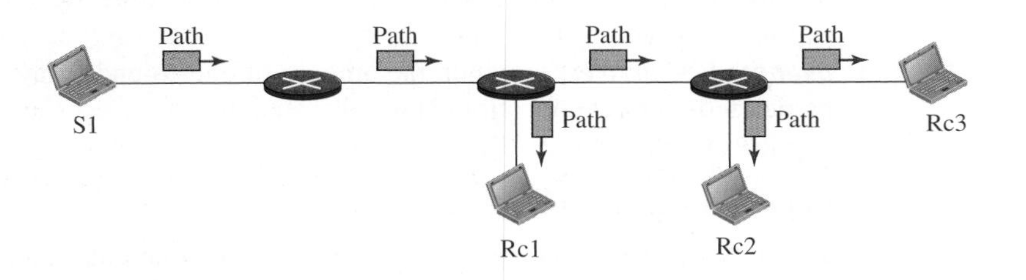

Resv Messages

After a receiver has received a Path message, it sends a Resv message. The Resv message travels toward the sender (upstream) and makes a resource reservation on the routers that support RSVP. If a router on the path does not support RSVP, it routes the packet based on the best-effort delivery methods we discussed before. Figure 30.8 shows the Resv messages.

Reservation Merging In RSVP, the resources are not reserved for each receiver in a flow; the reservation is merged. In Figure 30.9, Rc3 requests a 2-Mbps bandwidth while Rc2 requests a 1-Mbps bandwidth. Router R3, which needs to make a bandwidth reservation, merges the two requests. The reservation is made for 2 Mbps, the larger of the two, because a 2-Mbps input reservation can handle both requests. The same situation is true for R2. The reader may ask why Rc2 and Rc3, both belonging to a single flow, request different amounts of bandwidth. The answer is that, in a multimedia environment, different receivers may handle different grades of quality. For example, Rc2 may be able to receive video only at 1 Mbps (lower quality), while Rc3 may want to receive video at 2 Mbps (higher quality).

Figure 30.8 *Resv messages*

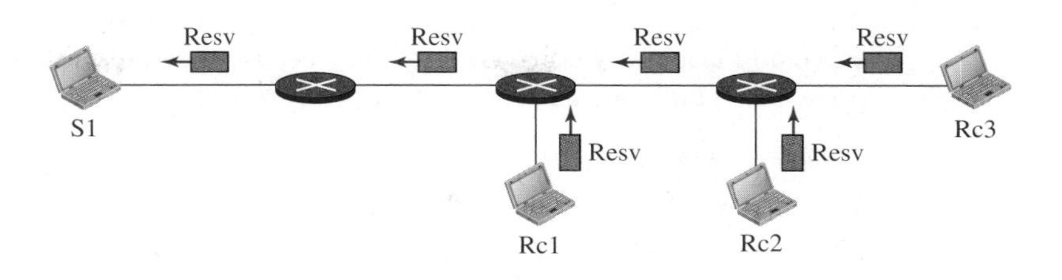

Figure 30.9 *Reservation merging*

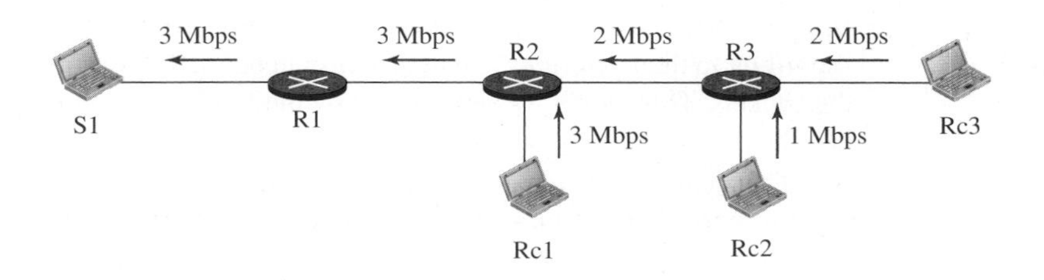

Reservation Styles When there is more than one flow, the router needs to make a reservation to accommodate all of them. RSVP defines three types of reservation styles: *wildcard filter* (WF), *fixed filter* (FF), and *shared explicit* (SE).

- ***Wild Card Filter Style.*** In this style, the router creates a single reservation for all senders. The reservation is based on the largest request. This type of style is used when the flows from different senders do not occur at the same time.
- ***Fixed Filter Style.*** In this style, the router creates a distinct reservation for each flow. This means that if there are *n* flows, *n* different reservations are made. This type of style is used when there is a high probability that flows from different senders will occur at the same time.
- ***Shared Explicit Style.*** In this style, the router creates a single reservation that can be shared by a set of flows.

Soft State The reservation information (state) stored in every node for a flow needs to be refreshed periodically. This is referred to as a *soft state,* as compared to the *hard state* used in other virtual-circuit protocols such as ATM, where the information about the flow is maintained until it is erased. The default interval for refreshing (soft state reservation) is currently 30 seconds.

30.3.5 Problems with Integrated Services

There are at least two problems with Integrated Services that may prevent its full implementation in the Internet: scalability and service-type limitation.

Scalability

The Integrated Services model requires that each router keep information for each flow. As the Internet is growing every day, this is a serious problem. Keeping information is especially troublesome for core routers because they are primarily designed to switch packets at a high rate and not to process information.

Service-Type Limitation

The Integrated Services model provides only two types of services, guaranteed and control-load. Those opposing this model argue that applications may need more than these two types of services.

30.4 DIFFERENTIATED SERVICES (DIFFSERV)

In this model, also called ***DiffServ,*** packets are marked by applications into classes according to their priorities. Routers and switches, using various queuing strategies, route the packets. This model was introduced by the IETF (Internet Engineering Task Force) to handle the shortcomings of Integrated Services. Two fundamental changes were made:

1. The main processing was moved from the core of the network to the edge of the network. This solves the scalability problem. The routers do not have to store information about flows. The applications, or hosts, define the type of service they need each time they send a packet.
2. The per-flow service is changed to per-class service. The router routes the packet based on the class of service defined in the packet, not the flow. This solves the service-type limitation problem. We can define different types of classes based on the needs of applications.

Differentiated Services is a class-based QoS model designed for IP.
In this Model packets are marked by applications according to their priority.

30.4.1 DS Field

In DiffServ, each packet contains a field called the DS field. The value of this field is set at the boundary of the network by the host or the first router designated as the boundary router. IETF proposes to replace the existing ToS (type of service) field in IPv4 or the priority class field in IPv6 with the DS field, as shown in Figure 30.10.

Figure 30.10 *DS field*

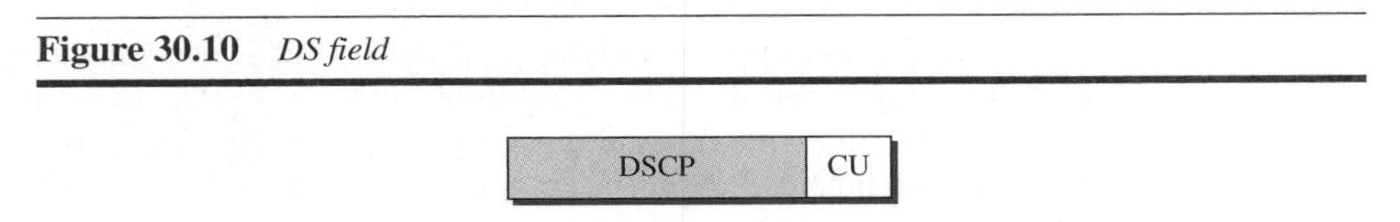

The DS field contains two subfields: DSCP and CU. The DSCP (Differentiated Services Code Point) is a 6-bit subfield that defines the **per-hop behavior (PHB).** The 2-bit CU (Currently Unused) subfield is not currently used.

The DiffServ capable node (router) uses the DSCP 6 bits as an index to a table defining the packet-handling mechanism for the current packet being processed.

30.4.2 Per-Hop Behavior

The DiffServ model defines per-hop behaviors (PHBs) for each node that receives a packet. So far three PHBs are defined: DE PHB, EF PHB, and AF PHB.

DE PHB

The DE PHB (default PHB) is the same as best-effort delivery, which is compatible with ToS.

EF PHB

The EF PHB (expedited forwarding PHB) provides the following services:

a. Low loss.

b. Low latency.

c. Ensured bandwidth.

This is the same as having a virtual connection between the source and destination.

AF PHB

The AF PHB (assured forwarding PHB) delivers the packet with a high assurance as long as the class traffic does not exceed the traffic profile of the node. The users of the network need to be aware that some packets may be discarded.

30.4.3 Traffic Conditioners

To implement DiffServ, the DS node uses traffic conditioners such as meters, markers, shapers, and droppers, as shown in Figure 30.11.

Figure 30.11 *Traffic conditioners*

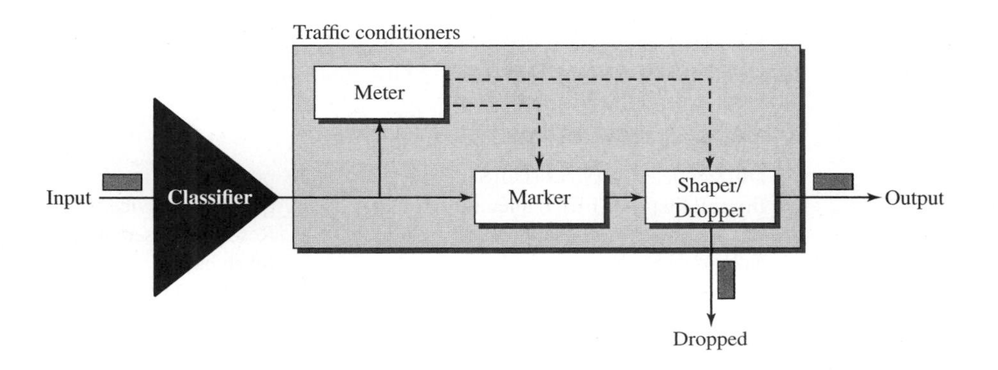

Meter

The meter checks to see if the incoming flow matches the negotiated traffic profile. The meter also sends this result to other components. The meter can use several tools such as a token bucket to check the profile.

Marker

A marker can re-mark a packet that is using best-effort delivery (DSCP: 000000) or down-mark a packet based on information received from the meter. Down-marking (lowering the class of the flow) occurs if the flow does not match the profile. A marker does not up-mark a packet (promote the class).

Shaper

A shaper uses the information received from the meter to reshape the traffic if it is not compliant with the negotiated profile.

Dropper

A dropper, which works as a shaper with no buffer, discards packets if the flow severely violates the negotiated profile.

30.5 END-CHAPTER MATERIALS

30.5.1 Recommended Reading

For more details about subjects discussed in this chapter, we recommend the following books and RFCs. The items enclosed in brackets refer to the reference list at the end of the book.

Books

Several books give some coverage of multimedia: [Com 06], [Tan 03], and [GW 04].

RFCs

Several RFCs show different updates on topics discussed in this chapter, including RFCs 2198, 2250, 2326, 2475, 3246, 3550, and 3551.

30.5.2 Key Terms

Differentiated Services (DiffServ)
first-in, first-out (FIFO) queuing
Integrated Services (IntServ)
leaky bucket
per-hop behavior (PHB)
priority queuing
quality of service (QoS)
Resource Reservation Protocol (RSVP)
token bucket
weighted fair queuing

30.5.3 Summary

To provide quality of service for an Internet application, we need to define the flow characteristics for the application: reliability, delay, jitter, and bandwidth. Common applications in the Internet have been marked with different levels of sensitivity to flow characteristics. A flow class is a set of applications with the same required level of flow characteristics. Traditionally, five flow classes have been defined by the ATM forum: CBR, VBR-NRT, VBR-RT, ABR, and UBR.

One way to improve QoS is to use flow control, which can be achieved using techniques such as scheduling, traffic shaping, resource reservation, and admission control. Scheduling uses FIFO queuing, priority queuing, and weighted fair queueing. Traffic shaping uses leaky or token bucket. Resource reservation can be made by creating a connection-oriented protocol at the top of the IP protocol to make the necessary allocation for the intended traffic. Admission control is a mechanism deployed by a router or switch to accept or reject a packet or a flow based on the packet class or the flow requirement.

Integrated Services (IntServ) is a flow-based architecture that tries to use flow specifications, admission control, and service classes to provide QoS in the Internet. The approach needs a separate protocol to create a connection-oriented service for this purpose. The protocol that provides connection-oriented service is called Resource Reservation Protocol (RSVP), which provides multicasting connection between the source and many destinations.

Differentiated Service (DiffServ) is an architecture that tries to handle traffic based on the class of packets, marked by the source. Each packet is marked by the source as belonging to a specific class; the packet, however, may be delayed or dropped if the traffic is busy with packets with a higher class level.

30.6 PRACTICE SET

30.6.1 Quizzes

A set of interactive quizzes for this chapter can be found on the book website. It is strongly recommended that the student take the quizzes to check his/her understanding of the materials before continuing with the practice set.

30.6.2 Questions

Q30-1. Which of the following applications are classified as VBR-RT in ATM?

a. HTTP **b.** SNMP **c.** SMTP **d.** VoIP

Q30-2. Which of the following applications are classified as UBR in ATM?

a. HTTP **b.** SNMP **c.** SMTP **d.** VoIP

Q30-3. Which of the following applications are classified as CBR in ATM?

a. HTTP **b.** SNMP **c.** SMTP **d.** VoIP

Q30-4. Is the flow label in IPv6 more appropriate for IntServ or DiffServ?

Q30-5. Why do we need Path and Resv messages in RSVP?

Q30-6. Distinguish between guaranteed services and controlled-load services in IntServ.

Q30-7. Define the parts of flow specification in IntServ.

Q30-8. Rank the following applications based on their sensitivity to bandwidth:

a. HTTP **b.** SNMP **c.** SMTP **d.** VoIP

Q30-9. Rank the following applications based on their sensitivity to reliability:

a. HTTP **b.** SNMP **c.** SMTP **d.** VoIP

Q30-10. How can multicasting be achieved using DiffServ?

Q30-11. List the components of a traffic conditioner in DiffServ.

Q30-12. DiffServ is normally called a source-based service. Explain the reason.

Q30-13. If a communication is unicast, how can we use RSVP, which is designed for multicast in IntServ?

Q30-14. Rank the following applications based on their sensitivity to delay:

a. HTTP **b.** SNMP **c.** SMTP **d.** VoIP

Q30-15. IntServ is normally called a destination-based service. Explain the reason.

Q30-16. Which of the following applications are classified as VBR-NRT in ATM?

a. HTTP **b.** SNMP **c.** SMTP **d.** VoIP

Q30-17. Which of the following technique(s) is (are) used for scheduling?

a. FIFO queuing **b.** Priority queuing **c.** Leaky Bucket

Q30-18. How many per-hop behaviors have been defined for DiffServ? Name them.

Q30-19. Rank the following applications based on their sensitivity to jitter:

a. HTTP **b.** SNMP **c.** SMTP **d.** VoIP

Q30-20. Which of the following technique(s) is (are) used for traffic shaping?

a. Token bucket **b.** Priority queuing **c.** Leaky Bucket

30.6.3 Problems

P30-1. To regulate its output flow, a router implements a weighted queueing scheme with three queues at the output port. The packets are classified and stored in one of these queues before being transmitted. The weights assigned to queues are w = 3, w = 2, and w = 1 (3/6, 2/6, and 1/6). The contents of each queue at time t_0 are shown in Figure 30.12. Assume packets are all the same size and that transmission time for each is 1 μs.

a. Using a time line, show the departure time for each packet.

b. Show the contents of the queues after 5, 10, 15, and 20 μs.

c. Find the departure delay of each packet with respect to the previous packet in the class w = 3. Has queuing created jitter in this class?

d. Find the departure delay of each packet with respect to the previous packet in the class w = 2. Has queuing created jitter in this class?

e. Find the departure delay of each packet with respect to the previous packet in the class w = 1. Has queuing created jitter in this class?

Figure 30.12 *Problem P30-1*

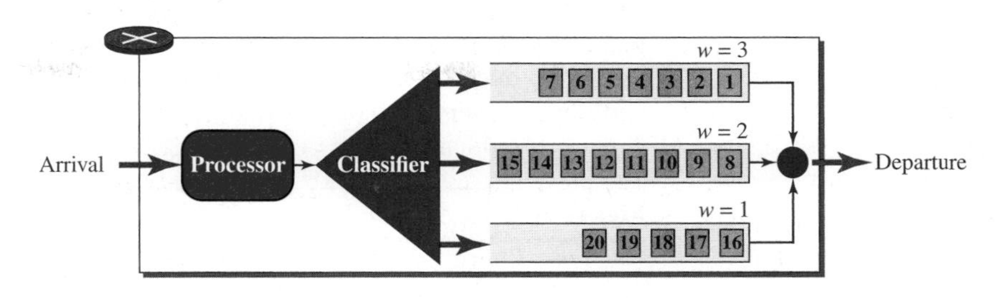

P30-2. Assume fixed-sized packets arrive at a router with a rate of three packets per second. Show how the router can use the leaky bucket algorithm to send out only two packets per second. What is the problem with this approach?

P30-3. Figure 30.13 shows a router using FIFO queuing at the input port.

Figure 30.13 *Problem P30-3*

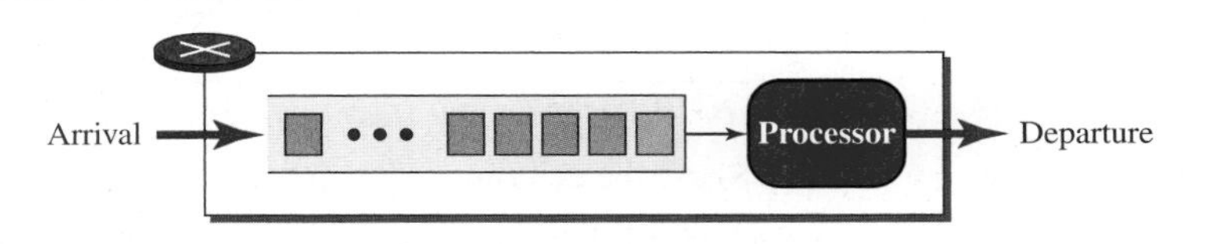

The arrival and required service times for seven packets are shown below; t_i means that the packet has arrived or departed *i* ms after a reference time. The values of required service times are also shown in ms. We assume the transmission time is negligible.

Packets	1	2	3	4	5	6	7
Arrival time	t_0	t_1	t_2	t_4	t_5	t_6	t_7
Required service time	1	1	3	2	2	3	1

a. Using time lines, show the arrival time, the process duration, and the departure time for each packet. Also show the contents of the queue at the beginning of each millisecond.

b. For each packet, find the time spent in the router and the departure delay with respect to the previously departed packet.

c. If all packets belong to the same application program, determine whether the router creates jitter for the packets.

P30-4. In Figure 30.3, assume the weight in each class is 4, 2, and 1. The packets in the top queue are labeled A, in the middle queue B, and in the bottom queue C. Show the list of packets transmitted in each of the following situations:

a. Each queue has a large number of packets.

b. The numbers of packets in queues, from top to bottom, are 10, 4, and 0.

c. The numbers of packets in queues, from top to bottom, are 0, 5, and 10.

P30-5. Assume an ISP uses three leaky buckets to regulate data received from three customers for transmitting to the Internet. The customers send fixed-size packets (cells). The ISP sends 10 cells per second for each customer and the maximum burst size of 20 cells per second. Each leaky bucket is implemented as a FIFO queue (of size 20) and a timer that extracts one cell from the queue and sends it every 1/10 of a second. (See Figure 30.14.)

Figure 30.14 *Problem P30-5*

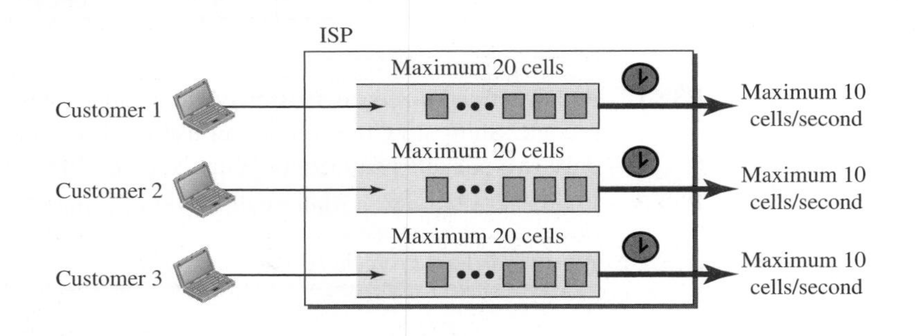

a. Show the customer rate and the contents of the queue for the first customer, which sends 5 cells per second for the first 7 seconds and 15 cells per second for the next 9 seconds.

b. Do the same for the second customer, which sends 15 cells per second for the first 4 seconds and 5 cells per second for the next 14 seconds.

c. Do the same for the third customer, which sends no cells for the first two seconds, 20 cells for the next two seconds, and repeats the pattern four times.

P30-6. An output interface in a switch is designed using the leaky bucket algorithm to send 8000 bytes/s (tick). If the following frames are received in sequence, show the frames that are sent during each second.

- Frames 1, 2, 3, 4: 4000 bytes each
- Frames 5, 6, 7: 3200 bytes each
- Frames 8, 9: 400 bytes each
- Frames 10, 11, 12: 2000 bytes each

P30-7. In a leaky bucket used to control liquid flow, how many gallons of liquid are left in the bucket if the output rate is 5 gal/min, there is an input burst of 100 gal/min for 12 s, and there is no input for 48 s?

P30-8. Assume that the ISP in Problem P30-5 decided to use token buckets (of capacity $c = 20$ and rate $r = 10$) instead of leaky buckets to give credit to the customer that does not send cells for a while but needs to send some bursts later. Each token bucket is implemented by a very large queue for each customer (no packet drop), a bucket that holds the token, and the timer that regulates dropping tokens in the bucket. (See Figure 30.15.)

a. Show the customer rate, the contents of the queue, and the contents of the bucket for the first customer, which sends 5 cells per second for the first 7 seconds and 15 cells per second for the next 9 seconds.

b. Do the same for the second customer, which sends 15 cells per second for the first 4 seconds and 5 cells per second for the next 14 seconds.

c. Do the same for the third customer, which sends no cells for the first 2 seconds, 20 cells for the next 2 seconds, and repeats the pattern 4 times.

Figure 30.15 *Problem P30-8*

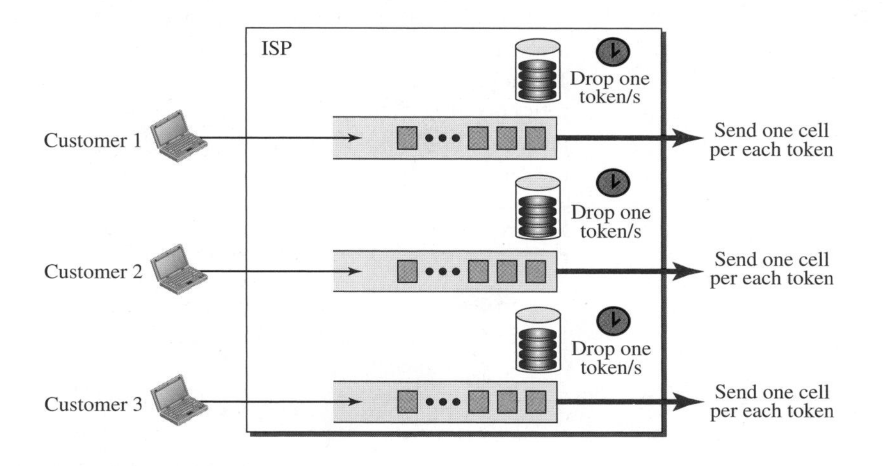

P30-9. Assume a router receives packets of size 400 bits every 100 ms, which means with the data rate of 4 kbps. Show how we can change the output data rate to less than 1 kbps by using a leaky bucket algorithm.

P30-10. In a switch using the token bucket algorithm, tokens are added to the bucket at a rate of $r = 5$ tokens/second. The capacity of the token bucket is $c = 10$. The switch has a buffer that can hold only eight packets (for the sake of example). The packets arrive at the switch at the rate of R packets/second. Assume that the packets are all the same size and need the same amount of time for processing. If at time zero the bucket is empty, show the contents of the bucket and the queue in each of the following cases and interpret the result.

a. $R = 5$ **b.** $R = 3$ **c.** $R = 7$

P30-11. To understand how the token bucket algorithm can give credit to the sender that does not use its rate allocation for a while but wants to use it later, let's repeat Problem P-10 with $r = 3$ and $c = 10$, but assume that the sender uses a variable rate, as shown in Figure 30.16. The sender sends only three packets per second for the first two seconds, sends no packets for the next two seconds, and sends seven packets for the next three seconds. The sender is allowed to send five packets per second, but, since it does not use this right

Figure 30.16 *Problem P30-11*

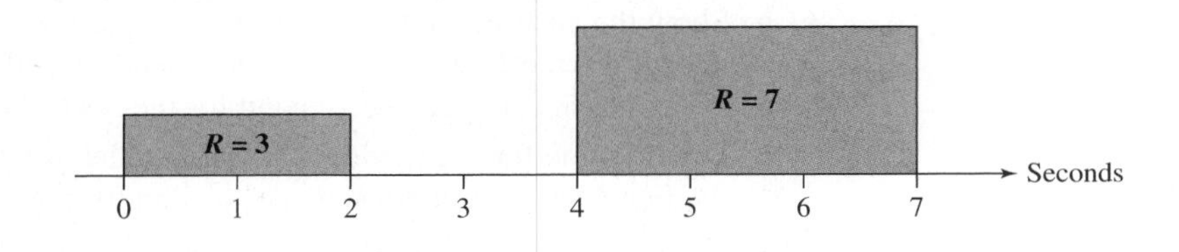

fully during the first four seconds, it can send more in the next three seconds. Show the contents of the token bucket and the buffer for each second to prove this fact.

P30-12. Figure 30.17 shows a router using priority queuing at the input port. The arrival and required service times (transmission time is negligible) for 10 packets are shown below; t_i means that the packet has arrived i ms after a reference time. The values of required service times are also shown in ms. The packets with higher priorities are packets 1, 2, 3, 4, 7, and 9 (shown in color); the other packets are packets with lower priorities.

Figure 30.17 *Problem P30-12*

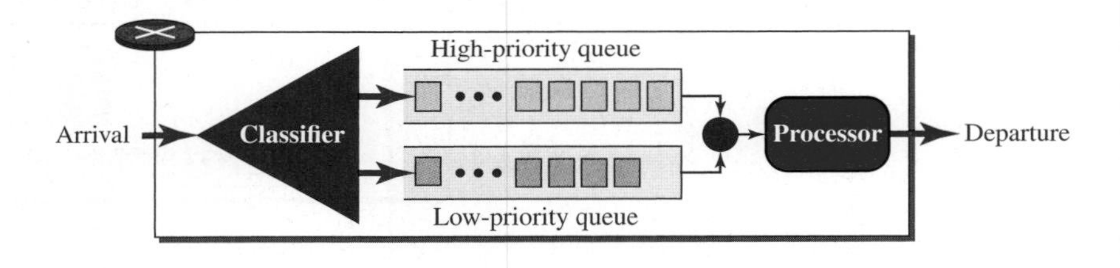

Packets	**1**	**2**	**3**	**4**	5	6	**7**	8	**9**	10
Arrival time	t_0	t_1	t_2	t_3	t_4	t_5	t_6	t_7	t_8	t_9
Required service time	2	2	2	2	1	1	2	1	2	1

a. Using time lines, show the arrival time, the processing duration, and the departure time for each packet. Also show the contents of the high-priority queue (Q1) and the low-priority queue (Q2) at each millisecond.

b. For each packet belonging to the high-priority class, find the time spent in the router and the departure delay with respect to the previously departed packet. Find if the router creates jitter for this class.

c. For each packet belonging to the low-priority class, find the time spent in the router and the departure delay with respect to the previously departed packet. Determine whether the router creates jitter for this class.

30.7 SIMULATION EXPERIMENTS

30.7.1 Applets

We have created some Java applets to show some of the main concepts discussed in this chapter. It is strongly recommended that the students activate these applets on the book website and carefully examine the protocols in action.

30.8 PROGRAMMING ASSIGNMENTS

Write the source code for, compile, and test the following programs in one of the programming language of your choice:

Prg30-21. A program to simulate a leaky bucket.

Prg30-22. A program to simulate a token bucket.

CHAPTER 31

Cryptography and Network Security

The topic of cryptography and network security is very broad and involves some specific areas of mathematics such as number theory. In this chapter, we try to give a very simple introduction to this topic to prepare the background for more study. We have divided this chapter into three sections.

- The first section introduces the subject. It first describes security goals such as confidentiality, integrity, and availability. The section shows how confidentiality is threatened by attacks such as snooping and traffic analysis. The section then shows how integrity is threatened by attacks such as modification, masquerading, replaying, and repudiation. The section mentions one attack that threatens availability, denial of service. This section ends with describing the two techniques used in security: cryptography and steganography. The chapter concentrates on the first.
- The second section discusses confidentiality. It first describes symmetric-key ciphers and explains traditional symmetric-key ciphers such as substitution and transposition ciphers. It then moves to modern symmetric-key ciphers and explains modern block and stream ciphers. The section then shows that denial of service is an attack to availability.
- The third section discusses other aspects of security: message integrity, message authentication, digital signature, entity authentication. These aspects today are part of the security system that complements confidentiality. The section also describes the topic of key management including the distribution of keys for both symmetric-key and asymmetric-key ciphers.

31.1 INTRODUCTION

We are living in the information age. We need to keep information about every aspect of our lives. In other words, information is an asset that has a value like any other asset. As an asset, information needs to be secured from attacks. To be secured, information needs to be hidden from unauthorized access (*confidentiality*), protected from unauthorized change (*integrity*), and available to an authorized entity when it is needed (*availability*).

During the last three decades, computer networks created a revolution in the use of information. Information is now distributed. Authorized people can send and retrieve information from a distance using computer networks. Although the three above-mentioned requirements—confidentiality, integrity, and availability—have not changed, they now have some new dimensions. Not only should information be confidential when it is stored; there should also be a way to maintain its confidentiality when it is transmitted from one computer to another.

In this section, we first discuss the three major goals of information security. We then see how attacks can threaten these three goals. We then discuss the security services in relation to these security goals. Finally we define two techniques to implement the security goals and prevent attacks.

31.1.1 Security Goals

Let us first discuss three security goals: confidentiality, integrity, and availability.

Confidentiality

Confidentiality is probably the most common aspect of information security. We need to protect our confidential information. An organization needs to guard against those malicious actions that endanger the confidentiality of its information. Confidentiality not only applies to the storage of information, it also applies to the transmission of information. When we send a piece of information to be stored in a remote computer or when we retrieve a piece of information from a remote computer, we need to conceal it during transmission.

Integrity

Information needs to be changed constantly. In a bank, when a customer deposits or withdraws money, the balance of her account needs to be changed. *Integrity* means that changes need to be done only by authorized entities and through authorized mechanisms. Integrity violation is not necessarily the result of a malicious act; an interruption in the system, such as a power surge, may also create unwanted changes in some information.

Availability

The third component of information security is *availability.* The information created and stored by an organization needs to be available to authorized entities. Information is useless if it is not available. Information needs to be constantly changed, which means it must be accessible to authorized entities. The unavailability of information is just as harmful for an organization as the lack of confidentiality or integrity. Imagine what would happen to a bank if the customers could not access their accounts for transactions.

31.1.2 Attacks

Our three goals of security—confidentiality, integrity, and availability—can be threatened by security *attacks.* Although the literature uses different approaches to categorizing the attacks, we divide them into three groups related to the security goals. Figure 31.1 shows the taxonomy.

Figure 31.1 *Taxonomy of attacks with relation to security goals*

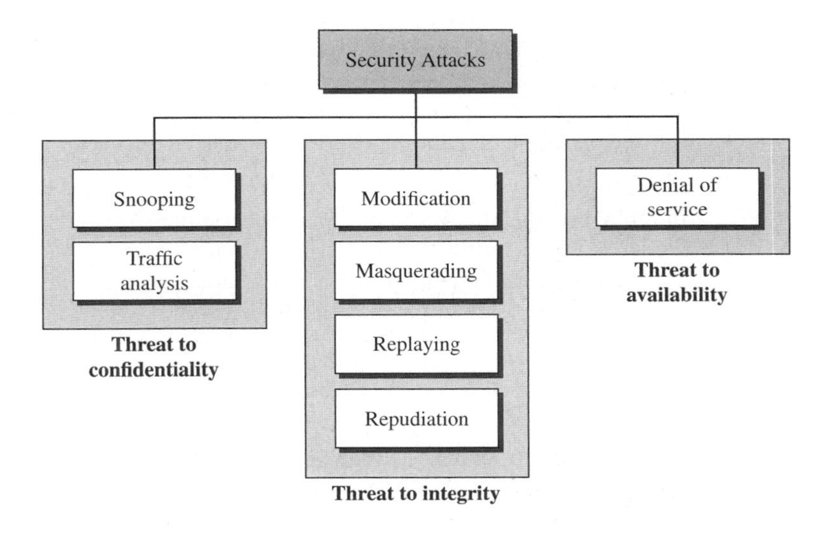

Attacks Threatening Confidentiality

In general, two types of attacks threaten the confidentiality of information: *snooping* and *traffic analysis.*

Snooping

Snooping refers to unauthorized access to or interception of data. For example, a file transferred through the Internet may contain confidential information. An unauthorized entity may intercept the transmission and use the contents for her own benefit. To prevent snooping, the data can be made nonintelligible to the intercepter by using encipherment techniques, discussed later.

Traffic Analysis

Although encipherment of data may make it nonintelligible for the intercepter, she can obtain some other types of information by monitoring online traffic. For example, she can find the electronic address (such as the e-mail address) of the sender or the receiver. She can collect pairs of requests and responses to help her guess the nature of the transaction.

Attacks Threatening Integrity

The integrity of data can be threatened by several kinds of attacks: *modification, masquerading, replaying, and repudiation.*

Modification

After intercepting or accessing information, the attacker modifies the information to make it beneficial to herself. For example, a customer sends a message to a bank to initiate some transaction. The attacker intercepts the message and changes the type of transaction to benefit herself. Note that sometimes the attacker simply deletes or delays the message to harm the system or to benefit from it.

Masquerading

Masquerading, or spoofing, happens when the attacker impersonates somebody else. For example, an attacker might steal the bank card and PIN of a bank customer and pretend that she is that customer. Sometimes the attacker pretends instead to be the receiver entity. For example, a user tries to contact a bank, but another site pretends that it is the bank and obtains some information from the user.

Replaying

In replaying, the attacker obtains a copy of a message sent by a user and later tries to replay it. For example, a person sends a request to her bank to ask for payment to the attacker, who has done a job for her. The attacker intercepts the message and sends it again to receive another payment from the bank.

Repudiation

This type of attack is different from others because it is performed by one of the two parties in the communication: the sender or the receiver. The sender of the message might later deny that she has sent the message; the receiver of the message might later deny that he has received the message. An example of denial by the sender would be a bank customer asking her bank to send some money to a third party but later denying that she has made such a request. An example of denial by the receiver could occur when a person buys a product from a manufacturer and pays for it electronically, but the manufacturer later denies having received the payment and asks to be paid.

Attacks Threatening Availability

We mention only one attack threatening availability: **denial of service.**

Denial of Service

Denial of service (DoS) is a very common attack. It may slow down or totally interrupt the service of a system. The attacker can use several strategies to achieve this. She might send so many bogus requests to a server that the server crashes because of the heavy load. The attacker might intercept and delete a server's response to a client, making the client believe that the server is not responding. The attacker may also intercept requests from the clients, causing the clients to send requests many times and overload the system.

31.1.3 Services and Techniques

ITU-T defines some security services to achieve security goals and prevent attacks. Each of these services is designed to prevent one or more attacks while maintaining security goals. The actual implementation of security goals needs some techniques. Two techniques are prevalent today: one is very general (cryptography) and one is specific (steganography).

Cryptography

Some security services can be implemented using cryptography. ***Cryptography,*** a word with Greek origins, means "secret writing." However, we use the term to refer to the science and art of transforming messages to make them secure and immune to attacks. Although in the past *cryptography* referred only to the **encryption** and **decryption** of messages using secret keys, today it is defined as involving three distinct mechanisms: symmetric-key encipherment, asymmetric-key encipherment, and hashing. We will discuss all these mechanisms later in the chapter.

Steganography

Although this chapter and the next are based on cryptography as a technique for implementing security mechanisms, another technique that was used for secret communication in the past is being revived at the present time: steganography. The word ***steganography,*** with origins in Greek, means "covered writing," in contrast with *cryptography,* which means "secret writing." *Cryptography* means concealing the contents of a message by enciphering; *steganography* means concealing the message itself by covering it with something else. We leave the discussion of steganography to some books dedicated to this topic.

31.2 CONFIDENTIALITY

We now look at the first goal of security, confidentiality. Confidentiality can be achieved using ciphers. Ciphers can be divided into two broad categories: symmetric-key and asymmetric-key.

31.2.1 Symmetric-Key Ciphers

A **symmetric-key cipher** uses the same key for both encryption and decryption, and the key can be used for bidirectional communication, which is why it is called *symmetric*. Figure 31.2 shows the general idea behind a symmetric-key cipher.

In Figure 31.2, an entity, Alice, can send a message to another entity, Bob, over an insecure channel with the assumption that an adversary, Eve, cannot understand the contents of the message by simply eavesdropping over the channel.

The original message from Alice to Bob is called ***plaintext;*** the message that is sent through the channel is called ***ciphertext.*** To create the ciphertext from the plaintext, Alice uses an **encryption algorithm** and a *shared secret key.*

To create the plaintext from ciphertext, Bob uses a **decryption algorithm** and the same secret key. We refer to encryption and decryption algorithms as ***ciphers.*** A *key* is a set of values (numbers) that the cipher, as an algorithm, operates on.

Figure 31.2 *General idea of a symmetric-key cipher*

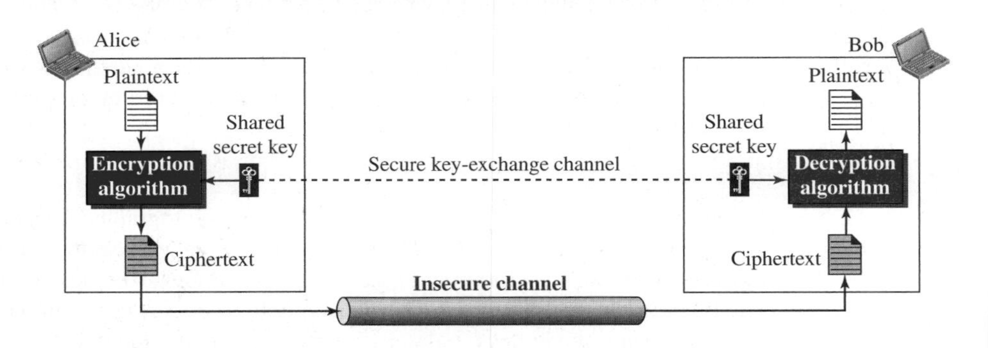

Symmetric-key ciphers are also called *secret-key ciphers*.

Note that the symmetric-key encipherment uses a single key (the key itself may be a set of values) for both encryption and decryption. In addition, the encryption and decryption algorithms are inverses of each other. If P is the plaintext, C is the ciphertext, and K is the key, the encryption algorithm $E_k(x)$ creates the ciphertext from the plaintext; the decryption algorithm $D_k(x)$ creates the plaintext from the ciphertext. We assume that $E_k(x)$ and $D_k(x)$ are inverses of each other: they cancel the effect of each other if they are applied one after the other on the same input. We have

Encryption: $C = E_k(P)$ **Decryption:** $P = D_k(C)$

in which, $D_k(E_k(x)) = E_k(D_k(x)) = x$. We need to emphasize that it is better to make the encryption and decryption public but keep the shared key secret. This means that Alice and Bob need another channel, a secured one, to exchange the secret key. Alice and Bob can meet once and exchange the key personally. The secured channel here is the face-to-face exchange of the key. They can also trust a third party to give them the same key. They can create a temporary secret key using another kind of cipher—asymmetric-key ciphers—which will be described later.

Encryption can be thought of as locking the message in a box; decryption can be thought of as unlocking the box. In symmetric-key encipherment, the same key locks and unlocks, as shown in Figure 31.3. Later sections show that the *asymmetric-key* encipherment needs two keys, one for locking and one for unlocking.

Figure 31.3 *Symmetric-key encipherment as locking and unlocking with the same key*

The symmetric-key ciphers can be divided into traditional ciphers and modern ciphers. Traditional ciphers are simple, character-oriented ciphers that are not secure based on today's standard. Modern ciphers, on the other hand, are complex, bit-oriented ciphers that are more secure. We briefly discuss the traditional ciphers to pave the way for discussing more complex modern ciphers.

Traditional Symmetric-Key Ciphers

Traditional ciphers belong to the past. However, we briefly discuss them here because they can be thought of as the components of the modern ciphers. To be more exact, we can divide traditional ciphers into substitution ciphers and transposition ciphers.

Substitution Ciphers

A **substitution cipher** replaces one symbol with another. If the symbols in the plaintext are alphabetic characters, we replace one character with another. For example, we can replace letter A with letter D and letter T with letter Z. If the symbols are digits (0 to 9), we can replace 3 with 7 and 2 with 6.

A substitution cipher replaces one symbol with another.

Substitution ciphers can be categorized as either monoalphabetic ciphers or polyalphabetic ciphers.

Monoalphabetic Ciphers In a **monoalphabetic cipher,** a character (or a symbol) in the plaintext is always changed to the same character (or symbol) in the ciphertext regardless of its position in the text. For example, if the algorithm says that letter A in the plaintext is changed to letter D, every letter A is changed to letter D. In other words, the relationship between letters in the plaintext and the ciphertext is one-to-one.

The simplest monoalphabetic cipher is the **additive cipher** (or **shift cipher**). Assume that the plaintext consists of lowercase letters (a to z), and that the ciphertext consists of uppercase letters (A to Z). To be able to apply mathematical operations on the plaintext and ciphertext, we assign numerical values to each letter (lowercase or uppercase), as shown in Figure 31.4.

Figure 31.4 *Representation of plaintext and ciphertext characters in modulo 26*

Plaintext →	a	b	c	d	e	f	g	h	i	j	k	l	m	n	o	p	q	r	s	t	u	v	w	x	y	z
Ciphertext →	A	B	C	D	E	F	G	H	I	J	K	L	M	N	O	P	Q	R	S	T	U	V	W	X	Y	Z
Value →	00	01	02	03	04	05	06	07	08	09	10	11	12	13	14	15	16	17	18	19	20	21	22	23	24	25

In Figure 31.4 each character (lowercase or uppercase) is assigned an integer in modulo 26. The secret key between Alice and Bob is also an integer in modulo 26. The encryption algorithm adds the key to the plaintext character; the decryption algorithm subtracts the key from the ciphertext character. All operations are done in modulo 26.

In additive cipher, the plaintext, ciphertext, and key are integers in modulo 26.

Historically, additive ciphers are called *shift ciphers* because the encryption algorithm can be interpreted as "shift *key* characters down" and the encryption algorithm can be interpreted as "shift *key* characters up." Julius Caesar used an additive cipher, with a key of 3, to communicate with his officers. For this reason, additive ciphers are sometimes referred to as the ***Caesar cipher.***

Example 31.1

Use the additive cipher with key = 15 to encrypt the message "hello".

Solution

We apply the encryption algorithm to the plaintext, character by character:

Plaintext: h → 07	Encryption: (07 + 15) mod 26	Ciphertext: 22 → W
Plaintext: e → 04	Encryption: (04 + 15) mod 26	Ciphertext: 19 → T
Plaintext: l → 11	Encryption: (11 + 15) mod 26	Ciphertext: 00 → A
Plaintext: l → 11	Encryption: (11 + 15) mod 26	Ciphertext: 00 → A
Plaintext: o → 14	Encryption: (14 + 15) mod 26	Ciphertext: 03 → D

The result is "WTAAD". Note that the cipher is monoalphabetic because two instances of the same plaintext character (*l*) are encrypted as the same character (*A*).

Example 31.2

Use the additive cipher with key = 15 to decrypt the message "WTAAD".

Solution

We apply the decryption algorithm to the plaintext character by character:

Ciphertext: W → 22	Decryption: (22 – 15) mod 26	Plaintext: 07 → h
Ciphertext: T → 19	Decryption: (19 – 15) mod 26	Plaintext: 04 → e
Ciphertext: A → 00	Decryption: (00 – 15) mod 26	Plaintext: 11 → l
Ciphertext: A → 00	Decryption: (00 – 15) mod 26	Plaintext: 11 → l
Ciphertext: D → 03	Decryption: (03 – 15) mod 26	Plaintext: 14 → o

The result is "hello". Note that the operation is in modulo 26, which means that we need to add 26 to a negative result (for example −15 becomes 11).

Additive ciphers are vulnerable to attacks using exhaustive key searches (brute-force attacks). The key domain of the additive cipher is very small; there are only 26 keys. However, one of the keys, zero, is useless (the ciphertext is the same as the plaintext). This leaves only 25 possible keys. Eve can easily launch a brute-force attack on the ciphertext. A better solution is to create a mapping between each plaintext character and the corresponding ciphertext character. Alice and Bob can agree on a table showing the mapping for each character. Figure 31.5 shows an example of such a mapping.

Example 31.3

We can use the key in Figure 31.5 to encrypt the message

Figure 31.5 *An example key for a monoalphabetic substitution cipher*

Plaintext →	a	b	c	d	e	f	g	h	i	j	k	l	m	n	o	p	q	r	s	t	u	v	w	x	y	z
Ciphertext →	N	O	A	T	R	B	E	C	F	U	X	D	Q	G	Y	L	K	H	V	I	J	M	P	Z	S	W

Plaintext: this message is easy to encrypt but hard to find the key
Ciphertext: ICFVQRVVNERFVRNVSIYRGAHSLIOJICNHTIYBFGTICRXRS

Polyalphabetic Ciphers In a **polyalphabetic cipher,** each occurrence of a character may have a different substitute. The relationship of a character in the plaintext to a character in the ciphertext is one-to-many. For example, "a" could be enciphered as "D" at the beginning of the text, but as "N" in the middle. Polyalphabetic ciphers have the advantage of hiding the letter frequency of the underlying language. Eve cannot use single-letter frequency statistics to break the ciphertext.

To create a polyalphabetic cipher, we need to make each ciphertext character dependent on both the corresponding plaintext character and the position of the plaintext character in the message. This implies that our key should be a stream of subkeys, in which each subkey depends somehow on the position of the plaintext character that uses that subkey for encipherment. In other words, we need to have a key stream $k = (k_1, k_2, k_3, \ldots)$ in which k_i is used to encipher the ith character in the plaintext to create the ith character in the ciphertext.

To see the position dependency of the key, let us discuss a simple polyalphabetic cipher called the **autokey cipher.** In this cipher, the key is a stream of subkeys, in which each subkey is used to encrypt the corresponding character in the plaintext. The first subkey is a predetermined value secretly agreed upon by Alice and Bob. The second subkey is the value of the first plaintext character (between 0 and 25). The third subkey is the value of the second plaintext character, and so on.

$$P = P_1P_2P_3 \ldots \qquad C = C_1C_2C_3\ldots \qquad k = (k_1, P_1, P_2, \ldots)$$
$$\text{Encryption: } C_i = (P_i + k_i) \bmod 26 \qquad \text{Decryption: } P_i = (C_i - k_i) \bmod 26$$

The name of the cipher, *autokey*, implies that the subkeys are automatically created from the plaintext cipher characters during the encryption process.

Example 31.4

Assume that Alice and Bob agreed to use an autokey cipher with initial key value $k_1 = 12$. Now Alice wants to send Bob the message "Attack is today". Enciphering is done character by character. Each character in the plaintext is first replaced by its integer value. The first subkey is added to create the first ciphertext character. The rest of the key is created as the plaintext characters are read. Note that the cipher is polyalphabetic because the three occurrences of "a" in the plaintext are encrypted differently. The three occurrences of "t" are also encrypted differently.

Plaintext:	a	t	t	a	c	k	i	s	t	o	d	a	y
P's Values:	00	19	19	00	02	10	08	18	19	14	03	00	24
Key stream:	***12***	*00*	*19*	*19*	*00*	*02*	*10*	*08*	*18*	*19*	*14*	*03*	*00*
C's Values:	12	19	12	19	02	12	18	00	11	7	17	03	24
Ciphertext:	**M**	**T**	**M**	**T**	**C**	**M**	**S**	**A**	**L**	**H**	**R**	**D**	**Y**

Transposition Ciphers

A **transposition cipher** does not substitute one symbol for another; instead it changes the location of the symbols. A symbol in the first position of the plaintext may appear in the tenth position of the ciphertext. A symbol in the eighth position in the plaintext may appear in the first position of the ciphertext. In other words, a transposition cipher reorders (transposes) the symbols.

A transposition cipher reorders symbols.

Suppose Alice wants to secretly send the message "Enemy attacks tonight" to Bob. The encryption and decryption is shown in Figure 31.6. Note that we added an extra character (z) to the end of the message to make the number of characters a multiple of 5.

Figure 31.6 *Transposition cipher*

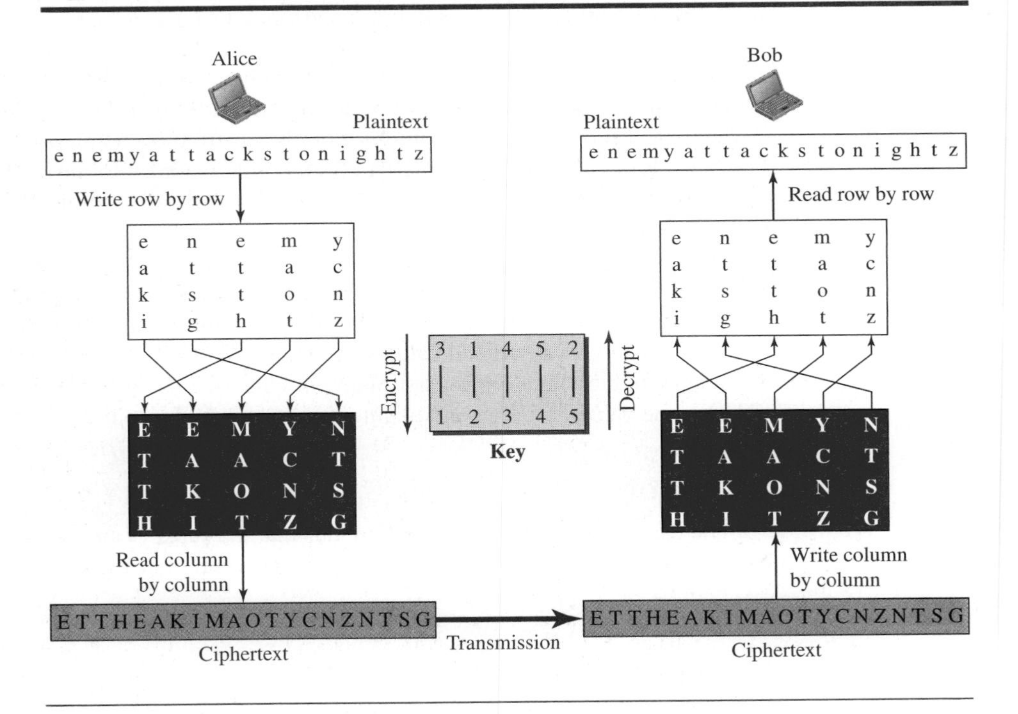

The first table is created by Alice writing the plaintext row by row. The columns are permuted using a key. The ciphertext is created by reading the second table column by column. Bob does the same three steps in the reverse order. He writes the ciphertext column by column into the first table, permutes the columns, and then reads the second table row by row. Note that the same key is used for encryption and decryption, but the algorithm uses the key in reverse order.

Stream and Block Ciphers

The literature divides the symmetric ciphers into two broad categories: stream ciphers and block ciphers.

Stream Cipher In a **stream cipher,** encryption and decryption are done one symbol (such as a character or a bit) at a time. We have a plaintext stream, a ciphertext stream, and a key stream. Call the plaintext stream P, the ciphertext stream C, and the key stream K.

$P = P_1P_2P_3, \ldots$	$C = C_1C_2C_3, \ldots$	$K = (k_1, k_2, k_3, \ldots)$
$C_1 = E_{k1}(P_1)$	$C_2 = E_{k2}(P_2)$	$C_3 = E_{k3}(P_3) \ldots$

Block Ciphers In a **block cipher,** a group of plaintext symbols of size m ($m > 1$) are encrypted together, creating a group of ciphertext of the same size. Based on the definition, in a block cipher, a single key is used to encrypt the whole block even if the key is made of multiple values. In a block cipher, a ciphertext block depends on the whole plaintext block.

Combination In practice, blocks of plaintext are encrypted individually, but they use a stream of keys to encrypt the whole message block by block. In other words, the cipher is a block cipher when looking at the individual blocks, but it is a stream cipher when looking at the whole message, considering each block as a single unit. Each block uses a different key that may be generated before or during the encryption process.

Modern Symmetric-Key Ciphers

The traditional symmetric-key ciphers that we have studied so far are *character-oriented ciphers.* With the advent of the computer, we need *bit-oriented ciphers.* This is because the information to be encrypted is not just text; it can also consist of numbers, graphics, audio, and video data. It is convenient to convert these types of data into a stream of bits, to encrypt the stream, and then to send the encrypted stream. In addition, when text is treated at the bit level, each character is replaced by 8 (or 16) bits, which means that the number of symbols becomes 8 (or 16) times larger. Mixing a larger number of symbols increases security. A modern cipher can be either a block cipher or a stream cipher.

Modern Block Ciphers

A symmetric-key *modern block cipher* encrypts an n-bit block of plaintext or decrypts an n-bit block of ciphertext. The encryption or decryption algorithm uses a k-bit key. The decryption algorithm must be the inverse of the encryption algorithm, and both operations must use the same secret key so that Bob can retrieve the message sent by Alice. Figure 31.7 shows the general idea of encryption and decryption in a modern block cipher.

If the message has fewer than n bits, padding must be added to make it an n-bit block; if the message has more than n bits, it should be divided into n-bit blocks and the appropriate padding must be added to the last block if necessary. The common values for n are 64, 128, 256, and 512 bits.

Figure 31.7 *A modern block cipher*

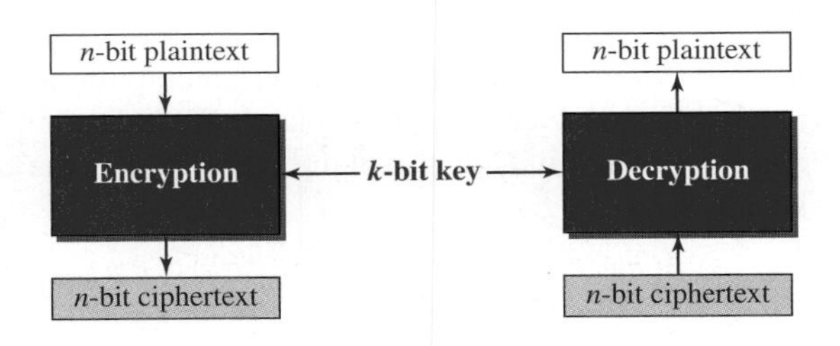

Components of a Modern Block Cipher Modern block ciphers are substitution ciphers when seen as a whole block. However, modern block ciphers are not designed as a single unit. To provide an attack-resistant cipher, a modern block cipher is made of a combination of transposition units (sometimes called *P-boxes*), substitution units (sometimes called *S-boxes*), and exclusive-OR (XOR) operations, as well as shifting elements, swapping elements, splitting elements, and combining elements. Figure 31.8 shows the components of a modern block cipher.

A **P-box** (permutation box) parallels the traditional transposition cipher for characters, but it transposes bits. We can find three types of P-boxes in modern block ciphers: straight P-boxes, expansion P-boxes, and compression P-boxes. An **S-box** (substitution box) can be thought of as a miniature substitution cipher, but it substitutes bits. Unlike the traditional substitution cipher, an S-box can have a different number of inputs and outputs. An important component in most block ciphers is the *exclusive-OR* operation, in which the output is 0 if the two inputs are the same, and the output is 1 if the two inputs are different. In modern block ciphers, we use n exclusive-OR operations to combine an n-bit data piece with an n-bit key. An exclusive-OR operation is normally the only unit where the key is applied. The other components are normally based on predefined functions.

Another component found in some modern block ciphers is the *circular shift operation*. Shifting can be to the left or to the right. The circular left-shift operation shifts each bit in an n-bit word k positions to the left; the leftmost k bits are removed from the left and become the rightmost bits. The *swap operation* is a special case of the circular shift operation where the number of shifted bits $k = n/2$.

Two other operations found in some block ciphers are split and combine. The *split operation* splits an n-bit word in the middle, creating two equal-length words. The *combine operation* normally concatenates two equal-length words, each of size $n/2$ bits, to create an n-bit word.

Data Encryption Standard (DES)

As an example of a modern block cipher, let us discuss the **Data Encryption Standard (DES).** Figure 31.9 shows the elements of a DES cipher at the encryption site.

At the encryption site, DES takes a 64-bit plaintext and creates a 64-bit ciphertext; at the decryption site, DES takes a 64-bit ciphertext and creates a 64-bit block of plaintext. The same 56-bit cipher key is used for both encryption and decryption.

Figure 31.8 *Components of a modern block cipher*

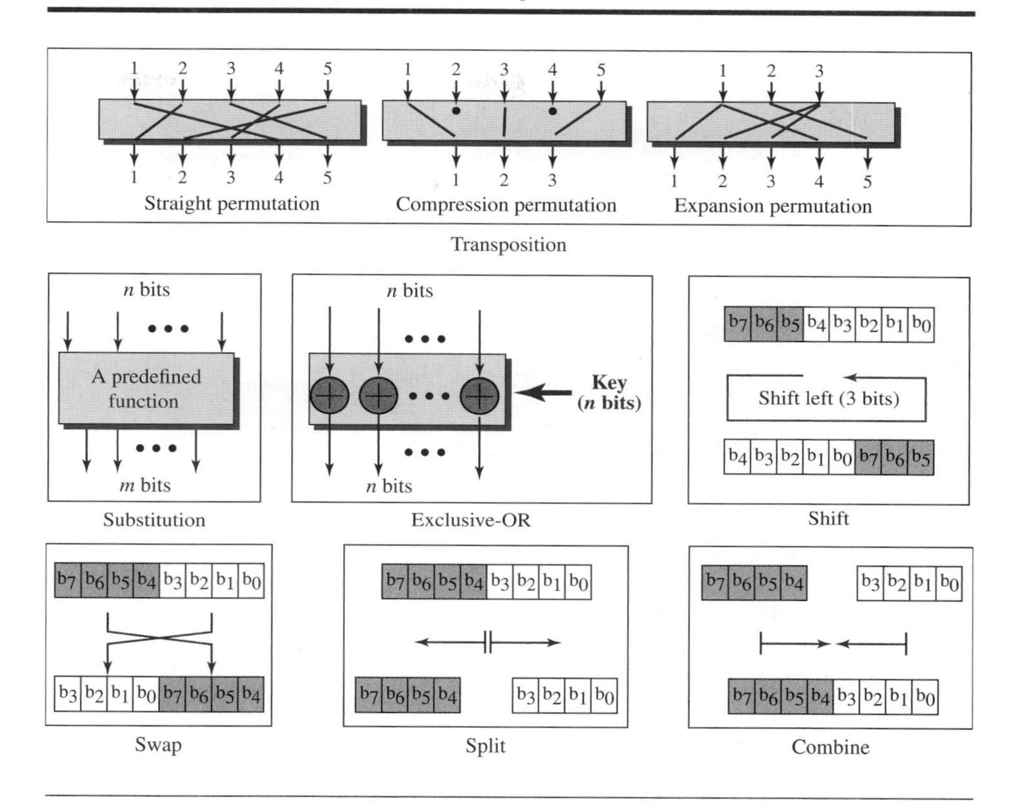

The initial permutation takes a 64-bit input and permutes them according to a predefined rule. The final permutation is the inverse of the initial permutation. These two permutations cancel the effect of each other. In other words, if the rounds are eliminated from the structures, the ciphertext is the same as the plaintext.

Rounds DES uses 16 rounds. Each round of DES is an invertible (Feistel) transformation, as shown in Figure 31.9. The round takes L_{i-1} and R_{i-1} from the previous round (or the initial permutation box) and creates L_i and R_i, which go to the next round (or final permutation box). Each round can have up to two cipher elements (mixer and swapper). Each of these elements is invertible. The swapper is obviously invertible. It swaps the left half of the text with the right half. The mixer is invertible because of the XOR operation. All noninvertible elements are collected inside the function $f(R_{i-1}, K_i)$.

DES Function The heart of DES is the DES function. The DES function applies a 48-bit key to the rightmost 32 bits (R_{i-1}) to produce a 32-bit output. This function is made up of four sections: an expansion P-box, a whitener (that adds a key), a group of S-boxes, and a straight P-box, as shown in Figure 31.10.

Since R_{i-1} is a 32-bit input and K_i is a 48-bit key, we first need to expand R_{i-1} to 48 bits. This expansion permutation follows a predetermined rule.

Figure 31.9 *General structure of DES*

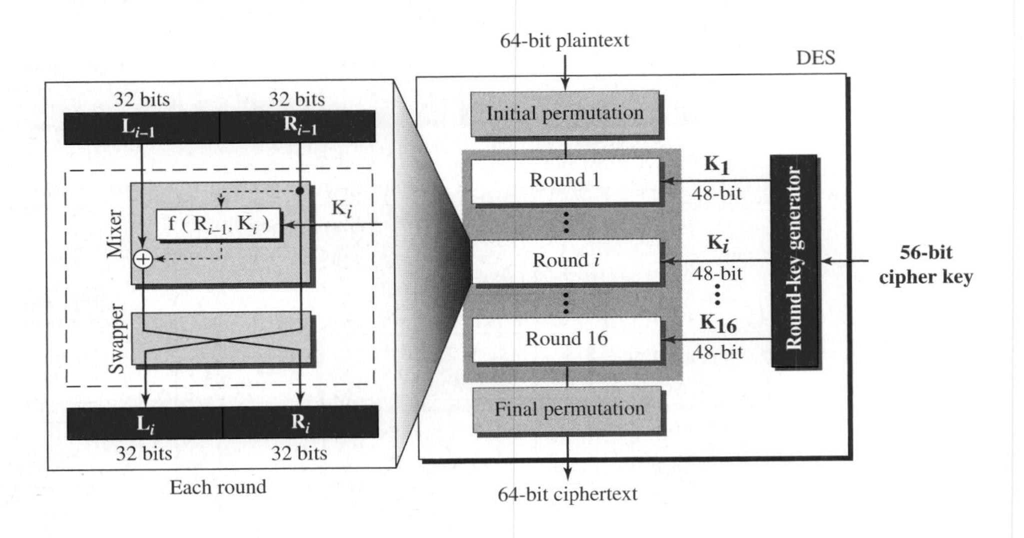

Figure 31.10 *DES function*

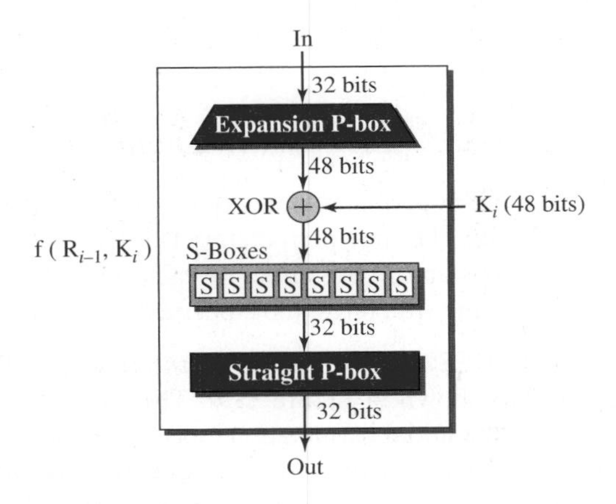

After the expansion permutation, DES uses the XOR operation on the expanded right section and the round key. The S-boxes do the real mixing. DES uses 8 S-boxes, each with a 6-bit input and a 4-bit output.The last operation in the DES function is a straight permutation with a 32-bit input and a 32-bit output.

Key Generation The round-key generator creates sixteen 48-bit keys out of a 56-bit cipher key. However, the cipher key is normally given as a 64-bit key in which 8 extra bits are the parity bits, which are dropped before the actual key-generation process, as shown in Figure 31.11.

Figure 31.11 *Key generation*

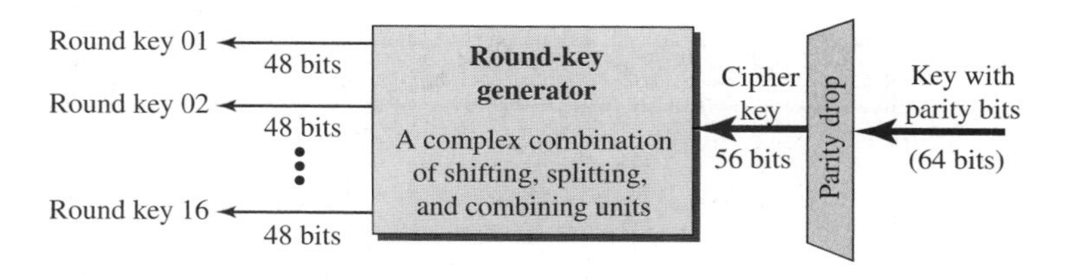

Example 31.5

We choose a random plaintext block and a random key, and determine (using a program) what the ciphertext block would be (all in hexadecimal) as shown below.

Plaintext:	Key:	CipherText:
123456ABCD132536	AABB09182736CCDD	C0B7A8D05F3A829C

Example 31.6

To check the effectiveness of DES when a single bit is changed in the input, we use two different plaintexts with only a single bit difference (in a program). The two ciphertexts are completely different without even changing the key.

Although the two plaintext blocks differ only in the rightmost bit, the ciphertext blocks differ in 29 bits.

Plaintext:	Key:	Ciphertext:
0000000000000000	22234512987ABB23	4789FD476E82A5F1
Plaintext:	Key:	Ciphertext:
0000000000000001	22234512987ABB23	0A4ED5C15A63FEA3

Modern Stream Ciphers

In addition to modern block ciphers, we can also use modern stream ciphers. The differences between modern stream ciphers and modern block ciphers are similar to the differences between traditional stream and block ciphers, which we explained in the previous section. In a *modern stream cipher,* encryption and decryption are done r bits at a time. We have a plaintext bit stream $P = p_n \cdots p_2 p_1$, a ciphertext bit stream $C = c_n \cdots c_2 c_1$, and a key bit stream $K = k_n \cdots k_2 k_1$, in which p_i, c_i, and k_i are r-bit words. Encryption is $c_i = E(k_i, p_i)$, and decryption is $p_i = D(k_i, c_i)$. Stream ciphers are faster than block ciphers. The hardware implementation of a stream cipher is also easier. When we need to encrypt binary streams and transmit them at a constant rate, a stream cipher is the better choice to use. Stream ciphers are also more immune to the corruption of bits during transmission.

The simplest and the most secure type of synchronous stream cipher is called the **one-time pad,** which was invented and patented by Gilbert Vernam. A one-time pad cipher uses a key stream that is randomly chosen for each encipherment. The encryption and decryption algorithms each use a single exclusive-OR operation. Based on properties of the exclusive-OR operation, the encryption and decryption algorithms are inverses of each other. It is important to note that in this cipher the exclusive-OR operation is used one bit at a time. Note also that there must be a secure channel so that Alice can send the key stream sequence to Bob (Figure 31.12).

Figure 31.12 *One-time pad*

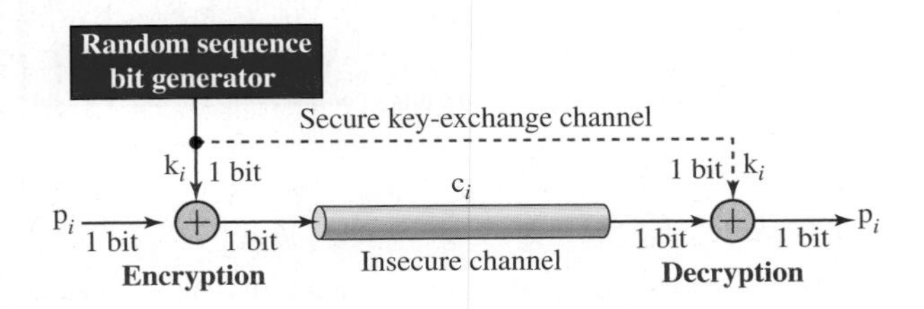

The one-time pad is an ideal cipher. It is perfect. There is no way that an adversary can guess the key or the plaintext and ciphertext statistics. There is no relationship between the plaintext and ciphertext, either. In other words, the ciphertext is a true random stream of bits even if the plaintext contains some patterns. Eve cannot break the cipher unless she tries all possible random key streams, which would be 2^n if the size of the plaintext is n bits. However, there is an issue here. How can the sender and the receiver share a one-time pad key each time they want to communicate? They need to somehow agree on the random key. So this perfect and ideal cipher is very difficult to achieve. However, there are some feasible, less secured, versions. One of the common alternatives is called a *feedback shift register* (FSR), but we leave the discussion of this interesting cipher to the books dedicated to the security topic.

31.2.2 Asymmetric-Key Ciphers

In previous sections we discussed symmetric-key ciphers. In this section, we start the discussion of **asymmetric-key ciphers.** Symmetric- and asymmetric-key ciphers will exist in parallel and continue to serve the community. We actually believe that they are complements of each other; the advantages of one can compensate for the disadvantages of the other.

The conceptual differences between the two systems are based on how these systems keep a secret. In symmetric-key cryptography, the secret must be shared between two persons. In asymmetric-key cryptography, the secret is personal (unshared); each person creates and keeps his or her own secret.

In a community of n people, $n\ (n - 1)/2$ shared secrets are needed for symmetric-key cryptography; only n personal secrets are needed in asymmetric-key cryptography. For a community with a population of 1 million, symmetric-key cryptography would require half a billion shared secrets; asymmetric-key cryptography would require 1 million personal secrets.

Symmetric-key cryptography is based on sharing secrecy; asymmetric-key cryptography is based on personal secrecy.

There are some other aspects of security besides encipherment that need asymmetric-key cryptography. These include authentication and digital signatures. Whenever an application is based on a personal secret, we need to use asymmetric-key cryptography.

Whereas symmetric-key cryptography is based on substitution and permutation of symbols (characters or bits), asymmetric-key cryptography is based on applying mathematical functions to numbers. In symmetric-key cryptography, the plaintext and ciphertext are thought of as a combination of symbols. Encryption and decryption permute these symbols or substitute one symbol for another. In asymmetric-key cryptography, the plaintext and ciphertext are numbers; encryption and decryption are mathematical functions that are applied to numbers to create other numbers.

In symmetric-key cryptography, symbols are permuted or substituted; in asymmetric-key cryptography, numbers are manipulated.

Asymmetric key cryptography uses two separate keys: one private and one public. If encryption and decryption are thought of as locking and unlocking padlocks with keys, then the padlock that is locked with a public key can be unlocked only with the corresponding private key. Figure 31.13 shows that if Alice locks the padlock with Bob's public key, then only Bob's private key can unlock it.

Figure 31.13 *Locking and unlocking in asymmetric-key cryptosystem*

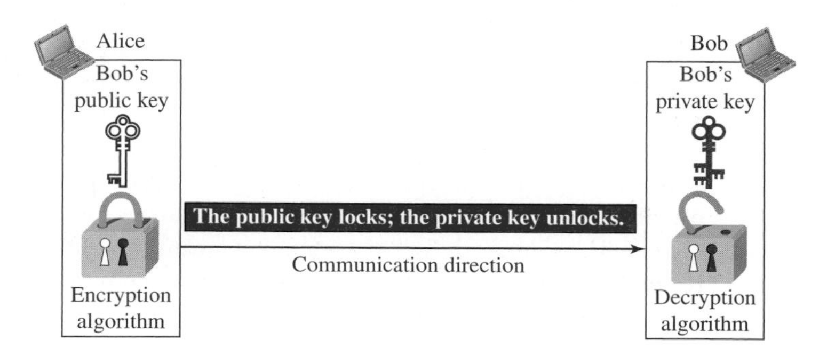

The figure shows that, unlike symmetric-key cryptography, there are distinctive keys in asymmetric-key cryptography: a **private key** and a **public key.** Although some books use the term *secret key* instead of *private key,* we use the term *secret key* only for symmetric-key cryptography and the terms *private key* and *public key* for asymmetric-key cryptography. We even use different symbols to show the three keys. In other words, we want to show that a *secret key* is not interchangeable with a *private key;* there are two different types of secrets.

Asymmetric-key ciphers are sometimes called public-key ciphers.

General Idea

Figure 31.14 shows the general idea of asymmetric-key cryptography as used for encipherment.

Figure 31.14 *General idea of asymmetric-key cryptosystem*

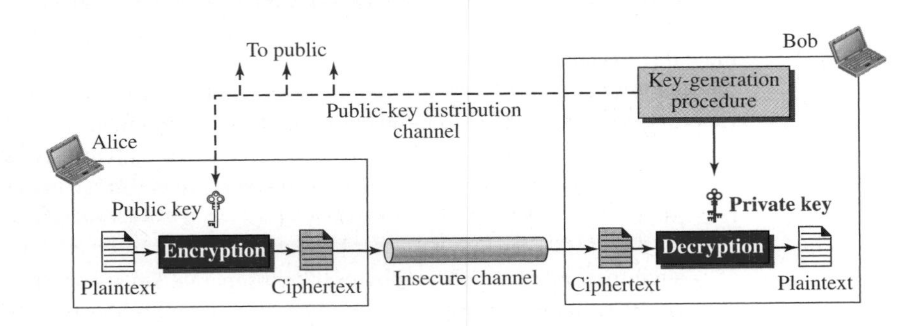

Figure 31.14 shows several important facts. First, it emphasizes the asymmetric nature of the cryptosystem. The burden of providing security is mostly on the shoulders of the receiver (Bob, in this case). Bob needs to create two keys: one private and one public. Bob is responsible for distributing the public key to the community. This can be done through a public-key distribution channel. Although this channel is not required to provide secrecy, it must provide authentication and integrity. Eve should not be able to advertise her public key to the community pretending that it is Bob's public key.

Second, asymmetric-key cryptography means that Bob and Alice cannot use the same set of keys for two-way communication. Each entity in the community should create its own private and public keys. Figure 31.14 shows how Alice can use Bob's public key to send encrypted messages to Bob. If Bob wants to respond, Alice needs to establish her own private and public keys.

Third, asymmetric-key cryptography means that Bob needs only one private key to receive all correspondence from anyone in the community, but Alice needs n public keys to communicate with n entities in the community, one public key for each entity. In other words, Alice needs a ring of public keys.

Plaintext/Ciphertext

Unlike in symmetric-key cryptography, plaintext and ciphertext in asymmetric-key cryptography are treated as integers. The message must be encoded as an integer (or a set of integers) before encryption; the integer (or the set of integers) must be decoded into the message after decryption. Asymmetric-key cryptography is normally used to encrypt or decrypt small pieces of information, such as the cipher key for a symmetric-key cryptography. In other words, asymmetric-key cryptography normally is used for ancillary goals instead of message encipherment. However, these ancillary goals play a very important role in cryptography today.

Asymmetric-key cryptography is normally used to encrypt or decrypt small pieces of information.

Encryption/Decryption

Encryption and decryption in asymmetric-key cryptography are mathematical functions applied over the numbers representing the plaintext and ciphertext. The ciphertext can be thought of as $C = f(K_{public}, P)$; the plaintext can be thought of as $P = g(K_{private}, C)$. The encryption function f is used only for encryption; the decryption function g is used only for decryption.

Need for Both

There is a very important fact that is sometimes misunderstood: the advent of asymmetric-key (public-key) cryptography does not eliminate the need for symmetric-key (secret-key) cryptography. The reason is that asymmetric-key cryptography, which uses mathematical functions for encryption and decryption, is much slower than symmetric-key cryptography. For encipherment of large messages, symmetric-key cryptography is still needed. On the other hand, the speed of symmetric-key cryptography does not eliminate the need for asymmetric-key cryptography. Asymmetric-key cryptography is still needed for authentication, digital signatures, and secret-key exchanges. This means that, to be able to use all aspects of security today, we need both symmetric-key and asymmetric-key cryptography. One complements the other.

RSA Cryptosystem

Although there are several asymmetric-key cryptosystems, one of the common public-key algorithms is the **RSA cryptosystem,** named for its inventors (Rivest, Shamir, and Adleman). RSA uses two exponents, e and d, where e is public and d is private. Suppose P is the plaintext and C is the ciphertext. Alice uses $C = P^e \bmod n$ to create ciphertext C from plaintext P; Bob uses $P = C^d \bmod n$ to retrieve the plaintext sent by Alice. The modulus n, a very large number, is created during the key generation process.

Procedure

Figure 31.15 shows the general idea behind the procedure used in RSA.

Bob chooses two large numbers, p and q, and calculates $n = p \times q$ and $\phi = (p - 1) \times (q - 1)$. Bob then selects e and d such that $(e \times d) \bmod \phi = 1$. Bob advertises e and n to the community as the public key; Bob keeps d as the private key. Anyone, including Alice, can encrypt a message and send the ciphertext to Bob, using $C = (P^e) \bmod n$; only Bob can decrypt the message, using $P = (C^d) \bmod n$. An intruder such as Eve cannot decrypt the message if p and q are very large numbers (she does not know d).

Example 31.7

For the sake of demonstration, let Bob choose 7 and 11 as p and q and calculate $n = 7 \times 11 = 77$. The value of $\phi(n) = (7 - 1)(11 - 1)$, or 60. If he chooses e to be 13, then d is 37. Note that $e \times d \bmod 60 = 1$. Now imagine that Alice wants to send the plaintext 5 to Bob. She uses the public exponent 13 to encrypt 5. This system is not safe because p and q are small.

Figure 31.15 *Encryption, decryption, and key generation in RSA*

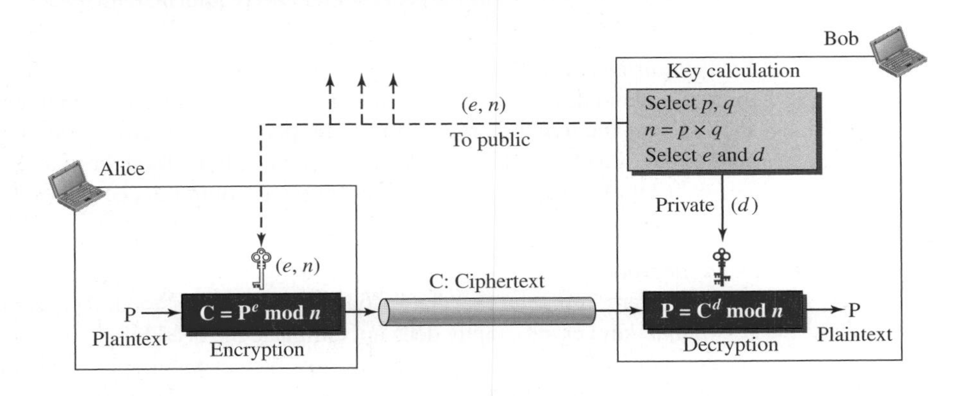

Plaintext: 5
$C = 5^{13} = 26 \bmod 77$
Ciphertext: 26

Ciphertext: 26
$P = 26^{37} = 5 \bmod 77$
Plaintext: 5

Example 31.8

Here is a more realistic example calculated using a computer program in Java. We choose a 512-bit p and q, calculate n and $\phi(n)$. We then choose e and calculate d. Finally, we show the results of encryption and decryption. The integer p is a 159-digit number.

$p =$	9613034531358350457419158128061542790930984559499621582258315087964794045505647063849125716018034750312098666606492420191808780667421096063354219926661209

The integer q is a 160-digit number.

$q =$	12060191957231446918276794204450896001555925054637033936061798321731482148483764659215389453209175225273226830107120695604602513887145524969000359660045617

The modulus $n = p \times q$. It has 309 digits.

$n =$	115935041739676149688925098646158875237714573754541447754855261376147885408326350817276878815968325168468849300625485764111250162414552339182927162507656772727460097082714127730434960500556347274566628060099924037102991424472292215772798531727033839381334692684137327622000966676671831831088373420823444370953

$\phi(n) = (p - 1)(q - 1)$ has 309 digits.

Bob chooses $e = 35535$ (the ideal is 65537). He then finds d.

$\phi(n) =$	115935041739676149688925098646158875237714573754541447754855261376147885408326350817276878815968325168468849300625485764111250162414552339182927162507656751054233608492916752034482627988117554787657013923444405716989581728196098226361075467211864612171359107358640614008885170265377277264467341066243857664128

$e =$	35535
$d =$	580083028600377639360936612896779175946690620896509621804228661113805938528223587317062869100300217108590443384021707298690876006115306202524959884448047568240966247081485817130463240644077704833134010850947385295645071936774061197326557424237217617674620776371642076003370853332885321447088595513667029483 1

Alice wants to send the message "THIS IS A TEST", which can be changed to a numeric value using the 00–26 encoding scheme (26 is the *space* character).

The ciphertext calculated by Alice is $C = P^e$, which is shown below.

P =	1907081826081826002619041819

C =	4753091236462268272063655506105451809423717960704917165232392430544529606131993285666178434183591141511974112520056829797945717360361012782188478927415660904800235071907152771859149751884658886321011483541033616578984679683867637337657774656250792805211481418440481418443081277305900469287424855916646210865 6

Bob can recover the plaintext from the ciphertext using $P = C^d$, which is shown below.

P =	1907081826081826002619041819

The recovered plaintext is "THIS IS A TEST" after decoding.

Applications

Although RSA can be used to encrypt and decrypt actual messages, it is very slow if the message is long. RSA, therefore, is useful for short messages. In particular, we will see that RSA is used in digital signatures and other cryptosystems that often need to encrypt a small message without having access to a symmetric key. RSA is also used for authentication, as we will see later in the chapter.

31.3 OTHER ASPECTS OF SECURITY

The cryptography systems that we have studied so far provide confidentiality. However, in modern communication, we need to take care of other aspects of security, such as integrity, message and entity authentication, nonrepudiation, and key management. We briefly discuss these issues in this section.

31.3.1 Message Integrity

There are occasions where we may not even need secrecy but instead must have integrity: the message should remain unchanged. For example, Alice may write a will to distribute her estate upon her death. The will does not need to be encrypted. After her death, anyone can examine the will. The integrity of the will, however, needs to be preserved. Alice does not want the contents of the will to be changed.

Message and Message Digest

One way to preserve the integrity of a document is through the use of a *fingerprint.* If Alice needs to be sure that the contents of her document will not be changed, she can put her fingerprint at the bottom of the document. Eve cannot modify the contents of this document or create a false document because she cannot forge Alice's fingerprint. To ensure that the document has not been changed, Alice's fingerprint on the document can be compared to Alice's fingerprint on file. If they are not the same, the document is not from Alice. The electronic equivalent of the document and fingerprint pair is the *message* and *digest* pair. To preserve the integrity of a message, the message is passed through an algorithm called a **cryptographic hash function.** The function creates a compressed image of the message, called a **digest,** that can be used like a fingerprint. To check the integrity of a message or document, Bob runs the cryptographic hash function again and compares the new digest with the previous one. If both are the same, Bob is sure that the original message has not been changed. Figure 31.16 shows the idea.

Figure 31.16 *Message and digest*

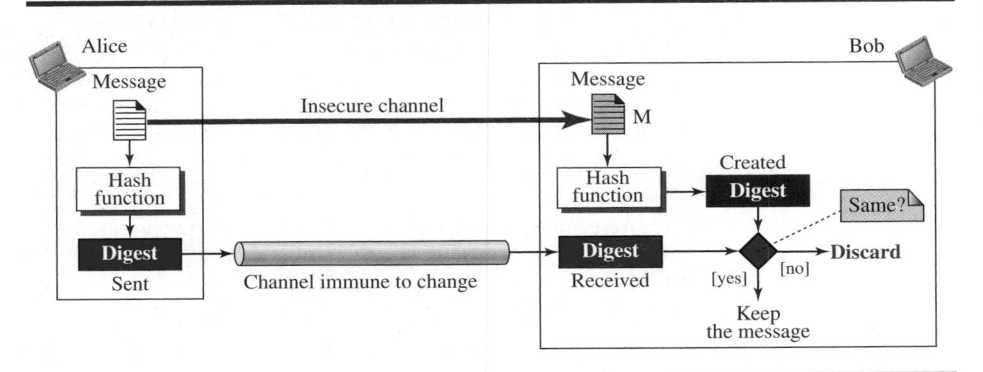

The two pairs (document/fingerprint) and (message/message digest) are similar, with some differences. The document and fingerprint are physically linked together. The message and message digest can be unlinked (or sent separately), and, most importantly, the message digest needs to be safe from change.

The message digest needs to be safe from change.

Hash Functions

A cryptographic hash function takes a message of arbitrary length and creates a message digest of fixed length. All cryptographic hash functions need to create a fixed-size digest out of a variable-size message. Creating such a function is best accomplished using iteration. Instead of using a hash function with variable-size input, a function with fixed-size input is created and is used a necessary number of times. The fixed-size input function is referred to as a *compression function*. It compresses an n-bit string to create an m-bit string where n is normally greater than m. The scheme is referred to as an *iterated cryptographic hash function*.

Several hash algorithms were designed by Ron Rivest. These are referred to as *MD2*, *MD4*, and *MD5*, where *MD* stands for ***Message Digest.*** The last version, MD5, is a strengthened version of MD4 that divides the message into blocks of 512 bits and creates a 128-bit digest. It turns out, however, that a message digest of size 128 bits is too small to resist attack.

In response to the insecurity of MD hash algorithms, the Secure Hash Algorithm was invented. The **Secure Hash Algorithm (SHA)** is a standard that was developed by the National Institute of Standards and Technology (NIST). SHA has gone through several versions.

31.3.2 Message Authentication

A digest can be used to check the integrity of a message—that the message has not been changed. To ensure the integrity of the message and the data origin authentication—that Alice, not somebody else, is the originator of the message—we need to include a secret shared by Alice and Bob (that Eve does not possess) in the process; we need to create a **message authentication code (MAC).** Figure 31.17 shows the idea.

Figure 31.17 *Message authentication code*

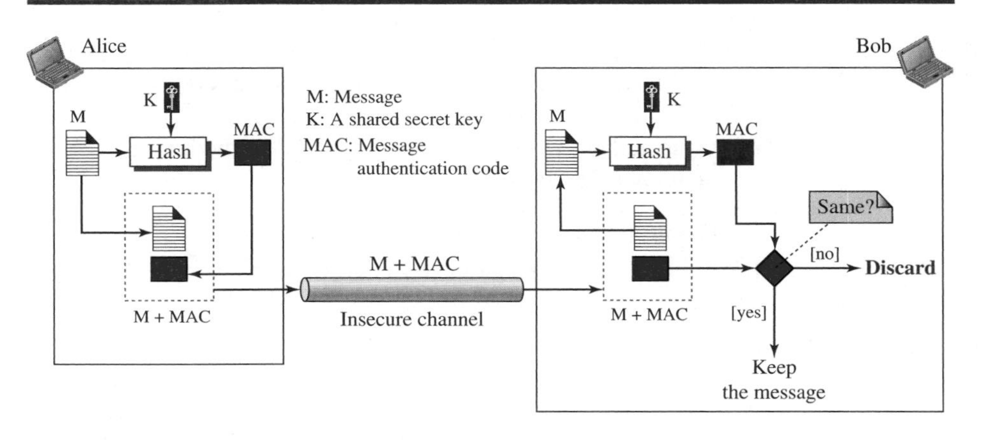

Alice uses a hash function to create a MAC from the concatenation of the key and the message, h(K + M). She sends the message and the MAC to Bob over the insecure channel. Bob separates the message from the MAC. He then makes a new MAC from the concatenation of the message and the secret key. Bob then compares the newly created MAC with the one received. If the two MACs match, the message is authentic and has not been modified by an adversary.

Note that there is no need to use two channels in this case. Both the message and the MAC can be sent on the same insecure channel. Eve can see the message, but she cannot forge a new message to replace it because Eve does not possess the secret key between Alice and Bob. She is unable to create the same MAC that Alice did.

A MAC provides message integrity and message authentication using a combination of a hash function and a secret key.

HMAC

The National Institute of Standards and Technology (NIST) has issued a standard for a nested MAC that is often referred to as ***HMAC (hashed MAC).*** The implementation of HMAC is much more complex than the simplified MAC and is not covered in this text.

31.3.3 Digital Signature

Another way to provide message integrity and message authentication (and some more security services, as we will see shortly) is a digital signature. A MAC uses a secret key to protect the digest; a digital signature uses a pair of private-public keys.

A digital signature uses a pair of private-public keys.

We are all familiar with the concept of a signature. A person signs a document to show that it originated from her or was approved by her. The signature is proof to the recipient that the document comes from the correct entity. When a customer signs a check, the bank needs to be sure that the check is issued by that customer and nobody else. In other words, a signature on a document, when verified, is a sign of authentication—the document is authentic. Consider a painting signed by an artist. The signature on the art, if authentic, means that the painting is probably authentic.

When Alice sends a message to Bob, Bob needs to check the authenticity of the sender; he needs to be sure that the message comes from Alice and not Eve. Bob can ask Alice to sign the message electronically. In other words, an electronic signature can prove the authenticity of Alice as the sender of the message. We refer to this type of signature as a **digital signature.**

Comparison

Let us begin by looking at the differences between conventional signatures and digital signatures.

Inclusion

A conventional signature is included in the document; it is part of the document. When we write a check, the signature is on the check; it is not a separate document. But when we sign a document digitally, we send the signature as a separate document.

Verification Method

The second difference between the two types of signatures is the method of verifying the signature. For a conventional signature, when the recipient receives a document, she compares the signature on the document with the signature on file. If they are the same, the document is authentic. The recipient needs to have a copy of this signature on file for comparison. For a digital signature, the recipient receives the message and the signature. A copy of the signature is not stored anywhere. The recipient needs to apply a verification technique to the combination of the message and the signature to verify the authenticity.

Relationship

For a conventional signature, there is normally a one-to-many relationship between a signature and documents. A person uses the same signature to sign many documents. For a digital signature, there is a one-to-one relationship between a signature and a message. Each message has its own signature. The signature of one message cannot be used in another message. If Bob receives two messages, one after another, from Alice, he cannot use the signature of the first message to verify the second. Each message needs a new signature.

Duplicity

Another difference between the two types of signatures is a quality called *duplicity.* With a conventional signature, a copy of the signed document can be distinguished from the original one on file. With a digital signature, there is no such distinction unless there is a factor of time (such as a timestamp) on the document. For example, suppose Alice sends a document instructing Bob to pay Eve. If Eve intercepts the document and the signature, she can resend it later to get money again from Bob.

Process

Figure 31.18 shows the digital signature process. The sender uses a *signing algorithm* to sign the message. The message and the signature are sent to the receiver. The receiver receives the message and the signature and applies the *verifying algorithm* to the combination. If the result is true, the message is accepted; otherwise, it is rejected.

Figure 31.18 *Digital signature process*

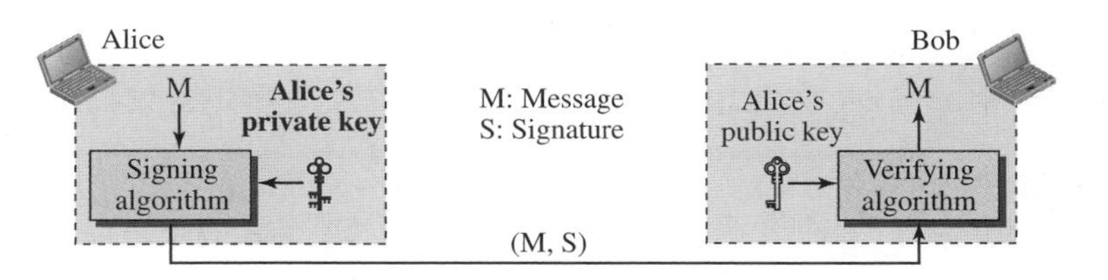

A conventional signature is like a private "key" belonging to the signer of the document. The signer uses it to sign documents; no one else has this signature. The copy of the signature on file is like a public key; anyone can use it to verify a document, to compare it to the original signature.

In a digital signature, the signer uses her private key, applied to a signing algorithm, to sign the document. The verifier, on the other hand, uses the public key of the signer, applied to the verifying algorithm, to verify the document.

Note that when a document is signed, anyone, including Bob, can verify it because everyone has access to Alice's public key. Alice must not use her public key to sign the document because then anyone could forge her signature.

Can we use a secret (symmetric) key to both sign and verify a signature? The answer is negative for several reasons. First, a secret key is known by only two entities (Alice and Bob, for example). So if Alice needs to sign another document and send it to

Ted, she needs to use another secret key. Second, as we will see, creating a secret key for a session involves authentication, which uses a digital signature. We have a vicious cycle. Third, Bob could use the secret key between himself and Alice, sign a document, send it to Ted, and pretend that it came from Alice.

A digital signature needs a public-key system.
The signer signs with her private key; the verifier verifies with the signer's public key.

We should make a distinction between private and public keys as used in digital signatures and public and private keys as used in a cryptosystem for confidentiality. In the latter, the public and private keys of the receiver are used in the process. The sender uses the public key of the receiver to encrypt; the receiver uses his own private key to decrypt. In a digital signature, the private and public keys of the sender are used. The sender uses her private key; the receiver uses the sender's public key.

A cryptosystem uses the public and private keys of the receiver;
a digital signature uses the private and public keys of the sender.

Signing the Digest

We said before that the asymmetric-key cryptosystems are very inefficient when dealing with long messages. In a digital signature system, the messages are normally long, but we have to use asymmetric-key schemes. The solution is to sign a digest of the message, which is much shorter than the message. A carefully selected message digest has a one-to-one relationship with the message. The sender can sign the message digest and the receiver can verify the message digest. The effect is the same. Figure 31.19 shows signing a digest in a digital signature system.

Figure 31.19 *Signing the digest*

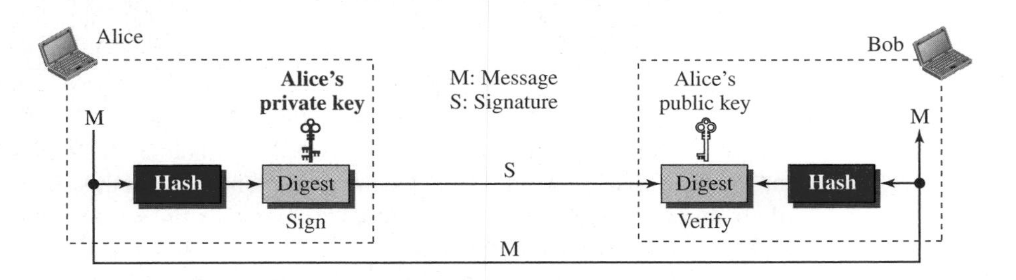

A digest is made out of the message at Alice's site. The digest then goes through the signing process using Alice's private key. Alice then sends the message and the signature to Bob.

At Bob's site, using the same public hash function, a digest is first created out of the received message. The verifying process is applied. If authentic, the message is accepted; otherwise, it is rejected.

Services

We discussed several security services in the beginning of the chapter including *message confidentiality, message authentication, message integrity*, and *nonrepudiation*. A digital signature can directly provide the last three; for message confidentiality we still need encryption/decryption.

Message Authentication

A secure digital signature scheme, like a secure conventional signature (one that cannot be easily copied) can provide message authentication (also referred to as data-origin authentication). Bob can verify that the message is sent by Alice because Alice's public key is used in verification. Alice's public key cannot verify the signature signed by Eve's private key.

Message Integrity

The integrity of the message is preserved if we sign the message or the digest of the message because we cannot get the same digest if any part of the message is changed. The digital signature schemes today use a hash function in the signing and verifying algorithms that preserves the integrity of the message.

Nonrepudiation

If Alice signs a message and then denies it, can Bob later prove that Alice actually signed it? For example, if Alice sends a message to a bank (Bob) and asks to transfer $10,000 from her account to Ted's account, can Alice later deny that she sent this message? With the scheme we have presented so far, Bob might have a problem. Bob must keep the signature on file and later use Alice's public key to create the original message to prove the message in the file and the newly created message are the same. This is not feasible because Alice may have changed her private or public key during this time; she may also claim that the file containing the signature is not authentic.

One solution is a trusted third party. People can create an established trusted party among themselves. Later in the chapter, we will see that a trusted party can solve many other problems concerning security services and key exchange. Figure 31.20 shows how a trusted party can prevent Alice from denying that she sent the message.

Alice creates a signature from her message (S_A) and sends the message, her identity, Bob's identity, and the signature to the center. The center, after checking that Alice's public key is valid, verifies through Alice's public key that the message came from Alice. The center then saves a copy of the message with the sender's identity, recipient's identity, and a timestamp in its archive. The center uses its private key to create another signature (S_T) from the message. The center then sends the message, the new signature, Alice's identity, and Bob's identity to Bob. Bob verifies the message using the public key of the trusted center.

If in the future Alice denies that she sent the message, the center can show a copy of the saved message. If Bob's message is a duplicate of the message saved at the center, Alice will lose the dispute. To make everything confidential, a level of encryption/decryption can be added to the scheme, as discussed in the next section.

Figure 31.20 *Using a trusted center for nonrepudiation*

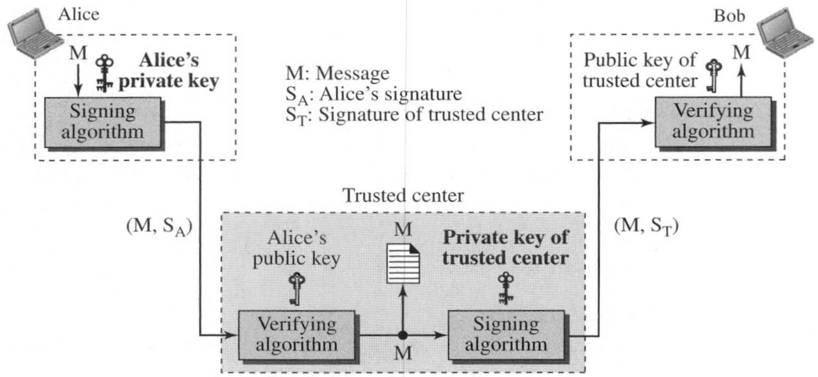

Confidentiality

A digital signature does not provide confidential communication. If confidentiality is required, the message and the signature must be encrypted using either a symmetric-key or an asymmetric-key cipher.

RSA Digital Signature Scheme

Several *digital signature schemes* have evolved during the last few decades. Some of them have been implemented. In this section, we briefly show one of them, RSA. In a previous section, we discussed how to use the RSA cryptosystem to provide privacy. The RSA idea can also be used for signing and verifying a message. In this case, it is called the *RSA digital signature scheme.* The digital signature scheme changes the roles of the private and public keys. First, the private and public keys of the sender, not the receiver, are used. Second, the sender uses her own private key to sign the document; the receiver uses the sender's public key to verify it. If we compare the scheme with the conventional way of signing, we see that the private key plays the role of the sender's own signature and the sender's public key plays the role of the copy of the signature that is available to the public. Obviously Alice cannot use Bob's public key to sign the message, because then any other person could do the same. The signing and verifying sites use the same function, but with different parameters. The verifier compares the message and the output of the function for equality in modulo arithmetic. If the result is true, the message is accepted. Figure 31.21 shows the scheme in which the signing and verifying is done on the digest of the message instead of the message itself because the public-key cryptography is not very efficient for use with long messages; the digest is much smaller than the message itself.

Alice, the signer, first uses an agreed-upon hash function to create a digest from the message, $D = \boldsymbol{h}(M)$. She then signs the digest, $S = D^d \bmod n$. The message and the signature are sent to Bob. Bob, the verifier, receives the message and the signature. He first uses Alice's public exponent to retrieve the digest, $D' = S^e \bmod n$. He then applies the hash algorithm to the message received to obtain $D = \boldsymbol{h}(M)$. Bob now

Figure 31.21 *The RSA signature on the message digest*

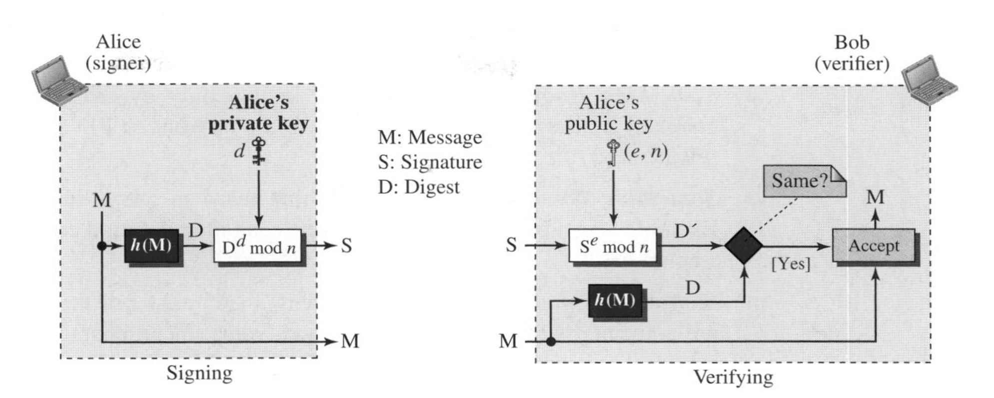

compares the two digests, D and D′. If they are equal (in modulo arithmetic), he accepts the message.

Digital Signature Standard (DSS)

The **Digital Signature Standard (DSS)** was adopted by NIST in 1994. DSS is a complicated, and more secure, digital signature scheme.

31.3.4 Entity Authentication

Entity authentication is a technique designed to let one party verify the identity of another party. An *entity* can be a person, a process, a client, or a server. The entity whose identity needs to be proven is called the *claimant;* the party that tries to verify the identity of the claimant is called the *verifier.*

Entity versus Message Authentication

There are two differences between *entity authentication* and *message authentication (data-origin authentication).*

1. Message authentication (or data-origin authentication) might not happen in real time; entity authentication does. In the former, Alice sends a message to Bob. When Bob authenticates the message, Alice may or may not be present in the communication process. On the other hand, when Alice requests entity authentication, there is no real message communication involved until Alice is authenticated by Bob. Alice needs to be online and to take part in the process. Only after she is authenticated can messages be communicated between Alice and Bob. Data-origin authentication is required when an e-mail is sent from Alice to Bob. Entity authentication is required when Alice gets cash from an automatic teller machine.
2. Message authentication simply authenticates one message; the process needs to be repeated for each new message. Entity authentication authenticates the claimant for the entire duration of a session.

Verification Categories

In entity authentication, the claimant must identify herself to the verifier. This can be done with one of three kinds of witnesses: *something known, something possessed,* or *something inherent.*

- ❑ ***Something Known.*** This is a secret known only by the claimant that can be checked by the verifier. Examples are a password, a PIN, a secret key, and a private key.
- ❑ ***Something Possessed.*** This is something that can prove the claimant's identity. Examples are a passport, a driver's license, an identification card, a credit card, and a smart card.
- ❑ ***Something Inherent.*** This is an inherent characteristic of the claimant. Examples are conventional signatures, fingerprints, voice, facial characteristics, retinal pattern, and handwriting.

In this section, we only discuss the first type of witness, *something known,* which is normally used for remote (online) entity authentication. The other two categories are normally used when the claimant is personally present.

Passwords

The simplest and oldest method of entity authentication is the use of a *password,* which is something that the claimant *knows*. A password is used when a user needs to access a system's resources (login). Each user has a user identification that is public, and a password that is private. Passwords, however, are very prone to attack. A password can be stolen, intercepted, guessed, and so on.

Challenge-Response

In password authentication, the claimant proves her identity by demonstrating that she knows a secret, the password. However, because the claimant sends this secret, it is susceptible to interception by the adversary. In **challenge-response authentication,** the claimant proves that she *knows* a secret without sending it. In other words, the claimant does not send the secret to the verifier; the verifier either has it or finds it.

> **In challenge-response authentication, the claimant proves that she knows a secret without sending it to the verifier.**

The *challenge* is a time-varying value such as a random number or a timestamp that is sent by the verifier. The claimant applies a function to the challenge and sends the result, called a *response,* to the verifier. The response shows that the claimant knows the secret.

Using a Symmetric-Key Cipher

Several approaches to challenge-response authentication use symmetric-key encryption. The secret here is the shared secret key, known by both the claimant and the verifier. The function is the encrypting algorithm applied on the challenge. Although there are several approaches to this method, we just show the simplest one to give an idea. Figure 31.22 shows this first approach.

Figure 31.22 *Unidirectional, symmetric-key authentication*

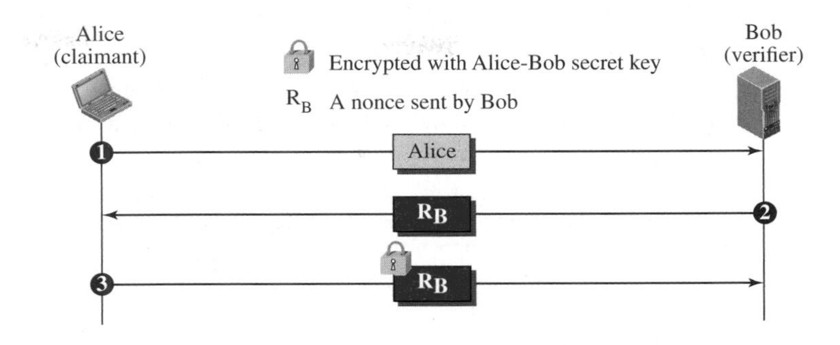

The first message is not part of challenge-response, it only informs the verifier that the claimant wants to be challenged. The second message is the challenge. R_B is the nonce (abbreviation for *number once*) randomly chosen by the verifier (Bob) to challenge the claimant. The claimant encrypts the nonce using the shared secret key known only to the claimant and the verifier and sends the result to the verifier. The verifier decrypts the message. If the nonce obtained from decryption is the same as the one sent by the verifier, Alice is granted access.

Note that in this process, the claimant and the verifier need to keep the symmetric key used in the process secret. The verifier must also keep the value of the nonce for claimant identification until the response is returned.

Using an Asymmetric-Key Cipher

Figure 31.23 shows this approach.

Figure 31.23 *Unidirectional, asymmetric-key authentication*

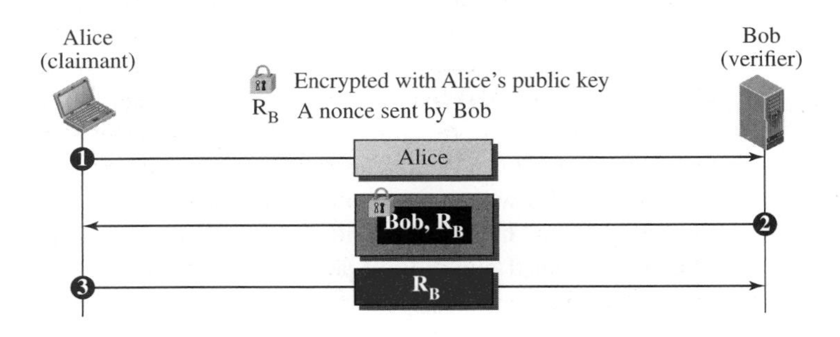

Instead of a symmetric-key cipher, we can use an asymmetric-key cipher for entity authentication. Here the secret must be the private key of the claimant. The claimant must show that she owns the private key related to the public key that is available to everyone. This means that the verifier must encrypt the challenge using the public key of the claimant; the claimant then decrypts the message using her private key. The response to the challenge is the decrypted message.

Using Digital Signatures

Entity authentication can also be achieved using a digital signature. When a digital signature is used for entity authentication, the claimant uses her private key for signing. In the first approach, shown in Figure 31.24, Bob uses a plaintext challenge and Alice signs the response.

Figure 31.24 *Digital signature, unidirectional authentication*

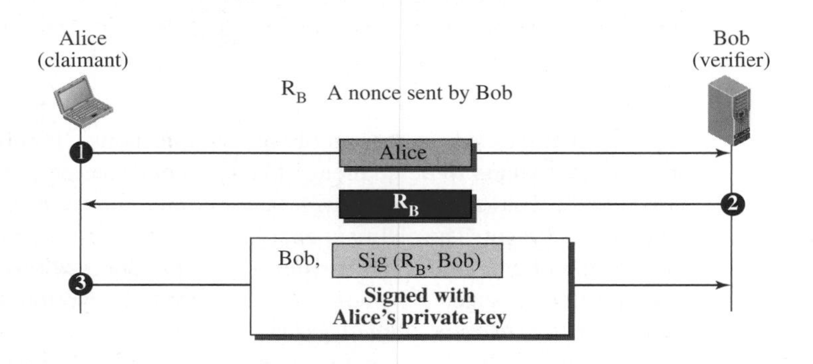

31.3.5 Key Management

We discussed symmetric-key and asymmetric-key cryptography in the previous sections. However, we have not yet discussed how secret keys in symmetric-key cryptography, and public keys in asymmetric-key cryptography, are distributed and maintained. This section touches on these two issues.

Symmetric-Key Distribution

Symmetric-key cryptography is more efficient than asymmetric-key cryptography for enciphering large messages. Symmetric-key cryptography, however, needs a shared secret key between two parties.

If Alice needs to exchange confidential messages with N people, she needs N different keys. What if N people need to communicate with each other? A total of $N(N-1)$ keys is needed if we require that two people use two keys for bidirectional communication; only $N(N-1)/2$ keys are needed if we allow a key to be used for both directions. This means that if one million people need to communicate with each other, each person has almost one million different keys; in total, half a trillion keys are needed. This is normally referred to as the N^2 problem because the number of required keys for N entities is close to N^2.

The number of keys is not the only problem; the distribution of keys is another. If Alice and Bob want to communicate, they need a way to exchange a secret key; if Alice wants to communicate with one million people, how can she exchange one million keys with one million people? Using the Internet is definitely not a secure method. It is obvious that we need an efficient way to maintain and distribute secret keys.

Key Distribution Center: KDC

A practical solution is the use of a trusted third party, referred to as a ***key distribution center*** **(KDC).** To reduce the number of keys, each person establishes a shared secret key with the KDC. A secret key is established between the KDC and each member. Now the question is how Alice can send a confidential message to Bob. The process is as follows:

1. Alice sends a request to the KDC stating that she needs a session (temporary) secret key between herself and Bob.
2. The KDC informs Bob about Alice's request.
3. If Bob agrees, a session key is created between the two.

The secret key between Alice and Bob that is established with the KDC is used to authenticate Alice and Bob to the KDC and to prevent Eve from impersonating either of them.

Multiple KDCs When the number of people using a KDC increases, the system becomes unmanageable and a bottleneck can result. To solve the problem, we need to have multiple KDCs. We can divide the world into domains. Each domain can have one or more KDCs (for redundancy in case of failure). Now if Alice wants to send a confidential message to Bob, who belongs to another domain, Alice contacts her KDC, which in turn contacts the KDC in Bob's domain. The two KDCs can create a secret key between Alice and Bob. There can be local KDCs, national KDCs, and international KDCs. When Alice needs to communicate with Bob, who lives in another country, she sends her request to a local KDC; the local KDC relays the request to the national KDC; the national KDC relays the request to an international KDC. The request is then relayed all the way down to the local KDC where Bob lives. Figure 31.25 shows a configuration of hierarchical multiple KDCs.

Figure 31.25 *Multiple KDCs*

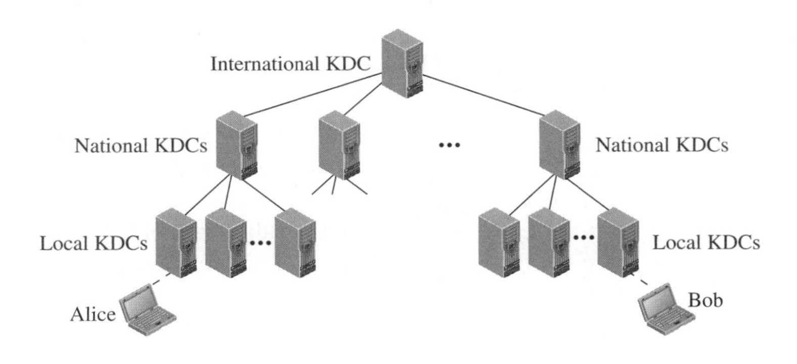

Session Keys A KDC creates a secret key for each member. This secret key can be used only between the member and the KDC, not between two members. If Alice needs to communicate secretly with Bob, she needs a secret key between herself and Bob. A KDC can create a *session key* between Alice and Bob, using their keys with the center.

The keys of Alice and Bob are used to authenticate Alice and Bob to the center and to each other before the session key is established. After communication is terminated, the session key is no longer useful.

A session symmetric key between two parties is used only once.

Several different approaches have been proposed to create the session key using ideas discussed in the previous sections. We show the simplest approach in Figure 31.26. Although this approach is very rudimentary, it helps to understand more sophisticated approaches in the literature.

Figure 31.26 *Creating a session key using KDC*

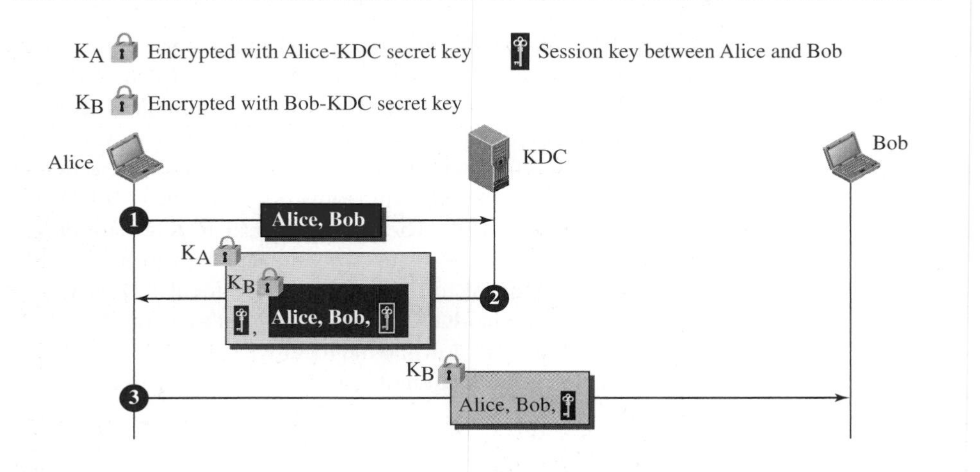

1. Alice sends a plaintext message to the KDC to obtain a symmetric session key between Bob and herself. The message contains her registered identity (the word *Alice* in the figure) and the identity of Bob (the word *Bob* in the figure). This message is not encrypted, it is public. The KDC does not care.
2. The KDC receives the message and creates what is called a **ticket.** The ticket is encrypted using Bob's key (K_B). The ticket contains the identities of Alice and Bob and the session key. The ticket with a copy of the session key is sent to Alice. Alice receives the message, decrypts it, and extracts the session key. She cannot decrypt Bob's ticket; the ticket is for Bob, not for Alice. Note that this message contains a double encryption—the ticket is encrypted, and the entire message is also encrypted. In the second message, Alice is actually authenticated to the KDC, because only Alice can open the whole message using her secret key with KDC.
3. Alice sends the ticket to Bob. Bob opens the ticket and knows that Alice needs to send messages to him using the session key. Note that in this message, Bob is authenticated to the KDC because only Bob can open the ticket. Because Bob is authenticated to the KDC, he is also authenticated to Alice, who trusts the KDC.

In the same way, Alice is also authenticated to Bob, because Bob trusts the KDC and the KDC has sent Bob the ticket that includes the identity of Alice.

Symmetric-Key Agreement

Alice and Bob can create a session key between themselves without using a KDC. This method of session-key creation is referred to as the *symmetric-key agreement*. Although there are several ways to accomplish this, we discuss only one method, Diffie-Hellman, which shows the basic idea used in more sophisticated (less prone to attack) methods.

Diffie-Hellman Key Agreement

In the **Diffie-Hellman protocol** two parties create a symmetric session key without the need of a KDC. Before establishing a symmetric key, the two parties need to choose two numbers p and g. These two numbers have some properties discussed in number theory, but that discussion is beyond the scope of this book. These two numbers do not need to be confidential. They can be sent through the Internet; they can be public. Figure 31.27 shows the procedure.

Figure 31.27 *Diffie-Hellman method*

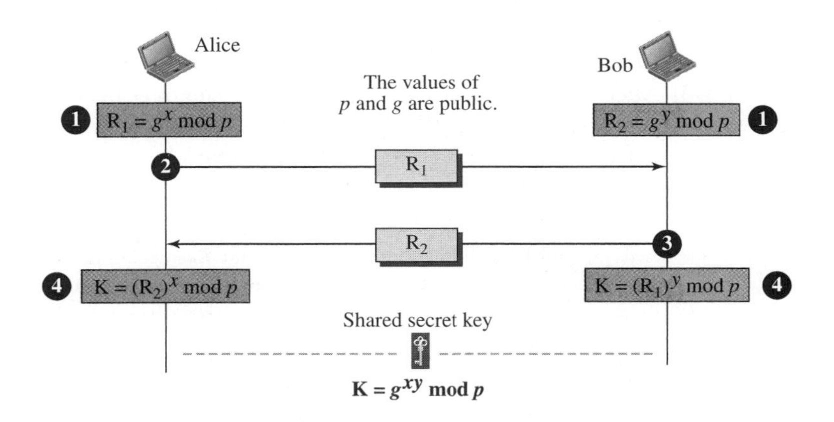

The steps are as follows:

1. Alice chooses a large random number x such that $0 \leq x \leq p - 1$ and calculates $R_1 = g^x \bmod p$. Bob chooses another large random number y such that $0 \leq y \leq p - 1$ and calculates $R_2 = g^y \bmod p$.
2. Alice sends R_1 to Bob. Note that Alice does not send the value of x; she sends only R_1.
3. Bob sends R_2 to Alice. Again, note that Bob does not send the value of y, he sends only R_2.
4. Alice calculates $K = (R_2)^x \bmod p$. Bob also calculates $K = (R_1)^y \bmod p$.

K is the symmetric key for the session.

$$K = (g^x \bmod p)^y \bmod p = (g^y \bmod p)^x \bmod p = g^{xy} \bmod p$$

Bob has calculated $K = (R_1)^y \bmod p = (g^x \bmod p)^y \bmod p = g^{xy} \bmod p$. Alice has calculated $K = (R_2)^x \bmod p = (g^y \bmod p)^x \bmod = g^{xy} \bmod p$. Both have reached the same value without Bob knowing the value of x and without Alice knowing the value of y.

The symmetric (shared) key in the Diffie-Hellman method is $K = g^{xy} \bmod p$.

Example 31.9

Let us give a trivial example to make the procedure clear. Our example uses small numbers, but note that in a real situation, the numbers are very large. Assume that $g = 7$ and $p = 23$. The steps are as follows:

1. Alice chooses $x = 3$ and calculates $R_1 = 7^3 \bmod 23 = 21$. Bob chooses $y = 6$ and calculates $R_2 = 7^6 \bmod 23 = 4$.
2. Alice sends the number 21 to Bob.
3. Bob sends the number 4 to Alice.
4. Alice calculates the symmetric key $K = 4^3 \bmod 23 = 18$. Bob calculates the symmetric key $K = 21^6 \bmod 23 = 18$.
   The value of K is the same for both Alice and Bob; $g^{xy} \bmod p = 7^{18} \bmod 23 = 18$.

Public-Key Distribution

In asymmetric-key cryptography, people do not need to know a symmetric shared key. If Alice wants to send a message to Bob, she only needs to know Bob's public key, which is open to the public and available to everyone. If Bob needs to send a message to Alice, he only needs to know Alice's public key, which is also known to everyone. In public-key cryptography, everyone shields a private key and advertises a public key.

In public-key cryptography, everyone has access to everyone's public key; public keys are available to the public.

Public keys, like secret keys, need to be distributed to be useful. Let us briefly discuss the ways public keys can be distributed.

Public Announcement

The naive approach is to announce public keys publicly. Bob can put his public key on his website or announce it in a local or national newspaper. When Alice needs to send a confidential message to Bob, she can obtain Bob's public key from his site or from the newspaper, or even send a message to ask for it. This approach, however, is not secure; it is subject to forgery. For example, Eve could make such a public announcement. Before Bob can react, damage could be done. Eve can fool Alice into sending her a message that is intended for Bob. Eve could also sign a document with a corresponding forged private key and make everyone believe it was signed by Bob. The approach is also vulnerable if Alice directly requests Bob's public key. Eve can intercept Bob's response and substitute her own forged public key for Bob's public key.

Certification Authority

The common approach to distributing public keys is to create **public-key certificates.** Bob wants two things; he wants people to know his public key, and he wants no one to accept a forged public key as his. Bob can go to a **certification authority (CA),** a federal or state organization that binds a public key to an entity and issues a certificate. Figure 31.28 shows the concept.

Figure 31.28 *Certification authority*

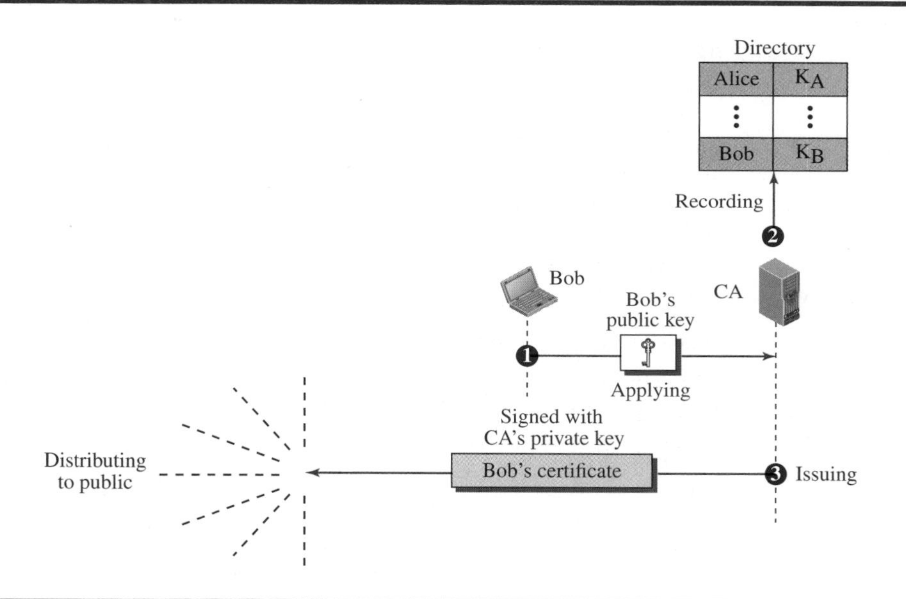

The CA itself has a well-known public key that cannot be forged. The CA checks Bob's identification (using a picture ID along with other proof). It then asks for Bob's public key and writes it on the certificate. To prevent the certificate itself from being forged, the CA signs the certificate with its private key. Now Bob can upload the signed certificate. Anyone who wants Bob's public key downloads the signed certificate and uses the authority's public key to extract Bob's public key.

X.509

Although the use of a CA has solved the problem of public-key fraud, it has created a side effect. Each certificate may have a different format. If Alice wants to use a program to automatically download different certificates and digests belonging to different people, the program may not be able to do this. One certificate may have the public key in one format and another certificate may have it in a different format. The public key may be on the first line in one certificate and on the third line in another. Anything that needs to be used universally must have a universal format. To remove this side effect, the ITU has designed **X.509,** a recommendation that has been accepted by the Internet with some changes. X.509 is a structured way to describe the certificate. It uses a well-known protocol called *ASN.1* (discussed in Chapter 27) that defines fields familiar to computer programmers.

31.4 END-CHAPTER MATERIALS

31.4.1 Recommended Reading

Several books give thorough coverage of cryptography and network security: [For 08], [Sta 06], [Bis 05], [Mao 04], [Sti 06]. [Res 01], [Tho 00], [DH 03], and [Gar 95].

31.4.2 Key Terms

additive cipher
asymmetric-key cipher
autokey cipher
block cipher
Caesar cipher
certification authority (CA)
challenge-response authentication
cipher
ciphertext
cryptographic hash function
cryptography
Data Encryption Standard (DES)
decryption
decryption algorithm
denial of service (DoS)
Diffie-Hellman protocol
digest
digital signature
Digital Signature Standard (DSS)
encryption
encryption algorithm
HMAC (hashed MAC)
key distribution center (KDC)
message authentication code (MAC)
Message Digest (MD)
monoalphabetic cipher
one-time pad
P-box
plaintext
polyalphabetic cipher
private key
public key
public-key certificate
RSA cryptosystem
S-box
Secure Hash Algorithm (SHA)
shift cipher
steganography
stream cipher
substitution cipher
symmetric-key cipher
ticket
transposition cipher
X.509

31.4.3 Summary

The three goals of security are confidentiality, integrity, and availability. These goals are threatened by attacks such as snooping, traffic analysis, modification, masquerading, replaying, repudiation, and denial of service. Cryptography is a technique described in this chapter to achieve security goals; the other technique, steganography, is left for more advanced books on security.

Confidentiality is achieved through asymmetric-key and symmetric-key ciphers. In a symmetric-key cipher the same key is used for encryption and decryption, and the key can be used for bidirectional communication. We can divide traditional symmetric-key ciphers into two broad categories: substitution ciphers and transposition ciphers. In asymmetric-key cryptography there are two separate keys: one private and one public. Asymmetric-key cryptography means that Bob and Alice cannot use the same set of keys for two-way communication.

Discussion of security is not limited to confidentiality; other aspects of security include integrity, message authentication, entity authentication, and key management. Message integrity is achieved using a hashing function to create a digest of the message. Message authentication is achieved using techniques such as message authentication code (MAC) or digital signature. Entity authentication is achieved using either personal identification, such as a password, or techniques such as the challenge-response process. To provide either confidentiality or other aspects of security, we need either secret keys or the combination of private-public keys. The distribution of secret keys can be done by a key distribution center (KDC) or through instantaneous methods such as Diffie-Hellman. The certification of public keys can be done through a certification authority (CA).

31.5 PRACTICE SET

31.5.1 Quizzes

A set of interactive quizzes for this chapter can be found on the book website. It is strongly recommended that the student take the quizzes to check his/her understanding of the materials before continuing with the practice set.

31.5.2 Questions

Q31-1. In Figure 31.10 in the text, why do we need an expansion P-box? Why can't we use a straight or a compression P-box?

Q31-2. If Alice and Bob need to communicate using asymmetric-key cryptography, how many keys do they need? Who needs to create these keys?

Q31-3. Alice needs to send a message to a group of fifty people. If Alice needs to use message authentication, which of the following schemes do you recommend?

a. MAC **b.** digital signature

Q31-4. In an asymmetric public key cipher, which key is used for encryption? Which key is used for decryption?

a. public key **b.** private key

Q31-5. In RSA, why can't Bob choose 1 as the public key e?

Q31-6. Which of the following services are not provided by digital signature?

a. message authentication **b.** confidentiality **c.** nonrepudiation

Q31-7. Which of the following attacks is a threat to confidentiality?

a. snooping **b.** masquerading **c.** repudiation

Q31-8. A certification authority (CA) is designed to solve the problem of distributing __________ keys.

a. secret **b.** public **c.** private

Q31-9. Alice has found a way to write secretly to Bob. Each time, she takes a new text, such as an article from the newspaper, but inserts one or two spaces between the words. A single space means a binary digit 0; a double space means a binary digit 1. Bob extracts the binary digits and interprets them using ASCII code. Is this an example of cryptography or steganography? Explain.

Q31-10. What is the role of the secret key added to the hash function in Figure 31.17 in the text (MAC)? Explain.

Q31-11. Which of the following attacks is a threat to availability?

a. repudiation **b.** denial of service **c.** modification

Q31-12. When a letter is sent from Bob to Alice in a language that only the two can understand, is this an example of cryptography or steganography?

Q31-13. A key distribution center (KDC) is designed to solve the problem of distributing __________ keys.

a. secret **b.** public **c.** private

Q31-14. Figure 31.9 in the text shows that DES creates 16 different 48-bit keys, one for each round. Why do we need 16 different keys? Why can't we use the same key in each round?

Q31-15. Assume Alice and Bob use an additive cipher in modulo 26 arithmetic. If Eve, the intruder, wants to break the code by trying all possible keys (brute-force attack), how many keys should she try on average?

Q31-16. A permutation block (P-box) in a modern block cipher is an example of a keyless transposition cipher. What does this statement mean? (See Figure 31.8 in the text.)

Q31-17. Assume we have a plaintext of 1000 characters. How many keys do we need to encrypt or decrypt the message in each of the following ciphers?

a. additive **b.** monoalphabetic **c.** autokey

Q31-18. In a club with 50 members, how many secret keys are needed to allow secret messages to be exchanged between any pair of members?

Q31-19. A permutation block (P-box) in a modern block cipher has five inputs and five outputs. This is a ______ permutation?

a. straight **b.** compression **c.** expansion

Q31-20. According to the definitions of stream and block ciphers, find which of the following ciphers is a stream cipher.

a. additive **b.** monoalphabetic **c.** autokey

Q31-21. Why do you think asymmetric-key cryptography is used only with small messages.

Q31-22. Alice signs the message she sends to Bob to prove that she is the sender of the message. Which of the following keys does Alice need to use?

a. Alice's public key **b.** Alice's private key

Q31-23. Assume Alice needs to send a confidential signed document to 100 people. How many keys does Alice need to use to prepare 100 copies if she uses asymmetric-key confidentiality? Explain.

Q31-24. Which of the following attacks is a threat to integrity?

a. modification **b.** replaying **c.** denial of service

Q31-25. In a cipher, all As in the plaintext have been changed to Ds in the ciphertext and all Ds in the plaintext have been changed to Hs in the ciphertext. Is this a monoalphabetic or polyalphabetic substitution cipher? Explain.

Q31-26. Alice and Bob exchange confidential messages. They share a very large number as the encryption and decryption key in both directions. Is this an example of symmetric-key or asymmetric-key cryptography? Explain.

Q31-27. Alice uses the same key when she encrypts a message to be sent to Bob and when she decrypts a message received from Bob. Is this an example of symmetric-key or asymmetric-key cryptography? Explain.

Q31-28. Which cipher can be broken more easily, monoalphabetic or polyalphabetic?

Q31-29. Distinguish message authentication and entity authentication.

Q31-30. In each round of DES, we have all components defined in Figure 31.8 in the text. Which components use a key and which components do not?

Q31-31. In a modern block cipher, we often need to use a component in the decryption cipher that is the inverse of the component used in the encryption cipher. What is the inverse of each of the following components?

a. swap **b.** shift right **c.** combine

Q31-32. Distinguish between a substitution cipher and a transposition cipher.

Q31-33. If the one-time pad cipher (Figure 31.12 in the text) is the simplest and most secure cipher, why is it not used all of the time?

Q31-34. Which of the following words means "secret writing"? Which one means "covered writing"?

a. cryptography **b.** steganography

Q31-35. When a sealed letter is sent from Alice to Bob, is this an example of using cryptography or steganography for confidentiality?

Q31-36. If we have a single integer key in Example 31.1 and 31.2 in the text, how many integer keys do we have in Example 31.3 in the text?

31.5.3 Problems

P31-1. Define the type of attack in each of the following cases:

a. A student breaks into a professor's office to obtain a copy of the next test.

b. A student gives a check for $10 to buy a used book. Later the student finds out that the check was cashed for $100.

c. A student sends hundreds of e-mails per day to the school using a phony return e-mail address.

P31-2. Assume we have a keyless substitution box (S-box) with three inputs (x_1, x_2, and x_3) and two outputs (y_1 and y_2). The relation between the inputs and outputs is defined as follows ($\oplus$ means XOR):

$y_1 = x_1 \oplus x_2 \oplus x_3$ $y_2 = x_1$

What is the output if the input is (110)? What is the output if the input is (001)?

P31-3. A substitution cipher does not have to be a character-to-character transformation. In a Polybius cipher, each letter in the plaintext is encrypted as two integers. The key is a 5×5 matrix of characters. The plaintext is the character in the matrix, the ciphertext is the two integers (each between 1 and 5) representing row and column numbers. Encipher the message "An exercise" using the Polybius cipher with the following key:

	1	2	3	4	5
1	z	q	p	f	e
2	y	r	o	g	d
3	x	s	n	h	c
4	w	t	m	i / j	b
5	v	u	l	k	a

P31-4. Another method used in a ciphertext attack is called the *statistical* approach, in which the intruder intercepts a long ciphertext and tries to analyze the statistics of the characters in the ciphertext. A simple cipher like the additive cipher does not change the statistics of the characters because encryption is one-to-one. Assume the intruder has intercepted the following ciphertext and the most common character in an English plaintext is the character "e". Use this knowledge to find the key of the cipher and decrypt the ciphertext.

XLILSYWIMWRSAJSVWEPIJSVJSYVQMPPMSRHSPPEVWMXMWASV XLQSVILYVVCFIJSVIXLIWIPPIVVIGIMZIWQSVISJJIVW

P31-5. Assume we have a very simple message digest. Our unrealistic message digest is just one number between 0 and 25. The digest is initially set to 0. The cryptographic hash function adds the current value of the digest to the value of the current character (between 0 and 25). Addition is in modulo 26. What is the value of the digest if the message is "HELLO"? Why is this digest not secure?

P31-6. In Figure 31.9 we have a swapper in each round. What is the use of this swapper?

P31-7. Change Figure 31.24 to allow bidirectional authentication. Alice needs to be authenticated for Bob and Bob for Alice.

P31-8. In RSA, given $p = 107$, $q = 113$, $e = 13$, and $d = 3653$, encrypt the message "THIS IS TOUGH" using 00 to 26 (A: 00 and space: 26) as the encoding scheme. Decrypt the ciphertext to find the original message.

P31-9. One of the attacks an intruder can apply to a simple cipher like an additive cipher is called the *ciphertext* attack. In this type of attack, the intruder intercepts the cipher and tries to find the key and eventually the plaintext. One of the methods used in a ciphertext attack is called the *brute-force* approach, in which the intruder tries several keys and decrypts the message until the message makes sense. Assume the intruder has intercepted the ciphertext "UVACLYZLJBYL". Try to decrypt the message by using keys beginning with 1 and continuing until a plaintext appears that makes sense.

P31-10. The circular shift operation is one of the components of the modern block ciphers.

a. Show the result of a 3-bit circular left shift on the word $(10011011)_2$.

b. Show the result of a 3-bit circular right shift on the result of part *a*.

c. Compare the result of part *b* with the original word in part *a* to show that shift right and shift left operations are inverses of each other.

P31-11. A cryptographic hash function needs to be *second preimage resistant,* which means that given the message M and the message digest *d,* we should not be able to find any other message, M′, whose digest is *d*. In other words, two different messages cannot have the same digest. Based on this requirement, show that a traditional checksum in the Internet cannot be used as a hash function.

P31-12. Alice can use only the additive cipher on her computer to send a message to a friend. She thinks that the message is more secure if she encrypts the message two times, each time with a different key. Is she right? Defend your answer.

P31-13. In Figure 31.9, we have two straight permutation operations: *initial permutation* and *final permutation*. Experts believe these operations are useless and do not help to make the cipher stronger. Can you find the reason for this statement?

P31-14. To understand the concept of secret-key distribution, assume a small private club has only 100 members (excluding the president). Answer the following questions:

a. How many secret keys are needed if all members of the club need to send secret messages to each other?

b. How many secret keys are needed if everyone trusts the president of the club? If a member needs to send a message to another member, she first sends it to the president; the president then sends the message to the other member.

c. How many secret keys are needed if the president decides that the two members who need to communicate should contact him first. The president then creates a temporary key to be used between the two. The temporary key is encrypted and sent to both members.

P31-15. Figure 31.22 shows a unidirectional authentication that authenticates Alice for Bob. Change this figure to provide bidirectional authentication: to authenticate Alice for Bob and then Bob for Alice.

P31-16. Use the additive cipher with k = 10 to encrypt the plaintext "book". Then decrypt the message to get the original plaintext.

P31-17. In a transposition cipher the encryption and decryption keys are often represented as two one-dimension tables (arrays) and the cipher is represented as a piece of software (a program).

a. Show the array for the encryption key in Figure 31.6 in the text. Hint: the value of each element can show the input-column number; the index can show the output-column number.

b. Show the array for the decryption key in Figure 31.6 in the text.

c. Explain, given the encryption key, how we can find the decryption key.

P31-18. The key in DES is 56 bits. Assume Eve, the intruder, tries to find the key using a brute-force attack (tries all of the keys one by one). If she can try one million keys (almost 2^{20}) in each second (using a powerful computer), how long does it take to break the code?

P31-19. Assume you want to write a program to simulate the permutation boxes in Figure 31.8 in the text.

a. Show how you represent each box as a table.

b. Show the inversion of each box as a table.

P31-20. Change Figure 31.23 to allow bidirectional authentication. Alice needs to be authenticated for Bob and Bob for Alice.

P31-21. Assume Bob, using the RSA cryptosystem, selects $p = 11$, $q = 13$, and $d = 7$, which of the following can be the value of public key e?

a. 11 **b.** 103 **c.** 19

P31-22. A very common operation in block ciphers is the XOR operation. Find the results of the following operations. Interpret the results.

a. $(01001101) \oplus (01001101)$

b. $(01001101) \oplus (00000000)$

P31-23. Encrypt the message "this is an exercise" using additive cipher with key = 20. Ignore the space between words. Decrypt the message to get the original plaintext.

P31-24. Explain why private-public keys cannot be used in creating a MAC.

P31-25. Each round in a block cipher should be invertible to make the whole block invertible. Modern block ciphers use two approaches to achieve this. In the first approach, each component is invertible; in the second approach some components are not invertible but the whole round is invertible using what is called a *Feistel cipher*. This approach is used in DES, described in the text. The trick in the Feistel cipher is to use the XOR operation as one of the components. To see the point, assume that a round is made of a noninvertible component, NI, and an XOR operation, as shown in Figure 31.29.

Figure 31.29 *Problem P31-25*

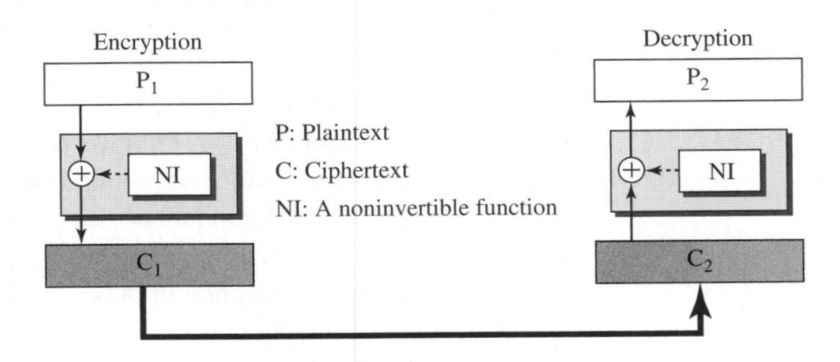

Prove that the whole round is invertible, which means that the plaintext can be recovered from the ciphertext. Hint: use XOR properties ($x \oplus x = 0$ and $x \oplus 0 = x$).

P31-26. Explain why encryption is used in the second message (from Bob to Alice) in Figure 31.23, but signing is done in the third message (from Alice to Bob) in Figure 31.24.

P31-27. The nonce in Figure 31.22 is to prevent a replay of the third message. Eve cannot replay the third message and pretend that it is a new request from Alice, because when Bob receives the response, the value of R_B is not valid anymore. This means that we can eliminate the first and the second message if we add a timestamp to the diagram. Show a new version of Figure 31.22 using a timestamp.

P31-28. You may have noticed that there is a flaw in Figure 31.26. Eve, the intruder, can replay the third message and, if she can somehow get access to the session key, can pretend to be Alice and exchange messages with Bob. The problem can be avoided if both Alice and Bob use two nonces. Remember that nonces have a lifetime and their main purpose is to prevent replaying. Modify Figure 31.26 to add two nonces.

P31-29. The swap operation is one of the components of the modern block ciphers.

a. Swap the word $(10011011)_2$.

b. Swap the word resulting from part *a*.

c. Compare the results of part *a* and part *b* to show that swapping is a self-invertible operation.

P31-30. Atbash was a popular cipher among Biblical writers. In Atbash, "A" is encrypted as "Z", "B" is encrypted as "Y", and so on. Similarly, "Z" is encrypted as "A", "Y" is encrypted as "B", and so on. Suppose that the alphabet is divided into halves and the letters in the first half are encrypted as the letters in the second and vice versa. Find the type of cipher and key. Encipher the plaintext "an exercise" using the Atbash cipher.

31.6 SIMULATION EXPERIMENTS

31.6.1 Applets

We have created some Java applets to show some of the main concepts discussed in this chapter. It is strongly recommended that the students activate these applets on the book website and carefully examine the protocols in action.

31.7 PROGRAMMING ASSIGNMENTS

Write the source code, compile, and test the following programs in one of the programming languages of your choice:

Prg31-1. A general program to implement substitution (additive) cipher (encryption and decryption). The input to the program is a flag that demands encryption or decryption, the symmetric key, and the plaintext or ciphertext. The output is the ciphertext or the plaintext, depending on the flag.

Prg31-2. A general program to implement transposition cipher (encryption and decryption). The input to the program is a flag that demands encryption or decryption, the symmetric key, and the plaintext or ciphertext. The output is the ciphertext or the plaintext, depending on the flag.

Prg31-3. A general program to implement RSA cryptosystem. The input to the program is a flag that demands encryption or decryption, the value of p and q, the value of e, and the plaintext or ciphertext. The output is the ciphertext or the plaintext, depending on the flag.

CHAPTER 32

Internet Security

After discussing cryptography and network security in the previous chapter, we concentrate on Internet security in this chapter. Internet security is normally applied at three layers in the Internet: the network layer, the transport layer, and the application layer. We also discuss firewalls, a technology that provides system security for an enterprise.

This chapter is divided into four sections.

- The first section discusses security at the network layer, IPSec. The section first describes the two modes of IPSec: transport mode and tunnel mode. It then describes the two versions of the protocol: AS and ESP. The section continues with listing and explaining the services provided by each protocol version. It next describes security association, a technique that changes the connectionless service of the network layer to a connection-oriented service for the purpose of applying security measures. The section ends with describing one application of IPSec, the virtual private networks.
- The second section discusses one of the security protocols at the transport layer, SSL (the other protocol, TLS, is similar). The section first describes the SSL architecture: services, algorithms, and parameter generation. It then explains the four protocols that SSL is made of: Handshake, ChangeCipherSpec, Alert, and Record.
- The third section discusses security at the application layer. At this layer, security is provided only for the e-mail application; other applications can use the security at the transport layer, but e-mail, because of its one-way communication, cannot do so. The section first describes Pretty Good Privacy (PGP), which provides e-mail security mostly for personal use. The section then describes S/MIME, a secured version of the MIME protocol that provides security mostly for an enterprise.
- The fourth section discusses firewalls, a technology that can protect an enterprise from the malicious intension of an intruder. The section describes two versions: packet-filter firewalls and proxy firewalls. The first gives protection only at the network layer; the second can provide protection at the application layer.

32.1 NETWORK-LAYER SECURITY

We start this chapter with the discussion of security at the network layer. At the network layer, security is applied between two hosts, two routers, or a host and a router. The purpose of network-layer security is to protect those applications that use the service of the network layer directly, such as routing protocols. Those applications that use the service of UDP can also benefit from this service because UDP is a connectionless protocol and transport-layer security protocols, as we discuss later, cannot be applied to UDP. The only application-layer security we discuss here is called *IPSec*. **IP Security (IPSec)** is a collection of protocols designed by the Internet Engineering Task Force (IETF) to provide security for a packet at the network level. IPSec helps create authenticated and confidential packets for the IP layer.

32.1.1 Two Modes

IPSec operates in one of two different modes: transport mode or tunnel mode.

Transport Mode

In **transport mode,** IPSec protects what is delivered from the transport layer to the network layer. In other words, transport mode protects the payload to be encapsulated in the network layer, as shown in Figure 32.1.

Figure 32.1 *IPSec in transport mode*

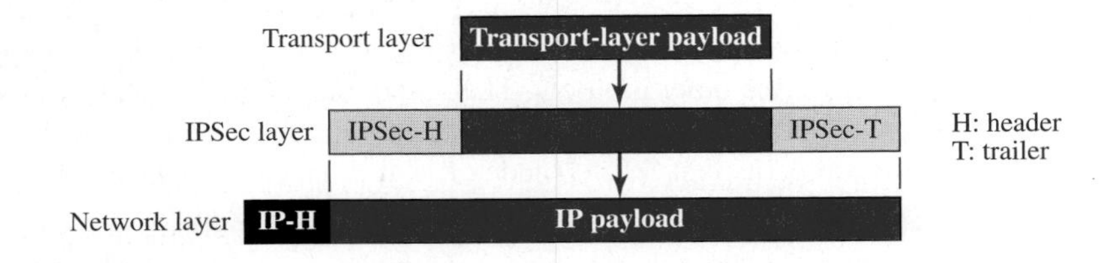

Note that transport mode does not protect the IP header. In other words, transport mode does not protect the whole IP packet; it protects only the packet from the transport layer (the IP-layer payload). In this mode, the IPSec header (and trailer) are added to the information coming from the transport layer. The IP header is added later.

IPSec in transport mode does not protect the IP header; it only protects the payload coming from the transport layer.

Transport mode is normally used when we need host-to-host (end-to-end) protection of data. The sending host uses IPSec to authenticate and/or encrypt the payload delivered from the transport layer. The receiving host uses IPSec to check the authentication and/or decrypt the IP packet and deliver it to the transport layer. Figure 32.2 shows this concept.

Figure 32.2 *Transport mode in action*

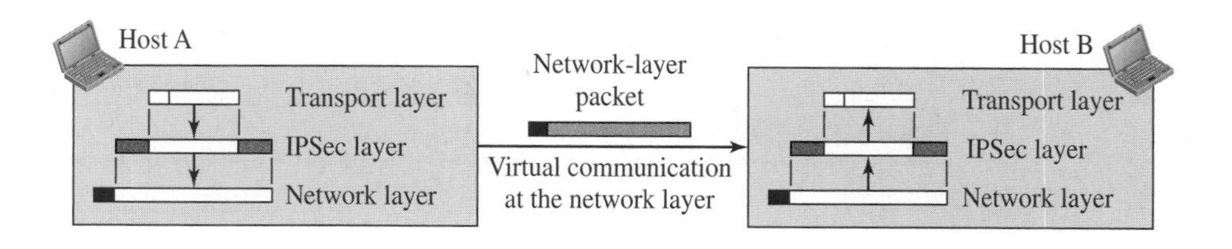

Tunnel Mode

In **tunnel mode,** IPSec protects the entire IP packet. It takes an IP packet, including the header, applies IPSec security methods to the entire packet, and then adds a new IP header, as shown in Figure 32.3.

Figure 32.3 *IPSec in tunnel mode*

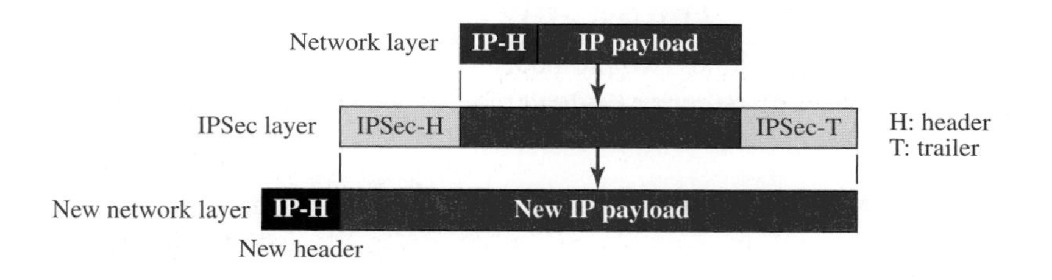

The new IP header, as we will see shortly, has different information than the original IP header. Tunnel mode is normally used between two routers, between a host and a router, or between a router and a host, as shown in Figure 32.4. The entire original packet is protected from intrusion between the sender and the receiver, as if the whole packet goes through an imaginary tunnel.

Figure 32.4 *Tunnel mode in action*

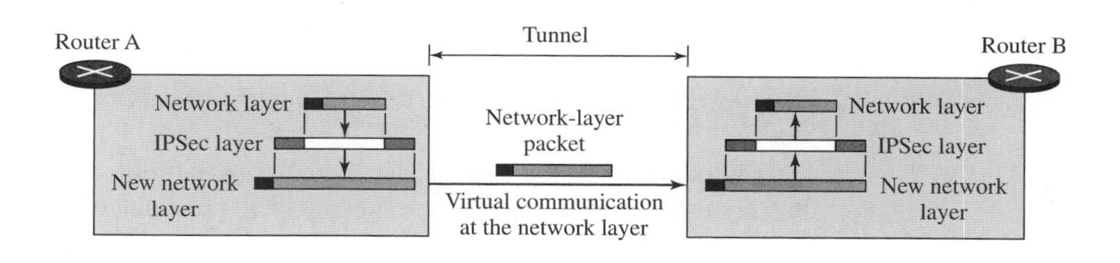

IPSec in tunnel mode protects the original IP header.

Comparison

In transport mode, the IPSec layer comes between the transport layer and the network layer. In tunnel mode, the flow is from the network layer to the IPSec layer and then back to the network layer again. Figure 32.5 compares the two modes.

Figure 32.5 *Transport mode versus tunnel mode*

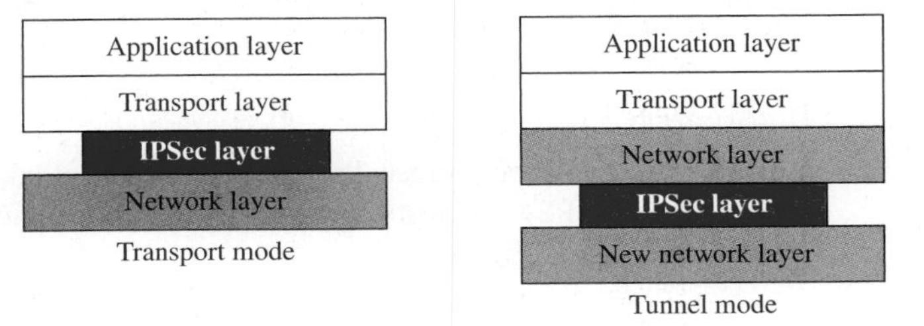

32.1.2 Two Security Protocols

IPSec defines two protocols—the Authentication Header (AH) Protocol and the Encapsulating Security Payload (ESP) Protocol—to provide authentication and/or encryption for packets at the IP level.

Authentication Header (AH)

The **Authentication Header (AH) protocol** is designed to authenticate the source host and to ensure the integrity of the payload carried in the IP packet. The protocol uses a hash function and a symmetric (secret) key to create a message digest; the digest is inserted in the authentication header (see MAC). The AH is then placed in the appropriate location, based on the mode (transport or tunnel). Figure 32.6 shows the fields and the position of the authentication header in transport mode.

When an IP datagram carries an authentication header, the original value in the protocol field of the IP header is replaced by the value 51. A field inside the authentication header (the next header field) holds the original value of the protocol field (the type of payload being carried by the IP datagram). The addition of an authentication header follows these steps:

1. An authentication header is added to the payload with the authentication data field set to 0.
2. Padding may be added to make the total length appropriate for a particular hashing algorithm.
3. Hashing is based on the total packet. However, only those fields of the IP header that do not change during transmission are included in the calculation of the message digest (authentication data).
4. The authentication data are inserted in the authentication header.

Figure 32.6 *Authentication Header (AH) protocol*

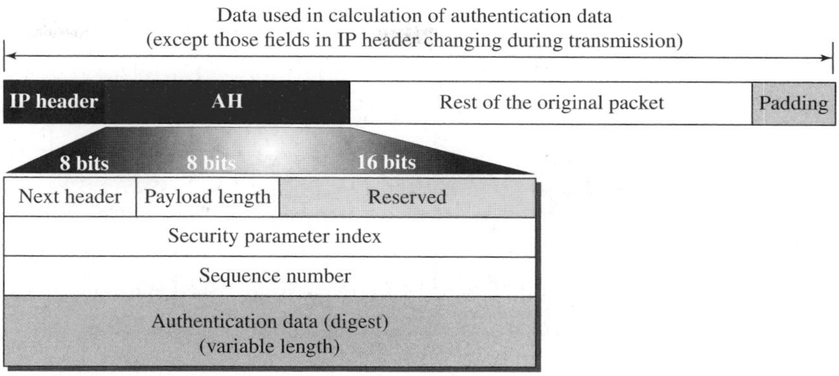

5. The IP header is added after changing the value of the protocol field to 51.

A brief description of each field follows:

- ***Next Header.*** The 8-bit next header field defines the type of payload carried by the IP datagram (such as TCP, UDP, ICMP, or OSPF).
- ***Payload Length.*** The name of this 8-bit field is misleading. It does not define the length of the payload; it defines the length of the authentication header in 4-byte multiples, but it does not include the first 8 bytes.
- ***Security Parameter Index.*** The 32-bit security parameter index (SPI) field plays the role of a virtual circuit identifier and is the same for all packets sent during a connection called a *Security Association* (discussed later).
- ***Sequence Number.*** A 32-bit sequence number provides ordering information for a sequence of datagrams. The sequence numbers prevent a playback. Note that the sequence number is not repeated even if a packet is retransmitted. A sequence number does not wrap around after it reaches 2^{32}; a new connection must be established.
- ***Authentication Data.*** Finally, the authentication data field is the result of applying a hash function to the entire IP datagram except for the fields that are changed during transit (e.g., time-to-live).

The AH protocol provides source authentication and data integrity, but not privacy.

Encapsulating Security Payload (ESP)

The AH protocol does not provide confidentiality, only source authentication and data integrity. IPSec later defined an alternative protocol, **Encapsulating Security Payload (ESP),** that provides source authentication, integrity, and confidentiality. ESP adds a header and trailer. Note that ESP's authentication data are added at the end of the packet, which makes its calculation easier. Figure 32.7 shows the location of the ESP header and trailer.

Figure 32.7 *Encapsulating Security Payload (ESP)*

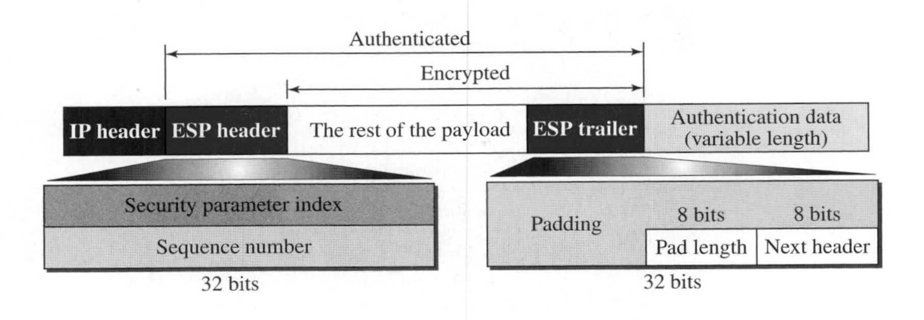

When an IP datagram carries an ESP header and trailer, the value of the protocol field in the IP header is 50. A field inside the ESP trailer (the next-header field) holds the original value of the protocol field (the type of payload being carried by the IP datagram, such as TCP or UDP). The ESP procedure follows these steps:

1. An ESP trailer is added to the payload.
2. The payload and the trailer are encrypted.
3. The ESP header is added.
4. The ESP header, payload, and ESP trailer are used to create the authentication data.
5. The authentication data are added to the end of the ESP trailer.
6. The IP header is added after changing the protocol value to 50.

The fields for the header and trailer are as follows:

- ***Security Parameter Index.*** The 32-bit security parameter index field is similar to the one defined for the AH protocol.
- ***Sequence Number.*** The 32-bit sequence number field is similar to the one defined for the AH protocol.
- ***Padding.*** This variable-length field (0 to 255 bytes) of 0s serves as padding.
- ***Pad Length.*** The 8-bit pad-length field defines the number of padding bytes. The value is between 0 and 255; the maximum value is rare.
- ***Next Header.*** The 8-bit next-header field is similar to that defined in the AH protocol. It serves the same purpose as the protocol field in the IP header before encapsulation.
- ***Authentication Data.*** Finally, the authentication data field is the result of applying an authentication scheme to parts of the datagram. Note the difference between the authentication data in AH and ESP. In AH, part of the IP header is included in the calculation of the authentication data; in ESP, it is not.

IPv4 and IPv6

IPSec supports both IPv4 and IPv6. In IPv6, however, AH and ESP are part of the extension header.

AH versus ESP

The ESP protocol was designed after the AH protocol was already in use. ESP does whatever AH does with additional functionality (confidentiality). We actually do not need AH. However, the implementation of AH is already included in some commercial products, which means that AH will remain part of the Internet until these products are phased out.

32.1.3 Services Provided by IPSec

The two protocols, AH and ESP, can provide several security services for packets at the network layer. Table 32.1 shows the list of services available for each protocol.

Table 32.1 *IPSec services*

Services	*AH*	*ESP*
Access control	Yes	Yes
Message authentication (message integrity)	Yes	Yes
Entity authentication (data source authentication)	Yes	Yes
Confidentiality	No	Yes
Replay attack protection	Yes	Yes

Access Control

IPSec provides access control indirectly, using a Security Association Database (SAD), as we will see in the next section. When a packet arrives at a destination and there is no Security Association already established for this packet, the packet is discarded.

Message Integrity

Message integrity is preserved in both AH and ESP. A digest of data is created and sent by the sender to be checked by the receiver.

Entity Authentication

The Security Association and the keyed-hash digest of the data sent by the sender authenticate the sender of the data in both AH and ESP.

Confidentiality

The encryption of the message in ESP provides confidentiality. AH, however, does not provide confidentiality. If confidentiality is needed, one should use ESP instead of AH.

Replay Attack Protection

In both protocols, the replay attack is prevented by using sequence numbers and a sliding receiver window. Each IPSec header contains a unique sequence number when the Security Association is established. The number starts from 0 and increases until the value reaches $2^{32} - 1$. When the sequence number reaches the maximum, it is reset to 0 and, at the same time, the old Security Association (see the next section) is deleted and a new one is established. To prevent processing duplicate packets, IPSec mandates the use of a fixed-size window at the receiver. The size of the window is determined by the receiver with a default value of 64.

32.1.4 Security Association

Security Association is a very important aspect of IPSec. IPSec requires a logical relationship, called a **Security Association (SA),** between two hosts. The security association changes the connectionless service provided by IP to a connection-oriented service upon which we can apply security. This section first discusses the idea and then shows how it is used in IPSec.

Idea of Security Association

A Security Association is a contract between two parties; it creates a secure channel between them. Let us assume that Alice needs to unidirectionally communicate with Bob. If Alice and Bob are interested only in the confidentiality aspect of security, they can get a shared secret key between themselves. We can say that there are two SAs between Alice and Bob; one outbound SA and one inbound SA. Each of them stores the value of the key in one variable and the name of the encryption/decryption algorithm in another. Alice uses the algorithm and the key to encrypt a message to Bob; Bob uses the algorithm and the key when he needs to decrypt the message received from Alice. Figure 32.8 shows a simple SA.

Figure 32.8 *Simple SA*

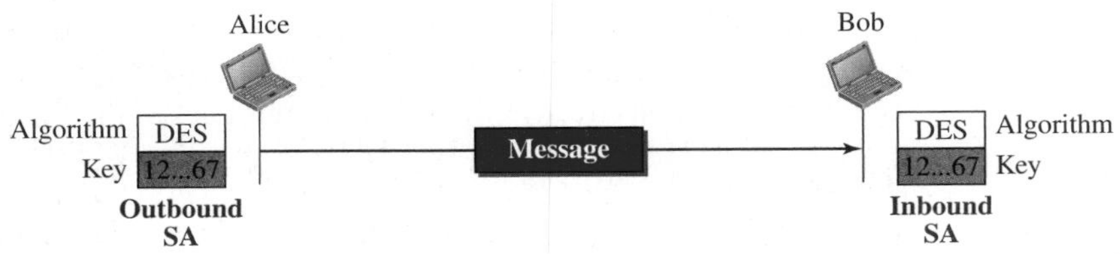

The Security Association can be more involved if the two parties need message integrity and authentication. Each association needs other data such as the algorithm for message integrity, the key, and other parameters. It can be much more complex if the parties need to use specific algorithms and specific parameters for different protocols, such as IPSec AH or IPSec ESP.

Security Association Database (SAD)

A Security Association can be very complex. This is particularly true if Alice wants to send messages to many people and Bob needs to receive messages from many people. In addition, each site needs to have both inbound and outbound SAs to allow bidirectional communication. In other words, we need a set of SAs that can be collected into a database. This database is called the **Security Association Database (SAD).** The database can be thought of as a two-dimensional table with each row defining a single

SA. Normally, there are two SADs, one inbound and one outbound. Figure 32.9 shows the concept of outbound or inbound SADs for one entity.

Figure 32.9 *SAD*

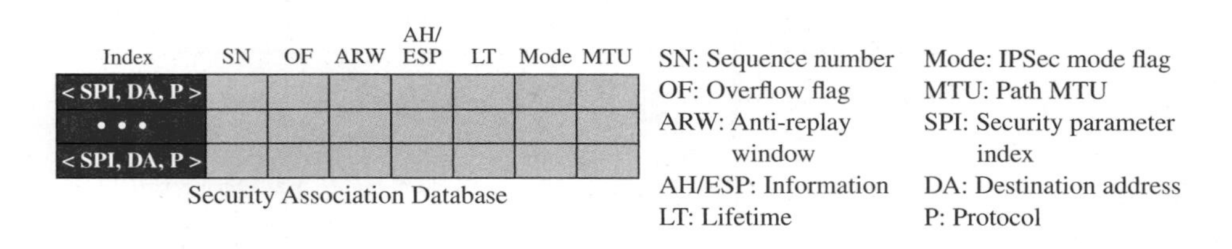

When a host needs to send a packet that must carry an IPSec header, the host needs to find the corresponding entry in the outbound SAD to find the information for applying security to the packet. Similarly, when a host receives a packet that carries an IPSec header, the host needs to find the corresponding entry in the inbound SAD to find the information for checking the security of the packet. This searching must be specific in the sense that the receiving host needs to be sure that correct information is used for processing the packet. Each entry in an inbound SAD is selected using a triple index: security parameter index (a 32-bit number that defines the SA at the destination), destination address, and protocol (AH or ESP).

Security Policy

Another important aspect of IPSec is the **Security Policy (SP),** which defines the type of security applied to a packet when it is to be sent or when it has arrived. Before using the SAD, discussed in the previous section, a host must determine the predefined policy for the packet.

Security Policy Database

Each host that is using the IPSec protocol needs to keep a **Security Policy Database (SPD).** Again, there is a need for an inbound SPD and an outbound SPD. Each entry in the SPD can be accessed using a sextuple index: source address, destination address, name, protocol, source port, and destination port, as shown in Figure 32.10. The name usually defines a DNS entity. The protocol is either AH or ESP.

Figure 32.10

Index	Policy
< SA, DA, Name, P, SPort, DPort >	
• • •	
< SA, DA, Name, P, SPort, DPort >	

SA: Source address
DA: Destination address
P: Protocol
SPort: Source port
DPort: Destination port

Outbound SPD

When a packet is to be sent out, the outbound SPD is consulted. Figure 32.11 shows the processing of a packet by a sender. The input to the outbound SPD is the sextuple index; the output is one of the three following cases: drop (packet cannot be sent), bypass (bypassing security header), and apply (applying the security according to the SAD; if no SAD, creating one).

Figure 32.11 *Outbound processing*

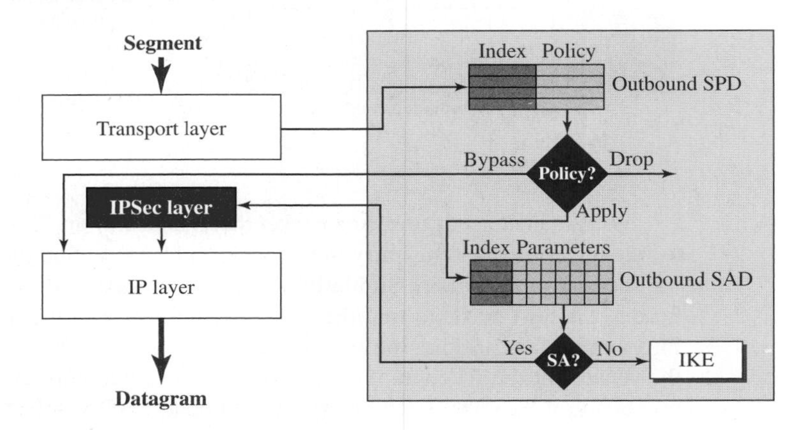

Inbound SPD

When a packet arrives, the inbound SPD is consulted. Each entry in the inbound SPD is also accessed using the same sextuple index. Figure 32.12 shows the processing of a packet by a receiver.

The input to the inbound SPD is the sextuple index; the output is one of the three following cases: discard (drop the packet), bypass (bypassing the security and delivering the packet to the transport layer), and apply (applying the policy using the SAD).

32.1.5 Internet Key Exchange (IKE)

The **Internet Key Exchange (IKE)** is a protocol designed to create both inbound and outbound Security Associations. As we discussed in the previous section, when a peer needs to send an IP packet, it consults the Security Policy Database (SPD) to see if there is an SA for that type of traffic. If there is no SA, IKE is called to establish one.

IKE is a complex protocol based on three other protocols: Oakley, SKEME, and ISAKMP, as shown in Figure 32.13.

The **Oakley** protocol was developed by Hilarie Orman. It is a key creation protocol. **SKEME,** designed by Hugo Krawcyzk, is another protocol for key exchange. It uses public-key encryption for entity authentication in a key-exchange protocol.

Figure 32.12 *Inbound processing*

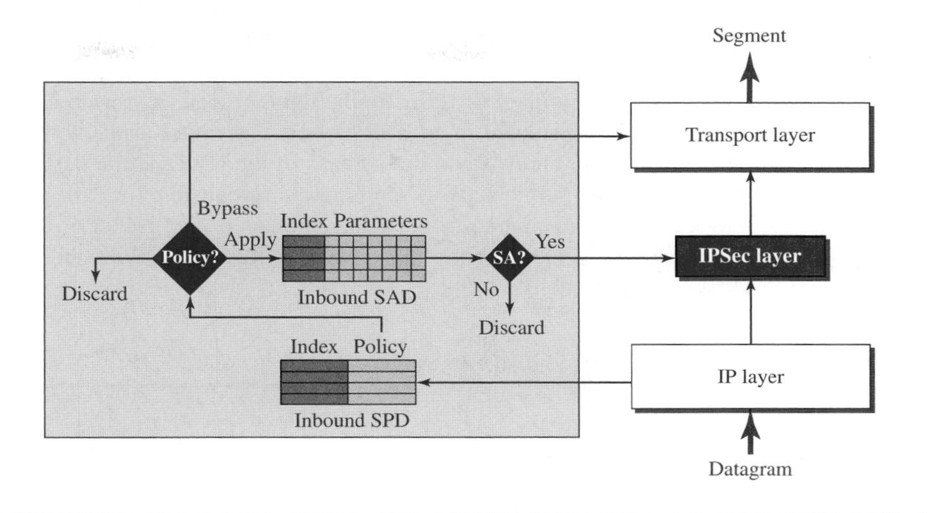

IKE creates SAs for IPSec.

Figure 32.13 *IKE components*

The **Internet Security Association and Key Management Protocol (ISAKMP)** is a protocol designed by the National Security Agency (NSA) that actually implements the exchanges defined in IKE. It defines several packets, protocols, and parameters that allow the IKE exchanges to take place in standardized, formatted messages to create SAs. We leave the discussion of these three protocols for books dedicated to security.

32.1.6 Virtual Private Network (VPN)

One of the applications of IPSec is in *virtual private networks*. A **virtual private network (VPN)** is a technology that is gaining popularity among large organizations that use the global Internet for both intra- and interorganization communication, but require privacy in their intraorganization communication. VPN is a network that is private but virtual. It is private because it guarantees privacy inside the organization. It is virtual because it does not use real private WANs; the network is physically public but virtually private. Figure 32.14 shows the idea of a virtual private network. Routers R1 and R2 use VPN technology to guarantee privacy for the organization. VPN technology uses the ESP protocol of IPSec in the tunnel mode. A private datagram, including the header, is encapsulated in an ESP packet. The router at the border of the sending site uses its own IP address and the address of the router at the destination site in the new datagram. The public network (Internet) is responsible for carrying the packet from R1 to R2. Outsiders cannot decipher the contents of the packet or the source and destination addresses. Deciphering takes place at R2, which finds the destination address of the packet and delivers it.

Figure 32.14 *Virtual private network*

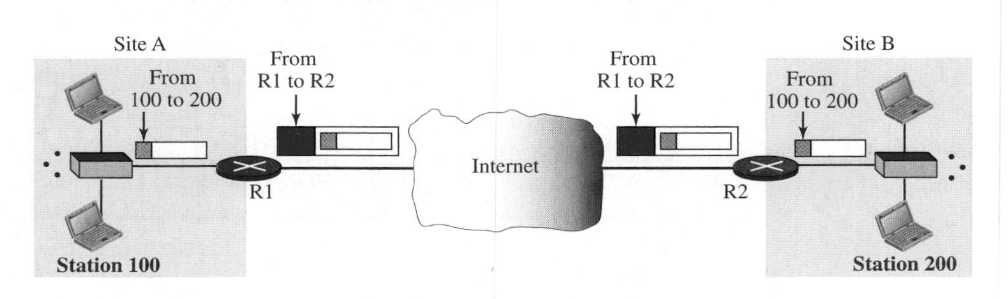

32.2 TRANSPORT-LAYER SECURITY

In fact, security at the transport layer provides security for the application layer, which uses the services of TCP (or SCTP) as a connection-oriented protocol. Before the messages of these applications are encapsulated in TCP, they are encapsulated in the security-protocol packets. Those applications that use the services of UDP cannot benefit from these security services because the nature of security requires connection establishment between the two entities. Anther application that cannot benefit from the transport-layer security is electronic mail (e-mail). This application provides one-way connection between the sender and the receiver; we need a special security provision for this application, as discussed in the next section.

Two protocols are dominant today for providing security at the transport layer: the **Secure Sockets Layer (SSL) protocol** and the **Transport Layer Security (TLS) protocol.** The latter is actually an IETF version of the former. We discuss SSL in this section; TLS is very similar. Figure 32.15 shows the position of SSL and TLS in the Internet model.

Some of the goals of these protocols are to provide server and client authentication, data confidentiality, and data integrity. Application-layer client/server programs,

Figure 32.15 *Location of SSL and TLS in the Internet model*

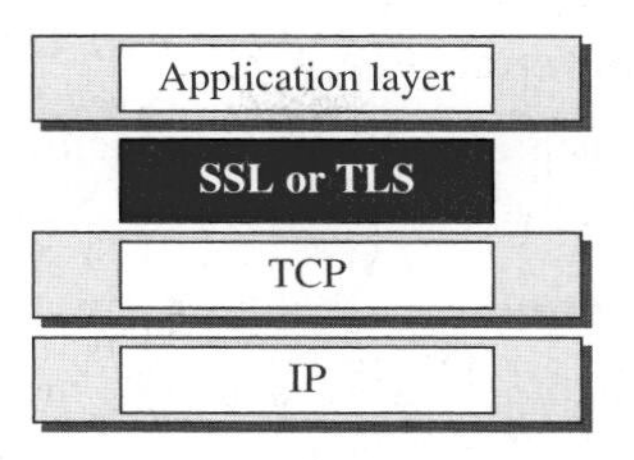

such as HTTP (see Chapter 26), that use the services of TCP can encapsulate their data in SSL packets (HTTPS). If the server and client are capable of running SSL (or TLS) programs, then the client can use the URL *https://...* instead of *http://...* to allow HTTP messages to be encapsulated in SSL (or TLS) packets. For example, credit card numbers can be safely transferred via the Internet for online shoppers.

32.2.1 SSL Architecture

SSL is designed to provide security and compression services to data generated from the application layer. Typically, SSL can receive data from any application-layer protocol, but usually the protocol is HTTP. The data received from the application is compressed (optional), signed, and encrypted. The data is then passed to a reliable transport-layer protocol such as TCP. Netscape developed SSL in 1994. Versions 2 and 3 were released in 1995. In this section, we discuss SSLv3.

Services

SSL provides several services on data received from the application layer.

- ***Fragmentation.*** First, SSL divides the data into blocks of 2^{14} bytes or less.
- ***Compression.*** Each fragment of data is compressed using one of the lossless compression methods negotiated between the client and server. This service is optional.
- ***Message Integrity.*** To preserve the integrity of data, SSL uses a keyed-hash function to create a MAC.
- ***Confidentiality.*** To provide confidentiality, the original data and the MAC are encrypted using symmetric-key cryptography.
- ***Framing.*** A header is added to the encrypted payload. The payload is then passed to a reliable transport-layer protocol.

Key Exchange Algorithms

To exchange an authenticated and confidential message, the client and the server each need a set of cryptographic secrets. However, to create these secrets, one pre-master secret must be established between the two parties. SSL defines several key-exchange methods to establish this pre-master secret.

Encryption/Decryption Algorithms

The client and server also need to agree to a set of encryption and decryption algorithms.

Hash Algorithms

SSL uses hash algorithms to provide message integrity (message authentication). Several hash algorithms have been defined for this purpose.

Cipher Suite

The combination of key exchange, hash, and encryption algorithms defines a **cipher suite** for each SSL session.

Compression Algorithms

Compression is optional in SSL. No specific compression algorithm is defined. Therefore a system can use whatever compression algorithm it desires.

Cryptographic Parameter Generation

To achieve message integrity and confidentiality, SSL needs six cryptographic secrets: four keys and two IVs (initialization vectors). The client needs one key for message authentication, one key for encryption, and one IV as the original block in calculation. The server needs the same. SSL requires that the keys for one direction be different from those for the other direction. If there is an attack in one direction, the other direction is not affected. The parameters are generated using the following procedure:

1. The client and server exchange two random numbers; one is created by the client and the other by the server.
2. The client and server exchange one *pre-master secret* using one of the predefined key-exchange algorithms.
3. A 48-byte *master secret* is created from the pre-master secret by applying two hash functions (SHA-1 and MD5), as shown in Figure 32.16.
4. The master secret is used to create variable-length *key material* by applying the same set of hash functions and prepending with different constants, as shown in Figure 32.17. The module is repeated until key material of adequate size is created. Note that the length of the key material block depends on the cipher suite selected and the size of keys needed for this suite.
5. Six different secrets are extracted from the key material, as shown in Figure 32.18.

Sessions and Connections

SSL differentiates a *connection* from a *session.* A session is an association between a client and a server. After a session is established, the two parties have common information such as the session identifier, the certificate authenticating each of them (if necessary), the compression method (if needed), the cipher suite, and a master secret that is used to create keys for message authentication encryption.

For two entities to exchange data, the establishment of a session is necessary, but not sufficient; they need to create a connection between themselves. The two entities exchange two random numbers and create, using the master secret, the keys and parameters needed for exchanging messages involving authentication and privacy.

Figure 32.16 *Calculation of master secret from pre-master secret*

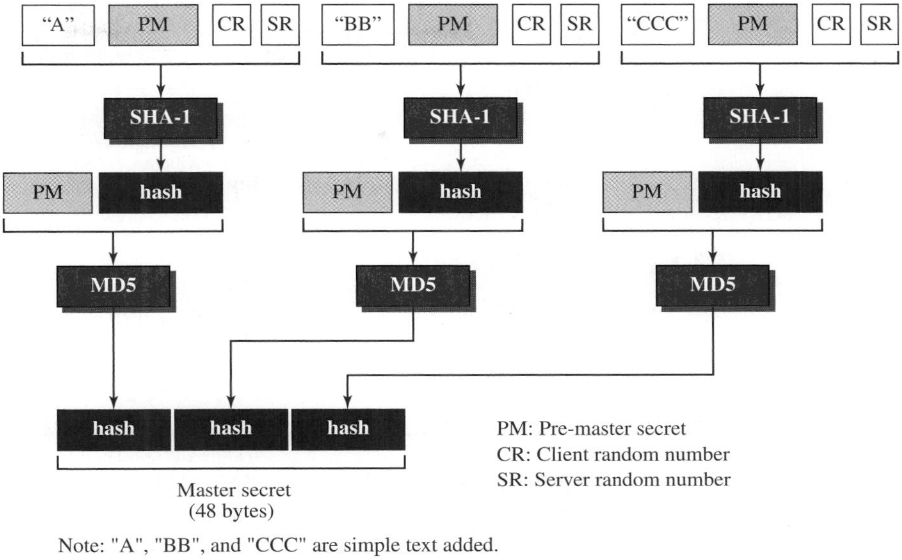

"A" PM CR SR "BB" PM CR SR "CCC" PM CR SR

SHA-1 SHA-1 SHA-1

PM hash PM hash PM hash

MD5 MD5 MD5

hash hash hash

Master secret
(48 bytes)

PM: Pre-master secret
CR: Client random number
SR: Server random number

Note: "A", "BB", and "CCC" are simple text added.

Figure 32.17 *Calculation of key material from master secret*

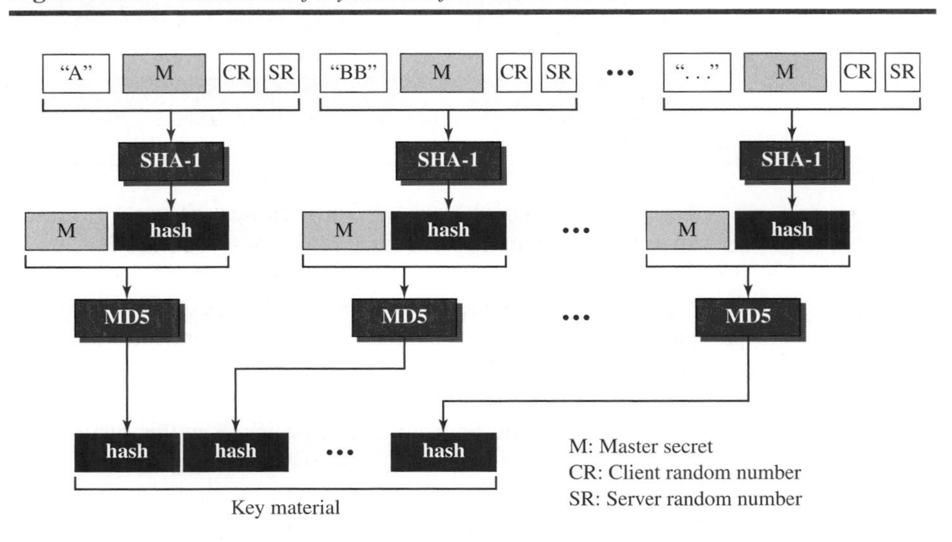

A session can consist of many connections. A connection between two parties can be terminated and reestablished within the same session. When a connection is terminated, the two parties can also terminate the session, but it is not mandatory. A session can be suspended and resumed.

Figure 32.18 *Extractions of cryptographic secrets from key material*

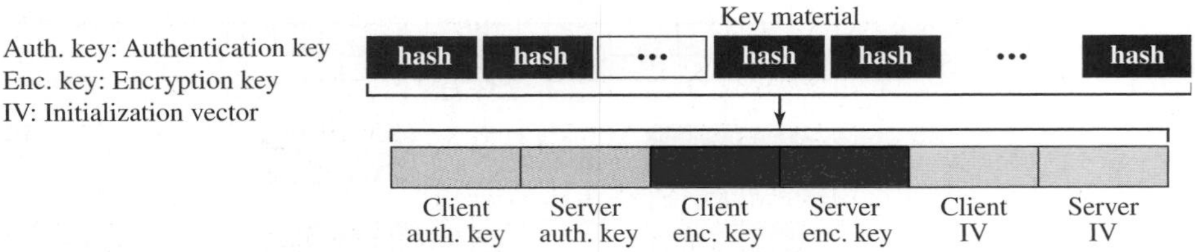

Auth. key: Authentication key
Enc. key: Encryption key
IV: Initialization vector

Key material
hash hash ... hash hash ... hash

Client auth. key | Server auth. key | Client enc. key | Server enc. key | Client IV | Server IV

32.2.2 Four Protocols

We have discussed the idea of SSL without showing how SSL accomplishes its tasks. SSL defines four protocols in two layers, as shown in Figure 32.19.

Figure 32.19 *Four SSL protocols*

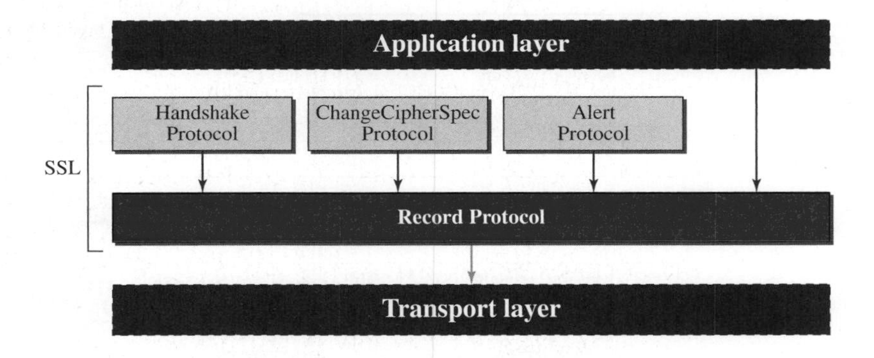

The Record Protocol is the carrier. It carries messages from three other protocols as well as the data coming from the application layer. Messages from the Record Protocol are payloads to the transport layer, normally TCP. The Handshake Protocol provides security parameters for the Record Protocol. It establishes a cipher set and provides keys and security parameters. It also authenticates the server to the client and the client to the server if needed. The ChangeCipherSpec Protocol is used for signaling the readiness of cryptographic secrets. The Alert Protocol is used to report abnormal conditions. We will briefly discuss these protocols in this section.

Handshake Protocol

The **Handshake Protocol** uses messages to negotiate the cipher suite, to authenticate the server to the client and the client to the server if needed, and to exchange information for building the cryptographic secrets. The handshaking is done in four phases, as shown in Figure 32.20.

Figure 32.20 *Handshake Protocol*

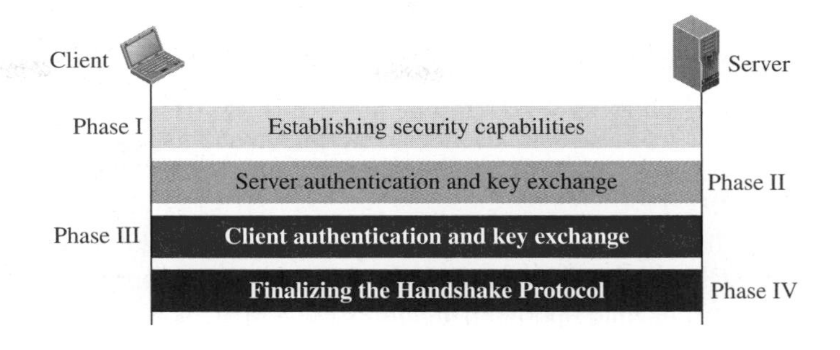

Phase I: Establishing Security Capabilities

In Phase I, the client and the server announce their security capabilities and choose those that are convenient for both. In this phase, a session ID is established and the cipher suite is chosen. The parties agree upon a particular compression method. Finally, two random numbers are selected, one by the client and one by the server, to be used for creating a master secret as we saw before. After Phase I, the client and server know the version of SSL, the cryptographic algorithms, the compression method, and the two random numbers for key generation.

Phase II: Server Authentication and Key Exchange

In Phase II, the server authenticates itself if needed. The server may send its certificate, its public key, and may also request certificates from the client. After Phase II, the server is authenticated to the client, and the client knows the public key of the server if required.

Phase III: Client Authentication and Key Exchange

Phase III is designed to authenticate the client. After Phase III, the client is authenticated for the server, and both the client and the server know the pre-master secret.

Phase IV: Finalizing and Finishing

In Phase IV, the client and server send messages to change cipher specifications and to finish the Handshake Protocol.

ChangeCipherSpec Protocol

We have seen that the negotiation of the cipher suite and the generation of cryptographic secrets are formed gradually during the Handshake Protocol. The question now is: When can the two parties use these parameters or secrets? SSL mandates that the parties cannot use these parameters or secrets until they have sent or received a special message, the ChangeCipherSpec message, which is exchanged during the Handshake Protocol and defined in the *ChangeCipherSpec Protocol.* The reason is that the issue is not just sending or receiving a message. The sender and the receiver need two states, not one. One state, the pending state, keeps track of the parameters and secrets. The other state, the active state, holds parameters and secrets used by the Record Protocol to sign/verify or encrypt/decrypt messages. In addition, each state holds two sets of values: *read* (inbound) and *write* (outbound).

Alert Protocol

SSL uses the *Alert Protocol* for reporting errors and abnormal conditions. It uses only one message that describes the problem and its level (warning or fatal).

Record Protocol

The *Record Protocol* carries messages from the upper layer (Handshake Protocol, ChangeCipherSpec Protocol, Alert Protocol, or application layer). The message is fragmented and optionally compressed; a MAC is added to the compressed message using the negotiated hash algorithm. The compressed fragment and the MAC are encrypted using the negotiated encryption algorithm. Finally, the SSL header is added to the encrypted message. Figure 32.21 shows this process at the sender. The process at the receiver is reversed.

Figure 32.21 *Processing done by the Record Protocol*

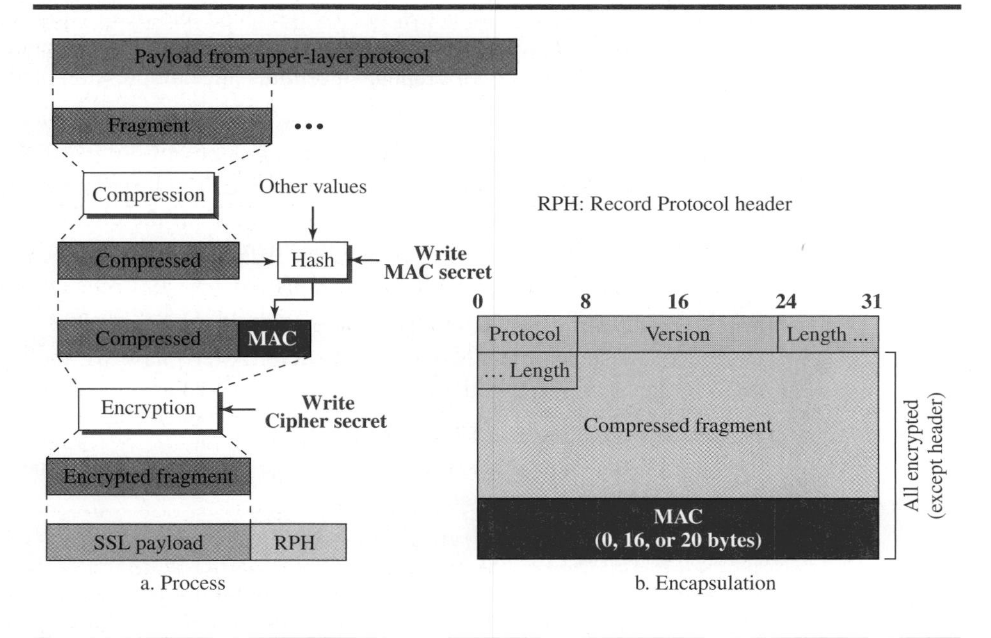

32.3 APPLICATION-LAYER SECURITY

This section discusses two protocols providing security services for e-mails: Pretty Good Privacy (PGP) and Secure/Multipurpose Internet Mail Extension (S/MIME).

32.3.1 E-mail Security

Sending an e-mail is a one-time activity. The nature of this activity is different from those we saw in the two previous sections: SSL or IPSec. In those protocols, we assume that the two parties create a session between themselves and exchange data in both directions. In e-mail, there is no session. Alice and Bob cannot create a session. Alice sends a message to Bob; sometime later, Bob reads the message and may or may not send a reply. We discuss the security of a unidirectional message because what Alice sends to Bob is totally independent from what Bob sends to Alice.

Cryptographic Algorithms

If e-mail is a one-time activity, how can the sender and receiver agree on a cryptographic algorithm to use for e-mail security? If there is no session and no handshaking to negotiate the algorithms for encryption/decryption and hashing, how can the receiver know which algorithm the sender has chosen for each purpose?

To solve the problem, the protocol defines a set of algorithms for each operation that the user used in his/her system. Alice includes the names (or identifiers) of the algorithms she has used in the e-mail. For example, Alice can choose DES for encryption/decryption and MD5 for hashing. When Alice sends a message to Bob, she includes the corresponding identifiers for DES and MD5 in her message. Bob receives the message and extracts the identifiers first. He then knows which algorithm to use for decryption and which one for hashing.

In e-mail security, the sender of the message needs to include the names or identifiers of the algorithms used in the message.

Cryptographic Secrets

The problem for the cryptographic algorithms also applies to the cryptographic secrets (keys). If there is no negotiation, how can the two parties establish secrets between themselves? The e-mail security protocols today require that encryption/decryption be done using a symmetric-key algorithm and a one-time secret key sent with the message. Alice can create a secret key and send it with the message she sends to Bob. To protect the secret key from interception by Eve, the secret key is encrypted with Bob's public key. In other words, the secret key itself is encrypted.

In e-mail security, the encryption/decryption is done using a symmetric-key algorithm, but the secret key to decrypt the message is encrypted with the public key of the receiver and is sent with the message.

Certificates

One more issue needs to be considered before we discuss any e-mail security protocol in particular. It is obvious that some public-key algorithms must be used for e-mail security. For example, we need to encrypt the secret key or sign the message. To encrypt the secret key, Alice needs Bob's public key; to verify a signed message, Bob needs Alice's public key. So, for sending a small authenticated and confidential message, two public keys are needed. How can Alice be assured of Bob's public key, and how can Bob be assured of Alice's public key? Each e-mail security protocol has a different method of certifying keys.

32.3.2 Pretty Good Privacy (PGP)

The first protocol discussed in this section is called **Pretty Good Privacy (PGP).** PGP was invented by Phil Zimmermann to provide e-mail with privacy, integrity, and authentication. PGP can be used to create secure e-mail messages.

Scenarios

Let us first discuss the general idea of PGP, moving from a simple scenario to a complex one. We use the term "Data" to show the message prior to processing.

Plaintext

The simplest scenario is to send the e-mail message in plaintext as shown in Figure 32.22. There is no message integrity or confidentiality in this scenario.

Figure 32.22 *A plaintext message*

Message Integrity

Probably the next improvement is to let Alice sign the message. Alice creates a digest of the message and signs it with her private key. Figure 32.23 shows the situation.

Figure 32.23 *An authenticated message*

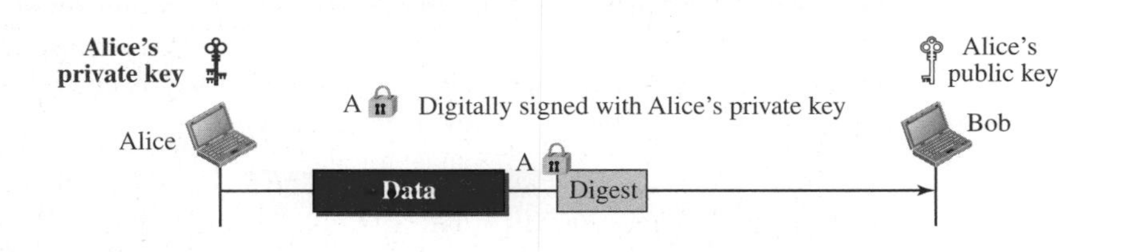

When Bob receives the message, he verifies the message by using Alice's public key. Two keys are needed for this scenario. Alice needs to know her private key; Bob needs to know Alice's public key.

Compression

A further improvement is to compress the message to make the packet more compact. This improvement has no security benefit, but it eases the traffic. Figure 32.24 shows the new scenario.

Figure 32.24 *A compressed message*

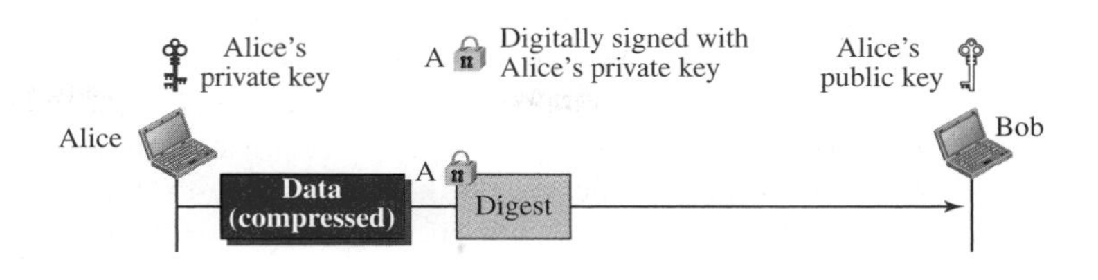

Confidentiality with One-Time Session Key

Figure 32.25 shows the situation. As we discussed before, confidentiality in an e-mail system can be achieved using conventional encryption with a one-time session key. Alice can create a session key, use the session key to encrypt the message and the digest, and send the key itself with the message. However, to protect the session key, Alice encrypts it with Bob's public key.

Figure 32.25 *A confidential message*

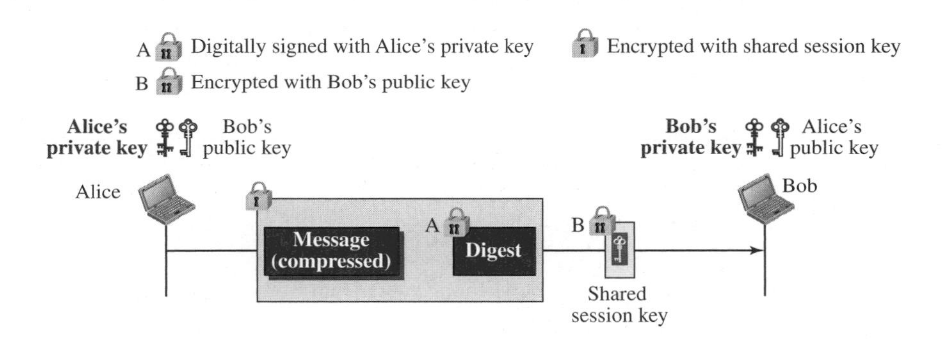

When Bob receives the packet, he first decrypts the session key, using his private key. He then uses the session key to decrypt the rest of the message. After decompressing the rest of the message, Bob creates a digest of the message and checks to see if it is equal to the digest sent by Alice. If it is, then the message is authentic.

Code Conversion

Another service provided by PGP is code conversion. Most e-mail systems allow the message to consist of only ASCII characters. To translate other characters not in the ASCII set, PGP uses Base-64 conversion (see Chapter 26).

Segmentation

PGP allows segmentation of the message after it has been converted to Radix-64 to make each transmitted unit the uniform size allowed by the underlying e-mail protocol.

Key Rings

In all previous scenarios, we assumed that Alice needs to send a message only to Bob. That is not always the case. Alice may need to send messages to many people; she needs *key rings.* In this case, Alice needs a ring of public keys, with a key belonging to each person with whom Alice needs to correspond (send or receive messages). In addition, the PGP designers specified a ring of private/public keys. One reason is that Alice may wish to change her pair of keys from time to time. Another reason is that Alice may need to correspond with different groups of people (friends, colleagues, and so on). Alice may wish to use a different key pair for each group. Therefore, each user needs to have two sets of rings: a ring of private keys and a ring of public keys of other people. Figure 32.26 shows a community of three people, each having a ring of pairs of private/public keys and, at the same time, a ring of public keys belonging to other people in the community.

Figure 32.26 *Key rings in PGP*

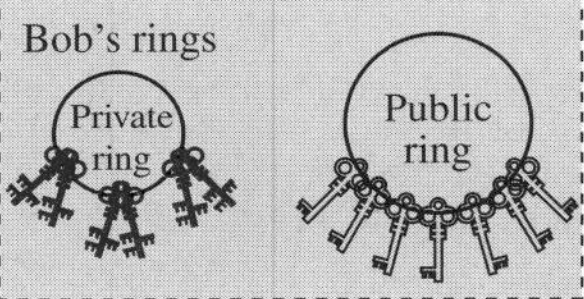

Alice, for example, has several pairs of private/public keys belonging to her and public keys belonging to other people. Note that everyone can have more than one public key. Two cases may arise.

1. Alice needs to send a message to another person in the community.
   a. She uses her private key to sign the digest.
   b. She uses the receiver's public key to encrypt a newly created session key.
   c. She encrypts the message and signed digest with the session key created.
2. Alice receives a message from another person in the community.
   a. She uses her private key to decrypt the session key.
   b. She uses the session key to decrypt the message and digest.
   c. She uses her public key to verify the digest.

PGP Algorithms

PGP defines a set of asymmetric-key and symmetric-key algorithms, cryptography hash functions, and compression methods. We leave the details of these algorithms to the books devoted to PGP. When Alice sends an e-mail to Bob, she defines the algorithm she has used for each purpose.

PGP Certificates and Trusted Model

PGP, like other protocols we have seen so far, uses certificates to authenticate public keys. However, the process is totally different, as explained below.

PGP Certificates

In PGP, there is no need for a certification authority (CA); anyone in the ring can sign a certificate for anyone else in the ring. Bob can sign a certificate for Ted, John, Anne, and so on. There is no hierarchy of trust in PGP; there is no tree. The lack of hierarchical structure may result in the fact that Ted may have one certificate from Bob and another certificate from Liz. If Alice wants to follow the line of certificates for Ted, there are two paths: one starts from Bob and one starts from Liz. An interesting point is that Alice may fully trust Bob, but only partially trust Liz. There can be multiple paths in the line of trust from a fully or partially trusted authority to a certificate. In PGP, the issuer of a certificate is usually called an *introducer.*

In PGP, there can be multiple paths from fully or partially trusted authorities to any subject.

- ***Trust and Legitimacy.*** The entire operation of PGP is based on trust in the introducer, trust in the certificate, and acceptance of the legitimacy of the public keys.
- ***Introducer Trust Levels.*** With the lack of a central authority, it is obvious that the ring cannot be very large if every user has to fully trust everyone else. (Even in real life we cannot fully trust everyone that we know.) To solve this problem, PGP allows different levels of trust. The number of levels is mostly implementation dependent, but for simplicity, let us assign three levels of trust to any introducer: *none, partial,* and *full*. The introducer trust level specifies the trust levels issued by the introducer for other people in the ring. For example, Alice may fully trust Bob, partially trust Anne, and not trust John at all. There is no mechanism in PGP to determine how to make a decision about the trustworthiness of the introducer; it is up to the user to make this decision.
- ***Certificate Trust Levels.*** When Alice receives a certificate from an introducer, she stores the certificate under the name of the subject (certified entity). She assigns a level of trust to this certificate. The certificate trust level is normally the same as the trust level for the introducer that issued the certificate. Assume that Alice fully trusts Bob, partially trusts Anne and Janette, and has no trust in John. The following scenarios can happen.

  1. Bob issues two certificates, one for Linda (with public key K1) and one for Lesley (with public key K2). Alice stores the public key and certificate for Linda under Linda's name and assigns a *full* level of trust to this certificate. Alice also stores the certificate and public key for Lesley under Lesley's name and assigns a full level of trust to this certificate.
  2. Anne issues a certificate for John (with public key K3). Alice stores this certificate and public key under John's name, but assigns a *partial* level for this certificate.
  3. Janette issues two certificates, one for John (with public key K3) and one for Lee (with public key K4). Alice stores John's certificate under his name and Lee's certificate under his name, each with a *partial* level of trust. Note that

John now has two certificates, one from Anne and one from Janette, each with a *partial* level of trust.

4. John issues a certificate for Liz. Alice can discard or keep this certificate with a signature trust of *none*.

- ***Key Legitimacy.*** The purpose of using trust in the introducer and certificate is to determine the legitimacy of a public key. Alice needs to know how legitimate the public keys of Bob, John, Liz, Anne, and so on are. PGP defines a very clear procedure for determining key legitimacy. The level of the key legitimacy for a user is the weighted trust levels of that user. For example, suppose we assign the following weights to certificate trust levels:
  1. A weight of 0 to a nontrusted certificate.
  2. A weight of 1/2 to a certificate with partial trust.
  3. A weight of 1 to a certificate with full trust.

Then to fully trust an entity, Alice needs one fully trusted certificate or two partially trusted certificates for that entity. For example, Alice can use John's public key in the previous scenario because both Anne and Janette have issued a certificate for John, each with a certificate trust level of 1/2. Note that the legitimacy of a public key belonging to an entity does not have anything to do with the trust level for that person. Although Bob can use John's public key to send a message to him, Alice cannot accept any certificate issued by John because, for Alice, John has a trust level of *none*.

Trust Model in PGP

As Zimmermann has proposed, we can create a trust model for any user in a ring with the user as the center of activity. Such a model can look like the one shown in Figure 32.27. The figure shows the trust model for Alice at some moment.

Figure 32.27 *Trust model*

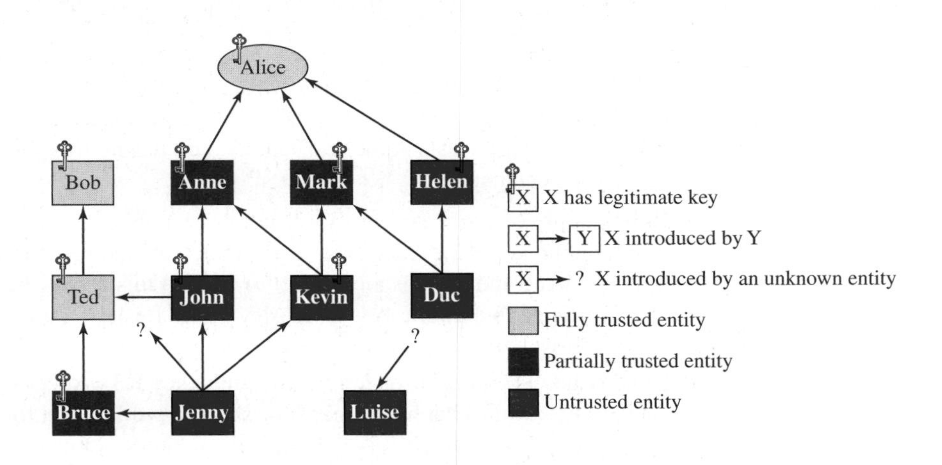

Let us elaborate on the figure. Figure 32.27 shows that there are three entities in Alice's ring with full trust (Alice herself, Bob, and Ted). The figure also shows three entities with partial trust (Anne, Mark, and Bruce). There are also six entities with no trust. Nine entities have a legitimate key. Alice can encrypt a message to any one of these entities or verify a signature received from one of these entities (Alice's key is never used in this model). There are also three entities that do not have any legitimate keys with Alice.

Bob, Anne, and Mark have made their keys legitimate by sending their keys by e-mail and verifying their fingerprints by phone. Helen, on the other hand, has sent a certificate from a CA because she is not trusted by Alice and verification on the phone is not possible. Although Ted is fully trusted, he has given Alice a certificate signed by Bob. John has sent Alice two certificates, one signed by Ted and one by Anne. Kevin has sent two certificates to Alice, one signed by Anne and one by Mark. Each of these certificates gives Kevin half a point of legitimacy; therefore, Kevin's key is legitimate. Duc has sent two certificates to Alice, one signed by Mark and the other by Helen. Since Mark is half-trusted and Helen is not trusted, Duc does not have a legitimate key. Jenny has sent four certificates, one signed by a half-trusted entity, two by untrusted entities, and one by an unknown entity. Jenny does not have enough points to make her key legitimate. Luise has sent one certificate signed by an unknown entity. Note that Alice may keep Luise's name in the table in case future certificates for Luise arrive.

- ***Web of Trust.*** PGP can eventually make a **web of trust** among a group of people. If each entity introduces more entities to other entities, the public key ring for each entity gets larger and larger and entities in the ring can send secure e-mail to each other.
- ***Key Revocation.*** It may become necessary for an entity to revoke his or her public key from the ring. This may happen if the owner of the key feels that the key is compromised (stolen, for example) or just too old to be safe. To revoke a key, the owner can send a revocation certificate signed by herself. The revocation certificate must be signed by the old key and disseminated to all the people in the ring that use that public key.

PGP Packets

A message in PGP consists of one or more packets. During the evolution of PGP, the format and the number of packet types have changed. We do not discuss the formats of these packets here.

Applications of PGP

PGP has been extensively used for personal e-mails. It will probably continue to be.

32.3.3 S/MIME

Another security service designed for electronic mail is **Secure/Multipurpose Internet Mail Extension (S/MIME).** The protocol is an enhancement of the Multipurpose Internet Mail Extension (MIME) protocol we discussed in Chapter 26.

Cryptographic Message Syntax (CMS)

To define how security services, such as confidentiality or integrity, can be added to MIME content types, S/MIME has defined **Cryptographic Message Syntax (CMS).** The syntax in each case defines the exact encoding scheme for each content type. The following describe the types of messages and different subtypes that are created from these messages. For details, the reader is referred to RFC 3369 and RFC 3370.

Data Content Type

This is an arbitrary string. The object created is called *Data*.

Signed-Data Content Type

This type provides only integrity of data. It contains any data type plus zero or more signature values. The encoded result is an *object* called *signedData*. Figure 32.28 shows the process of creating an object of this type. The following are the steps in the process:

1. For each signer, a message digest is created from the content using the specific hash algorithm chosen by that signer.
2. Each message digest is signed with the private key of the signer.
3. The content, signature values, certificates, and algorithms are then collected to create the *signedData* object.

Figure 32.28 *Signed-data content type*

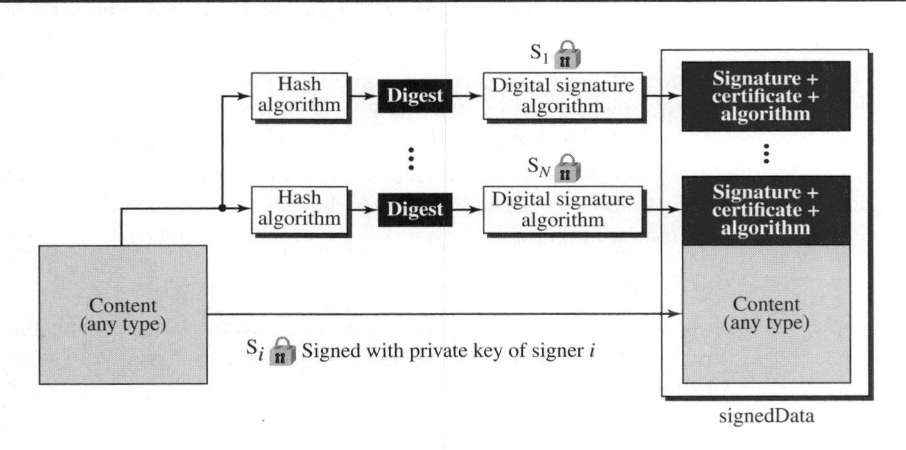

Note that, in this case, the content is not necessarily a personal message. It can be a document whose integrity needs to be preserved. The sender can collect the signatures and then send (or store) them with the message.

Enveloped-Data Content Type

This type is used to provide privacy for the message. It contains any message type plus zero or more encrypted keys and certificates. The encoded result is an *object* called *envelopedData*. Figure 32.29 shows the process of creating an object of this type.

Figure 32.29 *Enveloped-data content type*

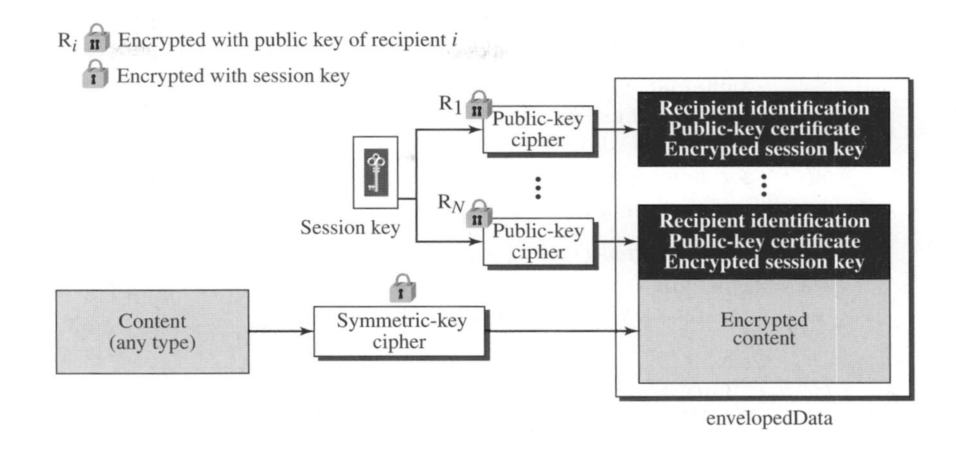

1. A pseudorandom session key is created for the symmetric-key algorithms to be used.
2. For each recipient, a copy of the session key is encrypted with the public key of that recipient.
3. The content is encrypted using the defined algorithm and created session key.
4. The encrypted contents, encrypted session keys, algorithm used, and certificates are encoded using Radix-64.

   Note that, in this case, we can have one or more recipients.

Digested-Data Content Type

This type is used to provide integrity for the message. The result is normally used as the content for the enveloped-data content type. The encoded result is an *object* called *digestedData*. Figure 32.30 shows the process of creating an object of this type.

Figure 32.30 *Digested-data content type*

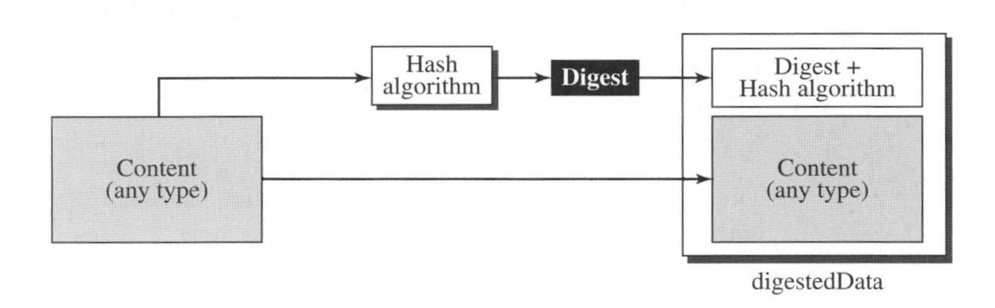

1. A message digest is calculated from the content.

2. The message digest, the algorithm, and the content are added together to create the *digestedData* object.

Encrypted-Data Content Type

This type is used to create an encrypted version of any content type. Although this looks like the enveloped-data content type, the encrypted-data content type has no recipient. It can be used to store the encrypted data instead of transmitting it. The process is very simple: the user employs any key (normally derived from the password) and any algorithm to encrypt the content. The encrypted content is stored without including the key or the algorithm. The object created is called *encryptedData*.

Authenticated-Data Content Type

This type is used to provide authentication of the data. The object is called *authenticatedData*. Figure 32.31 shows the process.

Figure 32.31 *Authenticated-data content type*

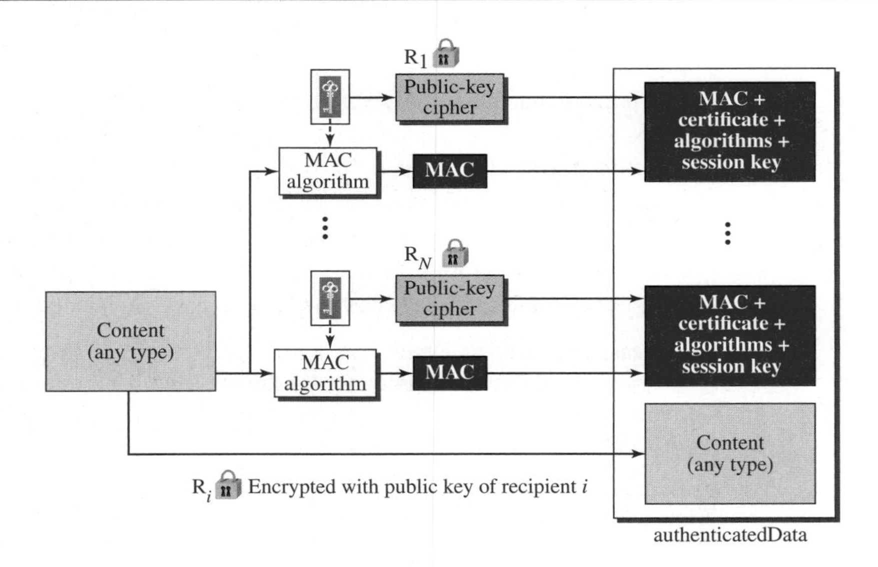

1. Using a pseudorandom generator, a MAC key is generated for each recipient.
2. The MAC key is encrypted with the public key of the recipient.
3. A MAC is created for the content.
4. The content, MAC, algorithms, and other information are collected to form the authenticatedData object.

Key Management

The key management in S/MIME is a combination of key management used by X.509 and PGP. S/MIME uses public-key certificates signed by the certificate authorities defined by X.509. However, the user is responsible for maintaining the web of trust to verify signatures as defined by PGP.

Cryptographic Algorithms

S/MIME defines several cryptographic algorithms. We leave the details of these algorithms to the books dedicated to security in the Internet.

Example 32.1

The following shows an example of an enveloped-data in which a small message is encrypted using triple DES.

Content-Type: application/pkcs7-mime; mime-type=enveloped-data
Content-Transfer-Encoding: Radix-64
Content-Description: attachment
name="report.txt";
cb32ut67f4bhijHU21oi87eryb0287hmnklsgFDoY8bc659GhIGfH6543mhjkdsaH23YjBnmN
ybmlkzjhgfdyhGe23Kjk34XiuD678Es16se09jy76jHuytTMDcbnmlkjgfFdiuyu678543m0n3hG
34un12P2454Hoi87e2ryb0H2MjN6KuyrlsgFDoY897fk923jljk1301XiuD6gh78EsUyT23y

Applications of S/MIME

It is predicted that S/MIME will become the industry choice to provide security for commercial e-mail.

32.4 FIREWALLS

All previous security measures cannot prevent Eve from sending a harmful message to a system. To control access to a system we need firewalls. A **firewall** is a device (usually a router or a computer) installed between the internal network of an organization and the rest of the Internet. It is designed to forward some packets and filter (not forward) others. Figure 32.32 shows a firewall.

Figure 32.32 *Firewall*

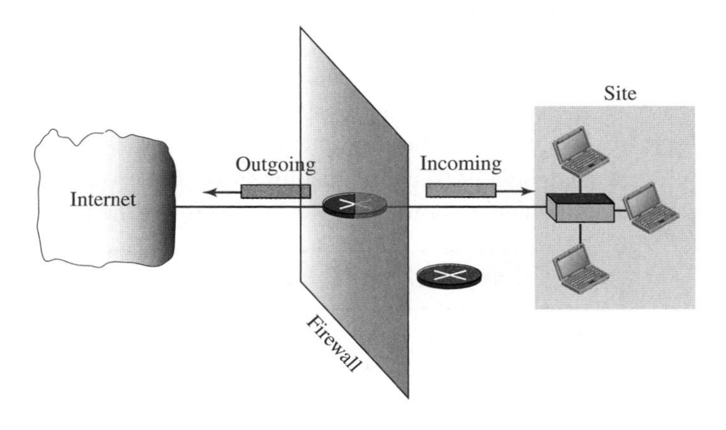

For example, a firewall may filter all incoming packets destined for a specific host or a specific server such as HTTP. A firewall can be used to deny access to a specific host or a specific service in the organization. A firewall is usually classified as a *packet-filter firewall* or a *proxy-based firewall*.

32.4.1 Packet-Filter Firewall

A firewall can be used as a packet filter. It can forward or block packets based on the information in the network-layer and transport-layer headers: source and destination IP addresses, source and destination port addresses, and type of protocol (TCP or UDP). A **packet-filter firewall** is a router that uses a filtering table to decide which packets must be discarded (not forwarded). Figure 32.33 shows an example of a filtering table for this kind of a firewall.

Figure 32.33 *Packet-filter firewall*

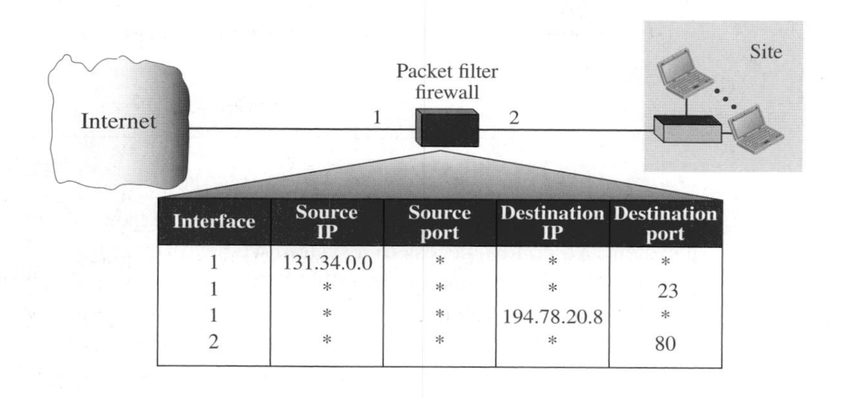

Interface	Source IP	Source port	Destination IP	Destination port
1	131.34.0.0	*	*	*
1	*	*	*	23
1	*	*	194.78.20.8	*
2	*	*	*	80

According to the figure, the following packets are filtered:

1. Incoming packets from network 131.34.0.0 are blocked (security precaution). Note that the * (asterisk) means "any."
2. Incoming packets destined for any internal TELNET server (port 23) are blocked.
3. Incoming packets destined for internal host 194.78.20.8 are blocked. The organization wants this host for internal use only.
4. Outgoing packets destined for an HTTP server (port 80) are blocked. The organization does not want employees to browse the Internet.

A packet-filter firewall filters at the network or transport layer.

32.4.2 Proxy Firewall

The packet-filter firewall is based on the information available in the network layer and transport layer headers (IP and TCP/UDP). However, sometimes we need to filter a message based on the information available in the message itself (at the application layer). As an example, assume that an organization wants to implement the following policies regarding its web pages: only those Internet users who have previously established business relations with the company can have access; access to other users must be blocked. In this case, a packet-filter firewall is not feasible because it cannot distinguish between different packets arriving at TCP port 80 (HTTP). Testing must be done at the application level (using URLs).

One solution is to install a **proxy firewall** (computer) (sometimes called an ***application gateway***), which stands between the customer computer and the corporation computer. When the user client process sends a message, the application gateway runs a server process to receive the request. The server opens the packet at the application level and finds out if the request is legitimate. If it is, the server acts as a client process and sends the message to the real server in the corporation. If it is not, the message is dropped and an error message is sent to the external user. In this way, the requests of the external users are filtered based on the contents at the application layer. Figure 32.34 shows an application gateway implementation for HTTP.

Figure 32.34 *Proxy firewall*

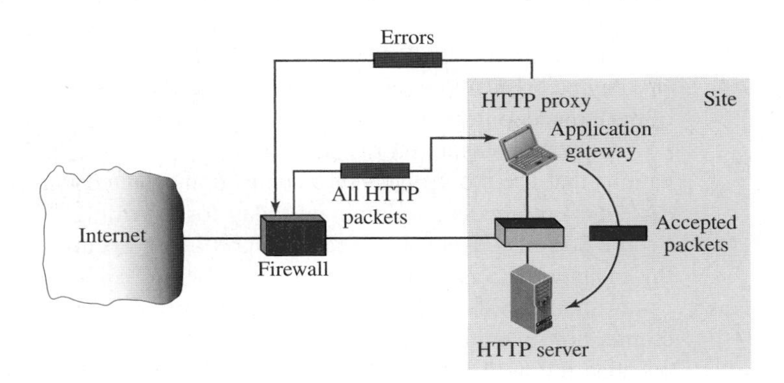

A proxy firewall filters at the application layer.

32.5 END-CHAPTER MATERIALS

32.5.1 Recommended Reading

For more details about subjects discussed in this chapter, we recommend the following books and RFCs. The items enclosed in brackets refer to the reference list at the end of the book.

Books

Several books give through coverage of cryptography and network security: [For 08], [Sta 06], [Bis 05], [Mao 04], [Sti 06]. [Res 01], [Tho 00], [DH 03], and [Gar 95].

RFCs

Two RFCs discuss Cryptographic Message Syntax: RFC 3369 and RFC 3370.

32.5.2 Key Terms

application gateway
Authentication Header (AH) protocol
Cryptographic Message Syntax (CMS)
Encapsulating Security Payload (ESP)
firewall
Handshake Protocol
Internet Key Exchange (IKE)
Internet Security Association and Key Management Protocol (ISAKMP)
IP Security (IPSec)
Oakley
packet-filter firewall
Pretty Good Privacy (PGP)
proxy firewall
Secure/Multipurpose Internet Mail Extension (S/MIME)
Secure Sockets Layer (SSL) protocol
Security Association (SA)
Security Association Database (SAD)
Security Policy (SP)
Security Policy Database (SPD)
SKEME
Transport Layer Security (TLS) protocol
transport mode
tunnel mode
virtual private network (VPN)
web of trust

32.5.3 Summary

IP Security (IPSec) is a collection of protocols designed by the IETF to provide security for a packet at the network level. IPSec operates in transport or tunnel mode. IPSec defines two protocols: Authentication Header (AH) Protocol and Encapsulating Security Payload (ESP) Protocol. IPSec creates a connection-oriented association at the top of the connectionless IP protocol to be able to provide security.

A transport-layer security protocol provides end-to-end security services for applications that use the services of a connection-oriented transport-layer protocol such as TCP. Two protocols are dominant today for providing security at the transport layer: Secure Sockets Layer (SSL) and Transport Layer Security (TLS). We discussed SSL in this chapter; TLS is similar.

Although SSL or TLS can provide security for applications that use the service of connection-oriented protocols such as TCP, the e-mail application is exceptional because the application uses a one-way communication. The Pretty Good Privacy (PGP), invented by Phil Zimmermann, provides e-mail with privacy, integrity, and authentication. Another security service designed for electronic mail is Secure/Multipurpose Internet Mail Extension (S/MIME).

A firewall is a device (usually a router or a computer) installed between the internal network of an organization and the rest of the Internet. It is designed to forward some packets and filter others. A firewall is usually classified as a packet-filter firewall or a proxy firewall.

32.6 PRACTICE SET

32.6.1 Quizzes

A set of interactive quizzes for this chapter can be found on the book website. It is strongly recommended that the student take the quizzes to check his/her understanding of the materials before continuing with the practice set.

32.6.2 Questions

Q32-1. What are the names of the protocols, discussed in this chapter, that provide security for e-mail?

Q32-2. What are the two types of firewalls?

Q32-3. What are the two protocols defined by IPSec?

Q32-4. What is IKE? What is its role in IPSec?

Q32-5. Why does IPSec need a security association?

Q32-6. How do LANs on a fully private internet communicate?

Q32-7. What are the two protocols discussed in this chapter that provide security at the transport layer?

Q32-8. What is the purpose of the Record Protocol in SSL?

Q32-9. What does ESP add to the IP packet?

Q32-10. How does IPSec create a set of security parameters?

Q32-11. What is the purpose of a firewall?

Q32-12. How does SSL create a set of security parameters?

Q32-13. What is a VPN and why is it needed?

Q32-14. How does PGP create a set of security parameters?

Q32-15. What is the difference between a session and a connection in SSL?

Q32-16. Are both AH and ESP needed for IP security? Why or why not?

Q32-17. What is the purpose of the Handshake Protocol in SSL?

Q32-18. What does AH add to the IP packet?

32.6.3 Problems

P32-1. When we talk about authentication in PGP (or S/MIME), do we mean *message authentication* or *entity authentication*? Explain.

P32-2. Compare and contrast PGP and S/MIME. What are the advantages and disadvantages of each?

P32-3. Host A and host B use IPSec in the transport mode. Can we say that the two hosts need to create a virtual connection-oriented service between them? Explain.

P32-4. When we talk about authentication in SSL, do we mean *message authentication* or *entity authentication*? Explain.

P32-5. If Alice and Bob are continuously sending messages to each other, can they create a security association once and use it for every packet exchanged? Explain.

P32-6. Assume Alice needs to send an e-mail to Bob. Explain how the confidentiality of the e-mail is achieved using PGP.

P32-7. Should the handshaking in SSL occur before or after the three-way handshaking in TCP? Can they be combined? Explain.

P32-8. Assume Alice needs to send an e-mail to Bob. Explain how the authentication of the e-mail is achieved using S/MIME.

P32-9. Assume Alice needs to send an e-mail to Bob. Explain how the integrity of the e-mail is achieved using S/MIME.

P32-10. Can we use SSL with UDP? Explain.

P32-11. Assume Alice needs to send an e-mail to Bob. Explain how the integrity of the e-mail is achieved using PGP.

P32-12. When we talk about authentication in IPSec, do we mean *message authentication* or *entity authentication*? Explain.

P32-13. Assume Alice needs to send an e-mail to Bob. Explain how the confidentiality of the e-mail is achieved using S/MIME.

P32-14. If cryptography algorithms in PGP or S/MIME cannot be negotiated, how can the receiver of the e-mail determine which algorithm has been used by the sender?

P32-15. Why is there no need for a Security Association with SSL?

P32-16. We defined two security services for e-mail (PGP and S/MIME). Explain why e-mail applications cannot use the services of SSL/TLS and need to use either PGP or S/MIME.

32.7 SIMULATION EXPERIMENTS

32.7.1 Applets

We have created some Java applets to show some of the main concepts discussed in this chapter. It is strongly recommended that the students activate these applets on the book website and carefully examine the protocols in action.

32.7.2 Lab Assignments

In this section, we use Wireshark to simulate two protocols: Secure Shell (SSH) and HyperText Transfer Protocol Secure (HTTPS). Full descriptions of these lab assignments are on the book website.

Lab32-1. In Chapter 19, we learned about IP. In this lab, we want to use IPsec to create a secure IP connection between two ends.

Lab32-2. In Chapter 26, we learned about HTTP, the protocol used to access web pages from the Internet. HTTP per se does not provide security. However, we can combine HTTP and SSL/TLS to add security to HTTP. The new protocol is called HyperText Transfer Protocol Secure (HTTPS). In this lab, we want to use HTTPS and capture the packet with Wireshark to examine the contents of the SSL/TSL packets when used with HTTPS.

Lab32-3. We learned about FTP and TELNET in Chapter 26. Transferring files using FTP and logging into a system using TELNET are not secured. In Chapter 26, we also learned that we can use Secure Shell (SSH) to simulate both FTP and TELNET. In this lab, we want to use SSH and capture the packet with Wireshark to learn how the Internet security protocols (SSL/TLS), which we learned in this chapter, can create secure file transfer and logging.

GLOSSARY

1-persistent method A CSMA persistence strategy in which a station sends a frame immediately if the line is idle.

4-dimensional, 5-level pulse amplitude modulation (4D-PAM5) An encoding scheme used by 1000Base-T.

10 Gigabit Ethernet The new implementation of Ethernet operating at 10 Gbps.

56K modem A modem technology using two different data rates: one for uploading and one for downloading from the Internet.

Abstract Syntax Notation One (ASN.1) A formal language using abstract syntax for defining the structure of a protocol data unit (PDU).

access point (AP) A central base station in a BSS.

acknowledgment (ACK) A response sent by the receiver to indicate the successful receipt of data.

acknowledgment number In TCP, the number in the acknowledgment field that defines the sequence number of the next byte expected.

active document In the World Wide Web, a document executed at the local site.

ad hoc network A self-configuring network connected by wireless link.

adaptive antenna system (AAS) A system that uses multiple antennas on both terminal and base station to increase performance.

adaptive delta modulation (ADM) A delta modulation technique in which the value of delta is adjusted in each step.

adaptive DPCM (ADPCM) A DPCM method in which the value of delta is adjusted in each step.

add/drop multiplexer A SONET device that removes and inserts signals in a path without demultiplexing and re-multiplexing.

additive cipher The simplest monoalphabetic cipher in which each character is encrypted by adding its value with a key.

additive increase In TCP, a congestion control strategy in which the window size is increased by just one segment instead of exponentially.

additive increase, multiplicative decrease (AIMD) Combination of additive increase and multiplicative decrease congestion control methods used in TCP.

address aggregation A mechanism in which the blocks of addresses for several organizations are aggregated into one larger block.

Address Resolution Protocol (ARP) In TCP/IP, a protocol for obtaining the link-layer address of a node when the Internet address is known.

address space The total number of addresses used by a protocol.

Advanced Encryption Standard (AES) An asynchronous block cipher adapted by NIST to replace DES.

Advanced Mobile Phone System (AMPS) A North American analog cellular phone system using FDMA.

Advanced Network and Services (ANS) A nonprofit organization created by IBM, Merit, and MCI to build a high-speed backbone.

Advanced Network and Services NET (ANSNET) The network created by ANS.

Advanced Research Projects Agency (ARPA) The government agency that funded ARPANET. The same agency later funded global Internet.

Advanced Research Projects Agency Network (ARPANET) The packet switching network that was funded by ARPA. It was used for internetworking research.

ALOHA The original random multiple access method in which a station can send a frame any time it has one to send (MA).

alternate mark inversion (AMI) A digital-to-digital bipolar encoding method in which the amplitude representing 1 alternates between positive and negative voltages.

American National Standards Institute (ANSI) A national standards organization that defines standards in the United States.

American Standard Code for Information Interchange (ASCII) A character code developed by ANSI and used extensively for data communication.

amplitude The strength of a signal, usually measured in volts.

amplitude modulation (AM) An analog-to-analog conversion method in which the carrier signal's amplitude varies with the amplitude of the modulating signal.

amplitude shift keying (ASK) A digital-to-analog conversion method in which the amplitude of the carrier signal is varied to represent binary 0 or 1.

analog data Data that are continuous and smooth and not limited to a specific number of values.

analog signal A continuous waveform that changes smoothly over time.

analog-to-analog conversion The representation of analog information by an analog signal.

analog-to-digital conversion The representation of analog information by a digital signal.

angle of incidence In optics, the angle formed by a light ray approaching the interface between two media and the line perpendicular to the interface.

anycast address An address that defines a group of computers in which the message is sent to the first member in the group.

aperiodic signal A signal that does not exhibit a pattern or repeating cycle.

applet A computer program for creating an active Web document. It is usually written in Java.

application adaptation layer (AAL) A layer in ATM protocol that breaks user data into 48-byte payloads.

application gateway In a proxy firewall, the computer that stands between the customer computer and corporation computer.

application layer The fifth layer in the Internet model; provides access to network resources.

application programming interface (API) A set of declarations, definitions, and procedures followed by programmers to write client-server programs.

area A collection of networks, hosts, and routers all contained within an autonomous system.

arithmetic coding A lossless coding in which the entire message is mapped to a small interval within the interval [0, 1). The small interval is then encoded as a binary pattern.

association A connection in SCTP.

asymmetric-key cipher A cipher using an asymmetric-key cryptosystem.

asynchronous balanced mode (ABM) In HDLC, a communication mode in which each station can be either primary or secondary.

asynchronous connectionless link (ACL) A link between a Bluetooth master and slave in which a corrupted payload is retransmitted.

Asynchronous Transfer Mode (ATM) A wide area protocol featuring high data rates and equal-sized packets (cells); ATM is suitable for transferring text, audio, and video data.

ATM adaptation layer (AAL) The layer in the ATM protocol that encapsulates the user data.

ATM layer A layer in ATM that provides routing, traffic management, switching, and multiplexing services.

attenuation The loss of a signal's energy due to the resistance of the medium.

audio Recording or transmitting of sound or music.

authentication Verification of the sender of a message.

Authentication Header (AH) Protocol A protocol defined by IPSec at the network layer that provides integrity to a message through the creation of a digital signature by a hashing function.

autokey cipher A stream cipher in which each subkey in the stream is the same as the previous plaintext character. The first subkey is the secret between two parties.

automatic repeat request (ARQ) An error-control method in which correction is made by retransmission of data.

autonegotiation A Fast Ethernet feature that allows two devices to negotiate the mode or data rate.

autonomous system (AS) A group of networks and routers under the authority of a single administration.

Backus-Naur Form (BNF) A meta language that specifies which sequences of symbols form a valid term.

band-pass channel A channel that can pass a range of frequencies.

bandwidth The difference between the highest and the lowest frequencies of a composite signal. It also measures the information-carrying capacity of a line or a network.

bandwidth-delay product A measure of the number of bits that can be sent while waiting for news from the receiver.

banyan switch A multistage switch with microswitches at each stage that route the packets based on the output port, represented as a binary string.

Barker sequence A sequence of 11 bits used for spreading.

baseband transmission Transmission of digital or analog signal without modulation, using a low-pass channel.

baseline wandering In decoding a digital signal, the receiver calculates a running average of the received signal power. This average is called the baseline. A long string of 0s or 1s can cause a drift in the baseline (baseline wandering) and make it difficult for the receiver to decode correctly.

Basic Encoding Rules (BER) A standard that encodes data to be transferred through a network.

basic service set (BSS) The building block of a wireless LAN as defined by the IEEE 802.11 standard.

Batcher-banyan switch A banyan switch that sorts the arriving packets based on their destination port.

baud rate The number of signal elements transmitted per second. A signal element consists of one or more bits.

Bayonet Neill-Concelman (BNC) connector A common coaxial cable connector.

beacon frame In Point Coordination Function of Project 802.11, a frame that starts the repetition interval.

Bellman-Ford An algorithm used to calculate routing tables in the distance vector routing method.

bidirectional frame (B-frame) An MPEG frame that is related to the preceding and following I-frame or P-frame.

bipolar encoding A digital-to-digital encoding method in which 0 amplitude represents binary 0 and positive and negative amplitudes represent alternate 1s.

bipolar with 8-zero substitution (B8ZS) A scrambling technique in which a stream of 8 zeros is replaced by a predefined pattern to improve bit synchronization.

bit Binary digit. The smallest unit of data (0 or 1).

bit length The length of a bit in transfer in meter.

bit rate The number of bits transmitted per second.

bit stuffing In a bit-oriented protocol, the process of adding an extra bit in the data section of a frame to prevent a sequence of bits from looking like a flag.

bit-oriented protocol A protocol in which the data frame is interpreted as a sequence of bits.

block cipher A type of cipher in which blocks of plaintext are encrypted one at a time using the same cipher key.

block coding A coding method in which blocks of n bits are encoded into blocks of m bits where $m > n$.

Bluetooth A wireless LAN technology designed to connect devices of different functions such as telephones and notebooks in a small area such as a room.

Bootstrap Protocol (BOOTP) The protocol that provides configuration information from a table (file).

Border Gateway Protocol (BGP) An inter-autonomous system routing protocol based on path vector routing.

bridge A network device operating at the first two layers of the Internet model with filtering and forwarding capabilities.

broadband transmission Transmission of signals using modulation of a higher frequency signal. The term implies a wide-bandwidth data combined from different sources.

broadcast address An address that allows transmission of a message to all nodes of a network.

broadcast link A link in which each station receives a sent packet.

broadcasting Transmission of a message to all nodes in a network.

browser An application program that displays a WWW document. A browser usually uses other Internet services to access the document.

BSS-transition mobility In a wireless LAN, a station that can move from one BSS to another but is confined inside one ESS.

bucket brigade attack See *man-in-the middle attack*.

buffer Memory set aside for temporary storage.

burst error Error in a data unit in which two or more bits have been altered.

bursty data Data with varying instantaneous transmission rates.

byte A group of 8 bits. An octet.

byte stuffing In a byte-oriented protocol, the process of adding an extra byte in the data section of a frame to prevent a byte from looking like a flag.

byte-oriented protocol A protocol in which the data section of the frame is interpreted as a sequence of bytes (characters).

cable modem A technology in which the TV cable provides Internet access.

cable modem transmission system (CMTS) A device installed inside the distribution hub that receives data from the Internet and passes them to the combiner.

cable TV network A system using coaxial or fiber optic cable that brings multiple channels of video programs into homes.

caching The storing of information in a small, fast memory.

Caesar cipher A shift cipher used by Julius Caesar with the key value of 3.

care-of address A temporary IP address used by a mobile host while visiting a foreign network.

carrier extension A technique in Gigabit Ethernet that increases the minimum length of the frame to achieve a higher maximum cable length.

carrier sense multiple access (CSMA) A contention access method in which each station listens to the line before transmitting data.

carrier sense multiple access with collision avoidance (CSMA/CA) An access method in wireless LANs that avoids collision by forcing the stations to send reservation messages when they find the channel is idle.

carrier sense multiple access with collision detection (CSMA/CD) An access method in which stations transmit whenever the transmission medium is available and retransmit when collision occurs.

carrier signal A high frequency signal used for digital-to-analog or analog-to-analog modulation. One of the characteristics of the carrier signal (amplitude, frequency, or phase) is changed according to the modulating data.

Cascading Style Sheets (CSS) A standard developed to use with HTML to define document presentation.

cell A small, fixed-size data unit; also, in cellular telephony, a geographical area served by a cell office.

cell network A network using the cell as its basic data unit.

cellular telephony A wireless communication technique in which an area is divided into cells. A cell is served by a transmitter.

Certification Authority (CA) An agency such as a federal or state organization that binds a public key to an entity and issues a certificate.

Challenge Handshake Authentication Protocol (CHAP) In PPP, a three-way handshaking protocol used for authentication.

challenge-response authentication An authentication method in which the claimant proves that she *knows* a secret without sending it.

channel A communications pathway.

channelization A multiple access method in which the available bandwidth of a link is shared in time.

character-oriented protocol See *byte-oriented protocol.*

checksum A value used for error detection. It is formed by adding data units using one's complement arithmetic and then complementing the result.

chip In CDMA, a number in a code that is assigned to a station.

choke point A packet sent by a router to the source to inform it of congestion.

Chord A P2P protocol that was published by Stoica et al. in 2001 in which the identifier space is made of 2^m points distributed in a circle in the clockwise direction.

chunk A unit of transmission in SCTP.

cipher A decryption and/or encryption algorithm.

cipher feedback mode (CFB) A DES and triple DES operation mode in which data is sent and received 1 bit at a time, with each bit independent of the previous bits.

cipher stream mode (CSM) A DES and triple DES operation mode in which data is sent and received 1 byte at a time.

cipher suite A list of possible ciphers.

ciphertext The message after being encrypted.

circuit switching A switching technology that establishes an electrical connection between stations using a dedicated path.

circuit-switched network A network in which circuit-switching technology is used. A good example is the old telephone voice network.

cladding Glass or plastic surrounding the core of an optical fiber; the optical density of the cladding must be less than that of the core.

Clark's solution A solution to prevent the silly window syndrome. An acknowledgment is sent as soon as the data arrive, but announces a window size of zero until either there is enough space to accommodate a segment of maximum size or until half of the buffer is empty.

classful addressing An addressing mechanism in which the IP address space is divided into five classes: A, B, C, D, and E.

classless addressing An addressing mechanism in which the IP address space is not divided into classes.

Classless InterDomain Routing (CIDR) A technique to reduce the number of routing table entries when supernetting is used.

client A program that can receive services from programs called servers.

client process A running application program on a local site that requests service from a running application program on a remote site.

client-server model The model of interaction between two application programs in which a program at one end (client) requests a service from a program at the other end (server).

client-server paradigm A paradigm in which two computers are connected by an internet and each must run a program, one to provide a service and one to request a service.

closed-loop congestion control A method to alleviate congestion after it happens.

coaxial cable A transmission medium consisting of a conducting core, insulating material, and a second conducting sheath.

code division multiple access (CDMA) A multiple access method in which one channel carries all transmissions simultaneously.

codeword The encoded dataword.

ColdFusion A dynamic web technology that allows the fusion of data items coming from a conventional database.

collision The event that occurs when two transmitters send at the same time on a channel designed for only one transmission at a time; data will be destroyed.

collision domain The length of the medium subject to collision.

collocated care-of address The care-of address for a mobile host that is acting as a foreign agent.

colon hexadecimal notation In IPv6, an address notation consisting of 32 hexadecimal digits, with every four digits separated by a colon.

committed burst size (Bc) The maximum number of bits in a specific time period that a Frame Relay network must transfer without discarding any frames.

committed information rate (CIR) The committed burst size divided by time.

common carrier A transmission facility available to the public and subject to public utility regulation.

common gateway interface (CGI) A standard for communication between HTTP servers and executable programs. CGI is used in creating dynamic documents.

community antenna TV (CATV) A cable network service that broadcasts video signals to locations with poor or no reception.

compatible address An IPv6 address consisting of 96 bits of zero followed by 32 bits of IPv4.

competitive local exchange carrier (CLEC) A telephone company that cannot provide main telephone services; instead, other services such as mobile telephone service and toll calls inside a LATA are provided.

complementary code keying (CCK) An HR-DSSS encoding method that encodes four or eight bits into one symbol.

composite signal A signal composed of more than one sine wave.

Computer Science Network (CSNET) A network sponsored by the National Science Foundation, originally intended for universities.

concurrent server A server that serves multiple clients simultaneously.

congestion Excessive network or internetwork traffic causing a general degradation of service.

congestion avoidance algorithm An algorithm used by TCP that tries to avoid congestion by slowing down the transmission.

congestion control The mechanism of eliminating or avoiding congestion in a network.

connecting device A device that connects computers or networks.

connection establishment The preliminary setup necessary for a logical connection prior to actual data transfer.

connection-oriented service A service for data transfer involving establishment and termination of a connection.

connectionless network A network that can provide only a connectionless service.

connectionless service A service for data transfer without connection establishment or termination.

constant bit rate (CBR) The data rate of an ATM service class that is designed for customers requiring real-time audio or video services.

constellation diagram A graphical representation of the phase and amplitude of different bit combinations in digital-to-analog modulation.

Consultative Committee for International Telegraphy and Telephony (CCITT) An international standards group now known as the ITU-T.

contention An access method in which two or more devices try to transmit at the same time on the same channel.

contention window In CSMA/CA, an amount of time divided into slots. Each slot is randomly selected by a station for transmission.

controlled access A multiple access method in which the stations consult one another to determine who has the right to send.

convergence sublayer (CS) In ATM protocol, the upper AAL sublayer that adds a header or a trailer to the user data.

cookie A string of characters that holds some information about the client and must be returned to the server untouched.

core The glass or plastic center of an optical fiber.

Core-Based Tree (CBT) In multicasting, a group-shared protocol that uses a center router as the root of the tree.

country domain A subdomain in the Domain Name System that uses two characters as the last suffix.

crossbar switch A switch consisting of a lattice of horizontal and vertical paths. At the intersection of each horizontal and vertical path, there is a crosspoint that can connect the input to the output.

crosspoint The junction of an input and an output on a crossbar switch.

crosstalk The noise on a line caused by signals traveling along another line.

cryptographic hash function A function that creates a much shorter output from an input. To be useful, the function must be resistant to image, preimage, and collision attacks.

Cryptographic Message Syntax (CMS) The syntax used in S/MIME that defines the exact encoding scheme for each content type.

cryptography The science and art of transforming messages to make them secure and immune to attacks.

customer premises equipment (CPE) In WiMAX, the customer premises equipment, or subscriber unit, is performing the same job as a modem in wired communication.

cyclic code A linear code in which the cyclic shifting (rotation) of each codeword creates another code word.

cyclic redundancy check (CRC) A highly accurate error-detection method based on interpreting a pattern of bits as a polynomial.

data element The smallest entity that can represent a piece of information. A bit.

Data Encryption Standard (DES) A symmetric-key block cipher using rounds of Feistel ciphers and standardized by NIST.

data rate The number of data elements sent in one second.

data transparency The ability to send any bit pattern as data without it being mistaken for control bits.

data-link control (DLC) The responsibilities of the data-link layer: flow control and error control.

data-link layer The second layer in the Internet model. It is responsible for node-to-node delivery.

data-transfer phase The intermediate phase in circuit-switched or virtual-circuit networks in which data transfer takes place.

datagram In packet switching, an independent data unit.

datagram network A packet-switched network in which packets are independent from each other.

dataword The smallest block of data in block coding.

deadlock A situation in which a task cannot proceed because it is waiting for an event that will never occur.

decapsulation A process, inverse of encapsulation, that extracts the payload of packets.

decibel (dB) A measure of the relative strength of two signal points.

decryption Recovery of the original message from the encrypted data.

decryption algorithm Algorithm to descramble the ciphertext to create the original plaintext.

default routing A routing method in which a router is assigned to receive all packets with no match in the routing table.

Defense Advanced Research Projects Agency (DARPA) A government organization, which, under the name of ARPA, funded ARPANET and the Internet.

delta modulation An analog-to-digital conversion technique in which the value of the digital signal is based on the difference between the current and the previous sample values.

demodulator A device that demodulates a modulated signal to get the original signal.

demultiplexing Inverse of multiplexing. To get the original signal or data from a multiplexed signal or data.

denial of service The only attack on the availability goal that may slow down or interrupt the system.

dense wave-division multiplexing (DWDM) A WDM method that can multiplex a very large number of channels by spacing channels closer together.

destination address The address of the receiver of the data unit.

differential Manchester encoding A digital-to-digital polar encoding method that features a transition at the middle of the bit interval as well as an inversion at the beginning of each 1 bit.

differential PCM (DPCM) DPCM is the generalization of delta modulation in which more than one previously reconstructed sample is used for prediction.

Differentiated Services (DS or DiffServ) A class-based QoS model designed for IP.

Diffie-Hellman protocol A key management protocol that provides a one-time session key for two parties.

digest A condensed version of a document.

digital AMPS (D-AMPS) A second-generation cellular phone system that is a digital version of AMPS.

digital data Data represented by discrete values or conditions.

digital data service (DDS) A digital version of an analog leased line with a rate of 64 kbps.

digital signal A discrete signal with a limited number of values.

digital signal (DS) service A telephone company service featuring a hierarchy of digital signals.

digital signature A security mechanism in which the sender can electronically sign the message and the receiver can verify the message to prove that the message is indeed signed by the sender.

Digital Signature Standard (DSS) The digital signature standard adopted by NIST under FIPS 186.

digital subscriber line (DSL) A technology using existing telecommunication networks to accomplish high-speed delivery of data, voice, video, and multimedia.

digital subscriber line access multiplexer (DSLAM) A telephone company site device that functions like an ADSL modem.

digital-to-analog conversion The representation of digital information by an analog signal.

digital-to-digital conversion The representation of digital data by a digital signal.

digitization Conversion of analog information to digital information.

Dijkstra's algorithm In link-state routing, an algorithm that finds the shortest path to other routers.

direct broadcast address The IP address in a block or subblock (with the suffix set all to 1s). The address is used by a router to send a message to all hosts in the block.

direct current (DC) A zero-frequency signal with a constant amplitude.

direct delivery A delivery in which the final destination of the packet is a host connected to the same physical network as the sender.

direct sequence spread spectrum (DSSS) distortion A wireless transmission method in which each bit to be sent by the sender is replaced by a sequence of bits called a chip code.

discard eligibility (DE) A bit that identifies a packet that can be discarded if there is congestion in the network.

discrete cosine transform (DCT) A compression technique in which the signal representation of data is changed from a time or space domain to the frequency domain.

discrete multitone technique (DMT) A modulation method combining elements of QAM and FDM.

Distance Vector Multicast Routing Protocol (DVMRP) A protocol based on distance vector routing that handles multicast routing.

distance vector routing A routing method in which each router sends its neighbors a list of networks it can reach and the distance to each network.

distortion Any change in a signal due to noise, attenuation, or other influences.

distributed coordination function (DCF) The basic access method in wireless LANs; stations contend with each other to get access to the channel.

distributed database Information stored in many locations.

distributed hash table (DHT) A DHT distributes data (or references to data) among a set of nodes according to some predefined rules. Each peer in a DHT-based network becomes responsible for a range of data items.

distributed interframe space (DIFS) In wireless LANs, a period of time that a station waits before sending a control frame.

distributed processing A strategy in which services provided for the network reside at multiple sites.

DNS server A computer that holds information about the name space.

domain A subtree of the domain name space.

domain name In the DNS, a sequence of labels separated by dots.

domain name space A method for organizing the name space in which the names are defined in an inverted-tree structure with the root at the top.

Domain Name System (DNS) A TCP/IP application service that converts user-friendly names to IP addresses.

dotted-decimal notation A notation devised to make the IP address easier to read; each byte is converted to its decimal equivalent and then set off from its neighbor by a decimal.

double crossing In mobile IP, double crossing occurs when a remote host communicates with a mobile host that has moved to the same network (or site) as the remote host.

downlink Transmission from a satellite to an earth station.

downloading Retrieving a file or data from a remote site.

dual stack Two protocols (IPv4 and IPv6) on the same station.

dynamic document A Web document created by running a program at the server site.

Dynamic Domain Name System (DDNS) A method to update the DNS master file dynamically.

Dynamic Host Configuration Protocol (DHCP) An extension to BOOTP that dynamically assigns configuration information.

dynamic mapping A technique in which a protocol is used for address resolution.

dynamic routing Routing in which the routing table entries are updated automatically by the routing protocol.

E lines The European equivalent of T lines.

eight-binary, six-ternary (8B6T) encoding A three-level line encoding scheme that encodes a block of 8 bits into a signal of 6 ternary pulses.

eight-binary, ten-binary (8B/10B) encoding A block coding technique in which 8 bits are encoded into a 10-bit code.

electromagnetic spectrum The frequency range occupied by electromagnetic energy.

Electronics Industries Association (EIA) An organization that promotes electronics manufacturing concerns. It has developed interface standards such as EIA-232, EIA-449, and EIA-530.

Encapsulating Security Payload (ESP) A protocol defined by IPSec that provides privacy as well as a combination of integrity and message authentication.

encapsulation The technique in which a data unit from one protocol is placed within the data field portion of the data unit of another protocol.

encryption Converting a message into an unintelligible form that is unreadable unless decrypted.

encryption algorithm An algorithm to convert a message into an unintelligible form that is unreadable unless decrypted.

end office A switching office that is the terminus for the local loops.

end system A sender or receiver of data.

entity authentication A technique designed to let one party prove the identity of another party.

ephemeral port number A port number used by the client.

error control The handling of errors in data transmission.

escape character A character that is used to change the meaning of the next character.

ESS transition mobility The movement of a wireless station from one ESS to another.

Ethernet A local area network using the CSMA/CD access method created by Xerox; has gone through four generations.

excess burst size (Be) In Frame Relay, the maximum number of bits in excess of Bc that the user can send during a predefined period of time.

extended binary coded decimal interchange code (EBCDIC) An 8-bit character code developed and used by IBM.

extended service set (ESS) A wireless LAN service composed of two or more BSSs with APs as defined by the IEEE 802.11 standard.

Extensible HyperText Markup Language (XHTML) HTML that conforms to the syntax of XML.

Extensible Markup Language (XML) A language that allows users to define representation of data.

Extensible Style Language (XSL) The style language of XML.

exterior routing Routing between autonomous systems.

extranet A private network that uses the TCP/IP protocol suite to allow authorized access from outside users.

Fast Ethernet Ethernet with a data rate of 100 Mbps.

fast-recovery algorithm An algorithm used in TCP when three duplicate ACKs arrive, which is interpreted as light congestion. Similar to additive increase but increases size of congestion window when a new duplicate ACK arrives.

fast retransmission Retransmission of a segment in the TCP protocol when three duplicate acknowledgments have been received that imply the loss or corruption of that segment.

Federal Communications Commission (FCC) A government agency that regulates radio, television, and telecommunications.

Feistel cipher A class of ciphers consisting of both invertible and noninvertible components.

fiber-optic cable A high-bandwidth transmission medium that carries data signals in the form of pulses of light. It consists of a thin cylinder of glass or plastic, called the core, surrounded by a concentric layer of glass or plastic called the cladding.

File Transfer Protocol (FTP) In TCP/IP, an application layer protocol that transfers files between two sites.

filtering A process in which a switch makes forwarding decisions.

finite state machine (FSM) A machine that goes through a limited number of states.

firewall A device (usually a router) installed between the internal network of an organization and the rest of the Internet to provide security.

first-in, first-out (FIFO) queue A queue in which the first item in is the first item out.

flag A bit pattern or a character added to the beginning and the end of a frame to separate the frames.

flat name space A name space in which there is no hierarchical structure.

flooding Saturation of a network with a message.

flow control A technique to control the rate of flow of frames (packets or messages).

footprint An area on Earth that is covered by a satellite at a specific time.

foreign agent In mobile IP, the foreign agent is a router or a host attached to the foreign network. The foreign agent receives and delivers packets sent by the home agent to the mobile host.

foreign network The network a mobile host is connected to that is not its home.

forward error correction (FEC) Correction of errors at the receiver without retransmission.

forwarding Placing the packet in its route to its destination.

four-binary, five-binary (4B/5B) encoding A block coding technique in which a four-bit block is encoded into a 5-bit code.

Fourier analysis The mathematical technique used to obtain the frequency spectrum of an aperiodic signal if the time-domain representation is given.

fragmentation The division of a packet into smaller units to accommodate a protocol's MTU.

frame A group of bits representing a block of data.

frame bursting A technique in CSMA/CD Gigabit Ethernet in which multiple frames are logically connected to each other to resemble a longer frame.

framing Grouping a set of bits or a set of bytes into a unit.

frequency The number of cycles per second of a periodic signal.

frequency masking Frequency masking occurs when a loud sound partially or totally masks a softer sound if the frequencies of the two are close.

frequency modulation (FM) An analog-to-analog modulation method in which the carrier signal's frequency varies with the amplitude of the modulating signal.

frequency shift keying (FSK) A digital-to-analog encoding method in which the frequency of the carrier signal is varied to represent binary 0 or 1.

frequency division multiple access (FDMA) An access method technique in which multiple sources use assigned bandwidth in a data communication band.

frequency division multiplexing (FDM) The combining of analog signals into a single signal.

frequency domain plot A graphical representation of a signal's frequency components.

frequency hopping spread spectrum (FHSS) A wireless transmission method in which the sender transmits at one carrier frequency for a short period of time, then hops to another carrier frequency for the same amount of time, hops again for the same amount of time, and so on. After N hops, the cycle is repeated.

full-duplex mode A transmission mode in which both parties can communicate simultaneously.

full-duplex switched Ethernet Ethernet in which each station, in its own separate collision domain, can both send and receive.

fully qualified domain name (FQDN) A domain name consisting of labels beginning with the host and going back through each level to the root node.

fundamental frequency The frequency of the dominant sine wave of a composite signal.

gatekeeper In the H.323 standard, a server on the LAN that plays the role of the registrar server.

generator polynomial The polynomial used as the divisor in CRC.

generic domain A subdomain in the domain name system that uses generic suffixes.

geographical routing A routing technique in which the entire address space is divided into blocks based on physical landmasses.

geostationary Earth orbit (GEO) A satellite orbit positioned above the upper Van Allen Belt. A satellite travelling in this orbit looks stationary to the people on Earth.

Gigabit Ethernet Ethernet with one gigabit per second (1000 Mbps) data rate.

Global Positioning System (GPS) An MEO public satellite system consisting of 24 satellites and used for land and sea navigation. GPS is not used for communications.

Global System for Mobile Communication (GSM) A second-generation cellular phone system used in Europe.

Globalstar A LEO satellite system with 48 satellites in six polar orbits with each orbit hosting eight satellites.

Go-Back-*N* protocol An error-control protocol in which the frame in error and all following frames must be retransmitted.

Graphic Interchange Format (GIF) A standard that normally uses a palette of only 256 different colors.

ground propagation Propagation of radio waves through the lowest portion of the atmosphere (hugging the earth).

group-shared tree A multicast routing feature in which each group in the system shares the same tree.

guard band A bandwidth separating two signals.

guided media Transmission media with a physical boundary.

H.323 A standard designed by ITU to allow telephones on the public telephone network to talk to computers (called terminals in H.323) connected to the Internet.

half-close In TCP, a type of connection termination in which one site stops sending data while it is still receiving data.

half-duplex mode A transmission mode in which communication can be two-way but not at the same time.

Hamming code A method that adds redundant bits to a data unit to detect and correct bit errors.

Hamming distance The number of differences between the corresponding bits in two codewords.

handoff Changing to a new channel as a mobile device moves from one cell to another.

handshake protocol The protocol used in connection-oriented networks to establish a connection or to tear down the connection.

harmonics Components of a digital signal, each having a different amplitude, frequency, and phase.

hash function An algorithm that creates a fixed-size digest from a variable-length message.

hashed MAC (HMAC) A MAC based on a hash function such as SHA-1.

hashing A cryptographic technique in which a fixed-length message digest is created from a variable-length message.

header Control information added to the beginning of a data packet.

hertz (Hz) Unit of measurement for frequency.

hexadecimal colon notation Same as colon hexadecimal notation.

hierarchical routing A routing technique in which the entire address space is divided into levels based on specific criteria.

high-density bipolar 3-zero (HDB3) A scrambling technique in which four consecutive zero-level voltages are replaced with one of the two predefined sequences.

High-level Data Link Control (HDLC) A bit-oriented data-link protocol defined by the ISO.

high-rate direct-sequence spread spectrum (HR-DSSS) A signal generation method similar to DSSS except for the encoding method (CCK).

home address The original address of a mobile host.

home agent Usually a router attached to the home network of the mobile host that receives and sends packets (for the mobile host) to the foreign agent.

home network A network that is the permanent home of the mobile host.

hop count The number of nodes along a route. It is a measurement of distance in routing algorithms.

hop-to-hop delivery Transmission of frames from one node to the next.

horn antenna A scoop-shaped antenna used in terrestrial microwave communication.

host A station or node on a network.

hostid The part of an IP address that identifies a host.

host-specific routing A routing method in which the full IP address of a host is given in the routing table.

hub A central device in a star topology that provides a common connection among the nodes.

Huffman coding A statistical compression method using variable-length codes to encode a set of symbols.

hybrid network A network with a private internet and access to the global Internet.

hybrid-fiber-coaxial (HFC) network The second generation of cable networks; uses fiber optic and coaxial cable.

hypermedia Information containing text, pictures, graphics, and sound that is linked to other documents through pointers.

hypertext Information containing text that is linked to other documents through pointers.

HyperText Markup Language (HTML) The computer language for specifying the contents and format of a Web document. It allows additional text to include codes that define fonts, layouts, embedded graphics, and hypertext links.

HyperText Transfer Protocol (HTTP) An application service for retrieving a web document.

inband signaling Using the same channel for data and control transfer.

indirect delivery A delivery in which the source and destination of a packet are in different networks.

infrared wave A wave with a frequency between 300 GHz and 400 THz; usually used for short-range communications.

initial sequence number (ISN) In TCP, the random number used as the first sequence number in a connection.

inner product A number produced by multiplying two sequences, element by element, and summing the products.

Institute of Electrical and Electronics Engineers (IEEE) A group consisting of professional engineers that has specialized societies whose committees prepare standards in members' areas of specialty.

Integrated Services (IntServ) A flow-based QoS model designed for IP.

interactive audio/video Real-time communication with sound and images.

interautonomous system routing protocol A protocol to handle transmissions between autonomous systems.

interdomain routing Routing among autonomous systems.

interface The boundary between two pieces of equipment. It also refers to mechanical, electrical, and functional characteristics of the connection. In network programming, a set of procedures available to the upper layer to use the services of the lower layer.

interference Any undesired energy that interferes with the desired signals.

interframe space (IFS) In wireless LANs, a time interval between two frames to control access to the channel.

Interim Standard 95 (IS-95) One of the dominant second-generation cellular telephony standards in North America.

interior routing Routing inside an autonomous system.

interleaved FDMA (IFDMA) The more efficient FDMA used in Universal Mobile Telecommunications System (UMTS).

interleaving In multiplexing, taking a specific amount of data from each device in a regular order.

International Organization for Standardization (ISO) A worldwide organization that defines and develops standards on a variety of topics.

International Telecommunication Union (ITU) An international telecommunication organization.

Internet A global internet that uses the TCP/IP protocol suite.

Internet address A 32-bit or 128-bit network-layer address used to uniquely define a host on an internet using the TCP/IP protocol.

Internet Architecture Board (IAB) The technical adviser to the ISOC; oversees the continuing development of the TCP/IP protocol suite.

Internet Assigned Numbers Authority (IANA) A group supported by the U.S. government that was responsible for the management of Internet domain names and addresses until October 1998.

Internet Control Message Protocol (ICMP) A protocol in the TCP/IP protocol suite that handles error and control messages. Two versions are used today, ICMPv4 and ICMPv6.

Internet Corporation for Assigned Names and Numbers (ICANN) A private, nonprofit corporation managed by an international board that assumed IANA operations.

Internet draft A working Internet document (a work in progress) with no official status and a six-month lifetime.

Internet Engineering Steering Group (IESG) An organization that oversees the activities of IETF.

Internet Engineering Task Force (IETF) A group working on the design and development of the TCP/IP protocol suite and the Internet.

Internet Group Management Protocol (IGMP) A protocol in the TCP/IP protocol suite that handles multicasting.

Internet Key Exchange (IKE) A protocol designed to create security associations in IPSec.

Internet Mail Access Protocol (IMAP) A complex and powerful protocol to pull e-mail messages from an e-mail server.

Internet Mobile Communication 2000 (ITM-2000) An ITU issued blueprint that defines criteria for third-generation cellular telephony.

Internet model A 5-layer protocol stack that dominates data communications and networking today.

Internet Network Information Center (INTERNIC) An agency responsible for collecting and distributing information about TCP/IP protocols.

Internet Protocol (IP) The network-layer protocol in the TCP/IP protocol suite governing connectionless transmission across packet-switched networks. Two versions commonly in use: IPv4 and IPv6.

Internet Protocol Control Protocol (IPCP) In PPP, the set of protocols that establish and terminate a network layer connection for IP packets.

Internet Protocol, next generation (IPng) Another term for the sixth version of the Internet Protocol, IPv6.

Internet Protocol, version 6 (IPv6) The sixth version of the Internet Protocol.

Internet Research Task Force (IRTF) A forum of working groups focusing on long-term research topics related to the Internet.

Internet Security Association and Key Management Protocol (ISAKMP) A protocol designed by the National Security Agency (NSA) that actually implements the exchanges defined in IKE.

Internet service provider (ISP) A company that provides Internet services.

Internet Society (ISOC) The nonprofit organization established to publicize the Internet.

Internet standard A thoroughly tested specification that is useful to and adhered to by those who work with the Internet. It is a formalized regulation that must be followed.

internetwork (internet) A network of networks.

internetworking Connecting several networks together using internetworking devices such as routers and gateways.

intracoded frame (I-frame) In MPEG, an I-frame is an independent frame that is not related to any other frame (not to the frame sent before or after). They are present at regular intervals.

intranet A private network that uses the TCP/IP protocol suite.

inverse domain A subdomain in the DNS that finds the domain name, given the IP address.

IP datagram The Internet Protocol data unit.

IP Security (IPSec) A collection of protocols designed by the IETF (Internet Engineering Task Force) to provide security for a packet carried on the Internet.

IrDA port A port that allows a wireless keyboard to communicate with a PC.

Iridium A 66-satellite network that provides communication from any Earth site to another.

iterative resolution Resolution of the IP address in which the client may send its request to multiple servers before getting an answer.

iterative server A server that can serve only one client at a time.

ITU Standardization Sector (ITU-T) A standards organization formerly known as the CCITT.

jamming signal In CSMA/CD, a signal sent by the first station that detects collision to alert every other station of the situation.

Java An object-oriented programming language.

jitter A phenomenon in real-time traffic caused by gaps between consecutive packets at the receiver caused by uneven delays.

Joint Photographic Experts Group (JPEG) A standard for compressing continuous-tone pictures.

Kademlia A DHT-based P2P network in which the distances between nodes are measured as the XOR of two identifiers.

Karn's Algorithm An algorithm that does not include the retransmitted segments in calculation of round-trip time.

keepalive timer A timer in TCP that checks to see if there is an active process at the other site.

key A set of values that the cipher, as an algorithm, operates on.

key-distribution center (KDC) In secret key encryption, a trusted third party that shares a key with each user.

label An identifier used in connection-oriented service to define the path.

leaky bucket algorithm An algorithm to shape bursty traffic.

least-cost tree In least-cost routing, a tree with the source at the root that spans the whole graph.

Lempel-Ziv-Welch (LZW) A group of compression methods based on dynamic creation of a dictionary (array) of strings in the text, which was invented by Lempel and Ziv and refined by Welch.

limited-broadcast address An address used to broadcast messages only to hosts inside a network (link).

line coding Converting binary data into signals.

line-of-sight propagation The transmission of very high frequency signals in straight lines directly from antenna to antenna.

linear block code A block code in which adding two codewords creates another codeword.

linear predictive coding (LPC) A predictive coding method in which, instead of sending quantized difference signals, the source analyzes the signals and determines their characteristics.

link The physical communication pathway that transfers data from one device to another.

Link Control Protocol (LCP) A PPP protocol responsible for establishing, maintaining, configuring, and terminating links.

link local address An IPv6 address that is used if a LAN is to use the Internet protocols but is not connected to the Internet for security reasons.

link-layer address The address of a device used at the data-link layer (MAC address).

link-state advertisement (LSA) In OSPF, a method to disperse information.

link-state database In link-state routing, a database common to all routers and made from LSP information.

link-state packet (LSP) In link-state routing, a small packet containing routing information sent by a router to all other routers.

link-state routing A routing method in which each router shares its knowledge of changes in its neighborhood with all other routers.

local area network (LAN) A network connecting devices inside a single building or inside buildings close to each other.

local logging Logging into a host using the terminal directly connected to the host.

local loop The link that connects a subscriber to the telephone central office.

logical address An address defined in the network layer.

logical link control (LLC) The upper sublayer of the data-link layer as defined by IEEE Project 802.2.

Logical Link Control and Adaptation Protocol (L2CAP) A Bluetooth layer used for data exchange on an ACL link.

logical tunnel The encapsulation of a multicast packet inside a unicast packet to enable multicast routing by nonmulticast routers.

longest mask matching The technique in CIDR in which the longest prefix is handled first when searching a routing table.

loopback address An address used by a host to test its internal software.

lossless compression A compression method in which the integrity of the data is preserved because compression and decompression algorithms are exact inverses of each other: no part of the data is lost in the process.

lossy compression A compression method in which the loss of some data is sacrificed to obtain a better compression ratio.

low-Earth-orbit (LEO) A polar satellite orbit with an altitude between 500 and 2000 km. A satellite with this orbit has a rotation period of 90 to 120 minutes.

low-pass channel A channel that passes frequencies between 0 and f.

magic cookie In DHCP, the number in the format of an IP address with the value of 99.130.83.99; indicates that options are present.

man-in-the-middle attack A key management problem in which an intruder intercepts and sends messages between the intended sender and receiver.

Management Information Base (MIB) The database used by SNMP that holds the information necessary for management of a network.

Manchester encoding A digital-to-digital polar encoding method in which a transition occurs at the middle of each bit interval to provide synchronization.

mapped address An IPv6 address used when a computer that has migrated to IPv6 wants to send a packet to a computer still using IPv4.

mask For IPv4, a 32-bit binary number that gives the first address in the block (the network address) when ANDed with an address in the block.

master secret In SSL, a 48-byte secret created from the *pre-master secret*.

maturity level The phases through which an RFC goes.

maximum transfer unit (MTU) The largest size data unit a specific network can handle.

media access control (MAC) sublayer (also called medium access control) The lower sublayer in the data-link layer defined by the IEEE 802 project. It defines the access method and access control in different local area network protocols.

media server A server used in streaming audio or video.

medium-Earth-orbit (MEO) A satellite orbit positioned between the two Van Allen belts. A satellite at this orbit takes six hours to circle the earth.

mesh topology A network configuration in which each device has a dedicated point-to-point link to every other device.

message access agent (MAA) A client-server program that pulls the stored email messages.

message authentication A security measure in which the sender of the message is verified for every message sent.

message authentication code (MAC) An MDC that includes a secret between two parties. A keyed hash function.

Message Digest (MD) A set of several hash algorithms designed by Ron Rivest and referred to as MD2, MD4, and MD5.

message transfer agent (MTA) An SMTP component that transfers the message across the Internet.

metafile In streaming audio or video, a file that holds information about the audio/video file.

metric A cost assigned for passing through a network.

metropolitan area network (MAN) A network that can span a geographical area the size of a city.

microwave Electromagnetic waves ranging from 2 GHz to 40 GHz.

Military Network (MILNET) A network for military use that was originally part of ARPANET.

minimum Hamming distance In a set of words, the smallest Hamming distance between all possible pairs.

mixer A device that mathematically adds signals coming from different sources to create a single signal.

mobile host A host that can move from one network to another.

mobile switching center (MSC) In cellular telephony, a switching office that coordinates communication between all base stations and the telephone central office.

mobile telephone switching office (MTSO) An office that controls and coordinates communication between all of the cell offices and the telephone control office.

modem A device consisting of a modulator and a demodulator. It converts a digital signal into an analog signal (modulation) and vice versa (demodulation).

modification detection code (MDC) The digest created by a hash function.

modular arithmetic Arithmetic that uses a limited range of integers (0 to $n - 1$).

modulation Modification of one or more characteristics of a carrier wave by an information-bearing signal.

modulator A device that modulates a signal to create another signal.

modulus The upper limit in modular arithmetic (n).

monoalphabetic cipher A substitution cipher in which a symbol in the plaintext is always changed to the same symbol in the ciphertext, regardless of its position in the text.

monoalphabetic substitution An encryption method in which each occurrence of a character is replaced by another character in the set.

Motion Picture Experts Group (MPEG) A method to compress videos.

MPEG audio layer 3 (MP3) A standard that uses perceptual coding to compress audio.

multi-carrier CDMA (MC-CDMA) An access method proposed for 4G wireless networks.

multi-user MIMO (MU-MIMO) A more sophisticated version of MIMO in which multiple users can communicate at the same time.

multicast address An address used for multicasting.

multicast backbone (MBONE) A set of internet routers supporting multicasting through the use of tunneling.

Multicast Open Shortest Path First (MOSPF) A multicast protocol that uses multicast link-state routing to create a source-based least-cost tree.

multicast router A router with a list of loyal members related to each router interface that distributes the multicast packets.

multicast routing Moving a multicast packet to its destinations.

multicasting A transmission method that allows copies of a single packet to be sent to a selected group of receivers.

multihoming service A service provided by SCTP protocol in which a host can be connected to more than one network.

multilevel multiplexing A technique in multiplexing that is used when the data rate of an input line is a multiple of others.

multiline transmission, three-level (MLT-3) encoding A line coding scheme featuring three levels of signals and transitions at the beginning of the 1 bit.

multimedia traffic Traffic consisting of data, video, and audio.

multimode graded-index fiber An optical fiber with a core having a graded index of refraction.

multimode step-index fiber An optical fiber with a core having a uniform index of refraction. The index of refraction changes suddenly at the core/cladding boundary.

multiple access (MA) A line access method in which every station can access the line freely.

multiple unicasting Sending multiple copies of a message, each with a different unicast destination address, from one source.

multiple-input and multiple-output (MIMO) antenna A branch of intelligent antenna proposed for 4G wireless systems that allows independent streams to be transmitted simultaneously from all the antennas to increase the data rate into multiple folds.

multiple-slot allocation A technique in multiplexing that allows an input to have more than one slot in the output.

multiplexer (MUX) A device used for multiplexing.

multiplexing The process of combining signals from multiple sources for transmission across a single data link.

multiplicative decrease A congestion avoidance technique in which the threshold is set to half of the last congestion window size, and the congestion window size starts from one again.

Multipurpose Internet Mail Extensions (MIME) A supplement to SMTP that allows non-ASCII data to be sent through SMTP.

multistage switch An array of switches designed to reduce the number of crosspoints.

multistream service A service provided by SCTP that allows data transfer to be carried using different streams.

Nagle's algorithm An algorithm that attempts to prevent silly window syndrome at the sender's site; both the rate of data production and the network speed are taken into account.

name space All the names assigned to machines on an internet.

name-address resolution Mapping a name to an address or an address to a name.

National Science Foundation Network (NSFNET) The network funded by the National Science Foundation.

National Security Agency (NSA) A U.S. intelligence-gathering security agency.

netid The part of an IP address that identifies the network.

network The interconnection of a set of devices capable of communication.

network access point (NAP) A complex switching station that connects backbone networks.

network address An address that identifies a network to the rest of the Internet; it is the first address in a block.

Network Address Translation (NAT) A technology that allows a private network to use a set of private addresses for internal communication and a set of global Internet addresses for external communication.

network allocation vector (NAV) In CSMA/CA, the amount of time that must pass before a station can check for an idle line.

Network Control Protocol (NCP) In PPP, a set of control protocols that allows the encapsulation of data coming from network-layer protocols.

Network Information Center (NIC) An agency responsible for collecting and distributing information about TCP/IP protocols.

network interface card (NIC) An electronic device, internal or external to a station, that contains circuitry to enable the station to be connected to the network.

network layer The third layer in the Internet model, responsible for the delivery of a packet to the final destination.

Network Virtual Terminal (NVT) A TCP/IP application protocol that allows remote logging.

network-specific routing Routing in which all hosts on a network share one entry in the routing table.

network-to-network interface (NNI) In ATM, the interface between two networks.

next-hop routing A routing method in which only the address of the next hop is listed in the routing table instead of a complete list of the stops the packet must make.

no-transition mobility *See* transition mobility.

node An addressable communication device (e.g., a computer or router) on a network.

node-to-node delivery Transfer of a data unit from one node to the next.

noise Random electrical signals that can be picked up by the transmission medium and cause degradation or distortion of the data.

noiseless channel An error-free channel.

noisy channel A channel that can produce error in data transmission.

nonce A large random number that is used once to distinguish a fresh authentication request from a used one.

nonperiodic (aperiodic) signal A signal that has no period; a signal that does not exhibit a pattern or repeating cycle.

nonpersistent connection A connection in which one TCP connection is made for each request/response.

nonpersistent method A random multiple access method in which a station waits a random period of time after a collision is sensed.

nonrepudiation A security aspect in which a receiver must be able to prove that a received message came from a specific sender.

non-return-to-zero (NRZ) A digital-to-digital polar encoding method in which the signal level is always either positive or negative, but never at zero level.

non-return-to-zero, invert (NRZ-I) An NRZ encoding method in which the signal level is inverted each time a 1 is encountered.

non-return-to-zero, level (NRZ-L) An NRZ encoding method in which the signal level is directly related to the bit value.

normal response mode (NRM) In HDLC, a communication mode in which the secondary station must have permission from the primary station before transmission can proceed.

Nyquist bit rate The data rate based on the Nyquist theorem.

Nyquist theorem A theorem that states that the number of samples needed to adequate represent an analog signal is equal to twice the highest frequency of the original signal.

Oakley A key creation protocol, developed by Hilarie Orman, which is one of the three components of the IKE protocol.

object identifier In MIB, an identifier for an object used in SNMP and some other network management protocols.

omnidirectional antenna An antenna that sends out or receives signals in all directions.

one-time pad A cipher invented by Vernam in which the key is a random sequence of symbols having the same length as the plaintext.

one's complement A representation of binary numbers in which the complement of a number is found by complementing all bits.

Open Shortest Path First (OSPF) An interior routing protocol based on link-state routing.

Open Systems Interconnection (OSI) model A seven-layer model for data communication defined by ISO.

open-loop congestion control Policies applied to prevent congestion.

optical carrier (OC) The hierarchy of fiber-optic carriers defined in SONET.

optical fiber A thin thread of glass or other transparent material to carry light beams.

orbit The path a satellite travels around the earth.

orthogonal-frequency-division-multiplexing (OFDM) A multiplexing method similar to FDM, with all the subbands used by one source at a given time.

orthogonal sequence A sequence with special properties between elements.

out-of-band signaling Using two separate channels for data and control.

P-box A component in a modern block cipher that transposes bits.

***p*-persistent method** A CSMA persistence strategy in which a station sends with probability p if it finds the line idle.

packet Synonym for data unit, mostly used in the network layer.

Packet Internet Groper (PING) An application program to determine the reachability of a destination using an ICMP echo request and reply.

packet switching Data transmission using a packet-switched network.

packet-filter firewall A firewall that forwards or blocks packets based on the information in the network-layer and transport-layer headers.

packet-switched network A network in which data are transmitted in independent units called packets.

parallel transmission Transmission in which bits in a group are sent simultaneously, each using a separate link.

parity check code An error-detection method using a parity bit.

partially qualified domain name (PQDN) A domain name that does not include all the levels between the host and the root node.

Password Authentication Protocol (PAP) A simple two-step authentication protocol used in PPP.

Pastry A DHT-based P2P network in which the identifiers are n-digit strings in base 2^b.

path vector routing A routing method on which BGP is based; in this method, the ASs through which a packet must pass are explicitly listed.

peak amplitude The maximum amplitude of an analog signal.

peer-to-peer (P2P) paradigm A paradigm in which two peer computers can communicate with each other to exchange services.

peer-to-peer process A process on a sending and a receiving machine that communicates at a given layer.

per-hop behavior (PHB) In the Diffserv model, a 6-bit field that defines the packet-handling mechanism for the packet.

perceptual coding The most common compression technique used to create CD-quality audio, based on the science of psychoacoustics. Algorithms used in perceptual coding first transform the data from time domain to frequency domain; the operations are then performed on the data in the frequency domain.

period The amount of time required to complete one full cycle.

periodic signal A signal that exhibits a repeating pattern.

permanent virtual circuit (PVC) A virtual circuit transmission method in which the same virtual circuit is used between source and destination on a continual basis.

persistence timer A timer in TCP that is used to prevent deadlock.

persistent connection A connection in which the server leaves the connection open for more requests after sending a response.

personal communication system (PCS) A generic term for a commercial cellular system that offers several kinds of communication services.

phase The relative position of a signal in time.

phase modulation (PM) An analog-to-analog modulation method in which the carrier signal's phase varies with the amplitude of the modulating signal.

phase shift keying (PSK) A digital-to-analog modulation method in which the phase of the carrier signal is varied to represent a specific bit pattern.

PHY sublayer The transceiver in Fast Ethernet.

physical address See link-layer address.

physical layer The first layer of the Internet model, responsible for the mechanical and electrical specifications of the medium.

piconet A Bluetooth network.

piggybacking The inclusion of acknowledgment on a data frame.

pipelining Sending several packets or frames before news is received concerning previous ones.

pixel A picture element of an image.

plain old telephone system (POTS) The conventional telephone network used for voice communication.

plaintext The message before encryption or after decryption.

playback buffer A buffer that stores the data until they are ready to be played.

point coordination function (PCF) In wireless LANs, an optional and complex access method implemented in an infrastructure network.

point of presence (POP) A switching office where carriers can interact with each other.

point-to-point link A dedicated transmission link between two devices.

Point-to-Point Protocol (PPP) A protocol for data transfer across a serial line.

poison reverse A variation of split horizons. In this method, information received by the router is used to update the routing table and then passed out to all interfaces. However, a table entry that has come through one interface is set to a metric of infinity as it goes out through the same interface.

polar encoding A digital-to-analog encoding method that uses two levels (positi negative) of amplitude.

policy routing A path vector routing feature in which the routing tables are based on rules set by the network administrator rather than on a metric.

poll In the primary/secondary access method, a procedure in which the primary station asks a secondary station if it has any data to transmit.

poll/final (P/F) bit A bit in the control field of HDLC; if the primary is sending, it can be a poll bit; if the secondary is sending, it can be a final bit.

poll/select An access method protocol using poll and select procedures. See *poll*. See *select*.

polling An access method in which one device is designated as a primary station and the others as the secondary stations. The access is controlled by the primary station.

polyalphabetic cipher A cipher in which each occurrence of a character may have a different substitute.

polyalphabetic substitution An encryption method in which each occurrence of a character can have a different substitute.

polynomial An algebraic term that can represent a CRC divisor.

port address In TCP/IP protocol, an integer that identifies a process (same as port numbers).

port forwarding A service provided by SSH to allow another application to secure channels using SSH.

port number An integer that defines a process running on a host (same as port address).

Post Office Protocol, version 3 (POP3) A popular but simple SMTP mail access protocol.

pre-master secret In SSL, a secret exchanged between the client and server before calculation of the master secret.

preamble The 7-byte field of an IEEE 802.3 frame consisting of alternating 1s and 0s that alert and synchronize the receiver.

predicted frame (P-frame) A predicted frame is related to the preceding I-frame or B-frame. In other words, each P-frame contains only the changes from the preceding frame.

predictive coding (PC) In audio compression, encoding only the differences between the samples.

prefix In an IP address, another name for the common part (similar to the netid).

presentation layer The sixth layer of the OSI model; responsible for translation, encryption, authentication, and data compression.

Pretty Good Privacy (PGP) A protocol invented by Phil Zimmermann to provide e-mail with privacy, integrity, and authentication.

primary station In primary/secondary access method, a station that issues commands to the secondary stations.

priority queuing A queuing technique in which there are two queues: one for regular packets, the other for the packet with priority.

privacy A security aspect in which the message makes sense only to the intended receiver.

private key In an asymmetric-key cryptosystem, the key used for decryption. In a digital signature, the key is used for signing.

private network A network that is isolated from the Internet.

process A running application program.

process-to-process communication Communication between two running application programs.

process-to-process delivery Delivery of a packet from the sending process to the destination process.

roject 802 The project undertaken by the IEEE in an attempt to solve LAN incompatibility.

pagation delay See propagation time.

propagation speed The rate at which a signal or bit travels; measured by distance/second.

propagation time The time required for a signal to travel from one point to another.

protocol Rules for communication.

Protocol Independent Multicast (PIM) A multicasting protocol family with two members, PIM-DM and PIM-SM; both protocols are unicast-protocol dependent.

Protocol Independent Multicast-Dense Mode (PIM-DM) A source-based routing protocol that uses RPF and pruning/grafting strategies to handle multicasting.

Protocol Independent Multicast-Sparse Mode (PIM-SM) A group-shared routing protocol that is similar to CBT and uses a rendezvous point as the source of the tree.

protocol layering The idea of using a set of protocols to create a hierarchy of rules for handling a difficult task.

protocol suite A stack or family of protocols defined for a complex communication system.

proxy ARP A technique that creates a subnetting effect; one server answers ARP requests for multiple hosts.

proxy firewall A firewall that filters a message based on the information available in the message itself (at the application layer).

proxy server A computer that keeps copies of responses to recent requests.

pruning Stopping the sending of multicast messages from an interface.

pseudoheader Information from the IP header used only for checksum calculation in the UDP and TCP packet.

pseudorandom noise (PN) A pseudorandom code generator used in FHSS.

pseudoternary A variation of AM encoding in which a 1 bit is encoded as zero voltage and a 0 bit is encoded as alternating positive and negative voltage.

psychoacoustics Psychoacoustics is the study of subjective human perception of sound. Perceptual coding takes advantage of flaws in the human auditory system.

public key In an asymmetric-key cryptosystem, the key used for encryption. In digital signature, the key is used for verification.

public key infrastructure (PKI) A hierarchical structure of CA servers.

public-key certificate A certificate that defines the owner of a public key.

public-key cryptography A method of encryption based on a nonreversible encryption algorithm. The method uses two types of keys: The public key is known to the public; the private key (secret key) is known only to the receiver.

pulse amplitude modulation (PAM) A technique in which an analog signal is sampled; the result is a series of pulses based on the sampled data.

pulse code modulation (PCM) A technique that modifies PAM pulses to create a digital signal.

pulse position modulation (PPM) The modulation technique used to modulate an infrared signal.

pulse stuffing A technique in multiplexing that adds dummy bits to an input to make its data rate the same as the others.

pure ALOHA The original ALOHA that does not use slots.

quadrature amplitude modulation (QAM) A digital-to-analog modulation method in which the phase and amplitude of the carrier signal vary with the modulating signal.

quality of service (QoS) An issue that refers to a set of techniques and mechanisms that guarantees the performance of a network.

quantization The assignment of a specific range of values to signal amplitudes.

quantization error Error introduced in the system during quantization (analog-to-digital conversion).

queue A waiting list.

quoted-printable An encoding scheme used when the data consist mostly of ASCII characters with a small non-ASCII portion.

Radio Government (RG) rating A government rating that defines the specification for the coaxial cable.

radio wave Electromagnetic energy in the 3-KHz to 300-GHz range.

random access A medium access category in which each station can access the medium without being controlled by any other station.

ranging In an HFC network, a process that determines the distance between the CM and the CMTS.

rate adaptive asymmetrical digital subscriber line (RADSL) A DSL-based technology that features different data rates depending on the type of communication.

raw socket A structure designed for protocols that directly use the services of IP and use neither stream sockets nor datagram sockets.

read-only memory (ROM) Permanent memory with contents that cannot be changed.

Real-Time Streaming Protocol (RTSP) An out-of-band control protocol designed to add more functionality to the streaming audio/video process.

Real-time Transport Control Protocol (RTCP) A companion protocol to RTP with messages that control the flow and quality of data and allow the recipient to send feedback to the source or sources.

Real-time Transport Protocol (RTP) A protocol for real-time traffic; used in conjunction with UDP.

recursive resolution Resolution of the IP address in which the client sends its request to a server that eventually returns a response.

redundancy The addition of bits to a message for error control.

Reed-Solomon code A complex, but efficient, cyclic code.

reflection The phenomenon related to the bouncing back of light at the boundary of two media.

refraction The phenomenon related to the bending of light when it passes from one medium to another.

regional ISP A small ISP that is connected to one or more backbones or international ISPs.

registered port A port number, ranging from 1024 to 49,151, not assigned or controlled by IANA.

registrar An authority to register new domain names.

registrar server In SIP, a server to which a user is registered at each moment.

relay agent For BOOTP, a router that can help send local requests to remote servers.

reliability A QoS flow characteristic; dependability of the transmission. A network is reliable when it does not corrupt, lose, or duplicate a packet.

remote bridge A device that connects LANs and point-to-point networks; often used in a backbone network.

remote logging (rlogin) The process of logging on to a remote computer from a terminal connected to a local computer.

rendezvous point (RP) A router used by PIM to distribute the multicast packets.

rendezvous router A router that is the core or center for each multicast group; it becomes the root of the tree.

rendezvous-point tree A group-shared tree method in which there is one tree for each group.

repeater A device that extends the distance a signal can travel by regenerating the signal.

replay attack The resending of a message that has been intercepted by an intruder.

Request for Comment (RFC) A formal Internet document concerning an Internet issue.

resolver The DNS client that is used by a host that needs to map an address to a name or a name to an address.

Resource Reservation Protocol (RSVP) A signaling protocol to help IP create a flow and make a resource reservation to improve QoS.

retransmission time-out (RTO) The expiration of a timer that controls the retransmission of packets.

return-to-zero (RZ) A digital-to-digital encoding technique in which the voltage of the signal is zero for the second half of the bit interval.

reuse factor In cellular telephony, the number of cells with a different set of frequencies.

Reverse Address Resolution Protocol (RARP) A TCP/IP protocol that allows a host to find its Internet address given its physical address.

reverse path broadcasting (RPB) In multicasting, a technique in which it is guaranteed that each destination receives one and only one copy of the packet.

reverse path forwarding (RPF) A technique in which the router forwards only the packets that have traveled the shortest path from the source to the router.

reverse path multicasting (RPM) A technique that adds pruning and grafting to RPB to create a multicast shortest path tree that supports dynamic membership changes.

ring topology A topology in which the devices are connected in a ring. Each device on the ring receives the data unit from the previous device, regenerates it, and forwards it to the next device.

Rivest, Shamir, Adleman (RSA) See *RSA cryptosystem.*

RJ45 A coaxial cable connector.

roaming In cellular telephony, the ability of a user to communicate outside of his own service provider's area.

root server In DNS, a server whose zone consists of the whole tree. A root server usually does not store any information about domains but delegates its authority to other servers, keeping references to those servers.

round-trip time (RTT) The time required for a datagram to go from a source to a destination and then back again.

route A path traveled by a packet.

router An internetworking device operating at the first three layers of the TCP/IP protocol suite. A router is attached to two or more networks and forwards packets from one network to another.

routing The process performed by a router; finding the next hop for a datagram.

Routing Information Protocol (RIP) A routing protocol based on the distance-vector routing algorithm.

routing table A table containing information a router needs to route packets. The information may include the network address, the cost, the address of the next hop, and so on.

RSA cryptosystem A popular public-key encryption method developed by Rivest, Shamir, and Adleman.

run-length coding A compression method for removing redundancy. The method replaces a repeated sequence, run, of the same symbol with two entities: a count and the symbol itself.

S-box An encryption device made of decoders, P-boxes, and encoders.

sample and hold A sampling method that samples the amplitude of an analog signal and holds the value until the next sample.

sampling The process of obtaining amplitudes of a signal at regular intervals.

sampling rate The number of samples obtained per second in the sampling process.

satellite network A combination of nodes that provides communication from one point on the earth to another.

scatternet A combination of piconets.

scrambling In digital-to-digital conversion, modifying part of the rules in a line coding scheme to create bit synchronization.

secondary station In the poll/select access method, a station that sends a response in answer to a command from a primary station.

secret-key encryption A security method in which the key for encryption is the same as the key for decryption; both sender and receiver have the same key.

Secure Hash Algorithm (SHA) A series of hash function standards developed by NIST and published as FIPS 180. It is mostly based on MD5.

Secure Key Exchange Mechanism (SKEME) A protocol for key exchange, designed by Hugo Krawcyzk, that uses public-key encryption for entity authentication.

Secure Shell (SSH) A client-server program that provides secure logging.

Secure Sockets Layer (SSL) protocol A protocol designed to provide security and compression services to data generated from the application layer.

Secure/Multipurpose Internet Mail Extensions (S/MIME) An enhancement to MIME designed to provide security for electronic mail.

Security Association (SA) An IPSec protocol that creates a logical connection between two hosts.

Security Association Database (SAD) A two-dimensional table with each row defining a single security association (SA).

security parameter index (SPI) A parameter that uniquely distinguishes one security association from the others.

Security Policy (SP) In IPSec, a set of predefined security requirements applied to a packet when it is to be sent or when it has arrived.

Security Policy Database (SPD) A database of security policies (SPs).

segment The packet at the TCP layer. Also, the length of the transmission medium shared by devices.

segmentation The splitting of a message into multiple packets; usually performed at the transport layer.

segmentation and reassembly (SAR) The lower AAL sublayer in the ATM protocol in which a header and/or trailer may be added to produce a 48-byte element.

select In the poll/select access method, a procedure in which the primary station asks a secondary station if it is ready to receive data.

selective-repeat (SR) protocol An error-control protocol in which only the frame in error is resent.

self-synchronization Synchronization of long strings of 1s or 0s through the coding method.

sequence number The number that denotes the location of a frame or packet in a message.

serial transmission Transmission of data one bit at a time using only a single link.

server A program that can provide services to other programs, called *clients*.

Session Initiation Protocol (SIP) In voice over IP, an application protocol that establishes, manages, and terminates a multimedia session.

session layer The fifth layer of the OSI model, responsible for the establishment, management, and termination of logical connections between two end users.

setup phase In virtual circuit switching, a phase in which the source and destination use their global addresses to help switches make table entries for the connection.

Shannon capacity The theoretical highest data rate for a channel.

shielded twisted-pair (STP) Twisted-pair cable enclosed in a foil or mesh shield that protects against electromagnetic interference.

shift cipher A type of additive cipher in which the key defines shifting of characters toward the end of the alphabet.

shift register A register in which each memory location, at a time click, accepts the bit at its input port, stores the new bit, and displays it on the output port.

short interframe space (SIFS) In CSMA/CA, a period of time that the destination waits after receiving the RTS.

shortest path tree A routing table formed by using Dijkstra's algorithm.

signal element The shortest section of a signal (timewise) that represents a data element.

signal rate The number of signal elements sent in one second.

signal-to-noise ratio (SNR) The ratio of average signal power to average noise power.

silly window syndrome A situation in which a small window size is advertised by the receiver and a small segment sent by the sender.

simple and efficient adaptation layer (SEAL) An AAL layer designed for the Internet (AAL5).

simple bridge A networking device that links two segments; requires manual maintenance and updating.

Simple Mail Transfer Protocol (SMTP) The TCP/IP protocol defining electronic mail service on the Internet.

Simple Network Management Protocol (SNMP) The TCP/IP protocol that specifies the process of management in the Internet.

Simple Protocol The simple protocol we used to show an access method without flow and error control.

simplex mode A transmission mode in which communication is one-way.

sine wave An amplitude-versus-time representation of a rotating vector.

single-bit error Error in a data unit in which only a single bit has been altered.

single-mode fiber An optical fiber with an extremely small diameter that limits beams to a few angles, resulting in an almost horizontal beam.

site local address An IPv6 address for a site having several networks but not connected to the Internet.

sky propagation Propagation of radio waves into the ionosphere and then back to earth.

slash notation A shorthand method to indicate the number of 1s in the mask.

sliding window protocol A protocol that allows several data units to be in transition before receiving an acknowledgment.

slotted ALOHA The modified ALOHA access method in which time is divided into slots and each station is forced to start sending data only at the beginning of the slot.

slow convergence A RIP shortcoming apparent when a change somewhere in the Internet propagates very slowly through the rest of the Internet.

slow start algorithm A congestion-control method in which the congestion window size increases exponentially at first.

socket An end point for a process; two sockets are needed for communication.

socket address A structure holding an IP address and a port number.

socket interface A set of system calls used in client-server paradigm.

Software Defined Radio (SDR) A radio communication system in which traditional hardware components are implemented in software.

source quench A method, used in ICMP for flow control, in which the source is advised to slow down or stop the sending of datagrams because of congestion.

source routing Explicitly defining the route of a packet by the sender of the packet.

source-based tree A tree used for multicasting by multicasting protocols in which a single tree is made for each combination of source and group.

source-to-destination delivery The transmission of a message from the original sender to the intended recipient.

space propagation A type of propagation that can penetrate the ionosphere.

space-division switching Switching in which the paths are separated from each other spatially.

spanning tree A tree with the source as the root and group members as leaves; a tree that connects all of the nodes.

spatial compression Compressing an image by removing redundancies.

spectrum The range of frequencies of a signal.

split horizon A method to improve RIP stability in which the router selectively chooses the interface from which updating information is sent.

spread spectrum (SS) A wireless transmission technique that requires a bandwidth several times the original bandwidth.

Standard Ethernet The original Ethernet operating at 10 Mbps.

star topology A topology in which all stations are attached to a central device (hub).

start bit In asynchronous transmission, a bit to indicate the beginning of transmission.

state transition diagram A diagram to illustrate the states of a finite state machine.

static document On the World Wide Web, a fixed-content document that is created and stored in a server.

static mapping A technique in which a list of logical and physical address correspondences is used for address resolution.

static routing A type of routing in which the routing table remains unchanged.

stationary host A host that remains attached to one network.

statistical TDM A TDM technique in which slots are dynamically allocated to improve efficiency.

status line In the HTTP response message, a line that consists of the HTTP version, a space, a status code, a space, a status phrase.

steganography A security technique in which a message is concealed by covering it with something else.

stop bit In asynchronous transmission, one or more bits to indicate the end of transmission.

Stop-and-Wait Protocol A protocol in which the sender sends one frame, stops until it receives confirmation from the receiver, and then sends the next frame.

store-and-forward switch A switch that stores the frame in an input buffer until the whole packet has arrived.

straight tip connector A type of fiber-optic cable connector using a bayonet locking system.

STREAM One of the interfaces that have been defined for network programming.

stream cipher A type of cipher in which encryption and decryption are done one symbol (such as a character or a bit) at a time.

Stream Control Transmission Protocol (SCTP) The transport-layer protocol designed to combine the features of UDP and TCP.

stream socket A structure designed to be used with a connection-oriented protocol such as TCP.

streaming live audio/video Broadcast data from the Internet that a user can listen to or watch.

streaming stored audio/video Data downloaded as files from the Internet that a user can listen to or watch.

strong collision Creating two messages with the same digest.

Structure of Management Information (SMI) In SNMP, a component used in network management.

structured data type A complex data type made of some simple or structured data types.

STS multiplexer/demultiplexer A SONET device that multiplexes and demultiplexes signals.

stub link A network that is connected to only one router.

subnet A subnetwork.

subnet address The network address of a subnet.

subnet mask The mask for a subnet.

subnetwork A part of a network.

substitution cipher A cipher that replaces one symbol with another.

suffix The varying part (similar to the hostid) of an IP address.

summary link to AS boundary router LSA An LSA packet that lets a router inside an area know the route to an autonomous boundary router.

summary link to network LSA An LSA packet that finds the cost of reaching networks outside of the area.

supergroup A signal composed of five multiplexed groups.

supernet A network formed from two or more smaller networks.

supernet mask The mask for a supernet.

switch A device connecting multiple communication lines together.

switched virtual circuit (SVC) A virtual circuit transmission method in which a virtual circuit is created and in existence only for the duration of the exchange.

switching office The place where telephone switches are located.

symmetric-key cipher A cipher using a symmetric-key cryptosystem.

symmetric-key cryptography A cipher in which the same key is used for encryption and decryption.

SYN flooding attack A serious security problem in the TCP connection establishment phase in which one or more malicious attackers send a large number of SYN segments.

synchronous connection-oriented (SCO) link In a Bluetooth network, a physical link created between a master and a slave that reserves specific slots at regular intervals.

Synchronous Digital Hierarchy (SDH) The ITU-T equivalent of SONET.

Synchronous Optical Network (SONET) A standard developed by ANSI for fiber-optic technology that can transmit high-speed data. It can be used to deliver text, audio, and video.

synchronous TDM A TDM technique in which each input has an allotment in the output even when it is not sending data.

synchronous transmission A transmission method that requires a constant timing relationship between the sender and the receiver.

synchronous transport signal (STS) A signal in the SONET hierarchy.

syndrome A sequence of bits generated by applying the error checking function to a codeword.

T-lines A hierarchy of digital lines designed to carry speech and other signals in digital forms.

TCP/IP protocol suite A group of hierarchical protocols used in an internet.

teardown phase In virtual circuit switching, the phase in which the source and destination inform the switch to erase their entry.

telecommunications Exchange of information over distance using electronic equipment.

teleconferencing Audio and visual communication between remote users.

Teledesic A system of satellites that provides fiber-optic communication (broadband channels, low error rate, and low delay).

temporal compression An MPEG compression method in which redundant frames are removed.

temporal masking A situation where a loud sound can numb our ears for a short time even after the sound has stopped.

Terminal Network (TELNET) A general purpose client-server program for remote logging.

three-way handshake A sequence of events for connection establishment or termination consisting of the request, then the acknowledgment of the request, and then confirmation of the acknowledgment.

throughput The number of bits that can pass through a point in one second.

ticket An encrypted message intended for entity B, but sent to entity A for delivery.

time to live (TTL) The lifetime of a packet.

time-division duplex TDMA (TDD-TDMA) In a Bluetooth network, a kind of half-duplex communication in which the slave and receiver send and receive data, but not at the same time (half-duplex).

time-division multiple access (TDMA) A multiple access method in which the bandwidth is just one time-shared channel.

time-division multiplexing (TDM) The technique of combining signals coming from low-speed channels to share time on a high-speed path.

time-division switching A circuit-switching technique in which time-division multiplexing is used to achieve switching.

time-domain plot A graphical representation of a signal's amplitude versus time.

time-space-time (TST) switch A switch that combines space-division and time-division technology to achieve better performance.

timestamp A field in the packet related to the absolute or relative time the packed is created or sent.

token A small packet used in the token-passing access method.

token bucket An algorithm that allows idle hosts to accumulate credit for the future in the form of tokens.

token passing An access method in which a token is circulated in the network. The station that captures the token can send data.

topology The structure of a network including physical arrangement of devices.

traffic control A method for shaping and controlling traffic in a wide area network.

traffic shaping A mechanism to improve QoS that controls the amount and the rate of the traffic sent to the network.

trailer Control information appended to a data unit.

transceiver A device that both transmits and receives.

transient link A network with several routers attached to it.

transition mobility In IEEE 802.11, a station with BSS-transition mobility can move from one BSS to another, but the movement is confined inside one ESS. A station with no-transition mobility is either stationary (not moving) or moving only inside a BSS.

Transmission Control Protocol (TCP) A transport-layer protocol in the TCP/IP protocol suite.

Transmission Control Protocol/Internet Protocol (TCP/IP) A five-layer protocol suite that defines the exchange of transmissions across the Internet.

transmission medium The physical path linking two communication devices.

transmission path (TP) In ATM, a physical connection between an end-point and a switch or between two switches.

transmission rate The number of bits sent per second.

transparency The ability to send any bit pattern as data without it being mistaken for control bits.

transport layer The fourth layer in the Internet and OSI model; responsible for reliable end-to-end delivery and error recovery.

Transport Layer Interface (TLI) A networking API provided by the UNIX system.

Transport Layer Security (TLS) protocol A security protocol at the transport level designed to provide security on the WWW. An IETF version of the SSL protocol.

transport mode Encryption in which a TCP segment or a UDP user datagram is first encrypted and then encapsulated in an IPv6 packet.

transposition cipher A cipher that transposes symbols in the plaintext to create the ciphertext.

Trap In SNMP, a PDU sent from an agent to the manager to report an event.

triangle routing In mobile IP, the less severe inefficiency case that occurs when the remote host communicates with a mobile host that is not attached to the same network (or site) as the mobile host.

triangulation The same as trilateration, but using three angles instead of three distances.

trilateration A two-dimensional method of finding a location given the distances from three different points.

trunk Transmission media that handle communications between offices.

tunnel mode A mode in IPSec that protects the entire IP packet. It takes an IP packet, including the header, applies IPSec security methods to the entire packet, and then adds a new IP header.

tunneling In multicasting, a process in which the multicast packet is encapsulated in a unicast packet and then sent through the network. In VPN, the encapsulation of an encrypted IP datagram in a second outer datagram. For IPv6, a strategy used when two computers using IPv6 want to communicate with each other when the packet must pass through a region that uses IPv4.

twisted-pair cable A transmission medium consisting of two insulated conductors in a twisted configuration.

two-binary, one quaternary (2B1Q) encoding A line encoding technique in which each pulse represents 2 bits.

two-dimensional parity check An error detection method in two dimensions.

type of service (TOS) A criteria or value that specifies the handling of the datagram.

unbalanced configuration An HDLC configuration in which one device is primary and the others secondary.

unguided media Transmission media with no physical boundaries (air).

unicast address An address belonging to one destination.

unicasting The sending of a packet to just one destination.

Unicode The international character set used to define valid characters in computer science.

unidirectional antenna An antenna that sends or receives signals in one direction.

uniform resource locator (URL) A string of characters (address) that identifies a page on the World Wide Web.

unipolar encoding A digital-to-digital encoding method in which one nonzero value represents either 1 or 0; the other bit is represented by a zero value.

Universal Mobile Telecommunication System (UMTS) One of the popular 3G technologies using a version of CDMA called Direct Sequence Wideband CDMA (DS-WCDMA).

unshielded twisted-pair (UTP) A cable with wires that are twisted together to reduce noise and crosstalk. See also *twisted-pair cable* and *shielded twisted-pair.*

unspecified bit rate (UBR) The data rate of an ATM service class specifying only best-effort delivery.

uplink Transmission from an earth station to a satellite.

uploading Sending a local file or data to a remote site.

user agent (UA) An SMTP component that prepares the message, creates the envelope, and puts the message in the envelope.

user authentication A security measure in which the sender identity is verified before the start of a communication.

user datagram The name of the packet in the UDP protocol.

User Datagram Protocol (UDP) A connectionless TCP/IP transport-layer protocol.

user-to-network interface (UNI) In ATM, the interface between an end point (user) and an ATM switch.

variable bit rate (VBR) The data rate of an ATM service class for users needing a varying bit rate.

video Recording or transmitting of a picture or a movie.

Vigenere cipher A polyalphabetic substitution scheme that uses the position of a character in the plaintext and the character's position in the alphabet.

virtual circuit (VC) A logical circuit made between the sending and receiving computers.

virtual circuit network A network that uses virtual circuit switching.

virtual circuit switching A switching technique used in switched WANs.

virtual link An OSPF connection between two routers that is created when the physical link is broken. The link between them uses a longer path that probably goes through several routers.

virtual local area network (VLAN) A technology that divides a physical LAN into virtual workgroups through software methods.

virtual path (VP) A combination of virtual circuits in ATM.

virtual private network (VPN) A technology that creates a network that is physically public, but virtually private.

virtual tributary (VT) A partial payload that can be inserted into a SONET frame and combined with other partial payloads to fill out the frame.

voice over IP A technology in which the Internet is used as a telephone network.

vulnerable time A time in which there is a possibility of collision between packets travelling in a network.

Walsh table In CDMA, a two-dimensional table used to generate orthogonal sequences.

wavelength The distance a simple signal can travel in one period.

wavelengh-division multiplexing (WDM) The combining of modulated light signals into one signal.

web of trust In PGP, the key rings shared by a group of people.

web page A unit of hypertext or hypermedia available on the Web.

weighted fair queuing A packet scheduling technique to improve QoS in which the packets are assigned to queues based on a given priority number.

well-known port number A port number that identifies a process on the server.

wide area network (WAN) A network that uses a technology that can span a large geographical distance.

window scale factor An option in TCP that allows for increasing the window size defined in the header.

working group An IETF committee concentrating on a specific Internet topic.

World Wide Web (WWW) A multimedia Internet service that allows users to traverse the Internet by moving from one document to another via links that connect them.

Worldwide Interoperability for Microwave Access (WiMAX) A family of IEEE 802.16 standards to deliver wireless data at the last mile (similar to cable or DLS networks for wired communication).

X.509 A recommendation devised by ITU and accepted by the Internet that defines certificates in a structured way.

zone In DNS, what a server is responsible for or has authority over.

REFERENCES

[AL 98] Albitz, P., and Liu, C. *DNS and BIND,* 3rd ed. Sebastopol, CA: O'Reilly, 1998.

[AZ 03] Agrawal, D., and Zeng, Q. *Introduction to Wireless and Mobile Systems.* Pacific Grove, CA: Brooks/Cole Thomson Learning, 2003.

[Bar 02] Barr, T., *Invitation to Cryptology.* Upper Saddle River, NJ: Prentice Hall, 2002.

[Bar et al. 05] Barrett, Daniel J., Silverman, Ricard E., and Byrnes, Robert G. *SSH: The Secure Shell: The Definitive Guide,* Sebastopol, CA: O'Reilly, 2005.

[BEL 01] Bellamy, J. *Digital Telephony.* New York, NY: Wiley, 2001.

[Ber 96] Bergman, J. *Digital Baseband Transmission and Recording.* Boston, MA: Kluwer, 1996.

[Bis 03] Bishop, D. *Introduction to Cryptography with Java Applets.* Sebastopol, CA: O'Reilly, 2003.

[Bis 05] Bishop, Matt. *Introduction to Computer Security.* Reading, MA: Addison-Wesley, 2005.

[Bla 00] Black, U. *QOS in Wide Area Networks.* Upper Saddle River, NJ: Prentice Hall, 2000.

[Bla 00] Black, U. *PPP and L2TP: Remote Access Communication.* Upper Saddle River, NJ: Prentice Hall, 2000.

[Bla 03] Blahut, R. *Algebraic Codes for Data Transmission.* Cambridge, UK: Cambridge University Press, 2003.

[BYL 09] Buford, J. F., Yu, H., and Lua, E. K. *P2P Networking and Applications.* San Francisco: Morgan Kaufmann, 2009.

[CBR 03] Cheswick, W., Bellovin, S., and Rubin, A. *Firewalls and Internet Security.* Reading, MA: Addison-Wesley, 2003.

[CD 08] Calvert, Kenneth L., and Donaho, Michael J. *TCP/IP Sockets in Java.* San Francisco, CA: Morgan Kaufmann, 2008.

[CER 89] Cerf, V. *A History of Arpanet, The Interoperability Report,* 1989.

[CHW 99] Crowcroft, J., Handley, M., Wakeman, I. *Internetworking Multimedia.* San Francisco, CA: Morgan Kaufmann, 1999.

[Com 00] Comer, D. *Internetworking with TCP/IP, Volume 1: Principles, Protocols, and Architecture.* Upper Saddle River, NJ: Prentice Hall, 2000.

[Com 04] Comer, D. *Computer Networks.* Upper Saddle River, NJ: Prentice Hall, 2004.

[Com 06] Comer, Douglas E. *Internetworking with TCP/IP,* vol. 1. Upper Saddle River, NJ: Prentice Hall, 2006.

[Cou 01] Couch, L. *Digital and Analog Communication Systems.* Upper Saddle River, NJ: Prentice Hall, 2001.

[DC 01] Donaho, Michael J., and Calvert, Kenneth L. *TCP/IP Sockets: C version.* San Francisco, CA: Morgan Kaufmann, 2001.

[DH 03] Doraswamy, H., and Harkins, D. *IPSec.* Upper Saddle River, NJ: Prentice Hall, 2003.

[Dro 02] Drozdek, A. *Elements of Data Compression.* Pacific Grove, CA: Brooks/Cole (Thomson Learning), 2002.

[Dut 01] Dutcher, D. *The NAT Handbook.* New York, NW: Wiley, 2001.

[Far 04] Farrel, A. *The Internet and Its Protocols.* San Francisco: Morgan Kaufmann, 2004.

[FH 98] Ferguson, P., and Huston, G. *Quality of Service*. New York: John Wiley and Sons, Inc., 1998.

[For 03] Forouzan, B. *Local Area Networks.* New York: McGraw-Hill, 2003.

[For 08] Forouzan, B., *Cryptography and Network Security.* New York: McGraw-Hill, 2008.

[For 10] Forouzan, B., *TCP/IP Protocol Suite*. New York: McGraw-Hill, 2010.

[Fra 01] Frankkel, S. *Demystifying the IPSec Puzzle*. Norwood, MA: Artech House, 2001.

[FRE 96] Freeman, R. *Telecommunication System Engineering*. New York, NW: Wiley, 1996.

[Gar 95] Garfinkel, S. *PGP: Pretty Good Privacy.* Sebastopol. CA: O'Reilly, 1995.

[Gar 01] Garret, P. *Making, Breaking Codes.* Upper Saddle River, NJ: Prentice Hall, 2001.

[Gas 02] Gast, M. *802.11 Wireless Networks.* Sebastopol, CA: O'Reilly, 2002.

[GGLLB 98] Gibson, J. D., Gerger, T., Lookabaugh, T., LindBerg, D., and Baker, R. L. *Digital Compression for Multimedia*. San Francisco: Morgan Kaufmann, 1998.

[GW 04] Garcia, A., and Widjaja, I. *Communication Networks.* New York, NY: McGraw-Hill, 2004.

[Hal 01] Halsall, F. *Multimedia Communication.* Reading, MA: Addison-Wesley, 2001.

[Ham 80] Hamming, R. *Coding and Information Theory.* Upper Saddle River, NJ: Prentice Hall, 1980.

[Har 05] Harol, Elliot R. *Java Network Programming.* Sebastopol, CA: O'Reilly, 2005.

[HM 10] Havaldar, P., and Medioni, G. *Multimedia Systems: Algorithms, Standards, and Industry Practices.* Boston: Course Technology (Cengage Learning), 2010.

[Hsu 03] Hsu, H. *Analog and Digital Communications.* New York, NY: McGraw-Hill, 2003.

[Hui 00] Huitema, C. *Routing in the Internet,* 2nd ed. Upper Saddle River, NJ: Prentice Hall, 2000.

[Izz 00] Izzo, P. *Gigabit Networks.* New York, NY: Wiley, 2000.

[Jam 03] Jamalipour, A. *Wireless Mobile Internet.* New York, NY: Wiley, 2003.

[Jen et al. 86] Jennings, D. M., Landweber, L. M., Fuchs, I. H., Farber, D. H., and Adrion, W. R. "Computer Networking for Scientists and Engineers," *Science* 231, no. 4741 (1986): 943–950.

[KCK 98] Kadambi, J., Crayford, I., and Kalkunte, M. *Gigabit Ethernet.* Upper Saddle River, NJ: Prentice Hall, 1998.

[Kei 02] Keiser, G. *Local Area Networks.* New York, NY: McGraw-Hill, 2002.

[Kes 02] Keshav, S. *An Engineering Approach to Computer Networking.* Reading, MA: Addison-Wesley, 2002.

[Kle 04] Kleinrock, L. *The Birth of the Internet.*

[KMK 04] Kumar, A., Manjunath, D., and Kuri, J. *Communication Network: An Analytical Approach.* San Francisco: Morgan Kaufmann, 2004.

[Koz 05] Kozierock, Charles M. *The TCP/IP Guide.* San Francisco: No Starch Press, 2005.

[KPS 02] Kaufman, C., Perlman, R., and Speciner, M. *Network Security.* Upper Saddle River, NJ: Prentice Hall, 2002.

[KR 05] Kurose, J., and Ross, K. *Computer Networking.* Reading, MA: Addison-Wesley, 2005.

[Lei et al. 98] Leiner, B., Cerf, V., Clark, D., Kahn, R., Kleinrock, L., Lynch, D., Postel, J., Roberts, L., and Wolff, S., *A Brief History of the Internet.* http://www.isoc.org/internet/history/brief.shtml.

[Los 04] Loshin, Pete. *IPv6: Theory, Protocol, and Practice.* San Francisco: Morgan Kaufmann, 2004.

[Mao 04] Mao, W. *Modern Cryptography.* Upper Saddle River, NJ: Prentice Hall, 2004.

[Max 99] Maxwell, K. *Residential Broadband.* New York, NY: Wiley, 1999.

[Mir 07] Mir, Nader F. *Computer and Communication Networks*. Upper Saddle River, NJ: Prentice Hall, 2007.

[MOV 97] Menezes, A., Oorschot, P., and Vanstone, S. *Handbook of Applied Cryptography.* New York, NY: CRC Press, 1997.

[Moy 98] Moy, John. *OSPF.* Reading, MA: Addison-Wesley, 1998.

[MS 01] Mauro, D., and Schmidt, K. *Essential SNMP.* Sebastopol, CA: O'Reilly, 2001.

[PD 03] Peterson, Larry L., and Davie, Bruce S. *Computer Networks,* 3rd ed. San Francisco: Morgan Kaufmann, 2003.

[Pea 92] Pearson, J. *Basic Communication Theory.* Upper Saddle River, NJ: Prentice Hall, 1992.

[Per 00] Perlman, Radia. *Interconnections,* 2nd ed. Reading, MA: Addison-Wesley, 2000.

[PHS 03] Pieprzyk, J., Hardjono, T., and Seberry, J., *Fundamentals of Computer Security.* Berlin, Germany: Springer, 2003.

[Pit 06] Pitt, E. *Fundamental Networking in Java.* Berlin: Springer-Verlag, 2006.

[PKA 08] Poo, D., Kiong, D., Ashok, S. *Object-Oriented Programming and Java.* Berlin: Springer-Verlag, 2008.

[Res 01] Rescorla, E. *SSL and TLS*. Reading, MA: Addison-Wesley, 2001.

[Rhe 03] Rhee, M, *Internet Security.* New York, NY: Wiley, 2003.

[Ror 96] Rorabaugh, C. *Error Coding Cookbook.* New York, NY: McGraw-Hill, 1996.

[RR 96] Robbins, Kay A., and Robbins, Steven. *Practical UNIX Programming,* Upper Saddle River, NJ: Prentice Hall, 1996.

[Sau 98] Sauders, S. *Gigabit Ethernet Handbook.* New York, NY: McGraw-Hill, 1998.

[Sch 03] Schiller, J. *Mobile Communications.* Reading, MA: Addison-Wesley, 2003.

[Sch 96] Schneier, B. *Applied Cryptography.* Reading, MA: Addison-Wesley, 1996.

[Seg 98] Segaller, S. *Nerds 2.0.1: A Brief History of the Internet*. New York: TV Books, 1998.

[Sna 00] Snader, J. C. *Effective TCP/IP Programming*. Reading, MA: Addison-Wesley, 2000.

[Sol 03] Solomon, D. *Data Privacy and Security.* Berlin, Germany: Springer, 2003.

[Spi 74] Spiegel, M. *Fourier Analysis.* New York, NY: McGraw-Hill, 1974.

[Spu 00] Spurgeon, C. *Ethernet.* Sebastopol, CA: O'Reilly, 2000.

[SSS 05] Shimonski, R., Steiner, R., Sheedy, S. *Network Cabling Illuminated.* Sudbury, MA: Jones and Bartlette, 2005.

[Sta 02] Stallings, W. *Wireless Communications and Networks.* Upper Saddle River, NJ: Prentice Hall, 2002

[Sta 04] Stallings, W. *Data and Computer Communications,* 7th ed. Upper Saddle River, NJ: Prentice Hall, 2004.

[Sta 06] Stallings, W. *Cryptography and Network Security,* 5th ed. Upper Saddle River, NJ: Prentice Hall, 2006.

[Sta 98] Stallings, W. *High Speed Networks.* Upper Saddle River, NJ: Prentice Hall, 1998.

[Ste 94] Stevens, W. Richard. *TCP/IP Illustrated,* vol. 1. Reading, MA: Addison-Wesley, 1994.

[Ste 95] Stevens, W. Richard. *TCP/IP Illustrated,* vol. 2. Reading, MA: Addison-Wesley, 1995.

[Ste 96] Stevens, W. *TCP/IP Illustrated, Volume 3.* Upper Saddle River, NJ: Prentice Hall, 2000.

[Ste et al. 04] Stevens, W. Richard, Fenner, Bill, and Rudoff, Andrew, M. *UNIX Network Programming: The Sockets Networking API*. Reading, MA: Addison-Wesley, 2004.

[Ste 99] Stewart, John W. III, J. *BGP4: Inter-Domain Routing in the Internet.* Reading, MA: Addison-Wesley, 1999.

[Sti 06] Stinson, D. *Cryptography: Theory and Practice.* New York: Chapman & Hall/CRC, 2006.

[Sub 01] Subramanian, M. *Network Management.* Reading, MA: Addison-Wesley, 2000.

[SW 05] Steinmetz, R., and Wehrle, K. *Peer-to-Peer Systems and Applications.* Berlin: Springer-Verlag, 2005.

[SWE 99] Scott, C., Wolfe, P, and Erwin, M. *Virtual Private Networks.* Sebastopol, CA: O'Reilly, 1998.

[SX 02] Stewart, Randall R., and Xie, Qiaobing. *Stream Control Transmission Protocol (STCP).* Reading, MA: Addison-Wesley, 2002.

[Tan 03] Tanenbaum, Andrew S. *Computer Networks,* 4th ed. Upper Saddle River, NJ: Prentice Hall, 2003.

[Tho 00] Thomas, S. *SSL and TLS Essentials.* New York: John Wiley & Sons, 2000.

[WV 00] Warland, J., and Varaiya, P. *High Performance Communication Networks.* San Francisco, CA: Morgan, Kaufmans, 2000.

[WZ 01] Wittmann, R., and Zitterbart, M. *Multicast Communication.* San Francisco: Morgan Kaufmann, 2001.

[YS 01] Yuan R. and Strayer, W. *Virtual Private Networks.* Reading, MA: Addison-Wesley, 2001.

[Zar 02] Zaragoza, R. *The Art of Error Correcting Coding.* Reading, MA: Addison-Wesley, 2002.

INDEX

Numerics

A

D

F

G

H

I

N

O

S